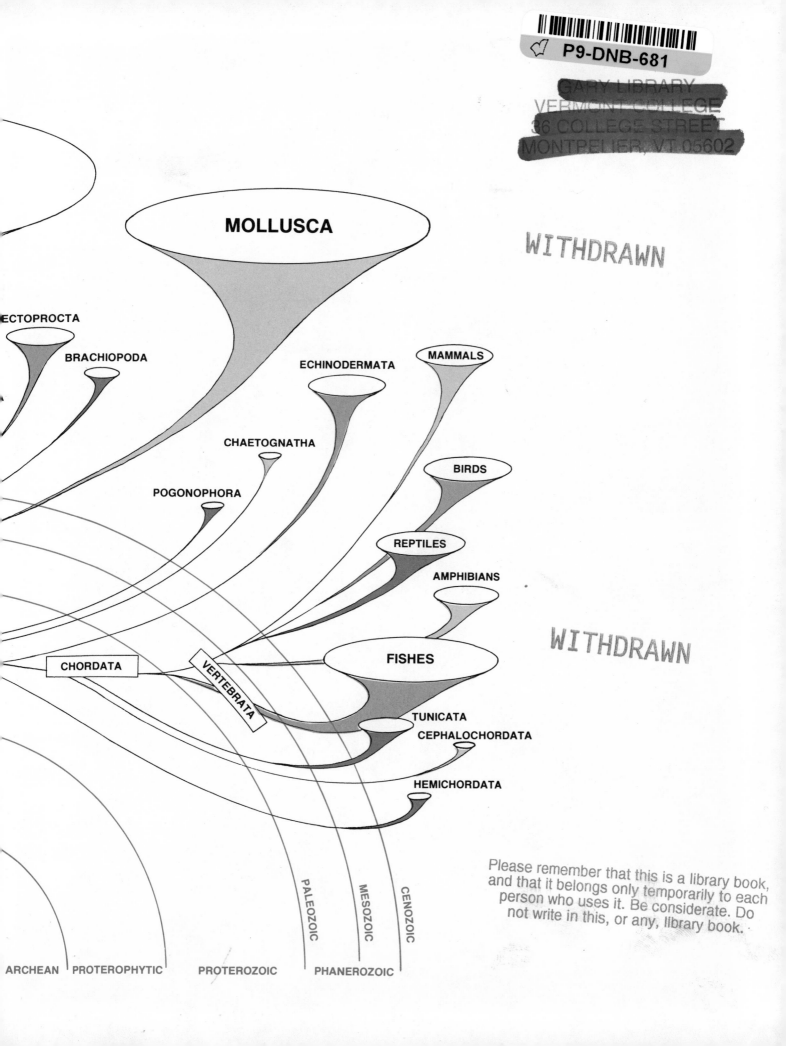

W.C. Ober.

INTEGRATED PRINCIPLES OF

ZOOLOGY

EIGHTH EDITION

CLEVELAND P. HICKMAN, Jr.

Professor of Biology
Washington and Lee University,
Lexington, Virginia

LARRY S. ROBERTS

Department of Biological Sciences,
Texas Tech University,
Lubbock, Texas

FRANCES M. HICKMAN

Emeritus, Department of Zoology
DePauw University,
Greencastle, Indiana

Original artwork by William C. Ober, M.D.
Crozet, Virginia

TIMES MIRROR/MOSBY COLLEGE PUBLISHING

ST. LOUIS • TORONTO • SANTA CLARA   1988

Editor: **David Kendric Brake**
Editorial assistant: **Mary Huggins**
Project manager: **Teri Merchant**
Production editor: **Mary Stueck**
Design: **Elizabeth Fett**

Cover art: "Master of the Herd—African Buffalo"© 1979 Robert Bateman.
Courtesy of the artist and Mill Pond Press, Inc., Venice, FL 34292-3505

EIGHTH EDITION

Copyright © 1988 by Times Mirror/Mosby College Publishing

A division of The C.V. Mosby Company
11830 Westline Industrial Drive, St. Louis, MO 63146

Previous editions copyrighted 1955, 1961, 1966, 1970, 1974, 1979, 1984

Printed in the United States of America

Library of Congress Cataloging-in-Publication Data

Hickman, Cleveland P.
   Integrated principles of zoology.

   Includes bibliographies and index.
   1. Zoology. I. Roberts, Larry S., 1935–
II. Hickman, Frances Miller. III. Title.
QL47.2.H54 1988   591   87-24001
ISBN 0-8016-2450-9

TS/VH/VH 9 8 7 6 5 4 3   02/C/221

W.C. Ober.

PREFACE

Integrated Principles of Zoology deals with the diversity and adaptations of the several million species that make up the Animal Kingdom. As it was with the first edition published 33 years ago, the aims of this eighth edition are to present the concepts and principles of zoology simply and directly and to instill in the student an enduring enthusiasm for the wonderful world of animals.

During the preparation of this edition it occurred to us more than once that this book had itself metamorphosed into a living organism or, perhaps more suitably, into an evolutionary lineage that, with each new edition, must adapt itself to an ever-changing environment. Fortunately, evolution has been conservative as well as inventive. So although a revision must respond to the relentless growth of biological information, there is much of zoological science that has permanence. Although it is sometimes said that biology today is in the midst of a revolution, this is in a sense untrue. More revolutionary by far were some of the advances made by great biologists of the past, such as Linnaeus, who showed us how to classify living things; Darwin, who explained the mechanism of evolution; and Mendel, who clarified variation and its inheritance. Discoveries such as these were true revolutions in biological thought. These individuals and many others bequeathed us the framework in which biology today rests, and it is these principles, summarized in Chapter 1, that form the central theme of this book.

This edition has been extensively revised. In addition to updating it throughout, we made important changes in presentation and content, guided in part by the suggestions and criticisms of eight zoologists, some of them long-time users of this text, who reviewed the entire book. We and our publishers were especially mindful of another hazard that attends the evolution of a textbook: the "dinosaur syndrome." Successive editions grow thicker and heavier, more encyclopedic, until usefulness to student and instructor alike is compromised. To avoid this fate, we have trimmed some portions of the text and in general followed the rule that for everything added something else must be cut out. At the same time we feel strongly that if a topic is to be included at all, it must treated with sufficient depth and rigor to be meaningful to the student.

W.C. Ober.

Changes in this edition

Those familiar with previous editions will notice that the organization remains unaltered (we admit to having flirted with other schemes that may have a special appeal for particular programs, but the present approach is clearly favored by the majority of users). Several important changes, however, have been made within the five parts of the text. The first two chapters of Part One were condensed and the discussion of nucleic acid chemistry was moved to Chapter 38 where it leads off the treatment of molecular genetics. Certain sections in Chapters 4 and 5 were condensed and the description of the cellular cytoskeleton (Chapter 4) and free energy (Chapter 5) were rewritten. Chapter 6 was extensively revised to provide a more comparative focus. The treatment of traditional and modern approaches to systematics in Chapter 7 was reworked to accord a more satisfactory description of cladistics, and the discussion of the origin of the metazoa was moved to Chapter 9.

The chapters on invertebrates have been updated. The coverage of Mesozoa was shortened, and that of the sponges was extensively rewritten (Chapter 10). Some additional information on Turbellaria was inserted in Chapter 11, and this was balanced by some abridgement of the section on tapeworms. Coverage of body wall muscles in nematodes and the function of the hydrostatic skeleton has been rewritten, and the relation of the lacunar system to the muscles of acanthocephalans has been clarified (Chapter 12). In Chapter 13 material on torsion and fouling in gastropods, evolution of the gills in bivalves, and internal features of cephalopods has been reworked. Chapter 14 has been reorganized, with the coverage of polychaetes preceding that of oligochaetes and leeches. The arthropod chapters (15, 16, 17) have been reorganized and rewritten as necessary to accord the Crustacea and the Uniramia the status of subphyla, rather than treating them as a single (probably polyphyletic) subphylum Mandibulata. The newly described class of crustaceans (Tantulocarida) has been included. The coverage of insects has been shortened somewhat, but we believe that adequate detail remains to give students a concept of this enormous group. Except for numerous minor rewrites and condensations, the remaining chapters on invertebrates are much as they were in the previous edition.

Chapter 22, the first of the six chordate chapters, was completely reorganized at the behest of the reviewers, with an expanded protochordate treatment placed before the treatment of vertebrate ancestry and evolution. The remaining vertebrate chapters were condensed in places and updated throughout, but in other respects their content has not been substantially modified. The classification of the living mammalian orders was updated and somewhat abridged.

For Part Three, The Activity of Life, the reviewers urged us to strengthen the chapters with a greater comparative representation and reduce the emphasis on human physiology. Hence we have weaved in many more comparative examples from the broad spectrum of animal life. Numerous vertebrate or human examples are retained, however, because many teachers endorse that approach. In some areas where there has been a rapid proliferation of knowledge as, for example, the defense mechanisms of animals (immunity), the discussion has been completely rewritten (Chapter 29). Other areas that received thorough revisions were ancestry and embryology of the vertebrate kidney (Chapter 30); countercurrent multiplier mechanism of the vertebrate kidney (Chapter 30); feeding mechanisms (Chapter 31); and resting and action potentials and synapse function (Chapter 32). Chapter 34 on animal behavior was somewhat condensed.

Part Four contains five chapters on the continuity and evolution of animal life. Several sections in Chapter 35 on reproduction were overhauled; these include the discussions of sexual reproduction and its significance, origin and migration of germ cells, and reproductive patterns. Chapters 36 and 39 on development and evolution were revised in areas that reflect our improved knowledge, such as fertilization and activation, and genetic drift. Human evolution, formerly a separate chapter, has been greatly condensed and included with Chapter 39 on organic evolution. This was done with some reluctance, but our reviewers convinced us that few courses in general zoology have the time to consider this dynamic field as a separate topic.

K. Sandved.

Inheritance and the burgeoning field of molecular genetics now require two chapters (37 and 38). The treatment of Mendelian genetics has been condensed by omitting subjects that we felt were unnecessary for understanding basic principles of inheritance. Molecular genetics, on the other hand, has been expanded with the inclusion of recent information and the addition of a new section on the molecular genetics of cancer.

The animal environment is the theme of Part Five. We condensed the treatment of the biomes and the aquatic environment (Chapter 40) and reworked the discussion of population growth (Chapter 41). We also separated the description of the logistic equation from the text by placing it in a box; this frees the instructor to include or omit the mathematics of the sigmoid growth curve, as determined by one's approach to the subject. The book ends, as before, with an updated discussion of the causes of animal extinction.

Special features

In-text learning aids

Key words are boldfaced and the derivations of generic names of animals are given where they first appear in the text. Each chapter ends with a concise summary, a list of review questions (new with this edition), and annotated selected references. The text has been augmented with marginal notes that provide interesting sidelights without interrupting the narrative.

J.L. Rotman.

Art program

Many new original drawings have been prepared by William C. Ober, whose knowledge of biology and medicine, in addition to his artistic skills, continues to enrich the excellent illustration program of this text. We have worked conscientiously with Dr. Ober in the research and design of the new illustrations for maximum accuracy, visual appeal, and value as learning aids.

Glossary and appendix

An extensive glossary of more than 1000 words provides pronunciation, derivation, and definition of each term. The historical appendix, unique to this textbook, lists key discoveries in zoology, and separately describes books and publications that have greatly influenced the development of zoology. Many readers have found this appendix an invaluable reference to be consulted long after completing their formal training in zoology.

Instructor's manual

The instructor's manual has been broadly revised and expanded for this edition. In addition to test questions, we have prepared a chapter outline, a commentary and lesson plan, and a listing of source materials for each chapter of the text. We trust this will be of particular value for first-time users of this text, but we think experienced teachers will also find much of value, particularly in the commentary and source material sections.

Laboratory manual

The laboratory manual by Hickman and Hickman, *Laboratory Studies in Integrated Zoology*, now in its seventh edition, has been extensively rewritten and reillustrated. Although it was written for a year-long course in zoology, it can conveniently be adapted for semester or term courses by judicious selection of exercises.

Acknowledgments

We are indebted especially to the following zoologists whose reviews of the entire text guided our approach to this revision: George B. Bourne, University of Calgary; David B. Campbell, University of New Hampshire; Walter J. Harman, Louisiana State University; John B. Hess, Central Missouri State University; Janann V. Jenner, New York University; David W. Phillips, University of California, Davis; Edwin C. Powell, Iowa State University; and Archie M. Waterbury, California Polytechnic and State University, San Luis Obispo. We are certain that the book has gained immeasurably from their insights. Although we have taken pains to follow their advice wherever possible, we know that they all understand that the instructional approaches of eight experienced zoologists often differ radically. In the end, we have to make the critical judgments and—how could it be otherwise—accept the responsibility for not following all the many fine suggestions offered.

We also thank Professor Alex L.A. Middleton, University of Guelph, for taking the time (as busy scientists seldom do), and without personal reward, to write a lengthy letter with numerous helpful suggestions. We are also grateful to Robin Leech, Edmonton, Alberta who, in the same spirit of unsolicited generosity, suggested many ways to improve the historical appendix. John Knox of Washington and Lee University helped to sharpen our views on certain aspects of organic evolution. And although we cannot mention them all by name, the perceptive comments from other colleagues at Washington and Lee and Texas Tech often provided inspiration and insights on occasions that are probably forgotten by them but remembered by us.

Finally, we are indebted to the able and conscientious staff of Times Mirror/Mosby who shepherded this manuscript through to the book now in the reader's hands. We especially thank our editor, David Brake, editorial assistant Mary Huggins, project editor Teri Merchant, production editor Mary Stueck, and design director Kay Kramer.

CLEVELAND P. HICKMAN, Jr.
LARRY S. ROBERTS
FRANCES M. HICKMAN

CONTENTS

DETAILED CONTENTS

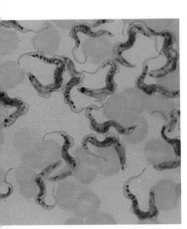

Manfred Kage/Peter Arnold, Inc.

PART TWO   THE DIVERSITY OF ANIMAL LIFE

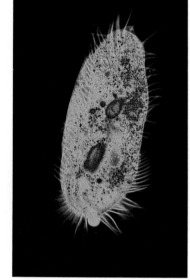

Manfred Kage/Peter Arnold, Inc.

C.P. Hickman, Jr.

C.P. Hickman, Jr.

C.P. Hickman, Jr.

C.P. Hickman, Jr.

C.P. Hickman, Jr.

C.P. Hickman, Jr.

PART FOUR   CONTINUITY AND EVOLUTION OF ANIMAL LIFE

35   THE REPRODUCTIVE PROCESS, 754

36   PRINCIPLES OF DEVELOPMENT, 775

C.P. Hickman, Jr.

C.P. Hickman, Jr.

C.P. Hickman, Jr.

PART ONE

INTRODUCTION TO THE LIVING ANIMAL

C H A P T E R   1

L I F E   General Considerations and Biological Principles

Zoology (Gr. *zōon*, animal, + *logos*, discourse on, study of) is the scientific study of animals. It is commonly considered a subdivision of an even broader science, biology (Gr. *bios*, life, + *logos*, discourse on, study of), the study of all life. The panorama of animal life, how animals function, live, reproduce, and interact with their environment, is exciting, fascinating, and awe inspiring. A complete understanding of all phenomena included in zoology is beyond the ability of any single person, perhaps of all humanity, but the satisfaction of knowing as much as possible is worth the effort. In the chapters to follow, we hope to give the student an introduction to this science and to share our excitement in the pursuit of it.

In this chapter we briefly discuss zoology as a science, the concept of organic evolution, and a number of principles that will be developed in the rest of the book. These principles will provide conceptual unity to the knowledge that has accumulated about animals.

Figure 1-1

Although both the living and the nonliving are subject to the same physical and chemical laws, living things are distinguished by a complex of unique structural and functional patterns that are unknown in the nonliving world. **A,** Bobwhite quail *Colinus virginianus.* **B,** Hermit crab *Coenbita compressus.* Photographs by C.P. Hickman, Jr.

ZOOLOGY: THE SCIENTIFIC STUDY OF ANIMALS
Biology as the Study of Life

We begin our discussion with two difficult questions: what is life, and what is an animal? At first these questions seem easy. Anyone, after all, can tell the difference between a rock and a robin, or between a toad and a tree. However, absolute definitions are not nearly as easy.

The question "What is life?" is one that many biologists have considered almost unanswerable. We must define life in terms of attributes that we associate with it. Several biological manifestations clearly distinguish the living from the nonliving. The essential differences between the two appear to be organization, metabolism, development, reproduction, interaction with the environment, and genetic control. As we have learned more about the structural and organizational properties of living matter, it has become increasingly apparent that the living and the nonliving share the same kind of chemical elements and that both are subject to the law of conservation of energy. Few biologists today hold to the vitalistic view, prevalent not so many years ago, that living things are endowed with an inexplicable vital force. The accumulating weight of scientific evidence indicates that life is governed by physical and chemical laws, even though we still may be far from understanding the complex fabric of life as a whole.

1. Organization. Nonliving matter is organized at least into atoms and molecules and often has a higher degree of organization as well. However, atoms and molecules are combined into patterns in living organisms that have no counterparts in the nonliving. Such combinations of common elements, which are highly improbable on thermodynamic grounds and are themselves nonliving, provide dynamic systems of coordinated chemical and physical activities when they are organized as they are in cells. Many of the molecules are organized into the various structures of the cell, where thousands of chemical reactions take place.

2. Metabolism. Metabolism is the collective name given the essential chemical processes that go on in living cells and organisms, including digestion, production of energy (respiration), and synthesis of molecules and structures. It is the sum of the constructive (anabolic) and destructive (catabolic) reactions.

3. Development. Even the simplest, one-celled organisms grow in size until they divide into two or more cells. Multicellular organisms undergo a process of change through their lives. They all begin as a single cell whose offspring repeatedly divide and progressively differentiate into the organized structures of the adult.

4. Reproduction. Some individual organisms may be unable to reproduce or replicate themselves, although they are undeniably alive. Nevertheless, the ability to reproduce is present in the population of those organisms and of course is necessary for the continued existence of the population.

5. Interaction with the environment. All organisms can respond to stimuli in their environment, a property called **irritability.** The stimulus and response may be simple—such as a bacterium moving away from or toward a light source or away from a noxious substance, or quite complex—such as a bird responding to a complicated series of signals in a courtship ritual. Life and the environment are inseparable. The evolutionary history of the organism has placed it in a specific environment that has determined the structural, functional, and behavioral properties of the organism. Indeed, all living organisms are products of evolution; that is, they have changed over the generations.

6. Genetic control. All living organisms have an information package in their genes that directs their organization, development, metabolism, reproduction, and environmental adaptations. The genes are inherited from one generation to the next, and changes in the genes and their combinations make evolution possible.

Figure 1-2

Mating green treefrogs *Hyla cinerea.* Reproduction, perhaps the ultimate objective of the life process, is indispensable for the survival of a species.
Photograph by C.P. Hickman, Jr.

• • •

Almost every criterion of life has its counterpart in the nonliving world: just as some living organisms may be unable to reproduce, some nonliving systems are even capable of a limited amount of metabolism or reproduction. However, only living things have combined these properties into unique structural and functional patterns.

Zoology as Biology

The question "What is an animal?" may be answered somewhat more definitely than "What is life?", but difficulty arises at the dividing line between animals and other living things.

Animals are organisms that comprise cells in which most of the material for inheritance is collected in membrane-enclosed structures (nuclei), which cannot

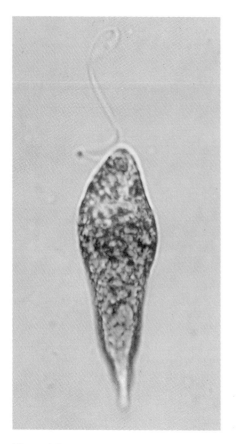

Figure 1-3

Some organisms, such as this *Euglena,* occupy the dividing line between animals and other living organisms.
Courtesy Claude Taylor, III.

use light energy to produce organic compounds, and in which the cells are not enclosed by cell walls (rigid, nonliving structures surrounding the cell membrane). This definition separates animals from bacteria, which do not have their genetic material enclosed in nuclei and have cell walls, and from fungi, which have cell walls. It also separates animals from plants, which use the energy of light (usually from the sun) to incorporate carbon dioxide from the atmosphere into organic compounds; these become building blocks for the cell and provide energy for other reactions. Animals are unable to do this and must depend on plants to capture the energy of the sun and make it available as organic compounds to the animals. The attributes on which these distinctions are based seem clear enough—certainly as applied to toads and trees—but cases exist in which the difference between plants and animals is not at all clear. For example, *Euglena* (Figure 1-3 and p. 145) exists in a transition zone, functioning as an animal in the dark and a plant in the light; some varieties have completely lost the ability to use light energy. Are *Euglena* plants or animals? Some authors prefer the five-kingdom arrangement (p. 126), with organisms such as *Euglena* placed with many other single-celled organisms in a separate kingdom, the Protista. Even so, we often find ourselves referring to the "plantlike" protistans and the "animal-like" protistans.

Principles of Science

We emphasized in the first sentence of this chapter that zoology is the scientific study of animals. Certain historical aspects of the development of modern science, particularly as it relates to zoology, are discussed in other parts of this book, but a basic understanding of zoology requires an understanding of what science is, what it is not, and how knowledge is gained by the scientific method.

Nature of science

What we call science, in modern sense, has arisen recently in human history, really within the last 200 years or so. Even in that span of time, the pace of discovery and accumulation of knowledge concerning our world and universe have been expanding as the number of scientists has grown and as the sophistication of our instruments and the power of our analyses have increased. Thus, scientific discoveries have changed human lives enormously from what they were 100, or even 20, years ago. Even the most primitive societies in the most remote locations have hardly remained untouched. Nevertheless, many, perhaps most, people in such highly developed countries as the United States and Canada have little or no notion of the real nature of science. Eloquent testimony to the truth of this is that on March 19, 1981, the governor of Arkansas signed into law the Balanced Treatment for Creation-Science and Evolution-Science Act (Act 590 of 1981). The enactment of this law led to a historic lawsuit tried in December 1981 in the court of Judge William R. Overton, U.S. District Court, Eastern District of Arkansas. The suit was brought by the American Civil Liberties Union on behalf of 23 plaintiffs, including a number of religious leaders and groups representing several denominations, individual parents, and educational associations. The plaintiffs contended that the law was a violation of the First Amendment to the U.S. Constitution, which prohibits "establishment of religion" by the government. This prohibition includes passing a law that would aid one religion or prefer one religion over another. On January 5, 1982, Judge Overton permanently enjoined the State of Arkansas from enforcing Act 590.

Considerable testimony during the trial dealt with the nature of science.

Some witnesses defined science simply, if not very informatively, as "what is accepted by the scientific community" and "what scientists do." However, on the basis of other testimony by scientists, Judge Overton was able to state explicitly these essential characteristics of science:

1. It is guided by natural law.
2. It has to be explanatory by reference to natural law.
3. It is testable against the empirical world.
4. Its conclusions are tentative, that is, are not necessarily the final word.
5. It is falsifiable.

The pursuit of scientific knowledge must be guided by the physical and chemical laws that govern the state of existence and interactions of atoms, sub-atomic particles, molecules, and so on. Scientific knowledge must explain what is observed by reference to natural law without requiring the intervention of any supernatural being or force. One may believe, as many scientists do, that the universe was brought into existence by the action of a supernatural being, but such a belief is neither within the realm of science nor contradictory to the tenets of science. We must be able to observe events in the real world, directly or indirectly, for them to have scientific value, and testing of hypotheses and theories must be accessible to our senses or to instruments that can measure the events. If we draw a conclusion relative to some event, we must always be ready to discard or modify our conclusion if it is inconsistent with further observations. As Judge Overton stated, "While anybody is free to approach a scientific inquiry in any fashion they choose, they cannot properly describe the methodology used as scientific, if they start with a conclusion and refuse to change it regardless of the evidence developed during the course of the investigation."

Scientific method

The manner in which a scientist seeks to gain new knowledge or explain natural phenomena is known as the scientific method. It has sometimes been described as ordinary common sense raised to a higher level and applied systematically. The first step is **observation.** The scientist observes a series of events, often indirectly by means of instruments. Frequently, the events are "observed" by reading descriptions of the observations made by other scientists as recorded in the scientific literature. By use of instruments or literature, the scientist may observe otherwise inaccessible events, such as those too small to be seen or those that may have taken place many years previously. After considering the observations, the scientist seeks to generalize about them, that is, to make a statement of explanation about the observations, such as their cause, mechanism, and relationship to each other. This statement becomes the **hypothesis.** To have any scientific value, the hypothesis must then be tested. On the basis of the hypothesis, the scientist must make a **prediction** about future observations. The scientist must say, "If my hypothesis is a valid explanation of past observations, then future observations ought to have certain characteristics." If the observations do have such characteristics, they constitute evidence in favor of the hypothesis and the hypothesis gains strength as an explanation of the events. In experimental science the "future observations" are in the form of experiments. An **experiment** is a manipulative process by which a prediction made on the basis of the hypothesis can be tested. A certain condition or manipulation is applied to an entity (such as a plant, an animal, a container of a substance, or a body of water), and the results are observed. If the results are as predicted, the hypothesis is supported (not proved), but if the results are otherwise, the hypothesis is invalidated. The results, usually called **data** (sing., **datum**), are evidence for or against the hypothesis. The condition or manipulation that has been applied and is being tested is referred to as the **experimental variable.** To have confidence that the experimental variable is responsible for the effect observed, another entity just like the first one must be subjected to all the conditions that prevailed during the experiment *except* the experimental variable. This part of the experiment is called the **control.** The difference observed between the experimental subject and the control is thus the effect of the experimental variable. In practice an experiment is almost never performed on a single entity, such as an individual plant or animal, but on a group of entities; therefore there are an experimental group and a control group. Furthermore, the experiment is usually repeated a number of times. The more data obtained, the more confidence can be placed in the conclusion about the hypothesis.

Not all scientific knowledge is gained by means of experiments. Much science is descriptive, but the scientific method is applied nevertheless and the steps are similar. Observations are made, a hypothesis is formed, predictions are made on the basis of the hypothesis as to what the results of further observations should be, and data are collected that may support or disprove the hypothesis. For example, suppose we examine several individual organisms and compare them with published descriptions of such organisms (initial observations). Based on these observations, we form a hypothesis that the organisms represent a new species, that is, a species that has never before been described in the literature. We collect and examine more of the same kinds of individuals from other locations, and we find that they are distinct from other known species. These data support our hypothesis. Alternatively, we may find that there are many intergrading individuals and that our specimens cannot be reliably distinguished from a species already known. In the latter case, we must discard the hypothesis.

If a hypothesis is supported by a great deal of data, and particularly if it is very powerful—explains a wide variety of related phenomena—the hypothesis may attain the status of a **theory.** We emphasize that the meaning of the word "theory," when used by scientists, is not "speculation" as it is in ordinary English usage. The failure to make this distinction has been prominent in the creationism versus evolution controversy. The creationists have spoken of evolution as "only a theory," as if it were little better than a guess. In fact, the theory of evolution is supported by such massive evidence that most biologists view repudiation of evolution as tantamount to repudiation of reality. Nonetheless, evolution, along with other theories in science, has not been *proved* in a mathematical sense but is testable, tentative, and falsifiable. It has been tested for more than 120 years, and to date there is no scientific evidence that it is false; indeed, organic evolution is accepted as the cornerstone of biology. On the other hand, although much has been learned about the *mechanisms* of evolution, they continue to be explored and clarified.

SOME IMPORTANT BIOLOGICAL PRINCIPLES AND CONCEPTS

A principle or generalization is a statement of fact that has a wide application and can be used to formulate other principles and concepts. Like the other sciences, biology has a number of fundamental principles and assumptions, although they may not lend themselves to the mathematical exactness of the physical sciences. Well-formulated principles are indispensable to help us organize our thinking, see important relationships, and form basic conclusions from the awesome mass of factual material with which we are confronted. These principles are products of the scientific method and have been tested by many workers over long periods of time. Inevitably, some will be subject to revision and new interpretation in the light of new knowledge.

Inevitably also, any such list compiled by other zoologists might have a different organization or emphasis, or even the addition or deletion of certain points, depending on the background and training of those individuals. Nonetheless, there would be significant agreement on the principles that demonstrate the essential unity of the organism and the integration of all biological systems.

Statements of principles are necessarily condensed, and many terms used here, although defined in the glossary, will be fully explained in later chapters. *The student is not expected to memorize the principles at this point, but rather should assimilate them gradually while progressing through the book and beginning to interrelate information.* References to the principles will be made in later chapters, where they will have significant application and will be better understood. Some are so well accepted that they form a part of the conceptual scheme of virtually all biologists. Others are of more recent development and have yet to be thoroughly integrated into the science.

Application of Physical and Chemical Principles

1. Living systems and their constituents obey physical and chemical laws. All living systems are subject to the same physical and chemical laws as are nonliving systems. Within the cells of any organism, the living substance comprises a multitude of nonliving constituents: proteins, nucleic acids, fats, carbohydrates, waste metabolites, crystalline aggregates, pigments, and many others, all of which are composed of molecules and their constituent atoms. Living substance (protoplasm) is alive because of the highly complex organization of these nonliving substances and the way they interact with one another, just as a watch is a timepiece only when all of its gears, springs, and bearings are organized in a particular way and interact with one another. Neither the gears of a watch nor the molecules in protoplasm can interact in any way that is contrary to universal physical laws. Consequently, the more completely we can understand the functioning of protoplasm and its constituents on the basis of chemical principles, the more completely we can understand the phenomena of life (Chapter 2).

2. All organisms capture, store, and transmit energy. Of the many physicochemical principles pertinent to the maintenance of life, the laws governing energy and its transformations (thermodynamics) are central. The first law of thermodynamics is the **law of conservation of energy.** This says that *energy can be neither created nor destroyed, although it may be transformed from one form to another.* All aspects of life require energy in some form, and the energy to support life on earth flows from the fusion reactions in our sun and reaches the earth in the form of light and heat. Sunlight is captured by green plants and transformed by the process of photosynthesis into chemical bond energy. Chemical bond energy is a form of potential energy that can be released when the bond is broken; the energy is then used to perform electrical, mechanical, and osmotic tasks in the cell. Energy made and stored in plants is used by animals that eat the plants, and these animals may in turn provide energy for other animals that eat them.

Ultimately, all the energy transformed and stored by the plants will again be transformed, little by little, and dissipated as heat. This is in accord with the second law of thermodynamics, which otherwise might seem to be violated by living systems. This law states that *there is a tendency in nature to proceed toward*

Figure 1-4

Fusion reactions in the sun support all life on earth.
Photograph by C.P. Hickman, Jr.

a state of greater molecular disorder or **entropy.** Thus, the high degree of molecular organization in living cells is attained and maintained *only* as long as energy fuels the organization. The ultimate fate of materials in the cells is degradation and dissipation of the chemical bond energy as heat, thereby approaching greater molecular disorder, or randomness (Chapters 5 and 41).

3. There is a biochemical and molecular unity of living systems. Given that living systems are subject to universal physicochemical laws and that a particular molecule or element has the same characteristics, whether it is in a bacterial cell or an elephant cell, it should not be surprising that living systems possess a high degree of molecular and biochemical unity. All cells require chemical constituents to serve the biochemical functions of food utilization and energy transformation and the hereditary functions of genetic replication. Despite the occurrence of more than 170 amino acids in nature, only 20 are used almost ubiquitously as building blocks of active proteins in living organisms. Deoxyribonucleic acid, the hereditary material of life, is constructed from only four particular nitrogenous bases, in addition to certain sugar phosphate units. The macromolecules used in energy transformations often show remarkable consistency in structure. The same sequences of metabolic reactions to release the stored energy in chemical bonds occur over an enormous variety of unrelated organisms, and the reactions often differ only in detail (Chapters 2 and 5).

___ Concept of Evolution

4. All modern organisms have arisen from preexisting organisms by organic evolution. This doctrine holds that all existing forms of life have descended over an enormous span of time from a single, simple, cellular protoplasmic mass that arose spontaneously, probably in the sea. Life may have arisen more than once. Organic evolution explains the diversity of modern organisms, their characteristics, and their distribution as the historical outcomes of change from previously existing forms (Chapter 39).

5. Animals evolve by the mechanism of natural selection. Charles Robert Darwin (Figure 1-5) and Alfred Russel Wallace simultaneously presented the first credible explanation of the mechanism of evolution, the **principle of natural selection.** The principle is founded on the observed facts that no two organisms are exactly alike; that some variations, at least, are inheritable; that all groups tend to overproduce their kind; and that because more individuals are reproduced than can survive, there is a struggle for existence among them. Organisms possessing variations that make them better adapted to the environment will survive and produce more offspring then those with less favorable variations; thus, their inheritable characteristics will appear in greater proportion in the next generation. Continuation of this natural selection will produce continued evolutionary change.

Evolution by means of natural selection allows us to ask two fundamental questions about any biological phenomenon or characteristic: "What is its function?" and "How did it evolve?"

The idea of function is unique to biology among the sciences. By function we do not mean "purpose" in the sense of consciously directed objective, but the way in which a particular characteristic helps an organism or a population to survive and reproduce. For example, the function of legs or limbs in particular animals may be locomotion, food gathering, food handling, copulation, respiration, or even a combination of these. The legs help the animal to survive and reproduce; hence they have **adaptive value,** that is, they are an **adaptation.** By the same token, the function of a particular enzyme may be to help derive the energy from a food molecule, or an entire metabolic sequence of reactions may have such a function,

Figure 1-5

Modern evolutionary theory is strongly identified with Charles Robert Darwin who, with Alfred Russel Wallace, provided the first credible explanation of evolution—natural selection. This photograph of Darwin was taken in 1854 when he was 45 years old.
Courtesy American Museum of Natural History.

and they can thus be viewed as adaptations. A living organism is a bundle of adaptations, some of which may be widely possessed by a great variety of organisms, and some of which are narrow, or specialized, adaptations, helping the organism to survive in its particular habitat (Chapter 39).

6. **All organisms are adapted to their habitat.** Environmental conditions vary from one habitat to another, and these conditions constitute **selective factors** or **pressures,** which favor the survival of organisms with certain characteristics over others; that is, they favor the survival of those organisms that are better adapted to a specific environment. Thus, over the generations, organisms with adaptations suitable for a particular habitat (special adaptations) evolve because of persistent selection by environmental pressures (Chapters 7 to 27 and 39).

7. **Animals that have many morphological characteristics in common share a common descent.** Since existing animals have evolved from preexisting ones, it is highly probable that those having many morphological characteristics in common share a common ancestor. The more characters organisms have in common, the more closely they are related, and the closer they are to their common ancestor. The phylogenetic scheme forms the basis for modern classification of animals. A few common characters shared by two groups may have limited significance because of the possibility of **convergent evolution;** similar selective pressures may have selected for the evolution of similar adaptations in unrelated groups. However, when common characters are homologous—similar in origin—the evidence for relationship is strong (Chapters 7 to 27).

8. **The embryos of animals tend to resemble the embryos of their ancestors.** Studies of the embryogenesis of animals in the context of evolution resulted in the observation that embryos often go through stages resembling their ancestors, which led to the formulation of the **biogenetic law.** As originally conceived, it was believed that the embryonic stages of an animal were similar to the adult stages of its phylogenetic ancestors: the **principle of recapitulation,** or the idea that ontogeny (the life history) repeats phylogeny (the ancestral history). This viewpoint assumed that all evolutionary advancements were added to the terminal stages of the life histories of organisms. Although subsequent observation has shown that this interpretation is incorrect, there is nevertheless a tendency for early developmental patterns to become more or less stabilized in successive ontogenies of later descendants (**paleogenesis**). Thus, a modern restatement of the biogenetic law would be that *embryos of animals tend to go through stages resembling the embryos of their ancestors.* This tendency is demonstrated mostly in the least specialized members of animal groups. It is not well demonstrated and may not be evident at all in specialized members for two reasons: embryos are just as subject to natural selection as are adults and so may show embryonic adaptations, and adult adaptations often begin their development in the embryonic stages. Therefore any similarity to ancestral embryos may be completely obscured. Paleogenesis is important, nevertheless, because study of the less highly specialized or modified members of a group may provide evidence of phylogenetic relationships with other groups (Chapters 7, 13 to 22, and 36).

9. **Evolution is irreversible.** An evolutionary generalization of wide applicability is sometimes known as **Dollo's law:** that with the possible exception of short intervals and in a very restricted sense, evolution is a one-way process, or is irreversible. An evolutionary path once taken cannot be retraced. A consequence of this fact is that the possible range of adaptations possessed by descendant groups is predetermined and limited by the array of adaptations possessed by their ancestors. Accordingly, each major group of animals usually has a basic adaptive pattern that may have determined its evolutionary divergence. For example, in the phylum Mollusca—a very large group of animals with over 100,000 living spe-

Figure 1-6

A living organism is a bundle of adaptations. An eastern chipmunk fills its pouches with food to be stored for winter use. Photograph by L.L. Rue, III.

Figure 1-7

Whelk *Busycon.* A living mollusc is an elaboration of an ancestral pattern. Photograph by C.P. Hickman, Jr.

cies—there is a certain array of characteristics that, if figuratively extrapolated backward, suggests a hypothetical ancestral mollusc. Most investigators of molluscs agree that the hypothetical ancestral mollusc must be very much like the actual ancestral mollusc and that all living molluscs, from the tiny snail to the giant squid, are but elaborations on the ancestral pattern. Thus, the adaptive radiation of present-day molluscs, although enormous, is strictly limited by the characteristics of their ancient ancestor (Chapters 8 to 27, 38).

___ Biogenesis

10. Life comes from life. Certainly a basic principle of the widest possible applicability is the concept that all organisms come from preexisting organisms, or the **principle of biogenesis.** This is in contrast to the notion that life can spring from inanimate matter, that is, abiogenesis or spontaneous generation—a concept that is now discredited. That living organisms could be spontaneously generated was widely accepted until the experiments of several workers, including Louis Pasteur, were carried out.

Modern biologists have recognized, however, that while the principle of biogenesis applies now, it was not always so. Life must have originated on earth from nonliving matter at least once, perhaps several times. But of course, conditions were far different in the primeval seas than they are now (Chapter 3).

11. Reproduction is a property of living systems. A corollary of the principle of biogenesis is the ability of living organisms to reproduce. Reproduction is a unique and ubiquitous property of living systems. This principle becomes self-evident when one considers the alternatives (Chapter 35).

___ Cell Theory

12. Cells are the fundamental and functional units of life. The cell theory, or **cell doctrine** as it is often called, is another of the great unifying concepts of biology. All animals and plants are composed of cells and cell products. New cells come from division of preexisting cells, and the activity of a multicellular organism as a whole is the sum of the activities of its constituent cells and their interactions.

Despite great variation in size (although most cells are between 0.5 and 40 μm in diameter), shape, and function, all cells are basically similar in structure. It is a remarkable fact that living forms, from amebas and unicellular algae to whales and giant redwood trees, are formed from this single type of building unit (Chapter 4).

13. Cells contain differentiated and functionally interdependent structures. The cell is surrounded by a membrane, and in plants also by a cell wall secreted by the cell. Within the cell are various structures referred to as **organelles.** In plant and animal cells a number of organelles are bounded by membranes, including a **nucleus,** a relatively large structure containing the hereditary material. The material and structures outside the nucleus are collectively called the **cytoplasm.** The cells of bacteria and blue-green algae lack membrane-bound organelles, and their nuclear area is not surrounded by a membrane (Chapter 4).

___ Gene Theory

14. All organisms inherit a structural and functional organization from their progenitors: the principle of hereditary transmission. This generalization involves the laws of heredity and applies to all living things. All organisms have the capacity for reproducing their own kind and transmitting their characteristics to their offspring. What is inherited by an offspring is not necessarily an exact copy of the parent, but rather a certain type of organization that, under the influence of developmental and environmental forces, gives rise to a certain expressed result. Many potentialities may be inherited, but those that are actually expressed are much more limited (Chapters 35 and 37).

15. The gene is the fundamental unit of inheritance, and inheritance obeys Mendel's laws. The gene is a bit of coded information that confers the potential for expression of a certain characteristic in the cell or organism. The genes are borne on structures called **chromosomes;** each chromosome bears many genes. With certain notable exceptions, the nucleus of each animal cell has two sets of chromosomes (**diploid condition**), with pairs of chromosomes bearing the genes for the same set of characteristics. The two genes for a particular characteristic, one on each member of a pair of homologous chromosomes, are described as allelic genes, or **alleles.** Only one of the alleles may result in the production of an expressed characteristic in the organism (the dominant allele), although both are present in each cell and either may be passed on to the progeny.

During the production of the sex cells (**gametes: sperm** and **ova**), the number of chromosomes is halved (**haploid condition**), with each gamete receiving one of each of the homologous pairs of chromosomes. Thus, each gamete has only one of each kind of gene. This is the **law of segregation,** first formulated by Gregor Mendel in 1866.

Mendel's second law, the **law of independent assortment,** implies that the likelihood of an individual progeny inheriting a particular allele is the result of chance alone. Although each gamete has a full haploid set of chromosomes, which chromosome of each of the homologous pairs is actually passed to a particular gamete is completely random; that is, it is independent of the other homologues passed to that gamete. Of course, all of the genes on the same chromosome tend to be inherited together and do not assort independently (Chapter 37).

16. Gene frequencies in a population can remain stable under certain conditions. When the female gamete (ovum) is fertilized by the male (sperm), a fusion of their nuclei restores the number of chromosomes to the diploid condition. Thus, the progeny receive one set of chromosomes from each parent and, consequently, have two genes for each trait. However, in the population as a whole, there may be several other alleles for the same trait, and the frequency of occurrence of a given gene can be expressed as a fraction of one (one equals the whole population, or the frequencies of all the alleles added together). It might be expected that if the frequency of a gene in a population were very low, the allele would eventually disappear. However, it can be shown that, provided certain conditions are met, the gene frequencies of all alleles in a population remain constant. This is the **Hardy-Weinberg law,** which operates under certain specific conditions (Chapter 39).

17. Gene frequencies in a population can change. The presence of numerous alleles, or combinations of alleles, in a population provides variability in the characteristics possessed by the individuals in that population. Clearly, if some of these characteristics are helpful to survival and reproduction of those individuals, they will provide a disproportionate share of their genes to the next generation. If selective pressures are strong, gene frequencies in a population may change markedly over time. Thus, *natural selection influences gene frequencies.*

Furthermore, the genes themselves may change; that is, the coded information in a gene may undergo a molecular alteration so that the trait produced by that gene is different. This is a **mutation.** Mutations are the ultimate source of biological variation. Mutations may produce genetic effects that are helpful in terms of natural selection, or they may be detrimental.

Genetic drift is a change in gene frequency by chance alone. This can occur

Figure 1-8

Walrus bulls. Gene frequencies within a population remain stable unless disturbed by natural selection, mutation, genetic drift, or migration.
Photograph by L.L. Rue, III.

Figure 1-9

James Watson and Francis Crick with one of their models of DNA.

From James D. Watson, *The Double Helix,* Atheneum, New York, 1968; copyright 1968 by James D. Watson; reprinted by permission of the publisher.

when the population is small, and "sampling errors" can accumulate, thus changing gene frequencies in the population over time (Chapter 39).

18. DNA is the substance that contains the inherited information. We have referred to the gene as a "bit of coded information." The discovery of the nature of the code and how the code is translated into the expression of a characteristic is among the great triumphs of modern biology. The genetic material on the chromosomes is **deoxyribonucleic acid (DNA).** This is a high–molecular weight substance composed of many units of a sugar phosphate (deoxyribose phosphate) and many units of four nitrogenous bases (adenine, guanine, thymine, and cytosine, abbreviated A, G, T, and C, respectively). In 1953 James Watson and Francis Crick suggested a **double helix model** for DNA, for which abundant evidence has subsequently accumulated. According to the Watson-Crick model, two very long strands of sugar phosphate units are connected together, and the strands twist around each other in a double helix form. Projecting toward each other between the two strands are the nitrogenous bases. When a precisely scaled molecular model was constructed, it was found that adenine would fit opposite thymine and cytosine would fit opposite guanine, but other combinations would not fit in the available space. This suggested a hypothesis for DNA replication: the strands could unwind, and each could serve as a template for the synthesis of the complementary strand, resulting in two double helices exactly the same as the preexisting one (Chapter 38).

19. The genetic code is in the linear order or sequence of bases on the DNA strand. The Watson-Crick model implied that the genetic code would lie in the linear order of the nitrogenous bases: A, G, T, and C. As mentioned previously, proteins are constructed of 20 different amino acid building blocks. Assuming that the order of bases in DNA represented a code for the order of amino acids in a protein, it was clear that some combination of bases must code for one protein, since there are only four bases but 20 amino acids. The minimum number of bases to provide a sufficient number of different combinations to specify each of the amino acids is three, as suggested by Francis Crick in 1961. Subsequent research has substantiated Crick's hypothesis. The genetic code is based on triplets of nitrogenous bases along the DNA molecule. Each triplet, called a **codon,** codes for a single amino acid, and the codons do not overlap. The genetic code is apparently universal; that is, the same codon or sequence of three nitrogenous bases codes for the same amino acid in all organisms (Chapter 38).

20. Transcription and translation of genetic information are mediated by RNA. Although the transmission of genetic information and the nature of the genetic code now seem to us at once ingenious and astonishingly simple, they fall far short of explaining how the cell reads the code and translates it into cell products and structures. Nevertheless, much progress has been made in elucidating these critical processes. They depend on molecules of **ribonucleic acid (RNA),** a substance somewhat similar to DNA except that ribose is present in its strand instead of deoxyribose, and uracil is present instead of thymine. There are three principal types of RNA: **messenger RNA (mRNA), transfer RNA (tRNA),** and **ribosomal RNA (rRNA).** Molecules of mRNA are synthesized alongside strands of DNA in the chromosome, using the DNA as a template and thus transcribing the code from the DNA to the mRNA. The mRNA moves out of the nucleus into the cytoplasm where it interacts with submicroscopic particles called **ribosomes.** Ribosomes are constructed of about 50% protein and 50% rRNA. The molecules of tRNA provide the translation function. The various types bind with specific amino acids. At a particular position in a specific tRNA molecule there is a triplet sequence of nitrogenous bases, the **anticodon,** which is complementary to the codon sequence in the mRNA. Ribosomes move along the strand of mRNA, positioning the molecules of tRNA according to the codon-anticodon specificity, and the amino acid attached to the tRNA is reattached to the chain of amino acids being assembled into a protein. The specificity or kind of protein is determined by the sequence of amino acids in it, and this depends on the sequence of codons in DNA as transcribed and translated by the RNA system (Chapter 38).

21. A gene constitutes the information needed to synthesize an enzyme: the "one gene–one enzyme" principle. A gene, then, may be redefined as a series of codons of DNA that result in the production of a particular protein. The physical and chemical properties of the protein depend on the ordinal sequence of its amino acids. The properties of many proteins allow them to function as **enzymes.** Enzymes are catalysts for the myriad chemical reactions that go on in a cell; that is, they facilitate reactions by participating in them but are not "used up" in the reactions—they are not substrates or products of the reactions. The hundreds of chemical reactions, collectively called **metabolism** or **metabolic reactions,** are each catalyzed by specific enzymes. These reactions would not occur, or would occur very slowly, if they were not mediated by enzymes. The reactions to derive energy from food molecules and synthesize cell structures and products, even the synthesis of DNA itself, require the participation of enzymes. In 1941 G.W. Beadle and E.L. Tatum suggested the one gene–one enzyme hypothesis: that a gene would constitute the genetic information necessary to synthesize one enzyme. The validity of the hypothesis, in a slightly modified form, is now widely accepted.

Since it is now known that a particular protein may be made of several chains of amino acids (polypeptides), the Beadle and Tatum hypothesis is expressed as **one gene–one polypeptide.** Thus, a mutation is a change in the order of the nitrogenous bases in the DNA, which codes for the order of amino acids in an enzyme, producing an enzyme with different properties. Mutations often involve a change in several nitrogenous bases in the gene but in some cases may result from the alteration of a single pair. When the properties of an enzyme are changed, the metabolic reaction it catalyzes is altered; therefore *metabolic reactions are under genic control* (Chapters 5 and 38).

___ Development

22. Growth is a fundamental characteristic of life. Its manifestation as the synthesis of protoplasm in every living organism has been referred to as the **growth law.** True growth is characterized by an increase in tissue or protoplasmic mass with cell division or cell enlargement or both. It thus depends on the incorporation of new materials from the environment and is subject to the availability of those materials (Chapters 6 and 36).

23. All organisms have a characteristic life cycle that they pass through as they grow. In the case of unicellular organisms, this life cycle may be as simple as growth and cell division, but the interspersion of resting stages, sexual reproduction, cyclical changes in body form, and other such events is common. In multicellular organisms, development begins with a single cell, usually a fertilized egg, or **zygote.** This initially undergoes division (**cleavage**) into smaller and smaller cells with no increase in total mass and no true growth, but the preliminary cell division subdivides the materials originally present in the zygote, which sets the stage for further development. Cleavage is followed by shaping of the embryo, called **morphogenesis,** and a progressive specialization of the cells, tissues, and structures into the organization of the whole organism, called **differentiation.** Thus begins the life cycle of the organism, which may be quite complex, including

Figure 1-10

All organisms have a characteristic life cycle, which may be simple and direct, as in most unicellular organisms, or quite complex in advanced multicellular organisms such as these broad-tailed hummingbirds.

Photograph by L.L. Rue, III.

embryonic and juvenile development, adolescence, adult equilibrium, and senescence (Chapters 5, 6, 35, and 36).

24. All organisms develop a characteristic body plan. The patterns of growth are genetically directed and are usually constant for each stage of the life cycle from generation to generation. The inherited body plan may be described in terms of broadly inclusive characteristics, such as presence and type of internal body cavities, symmetry, and type of nervous system—all of which would be possessed by organisms with a common ancestor—or the plan may be described in terms of narrow characteristics possessed only by individuals of a certain species or population (Chapter 6).

25. Nuclear equivalence and cell differentiation are characteristics of cells in multicellular organisms. Since the body plan and all organization of the body are inherited, it is clear that *all* genetic information (the **genome**) for the attainment of that body plan must be present in the zygote, which replicates its DNA and confers that information on all its progeny cells. Thus, the cell nuclei of a multicellular organism, with few exceptions, are genetically alike. This is the **principle of nuclear equivalence.** However, it is also clear that *differentiated* cells are not all alike. Therefore only some of the information is transcribed and translated in a given cell. Cell differentiation results from the differential activity of the same set of genes in different cells. In the fully differentiated cell, most of the genome is fully repressed and silent. A muscle cell, for example, never translates the parts of its DNA instructions that would produce proteins unique to a nerve cell. During embryonic development, nuclei in different regions of the embryo provide different genetic information, resulting in regional differentiation of cells and tissue. The control of differentiation remains one of the great unsolved problems of biology. Whatever the molecular mechanisms, there is abundant evidence that differentiation is one of the responses to conditions or substances in the environment of a cell (Chapters 36 and 38).

____ Responses and Relations to the Environment of a Cell or Organism

26. Irritability is a characteristic of life. One of the fundamental characteristics of life is irritability, that is, the ability to react to an environmental stimulus. This is perhaps the most widely occurring of all adaptations. Broadly speaking, it can be interpreted as the ability to react to a stimulus in such a way as to promote the continued life of a cell or organism. The reaction may be, for example, to consume a food molecule or particle when such is present, to seek or avoid light, to pursue a prey organism, or to avoid a noxious substance. When such reactions can no longer be accomplished, the organism perishes. Throughout the life of an animal, matter and energy pass through the body, providing perturbations of the internal physiological state. Many physiological and metabolic mechanisms exist that function to compensate for such disturbances and to maintain conditions compatible with continued life within the organism. This tendency toward internal stabilization was recognized by Claude Bernard in the nineteenth century, and the concept was developed in the twentieth century by Walter B. Cannon, who called it the **principle of homeostasis.** Homeostasis is achieved at the cell, organ, and system levels by material and energy transport and in many instances is controlled by feedback mechanisms (Chapters 30 and 32).

27. Cells in a multicellular animal communicate with and affect each other. Reaction to stimuli and maintenance of homeostasis require mechanisms for communication between cells and organs in multicellular animals. Increasingly complex levels of organization in animals require correspondingly increasing complexities

Figure 1-11

Wolf spider with egg case. All organisms develop a characteristic, genetically directed body plan.
Photograph by C.P. Hickman, Jr.

in communication between cells and organs. Communication between parts of an animal is by two primary means: **neural** and **hormonal.** Relatively rapid communication is by neural mechanisms and involves propagated electrochemical changes in cell membranes. Highly organized and complex nervous systems carry out this function in higher animals. Relatively less rapid or long-term adjustments in an animal are by hormonal or endocrine mechanisms. Hormones are substances produced by cells in one part of the body that regulate cell processes elsewhere in the organism when carried there by body fluids. Many instances are now known in which substances produced by some individual animals reach other individuals in the population and function as hormones in the other individuals, producing adjustments in their physiological processes or behavior. Such substances are called **pheromones** (Chapters 32 to 34).

28. All animals interact with their environment. Whether able to react to a given environmental stimulus or not, all organisms in a given area interact with both the biological and physical factors in their environment. The earth's biomass is organized into a hierarchy of interacting units: the individual organism, the population, the community, and the ecosystem. Among the important generalizations in ecology are the concepts of **food chains** (producers, primary and secondary consumers, decomposers), **habitats,** and **niches.** The habitat is the spatial location where an animal lives. It is always physically circumscribed, although the space may be very large, such as an entire ocean, or tiny, such as the intestine of an insect. What the organism does—its role—in its habitat is its niche. The niche of an animal must be described in terms of the effects the organism has on other organisms in its habitat, and vice versa, as well as the effects on and by nonliving resources in the habitat. The possible effects on other animals include an array, such as whether the organism interacts with them as a predator, a prey, a mutual, a parasite, a competitor, or others. Insofar as individuals or populations compete for food, space, or other resources in a habitat, their niches overlap. If the competition is too strong, one of the individuals or populations will perish, be driven out of the habitat, or be forced to use other resources—to occupy a different niche. It has been found that no two species can occupy the same ecological niche at the same time and place (**principle of competitive exclusion; Gause's rule**) (Chapters 39 and 41).

SUMMARY

Zoology is the scientific study of animals and is part of biology, the scientific study of life. Animals and life can be defined only in terms of attributes that are associated with them. Characteristics of living organisms include organization, metabolism, development, reproduction, interaction with the environment, and genetic control.

Science is characterized by a particular approach to the acquisition of human knowledge. It is guided by, and is explanatory with reference to, natural law, and it is testable, tentative, and falsifiable. The scientific method is the manner in which scientific knowledge is gained, and it has proved to be a powerful tool. Its essential constituents are observing events, forming a hypothesis on the basis of the observations, testing the hypothesis, and drawing a conclusion about the hypothesis on the basis of the tests. Tests may take the form of experiments. Every experiment must have two components: an experimental group and a control group. The control group is exactly like the experimental group and is treated exactly the same, except that the experimental variable is not applied to the control group. Results of the tests of the hypothesis are the data. A hypothesis for

Figure 1-12

Club-headed hydroids *Clava squamata* on seaweed. The concept of niche—the position or status or functional relationship of a species within a community—is an important ecological generalization. No two species can occupy precisely the same ecological niche at the same time and place. Photograph by D.P. Wilson.

which there is a great deal of supporting data, particularly one that explains a very large number of observations, may be elevated to the status of a theory. Evolution based primarily on the principle of natural selection is such a theory.

The principles given in this chapter illustrate the unity of biological science. All the cellular constituents, like the cells themselves, are governed by natural laws. Living organisms can come only from other living organisms, just as new cells can be produced only from preexisting cells. The nucleus of the cell carries the hereditary material, which confers a certain set of characteristics on its cellular progeny, which themselves may become new organisms on which natural selection may act. Living organisms have mechanisms that tend to compensate through internal and external changes to keep conditions within limits compatible with continued life.

Review questions

1. Name and briefly explain the characteristics of life.
2. Explain the difference between science and, for example, a religious belief.
3. Name some common, everyday event, and tell how you would investigate it by using the scientific method.
4. What is the relationship of living organisms to physical and chemical laws?
5. What is the doctrine of organic evolution?
6. What is the principle of natural selection?
7. How can we interpret the *function* of any characteristic of a living organism in the context of natural selection?
8. Give a modern restatement of the biogenetic law.
9. What is biogenesis?
10. What is the cell doctrine?
11. Genes are the fundamental units of inheritance. Answer the following with respect to genes:
    a. Name the structures that bear the genes in cells.
    b. What are two factors that can change gene frequencies?
    c. What is the name of the substance that contains the inherited information in a coded form?
    d. What is the name of the substance that mediates translation of the coded information?
12. Distinguish growth, morphogenesis, and differentiation.
13. What is the principle of nuclear equivalence?
14. Define irritability.
15. What is the principle of homeostasis?
16. What are two means of cellular communication within an organism?
17. What is the principle of competitive exclusion?

Selected references

Beveridge, W.I.B. 1957. The art of scientific investigation. New York, W.W. Norton & Co., Inc. *A treatise on how knowledge is gained in science.*

de Camp, L.S. 1969. The end of the monkey war. Sci. Am. **220**:15-21 (Feb.). *In 1968 the U.S. Supreme Court struck down the Arkansas statute that forbade the teaching of evolution in public schools and colleges. The conclusion that this marked the "end of the monkey war" was somewhat premature.*

Hailman, J.P. 1977. Optical signals. Animal communication and light. Bloomington, Indiana University Press. *The introduction has a concise statement of the scientific method.*

McCarty, M. 1985. The transforming principle. New York, W.W. Norton & Co. *An engrossing account of the discovery that genes are made of DNA and of the author's role in the discovery.*

Overton, W.R. 1982. Judgment, injunction, and memorandum opinion in the case of McLean vs. Arkansas Board of Education. Science **215**:934-943. *Judge Overton's opinion is reprinted verbatim. It is highly recommended reading.*

Stebbins, G.L., and F.J. Ayala. 1985. The evolution of Darwinism. Sci. Am. **253**:72-82 (July). *Recent developments in evolutionary theory are reviewed. It is stressed that, contrary to the claims of the creationists, these developments do not deny that Darwinian natural selection plays an important part in the process.*

CHAPTER 2

MATTER AND LIFE

__BASIC STRUCTURE OF MATTER

As we noted in Chapter 1, living systems and their constituents obey physical and chemical laws. Within the cells of any organism, the living substance is composed of a multitude of nonliving constituents: proteins, nucleic acids, fats, carbohydrates, waste metabolites, crystalline aggregates, pigments, and many others. Physical and chemical interactions of such substances account for the many processes essential to life, including digestion and absorption of nutrients, derivation of energy, removal of waste, communication of cells with each other, conduction of nerve impulses, and transmission of genetic information from one generation to the next. Because these phenomena will be discussed in later pages, we must present some basic information on chemistry and biochemistry here.

___ Elements and Atoms

All matter is composed of **elements,** which are substances that cannot be subdivided further by ordinary chemical reactions. Only 92 elements occur naturally, but the elements may be combined by chemical bonds into a vast number of different compounds. The elements are designated by one or two letters derived from their Latin or English names (Table 2-1). The elements are composed of discrete units called **atoms,** which are the smallest components into which an element can be subdivided by normal chemical means. Combination of the atoms of an element with each other or with those of other elements by chemical bonds creates **molecules.** In a chemical formula, the symbol for an element stands for one atom of the element, with additional atoms indicated by appropriately placed numbers. Thus atmospheric nitrogen is N_2 (each molecule is composed of 2 atoms of nitrogen), and water is H_2O (2 atoms of hydrogen and 1 of oxygen in each molecule), and so on.

Subatomic particles

Each atom is composed of subatomic particles, of which there are three with which we need concern ourselves: protons, neutrons, and electrons. Every atom consists of a positively charged nucleus surrounded by a negatively charged system of electrons (Figure 2-1). The nucleus, containing most of the atom's mass, is made up of protons and neutrons clustered together in a very small volume. These two particles have about the same mass, each being about 2000 times heavier than an electron. The protons bear positive charges, and the neutrons are uncharged (neutral). Although the number of protons in the nucleus is the same as the num-

Figure 2-1

Structure of carbon atom. A planetary system of 6 negatively charged electrons revolves around a dense nucleus of 6 positive protons and 6 uncharged neutrons.

Table 2-1 Some of the Most Important Elements in Living Organisms

Element	Symbol	Atomic number	Approximate atomic weight
Carbon	C	6	12
Oxygen	O	8	16
Hydrogen	H	1	1
Nitrogen	N	7	14
Phosphorus	P	15	31
Sodium	Na	11	23
Sulfur	S	16	32
Chlorine	Cl	17	35
Potassium	K	19	39
Calcium	Ca	20	40
Iron	Fe	26	56
Iodine	I	53	127

ber of electrons around the nucleus, the number of neutrons may vary. For every positively charged proton in the nucleus, there is a negatively charged electron; the total charge of the atom is thus neutral.

The **atomic number** of an element is equal to the number of protons in the nucleus, whereas the **atomic mass** is nearly equal to the number of protons plus the number of neutrons (explanation of why atomic mass is not exactly equal to protons plus neutrons can be found in any introductory chemistry text). The mass of the electrons may be neglected.

Isotopes

It is possible for two atoms of the same element to have the same number of protons in their nuclei but have a different number of neutrons. Such different forms, having the same number of protons but different atomic masses, are called **isotopes.** For example, the predominant form of hydrogen in nature has 1 proton and no neutron (^{1}H) (Figure 2-2). Another form (deuterium [^{2}H]) has 1 proton and 1 neutron. Tritium (^{3}H) has 1 proton and 2 neutrons. Some isotopes are unstable, undergoing a spontaneous disintegration with the emission of one or more of three types of particles, or rays: gamma rays (a form of electromagnetic radiation), beta rays (electrons), and alpha rays (positively charged helium nuclei stripped of their electrons). These unstable isotopes are said to be **radioactive.** Using radioisotopes, biologists are able to trace movements of elements and tagged compounds through organisms. Our present understanding of metabolic pathways in animals and plants is in large part the result of this powerful analytical tool. Among the commonly used radioisotopes are carbon 14 (^{14}C), tritium, and phosphorus 32 (^{32}P).

Electron "shells" of atoms

According to Niels Bohr's planetary model of the atom, the electrons revolve around the nucleus of an atom in circular orbits of precise energy and size. All of the orbits of any one energy and size comprise an electron shell. This simplified picture of the atom has been greatly modified by more recent experimental evidence; definite electron pathways are no longer hypothesized, and an electron

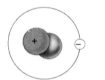

Deuterium (^{2}H)   Hydrogen (H)

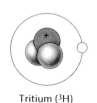

Tritium (^{3}H)

Figure 2-2

Three isotopes of hydrogen. Of the three isotopes, hydrogen 1 makes up about 99.98% of all hydrogen, and deuterium (heavy hydrogen) makes up about 0.02%. Tritium is radioactive and is found only in traces in water. Numbers indicate approximate atomic weights. Most elements are mixtures of isotopes. Some elements (for example, tin) have as many as 10 isotopes. From Raven, P.H., and G.B. Johnson. 1986. Biology. St. Louis, Times Mirror/Mosby.

shell is more vaguely understood as a thick region of space around the nucleus rather than a narrow shell of a particular radius out in space.

However, the old planetary model with the idea of electronic shells is still useful in interpreting chemical phenomena. The number of concentric shells required to contain an element's electrons varies with the element. Each shell can hold a maximum number of electrons. The first shell next to the atomic nucleus can hold a maximum of 2 electrons, and the second shell can hold 8; other shells also have a maximum number, but no atom can have more than 8 electrons in its outermost shell. Inner shells are filled first, and if there are not enough electrons to fill all the shells, the outer shell is left incomplete. Hydrogen has 1 proton in its nucleus and 1 electron in its single orbit but no neutron. Since its shell can hold 2 electrons, it has an incomplete shell. Helium has 2 electrons in its single shell, and its nucleus is made up of 2 protons and 2 neutrons. Since the 2-electron arrangement in helium's shell is the maximum number for this shell, the shell is closed and precludes all chemical activity. There is no known compound of helium. Neon is another inert (chemically inactive) gas because its outer shell contains 8 electrons, the maximum number (Figure 2-3). However, stable compounds of xenon (an inert gas) with fluorine and oxygen are formed under special conditions. Oxygen has an atomic number of 8. Its 8 electrons are arranged with 2 in the first shell and 6 in the second shell (Figure 2-3). It is active chemically, forming compounds with almost all elements except inert gases.

____ Chemical Bonds

As noted above, atoms joined to each other by chemical bonds form molecules, and atoms of each element form molecules with each other or with atoms of other elements in particular ways, depending on the number of electrons in their outer orbits.

Ionic bonds

Elements react in such a way as to gain a stable configuration of electrons in their outer shells. The number of electrons in the outer shell varies from 0 to 8. With either 0 or 8 in this shell, the element is chemically inactive. When there are fewer than 8 electrons in the outer shell, the atom will tend to lose or gain electrons to have an outer shell of 8, which will result in a charged ion. Atoms with 1 to 3 electrons in the outer shell tend to lose them to other atoms and to become positively charged ions because of the excess protons in the nucleus. Atoms with 5 to 7 electrons in the outer orbit tend to gain electrons from other atoms and to become negatively charged ions because of the greater number of electrons than protons. Positive and negative ions tend to unite.

Every atom has a tendency to complete its outer shell to increase its stability in the presence of other atoms. Let us examine how 2 atoms with incomplete outer shells, sodium and chloride, can interact to fill their outer shells. Sodium, with 11 electrons, has 2 electrons in its first shell, 8 in its second shell, and only 1 in the third shell. The third shell is highly incomplete; if this third-shell electron were lost, the second shell would be the outermost shell and would produce a stable atom. Chlorine, with 17 electrons, has 2 in the first shell, 8 in the second, and 7 in the incomplete third shell. Chlorine must gain an electron to fill the outer shell and become a stable atom. Clearly, the transfer of the third-shell sodium electron to the incomplete chlorine third shell would yield simultaneous stability to both atoms.

Sodium, now with 11 protons but only 10 electrons, becomes electropositive

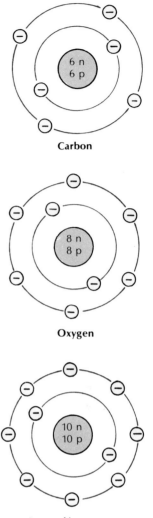

Figure 2-3

Electron shells of three common atoms. Since no atom can have more than 8 electrons in its outermost shell and 2 electrons in its innermost shell, neon is chemically inactive. However, the second shells of carbon and oxygen, with 4 and 6 electrons, respectively, are open so that these elements are electronically unstable and react chemically whenever appropriate atoms come into contact. Chemical properties of atoms are determined by their outermost electron shells.

Figure 2-4

Ionic bond. When 1 atom of sodium and 1 of chlorine react to form a molecule, a single electron in the outer shell of sodium is transferred to the outer shell of chlorine. This causes the outer or second shell (third shell is empty) of sodium to have 8 electrons and also chlorine to have 8 electrons in its outer or third shell. The compound thus formed is sodium chloride (NaCl). By losing 1 electron, sodium becomes a positive ion, and by gaining 1 electron, chlorine (chloride) becomes a negative ion. This ionic bond is the strong electrostatic force acting between positively and negatively charged ions.

From Raven, P.H., and G.B. Johnson. 1986. Biology. St. Louis, Times Mirror/Mosby.

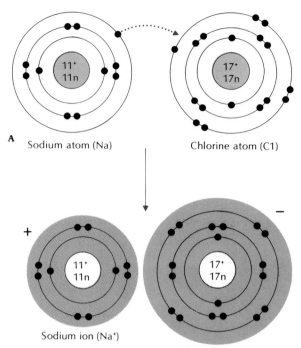

A   Sodium atom (Na)   Chlorine atom (C1)

Sodium ion (Na⁺)

Chloride ion (C1⁻)

(Na^+). In gaining an electron from sodium, chlorine contains 18 electrons but only 17 protons and thus becomes an electronegative chloride ion (Cl^-). Since unlike charges attract, a strong electrostatic force, called an **ionic bond** (Figure 2-4), is formed. The ionic compound formed, sodium chloride, can be represented in electron dot notation ("fly-speck formulas") as:

$$Na\cdot + \cdot \overset{\cdot\cdot}{\underset{\cdot\cdot}{Cl}} : \;\to\; Na^+ + (:\overset{\cdot\cdot}{\underset{\cdot\cdot}{Cl}} :)^-$$

The number of dots shows the number of electrons present in the outer shell of the atom: 7 in the case of the neutral chlorine atom and 8 for the chloride ion; 1 in the case of the neutral sodium atom and none for the sodium ion.

If an element with 2 electrons in its outer shell, such as calcium, reacts with chlorine, it must give them both up, one to each of 2 chlorine atoms, and calcium becomes doubly positive:

$$Ca: + 2 \cdot \overset{\cdot\cdot}{\underset{\cdot\cdot}{Cl}} : \;\to\; Ca^{2+} + 2(:\overset{\cdot\cdot}{\underset{\cdot\cdot}{Cl}} :)^-$$

Processes that involve the **loss of electrons** are called **oxidation** reactions; those that involve the **gain of electrons** are **reduction** reactions. Since oxidation and reduction always occur simultaneously, each of these processes is really a "half-reaction." The entire reaction is called an **oxidation-reduction** reaction, or simply a **redox** reaction. The terminology is confusing because oxidation-reduction reactions involve electron transfers, rather than (necessarily) any reaction with oxygen. However, it is easier to learn the system than to try to change accepted usage.

Covalent bonds

Stability can also be achieved when 2 atoms share electrons. Let us again consider the chlorine atom, which, as we have seen, has an incomplete 7-electron outer shell. Stability is attained by gaining an electron. One way this can be done is for 2

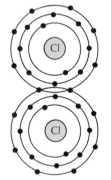

Figure 2-5

Covalent bond. Each chlorine atom has 7 electrons in its outer shell, and by sharing one pair of electrons, each atom acquires a complete outer shell of 8 electrons, thus forming a molecule of chlorine (Cl_2). Such a reaction is called a molecular reaction, and such bonds are called covalent bonds.

chlorine atoms to *share* one pair of electrons (Figure 2-5). To do this, the 2 chlorine atoms must *overlap* their third shells so that the electrons in these shells can now spread themselves over both atoms, thereby completing the filling of both shells. Many other elements can form covalent (or electron-pair) bonds. Examples are hydrogen (H_2)

$$H\cdot + H\cdot \rightarrow H\!:\!H$$

and oxygen (O_2)

$$\ddot{O}\!: + :\!\ddot{O} \rightarrow \ddot{O}\!::\!\ddot{O}$$

In this case oxygen must share two pairs of electrons to achieve stability. Each atom now has 8 electrons available to its outer shell, the stable number.

Covalent bonds are of great significance to living systems, since the major elements of living matter (carbon, oxygen, nitrogen, hydrogen) almost always share electrons in strong covalent bonds. The stability of these bonds is essential to the integrity of DNA and other macromolecules, which, if easily dissociated, would result in biological disorder.

The outer shell of carbon contains 4 electrons. This element is endowed with great potential for forming a variety of atomic configurations with itself and other molecules. It can, for example, share its electrons with hydrogen to form methane (Figure 2-6). Carbon now achieves stability with 8 electrons, and each hydrogen atom becomes stable with 2 electrons. Carbon can also bond with itself (and hydrogen) to form, for example, ethane:

$$
\begin{array}{ccc}
H & H & \\
H\!:\!\overset{\cdot\cdot}{\underset{\cdot\cdot}{C}}\!:\!\overset{\cdot\cdot}{\underset{\cdot\cdot}{C}}\!:\!H & \text{or} & H\!-\!\overset{\displaystyle H}{\underset{\displaystyle H}{C}}\!-\!\overset{\displaystyle H}{\underset{\displaystyle H}{C}}\!-\!H \\
H & H &
\end{array}
$$

Carbon also forms covalent bonds with oxygen:

$$\cdot\dot{C}\cdot + 2\,\ddot{O}\!: \rightarrow \ddot{O}\!::\!C\!::\!\ddot{O}$$

This is a "double-bond" configuration usually written as $O\!=\!C\!=\!O$. Carbon can even form triple bonds as, for example, in acetylene:

$$H\!:\!C\!:\!:\!:\!C\!:\!H \text{ or } H\!-\!C\!\equiv\!C\!-\!H$$

The significant aspect of each of these molecules is that each carbon gains a share in 4 electrons from atoms nearby, thus attaining the stability of 8 electrons. The sharing may occur between carbon and other elements or other carbon atoms, and in many instances the 8-electron stability is achieved by means of multiple bonds.

These examples only begin to illustrate the amazing versatility of carbon. It is a part of virtually all compounds comprising living substance, and without carbon, life as we know it would not exist.

Hydrogen bonds

Hydrogen bonds are described as "weak" bonds because they require little energy to break. They do not form by transfer or sharing of electrons, but result from unequal charge distribution on a molecule, so that the molecule is polar. For example, the 2 hydrogen atoms that share electrons with an oxygen atom to form water (H_2O) are not 180 degrees away from each other around the oxygen, but

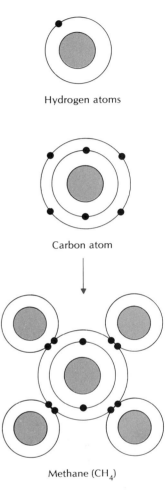

Hydrogen atoms

Carbon atom

Methane (CH_4)

Figure 2-6

In methane the 4 hydrogen atoms each share an electron with a carbon atom. They are arranged symmetrically around the carbon atom and form a pyramid-shaped tetrahedron in which each of the hydrogen atoms is equally distant from the others. From Raven, P.H., and G.B. Johnson. 1986. Biology. St. Louis, Times Mirror/Mosby.

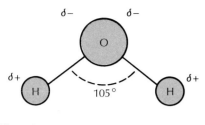

Figure 2-7

Molecular structure of water. The 2 hydrogen atoms bonded covalently to an oxygen atom are arranged at an angle of about 105 degrees. Since the electrical charge is not symmetrical, the molecule is polar with positively charged and negatively charged ends (Σ denotes partial charge).

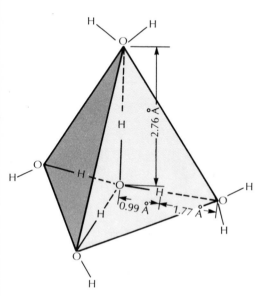

Figure 2-8

Geometry of water molecules. Each water molecule is linked by hydrogen bonds *(dashed lines)* to 4 other water molecules. If imaginary lines are used to connect the divergent oxygen atoms, a tetrahedron is obtained. In ice, the individual tetrahedrons associate to form an open lattice structure.

form an angle of about 105 degrees (Figure 2-7). Thus the side of the molecule away from the hydrogen atoms is more negative, and the hydrogen side is more positive (contrast the methane molecule [Figure 2-6], in which the equidistant placement of hydrogen atoms cancels out charge displacements). The electrostatic attraction between the electropositive part of one molecule forms a hydrogen bond with the electronegative part of an adjacent molecule. The ability of water molecules to form hydrogen bonds with each other (Figure 2-8) accounts for many unusual properties of this unique substance (p. 24). Hydrogen bonds are important in the formation and function of other biologically active substances, such as proteins and nucleic acids (pp. 29-31).

____ Acids, Bases, and Salts

The hydrogen ion (H^+) is one of the most important ions in living organisms. The hydrogen atom contains a single electron. When this electron is completely transferred to another atom (not just shared with another atom as in the covalent bonds with carbon), only the hydrogen nucleus with its positive proton remains. Any molecule that dissociates in solution and gives rise to a hydrogen ion is an **acid.** An acid is classified as strong or weak, depending on the extent to which the acid molecule is dissociated in solution. Examples of strong acids that dissociate completely in water are hydrochloric acid ($HCl \rightarrow H^+ + Cl^-$) and nitric acid ($HNO_3 \rightarrow H^+ + NO_3^-$). Weak acids, such as carbonic acid ($H_2CO_3 \rightleftharpoons H^+ + HCO_3^-$), dissociate only slightly. A solution of carbonic acid is mostly undissociated carbonic acid molecules with only a small number of bicarbonate (HCO_3^-) and hydrogen ions (H^+) present.

A **base** contains negative ions called hydroxide ions and may be defined as a molecule or ion that will accept a proton (hydrogen ion). Bases are produced when compounds containing them are dissolved in water. Sodium hydroxide ($NaOH$) is a strong base because it will dissociate completely in water into sodium (Na^+) and hydroxide (OH^-) ions. Among the characteristics of bases is their ability to combine with hydrogen ions, thus decreasing the concentration of the hydrogen ions. Like acids, bases vary in the extent to which they dissociate in aqueous solutions into hydroxide ions.

A **salt** is a compound resulting from the chemical interaction of an acid and a base. Common salt, sodium chloride ($NaCl$), is formed by the interaction of hydrochloric acid (HCl) and sodium hydroxide ($NaOH$). In water the HCl is dissociated into H^+ and Cl^- ions. The hydrogen and hydroxide ions combine to form water (H_2O), and the sodium and chloride ions remain as a dissolved form of salt (Na^+Cl^-):

$$H^+Cl^- + Na^+OH^- \rightarrow Na^+Cl^- + H_2O$$

$$\text{Acid} \qquad \text{Base} \qquad \text{Salt}$$

Organic acids are usually characterized by having in their molecule the carboxyl group (—COOH). They are weak acids because a relatively small proportion of the H^+ reversibly dissociates from the carboxyl:

$$\begin{array}{ccc} R—C=O & & R—C=O \\ | & \rightleftharpoons & | \qquad + H^+ \\ O—H & & O— \end{array}$$

R refers to an atomic grouping unique to the molecule. Some common organic acids are acetic, citric, formic, lactic, and oxalic. Many more of these will be encountered later in discussions of cellular metabolism.

Hydrogen ion concentration (pH)

Solutions are classified as acid, basic, or neutral according to the proportion of hydrogen (H^+) and hydroxide (OH^-) ions they possess. In acid solutions there is an excess of hydrogen ions; in alkaline, or basic, solutions the hydroxide ion is more common; and in neutral solutions both hydrogen and hydroxide ions are present in equal numbers.

To express the acidity or alkalinity of a substance, a logarithmic scale, a type of mathematical shorthand, is employed that uses the numbers 1 to 14. This is the pH, defined as:

$$pH = \log_{10}\frac{1}{[H^+]}$$

or

$$pH = -\log_{10}[H^+]$$

Thus pH is the negative logarithm of the hydrogen ion concentration. In other words, when the hydrogen ion concentration is expressed exponentially, pH is the exponent, but with the *opposite* sign; if $[H^+] = 10^{-2}$, then $pH = -(-2) = +2$. Unfortunately, pH can be a confusing concept because, as the $[H^+]$ decreases, the pH increases. Numbers below 7 indicate an acid range, and numbers above 7 indicate alkalinity. The number 7 indicates neutrality, that is, the presence of equal numbers of H^+ and OH^- ions. According to this logarithmic scale, a pH of 3 is 10 times more acid than one of 4; a pH of 9 is 10 times more alkaline than one of 8.

Buffer action

The hydrogen ion concentration in the extracellular fluids must be regulated so that metabolic reactions within the cell will not be adversely affected by a constantly changing hydrogen ion concentration, to which they are extremely sensitive. A change in pH of only 0.2 from the normal blood pH of about 7.35 can cause serious metabolic disturbances. To maintain pH within physiological limits, there are certain substances in cells and organisms that tend to compensate for any change in the pH when acids or alkalies are produced in metabolic reactions or are added to the body fluids. These are called **buffers.** A buffer is a mixture of slightly ionized weak acid and its completely ionized salt. In such a system, added H^+ combines with the anion of the salt to form undissociated acid, and added OH^- combines with H^+ to form water. The most important buffers in blood and other extracellular fluids are the bicarbonates and phosphates, and organic molecules such as amino acids and proteins are important buffers within cells. The bicarbonate buffer system consists of carbonic acid (H_2CO_3, a weak acid) and its salt, sodium bicarbonate ($NaHCO_3$). Sodium bicarbonate is strongly ionized into sodium ions (Na^+) and bicarbonate ions (HCO_3^-). When a strong acid (for example, HCl) is added to the fluid, the H^+ ions of the dissociated acid will react with the bicarbonate ion (HCO_3^-) to form a very weak acid, carbonic acid, which dissociates only slightly. Thus the H^+ ions from the HCl are removed and the pH is little altered. When a strong base (for example, NaOH) is added to the fluid, the OH^- ions of the strong base will react with the carbonic acid by removing H^+ ions from the H_2CO_3, to form water and bicarbonate ions. Again the H^+ ion concentration in solution is little altered and the pH remains nearly unchanged.

___CHEMISTRY OF LIFE
___Water and Life

Water is the most abundant of all compounds in cells, making up about 60% to 90% of most living organisms. The maintenance of a constant aqueous internal environment is a major physiological task for all organisms, both terrestrial and aquatic.

Water has several extraordinary properties that make it especially fit for its essential role in living systems. It is the most stable yet versatile of all solvents, and it is the only substance that occurs in nature in the three phases of solid, liquid, and vapor within the ordinary range of earth's temperatures. We now know that the remarkable properties of water can be explained in large part on the basis of the hydrogen bonds that form between its molecules. Although hydrogen bonds are much weaker than the covalent bonds within a water molecule, they require substantial energy for breakage.

Water has a **high specific heat capacity:** 1 calorie is required to elevate the temperature of 1 g of water 1° C (such as from 15° to 16° C). Every other liquid but ammonia requires less heat to accomplish the same temperature increase. When water is heated, much of the heat energy is used to rupture some of the hydrogen bonds, leaving less heat to increase the kinetic energy (molecular movement), and thus the temperature, of the water. The high thermal capacity of water has a great moderating effect on environmental temperature changes and is a great protective agent for all life.

Water also has a **high heat of vaporization.** More than 500 calories is required to change 1 g of liquid water into water vapor. This is so because all of the hydrogen bonds between a water molecule and its neighbors must be ruptured before water can escape the surface and enter the air. For terrestrial animals (and plants), cooling produced by the evaporation of water is an important means of getting rid of excess heat.

Another important property of water from a biological standpoint is its **unique density behavior** during changes of temperature. Most liquids become denser with decreasing temperature. Water, however, reaches its maximum density at 4° C *while still a liquid*, then becomes less dense with further cooling. Therefore, ice *floats* rather than forming on the bottom of lakes and ponds. If it were not for this property, bodies of water would freeze solid from the bottom up in winter, and, except in warmer climates, would not necessarily completely melt in summer. Under these conditions, aquatic life would be severely limited. In ice, all molecules form hydrogen bonds with others (Figure 2-9). The molecules form an extensive, open, crystallike network held together by hydrogen bonds. The molecules in this latticelike form are further apart, and thus less dense, than when some of the molecules have not formed hydrogen bonds at 4° C.

Water has a **high surface tension,** greater than that of any other liquid but mercury. This property is an aspect of the great cohesiveness of water molecules: their tendency to be held together by hydrogen bonds between them. Cohesiveness is important in the maintenance of protoplasmic form and movement, and the high surface tension creates a unique ecological niche (see p. 909) for insect forms, such as water striders (Figure 2-10) and whirligig beetles, that skate on the surfaces of ponds. Despite its high surface tension, water has **low viscosity,** a property that favors the movement of blood through minute capillaries and of cytoplasm inside cellular boundaries.

Finally, water is an excellent **solvent** for the ions of salts, which are so

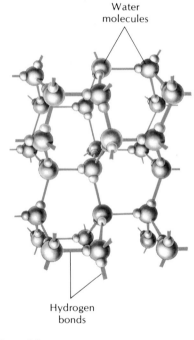

Water
molecules

Hydrogen
bonds

Figure 2-9

When water freezes at 0° C, the four partial charges of each atom in the molecule form interact with the opposite charges of atoms in other water molecules. The hydrogen bonds between all the molecules form a crystal-like lattice structure, and the molecules are farther apart (less dense) than when some of the molecules have not formed hydrogen bonds at 4° C.
From Raven, P.H., and G.B. Johnson. 1986. Biology. St. Louis, Times Mirror/Mosby.

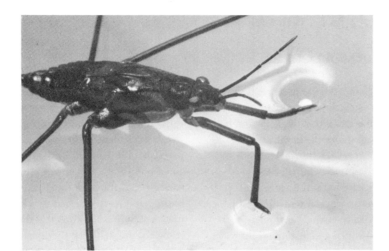

Figure 2-10

Because of hydrogen bonds between water molecules at the water-air interface, the water molecules cling together and create a high surface tension. Thus some insects, such as this water strider, can literally walk on water.
From Raven, P.H., and G.B. Johnson. 1986. Biology. St. Louis, Times Mirror/Mosby

important to life processes. Salts dissolve to a much greater degree in water than they do in any other solvent. This results from the dipolar nature of water, which causes it to orient itself around charged particles dissolved in it. When, for example, NaCl is dissolved in water, the Na^+ and Cl^- ions present in the solid salt rapidly separate into independent Na^+ and Cl^- ions. The negative zones of the water dipoles align themselves around the Na^+ ions while the positive zones align themselves around the Cl^- ions (Figure 2-11). This keeps the ions separated, promoting a high degree of dissociation. Other solvents having less or no dipolar character are less able to align effectively around such ions and are therefore less able to dissolve the salt.

Water and life are part and parcel of each other. The special conditions on earth resulting from its ideal size, element composition, and nearly circular orbit at a perfect distance from a long-lived star, the sun, made possible the accumulation of water on the earth's surface. It is difficult even to imagine the origin of life without water.

Salt crystal

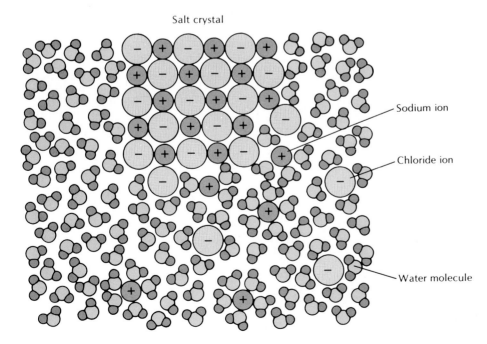

Sodium ion

Chloride ion

Water molecule

Figure 2-11

When a crystal of sodium chloride dissolves in water, the negative ends of the dipolar molecules of water surround the Na^+ ions, while the positive ends of water molecules face the Cl^- ions. The ions are thus separated and do not reenter the salt lattice.
From Raven, P.H., and G.B. Johnson. 1986. Biology. St. Louis, Times Mirror/Mosby.

___ Organic Molecules

The term *organic compounds* has been applied to substances derived from plants and animals. All organic compounds contain carbon, but many also contain hydrogen, oxygen, nitrogen, sulfur, phosphorus, salts, and other elements. Organic compounds specifically are those carbon compounds in which the principal bonds are carbon-to-carbon and carbon-to-hydrogen.

Carbon has a great ability to bond with other carbon atoms in chains of varying lengths and configurations. More than a million organic compounds have been identified; more are being added daily. Carbon-to-carbon combinations introduce the possibility of enormous complexity and variety into molecular structure. Examples will be found in the pages to follow.

Carbohydrates: nature's most abundant organic substance

Carbohydrates are compounds of carbon, hydrogen, and oxygen. They are usually present in the ratio of $1\,C:2\,H:1\,O$ and are grouped as H—C—OH. Familiar examples of carbohydrates are sugars, starches, and cellulose (the woody structure of plants). There is more cellulose on earth than all other organic materials combined. Carbohydrates are made synthetically from water and carbon dioxide by green plants, with the aid of the sun's energy. This process, called **photosynthesis,** is a reaction on which all life depends, for it is the starting point in the formation of food.

Carbohydrates are usually divided into the following three classes: (1) **monosaccharides,** or simple sugars; (2) **disaccharides,** or double sugars; and (3) **polysaccharides,** or complex sugars. Simple sugars are composed of carbon chains containing 4 carbons (tetroses), 5 carbons (pentoses), or 6 carbons (hexoses). Other simple sugars have up to 10 carbons, but these are not biologically important. Simple sugars, such as glucose, galactose, and fructose, all contain a free sugar group,

in which the double-bonded O may be attached to the terminal C of a chain or to a nonterminal C. The hexose **glucose** (also called dextrose) is the most important carbohydrate in the living world. Glucose is often shown as a straight chain (Figure 2-12, *A*), but in water it tends to form a cyclic compound (Figure 2-12, *B*). The "chair" diagram (Figure 2-13) of glucose best represents its true configuration, but we must remember that all forms of glucose, however represented, are the same molecule.

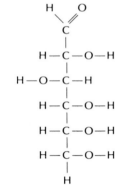

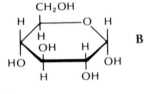

Figure 2-12

Two ways of depicting the structural formula of the simple sugar glucose. In **A** the carbon atoms are shown in open-chain form. When dissolved in water, glucose tends to assume a ring form as in **B**. In this ring model the carbon atoms located at each turn in the ring are usually not shown.

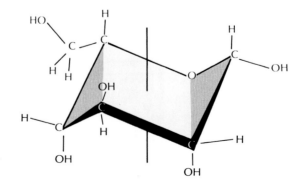

Figure 2-13

"Chair" representation of glucose molecule.

Other hexoses of biological significance are galactose and fructose. Their straight-chain structures are compared with that of glucose in Figure 2-14.

Disaccharides are double sugars formed by the bonding of two simple sugars. An example is maltose (malt sugar), composed of 2 glucose molecules. As shown in Figure 2-15, the 2 glucose molecules are condensed together by the removal of a molecule of water. This dehydration reaction, with the sharing of an oxygen atom by the two sugars, characterizes the formation of all disaccharides. Two other common disaccharides are sucrose (ordinary cane, or table, sugar), formed by the linkage of glucose and fructose, and lactose (milk sugar), composed of glucose and galactose.

Polysaccharides are made up of many molecules of simple sugars (usually glucose) linked together in long chains and are referred to by the chemist as polymers. Their empirical formula is usually written $(C_6H_{10}O_5)_n$, where n stands for the unknown number of simple sugar molecules of which they are composed. Starch is the common storage form of sugar in most plants and is an important food constituent for animals. **Glycogen** is an important storage form for sugar in animals. It is found mainly in liver and muscle cells in vertebrates. When needed, glycogen is converted into glucose and is delivered by the blood to the tissues. Another polymer is **cellulose,** which is the principal structural carbohydrate of plants.

The main role of carbohydrates in protoplasm is to serve as a source of chemical energy. Glucose is the most important of these energy carbohydrates. Some carbohydrates become basic components of protoplasmic structure, such as the pentoses that form constituent groups of nucleic acids and of nucleotides.

Lipids: fuel storage and building material

Lipids are fats and fatlike substances. They are composed of molecules of low polarity; consequently, they are virtually insoluble in water but are soluble in organic solvents such as acetone and ether. Three principal groups of lipids are neutral fats, phospholipids, and steroids.

Neutral fats

The neutral or "true" fats are major fuels of animals. Stored fat may be derived directly from dietary fat or indirectly from dietary carbohydrates that are converted to fat for storage. Fats are oxidized and released into the bloodstream as needed to meet tissue demands, especially for muscles.

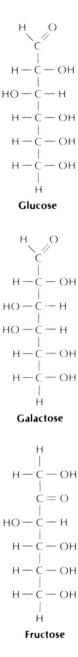

Glucose

Galactose

Fructose

Figure 2-14

These three hexoses are the most common monosaccharides. Glucose and galactose are aldehyde sugars; fructose is a ketone sugar.

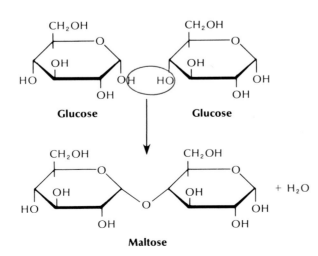

Maltose

Figure 2-15

Formation of a double sugar (disaccharide maltose) from 2 glucose molecules with the removal of 1 molecule of water.

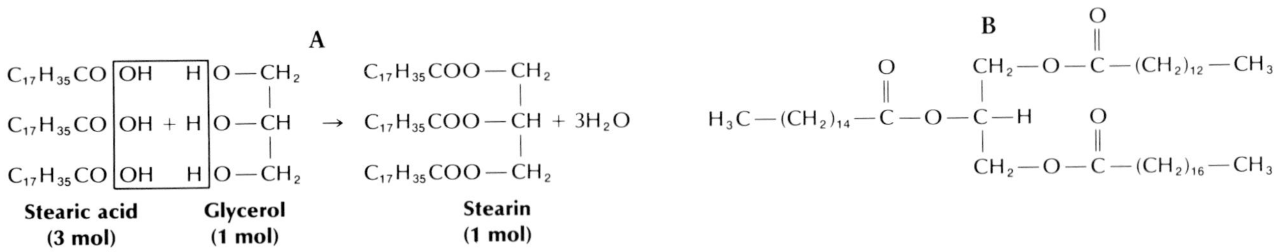

A

$$C_{17}H_{35}CO\boxed{OH \quad H}O-CH_2 \qquad C_{17}H_{35}COO-CH_2$$
$$C_{17}H_{35}CO\boxed{OH + H}O-CH \quad \rightarrow \quad C_{17}H_{35}COO-CH + 3H_2O$$
$$C_{17}H_{35}CO\boxed{OH \quad H}O-CH_2 \qquad C_{17}H_{35}COO-CH_2$$

Stearic acid   **Glycerol**   **Stearin**
(3 mol)   (1 mol)   (1 mol)

B

$$H_3C-(CH_2)_{14}-\overset{\overset{\displaystyle O}{\|}}{C}-O-\overset{\displaystyle CH_2-O-\overset{\overset{\displaystyle O}{\|}}{C}-(CH_2)_{12}-CH_3}{\underset{\displaystyle CH_2-O-\overset{\overset{\displaystyle O}{\|}}{C}-(CH_2)_{16}-CH_3}{C-H}}$$

Figure 2-16

Neutral fats. **A,** Formation of a neutral fat from 3 molecules of stearic acid (a fatty acid) and glycerol. **B,** A neutral fat bearing three different fatty acids.

Neutral fats are triglycerides, which are molecules consisting of glycerol and 3 molecules of fatty acids. Neutral fats are therefore esters, that is, a combination of an alcohol (glycerol) and an acid. Fatty acids in triglycerides are simply long-chain monocarboxylic acids; they vary in size but are commonly 14 to 24 carbons long. The production of a typical fat by the union of glycerol and stearic acid is shown in Figure 2-16, *A.* In this reaction it can be seen that the 3 fatty acid molecules have united with the OH group of the glycerol to form stearin (a neutral fat), with the production of 3 molecules of water.

Most triglycerides contain two or three different fatty acids attached to glycerol, bearing ponderous names such as myristoyl stearoyl glycerol (Figure 2-16, *B*). The fatty acids in this triglyceride are **saturated**; that is, every carbon within the chain holds 2 hydrogen atoms. Saturated fats, more common in animals than in plants, are usually solid at room temperature. **Unsaturated** fatty acids, typical of plant oils, have 2 or more carbon atoms joined by double bonds; that is, the carbons are not "saturated" with hydrogen atoms and are able to form additional bonds with other atoms. Two common unsaturated fatty acids are oleic acid and linoleic acid (Figure 2-17). Plant fats such as peanut oil and corn oil tend to be liquid at room temperature.

Figure 2-17

Unsaturated fatty acids: oleic acid having one double bond and linoleic acid having two double bonds. The remainder of the hydrocarbon chains of both acids is saturated.

$$CH_3-(CH_2)_7-CH=CH-(CH_2)_7-COOH$$
Oleic acid

$$CH_3-(CH_2)_4-CH=CH-CH_2-CH=CH-(CH_2)_7-COOH$$
Linoleic acid

Phospholipids

Unlike the fats that are fuels and serve no structural roles in the cell, phospholipids are important components of the molecular organization of tissues, especially membranes. They resemble triglycerides in structure, except that 1 of the 3 fatty acids is replaced by phosphoric acid and an organic base. An example is lecithin, an important phospholipid of nerve membrane (Figure 2-18). Because the phosphate group on phospholipids is charged and polar and therefore soluble in water and the remainder of the molecule is nonpolar, phospholipids can bridge two environments and bind water-soluble molecules such as proteins to water-insoluble materials.

Steroids

Steroids are complex alcohols, structurally unlike fats but have fatlike properties. The steroids are a large group of biologically important molecules, including cholesterol (Figure 2-19), vitamin D, many adrenocortical hormones, and the sex hormones.

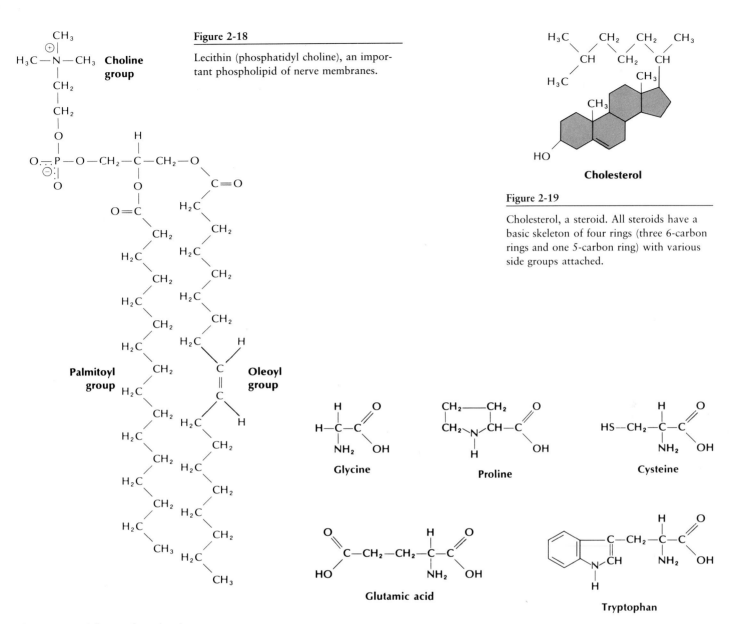

Choline group

Palmitoyl group

Oleoyl group

Figure 2-18

Lecithin (phosphatidyl choline), an important phospholipid of nerve membranes.

Cholesterol

Figure 2-19

Cholesterol, a steroid. All steroids have a basic skeleton of four rings (three 6-carbon rings and one 5-carbon ring) with various side groups attached.

Glycine

Proline

Cysteine

Glutamic acid

Tryptophan

Figure 2-20

Five of the 20 naturally occurring amino acids.

Amino acids and proteins

Proteins are large, complex molecules composed of 20 commonly occurring amino acids (Figure 2-20). The amino acids are linked together by **peptide bonds** to form long, chainlike polymers. In the formation of a peptide bond, the carboxyl group of one amino acid is linked by a covalent bond to the amino group of another, with the elimination of water, as follows:

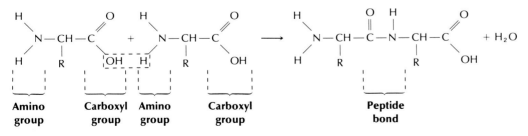

Amino group   **Carboxyl group**   **Amino group**   **Carboxyl group**   **Peptide bond**

The combination of two amino acids by a peptide bond forms a dipeptide, and, as is evident, there is still a free amino group on one and a free carboxyl group on the other; therefore additional amino acids can be joined to both ends until a long chain is produced. The 20 different kinds of amino acids can be arranged in an enormous variety of sequences of up to several hundred amino acid units; therefore it is not difficult to account for practically countless varieties of proteins among living organisms.

A protein is not just a long string of amino acids; it is a highly organized molecule. For convenience, biochemists have recognized four levels of protein organization called primary, secondary, tertiary, and quaternary.

The **primary structure** of a protein is determined by the kind and sequence of amino acids making up the polypeptide chain. Because the bonds between the amino acids in the chain are characterized by a limited number of stable angles, certain recurrent structural patterns are assumed by the chain. This is called the **secondary structure,** and it is often that of an **alpha-helix,** that is, helical turns in a clockwise direction like a screw (Figure 2-21). The spirals of the chains are stabilized by hydrogen bonds, usually between a hydrogen atom of one amino acid and the peptide-bond oxygen of another in an adjacent turn of the helix.

Not only does the polypeptide chain (primary structure) spiral into helical configurations (secondary structure), but also the helices themselves bend and fold, giving the protein its complex, yet stable, three-dimensional **tertiary structure** (Figure 2-22). The folded chains are stabilized by the interactions between side groups of amino acids. One of these interactions is the **disulfide bond,** a covalent bond between the sulfur(s) atoms in pairs of cysteine (sis'tee-in) units that are brought together by folds in the polypeptide chain. Other kinds of bonds that help stabilize the tertiary structure of proteins are hydrogen bonds, ionic bonds, and hydrophobic bonds.

The term **quaternary structure** describes those proteins that contain more than one polypeptide chain unit. For example, hemoglobin (the oxygen-carrying substance in blood) of higher vertebrates is composed of four polypeptide subunits nested together into a single protein molecule.

Proteins as enzymes

Proteins perform many functions in living things. They serve as the structural framework of protoplasm and form many cell components. However, the most important role of proteins by far is as **enzymes,** the biological catalysts required for almost every reaction in the body.

Enzymes lower the activation energy required for specific reactions and enable life processes to proceed at moderate temperatures. They control the reactions by which food is digested, absorbed, and metabolized. They promote the synthesis of structural materials for growth and to replace the wear and tear on the body. They determine the release of energy used in respiration, growth, muscle

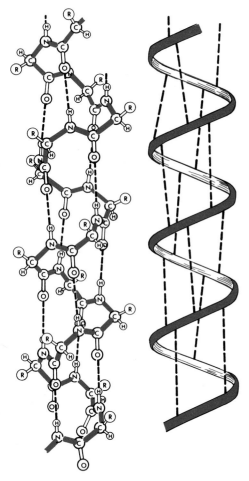

Figure 2-21

Alpha-helix pattern of a polypeptide chain. Hydrogen bonds *(dashed lines)* stabilize adjacent turns of the helix. *R,* Amino acid side chains.

Modified from Green D. 1956. Currents of biochemical research. New York, Interscience Publishers, Inc.

Figure 2-22

Three-dimensional tertiary structure of the protein myoglobin. Adjacent folds of the polypeptide chain are held together by disulfide bonds that form between pairs of cysteine molecules. In the upper center of the molecule is the heme group, which combines with oxygen.

From Neurath, H. 1964. The proteins, vol. 2, ed. 2. New York, Academic Press, Inc.

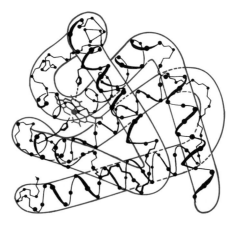

contraction, physical and mental activities, and many other activities. Enzyme action is described in Chapter 5.

Nucleic acids

Nucleic acids are complex substances of high molecular weight that represent a basic manifestation of life. The sequence of nitrogenous bases in these polymeric molecules encodes the genetic information necessary for all aspects of biological inheritance. They not only direct the synthesis of enzymes and other proteins, but are also the only molecules that have the power (with the help of the right enzymes) to replicate themselves. The two kinds of nucleic acids in cells are **deoxyribose nucleic acid (DNA)** and **ribose nucleic acid (RNA)**. They are polymers of repeated units called **nucleotides,** each containing a sugar, nitrogenous base, and phosphate group. The structure of nucleic acids is crucial to the mechanism of inheritance and protein synthesis, so this subject is discussed further in Chapter 38.

⎯ SUMMARY

All matter is composed of elements, which are themselves composed of discrete units called atoms. Atoms consist of a positively charged nucleus of very small volume, surrounded by negatively charged electrons of almost no mass. The nucleus is made up of 1 or more positively charged protons (the number of protons is the atomic number) and uncharged neutrons. The total number of protons and neutrons is nearly equal to the atomic mass of the element. Forms of an element that differ in the number of neutrons in the nucleus are isotopes, and isotopes that spontaneously disintegrate are radioactive isotopes.

The innermost "shell" of an atom can hold only 2 electrons, and each outer shell can hold a maximum of 8 electrons. Inner shells are filled first, and atoms tend to complete their shells by gaining or losing electrons (to form ionic bonds) or by sharing electrons with other atoms (to form covalent bonds). Molecules are formed when 2 or more atoms are joined together by chemical bonds. Loss of electrons is oxidation, and gain of electrons is reduction. Hydrogen bonds are weak bonds resulting from unequal charge distribution; they are important in biochemical processes.

Acids are molecules that dissociate in solution and give rise to a hydrogen ion (H^+); bases accept hydrogen ions and contain hydroxide ions (OH^-). Salts are compounds resulting from chemical reactions between acids and bases. The acidity or alkalinity of solutions is expressed by the pH scale, defined as the negative logarithm of the hydrogen ion concentration. Undesirable changes in pH in living systems are prevented by the action of buffers.

The unique structure of water and its ability to form hydrogen bonds between adjacent water molecules are responsible for its special properties; solvency for ionic and polar substances; high heat capacity, boiling point, and surface tension; and less density as a solid than as a liquid. Life on earth could not have appeared without water.

Carbon is especially versatile in bonding with itself or with other atoms and is the only element capable of forming the variety of molecules found in living things. Carbohydrates are composed primarily of carbon, hydrogen, and oxygen grouped as H—C—OH. Sugars serve as immediate sources of energy in living systems. Monosaccharides, or simple sugars, may bond together to form disaccharides or polysaccharides, which serve as storage forms of sugar or perform structural roles. Lipids exist principally as fats, phospholipids, and steroids.

Proteins are large molecules composed of amino acids linked together by peptide bonds. Proteins have a primary, secondary, tertiary, and often, quaternary structure. Proteins perform many functions, especially as enzymes (biological catalysts).

Nucleic acids are polymers of nucleotide units, each composed of a sugar, a nitrogenous base, and a phosphate group. They contain the material of inheritance and function in protein synthesis.

Review questions

1. Distinguish among the following: element, atom, proton, neutron, electron.
2. How do isotopes of elements differ from each other?
3. Write the names of the elements for each of the following symbols: P, Cl, K, Fe, H, N, C, S, Ca, O, Na, I.
4. Distinguish among ionic bonds, covalent bonds, and hydrogen bonds.
5. In the reaction below, name the following: oxidizing agent, reducing agent, electron donor, electron acceptor.

$$K + Cl \rightarrow K^+ + Cl^-$$

6. Match the items in the left column with the most appropriate terms in the right column:

   _____$H_2^{2+}SO_4^{2-}$ a. Acid
   _____K^+OH^- b. Base
   _____K^+Cl^- c. Salt
   _____$CH_3-\underset{\underset{O^-H^+}{|}}{C}=O$
   _____$Ca^{2+}SO_4^{2-}$

7. A solution with a pH of 6 is 10 times more acid than one with a pH of 7. Explain why this is so.
8. Explain why water molecules tend to form hydrogen bonds with other water molecules.
9. Explain each of the following properties of water, and tell how each is conferred by the dipolar nature of the water molecule: High specific heat capacity; high heat of vaporization; unique density behavior; high surface tension; good solvent for ions of salts.
10. Name two simple carbohydrates, two storage carbohydrates, and a structural carbohydrate.
11. What are characteristic differences in molecular structure between lipids and carbohydrates?
12. Explain the difference between primary, secondary, tertiary, and quaternary structure of a protein.
13. What are the important nucleic acids in a cell, and of what units are they constructed?

Selected references

Standard texts in freshman chemistry, and in biochemistry and cellular biology are good sources of additional information on the topics in this chapter. The following selection is by no means exhaustive.

Brady, J.E., and J.R. Holum. 1984. Fundamentals of chemistry, ed. 2. New York, John Wiley & Sons, Inc.

Jones, M.M., D.O. Johnston, J.T. Netterville, and J.L. Wood. 1983. Chemistry, man, and society. Philadelphia, Saunders College Publishing.

Lehninger, A.L. 1982. Principles of biochemistry. New York, Worth Publishers, Inc. *Clearly presented advanced text.*

Phillips, D.C., and A.C.T. North. 1978. Protein structure. Carolina Biology Readers #34.

Burlington, North Carolina, Carolina Biological Supply Co. *Comes with pair of spectacles for three-dimensional viewing of structural diagrams.*

Sheeler, P., and D.E. Bianchi. 1980. Cell biology: structure, biochemistry, and function. New York, John Wiley & Sons, Inc. *Clear treatment of chemistry of macromolecules.*

Weinberg, R.A. 1985. The molecules of life. Sci. Am. 253:48-57 (Oct.). *This entire number of Scientific American is devoted to the modern findings of molecular biology.*

CHAPTER 3

ORIGIN OF LIFE

Where did life on earth come from? This is an ancient, intriguing question that we must suppose has aroused human curiosity since the dawn of cultural development. Although the great religions have sought to satisfy this curiosity, they have not provided, nor were they ever intended to provide, a scientific explanation of the detailed sequence of events that culminated in the first appearance of life. To most biologists the question of life's beginnings is one of profound interest. The biologist is struck by the remarkable unity of nature (Principle 3, p. 8). As more concerning the identifiable components of life is learned, an evolutionary pattern in the structure and function of living things can be seen.

All organisms, from humans to the smallest microbes that transcend the rather arbitrary boundary between life and nonlife, share two kinds of basic biomolecules: nucleic acid and protein. As we have seen in the previous chapter, both molecules are large and complex in form. Except in some viruses, DNA is the material of inheritance; it provides the code by which inheritable traits are passed to the next generation of cells or organisms. RNA functions in several ways to translate the genetic code into materials characteristic of the cell. Proteins perform structural functions and, more important, catalyze or facilitate the myriad chemical reactions necessary to life. Yet despite the chemical complexity of these molecules and the diversity of functions they perform, in all organisms they are composed of relatively few main building blocks: 20 amino acids, five bases (adenine, guanine, cytosine, uracil, and thymine), two sugars (ribose and deoxyribose), and phosphate.

The remarkable uniformity of life extends also to cell function. Certain metabolic processes that convert foodstuffs into a usable form of energy consistently occur in a wide range of organisms, from the simplest to the most complex. This example, along with other examples of molecular and functional identity, suggests that all life must have had a common beginning.

Even though we acknowledge the kinship of living things, we must admit at the beginning that we do not know how life on earth originated. Until recently the study of life's origins was not considered worthy of serious speculation by biologists because, it was argued, the absence of a geological record made the course of events resulting in the appearance of life unknowable. This situation has changed.

Since 1950 several laboratories around the world have been devoting fulltime research to origin-of-life studies. It is a multidisciplinary effort that requires the contributions of scientists of several specialties: biologists, chemists, physicists, geologists, and astronomers. From such studies it has been possible to reconstruct a scenario of ancient events in which the earliest life form could have evolved more than 4 billion years BP (before the present) from inorganic constituents present on the surface of the earth. These studies are not attempts to prove or disprove any religious or philosophical belief, but rather they are endeavors to provide an intellectually satisfying account of how life on earth could have arisen by natural means.

___HISTORICAL PERSPECTIVE

From ancient times it was commonly believed that life could arise by spontaneous generation from dead material, in addition to arising from parental organisms by reproduction (biogenesis). Frogs appeared to arise from damp earth, mice from putrefied matter, insects from dew, maggots from decaying meat, and so on. Warmth, moisture, sunlight, and even starlight were often mentioned as beneficial factors that encouraged spontaneous generation.

These ideas were elaborately developed by the Greeks and appeared frequently in the writings of Aristotle (384 to 322 BC). They became firmly entrenched in virtually all cultures, including those of the Far East, and were unquestioningly accepted by even relatively recent great figures such as Copernicus, Bacon, Galileo, Harvey, Descartes, Goethe, and Schelling. Spontaneous generation was also supported by Christian philosophers who pointed out that, according to the first chapter of Genesis, God did not create plants and animals directly but bade the waters to bring them forth.

Inevitably the question of spontaneous generation fell under the scrutiny of experimental science in the sixteenth, seventeenth, and eighteenth centuries. At first such studies were ill-conceived efforts to supplement natural observations of spontaneous generation by artificially producing various organisms in the laboratory.

The first attack on the doctrine of spontaneous generation occurred in 1668 when the Italian physician Francesco Redi exposed meat in jars, some uncovered and some covered with parchment or wire gauze. The meat in all the vessels spoiled, but only the open vessels had maggots, and he noticed that flies were constantly entering and leaving these vessels. He concluded that, if flies had no access to the meat, no worms would be found.

Although Redi's refutation of spontaneous generation became widely known, the doctrine was too firmly entrenched to be abandoned. In 1748 the English Jesuit priest John T. Needham boiled mutton broth and put it in corked containers. After a few days the medium was swarming with microscopic organisms. He concluded that spontaneous generation was real because he believed that he had killed all living organisms by boiling the broth and that he had excluded the access of others by sealing the tubes.

However, an Italian investigator, Abbé Lazzaro Spallanzani (1767), was critical of Needham's experiments and conducted experiments that dealt another blow against the theory of spontaneous generation. He thoroughly boiled extracts of vegetables and meat, placed these extracts in clean vessels, and sealed the necks of the flasks hermetically in flame. He then immersed the sealed flasks in boiling water for several minutes to make sure that all germs were destroyed. As controls, he left some tubes open to the air. At the end of 2 days, he found the open flasks swarming with organisms; the others contained none.

This experiment still did not settle the issue, for the advocates of spontaneous generation maintained either that air, which Spallanzani had excluded, was necessary for the production of new organisms or that the method he used had destroyed the vegetative power of the medium. When oxygen was discovered (1774), the opponents of Spallanzani seized on this as the vital element that he had destroyed in his experiments.

Among the accounts of early efforts to produce organisms spontaneously in the laboratory is a recipe for making mice, given by the Belgian plant nutritionist Jean Baptiste van Helmont (1648). "If you press a piece of underwear soiled with sweat together with some wheat in an open jar, after about 21 days the odor changes and the ferment . . . changes the wheat into mice. But what is more remarkable is that the mice which came out of the wheat and underwear were not small mice, not even miniature adults or aborted mice, but adult mice emerge!"

Figure 3-1

Louis Pasteur, holding in his left hand one of the swan-necked flasks used to demonstrate the absence of spontaneous generation.
Courtesy Parke-Davis.

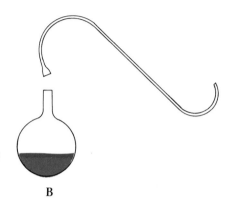

A   B

Figure 3-2

Louis Pasteur's swan-neck flask experiment. **A,** Sugared yeast water boiled in swan-neck flask remains sterile until neck is broken. **B,** Within 48 hours, flask is swarming with life.

It remained for the great French scientist Louis Pasteur (Figure 3-1) to silence all but the most stubborn proponents of spontaneous generation with an elegant series of experiments with his famous "swan-neck" flasks. Pasteur (1861) answered the objection to the lack of air by introducing fermentable material into a flask with a long S-shaped neck that was open to the air (Figure 3-2). The flask and its contents were then boiled for a long time. Afterwards the flask was cooled and left undisturbed. No fermentation occurred because all organisms that entered the open end were deposited on the floor of the neck and did not reach the flask contents. When the neck of the flask was cut off, the organisms in the air could fall directly on the fermentable mass and fermentation occurred within it in a short time. Pasteur concluded that, if suitable precautions were taken to keep out the germs and their reproductive elements, such as eggs and spores, no fermentation or putrefaction could take place.

Pasteur brought an end to the long and tenacious career of the concept of spontaneous generation. Pasteur's work showed that no living organisms come into existence except as descendants of similar organisms (Principle 10, p. 10). In announcing his results before the French Academy, Pasteur proclaimed, "Never will the doctrine of spontaneous generation arise from this mortal blow." Paradoxically, in showing that spontaneous generation did not occur as previously claimed (production of mice, maggots, frogs, and others), Pasteur also ended for a time further inquiry into the spontaneous origins of life. A lengthy period of philosophical speculation followed, but virtually no experimentation on life's origins was performed for 60 years.

Renewal of Inquiry: Oparin-Haldane Hypothesis

The rebirth of interest in the origins of life occurred in the 1920s. In that decade the Russian biochemist Alexander I. Oparin and the British biologist J.B.S. Haldane independently proposed that life originated on earth after an inconceivably long period of "abiogenic molecular evolution." Rather than arguing that the first living organisms miraculously originated all at once, a notion that had constrained fresh thinking for so long, Oparin and Haldane suggested that the simplest living units (for example, bacteria) came into being gradually by the progressive assembly of inorganic molecules into more complex organic molecules. These molecules would react with each other to form living microorganisms.

Haldane proposed that the earth's primitive atmosphere consisted of water, carbon dioxide, and ammonia. When such a gas mixture is exposed to ultraviolet radiation, many organic substances such as sugars and amino acids are formed. Ultraviolet light must have been very intense on the primitive earth before the production of oxygen (by photosynthetic organisms), which reacted with ultraviolet rays to form ozone, a 3-atom form of oxygen. Today ozone serves as a protective screen to prevent such intense ultraviolet radiation from reaching the earth's surface. Haldane believed that the early organic molecules could accumulate in the primitive oceans to form a "hot dilute soup." In this primordial broth carbohydrates, fats, proteins, and nucleic acids might have been assembled to form the earliest microorganisms.

Oparin, too, suggested that the earth's primitive atmosphere lacked oxygen and instead contained hydrogen, methane, ammonia, and other reducing compounds. He proposed that the organic compounds required for life were formed spontaneously in such a reducing atmosphere under the influence of sunlight, lightning, and the intense heat of volcanoes.

The Oparin-Haldane hypothesis greatly influenced theoretical speculation on the origins of life during the 1930s and 1940s. Finally in 1953 Stanley Miller, working with Harold Urey in Chicago, made the first successful attempt to simulate with laboratory apparatus the conditions thought to prevail on the primitive earth. This experiment, described in more detail later in the chapter, demonstrated that important biomolecules are formed in surprisingly large amounts when an electrical discharge is passed through a reducing atmosphere of the kind proposed by Oparin. The realization that it was possible to simulate a prebiotic milieu in the laboratory ushered in a new era in origin-of-life studies. It coincided with the dawn of the space age and a new public interest in the question of life's origins.

PRIMITIVE EARTH

Most astronomers accept the "Big Bang" model, according to which the universe originated from a primeval fireball and has been expanding and cooling since its inception 10 to 20 billion years ago. The sun and the planets are believed to have been formed approximately 4.6 billion years ago out of a spherical cloud of cosmic dust and gases that had some angular momentum. The cloud collapsed under the influence of its own gravity into a rotating disc. As the material in the central part of the disc condensed to form the sun, a substantial amount of gravitational energy was released as radiation. The pressure of this outwardly directed radiation prevented the complete collapse of the nebula into the sun. The material left behind began to cool and eventually gave rise to the planets (Figure 3-3).

Origin of the Earth's Atmosphere

Evolution of the primeval reducing atmosphere

While the earth was condensing, the various atoms were sorted out according to weight. Heavy elements (silicon, aluminum, nickel, and iron) gravitated toward the center, whereas the lighter elements (hydrogen, oxygen, carbon, and nitrogen)

Figure 3-3

Solar system showing narrow range of conditions suitable for life.

remained in the surface gas. Hydrogen and helium, because of their volatility, escaped as the earth condensed, and these elements became severely depleted from the primeval atmosphere. Neon and argon were almost completely lost from the atmosphere. Oxygen, nitrogen, and water, which are the major constituents of our present oxidizing atmosphere, could not escape from the earth because they were present in nonvolatile, chemically combined forms trapped in dust particles. Later, as the earth condensed, water, carbon compounds, and nitrogen were released into the atmosphere from the earth's interior by volcanic activity, which was much more extensive then than it is now.

It is generally agreed that the earth's primeval atmosphere contained no more than a trace of oxygen. It was a **reducing atmosphere;** that is, it contained a predominance of molecules having less oxygen than hydrogen. Methane (CH_4) and ammonia (NH_3) are examples of fully reduced compounds. According to one view, such compounds, together with water, composed the early atmosphere of the earth. Carbon dioxide (CO_2) nitrogen (N_2), and traces of hydrogen (H_2) may also have been present. Another view is that the earth's early atmosphere was chiefly carbon dioxide and water, with lesser amounts of H_2, N_2, ammonia, hydrogen sulfide, sulfur dioxide, and methane. According to this concept, the early atmosphere was only mildly reducing rather than strongly reducing as it would have been were methane and ammonia its principal components.

The character of the primeval atmosphere is important in any discussion of the origins of life because the organic compounds of which living organisms are made are not stable in an oxidizing atmosphere. Organic compounds are not synthesized nonbiologically in our oxidizing atmosphere today, and they are not stable if they are introduced into it. According to Oparin, life could not have originated without the basic organic building blocks from which the first organisms were assembled. Consequently, modern origin-of-life hypotheses assume that the primeval atmosphere was reducing because the synthesis of compounds of biological importance occurs only under reducing conditions. Geochemical evidence tends to support the belief that the early atmosphere was a reducing one that arose by degassing of the earth's interior.

Appearance of oxygen

As water, nitrogen, and carbon dioxide continued to enter the atmosphere from volcanoes, the atmosphere became saturated with water vapor, and rain began to fall. Small lakes formed, enlarging into oceans. The oceans gradually became salted as rocks weathered. During this early period, which lasted 1.5 billion years, the atmosphere probably was still reducing well after the first living protocells such as algae and bacteria had evolved.

Our atmosphere today is strongly oxidizing. It contains 78% molecular nitrogen, approximately 21% free oxygen, 1% argon, and 0.03% carbon dioxide. Although the time course for its development is much disputed, at some point oxygen began to appear in significant amounts in the atmosphere. In the primitive reducing atmosphere, oxygen was formed by the decomposition of water in either of two ways.

The first is the photolytic action of ultraviolet light from the sun on water in the upper atmosphere.

$$2\,H_2O \xrightarrow{\text{UV light}} 2\,H_2 + O_2$$

The hydrogen produced escapes from the earth's gravitational field, leaving free oxygen to accumulate in the atmosphere. The amount produced by the photodissociation of water is now quite small, although it may have been significant over the immense span of time in which it has occurred.

The second and probably more important source of oxygen is photosynthesis. Almost all oxygen produced at the present time is produced by cyanobacteria (blue-green algae), eukaryotic algae, and land plants. Each day the earth's plants combine approximately 400 million tons of carbon with 70 million tons of hydrogen to set free 1.1 billion tons of oxygen. Oceans are a major source of oxygen. Almost all oxygen produced today is consumed by organisms oxidizing their food to carbon dioxide; if this did not occur, the amount of oxygen in the atmosphere would double in approximately 3000 years. Since Precambrian fossil cyanobacteria resemble modern cyanobacteria, it seems probable that most of the oxygen in the early atmosphere was produced by photosynthesis.

CHEMICAL EVOLUTION
Sources of Energy

According to the Oparin-Haldane hypothesis, a variety of carbon compounds gradually accumulated on the surface of the earth during a lengthy period of prebiotic chemical evolution. The primitive reducing atmosphere contained simple gaseous compounds of carbon, nitrogen, oxygen, and hydrogen, such as carbon dioxide, molecular nitrogen, water vapor, and perhaps methane and ammonia. These were the starting materials from which organic compounds were made. However, if these gaseous compounds are mixed together in a closed glass system and allowed to stand at room temperature, they never chemically react with each other. To promote a chemical reaction, a continuous source of **free energy** sufficient to overcome reaction-activation barriers must be supplied.

For example, one of the simplest and most important prebiotic chemical reactions is the formation of hydrogen cyanide (HCN) from nitrogen and methane. When an electrical discharge is passed through an atmosphere of nitrogen, some of the molecules absorb enough energy to dissociate into atoms.

$$N_2 \rightarrow 2\,N$$

Dissociated nitrogen atoms are highly reactive. If methane is present in the atmosphere, nitrogen atoms react to form hydrogen cyanide and hydrogen.

$$N + CH_4 \rightarrow HCN + \frac{3}{2}\,H_2$$

Other sources of energy such as ultraviolet light or heat could equally well cause the formation of hydrogen cyanide from nitrogen and methane. Once formed, hydrogen cyanide dissolves in rain and is carried into lakes and oceans where it and other reactive molecules can form organic compounds.

The sun is by far the most powerful source of free energy for the earth. Each year each square centimeter of the earth receives an average 260,000 calories of radiant energy. The way in which this energy is distributed over the infrared, visible, and ultraviolet regions of the spectrum is shown in Table 3-1. Some investigators have suggested that life arose in vast clouds of water and dust particles surrounding the primitive earth. Such particles could have absorbed considerable energy in the visible and near infrared wavelengths. Since the solar energy avail-

Table 3-1 Present Sources of Energy Averaged over the Earth

Source	Energy (cal/cm^2/year)
Total radiation from sun	260,000
Infrared (above 700 nm)	143,000
Visible (350-700 nm)	113,600
Ultraviolet	
250-350 nm	2837
200-250 nm	522
150-200 nm	39
<150 nm	1.7
Electrical discharges	4
Shock waves	1.1
Radioactivity (to 1 km depth)	0.8
Volcanoes	0.13
Cosmic rays	0.0015

Modified from Miller, S.L., and L.E. Orgel. 1974. The origins of life on the earth. Englewood Cliffs, N.J., Prentice-Hall, Inc.

able falls off rapidly in the ultraviolet region, only approximately 0.2% of the total energy is at wavelengths shorter than 200 nm, where it can be absorbed by molecules such as methane, water, and ammonia. Nevertheless, ultraviolet radiation could have been an important source of energy for photochemical reactions in the primitive atmosphere.

Electrical discharges could have provided another source of energy for chemical evolution. Although the total amount of electrical energy released by lightning is small compared with solar energy, nearly all of the energy of lightning is effective in synthesizing organic compounds in a reducing atmosphere. A single flash of lightning through a reducing atmosphere generates a large amount of organic matter. Thunderstorms may have been one of the most important sources of energy for organic synthesis.

Volcanoes and hot springs were available on the primitive earth as sources of energy, but it is doubtful that they were a major site of prebiotic organic synthesis. Cosmic rays, radioactivity, and sonic energy generated by ocean waves were also available sources of energy, but their contribution cannot be evaluated and was probably small in any case. Of the other sources of energy, only shock waves generated by meteorites passing through the primitive atmosphere may have produced a significant amount of organic matter. Meteorites generate intense temperatures as high as 20,000° C and pressures exceeding 15,000 atmospheres. A large meteorite could generate millions of tons of organic matter in its wake.

Prebiotic Synthesis of Small Organic Molecules

Earlier in this chapter we referred to Stanley Miller's pioneering simulation of primitive earth conditions in the laboratory. Miller built an apparatus designed to circulate a mixture of methane, hydrogen, ammonia, and water past an electric spark (Figure 3-4). Water in the flask was boiled to produce steam that helped to circulate the gases. The products formed in the electrical discharge (representing

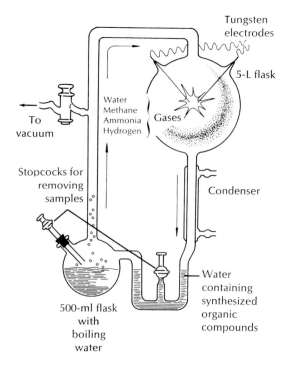

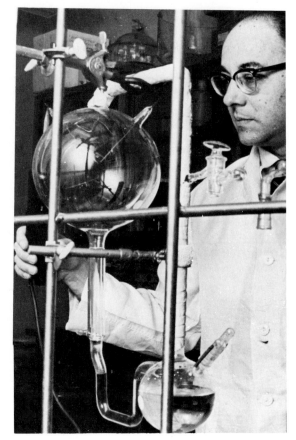

Figure 3-4

Dr. S.L. Miller and apparatus used in experiment on the synthesis of amino acids with an electric spark in a strongly reducing atmosphere.
Courtesy S.L. Miller.

lightning) were condensed in the condenser and collected in the U-tube and small flask (representing ocean).

After a week of continuous sparking, the water containing the products was analyzed. The results were surprising. Approximately 15% of the carbon that was originally in the reducing "atmosphere" had been converted into organic compounds that collected in the "ocean." The most striking finding was that many compounds related to life were synthesized. These included four amino acids commonly found in proteins, urea, and several simple fatty acids.

We can appreciate the astonishing nature of this synthesis when we consider that there are thousands of known organic compounds with structures no more complex than those of the amino acids formed. Yet in Miller's synthesis most of the relatively few substances formed were compounds found in living organisms. This was surely no coincidence, and it suggests that prebiotic synthesis on the primitive earth may have occurred under conditions that were not greatly different from those that Miller chose to simulate.

Miller's experiments recently have been criticized by geochemists who believe that the early atmosphere of the earth was quite different from Miller's strongly reducing simulated atmosphere. Nevertheless, Miller's work stimulated many other investigators to repeat and extend his experiment. It was soon found that amino acids could be synthesized in many different kinds of gas mixtures that were heated (volcanic heat), irradiated with ultraviolet light (solar radiation), or subjected to electrical discharge (lightning). All that was required to produce amino acids was that the gas mixture be reducing and that it be subjected violently to some energy source. In recent experiments, electrical discharges have been passed through mixtures of carbon monoxide, nitrogen, and water, yielding amino acids and nitrogenous bases. Although reaction rates were much slower than in atmospheres containing methane and ammonia, and yields were poor in comparison, these experiments support the hypothesis that the chemical beginnings of life can occur in atmospheres that are only mildly reducing. It has also been confirmed that no amino acids can be produced in an atmosphere containing oxygen.

Thus, the experiments of many scientists have shown that highly reactive intermediate molecules such as hydrogen cyanide, formaldehyde, and cyanoacetylene are formed when a reducing mixture of gases is subjected to a violent energy source. These react with water and ammonia or nitrogen to form more complex organic molecules, including amino acids, fatty acids, urea, aldehydes, sugars, purine and pyrimidine bases—indeed all the building blocks required for the synthesis of the most complex organic compounds of living matter.

___ Formation of Polymers

Need for concentration

The next stage in chemical evolution involved the condensation of amino acids, purines, pyrimidines, and sugars to yield larger molecules that resulted in proteins and nucleic acids. Such condensations do not occur easily in dilute solutions because the presence of excess water tends to drive reactions toward decomposition (hydrolysis). Although the primitive ocean has been called a primordial soup, it was probably a rather dilute one containing organic material that was approximately one tenth to one third as concentrated as chicken bouillon.

Prebiotic synthesis must have occurred in restricted regions where concentrations were higher. The primordial soup might have been concentrated by evaporation in lakes, ponds, or tide pools. Dilute aqueous solutions could also have been concentrated effectively by freezing. As ice freezes, organic solutes are concentrated in the solution that separates from the pure ice. This technique is employed to produce applejack from cider in the northern United States and Canada. When a barrel of cider is allowed to freeze, a liquid residue remains that contains most of the alcohol and flavoring materials.

Because the earliest life form may have originated 4 billion years ago, the earth's surface may have been still too warm to allow the formation of oceans. Violent weather on the primitive earth would have created enormous dust storms, and the dust particles could have become foci of water droplets. Salt concentration in the particles could have been high because of the large quantities of minerals swept up by the dust storms. An alternative hypothesis to account for life's origin under conditions of a hot earth exists: perhaps the earth was too warm to have oceans, but not too hot to have a damp surface. This would have resulted from constant rain and rapid evaporation. Thus the earth's surface could have become coated with organic molecules, an "incredible scum." Prebiotic molecules might have been concentrated by adsorption on the surface of clay and other minerals. Clay has the capacity to concentrate and condense large amounts of organic molecules from an aqueous solution.

Thermal condensations

Most biological polymerizations are dehydration reactions; that is, monomers are linked together by the removal of water. The peptide bond is a familiar example. In living systems dehydration reactions always take place in an aqueous (cellular) environment in the presence of the appropriate dehydrating enzymes. Without enzymes and energy supplied by ATP, the macromolecules (proteins and nucleic acids) of living systems soon break down into their constituent monomers.

One of the ways in which dehydration reactions could have occurred in primitive earth conditions without enzymes is by thermal condensation. The simplest dehydration is accomplished by driving off water from solids by direct heating. For example, if a mixture of all 20 amino acids is heated to 180° C, a good yield of polypeptides is obtained.

The thermal synthesis of polypeptides to form "proteinoids" has been studied extensively by the American scientist Sidney Fox. He showed that heating dry mixtures of amino acids and then mixing the resulting polymers with water will form small spherical bodies. These proteinoid microspheres (Figure 3-5) possess certain characteristics of living systems. Each is not more than 2 μm in diameter and is comparable in size and shape to spherical bacteria. Their outer walls appear to have a double layer, and they show osmotic and selective diffusion properties. They may grow by accretion or proliferate by budding like bacteria. There is no way to know whether proteinoids may have been the ancestors of the first cells or whether they are just interesting creations of the chemist's laboratory. They must be formed under conditions that would have been found only in volcanoes. Possibly organic polymers might have condensed on or in volcanoes and then, wetted by rain or dew, reacted further in solution to form polypeptides or polynucleotides.

ORIGIN OF LIVING SYSTEMS

We now have evidence from the fossil record that life existed by 3.8 billion years ago; therefore, the origin of the earliest life form can be estimated at 4 billion years BP. The first living organisms were protocells: autonomous membrane-bound units with a complex functional organization that permitted the essential activity of self-reproduction. The primitive chemical systems we have described lack this essential property. The principal problem in understanding the origin of life is explaining how primitive chemical systems could have become organized into living, autonomous, self-reproducing cells.

As we have seen, a lengthy chemical evolution on the primitive earth produced several molecular components of living forms. In a later stage of evolution, nucleic acids (DNA and RNA) began to behave as simple genetic systems that directed the synthesis of proteins, especially enzymes. However, this has led to a troublesome chicken-egg paradox: (1) How could nucleic acids have appeared without enzymes to synthesize them? (2) How could enzymes have evolved without nucleic acids to direct their synthesis? These questions are predicated on a long-accepted dogma that only proteins could act as enzymes. Startling new evidence has suggested that RNA in some instances may have enzymatic activity! Therefore the earliest enzymes could have been RNA. Nevertheless, proteins have several important advantages over RNA as catalysts, and the first protocells with protein enzymes would have had a powerful selective advantage over those with only RNA.

Once this stage of organization was reached, natural selection (Principle 5, p. 8) began acting on these primitive self-replicating systems. This was a critical point. Before this stage, biogenesis was shaped by the favorable environmental conditions on the primitive earth and by the nature of the reacting elements themselves. When self-replicating systems became responsive to the forces of natural

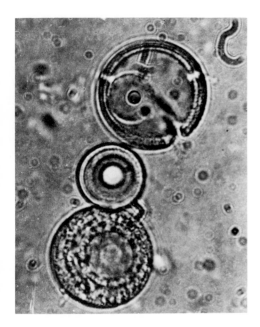

Figure 3-5

Electron micrograph of proteinoid microspheres. These proteinlike bodies can be produced in the laboratory from polyamino acids and may represent precellular forms. They have definite internal ultrastructure. (×1700.)

Courtesy R.M. Syren and S.W. Fox, Institute of Molecular Evolution, University of Miami, Coral Gables, Fla.

selection, they began to evolve. The more rapidly replicating and more successful systems were favored, and they replicated even faster. In short, the most efficient forms survived. From this evolved the genetic code and fully directed protein synthesis. The system was a protocell, and it could be called a living organism.

___ Origin of Metabolism

Living cells today are organized systems that possess complex and highly ordered sequences of enzyme-mediated reactions. Some cells trap solar energy and convert it into chemical bond energy, which is stored in glucose, ATP, and other molecules. Other cells are able to use these sources of bond energy to grow, divide, and maintain their internal integrity. Indeed, the attributes of life, involving energy conversion, assimilation, secretion, excretion, responsiveness to stimuli, and capacity to reproduce, all depend on the complex metabolic patterns characteristic of contemporary cells. How did such vastly complex metabolic schemes develop?

Organisms that depend on nutrient molecules they have not synthesized for their food supplies are known as **heterotrophs** (Gr., *heteros*, another, + *trophos*, feeder), whereas organisms that can synthesize their food from inorganic sources using light or another source of energy are called **autotrophs** (Gr., *autos*, self, + *trophos*, feeder) (Figure 3-6). The earliest microorganisms are sometimes referred to as **primary heterotrophs** because they existed before there were any autotrophs. They were probably anaerobic, bacterium-like organisms similar to modern *Clostridium,* and they obtained all their nutrients directly from the environment. Chemical evolution had already supplied generous stores of nutrients in the prebiotic soup. There would be neither advantage nor need for the earliest organisms to synthesize their own compounds, as long as they were freely available from the environment. But as soon as deficiencies of certain essential nutrients occurred, alternative sources had to be found. At that point, microorganisms that could synthesize these essential compounds from other accessible compounds would clearly have a greater advantage for surviving than those that could not.

For example, ATP is the immediate energy coinage of all living organisms. Since it has been formed experimentally in simulated prebiotic conditions, there is good reason to believe that it was present in the primitive environment and available to and used by protocells. Thus, early organisms would have depended on the environmental supply for their ATP requirements.

Once the supply was exhausted or became precarious, perhaps because of an increase in the numbers of organisms using it, protocells able to convert a precursor such as phosphoenolpyruvate to pyruvate and capture the energy released in the form of ATP would have had a tremendous advantage over those that lacked this capability. They would be selected for survival and would thrive. In the same way, when the supply of phosphoenolpyruvate became limiting, it would be necessary to synthesize it from another precursor (such as 2-phosphoglycerate) supplied by the environment. Again, the most successful organisms would have been those that chanced to develop this metabolic capability. Long sequences of reactions could have arisen in this manner.

An enzyme is required to catalyze each of these reactions. So, when we say that early protocells developed a reaction sequence as we have described (A made from B, B from C, and so on), we are really assuming that the appropriate enzymes appeared to catalyze these reactions. The numerous enzymes of cellular metabolism appeared when cells became able to utilize proteins for catalytic functions

We present here the traditional view that the first organisms were primary heterotrophs. This concept has been questioned by Carl Woese, who finds it easier to visualize membrane-associated molecular aggregates that absorbed visible light and converted it with some efficiency into chemical energy. Thus the first organisms would have been autotrophs. Woese has also challenged the notion that metabolism has evolved step by step backward, suggesting instead that the earliest "metabolism" may have consisted of numerous chemical reactions catalyzed by nonprotein cofactors (substances necessary for the function of many of the protein enzymes in living cells). These cofactors would also have been associated with membranes.

Figure 3-6

Koala, a heterotroph, feeding on a eucalyptus tree, an autotroph. All heterotrophs depend for their nutrients directly or indirectly on autotrophs that capture the sun's energy to synthesize their own nutrients. Photograph by C.P. Hickman, Jr.

and thereby gained a selective advantage. No planning was required; the results were achieved through natural selection.

Appearance of Photosynthesis and Oxidative Metabolism

Eventually, almost all usable energy-rich nutrients of the prebiotic soup were consumed. This ushered in the next stage of biochemical evolution: the use of readily available solar radiation to provide metabolic energy. Photosynthesis, the production of organic compounds from sunlight and atmospheric carbon dioxide, is the only process that restores free energy to the biosphere. Plant photosynthesis makes possible the richness of life on earth as we know it today. The self-nourishing photoautotrophic organisms, mainly green plants, capture solar energy and use it to convert simple inorganic substances into organic materials. The energy-rich compounds they produce provide not only for their own functioning but for the secondary heterotrophic organisms, mainly animals, that feed on autotrophs. The heterotrophs in turn release important raw materials for autotrophs. This is the energy cycle of the biosphere that is powered by a steady supply of energy from the sun (Principle 2, p. 7).

The appearance of photosynthesis was of enormous consequence for evolution, but like other metabolic events it did not appear all at once. In plant photosynthesis, water is the source of the hydrogen that is used to reduce carbon dioxide to sugars. Oxygen is liberated into the atmosphere.

$$6 \ CO_2 + 6 \ H_2O \xrightarrow{\text{light}} C_6H_{12}O_6 + O_2$$

However, the first steps in the development of photosynthesis almost certainly did not involve the splitting of water because a large input of energy is required. Hydrogen sulfide is thought to have been abundant in the primitive earth and was probably the first reducing agent used in photosynthesis.

$$6 \ CO_2 + 12 \ H_2S \xrightarrow{\text{light}} C_6H_{12}O_6 + 12 \ S + 6 \ H_2O$$

Later, as hydrogen sulfide and other reducing agents except water were used up, oxygen-evolving photosynthesis appeared. This may have occurred as early as 3.5 billion years BP. Gradually oxygen began to accumulate in the atmosphere. When atmospheric oxygen reached approximately 1% of its present level, ozone began to accumulate and ultraviolet radiation was screened out. Now land and surface waters could be occupied, and oxygen production probably increased sharply.

But at this important juncture, accumulating atmospheric oxygen began to interfere with cellular metabolism, which up to this point had evolved under strictly reducing conditions. As the atmosphere slowly changed from a reducing to an oxidizing one, a new and highly efficient kind of energy metabolism appeared: **oxidative (aerobic) metabolism.** By using the available oxygen to oxidize glucose to carbon dioxide and water, much of the bond energy stored by photosynthesis could be recovered. Most living forms became wholly dependent on oxidative metabolism, and oxygen-evolving photosynthesis became essential for the continuation of life on earth.

The final phase in life's evolution followed. Although a vast span of time—more than 2 billion years—passed before multicellular organisms appeared, living cells, much as we know them today, surrounded by semipermeable membranes and supporting an efficient oxygen-consuming form of metabolism, were flourishing on earth.

PRECAMBRIAN LIFE

As depicted on the inside back cover of this book, the Precambrian period spanned the geological time before the beginning of the Cambrian period 600 million years BP. At the beginning of the Cambrian period, most of the major phyla of invertebrate animals made their appearance within a few million years. This has been called the "Cambrian explosion" because before this time fossil deposits were rare and almost devoid of anything more complex than single-celled bacteria. We now recognize that the apparent rarity of Precambrian fossils was because they escaped notice owing to their microscopic size. What forms of life existed on earth before the burst of evolutionary activity in the early Cambrian world, and what organisms were responsible for the momentous change from a reducing to an oxidizing atmosphere?

Prokaryotes and the Age of Cyanobacteria (Blue-Green Algae)

The earliest bacterium-like organisms proliferated, giving rise to a great variety of bacterial forms, some of which were capable of photosynthesis. From these arose the oxygen-evolving **cyanobacteria** some 3 billion years ago.

Bacteria are called **prokaryotes,** meaning literally "before the nucleus." They contain a single chromosome comprising a single, large molecule of DNA not located in a membrane-bound nucleus, but found in a nuclear region, or **nucleoid.** The DNA is not complexed with histone proteins, and prokaryotes lack membranous organelles such as mitochondria, plastids, Golgi apparatus, and endoplasmic reticulum (Chapter 4). During cell division, the nucleoid divides without visible chromosomes, never by true chromosomal (mitotic) division.

Bacteria and especially cyanobacteria ruled the earth's oceans unchallenged for some 1½ to 2 billion years. The cyanobacteria reached the zenith of their success approximately 1 billion years BP, when filamentous forms produced great floating mats on the ocean surface. This long period of cyanobacterial dominance, encompassing approximately two thirds of the history of life, has been called with justification the "age of blue-green algae" (Schopf, 1974). Bacteria and cyanobacteria are so completely different from forms of life that evolved later that they have been placed in a separate kingdom, Monera (p. 127).

Within the last decade, however, Carl Woese and his colleagues at the University of Illinois have discovered that the prokaryotes actually comprise two distinct lines of descent: the eubacteria ("true" bacteria) and the archaebacteria. Although these two groups of bacteria look very much alike when viewed with the electron microscope, they are biochemically distinct. The cell walls of the archaebacteria do not contain muramic acid, as do those of all other bacteria, and there are fundamental differences in their metabolism. But the most compelling evidence for differentiating these two groups comes from the use of one of the newest and most powerful tools at the disposal of the evolutionist, the molecular sequencing technique (see marginal note). Woese found that the sequence of bases in one kind of RNA, ribosomal RNA, is sharply different from that of all other bacteria as well as from those found in the eukaryotes (see the following discussion). Woese believes that the archaebacteria are so distinctly different from the true bacteria that they should be considered as a separate kingdom, Archaebacteria. The Monera would then comprise only the true bacteria.

The name "algae" is misleading because it suggests a relationship to the eukaryotic algae, and many scientists prefer the alternative name "cyanobacteria" rather than "blue-green algae." These were the organisms responsible for producing oxygen released into the atmosphere. Study of the biochemical reactions in present cyanobacteria suggests that they evolved in a time of fluctuating oxygen concentration. For example, although they can tolerate atmospheric concentrations of oxygen (21%), the optimum concentration for many of their metabolic reactions is only 10%.

Molecular sequencing has emerged as the most successful approach to unraveling the genealogies of very ancient forms of life. An organism contains a living record of its past in the amino acid sequences of its proteins and in the nucleotide sequences of its nucleic acids, DNA and RNA. The sequences of nucleotides in the DNA of an organism's genes are an especially faithful record of evolutionary relationship because every gene that exists today is an evolved copy of a gene that existed millions, even billions, of years ago. Genes become altered by mutations through the course of time, but vestiges of the original gene usually persist. Using a technique that has become available only in the last few years, one can determine the sequence of nucleotides in the entire molecule or in short segments of the molecule that have been cut apart with a specific enzyme. A catalog of the nucleotide sequences characteristic of each organism is then compiled. If genes for the same function in two different organisms are compared, the extent to which they differ is correlated with the time elapsed since the two organisms diverged from a common ancestor. Molecular phylogeny based on the sequencing technique has permitted biologists to move from speculation about relationships to the construction of actual phylogenies based on information locked within the cell.

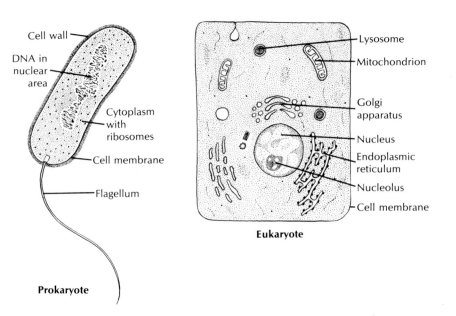

Prokaryote

Eukaryote

Figure 3-7

Comparison of prokaryote and eukaryote cells. The prokaryote cell is about one-tenth the size of the eukaryote cell.

___ Appearance of the Eukaryotes

Approximately 1.5 billion years ago, after the accumulation of an oxygen-rich atmosphere, organisms with nuclei appeared. The **eukaryotes** ("true nucleus") have cells with membrane-bound nuclei containing **chromosomes** composed of **chromatin**. In contrast to the prokaryote chromosome, constituents of chromatin include proteins called **histones** and RNA, in addition to the DNA. Both prokaryote and eukaryote chromosomes include some nonhistone proteins. Eukaryotes are generally larger than prokaryotes, contain much more DNA, and usually divide by some form of mitosis. Within their cells are numerous membranous organelles, including mitochondria in which the enzymes for oxidative metabolism are packaged. Protozoa, fungi, green and other algae, higher plants, and multicellular animals are composed of eukaryotic cells.

Prokaryotes and eukaryotes are profoundly different from each other (Figure 3-7) and clearly represent a marked dichotomy in the evolution of life. The ascendancy of the eukaryotes resulted in a rapid decline in the dominance of cyanobacteria as the eukaryotes proliferated and fed on them.

Why were the eukaryotes immediately so successful? Probably because they developed an important process facilitating rapid evolution: sex. Sex promotes great genetic variability in populations by mixing the genes of each two individuals that mate. By preserving favorable genetic variants, natural selection encourages rapid evolutionary change (Principle 5, p. 8). Prokaryotes propagate effectively and efficiently, but their mechanisms for interchange of genes, which does occur in some cases, lack the systematic genetic recombination characteristic of sexual reproduction.

The organizational complexity of the eukaryotes is so much greater than that of the prokaryotes that it is difficult to visualize how a eukaryote could have arisen from any known prokaryote. Margulis and others have proposed that eukaryotes did not in fact arise from any single prokaryote but were derived from a symbiosis ("life together") of two or more types. Mitochondria and plastids, for example, each contain their own complement of DNA (apart from the nucleus of the cell), which has some prokaryote characteristics. Mitochondria contain the enzymes of oxidative metabolism, and plastids (a plastid with chlorophyll is a chloroplast) carry out photosynthesis. It is easy to see how a host cell that was able

In addition to maintaining that mitochondria and plastids originated as bacterial symbionts, Margulis argues that eukaryote flagella, cilia (locomotory structures), and even the spindle of mitosis came from a kind of bacterium like a spirochete. Indeed, she suggests that this association (the spirochete with its new host cell) was what made the evolution of mitosis possible.

to accommodate such guests in its cytoplasm would have had enormous competitive advantages.

Eukaryotes may have originated more than once. They were no doubt unicellular, and many were photosynthetic autotrophs. Some of these lost their photosynthetic ability and became heterotrophs, feeding on the autotrophs and the prokaryotes. As the cyanobacteria were cropped, their dense filamentous mats began to thin, providing space for other species. Carnivores appeared and fed on the herbivores. Soon a balanced ecosystem of carnivores, herbivores, and primary producers appeared. This was ideal for evolutionary diversity. By freeing space, cropping herbivores encouraged a greater diversity of producers, which in turn promoted the evolution of new and more specialized croppers. An ecological pyramid developed with carnivores at the top.

The burst of evolutionary activity that followed at the end of the Precambrian period and beginning of the Cambrian period was unprecedented, and nothing approaching it has occurred since. Nearly all animal and plant phyla appeared and established themselves within a relatively brief period of a few million years. The eukaryotic cell made possible the richness and diversity of life on earth today.

SUMMARY

A remarkable uniformity in the chemical constituents of living things and many of the reactions that go on in their cells suggests that life on earth may have had a common origin. Historically, it was believed that life could arise anytime, spontaneously, as long as the conditions were right. Some 60 years after Louis Pasteur disproved spontaneous generation, A.I. Oparin and J.B.S. Haldane proposed a long "abiogenic molecular evolution" on earth in which organic molecules slowly accumulated in a "primordial soup." The atmosphere of the primitive earth was reducing, and little or no free oxygen was present. Ultraviolet radiation reached the earth's surface with much greater intensity then because there was no ozone layer in the atmosphere. This, plus the electrical discharges of lightning, provided energy for the formation of organic molecules from hydrogen, methane, ammonia, water, and carbon dioxide. Some energy may have been provided by the shock waves of meteorites passing through the atmosphere. Stanley Miller and Harold Urey showed the plausibility of the Oparin-Haldane hypothesis by simple but ingenious experiments. Thus, amino acids, purines, fatty acids, sugars, and other molecules have been synthesized under conditions that may have prevailed on the primitive earth. Although the compounds could have accumulated in rather dilute concentrations, dehydration reactions could have occurred after concentration in protein microspheres, during freezing, within droplets in the atmosphere, or by adsorption onto clay particles. Enzymes and the genetic coding of nucleic acids probably arose together, rather than one before the other. When self-replicating systems became responsive to the forces of natural selection, evolution proceeded more rapidly.

The first organisms were the primary heterotrophs, living on the energy stored in molecules dissolved in the primordial soup. As such molecules were used up, autotrophs had a great selective advantage. Molecular oxygen began to accumulate in the atmosphere as an end product of photosynthesis, and finally the atmosphere became oxidizing. The organisms responsible for this were apparently cyanobacteria. All bacteria are prokaryotes, organisms that lack a membrane-bound nucleus and other organelles in their cytoplasm. The prokaryotes consist of two genetically distinct groups that some believe should be considered separate kingdoms, Archaebacteria and Monera.

The eukaryotes originated after the atmosphere was oxidizing and apparently arose from symbiotic unions of two or more types of prokaryotes. Eukaryotes have most of their genetic material (DNA) borne in a membrane-bound nucleus and have mitochondria and other features. They include the eukaryotic algae, fungi, other plants, and all animals. Their evolutionary success results in great degree from the variability conferred by sexual reproduction.

Review questions

1. In regard to the experiments of Louis Pasteur and Stanley Miller described in this chapter, explain what constituted the following in each case: observations, induction, hypothesis, deduction, prediction, data, control.
2. What are two views on the composition of the earth's early atmosphere? On what do both views agree?
3. What is the most important way in which the earth's early atmosphere differed from that of today?
4. Name four different sources of energy that could have powered reactions on early earth to form organic compounds.
5. Explain the significance of the Miller-Urey experiments.
6. What are several mechanisms by which organic molecules in the prebiotic world could have been concentrated so that further reactions could occur?
7. Distinguish among the following: primary heterotroph, autotroph, secondary heterotroph.
8. What is the origin of the oxygen in the present-day atmosphere, and what is its metabolic significance to most organisms living today?
9. Distinguish between prokaryotes and eukaryotes as completely as you can.
10. Describe Margulis' view on the origin of eukaryotes from prokaryotes.

Selected references

Cairns-Smith, A.G. 1985. The first organisms. Sci. Am. **252**:90-100 (June). *Argues that a genetic system based on nucleic acids was too complex for the earliest organisms to have evolved. Perhaps the earliest organisms depended on clay crystals with genetic properties.*

Holzman, D. 1984. RNA: messenger, self-splicer, catalyst. Mosaic **15**(2):16-21. *Recounts some of the current investigations of RNA as a catalyst.*

Margulis, L. 1982. Early life. Boston, Science Books International. *A simplified account of the development of life on earth. Well-illustrated; undergraduate level.*

Trachtman, P. 1984. Searching for the origins of life. Smithsonian **15**(3):42-51 (June). *Current research on the origin of life.*

Vidal, G. 1984. The oldest eukaryotic cells. Sci. Am. **250**:48-57 (Feb.). *Careful examination of microscopic fossils shows that eukaryotes evolved in the form of unicellular plankton about 1.4 billion years ago.*

Woese, C.R. 1984. The origin of life. Carolina Biology Readers, no. 13. Burlington, North Carolina, Carolina Biological Supply Co. *Thought-provoking account of the author's views, critique of traditional concepts, focus on problems.*

CHAPTER 4

THE CELL AS THE UNIT
OF LIFE

━━CELL CONCEPT

More than 300 years ago the English scientist and inventor Robert Hooke, using a primitive compound microscope, observed boxlike cavities in slices of cork and leaves. He called these compartments "little boxes or cells." In the years that followed Hooke's first demonstration of the remarkable powers of the microscope before the Royal Society of London in 1663, biologists gradually began to realize that cells were far more than simple containers filled with "juices."

Cells are the fabric of life. Even the most primitive cells are enormously complex structures that form the basic units of all living matter. All tissues and organs are composed of cells. In a human an estimated 60 trillion cells interact, each performing its specialized role in an organized community. In single-celled organisms all the functions of life are performed within the confines of one microscopic package. There is no life without cells. The idea that the cell represents the basic structural and functional unit of life is an important unifying concept of biology (Principle 12, p. 10).

With the exception of eggs, which are the largest cells (in volume) known, cells are small and mostly invisible to the unaided eye. Consequently, our understanding of cells paralleled technical advances in the resolving power of microscopes. The Dutch microscopist A. van Leeuwenhoek sent letters to the Royal Society of London containing detailed descriptions of the numerous organisms he had observed using high-quality single lenses that he had made (1673 to 1723). In the early nineteenth century, the improved design of the microscope permitted biologists to see separate objects only 1 μm apart. This advance was quickly followed by new discoveries that laid the groundwork for **modern cell theory**—a theory stating that all living organisms are composed of cells.

In 1838 Matthias Schleiden, a German botanist, announced that all plant tissue was composed of cells. A year later one of his countrymen, Theodor Schwann, described animal cells as being similar to plant cells, an understanding that had been long delayed because the animal cell is bounded only by a nearly invisible plasma membrane rather than the distinct cell wall characteristic of the plant cell. Schleiden and Schwann are thus credited with the unifying cell theory that ushered in a new era of productive exploration in cell biology.

In 1840 J. Purkinje introduced the term **protoplasm** to describe the cell contents. Protoplasm was at first thought to be a granular, gel-like mixture with special and elusive life properties of its own; the cell was viewed as a bag of thick

Units of measurement commonly used in microscopic study are micrometers, nanometers, and angstroms: 1 micrometer (μm) = 0.000001 meter (about $\frac{1}{25,000}$ inch); 1 nanometer (nm) = 0.000000001 meter; (1 angstrom (Å) = 0.0000000001 meter. Thus, 1 m = 10^3 mm = 10^6 μm = 10^9 nm = 10^{10} Å. Use of the angstrom is being discontinued.

soup containing a nucleus. Later the interior of the cell became increasingly visible as microscopes were improved and better tissue-sectioning and staining techniques were introduced. Rather than being a uniform granular soup, the cell interior is composed of numerous **cell organelles,** each performing a specific function in the life of the cell (Principle 13, p. 10). Today we realize that the components of a cell are so highly organized, structurally and functionally, that describing its contents as "protoplasm" is a bit like describing the contents of an automobile engine as "autoplasm."

___ How Cells are Studied

The light microscope, with all its variations and modifications, has contributed more to biological investigation than any other instrument developed by humans. It has been a powerful exploratory tool for 300 years and continues to be so more than 50 years after the invention of the electron microscope. However, the electron microscope has vastly enhanced our understanding of the delicate internal organization of cells, and modern biochemical, immunological, and physical techniques have contributed enormously to our understanding of cell structure and function.

The electron microscope employs high voltages to direct a beam of electrons through the object to be examined. The wavelength of the electron beam is approximately 0.00001 that of ordinary white light, thus permitting far greater magnification and resolution (compare *A* and *B* of Figure 4-1). In preparation for viewing, specimens are cut into extremely thin sections and treated with "electron stains" (ions of elements such as osmium, lead, and uranium) to increase contrast between different structures. Images are seen on a fluorescent screen and photographed (Figure 4-2). Because electrons pass through the specimen to the photographic plate, the instrument is called a transmission electron microscope.

In contrast, specimens prepared for the scanning electron microscope are not sectioned, and electrons do not pass through them. The whole specimen is bombarded with electrons, causing secondary electrons to be emitted. An apparent three-dimensional image is recorded in the photograph. Although the magnifica-

Figure 4-1

Liver cells of rat. **A,** Magnified approximately 600 times through light microscope (scale bar, 34 µm). Note prominently stained nucleus in each polyhedral cell. **B,** Portion of single liver cell, magnified approximately 5000 times through electron microscope (scale bar, 4 µm). Single large nucleus dominates field; mitochondria *(M),* rough endoplasmic reticulum *(RER),* and glycogen granules *(G)* are also seen.
From Morgan, C.R., and R.A. Jersild, Jr. 1970. Anat. Record **166:**575-586.

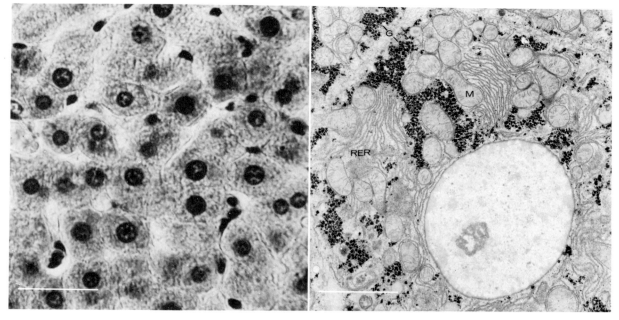

A   B

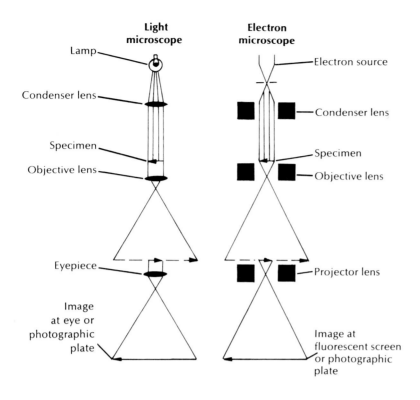

Light microscope

Lamp

Condenser lens

Specimen

Objective lens

Eyepiece

Image at eye or photographic plate

Electron microscope

Electron source

Condenser lens

Specimen

Objective lens

Projector lens

Image at fluorescent screen or photographic plate

Figure 4-2

Comparison of optical paths of light and electron microscopes. To facilitate comparison, the scheme of the light microscope has been inverted from its usual orientation with light source below and image above. In an electron microscope the lenses are magnets to focus the beam of electrons.

tion capability of the scanning instrument is not as great as the transmission microscope, a great deal has been learned about the surface features of organisms and cells. Examples of scanning electron micrographs are shown on pp. 615 and 778.

Advances in the techniques of cell study were not limited to improvements in microscopes but have included new methods of tissue preparation, staining for microscopic study, and the great contributions of modern biochemistry. For example, the various organelles of cells have differing, characteristic densities. Cells can be broken up with most of the organelles remaining intact, then centrifuged in a density gradient (Figure 4-3), and relatively pure preparations of each organelle may be recovered. Thus the biochemical functions of the various organelles may be studied separately. The DNA and various types of RNA can be extracted and studied. Many enzymes can be purified and their characteristics determined. The use of radioactive isotopes has allowed elucidation of many metabolic reactions and pathways in the cell. Modern chromatographic techniques can separate chemically similar intermediates and products. A particular protein in cells can be extracted, purified, and specific antibodies (see p. 617) against the protein can be prepared. When the antibody is complexed with a fluorescent substance and the complex is used to stain cells, the complex binds to the protein of interest, and its precise location in the cells can be determined. Many more examples could be cited, and these have contributed enormously to our present understanding of cell stucture and function.

Figure 4-3

A rotor containing samples is being placed in an ultracentrifuge. Spinning at high speeds, such devices exert forces many thousands of times the force of gravity on the samples.

Photograph courtesy Department of Biological Sciences, Texas Tech University.

ORGANIZATION OF CELLS

If we were to restrict our study of cells to fixed and sectioned tissues, we would be left with the erroneous impression that cells are static, quiescent, rigid structures. In fact, the cell interior is in a constant state of upheaval. Most cells are continually changing shape, pulsing, and heaving; their organelles twist and regroup in a

Table 4-1 Comparison of Prokaryotic and Eukaryotic Cells

Characteristic	Prokaryotic cell	Eukaryotic cell
Cell size	Mostly small (1-10 μm)	Mostly large (10-100 μm)
Genetic system	DNA with some nonhistone protein; simple, circular chromosome in nucleoid; nucleoid not membrane bound	DNA complexed with histone and nonhistone proteins in complex chromosomes within nucleus with membranous envelope
Cell division	Direct by binary fission or budding; no mitosis	Some form of mitosis; centrioles in many; mitotic spindle present
Sexual system	Absent in most; highly modified if present	Present in most; male and female partners; gametes that fuse
Nutrition	Absorption by most; photosynthesis by some	Absorption, ingestion, photosynthesis by some
Energy metabolism	Mitochondria absent; oxidative enzymes bound to cell membrane, not packaged separately; great variation in metabolic pattern	Mitochondria present; oxidative enzymes packaged therein; more unified pattern of oxidative metabolism
Intracellular movement	None	Cytoplasmic streaming, phagocytosis, pinocytosis

cytoplasm teeming with starch granules, fat globules, and vesicles of various sorts. This description is derived from studies of living cell cultures with time-lapse photography and video. If we could see the swift shuttling of molecular traffic through gates in the cell membrane and the metabolic energy transformations within cell organelles, we would have an even stronger impression of internal turmoil. However, the cell is anything but a bundle of disorganized activity. There is order and harmony in the cell's functioning that represents the elusive phenomenon we call life. Studying this dynamic miracle of evolution through the microscope, we realize that, as we gradually comprehend more and more about this unit of life and how it operates, we are gaining a greater understanding of the nature of life itself.

___ Prokaryotic and Eukaryotic Cells

The radically different cell plan of prokaryotes and eukaryotes is described previously (p. 46). A fundamental distinction, expressed in their names, is that prokaryotes lack the membrane-bound nucleus present in all eukaryotic cells. Other major differences are summarized in Table 4-1.

Despite these differences, which are of paramount importance in cell studies, prokaryotes and eukaryotes have much in common. Both have DNA, use the same genetic code, and synthesize proteins. Many specific molecules such as ATP perform similar roles in both. These fundamental similarities imply common ancestry (Principle 3, p. 8).

Prokaryotic organisms are bacteria separated into the kingdoms Archaebacteria and Monera. The most complex of these are the filamentous forms of cyanobacteria and certain other bacteria. All other organisms are eukaryotes distributed among four kingdoms: the unicellular kingdom Protista (protozoa and nucleated algae) and three multicellular kingdoms—Plantae (green plants), Fungi (true fungi), and Animalia (multicellular animals). The kingdom classifications are discussed in Chapter 7 (p. 126). The following discussion is restricted to eukaryotic cells, of which all animals are composed.

___ Components of Eukaryotic Cells and Their Functions

If the inside of the cheek is gently scraped with a blunt instrument and the scrapings are put on a slide in a drop of physiological salt solution and examined unstained with a microscope, living cells can be seen. The flat circular cells with small nuclei are the squamous epithelial cells that line much of the mouth.

Flat epithelial cells are only one of many different shapes assumed by cells. Although many cells, because of surface tension forces, assume a spherical shape when freed from restraining influences, others retain their shape under most conditions because of their characteristic cytoskeleton (Figure 4-4).

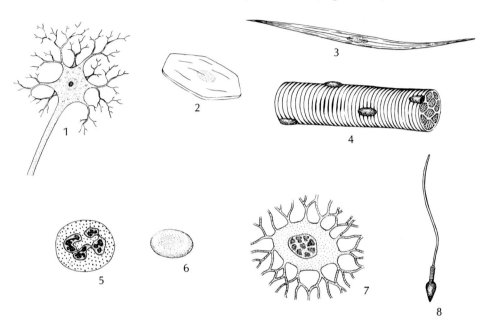

Figure 4-4

Types of cells. **1,** Nerve cell, showing cell body (soma) surrounded by numerous dendritic extensions and a portion of the axon extending below. **2,** Epithelial cell from lining of the mouth. **3,** Smooth muscle cell from intestinal wall. **4,** Striated muscle cell from skeletal muscle. **5,** White blood corpuscle. **6,** Red blood corpuscle (erythrocyte). **7,** Bone cell. **8,** Human spermatozoon. (Not drawn to the same scale.)

Typically the eukaryotic cell is enclosed within a thin, sturdy, differentially permeable **plasma membrane** (Figures 4-5 and 4-6). This structure regulates the flow of materials between the cell and its surroundings. In some cells, such as nerve cells, the plasma membrane also is involved in intercellular communication. In other cells, such as intestinal epithelium, the plasma membrane is modified into numerous, small, fingerlike projections called **microvilli** (sing. **microvillus**) that increase the surface area of the cell (Figure 4-7).

The most prominent organelle is the spherical or ovoid nucleus enclosed within *two* membranes to form the double-layered **nuclear envelope** (Figure 4-8). At intervals the nuclear envelope is perforated by pores, permitting some continuity between the nuclear contents and the cytoplasm surrounding the nucleus. The nucleus contains chromatin and one or more dense, granular structures called **nucleoli** (sing., **nucleolus**) (Figure 4-8). The chromatin is a complex of DNA and histone and nonhistone protein and carries the genetic information of the cell. Nucleoli are specialized parts of certain chromosomes that carry multiple copies of the DNA information to synthesize ribosomal RNA. After transcription from the nucleolar DNA, the ribosomal RNA combines with several different proteins to form a **ribosome** (Figure 4-8), detaches from the nucleolus, and passes through nuclear pores to the cytoplasm.

The cytoplasm contains many organelles such as mitochondria, Golgi complexes, and centrioles. Plant cells typically contain **plastids,** the photosynthetic organelles, and bear a cell wall containing cellulose outside the plasma membrane. Animal cells lack plastids and a cell wall.

Figure 4-5

Generalized cell with principal organelles, as might be seen with the electron microscope. Each of the major organelles is shown enlarged. Membranes of organelles are believed to be continuous with, or derived from, the plasma membrane by an infolding process. Structure of other membranes (of nucleus, endoplasmic reticulum, mitochondria, and others) is probably similar to that of plasma membrane, shown enlarged at lower left.

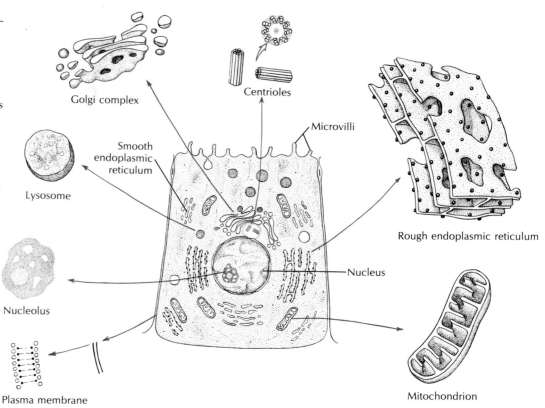

Golgi complex

Centrioles

Microvilli

Smooth endoplasmic reticulum

Lysosome

Rough endoplasmic reticulum

Nucleus

Nucleolus

Plasma membrane

Mitochondrion

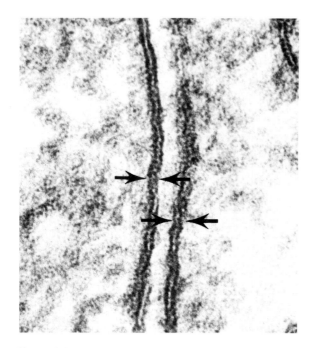

Figure 4-6

Plasma membranes of two adjacent cells. Each membrane *(between arrows)* shows a typical dark-light-dark staining pattern. (×325,000.)
Courtesy A. Wayne Vogl.

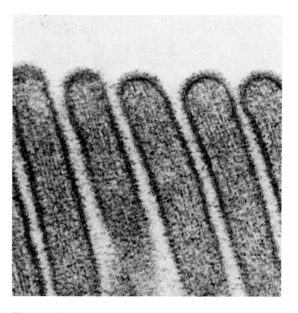

Figure 4-7

Microvilli of small intestine. (×85,000.)
Courtesy J.D. Berlin.

Nucleolus   Nucleus   Mitochondrion

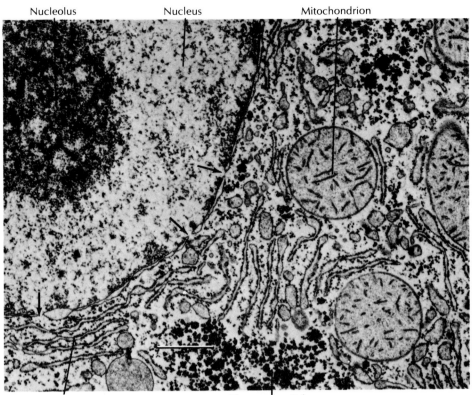

Endoplasmic reticulum   Glycogen particles

Figure 4-8

Electron micrograph of part of hepatic cell of rat showing portion of nucleus *(left)* and surrounding cytoplasm. Endoplasmic reticulum and mitochondria are visible in cytoplasm, and pores *(arrows)* can be seen in nuclear envelope. (×14,000; scale bar 1 μm)
Courtesy G.E. Palade, The Rockefeller University, New York.

The space between the membranes comprising the nuclear envelope connects at some points with the space (channels, or **cisternae** [sing. **cisterna**]) within the membranes of the **endoplasmic reticulum (ER)**. The ER (Figures 4-5 and 4-8) is a complex of membranes that separates some of the products of the cell from the synthetic machinery that produces them. The cisternae of the ER apparently function as routes for transport of certain substances within the cell. Often the membranes of the ER are lined on their outer surfaces with ribosomes and are thus designated **rough ER** (contrasted with **smooth ER,** without ribosomes) (Figure 4-9). In some instances it has been shown that the protein synthesized by the ribosomes on the rough ER enters the cisternae and from there is transported to the Golgi apparatus or complex (Figure 4-10). The **Golgi complex** is a stack of smooth, membranous cisternae (Figure 4-11) that functions in the storage, modification, and packaging of protein products, especially secretory products. It does not synthesize protein but may add polysaccharide to the complex. As its products mature, parts of the cisternae pinch off and become membrane-bound **vesicles** in the cytoplasm. The contents of some of these vesicles may be expelled to the outside of the cell as secretory products destined to be exported from a glandular cell. Others may contain digestive enzymes that remain in the cell that produces them. Such vesicles are called **lysosomes** (literally "releasing body," a body capable of causing lysis, or disintegration). The enzymes they contain are involved in the breakdown of foreign material, including bacteria engulfed by the cell. Lysosomes also are capable of breaking down injured or diseased cells and worn-out cellular components, since the enzymes they contain are so powerful that they kill the cell that formed them if the lysosome membrane ruptures. In normal cells the enzymes remain safely enclosed within the protective membrane.

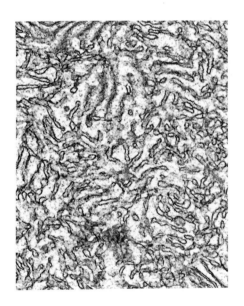

Figure 4-9

Extensive smooth endoplasmic reticulum in a Sertoli cell from the testis of a ground squirrel. This kind of endoplasmic reticulum lacks ribosomes. (×22,800.)
Courtesy A. Wayne Vogl.

Figure 4-10

System for assembling, isolating, and secreting proteins for export from, and use inside, a eukaryotic cell.

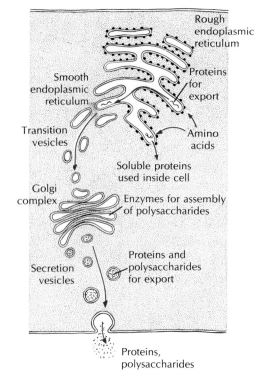

Rough endoplasmic reticulum

Smooth endoplasmic reticulum

Proteins for export

Transition vesicles

Amino acids

Soluble proteins used inside cell

Golgi complex

Enzymes for assembly of polysaccharides

Secretion vesicles

Proteins and polysaccharides for export

Proteins, polysaccharides

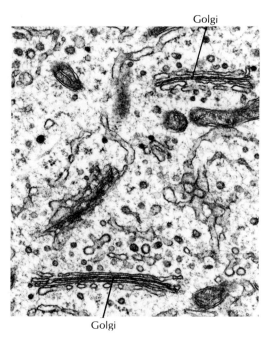

Golgi

Golgi

Figure 4-11

Golgi complexes in a Sertoli cell of ground squirrel testis. (×26,700.) Courtesy A. Wayne Vogl.

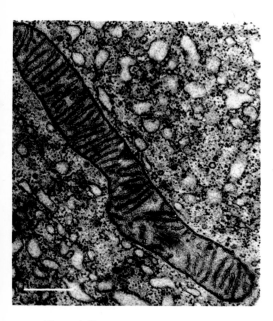

Figure 4-12

Electron micrograph of elongated mitochondrion in pancreatic exocrine cell of guinea pig. The mitochondrion has two membranes, an outer smooth membrane and an inner membrane that is folded inward into invaginations called cristae. These appear as incomplete transverse septa inside the mitochondrion. (×30,000; scale bar 6.5 μm) Courtesy G.E. Palade.

Mitochondria (sing., **mitochondrion**) (Figure 4-12) are conspicuous organelles present in nearly all eukaryotic cells. They are diverse in size, number, and shape; some are rodlike, and others are more or less spherical. They may be scattered uniformly through the cytoplasm, or they may be localized near cell surfaces and other regions where there is high metabolic activity. The mitochondrion is composed of a double membrane. The outer membrane is smooth, whereas the inner membrane is folded into numerous platelike or fingerlike projections called **cristae** (Figures 4-5 and 4-12). These characteristic features make mitochondria easy to identify among the organelles. Mitochondria are often called "powerhouses of the cell" because enzymes located on the cristae carry out the energy-yielding steps of aerobic metabolism. ATP, the most important energy-storage molecule of all cells, is produced in this organelle. Mitochondria are self-replicating. They have a tiny, circular chromosome, resembling the chromosomes of prokaryotes but much smaller. The chromosome contains DNA that specifies some, but not all, of the proteins of the mitochondrion.

Eukaryotic cells characteristically have a system of tubules and filaments that form the **cytoskeleton** (Figures 4-13 and 4-14). These provide support and maintain the form of the cell, and in many cells, they provide a means of locomotion and translocation of organelles within the cell. **Microfilaments** are thin, linear structures, first observed distinctly in muscle cells, where they are responsible for the ability of the cell to contract. They are made of a protein called **actin.** Several dozen other proteins are known that bind with actin and determine its configuration and behavior in particular cells. One of these is **myosin,** whose interaction with actin causes contraction in muscle and other cells. **Microtubules,** somewhat larger than microfilaments, are tubular structures composed of a protein called **tubulin** (Figure 4-14). They play a vital role in moving the chromosomes toward the daughter cells during cell division as will be seen later, and they

Microtubules   Intermediate filaments

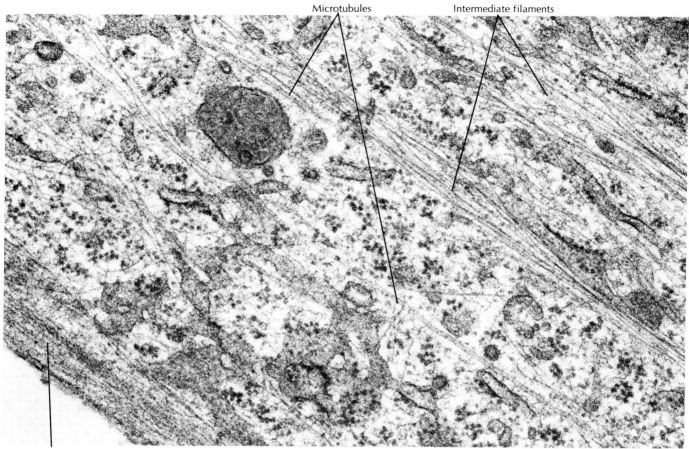

Microfilaments

Figure 4-13

Cytoskeleton of a cell, showing its complex nature. Three visible cytoskeletal elements, in order of increasing diameter, are microfilaments, intermediate filaments, and microtubules. (×66,600.)
Courtesy A. Wayne Vogl.

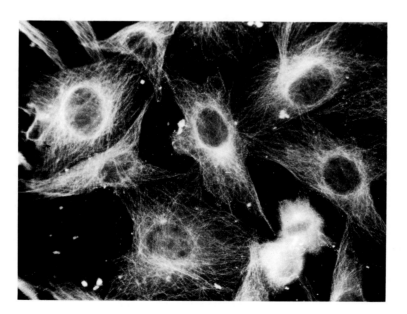

Figure 4-14

The microtubules in kidney cells of a baby hamster have been rendered visible by treating them with a preparation of fluorescent proteins that specifically bind to tubulin.
Photograph by K.G. Murti, Visuals Unlimited.

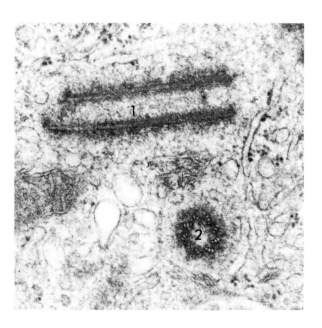

Figure 4-15

Electron micrograph of centrioles.
Photograph courtesy Charles J. Flickinger, M.D.

are important in intracellular architecture, organization, and transport. In addition, microtubules form essential parts of the structures of cilia and flagella. Microtubules radiate out from a microtubule organizing center near the nucleus, sometimes called the **centrosome**. Within the centrosome are found a pair of **centrioles** (Figures 4-5 and 4-15), which are themselves composed of microtubules. Each centriole of a pair lies at right angles to the other and is a short cylinder of nine triplets of microtubules. They replicate before cell division. Although cells of higher plants do not have centrioles, a microtubule organizing center is present. **Intermediate filaments** are larger than microfilaments and smaller than microtubules. There are five biochemically distinct types of intermediate filaments, and their composition and arrangement depend on the cell type in which they are found.

___ Surfaces of Cells and Their Specializations

The free surface of epithelial cells (cells that cover the surface of a structure or line a tube or cavity) sometimes bears either **cilia** or **flagella** (sing., **cilium, flagellum**). These are motile extensions of the cell surface that sweep materials past the cell. In single-celled animals (many of the protozoa) and some small multicellular forms, they propel the entire animal through a liquid medium. Flagella provide the means of locomotion for the sperm of most animals and many plants.

Cilia occur in large numbers on each cell and are relatively short (5 to 10 μm). Flagella typically, although not always, occur singly or in a small number per cell and are long, whiplike structures that may reach 150 μm in length. The distinction between cilia and flagella is that a cilium propels water parallel to the surface to which the cilium is attached, whereas a flagellum propels water parallel to the main axis of the flagellum (Fig. 28-7, p. 602). Their internal structure is the same. With few exceptions, the internal structures of locomotory cilia and flagella are composed of a long cylinder of nine pairs of microtubules enclosing a central pair. At the base of each cilium or flagellum is a **basal body (kinetosome),** which is identical in structure to a centriole.

Many cells move neither by cilia or flagella but by **ameboid movement** using **pseudopodia.** Some groups of protozoa (p. 138), migrating cells in embryos of multicellular animals, and some cells of adult multicellular animals, such as white blood cells, show ameboid movement. Cytoplasmic streaming through the action

Indeed, cilia and flagella are so alike in the details of their structure that it seems highly likely that they had a common evolutionary origin. Whether their origin was the symbiosis of a spirochete-like bacterium and host cell, as suggested by Margulis, is more conjectural. Margulis and others prefer the term undulipodia to include both cilia and flagella, and it is less awkward to use one word for structures that are alike in structure and origin. However, the terms "cilia" and "flagella" are so common and widely used that the student should be familiar with them.

of actin microfilaments extends a lobe (pseudopodium) outward from the surface of the cell. Continued streaming in the direction of the pseudopodium brings the cytoplasmic organelles into the lobe and accomplishes movement of the entire cell. Some specialized pseudopodia have cores of microtubules (p. 139), and movement is effected by assembly and disassembly of the tubular rods.

Cells covering the surface of a structure or cells packed together in a tissue may have specialized junctional complexes between them. Nearest the free surface, the two apposing cell membranes appear to fuse, forming a **tight junction** (Figure 4-16). At various points small ellipsoid discs occur, just beneath the cell membrane in each cell. These discs are known as **desmosomes.** Specialized proteins extend through the cell membrane of each cell and project a **glycoprotein** (a protein with a sugar molecule attached) that binds with the glycoprotein of the other cell. Intermediate filaments are attached to the desmosome within each cell and extend into the cytoplasm. **Gap junctions,** rather than serving as points of attachment, provide a means of intercellular communication. They form tiny canals between cells so that the cytoplasm becomes continuous, and small molecules can pass from one cell to another.

Another specialization of the cell surface is the lacing together of adjacent cell surfaces where the plasma membranes of the cells infold and interdigitate much like a zipper. They are especially common in the epithelium of kidney tubules. Microvilli, mentioned earlier, are another type of cell surface specialization (Figures 4-7 and 4-16). With the electron microscope they can be seen clearly in the lining of the intestine where they greatly increase the absorptive and digestive surface. Such specializations appear as brush borders when viewed with the light microscope.

___MEMBRANE STRUCTURE AND FUNCTION
___Structure of the Cell Membrane

The incredibly thin, yet sturdy, plasma membrane that encloses every cell is vitally important in maintaining cellular integrity. Once believed to be a rather static entity that defined cell boundaries and kept cell contents from spilling out, the plasma membrane (also called the plasmalemma) has proved to be a dynamic structure having remarkable activity and selectivity. It is a permeability barrier that separates the interior from the external environment of the cell, regulates the vital flow of molecular traffic into and out of the cell, and provides many of the unique functional properties of specialized cells.

Membranes inside the cell surround a variety of organelles. Indeed, the cell is a system of membranes that divide it into numerous compartments. Someone has estimated that if all the membranes present in 1 g of liver tissue were spread out flat, they would cover 30 square meters! Internal membranes share many of the structural features of the plasma membrane and are the site for many, perhaps most, of the cell's enzymatic reactions.

The basic structure of all biological membranes is a phospholipid bilayer. (The structure of phospholipids is given on p. 29.) One end of the molecule is water soluble (hydrophilic, or "water loving"), whereas the other end, consisting of hydrocarbon chains of fatty acids that are insoluble in water, is hydrophobic ("water fearing"). In membranes the hydrophobic ends of the molecules in each layer point toward each other, and the hydrophilic ends are directed toward the water phase (Figure 4-17). On mixing phospholipids with water, the phospholipid molecules spontaneously arrange themselves in a bilayer, which is the state of lowest free energy for these molecules in water. Because ions and most biological

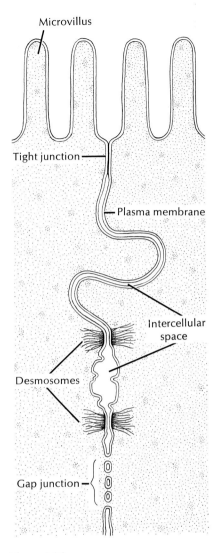

Figure 4-16

Two opposing plasma membranes forming the boundary between two epithelial cells. Various kinds of junctional complexes are found. The tight junction is a firm, adhesive band completely encircling the cell. Desmosomes are isolated "spot-welds" between cells. Gap junctions serve as sites of intercellular communication. Intercellular space may be greatly expanded in epithelial cells of some tissues.

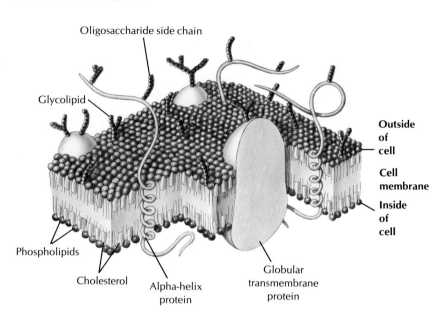

Oligosaccharide side chain

Glycolipid

**Outside
of
cell**

**Cell
membrane**

**Inside
of
cell**

Phospholipids

Cholesterol   Alpha-helix
protein

Globular
transmembrane
protein

Figure 4-17

Fluid-mosaic model of a cell membrane.

molecules are water soluble, the hydrocarbon forms a barrier between the cell's interior and its environment. An important characteristic of the phospholipid bilayer is that it is a liquid. This gives the membrane flexibility and allows the phospholipid molecules to diffuse sideways freely within their own monolayer.

Another important lipid in membranes is cholesterol (see p. 29). There are about equal numbers of cholesterol and phospholipid molecules in a membrane. The cholesterol in the membrane makes it even less permeable and decreases the flexibility.

Glycoproteins are essential components of plasma membranes. The glycoproteins fall into two broad categories according to their shape within the hydrocarbon region of the membrane (Figure 4-17). The hydrophilic, carbohydrate portions of both types are located on the extracellular ends of the protein. One type has a substantial globular portion within the hydrocarbon core. Some of these proteins catalyze the transport of substances such as negatively charged ions across the membrane. The other type has a tightly coiled alpha helix (p. 30) spanning the hydrophobic core of the membrane, and the hydrophobic side chains of the amino acids project outward from the helix. A hydrophilic end anchors the protein within the cytoplasm. Important functions of this type of protein include action as specific receptors for various molecules or as highly specific markings. For example, the self-nonself recognition that enables the immune system to react to invaders (Chapter 29) is based on proteins of this type.

Like the phospholipid molecules, most of the glycoproteins can move laterally in the membrane, although more slowly. This view of the membrane, in which a variety of protein molecules are embedded in a phospholipid bilayer, and in which they can move about, is known as the **fluid-mosaic model** and is now widely accepted.

Function of the Cell Membrane

The plasma membrane acts as a gatekeeper for the entrance and exit of the many substances involved in cell metabolism. Some substances can pass through with ease, others enter slowly and with difficulty, and still others cannot enter at all. This is called the **selective behavior** of the cell membrane. Because conditions outside the cell are different from and more variable than conditions within the cell, it is necessary that the passage of substances across the membrane be rigorously controlled.

We recognize three principal ways that a substance may traverse the cell membrane: (1) by **diffusion** along a concentration gradient; (2) by a **mediated transport system,** in which the substance binds to a specific site that in some way assists it across the membrane; and (3) by **endocytosis,** in which the substance is enclosed within a vesicle that forms on and detaches from the membrane surface to enter the cell.

Diffusion and osmosis

If a living cell surrounded by a membrane is immersed in a solution having more solute molecules than the fluid inside the cell, a **concentration gradient** instantly exists between the two fluids. Assuming that the membrane is **permeable** to the solute, there is a net movement of solute toward the inside, the side having the lower concentration. The solute diffuses "downhill" across the membrane until its concentrations on each side are equal.

Most cell membranes are **selectively permeable,** that is, permeable to water but selectively permeable or impermeable to solutes. In free diffusion it is this selectiveness that regulates molecular traffic. As a rule, gases (such as oxygen and carbon dioxide), urea, and lipid-soluble solutes (such as hydrocarbons and alcohol) are the only solutes that can diffuse through biological membranes with any degree of freedom. Since many water-soluble molecules readily pass through membranes, such movements cannot be explained by simple diffusion. Instead sugars, as well as many electrolytes and macromolecules, are moved across membranes by carrier-mediated processes, described in the next section.

If a membrane is placed between two unequal concentrations of solutes to which the membrane is impermeable or only weakly permeable, water flows through the membrane from the more dilute to the more concentrated solution. In effect the water molecules move down a concentration gradient from an area where the water molecules are more concentrated to an area where they are less concentrated. This is **osmosis.** To understand why this happens, we must view the system from the standpoint of the state of the water on each side.

Osmosis can be demonstrated by a simple experiment in which a selectively permeable membrane such as cellophane is tied tightly over the end of a funnel. The funnel is filled with a sugar solution and placed in a beaker of pure water so that the water levels inside and outside the funnel are equal. In a short time the water level in the glass tube of the funnel rises, indicating that water is passing through the cellophane membrane into the sugar solution (Figure 4-18).

Inside the funnel are sugar molecules, as well as water molecules. In the beaker outside the funnel are only water molecules. Thus the concentration of water is less on the inside because some of the available space is occupied by the larger, nondiffusible sugar molecules. A concentration gradient is said to exist for water molecules in the system. Water diffuses from the region of greater concentration of water (pure water outside) to the region of lesser concentration (sugar solution inside).

As water enters the sugar solution, the fluid level in the funnel rises, creating a hydrostatic pressure because of gravity inside the osmometer. Eventually the pressure produced by the increasing weight of solution in the funnel pushes water molecules out as fast as they enter. The level in the funnel becomes stationary and the system is in equilibrium. The **osmotic pressure** of the solution is equivalent to the hydrostatic pressure necessary to prevent further net entry of water.

The concept of osmotic pressure is not without problems. A solution reveals an osmotic "pressure" only when it is separated from solvent by a differentially permeable membrane. It can be disconcerting to think of an isolated bottle of

Figure 4-18

Simple membrane osmometer.

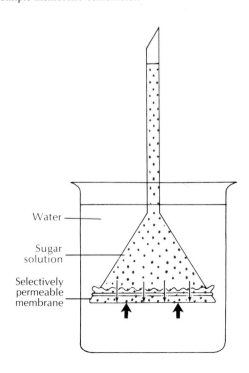

Water

Sugar
solution

Selectively
permeable
membrane

sugar solution as having osmotic "pressure" much as compressed gas in a bottle would have. Furthermore, the osmotic pressure is really the hydrostatic pressure that must be applied to a solution to keep it from gaining water. Consequently, biologists frequently use the term **osmotic potential** rather than osmotic pressure. However, since the term "osmotic pressure" is so firmly fixed in our vocabulary, it is necessary to understand the usage despite its potential confusion.

The *direct* measurement of osmotic pressure in biological solutions is seldom done today because the osmotic pressures of most biological solutions are so great that it would be impractical if not impossible to measure them with the simple membrane osmometer described. The osmotic pressure of human blood plasma would lift a fluid column more than 250 feet—if we could construct such a long, vertical tube and find a membrane that would not rupture from the pressure.

Indirect methods of measuring osmotic pressure are more practical. By far the most widely used measurement is the **freezing point depression.** This is a much faster and more accurate determination than is the direct measurement of osmotic pressure by the collodion membrane osmometer. Pure water freezes at exactly 0° C. As solutes are added, the freezing point is lowered; the greater the concentration of solutes, the lower the freezing point. Human blood plasma freezes at approximately −0.56° C; seawater freezes at approximately −1.80° C. Although the lowering of the freezing point of water by the presence of solutes is small, great accuracy of measurement is possible because the instruments used by biologists can detect differences of as little as 0.001° C.

Carrier-mediated transport

We have seen that the cell membrane is an effective barrier to the free diffusion of most molecules of biological significance. Yet it is essential that such materials enter and leave the cell. Nutrients such as sugars and materials for growth such as amino acids must enter the cell, and the wastes of metabolism must leave. Such molecules are moved across the membrane by special proteins called **carriers,** built into the structure of the membrane. It has been assumed that the carrier molecule captures a solute molecule to be transported, forming a temporary carrier-solute complex, much as enzymes form a temporary enzyme-substrate complex (p. 76). The carrier molecule, with its fare, would move or rotate to the opposite surface, where the solute would detach and leave the membrane. The carrier would move back, again presenting its attachment site for the pickup and transport of another solute molecule. In some cases, however, it has been shown that the carrier protein forms a small passageway through the protein, enabling the solute molecule to cross the phospholipid bilayer. Protein carriers are usually quite specific, recognizing and transporting only a limited group of chemical substances or perhaps even a single substance.

Figure 4-19

A, Diffusion through membrane channels. **B,** Carrier-mediated transport. Simple diffusion through membrane channels differs from carrier-mediated transport in that the latter shows saturation at high concentration. Simple diffusion does not exhibit saturation at high concentration.

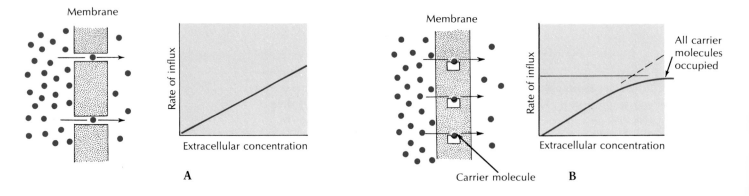

Membrane

Rate of influx

Extracellular concentration

A

Membrane

Rate of influx

All carrier molecules occupied

Extracellular concentration

Carrier molecule   **B**

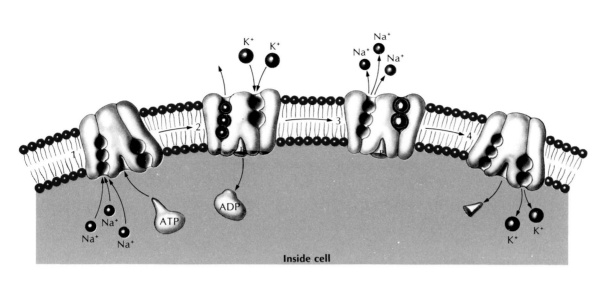

Inside cell

The carrier hypothesis is consistent with the observation that at high concentrations some solutes show a saturation effect. This means simply that the rate of influx reaches a plateau beyond which increasing the solute concentration has no further effect on influx rate (Figure 4-19, *B*). This is evidence that the number of carriers available in the membrane is limited. When all become occupied by solutes, the rate of transport is at a maximum and cannot be increased. Simple diffusion shows no such limitation; the greater the difference in solute concentration on the two sides of the membrane, the faster the influx (Fig. 4-19, *A*).

At least two distinctly different kinds of carrier-mediated transport mechanisms are recognized: (1) **facilitated transport,** in which the carrier assists a molecule to diffuse through the membrane that it cannot otherwise penetrate, and (2) **active transport,** in which energy is supplied to the carrier systems to transport molecules in the direction opposite to the gradient. Facilitated transport therefore differs from active transport in that it sponsors movement in a downhill direction (in the direction of the concentration gradient) only and requires no metabolic energy to drive the carrier system.

In higher animals facilitated transport aids in the transport of glucose (blood sugar) into body cells that burn it as a principal energy source for the synthesis of ATP. The concentration of glucose is greater in the blood than in the cells that consume it, favoring inward diffusion, but glucose is a polar molecule that does not, by itself, penetrate the membrane rapidly enough to support the metabolism of many cells; the carrier system increases the inward flow of glucose.

In active transport, molecules are moved uphill against the forces of passive diffusion. Active transport always involves the expenditure of energy (from ATP) because materials are pumped against a concentration gradient. Among the most important active transport systems in all animals are those that maintain sodium and potassium gradients between cells and the surrounding extracellular fluid or external environment. Most animal cells require a high internal concentration of potassium for protein synthesis at the ribosome and for certain enzymatic functions. The potassium concentration may be 20 to 50 times greater inside the cell than outside. Sodium, on the other hand, may be 10 times more concentrated outside the cell than inside. Both of these electrolyte gradients are maintained by the active transport of potassium into and sodium out of the cell. In many cells one kind of molecule on one side of the membrane is transported simultaneously with another kind of molecule or ion on the opposite side. The **sodium-potassium pump** that maintains the sodium and potassium gradients across nerve cell membranes is an example (Figure 4-20). There is evidence that 10% to 40% of all energy produced by some cells is used to power the sodium-potassium pump.

Figure 4-20

Sodium-potassium pump, powered by bond energy of ATP, maintains the normal gradients of these ions across the cell membrane. The pump works by a series of conformational changes in the transmembrane protein: *Step 1.* Three ions of Na^+ bind to the interior end of the protein, producing a conformational change in the protein complex. *Step 2.* The complex binds a molecule of ATP and cleaves it. *Step 3.* The binding of the phosphate group to the complex induces a second conformational change, passing the three Na^+ ions across the membrane, where they are now positioned facing the exterior. This new conformation has a very low affinity for the Na^+ ions, which dissociate and diffuse away, but it has a high affinity for K^+ ions and binds two of them as soon as it is free of the Na^+ ions. *Step 4.* Binding of the K^+ ions leads to another conformational change in the complex, this time leading to dissociation of the bound phosphate. Freed of the phosphate, the complex reverts to its original conformation, with the two K^+ ions exposed on the interior side of the membrane. This conformation has a low affinity for K^+ ions so that they are now released, and the complex has the conformation it started with, having a high affinity for Na^+ ions.

From Raven, P., and G.B. Johnson. 1986. Biology. St. Louis, Times Mirror/Mosby.

Figure 4-21

Three types of endocytosis. In phagocytosis the cell membrane binds to a large particle and extends to engulf it. In pinocytosis a small area of cell membrane invaginates to finally close off a droplet of surrounding medium. Receptor-mediated endocytosis is a mechanism for highly selective uptake of large molecules or particles. Binding of the ligand to the specific receptor on the surface membrane stimulates invagination of the receptor-ligand complex.

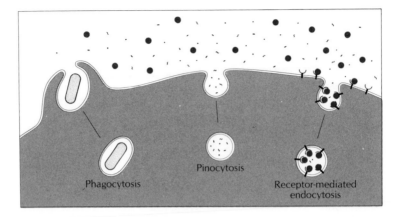

Phagocytosis

Pinocytosis

Receptor-mediated endocytosis

Endocytosis

The ingestion of solid or fluid material by cells was observed by microscopists nearly 100 years before phrases such as "active transport" and "protein carrier mechanism" were a part of the biologist's vocabulary. Endocytosis is a collective term that describes three similar processes, **phagocytosis, pinocytosis,** and **receptor-mediated endocytosis** (Figure 4-21). All require energy and thus may be considered forms of active transport.

Phagocytosis, which literally means "cell eating," is a common method of feeding among the protozoa and lower metazoa. It is also the way in which white blood cells (leukocytes) engulf cellular debris and uninvited microbes in the blood. By phagocytosis, the cell membrane forms a pocket that engulfs the solid material. The membrane-enclosed vesicle then detaches from the cell surface and moves into the cytoplasm where its contents are digested by intracellular enzymes (Figure 31-5, p. 666).

Pinocytosis, or "cell drinking," is similar to phagocytosis except that drops of fluid are sucked discontinuously through tubular channels into cells to form tiny vesicles. These may combine to form larger vacuoles. When a cell engages in pinocytosis, it brings within itself whatever molecules are contained in the fluid droplets. Thus it is nonspecific.

In contrast, receptor-mediated endocytosis is highly specific. Proteins of the plasma membrane specifically bind particular molecules (referred to as **ligands** in this process), which may be present in the extracellular fluid in very low concentrations. The area of the cell surface that contains the receptors and bound ligands then invaginates and is brought within the cell as in phagocytosis and pinocytosis. Within the vesicle, the receptor and the ligand are dissociated, and the receptor and membrane material are recycled back to the surface membrane. A variety of important molecules are brought into cells in this manner.

Exocytosis

Just as materials can be brought into the cell by invagination and formation of a vesicle, the membrane of a vesicle can fuse with the plasma membrane and extrude its contents to the surrounding medium. This is the process of **exocytosis.** This process occurs in various cells to remove undigestible residues of substances brought in by endocytosis, to secrete substances such as hormones, (Figure 4-10), and to transport a substance completely across a cellular barrier. For example, a substance may be picked up on one side of the wall of a blood vessel by endocytosis, moved across the cell, and released by exocytosis.

___MITOSIS AND CELL DIVISION

All cells of the body arise from the division of preexisting cells (Principle 12, p. 10). All the cells found in most multicellular organisms have originated from the division of a single cell, the **zygote,** which is formed from the union (fertilization) of an **egg** and a **sperm.** Cell division provides the basis for one form of growth, for both sexual and asexual reproduction, and for the transmission of hereditary qualities from one cell generation to another cell generation.

In the formation of **body cells** (somatic cells) the process of nuclear division is referred to as **mitosis.** By mitosis each "daughter cell" is assured of receiving a complete set of genetic instructions. Mitosis is a delivery system for distributing the chromosomes and the DNA they contain to continuing cell generations. Thus, all somatic cells, which number hundreds of billions in large animals, have the same genetic content, since all are descended by faithful reproduction of the original fertilized egg. As an animal grows, its somatic cells differentiate and assume different functions and appearances because of differential gene action. Even though most of the genes in specialized cells remain silent and unexpressed throughout the lives of those cells, every cell possesses a complete genetic complement. Mitosis ensures equality of genetic potential; later, other processes direct the orderly expression of genes during embryonic development by selecting from the genetic instructions that each cell contains. (These fundamental properties of cells of multicellular organisms are stated in Principles 22, 23, and 25, pp. 13 and 14, and are discussed further in Chapter 36.)

In animals that reproduce **asexually,** mitosis is the only mechanism for the faithful transfer of genetic information from parent to progeny. In animals that reproduce **sexually,** the parents must produce **sex cells** (gametes or germ cells) that contain only half the usual number of chromosomes, so that the offspring formed by the union of the gametes will not contain double the number of parental chromosomes. This requires a special type of *reductional* division called **meiosis,** described in Chapter 35.

___ Cell Cycle

Cycles are conspicuous attributes of life. The descent of a species through time is in a very real sense a sequence of life cycles (Principle 23, p. 13). Similarly, cells undergo cycles of growth and replication as they repeatedly divide. A cell cycle is a mitosis-to-mitosis cycle, that is, the interval between one cell generation and the next (Figure 4-22).

Actual nuclear division occupies only about 5% of the cell cycle; the rest of the cell's time is spent in **interphase,** the stage between nuclear divisions. For many years it was thought that interphase was a period of rest because the nucleus appeared inactive when observed with the ordinary light microscope. In the early 1950s new techniques for revealing DNA replication in nuclei were introduced at the same time that biologists came to appreciate fully the significance of DNA as the genetic material. It was then discovered that DNA replication occurred during the interphase stage. Further studies revealed that many other protein and nucleic acid components essential to normal cell growth and division were synthesized during the seemingly quiescent interphase period.

Replication of DNA occurs during a phase called the S period (period of synthesis). In mammalian cells in tissue culture, the S period lasts about 6 of the 18 to 24 hours required to complete one cell cycle. In this period both strands of DNA must replicate; that is, new complementary partners are synthesized for

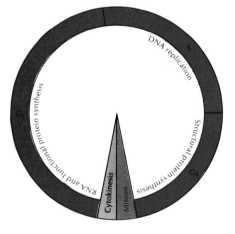

Figure 4-22

Cell cycle, showing relative duration of recognized periods. S, G_1, and G_2 are periods within interphase; S, synthesis of DNA; G_1, presynthetic period; G_2, postsynthetic period. Actual duration of the cycle and the different periods varies considerably in different cell types.

both strands so that two identical molecules are produced from the original strand.

The S period is preceded and succeeded by G_1 and G_2 periods respectively (G stands for "gap"), during which no DNA synthesis is occurring. For most cells, G_1 is an important preparatory stage for the replication of DNA that follows. During G_1, transfer RNA, ribosomes, messenger RNA, and several enzymes are synthesized. During G_2, spindle and aster proteins are synthesized in preparation for chromosome separation during mitosis. G_1 is typically of longer duration than G_2, although there is much variation in different cell types. Embryonic cells divide very rapidly because there is no cell growth between divisions, only subdivision of mass. DNA synthesis may proceed a hundred times more rapidly in embryonic cells than in adult cells, and the G_1 period is very shortened.

Structure of Chromosomes

As mentioned earlier, DNA in the eukaryotic cell occurs in chromatin, a complex of DNA with histone and nonhistone protein. The chromatin is organized into a number of discrete bodies called **chromosomes** (color bodies), so named because they stain deeply with certain biological dyes. In cells that are not dividing, the chromatin is loosely organized and dispersed, so that the individual chromosomes cannot be distinguished (Chapter 38, p. 822). Before division the chromatin condenses, and the chromosomes can be recognized and their individual morphological characteristics determined. They are of varied lengths and shapes, some bent and some rodlike. Their number is constant for the species, and every body cell (but not the germ cells) has the same number of chromosomes regardless of the cell's function. A human, for example, has 46 chromosomes in each body (somatic) cell.

During mitosis (nuclear division) the chromosomes shorten and become increasingly condensed and distinct, and each assumes a characteristic shape. At some point on the chromosome is a **centromere,** or constriction, to which are attached several spindle fibers that move the chromosome toward the pole during mitosis.

The formidable problem of packaging the cell's DNA so that the genetic instructions are accessible during the transcription process (the formation of messenger RNA from nuclear DNA) is described in Chapter 38, p. 828.

In 1974 it was discovered that the chromatin is composed of repeating subunits, called **nucleosomes.** Each nucleosome is a narrow "spool" or disc of histone proteins around which one and three fourths turns of double-helical DNA are wound to form a superhelix. The nucleosomes are linked together by the continuous DNA strand much like beads on a string. This arrangement is thought to explain the knobby appearance of chromatin fibers as revealed by high-resolution electron micrographs. Since the DNA is wound around the *outside* of the nucleosome beads, all base pairs of the molecule are accessible for transcription.

Phases in Mitosis

There are two distinct stages of cell division: the division of the nuclear chromosomes (**mitosis**) and the division of the cytoplasm (**cytokinesis**). Mitosis, the division of the nucleus (that is, chromosomal segregation), is certainly the most obvious and complex part of cell division and that of greatest interest to the cytologist. Cytokinesis normally immediately follows mitosis, although occasionally the nucleus may divide a number of times without a corresponding division of the cytoplasm. In such a case the resulting mass of protoplasm containing many nuclei is referred to as a **multinucleate cell.** An example is the giant resorptive cell type of bone (osteoclast), which may contain 15 to 20 nuclei. Sometimes a multinucleate mass is formed by cell fusion rather than nuclear proliferation. This arrangement is called a **syncytium.** An example is vertebrate skeletal muscle, which is composed of multinucleate fibers formed by the fusion of numerous embryonic cells.

The process of mitosis is arbitrarily divided for convenience into four successive stages or phases, although one stage merges into the next without sharp lines of transition. These phases are prophase, metaphase, anaphase, and telophase (Figure 4-23). When the cell is not actively dividing, it is in interphase.

Figure 4-23

Stages of mitosis, showing division of a cell with two pairs of chromosomes. One chromosome of each pair is shown in red.

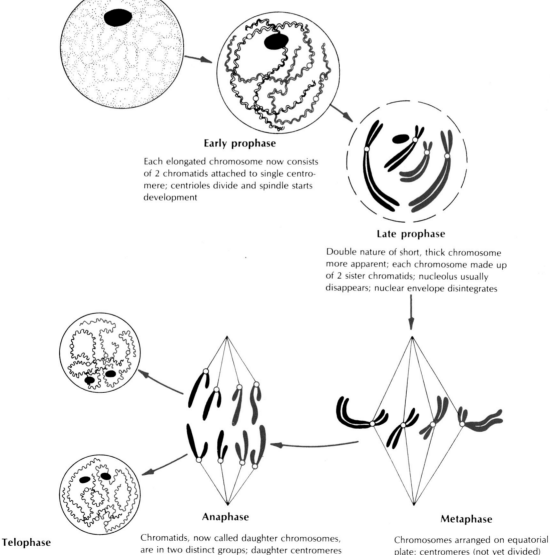

Interphase

Chromatin material appears granular; each chromosome reaches its maximum length and minimum thickness; duplication of chromosomes occurs at this stage

Early prophase

Each elongated chromosome now consists of 2 chromatids attached to single centromere; centrioles divide and spindle starts development

Late prophase

Double nature of short, thick chromosome more apparent; each chromosome made up of 2 sister chromatids; nucleolus usually disappears; nuclear envelope disintegrates

Metaphase

Chromosomes arranged on equatorial plate; centromeres (not yet divided) anchored to equator of spindle

Anaphase

Chromatids, now called daughter chromosomes, are in two distinct groups; daughter centromeres move apart and pull daughter chromosomes toward respective poles

Telophase

Chromosomes become longer and thinner; chromosomes may lose identity; nuclear membrane reappears and spindle-astral fibers fade away; cell body divides into 2 daughter cells, each of which now enters interphase

Because the DNA has been replicated during interphase, it already has a double set of chromosomes when mitosis begins.

Prophase

At the beginning of prophase, the centrioles replicate, and pairs migrate toward opposite sides of the nucleus. At the same time, fine fibers (microtubules) appear between the two pairs of centrioles to form a football-shaped **spindle,** so named because of its resemblance to nineteenth-century wooden spindles, used to twist thread together in spinning. Other fibers radiate outward from each pair of centrioles to form **asters.** The entire structure is the **mitotic apparatus,** and it increases in size and prominence as the centrioles move farther apart (Figure 4-24).

At the same time, the diffuse nuclear chromatin condenses to form visible chromosomes. These actually consist of two identical sister **chromatids** formed during interphase. The sister chromatids are joined together at their centromere. At the end of prophase, the nuclear envelope quickly disappears.

Metaphase

During metaphase the condensed sister chromatids rapidly migrate to the middle of the nuclear region to form a **metaphasic plate** (Figure 4-25). The centromeres line up precisely on the plate with the arms of the chromatids trailing off randomly in various directions. Spindle fibers now attach to each centromere.

Figure 4-24

Fine structure of mitotic apparatus. At each pole of the spindles is a clear zone occupied by a pair of centrioles and surrounded by short microtubules. Other microtubules form spindle fibers, some of which extend from pole to pole while others attach to chromosomes.

From DuPraw, E.J. 1968. Cell and molecular biology. New York, Academic Press, Inc.

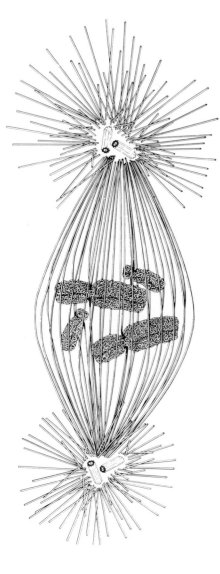

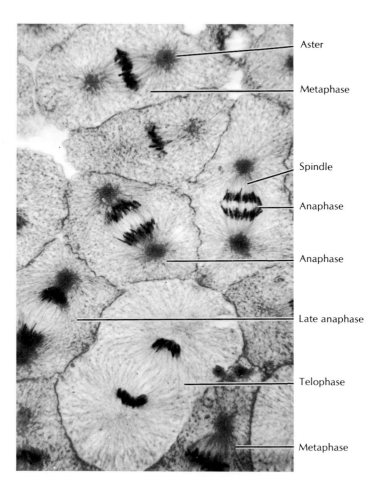

Aster

Metaphase

Spindle

Anaphase

Anaphase

Late anaphase

Telophase

Metaphase

Figure 4-25

Stages of mitosis in whitefish.

Courtesy General Biological Supply House, Inc., Chicago.

Anaphase

The two chromatids of each double chromosome thicken and separate. The single centromere that has held the two chromatids together now splits so that two independent chromosomes, each with its own centromere, are formed. The chromosomes part more, evidently pulled by the spindle fibers attached to the centromeres. The arms of each chromosome trail along behind as the microtubules of the spindle fibers shorten to drag the chromosomes toward their respective poles.

Telophase

When the daughter chromosomes reach their respective poles, telophase has begun. The daughter chromosomes are crowded together and stain intensely with histological stains. The spindle fibers disappear and the chromosomes lose their identity, reverting to the diffuse chromatin network characteristic of the interphase nucleus. Finally, the nuclear membranes reappear around the two daughter nuclei.

Cytokinesis: Cytoplasmic Division

During the final stages of nuclear division a **cleavage furrow** appears on the surface of the dividing cell and encircles it at the midline of the spindle. The cleavage furrow deepens and pinches the plasma membrane as though it were being tightened by an invisible rubber band. Microfilaments of actin are present just beneath the surface in the furrow between the cells. Interaction with myosin, similar to that which occurs when muscle cells contract (p. 605) draws the furrow inward. Finally, the infolding edges of the plasma membrane meet and fuse, completing the cell division.

Flux of Cells

Cell division is important for growth, for replacement of cells lost to natural attrition and wear and tear, and for wound healing. Cell division is especially rapid during the early development of the organism. At birth the human infant has about 2 trillion cells from repeated division of a single fertilized egg. This immense number could be attained by just 42 cell divisions, with each generation dividing once every 6 to 7 days. With only five more cell divisions, the cell number would increase to approximately 60 trillion, the number of cells in a mature man weighing 75 kg. But of course no organism develops in this machinelike manner. Cell division is rapid during early embryonic development, then slows with age. Furthermore, different cell populations divide at widely different rates. In some the average period between divisions is measured in hours, whereas in others it is measured in days, months, or even years. Cells in the central nervous system stop dividing altogether after the early months of fetal development and persist without further division for the life of the individual. Muscle cells also stop dividing during the third month of fetal development, and future growth depends on enlargement of fibers already present.

In other tissues that are subject to wear and tear, lost cells must be constantly replaced. It has been estimated that in humans about 1% to 2% of all body cells— a total of 100 billion—are shed daily. Mechanial rubbing wears away the outer cells of the skin, and emotional stress can result in physical stress that affects many cells. Food in the alimentary canal rubs off lining cells, the restricted life cycle of

The spindle fibers are microtubules—long, hollow cylinders composed of the protein tubulin. Each tubulin molecule, actually a doublet composed of two globular proteins, is attached end to end to form a strand. Thirteen strands aggregate to form a microtubule. There is evidence that microtubules are able to shift their position in cells by a peculiar treadmill process; protein doublets are disassembled at one end of the tubule, are added again at the opposite end, then travel down the microtubule treadmill to be dropped off once again at the disassembly end. Energy for this process is provided by the nucleotide guanosine triphosphate (GTP). It is thought that the spindle fibers that attach the chromosome by its centromere to the centriole shorten by losing tubulin doublets at the polar end, while assembly at the chromosome end is blocked. Thus, the chromosomes are dragged toward their poles.

The usual constraints that operate to switch off normal cell division when tissue growth is complete are lacking in uncontrolled cancer cell growth. Whereas normal cells "know" when to stop dividing, cancer cells not only divide rapidly but also grow between divisions. Part of the current research on cancer is focused on how the cell division switch is controlled and why it becomes overridden by internal and external influences during tumor growth.

blood corpuscles must involve enormous numbers of replacements, and during active sex life of males many millions of sperm are produced each day. Such losses of cells are made up by mitosis.

Cells undergo a senescence with aging. At some point in the life cycle of most cells, the cell substance breaks down, inert material is formed, metabolic processes slow down, and the synthetic power of enzymes decreases. These factors lead eventually to the death of the cell. In certain cases, products and secretions such as scales, feathers, and bony structures may persist after the death of the cell.

─ SUMMARY

Cells are the basic structural and functional units of all living organisms. Cells and their highly organized components are studied by light and electron microscopy and by biochemical methods. The eukaryotic cells of animals and other higher organisms differ from the prokaryotic cells of bacteria in several respects, the most distinctive of which is the presence of a membrane-bound nucleus containing chromosomes that carry the hereditary material.

Cells are surrounded by a plasma membrane that regulates the flow of molecular traffic between the cell and its surroundings. The nucleus, enclosed by a double membrane, contains chromatin, which condenses into rodlike chromosomes during cell division, and one or more nucleoli. Outside the nuclear envelope is the cell cytoplasm, subdivided by a membranous network, the endoplasmic reticulum, which is often associated with ribosomes and probably functions in the transport of materials within the cell. Among the organelles within the cell is the Golgi complex, which functions in storage and packaging of proteins. The mitochondria are especially conspicuous organelles that contain the enzymes of oxidative energy metabolism. Lysosomes and other membrane-bound vesicles are found in the cytoplasm. The cytoskeleton is composed of microfilaments (actin), microtubules (tubulin), and intermediate filaments (five biochemical types). Cilia and flagella are hairlike, motile appendages that contain microtubules and are associated with cell movement. Ameboid movement by pseudopodia operates by means of actin microfilaments. Tight junctions, desmosomes, and gap junctions are structurally and functionally distinct connections between cells.

Membranes in the cell are composed of a phospholipid bilayer and other materials including cholesterol. The hydrophilic ends of the phospholipid molecules are on the outer and inner surface of membranes, and the fatty acid portions are directed inward, toward each other, to form a hydrophobic core. Proteins anchored in the membrane forming the cell surface perform a variety of vital functions, including governing entry and exit of substances from the cell.

Substances can enter cells by diffusion, mediated transport, and endocytosis. Diffusion is the movement of molecules from an area of higher concentration to one of lower concentration, and osmosis is diffusion of water through a differentially permeable membrane as a result of osmotic pressure. Osmotic pressure is not the same as hydrostatic pressure but is defined in terms of equilibrium hydrostatic pressure. Solutes to which the membrane is impermeable require a carrier molecule to traverse it; carrier-mediated systems include facilitated transport (in the direction of a concentration gradient) and active transport (against a concentration gradient, which requires energy). Endocytosis includes bringing droplets (pinocytosis) or particles (phagocytosis) into the cell. Receptor-mediated endocytosis resembles pinocytosis except that it is much more specific; particular molecules bind to proteins on the cell surface, then the area invaginates and becomes a vesicle. In exocytosis the process of endocytosis is reversed.

The capacity to grow by cell multiplication is a fundamental characteristic of living systems. Cell division in eukaryotes includes mitosis, the division of the nuclear chromosomes, and cytokinesis, the division of the cytoplasm. Ordinary somatic cells replicate during mitosis and provide each daughter cell with a complete set of genetic instructions. Mitosis itself is only a small part of the total cell cycle. In interphase, G_1, S, and G_2 periods are recognized, and the S period is the time when DNA is synthesized (the chromsomes are replicated).

After replication, the chromosomes are held together by a centromere. In prophase, the first stage of mitosis, the replicated chromosomes condense into recognizable bodies. A spindle forms between the centrioles as they separate to opposite poles of the cell. At the end of prophase the nuclear envelope disintegrates. At metaphase the sister chromatids migrate to the center of the cell. At anaphase the centromeres divide and one of each kind of chromosome is pulled toward the centriole by the attached spindle fiber. At telophase the chromosomes gather in the position of the nucleus in each cell and revert to a diffuse chromatin network. The nuclear membranes reappear. Cell division is completed by separation of the cytoplasm between the daughter cells (cytokinesis).

Cells divide rapidly during embryonic development, then more slowly with age. Some cells continue to divide throughout the life of the animal to replace cells lost by attrition and wear, whereas others, such as nerve and muscle cells, complete their division during early development and never divide again.

Review questions

1. Explain the difference (in principle) between a light microscope and an electron microscope.
2. Give a one-sentence definition of each of the following: plasma membrane, chromatin, nucleus, nucleolus, rough endoplasmic reticulum (rough ER), Golgi complex, lysosomes, mitochondria, microfilaments, microtubules, intermediate filaments, centrioles, basal body (kinetosome), tight junction, gap junction, desmosome, glycoprotein, microvilli.
3. Name two functions each for actin and for tubulin.
4. Distinguish between cilia, flagella, and pseudopodia.
5. Name the types of chemical components found in a plasma membrane and briefly indicate the function of each.
6. State the fluid-mosaic model of the plasma membrane.
7. You place some red blood cells in a solution and observe that they swell and burst. You place some cells in another solution, and they shrink and become wrinkled. Explain what has happened in each case.
8. Explain the difference between osmotic and hydrostatic pressure.
9. Distinguish between two kinds of carrier-mediated transport.
10. Distinguish between phagocytosis, pinocytosis, receptor-mediated endocytosis, and exocytosis.
11. Distinguish between three kinds of endocytosis.
12. Define the following: chromosome, nucleosome, centromere, centrosome, mitosis, cytokinesis, syncytium.
13. When does the S period of the cell cycle occur, and what is happening at that period?
14. Name the phases of mitosis in order, and describe the behavior of the chromosomes at each stage.

Selected references

Bretscher, M.S. 1985. The molecules of the cell membrane. Sci. Am. **253**:100-108 (Oct.). *Good presentation of molecular structure of cell membranes, junctions, and mechanism of receptor-mediated endocytosis.*

Darnell, J., H. Lodish, and D. Baltimore. 1986. Molecular cell biology. New York, Scientific American Books, W.H. Freeman & Co. *Up-to-date, thorough, and readable. Includes both cell biology and molecular biology. Advanced, but highly recommended.*

Dautry-Varsat, A., and H.F. Lodish. 1984. How receptors bring proteins and particles into cells. Sci. Am. **250**:52-58 (May). *Good coverage of receptor-mediated endocytosis.*

Rothman, J.E. 1985. The compartmental organization of the Golgi apparatus. Sci. Am. **253**:74-89 (Sept.). *Modern concepts of the Golgi, describes functional distinctions between compartments.*

Weber, K., and M. Osborn. 1985. The molecules of the cell matrix. Sci. Am. **253**:110-120 (Oct.). *Fine account of microfilaments, microtubules, and intermediate filaments in cytoskeleton.*

Wolfe, S.L. 1985. Cell ultrastructure. Belmont, California, Wadsworth Publishing Co. *Brief but up-to-date atlas and functional account.*

CHAPTER 5

PHYSIOLOGY OF THE CELL

In protozoa all life processes occur within the borders of a single cell. In metazoa they are distributed among groups of specialized cells that have specific functions, such as nervous system coordination, digestion, and excretion, that benefit the whole organism—a "division of labor" that has contributed to the evolutionary success of multicellular animals. Nevertheless, certain activities are performed by every animal cell, whether a liver cell, a nerve cell, or a muscle cell. All cells must produce energy, synthesize their own internal structure, control much of their own activity, and guard their boundaries. In this chapter we consider these shared activities.

ENERGY AND THE LAWS OF THERMODYNAMICS

In this section we will introduce some general principles that apply to all energy exchanges in cells (Principle 2, p. 7).

The **first law of thermodynamics** states that energy cannot be created or destroyed. It can change from one form to another, but the total amount of energy in a system remains the same. In short, energy is conserved. Thus, if we burn gasoline in an engine, we do not create new energy but merely convert the chemical energy in the gasoline to another form, in this example mechanical energy and heat. The first law explains why a perpetual motion machine could never work, no matter how ingeniously it is constructed. In such a machine the energy is always transformed into heat because of friction between its parts.

Heat can be made to do work, of course, as it does in steam engines and in numerous industrial applications. However, living systems have limited applications for heat as a form of energy, although endothermic mammals and birds use it to elevate and maintain a constant body temperature. But for the most part, heat is a useless commodity to cells because it is a nonspecific form of energy that cannot be recaptured and redistributed to power metabolic processes. Heat can do work only when there is a temperature gradient, and since all cells in an organism are at virtually the same temperature, there is no heat flow and no chance of accomplishing work.

Why Cells Use Energy Stored in Chemical Bonds

Bond energy is a form of potential energy, or energy of position, as opposed to kinetic energy, or energy of motion. Chemical bonds in organic molecules are storehouses of chemical potential energy that the cell gradually liberates and couples to a variety of energy-consuming processes. As living organisms convert

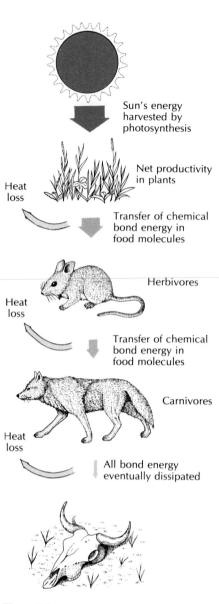

Figure 5-1

Solar energy sustains virtually all life on earth. With each energy transfer, however, much energy is lost as heat.

stored energy into other forms, and eventually into heat, less and less remains in reserve. How is it replenished? Our common experience tells us that the ultimate source of energy for life on earth is the sun. Sunlight is captured by green plants, which transform a portion of this energy into chemical bond energy (food energy). The energy accumulated by plants supports virtually all the rest of life on earth. Herbivorous animals eat plants and convert a small part of the potential chemical energy into animal tissue and a large part into heat that is invisibly dissipated into space. Carnivorous animals eat the herbivores, and they in turn are eaten by other carnivores. At each step in the transfer of energy in such a food chain, a large part of the energy is lost as heat (Figure 5-1).

At the cellular level, animals and plants use chemical energy stored in food to carry out hundreds of activities. Thus animals are totally dependent on plants, which fortunately accumulate enough energy to sustain both themselves and the animals that feed on them. Plant photosynthesis and animal respiration are inextricably woven together in a vast cycle that is driven by the constant flow of energy from the sun.

Entropy and Living Systems

Another reason why living organisms require constant replenishment of energy from the outside is expressed in the **second law of thermodynamics:** all energy of the universe is steadily moving toward increasing disorder and randomness. In effect, all forms of energy will inevitably be degraded to heat. In thermodynamics, disorder or randomness is termed **entropy.** Since entropy (disorder) increases in any given system (Figure 5-2), the change in entropy is positive.

If this is true, do not living systems disobey the second law by increasing the molecular orderliness of their structure? Certainly an organism becomes vastly more complex during its development from fertilized egg to adult. But the process of growth and maintenance is achieved by breaking down complex food molecules into simple inorganic waste products. In other words, an animal increases the entropy of its food while growing or maintaining a constant body weight. If we balance the increase in complexity of the adult against the increase in entropy of waste products, we find that the total yield is greater disorder as predicted by the second law. Even the orderly structure of the individual is not permanent but will be dissipated when the animal dies.

Entropy and Evolution

The evolution of life forms seems at first to present a more difficult problem. Has not evolution over the ages seen the appearance of life forms of ever-increasing complexity and order and requiring ever more energy for their sustenance? Does evolution disprove the second law of thermodynamics? Or does the second law disprove evolution? The argument that all forms of evolution violate the second law is a favorite of creationists. Since evolution proceeds from randomness to order and entropy proceeds from order to randomness, how does evolution defy the randomizing force of the second law?

There is an answer to this conundrum. The second law of thermodynamics applies to closed systems that no new energy can enter. The sun—earth—outer space system as a whole is closed in this sense and is trending toward increasing entropy in accordance with the second law. But the biosphere is an open system. Life on earth uses the continuous flow of solar energy within the system to remain a steady-state system of high internal order that opposes the tendency for entropic decay, at least for the period of time that life exists on earth. Hence evolution is possible and has been and is occurring. But life on earth is on borrowed time. Eventually, long after our brief walk upon this flowering earth, the sun will burn out and life on earth will end.

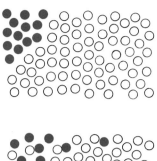

___ Free Energy

Instead of trying to measure entropic changes in biochemical reactions, biochemists use the concept of **free energy.** Free energy is simply the energy in a system available for doing work. In a molecule, free energy equals the energy present in chemical bonds minus the energy that cannot be used because of disorder (entropy). The great majority of reactions in cells release free energy and are said to be **exergonic** (Gr. *ex,* out, + *ergon,* work). Such reactions are spontaneous and always proceed "downhill" since free energy is lost from the system. Thus,

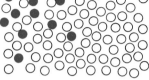

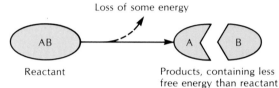

Loss of some energy

Reactant

Products, containing less
free energy than reactant

However, some important reactions in cells require the addition of free energy and are said to be **endergonic** (Gr. *endon,* within, + *ergon,* work). Such reactions have to be "pushed uphill" because they end up with more energy than they started with:

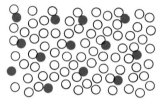

Reactants

Product, containing more
free energy than reactants

Energy input

Figure 5-2

Diffusion of solute through a solution, an example of entropy.

As we will see in a following section, ATP is the ubiquitous, energy-rich intermediate used by organisms to power important uphill reactions such as those required for active transport and in cellular synthesis.

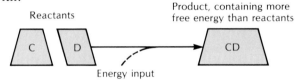

___ THE ROLE OF ENZYMES
___ Enzymes and Activation Energy

For any reaction to occur, even exergonic ones that tend to proceed spontaneously, the chemical bonds must first be destabilized. For example, if the reaction involves splitting a covalent bond, the atoms forming the bond must first be stretched apart to make them less stable. Some energy, termed the **activation energy,** must be supplied before the bond will be stressed enough to break. Only then will there be an overall loss of free energy and the formation of reaction products. This requirement can be likened to the energy needed to push a cart over the crest of a hill before it will roll spontaneously down the other side, liberating its potential energy as the cart descends.

One way to activate chemical reactants is to raise the temperature. By increasing the rate of molecular collisions and pushing chemical bonds apart, heat can impart the necessary activation energy to make a reaction proceed. But metabolic reactions must occur at biologically tolerable temperatures, temperatures too low to allow reactions to proceed beyond imperceptible rates. Instead, living systems have evolved a different strategy: they employ **catalysts.**

Catalysts are chemical substances that accelerate reaction rates without affecting the products of the reaction and without being altered or destroyed as a result of the reaction. A catalyst cannot make an energetically impossible reaction

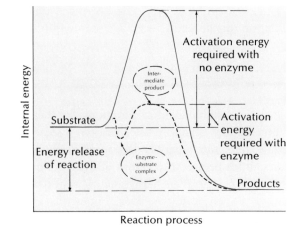

Figure 5-3

Energy changes during enzyme catalysis of a substrate. The overall reaction proceeds with a net release of energy. In the absence of an enzyme, substrate is stable because of the large amount of activation energy needed to disrupt strong chemical bonds. The enzyme reduces the energy barrier by forming a chemical intermediate with a much lower internal energy state.

happen; it simply accelerates a reaction that would have proceeded at a very slow rate otherwise.

Enzymes are the catalysts of the living world. The special catalytic talent of an enzyme is its power to reduce the amount of activation energy required for a reaction. In effect, an enzyme steers the reaction through one or more intermediate steps, each of which requires much less activation energy than that required for a single-step reaction (Figure 5-3). Note that enzymes do not supply the activation energy, but rather lower the activation energy barrier, making a reaction more likely to proceed. Enzymes affect only the reaction rate; they do not in any way alter the free energy change of a reaction, nor do they change the proportions of reactants and products in a reaction.

Nature of Enzymes

Enzymes are complex molecules that vary in size from small, simple proteins with a molecular weight of 10,000 to highly complex molecules with molecular weights up to 1 million. Many enzymes are pure proteins—delicately folded and inter-linked chains of amino acids. Other enzymes require the participation of small nonprotein groups called **cofactors** to perform their enzymatic function. In some cases these cofactors are metallic ions (such as ions of iron, copper, zinc, magnesium, potassium, and calcium) that form a functional part of the enzyme. Examples are carbonic anhydrase, which contains zinc; the cytochromes, which contain iron; and troponin (a muscle contraction enzyme), which contains calcium. Another class of cofactors, called **coenzymes,** is organic. All coenzymes contain groups derived from vitamins, compounds that must be supplied in the diet. All of the B complex vitamins are coenzyme compounds. Since animals have lost the ability to synthesize the vitamin components of coenzymes, it is obvious that a vitamin deficiency can be serious. However, unlike dietary fuels and nutrients that must be replaced after they are burned or assembled into structural materials, vitamins are recovered in their original form and are used repeatedly. Examples of enzymes that contain vitamins are nicotinamide adenine dinucleotide (NAD), which contains nicotinic acid; coenzyme A, which contains pantothenic acid; and flavin adenine dinucleotide (FAD), which contains riboflavin.

Action of Enzymes

An enzyme functions by combining in a highly specific way with its **substrate,** the molecule on which it acts. The enzyme bears an active site located within a cleft or pocket and contains a unique molecular configuration. The active site has a flex-

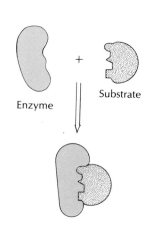

Figure 5-4

Interaction of substrate and enzyme. The enzyme changes shape when it binds the substrate.

ible surface that enfolds and conforms to the substrate (Figure 5-4). The binding of enzyme to substrate forms an **enzyme-substrate complex (ES complex)** in which the substrate is secured by covalent bonds to several points in the active site of the enzyme. The ES complex is not strong and will quickly dissociate, but during this fleeting moment the enzyme provides a unique chemical environment that stresses certain chemical bonds in the substrate so that much less energy is required to complete the reaction.

Enzymes that engage in important main-line sequences—such as the crucial energy-providing reactions of the cell that go on constantly—seem to operate in enzyme sets rather than in isolation. For example, the conversion of glucose to carbon dioxide and water proceeds through 19 reactions, each requiring a specific enzyme. Main-line enzymes are found in relatively high concentrations in the cell, and they may implement quite complex and highly integrated enzymatic sequences. One enzyme carries out one step, then passes the product to another enzyme that catalyzes another step, and so on. The reactions may be said to be coupled.

Specificity of Enzymes

One of the most distinctive attributes of enzymes is their high specificity. This is a consequence of the exact molecular fit that is required between enzyme and substrate. Furthermore, an enzyme catalyzes only one reaction; unlike reactions carried out in the organic chemist's laboratory, no side reactions or by-products result. Specificity of both substrate and reaction is obviously essential to prevent a cell from being swamped with useless by-products.

However, there is some variation in degree of specificity. Some enzymes catalyze the oxidation (dehydrogenation) of only one substrate; for example, succinic dehydrogenase catalyzes the oxidation of succinic acid only. Others, such as proteases (for example, pepsin and trypsin), will act on almost any protein, but each protease has its particular point of attack in the protein (Figure 5-5). Usually an enzyme will take on one substrate molecule at a time, catalyze its chemical change, release the product, and then repeat the process with another substrate molecule. The enzyme may repeat this process billions of times until it is finally worn out (after a few hours to several years) and is broken down by scavenger enzymes in the cell. Some enzymes undergo successive catalytic cycles at dizzying speeds of up to a million cycles per minute, but most operate at slower rates.

Enzyme-Catalyzed Reactions

Enzyme-catalyzed reactions are theoretically reversible. This is signified by the double arrows between substrate and products. For example:

$$\text{Fumaric acid} + H_2O \rightleftharpoons \text{Malic acid}$$

However, for various reasons the reactions catalyzed by most enzymes tend to go only in one direction. For example, the proteolytic enzyme pepsin can degrade proteins into amino acids, but it cannot accelerate the rebuilding of amino acids into any significant amount of protein. The same is true of most enzymes that catalyze the hydrolysis of large molecules such as nucleic acids, polysaccharides, lipids, and proteins. There is usually one set of reactions and enzymes that break them down, but they must be resynthesized by a different set of reactions that are catalyzed by different enzymes. This apparent irreversibility exists because the chemical equilibrium usually favors the formation of the smaller degradation products.

How can biochemists be certain that an enzyme-substrate (ES) complex exists? The original evidence offered by Leonor Michaelis in 1913 is that, when the substrate concentration is increased while the enzyme concentration is held constant, the reaction rate reaches a maximum velocity. This *saturation effect* is interpreted to mean that all catalytic sites become filled at high substrate concentration. It is not seen in uncatalyzed reactions. Other evidence includes the observation that the ES complex displays unique spectroscopic characteristics not displayed by either the enzyme or the substrate alone. Furthermore, some ES complexes can be isolated in pure form, and at least one kind (nucleic acids and their polymerase enzymes) has been directly visualized with the electron microscope.

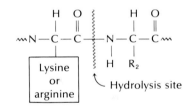

Figure 5-5

High specificity of trypsin. It splits only peptide bonds adjacent to lysine or arginine.

The net **direction** of any chemical reaction is dependent on the relative energy contents of the substances involved. If there is little change in the chemical bond energy of the substrate and the products, the reaction is more easily reversible. However, if large quantities of energy are released as the reaction proceeds in one direction, more energy must be provided in some way to drive the reaction in the reverse direction. Thus, many if not most enzyme-catalyzed reactions are in practice irreversible unless the reaction is coupled to another that makes energy available. In the cell both reversible and irreversible reactions are combined in complex ways to make possible both synthesis and degradation.

CHEMICAL ENERGY TRANSFER BY ATP

We have seen that endergonic reactions are those that will not proceed spontaneously by themselves because the products require an input of free energy. However, an endergonic reaction may be driven by coupling the energy-requiring reaction with an energy-yielding reaction. ATP is the intermediate in **coupled reactions,** and because it can drive such energetically unfavorable reactions, it is of central importance in metabolic processes.

The ATP molecule consists of adenosine (the purine adenine and the 5-carbon sugar ribose) and a triphosphate group (Figure 5-6). Most of the free energy in ATP resides in the triphosphate group, especially in two **phosphoanhydride bonds** between the three phosphate groups. These two bonds are called **high-energy bonds** because a great deal of free energy in the bonds is liberated when ATP is hydrolyzed to adenosine diphosphate (ADP) and inorganic phosphate. The high-energy groups in ATP are designated by the "tilde" symbol ~. A high-energy phosphate bond is shown as ~P and a low-energy bond (such as the bond linking the triphosphate group to adenosine) as —P. ATP may be symbolized as A— P~P~P and ADP as A—P~P.

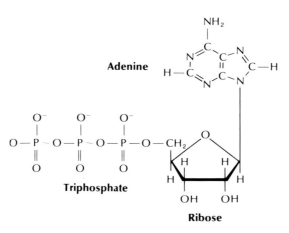

Figure 5-6

Adenosine triphosphate (ATP).

The way that ATP can act to drive a coupled reaction is shown in Figure 5-7. A coupled reaction is really a system involving two reactions linked by an energy shuttle (ATP). The conversion of substrate A to product A is endergonic because the product contains more free energy than the substrate. Therefore energy must be supplied by coupling the reaction to one that is exergonic, the conversion of substrate B to product B. Substrate B in this reaction is commonly called a **fuel** (for example, glucose or a lipid). The bond energy that is released in reaction B is transferred to ADP, which in turn is converted to ATP. ATP now contributes its phosphate-bond energy to reaction A, and ADP is produced again.

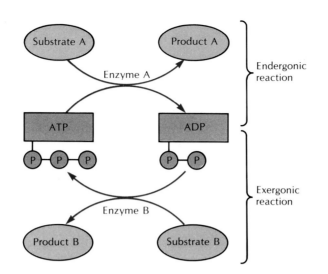

Figure 5-7

A coupled reaction. The endergonic conversion of substrate A to product A will not occur spontaneously but requires an input of energy from another reaction involving a large release of energy. ATP is the intermediate through which the energy is shuttled.

Note that ATP is an **energy-coupling agent** and *not* a fuel. It is not a storehouse of energy set aside for some future need. Rather it is produced by one set of reactions and is almost immediately consumed by another. ATP is formed as it is needed, primarily by oxidative processes in the mitochondria. Oxygen is not consumed unless ADP and phosphate molecules are available, and these do not become available until ATP is hydrolyzed by some energy-consuming process. *Metabolism is therefore mostly self-regulating.*

____CELLULAR METABOLISM

Cellular metabolism refers to the collective total of chemical processes that occur within living cells. It is often called **intermediary metabolism** because the exchange of matter and energy between the cell and its environment proceeds in a stepwise manner through chemical pathways composed of numerous intermediates, or **metabolites.** Although intermediary metabolism appears hopelessly complex as depicted in detailed metabolic charts that often grace the walls of biochemists' laboratories, the central metabolic routes through which matter and energy are channeled are not difficult to understand. Biochemists are vastly furthered in their research by the fact that the same kinds of reaction sequences in metabolism occur in a great variety of life forms, from bacteria to humans.

Animal cells tap the stored energy of organic fuels (for example, simple sugars, fatty acids, and amino acids) through a series of controlled degradative steps. This process commonly makes use of molecular oxygen from the atmosphere. In return animal cells give off carbon dioxide as an end product, which is used by plant cells in making glucose and the more complex molecules. In this way the cellular energy cycle of life involves the harnessing of sunlight energy by green plants directly and by animal cells indirectly.

____Respiration: Generating ATP in the Presence of Oxygen

How electron transport is used to trap chemical bond energy

Having seen that ATP is the one common energy denominator by which all cellular machines are powered, we are in a position to ask how this energy is captured from fuel substrates. This question directs us to an important generalization: *all cells obtain their chemical energy requirements from oxidation-reduction reac-*

tions. This means simply that in the degradation of fuel molecules, hydrogen atoms (electrons and protons) are passed from reducing agents to oxidizing agents with a release of energy. A portion of this energy is trapped and used to form the high-energy bonds of ATP. The release of energy during electron transfer and its conservation as ATP is the mainspring of cell activity and was a crucial evolutionary achievement.

Before we consider the important oxidation-reduction reactions in more detail, let us see what happens when a fuel molecule is degraded to liberate its free (chemical bond) energy. If we completely burn 1 mole (180 g) of glucose in a bomb calorimeter to carbon dioxide and water, we can show the combustion as

$$C_6H_{12}O_6 + 6\ O_2 \rightarrow 6\ CO_2 + 6\ H_2O \quad \Delta G = -686\ kcal$$

Burned in this way, all of the 686 kcal (686,000 calories) of free energy locked up in the structure of the glucose molecule is released as 686 kcal of heat. But in cell respiration a significant portion of this free energy is channeled into useful chemical energy as ATP. Thus, in the cell the reaction can be summarized as

$$C_6H_{12}O_6 + 38\ P + 38\ ADP + 6\ O_2 \rightarrow$$
$$6\ CO_2 + 6\ H_2O + 38\ ATP \quad \Delta G = -409\ kcal\ (as\ heat)$$

The difference between the 409 kcal lost as heat and the original 686 kcal of free energy in glucose is 277 kcal in 38 moles of ATP (about 7.3 kcal/mole). The passing of this free energy to ATP is accomplished, as we have already suggested, by electron transfer. Chemical energy is liberated when electrons are transferred from one compound to another.

Because they are so important, and to avoid later confusion, let us review what we mean by oxidation-reduction ("redox") reactions (p. 20). In these reactions there is a transfer of electrons from an electron donor (the reducing agent) to an electron acceptor (the oxidizing agent). As soon as the electron donor loses its electrons, it becomes oxidized. As soon as the electron acceptor accepts electrons, it becomes reduced. In other words, a reducing agent becomes oxidized when it reduces another compound, and an oxidizing agent becomes reduced when it oxidizes another compound. Thus, for every oxidation there must be a corresponding reduction.

In an oxidation-reduction reaction the electron donor and electron acceptor form a redox pair:

$$\text{Electron donor} \quad \rightleftharpoons e^- + \text{Electron acceptor}$$
$$\text{(reducing agent)} \qquad\qquad \text{(oxidizing agent)}$$

When electrons are accepted by the oxidizing agent, energy is liberated because the electrons move to a more stable position. The amount of free energy liberated depends on the difference in tendency of the two compounds to donate electrons. This is termed their **electron pressure.** In the cell, electrons move from compounds of high electron pressure to compounds of lower electron pressure. By transferring electrons stepwise in this manner, energy is gradually released, and a maximum yield of ATP is realized.

Aerobic versus anaerobic metabolism

Ultimately, the electrons are transferred to a **final electron acceptor.** The nature of this final acceptor is the key that determines the overall efficiency of cellular metabolism. The heterotrophs can be divided into two great groups: **aerobes,**

A calorie (cal) is the amount of heat required to heat 1 g of water from 14.5° to 15.5° C. One kilocalorie (kcal) = 1000 cal. Although the calorie is almost universally used in books and tables, it is no longer a part of the recently adopted International System of Units (SI system). The unit of energy in the SI system is the joule (J), defined as 4.184 J = 1 cal.

those that use molecular oxygen as the final electron acceptor, and **anaerobes,** those that employ some other molecule as the final electron acceptor.

Because humans and all of the other animals with which we are most familiar depend on oxygen, it may seem surprising that many primitive organisms live successfully under oxygen-free conditions. But we must recall that life on earth had its origins under reducing conditions; there was no oxygen in the primitive atmosphere for perhaps the first billion years of life on the primeval earth. During this formative period, the basic pattern of cellular metabolism evolved under strictly anaerobic conditions. Only later, as evolving photosynthetic organisms began to produce oxygen, did metabolic reactions appear in which the generation of ATP was coupled to the use of oxygen.

Today most animals use oxygen to generate high-energy phosphate. Strictly anaerobic organisms still exist and indeed occupy some important ecological niches. These organisms for the most part are bacteria that are restricted to the relatively few oxygen-free environments that remain on earth. Although some organisms are facultatively anaerobic, for the majority of animals oxygen is a necessity of life. Evolution has favored aerobic metabolism, not only because oxygen became available, but also because aerobic metabolism is vastly more efficient than anaerobic metabolism. In the absence of oxygen, only a very small fraction of the bond energy present in foodstuffs can be released. For example, when an anaerobic microorganism degrades glucose, the final electron acceptor (such as pyruvic acid) still contains most of the energy of the original glucose molecule. On the other hand, an aerobic organism, using oxygen as the final electron acceptor, can completely oxidize glucose to carbon dioxide and water. Almost 20 times as much energy is released when glucose is completely oxidized as when it is degraded only to the stage of lactic acid. An obvious advantage of aerobic metabolism is that a much smaller quantity of foodstuff is required to maintain a given rate of metabolism.

General description of respiration

Aerobic metabolism is more familiarly known as true **cellular respiration,** defined as the oxidation of fuel molecules with molecular oxygen as the final electron acceptor. As mentioned previously, the oxidation of fuel molecules describes the *removal of electrons* and *not* the direct combination of molecular oxygen with fuel molecules. Let us look at this process in general before considering it in more detail.

Hans Krebs, the British biochemist who contributed so much to our understanding of respiration, described three stages in the complete oxidation of fuel molecules to carbon dioxide and water (Figure 5-8). In stage I, foodstuffs are digested into small molecules that can be absorbed into the circulation. There is no useful energy yield during digestion, which is discussed in Chapter 31. In stage II, most of the degraded foodstuffs are converted into a 2-carbon acetyl group, acetyl coenzyme A. This stage occurs in the cytoplasmic matrix (cytosol). Some ATP is generated in stage II, but the yield is small compared with that obtained in the final stage of respiration. In stage III the final oxidation of fuel molecules occurs, with a large yield of ATP. This stage takes place in the mitochondria. Acetyl coenzyme A is channeled into the Krebs cycle where the acetyl group is completely oxidized to carbon dioxide. Electrons released from the acetyl groups are transferred to special carriers that pass them to electron acceptor compounds in the electron transport chain. At the end of the chain the electrons (and the protons accompanying them) are accepted by molecular oxygen to form water.

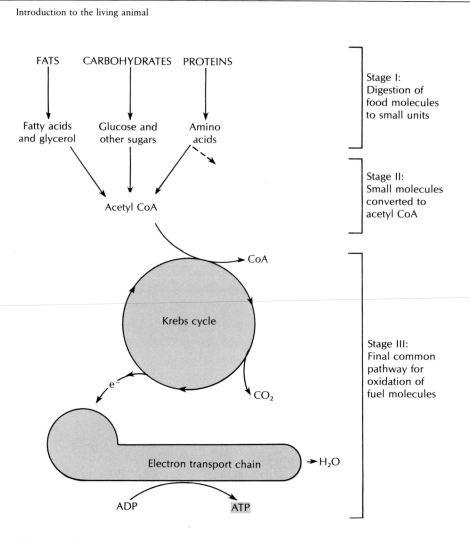

FATS   CARBOHYDRATES   PROTEINS

Fatty acids   Glucose and   Amino
and glycerol   other sugars   acids

Acetyl CoA

CoA

Krebs cycle

e^-

CO_2

Electron transport chain   → H_2O

ADP   ATP

Stage I:
Digestion of
food molecules
to small units

Stage II:
Small molecules
converted to
acetyl CoA

Stage III:
Final common
pathway for
oxidation of
fuel molecules

Figure 5-8

Overview of respiration, showing the three stages in the complete oxidation of food molecules to carbon dioxide and water.

Glycolysis

We begin our journey through the stages of respiration with glycolysis, a nearly universal pathway in living organisms that converts glucose into pyruvic acid. In a series of reactions, glucose and other 6-carbon monosaccharides are split into 3-carbon fragments, pyruvic acid. Pyruvic acid is then enzymatically stripped of carbon dioxide to form acetyl coenzyme A. A single oxidation occurs during glycolysis, and each molecule of glucose yields 2 molecules of ATP.

The metabolism of glucose begins with its phosphorylation by ATP to form **glucose-6-phosphate.** Glucose-6-phosphate is an important intermediate because it is a stem compound that can lead into any of several metabolic pathways. However, the predominant metabolic fate for glucose-6-phosphate is entry into the glycolytic sequence. The glucose-6-phosphate is then rearranged at the expense of another molecule of ATP to form **fructose-1,6-diphosphate** (Figure 5-9). The fuel has now been "primed" with phosphate groups and is sufficiently

Figure 5-9

Phosphorylation of glucose.

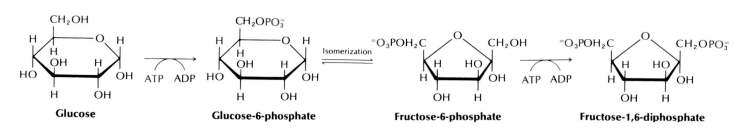

Glucose   Glucose-6-phosphate   Fructose-6-phosphate   Fructose-1,6-diphosphate

reactive to enable subsequent reactions to proceed. Obviously, the fuel molecule must release more energy—it eventually provides *much* more energy—than is loaned to it by ATP at the start. This is a kind of deficit financing that is required for an ultimate energy return many times greater than the original energy investment.

Fructose-1,6-diphosphate is next cleaved into two 3-carbon sugars, called **triose phosphates** (Figure 5-10). These high-energy derivatives of glucose are then oxidized (the electrons are removed) in an important energy-yielding step. The electrons removed are accepted by **nicotinamide adenine dinucleotide** (NAD), a derivative of the vitamin niacin. NAD serves as a carrier molecule to convey high-energy electrons to the final electron transport chain. NAD and its reduced form NADH exist as a redox pair:

$$\text{Reduced substrate} + \text{NAD}^+ \rightleftharpoons \text{NADH} + \text{H}^+ + \text{Oxidized substrate}$$

In this oxidation, NAD^+ accepts a hydrogen ion and two electrons to form NADH. The remaining proton is released as free H^+ in the medium. This transfer produces a structural change in NAD having a free energy change of about -52 kcal/mole. This is a very large proportion of the total free energy available from the oxidation of glucose (686 kcal). As we shall see, several more molecules of reduced NAD are formed as glucose oxidation proceeds. Altogether about 75% of the free energy in glucose is transferred to NAD. This reduced NAD and the free energy it contains are eventually used to synthesize ATP in the electron transport chain.

The two triose phosphates next enter a four-step reaction sequence ending with the formation of 2 molecules of pyruvic acid (Figure 5-10). In two of these steps, each molecule of triose phosphate yields a molecule of ATP. In other words, each triose phosphate yields 2 ATP molecules, and since there are 2 triose phosphate molecules 4 ATP molecules are generated. Recalling that 2 ATP molecules were used to prime the glucose initially, the net yield up to this point is 2 ATP molecules.

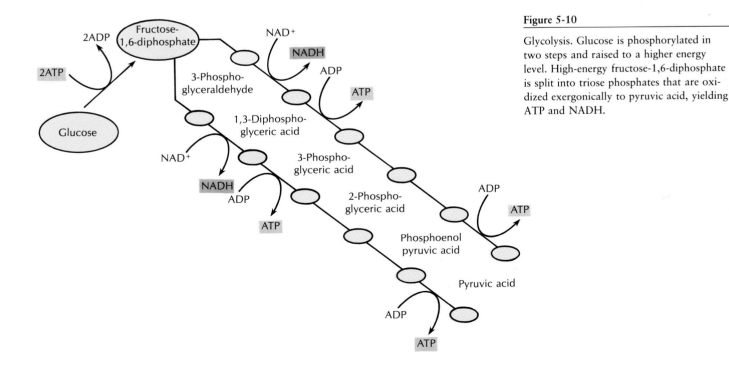

Figure 5-10

Glycolysis. Glucose is phosphorylated in two steps and raised to a higher energy level. High-energy fructose-1,6-diphosphate is split into triose phosphates that are oxidized exergonically to pyruvic acid, yielding ATP and NADH.

Acetyl coenzyme A: strategic intermediate in respiration

In aerobic metabolism the 2 molecules of pyruvic acid formed during glycolysis enter the mitochondrion. Once inside, each pyruvic acid molecule is oxidized and the 2 electrons are accepted by NAD^+ to form NADH. In the same reaction the third carbon of pyruvic acid is released as carbon dioxide. This allows the 2-carbon residue from pyruvic acid to condense with coenzyme A to form acetyl coenzyme A (Figure 5-11).

Acetyl coenzyme A is a critically important compound. Some two thirds of all the carbon atoms in foods eaten by animals appear as acetyl coenzyme A at some stage. It is the final oxidation of acetyl coenzyme A that provides energized electrons used to generate ATP. Acetyl coenzyme A is also the source of nearly all of the carbon atoms found in the body's fats. A part of the molecule is a coenzyme containing the vitamin pantothenic acid, another example of how vitamins play important structural roles in critical cellular functions.

Krebs cycle: oxidation of acetyl coenzyme A

The degradation (oxidation) of the 2-carbon acetyl group of acetyl coenzyme A occurs in a cyclic sequence called the Krebs cycle (Figure 5-12) in honor of Sir Hans A. Krebs, who worked out the sequence in the 1930s. The cycle is also often referred to as the tricarboxylic acid (TCA) cycle or the citric acid cycle. The reactions in this final common pathway for the oxidation of fuel molecules occur in the matrix of the mitochondrion (see Figure 5-15). Fuel molecules that enter the cycle as acetyl coenzyme A are first condensed with 4-carbon oxaloacetic acid to form 6-carbon citric acid. Coenzyme A then recycles to pick up another acetyl

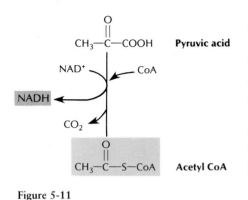

Figure 5-11

Formation of acetyl coenzyme A from pyruvic acid.

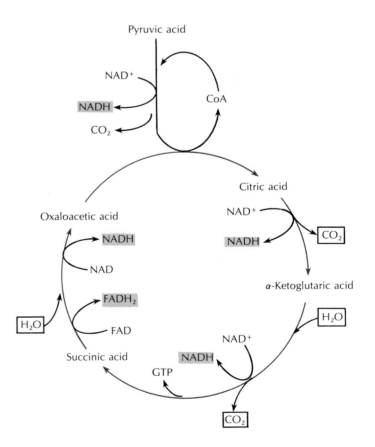

Figure 5-12

Krebs cycle in outline form showing the production of reduced NAD, reduced FAD, guanosine triphosphate (GTP), and carbon dioxide.

group, thus repeatedly transferring acetyl groups from pyruvic acid to oxaloacetic acid. Citric acid is then degraded through a series of reactions that end in the regeneration of oxaloacetic acid. The 2 carbon atoms that enter the cycle as acetyl coenzyme A leave the cycle as 2 molecules of carbon dioxide.

As the acetyl group is oxidized, carbon atom by carbon atom, three pairs of electrons are transferred to NAD^+ and one pair to flavine adenine dinucleotide (FAD), an electron carrier that functions like NAD. Thus, with each completion of the Krebs cycle, 8 hydrogen atoms are removed in pairs by the electron carriers NAD and FAD. These reduced electron carriers are subsequently oxidized by the electron transport chain, where the electron energy is used to generate ATP. The 2 molecules of carbon dioxide diffuse out of the mitochondrion and into the circulation to be finally eliminated via the respiratory system. At another point in the cycle a molecule of guanosine triphosphate (GTP) is formed; its high-energy phosphate is readily transferred to ADP to form ATP. It is noteworthy that this is the only place in the Krebs cycle were ATP is formed *directly*. All of the remaining energy is transferred to the electron transport chain.

Electron transport chain

The transfer of electrons from reduced NAD and FAD to the final electron acceptor, molecular oxygen, is accomplished in an elaborate electron transport chain. The function of this chain is to permit the controlled release of free energy to drive the synthesis of ATP. At three points along the chain, ATP production occurs by the phosphorylation of ADP. This method of energy capture is called oxidative phosphorylation because the formation of high-energy phosphate is coupled to oxygen consumption, and this depends on the demand for ATP by other metabolic activities within the cell. The actual mechanism of ATP formation by oxidative phosphorylation is not known with certainty; we can only say that the transfer of electrons does something that is translated into the production of high-energy phosphate bonds. Currently, the most widely accepted explanation of this mechanism is the chemiosmotic coupling theory (Hinkle and McCarty, 1978), described later in the chapter.

Localization and function of electron carriers

Oxidative phosphorylation, a complex process, would be unable to function efficiently, if at all, were the enzymes just floating freely in the cytoplasm of the cell. There is now abundant evidence that the oxidative enzymes and electron carriers are arranged in a highly ordered state on the inner membranes of the mitochondria.

Remember that mitochondria are composed of two membranes. The outside membrane forms a smooth sac enclosing the inner membrane, which is turned into numerous ridges called **cristae.** The inner membrane is studded with enormous numbers of minute, stalked particles called **inner membrane spheres.** The electron carriers of the respiratory chain are restricted to the inner membrane where, presumably, they are located on the inner membrane spheres. An artist's rendering of a section of mitochondrion as it might appear under the high-resolution electron microscope is shown in Figure 5-13.

Electrons are transported from NADH to oxygen through a series of electron carriers (Figure 5-14). These are, in their order in the chain, **flavin mononucleotide (FMN)**, the prosthetic group of a protein (flavoprotein); **coenzyme Q,** a nonprotein chemically related to vitamin K; and a series of **cytochromes,** all pro-

Figure 5-13

Representation of a section of mitochondrion as seen through a high-resolution electron microscope, showing the inner membrane spheres where ATP is generated. The density of the spheres is actually many times greater than depicted.

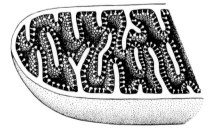

A prosthetic group is a nonpolypeptide unit, tightly bound to a protein, that is essential to the protein's activity. Many enzymes have prosthetic groups.

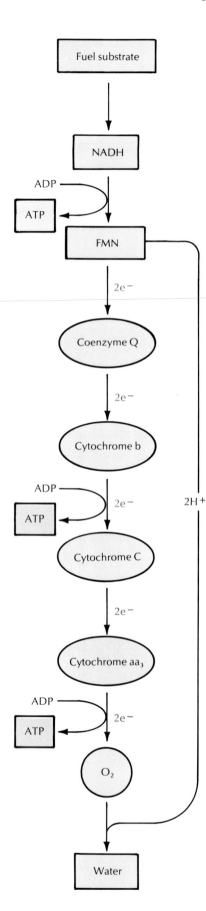

teins containing a heme prosthetic group and similar in structure to hemoglobin of vertebrate red blood cells. The step-by-step transfer of electrons down this chain happens because each successive enzyme has a lower electron pressure than its predecessor. With each step a portion of the energy is lost as heat, but the rest of the energy is conserved in the molecules of ATP formed during the electron transfer. Only the last member of the chain, cytochrome a-a$_3$ (also called cytochrome oxidase) can transfer its electrons to oxygen. Thus, the 2 electrons from NADH pass through an electron cascade, a series of discrete steps that permits the gradual harvesting of electron energy.

The advantage of an electron cascade becomes apparent when we compare the free energy present in NADH (53 kcal/mole) with the energy required to synthesize ATP from ADP. The formation of ATP requires at least 7.3 kcal/mole, the energy necessary for the hydrolysis of ATP. Clearly, there is enough energy in NADH to form several molecules of ATP. Since all exergonic energy transfers are accompanied by loss of free energy, it would be impossible to realize the theoretical ATP harvest of 7 ATP molecules (52 divided by 7.3). In fact there are three sites along the respiratory chain where there is enough free energy change (9 to 10 kcal/mole) to make ATP synthesis possible. These sites are shown in Figure 5-14. Thus, the oxidation of NADH yields 3 ATP molecules. The other electron carrier bringing energy from the Krebs cycle is FADH$_2$. It enters the electron transport chain at coenzyme Q, which is at a lower energy level in the sequence and so yields only 2 ATP molecules.

Coupling mechanisms for ATP synthesis

How is the oxidation of NADH coupled to ATP synthesis? This question has been intensively investigated. The currently favored hypothesis, now supported by a wealth of evidence, is the **chemiosmotic mechanism** first proposed by the British biochemist Peter Mitchell in 1961. The model states that ATP synthesis is driven by a gradient of hydrogen ions created by the energy released during electron transport. The mechanism is easiest to understand if it is divided into two broad parts: (1) establishing the hydrogen ion gradient, and (2) synthesis of ATP.

There is abundant evidence that the transport of electrons through the respiratory chain results in the pumping of hydrogen ions from the mitochondrial matrix to the intermembrane space between the outer and inner mitochondrial membranes (Figure 5-15). The electron transport carriers are located *in* the inner membrane. Each pair of electrons donated to the respiratory chain by NADH produces three pairs of hydrogen ions, which are ejected to the intermembrane space by enzymatic transfer. Because NADH carries only 1 hydrogen, the additional hydrogen ions are absorbed from the matrix as the electrons flow down the respiratory chain. At the end of the chain an additional 2 hydrogen ions are removed from the matrix and combined with one half of an oxygen molecule to form H$_2$O. Thus the hydrogen ion concentration increases in the intermembrane compartment and a membrane potential is created, with the intermembrane side

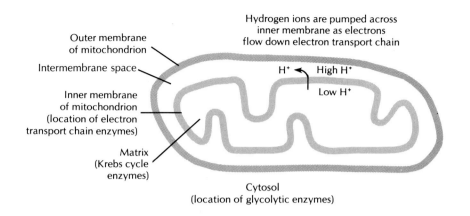

Outer membrane
of mitochondrion

Intermembrane space

Inner membrane
of mitochondrion
(location of electron
transport chain enzymes)

Matrix
(Krebs cycle
enzymes)

Hydrogen ions are pumped across
inner membrane as electrons
flow down electron transport chain

H⁺   High H⁺

Low H⁺

Cytosol
(location of glycolytic enzymes)

Figure 5-15

Mitchell's hypothesis for pumping hydrogen ions from matrix to intermembrane space of mitochondrion.
Modified from Stryer, L. 1981. Biochemistry, ed 2. San Francisco, W.H. Freeman & Co. Publishers.

positive. The model also postulates that the electron carriers are very precisely positioned in the inner membrane so that hydrogen ions can move in *only* one direction, from inside to outside. Furthermore, the inner membrane must be impermeable to hydrogen ions to prevent the hydrogen gradient from running down as fast as it is created.

The hydrogen gradient is believed to drive the synthesis of ATP in an **ATP synthesis complex** that consists of two parts. High-resolution electron micrographs of the inner mitochondrial membrane show it to be studded on the matrix side with numerous spherical projections (Figure 5-16). These contain **adenosine triphosphatase** (ATPase), an enzyme capable of catalyzing the formation of ATP from ADP and inorganic phosphate. The other portion of the complex consists of a channel in the inner membrane lying directly beneath the spherical projections. Through these channels hydrogen ions can approach from the intermembrane compartment to the active enzyme site in the spheres. The hydrogen ions flowing into this channel, following an osmotic gradient, interact with ATPase and its substrates at the active site, resulting in the synthesis of ATP. These are the major features of the hypothesis for which Mitchell received the Nobel Prize in 1978. As with any hypothesis of such complexity, and dealing with elements that cannot be seen but must be measured indirectly by ingenious experiments, certain details remain controversial. At first largely rejected by the scientific community that now embraces it, the model is a brilliant example of deductive biochemistry.

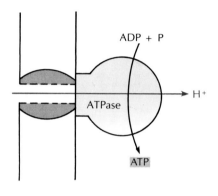

ADP + P

ATPase

H⁺

ATP

Figure 5-16

Tentative mechanism for generating ATP in mitochondrion inner membrane spheres.
Modified from Hinkle. P.C., and R.E. McCarty. 1978. Sci. Am. **238**:104-123 (Mar.).

Efficiency of oxidative phosphorylation

We are now in a position to calculate the ATP yield from the complete oxidation of glucose (Figure 5-17). The overall reaction is:

$$\text{Glucose} + 36\ \text{ADP} + 36\ \text{P} + 6\ O_2 \rightarrow 6\ CO_2 + 36\ \text{ATP} + 6\ H_2O$$

ATP has been generated at several points along the way (Table 5-1). The total yield is 38 molecules of ATP, but because 2 molecules were used up in the initial glucose phosphorylation, the net yield is 36 ATP molecules.

The free energy yield from glucose is 686 kcal/mole. The free energy stored in the net yield of 36 molecules of ATP is 263 kcal/mole. Thus, the thermodynamic efficiency of ATP formation is $^{263}/_{686}$, or 38.3%.

We must note, however, that energy yield calculations are not nearly as precise as our neat figures suggest. The free energy change for the hydrolysis of ATP rises as the concentration of reactants and products is reduced. So the efficiency for the total oxidation of glucose could be higher than 38%. On the other hand, the stated yield of 3 ATP molecules for every NADH molecule entering the

Figure 5-17

Pathway for oxidation of glucose
and other carbohydrates. Glucose
is degraded to pyruvate by cytoplasmic enzymes (glycolytic
pathway). Acetyl coenzyme A is
formed from pyruvate and is fed
into the Krebs cycle. An acetyl
group (2 carbons) is oxidized to 2
molecules of carbon dioxide with
each turn of the cycle. Pairs of
electrons (2H) are removed from
the carbon skeleton of the substrate at several points in the
pathway and are carried by oxidizing agents (NADH or $FADH_2$,
not shown) to the electron transport chain where 32 molecules of
ATP are generated. Four molecules of ATP are also generated
by substrate phosphorylation in
the glycolytic pathway, and two
molecules of ATP (initially GTP)
are formed in the Krebs cycle.
This yields a total of 38 molecules of ATP (36 moles net) per
glucose molecule. Molecular oxygen is involved only at the very
end of the pathway.

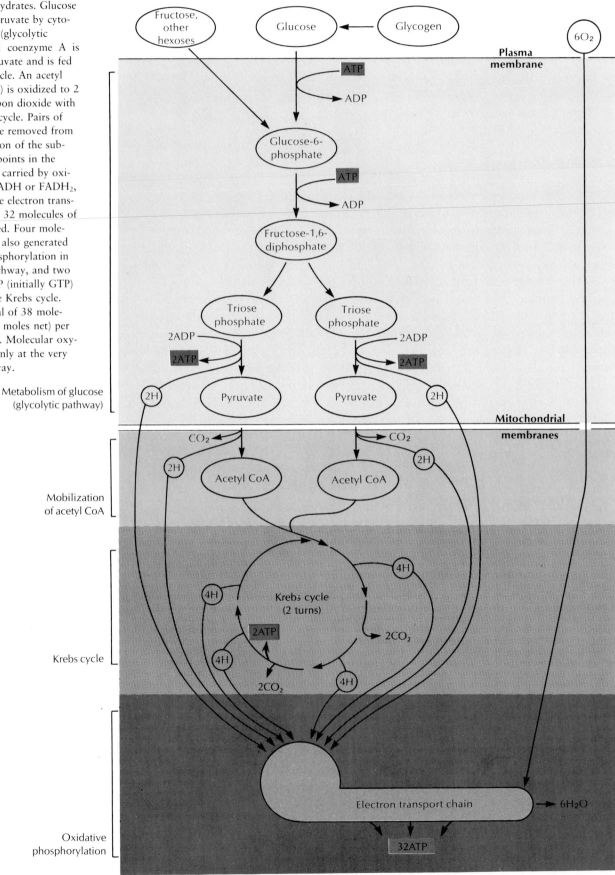

electron transport chain was never better than an approximation because it was difficult to measure the true energy cost of making the process work. Studies by Hinkle and McCarty in 1978 suggest that only about 2 ATP molecules are synthesized per electron pair. The energy loss is believed to be caused by the cost of concentrating ADP in the mitochondria and of moving ATP back out of the mitochondria as it is formed. If these values are correct, the ATP yield for the complete oxidation of glucose should be revised sharply downward from 36 molecules of ATP per molecule of glucose to about 25. Still, the overall efficiency of glucose oxidation would be about 27%, comparing favorably with human-designed energy conversion systems, which seldom exceed 5% to 10% efficiency.

___ Anaerobic Glycolysis: Generating ATP without Oxygen

Up to this point we have been describing aerobic metabolism, or respiration. The complete oxidation of glucose involves the formation of pyruvate by cytoplasmic enzymes followed by the oxidation of pyruvate in the mitochondria. It is in the latter, mitochondrial stage of respiration that the energies of electrons are tapped and used to generate ATP by oxidative phosphorylation. The important feature of aerobic metabolism is that oxygen is the final electron acceptor, and the complete oxidation of fuel substrates is possible only in the presence of oxygen.

We will now consider how animals generate ATP without oxygen, that is, anaerobically. Anaerobic organisms break down carbon compounds by **fermentation,** an ancient metabolic device for obtaining energy in an atmosphere devoid of oxygen. The term "fermentation," meaning "cause to rise," was originally used to describe the action of yeasts that break down glucose into alcohol (ethanol) and carbon dioxide, end products that long ago created occupations for both brewers and bakers. Alcohol (or in animals, lactic acid) is the major end product of **glycolytic fermentation,** or simply **anaerobic glycolysis,** defined as the anaerobic degradation of glucose to yield a reduced end product not requiring molecular oxygen.

In anaerobic glycolysis, glucose and other 6-carbon sugars are first broken down stepwise to a pair of 3-carbon pyruvic acid molecules, yielding 2 molecules of ATP and 2 atoms of hydrogen. This pathway, shown in Figure 5-18, is precisely the same glycolytic pathway that in aerobic metabolism directs glucose into the Krebs cycle (compare Figure 5-17 with Figure 5-18). But in the absence of molecular oxygen, further oxidation of pyruvic acid cannot occur. Both pyruvic acid and carrier-bound hydrogen accumulate in the cytoplasm because neither can proceed in oxidative channels without oxygen. The problem is neatly solved by the formation of lactic acid from pyruvic acid. Pyruvate becomes the final electron acceptor and lactate the end product of anaerobic glycolysis.

Anaerobic glycolysis is a primitive and inefficient metabolic pathway that yields only 2 moles of ATP per mole of glucose; by comparison oxidative phosphorylation nets 36 moles. Despite its inefficiency, it has not been discarded by evolution—its key virtue being that it provides *some* high-energy phosphate in situations where oxygen is absent or in short supply. Many microorganisms have the opportunistic capacity to metabolize anaerobically when the oxygen supply is limited, then to switch to aerobic metabolism when oxygen is plentiful (facultative anaerobes). Most animals have retained the glycolytic pathway that serves as an important backup system capable of providing short-term generation of ATP during brief periods of heavy energy expenditure, when the slow rate of oxygen

Table 5-1 Calculation of Total ATP Molecules Generated in Respiration

ATP generated	Source
4	Directly in glycolysis
2	As GTP in Krebs cycle
4	From NADH in glycolysis
6	From NADH produced in pyruvic acid to acetyl coenzyme A reaction
4	From reduced FAD in Krebs cycle
18	From NADH produced in Krebs cycle
38	TOTAL
−2	Used in priming reactions in glycolysis
36	NET

The term "glycolysis" ("sweet releasing") was coined to describe the anaerobic conversion of glucose to lactic acid. However, many biochemists today use "glycolysis" to refer to the breakdown of sugar to pyruvate, the initial sequence of reactions leading into the Krebs cycle. This is the sense in which we use the term here. The anaerobic breakdown of sugar to lactic acid is referred to as anaerobic glycolysis. The distinction between "glycolysis" and "anaerobic glycolysis" is one of convenience, because the enzyme sequence from glucose to pyruvate is the same under both aerobic and anaerobic conditions.

Figure 5-18

Anaerobic glycolysis, a process that proceeds in the absence of oxygen. Glucose is broken down to 2 molecules of pyruvic acid, generating 4 molecules of ATP and yielding 2, since 2 molecules of ATP are used to produce fructose-1,6-diphosphate. Pyruvic acid, the final electron acceptor for the hydrogen atoms and electrons released during pyruvic acid formation, is converted to lactic acid.

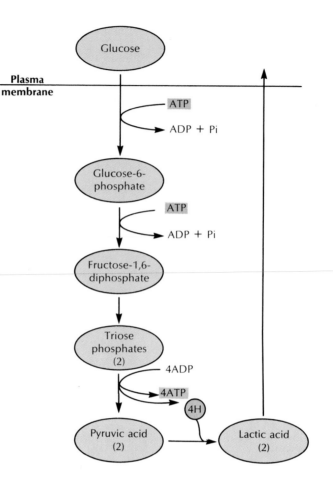

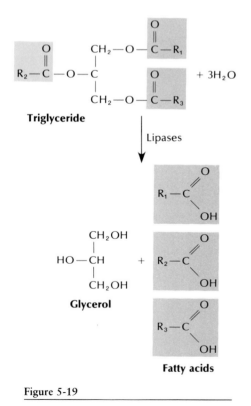

Figure 5-19

Hydrolysis of a triglyceride (neutral fat) by intracellular lipase. The R groups of each fatty acid represent a hydrocarbon chain.

delivery would be a limiting factor. Vertebrate skeletal muscle may rely heavily on glycolysis during short bursts of activity when contraction is too rapid and too powerful to be sustained by oxidative phosphorylation. This may mean the difference between survival and death in emergency situations. The lactic acid that accumulates in muscle during anaerobic glycolysis diffuses out into the blood and is carried to the liver where it is oxidized.

Many intertidal animals such as oysters and polychaete worms metabolize anaerobically when the tide is out and the beach is exposed. Diving birds and mammals fall back on glycolysis almost entirely to give them the energy needed to sustain a long dive. And salmon would never reach their spawning grounds were it not for anaerobic glycolysis providing almost all the ATP used in the powerful muscular bursts required to carry them up rapids and over falls. Many parasitic worms (e.g., nematodes, trematodes, cestodes) have dispensed with oxidative phosphorylation entirely. Living as they do in oxygen-deprived environments, they have evolved modified anaerobic pathways that are somewhat more efficient than classical anaerobic glycolysis.

Metabolism of Lipids

Animal fats are **triglycerides** (neutral fats), molecules composed of glycerol and 3 molecules of fatty acids. These fuels are important sources of energy for many metabolic processes in most animals. There is far more fat than glycogen in a mammalian body.

The first step in the breakdown of a triglyceride is the splitting of glycerol from the 3 fatty acid molecules (Figure 5-19). Glycerol and the fatty acids then proceed through separate pathways. Glycerol, a 3-carbon carbohydrate, is phosphorylated and enters the glycolytic pathway as a triose phosphate. From that point it is oxidized in the aerobic pathway like any other carbohydrate.

The remainder of the triglyceride molecule is fatty acids, carboxylic acids with long hydrocarbon chains. One of the abundant naturally occurring fatty acids is **stearic acid.**

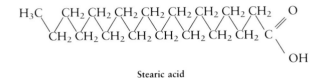

Stearic acid

We know that fats enter the mitochondrial processes through acetyl coenzyme A. What happens in brief is that the long hydrocarbon chain of a fatty acid is sliced up by oxidation, 2 carbons at a time; these are released from the end of the molecule as acetyl coenzyme A. The process is repeated until the entire chain has been reduced to several 2-carbon acetyl units. All of the acetyl coenzyme A molecules are then fed into the Krebs cycle (Figure 5-8).

How much ATP is gained by fatty acid oxidation? ATP is generated in the electron transport chain by the hydrogens stripped from the carbon chain during fatty acid breakdown, as well as by metabolism of the acetyl coenzyme A units. With allowance for the 2 high-energy phosphate bonds expended to attach the first acetyl coenzyme A molecule, it has been calculated that the complete oxidation of 18-carbon stearic acid will net 146 ATP molecules. By comparison, 3 molecules of glucose (also totaling 18 carbons) yield 108 ATP molecules. Since there are three fatty acids in each triglyceride molecule, a total of 440 ATP molecules is formed. An additional 22 molecules of ATP are generated in the breakdown of glycerol, giving a grand total of 462 molecules of ATP. Little wonder that fat is considered the king of animal fuels! Fats are more concentrated fuels than carbohydrates because fats are almost pure hydrocarbons; they contain more hydrogen per carbon atom than sugars do, and it is the energized electrons of hydrogen that generate high-energy bonds, when they are carried through the mitochondrial electron transport system.

Fat stores are derived principally from surplus fats and carbohydrates in the diet. Acetyl coenzyme A is the source of carbon atoms used to build fatty acids. Since all major classes of organic molecules (carbohydrates, fats, and proteins) can be degraded to acetyl coenzyme A, all can be converted into stored fat. The biosynthetic pathway for fatty acids resembles a reversal of the catabolic pathway already described but requires an entirely different set of enzymes. From acetyl coenzyme A, the fatty acid chain is assembled 2 carbons at a time. Because fatty acids release energy when they are oxidized, they obviously require an input of energy for their synthesis. This is provided principally by electron energy from glucose degradation. Thus the total ATP derived from oxidation of a molecule of triglyceride is not as great as previously calculated, because varying amounts of energy are required for synthesis and storage.

Metabolism of Proteins

Because proteins are composed of amino acids, 20 in all (Figure 2-20, p. 29), the central topic of our consideration is amino acid metabolism. Amino acid metabolism is complex. For one thing each of the 20 amino acids requires a separate pathway of biosynthesis and degradation. For another, amino acids are precursors

Goldfish are remarkably tolerant of oxygen-poor water and, in winter, can survive for days without any oxygen at all. Under such conditions they rely on glycolysis for their energy requirements. To avoid excessive buildup of lactic acid, the end product of glycolysis, they convert much of the lactic acid into alcohol, which diffuses out of the body. The Canadian team that made the discovery points out that the novel pathway prevents lactic acid from making the fish acidotic by swamping the fish's limited blood buffering system. Ethanol is neutral, and excessive acid accumulation is avoided.

Stored fats are the greatest reserve fuel in the body. Most of the usable fat resides in adipose tissue that is composed of specialized cells packed with globules of triglycerides. Adipose tissue is widely distributed in the abdominal cavity, in muscles, around deep blood vessels, and especially under the skin. Women average about 30% more fat than men, and this is responsible in no small measure for the curved contours of the female figure. However, its aesthetic contribution is strictly subsidiary to its principal function as an internal fuel depot. Indeed, humans can only too easily deposit large quantities of fat, generating personal unhappiness and hazards to health.

The physiological and psychological aspects of obesity are now being investigated by many researchers. There is increasing evidence that body fat deposition is regulated by a feeding control center located in the lateral and ventral regions of the hypothalamus, an area in the floor of the forebrain. The set point of this regulator determines the normal weight for the individual, which may be rather persistently maintained above or below what is considered normal for the human population. Thus, obesity is not always the result of overindulgence and lack of self-control, despite popular notions to the contrary.

to tissue proteins, enzymes, nucleic acids, and other nitrogenous constituents that form the very fabric of the cell. The central purpose of carbohydrate and fat oxidation is to provide energy to construct and maintain these vital macromolecules.

Let us begin with the **amino acid pool** in the blood and extracellular fluid from which the tissues draw their requirements. When animals eat proteins, these are digested in the gut, releasing the constituent amino acids, which are then absorbed (Figure 5-20). Tissue proteins also are hydrolyzed during normal growth, repair, and tissue restructuring; their amino acids join those derived from protein foodstuffs to enter the amino acid pool. A portion of the amino acid pool is used to rebuild tissue proteins, but most animals ingest a protein surplus. Since amino acids are not excreted as such in any significant amounts, they must be disposed of in some way. In fact, amino acids can be and are metabolized through oxidative pathways to yield high-energy phosphate. In short, excess proteins serve as fuel as do carbohydrates and fats. Their importance as fuel obviously depends on the nature of the diet. In carnivores that ingest a diet of almost pure protein and fat, nearly half of their high-energy phosphate is derived from amino acid oxidation.

Before entering the fuel depot, nitrogen must be removed from the amino acid molecule. This is done in either of two ways. Some amino acids are **deaminated** (their amino group splits off) to yield ammonia and a keto acid:

$$R-\underset{\underset{NH_2}{|}}{CH}-COOH + H_2O \rightarrow R-\overset{\overset{O}{\|}}{C}-COOH + NH_3 + H_2$$

Amino acid   Keto acid   Ammonia

Most amino acids, however, undergo **transamination** in which the amino group is transferred to a keto acid to yield a new amino acid, often glutamic acid:

$$R-\underset{\underset{NH_2}{|}}{CH}-COOH + HOOC-CH_2-CH_2-\overset{\overset{O}{\|}}{C}-COOH \rightleftharpoons R-\overset{\overset{O}{\|}}{C}-COOH + HOOC-CH_2-CH_2-\underset{\underset{NH_2}{|}}{CH}-COOH$$

Amino Acid   α-Ketoglutaric acid   Keto acid   Glutamic acid

Glutamic acid can then be oxidized to liberate ammonia. Thus, amino acid degradation yields two main products, ammonia and carbon skeletons, which are handled in different ways.

Once the nitrogen atoms are removed, the carbon skeletons of amino acids can be completely oxidized, usually by way of pyruvate or acetate. These residues then enter regular routes used by carbohydrate and fat metabolism. This brings us finally to the disposal of ammonia, a complicated story both biochemically and evolutionarily that we can examine only in a general way in this chapter.

Ammonia is a highly toxic waste product. Its disposal offers little problem to aquatic animals because it is soluble and readily diffuses through the respiratory surfaces into the surrounding medium. Terrestrial forms cannot get rid of ammonia so conveniently and must detoxify it by converting it to a relatively nontoxic compound. The two principal compounds formed are **urea** and **uric acid,** although a variety of other detoxified forms of ammonia are excreted by different invertebrate and vertebrate groups. Among the vertebrates, amphibians and especially mammals produce urea. Reptiles and birds, as well as many terrestrial invertebrates, produce uric acid.

The key feature that seems to determine the choice of nitrogenous waste is the availability of water in the environment. When water is abundant, the chief nitrogenous waste is ammonia. When water is restricted, it is urea. For animals living in truly arid habitats (for example, insects, pulmonate snails, many birds, and lizards) it is uric acid. The distinguishing attribute of uric acid is that it is highly insoluble and easily precipitates from solution, allowing its removal in solid form. Among vertebrates, it is not so much the adults that benefit from uric acid's insolubility (since many forms excreting urea also live successfully in dry habitats) as it is their embryos. Birds and reptiles lay eggs enclosed in water-impermeable shells containing not only the embryo but also all the supportive supplies that it will require for its development: nutrients, fuel, and a small amount of water that must be used with the greatest economy. Furthermore, the embryo has no way to jettison its wastes. As the embryo develops, amino acids are metabolized, yielding ammonia that must be immediately detoxified. Urea is an unsuitable product because it requires too much water for its storage. The solution is uric acid, which is retained in harmless, solid form in the allantois, one of the embryonic membranes (p. 786). When the hatchling emerges into its new world, the accumulated uric acid, along with the shell and membranes that supported development, is discarded.

___ Management of Metabolism

The complex pattern of enzyme reactions that constitutes metabolism cannot be explained entirely in terms of physicochemical laws or chance happenings. Although some enzymes do indeed "flow with the tide," the activity of others is rigidly controlled. In the former case, suppose the purpose of an enzyme is to convert A to B. If B is removed by conversion into another compound, the enzyme will tend to restore the original ratio of B to A. Since many enzymes act reversibly, this can result, according to the metabolic situation prevailing, in synthesis or degradation. For example, an excess of a Krebs cycle intermediate would result in its contribution to glycogen synthesis; a depletion of such a metabolite would lead to glycogen breakdown. This automatic compensation (equilibration) is not, however, sufficient to explain all that actually takes place in an organism, as for example, what happens at branch points in a metabolic pathway.

Mechanisms have been discovered that critically regulate enzymes in both quantity and activity. Enzyme induction in bacteria is an example of quantity regulation. The genes leading to synthesis of the enzyme are switched on or off, depending on the presence or absence of a substrate molecule. In this way the amount of an enzyme can be controlled. It is a relatively slow process.

There are also more subtle mechanisms of control that operate by affecting the *activity* of already existing enzymes. The systems producing ATP must not be excessively active when the organism is at rest but must be called into action when needed. Several enzymes, occurring at certain critical positions on the pathway from glycogen to carbon dioxide, can be "switched on or off." A good example of this is the enzyme phosphofructokinase, which acts between glucose-6-phosphate and fructose-1,6-diphosphate (Figure 5-9). The activity of this enzyme is depressed by high concentrations of ATP (Figure 5-21) or citric acid, because their presence means that a sufficient amount of precursors has reached the Krebs cycle and additional glucose is not needed. Alternatively, the activity of the enzyme is *enhanced* by AMP because its presence means that ATP is rapidly being consumed for some cellular activity. This common form of enzyme regulation is character-

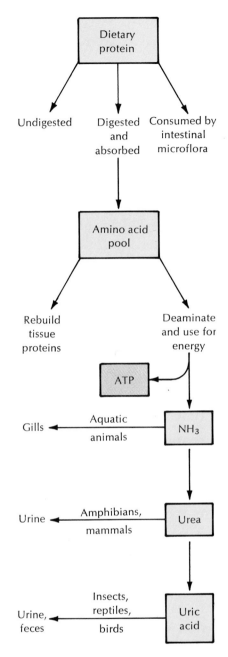

Figure 5-20

Fate of dietary protein.

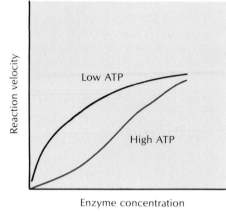

Figure 5-21

Regulation by ATP of the activity of phosphofructokinase as measured by reaction velocity. A high ATP level depresses enzyme activity by binding to a regulatory site on the enzyme molecule.

istically brought about by signal molecules that are normal metabolites (in contrast to hormones that are solely for communication). They act by altering the shape (or conformation) of the enzyme protein and thus improve or diminish the effectiveness of the enzyme as a catalyst (Figure 5-22).

As well as being subject to alteration in physical shape, some enzymes exist in two forms. These may be chemically different. Glycogen phosphorylase, which breaks glucose-6-phosphate units off a glycogen molecule, can exist as a relatively inert *b* form, but the *b* form can be converted reversibly to a more active *a* form. The hormone epinephrine, reaching a liver cell by way of the blood, affects the cell membrane in such a way that a complicated train of events is set in motion, resulting in a phosphate group being attached to a serine in the enzyme. This phosphorylated enzyme is extremely active and enables a rapid flow of sugar toward the Krebs cycle to take place. Epinephrine also has the reverse effect on the enzymes that work in the other direction, such as glycogen synthase, leading to the synthesis of glycogen, and it switches the synthase off at the critical time when glucose is needed. The messenger for this kind of activation/inactivation can be either a hormone or a particular metabolite.

Many cases of enzyme regulation are known, but these selected examples must suffice to illustrate the importance of enzyme regulation in the integration of metabolism.

___ SUMMARY

Living systems are subject to the same laws of thermodynamics that govern non-living systems. The first law states that energy cannot be destroyed, although it may change form. The second law states that all energy proceeds toward total randomness, or increasing entropy. Solar energy trapped by photosynthesis as chemical bond energy is passed through the food chain where it is used for biosynthesis, active transport, and motion, before finally being degraded to heat energy. Living organisms are able to decrease their entropy and maintain high internal order because the biosphere is an open system from which energy can be captured and used. Energy available for use in biochemical reactions is termed free energy.

Enzymes are pure proteins or proteins associated with complex organic groups (coenzymes) that vastly accelerate reaction rates in living systems. An enzyme does this by temporarily binding its active site to a reactant (substrate). In this configuration, internal activation energy barriers are lowered enough to disrupt and split the substrate, and the enzyme is restored to its original form.

The coupling of biochemical reactions is achieved primarily by adenosine triphosphate (ATP). Free energy held in chemical bonds can be transferred through the high-energy phosphate bonds of ATP to molecules having lower bond energy. ATP is formed as it is needed from fuels such as glucose and is used to power various cellular processes.

Animals liberate the stored energy of fuels (sugars, fatty acids, amino acids) through a series of discrete degradative steps. Energy is released as electrons, which are transferred by oxidation-reduction reactions that end in the generation of ATP.

Most animals are aerobic, meaning that oxygen from the environment is used as the final electron acceptor in cellular metabolism. Organisms that can survive without oxygen, and that use some other molecule (such as pyruvic acid)

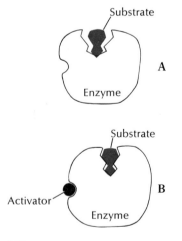

Figure 5-22

Enzyme regulation. **A,** The active site of an enzyme may only loosely fit its substrate in the absence of an activator. **B,** With the regulatory site of the enzyme occupied by an activator, the enzyme binds the substrate, and the site becomes catalytically active.

as the final electron acceptor, are termed anaerobic. Aerobic metabolism is much more efficient than anaerobic metabolism.

Glycolysis, the first stage in the oxidation of fuel molecules, is a sequence of reactions that converts glucose to pyruvic acid. The net yield is 2 molecules of ATP and 2 molecules of reduced NAD, an electron acceptor that, in aerobic organisms, transfers its electrons to the electron transport chain.

In aerobic metabolism, pyruvic acid is decarboxylated to form acetyl coenzyme A, which enters the Krebs cycle. The Krebs cycle is a series of reactions occurring inside the mitochondria where acetyl coenzyme A is broken down to carbon dioxide. In the course of the reactions, energized electrons are passed to electron acceptors NAD and FAD. A molecule of ATP also is formed.

The final stage in respiration is the electron transport chain, composed of a complex series of enzymes located in the inner membrane of the mitochondrion. ATP is formed at three sites as electrons are passed along this chain of electron carriers from reduced NAD and FAD to oxygen. This is called oxidative phophorylation. The currently favored hypothesis to explain the coupling of electron flow to ATP synthesis is the Mitchell chemiosmotic theory. It proposes that ATP synthesis is driven by a gradient of hydrogen ions across the inner mitochondrial membrane. The complete degradation of a molecule of glucose by oxidative phosphorylation yields 36 molecules of ATP. However, recent calculations suggest that the yield of ATP available for cellular work may be less than this.

Triglycerides (neutral fats) are especially rich depots of metabolic energy because the fatty acids of which they are composed are highly reduced and anhydrous. Fatty acids are degraded by sequential removal of 2-carbon units, which enter the Krebs cycle through acetyl coenzyme A.

Amino acids in excess of requirements for synthesis of proteins and other biomolecules are used as fuel. They are degraded by deamination or transamination to yield ammonia and carbon skeletons. The latter enter the Krebs cycle to be oxidized. Ammonia is a highly toxic waste product that aquatic animals quickly dispose of through respiratory surfaces. Terrestrial animals, however, convert ammonia into much less toxic compounds, urea or uric acid, for disposal.

The integration of metabolic pathways is rigidly regulated by mechanisms that control both the amount and activity of enzymes. The quantity of many enzymes is regulated by certain molecules that switch on or off enzyme synthesis in the nucleus. Enzyme activity may be altered by the presence or absence of metabolites that cause conformational changes in enzymes and thus improve or diminish their effectiveness as catalysts. Certain hormones may activate inactive enzymes.

Review questions
1. State the first and second laws of thermodynamics. How do these laws relate to living systems?
2. Living things maintain a high degree of organization despite a universal trend toward increasing disorganization and thus they appear to violate the second law of thermodynamics. What is the explanation for this paradox?
3. Explain what is meant by "free energy" in a system. Will a reaction that proceeds spontaneously have a positive or a negative change in free energy?
4. Many biochemical reactions proceed slowly unless the energy barrier to the change is lowered. How is this accomplished in living systems?
5. What happens in the formation of an enzyme-substrate complex that favors the disruption of substrate bonds?
6. What is meant by a "high-energy bond?"

7. If ATP is able to supply energy to an endergonic reaction, why should it not be considered a fuel?

8. What is an oxidation-reduction reaction and why are such reactions considered so important in cellular metabolism?

9. Give an example of the final electron acceptor found in aerobic and anaerobic organisms.

10. Why is it necessary for glucose to be "primed" with a high-energy phosphate bond before it can be degraded in the glycolytic pathway?

11. What happens to the electrons that are removed during the oxidation of triose phosphates during glycolysis?

12. Why is acetyl coenzyme A considered to be a "strategic intermediate" in respiration?

13. Study Figure 5-17. Then, without reference to the illustration, diagram the generation of ATP from one glucose molecule sent down the aerobic pathway, from glycolysis through the Krebs cycle and the electron transport chain. Do the same for electron pairs (shown as 2H in the diagram).

14. The currently favored hypothesis to explain the coupling of ATP synthesis to the oxidation of NADH is the chemiosmotic mechanism. Explain how the movement of hydrogen ions across mitochondrial membranes can lead to ATP formation.

15. Why are oxygen atoms important in oxidative phosphorylation?

16. Explain how animals can generate ATP *without* oxygen. Since anaerobic glycolysis is much less efficient than oxidative phosphorylation, why has anaerobic glycolysis not been discarded during animal evolution?

17. Why are animal fats sometimes referred to as "the king of animal fuels?"

18. The breakdown of amino acids yields two products: ammonia and carbon skeletons. What happens to these products?

19. Explain the relationship between the amount of water in an animal's environment and the kind of nitrogenous waste it produces.

20. Explain three ways that enzymes may be regulated in cells.

Selected references

Alberts, B., D. Bray, J. Lewis, M. Raff, K. Roberts, and J.D. Watson. 1983. Molecular biology of the cell. New York, Garland Publishing, Inc. *Chapter 2 presents an excellent summary of cellular energy and intermediary metabolism.*

Dickerson, R.E. 1980. Cytochrome c and the evolution of energy metabolism. Sci. Am. **242**:136-153 (Mar.). *How the metabolism of modern organisms evolved.*

Hinkle, P., and R. McCarty. 1978. How cells make ATP. Sci. Am. **238**:104-123 (Mar.). *Good description of the chemiosmotic hypothesis of ATP formation.*

Rawn, J.D. 1983. Biochemistry. New York, Harper & Row, Publishers. *Introductory text that assumes a strong background in chemistry. A novel feature is a stereo viewer that provides stunning three-dimensional views of macromolecules.*

Stryer, L. 1981. Biochemistry, ed. 2. San Francisco, W.H. Freeman & Co. Publishers. *One of the best of undergraduate biochemistry texts with lucid explanations and good diagrams.*

CHAPTER 6

ARCHITECTURAL PATTERN OF AN ANIMAL

Animals exist in an enormous variety of forms and our knowledge of fossils suggests that even greater diversity existed in the past. Yet although the possibilities for designs for living seem unlimited, they actually are not. This is because groups of animals have single evolutionary origins that have provided each group with a basic body plan that is a primary determinant of body form. No matter how much that body plan is altered and adapted to different ways of life, the evolution of new forms within a group must always develop within the constraints imposed by the group's architectural blueprint. So we do not find molluscs that fly or protozoans as large as garden snails. Nor are there giant insects or microscopic reptiles. However, such restrictions do not prevent animals of completely different groups from evolving convergently similar solutions to common problems. There *are* fishes that fly and birds that swim.

Animals are also shaped by their particular habitat and different ways of life. A worm that adopts a parasitic life in a vertebrate's intestine will look and function very differently from a free-living member of the same group, yet both will share the same distinguishing hallmarks. In this chapter we will consider the limited number of body plans that underlie the apparent diversity of animal form and examine some of the common architectural themes that animals share.

——LEVELS OF ORGANIZATION IN ANIMAL COMPLEXITY

Increasing complexity of organization is the most evident feature of animal phylogeny. The simplest animals are the unicellular protozoa and, being small, have much less scope for complexity than do larger creatures. Size then is an important determinant of complexity. Nevertheless, unicellular forms are complete organisms that carry on all the basic life functions of more complex animals. Within the confines of their cell they often show remarkable organization and division of labor, such as skeletal elements, locomotor devices, fibrils, and beginnings of sensory structures. The **metazoa,** or multicellular animals, on the other hand, have a much greater capacity for internal specialization. Different cells can take on different functions, becoming increasingly specialized and increasingly efficient in carrying out particular tasks. Consequently, the metazoan cell is not the equivalent of a protozoan cell; it is only a specialized part of the whole organism and is incapable of independent existence.

We can recognize five grades of organization. Each grade is more complex than the one before, and, as a general rule, it is a more advanced and more recent evolutionary product.

1. *Protoplasmic grade of organization.* Protoplasmic organization is found in protozoa and other unicellular organisms. All life functions are confined within the boundaries of a single cell, the fundamental unit of life. Within the cell, protoplasm is differentiated into organelles capable of carrying on specialized functions.

2. *Cellular grade of organization.* Cellular organization is an aggregation of cells that are functionally differentiated. A division of labor is evident, so that some cells are concerned with, for example, reproduction, others with nutrition. Such cells have little tendency to become organized into tissues (a tissue is a group of similar cells organized to perform a common function). Some protozoan colonial forms having somatic and reproductive cells might be placed at the cellular level of organization. Many authorities also place sponges at this level.

3. *Cell-tissue grade of organization.* A step beyond the preceding is the aggregation of similar cells into definite patterns or layers, thus becoming a **tissue.** Sponges are considered by some authorities to belong to this grade, although the jellyfish and their relatives (Cnidaria) are usually referred to as the beginning of the tissue plan. Both groups are still largely of the cellular grade of organization because most of the cells are scattered and not organized into tissues. An excellent example of a tissue in cnidarians is the **nerve net,** in which the nerve cells and their processes form a definite tissue structure, with the function of coordination.

4. *Tissue-organ grade of organization.* The aggregation of tissues into organs is a further step in advancement. Organs are usually made up of more than one kind of tissue and have a more specialized function than tissues. The first appearance of this level is in the flatworms (Platyhelminthes), in which there are a number of well-defined organs such as eyespots, proboscis, and reproductive organs. In fact, the reproductive organs are well organized into a reproductive system.

5. *Organ-system grade of organization.* When organs work together to perform some function, we have the highest level of organization—the organ system. The systems are associated with the basic body functions—circulation, respiration, digestion, and the others. Typical of all the higher forms, this type of organization is first seen in the nemertean worms, which have a complete digestive system distinct from the circulatory system.

___ SIZE AND COMPLEXITY

It is evident from the preceding discussion that as animals become more complex, they tend to become larger. An ameba can be small because it can move without muscles, digest food without a gut, excrete its wastes without a kidney, coordinate its activities without a brain, and breathe without gills. More advanced multicellular animals have specialized tissues and organs for these functions, and consequently must be larger to accommodate their increased complexity.

But if the single-celled protozoa are so successful, why has there been a tendency in animal lineages for increasingly larger multicellular forms to evolve? Certainly a prime selective force was the availability on earth of scores of habitats suitable for animals larger than a single cell. Small organisms, successful as they were (and are), long ago occupied all the available small ecological niches, and the only way a new species could succeed was to displace an organism from an existing niche or adapt to a new and larger one. Since single cells, excepting eggs, cannot become very large, multicellularity became a reasonably simple and highly adaptive path toward increased body size.

Once embarked on this path and as organisms became bulkier, increasing complexity was inevitable. As animals grow larger, the body surface (length^2) increases much more slowly than body volume (length^3) making it increasingly difficult for activities served from the surface to provision the mass of cells within. Systems had to evolve to supply nutrients and gases and to serve as an avenue for waste removal. Cilia and flagella are inadequate for moving anything larger than a flatworm, so muscles for locomotion appeared. And muscles in turn required some kind of skeletal support to work against.

If in becoming larger, animals lose the advantage of architectural simplicity, what is to be gained by being big? Perhaps the most obvious advantage is that in the predator-prey contest, predators are almost always larger than their prey. Exceptions are few and are usually related to an especially aggressive behavior on the part of the predator that compensates for small size. Since the predator must first catch its prey, whereas the prey must escape its predator if it is to survive, both predator and prey will experience selective pressures that encourage an evolutionary drift toward larger size.

Another advantage of large body size that is usually overlooked is that large animals can move themselves about at a lesser energy cost than small animals. This becomes clear when we look at the cost of running for a given group of animals (for example, mammals) of various body sizes (Figure 6-1). A large mammal consumes more oxygen than a small mammal, of course, but the cost of moving 1 g of its body over a given distance is much less for the large animal than for a small one. This makes sense when we compare the relative ease with which a horse can run a mile with the herculean task that running the same distance would present to a mouse!

Other advantages to large size have been cited by biologists. Nevertheless, every animal, whether large or small, is a success in its ecological niche. In fact, the most abundant groups of animals on earth today are those with short generation times and, consequently, small body size. To these countless creatures, "small is beautiful."

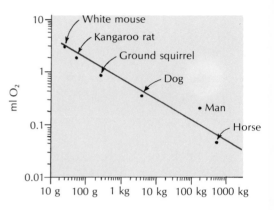

Figure 6-1

Net cost of running for mammals of various sizes. Each point represents the cost (measured in rate of oxygen consumption) of moving 1 g of body over 1 km. The cost decreases with increasing body size. Modified from Schmidt-Nielsen, K. 1972. How animals work. New York, Cambridge University Press.

____ PATTERNS OF ANIMAL DEVELOPMENT

If it is true, as we have just seen, that large body size and increased complexity have been encouraged in the course of evolution, it is also true that every metazoan animal, no matter how large it may become, must begin its individual life small. The final adult architecture of a multicellular animal is the result of progressive growth and differentiation from two critical components in sexual reproduction, the sperm and the egg. These are the **sex cells,** or **gametes,** typically produced in testes and ovaries of separate male and female sexes (exceptions to this generalization are described in Chapter 35).

The uniting of the germ cells (sperm and egg) by **fertilization** begins the life cycle of the organism (Principle 23, p. 13). The fertilized egg, now called a **zygote,** is really a one-celled organism, and from it develops a complete animal by the processes of **morphogenesis** and **differentiation.** How this occurs is only beginning to become clear. All the information necessary for development is contained within the zygote, principally within the genes of its nucleus. The actual blueprint for development is coded within the DNA molecules of the genes. The heredity of the

Figure 6-2

Cell division in developing embryos of *Echinus esculentus,* a common sea urchin. Cells become progressively smaller with no growth between divisions. Present are four- and eight-cell stages, many-celled stage, and early blastula. (×165.)
Photograph by D.P. Wilson.

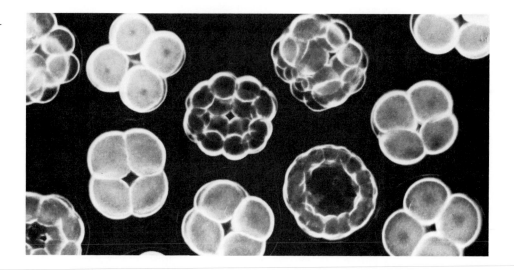

Isolecithal   Centrolecithal

Telolecithal-holoblastic

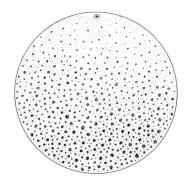

Telolecithal-meroblastic

Figure 6-3

Types of eggs. Isolecithal (echinoderms and amphioxus) have small yolk evenly distributed. Centrolecithal (insects and other arthropods) have cytoplasm centered but migrating out at cleavage, leaving yolk centered. Telolecithal with holoblastic cleavage (amphibians and bony fishes) have yolk concentrated at vegetal pole. Telolecithal with meroblastic cleavage (birds and reptiles) have cytoplasm concentrated in germinal disc at animal pole.

organism stabilizes development, making it a well-knit, smoothly functioning, and highly predictable process. Nevertheless, variations in development do occur, and this provides the material that makes evolutionary change possible (the sources of biological variation are considered in Chapter 38).

Every species develops a characteristic body plan (Principle 24, p. 14). As the body plan takes shape during development, certain distinguishing hallmarks of the phylum appear before the specific qualities of the species. Such basic qualities as symmetry and a longitudinal axis appear first, then a gut, sensory structures, and so on. As development continues, the organism first acquires the morphological characters of its class, then of its order, then its family, then its genus and finally of its own species. Thus development proceeds from the general to the specific.

___ Morphogenesis

Morphogenesis is the term used to describe the formation of shape in the embryo. The initial stages of morphogenesis are remarkably similar in nearly all groups, except for sponges, which develop in a peculiar way unlike that of any other metazoa.

Cleavage

The unicellular zygote begins to divide, first into two cells, those two into four cells, those four into eight. Repeated again and again, these cell divisions soon convert the zygote into a ball of cells. This process, called **cleavage,** occurs by mitosis. But unlike ordinary body-cell mitosis, there is no true growth (that is, increase in protoplasmic mass) (Figure 6-2). With each subsequent division, the cells are reduced in size by one half. The cleavage process converts a single, very large, unwieldy egg into many small, more maneuverable, ordinary-sized cells called blastomeres (Gr. *blastos,* germ, + *meros,* part).

Cleavage patterns are greatly affected by the amount of yolk in the egg. Eggs with very little yolk that is evenly distributed in the egg are called **isolecithal** (Gr. *isos,* equal, + *lekithos,* yolk) (Figure 6-3). In such eggs, which are typical of many invertebrates as well as marsupial and placental mammals (including humans), cleavage is complete (**holoblastic** [Gr. *holo,* whole, + *blastos,* germ]), and the daughter cells formed at each cleavage are usually approximately equal in size (Figure 6-2). However, eggs of many annelids and molluscs exhibit unequal cleavage; one blastomere is much larger than the others and holds most of the yolk.

Eggs with larger amounts of yolk are called **telolecithal** (Gr. *telos,* end, + *lekithos,* yolk) because the yolk tends to concentrate in the so-called **vegetal** pole of the egg (Figure 6-3). The opposite, or **animal,** pole contains mostly cytoplasm and very little yolk. Some telolecithal eggs, such as those of squid, marine bony fish, and amphibians, cleave completely (holoblastically), but the higher concentration of yolk in the vegetal half of the egg retards cleavage in that region. The result after a few cleavages is a greater number of small cells in the animal pole and fewer large cells in the vegetal half. A good example is cleavage in the frog egg (Figure 6-6).

Birds and reptiles produce the largest eggs of all animals, containing so much yolk that the actively dividing cytoplasm is confined to a narrow disc-shaped mass lying on top of the yolk. Cleavage is partial, or **meroblastic** (Gr., *meros,* part, + *blastos,* germ) (Figure 6-3) because the cleavage furrows cannot cut through the whole egg mass, but instead stop at the border between the cytoplasm and the yolk below.

The eggs of advanced arthropods, especially insects, have a completely different cleavage pattern. The centrally located egg nucleus is surrounded by a small island of cytoplasm that is in turn surrounded by yolk. This type of egg is called **centrolecithal** (Gr. *kentron,* center, + *lekithos,* yolk) (Figure 6-3). The nucleus divides repeatedly by mitosis without being accompanied by cytokinesis until hundreds or thousands of nuclei lie in the yolky center, each surrounded by a minute islet of cytoplasm. Only then, at some common signal, do these free nuclei migrate to the surface cytoplasm where membranes rapidly appear to partition off the nuclei into cells.

Radial and spiral cleavage

Although several different cleavage patterns among animal groups have been described, the eggs of the great majority of invertebrates cleave by one of two patterns—radial or spiral. In **radial cleavage,** the cleavage planes are symmetrical to the polar axis and produce tiers, or layers, of cells on top of each other (Figure 6-4). Radial cleavage is also said to be **regulative** (or **indeterminate**) because each blastomere of the early embryo, if separated from the others, can adjust or "regulate" its development into a complete and well-proportioned embryo. This happens because the blastomeres are equipotential in the sense that there is no definite relation between the position of any of the *early* blastomeres and the specific tissue it will form in the developing embryo.

Spiral cleavage is different from radial. The cleavage planes are oblique to the polar axis and typically produce quartets of unequal cells that come to lie, not on top of each other, but in the furrows between the cells (Figure 6-4). Spirally cleaving embryos also differ from radial embryos in having a **mosaic** (or **determinate**) form of development. This means that the organ-forming regions of the egg cytoplasm become strictly localized in the egg, even before the first cleavage division. The result is that if the early blastomeres are separated, each will continue to develop for a time as though it were still part of the whole. Each forms a defective, partial embryo. A curious feature of most spirally cleaving embryos is that at about the 29-cell stage a blastomere called the 4d cell is formed that will give rise to all the mesoderm of the embryo.

The importance of these two cleavage patterns extends well beyond the

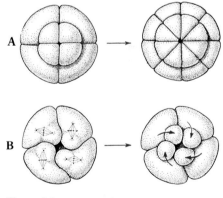

Figure 6-4

Radial and spiral cleavage. **A,** Radial cleavage shown at eight- and 16-cell stages. **B,** Spiral cleavage, showing the transition from four- to eight-cell stage. Arrows indicate clockwise movement of small cells (micromeres) following division from large cells (macromeres).

differences we have described. They are signals of a fundamental evolutionary dichotomy, the divergence of bilaterial metazoan animals into two separate lineages. Spiral cleavage is found in the Annelida, most Mollusca, Nemertea, turbellarian Platyhelminthes, some Brachiopoda and Echiurida. These and several other phyla are included in the **Protostomia** ("primary mouth") division of the Animal Kingdom (see the illustration inside the front cover of this book). Radial cleavage (and a modified pattern called biradial) is characteristic of the **Deuterostomia** ("secondary mouth") division of the Animal Kingdom that includes the Echinodermata, Chaetognatha, Hemichordata, and Chordata. Other distinguishing hallmarks of these two divisions are summarized in the next chapter (p. 128).

Blastulation

Cleavage, however modified by different cleavage patterns and by the presence of varying amounts of yolk, results in a cluster of cells called a **blastula** (Figures 6-5 and 6-6). In many animals the cells arrange themselves around a central fluid-filled cavity called the **blastocoel**. At this point the embryo consists of a few hundred to several thousand cells poised for further development. There has been a great increase in total DNA content, since each of the many daughter cell nuclei, by chromosomal replication at mitosis, contains as much DNA as the original zygote nucleus. The original zygote, however, has not increased in size; it has been subdivided into smaller and smaller cells.

Figure 6-5

Early embryology of a nemertean worm, a protostome with spiral cleavage.

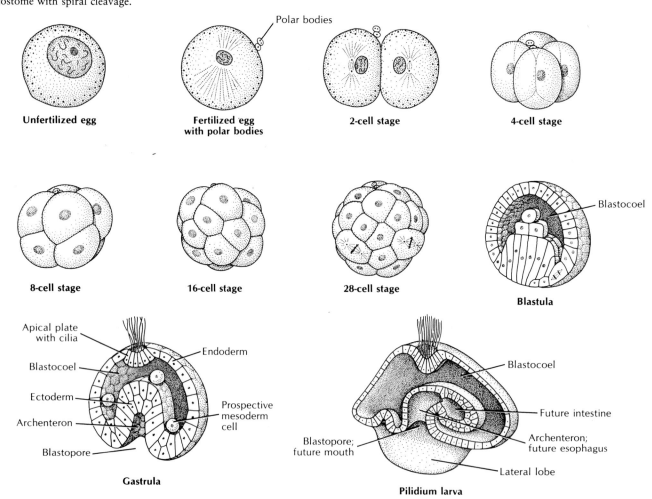

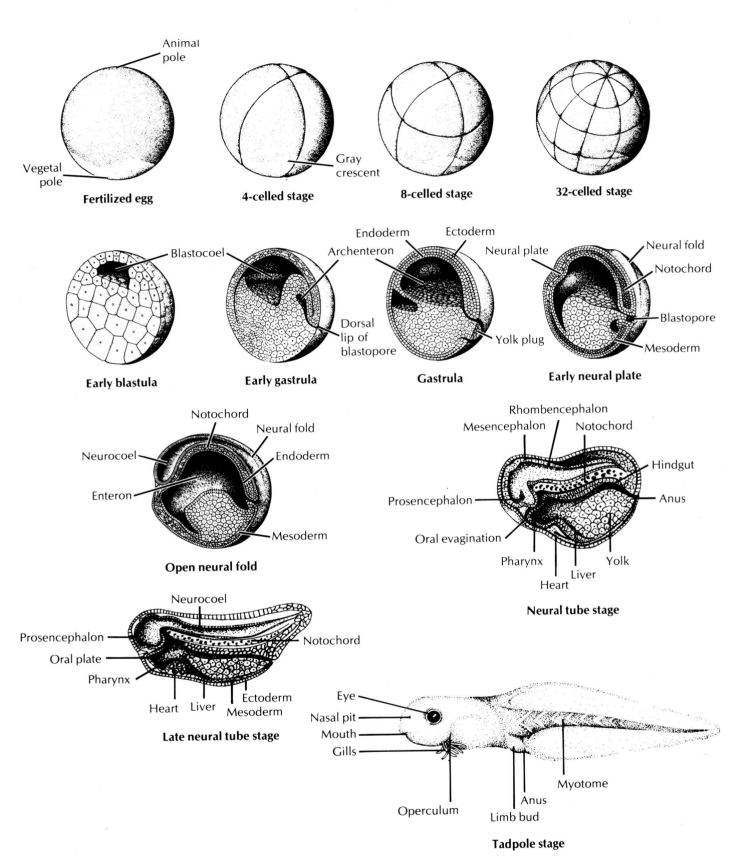

Figure 6-6

Early embryology of a frog, a deuterostome
with radial cleavage.

Gastrulation and the formation of germ layers

Gastrulation is a regrouping process in which new and important cell associations are formed. Up to this point the embryo has divided itself into a multicellular complex; the cytoplasm of these numerous cells is nearly in the same position as in the original undivided egg. In other words, there has been no significant movement or displacement of the cells from their place of origin. As gastrulation begins, the cells become rearranged in an orderly way by morphogenetic movements.

The early cell movements in gastrulation are remarkably similar in most protostomes and deuterostomes. In the nemertean worm, a protostome with spiral cleavage (Figure 6-5), cells at the vegetal pole indent, then migrate inward by a process called **invagination**. In a way similar to pushing in the side of a soft tennis ball, an internal cavity is created called the **archenteron**. The archenteron is the primitive gut and its opening to the outside, the **blastopore**, will become the future mouth of the worm. The gastrula is now an embryo of two **germ layers**. The outer layer is the **ectoderm**; it will give rise to the epithelium of the body surface and the nervous system. The inner layer that forms the archenteron is the **endoderm**; it will give rise to the epithelial lining of the digestive tube. The cavity between these two layers is the old blastocoel. A third germ layer, the **mesoderm**, arises from cells in the blastocoel that pinch off from either side of the blastopore. The mesoderm will later differentiate into muscles, blood vessels, and the reproductive system of the future adult body.

In the frog, a deuterostome with radial cleavage (Figure 6-6), the morphogenetic movements of gastrulation are greatly influenced by the mass of inert yolk in the vegetal half of the embryo. Cleavage divisions are slowed in this half so that the resulting blastula consists of many small cells in the animal half and a few large cells in the vegetal half (Figure 6-6). Cells on the surface begin to sink inward (invaginate) at the blastopore and continue to move inward as a sheet to form a flattened cavity, the archenteron. Although the cell movements are different, the result of gastrulation is the same in the frog and the nemertean worm: an embryo of two germ layers (endoderm and ectoderm) is formed. As invagination proceeds, the third germ layer, the mesoderm, rolls over the lateral and ventral lip of the blastopore and, once inside, penetrates between the endoderm on the inside and the ectoderm on the outside. The three germ layers now formed are the primary structural layers that play crucial roles in the further differentiation of the embryo.

In the most primitive of the true metazoa, the Cnidaria and the Ctenophora, only two germ layers are formed, the endoderm and ectoderm. These animals are **diploblastic**. In all more advanced metazoa, the mesoderm also appears, either from pouches of the archenteron or from other cells associated with endoderm formation. This three-layered condition is called **triploblastic**.

Formation of the coelom

The coelom, or true body cavity that contains the viscera, may be formed by one of two methods—**schizocoelous** or **enterocoelous** (Figure 6-7)—or by modification of these methods. (The two terms are descriptive, for *schizo* comes from the Greek *schizein*, to split; *entero* is a Greek form from *enteron*, gut; *coelous* comes from the Greek *koilos*, hollow or cavity.) In schizocoelous formation the coelom arises, as the word implies, from the splitting of mesodermal bands that originate from the blastopore region and grow between the ectoderm and endoderm; in

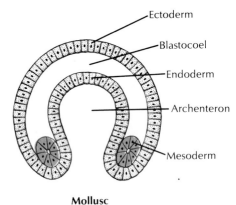

Ectoderm

Blastocoel

Endoderm

Archenteron

Mesoderm

**Mollusc
Schizocoelous**

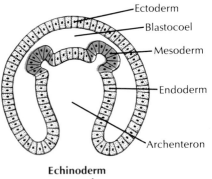

Ectoderm

Blastocoel

Mesoderm

Endoderm

Archenteron

**Echinoderm
Enterocoelous**

Figure 6-7

Two types of mesoderm and coelom formation: schizocoelous, in which mesoderm originates from wall of archenteron near lips of blastopore, and enterocoelous, in which mesoderm and coelom develop from endodermal pouches.

enterocoelous formation the coelom comes from pouches of the archenteron, or primitive gut.

These two quite different origins for the coelom are another expression of the deuterostome-protostome dichotomy of bilateral animals. The coelom of protostome animals develops by the schizocoelous method. The deuterostomes mostly follow the enterocoelous plan. Advanced chordates, however, are exceptions to this distinction because their coelom is formed by mesodermal splitting (schizocoelous), although in other respects they develop as deuterostomes, the division to which they are assigned. Other characteristics that distinguish these two phylogenetic divisions of bilateral animals are radial, regulative cleavage and spiral, mosaic cleavage, mentioned before.

Differentiation

With formation of the three primary germ layers, cells continue to regroup and rearrange themselves into primordial cell masses. As masses develop, they become increasingly committed to specific directions of differentiation. Cells that previously had the potential to develop into a variety of structures now lose this diverse potential and assume commitments to become, for example, kidney cells, intestinal cells, or brain cells. Differentiation is discussed in more detail in Chapter 36.

ORGANIZATION OF THE BODY

The body consists of three elements: (1) cells (discussed earlier), (2) body fluids, and (3) extracellular structural elements.

The animal body is separated from the outside world by a continuous layer of cells. This is the body's frontier that protects the controlled internal environment from the uncontrolled environment outside. Within this protective wrapping are the cells of the body's interior. These are of great variety and serve numerous functions, for example blood cells, muscle cells for movement, supportive cells, nervous and other cells for communication, and cells concerned with internal defense (Figure 6-8). All these functional types are treated in chapters to follow.

Body fluids permeate all tissues and spaces in the body but are naturally separated into certain fluid "compartments." In all metazoa, the two major body fluid compartments are the **intracellular space,** within the body's cells, and the **extracellular space,** outside the cells. In animals with closed vascular systems (such as segmented worms and vertebrates), the extracellular space can be further subdivided into the **blood plasma** (the fluid portion of the blood outside the cells; blood cells are really part of the intracellular compartment) and **interstitial fluid.** The interstitial fluid, or tissue fluid, occupies the spaces surrounding the cells. Unlike the vertebrates, many invertebrate groups have open blood systems with no true separation of blood plasma from interstitial fluid. However, all eumetazoan invertebrates share with the vertebrates the basic subdivision of body fluids between the intracellular and extracellular compartments. These relationships will be explored further in Chapter 29.

If we were to remove all the specialized cells and body fluids from the interior of the body, we would be left with the third element of the animal body: extracellular structural elements. This is the supportive material of the organism,

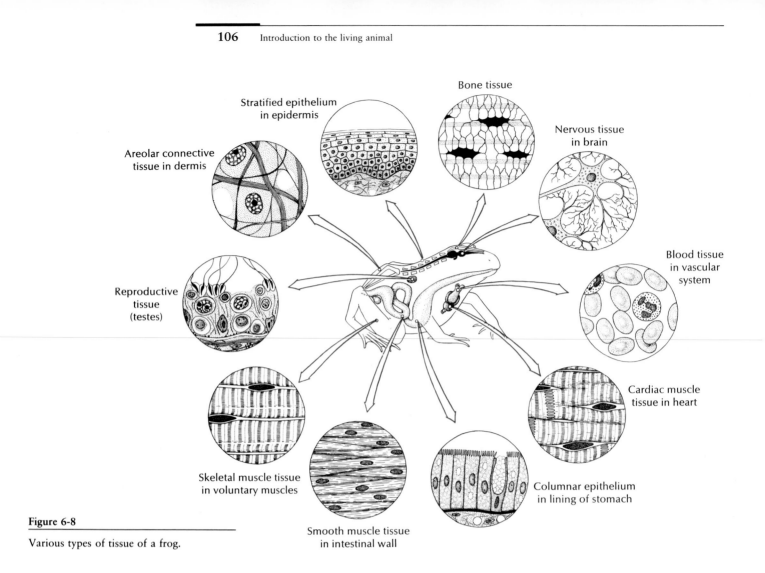

Figure 6-8

Various types of tissue of a frog.

such as loose connective tissue (especially well developed in vertebrates but present in all metazoa), cartilage (molluscs and chordates), bone (vertebrates), and cuticle (arthropods, nematodes, annelids, and others). These elements provide mechanical stability, protection, and in some instances act as a depot of materials for exchange and serve as a medium for extracellular reactions.

Two types of extracellular materials are recognized: **formed** and **amorphous** (meaning without definite form). One type of formed element that is extremely plentiful in the body is **collagen,** a white, fibrous protein material with great tensile strength. Collagen is the most abundant protein in the animal kingdom, found in animal bodies wherever both flexibility and resistance to stretching are required, such as in the integumentary system and where tissues are attached to skeletal structures. It is also found abundantly in vertebrate cartilage and bone, but these tissues have special mechanisms to prevent the collagen from crumpling.

Elastic fibers are another type of formed element. Unlike collagen, elastic fibers can be stretched but will spring back to their original length when the tension is released. Elastic fibers are found where springiness is required, such as in the walls of blood vessels and in the vertebrate lungs.

Amorphous extracellular substance is a jellylike, structureless material often called "ground substance." It is composed of complex, hydrated mucopolysaccharides and serves as a medium between cells through which nutrients, respiratory gases, wastes, and other materials pass.

___ Tissues

A **tissue** is a group of similar cells (together with associated cell products) specialized for the performance of a common function. The study of tissues is called **histology.** All cells in metazoan animals take part in the formation of tissues. Sometimes the cells of a tissue may be of several kinds, and some tissues have a great many intercellular* materials.

During embryonic development, the germ layers become differentiated into four kinds of tissues. These are epithelial, connective, muscular, and nervous tissues. This is a surprisingly short list of basic tissue types that are able to meet the diverse requirements of animal life.

Epithelial tissue

An **epithelium** is a sheet of cells that covers an external or internal surface. On the outside of the body, the epithelium forms a protective covering. Inside, the epithelium lines all the organs of the body cavity, as well as ducts and passageways through which various materials and secretions move. On many surfaces the epithelial cells are often modified into glands that produce lubricating mucus or specialized products such as hormones or enzymes.

Epithelia are classified on the basis of cell form and number of cell layers. A **simple epithelium** is one layer thick (Figure 6-9), and its cells may be **squamous** (flat), as in endothelium of blood vessels; **cuboidal** (short prisms), as in glands and ducts; or **columnar** (tall), as in the intestinal tract of most animals. In addition to simple epithelia that are present in nearly all metazoa, vertebrate animals often have epithelia arranged in several layers. This is called **stratified epithelium,** found for example in skin, sweat glands, and the urethra (Figure 6-10). Some stratified epithelia can change the number of their cell layers by being stretched out (**transitional,** such as the urinary bladder). Others have cells of different heights and give the appearance of stratified epithelia (**pseudostratified,** such as the trachea). Many epithelia may be **ciliated** at their free surfaces (such as the oviduct).

*The term "intercellular," meaning "between cells," should not be confused with the term "intracellular," meaning "within cells."

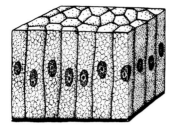

Simple columnar

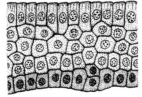

Simple squamous

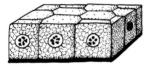

Simple cuboidal

Figure 6-9

Types of simple epithelium.

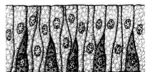

Pseudostratified

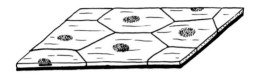

Stratified columnar

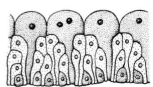

Transitional (relaxed)

Transitional (stretched)

Figure 6-10

Types of stratified and transitional epithelial tissue.

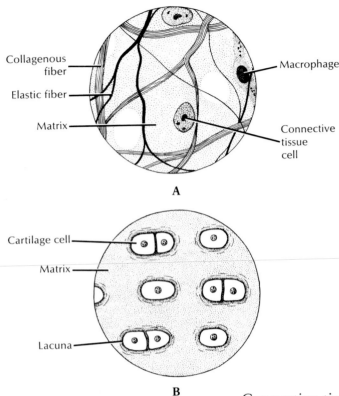

Collagenous fiber

Elastic fiber

Matrix

Macrophage

Connective tissue cell

A

Cartilage cell

Matrix

Lacuna

B

Figure 6-11

A, Areolar, a type of loose connective tissue. **B,** Hyaline cartilage, most common form of cartilage in body and a type of dense connective tissue.

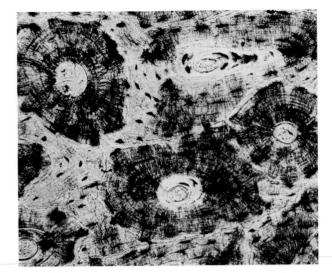

Figure 6-12

Section of bone, a type of dense connective tissue, showing several cylindrical osteons typical of bone. (×180.)

Connective tissue

Connective tissues bind together and support all other structures. They are so common that the removal of all other tissues from the body would still leave the complete form of the body clearly apparent. Connective tissue is made up of relatively few **cells** and a great deal of formed material such as **fibers** and ground substance (**matrix**) in which the fibers are embedded. There are three types of fibers: white or collagenous (the most common type), yellow or elastic, and branching or reticular. Connective tissue may be classified in various ways, but all the types fall under either **loose connective tissue** (reticular, areolar, adipose) or **dense connective tissue** (sheaths, ligaments, tendons, cartilage, bone) (Figures 6-11 and 6-12). The connective tissue of invertebrates, as in vertebrates, consists of cells, fibers, and ground substance, but it is not as elaborately developed.

Vascular tissue is considered by many histologists to be a type of fluid connective tissue. Vascular tissue includes blood, lymph, and tissue fluid. In vertebrates, blood is composed of **white blood cells** (leukocytes), **red blood cells** (erythrocytes), and **platelets,** all suspended in a liquid matrix, the **plasma.** Traveling through blood vessels, the blood carries to the tissue cells the materials necessary for the life processes. **Lymph** and **tissue fluids** arise from blood by filtration and serve in the exchange between cells and blood.

Erythrocytes are rare among invertebrates because when respiratory pigments such as hemoglobin are present, they usually circulate as giant molecules in the plasma. Most invertebrates have phagocytic blood cells, often called amebocytes, that correspond to the white blood cells of vertebrates.

Muscular tissue

Muscle is the most common tissue in the body of most animals. It is made up of elongated cells or fibers specialized for contraction. It originates (with few exceptions) from the mesoderm, and its unit is the cell or **muscle fiber.** The unspecialized cytoplasm of muscles is called **sarcoplasm,** and the contractile elements within the

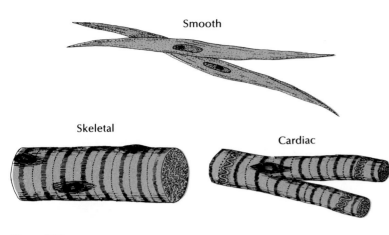

Smooth

Skeletal

Cardiac

Figure 6-13

Three kinds of vertebrate muscle fibers, as they appear when viewed with the light microscope.

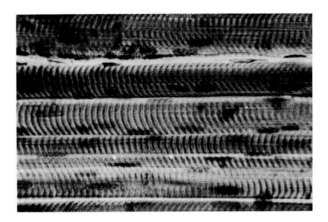

Figure 6-14

Photomicrograph of skeletal muscle showing several striated fibers lying side by side. (×600.)
Courtesy J.W. Bamberger.

fiber are the **myofibrils.** Structurally, muscles are either **smooth** (fibers unstriped) or **striated** (fibers cross-striped) (Figure 6-13).

Smooth muscle is the most common type of invertebrate muscle, serving as body wall musculature, lining ducts and sphincters, and functioning in other contractile activities. Smooth muscle fibers are often present in epithelial cells (**epitheliomuscular cells**), which combine the functions of motility with those of digestion or protection. In vertebrates, smooth muscle lines the walls of blood vessels and surrounds internal organs such as the intestine and uterus. It is often called involuntary muscle, since its contraction is generally not consciously controlled.

Striated muscle appears in several invertebrate groups but is most highly developed in the arthropods. Insects have only striated muscle, found in the walls of the heart, intestine, and other internal organs and, with the exoskeleton to which it is attached, forms the locomotory apparatus. Two types of striated muscle are found in vertebrates. **Skeletal muscle** (Figure 6-14), composed of extraordinarily long fibers, moves the skeleton and is responsible for posture. It is voluntary muscle innervated by spinal and some cranial nerves. **Cardiac muscle,** the involuntary muscle of the vertebrate heart, is composed of shorter fibers arranged in sheets and is controlled by the autonomic nervous system.

Nervous tissue

Nervous tissue is specialized for the reception of stimuli and the conduction of impulses. The structural and functional unit of the nervous system is the **neuron** (Figure 6-15), a nerve cell made up of a body containing the nucleus and its processes or fibers. In most animals the bodies of nerve cells are restricted to the central nervous system and ganglia, but the fibers may extend long distances through the body. Neurons are arranged in chains, and the point of contact between neurons is a **synapse.** In vertebrates, some fibers are wrapped with an insulating myelin sheath. Insulating sheaths occur in some invertebrates, especially arthropods, but most invertebrate nerve fibers are naked.

Sensory neurons are concerned with conducting impulses from sensory **receptors** in the skin or sense organs to nerve centers (brain or spinal cord). **Motor neurons** carry impulses from the nerve centers to muscles or glands (**effectors**) that are thus stimulated to respond. **Association neurons** form various connections between other neurons.

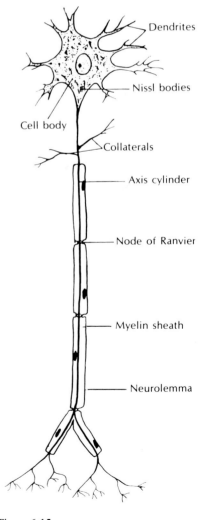

Dendrites

Nissl bodies

Cell body

Collaterals

Axis cylinder

Node of Ranvier

Myelin sheath

Neurolemma

Figure 6-15

Structure of a neuron.

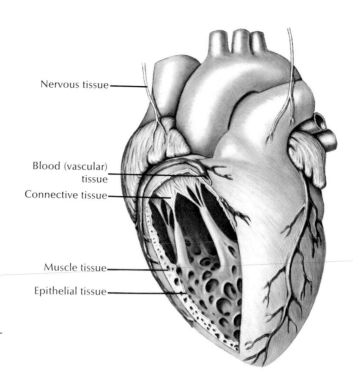

Nervous tissue

Blood (vascular) tissue

Connective tissue

Muscle tissue

Epithelial tissue

Figure 6-16

Heart showing various types of tissue in its structure.

Organs and Systems

An organ is a group of tissues organized into a larger functional unit. In more complex forms an organ may have most or all of the four basic tissue types. For example, the vertebrate heart (Figure 6-16) has epithelial tissue for covering and lining, connective tissue for framework and vascular transportation, muscular walls for contraction, and nervous elements for coordination.

All organs have a characteristic structural plan. Usually one tissue carries the burden of the organ's chief function, as muscle does in the heart; the other tissues perform supportive roles. The chief functional cells of an organ are called its **parenchyma;** the supporting tissues are its **stroma.** For instance, in the pancreas the secreting cells are the parenchyma; the capsule and connective tissue framework represent the stroma.

Organs are, in turn, associated in groups to form **systems,** with each system concerned with one of the basic functions. The higher metazoa have 11 organ systems: skeletal, muscular, integumentary, digestive, respiratory, circulatory, excretory, nervous, special sensory, endocrine, and reproductive. However, all living organisms perform the same basic functions. The need for procuring and using food and for movement, protection, perception, and reproduction are equally basic to an ameba, a clam, an insect, or a human. Obviously, because of differences in size, structure, and environment, each must meet these problems in a different manner.

ANIMAL BODY PLANS

Thus far in this chapter we have considered the characteristics of body design that animals share. We are ready now to consider the various architectural plans that distinguish major groups of animals. These appear enormously diverse, as even a cursory examination of the different invertebrate and vertebrate groups will reveal. However, it is possible to resolve them into four "master plans" (Gardiner,

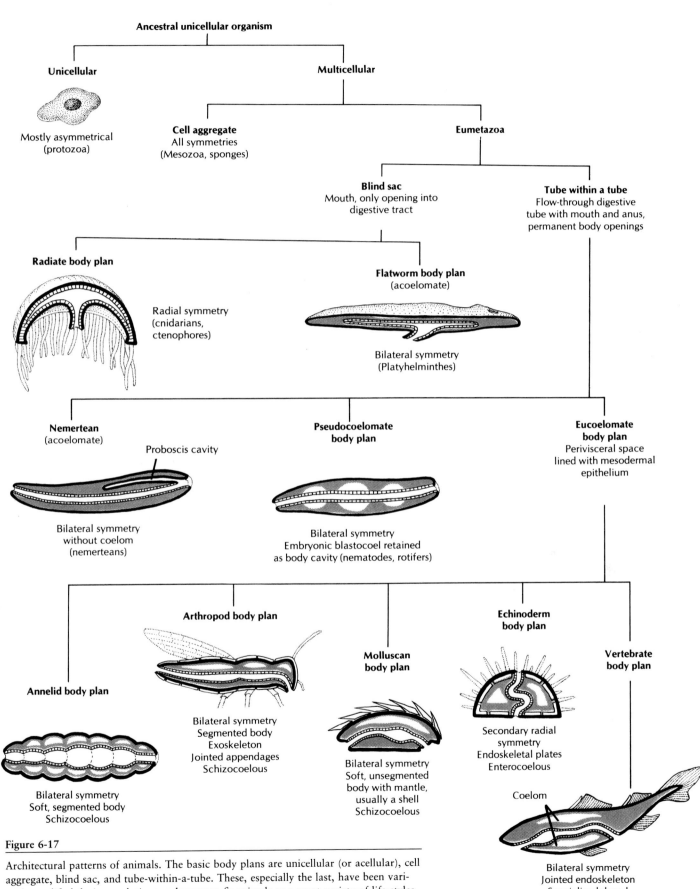

Figure 6-17

Architectural patterns of animals. The basic body plans are unicellular (or acellular), cell aggregate, blind sac, and tube-within-a-tube. These, especially the last, have been variously modified during evolutionary descent to fit animals to a great variety of life-styles. Ectoderm is shown in solid black, endoderm as open blocks, mesoderm in red.

1972): the unicellular plan, cell-aggregate plan, blind-sac plan, and tube-within-a-tube plan. These, together with some of their most important modifications, are shown in Figure 6-17. All but the first are multicellular plans. Four of the most important determinants of multicellular body plans are symmetry, presence or absence of a body cavity, presence or absence of segmentation, and cephalization.

___ Animal Symmetry

Symmetry refers to balanced proportions, or the correspondence in size and shape of parts on opposite sides of a median plane. **Spherical symmetry** means that any plane passing through the center divides the body into equivalent, or mirrored, halves. This type of symmetry is found chiefly among some of the protozoa and is rare in other groups of animals. Spherical forms are best suited for floating and rolling.

    Radial symmetry (Figure 6-17) applies to forms having a central axis around which similar body parts are concentrically arranged. These are the tubular, vase, or bowl shapes found in some sponges and in the hydras, jellyfish, sea urchins, and the like, in which one end of the central axis is usually the mouth. A variant form is **biradial symmetry,** in which, because of some part that is single or paired rather than radial, only two planes passing through the central axis produce mirrored halves. Sea walnuts, which are more or less globular in form but have a pair of tentacles, are an example. Radial and biradial animals are usually sessile, freely floating, or weakly swimming. The two phyla that are primarily radial, Cnidaria and Ctenophora, are called the **Radiata.**

    In **bilateral symmetry** only a sagittal plane can divide the animal into two mirrored portions—right and left halves (Figure 6-18). Bilateral animals make up all of the higher phyla and are collectively called the **Bilateria.** They are better fitted for directional movement (forward) than radially symmetrical animals.

    Let us review some of the convenient terms used for locating regions of animal bodies (Figure 6-18). **Anterior** is used to designate the head end; **posterior,** the opposite or tail end; **dorsal,** the back side; and **ventral,** the front or belly side. **Medial** refers to the midline of the body, **lateral** to the sides. **Distal** parts are farther from the middle of the body than some point of reference; **proximal** parts

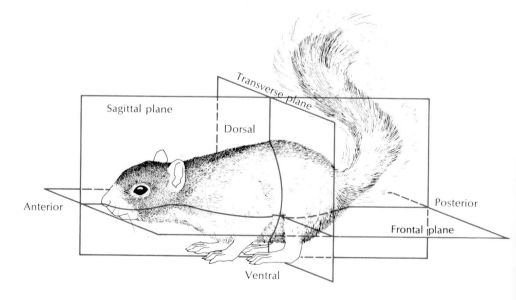

Figure 6-18

The planes of symmetry as illustrated by a bilateral animal.

are nearer. **Pectoral** refers to the chest region or the area supported by the forelegs, and **pelvic** refers to the hip region or the area supported by the hindlegs. A **frontal plane** divides a bilateral body into dorsal and ventral halves by running through the anteroposterior axis and the right-left axis at right angles to the **sagittal plane,** the plane dividing an animal into right and left halves. A **transverse plane** would cut through a dorsoventral and a right-left axis at right angles to both the sagittal and frontal planes and would result in anterior and posterior portions.

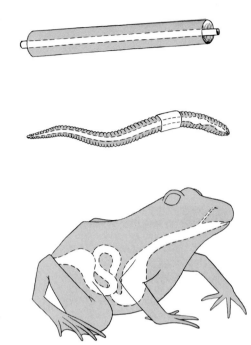

____ Body Cavities

The bilateral animals can be grouped according to their body-cavity type or lack of body cavity (Figure 6-17). In higher animals the main body cavity is the **coelom,** a fluid-filled space that surrounds the gut. The two methods of coelom formation—schizocoelous and enterocoelous—were described on p. 104. The coelom provides coelomic animals with a "tube-within-a-tube" arrangement (Figure 6-19). The true coelom develops within the mesoderm and is thus lined with mesodermal epithelium called the **peritoneum.** The coelom is of great significance in animal evolution (p. 261). It provides increased body flexibility and space for visceral organs and permits greater size and complexity by exposing more cells to surface exchange. The fluid-filled space also serves as a hydrostatic skeleton in some forms, aiding in such functions as movement and burrowing.

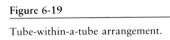

Figure 6-19

Tube-within-a-tube arrangement.

Acoelomate Bilateria

The more primitive bilateral animals do not have a true coelom. In fact, the flatworms and a few others have *no body cavity* surrounding the gut. The region between the ectodermal epidermis and the endodermal digestive tract is completely filled with mesoderm in the form of parenchyma.

Pseudocoelomate Bilateria

Nematodes and several other phyla have a cavity surrounding the gut, but it is not lined with mesodermal peritoneum. It is derived from the blastocoel of the embryo and represents a persistent blastocoel. This type of body cavity is called a **pseudocoel,** and its possessors also have a "tube-within-a-tube" arrangement.

Eucoelomate Bilateria

The remainder of the bilateral animals possess a **true coelom** lined with mesodermal peritoneum.

____ Metamerism (Segmentation)

Metamerism is the serial repetition of similar body segments along the longitudinal axis of the body. Each segment is called a **metamere,** or **somite.** In forms such as the earthworm and other annelids, in which metamerism is most clearly represented, the segmental arrangement includes both external and internal structures of several systems. There is repetition of muscles, blood vessels, nerves, and the setae of locomotion. Some other organs, such as those of sex, are repeated in only a few somites. In higher animals much of the segmental arrangement has become obscure.

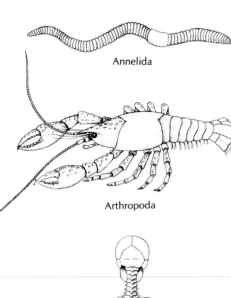

Annelida

Arthropoda

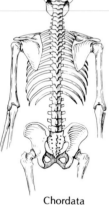

Chordata

Figure 6-20

Segmented phyla. These three phyla have all made use of an important principle in nature—metamerism, or repetition of structural units. Annelids and arthropods are definitely related, but chordates have derived their segmentation independently. Segmentation brings more varied specialization because segments, especially in arthropods, have become modified for different functions.

True metamerism is found in only three phyla—Annelida, Arthropoda, and Chordata (Figure 6-20)—although superficial segmentation of the ectoderm and the body wall may be found among many diverse groups of animals.

Cephalization

The differentiation of a head end is called **cephalization** and is found chiefly in bilaterally symmetrical animals. The concentration of nervous tissue and sense organs in the head bestows obvious advantages to an animal moving through its environment head first. This is the most efficient positioning of instruments for sensing the environment and responding to it. Usually the mouth of the animal is located on the head as well, since so much of an animal's activity is concerned with procuring food. Cephalization is always accompanied by differentiation along an anteroposterior axis (**polarity**). Polarity usually involves gradients of activities between limits, such as between the anterior and posterior ends.

HOMOLOGY AND ANALOGY

In comparative studies of animals the concepts of **homology** (similarity in origin) and **analogy** (similarity in function or appearance, but not in origin) are frequently used to express the mechanisms responsible for similar patterns of morphology or function in different animals. For example, the bones in a whale flipper are homologous to the bones of a human arm and of a bat wing (Figure 39-10, p. 851), although in appearance and function they are quite different. They are homologous because all share a common ancestry and embryonic origin. All develop similarly from a limb bud within which forms a single proximal bone (humerus), two parallel bones located more distally (radius and ulna), and several small bones more distal to the latter (wrist bones and phalanges). Later in development, significant modifications appear that shape the limb for the totally different function each will perform in these three different mammals.

Analogy denotes similarity of function and often of appearance as well. Just as different functions make homologous structures dissimilar in form, similar functions can make nonhomologous structures resemble each other. The wings of a bird and a butterfly obviously perform similar flight functions, but they are not homologous because they have totally different origins. Their resemblance is only superficial. The skeletal framework of the bird's wing with its feathered surface corresponds to the basic vertebrate pentadactyl limb plan. The wing of a butterfly is supported by cuticular thickenings ("veins"), and the surface is a cuticular membrane composed of two cell layers. They are analogous structures.

The principle of homology has wide application in zoology. Homology is a powerful argument for evolution and, because it is premised on the idea of inheritance from common ancestors, it is of cardinal importance in working out ancestral lineages in animals. Incorrect conclusions about homology will invariably lead to incorrect genealogical classifications. The problem now, as it was in the nineteenth century when evolutionary pathways were first being worked out, is to recognize homology. The best criterion of homology would be the identification of homologous genes. This is not yet possible, although recent advances in molecular sequencing suggest that this criterion of homology may soon be successfully applied. Genealogies can than be worked out, not only from morphology, but from biochemical resemblances.

___ SUMMARY

From the relatively simple organisms that made up the beginnings of life on earth, animal evolution has progressed through a history of ever more intricately organized forms. Organelles became integrated into cells, cells into tissues, tissues into organs, and organs into systems. Whereas a unicellular animal carries out all life functions within the confines of a single cell, an advanced multicellular animal is an organization of subordinate units that are united at successive levels.

During the course of evolution the upper limit of body size in animal lineages has tended to increase; this led inevitably to the division of labor among body parts and increased complexity.

All metazoa pass through a characteristic life cycle that usually begins with the union of male and female germ cells in fertilization. This fusion of sperm and egg pronuclei is followed by a series of developmental stages—cleavage, blastulation, gastrulation, differentiation of tissues and organs, and growth—that adheres to a predictable pattern within a species and among related species within a taxon.

Two quite different patterns of cleavage and coelom formation are recognized among multicellular animals and signal a fundamental division in their evolution. Protostome embryos typically show spiral cleavage with mosaic development, and the coelom is formed by splitting of mesoderm (schizocoelous). Deuterostome embryos show radial (or biradial) cleavage with regulative embryonic development. The coelom of most deuterostomes develops from mesodermal pouches off the primitive gut (enterocoelous) but advanced chordates form a chizocoel. During gastrulation, three germ layers—ectoderm, endoderm, and mesoderm—are formed that are destined to differentiate into the major organ systems of the body.

The metazoan body consists of cells, most of which are functionally specialized; body fluids, divided into intracellular and extracellular fluid compartments; and extracellular structural elements, which are fibrous or formless elements that serve various structural functions in the extracellular space. The cells of metazoa develop into various tissues made up of similar cells performing common functions. The basic tissue types are epithelial, connective, muscular, and nervous. Tissues are organized into larger functional units called organs, and organs are associated to form systems.

Every organism has an inherited body plan that may be described in terms of broadly inclusive characteristics, such as symmetry, presence or absence of body cavities, partitioning of body fluids, presence or absence of segmentation, degree of cephalization, and type of nervous system.

The concept of homology expresses the idea of similarity resulting from common embryological origin and evolutionary ancestry. Analogy means similarity of function and/or appearance, but without common origin. Homologous structures and behaviors are accepted as evidence for evolution and are used in animal classification.

Review questions

1. Name the five levels of organization in animal complexity and explain how each successive level is more advanced than the one preceding it.
2. Can you suggest why, during the evolution of separate animal lineages, there has been a tendency for the maximum body size to increase? Why should it be inevitable that complexity tends to increase along with body size?
3. Explain the event or feature that distinguishes each of the following developmental stages: fertilization, zygote, cleavage, blastula, gastrula.
4. What is different about isolecithal, telolecithal, and centrolecithal eggs?
5. Describe the two fundamentally different patterns of cleavage and coelom formation. With which group of bilateral metazoan animals is each pattern associated?
6. Body fluids of eumetazoan animals are separated into fluid "compartments." Name these compartments and explain how compartmentalization may differ in animals with open and closed circulatory systems.
7. What are the major kinds of extracellular structural elements in the metazoan body?
8. What are the four major tissue types in the body of a metazoan? For what function(s) is each tissue type specialized?
9. What is the difference between tissues, organs, and systems?
10. Distinguish among spherical, radial, biradial, and bilateral symmetry.
11. Match the animal group with its body plan:
    ___Unicellular   a. Nematode
    ___Cell aggregate   b. Vertebrate
    ___Blind sac, acoelomate   c. Protozoan
    ___Tube-within-a-tube, pseudocoelomate   d. Flatworm
    ___Tube-within-a-tube, eucoelomate   e. Sponge
12. Use the following terms to identify regions on your body and on the body of a frog: anterior, posterior, dorsal, ventral, lateral, distal, and proximal.
13. How would frontal, sagittal, and transverse planes divide your body?
14. What is meant by metamerism? Name three phyla showing metamerism.
15. Define the term homology and give an example of homologous structures among the vertebrates. Why are only homologous, and not analogous, structures of any significance in working out evolutionary lineages?

Selected references

Gardiner, M.S. 1972. The biology of invertebrates. New York, McGraw-Hill Book Co. *Functional approach.*

Grant, P. 1978. Biology of developing systems. New York, Holt, Rinehart & Winston. *Developmental biology text with good treatment of cleavage patterns.*

McMahon, T.A., and J.T. Bonner. 1983. On size and life. New York, Scientific American Books, Inc. *Wide ranging, artfully written, and colorfully produced introductory book on the biology and physics of size.*

von Bertalanffy, L. 1952. Problems of life. London, Watts & Co. *Levels of organization are among several biological concepts discussed in this classic.*

Welsch, U., and V. Storch. 1976. Comparative animal cytology and histology. London, Sidgwick & Jackson. *Modern comparative histology with good treatment of the invertebrates. Translation of German text.*

CHAPTER 7

CLASSIFICATION AND PHYLOGENY OF ANIMALS

ANIMAL CLASSIFICATION

Zoologists have named more than 1.5 million species of animals, and thousands more are added to the list each year. Yet some zoologists believe that species named so far make up less than 20% of all living animals and less than 1% of all those that have existed in the past.

To communicate with each other about the diversity of life, biologists have found it a practical necessity not only to name living organisms but to classify them. It is not just that the desire to put things into some kind of order is a fundamental activity of the human mind. A system of classification is a storage, retrieval, and communication mechanism for biological information. This is the science of **taxonomy.** It is concerned with the naming of each kind of organism by a uniformly adopted system that best expresses the degree of similarity of organisms. The science of **systematics** is somewhat broader in that it embraces taxonomy, which is concerned with classification, and evolutionary biology, which is the study of organismic diversity and orderliness of nature. Systematics is really comparative biology, because it uses everything that is known about animals to understand their relationships and evolutionary history.

Linnaeus and the Development of Classification

Although the history of human efforts to distinguish and name plants and animals must have been rooted in the beginnings of language, the great Greek philosopher and biologist Aristotle was the first to attempt seriously the classification of organisms on the basis of structural similarities. Following the Dark Ages in Europe, the English naturalist John Ray (1627-1705) introduced a more comprehensive classification system and a modern concept of species. The flowering of systematics in the eighteenth century culminated in the work of Carolus Linnaeus (1707-1778) (Figure 7-1), who gave us our modern scheme of classification.

Linnaeus was a Swedish botanist at the University of Uppsala. He had a great talent for collecting and classifying objects, especially flowers. Linnaeus worked out a fairly extensive system of classification for both plants and animals. His scheme of classification, published in his great work *Systema Naturae,* emphasized morphological characters as a basis for arranging specimens in collections. Actually his classification was largely arbitrary and artificial, and he believed strongly in the constancy of species. He divided the animal kingdom down to

Figure 7-1

Carolus Linnaeus (1707-1778). This portrait was made of Linnaeus at age 68, 3 years before his death.
Courtesy Library of Congress.

Table 7-1 Examples of Classification of Animals

	Human	Gorilla	Southern leopard frog	Katydid
Phylum	Chordata	Chordata	Chordata	Arthropoda
Subphylum	Vertebrata	Vertebrata	Vertebrata	Uniramia
Class	Mammalia	Mammalia	Amphibia	Insecta
Subclass	Eutheria	Eutheria		
Order	Primates	Primates	Salientia	Orthoptera
Suborder	Anthropoidea	Anthropoidea		
Family	Hominidae	Pongidae	Ranidae	Tettigoniidae
Subfamily			Raninae	
Genus	*Homo*	*Gorilla*	*Rana*	*Scudderia*
Species	*Homo sapiens*	*Gorilla gorilla*	*Rana pipiens*	*Scudderia furcata*
Subspecies			*Rana pipiens sphenocephala*	*Scudderia furcata furcata*

Linnaeus wrote, "The Author of Nature, when He created species, imposed on His Creations an eternal law of reproduction and multiplication within the limits of their proper kinds. He did indeed in many instances allow them the power of sporting in their outward appearances, *but never that of passing from one species to another*" (italics ours). It is interesting too that, whereas Linnaeus was working from a creationist model, the same system is used today to show phylogenetic relationships. This is an example of facts supporting different interpretations, rather than speaking for themselves. As new evidence accumulates, inadequate interpretations must be discarded.

species, and according to his scheme each species was given a distinctive name. He recognized four classes of vertebrates and two classes of invertebrates. These classes were divided into orders, the orders into genera, and the genera into species. Since his knowledge of animals was limited, his lower groups, such as the genera, were very broad and included animals that we now realize are only distantly related. As a result, much of his classification has been drastically altered, yet the basic principle of his scheme is followed at the present time.

Linnaeus' scheme of arranging organisms into an ascending series of groups of ever-increasing inclusiveness is the **hierarchical system** of classification. Species were grouped into genera, genera into orders, and orders into classes. This taxonomic hierarchy has been considerably expanded since Linnaeus' time. The major categories, or **taxa** (sing., **taxon**), now used are as follows, in descending series: kingdom, phylum, class, order, family, genus, and species. This hierarchy of seven ranks can be subdivided into finer categories, such as superclass, subclass, infraclass, superorder, and suborder. In all, more than 30 taxa are recognized. For very large and complex groups, such as the fishes and insects, these additional ranks are required to express recognized degrees of evolutionary divergence. Unfortunately, they also contribute complexity to the system.

Linnaeus's system for naming species is known as **binomial nomenclature.** Each species has a latinized name composed of two words (hence binomial). The first word is the **genus,** written with a capital initial letter; the second word is the **species epithet** that is peculiar to the species and is written with a small initial letter (Table 7-1). The genus name is always a noun, and the species epithet is usually an adjective that must agree in gender with the genus. For instance, the scientific name of the common robin is *Turdus migratorius* (L. *turdus,* thrush; *migratorius,* of the migratory habit).

There are times when a species is divided into subspecies, in which case a **trinomial nomenclature** is employed (see katydid example, Table 7-1). Thus to distinguish the southern form of the robin from the eastern robin, the scientific term *Turdus migratorius achrustera* (duller color) is employed for the southern type. The trinomial nomenclature is really an addition to the Linnaean system, which is basically binomial. The generic, specific, and subspecific names are printed in italics (underlined if handwritten or typed).

It is important to recognize that *only* the species is binomial. All ranks above the species are uninomial nouns, written with a capital initial letter. We must also note that the second word of a species is an epithet that has no meaning by itself. The scientific name of the white-breasted nuthatch is *Sitta carolinensis*. The "carolinensis" may be and is used in combination with other genera to mean "of Carolina," as for instance *Parus carolinensis* (Carolina chickadee) and *Anolis carolinensis* (green anole, a lizard). The genus name, on the other hand, may stand alone to designate a taxon that may include several species.

___ Species

Despite the central importance of the species concept in biology, biologists do not agree on a single rigid definition that applies to all cases. Before Darwin's time, the species was considered a primeval pattern, or archetype, divinely created. With gradual acceptance of the concept of organic evolution, scientists realized that species were not fixed, immutable units but have evolved one from another. Sometimes the gaps between species are so subtle that they can be distinguished only by the most careful examination. In other instances, a species is so distinctive in every way that it is clearly unique and only remotely related to other species. Consequently, the criteria of taxonomy have undergone gradual changes.

At first each species was supposed to have been represented by a **type** that was used as a fixed standard. The **typological specimen** was duly labeled and deposited in some prestigious center such as a museum (Figure 7-2). Anyone classifying a particular group would always take the pains to compare specimens with the available typological specimens. Since variations from the type specimen nearly always occurred, these differences were supposed to be attributable to imperfections during embryonic development and were considered of minor significance.

The **typological** (or **morphological**) **species concept** of classifying persisted for a long period (and still does to some extent). During this time, though, the idea

The person who first describes and publishes the name of a species is called the authority, and this person's name and date of publication is often written after the species name. Thus *Didelphis marsupialis* Linnaeus, 1758, tells us that Linnaeus was the first person to publish the species name of the opossum. The authority citation is not part of the scientific name but rather is an abbreviated bibliographic reference.

Figure 7-2

Type specimens of crustaceans in the Smithsonian Institution's Natural History Museum in Washington, D.C.
Photograph by C.P. Hickman, Jr.

that species represent lineages in evolutionary descent was becoming more firmly established. Gradually taxonomists began to think of species as *groups of interbreeding natural populations that are reproductively isolated from other such groups.* This is the **biological species concept,** in which a species is considered a population composed of unique individuals that may change to a greater or lesser extent when placed in a different environment. It is still important, nevertheless, for a person who describes a species to deposit a specimen in a museum so that later taxonomists can examine an actual specimen that the original author believed was a member of that species.

The modern concept of species is of course the antithesis of that of the typologist, to whom variations from the type specimen are illusions caused by small mistakes during embryonic development. In population studies the type is considered an abstract average of *real* variations within the interbreeding population. Thus, the species must be regarded as an **interbreeding population** made up of individuals of common descent and sharing intergrading characteristics.

The biological species concept is not without difficulties. For one thing, it does not apply to organisms that reproduce asexually because there is no way to test the interbreeding criterion in uniparental species. Even in species that reproduce sexually it is impractical (and often impossible) to conduct interbreeding experiments to test for reproductive isolation. Furthermore the criterion of reproductive isolation may not be absolute. Not infrequently, populations are found that are in an intermediate stage of differentiation between races, which can interbreed, and species, which cannot. They cannot be classified as definite races or as definite species, and so are sometimes called semispecies. Thus, as a practical matter, taxonomists still use mainly morphological criteria in describing a new species, supplemented by genetic, behavioral, physiological, and biochemical criteria when these can be applied.

Despite these problems, the biological species concept works satisfactorily most of the time. Reproductive barriers do exist. Sometimes pairs of species can be made to hybridize, but then produce sterile offspring. A well-known example is the sterile mule, the offspring of a mare and a male ass. Even when fertile hybrids are possible, they are normally prevented by various external barriers, such as anatomical incompatibilities and, especially, behavioral patterns that create aversions to mating with the wrong species.

Basis for Formation of Taxa

Classification emphasizes the natural relationships of animals. Descent from a common ancestor explains similarity in character; the more recent the descent, the more closely the animals are grouped in taxonomic units. The genera of a particular family show less diversity than do the families of an order. This is because the common ancestor of families within an order is more remote than the common ancestor of different genera within a family. The same applies to higher categories. The common ancestor of the various vertebrate classes, for example, must be much older than the common ancestor of orders of mammals within the class Mammalia.

It is apparent that this criterion of evolutionary ancestry is not a very definite one for use in setting up taxa. There is no way to define a class that does not apply equally well to an order or a family. The genera of certain ancient groups, such as the molluscs, may actually be much older than orders and classes of other groups. The taxonomist must arrange the classifications so that all members of a taxon, as far as can be determined, are related to one another more than they are to the members of any other taxon of the same rank. Obviously, the assignment of

To avoid problems with the biological species concept, the evolutionary species concept has been proposed. *An evolutionary species is a single lineage of ancestor-descendant populations that maintains its identity from other such lineages and that has its own evolutionary tendencies and historical fate* (Wiley, 1981). In this definition, "identity" means any quality a species uses to keep its line separated from other species. Each species "knows" its identity and has recognition systems in appearance and behavior and ecological roles to maintain reproductive isolation. The evolutionary species concept has the advantage of embracing both asexual and sexual organisms.

taxonomic rank depends on the opinion of the taxonomist making the study, and this is one reason classification has been called as much an art as a science (G.G. Simpson, 1961*). As more is learned about animals and their relationships, changes in classification are required. This brings instability to biological nomenclature and reduces its efficiency as a reference system.

The **law of priority** also brings about frequent changes. The first name proposed for a taxonomic unit that is published and meets other proper specifications has priority over all subsequent names proposed. The rejected duplicate names are called **synonyms.** It is disturbing sometimes to find that a species that has been well established for years must undergo a change in terminology when some industrious systematist discovers that on the basis of priority or for some other reason the species is, according to this "law," misnamed.

Despite such difficulties, the hierarchical system of classification is the accepted system, and it is a very useful one. To reduce confusion in the field of taxonomy and to lay down a uniform code of rules for the classification of animals, the International Commission on Zoological Nomenclature was established in 1898. This commission meets from time to time to formulate rules and make decisions in connection with taxonomic work.

___ Traditional and Modern Approaches to Systematics

The science of systematics is charged with two tasks. On the one hand it is required to name organisms and place them in some kind of order; this is taxonomy. On the other hand, systematics is supposed to help unravel the course of evolution. It is an unsteady mix of the practical and the theoretical. Most biologists firmly believe that systematics should legitimately concern itself with both these functions, but the difficulty of doing so has placed stresses on established taxonomic procedures.

Evolutionary taxonomy

Since Darwin's time, **evolutionary taxonomy** has been considered the traditional approach to classification. It bases taxonomy on evolutionary theory, and its goal is to reconstruct evolutionary history as closely as possible. Ancestor-descendant relationships, established from the fossil record and from morphological and biochemical homologies (see p. 851) are often used to construct phylogenetic trees showing the pattern of evolutionary descent. The lines in the phylogenetic tree portray evolutionary species and the branches represent speciation events, that is, the appearance of new species. A geological time scale is usually included, and sometimes the number of species in a group at any time is suggested by the widening and thinning of the branch lines (see Figure 27-1, pp. 566-567).

Ideally, the taxon depicted in a phylogenetic tree is **monophyletic** ("one tribe"); that is, all the organisms in the taxon have descended from a common ancestor. For example, all birds are believed to have descended from a common thecodont (reptile) ancestor; birds are therefore a monophyletic group (Figure 26-2, pp. 536-537). Sometimes, however, a taxon is suspected to have arisen from more than one common ancestor. Such a group is called **polyphyletic** ("many tribes"). For example, mammals were once thought to have arisen independently as many as nine different times. If this were true (and this view is no longer in favor), the mammals would be an artificial taxon composed of animals of more

---

* Simpson, G.G. 1961. Principles of animal taxonomy, New York, Columbia University Press.

than one origin that share characters (such as hair and mammary glands) that were independently acquired.

The traditional evolutionary approach to taxonomy has been criticized for the sometimes arbitrary and subjective way in which classification schemes had been applied. Evolutionary taxonomy is also vulnerable to circular reasonings: hypotheses about taxonomic relationships are used to establish phylogeny, and phylogeny is then used to determine taxonomy. Out of the search for objective remedies to these problems grew two new methodologies: **numerical taxonomy** and **cladistics.**

Numerical taxonomy

In all taxonomic work it is first necessary to recognize **characteristics** that can be used to describe an organism adequately and place it in the proper taxon. A characteristic is any feature or attribute that can be described, measured, weighed, pictured, counted, scored, or otherwise defined about an organism. In numerical taxonomy as many arbitrarily chosen, equally weighted characters as possible (well over 100) are selected, coded, and fed into a computer, and an analysis is made of the calculations. Species or groups are then clustered by similarity. This approach aims simply at producing meaningful groupings or organisms and makes no attempt to reconstruct the evolutionary history of groups.

Cladistics

An alternative to numerical taxonomy is the method of cladistics (Gr. *cladus,* branch). Unlike numerical taxonomy this approach bases classification exclusively on phylogeny. With it one can set up a kind of branching pattern called a **cladogram,** which resembles a phylogenetic tree but is not one (Figure 7-3). The cladogram is a series of branches, each branch representing the splitting of a parental species into two daughter species.

Although the theory of cladistics may seem complicated, its rationale is fairly simple. Suppose that one is interested in studying the relationships of certain groups of animals, for example, species within a genus or genera within a family. Another genus is selected, which is believed to be only distantly related to the groups of interest; this distantly related taxon is called the **outgroup.**

In the simplified example of four vertebrate animals shown in Figure 7-3, the bass would be considered the outgroup because we would suspect that it is only

Figure 7-3

A, A simple cladogram. The bass is the outgroup, and lizard, horse, and monkey are closely related groups. Note that a cladogram, unlike traditional phylogenetic trees, carries no time scale because it is meant to show only relative degree of relationship and not actual historical events. **B,** The "nested sets" of the cladistic hierarchy, constructed from the cladogram.

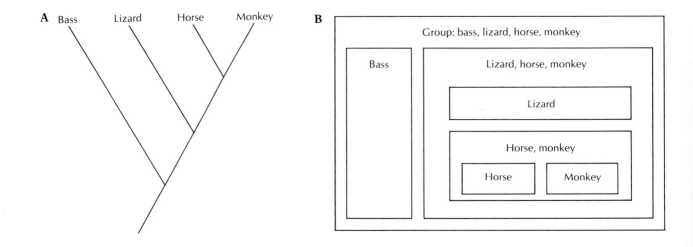

distantly related to the other three vertebrates. Then a variety of characteristics are chosen that will be used to compare among the more closely related animals and the outgroup. Any characteristic that the bass shares with the lizard, horse, and monkey—vertebrae or jaws, for example—would be considered **primitive** (or ancestral) because it is a characteristic possessed by the common ancestor of all four vertebrates. However, characteristics shared by the related group, but not possessed by the outgroup, are **derived.** In the example, the horse and monkey are most closely related because they have the greatest number of *shared* derived characters. The lizard would have a smaller number of derived characters shared with the horse and monkey and so we could conclude that it branched off from the common ancestral line of all three at an earlier stage. For example, all three have tetrapod limbs and amniotic eggs, but only the horse and monkey have hair and mammary glands.

From this sort of analysis a branching cladogram is constructed that shows successive splits of phyletic lines. Unlike a traditional phylogenetic tree, a cladogram does not attempt to show actual historical events, and the branch points do not identify particular ancestors. The advantage of cladistics is that it uses strictly defined, objective criteria, and yields a cladogram that gives an immediate reading of group phylogeny. Critics of cladistics, however, argue that it disregards subsequent evolutionary changes that occur in split lines. This can be seen most clearly in the "nested sets" formed in a cladistic hierarchy (Figure 7-3, *B*), which results in the bass being placed in a taxon equal in rank to one containing the lizard, horse, and monkey.

ANIMAL PHYLOGENY

The different approaches to classification that we have just described tend to obscure a basic agreement: order in the biological world derives from evolution. Hierarchical classification schemes are possible because relationships between species are the result of descent with modification from ancestral species. Reconstructing lines of descent leading to living animals is not unlike attempts to ferret out the ancestral history (genealogy) of one's own family. Neither plant or animal species nor members of one's family arose spontaneously, but rather they represent branching of ancestral forms. The phylogeny of an animal group, then, represents our concept of the path its evolution has taken. It is the evolutionary history of the group.

Unfortunately, the origin of most animal phyla is shrouded in the obscurity of Precambrian times. Because fossil records are fragmentary, our reconstructions of patterns of evolutionary relationships must have relied to a large extent on evidence from comparative morphology and embryology. Relationships are naturally more easily established within the smaller taxonomic units (such as species, genera, orders) than in the classes and phyla.

Other techniques are now in use in systematics that offer great promise in solving problems of phylogenetic relationships and evolution. Research is being carried out in animal behavior, comparative biochemistry, serology, cytology, genetic homology, and comparative physiology with this aim in view.

For example, it has been shown that there are certain homologies among polynucleotide sequences in the DNA molecules of such different forms as fish and humans. These sequences appear to be genes that have been retained with little change throughout vertebrate evolution. Possible phenotypical expressions of these homologous sequences are bilateral symmetry, notochord, and hemoglobin, as well as others. By using a single strand of DNA from one species and short radioactive pieces of a DNA strand from another species and mixing the strands together, it was found that some of the smaller strands paired with similar regions on the large strand, indicating that the paired parts had common genes.

Another recent technique involves the recognition of RNA codons by transfer RNA of another species. These new biochemical methods of classification are used to complement, rather than replace, the older, more established methods. In general, molecular evidences have not agreed very well with the fossil evidence. Nevertheless, the new molecular approach provides a potentially powerful tool for the systematist.

Although the representation of ancestral relationships of animals in the form of a "family tree" seems obvious today, it was not apparent to biologists before classification became founded on evolutionary theory. Even Darwin never attempted a pictorial diagram of animal relationships. Yet, despite all the shortcomings that any phylogenetic tree must possess, especially the danger of depicting and accepting highly tentative relationships as dogmatic certainty, such a scheme is a valuable tool. A family tree ties the taxa together in an evolutionary blueprint. Constructing a family tree is a creative activity based on judgment and experience and as such is always subject to modification as new information becomes available.

In each of the following chapters on animal phyla, we include a summary of that group's origins and the relationships within the group. The student is encouraged to make frequent reference to the geological time scale on the inside back cover of this book and to the phylogenetic tree of multicellular animals on the inside front cover. We have based conclusions about group histories on recent informed opinion. Although we recognize that family trees can give false impressions, we have nonetheless used them for lack of a better alternative. They may be viewed as educated speculations and as such with a certain measure of skepticism; at the same time they are not science fiction. They are derived from close studies of morphology and a thorough understanding of general biological principles by scientists who have devoted their lives to this form of detective work. Evolution, with its idea of life transforming itself through the ages, is supported by a vast wealth of fossil and living evidence. It is after all the framework of biology.

____ Some Helpful Definitions

In the discussions on evolutionary relationships in ensuing chapters, terms such as lower, higher, primitive, advanced, specialized, generalized, adaptation, and fitness will be used frequently, and an understanding of their meaning may be critical to an understanding of the discussion to follow.

The terms **lower** and **higher** usually refer to a group's relative position on a phylogenetic tree; that is, the level at which it is believed to have branched off a main stem of evolution. For example, sponges and jellyfish are usually considered to belong to "lower" phyla because they are believed to have been among the earliest metazoa. In other words, they originated near the base of the family tree of the animal kingdom.

The terms **primitive** and **advanced** are often used when discussing relationships within a particular group. A primitive species is one that possesses a great many of the same characteristics believed to have belonged to the ancestral stock from which it evolved. An advanced species is one that has undergone considerable change from the primitive condition, usually because of adaptation to a changed environment or to a different mode of living. Among the molluscs, for example, the caudofoveates are considered to be more primitive (that is, more like the hypothetical mollusc ancestor) than the snails, clams, or octopuses. A primitive species is not "less perfect" than an advanced species, since it may be as well adapted to its own type of environment as the advanced one is to the environment for which it has become especially adapted. Also, a species or group may be primitive in some respects and advanced in others.

Specialized might refer either to an organism or to one or more of its parts that has become adapted to a particular ecological niche or to a particular function. A more **generalized** species or structure may share the characteristics of two or more distinct groups or structures. For example, many of the small aquatic crustaceans have a number of similar feathery trunk appendages, all serving several functions, such as swimming, respiration, filter feeding, or egg bearing. Such multipurpose appendages would be considered generalized in comparison with the highly specialized defensive chelipeds or sensory antennae of the crayfish, the pollen-collecting legs of the honeybee, or the digging forelegs of the mole cricket. Each of these specialized appendages is adapted for one primary function.

An **adaptation** is any characteristic of an organism taken in the context of how the characteristic helps the organism survive and reproduce. Many adaptations are possessed in common by a wide variety of animals, but the word is most often used in reference to special adaptations for a particular habitat or environment. Indeed, the more *special adaptations* the animal has, the more *specialized* the animal is. Any characteristic or any change in a characteristic is said to have **adaptive value,** if it tends toward greater fitness of the organism for a particular niche or habitat.

An organism is **fitted** to its environment when it is adjusted or adapted to it. Its **fitness** for any particular place or mode of living is the degree of its adjustment, suitability, or adaptation to that particular environment or niche.

ONTOGENY, PHYLOGENY, AND RECAPITULATION

Phylogeny, as we have seen, represents the evolutionary history of any taxon. **Ontogeny,** on the other hand, refers to the history of the development of an individual through its entire life. All metazoa pass through certain common developmental stages. The stages, in succession, are the zygote, cleavage stages, blastula, and gastrula. The German zoologist Ernst Haeckel believed that each of these successive stages in the development of an individual represented one of the adult forms that appeared in its evolutionary history. As he phrased it, "Ontogeny is the short and rapid recapitulation of phylogeny. During its own development . . . an individual repeats the most important changes in form evolved by its ancestors during their long and slow paleontological development." According to this statement, the zygote of a developing individual represented the protozoan or protistan stage of its evolutionary history; the blastula represented the hollow colonial protozoa; and the gastrula, the adult cnidarians. The human embryo with gill depressions in the neck was believed to represent the stage when our adult ancestors were aquatic. On this basis he gave his generalization: *ontogeny (individual development) recapitulates (repeats) phylogeny (evolutionary descent).* This notion later became known as the **biogenetic law** (Principle 8, p. 9).

Haeckel's hypothesis of **recapitulation** was based on the flawed premise that evolutionary change occurs by the successive addition of stages to the end of an unaltered ancestral ontogeny. This was an outgrowth of his belief in Lamarck's concept of the inheritance of acquired characteristics (p. 44).

K.E. von Baer, a nineteenth-century embryologist, had noticed long before Haeckel the general similarity between the embryonic stages and the adults of certain animals, but von Baer had arrived at a different and sounder interpretation. According to his view, the earlier stages of all embryos tend to look alike, but as development proceeds, the embryos become more dissimilar. In other words, he believed that the embryos of higher and lower forms resemble each other more the earlier in their development that they are compared and *not* that embryos of higher forms resemble the adults of lower forms.

The fascination with and debates over recapitulation gradually waned as scientists became more absorbed in Mendelian genetics, experimental embryology, and cell and molecular biology, and the whole question fell into a mild state of disrepute. S.J. Gould (1977) has brought the matter again into the limelight. He contends that there *are* parallels between the stages of ontogeny and phylogeny but that features appearing at one stage of an ancestral ontogeny may be shifted to earlier or later stages in descendants. If a feature that appears at a specific point in the ontogeny of an ancestor arises earlier and earlier in descendants, we have one parallel producing recapitulation (the descendant's ontogeny repeats an evolutionary sequence of stages that characterized its ancestors at that specific point). If a feature that appears at a specific point in the ontogeny of an ancestor arises later and later in the descendants, we have an inverse parallel producing **paedomorphosis** (the retention of ancestral juvenile characters by later stages in the ontogeny of descendants). In other words, the rate of development of a feature in a descendant's ontogeny may be faster or slower than the rate of development of that same feature in an ancestor's ontogeny. This phyletic change in timing of development Gould calls **heterochrony.**

KINGDOMS OF LIFE

Since Aristotle's time, it has been traditional to assign every living organism to one of two kingdoms: plant or animal. However, the two-kingdom system has outlived its usefulness. Although it is easy to place rooted, photosynthetic organisms such as trees, flowers, mosses, and ferns among the plants and to place food-ingesting, motile forms such as worms, fishes, and mammals among the animals, unicellular organisms present difficulties (Chapter 1). Some forms are claimed both for the plant kingdom by botanists and for the animal kingdom by zoologists. An example is *Euglena* (p. 145) and its phytoflagellate kin, which are motile, like animals, but have chlorophyll and photosynthesize, like plants. Other groups, such as the bacteria, were rather arbitrarily assigned to the plant kingdom.

It was inevitable that biologists would try to resolve these problems by separating problem groups into new kingdoms. This was first done in 1866 by Haeckel, who proposed the new kingdom Protista to include all single-celled organisms. At first the bacteria and cyanobacteria (blue-green algae) forms that lack nuclei bounded by a membrane, were included with nucleated unicellular organisms. Finally the important differences between the anucleate bacteria and cyanobacteria (prokaryotes) and all other organisms that have cells with membrane-bound nuclei (eukaryotes) were recognized. The prokaryote-eukaryote distinction is actually much more profound than the plant-animal distinction of the traditional system. The many differences between prokaryotes and eukaryotes are summarized on p. 52.

von Baer was the greatest descriptive embryologist of the nineteenth century and is considered the father of modern embryology. Although it seems obvious to us today that von Baer's laws of progressive differentiation clearly support the concept of evolution, he never put an evolutionary interpretation to his observations and in fact remained implacably opposed to Darwin's theory. Toward the end of his long life (1792-1876) he wrote essays attacking the new Darwinism.

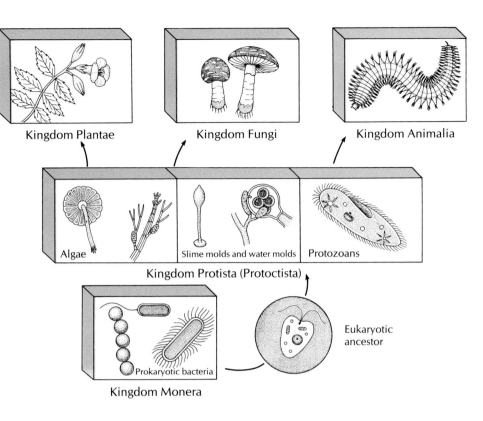

Figure 7-4

Five-kingdom system of classification, showing postulated evolutionary relationships.

In 1969 R.H. Whittaker proposed a five-kingdom system that incorporated the basic prokaryote-eukaryote distinction (Figure 7-4). The kingdom Monera contains the prokaryotes; the eukaryotes are divided among the remaining four kingdoms. The kingdom Protista contains the unicellular eukaryotic organisms (protozoa and unicellular eukaryotic algae). The multicellular organisms are split into three kingdoms on the basis of mode of nutrition and other fundamental differences in organization. The kingdom Plantae includes multicellular photosynthesizing organisms, higher plants, and multicellular algae. Kingdom Fungi contains the molds, yeasts, and fungi that obtain their food by absorption. The invertebrates (except the protozoa) and the vertebrates make up the kingdom Animalia. Most of these forms ingest their food and digest it internally, although some parasitic forms are absorptive. The supposed evolutionary relationships of the five kingdoms are shown in Figure 7-4. The Protista are believed to have given rise to all three multicellular kingdoms, which therefore have evolved independently.

Despite the appeal of Whittaker's five-kingdom system and its growing acceptance by biologists, the exclusion of the protozoa from the animal kingdom, though of no concern to the protozoa, does concern many zoologists whose scientific endeavor is the study of these organisms. The protozoa share many animal-like characteristics with the metazoa. Most ingest their food; many have specialized organelles and advanced locomotory systems, portending tissue differentiation in multicellular forms; many reproduce sexually; and some flagellate protozoa are colonial with division of labor among cell types, again suggestive of the metazoan pattern. Indeed, there is little doubt that the metazoan animals evolved from one or more protozoan groups. Thus zoologists, although acknowledging that the protozoa are, according to the five-kingdom system, eukaryotic protists and not animals, continue to claim the protozoa as their own. The kinship of the protozoan and the metazoan phyla is depicted in the family tree of animals on the inside front cover of this book.

Table 7-2   Basis for Distinction between Divisions of Bilateral Animals*

Protostomes	Deuterostomes
Mouth from, at, or near blastopore	New mouth from stomodeum
Anus new formation	Anus from or near blastopore
Cleavage mostly spiral	Cleavage mostly radial
Embryology mostly determinate (mosaic)	Embryology usually indeterminate (regulative)
In coelomate protostomes coelom forms as split in mesodermal bands (schizocoelous)	All coelomate, coelom from fusion of enterocoelous pouches (except chordates, which are schizocoelous)
Endomesoderm usually from a particular blastomere designated 4d	Endomesoderm from enterocoelous pouching (except chordates)
Includes phyla Platyhelminthes, Nemertea, Annelida, Mollusca, Arthropoda, minor phyla	Includes phyla Echinodermata, Hemichordata, Chaetognatha, and Chordata

*Embryos of many of the advanced members of each group may have some or all of these characteristics obscured or lost.

LARGER DIVISIONS OF THE ANIMAL KINGDOM

Although the phylum is often considered to be the largest and most distinctive taxonomic unit, zoologists often find it convenient to combine phyla under a few large groups because of certain common embryological and anatomical features. Such large divisions may have a logical basis, because not only are the members of some of these arbitrary groups united by common traits, but evidence also indicates some relationship in phylogenetic descent.

Subkingdom Protozoa (unicellular): protozoan phyla
Subkingdom Metazoa (multicellular): all other phyla
  Branch A (Mesozoa): phylum Mesozoa, the mesozoa
  Branch B (Parazoa): phylum Porifera, the sponges, and phylum Placozoa
  Branch C (Eumetazoa): all other phyla
    Grade I (Radiata): phyla Cnidaria, Ctenophora
    Grade II (Bilateria): all other phyla
      Division A (Protostomia): characteristics in Table 7-2
        Acoelomates: phyla Platyhelminthes, Nemertea
        Pseudocoelomates: phyla Rotifera, Gastrotricha, Kinorhyncha, Gnathostomulida, Nematoda, Nematomorpha, Acanthocephala, Entoprocta
        Eucoelomates: phyla Mollusca, Annelida, Arthropoda, Priapulida, Echiurida, Sipunculida, Tardigrada, Pentastomida, Onychophora, Pogonophora, Phoronida, Ectoprocta, Brachiopoda
      Division B (Deuterostomia): characteristics in Table 7-2
        Phyla Echinodermata, Chaetognatha, Hemichordata, Chordata

As in the outline, the bilateral animals are customarily divided into protostomes and deuterostomes on the basis of their embryological development (Table 7-2). However, some of the phyla are difficult to place into one of these two categories because they possess some of the characteristics of each group (Chapter 19).

▬ SUMMARY

Systematics is concerned with the speciation, classification, and phylogeny of animals. It includes a system of uniform naming of kinds of animals and the erection of a classification scheme that reflects evolutionary relationships. The scheme is hierarchical, and the most widely used groupings (taxa) are as follows, in order of increasing inclusiveness: species, genus, family, order, class, phylum, and kingdom. Subdivisions of all of these are commonly used as necessary. The basic binomial system of nomenclature, giving each kind of organism a generic name and a species epithet, originated with Carolus Linnaeus and is still in use today. The biological species concept has replaced the older typological concept of a species. A biological species is an interbreeding (or potentially interbreeding) population that is reproductively isolated from other such groups. It is not immutable through time but changes during the course of evolution. If a species is given more than one name, the earliest name published with an appropriate description of the organism is considered the correct name, and other, later names become synonyms.

There are currently three major schools of taxonomy. The traditional approach, called evolutionary taxonomy, attempts to reconstruct actual ancestor-descendant relationships, which may be depicted in the form of a phylogenetic tree. A second approach, called numerical taxonomy, strives to group taxa according to a statistical analysis of their total similarities and dissimilarities. Numerical taxonomy does not infer evolutionary relationships. A third approach is cladistics, whose proponents attempt to achieve phylogenetically significant groupings. Cladistics discriminates between primitive and derived characteristics of organisms.

Although morphology, embryology, and the fossil record are still widely used as criteria for assessing evolutionary relationships, other characteristics such as behavioral, serological, biochemical, cytological, and physiological characteristics are increasingly used in modern systematics. A highly useful device for visualizing evolutionary relationships and derivations from ancestral groups is the phylogenetic tree.

The following terms have specific meanings in evolutionary biology, each of which should be carefully distinguished and understood: higher, lower, primitive, advanced, specialized, generalized, adaptation, and fitness.

The earlier view that embryonic stages of an animal were similar to adult stages of its ancestor (recapitulation) has been replaced by the idea that animal embryos often tend to resemble the embryos of their ancestors.

The broadest, most inclusive categories of living organisms have been traditionally considered the plant and animal kingdoms. However, most biologists now prefer the five-kingdom scheme: Monera (prokaryotes), Protista (protozoa and eukaryotic algae), Fungi, Plantae, and Animalia. As considered in this book, the Animalia contains the subkingdoms Protozoa (unicellular phyla) and Metazoa (multicellular phyla). Among the metazoa, important groupings are the radial and the bilateral animals, and among the bilateral, the Protostomia and the Deuterostomia.

Review questions

1. Distinguish between systematics and taxonomy.
2. List in order, from most inclusive to least inclusive, the principle categories (taxa) in Carolus Linnaeus' system of classification.
3. Explain why the system for naming species that originated with Linnaeus is "binomial."
4. How does the biological species concept differ from the earlier typological concept of a species? Why is it considered to be a more modern concept of species?
5. What should be done if a species is discovered to have been named more than once?
6. What is the difference between a monophyletic and a polyphyletic taxon?
7. If you were to classify groups of animals by the method of cladistics, how should you proceed?
8. If you wished to reconstruct lines of descent for an animal group having only a fragmentary fossil record, what other kinds of information might you use?
9. What do you understand the following terms to mean in the context of evolutionary biology: higher, lower; primitive, advanced; specialized, generalized; adaptation, adaptive value; fitness?
10. What did Ernst Haeckel mean by the generalization, "Ontogeny recapitulates phylogeny?" What is the more modern interpretation of Haeckel's ideas?
11. What are the five kingdoms distinguished by Whittaker, and on what basis did he distinguish them?

References

Gould, S.J. 1977. Ontogeny and phylogeny. Cambridge, Mass., Harvard University Press. *A review of the historical development and collapse of the theory of recapitulation, and a discussion of modern concepts, such as heterochrony (changes in developmental timing, producing parallels between ontogeny and phylogeny) and neoteny, which Gould terms the most important determinant of human evolution.*

Jeffrey, C. 1973. Biological nomenclature. London, Edward Arnold, Ltd. *A concise, practical guide to the principles and practice of biological nomenclature and a useful interpretation of the Codes of Nomenclature.*

Margulis, L., and K.V. Schwartz. 1987. Five kingdoms: an illustrated guide to the phyla of life on earth, ed. 2. San Francisco, W.H. Freeman & Co. Publishers. *Illustrated catalog and descriptions of all major groups with bibliography and glossary.*

Ross, H.H. 1974. Biological systematics. Reading, Mass., Addison-Wesley Publishing Co., Inc. *Theory and practice of systematics are presented and exemplified from animals, plants, and microorganisms. Very comprehensive and useful.*

Wiley, E.O. 1981. Phylogenetics: the theory and practice of phylogenetic systematics. New York, John Wiley & Sons, Inc. *Excellent, thorough presentation of cladistic theory.*

PART TWO

THE DIVERSITY OF ANIMAL LIFE

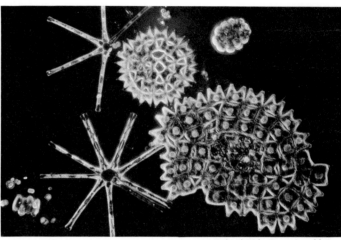

Roland Birke/Peter Arnold, Inc.

CHAPTER 8

THE UNICELLULAR ANIMALS

Protozoan Phyla

Position in Animal Kingdom

The protozoan is a complete organism in which all life activities are carried on within the limits of a single plasma membrane. Because their protoplasmic mass is not subdivided into cells, protozoa are sometimes termed "acellular," but many people prefer "unicellular" to emphasize the many structural similarities to the cells of multicellular animals (see Principle 12, p. 10).

The term "Protista" is used by some biologists to refer to all unicellular organisms, whether they are plant or animal. When applied in this way, the category includes the unicellular algae, protozoa, bacteria, yeasts, and so on. Most biologists prefer to use "Protista" to denote only those unicellular organisms whose nucleus is bounded by a nuclear membrane (eukaryotic) and the term "Monera" to include those unicellular organisms without a nuclear membrane (prokaryotic), such as the bacteria and the cyanobacteria (see p. 126).

The protozoa are eukaryotic protistans. The eukaryote-prokaryote distinction was summarized in Chapter 4 (p. 52) and mentioned again in Chapter 7 (p. 126).

Biological Contributions

1. **Intracellular specialization** (division of labor within the cell) involves the organization of functional organelles in the cell.
2. The earliest indication of **division of labor between cells** is seen in certain colonial protozoa that have both somatic and reproductive zooids (individuals) in the colony.
3. **Asexual reproduction** by mitotic division is first developed in the protists.
4. **True sexual reproduction** with zygote formation is found in some protozoa.
5. The responses (taxes) of protozoa to stimuli represent the **beginning of reflexes and instincts** as we know them in metazoans.
6. The most primitive animals with an **exoskeleton** are certain shelled protozoa.
7. **All types of nutrition** are developed in the protozoa: autotrophic, saprozoic, and holozoic. **Basic enzyme systems** to accomplish these types of nutrition are developed.
8. Means of **locomotion** in aqueous media are developed.

—— THE PROTOZOAN PHYLA

The organisms referred to as protozoa are united only on the basis of a single, negative characteristic: they are not multicellular. This concept was recognized, in a way, by the American zoologist Libbie Hyman (1940), who preferred the term "acellular" rather than the traditional "unicellular" to describe protozoa. She distinguished them as "animals whose body substance is not partitioned into cells." Although most zoologists have returned to describing protozoa as unicellular because of electron microscope studies subsequent to Hyman's book, the concept of acellularity is still a valuable one. It reminds us that the traditionally recognized phylum Protozoa was not a natural phylogenetic grouping. An enormous amount of information on protozoan structure, life histories, and physiology has accumulated in recent years, and the Society of Protozoologists published a new classification of protozoa in 1980, recognizing *seven* separate phyla. We will adopt the new classification because it comes closer to reflecting real evolutionary relationships than the older, simpler systems, but there is no way to give adequate treatment to all groups, even all phyla, of protozoa in a book of this size.

We can say, nevertheless, that the protozoan phyla demonstrate a basic body plan—the single eukaryotic cell—and that they amply demonstrate the enormous adaptive potential of that plan. Over 64,000 species have been named, and over half of these are fossil. Although they are unicellular, protozoa are not simple. They are functionally complete organisms with many complicated structures. Their various organelles tend to be more specialized than those of the average cell in a multicellular organism. Particular organelles may perform as skeletons, sensory structures, conducting mechanisms, and so on.

Protozoa are found wherever life exists. They are highly adaptable and easily distributed from place to place. They require moisture, whether they live in marine or freshwater habitats, soil, decaying organic matter, or plants and animals. They may be sessile or free swimming, and they form a large part of the floating plankton. The same species are often found widely separated in time as well as in space. Some species may have spanned geological eras of more than 100 million years.

Despite their wide distribution, many protozoa can live successfully only within narrow environmental ranges. Species adaptations vary greatly, and successions of species frequently occur as environmental conditions change. These changes may be brought about by physical factors, such as the drying up of a pond or seasonal changes in temperature, or by biological changes, such as predator pressure.

Protozoa play an enormous role in the economy of nature. Their fantastic numbers are attested by the gigantic ocean and soil deposits formed by their skeletons. About 10,000 species of protozoa are symbiotic in or on other animals or plants, sometimes even other protozoa. (See the discussion of population interactions in Chapter 41.) The relationship may be mutualistic, commensalistic, or parasitic, depending on the species involved. Some of the most important diseases of humans and domestic animals are caused by parasitic protozoa.

A number of species are colonial and some have multicellular stages in their life cycles, which may lead one to wonder why such protozoa are not considered metazoa. The reasons are that they usually have clearly recognizable, noncolonial relatives and, more arbitrarily, that they do not have more than one kind of nonreproductive cell and they do not undergo embryonic development. By definition, metazoa have more than one kind of nonreproductive cell in their bodies and undergo embryogenesis.

CHARACTERISTICS

1. **Unicellular;** some colonial, and some with multicellular stages in their life cycles
2. **Mostly microscopic,** although some are large enough to be seen with the unaided eye
3. All symmetries represented in the group; shape variable or constant (oval, spherical, or other)
4. **No germ layer present**
5. No organs or tissues, but **specialized organelles** are found; nucleus single or multiple
6. Free living, mutualism, commensalism, parasitism all represented in the group

7. Locomotion by **pseudopodia, flagella, cilia,** and direct cell movements; some sessile
8. Some provided with a **simple endoskeleton** or **exoskelton,** but mostly naked
9. **Nutrition of all types:** autotrophic (manufacturing own nutrients by photosynthesis), heterotrophic (depending on other plants or animals for food), saprozoic (using nutrients dissolved in the surrounding medium)
10. Aquatic or terrestrial habitat; free-living or symbiotic mode of life.
11. Reproduction **asexually** by fission, budding, and cysts and **sexually** by conjugation or by syngamy (union of male and female gametes to form a zygote)

CLASSIFICATION

The four main groups of protozoa traditionally recognized were the flagellates, amebas, spore formers, and ciliates. The system that follows reflects a much more phylogenetic arrangement, including the recognition that amebas and flagellates are more closely related to each other than they are to other groups, and that the "spore formers" represent several completely unrelated forms.

Phylum Sarcomastigophora (sar'ko-mas-ti-gof'o-ra) (Gr. *sarkos,* flesh, + *mastix,* whip, + *phora,* bearing). Flagella, pseudopodia, or both types of locomotory organelles; usually with only one type of nucleus; typically no spore formation; sexuality, when present, essentially syngamy.

Subphylum Mastigophora (mas-ti-gof'o-ra) (Gr. *mastix,* whip, + *phora,* bearing). One or more flagella typically present in adult stages; autotrophic or heterotrophic or both; reproduction usually asexual by fission.

Class Phytomastigophorea (fi'to-mas-ti-go-for'e-a) (Gr. *phyton,* plant, + *mastix,* whip, + *phora,* bearing). Plantlike flagellates, usually bearing chromoplasts, which contain chlorophyll. Examples: *Chilomonas, Euglena, Volvox, Ceratium, Peranema, Noctiluca.*

Class Zoomastigophorea (zo'o-mas-ti-go-for'e-a) (Gr. *zōon,* animal, + *mastix,* whip, + *phora,* bearing). Flagellates without chromoplasts; one to many flagella; ameboid forms with or without flagella in some groups; species predominantly symbiotic. Examples: *Trichomonas, Trichonympha, Trypanosoma, Leishmania, Dientamoeba.*

Subphylum Opalinata (o'pa-lin-a'ta) (N.F. *opaline,* like opal in appearance, + *ata,* group suffix). Body covered with longitudinal rows of cilium-like organelles; parasitic; cytostome (cell mouth) lacking; two to many nuclei of one type. Examples: *Opalina, Protoopalina.*

Subphylum Sarcodina (sar-ko-di'na) (Gr. *sarkos,* flesh, + *ina,* belonging to). Pseudopodia typically present; flagella present in developmental stages of some; free living or parasitic.

Superclass Rhizopoda (ri-zop'o-da) (Gr. *rhiza,* root, + *pous, podos,* foot). Locomotion by lobopodia, filopodia, or reticulopodia, or by cytoplasmic flow without production of discrete pseudopodia. Composed of eight classes, some of which are listed below.

Class Lobosea (lo-bo'se-a) (Gr. *lobos,* lobe). Pseudopodia lobose or more or less filiform but produced from broader lobe; usually uninucleate; no fruiting bodies. Examples: *Amoeba, Entamoeba, Acanthamoeba, Naegleria, Pelomyxa, Arcella, Difflugia.*

Class Eumycetozoea (yu'mi-set-o-zo'e-a) (Gr. *eu,* good, true, + *mykes,* fungus, + *zōon,* animal). Ameboid feeding stage, flagellated stage present or absent; produce aerial fruiting bodies with one to thousands of spores. Examples: *Dictyostelium, Physarum.*

Class Filosea (fi-los'e-a) (L. *filum,* thread). Hyaline, filiform pseudopodia, often branching, sometimes rejoining; no spores or flagellated stages known. Examples: *Vampyrella, Euglypha, Gromia.*

Class Granuloreticulosea (gran'yu-lo-re-tik'yu-los'e-a) (L. *granulum,* dim. of *granum,* grain, + *reticulum,* dim. of *rete,* net). Delicate, finely granular or hyaline reticulopodia or, rarely, finely pointed, granular but nonrejoining pseudopodia. Examples: *Allogromia, Fusulina, Textularia, Elphidium, Globigerina,* other foraminiferans.

Superclass Actinpoda (ak'ti-nop'o-da) (Gr. *aktis, aktinos,* ray, + *pous, podos,* foot). Often spherical, usually planktonic; pseudopodia in form of axopodia, with microtubular supporting structure.

Class Acantharea (a'kan-thar'e-a) (Gr. *akantha,* spine or thorn). Strontium sulfate skeleton composed of 20 or more radiating spines more or less joined in cell center; marine, usually planktonic. Examples: *Acanthometra, Lithoptera.*

Class Polycystinea (pol'e-sis-tin'e-a) (Gr. *polys,* many, + *kystis,* bladder). Siliceous skeleton in most species, usually of solid elements, consisting of one or more latticed shells with or without radial spines, or of spicules; capsular membrane usually of grossly polygonal plates with many more than three pores; marine, planktonic. Example: *Thalassicolla.*

Class Phaeodarea (fe'o-dar'e-a) (Gr. *phaios,* dusky, + *daria,* suffix). Skeleton of mixed silica and organic matter, consisting of usually hollow spines and shells; very thick capsular membrane with three pores; marine, planktonic. Examples: *Aulacantha, Challengeron.*

Class Heliozoea (he'le-o-zo'e-a) (Gr. *helios,* sun, + *zōon,* animal). Without central capsule; skeletal structures, if present, siliceous or organic; axopodia radiating on all sides; most species freshwater. Examples: *Clathrulina, Actinophrys, Actinosphaerium.*

Phylum Labyrinthomorpha (la'bi-rinth-o-morf'a) (Gr. *labyrinth,* maze, labyrinth, + *morph,* form, + *a,* suffix). Small group living on algae; mostly marine or estuarine. Example: *Labyrinthula.*

Phylum Apicomplexa (a'pi-com-plex'a) (L. *apex,* tip or summit, + *complex,* twisted around, + *a,* suffix). Characteristic set of organelles (apical complex) associated with anterior end present in some developmental stages; cilia and flagella absent except for flagellated microgametes in some groups; cysts often present; all species parasitic.

Class Perkinsea (per-kin'se-a). Small group parasitic in oysters.

Class Sporozoea (spor'o-zo'e-a) (Gr. *sporos,* seed, + *zōon,* animal). Spores or oocysts typically present that contain infective sporozoites; flagella present only in microgametes of some groups; pseudopods ordinarily absent, if present they are used for feeding, not locomotion; one or two host life cycles. Examples: *Monocystis, Gregarina, Eimeria, Plasmodium, Toxoplasma, Babesia.*

Phylum Myxozoa (mix-o-zo'a) (Gr. *myxa,* slime, mucus, + *zōon,* animal). Parasites of lower vertebrates, especially fishes, and invertebrates.

Phylum Microspora (mi-cros'por-a) (Gr. *micro,* small, + *sporos,* seed). Parasites of invertebrates, especially arthropods, and lower vertebrates.

Phylum Ascetospora (as-e-tos'por-a) (Gr. *asketos,* curiously wrought, + *sporos,* seed). Small group that is parasitic in invertebrates and a few vertebrates.

Phylum Ciliophora (sil-i-of'or-a) (L. *cilium,* eyelash, + Gr. *phora,* bearing). Cilia or ciliary organelles in at least one stage of life cycle; two types of nuclei, with rare exception; binary fission across rows of cilia, budding and multiple fission also occur; sexuality involving conjugation, autogamy, and cytogamy; nutrition heterotrophic; contractile vacuole typically present; most species free living, but many commensal, some parasitic. (This is a very large group, now divided by the Society of Protozoologists classification into three classes and numerous orders and suborders. The classes are separated on the basis of technical characteristics of the ciliary patterns, especially around the cytostome, the development of the cystostome, and other characteristics.) Examples: *Paramecium, Colpoda, Tetrahymena, Balantidium, Stentor, Blepharisma, Epidinium, Euplotes, Vorticella, Carchesium, Trichodina, Podophrya, Ephelota.*

Form and Function

Inasmuch as protozoa are cells, in many aspects their structure and physiology are the same as those of cells of multicellular organisms. However, because they must carry on all the functions of life as individual organisms, and because they show such enormous diversity in form, habitat, feeding, and so on, many features are unique to various protozoan cells.

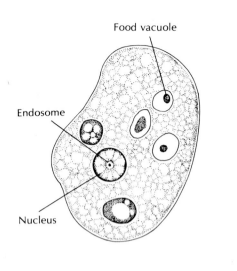

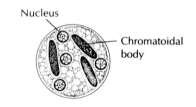

Figure 8-1

Entamoeba histolytica, the cause of amebic dysentery in humans. The trophozoite *(above)* is the actively moving and feeding form. It contains a single nucleus and several food vacuoles. The cyst *(below)* can tolerate conditions outside the body and is infective to the new host. It contains four nuclei and several chromatoidal bodies. The chromatoidal bodies are an organized form of RNA. On excystment in the small intestine of the host, the nuclei will divide again, and the cytoplasm divides to produce eight small amebas. The nuclei of both the trophozoite and the cyst are vesicular and have a central endosome (see text).
Drawing by Jeanne Robertson. From Schmidt, G.D., and L.S. Roberts. 1981. Foundations of parasitology, ed. 2. St. Louis, The C.V. Mosby Co.

Nucleus and cytoplasm

As in other eukaryotes, the nucleus is a membrane-bound structure whose interior communicates with the cytoplasm by small pores (Figure 4-8, p. 55). Within the nucleus the genetic material (DNA) is borne on chromosomes. Except during cell division, the chromosomes are not usually condensed in a form that can be distinguished, although during fixation of the cells for light microscopy, the chromosome material (chromatin) often clumps together irregularly, leaving some areas within the nucleus relatively clear. The appearance is described as **vesicular** and is characteristic of many protozoan nuclei (Figure 8-1). Condensations of chromatin may be distributed around the periphery of the nucleus or internally in distinct patterns. In some flagellates the chromosomes are visible through the interphase as they would appear during prophase of mitosis.

Also within the nucleus, one or more **nucleoli** are often present. **Endosomes** are nucleoli that remain as discrete bodies during mitosis; they are characteristic of phytoflagellates, parasitic amebas, and trypanosomes (Figures 8-1, 8-8, and 8-11).

The **macronuclei** of ciliates are described as **compact** or **condensed** because the chromatin material is more finely dispersed and clear areas cannot be observed with the light microscope (Figure 8-21).

Cellular organelles like those in the cells of multicellular animals can be distinguished in the cytoplasm of many protozoa. These include mitochondria, endoplasmic reticulum, Golgi apparatus, and various vesicles. Chloroplasts, the membrane-bound organelles in which photosynthesis takes place, are found in most phytoflagellates (Figure 8-9).

Sometimes the peripheral and the central areas of the cytoplasm can be distinguished as **ectoplasm** and **endoplasm** (Figure 8-3). The endoplasm appears more granular and contains the nucleus and cytoplasmic organelles. The ectoplasm appears more transparent (hyaline) under the light microscope, and it bears the bases of the cilia or flagella. The ectoplasm is often more rigid and is in the gel state of a colloid, whereas the more fluid endoplasm is in the sol state.

Locomotor organelles

The chief means by which protozoa move are by cilia and flagella and by pseudopodial movement. These mechanisms are extremely important in the biology of higher animals as well.

Cilia and flagella

Many small metazoans use cilia not only for locomotion but also to create water currents for their feeding and respiration. Ciliary movement is vital to many species in such functions as handling food, reproduction, excretion, and osmoregulation (as in flame cells, p. 644).

There is no real morphological distinction between cilia and flagella, and some investigators have preferred to call them both undulipodia. However, a cilium propels water parallel to the surface to which the cilium is attached, whereas a flagellum propels water parallel to the main axis of the flagellum. Thus there are important effects on the mechanics and speed of propulsion. Each flagellum or cilium contains nine pairs of longitudinal microtubules arranged in a circle around a central pair (Figure 8-2), and this is true for all flagella and cilia in the animal kingdom, with a few notable exceptions. This "9 + 2" tube of microtubules in the flagellum or cilium is its **axoneme;** the axoneme is covered by a

A

B

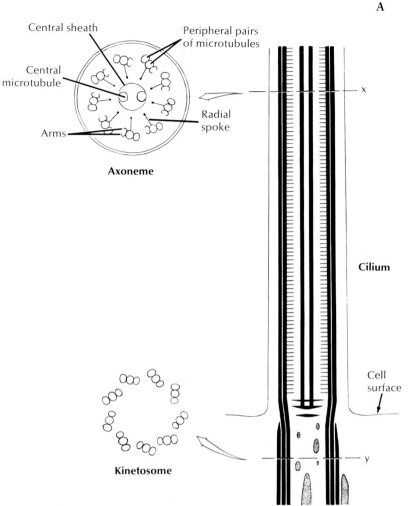

Central sheath

Peripheral pairs
of microtubules

Central
microtubule

Radial
spoke

Arms

Axoneme

Kinetosome

Cilium

Cell
surface

x

y

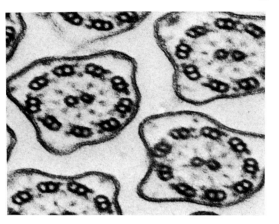

Figure 8-2

A, The axoneme is composed of nine pairs
of microtubules plus a central pair, and it is
enclosed within the cell membrane. The
central pair ends at about the level of the
cell surface in a basal plate (axosome). The
peripheral microtubules continue inward for
a short distance to compose two of each of
the triplets in the kinetosome (at level *y* in
A). **B,** Electron micrograph of section
through several cilia, corresponding to
section at *x* in **A**. (×133,000.)
Electron micrograph courtesy I.R. Gibbons, Harvard University.

membrane continuous with the cell membrane covering the rest of the organism.
At about the point where the axoneme enters the cell proper, the central pair of
microtubules ends at a small plate within the circle of nine pairs (Figure 8-2, *A*).
Also at about that point, another microtubule joins each of the nine pairs, so that
these form a short tube extending from the base of the flagellum into the cell and
consisting of nine *triplets* of microtubules. The short tube of nine triplets is the
kinetosome and is exactly the same in structure as the **centriole** (see p. 58 and
Figure 4-24, p. 68). The centrioles of some flagellates may give rise to the kinetosomes, or the kinetosomes may function as centrioles. All typical flagella and cilia
have a kinetosome at their base, regardless of whether they are borne by a protozoan or metazoan cell. The kinetosomes of protozoa have older, traditional
names (**blepharoplast, basal body, basal granule**) that are still commonly used.

The current explanation for ciliary and flagellar movement is the **sliding
microtubule hypothesis.** The movement is powered by the release of chemical
bond energy in ATP (p. 78). Two little arms are visible in electron micrographs on
each of the pairs of peripheral tubules in the axoneme (Figure 8-2), and these bear
the enzyme adenosine triphosphatase (ATPase), which cleaves the ATP. When the
bond energy in ATP is released, the arms "walk along" one of the filaments in the
adjacent pair, causing it to slide relative to the other filament in the pair. Shear
resistance, causing the axoneme to bend when the filaments slide past each other,
is provided by "spokes" from each doublet to the central pair of fibrils. These are
also visible in electron micrographs.

Figure 8-3

Ameba in active locomotion. Arrows indicate the direction of streaming protoplasm. The first sign of new pseudopodium is thickening of the ectoplasm to form a clear hyaline cap, into which the fluid endoplasm flows. As the endoplasm reaches the forward tip, it fountains out and is converted into ectoplasm, forming a stiff outer tube that lengthens as the forward flow continues. Posteriorly the ectoplasm is converted into fluid endoplasm, replenishing the flow. Substratum is necessary for ameboid movement.

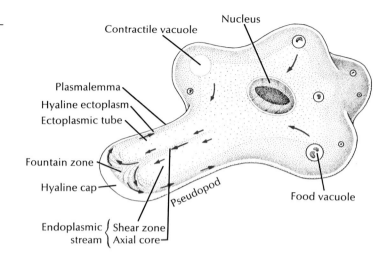

Pseudopodia

Although pseudopodia are the chief means of locomotion in the Sarcodina, they can be formed by a variety of flagellate protozoa, as well as by ameboid cells of many invertebrates. In fact, much of the defense against disease in the human body depends on ameboid white blood cells, and ameboid cells in many other animals, vertebrate and invertebrate, play similar roles.

In the protozoa, pseudopodia exist in several forms. The most familiar are the **lobopodia** (Figure 8-3), which are rather large, blunt extensions of the cell body containing both endoplasm and ectoplasm. Some amebas characteristically do not extend individual pseudopodia, but the whole body moves with pseudopodial movement; this is known as the **limax** form (for a genus of slugs, *Limax*). **Filipodia** are thin extensions, usually branching, and containing only ectoplasm. They are found in members of the sarcodine class Filosea, such as *Euglypha* (Figure 8-7, *B*). **Reticulopodia** (Figure 8-14) are distinguished from filipodia in that reticulopodia repeatedly rejoin to form a netlike mesh, although some protozoologists believe that the distinction between filipodia and reticulopodia is artificial. Members of the superclass Actinopoda have **axopodia** (Figure 8-14), which are long, thin pseudopodia supported by axial rods of microtubules (Figure 8-5). The microtubules are arranged in a definite spiral or geometrical array, depending on the species, and constitute the axoneme of the axopod. Axopodia can be extended

Figure 8-4

Ameboid movement. Series photographed at approximately 30-second intervals. *Right,* Pseudopodium is extending toward escaping rotifer.

Photograph by F.M. Hickman.

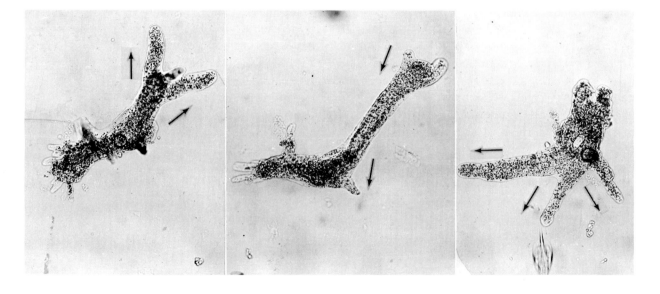

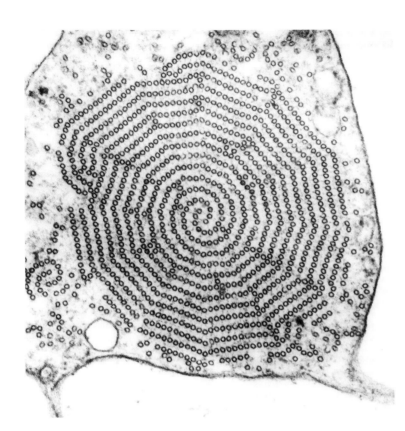

Figure 8-5

Electron micrograph of a section through an axopodium of *Actinosphaerium nucleofilum*. The axoneme of the axopodium is composed of an array of microtubules, which may vary from three to many in number depending on the species. Some species can extend or retract their axopodia quite rapidly. (×99,000.) Courtesy L. Evans Roth.

or retracted, apparently by addition or removal of microtubular material. Since the tips can adhere to the substrate, the animal can progress by a rolling motion, shortening the axonemes in front and extending those in the rear. Cytoplasm can flow along the axonemes, toward the body on one side and in the reverse direction on the other.

How pseudopodia work has long attracted the interest of zoologists, but we have only recently gained some insight into the phenomenon. When a typical lobopodium begins to form, an extension of ectoplasm called the hyaline cap appears, and endoplasm begins to flow toward and into the hyaline cap (Figure 8-3). As the endoplasmic material flows into the hyaline cap, it fountains out to the periphery and changes from the sol to the gel state; that is, it becomes ectoplasm. Thus, the ectoplasm is a tube through which the endoplasm flows as the pseudopodium extends. On the trailing (uroid) side of the animal, ectoplasm becomes endoplasm. At some point the pseudopodium becomes anchored to the substrate, and the animal is drawn forward. Although these events can be easily observed under the light microscope, the force that causes the endoplasm to flow through has been obscure. It now appears that ameboid movement, like that of muscle, involves microfilaments of actin (p. 605) sliding past each other. Microfilaments have been shown in the gel phase by electron microscopy, and myosin has been identified in *Amoeba proteus*. A similar mechanism apparently accounts for cytoplasmic streaming in all types of pseudopodia.

Excretion and osmoregulation

Vacuoles can be seen in the cytoplasm of many protozoa under the light microscope, and some of these vacuoles periodically fill with a fluid substance that is then expelled. Evidence is strong that these **contractile vacuoles** (Figures 8-3, 8-9, and 8-21) function principally in osmoregulation. They are more prevalent and fill

Amoeba

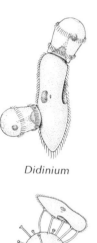

Leidyopsis

Didinium

Podophrya

Figure 8-6

Some feeding methods among protozoa. *Amoeba* surrounds a small flagellate with pseudopodia. *Leidyopsis*, a flagellate living in the intestine of termites, forms pseudopodia and ingests wood chips. *Didinium*, a ciliate, feeds only on *Paramecium*, which it swallows through a temporary cytostome in its anterior end. Sometimes more than one *Didinium* feed on the same *Paramecium*. *Podophrya* is a suctorian ciliophoran. Its tentacles attach to its prey and suck prey cytoplasm into the body of the *Podophrya*, where it is pinched off to form food vacuoles. Technically, all of these methods are types of phagocytosis.

and empty more frequently in freshwater protozoa than in marine and endosymbiotic species, where the surrounding medium would be more nearly isosmotic (having the same osmotic pressure) to the cytoplasm. Smaller species, which have a greater surface-to-volume ratio, generally have more rapid filling and expulsion rates in their contractile vacuoles. Excretion of metabolic wastes, on the other hand, is almost entirely by diffusion. The main end product of nitrogen metabolism is ammonia, which readily diffuses out of the small bodies of protozoa.

The contractile vacuoles, sometimes called water expulsion vesicles, differ considerably in complexity among the various types of protozoa. In amebas the contractile vacuoles are not usually found at any particular site, being passively carried around in the endoplasm. Small vesicles join the contractile vacuole, emptying their contents into it as the vacuole fills. The vacuole finally joins its membrane to the surface membrane and empties its contents to the outside.

Some ciliates, such as *Blepharisma*, have contractile vacuoles with filling mechanisms similar to that described for amebas. Others, such as *Paramecium*, have more complex contractile vacuoles. In these the contractile vacuoles are located in a specific position beneath the cell membrane, with an "excretory" pore leading to the outside, and surrounded by the ampullae of about six feeder canals (Figure 8-21). The feeder canals, in turn, are surrounded by fine tubules about 20 nm in diameter, which connect with the canals during filling of the ampullae and at their lower ends connect with the tubular system of the endoplasmic reticulum. The ampullae and the contractile vacuole are surrounded by bundles of fibrils, which may play a role in the contraction of these structures. Contraction of the ampullae fills the vacuole. When the vacuole contracts to discharge its contents to the outside, the ampullae become disconnected from the vacuole, so that backflow is prevented.

Nutrition

Protozoa can be categorized broadly into autotrophs and heterotrophs (see p. 43) according to whether they can synthesize their own organic constituents from inorganic substrates or must have organic molecules synthesized by other organisms. Another kind of classification, usually applied to heterotrophs, involves those that ingest visible particles of food (**phagotrophs,** or **holozoic** feeders) as contrasted with those ingesting food in a soluble form (**osmotrophs,** or **saprozoic** feeders). However, reality is not so simple, even among the one-celled animals. Autotrophic protozoa use light energy to synthesize their organic molecules (phototrophs), but they often practice phagotrophy and osmotrophy as well. Even among the heterotrophs, few are exclusively either phagotrophic or osmotrophic. The single order Euglenida (class Phytomastigophorea) contains some forms that are mainly phototrophs, some that are mainly osmotrophs, and some that are primarily phagotrophs. The species of *Euglena* show considerable variety in nutritional capability. Some require certain preformed organic molecules, even though they are autotrophs, and some lose their chloroplasts if maintained in darkness, thus becoming permanent osmotrophs.

Holozoic nutrition implies phagocytosis (Figure 8-6), in which there is an infolding or invagination of the cell membrane around the food particle. As the invagination extends farther into the cell, it is pinched off at the surface (p. 64). The food particle is thus contained in an intracellular, membrane-bound vesicle, the **food vacuole** or **phagosome.** Lysosomes, small vesicles containing digestive habitats frequently subjected to extremely harsh conditions. This is surely related to their ability to form cysts: dormant forms marked by the possession of resistant external coverings and a more or less complete shutdown of metabolic machinery. Cyst formation is also important to many parasitic forms that must survive a harsh environment between hosts (Figure 8-1). However, some parasites do not form cysts, apparently depending on direct transfer from one host to another. Reproductive phases such as fission, budding, and syngamy may occur in the cysts of some species (Figure 8-1). Encystment has not been found in *Paramecium*, and it is rare or absent in marine forms.

The conditions stimulating encystment are incompletely understood, although in some cases cyst formation is cyclic, occurring at a certain stage in the life cycle. In most free-living forms, adverse change in the environment favors encystment. Such conditions may include food deficiency, desiccation, increased tonicity of the environment, decreased oxygen concentration, or pH or temperature change.

During encystment a number of organelles, such as cilia or flagella, are resorbed, and the Golgi apparatus secretes the cyst wall material, which is carried to the surface in vesicles and extruded.

Although the exact stimulus for excystation (escape from cysts) is usually unknown, a return of favorable conditions initiates excystment in those protozoa in which the cysts are a resistant stage. In parasitic forms the excystment stimulus may be more specific, requiring conditions similar to those found in the host.

___ Representative Types

This section describes some representatives of each large group of protozoa to give the student a basis for comparing the groups and an idea of the diversity of protozoa. Forms such as *Amoeba* and *Paramecium*, although large and easy to obtain for study, are not wholly representative because their life histories are somewhat simpler than those of other members of the respective groups.

Phylum Sarcomastigophora

The Sarcomastigophora includes both protozoa that move by flagella (Mastigophora) and those that move by pseudopodia (Sarcodina). These characteristics are not mutually exclusive; some mastigophorans (flagellates) can form and use pseudopodia, and a number of sarcodines have flagellated stages in their life cycles.

Subphylum Mastigophora: the flagellated protozoa

Although some flagellates can form pseudopodia, their primary means of locomotion is by one or more flagella. What are probably the most primitive protozoa are found in this subphylum. The group is divided into the phytoflagellates (class Phytomastigophorea), which usually have chlorophyll and are thus plantlike, and the zooflagellates (class Zoomastigophorea), which do not have chlorophyll, are either holozoic or saprozoic, and thus are animal-like.

Phytoflagellates

Phytoflagellates usually have one or two (sometimes four) flagella and chromoplasts (also called chloroplasts), which contain the pigments used in photosynthesis. They are mostly free living and include such familiar forms as *Euglena, Chlamydomonas, Peranema, Volvox,* and the dinoflagellates. *Peranema* (Figure 8-8) is related to *Euglena* but is a colorless phytoflagellate with holozoic nutrition.

Cysts of some soil-inhabiting and freshwater protozoa have amazing durability. The cysts of the soil ciliate *Colpoda* can survive 7 days in liquid air and 3 hours at 100° C. Survival of *Colpoda* cysts in dried soil has been shown for up to 38 years, and those of a certain small flagellate *(Podo)* can survive up to 49 years! Not all cysts are so sturdy, however. Those of *Entamoeba histolytica* will tolerate gastric acidity but not desiccation, temperature above 50° C, or sunlight.

Consideration of the phytoflagellates as animals by the zoologists and as plants by the botanists has led to a curious taxonomic anomaly. According to the zoologists (Society of Protozoologists classification), they comprise a class of one phylum, but according to the botanists they should be placed in several different divisions (a division is a taxon of plants equivalent in rank to a phylum of animals) and classes within the divisions. In our acceptance of the Sarcomastigophora as including the plantlike flagellates, we tend to concur with Dodson (1971) that "the blurred areas then remain, not as a defect, but as evidence that evolution from common ancestors is one of the basic facts of life, that all eukaryotes are in fact related."

Figure 8-12

Giant ameba (*Pelomyxa* sp.) is easily visible
to the unassisted eye. At bottom is a nor-
mal-sized *Amoeba proteus;* above and at
left are several paramecia.
Courtesy Carolina Biological Supply Co., Burlington, N.C.

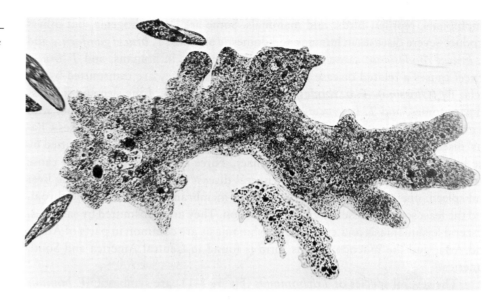

Figure 8-13

A, Living foraminiferans. **B,** Glass model of
Globigerina bulloides. Foraminiferans (class
Granuloreticulosea) are ameboid marine
protozoans that secrete a calcareous, many-
chambered test in which to live and then
extrude protoplasm through pores to form a
layer over the outside. The animal begins
with one chamber, and as it grows, it secretes a succession of new and larger chambers, continuing this process throughout
life. Foraminiferans are planktonic animals,
and when they die, their shells are added to
the ooze on the ocean's bottom.
A, Courtesy R. Vishniac; **B,** courtesy American
Museum of Natural History.

and cause abscesses there. Many infected persons show few or no symptoms but
are carriers, passing cysts in their feces. Infection is spread by contaminated water
or food containing the cysts.

Other species of *Entamoeba* found in humans are *E. coli* in the intestine and
E. gingivalis in the mouth. Neither of these species is known to cause disease.

Not all rhizopods are "naked" as are the amebas. Some have their delicate
plasma membrane covered with a protective **test** or shell. *Arcella* and *Difflugia*
(Figure 8-14) are common sarcodines. They have a test of secreted siliceous or
chitinoid material that may be reinforced with grains of sand. They move by
means of pseudopodia that project from openings in the shell.

The **foraminiferans** (class Granuloreticulosea) are an ancient group of
shelled rhizopods found in all oceans, with a few in fresh and brackish water.
Most foraminiferans live on the ocean floor in incredible numbers, having perhaps
the largest biomass of any animal group of earth. Their tests are of numerous types
(Figures 8-13 and 8-14). Most tests are many chambered and are made of calcium

A   B

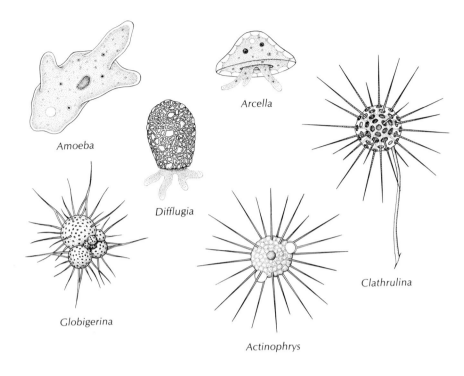

Amoeba

Arcella

Difflugia

Globigerina

Actinophrys

Clathrulina

Figure 8-14

Diversity among the Sarcodina. *Difflugia*, *Arcella*, and *Amoeba* belong to the rhizopod class Lobosea and have lobopodia. The foraminiferan *Globigerina* belongs to the class Granuloreticulosea and shows reticulopodia. *Actinophrys* and *Clathrulina* are actinopod heliozoeans. They have axopodia.

carbonate, although silica, silt, and other foreign materials are sometimes used. Slender pseudopodia extend through openings in the test, then branch and run together to form a protoplasmic net (**reticulopodia**) in which they ensnare their prey. Here the captured prey is digested, and the digested products are carried into the interior by the flowing protoplasm. The life cycles of foraminiferans are complex, for they have multiple division and alternation of haploid and diploid generations (intermediary meiosis).

Some of the slime molds (class Eumycetozoa), especially *Dictyostelium discoideum,* have been studied intensively because of their fascinating developmental cycle. Under natural conditions this species lives in forest detritus throughout the world. It feeds on bacteria and reproduces by binary fission as long as the food supply is plentiful. When food runs short, however, the amebas are attracted to each other, streaming toward a central point to form a **pseudoplasmodium** (large mass of discrete cells). Under the same conditions, some species actually fuse to become a large multinucleate individual (**plasmodium**). The pseudoplasmodium of *Dictyostelium* may migrate some distance to a favorable location, where it forms a stalk with a fruiting body on top. Within the fruiting body are formed resistant cysts, which are widely dispersed upon rupture of the fruiting body. Many details about the development, genetics, and biochemistry of these organisms are known.

Superclass Actinopoda

The Actinopoda is composed of the mostly freshwater class Heliozoea and the three marine classes Acantharea, Phaeodarea, and Polycystinea. Members of the marine classes are commonly known as **radiolarians.** All have axopodia, and except for some of the heliozoeans, they have tests (Figure 8-15). These protozoa are beautiful little animals.

The biological characteristics of the freshwater Heliozoea are somewhat better known than those of the other actinopods. Examples are *Actinosphaerium,* which is about 1 mm in diameter and can be seen with the unassisted eye, and *Actinophrys* (Figure 8-14), only 50 μm in diameter; neither has a test. *Clathrulina* (Figure 8-14) secretes a latticed test.

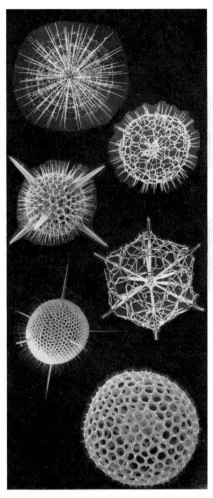

Figure 8-15

Types of radiolarian tests (class Polycystinea). In his study of these beautiful forms collected on the famous *Challenger* expedition, Haeckel worked out our present concepts of symmetry.

Photograph courtesy J. and M. Cachon. From Lee, J.J., S.H. Hutner, and E.C. Bovee. (eds.) 1985. An illustrated guide to the protozoa. Lawrence, Kansas, Society of Protozoologists. Allen Press.

The oldest known animals are found among the radiolarians. Radiolarians are nearly all pelagic (live in open water). Most of them are planktonic in shallow water, although some live in deep water. Their highly specialized skeletons are intricate in form and of great beauty (Figure 8-15). The body is divided by a central capsule that separates inner and outer zones of cytoplasm. The central capsule, which may be spherical, ovoid, or branched, is perforated to allow cytoplasmic continuity. The skeleton is made of silica, strontium sulfate, or a combination of silica and organic matter and usually has a radial arrangement of spines that extend through the capsule from the center of the body. At the surface a shell may be fused with the spines. Around the capsule is a frothy mass of cytoplasm from which axopodia arise (p. 138). These are sticky to catch the prey, which are carried by the streaming protoplasm to the central capsule to be digested. The ectoplasm on one side of the axial rod moves outward, or toward the tip, while on the other side it moves inward, or toward the test.

Radiolarians may have one or many nuclei. Their life history is not completely known, but binary fission, budding, and sporulation have been observed in them.

Role of Sarcodina in building earth deposits

The foraminiferans and the radiolarians have existed since Precambrian times and have left excellent fossil records. In many instances their hard shells have been preserved unaltered. Many of the extinct species closely resemble present-day ones. They were especially abundant during the Cretaceous and Tertiary periods. Some of them were among the largest protozoa that have ever existed, measuring up to 100 mm or more in diameter.

For untold millions of years the tests of dead foraminiferans have been sinking to the bottom of the ocean, building up a characteristic ooze rich in lime and silica. Much of this ooze is made up of the shells of the genus *Globigerina* (Figure 8-13, *B*). About one third of the sea bottom is covered with *Globigerina* ooze. This ooze is especially abundant in the Atlantic Ocean.

The radiolarians (Figures 8-14 and 8-15), with their less soluble siliceous shells, are usually found at greater depths (15,000 to 20,000 feet), mainly in the Pacific and Indian oceans. Radiolarian ooze probably covers about 2 to 3 million square miles. Under certain conditions, radiolarian ooze forms rocks (chert). Many fossil radiolarians are found in the Tertiary rocks of California.

The thickness of these deep-sea sediments has been estimated at 700 to 4000 m. Although the average rate of sedimentation must vary greatly, it is always very slow. *Globigerina* ooze has probably increased 1 to 12.5 mm in 1000 years. As many as 50,000 shells of foraminiferans may be found in a single gram of sediment, which gives some idea of the magnitude of numbers of these microorganisms and the length of time it has taken them to form the sediment carpet on the ocean floor.

Of equal interest and of greater practical importance are the limestone and chalk deposits that were laid down by the accumulation of these microorganisms when sea covered what is now land. Later, through a rise in the ocean floor and other geological changes, this sedimentary rock emerged as dry land. The chalk deposits of many areas of England, including the White Cliffs of Dover, were formed in this way. The great pyramids of Egypt were made from stone quarried from limestone beds that were formed by a very large foraminiferan population that flourished during the early Tertiary period.

Since fossil foraminiferans and radiolarians can be brought up in well drillings, their identification is often important to oil geologists for correlation of rock strata.

Phylum Apicomplexa

All apicomplexans are endoparasites, and their hosts are found in many animal phyla. The presence of a certain combination of organelles, the **apical complex,** distinguishes this subphylum (Figure 8-16, *A*). The apical complex is usually present only in certain developmental stages of the organisms; for example, **merozoites** and **sporozoites** (Figure 8-17). Some of the structures, especially the **rhoptries** and **micronemes,** apparently aid in penetrating the host's cells or tissues.

Locomotor organelles are less obvious in this group than in other protozoa. Pseudopodia occur in some intracellular stages, and gametes of some species are flagellated. Tiny contractile fibrils can form waves of contraction across the body surfaces to propel the organism through a liquid medium.

The life cycle usually includes both asexual and sexual reproduction, and there is sometimes an invertebrate intermediate host. At some point in the life cycle, the organisms develop a **spore (oocyst),** which is infective for the next host and is often protected by a resistant coat.

Class Sporozoea

The most important class of the phylum Apicomplexa, the Sporozoea, contains three subclasses: the Gregarinia, the Coccidia, and the Piroplasmia. The gregarines are common parasites of invertebrates, but they are of little economic significance. Piroplasms are of some veterinary importance; for example, *Babesia bigemina* causes Texas red-water fever in cattle.

Subclass Coccidia

The Coccidia are intracellular parasites in invertebrates and vertebrates, and the group includes species of very great medical and veterinary importance.

Eimeria. The name "coccidiosis" is generally applied only to infections with *Eimeria* or *Isospora.* Humans are occasionally infected with species of *Isospora,* but apparently there is little disease. Some species of *Eimeria* may cause serious disease in some domestic animals. The symptom is usually severe diarrhea or dysentery.

E. tenella is often fatal to young fowl, producing severe pathogenesis in the intestine. The organisms undergo schizogony (p. 141) in the intestinal cells, finally producing gametes. After fertilization the zygote forms an oocyst that passes out of the host in the feces (Figure 8-16, *B*). Sporogony occurs within the oocyst outside the host, producing eight sporozoites in each oocyst. Infection occurs when a new host accidentally ingests a sporulated oocyst and the sporozoites are released by digestive enzymes.

Toxoplasma. A similar life cycle occurs in *Toxoplasma gondii,* a parasite of cats, but this species produces extraintestinal stages as well. The extraintestinal stages can develop in a wide variety of animals other than cats—for example, rodents, cattle, and humans. Gametes and oocysts are not produced by the extraintestinal forms, but they can initiate the intestinal cycle in a cat that eats infected prey. In humans *Toxoplasma* causes little or no ill effects except in a woman infected during pregnancy, particularly in the first trimester. Such infection greatly increases the chances of a birth defect in the baby; it is now believed that 2% of all mental retardation in the United States is the result of congenital toxoplasmosis. The normal route of infection for humans is apparently the consumption of infected meat that is insufficiently cooked.

Plasmodium: the malarial organism. The best known of the coccidians is *Plasmodium,* the causative organism of the most important infectious disease of

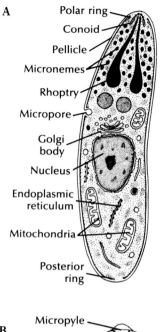

A

Polar ring
Conoid
Pellicle
Micronemes
Rhoptry
Micropore
Golgi body
Nucleus
Endoplasmic reticulum
Mitochondria
Posterior ring

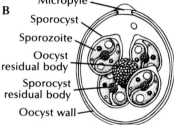

B

Micropyle
Sporocyst
Sporozoite
Oocyst residual body
Sporocyst residual body
Oocyst wall

Figure 8-16

A, Diagram of an apicomplexan sporozoite or merozoite at the electron microscope level, illustrating the apical complex. The polar ring, conoid, micronemes, rhoptries, subpellicular microtubules, and micropore (cytostome) are all considered components of the apical complex. **B,** Infective oocyst of *Eimeria.* The oocyst is the resistant stage and has undergone multiple fission after zygote formation (sporogony).

Another important source of infection with *Toxoplasma* is domestic cats. The oocysts are passed in the cat's feces, and a pregnant woman should not empty the litter box. If such a chore cannot be avoided, daily clean-up should be acceptable because it takes 3 days for the oocysts to sporulate and become infective.

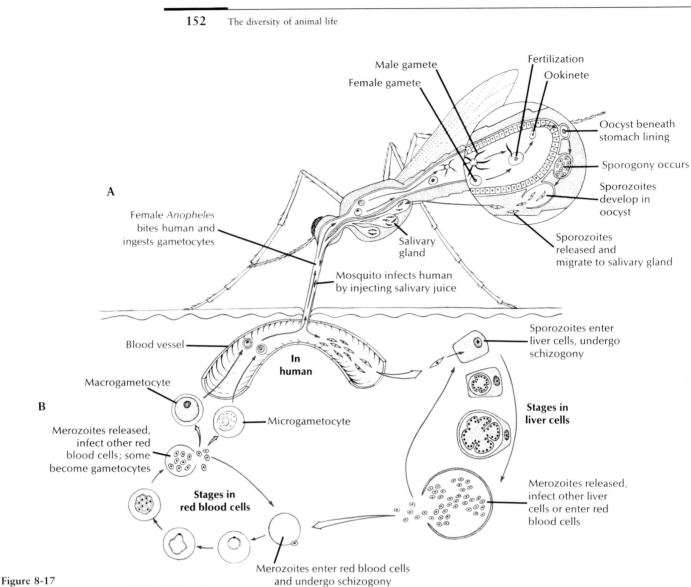

A, Female *Anopheles* bites human and ingests gametocytes

Male gamete
Female gamete
Fertilization
Ookinete
Oocyst beneath stomach lining
Sporogony occurs
Sporozoites develop in oocyst
Sporozoites released and migrate to salivary gland

Salivary gland

Mosquito infects human by injecting salivary juice

Blood vessel

In human

Sporozoites enter liver cells, undergo schizogony

Macrogametocyte

Microgametocyte

Stages in liver cells

Merozoites released, infect other red blood cells; some become gametocytes

B,

Merozoites released, infect other liver cells or enter red blood cells

Stages in red blood cells

Merozoites enter red blood cells and undergo schizogony

Figure 8-17

Life cycle of *Plasmodium vivax*, one of the protozoa (class Sporozoea) that causes malaria in humans. **A,** Sexual cycle produces sporozoites in body of mosquito. Meiosis occurs just after zygote formation (zygotic meiosis). **B,** Sporozoites infect a human and reproduce asexually, first in liver cells and then in red blood cells. Malaria is spread by *Anopheles* mosquito, which sucks up gametocytes along with human blood, then, when biting another victim, leaves sporozoites in new wound.

humans: **malaria.** Malaria is a very serious disease, difficult to control and widespread, particularly in tropical and subtropical countries. Four species of *Plasmodium* infect humans. Although each produces its own peculiar clinical picture, all four have similar cycles of development in their hosts (Figure 8-17).

The parasite is carried by mosquitoes *(Anopheles),* and sporozoites are injected into the human with the insect's saliva during its bite. The sporozoites penetrate liver cells and initiate schizogony. The products of this division then enter other liver cells to repeat the schizogonous cycle, or in *P. falciparum* they penetrate the red blood cells after only one cycle in the liver. The period when the parasites are in the liver is the **incubation period,** and it lasts from 6 to 15 days, depending on the species of *Plasmodium.*

Merozoites released as a result of the liver schizogony enter red blood cells, where they begin a series of schizogonous cycles. When they enter the cells, they become ameboid **trophozoites,** feeding on hemoglobin. The end product of the parasite's digestion of hemoglobin is a dark, insoluble pigment: **hemozoin.** The hemozoin accumulates in the host cell, is released when the next generation of merozoites is produced, and eventually accumulates in the liver, spleen, or other organs. The trophozoite within the cell grows and undergoes schizogony, producing six to 36 merozoites, depending on the species, which burst forth to infect new red cells. When the red blood cell containing the merozoites bursts, it releases the parasite's metabolic products, which have accumulated there. The release of these foreign substances into the patient's circulation causes the chills and fever characteristic of malaria.

Since the populations of schizonts maturing in the red blood cells are synchronized to some degree, the episodes of chills and fever have a periodicity characteristic of the particular species of *Plasmodium*. In *P. vivax* (benign tertian) malaria, the episodes occur every 48 hours; in *P. malariae* (quartan) malaria, every 72 hours; in *P. ovale* malaria, every 48 hours; and in *P. falciparum* (malignant tertian) malaria, about every 48 hours, although the synchrony is less well defined in this species. People usually recover from infections with the first three species, but mortality may be high in untreated cases of *P. falciparum* infection. Unfortunately, *P. falciparum* is the most common species, accounting for 50% of all malaria in the world.

After some cycles of schizogony in the red blood cells, infection of new cells by some of the merozoites results in the production of **microgametocytes** and **macrogametocytes** rather than another generation of merozoites. When the gametocytes are ingested by a mosquito feeding on the patient's blood, they mature into **gametes,** and fertilization occurs. The zygote becomes a motile **ookinete,** which penetrates the stomach wall of the mosquito and becomes the **oocyst.** Within the oocyst sporogony occurs, and thousands of **sporozoites** are produced. The oocyst ruptures, and the sporozoites migrate to the salivary glands, from which they are transferred to a human by the bite of the mosquito. Development in the mosquito requires 7 to 18 days but may be longer in cool weather.

The elimination of mosquitoes and their breeding places by insecticides, drainage, and other methods has been effective in controlling malaria in some areas. However, the difficulties in carrying out such activities in remote areas and the acquisition of resistance to insecticides by mosquitoes and to antimalarial drugs by *Plasmodium* (especially *P. falciparum*) mean that malaria will be a serious disease of humans for a long time to come.

Other species of *Plasmodium* parasitize birds, reptiles, and mammals. Those of birds are transmitted chiefly by the *Culex* mosquito.

Phylum Ciliophora

The ciliates are a large and interesting group, with a great variety of forms living in all types of freshwater and marine habitats. They are the most structurally complex and diversely specialized of all the protozoa. The majority are free living, but some are commensal or parasitic. They are usually solitary and motile, but some are sessile and some colonial. There is great diversity of shape and size. In general, they are larger than most other protozoa, but they range from very small (10 to 12 μm) up to 3 mm long. All have cilia that beat in a coordinated rhythmical manner, although the arrangement of the cilia may vary and some lack cilia as adults.

Ciliates are always multinucleate, possessing at least one **macronucleus** and one **micronucleus,** but varying from one to many of either type. The macronuclei are apparently responsible for metabolic and developmental functions and for maintaining all the visible traits, such as the pellicular apparatus. Macronuclei vary in shape among the different species (Figures 8-18 and 8-21). The micronuclei participate in sexual reproduction and give rise to macronuclei after exchange of micronuclear material between individuals. The micronuclei divide mitotically, and the macronuclei divide amitotically (see p. 141).

The **pellicle** of ciliates may consist only of the cell membrane or in some species may form a thickened armor. The cilia are short and usually arranged in

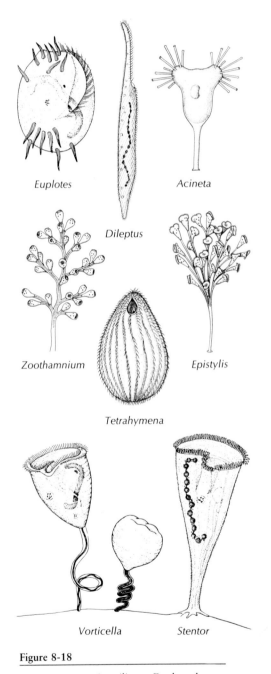

Euplotes

Acineta

Dileptus

Zoothamnium

Epistylis

Tetrahymena

Vorticella   Stentor

Figure 8-18

Some representative ciliates. *Euplotes* have stiff cirri used for crawling about. Contractile fibrils in ectoplasm of *Stentor* and in stalks of *Vorticella* allow great expansion and contraction. Note the macronuclei, long and curved in *Euplotes* and *Vorticella*, shaped like a string of beads in *Stentor*.

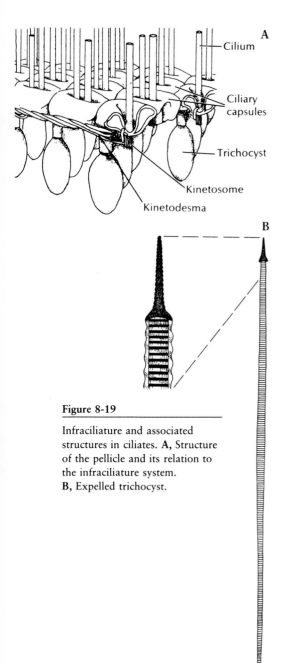

Figure 8-19

Infraciliature and associated structures in ciliates. **A,** Structure of the pellicle and its relation to the infraciliature system. **B,** Expelled trichocyst.

longitudinal or diagonal rows. Cilia may cover the surface of the animal or may be restricted to the oral region or to certain bands. In some forms the cilia are fused into a sheet called an **undulating membrane** or into smaller **membranelles,** both used to propel food into the **cytopharynx** (gullet). In other forms there may be fused cilia forming stiffened tufts called **cirri,** often used in locomotion by the creeping ciliates (Figure 8-18).

An apparently structural system of fibers, in addition to the kinetosomes, makes up the **infraciliature,** just beneath the pellicle (Figure 8-19). Each cilium terminates beneath the pellicle in its kinetosome, and from each kinetosome a **kinetodesmal fibril** arises and passes along beneath the row of cilia, joining with the other fibrils of that row. The kinetosomes and fibrils (**kinetodesmata**) of that row make up what is known as a **kinety** (Figure 8-19). All ciliates seem to have kinety systems, even those that lack cilia at some stage. The infraciliature apparently does not coordinate the ciliary beat, as formerly thought. Coordination of the ciliary movement seems to be by waves of depolarization of the cell membrane moving down the animal, similar to the phenomenon in a nerve impulse (p. 682).

Most ciliates are holozoic. Most of them possess a cytostome (mouth) that in some forms is a simple opening and in others is connected to a gullet or ciliated groove. The mouth in some is strengthened with stiff, rodlike trichites for swallowing larger prey; in others, such as the paramecia, ciliary water currents carry microscopic food particles toward the mouth. *Didinium* has a proboscis for engulfing the paramecia on which it feeds (Figure 8-6). Suctorians paralyze their prey and then ingest the contents through tubelike tentacles by a complex feeding mechanism that apparently combines phagocytosis with a sliding filament action of microtubules in the tentacles (Figure 8-6).

Some ciliates have curious small bodies in their ectoplasm between the bases of the cilia. Examples are **trichocysts** (Figures 8-19 and 8-21) and **toxicysts.** Upon mechanical or chemical stimulation, these bodies explosively expel a long, threadlike structure. The mechanism of expulsion is unknown. The function of trichocysts is thought to be defensive, although this is unclear. When a paramecium is attacked by a *Didinium,* it expels its trichocysts but to no avail. Toxicysts, however, release a poison that paralyzes the prey of carnivorous ciliates. Toxicysts are structurally quite distinct from trichocysts. Many dinoflagellates have structures very similar to trichocysts.

Among the more striking and familiar of the ciliates are *Stentor* (Gr. herald with a loud voice), trumpet shaped and solitary, with a bead-shaped macronucleus (Figure 8-18); *Vorticella* (L. dim. of *vortex,* a whirlpool), bell shaped and attached by a contractile stalk (Figure 8-18); *Euplotes* (Gr. *eu,* true, good, + *ploter,* swimmer) and *Stylonychia,* with flattened bodies and groups of fused cilia (cirri) that function as legs (Figure 8-18); *Blepharisma,* slender and pink; and *Spirostomum,* very long and wormlike in appearance.

Paramecium: a representative ciliate

Paramecia are usually abundant in ponds or sluggish streams containing aquatic plants and decaying organic matter.

Form and function

The paramecium is often described as slipper shaped. *Paramecium caudatum* is 150 to 300 μm in length and is blunt anteriorly and somewhat pointed posteriorly (Figure 8-20). The animal has an asymmetrical appearance because of the **oral groove,** a depression that runs obliquely backward on the ventral side.

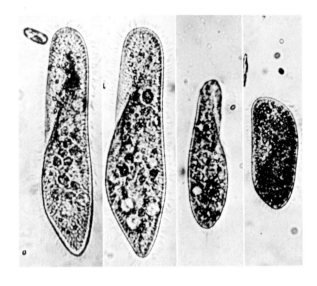

Figure 8-20

Comparison of four common species of *Paramecium* photographed at the same magnification. *Left to right: P. multimicronucleatum, P. caudatum, P. aurelia,* and *P. bursaria.*
Courtesy Carolina Biological Supply House Co., Burlington, N.C.

The **pellicle** is a clear, elastic membrane divided into hexagonal areas by tiny elevated ridges (Figure 8-19), and its entire surface is covered with cilia arranged in lengthwise rows. Just below the pellicle is the thin clear **ectoplasm** that surrounds the larger mass of granular **endoplasm** (Figure 8-21). Embedded in the ectoplasm just below the surface are the spindle-shaped **trichocysts,** which alternate with the bases of the cilia. The infraciliature can be seen only with special fixing and staining methods.

The **cytostome** at the end of the oral groove leads into a tubular **cytopharynx,** or **gullet.** Along the gullet an undulating membrane of modified cilia keeps food moving. Fecal material is discharged through a **cytoproct** posterior to the oral groove (Figure 8-21). Within the endoplasm are food vacuoles containing

Figure 8-21

Left, enlarged section of a contractile vacuole (water expulsion vesicle) of *Paramecium.* Water is believed to be collected by endoplasmic reticulum, emptied into feeder canals and then into the vesicle. The vesicle contracts to empty its contents to the outside, thus serving as an osmoregulatory organelle. *Right, Paramecium,* showing gullet, food vacuoles, and nuclei.

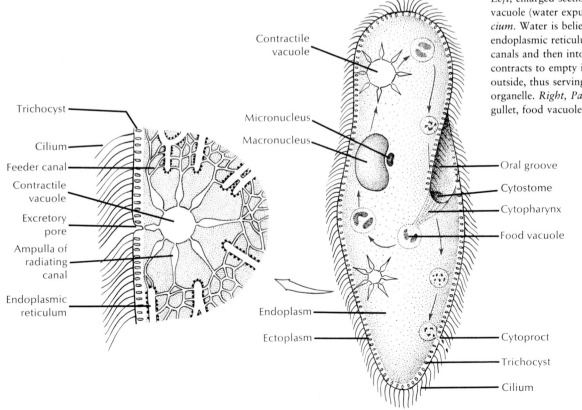

Locomotor responses, by which an animal more or less continuously orients itself with respect to a stimulus, are called taxes (sing, taxis). Movement toward the stimulus is a positive taxis; movement away is a negative taxis. Some examples are thermotaxis, response to heat; phototaxis, response to light; thigmotaxis, response to contact; chemotaxis, response to chemical substances; rheotaxis, response to currents of air or water; galvanotaxis, response to constant electric current; and geotaxis, response to gravity. Some stimuli do not cause an orienting response but simply a change in movement: more rapid movement, more frequent random turning, or slowing or cessation of movement. Such responses are known as kineses. Is the avoiding reaction of a paramecium a taxis or a kinesis?

food in various stages of digestion. There are two **contractile vacuoles**, each consisting of a central space surrounded by several **radiating canals** that collect fluid and empty it into the central vacuole. Excretion and osmoregulation were described on p. 139.

P. caudatum has two nuclei: a large kidney-shaped **macronucleus** and a smaller **micronucleus** fitted into the depression of the former. These can usually be seen only in stained specimens. The number of micronuclei varies in different species. *P. multimicronucleatum* may have as many as seven.

Paramecia are holozoic, living on bacteria, algae, and other small organisms. The cilia in the oral groove sweep food particles in the water into the cytostome, from which point they are carried into the cytopharynx by the undulating membrane. From the cytopharynx the food is collected into a food vacuole that is constricted into the endoplasm. The food vacuoles circulate in a definite course through the protoplasm (cyclosis) while the food is being digested by enzymes from the endoplasm. The indigestible part of the food is ejected through the cytoproct.

The body of the paramecium is elastic, allowing it to bend and squeeze its way through narrow places. Its cilia can beat either forward or backward, so that the animal can swim in either direction. The cilia beat obliquely, causing the animal to rotate on its long axis. In the oral groove the cilia are longer and beat more vigorously than the others so that the anterior end swerves aborally. As a result of these factors, the animal moves forward in a spiral path (Figure 8-22, *A*).

When a ciliate, such as a paramecium, comes in contact with a barrier or a disturbing chemical stimulus, it reverses its cilia, backs up a short distance, and swerves the anterior end as it pivots on its posterior end. This is called an **avoiding reaction** (Figure 8-22, *B*). It may continue to change its direction to keep itself some distance away from the noxious stimulus, and it may react in a similar fashion to keep itself within the zone of an attractant. The paramecium may also change its swimming speed. How does the paramecium "know" when to change directions or swimming speed? Interestingly, the reactions of the organism depend on the effects of the stimulus on the electrical potential difference across its cell membrane (see p. 682). The paramecia slightly hyperpolarize in attractants and depolarize in repellents that depend on the avoiding reaction, and they more strongly hyperpolarize in repellents and depolarize in attractants that depend on changes in swimming speed.

Reproduction

Paramecia reproduce only by binary fission across kineties but have certain forms of sexual phenomena called conjugation and autogamy.

In **binary fission** the micronucleus divides mitotically into two daughter

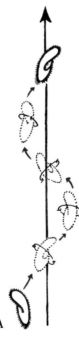

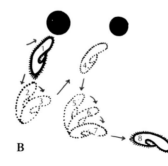

Figure 8-22

A, Spiral path of swimming *Paramecium.*
B, Avoidance reaction of *Paramecium.*

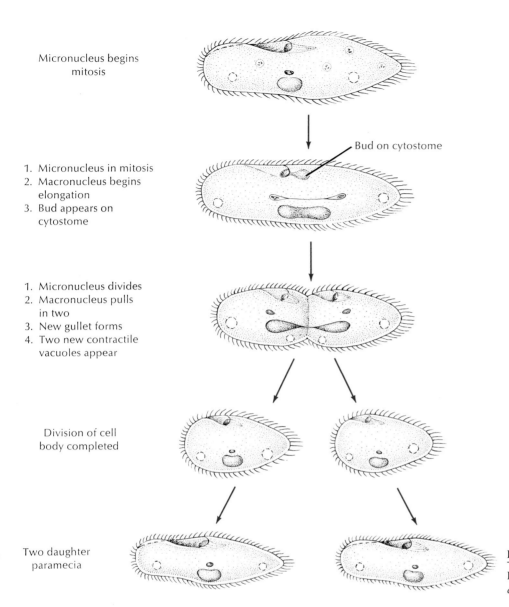

Micronucleus begins
mitosis

Bud on cytostome

1. Micronucleus in mitosis
2. Macronucleus begins
   elongation
3. Bud appears on
   cytostome

1. Micronucleus divides
2. Macronucleus pulls
   in two
3. New gullet forms
4. Two new contractile
   vacuoles appear

Division of cell
body completed

Two daughter
paramecia

Figure 8-23

Binary fission in a ciliophoran *(Paramecium)*. Division is across rows of cilia.

micronuclei, which move to opposite ends of the cell (Figure 8-23). The macronucleus elongates and divides amitotically.

Conjugation occurs at intervals in ciliates. Conjugation is the temporary union of two individuals to exchange chromosomal material (Figure 8-24). During the union the macronucleus disintegrates and the micronucleus of each individual undergoes meiosis (p. 762), giving rise to four haploid micronuclei, three of which degenerate (Figure 8-24, *A* to *D*). The remaining micronucleus then divides into two haploid pronuclei, one of which crosses over into the conjugant partner. When the exchanged pronucleus unites with the pronucleus of the partner, the diploid number of chromosomes is restored (Figure 8-24, *E* and *F*).

The two paramecia next separate, and in each paramecium the fused micronucleus, which is comparable to a zygote in higher forms, divides by mitosis into two, four, and eight micronuclei. Four of these micronuclei enlarge and become macronuclei, and three of the other four micronuclei disappear (Figure 8-24, *G* and *J*). Then the paramecium itself divides twice, resulting in four paramecia, each with one micronucleus and one macronucleus. Following this complicated pro-

Figure 8-24

Sexual reproduction, or conjugation, in *Paramecium caudatum*. **A,** Similar individuals come in contact on oral surface; micronuclei prepare to divide. **B** and **C,** Micronuclei divide twice (meiosis), resulting in four haploid nuclei in each partner. **D,** Three nuclei degenerate; remaining one divides to form "male" and "female" pronuclei. **E,** Male pronuclei are exchanged between conjugants. **F,** In each conjugant, male and female pronuclei fuse to form synkaryon pronucleus (diploid); conjugants now separate. **G** to **I,** In each exconjugant the synkaryon divides three times to form eight micronuclei; the old macronucleus is gradually resorbed. **J,** Four of the micronuclei become macronuclei; three are resorbed; one prepares to divide as the cytosome divides. **K** and **L,** Micronucleus divides twice as the cytosome divides twice; each of the four resulting daughter cells receives a micronucleus and one of the four macronuclei.

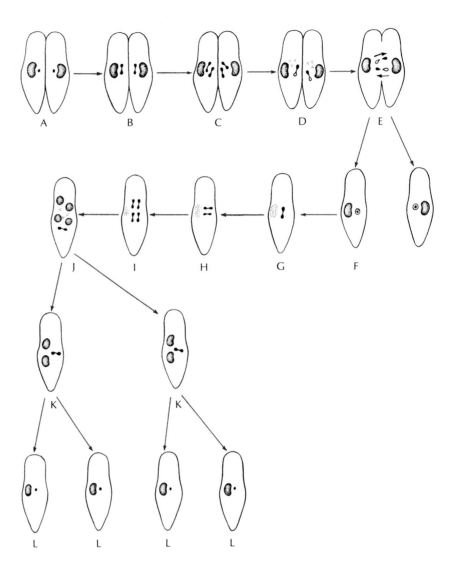

Within each species of *Paramecium* the individuals exhibit morphological and physiological differences. Since these differences are usually more minor and more superficial than those that distinguish species, the groups within a species are referred to as strains, biotypes, or varieties. Most species of protozoa can be divided into a number of these groups.

cess, the animals may continue to reproduce by binary fission without the necessity of conjugation.

The result of conjugation is similar to that of zygote formation, for each exconjugant contains hereditary material from two individuals. The advantage of sexual reproduction is that it permits gene recombinations, thus increasing genetic variation in the population. Although ciliates in clone cultures can apparently reproduce repeatedly and indefinitely without conjugation, the stock seems eventually to lose vigor. Conjugation restores vitality to a stock. Seasonal changes or a deteriorating environment will usually stimulate sexual reproduction.

In 1937 it was discovered that not every paramecium would conjugate with any other paramecium of the same species. T.M. Sonneborn (1957) found that there were physiological differences between individuals that set them off into **mating types.** Ordinarily conjugation will not occur between individuals of the same mating type but only with an individual of another (complimentary) mating type. It was also found that within a single species there are a number of varieties, each of which has mating types that conjugate among themselves but not with the mating types of other varieties. In *Paramecium aurelia,* for instance, each of six varieties has two mating types; conjugation, however, will occur only between members of opposite or complementary mating types within their own variety. With few exceptions, each variety has only two interbreeding mating types. There is no morphological basis for distinguishing mating types within a variety; such differences that exist must be physiological. Some varieties, however, can be distinguished from each other morphologically.

Mating types are also found in other species of paramecia and in other ciliates.

Autogamy is a process of self-fertilization that is similar to conjugation except that there is no exchange of nuclei. After the disintegration of the macronucleus and the meiotic divisions of the micronucleus, the two haploid pronuclei fuse to form a synkaryon that is homozygous (Chapter 37, p. 803), rather than heterozygous as in the case of exconjugants.

Symbiotic ciliates

Many symbiotic ciliates live as commensals, but some can be harmful to their hosts. *Balantidium coli* lives in the large intestine of humans, pigs, rats, and many other mammals (Figure 8-25). There seem to be host-specific strains, and the organism is not easily transmitted from one species to another. Transmission is by fecal contamination of food or water. Usually the organisms are not pathogenic, but in humans they sometimes invade the intestinal lining and cause a dysentery similar to that caused by *Entamoeba histolytica*. The disease can be serious and even fatal.

Other species of ciliates live in other hosts. *Entodinium* (Figure 8-25) belongs to a group that has very complex structure and lives in the digestive tract of ruminants, where they may be very abundant. *Nyctotherus* live in the colon of frogs and toads. In aquarium and wild freshwater fish, *Ichthyophthirius* causes a disease known to many fish culturists as "ick."

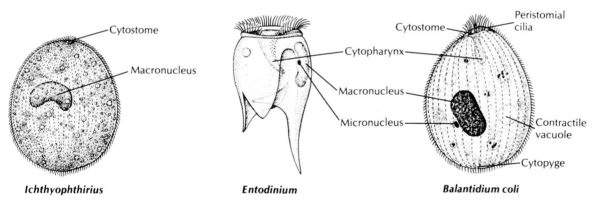

Ichthyophthirius   **Entodinium**   **Balantidium coli**

Figure 8-25

Some symbiotic ciliates. *Balantidium coli* is a parasite of humans and other mammals. This ciliate is common in pigs, in which it does little damage. In humans it may produce intestinal ulcers and severe chronic dysentery. Infections are common in parts of Europe, Asia, and Africa but are rare in the United States. *Ichthyophthirius* causes a common disease in aquarium and wild freshwater fish, known as "ick" to many fish culturists. Untreated, it can cause much loss of exotic fish. *Entodinium* is a complex ciliate found in the rumen of cows and sheep.

Suctorians

Suctorians are ciliates in which the young possess cilia and are free swimming, and the adults grow a stalk for attachment, become sessile, and lose their cilia. They have no cytostome but feed by long, slender, tubelike tentacles. The suctorian captures living prey, usually a ciliate, by the tip of one or more tentacles and paralyzes it. The cytoplasm of the prey then flows through the attached tentacles, forming food vacuoles in the feeding suctorian (Figure 8-6).

One of the best places to find freshwater suctorians is in the algae that grow on the carapace of turtles. Common genera of suctorians found there are *Anarma* (without stalk or test) and *Squalorophrya* (with stalk and test). Other freshwater representatives are *Podophrya* (Figure 8-6) and *Dendrosoma*. *Acinetopsis* and *Ephelota* are saltwater forms.

Suctorian parasites include *Trichophrya*, the species of which are found on a variety of invertebrates and freshwater fish; *Allantosoma*, which live in the intestine of certain mammals; and *Sphaerophrya*, which are found in *Stentor*.

___ Phylogeny and Adaptive Radiation

Phylogeny

Protozoa are usually placed at the base of the phylogenetic tree of animals. Doubtless, multicellular animals were derived from a protozoan or a protozoan-like ancestor, perhaps more than once. Sponges, for example, may well have been derived separately from other metazoa. Colonial protozoa, particularly among flagellates (Figure 8-10), show various degrees of cell aggregation and some differentiation that suggest the body plans of early metazoa.

The mechanism of ciliary coordination and kineses shown by ciliates, depending on establishment of a membrane potential, polarization, and depolarization, suggests the antecedent among the protozoa of the mechanism of neural transmission, coordination, and cell-to-cell communication among the metazoa.

With the exception of certain shell-bearing Sarcodina, such as foraminiferans and radiolarians, protozoa have left no fossil records. Mastigophorans may be the oldest of all protozoa, perhaps having arisen from bacteria and spirochetes, but the group is probably polyphyletic. Some of the phytoflagellates are more closely related to the green algae than they are to zooflagellates. Some colorless phytoflagellates have chlorophyll-bearing relatives, and some autotrophic forms are facultatively saprophytic in darkness. Hence, the common origin of animals and plants may lie in the Phytomastigophorea. That the amebas were derived from the flagellates is indicated by the pseudopodia of some flagellates and the flagellated states of some amebas. However, the different classes of Sarcodina may have arisen independently from different kinds of flagellates.

Sporozoea, which are all specialized parasites, probably came from flagellated ancestors: they often have ameboid feeding stages and flagellated gametes. Aggregation and fruiting body formation in the Eumycetozoa suggest affinities to the fungi, as is indicated by their common name, slime molds, and some scientists do consider them fungi. The origin of the ciliates is somewhat obscure, but the basic structural similarity of the flagellum to the cilium is undeniable.

Adaptive radiation

Some of the wide range of adaptations of protozoa have been described in the preceding pages. The Sarcodina range from bottom-dwelling, naked species to planktonic forms such as the foraminiferans and radiolarians with beautiful, intricate tests. There are many symbiotic species of amebas. Flagellates likewise show adaptations for a similarly wide range of habitats, with the added variation of photosynthetic ability in many species of Phytomastigophorea. The fine line between plants and animals at this level is shown by our ability to turn a "plant" into an "animal" by experimentally destroying the chloroplasts of *Euglena*.

Within a single-cell body plan, the division of labor and specialization of organelles are carried furthest in the ciliates. These have become the most complex of all protozoa. Specializations for intracellular parasitism have been adopted by the Sporozoea, Microspora, and Myxozoa.

Ciliophrys marina is a small marine heliozoan (subphylum Sarcodina, superclass Actinopoda) that can change *rapidly* and *reversibly* from a rapidly swimming flagellate with no axopodia to a more typical heliozoan with axopodia and slowly beating flagellum. Davidson (1982) suggested that the structure of *C. marina* shows derivation from the chrysomonad phytoflagellates.

SUMMARY

The assemblage of animals known as protozoa is a large, heterogeneous group now recognized as being composed of seven phyla. The largest and most important of the phyla are the Sarcomastigophora (flagellates and amebas), the Apicomplexa (coccidians, malaria-causing organisms, and others), and the Ciliophora (ciliates). They demonstrate the great adaptive potential of the basic body plan, the single eukaryotic cell. They occupy a vast array of niches and habitats, and many species have complex and specialized organelles.

All protozoa have one or more nuclei, and these often appear vesicular under the light microscope. Macronuclei of ciliates are compact. Endosomes are often present in the nuclei. Many protozoa have organelles similar to those found in metazoan cells.

Pseudopodial or ameboid movement is a locomotory and food-gathering mechanism in protozoa and plays a vital role as a defense mechanism in metazoa. It is accomplished by microfilaments moving past each other, and it requires expenditure of energy from ATP. Ciliary movement is likewise important in both protozoa and metazoa. Currently, the most widely accepted mechanism to account for ciliary movement is the sliding microtubule hypothesis.

Various protozoa feed by holophytic, holozoic, or saprozoic means. The excess water that enters their bodies is expelled by contractile vacuoles (water-expulsion vesicles). Respiration and waste elimination are through the body surface. Protozoa can reproduce asexually by binary fission, multiple fission, and budding; sexual processes are common. Cyst formation to withstand adverse environmental conditions is an important adaptation in many protozoa.

Most phytoflagellates are photosynthetic, and many zooflagellates are important parasites. They move by beating one or more flagella. Sarcodines move by pseudopodia; many are important members of planktonic communities, and some are parasites. Many have a test, or shell. All apicomplexans are parasitic, and they include *Plasmodium,* which causes malaria. The Ciliophora move by means of cilia or ciliary organelles. They are a large and diverse group, and many are complex in structure.

Review questions
1. Explain why a protozoan may be very complex, even though it is composed of only one cell.
2. What are eight characteristics of protozoa?
3. Distinguish among the following protozoan phyla: Sarcomastigophora, Apicomplexa, Ciliophora.
4. Distinguish vesicular and compact nuclei.
5. Explain the transitions of endoplasm and ectoplasm in ameboid movement.
6. Distinguish lobopodia, filipodia, reticulopodia, and axopodia.
7. Contrast the structure of an axoneme of a cilium with that of a kinetosome.
8. What is the sliding-microtubule hypothesis?
9. Explain how protozoa eat, digest their food, osmoregulate, and respire.
10. Distinguish the following: binary fission, budding, multiple fission, and sexual and asexual reproduction.
11. Distinguish gametic meiosis, zygotic meiosis, and intermediary meiosis.
12. What is the survival value of encystment?
13. Contrast and give an example of phytoflagellates and zooflagellates.
14. Name three kinds of sarcodines, and tell where they are found (their habitats).
15. Outline the general life cycle of malaria organisms.

16. What is the public health importance of *Toxoplasma,* and how do humans become infected with it?
17. Define the following with references to ciliates: macronucleus, micronucleus, pellicle, undulating membrane, cirri, infraciliature, trichocysts, conjugation.
18. Outline the steps in conjugation of ciliates.
19. What are indications that the Sarcodina, Apicomplexa, and Ciliophora may have been derived from ancestral Phytomastigophorea?

Selected references

See also general references for Part Two, p. 590.

Bonner, J.T. 1983. Chemical signals of social amoebae. Sci. Am. **248:**114-120 (April). *Social amebas of two different species secrete different chemical compounds that act as aggregation signals. The evolution of the aggregation signals in these protozoa may provide clues to the origin of the diverse chemical signals (neurotransmitters and hormones) in more complex organisms.*

Davidson, L.A. 1982. Ultrastructure, behavior, and algal flagellate affinities of the helioflagellate *Ciliophrys marina,* and the classification of the helioflagellates (Protista, Actinopoda, Heliozoea). J. Protozool. **29:**19-29. *Hypothesizes derivation of this heliozoan from chrysomonad phytoflagellates.*

Dodson, E.O. 1971. The kingdom of organisms. Syst. Zool. **20:**265-281.

Farmer, J.N. 1980. The protozoa: introduction to protozoology. St. Louis, The C.V. Mosby Co.

Harrison, G. 1978. Mosquitoes, malaria and man: a history of the hostilities since 1880. New York, E.P. Dutton. *A fascinating story, well told.*

Lee, J.J., S.H. Hutner, and E.C. Bovee (eds.). 1985. An illustrated guide to the protozoa. Lawrence, Kansas, Society of Protozoologists. *A comprehensive guide and essential reference for students of the protozoa.*

Levine, N.D., and others. 1980. A newly revised classification of the protozoa. J. Protozool. **27:**37-58. *This classification is a revision of that published in 1964 by the Committee on Systematics and Evolution of the Society of Protozoologists. It is the most authoritative classification currently available.*

T H E   L O W E S T   M E T A Z O A

Phylum Mesozoa

Phylum Placozoa

Phylum Porifera: Sponges

Position in Animal Kingdom

The multicellular animals, or metazoa, are typically divided into three branches: (1) Mesozoa (a single phylum), (2) Parazoa (phylum Porifera, the sponges; and phylum Placozoa), and (3) Eumetazoa (all other phyla).

Although Mesozoa and Parazoa are multicellular, neither fits into the general plan of organization of the other phyla. Such cellular layers as they possess are not homologous to the germ layers of the Eumetazoa, and neither group has developmental patterns in line with the other metazoa. The poriferans are considered to be aberrant, that is, deviating widely from standard patterns. This is the reason for the name Parazoa, which means the "beside-animals." Placozoans may be the most primitive free-living animals. They show relationships to the Porifera.

Biological Contributions

1. Although the simplest in organization of all the metazoa, these groups do compose a higher level of morphological and physiological integration than that found in protozoan colonies. The Mesozoa and Parazoa may be said to belong to a **cellular level of organization.**

2. The mesozoans, although composed simply of an outer layer of somatic cells and an inner layer of reproductive cells, nevertheless have a very complex reproductive cycle somewhat suggestive of that of the trematodes (flukes). Mesozoans are entirely parasitic.

3. Placozoans are essentially composed of two epithelia with fluid and some fibrous cells between them.

4. The poriferans are more complex, with several types of cells differentiated for various functions, some of which are organized into **incipient tissues** of a low level of integration.

5. The developmental patterns of the phyla are different from those of other phyla, and their embryonic layers are not homologous to the germ layers of other phyla.

6. The sponges have developed a unique system of **water currents** on which they depend for food and oxygen.

ORIGIN OF METAZOA

Unraveling the origin of the multicellular animals (metazoans) has presented many problems for zoologists. It is generally believed that they evolved from unicellular organisms, but there is much disagreement over which group of unicellular organisms gave rise to them, and how. The three prevalent hypotheses in current use are (1) that the metazoans arose from a syncytial (multinucleate) ciliate in which cell boundaries later evolved, (2) that they arose from a colonial flagellate in which the cells gradually became more specialized and interdependent, and (3) that the origin of metazoans was polyphyletic, or derived from more than one group of unicellular organisms.

Proponents of the **syncytial ciliate hypothesis** believe that metazoans arose from primitive single-celled ciliates, which at first were multinucleated (having more than one nucleus) but later, by acquiring cell membranes around the nuclei, became compartmentalized into the multicellular condition. The ancestral ciliates, like many modern ones, tended toward bilateral symmetry, so that the earliest metazoans were bilateral and similar to the present primitive flatworms. There are several objections to this hypothesis. It ignores the embryology of the flatworms in which nothing similar to cellularization occurs; it does not explain the presence of flagellated sperm in the metazoans; and, perhaps more important, it infers that the radially symmetrical cnidarians are derived from the flatworms, thus making bilateral symmetry more primitive than radial symmetry.

The **colonial flagellate hypothesis**—first proposed by Haeckel in 1874—is the classical scheme, which, with various revisions, still has many followers.

Figure 9-1

Two methods of reproduction by mesozoans. **A,** Asexual development of vermiform larvae from reproductive cells in the axial cell of the adult. **B,** Under crowded conditions in the host kidney, reproductive cells develop into gonads with gametes that produce infusoriform dispersal larvae that are shed in the host urine.
Modified from Lapan, E.A., and H. Morowitz. 1972. Sci. Am. **227:**94-101 (Dec).

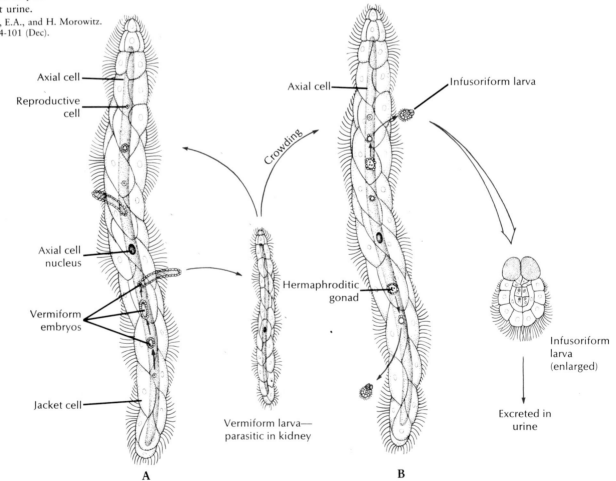

Axial cell

Reproductive cell

Axial cell nucleus

Vermiform embryos

Jacket cell

Axial cell

Crowding

Hermaphroditic gonad

Vermiform larva—parasitic in kidney

Infusoriform larva

Infusoriform larva (enlarged)

Excreted in urine

A   **B**

According to this hypothesis, the metazoans were derived from a hollow, spherical, colonial flagellate (protozoa). The individual cells within the colony became differentiated for specific functional roles (reproductive cells, nerve cells, somatic cells, and so on), thus subordinating cellular independence to the welfare of the colony as a whole. The colonial ancestral form was at first radially symmetrical, similar perhaps to the free-swimming planula larvae of the cnidarians (jellyfishes and others). This larva is radially symmetrical and has no mouth. The cnidarians with their radial symmetry could have evolved from this form.

Bilateral symmetry could have evolved later when some of these planula-like ancestors became adapted for a creeping form of locomotion on the ocean floor. Dorsal and ventral surfaces would have differentiated, a ventral mouth would have appeared, and a start would have been made toward cephalization (a concentration of neurons and sensory structures at the anterior). This would have led to bilateral symmetry and the evolution of the primitive flatworms, which later gave rise to all other bilateral forms.

Some zoologists prefer the idea that the metazoans had a **polyphyletic origin** and suggest that the sponges, cnidarians, ctenophores, and flatworms each evolved independently—the sponges and cnidarians probably evolving from colonial flagellates, and the ctenophores and flatworms from ciliates. This is of course a compromise between the other two hypotheses.

In any case, by whatever route, the primitive flatworms (acoels) are believed to be the most primitive of the bilateral animals.

PHYLUM MESOZOA

The name Mesozoa (mes-o-zo'a) (Gr. *mesos*, in the middle, + *zōon*, animal) was coined by an early investigator (van Beneden, 1876) who believed the group to be a "missing link" between protozoa and metazoa. These minute, ciliated, wormlike animals represent an extremely simple level of organization. All mesozoans live as parasites in marine invertebrates, and the majority of them are only 0.5 to 7 mm in length. Most are made up of only 20 to 30 cells arranged basically in two layers. The layers are not homologous to the germ layers of higher metazoans.

There are two classes of mesozoans, the Rhombozoa and the Orthonectida, but they differ so much from each other that some authorities believe they should be placed in separate phyla.

The rhombozoans (Gr. *rhombos*, a spinning top, + *zōon*, animal) live in the kidneys of benthic cephalopods (bottom-dwelling octopuses, cuttlefishes, and squids). The adults, called **vermiforms** (or nematogens), are long and slender (Figure 9-1). Their inner, reproductive cells give rise to vermiform larvae that grow and then reproduce. When the population becomes crowded, the reproductive cells of some adults develop into gonadlike structures producing male and female gametes. The zygotes grow into minute (0.04 mm) ciliated infusoriform larvae (Figure 9-1, *B*), quite unlike the parent. These are shed with the urine into the seawater. The next part of the life cycle is unknown because the infusoriform larvae are not immediately infective to a new host.

Orthonectids (Gr. *orthos*, straight, + *nektōs*, swimming) (Figure 9-2) parasitize a variety of invertebrates, such as brittle stars, bivalve molluscs, polychetes, and nemerteans. The life cycles involve sexual and asexual phases, and the asexual stage is quite different from that of the dicyemids. It consists of a multinucleated mass called a **plasmodium,** which by division ultimately gives rise to males and females.

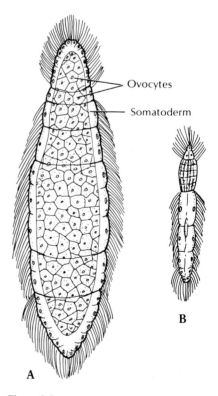

Ovocytes

Somatoderm

A   **B**

Figure 9-2

A, Female and, **B,** male orthonectid (*Rhopalura*). This mesozoan parasitizes such forms as flatworms, molluscs, annelids, and brittle stars. The structure consists of a single layer of ciliated epithelial cells surrounding an inner mass of sex cells.

Phylogeny of Mesozoans

There is still much to learn about these mysterious little parasites, but probably one of the most intriguing questions is the place of mesozoans in the evolutionary picture. Some investigators believe them to be primitive or degenerate flatworms and even believe that they should be classed with the Platyhelminthes. Others place them close to protozoa as truly primitive forms, possibly related to the ciliates. Whether metazoans and mesozoans derived independently from protozoan beginnings or whether mesozoans are indeed degenerate flatworms is still an enigma.

PHYLUM PLACOZOA

The phylum Placozoa (Gr. *plax, plakos,* tablet, plate, + *zōon,* animal) was proposed in 1971 by K.G. Grell to contain a single species, *Trichoplax adhaerens* (Figure 9-3, *A*), a tiny (2 to 3 mm) marine form that had been considered to be either a mesozoan or a cnidarian larva by various workers in the past. The body is platelike and has no symmetry, no organs, and no muscular or nervous system. It is composed of a dorsal epithelium or flagellated cover cells and shiny spheres, a thick ventral epithelium containing flagellated cells and nonflagellated gland cells, and a space between the epithelia containing fluid and fibrous cells (Figure 9-3, *B*). The organisms glide over their food, secrete digestive enzymes upon it, and then absorb the products. Grell considers *Trichoplax* diploblastic (see p. 104), with the dorsal epithelium representing ectoderm and the ventral epithelium representing endoderm because of its nutritive function. The phylogenetic position of placozoans is uncertain, although they are certainly very primitive. They seem to be closest to the Porifera.

Figure 9-3

A, *Trichoplax adhaerens* is a marine, platelike animal only 2 to 3 mm in diameter. The only member of the phylum Placozoa, it is one of the most primitive multicellular animals known. **B,** Section through *Trichoplax adhaerens,* showing histological structure.

Redrawn from Grell, K.G. 1972. Z. Morph. Tiere 73:297-314.

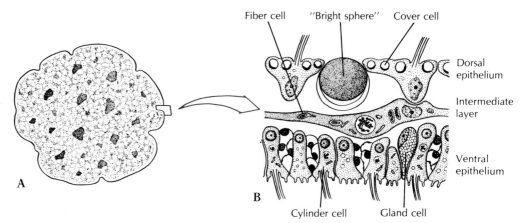

PHYLUM PORIFERA: SPONGES

Sponges belong to phylum Porifera (po-rif'-er-a) L. *porus,* pore, + *fera,* bearing). The bodies of sponges bear myriads of tiny pores and canals that comprise a filter-feeding system adequate for their inactive life-style, for they are sessile animals. They depend on the water currents carried through their unique canal systems to bring them food and oxygen and to carry away their body wastes. Their bodies are little more than masses of cells embedded in a gelatinous matrix and stiffened by a skeleton of minute **spicules** of calcium or silica or by "spongy" fibers of a collagenous substance called **spongin.** They have no organs or true tissues, and even their cells show a certain degree of independence. As sessile animals with only negligible body movement, they have not evolved a nervous system or sense organs and have only the simplest of contractile elements.

So, although they are multicellular, sponges share few of the characteristics of other metazoan phyla. They seem to be outside the line of evolution leading from the protozoa to the other metazoa: a dead-end branch. It is for this reason that they are often called the Parazoa (Gr. *para*, beside or alongside of, + *zōon*, animal).

Most sponges are colonial, and they vary in size from a few millimeters to the great loggerhead sponges, which may reach 2 m or more across. Many sponge species are brightly colored because of pigments in the dermal cells. Red, yellow, orange, green, and purple sponges are not uncommon. However, the color fades quickly when the sponges are removed from the water. Some sponges, including the simplest and most primitive, are radially symmetrical, but many are quite irregular in shape. Some stand erect, some are branched or lobed, and others are low, even encrusting, in form. Some bore holes into shells or rocks.

Most of the 5000 or more sponge species are marine, although some 150 species live in fresh water. Marine sponges are abundant in all seas and at all depths, and a few even exist in brackish water. Although the embryos are free swimming, the adults are always attached, usually to rocks, shells, corals, or other submerged objects. Some benthic forms even grow on sand or mud buttoms. Their growth patterns often depend on the shape of the substratum, the direction and speed of the water currents, and the availability of space, so that the same species may differ markedly in appearance under different environmental circumstances. Sponges in calm waters may grow taller and straighter than those in rapidly moving waters.

Many animals (crabs, nudibranchs, mites, bryozoans, and fish) live as commensals or parasites in or on sponges. The larger sponges particularly tend to harbor a large variety of invertebrate commensals. On the other hand, sponges grow on many other living animals, such as molluscs, barnacles, brachiopods, corals, or hydroids. Some crabs attach pieces of sponge to their carapaces for camouflage and for protection, since most predators seem to find sponges distasteful. Some reef fishes, however, are known to graze on shallow-water sponges.

The sponges are an ancient group, with an abundant fossil record extending back to the Cambrian period and even, according to some claims, the Precambrian.

Certainly one reason for the success of sponges as a group is that they have few enemies. Because of a sponge's elaborate skeletal framework and often noxious odor, most potential predators find sampling a sponge about as pleasant as eating a mouthful of glass splinters embedded in fibrous gelatin.

CHARACTERISTICS

1. Multicellular; body a loose aggregation of cells of mesenchymal origin
2. Body with pores (ostia), canals, and chambers that serve for passage of water
3. Mostly marine; all aquatic
4. Radial symmetry or none
5. Epidermis of flat pinacocytes; most interior surfaces lined with flagellated collar cells (choanocytes) that create water currents; a gelatinous protein matrix called mesohyl (mesoglea) contains amebocytes of various types and skeletal elements
6. Skeletal structure of fibrillar collagen (a protein) and calcareous or siliceous crystalline spicules, often combined with variously modified collagen (spongin)
7. No organs or true tissues; digestion intracellular; excretion and respiration by diffusion
8. Reactions to stimuli apparently local and independent; nervous system probably absent
9. All adults sessile and attached to substratum
10. Asexual reproduction by buds or gemmules and sexual reproduction by eggs and sperm; free-swimming ciliated larvae

CLASSIFICATION

Class Calcarea (cal-ca're-a) (L. *calcis*, lime, + Gr. *spongos*, sponge) (**Calcispongiae**). Have spicules of calcium carbonate that often form a fringe around the osculum (main water outlet); spicules needle shaped or three or four rayed; all three typles of canal systems (asconoid, syconoid, leuconoid) represented; all marine. Examples: *Sycon, Leucosolenia.*

Class Hexactinellida (hex-ak-tin-el'i-da) (Gr. *hyalos*, glass, + *spongos*, sponge) (**Hyalospongiae**). Have six-rayed, siliceous spicules extending at right angles from a central point; spicules often united to form network; body often cylindrical or funnel shaped; flagellated chambers in simple syconoid or leuconoid arrangement; habitat mostly deep water; all marine. Examples: Venus' flower basket *(Euplectella), Hyalonema.*

Class Demospongiae (de-mo-spun'je-e) (Gr. *demos*, people, + *spongos*, sponge). Have siliceous spicules that are not six rayed, or spongin, or both; leuconoid-type canal systems; one family found in fresh water; all others marine. Examples: *Thenea, Cliona, Spongilla, Myenia,* and all bath sponges.

Class Sclerospongiae (skler'o-spun'je-e) (Gr. *skleros*, hard, + *spongos*, sponge). Secrete massive basal skeleton of calcium carbonate, with living tissue extending into the skeleton from 1 mm to 3 cm or more, extending above skeleton less than 1 mm; have siliceous spicules similar to Demospongiae (sometimes absent), and spongin fibers; leuconoid organization; inhabit caves, crevices, tunnels, and deep water on coral reefs. Examples: *Astrosclera, Calcifibrospongia, Ceratoporella, Merlia.*

___ Form and Function

The only body openings of these unusual animals are pores, usually many tiny ones called **ostia** for incoming water, and a few large ones called **oscula** (sing., **osculum**) for water outlet. These openings are connected by a system of canals, some of which are lined with peculiar flagellated collar cells called **choanocytes,** whose flagella maintain a current of environmental water through the canals. Water enters the canals through a multitude of tiny incurrent pores (**dermal ostia**) and leaves by way of one or more large oscula. The choanocytes not only keep the water moving but also trap and phagocytize food particles that are carried in the water. The cells lining the passageways are very loosely organized. Collapse of the canals is prevented by the skeleton, which, depending on the species, may be made up of needlelike calcareous or siliceous spicules, a meshwork of organic spongin fibers, or a combination of the two.

Sessile animals make few movements and therefore need little in the way of nervous, sensory, or locomotor parts. Sponges apparently have been sessile from their earliest appearance and have never acquired specialized nervous or sensory structures, and they have only the very simplest of contractile systems.

Types of canal systems

Most sponges fall into one of three types of canal systems: asconoid, syconoid, or leuconoid (Figure 9-4).

Asconoids: flagellated spongocoels

The asconoid sponges have the simplest type of organization. They are small and tube shaped. Water enters through microscopic dermal pores into a large cavity called the **spongocoel,** which is lined with choanocytes. The choanocyte flagella pull the water through the pores and expel it through a single large osculum (Figure 9-4). *Leucosolenia* (Gr. *leukos*, white, + *sōlen*, pipe) is an asconoid type of sponge. Its slender, tubular individuals grow in groups attached by a common stolon, or stem, to objects in shallow seawater (Figure 9-5). Asconoids are found only in class Calcarea.

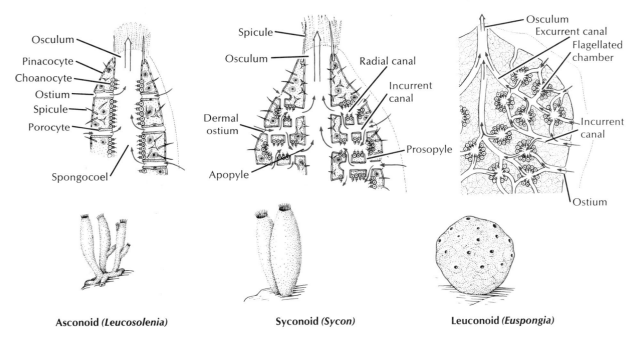

Asconoid *(Leucosolenia)*   Syconoid *(Sycon)*   Leuconoid *(Euspongia)*

Figure 9-4

Three types of sponge structure. The degree of complexity from simple asconoid to complex leuconoid type has involved mainly the water-canal and skeletal systems, accompanied by outfolding and branching of the collar cell layer. The leuconoid type is considered the major plan for sponges, for it permits greater size and more efficient water circulation.

Figure 9-5

A cluster of sponges removed from a dock where it was attached by a slender stalk, perhaps a bit of algae. Around the stalk *(above, encrusting)* is a bit of breadcrumb sponge *Halichondria,* a leuconoid-type sponge. To the right is a mass of branching *Leucosolenia,* an asconoid sponge, and at left are three large and two small *Sycon,* syconoid sponges.

Photograph by D.P. Wilson.

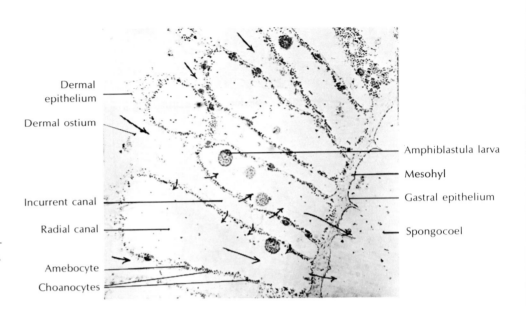

Dermal epithelium

Dermal ostium

Incurrent canal

Radial canal

Amebocyte

Choanocytes

Amphiblastula larva

Mesohyl

Gastral epithelium

Spongocoel

Figure 9-6

Cross section through wall of sponge *Sycon*, showing canal system. Photomicrograph of stained slide.

Micrograph by F.M. Hickman.

Syconoids: flagellated canals

Syconoid sponges look somewhat like larger editions of asconoids, from which they were derived. They have the tubular body and single osculum, but the body wall, which is thicker and more complex than that of asconoids, contains choanocyte-lined radial canals that empty into the spongocoel (Figure 9-4). The spongocoel in syconoids is lined with epithelial-type cells rather than flagellated cells as in asconoids. Water enters through a large number of dermal ostia into **incurrent canals** and then filters through tiny openings called **prosopyles** into the **choanocyte chambers** (Figure 9-6). There food is ingested by the choanocytes, whose flagella force the water on through internal pores (**apopyles**) into the spongocoel. From there it emerges through the osculum. Syconoids do not usually form highly branched colonies as do the asconoids. During development, syconoid sponges pass through an asconoid stage; the flagellated canals form by evagination of the body wall. This is one evidence that syconoid sponges were derived from asconoid ancestral stock. Syconoids are found in classes Calcarea and Hexactinellida. *Sycon* (Gr. *sykon*, a fig) is a commonly studied example of the syconoid type of sponge (Figure 9-5).

Leuconoids: flagellated chambers

Leuconoid organization is the most complex of the sponge types and the best adapted for increase in sponge size. Most leuconoids form large colonial masses, each member of the mass having its own osculum, but individual members are poorly defined and often impossible to distinguish (Figure 9-7). Clusters of flagellated chambers are filled from incurrent canals and discharge water into excurrent canals that eventually lead to the osculum (Figure 9-4). Most sponges are of the leuconoid type, which occurs in most Calcarea and in all other classes.

These three types of canal systems—asconoid, syconoid, and leuconoid—demonstrate an increase in complexity and efficiency of the water pumping system, but they do not imply an evolutionary or development sequence. The leuconoid grade of construction has evolved independently many times in sponges. Possession of the leuconoid plan is of clear adaptive value; it increases the proportion of flagellated surfaces compared with the volume, thus providing more collar cells to meet food demands. It makes possible a much larger body size than asconoid or syconoid grades.

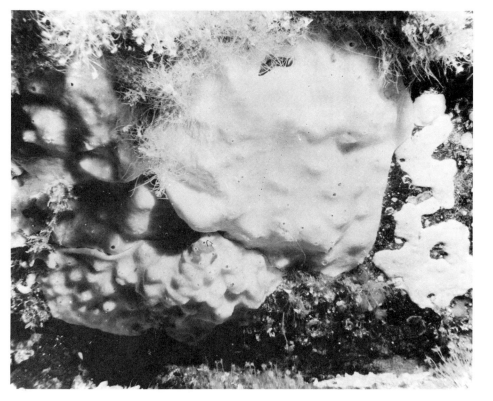

Figure 9-7

Large marine sponge (class Demospongiae) growing on underside of overhanging rock, surrounded by clusters of flowerlike Hydrozoa. Note the many small oscular openings. Photograph by T. Lündalvy.

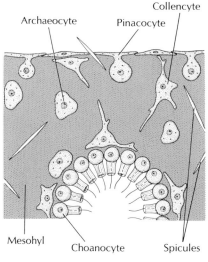

Figure 9-8

Small section through sponge wall, showing four types of sponge cells. Pinacocytes are protective and contractile; choanocytes create water currents and engulf food particles; archaeocytes have a variety of functions; collencytes secrete collagen.

Types of cells

Sponge cells are loosely arranged in a gelatinous matrix called **mesohyl** (mesoglea, mesenchyme) (Figures 9-6 and 9-8). The mesohyl is the "connective tissue" of the sponges; in it are found various ameboid cells, fibrils, and skeletal elements. There are several types of cells in sponges.

Pinacocytes

The nearest approach to a true tissue in sponges is found in the arrangement of the **pinacocyte** cells of the **pinacoderm** (Figure 9-8). These are thin, flat, epithelial-type cells that cover the exterior surface and some interior surfaces. Some are T-shaped, with their cell bodies extending into the mesohyl. Pinacocytes are somewhat contractile and help regulate the surface area of the sponge. Some of the pinacocytes are modified as contractile **myocytes**, which are usually arranged in circular bands around the oscula or pores, where they help regulate the rate of water flow.

Choanocytes

The choanocytes, which line the flagellated canals and chambers, are ovoid cells with one end embedded in the mesohyl and the other exposed. The exposed end bears a flagellum surrounded by a collar (Figures 9-8 and 9-9). The electron microscope shows the collar to be made up of adjacent microvilli, connected to each other by delicate microfibrils, so that the collar forms a fine filtering device for straining food particles from the water (Figure 9-9, *B* and *C*). The beat of the flagellum pulls water through the sievelike collar and forces it out through the open top of the collar. Particles too large to enter the collar become trapped in secreted mucus and slide down the collar to the base where they are phagocytized by the cell body. Larger particles have already been screened out by the small size of the dermal pores and prosopyles. The food engulfed by the cells is passed on to a neighboring archaeocyte for digestion.

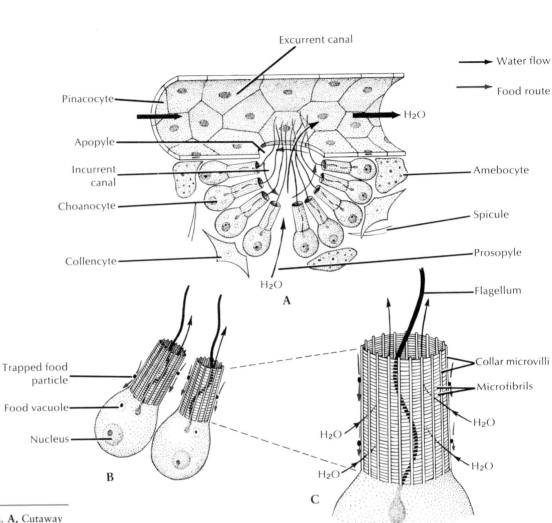

Figure 9-9

Food trapping by sponge cells. **A,** Cutaway section of canals showing cellular structure and direction of water flow. **B,** Two choanocytes and, **C,** structure of the collar. Small red arrows indicate movement of food particles.

Archaeocytes

Archaeocytes are ameboid cells that move about in the mesohyl (Figure 9-8) and carry out a number of functions. They can phagocytose particles at the pinacoderm and receive particles for digestion from the choanocytes. Archaeocytes apparently can differentiate into any of the other types of more specialized cells in the sponge. Some, called **sclerocytes**, secrete spicules. Others, called **spongocytes**, secrete the spongin fibers of the skeleton, and **collencytes** secrete fibrillar collagen. **Lophocytes** secrete large quantities of collagen but are distinguishable morphologically from collencytes.

Types of skeletons

The skeleton gives support to the sponge, preventing collapse of the canals and chambers. The major structural protein in the animal kingdom is collagen, and fibrils of collagen are found throughout the intercellular matrix of all sponges. In addition, various Demospongiae secrete a form of collagen traditionally known as spongin. There are several types of spongin, differing in chemical composition and form (fibers, spicules, filaments, spongin surrounding spicules, and so on) that are found in the various demosponges. Demospongiae also secrete siliceous spicules, as do Sclerospongiae. The siliceous spicules and organic collagen skeleton of the sclerosponges is limited to a thin layer of living tissue above their massive skeleton of calcium carbonate. Calcareous sponges secrete spicules composed mostly of crystalline calcium carbonate that have one, three, or four rays (Figure 9-10). Glass sponges have siliceous spicules with six rays arranged in three planes at right angles to each other. There are many variations in the shape of spicules, and these structural variations are of taxonomic importance.

Sponge physiology

All the life activities of the sponge depend on the current of water flowing through the body. A sponge pumps a remarkable amount of water. *Leuconia* (Gr. *leukos*, white), for example, is a small leuconoid sponge about 10 cm tall and 1 cm in diameter. It is estimated that water enters through some 81,000 incurrent canals at a velocity of 0.1 cm/second. However, *Leuconia* has more than 2 million flagellated chambers whose combined diameter is much greater than that of the canals, so that in the chambers the water slows down to 0.001 cm/second, allowing ample opportunity for food capture by the collar cells. All of the water is expelled through a single osculum at a velocity of 8.5 cm/second: a jet force capable of carrying waste products some distance away from the sponge. Some large sponges have been found to filter 1500 liters of water a day.

Sponges feed primarily on particles suspended in the water pumped through their canal systems. Detritus particles, planktonic organisms, and bacteria are consumed nonselectively in the size range from 50 μm (average diameter of ostia) to 0.1 μm (width of spaces between microvilli of choanocyte collar). Pinacocytes may phagocytose particles at the surface, but most of the larger particles are consumed in the canals by archaeocytes that move close to the lining of the canals. The smallest particles, accounting for about 80% of the particulate organic carbon, are phagocytosed by the choanocytes. Sponges can also absorb dissolved nutrients from the water passing through the system. Protein molecules are taken into the choanocytes by pinocytosis.

Digestion is entirely **intracellular** (occurs within cells), and present evidence indicates that this chore is performed by the archaeocytes. Particles taken in by the choanocytes are passed on to archaeocytes for digestion.

There are no respiratory or excretory organs; both functions are apparently carried out by diffusion in individual cells. Contractile vacuoles have been found in archaeocytes and choanocytes of freshwater sponges.

The only visible activities and responses in sponges, other than the propulsion of water, are slight alterations in shape and the closing and opening of the incurrent and excurrent pores, and these movements are very slow. The most common response is closure of the oscula. Apparently excitation spreads from cell to cell, although some zoologists point to the possibility of coordination by means of substances carried in the water currents, and some have tried, not very successfully, to demonstrate the presence of nerve cells.

Reproduction

Sponges reproduce both asexually and sexually. **Asexual reproduction** occurs by means of bud formation and by regeneration following fragmentation. **External buds,** after reaching a certain size, may become detached from the parent and float away to form new sponges, or they may remain to form colonies. **Internal buds,** or **gemmules,** are formed in freshwater sponges and some marine sponges. Here, archaeocytes are collected together in the mesohyl and become surrounded by a tough spongin coat incorporating siliceous spicules. When the parent animal dies, the gemmules survive and remain dormant, preserving the life of the species dur-

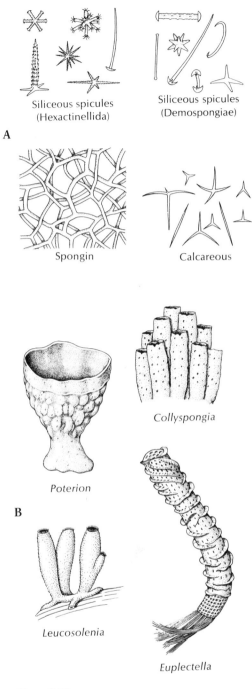

Figure 9-10

A, Types of spicules found in sponges. There are amazing diversity, complexity, and beauty of form among the many types of spicules. **B,** Some sponge body forms.

It is interesting to note that investigators studying *Ephydatia fluviatilis* (Gr. *ephydatios,* of the water), a freshwater sponge, have discovered that there are several strains of the species and that sponges hatched from gemmules of one strain can fuse with others of the same strain, but not with products of gemmules of any other strain. This characteristic is probably the result of strain-specific markers or receptors on the plasma membrane of the cells, which are under genetic control.

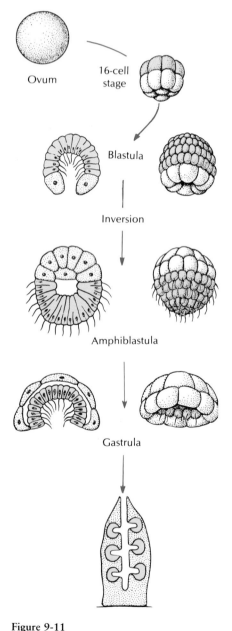

Figure 9-11

Development of the sponge *Sycon.*

ing periods of freezing or severe drought. Later the cells in the gemmules escape through a special opening, the **micropyle,** and develop into new sponges. Gemmulation in freshwater sponges (Spongillidae) is thus an adaptation to the changing seasons. Gemmules are also a means of colonizing new habitats, since they can be spread by streams or by animal carriers. What prevents the gemmules from hatching during the season of formation rather than remaining dormant? Some species secrete a substance that inhibits early germination of the gemmules, and the gemmules do not germinate as long as they are held in the body of the parent. Other species undergo a period of maturation at low temperatures (as in winter) before they germinate. Gemmules in marine sponges also seem to be an adaptation to pass the cold of winter; they are the only form in which *Haliclona loosanoffi* exists during the colder parts of the year in the northern part of its range.

In **sexual reproduction** most sponges are **monoecious** (have both male and female sex cells in one individual, although not at the same time). Sperm arise from transformation of choanocytes. In Calcarea and at least some Demospongiae, oocytes also develop from choanocytes; in other demosponges oocytes apparently are derived from archaeocytes. Most sponges are viviparous, that is, after fertilization the zygote is retained in and derives nourishment from the parent, and a ciliated larva is released. In such sponges, sperm are released into the water by one individual and are taken into the canal system of another. There they are phagocytosed by choanocytes, which transform into carrier cells and carry the sperm through the mesohyl to the oocytes. Other sponges are oviparous, and both oocytes and sperm are shed free into the water. The free-swimming larva of most sponges is a solid-bodied **parenchymella.** The outwardly directed, flagellated cells migrate to the interior after the larva settles and become the choanocytes in the flagellated chambers. The Calcarea and a few Demospongiae have a very strange developmental pattern. A hollow blastula, called an **amphiblastula** (Figure 9-11), develops, with flagellated cells toward the interior. The blastula then turns *inside out* (**inversion**), the flagellated ends of the cells becoming directed to the outside! Its flagellated cells (**micromeres**) are at one end, and the larger, nonflagellated cells (**macromeres**) are at the other. In contrast to other metazoan embryos, the micromeres invaginate into and are overgrown by the macromeres. The flagellated micromeres become the choanocytes, archeocytes, and collencytes of the new sponge, and the nonflagellated cells give rise to pinacorderm and sclerocytes.

Regeneration and somatic embryogenesis

Sponges have a tremendous ability to repair injuries and to restore lost parts, a process called **regeneration.** Regeneration does not imply a reorganization of the entire animal, but only of the wounded portion.

On the other hand, if a sponge is cut into small fragments, or if the cells of a sponge are entirely dissociated and are allowed to fall into small groups, or aggregates, entire new sponges can develop from these fragments or aggregates of cells. This process has been termed **somatic embryogenesis.** Somatic embryogenesis involves a complete reorganization of the structure and functions of the participating cells or bits of tissue. Isolated from the influence of adjoining cells, they can realize their own potential to change in shape or function as they develop into a new organism.

A great deal of experimental work has been done in this field. The process of reorganization appears to differ in sponges of differing complexity. There is still some controversy concerning just what mechanisms lead to the adhesion of the cells and the share that each type of cell plays in the formative process.

___ Class Calcarea (Calcispongiae)

The Calcarea (also called Calcispongiae) are the calcareous sponges, so called because their spicules are composed of calcium carbonate. The spicules are straight (monaxons) or have three or four rays. These sponges tend to be small—10 cm or less in height—and tubular or vase shaped. They may be asconoid, syconoid, or leuconoid in structure. Though many are drab in color, some are bright yellow, red, green, or lavender. *Leucosolenia* and *Sycon* (*Scypha, Grantia* of supply houses) are marine shallow-water forms commonly studied in the laboratory. *Leucosolenia* is a small asconoid sponge that grows in branching colonies, usually arising from a network of horizontal, stolonlike tubes (Figure 9-5). *Sycon* is a solitary sponge that may live singly or form clusters by budding (Figure 9-5). The vase-shaped, typically syconoid animal is 1 to 3 cm long, with a fringe of straight spicules around the osculum that discourages small animals from entering.

Leucosolenia

___ Class Hexactinellida (Hyalospongiae): Glass Sponges

The glass sponges make up the class Hexactinellida (or Hyalospongiae). They are nearly all deep-sea forms that are collected by dredging. Most of them are radially symmetrical, with vase- or funnel-shaped bodies that are usually attached by stalks of root spicules to a substratum (Figure 9-10, *Euplectella*) (N.L. from Gr. *euplektos,* well plaited). They range from 7.5 cm to more than 1.3 m in length. Their distinguishing features are the skeleton of six-rayed siliceous spicules that are commonly bound together into a network forming a glasslike structure and the **trabecular net** of living tissue produced by the fusion of the pseudopodia of archaeocytes. Within the trabecular net are elongated, finger-shaped chambers lined with choanocytes and opening into the spongocoel. The osculum is unusually large and may be covered over by a sievelike plate of silica. There is no pinacoderm or gelatinous mesohyl, and both the external surface and the spongocoel are lined with the trabecular net. The skeleton is rigid, and muscular elements (myocytes) appear to be absent. The general arrangement of the chambers fits glass sponges into both syconoid and leuconoid types. Their structure is adapted to the slow, constant currents of sea bottoms, for the channels and pores of the sponge wall are relatively large and uncomplicated and permit an easy flow of water. Little, however, is known about their physiology, doubtless because of their deep-water habitat.

Euplectella

The latticelike network of spicules found in many glass sponges is of exquisite beauty, such as that of *Euplectella,* or Venus' flower basket (Figure 9-10), a classic example of the Hexactinellida.

___ Class Demospongiae

Class Demospongiae contains 95% of living sponge species, including most of the larger sponges. The spicules are siliceous but are not six rayed, and they may be bound together by spongin or may be absent altogether. All members of the class are leuconoid, and all are marine except one family, the Spongillidae, or freshwater sponges.

Freshwater sponges are widely distributed in well-oxygenated ponds and streams, where they encrust plant stems and old pieces of submerged wood. They may resemble a bit of wrinkled scum, be pitted with pores, and be brownish or greenish in color. Common genera are *Spongilla* (L. *spongia,* from Gr. *spongos,*

A   B   C

Figure 9-12

Marine Demospongiae. **A,** *Dasychalina cyathina* is a vaselike sponge that reaches 30 cm in height. The large opening at the top is not an osculum; rather, the oscula open into the interior cavity. **B,** *Tethya aurantia* is orange and shaped like a flattened sphere. **C,** *Callispongia plicifera* is another vasiform species. Dull purple, it may show a light blue fluorescence. This is a view of a colony from above, looking into the "vase."
A and C, Photographs by L.S. Roberts; B, photograph by D.W. Gotshall.

sponge) and *Myenia.* Freshwater sponges are most common in midsummer, although some are more easily found in the fall. They die and disintegrate in late autumn, leaving the gemmules (already described) to produce the next year's population. They also reproduce sexually. When examined closely, freshwater sponges reveal a thin dermis overlying large subdermal spaces (separated by columns of spicules) with many water channels in the interior. There are usually several oscula, each of which (at least in *Myenia*) is mounted on a small chimneylike tube. Their spiculation also includes a spongin network.

The marine Demospongiae are quite varied and may be quite striking in color and shape (Figure 9-12). Some are encrusting; some are tall and fingerlike; some are low and spreading; some bore into shells; and some are shaped like fans, vases, cushions, or balls (Figure 9-12). Loggerhead sponges may grow several meters in diameter.

The so-called bath sponges *(Spongia, Hippospongia)* belong to the group called horny sponges, which have spongin skeletons and lack siliceous spicules entirely.

___ Class Sclerospongiae

The Sclerospongiae are a small group of sponges that secrete a massive calcareous skeleton and are thus often called coralline sponges. Living tissue extends from 1 mm to 3 cm or more into the skeleton but only 1 mm above it. Their organization is leuconoid, as in the Demospongiae, and siliceous spicules and spongin are present in three of the four orders. Sclerosponges live in cryptic habitats (deeply shaded or completely dark locations) on coral reefs, such as crevices, caves, undersurfaces of overhangs, and deep water. They are believed to be relict representatives of ancient groups with a geological history extending from the Cambrian period. When modern corals and their symbiosis with algal cells (see p. 204) arose in the Mesozoic era, they were very successful and came to dominate the reefs. However, because the corals depend physiologically on the algal cells growing in their tissues, they must have adequate light. Thus, sclerosponges and certain other organisms were able to exploit the unoccupied cryptic habitats.

___ Phylogeny and Adaptive Radiation

Phylogeny

The sponges originated before the Cambrian period. Two groups of calcareous spongelike organisms occupied early Paleozoic reefs. The Devonian period saw the rapid development of many glass sponges. That sponges are related to protozoa is shown by their phagocytic method of nutrition and the resemblance of their flagellated larvae to colonial protozoa. The possibility that sponges arose from choanoflagellates (protozoa that bear collars and flagella) earned support for a time. However, many zoologists object to that hypothesis because sponges do not acquire collars until later in their embryological development. The outer cells of the larvae are flagellated but not collared, and they do not become collar cells until they become internal. Also, collar cells are found in certain corals and echinoderms, so they are not unique to the sponges.

Another hypothesis is that sponges are derived from a hollow, free-swimming colonial flagellate, such as gave rise to the ancestral stocks of other metazoans. Certainly the sponge larvae resemble such flagellate colonies. The curious process of inversion occurs in the colonial phytoflagellate *Volvox*. The development of the water canals and the movement of the flagellated cells to the interior to become choanocytes may have occurred as the sponges began to assume a sessile existence. Whatever the origin, it is obvious that the sponges diverged early from the main line leading to other metazoans. Their phylogenetic remoteness from other metazoans is shown by their low level of organization, the independent nature of their cells, the absence of organs, and their body structure built around a system of water canals. They became a "dead-end" phylum.

Adaptive radiation

The Porifera have been a highly successful group that has branched out into several thousand species and a variety of marine and freshwater habitats. Their diversification centers largely on their unique water-current system and its various degrees of complexity. The proliferation of the flagellated chambers in the leuconoid sponges was more favorable to an increase in body size than that of the asconoid and syconoid sponges because facilities for feeding and gaseous exchange were greatly enlarged.

___ SUMMARY

Members of the phylum Mesozoa are very simply organized animals that are parasitic in the kidneys of cephalopod molluscs (class Rhombozoa) and in several other invertebrate groups (class Orthonectida). They have only two cell layers, but these are not homologous to the germ layers of higher metazoans. They have a complicated life history that is still incompletely known. Whether their simple organization is primitive or derived from a more advanced group is unknown.

The phylum Placozoa has only one member, a small platelike marine organism. It too has only two cell layers, but some workers believe that these layers are homologous to ectoderm and endoderm of higher metazoans. The closest relatives of the placozoans seem to be the sponges.

The sponges (phylum Porifera) are an abundant marine group with some freshwater representatives. They have various specialized cells, but these are not organized into tissues or organs. They depend on the flagellar beat of their choanocytes to circulate water through their bodies for food gathering and respiratory gas exchange. They are supported by secreted skeletons of fibrillar collagen, collagen in the form of large fibers or filaments (spongin), calcareous or siliceous spicules, or a combination of spicules and spongin in most species.

Sponges reproduce asexually by budding, fragmentation, and gemmules (internal buds). Most sponges are monoecious but produce sperm and oocytes at different times. Embryogenesis is unusual, with a migration of flagellated cells at the surface to the interior (parenchymella) or the production of an amphiblastula with inversion and growth of macromeres over micromeres. Sponges have great regenerative abilities.

Calcarea have calcareous spicules of monaxons or are three or four rayed. They are usually small and may be asconoid, syconoid, or leuconoid in structure.

Hexactinellida are mostly deep-sea forms and have a skeleton of six-rayed siliceous spicules bound together in a trabecular net. The arrangement of the flagellated chambers shows both syconoid and leuconoid types.

Demospongiae are the largest and most abundant class of sponges. One family occurs in fresh water. Their spicules are siliceous (but not six rayed) but may be absent in some species. A protein called spongin may be present. All species are leuconoid, and some may reach great size (up to several meters). They show great diversity in size, color, form, and habitat.

Sclerospongiae are a small group of sponges with a massive, basal calcareous skeleton, of which they inhabit the outer few millimeters or centimeters. Siliceous spicules and spongin may be present. They occupy cryptic habitats on coral reefs.

Sponges are an ancient group, remote phylogenetically from other metazoa. They have been successful, and their adaptive radiation is centered on elaboration of the water circulation and filter feeding system.

Review questions

1. Briefly describe and contrast the syncytial ciliate hypothesis, the colonial flagellate hypothesis, and the polyphyletic origin of the metazoa.
2. Describe the body plan of Mesozoa and Placozoa.
3. Give eight characteristics of sponges.
4. Briefly describe asconoid, syconoid, and leuconoid body types in sponges.
5. What sponge body type is most efficient and makes possible the largest body size?
6. Define the following: ostia, osculum, spongocoel, apopyles, prosopyles.
7. Define the following: pinacocytes, choanocytes, archaeocytes, sclerocytes, spongocytes, collencytes.
8. What material is found in the skeleton of all sponges?
9. Describe the skeletons of each of the classes of sponges.
10. Describe how sponges feed, respire, and excrete.
11. What is a gemmule?
12. Describe how gametes are produced and the process of fertilization in most sponges.
13. Contrast embryogenesis in most Demospongiae with that in the Calcarea.
14. What is the largest class of sponges, and what is its body type?
15. What are possible ancestors to sponges? Justify your answer.

Selected references

See also general references for Part Two, p. 590.

Bergquist, P.R. 1978. Sponges. Berkeley, University of California Press. *Excellent monograph on sponge structure, classification, evolution, and general biology.*

Grell, K.G. 1982. Placozoa. In S.P. Parker (ed.). Synopsis and classification of living organisms, vol. 1. New York, McGraw-Hill Book Co. *Synopsis of placozoan characteristics.*

Hartman, W.D. 1982. Porifera. In S.P. Parker (ed.). Synopsis and classification of living organisms, vol. 1. New York, McGraw-Hill Book Co. *Review of sponge classification.*

Long, M.E., and D. Doubelet. 1977. Consider the sponge. Nat. Geogr. **151**(3):392-407. *Beautiful color photographs of sponges.*

Simpson, T.L. 1984. The cell biology of sponges. New York, Springer-Verlag. *A review and synthesis, points out many problems yet to be solved.*

C H A P T E R   1 0

T H E   R A D I A T E   A N I M A L S

Phylum Cnidaria

Phylum Ctenophora

Position in Animal Kingdom

The two phyla Cnidaria and Ctenophora make up the radiate animals, which are characterized by **primary radial** or **biradial symmetry** and which represent the most primitive of the eumetazoans. Radial symmetry, in which the body parts are arranged concentrically around the oral-aboral axis, is particularly suitable for **sessile** or sedentary animals and for free-floating animals. Biradial symmetry is basically a type of radial symmetry in which only two planes through the oral-aboral axis divide the animal into mirror images because of the presence of some part that is single or paired. All other eumetazoans have a primary bilateral symmetry; that is, they are bilateral or were derived from an ancestor that was bilateral.

Neither phylum has advanced generally beyond the **tissue level of organization,** although a few organs occur. In general, the ctenophores have a more complex structure than that of the cnidarians.

Biological Contributions

1. Both phyla have developed two well-defined **germ layers,** ectoderm and endoderm; a third, or mesodermal, layer, which is derived embryologically from the ectoderm, is present in some. The body plan is saclike, and the body wall is composed of two distinct layers, epidermis and gastrodermis, derived from the ectoderm and endoderm, respectively. The gelatinous matrix, mesoglea, between these layers may be structureless, may contain a few cells and fibers, or may be composed largely of mesodermal connective tissue and muscle fibers.

2. An internal body cavity, the **gastrovascular cavity,** is lined by the gastrodermis and has a single opening, the mouth, which also serves as the anus.

3. **Extracellular digestion** occurs in the gastrovascular cavity, and intracellular digestion takes place in the gastrodermal cells.

4. Most radiates have **tentacles,** or extensible projections around the oral end, that aid in food capture.

5. The first true **nerve cells** (protoneurons) occur in the radiates, but the nerves are arranged as a nerve net, with no central nervous system.

6. **Sense organs** appear first in the radiates and include well-developed statocysts (organs of equilibrium) and ocelli (photosensitive organs).

7. Locomotion in the free-moving forms is achieved by either **muscular contractions** (cnidarians) or **ciliary comb plates** (ctenophores). However, both groups are still better adapted to floating or being carried by currents than to strong swimming.

8. **Polymorphism** in the cnidarians has widened their ecological possibilities. In many species the presence of both a polyp (sessile and attached) stage and a medusa (free-swimming) stage permits occupation of a benthic (bottom) and a pelagic (open-water) habitat by the same species.

Polymorphism also widens the possibilities of structural complexity.

9. Some unique features are found in these phyla, such as **nematocysts** (stinging organelles) in cnidarians and **colloblasts** (adhesive organelles) and **ciliary comb plates** in ctenophores.

▬ PHYLUM CNIDARIA

The phylum Cnidaria (ny-dar'e-a) (Gr. *knide*, nettle, + L. *aria* [pl. suffix]; like or connected with) is an interesting group of more than 9000 species. It takes its name from cells called **cnidocytes,** which contain the stinging organelles (**nematocysts**) characteristic of the phylum. Nematocysts are *formed and used* only by cnidarians and by one species of ctenophore. Another name for the phylum, Coelenterata (se-len'te-ra'ta) (Gr. *koilos*, hollow, + *enteron*, gut, + L. *ata* [pl. suffix], characterized by), is used less commonly than formerly, and it sometimes is now employed to refer to both radiate phyla, since its meaning is equally applicable to both.

The cnidarians are generally regarded as being close to the basic stock of the metazoan line. Although their organization has a structural and functional simplicity not found in other metazoans, they are a rather successful phylum, forming a significant proportion of the biomass in some locations. They are widespread in marine habitats, and there are a few in fresh water. Although they are sessile or, at best, fairly slow moving or slow swimming, they are quite efficient predators of organisms that are much swifter and more complex. The phylum includes some of nature's strangest and loveliest creatures: the branching, plantlike hydroids; the flowerlike sea anemones; the jellyfishes; and those architects of the ocean floor, the horny corals (sea whips, sea fans, and others), and all the hard corals whose thousands of years of calcareous house-building have produced great reefs and coral islands (p. 203).

Cnidarians are found most abundantly in shallow marine habitats, especially in warm temperature and tropical regions. There are no terrestrial species. Colonial hydroids are usually found attached to mollusc shells, rocks, wharves, and other animals in shallow coastal water, but some species are found at great depths. Floating and free-swimming medusae are found in open seas and lakes, often far from the shore. Floating colonies such as the Portuguese man-of-war and *Velella* (L. *velum*, veil, + *ellus*, dim. suffix) have floats or sails by which the wind carries them.

Some molluscs and flatworms eat hydroids bearing nematocysts and use these stinging structures for their own defense. Some other animals feed on cnidarians, but cnidarians rarely serve as food for humans.

Cnidarians sometimes live symbiotically with other animals, often as commensals on the shell or other surface of their host. Algae frequently live as mutuals in the tissues of cnidarians, notably in some freshwater hydras and in reef-building corals. The presence of the algae in reef-building corals limits the occurrence of coral reefs to relatively shallow, clear water where there is sufficient light for the photosynthetic requirements of the algae. These kinds of corals are an essential component of coral reefs, and reefs are extremely important habitats in tropical waters. Coral reefs are discussed further later in the chapter.

Although many cnidarians have little economic importance, reef-building corals are an important exception. Fish and other animals associated with reefs provide substantial amounts of food for humans, and reefs are of economic value as tourist attractions. Precious coral is used for jewelry and ornaments, and coral rock serves for building purposes.

Planktonic medusae may be of some importance as food for fish that are of commercial value; the reverse is also true—the young of the fish fall prey to cnidarians.

CHARACTERISTICS

1. Entirely aquatic, some in fresh water but mostly marine
2. **Radial symmetry** or biradial symmetry around a longitudinal axis with **oral** and **aboral** ends; no definite head
3. Two basic types of individuals: **polyps** and **medusae**
4. Exoskeleton or endoskeleton of chitinous, calcareous, or protein components in some
5. Body with two layers, epidermis and gastrodermis, with mesoglea (**diploblastic**); mesoglea with cells and connective tissue (ectomesoderm) in some (**triploblastic**)
6. **Gastrovascular cavity** (often branched or divided with septa) with a single opening that serves as both mouth and anus; extensible tentacles usually encircling the mouth or oral region
7. Special stinging cell organelles called **nematocysts** in either epidermis or gastrodermis or in both; nematocysts abundant on tentacles, where they may form batteries or rings
8. **Nerve net** with symmetrical and asymmetrical synapses; with some sensory organs; diffuse conduction
9. Muscular system (epitheliomuscular type) of an outer layer of longitudinal fibers at base of epidermis and an inner one of circular fibers at base of gastrodermis; modifications of this plan in higher cnidarians, such as separate bundles of independent fibers in the mesoglea
10. Asexual reproduction by budding (in polyps) or sexual reproduction by gametes (in all medusae and some polyps); sexual forms monoecious or dioecious; **planula larva**; holoblastic indeterminate cleavage
11. No excretory or respiratory system
12. No coelomic cavity

CLASSIFICATION

Class Hydrozoa (hy-dro-zo′a) (Gr. *hydra*, water serpent, + *zōon*, animal). Solitary or colonial; asexual polyps and sexual medusae, although one type may be suppressed; hydranths with no mesenteries; medusae (when present) with a velum; both freshwater and marine. Examples: *Hydra, Obelia, Physalia, Tubularia.*

Class Scyphozoa (sy-fo-zo′a) (Gr. *skyphos*, cup, + *zōon*, animal). Solitary; polyp stage reduced or absent; bell-shaped medusae without velum; gelatinous mesoglea much enlarged; margin of bell or umbrella typically with eight notches that are provided with sense organs; all marine. Examples: *Aurelia, Cassiopeia, Rhizostoma.*

Class Cubozoa (ku′bo-zo′a) (Gr. *kybos*, a cube, + *zōon*, animal). Solitary; polyp stage reduced; bell-shaped medusae square in cross section, with tentacle or group of tentacles hanging from a bladelike pedalium at each corner of the umbrella; margin of umbrella entire, without velum but with velarium; all marine. Examples: *Tripedalia, Carybdea, Chironex, Chiropsalmus.*

Class Anthozoa (an-tho-zo′a) (Gr. *anthos*, flower, + *zōon*, animal). All polyps; no medusae; solitary or colonial; enteron subdivided by at least eight mesenteries or septa bearing nematocysts; gonads endodermal; all marine.

  Subclass Zoantharia (zo′an-tha′ri-a) (N.L. from Gr. *zōon*, animal, + *anthos*, flower, + L., *aria*, like or connected with). With simple unbranched tentacles; mesenteries in pairs; sea anemones, hard corals, and others. Examples: *Metridium, Anthopleura, Tealia, Astrangia, Acropora.*

Subclass Ceriantipatharia (se-ri-ant'i-pa-tha'ri-a) (N.L. combination of Ceriantharia and Antipatharia). With simple unbranched tentacles; mesenteries unpaired; tube anemones and black or thorny corals. Examples: *Cerianthus, Antipathes, Stichopathes.*

Subclass Alcyonaria (al'cy-o-na'ri-a) (Gr. *alkonion,* kind of sponge resembling nest of kingfisher [*alkyōn,* kingfisher], + L. *aria,* like or connected with). With eight pinnate tentacles; eight complete, unpaired mesenteries; soft and horny corals. Examples: *Tubipora, Alcyonium, Gorgonia, Plexaura, Renilla.*

____ Form and Function

Polymorphism in cnidarians

One of the most interesting—and sometimes puzzling—aspects of this phylum is the dimorphism and often polymorphism displayed by many of its members. In general, all cnidarian forms fit into one of two morphological types: the **polyp,** or hydroid form, which is adapted to a sedentary or sessile life, and the **medusa,** or jellyfish form, which is adapted for a floating or free-swimming existence (Figure 10-1).

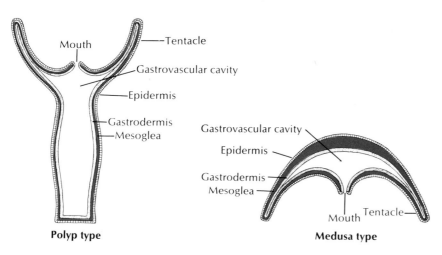

Figure 10-1

Comparison between the polyp and medusa types of individuals.

    Most polyps have tubular bodies with a mouth at one end surrounded by tentacles. The aboral end is usually attached to a substratum by a pedal disc or other device. Polyps may live singly or in colonies. Colonies of some species include more than one kind of individual, each specialized for a certain function, such as feeding, reproduction, or defense (Figure 10-2).

    Medusae are usually free swimming and have bell-shaped or umbrella-shaped bodies and tetramerous symmetry (body parts arranged in fours). The mouth is usually centered on the concave side, and tentacles extend from the rim of the umbrella.

    The sea anemones and corals (class Anthozoa) are all polyps, and the true jellyfishes (class Scyphozoa) are all medusae but may have a polypoid larval stage. The colonial hydroids of class Hydrozoa, however, sometimes have life histories that feature both the polyp, or hydroid, stage and the free-swimming medusa stage—rather like a Jeckyll-and-Hyde existence. A species that has both the attached polyp and the floating medusa within its life history can take advantage of the feeding and distribution possibilities of both the pelagic (open-water) and the benthic (bottom) types of environment.

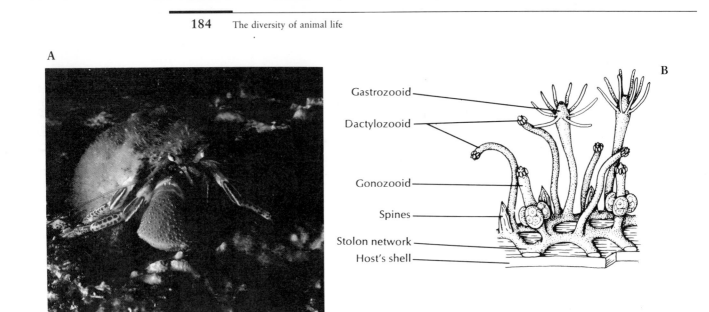

A

B

Gastrozooid

Dactylozooid

Gonozooid

Spines

Stolon network

Host's shell

Figure 10-2

A, A hermit crab with its cnidarian commensals. The shell is blanketed with polyps of the hydrozoan *Hydractinia milleri*. The crab is camouflaged by the cnidarians, and the cnidarians get a free ride and bits of food from their host's meals. **B,** Portion of a colony of *Hydractinia,* showing the types of zooids, and the stolon (hydrorhiza) from which they grow.

A, Photograph by R. Harbo.

Superficially the polyp and medusa seem very different. But actually each has retained the saclike body plan that is basic to the phylum (Figure 10-1). The medusa is essentially an unattached polyp with the tubular portion widened and flattened into the bell shape.

Both the polyp and the medusa possess the three body wall layers typical of the cnidarians, but the jellylike layer of mesoglea is much thicker in the medusa, constituting the bulk of the animal and making it more buoyant. It is because of this mass of mesogleal "jelly" that the medusae are commonly called jellyfishes.

Nematocysts: the stinging organelles

One of the most characteristic structures in the entire cnidarian group is the stinging organelle called the **nematocyst** (Figure 10-3). Over 20 different types of nematocysts (Figure 10-4) have been described in the cnidarians so far; they are important in taxonomic determinations. The nematocyst is a tiny capsule composed of material similar to chitin and containing a coiled tubular "thread" or

Figure 10-3

At left, structure of a stinging cell. At right, portion of the body wall of a hydra. Cnidocytes, which contain the nematocysts, arise in the epidermis from interstitial cells.

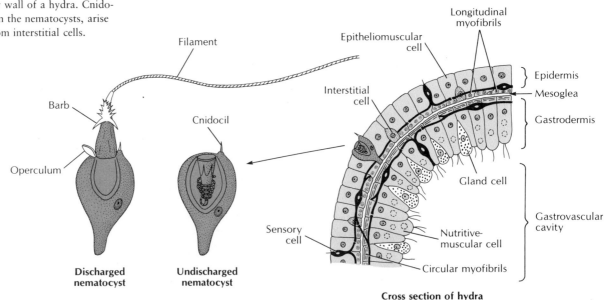

Filament

Barb

Cnidocil

Operculum

Epitheliomuscular cell

Longitudinal myofibrils

Interstitial cell

Epidermis

Mesoglea

Gastrodermis

Gland cell

Sensory cell

Nutritive-muscular cell

Circular myofibrils

Gastrovascular cavity

Discharged nematocyst

Undischarged nematocyst

Cross section of hydra

filament, which is a continuation of the narrowed end of the capsule. This end of the capsule is covered by a little lid, or **operculum.** The inside of the undischarged thread may bear tiny barbs, or spines.

The nematocyst is enclosed in the cell that has secreted it, the **cnidocyte** (during its development, the cnidocyte is properly called the **cnidoblast**). Most cnidocytes are provided with a triggerlike **cnidocil,** which is a modified flagellum with a kinetosome at its base. Contact of the cnidocil with an object such as prey provides tactile stimulation for the nematocyst to discharge. Cnidocytes with their nematocysts are borne in invaginations of ectodermal cells and, in some forms, in gastrodermal cells, and they are especially abundant on the tentacles. When a nematocyst is discharged, its cnidocyte is absorbed and a new one replaces it. Nematocysts used in defense and food capture require chemical stimulation (presence of organic compounds from other animals) to discharge, as well as tactile stimulation. Not all nematocysts have barbs or inject poison. Some, for example, do not penetrate the prey but rapidly recoil like a spring after discharge, grasping and holding any part of the prey caught in the coil (Figure 10-4). Adhesive nematocysts usually do not discharge in food capture.

The mechanism of nematocyst discharge is remarkable. Inside the capsule the osmotic pressure is 140 atmospheres. When stimulated to discharge, the nematocyst membrane becomes permeable to water, and the high internal osmotic pressure causes water to rush into the capsule. The operculum opens, the increase in *hydrostatic pressure* within the capsule pushes the thread out with great force, and the thread turns inside out as it goes. At the everting end of the thread, the barbs flick to the outside like tiny switchblades. This minute but awesome weapon then injects poison when it penetrates the prey.

The nematocysts of most cnidarians are not harmful to humans, but the stings of the Portuguese man-of-war (Figure 10-15) and certain jellyfish are quite painful and in some cases may be dangerous.

Nerve net

The nerve net of the cnidarians is one of the best examples of a diffuse nervous system in the animal kingdom. This plexus of nerve cells is found both at the base of the epidermis and at the base of the gastrodermis, forming two interconnected nerve nets. Nerve processes (axons) end on other nerve cells at synapses or at junctions with sensory cells or effector organs (nematocysts or epitheliomuscular cells). Nerve impulses are transmitted from one cell to another by release of a neurotransmitter from small vesicles on one side of the synapse or junction (p. 684). One-way transmission between nerve cells in higher animals is ensured because the vesicles are located on only one side of the synapse. However, cnidarian nerve nets are peculiar in that many of the synapses have vesicles of neurotransmitters on both sides, allowing transmission across the synapse in either direction. Another peculiarity of cnidarian nerves is the absence of any sheathing material (myelin) on the axons.

There is no concentrated grouping of nerve cells to suggest a "central nervous system." Nerves are grouped, however, in the "ring nerves" of hydrozoan medusae and in the marginal sense organs of scyphozoan medusae. In some cnidarians the nerve nets form two or more systems: in Scyphozoa there is a fast conducting system to coordinate swimming movements and a slower one to coordinate movements of tentacles.

The nerve cells of the net have synapses with slender sensory cells that receive external stimuli, and the nerve cells have junctions with epitheliomuscular

Figure 10-4

Several types of nematocysts shown after discharge. At bottom are two views of a type that does not impale the prey, rather it recoils like a spring, catching any small part of the prey in the path of the recoiling thread.

Note again the distinction between osmotic and hydrostatic pressure (p. 61). The nematocyst is never required actually to contain 140 atmospheres of hydrostatic pressure within itself; such a hydrostatic pressure would doubtless cause it to explode. As the water rushes in during discharge, the osmotic pressure falls rapidly, while the hydrostatic pressure rapidly increases.

Note that there is little adaptive value for a radially symmetrical animal to have a central nervous system with a brain. The environment is approached from all sides equally, and there is no control over the direction of approach to a prey organism.

the prey and inject poison (penetrants, Figure 10-3); those that recoil and entangle the prey (volvents); and those that secrete an adhesive substance used in locomotion and attachment (glutinants).

Sensory cells are scattered among the other epidermal cells, especially near the mouth and tentacles and on the pedal disc. The free end of each sensory cell bears a flagellum, which is the sensory receptor for chemical and tactile stimuli. The other end branches into fine processes, which synapse with the nerve cells.

Nerve cells of the epidermis are generally multipolar (have many processes), although in more highly organized cnidarians the cells may be bipolar (with two processes). Their processes (axons) form synapses with sensory cells and other nerve cells and junctions with epitheliomuscular cells and cnidocytes. There are both one-way (morphologically asymmetrical) and two-way synapses with other nerve cells.

Gastrodermis

The gastrodermis, a layer of cells lining the gastrovascular cavity, is made up chiefly of large, flagellated, columnar epithelial cells with irregular flat bases. The cells of the gastrodermis include nutritive-muscular, interstitial, and gland cells.

Nutritive-muscular cells are usually tall columnar cells and have laterally extended bases containing myofibrils. The myofibrils run at right angles to the body or tentacle axis and so form a circular muscle layer. However, this muscle layer in hydras is very weak, and longitudinal extension of the body and tentacles is brought about mostly by increasing the volume of water in the gastrovascular cavity. The water is brought in through the mouth by the beating of the flagella on the nutritive-muscular cells. Thus, the water in the gastrovascular cavity serves as a **hydrostatic skeleton.** The two flagella on the free end of each cell also serve to circulate food and fluids in the digestive cavity. The cells often contain large numbers of food vacuoles. Gastrodermal cells in the green hydra *(Chlorohydra* [Gr. *chloros,* green, + *hydra,* a mythical nine-headed monster slain by Hercules]) bear green algae (zoochlorella), which give the hydras their color. This is probably a case of symbiotic mutualism, since the algae use the respiratory carbon dioxide from the hydra to form organic compounds useful to the host and secrete oxygen as a by-product of their photosynthesis. They receive shelter and probably other physiological requirements in return.

Interstitial cells are scattered among the bases of the nutritive cells. They may transform into other types of cells when the need arises.

Gland cells in the hypostome and in the column secrete digestive enzymes. Mucous glands about the mouth aid in ingestion.

Nematocysts are not found in the gastrodermis because cnidocytes are lacking in this layer.

Mesoglea

The mesoglea lies between the epidermis and gastrodermis and is attached to both layers. It is gelatinous, or jellylike, and has no fibers or cellular elements. It is a continuous layer that extends over both body and tentacles, thickest in the stalk portion and thinnest in the tentacles. This arrangement allows the pedal region to withstand great mechanical strain and gives the tentacles more flexibility. The mesoglea helps to support the body and acts as a type of elastic skeleton.

Locomotion

Unlike colonial polyps, which are permanently attached, the hydra can move about freely by gliding on its basal disc, aided by mucus secretions. Or using a "inch worm" movement, it can loop along by bending over and attaching its

Over 230 years ago, Abraham Trembley was astonished to discover the isolated sections of the stalk of hydra could regenerate and each become a complete animal. Since then, over 2000 investigations of hydra have been published, and the organism has become a classic model for the study of morphological differentiation. The mechanisms governing morphogenesis have great practical importance, and the simplicity of hydra lends itself to these investigations. Substances controlling development (morphogens), such as those determining which end of a cut stalk will develop a mouth and tentacles, have been discovered, and they may be present in the cells in extremely low concentrations (10^{-10} M).

tentacles to the substratum. It may even turn end over end or detach itself and, by forming a gas bubble on its basal disc, float to the surface.

Feeding and digestion

Hydras feed on a variety of small crustacea, insect larvae, and annelid worms. The hydra awaits its prey with tentacles extended (Figure 10-8). The food organism that brushes against its tentacles may find itself harpooned by scores of nematocysts that render it helpless, even though it may be larger than the hydra. The tentacles move toward the mouth, which slowly widens. Well moistened with mucus secretions, the mouth glides over and around the prey, totally engulfing it.

The activator that actually causes the mouth to open is the reduced form of glutathione, which is found to some extent in all living cells. Glutathione is released from the prey through the wounds made by the nematocysts, but only animals releasing enough of the chemical to activate the feeding response are eaten by the hydra. This explains how a hydra distinguishes between *Daphnia,* which it relishes, and some other forms that it refuses. When glutathione is placed in water containing hydras, each hydra will go through the motions of feeding even though no prey is present.

Inside the gastrovascular cavity, gland cells discharge enzymes on the food. The digestion is started in the gastrovascular cavity (extracellular digestion), but many of the food particles are drawn by pseudopodia into the nutritive-muscular cells of the gastrodermis, where intracellular digestion occurs. Ameboid cells may carry undigested particles to the gastrovascular cavity, where they are eventually expelled with other indigestible matter.

Reproduction

The hydra reproduces sexually and asexually. In asexual reproduction, buds appear as outpocketings of the body wall and develop into young hydras that eventually detach from the parent. Most species are dioecious. Temporary gonads (Figure 10-6) usually appear in the autumn, stimulated by the lower temperatures and perhaps also by the reduced aeration of stagnant waters. Eggs in the ovary usually mature one at a time and are fertilized by sperm shed into the water.

The zygotes undergo holoblastic cleavage to form a hollow blastula. The inner part of the blastula delaminates to form the endoderm (gastrodermis), and the mesoglea is laid down between the ectoderm and endoderm. A cyst forms around the embryo before it breaks loose from the parent, enabling it to survive the winter. Young hydras hatch out in spring when the weather is favorable.

Hydroid colonies

Far more representative of class Hydrozoa than the hydras are hydroids that have a medusa stage in their life cycle. *Obelia* is often used in laboratory exercises for beginning students to illustrate the hydroid type (Figure 10-9).

A typical hydroid has a base, a stalk, and one or more terminal zooids. The base by which the colonial hydroids are attached to the substratum is a rootlike stolon, or **hydrorhiza,** which gives rise to one or more stalks called **hydrocauli.** The living cellular part of the hydrocaulus is the tubular **coenosarc,** composed of the three typical cnidarian layers surrounding the coelenteron (gastrovascular cavity). The protective covering of the hydrocaulus is a nonliving chitinous sheath, or **perisarc.** Attached to the hydrocaulus are the individual polyp animals, or zooids. Most of the zooids are feeding polyps called **hydranths,** or **gastrozooids.** They may be tubular, bottle shaped, or vaselike, but all have a terminal mouth and a circlet

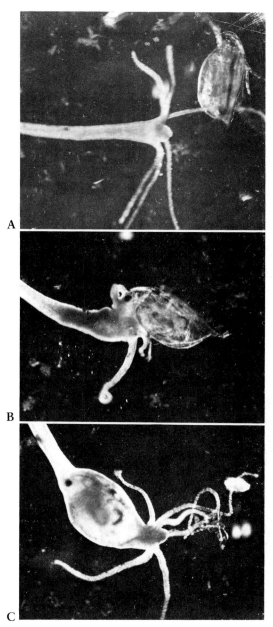

A

B

C

Figure 10-8

A, Hungry hydra catches an unwary water flea with the nematocysts of its tentacles and, **B,** swallows it whole. **C,** Hydra is full, but not too full to capture a protozoan for dessert.

Photographs by F.M. Hickman.

Figure 10-13

Life cycle of *Craspedacusta*, a freshwater hydrozoan. The polyp has three methods of sexual reproduction: by budding off new individuals, which may remain attached to the parent (colony formation); by constricting off nonciliated planula-like larvae (frustules), which can move around and give rise to new polyps; and by producing medusa buds, which develop into sexual jellyfish.

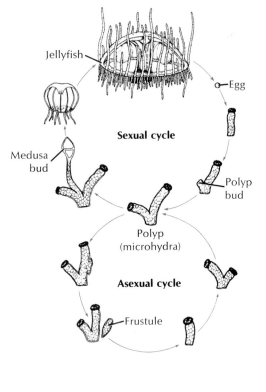

and four radial canals that connect with a ring canal around the margin. This in turn connects with the hollow tentacles. Thus, the coelenteron is continuous from mouth to tentacles, and the entire system is lined with gastrodermis. Nutrition is similar to that of the hydranths.

The nerve net is usually concentrated into two nerve rings at the base of the velum. The bell margin is liberally supplied with sensory cells. It usually also bears two kinds of specialized sense organs: **statocysts,** which are small organs of equilibrium (Figure 10-12, *B*), and **ocelli,** which are light-sensitive organs.

Freshwater medusae

The freshwater medusa *Craspedacusta sowerbyi* (Figure 10-13) (order Hydroida) may have evolved from marine ancestors in the Yangtze River of China. Probably introduced with shipments of aquatic plants, this interesting form has now been found in many parts of Europe, all over the United States, and in parts of Canada. The medusa may attain a diameter of 20 mm.

The polyp phase of this animal is tiny (2 mm) and appears to be more or less degenerate, for it has no perisarc and no tentacles. It occurs in colonies of a few polyps. For a long time its relation to the medusa was not recognized, and thus the polyp was given a name of its own, *Microhydra ryderi*. On the basis of its relationship to the jellyfish and the law of priority, both the polyp and the medusa should be called *Craspedacusta* (N.L. *craspedon*, velum, + Gr. *kystis*, bladder).

The polyp has three methods of asexual reproduction, as shown in Figure 10-13.

Figure 10-14

These hydrozoans form calcareous skeletons that resemble true coral. **A,** *Stylaster roseus* (order Stylasterina) occurs commonly in caves and crevices in coral reefs. These fragile colonies branch in only a single plane and may be white, pink, purple, red, or red with white tips. **B,** Species of *Millepora* (order Milleporina) form branching or platelike colonies and often grow over the horny skeleton of gorgonians (see p. 203), as is shown here. They are generously supplied with powerful nematocysts that produce a burning sensation on human skin, justly earning the common name fire coral.

Photographs by L.S. Roberts.

Figure 10-15

Three Portuguese man-of-war colonies, *Physalia physalis* (order Siphonophora, class Hydrozoa), lie stranded in shallow water. Colonies often drift onto southern ocean beaches, where they are a hazard to bathers. Each colony of medusa and polyp types is integrated to act as one individual. As many as a thousand zooids may be found in one colony. The nematocysts secrete a powerful neurotoxin.
Photograph by C. Lane.

Floating colonies

Members of the orders Siphonophora and Chondrophora are among the most specialized of the Hydrozoa. They form polymorphic swimming or floating colonies made up of several types of modified medusae and polyps.

Physalia (Gr. *physallis*, bladder), or the Portuguese man-of-war (Figure 10-15), is one such colony with a rainbow-hued float of blues and pinks that carries it along on the surface waters of tropical seas. Many are blown to shore on the eastern coast of the United States. The long, graceful tentacles, actually zooids, are laden with nematocysts and are capable of inflicting painful stings. The float, called a **pneumatophore,** is believed to have expanded from the original larval polyp. It contains an air sac arising from the body wall and filled with a gas similar to air. The float acts as a type of nurse-carrier for future generations of individuals that bud from it and hang suspended in the water. Some of the siphonophores, such as *Stephalia* and *Nectalia,* possess swimming bells as well as a float.

There are several types of polyp individuals. The **gastrozooids** are feeding polyps with a single long tentacle arising from the base of each. Some of these long, stinging tentacles become separated from the feeding polyp and are called **dactylozooids,** or fishing tentacles. These sting the prey and lift them to the lips of the feeding polyps. Among the modified medusoid individuals are the **gonophores,** which are little more than sacs containing either ovaries or testes.

____ Class Scyphozoa

Class Scyphozoa (si-fo-zo′a) (Gr. *skyphos,* cup) includes most of the larger jellyfishes, or "cup animals." A few, such as *Cyanea* (Gr. *kyanos,* dark-blue substance), may attain a bell diameter exceeding 2 m and tentacles 60 to 70 m long (Figure 10-16). Most scyphozoa, however, range from 2 to 40 cm in diameter. Most are found floating in the open sea, some even at depths of 3000 m, but one unusual order is sessile and attaches by a stalk to seaweeds and other objects on the sea bottom (Figure 10-17). Their coloring may range from colorless to striking orange and pink hues.

An interesting mutualistic relationship exists between *Physalia* and a small fish called *Nomeus* (Gr. herdsman) that swims among the tentacles with perfect safety. Other larger fish trying to catch *Nomeus* are caught by the deadly tentacles. *Nomeus* probably feeds on bits of *Physalia*'s prey and certainly gains a measure of protection from predation by its refuge among the *Physalia* tentacles. Why the fish is not stung to death by its host's nematocysts is unclear, but like the anemone fish to be discussed later, *Nomeus* is probably protected by a skin mucus that does not stimulate nematocyst discharge.

Figure 10-16

Giant jellyfish, *Cyanea capillata* (order Se-maeostomeae, class Scyphozoa). A North Atlantic species of *Cyanea* reaches a bell diameter exceeding 2 m. It is known as the "sea blubber" by fishermen.
Photograph by R. Harbo.

Figure 10-17

Haliclystus auricula (order Stauromedusae, class Scyphozoa). Members of this order are unusual scyphozoans in that the medusae are sessile and attached to seaweed or other objects.
Photograph by K. Sandved.

The bells of different species vary in depth from a shallow saucer shape to a deep helmet or goblet shape. The jelly (mesoglea) layer is unusually thick, giving the bell a fairly firm consistency. The jelly is 95% to 96% water. Unlike the hydromedusae, this layer in the scyphomedusae also contains ameboid cells and fibers, so that it is called a **collenchyme.** Movement is by rhythmical pulsations of the umbrella. There is no velum as in the hydromedusae. There may be many tentacles or few, and they may be short as in *Aurelia* (L. *aurum*, gold) or long as in *Cyanea. Aurelia aurita* (Figure 10-18) is a familiar species 7 to 10 cm in diameter, commonly found in the waters off both the east and west coasts of the United States, and widely used for the study of Scyphozoa.

The margin of the umbrella is scalloped, usually with each indentation bearing a pair of **lappets,** and between them is a sense organ called a **rhopalium** (tentaculocyst). *Aurelia* has eight such notches. Some scyphozoans have four, others 16. Each rhopalium is club shaped and contains a hollow statocyst for equilibrium and one or two pits lined with sensory epithelium. In some species the rhopalia also bear ocelli.

The mouth is centered on the subumbrella side. The manubrium is usually drawn out to form four frilly **oral arms** that are used in capturing and ingesting prey.

The tentacles, the manubrium, and often the entire body surface are well supplied with nematocysts that can give painful stings. However, the primary function of scyphozoan nematocysts is not to attack humans but to paralyze prey animals, which are conveyed to the mouth lobes with the help of the other tentacles or by the bending of the umbrella margin.

Aurelia, which has comparatively short tentacles, feeds on small planktonic animals. These are caught in the mucus of the umbrella surface, are carried to "food pockets" on the umbrella margin by cilia, and are picked up from the pockets by the oral lobes whose cilia carry the food to the gastrovascular cavity. Flagella in the gastrodermis layer keep a current of water moving to bring food and oxygen into the stomach and carry out wastes.

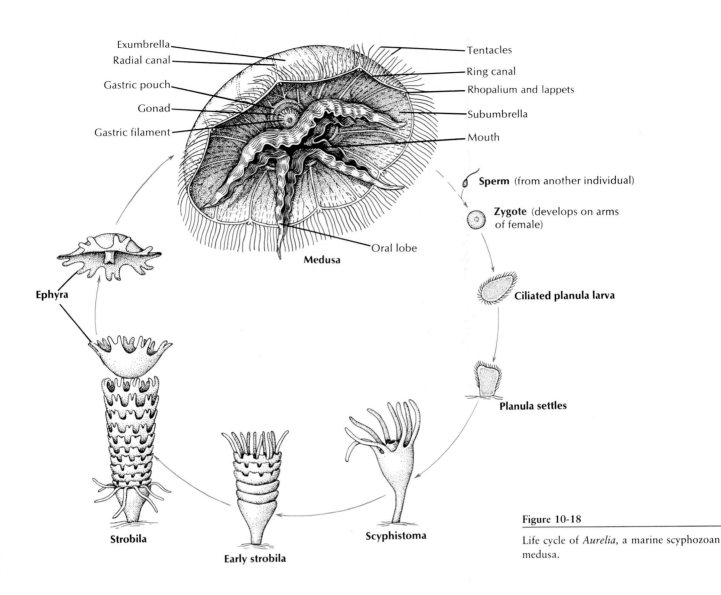

Exumbrella
Radial canal
Gastric pouch
Gonad
Gastric filament

Tentacles
Ring canal
Rhopalium and lappets
Subumbrella
Mouth

Oral lobe

Medusa

Sperm (from another individual)

Zygote (develops on arms of female)

Ciliated planula larva

Planula settles

Ephyra

Strobila

Early strobila

Scyphistoma

Figure 10-18

Life cycle of *Aurelia*, a marine scyphozoan medusa.

Cassiopeia (L. mythical queen of Ethiopia), a jellyfish common to Florida waters, and *Rhizostoma* (Gr. *rhiza*, root, + *stoma*, mouth), which can be found in colder waters, belong to a group differing from that of *Aurelia* both in their lack of tentacles on the umbrella margin and in the structure of the oral arms. During development, the edges of the oral lobes fold over and fuse, forming canals (**arms** or **brachial canals**) that become highly branched. These open to the surface at frequent intervals by pores called "mouths"; the original mouth is obliterated in the fusion of the oral lobes. Planktonic organisms caught in the mucus of the frilly oral arms are transported to the mouths and then up the brachial canals to the gastric cavity by cilia. In contrast to the usual swimming habit of medusae, *Cassiopeia* is usually found lying on its "back" in shallow lagoons. Its umbrella margin contracts about 20 times a minute, creating water currents to bring plankton into contact with the mucus and nematocysts of its oral lobes. Its tissues are abundantly supplied with symbiotic algal cells (zooxanthellae).

Internally, extending out from the stomach of scyphozoa are four **gastric pouches** in which gastrodermis extends down in little tentacle-like projections called **gastric filaments**. These are covered with nematocysts to quiet further any prey that may still be struggling. Gastric filaments are lacking in the hydromedusae. A complex system of **radial canals** branches out from the pouches to a **ring canal** in the margin and makes up a part of the gastrovascular cavity.

Figure 10-24

A sea anemone that swims. When attacked by a predatory sea star *Dermasterias*, the anemone *Stomphia didemon* detaches from the bottom and rolls or swims spasmodically to a safer location.
Photographs by R. Harbo.

When disturbed, sea anemones contract and draw in their tentacles and oral discs. Some anemones are able to swim, to a limited extent, by rhythmical bending movements, which may be a mechanism for escape from enemies such as sea stars and nudibranchs. *Stomphia*, for example, at the touch of a predatory sea star, will detach its pedal disc and make creeping or swimming movements to escape (Figure 10-24). This escape reaction is caused not only by the touch of the star but also by exposure to drippings exuded by the star or to crude extracts made from its tissues. The sea drippings contain steriod saponins that are toxic and irritating to most invertebrates. Extracts given off by nudibranchs can also provoke this reaction in sea anemones.

Anemones form some interesting mutualistic relationships with other organisms. Many species harbor symbiotic algae (zooxanthellae) within their tissues, similar to the hard coral—zooxanthellae association (described later in the chapter), and the anemones profit from the products of algal photosynthesis (p. 204). Some anemones habitually attach to the shells occupied by certain hermit crabs. The hermit encourages the relationship and, finding its favorite species, which it recognizes by touch, it massages the anemone until it detaches. The hermit crab holds the anemone against its own shell until the anemone is firmly attached. The crab probably derives some protection and camouflage from the anemone. The anemone gets free transportation and particles of food dropped by the hermit crab.

Certain damselfishes (clown fishes) (family Pomacentridae) form associations with large anemones, especially in tropical Indo-Pacific waters. An unknown property of the skin mucus of the fish causes the anemone's nematocysts not to discharge, but if some other fish is so unfortunate as to brush the anemone's tentacles, it is likely to become a meal. The anemone obviously provides shelter for the anemone fish, and the fish may help ventilate the anemone by its movements, keep the anemone free of sediment, and even lure an unwary victim to seek the same shelter.

The sexes are separate in some sea anemones, and some are hermaphroditic. Monoecious species are **protandrous** (produce sperm first, then eggs). Gonads are arranged on the margins of the mesenteries, and fertilization takes place externally or in the gastrovascular cavity. The zygote develops into a ciliated larva. Asexual reproduction commonly occurs by **pedal laceration** or by longitudinal fission, occasionally by transverse fission or by budding. In pedal laceration, small pieces of the pedal disc break off as the animal moves, and each of these regenerates a small anemone.

Zoantharian corals

The zoantharian corals belong to the order Scleractinia, sometimes known as the true or stony corals. The stony corals might be described as miniature sea anemones that live in calcareous cups they themselves have secreted (Figures 10-25 and 10-26). Like that of the anemones, the coral polyp's gastrovascular cavity is subdivided by mesenteries arranged in multiples of six (hexamerous) and its hollow tentacles surround the mouth, but there is no siphonoglyph.

Instead of a pedal disc, the epidermis at the base of the column secretes the limy skeletal cup, including the sclerosepta, which project up into the polyp between the true mesenterial septa (Figure 10-26). The living polyp can retract into the safety of the cup when not feeding. Since the skeleton is secreted below the living tissue rather than within it, the calcareous material is an exoskeleton. In the colonial corals, the skeleton may become massive, building up over many years, with the living coral forming a sheet of tissue over the surface. The gastrovascular cavities of the polyps are all connected through this sheet of tissue.

Three other small orders of Zoantharia are recognized.

Tube anemones and thorny corals

Members of the subclass Ceriantipatharia have coupled but unpaired mesenteries. The tube anemones (order Ceriantharia) are solitary and live in soft bottom sediments, buried to the level of the oral disc. They occupy tubes constructed of secreted mucus and threads of nematocyst-like organelles, into which they can

Figure 10-25

A, Cup coral *Tubastrea* sp. The polyps form clumps resembling groups of sea anemones. Although often found on coral reefs, *Tubastrea* is not a reef-building coral (ahermatypic) and has no symbiotic zooxanthellae in its tissues. **B,** The polyps of *Montastrea cavernosa* are tightly withdrawn in the daytime but open to feed at night, as in C.
A, Photograph by C.P. Hickman, Jr.; **B** and **C,** photographs by L.S. Roberts.

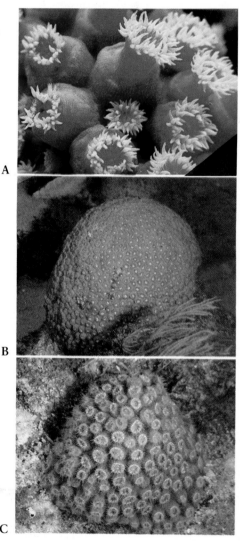

A

B

C

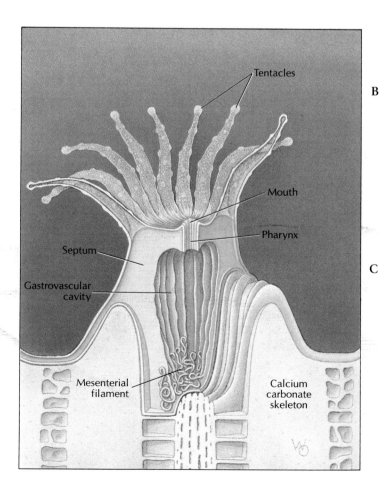

Figure 10-26

Polyp of a zoantharian coral (order Scleractinia) showing calcareous cup (exoskeleton), gastrovascular cavity, sclerosepta, mesenterial septa, and mesenterial filaments.

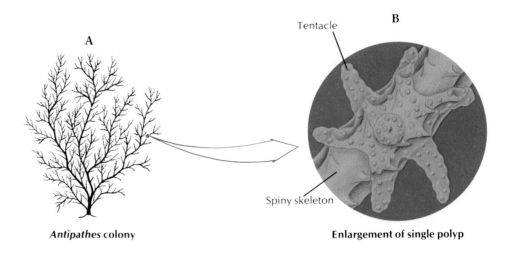

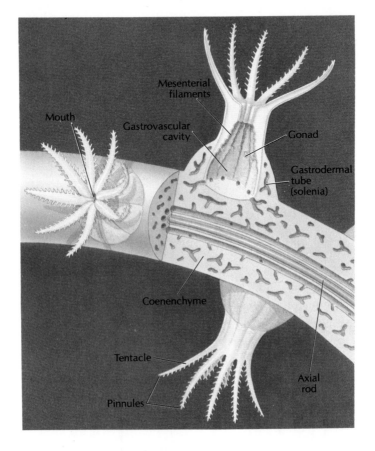

Figure 10-27

A, Colony of *Antipathes*, a black or thorny coral (order Antipatharia, class Anthozoa). Most abundant in deep waters in the tropics, black corals secrete a tough, proteinaceous skeleton that can be worked into jewelry. B, The polyps of Antipatharia have six simple, nonretractile tentacles. The spiny processes on the skeleton are the origin of the common name thorny corals.

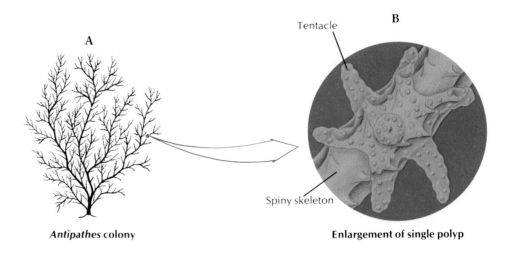

Antipathes colony   Enlargement of single polyp

withdraw. The thorny or black corals (order Antipatharia) (Figure 10-27) are colonial and attached to a firm substratum. Their skeleton is of a horny material and has thorns. Both of these orders are small in numbers of species and are limited to warmer waters of the sea.

Alcyonarian corals

Alcyonarians are often referred to as octocorals because of their strict octomerous symmetry, with eight pinnate tentacles and eight unpaired, complete mesenteries (Figure 10-22). They are all colonial, and the gastrovascular cavities of the polyps communicate through a system of gastrodermal tubes called **solenia** (Figure 10-28). The solenia run through an extensive mesoglea (**coenenchyme**) in most alcyonarians, and the surface of the colony is covered by epidermis. The skeleton is

Figure 10-28

Polyps of an alcyonarian coral (octocoral). Note the eight, pinnate tentacles, coenenchyme, and solenia. They have an endoskeleton of limy spicules often with a horny protein, which may be in the form of an axial rod.

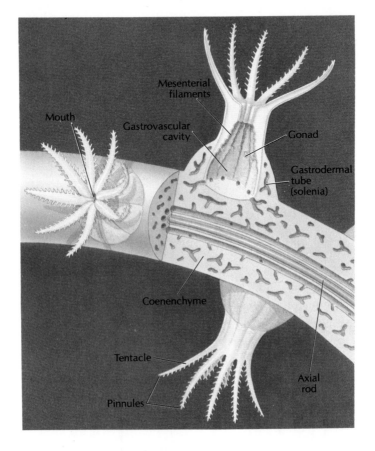

A

B

C

Figure 10-29

Colonial gorgonian, or horny, corals (order Gorgonacea, class Anthozoa) are conspicuous components of reef faunas, especially of the West Indies. **A,** Red gorgonian *Lophogorgia chilensis.* **B,** A sea plume, *Pseudopterogorgia* sp. **C,** A sea fan *Gorgonia ventalina.*

A, Photograph by D.W. Gotshall; **B,** photograph by L.S. Roberts; **C,** photograph by W.C. Ober.

secreted in the coenenchyme and consists of limy spicules, fused spicules, or a horny protein, often in combination. Thus, the skeletal support of most alcyonarians is an endoskeleton. The variation in pattern among the species of alcyonarians lends great variety to the form of the colonies: from the soft coral *Gersemia,* with its spicules scattered through the coenenchyme, to the tough, axial supports of the sea fans and other gorgonian corals (Figure 10-29), to the fused spicules of the organ-pipe coral. *Renilla* (L. *ren,* kidney, + *illa,* suffix), the sea pansy, is a colony reminiscent of a pansy flower. Its polyps are embedded in the fleshy upper side and a short stalk that supports the colony is embedded in the sea floor (Figure 10-30). *Ptilosarcus* (Gr. *ptilon,* feather, + *sarkos,* flesh), a sea pen, is a member of the same order and may reach a length of 50 cm (Figure 10-22).

The graceful beauty of the alcyonarians—in hues of yellow, red, orange, and purple—helps create the "submarine gardens" of the coral reefs.

Coral reefs

Most students will have seen photographs or movies giving a glimpse of the vibrant color and life found on coral reefs, and some may have been fortunate enough to visit a reef. Coral reefs are among the most productive of all ecosystems, and they have a diversity of life forms rivaled only by the tropical rain forest. They are large formations of calcium carbonate (limestone) in shallow tropical seas laid down by living organisms over thousands of years; living plants and animals are confined to the top layer of reefs where they add more and more calcium carbonate to that deposited by their predecessors. The most important organisms that precipitate calcium carbonate from seawater to form reefs are the scleractinian, **hermatypic** (reef-building) **corals** (Figure 10-25) and **coralline algae.** Not only do the coralline algae contribute to the total mass of calcium carbonate, but their precipitation of the substance helps to hold the reef together. Some alcyonarians

Figure 10-30

Sea pansy *Renilla reniformis* (order Pennatulacea, class Anthozoa). The eight pinnate tentacles characteristic of the subclass Alcyonaria can be distinguished.

Photograph by K. Sandved.

and hydrozoa (especially *Millepora* [L. *mille,* thousand, + *porus,* pore] spp., the "fire coral," Figure 10-14, *B*) contribute in some measure to the calcareous material, and an enormous variety of other organisms contribute small amounts. However, hermatypic (Gr. *herma,* support, mound, + *typos,* type) corals seems essential to the formation of large reefs, since such reefs do not occur where these corals cannot live.

Hermatypic corals require warmth, light, and the salinity of undiluted seawater. This limits coral reefs to shallow waters between 30 degrees north and 30 degrees south latitude and excludes them from areas with upwelling of cold water or areas near major river outflows with attendant low salinity and high turbidity. These corals require light because they have mutualistic algae (zooxanthellae) living in their tissues. The microscopic zooxanthellae are very important to the corals: their photosynthesis and fixation of carbon dioxide furnish food molecules for their hosts, they recycle phosphorus and nitrogenous waste compounds that otherwise would be lost, and they enhance the ability of the coral to deposit calcium carbonate.

Several types of reefs are commonly recognized. The **fringing reef** is close to a land mass with either no lagoon or a narrow lagoon between it and the shore. A **barrier reef** runs roughly parallel to shore and has a wider and deeper lagoon than does a fringing reef. **Atolls** are reefs that encircle a lagoon but not an island. These types of reefs typically slope rather steeply into deep water at their seaward edge. **Patch** or **bank reefs** occur some distance back from the steep, seaward slope in the lagoons of barrier reefs or atolls. The so-called Great Barrier Reef, extending 1260 miles long and up to 90 miles from the shore off the northeast coast of Australia, is actually a complex of reef types.

Fringing, barrier, and atoll reefs all have distinguishable zones that are characterized by different groups of corals and other animals. The side of the reef facing the sea is the **reef front** or **fore reef slope** (Figure 10-31). The reef front is more or less parallel to the shore and perpendicular to the predominant direction of wave travel. It slopes downward into deeper water, sometimes gently at first, then precipitously. Characteristic assemblages of scleratinian corals grow deep on the slope, high near the crest, and in zones between. In shallow water or slightly emergent at the top of the reef front is the **reef crest.** The upper front and the crest bear the greatest force of the waves and must absorb great energy during storms. Pieces of coral and other organisms are broken off at such times and thrown shoreward onto the **reef flat,** which slopes down into the lagoon. The reef flat thus receives a supply of calcareous material that is eventually broken down into coral sand. The sand is stabilized by the growth of plants such as turtle grass and coralline algae and is ultimately cemented into the mass of the reef by precipitation of carbonates. A reef is not an unbroken wall facing the sea but is highly irregular, with grooves, caves, crevices, channels through from the flat to the front, and deep, cup-shaped holes ("blue holes"). Alcyonarians tend to grow in these areas that are more protected from the full force of the waves, as well as on the flat and the deeper areas of the fore reef slope. Many other kinds of organisms inhabit the cryptic locations such as caves and crevices.

Enormous numbers of species and individuals of invertebrate groups and fishes populate the reef ecosystem. For example, there are 300 *common* species of fishes on Caribbean reefs. It is marvelous that such diversity and productivity can be maintained, since the reefs are washed by the nutrient-poor waves of the open ocean. Although relatively little nutrient enters the ecosystem, little is lost because the interacting organisms of the reef are so efficient in recycling. Even the feces released by the fish are fed upon by the corals!

Because zooxanthellae are vital to hermatypic corals, and water absorbs light, hermatypic corals rarely live below a depth of 100 feet. Interestingly, some deposits of coral reef limestone, particularly around Pacific islands and atolls, reach great thickness—even thousands of feet. Clearly, the corals and other organisms could not have grown from the bottom in the abyssal blackness of the deep sea and reached shallow water where light could penetrate. Charles Darwin was the first to realize that such reefs began their growth in *shallow* water around volcanic islands; then as the islands slowly sank beneath the sea, the growth of the reefs kept up with the rate of sinking, thus accounting for the depth of the deposits.

A

B

Figure 10-31

A, Profile of a barrier reef. B, Portion of an atoll from the air. Reef slope plunges into deep water at left (dark blue), lagoon at right.

B, Photograph by C.P. Hickman, Jr.

▬ PHYLUM CTENOPHORA

Ctenophora (te-nof'o-ra) (Gr. *kteis, ktenos,* comb, + *phora,* pl. of bearing) is composed of fewer than 100 species. All are marine forms. They take their name from the eight rows of comblike plates they bear for locomotion. Common names for ctenophores are "sea walnuts" and "comb jellies." Ctenophores, along with cnidarians, represent the only two phyla having primary radial symmetry, in contrast to other metazoans, which have primary bilateral symmetry.

Ctenophores do not have nematocysts, except in one species (*Haeckelia rubra,* after Ernst Haeckel, nineteenth-century German zoologist) that is provided with nematocysts on certain regions of its tentacles but lacks colloblasts. These nematocysts are apparently a part of this ctenophore and are not "appropriated" by eating a cnidarian.

In common with the cnidarians, ctenophores have not advanced beyond the tissue grade of organization. There are no definite organ systems in the strict meaning of the term.

The ctenophores are strictly marine animals and are all free swimming except for a few creeping and sessile forms. They occur in all seas but especially in warm waters.

Although they are feeble swimmers and are more common in surface waters, ctenophores are sometimes found at considerable depths. They are often at the mercy of tides and strong currents, but they avoid storms by swimming downward in the water. In calm water they may rest vertically with little movement, but when moving they use their ciliated comb plates to propel themselves mouth-end forward. Highly modified forms such as *Cestum* (L. *cestus,* girdle) use sinuous body movements as well as their comb plates in locomotion.

The fragile, transparent bodies of ctenophores are easily seen at night when they emit light (luminesce).

CHARACTERISTICS

1. Symmetry **biradial;** arrangement of internal canals and the position of the paired tentacles change the radial symmetry into a combination of the two (**radial + bilateral**)
2. Usually ellipsoidal or spherical in shape, **with radially arranged rows of comb plates for swimming**
3. Ectoderm, endoderm, and a mesoglea (ectomesoderm) with scattered cells and muscle fibers; may be considered **triploblastic**
4. Nematocysts absent (except in one species), but **adhesive cells (colloblasts)** present
5. Digestive system consisting of mouth, pharynx, stomach, a series of canals, and anal pores
6. Nervous system consisting of a subepidermal plexus concentrated around the mouth and beneath the comb plate rows; an **aboral sense organ** (statocyst)
7. No polymorphism
8. Reproduction monoecious; gonads (endodermal origin) on the walls of the digestive canals, which are under the rows of comb plates; determinate cleavage; cydippid larva
9. Luminescence common

COMPARISON WITH CNIDARIA

Ctenophores resemble the cnidarians in the following ways:
1. Form of radial symmetry; together with the cnidarians, they form the group Radiata
2. Aboral-oral axis around which the parts are arranged
3. Well-developed gelatinous ectomesoderm (collenchyme)
4. No coelomic cavity
5. Diffuse nerve plexus
6. Lack of organ systems

They differ from the cnidarians in the following ways:
1. No nematocysts except in *Haeckelia*
2. Development of muscle cells from mesenchyme
3. Presence of comb plates and colloblasts
4. Mosaic, or determinate type of development
5. Presence of pharynx generally
6. No polymorphism
7. Never colonial
8. Presence of anal openings

CLASSIFICATION

Class Tentaculata (ten-tak-yu-láta) (L. *tentaculum*, feeler, + *ata*, group suffix). With tentacles; tentacles may or may not have sheaths into which they retract; some types flattened in oral-aboral axis for creeping; others compressed in tentacular plane to a bandlike form; in some the comb plates may be confined to the larval form. Examples: *Pleurobrachia, Cestum.*

Class Nuda (núda) (L. *nudus*, naked). Without tentacles, but flattened in tentacular plane; wide mouth and pharynx; gastrovascular canals much branched. Example: *Beroe.*

____ Class Tentaculata

Representative type: Pleurobrachia

Pleurobrachia (Gr. *pleuron*, side, + L. *brachia*, arms) is often used as a representative of this group of ctenophores. Its transparent body is about 1.5 to 2 cm in diameter (Figure 10-32, *A*). The oral pole bears the mouth opening, and the aboral pole has a sensory organ, the **statocyst.**

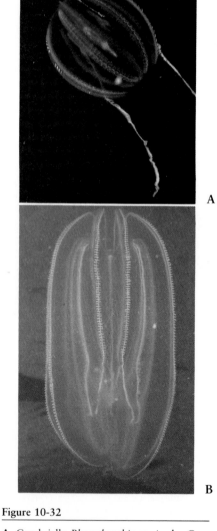

Figure 10-32

A, Comb jelly *Pleurobrachia* sp. (order Cydippida, class Tentaculata). Its fragile beauty is especially evident at night when it luminesces from its comb rows. **B,** *Mnemiopsis* sp. (order Lobata, class Tentaculata). **A,** Photograph by J.L. Rotman; **B,** photograph by K. Sandved.

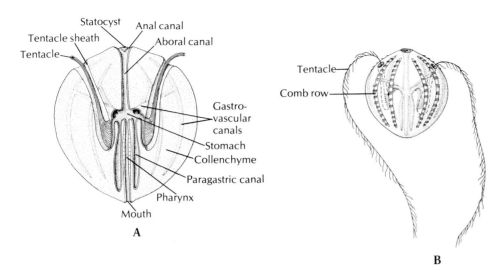

Figure 10-33

The comb jelly *Pleurobrachia*, a ctenophore. **A,** Hemisection. **B,** External view.

Comb plates

On the surface are eight equally spaced bands called **comb rows** that extend as meridians from the aboral pole and end before reaching the oral pole (Figure 10-33). Each band is made up of transverse plates of long fused cilia called **comb plates** (Figure 10-34, *B*). Ctenophores are propelled by the beating of the cilia on the comb plates. The beat in each row starts at the aboral end and proceeds successively along the combs to the oral end. All eight rows normally beat in unison. The animal is thus driven forward with the mouth in advance. The animal can swim backward by reversing the direction of the wave.

Tentacles

The two **tentacles** are long, solid and very extensible, and they can be retracted into a pair of **tentacle sheaths.** When completely extended, they may measure 15 cm in length. The surface of the tentacles bears **colloblasts,** or glue cells (Figure 10-34, *A*), which secrete a sticky substance that is used for catching and holding small animals.

Body wall

The cellular layers of ctenophores are generally similar to those of cnidarians. Between the epidermis and gastrodermis is a gelatinous **collenchyme** that makes up most of the interior of the body and contains muscle fibers and ameboid cells. Although they are derived from ectodermal cells, the muscle cells are distinct and are not contractile portions of epitheliomuscular cells (in contrast to the Cnidaria).

Digestive system and feeding

The **gastrovascular system** consists of a mouth, a pharynx, a stomach, and a system of gastrovascular canals that branch through the jelly to extend to the comb plates, tentacular sheaths, and elsewhere (Figure 10-33). There are two blind canals that terminate near the mouth, and an aboral canal that passes near the statocyst and then divides into two small **anal canals** through which undigested material is expelled.

Ctenophores live on small planktonic organisms such as copepods. The glue cells on the tentacles stick to the small prey and enable the tentacles to carry the prey to the ctenophore's mouth. Digestion is both extracellular and intracellular.

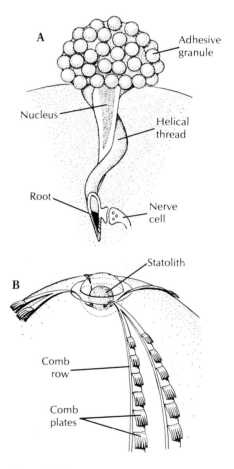

Figure 10-34

A, Colloblast, an adhesive cell characteristic of ctenophores. **B,** Portion of comb rows showing comb plates, each composed of transverse rows of long fused cilia.

Respiration and excretion

Respiration and excretion occur through the body surface.

Nervous and sensory systems

Ctenophores have a nervous system similar to that of the cnidarians. It is made up of a subepidermal plexus, which is concentrated under each comb plate, but there is no central control as is found in higher animals.

The **sense organ** at the aboral pole is a statocyst. Tufts of cilia support a calcareous statolith, with the whole being enclosed in a bell-like container. Alterations in the position of the animal change the pressure of the statolith on the tufts of cilia. The sense organ is also concerned in coordinating the beating of the comb rows but does not trigger their beat.

The epidermis of ctenophores is abundantly supplied with sensory cells, so the animals are sensitive to chemical and other forms of stimuli. When a ctenophore comes in contact with an unfavorable stimulus, it often reverses the beat of its comb plates and backs up. The comb plates are very sensitive to touch, which often causes them to be withdrawn into the jelly.

Reproduction

Pleurobrachia, like other ctenophores, are monoecious. The gonads are located on the lining of the gastrovascular canals under the comb plates. Fertilized eggs are discharged through the epidermis into the water.

Cleavage in the ctenophores is determinate (mosaic), since the various parts of the animal that will be formed by each cleavage cell are determined early in embryogenesis. If one of the cells is removed in the early stages, the resulting embryo will be deficient. This type of development differs from that of cnidarians, which is regulative (indeterminate). The free-swimming cydippid larva is superficially similar to the adult ctenophore and develops directly into an adult.

____ Other Ctenophores

Ctenophores are fragile and beautiful creatures. Their transparent bodies glisten like fine glass, brilliantly iridescent during the day and luminescent at night.

One of the most striking ctenophores is *Beroe* (L. a nymph), which may be more than 100 mm in length and 50 mm in breadth (Figure 10-35, *A*). It is conical or thimble shaped and is flattened in the tentacular plane. The tentacular plane in *Beroe* is defined as where the tentacles would have been, because it has a large mouth but no tentacles. It is pink. The body wall is covered with an extensive network of canals formed by the union of the paragastric and meridional canals. Venus' girdle (*Cestum*, Figure 10-35, *B*) is highly compressed in the tentacular plane. Bandlike, it may be more than 1 m long and presents a graceful appearance as it swims in the oral direction. The highly modified *Ctenoplana* (Gr. *ktenos*, comb, + L. *planus*, flat) and *Coeloplana* (Gr. *koilos*, hollow, + L. *planus*, flat) (Figure 10-35, *C*) are rare but are interesting because they have disc-shaped bodies flattened in the oral-aboral axis and are adapted for creeping rather than swimming. A common ctenophore along the Atlantic and Gulf coasts is *Mnemiopsis* (Gr. *mnēmē*, memory, + *opsis*, appearance) (Figure 10-32, *B*), which has a laterally compressed body with two large oral lobes and unsheathed tentacles.

Nearly all ctenophores give off flashes of luminescence at night, especially such forms as *Mnemiopsis* (Figure 10-32, *B*). The vivid flashes of light seen at night in southern seas are often caused by members of this phylum.

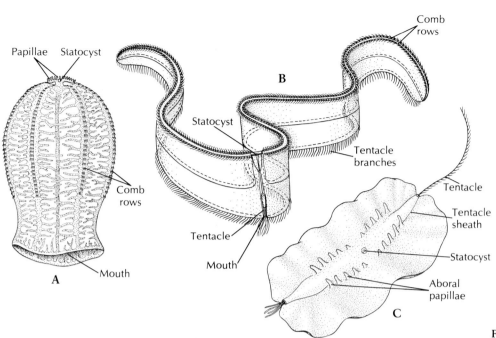

Papillae Statocyst

Comb rows

Comb rows

Statocyst

Tentacle branches

Tentacle

Tentacle sheath

Statocyst

Aboral papillae

Comb rows

Tentacle

Mouth

Mouth

A

C

Figure 10-35

Diversity among the phylum Ctenophora. **A,** *Beroe* sp. (order Beroida, class Nuda). **B,** *Cestum* sp. (order Cestida, class Tentaculata). **C,** *Coeloplana* sp. (order Platyctenea, class Tentaculata).

PHYLOGENY AND ADAPTIVE RADIATION
Phylogeny

The origin of the cnidarians and ctenophores is obscure, although the most widely supported hypothesis today is that the radiate phyla arose from a radially symmetrical, planula-like ancestor. Such an ancestor could have been common to the radiates and to the higher metazoans, the latter having been derived from a branch whose members habitually crept about on the sea bottom. Such a habit would select for bilateral symmetry. Others became sessile or free floating, conditions for which radial symmetry is a selective advantage. A planula larva in which an invagination formed to become the gastrovascular cavity would correspond roughly to a cnidarian with an ectoderm and an endoderm.

The trachyline medusae (an order of class Hydrozoa) are sometimes considered the most primitive of modern cnidarians because of their direct development from the planula and actinula larvae to the medusa. Such a group could have given rise to the three classes of cnidarians, with further development of the polyp and subsequent loss of the medusa occurring in the anthozoan line. Some zoologists now believe that the Scyphozoa are the most primitive of the cnidarian classes, with the Cubozoa providing a link between the Scyphozoa and the Hydrozoa. Colonial hydrozoans could then have arisen from an ancestor like *Tripedalia* whose polyp buds failed to separate.

In the past it has been assumed that the ctenophores arose from a medusoid cnidarian, but this assumption has been questioned recently. The similarities between the groups are mostly of a general nature and do not seem to indicate a close relationship.

Some biologists have regarded the ctenophores and some advanced cnidarians (for example, advanced anthozoans) as triploblastic because the highly cellular nature of the mesoglea would constitute a mesoderm. However, others define mesoderm strictly as a layer derived from endoderm; thus both cnidarians and ctenophores would be diploblastic.

___ Adaptive Radiation

In their evolution neither phylum has deviated far from its basic plan of structure. In the Cnidaria, both the polyp and medusa are constructed on the same scheme. Likewise, the ctenophores have adhered to the arrangement of the comb plates and their biradial symmetry.

Nonetheless, the cnidarians are a successful phylum in terms of numbers of individuals and species, demonstrating a surprising degree of diversity considering the simplicity of their basic body plan. They are efficient predators, many feeding on prey quite large in relation to themselves. Some are adapted for feeding on small particles. The colonial form of life is well explored, with some colonies growing to great size among the corals, and others, such as the siphonophores, showing astonishing polymorphism and specialization of individuals within the colony.

___ SUMMARY

The phyla Cnidaria and Ctenophora have a primary radial symmetry; radial symmetry is an advantage for sessile or free-floating organisms. The Cnidaria are surprisingly efficient predators because they possess stinging organelles called nematocysts. Both phyla are essentially diploblastic (some triploblastic, according to definition of mesoderm), with a body wall composed of epidermis and gastrodermis and a mesoglea between. The digestive-respiratory (gastrovascular) cavity has a mouth and no anus. Cnidarians are at the tissue level of organization. They have two basic body types (polypoid and medusoid), and in many hydrozoans and scyphozoans the life cycle involves both the asexually reproducing polyp and the sexually reproducing medusa.

That unique organelle, the nematocyst, is secreted by a cnidoblast (which becomes the cnidocyte) and is contained coiled within a capsule. When discharged, some types of nematocysts penetrate the prey and inject poison. Discharge is effected by a change in permeability of the capsule and an increase in internal hydrostatic pressure because of the high osmotic pressure within the capsule.

There is no concentrated central nervous system in cnidarians; the nerves are spread throughout the body in a netlike arrangement. Some synapses can transmit impulses in either direction.

Most hydrozoans are colonial and marine, but the freshwater hydras are commonly demonstrated in class laboratories and have the typical polypoid structure. The body is approximately cylindrical in shape, with the mouth surrounded by nematocyst-bearing tentacles at one end. The various cell types are also typical: epitheliomuscular, interstitial, gland, sensory, nerve, and nutritive-muscular cells. Hydras reproduce asexually by budding and sexually after formation of gonads in the cylinder wall; they have no medusoid stage. The more typical form, found in most marine hydrozoans, is that of a branching colony containing many polyps (hydranths), and a free-swimming medusa stage. An example of a floating, colonial hydrozoan with numerous body forms (both polypoid and medusoid) in the colony is the Portuguese man-of-war, *Physalia*.

The scyphozoans are the typical jellyfish, in which the medusoid is the dominant body form. Many scyphozoans have an inconspicuous polypoid stage that reproduces asexually (scyphistoma). Scyphozoan medusae differ from hydrozoan

Sea anemones *Metridium senile* in a tide pool.
Photograph by H.W. Pratt/BPS.

medusae in that scyphozoans have complex sensory structures called rhopalia and do not have a velum.

The cubozoans are a small class of cnidarians that are predominantly medusoid. They include the dangerous sea wasps. Cubomedusae have rhopalia and a velarium, which is a velumlike structure.

The anthozoans are all marine and are polypoid; there is no medusoid stage. They include two large and important subclasses: Zoantharia (with hexamerous or polymerous symmetry and paired mesenteries) and Alcyonaria (with octomerous symmetry). A small subclass of the Anthozoa is the Ceriantipatharia (with symmetry as in Zoantharia but with unpaired mesenteries). The largest zoantharian orders contain the sea anemones, which are mostly solitary and do not have a secreted skeleton, and the stony corals, which are mostly colonial and secrete a calcareous exoskeleton. Stony corals are the critical component in coral reefs, which are habitats of great beauty, productivity, and ecological and economic value. The Alcyonaria contain the soft and horny corals, many of which are important and beautiful components of coral reefs.

The Ctenophora are biradial organisms that swim by means of eight comb rows. The combs are modified from cilia. Only one species has nematocysts, but adhesive cells (colloblasts) are characteristic of the phylum. These weakly swimming animals capture small prey with the aid of the colloblasts.

Cnidaria and Ctenophora are probably derived from an ancestor that resembled the planula larva of the cnidarians. In spite of their relatively simple level of organization, the cnidarians are a successful and important phylum.

Review questions

1. Explain the selective advantage of radial symmetry for sessile and free-floating animals.
2. Give ten characteristics of the phylum Cnidaria.
3. Name and distinguish the classes in the phylum Cnidaria.
4. Distinguish between the polyp and medusa forms.
5. Explain the mechanism of nematocyst discharge.
6. What is an unusual feature of the nervous system of cnidarians?
7. Diagram a hydra and label the main body parts.
8. Name and give the functions of the main cell types in the epidermis and in the gastrodermis of hydra.
9. What stimulates feeding behavior in hydras?
10. Define the following with regard to hydroids: hydrorhiza, hydrocaulus, coensosarc, perisarc, hydranth, gonangium, manubrium, statocyst, ocellus.
11. Give an example of a highly polymorphic, floating, colonial hydrozoan.
12. Distinguish the following from each other: statocyst and rhopalium; scyphomedusae and hydromedusae; scyphistoma, strobila, and ephyrae; velum, velarium, and pedalium; Zoantharia and Alcyonaria.
13. Define the following with regard to sea anemones: siphonoglyph; primary septa or mesenteries; incomplete mesenteries; septal filaments; acontia threads; pedal laceration.
14. Describe three specific interactions of anemones with nonprey organisms.
15. Contrast the skeletons of zoantharian and alcyonarian corals.
16. Why do hermatypic corals grow only in relatively shallow water?
17. Specifically, what kinds of organisms are most important in deposition of calcium carbonate on coral reefs?
18. How do zooxanthellae contribute to the welfare of hermatypic corals?
19. Distinguish each of the following from each other: fringing reefs; barrier reefs; atolls; patch or bank reefs.
20. Name six characteristics of ctenophores.

21. How do ctenophores swim, and how do they obtain food?
22. Compare cnidarians and ctenophores, giving five ways in which they resemble each other and five ways in which they differ.
23. What is the prevalent hypothesis on the origin of the radiate phyla?

Selected references

See also general references for Part Two, p. 590.

Barnes, J.H. 1966. Studies on three venomous cubomedusae. In W.J. Rees (ed.). The Cnidaria and their evolution. Symposia of the Zoological Society of London, no. 16. London, Academic Press, Inc., Ltd. *Presents morphological studies of cubomedusae producing significant human injury along the Australian coast, clinical details, treatment, and evidence that fatal stingings have been due to* Chironex fleckeri.

Bennett, I. 1971. The Great Barrier Reef. Sydney, Lansdowne Press. *Good introduction to this 1200 mile-long complex of islands and reefs. Many color photographs.*

Colin, P.I. 1978. Caribbean reef invertebrates and plants. Neptune City, N.J., T.F.H. Publications, Inc. *Many color photographs, very good for sight recognition of Caribbean reef corals and other invertebrates.*

Faulkner, D., and R. Chesher. 1979. Living corals. New York, Clarkson N. Potter, Inc. *A volume of color photographs of Pacific corals, including considerable information about them; a beautiful book and a feast for the eyes.*

Muscatine, L., and H.M. Lenhoff (eds.). 1974. Coelenterate biology: reviews and new perspectives. New York, Academic Press, Inc. *Highly recommended for further reading; it has chapters on histology, nematocysts, neurobiology, development, symbioses, bioluminescence, and Ctenophora.*

Wood, E.M. 1983. Reef corals of the world, biology and field guide. Neptune City, N.J., T.F.H. Publications, Inc. *Good guide for identification of corals down to genus, grouped according to whether they occur in the Western Atlantic or Indo-Pacific regions. Many line drawings, black and white photographs, and some color illustrations.*

CHAPTER 11

THE ACOELOMATE ANIMALS

Phylum Platyhelminthes

Phylum Nemertea

Phylum Gnathostomulida

Position in Animal Kingdom

1. The Platyhelminthes, or flatworms, the Nemertea or ribbon worms, and the Gnathostomulida, or jaw worms, are the most primitive groups of animals to have **primary bilateral symmetry,** the type of symmetry assumed by all higher animals.
2. These phyla have only one internal space, the digestive cavity, with the region between the ectoderm and endoderm filled with mesoderm in the form of muscle fibers and mesenchyme (parenchyma). Since they lack a coelom or a pseudocoelom, they are termed **acoelomate animals,** and because they have three well-defined germ layers, they are termed triploblastic.
3. Acoelomates show more specialization and division of labor among their organs than do the radiate animals because the mesoderm makes more elaborate organs possible. Thus, the acoelomates are said to have reached the **organ-system level of organization.**
4. They belong to the protostome division of the Bilateria and have spiral cleavage, and at least the platyhelminths and nemerteans have determinate (mosaic) cleavage.

Biological Contributions

1. The acoelomates developed the basic **bilateral** plan of organization that has been widely exploited in the animal kingdom.
2. The **mesoderm** developed into a well-defined embryonic germ layer (**triploblastic**), making available a great source of tissues, organs, and systems.
3. Along with bilateral symmetry, **cephalization** was established. There is some centralization of the nervous system evident in the **ladder type of system** found in flatworms.
4. Along with the subepidermal musculature, there is also a mesenchymal system of muscle fibers.
5. They are the most primitive animals with an **excretory system.**
6. The nemerteans are the most primitive animals to have a **circulatory system** with blood and a **one-way alimentary canal.** Although not stressed by zoologists, the rhynchocoel cavity in ribbon worms is technically a true coelom, but as it is merely a part of the proboscis mechanism, it is not of evolutionary significance.
7. Unique and specialized structures occur in all three phyla. The parasitic habit of many flatworms has led to many specialized adaptations, such as organs of adhesion.

The Platyhelminthes (Gr. *platys,* flat, + *helmins,* worm), or flatworms, the Nemertea (Gr. *Nemertes,* one of the nereids, unerring one), or ribbon worms, and the Gnathostomulida (Gr. *gnathos,* jaw, + *stoma,* mouth, + L. *ulus,* dim.), or jaw worms, are the most primitive groups of animals to have primary bilateral symmetry, the type of symmetry assumed by all higher animals. Bilateral symmetry is much more suitable for active movement than is radial symmetry. Active crawling or swimming is itself a selective pressure for better sensory organs and nervous control and coordination. It is clearly more efficient to have such organs and centers concentrated in the area of the body that meets the environment first: the anterior end. Selection for an anterior end with its concentrated sensory and nervous organs creates bilateral symmetry and, concurrently, the evolution of a head, or cephalization. All animal groups higher in the evolutionary tree than the Platyhelminthes are bilaterally symmetrical or have been derived from bilateral ancestors.

These phyla have only one internal space, the digestive cavity, with the region between the ectoderm and endoderm filled with mesoderm in the form of muscle fibers and mesenchyme (parenchyma). Since they lack a coelom or a pseudocoel, they are termed **acoelomate** animals (Figure 11-1), and because they have three well-defined germ layers, they are **triploblastic.** Acoelomates show more specialization and division of labor among their organs than do the radiate animals because the mesoderm makes more elaborate organs possible; thus, the acoelomates are said to have reached the organ-system level of organization.

These phyla belong to the protostome division of the Bilateria and typically have spiral cleavage. They have some centralization of the nervous system, with a concentration of nerves anteriorly and a ladder-type arrangement of trunks and connectives down the body. They have an excretory (or osmoregulatory) system, and the rhynchocoels are the most primitive phylum with a circulatory system. They also have a one-way digestive system, with an anus as well as a mouth.

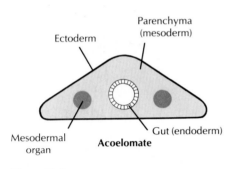

Figure 11-1

Acoelomate body plan.

PHYLUM PLATYHELMINTHES

The term "worm" has been loosely applied to elongated, bilateral invertebrate animals without appendages. At one time zoologists considered worms (Vermes) to be a group in their own right. Such a group included a highly diverse assortment of forms. This unnatural assemblage has been reclassified into various phyla. By tradition, however, zoologists still refer to the various groups of these animals as flatworms, ribbon worms, roundworms, segmented worms, and the like.

The Platyhelminthes were derived from an ancestor that probably had many cnidarian-like characteristics, perhaps a common ancestor with the cnidarians. Nonetheless, replacement of the gelatinous mesoglea of cnidarians with a cellular, mesodermal parenchyma laid the basis for a more complex organization. Parenchyma is a form of "packing" tissue containing more cells and fibers than the mesoglea of the cnidarians. In at least some platyhelminths, the parenchyma is made up of noncontractile cell bodies of muscle cells, that is, the cell body containing the nucleus and other organelles is connected to an elongated contractile portion in somewhat the same manner as the epitheliomuscular cells of the cnidarians (see Figure 10-7).

Flatworms range in size from a millimeter or less to some of the tapeworms that are many meters in length. Their flattened bodies may be slender, broadly leaflike, or long and ribbonlike.

The flatworms include both free-living and parasitic forms, but the freeliving members are found exclusively in the class Turbellaria. A few turbellarians are symbiotic or parasitic, but the majority are adapted as bottom dwellers in marine or fresh water or live in moist places on land. Many, especially of the larger species, are found on the underside of stones and other hard objects in freshwater streams or in the littoral zones of the ocean.

Relatively few turbellarians live in fresh water. Planarians (Figure 11-2) and some others frequent streams and spring pools; others prefer flowing water of mountain streams. Some species occur in fairly hot springs. Terrestrial turbellarians are found in fairly moist places under stones and logs.

All members of the classes Monogenea and Trematoda (the flukes) and the class Cestoda (the tapeworms) are parasitic. Most of the Monogenea are ectoparasites, but all the trematodes and cestodes are endoparasitic. Many species have indirect life cycles with more than one host; the first host is often an invertebrate, and the final host is usually a vertebrate. Humans serve as hosts for a number of species. Certain larval stages may be free living.

CHARACTERISTICS

1. Three germ layers (**triploblastic**)
2. **Bilateral symmetry**; definite polarity of anterior and posterior ends
3. **Body flattened dorsoventrally**; oral and genital apertures mostly on ventral surface
4. Epidermis may be cellular or syncytial (ciliated in some); **rhabdites** in epidermis of most Turbellaria; epidermis a syncytial **tegument** in Monogenea, Trematoda, and Cestoda
5. Muscular system primarily of a sheath form and of mesodermal origin; layers of circular, longitudinal, and sometimes oblique fibers beneath the epidermis
6. No internal body space other than digestive tube (acoelomate); spaces between organs filled with parenchyma, a form of connective tissue or mesenchyme
7. Digestive system incomplete (gastrovascular type); absent in some
8. **Nervous system consisting of a pair of anterior ganglia with longitudinal nerve cords connected by transverse nerves and located in the mesenchyme** in most forms; similar to cnidarians in primitive forms
9. Simple sense organs; eyespots in some
10. Excretory system of two lateral canals with branches bearing **flame cells** (**protonephridia**); lacking in some primitive forms
11. Respiratory, circulatory, and skeletal systems lacking; lymph channels with free cells in some trematodes
12. Most forms monoecious; reproductive system complex, usually with well-developed gonads, ducts, and accessory organs; internal fertilization; development direct in free-swimming forms and those with a single host in the life cycle; usually indirect in internal parasites in which there may be a complicated life cycle often involving several hosts
13. Class Turbellaria mostly free living; classes Monogenea, Trematoda, and Cestoda entirely parasitic

CLASSIFICATION

Class Turbellaria (tur′bel-lar′e-a) (L. *turbellae* [pl.], stir, bustle, + *aria*, like or connected with): the turbellarians. Usually free-living forms with soft, flattened bodies; covered with ciliated epidermis containing secreting cells and rodlike bodies (rhabdites); mouth usually on ventral surface sometimes near center of body; no body cavity except intercellular lacunae in parenchyma; mostly hermaphroditic, but some have asexual fission. Examples: *Dugesia* (planaria), *Microstomum, Planocera.*

Class Monogenea (mon′o-gen′e-a) (Gr. *mono*, single, + *gene*, origin, birth): the monogenetic flukes. Body covered with a syncytial tegument without cilia; body usually leaflike to cylindrical in shape; posterior attachment organ with hooks, suckers, or clamps, usually in combination; monoecious; development direct, with single host and usually with free-swimming, ciliated larva; all parasitic, mostly on skin or gills of fish. Examples: *Dactylogyrus, Polystoma, Gyrodactylus.*

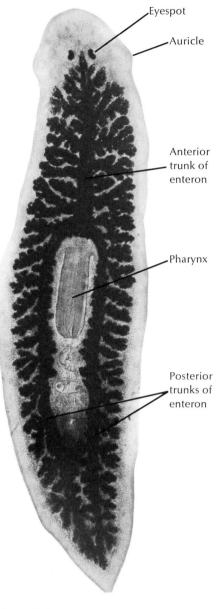

Figure 11-2

Stained planarian.
Courtesy Carolina Biological Supply Co., Burlington, N.C.

Class Trematoda (trem′a-to′da) (Gr. *trematodes*, with holes, + *eidos*, form): the digenetic flukes. Body covered with a syncytial tegument without cilia; leaflike or cylindrical in shape; usually with oral and ventral suckers, no hooks; alimentary canal usually with two main branches; mostly monoecious; development indirect, with first host a mollusc, final host usually a vertebrate; parasitic in all classes of vertebrates. Examples: *Fasciola, Clonorchis, Schistosoma.*

Class Cestoda (ses-to′da) (Gr. *kestos*, girdle, + *eidos*, form): the tapeworms. Body covered with nonciliated, syncytial tegument; general form of body tapelike; scolex with suckers or hooks, sometimes both, for attachment; body usually divided into series of proglottids; no digestive organs; usually monoecious; parasitic in digestive tract of all classes of vertebrates; development indirect with two or more hosts; first host may be vertebrate or invertebrate. Examples: *Diphyllobothrium, Hymenolepis, Taenia.*

_____ Class Turbellaria

Turbellarians are mostly free-living worms that range in length from 5 mm or less to 50 cm. Usually covered with ciliated epidermis, these are mostly creeping worms that combine muscular with ciliary movements to achieve locomotion. The mouth is on the ventral side. Unlike the trematodes and cestodes, they have simple life cycles.

The orders of Turbellaria can be divided into two groups based on specialization of the female reproductive system and embryogenesis. In the more primitive group, the yolk for nutrition of the developing embryo is contained in the egg cell itself (**endolecithal**), and the embryogenesis shows the spiral determinate cleavage typical of protostomes (p. 101). In the more advanced group, the egg cell contains little or no yolk, and the yolk is contributed by cells released from organs called **vitellaria**. Usually a number of yolk cells surround the zygote within the eggshell (**ectolecithal**), affecting cleavage in such a way that the spiral pattern cannot be distinguished. Other characteristics of value in distinguishing orders of Turbellaria are the form of the gut (present or absent; simple or branched; pattern of branching) and the pharynx (simple; folded; bulbous). Except for the order Polycladida (Gr. *poly*, many, + *klados*, branch), the turbellarians with endolecithal eggs have a simple gut or no gut and a simple pharynx. In a few there is no recognizable pharynx. The polyclads have a folded pharynx and a gut with many branches. The polyclads include many marine forms of moderate to large size (3 to 20 mm), and a highly branched intestine is correlated with larger size in Turbellaria. Members of the order Tricladida (Gr. *treis*, three, + *klados*, branch), which are in the ectolecithal group and include the freshwater planaria, have a three-branched intestine.

Of the orders in the endolecithal group, the order Acoela (Gr. *a*, without, + *koilos*, hollow) is often regarded as having changed least from the ancestral form. Its members are small and have a mouth but no gastrovascular cavity or excretory system. Food is merely passed through the mouth or pharynx into temporary spaces that are surrounded by mesenchyme where gastrodermal phagocytic cells digest the food intracellularly. The Acoela have a diffuse nervous system.

Form and function

The freshwater planarians, such as *Dugesia* (formerly called *Euplanaria* but changed by priority to *Dugesia* after Dugès, who first described the form in 1830), belong to the Tricladida and are used extensively in introductory laboratory courses. The outer covering is a ciliated epidermis resting on a basement membrane. It contains rod-shaped **rhabdites** that, when discharged with water, swell

Based on observations with the light microscope, the Acoela were long thought to be syncytial. However, cell membranes have been shown in all species examined by electron microscopy.

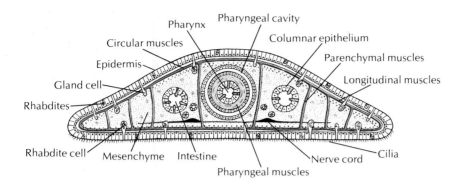

Figure 11-3

Cross section of planarian through pharyngeal region, showing relationships of body structures.

and form a protective mucous sheath around the body. Single-cell mucous glands open on the surface of the epidermis (Figure 11-3). Tyler (1976) described dual-gland adhesive organs in the epidermis of most turbellarian orders. These consist of three cell types: viscid and releasing gland cells and anchor cells (Figure 11-4). Secretions of the viscid gland cells apparently fasten the microvilli of the anchor cells to the substrate, and the secretions of the releasing gland cells provide a quick, chemical detaching mechanism.

In the body wall below the basement membrane of tubellarians are layers of **muscle fibers** that run circularly, longitudinally, and diagonally. A meshwork of **parenchyma** cells, developed from mesoderm, fills the spaces between muscles and visceral organs. Parenchyma cells in some, perhaps all, flatworms are not a separate cell type but are the noncontractile portions of muscle cells (Lumsden and Specian, 1980).

Very small planaria swim by means of their cilia. Others move by gliding, head slightly raised, over a slime track secreted by the marginal adhesive glands. The beating of the epidermal cilia in the slime track drives the animal along, while rhythmical muscular waves can be seen passing backward from the head. Large polyclads and terrestrial turbellarians crawl by muscular undulations, much in the manner of a snail.

Nutrition and digestion

The digestive system includes a mouth, a pharynx, and an intestine. The pharynx, enclosed in a **pharyngeal sheath** (Figure 11-5), opens posteriorly just inside the mouth, through which it can extend (Figure 11-5). The intestine has three many-branched trunks, one anterior and two posterior. The whole forms a **gastrovascular cavity** lined with columnar epithelium (Figure 11-5).

Planarians are mainly carnivorous, feeding largely on small crustaceans, nematodes, rotifers, and insects. They can detect food from some distance by means of chemoreceptors. They entangle their prey in mucus secretions from the mucous glands and rhabdites. The planarian grips its prey with its anterior end, wraps its body around the prey, extends its proboscis, and sucks up minute bits of the food. Intestinal secretions contain proteolytic enzymes for some **extracellular digestion.** Bits of food are sucked up into the intestine, where the phagocytic cells of the gastrodermis complete the digestion (**intracellular**). The gastrovascular cavity ramifies to most parts of the body, and food is absorbed through its walls into the body cells. Undigested food is egested through the pharynx.

Excretion and osmoregulation

Except in the Acoela, the osmoregulatory system of turbellarians consists of **protonephridia** (excretory or osmoregulatory organs closed at the inner end) with **flame cells** (Figure 11-5, A). The flame cell is cup shaped with a tuft of cilia

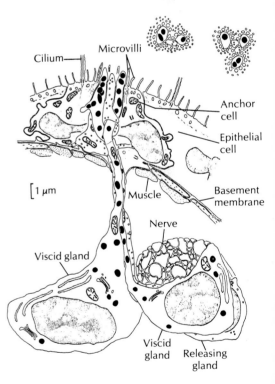

Figure 11-4

Reconstruction of dual-gland adhesive organ of the turbellarian *Haplopharynx* sp. There are two viscid glands and one releasing gland, which lie beneath the body wall. The anchor cell lies within the epidermis, and one of the viscid glands and the releasing gland are in contact with a nerve.
From Tyler, S. 1976. Zoomorphologic **84**:1-76.

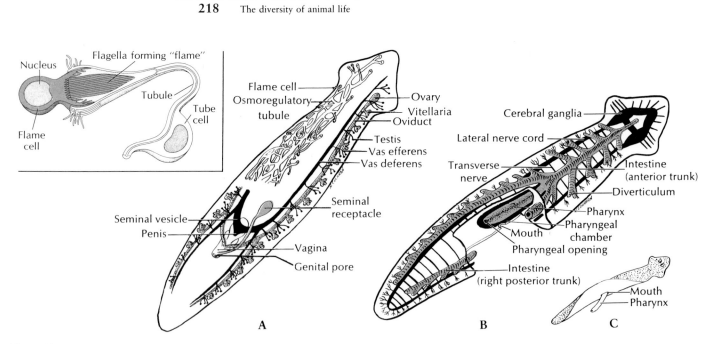

Figure 11-5

A, Reproductive and osmoregulatory systems in planaria. Portions of male and female organs omitted to show part of osmoregulatory system. Inset at left is enlargement of flame cell. B, Diagrammatic view of digestive system and ladder-type nervous system of planaria. Cut section shows relation of pharynx, in resting position, to digestive system and mouth on ventral surface. C, Pharynx extended through ventral mouth.

extending from the inner face of the cup. The rim of the cup is elongated into fingerlike projections that interdigitate with similar projections of a tubule cell. The space (lumen) enclosed by the tubule cell continues into collecting ducts that finally open to the outside by pores. The beat of the cilia (resembling a flickering flame) provides a negative pressure to draw fluid through the delicate interdigitations between the flame cell and the tubule cell. The wall of the duct beyond the flame cell commonly bears folds or microvilli that probably function in resorption of certain ions or molecules.

Among the various flatworms, there may be a single protonephridium or from one to four pairs. In planarians they anastomose into a network along each side of the animal (Figure 11-5) and may empty through many nephridiopores. That this system is mainly osmoregulatory is indicated by the observation that it is reduced or absent in marine turbellarians, which do not have to expel excess water.

Metabolic wastes are removed largely by diffusion through the body wall.

Respiration

There are no respiratory organs. Exchange of gases takes place through the body surface.

Nervous system

The most primitive flatworm nervous system, found in some of the acoels, is a **subepidermal nerve plexus** resembling the nerve net of the cnidarians. Other flatworms have, in addition to a nerve plexus, one to five pairs of **longitudinal nerve cords** lying under the muscle layer. The more advanced flatworms tend to have the lesser number of nerve cords. Freshwater planarians have one ventral pair (Figure 11-5, B). Connecting nerves form a "ladder-type" pattern. The brain is a bilobed mass of ganglion cells arising anteriorly from the ventral nerve cords. Except in the acoels, which have a diffuse system, the neurons are organized into sensory, motor, and association types—an important advance in the evolution of the nervous system.

Sense organs

Active locomotion in flatworms has favored not only cephalization in the nervous system but also advancements in the development of sense organs. **Ocelli,** or light-sensitive eyespots, are common in the turbellarians (Figure 11-2).

Tactile cells and chemoreceptive cells are abundant over the body, and in planarians they form definite organs on the auricles (the earlike lobes on the sides of the head). Some planarians also have statocysts for equilibrium and rheoreceptors for sensing water current direction.

Reproduction and regeneration

Many turbellarians reproduce both asexually (by fission) and sexually. Asexually, the freshwater planarian merely constricts behind the pharynx and separates into two animals, each of which regenerates the missing parts—a quick means of population increase. There is evidence that a reduced population density results in an increase in the rate of fissioning. In some forms, such as *Stenostomum* (Gr. *stenos,* narrow, + *stoma,* mouth) and *Microstomum* (Gr. *mikros,* small, + *stoma,* mouth), in which fissioning occurs, the individuals do not separate at once but remain attached, forming chains of zooids (Figure 11-6, *B* and *C*).

The considerable powers of regeneration in planarians have provided an interesting system for experimental studies of development. For example, a piece excised from the middle of the planarian can regenerate both a new head and a new tail. However, the piece retains its original polarity: the head grows at the anterior end and the tail at the posterior end. An extract of heads added to a culture medium containing headless worms will prevent regeneration of new heads, suggesting that substances in one region will suppress the regeneration of the same region at another level of the body. Many other experiments could be cited.

Turbellarians are monoecious (hermaphroditic) but practice cross-fertilization. During the breeding season each individual develops both male and female organs, which usually open through a common genital pore (Figure 11-5, *A*).

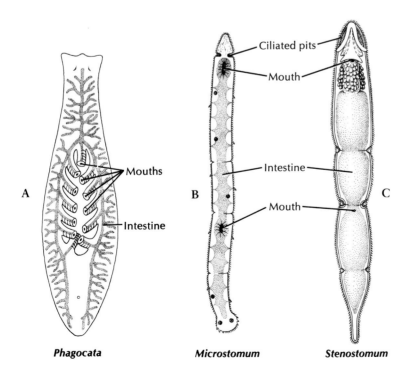

Figure 11-6

Some small freshwater turbellarians. **A,** *Phagocata* has numerous pharynges. **B** and **C,** Incomplete fission results for a time in a series of attached zooids.

Phagocata   *Microstomum*   *Stenostomum*

After copulation one or more fertilized eggs and some yolk cells become enclosed in a small cocoon. The cocoons are attached by little stalks to the underside of stones or plants. Embryos emerge as juveniles that resemble mature adults. In some marine forms the egg develops into a ciliated free-swimming larva.

Class Monogenea

The monogenetic flukes traditionally have been placed as an order of the Trematoda, but they are sufficiently different to deserve a separate class. They are all parasites, primarily of the gills and external surfaces of fish. A few are found in the urinary bladders of frogs and turtles, and one has been reported from the eye of a hippopotamus. Although widespread and common, monogeneans seem to cause little damage to their hosts under natural conditions. However, like numerous other fish pathogens, they become a serious threat when their hosts are crowded together, as in fish farming.

The life cycles of monogeneans are direct, with a single host. The egg hatches a ciliated larva, the **oncomiracidium,** that attaches to the host or swims around awhile before attachment. The oncomiracidium bears hooks on its posterior, which in many species become the hooks on the large posterior attachment organ (**opisthaptor**) of the adult. Because the monogenean must cling to the host and withstand the force of water flow over the gills or skin, adaptive radiation has produced a wide array of opisthaptors in different species. Opisthaptors may bear large and small hooks, suckers, and clamps, often in combination with each other.

Common genera are *Gyrodactylus* (L. *gyro,* a circle, + Gr. *daktylos,* toe, finger) (Figure 11-7) and *Dactylogyrus* (Gr. *daktylos,* toe, finger, + L. *gyro,* a circle), both of economic importance to fish culturists, and *Polystoma* (Gr. *polys,* many, + *stoma,* mouth), found in the urinary bladder of frogs.

Class Trematoda

Trematodes are all parasitic flukes, and as adults they are almost all found as endoparasites of vertebrates. They are chiefly leaflike in form and are structurally similar in many respects to the ectolecithal Turbellaria. A major difference is found in the body covering, or **tegument,** which does not bear cilia in the adult. Furthermore, in common with Monogenea and Cestoda, cell bodies are sunk beneath the outer layer (Figure 11-8) and superficial muscle layers and they communicate with the outer layer (distal cytoplasm) by processes extending between the muscles. Because the distal cytoplasm is continuous, with no intervening cell membranes, the tegument is **syncytial.** This peculiar epidermal arrangement is probably related to adaptations for parasitism in ways that are still unclear.

Other structural adaptations for parasitism are more apparent: various penetration glands or glands to produce cyst material; organs for adhesion such as suckers and hooks; and increased reproductive capacity. Otherwise, trematodes retain several turbellarian characteristics, such as a well-developed alimentary canal (but with the mouth at the anterior, or cephalic, end) and similar reproductive, excretory, and nervous systems, as well as a musculature and parenchyma that are only slightly modified from those of the Turbellaria. Sense organs are poorly developed.

Of the subclasses of Trematoda, Aspidogastrea and Didymozoidea are small and poorly known groups, but Digenea (Gr. *dis,* double, + *genos,* race) is a large group with many species of medical and economic importance.

We feel that there are ample reasons for considering the Monogenea in a class separate from the digenetic trematodes (see Schmidt and Roberts [1985] in references for Part Two, p. 590). However, this separation is not universally accepted (see Barnes [1987] in references for Part Two).

Figure 11-7

Gyrodactylus cylindriformis, ventral view. From Mueller, J.F., and H.J. Van Cleave. 1932. Roosevelt Wildlife Animals.

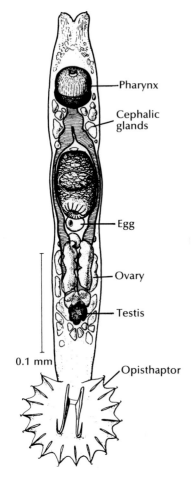

Pharynx

Cephalic glands

Egg

Ovary

Testis

0.1 mm

Opisthaptor

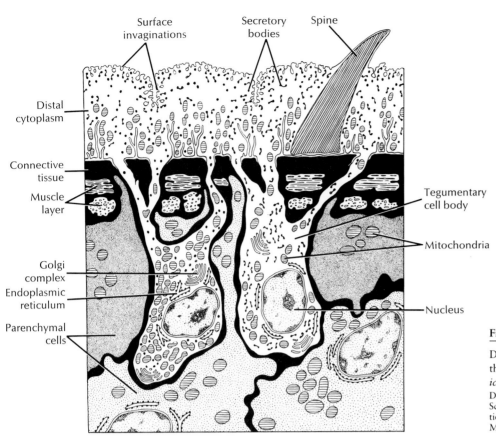

Distal cytoplasm

Connective tissue

Muscle layer

Golgi complex

Endoplasmic reticulum

Parenchymal cells

Surface invaginations

Secretory bodies

Spine

Tegumentary cell body

Mitochondria

Nucleus

Figure 11-8

Diagrammatic drawing of the structure of the tegument of a trematode *Fasciola hepatica.*

Drawing by L.T. Threadgold. Modified from Schmidt G.D., and L.S. Roberts. 1985. Foundations of parasitology, ed. 3, St. Louis, The C.V. Mosby Co.

Subclass Digenea

With rare exceptions, digenetic trematodes have an indirect life cycle, the first (**intermediate**) host being a mollusc and the **definitive** host (the host in which sexual reproduction occurs, sometimes called the **final** host) being a vertebrate. In some species a second, and sometimes even a third, intermediate host intervenes. The group has been very successful and they inhabit, according to species, a wide variety of sites in their hosts: all parts of the digestive tract, respiratory tract, circulatory system, urinary tract, and reproductive tract.

One of the world's most amazing biological phenomena is the digenean life cycle. Although the cycles of different species vary widely in detail, a typical example would include the adult, egg, miracidium, sporocyst, redia, cercaria, and metacercaria stages (Figure 11-9). The **egg** usually passes from the definitive host in the excreta and must reach water to develop further. There, it hatches to a free-swimming, ciliated larva, the **miracidium.** The miracidium penetrates the tissues of a snail, where it transforms into a **sporocyst.** The sporocyst reproduces asexually to yield either more sporocysts or a number of **rediae.** The rediae, in turn, reproduce asexually to produce more rediae or to produce **cercariae.** In this way a single egg can give rise to an enormous number of progeny. The cercariae emerge from the snail and penetrate a second intermediate host or encyst on vegetation or other objects to become **metacercariae,** which are juvenile flukes. The adult grows from the metacercaria when that stage is eaten by the definitive host.

Some of the most serious parasites of humans and domestic animals belong to the Digenea. The first digenetic life cycle to be worked out was that of *Fasciola hepatica* (L. *fasciola,* a small bundle, band), which causes "liver rot" in sheep and

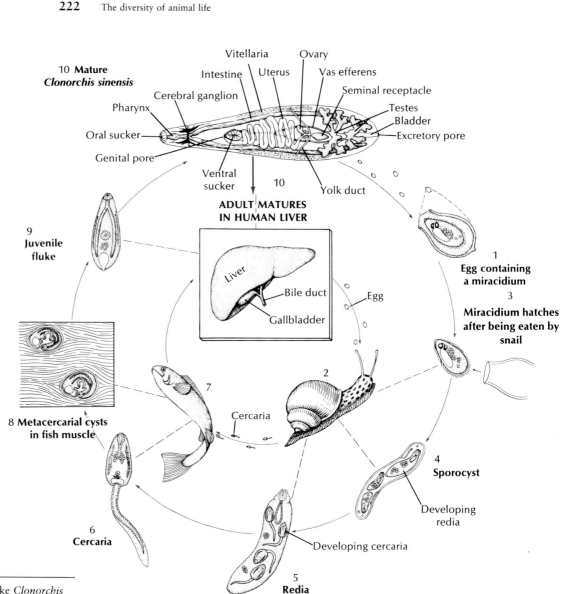

10 **Mature**
Clonorchis sinensis

Vitellaria
Ovary
Intestine   Uterus   Vas efferens
Cerebral ganglion   Seminal receptacle
Pharynx   Testes
Bladder
Oral sucker   Excretory pore
Genital pore   Yolk duct
Ventral sucker   **10**

**ADULT MATURES
IN HUMAN LIVER**

Liver
Bile duct
Gallbladder

Egg

1
**Egg containing
a miracidium**

3
**Miracidium hatches
after being eaten by
snail**

9
**Juvenile
fluke**

7

Cercaria

2

4
Sporocyst

Developing
redia

8 **Metacercarial cysts
in fish muscle**

6
Cercaria

Developing cercaria

5
Redia

Figure 11-9

Life cycle of human liver fluke *Clonorchis sinensis*. Egg, *1*, shed from adult trematode, *10*, in bile ducts of human, is carried out of body in feces and is ingested by snail *(Parafossarulus), 2,* in which miracidium, *3,* hatches and becomes mother sporocyst. *4,* Young rediae are produced in sporocyst, grow, *5,* and in turn produce young cercariae. Cercariae then leave snail, *6,* find a fish host, *7,* and burrow under scales to encyst in muscle, *8.* When raw or improperly cooked fish containing cysts is eaten by a human, metacercaria is released, *9,* and enters bile duct, where it matures, *10,* to shed eggs into feces, *1,* thus starting another cycle.

other ruminants. The adult fluke lives in the bile passage of the liver, and the eggs are passed in the feces. After hatching, the miracidium penetrates a snail to become a sporocyst. There are two generations of rediae, and the cercaria encysts on vegetation. When the infested vegetation is eaten by the sheep or other ruminant (or sometimes humans), the metacercariae excyst and grow into young flukes.

Clonorchis sinensis: liver fluke in humans

Clonorchis (Gr. *clon*, branch, + *orchis*, testis) (Figure 11-9) is the most important liver fluke of humans and is common in many regions of the Orient, especially in China, southern Asia, and Japan. Cats, dogs, and pigs are also often infected.

Structure

The worms vary from 10 to 20 mm in length (Figure 11-9). There structure is typical of many trematodes in most respects. They have an **oral sucker** and a **ventral sucker.** The **digestive system** consists of a pharynx, a muscular esophagus, and two long, unbranched intestinal ceca. The **excretory system** consists of two protonephridial tubules, with branches provided with flame cells or bulbs. The two tubules unite to form a single median tubule that opens to the outside. The **nervous system,** like that of turbellarians, is made up of two cerebral ganglia connected to longitudinal cords that have transverse connectives.

The **reproductive system** is hermaphroditic and complex. The male system is made up of two branched **testes** and two **vasa efferentia** that unite to form a single **vas deferens,** which widens into a **seminal vesicle.** The seminal vesicle leads into an **ejaculatory duct,** which terminates at the genital opening. Unlike most trematodes, *Clonorchis* does *not* have a protrusible copulatory organ, the cirrus. The female system contains a branched **ovary** with a short **oviduct,** which is joined by ducts from the **seminal receptacle** and the **vitellaria** at the **ootype.** The ootype is surrounded by a glandular mass, **Mehlis' gland,** of uncertain function. From Mehlis' gland the much-convoluted **uterus** runs to the genital pore. Cross-fertilization between individuals is usual, and sperm are stored in the seminal receptacle. When an ovum is released from the ovary, it is joined by a sperm and a group of vitelline cells and is fertilized. The vitelline cells release a proteinaceous shell material, which is stabilized by a chemical reaction; the Mehlis' gland secretions are added; and the egg passes into the uterus.

Life cycle

The normal habitat of the adults is in the bile passageways of humans and other fish-eating mammals (Figure 11-9). The eggs, each containing a complete miracidium, are shed into the water with the feces but do not hatch until they are ingested by the snail *Parafossarulus* or related genera. The eggs, however, may live for some weeks in water. In the snail the miracidium enters the tissues and transforms into the sporocyst (a baglike structure with embryonic germ cells), which produces one generation of rediae. The redia is elongated, with an alimentary canal, a nervous system, an excretory system, and many germ cells in the process of development. The rediae pass into the liver of the snail where the germ cells continue embryonation and give rise to the tadpolelike cercariae.

The cercariae escape into the water, swim about until they meet with fish of the family Cyprinidae, and then bore into the muscles or under the scales. Here the cercariae lose their tails and encyst as metacercariae. If a mammal eats raw infected fish, the metacercarial cyst dissolves in the intestine, and the young flukes apparently migrate up the bile duct, where they become adults. There the flukes may live for 15 to 30 years.

The effect of the flukes on humans depends mainly on the extent of the infection. A heavy infection can cause a pronounced cirrhosis of the liver and can result in death. Cases are diagnosed through fecal examinations. To avoid infection, all fish used as food should be thoroughly cooked. Destruction of the snails that carry larval stages would be a method of control.

Schistosoma: blood flukes

Schistosomiasis, infection with blood flukes of the genus *Schistosoma* (Gr. *schistos,* divided, + *soma,* body), ranks as one of the major infectious diseases in the world, with 200 million people infected. The disease is widely prevalent over much of Africa and parts of South America, the West Indies, the Middle East, and the Far East. The old generic name for the worms was *Bilharzia* (from Theodor Bilharz, German parasitologist who discovered *Schistosoma haematobium*), and the infection was called bilharziasis, a name still used in many areas.

The blood flukes differ from most other flukes in being dioecious and having the two branches of the digestive tube united into a single tube in the posterior

Unfortunately, some projects intended to raise the standard of living in some tropical countries, such as the Aswan High Dam in Egypt, have increased the prevalence of schistosomiasis by creating more habitats for the snail intermediate hosts. Before the dam was constructed, the 500 miles of the Nile River between Aswan and Cairo was subjected to annual floods; alternate flooding and drying killed many snails. Four years after dam completion, the prevalence of schistosomiasis had increased sevenfold along that segment of the river. The prevalence in fishermen around the lake above the dam increased from a very low level to 76%.

Figure 11-10

Adult male and female *Schistosoma mansoni* in copulation. Male has long gynecophoric canal that holds female (the darkly stained individual) during insemination and oviposition. Humans are usually hosts of adult parasites, found mainly in Africa but also in South America and elsewhere. Humans become infected by wading or bathing in cercaria-infested waters.
AFIP No. 56-3334.

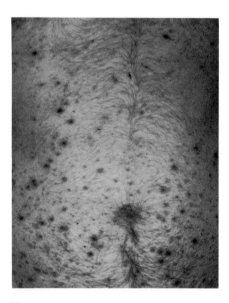

Figure 11-11

Human abdomen, showing schistosome dermatitis caused by penetration of schistosome cercariae that are unable to complete development in humans. Sensitization to allergenic substances released by cercariae results in rash and itching.
Original photograph courtesy R.E. Kuntz. From H. Zaiman (ed.). A pictorial presentation of parasites.

part of the body. The male is broader and heavier and has a large, ventral groove, the **gynecophoric canal,** posterior to the ventral sucker. The gynecophoric canal embraces the long, slender female (Figure 11-10).

Three species account for most of the schistosomiasis in humans: *S. mansoni,* which lives primarily in the venules draining the large intestine; *S. japonicum,* which is found mostly in the venules of the small intestine; and *S. haematobium,* which lives in the venules of the urinary bladder. *S. mansoni* is common in parts of Africa, Brazil, northern South America, and the West Indies; species of *Biomphalaria* are the principal snail intermediate hosts. *S. haematobium* is widely prevalent in Africa, using snails of the genera *Bulinus* and *Physopsis* as the main intermediate hosts. *S. japonicum* is confined to the Far East, and its hosts are several species of *Oncomelania.*

The life cycle of blood flukes is similar in all species. Eggs are discharged in human feces or urine; if they get into water, they hatch out as ciliated miracidia, which must contact the required kind of snail within a few hours to survive. In the snail, they transform into sporocysts, which produce another generation of sporocysts. The daughter sporocysts give rise to cercariae directly, without the formation of rediae. The cercariae escape from the snail and swim about until they come in contact with the bare skin of a human. They penetrate the skin, shedding their tails in the process, and reach a blood vessel where they enter the circulatory system. There is no metacercarial stage. The young schistosomes make their way to the hepatic portal system of blood vessels and undergo a period of development in the liver before migrating to their characteristic sites. As eggs are released by the adult female, they are somehow extruded through the wall of the venule and through the gut or bladder lining, to be voided with the feces or urine, according to species. Many eggs do not make this difficult transit and are swept by the blood flow back to the liver or other areas, where they become centers of inflammation and tissue reaction.

The main ill effects of schistosomiasis result from the eggs. With *S. mansoni* and *S. japonicum,* eggs in the intestinal wall cause ulceration, abscesses, and bloody diarrhea with abdominal pain. Similarly, *S. haematobium* causes ulceration of the bladder wall with bloody urine and pain on urination. Eggs swept to the liver or other sites cause symptoms associated with the organs where they lodge. When they are caught in the capillary bed of the liver, they impede circulation and cause cirrhosis, a fibrotic reaction that interferes with liver function. Of the three species, *S. haematobium* is considered the least serious and *S. japonicum* the most severe. The prognosis is poor in heavy infections of *S. japonicum* without early treatment.

Control is best carried out by educating people to dispose of their body wastes hygienically, a difficult problem with poor people under primitive conditions.

Schistosome dermatitis (swimmer's itch)

Various species of schistosomes in several genera are known to cause a rash or dermatitis when their cercariae penetrate hosts that are unsuitable for further development (Figure 11-11). The cercariae of several genera whose normal hosts are North American birds cause dermatitis in bathers in northern lakes. The severity of the rash increases with an increasing number of contacts with the organisms, or sensitization. After penetration, the cercariae are attacked and killed by the host's immune mechanisms, and they release allergenic substances, causing itching. The condition is more an annoyance than a serious threat to health, but there may be economic losses to persons depending on vacation trade around infested lakes.

Paragonimus: lung flukes

Several species of *Paragonimus* (Gr. *para*, beside, + *gonimos*, generative), a fluke that lives in the lungs of its host, are known from a variety of mammals. *Paragonimus westermani* (Figure 11-12), found in the Orient, Southwest Pacific, and some parts of South America, parasitizes a number of wild carnivores, humans, pigs, and rodents. Its eggs are coughed up in the sputum, swallowed, then eliminated with the feces. The metacercariae are found in freshwater crabs, and the infection is acquired by eating uncooked crab meat. The infection causes respiratory symptoms, with breathing difficulties and chronic cough. Fatal cases are common. A closely related species, *P. kellicotti*, is found in mink and similar animals in North America, but only one human case has been recorded. Its metacercariae are in crayfish.

Some other trematodes

Fasciolopsis buski (L. *fasciola*, small bundle, + Gr. *opsis*, appearance) (intestinal fluke of humans) parasitizes humans and pigs in India and China. Larval stages occur in several species of planorbid snails, and the cercariae encyst on water chestnuts, an aquatic vegetation eaten raw by humans and pigs.

Leucochloridium is noted for its remarkable sporocysts. Snails *(Succinea)* eat vegetation infected with eggs from bird droppings. The sporocysts become much enlarged and branched, and the cercariae encyst within the sporocyst. The sporocysts enter the snail's head and tentacles, become brightly striped with orange and green bands, and pulsate at frequent intervals. Birds are attracted by the enlarged and pulsating tentacles, eat the snails, and so complete the life cycle.

____ Class Cestoda

Cestoda, or tapeworms, differ in many respects from the preceding classes. They usually have long flat bodies in which there is a linear series of sets of reproductive organs. Each set is called a **proglottid** and is usually set off at its anterior and posterior ends by zones of muscle weakness, marked externally by grooves. There is a complete lack of a digestive system. As in Monogenea and Trematoda, there are no external, motile cilia in the adult, and the tegument is of a distal cytoplasm with sunken cell bodies beneath the superficial muscle layer (Figure 11-13). In contrast to the monogenes and trematodes, however, their entire surface is covered with minute projections called **microtriches.** The microtriches greatly enlarge the surface area of the tegument, which is a vital adaptation of the tapeworm since it must absorb all its nutrients across the tegument.

Tapeworms are nearly all monoecious. They have well-developed muscles, and their excretory system and nervous system are somewhat similar to those of other flatworms. They have no special sense organs but do have sensory endings in the tegument that are modified cilia (Figure 11-13). One of their most specialized structures is the **scolex,** or holdfast, which is the organ of attachment. It is usually provided with suckers or suckerlike organs and often with hooks or spiny tentacles (Figure 11-14).

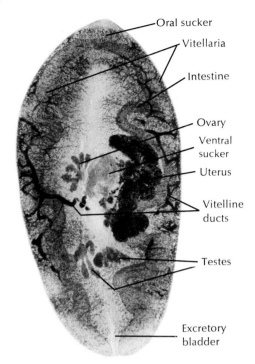

Oral sucker
Vitellaria
Intestine
Ovary
Ventral sucker
Uterus
Vitelline ducts
Testes
Excretory bladder

Figure 11-12

Pulmonary fluke *Paragonimus westermani* produces paragonimiasis in human lung. Adults are up to 2 cm long. Eggs discharged in sputum or feces hatch into free-swimming miracidia that enter snails. Cercariae from snails enter freshwater crabs and encyst in soft tissues. Humans are infected by eating poorly cooked crabs or by drinking water containing larvae freed from dead crabs. Photograph by R.E. Kuntz and J.A. Moore. From Schmidt, G.D., and L.S. Roberts. 1985. Foundations of parasitology, ed. 3. St. Louis, The C.V. Mosby Co.

Gutless Wonder

Though lacking skeletal strengths
Which we associate with most
Large forms, tapeworms go to
    great lengths
To take the measure of a host.

Monotonous body sections
In a limp mass-production line
Have nervous and excretory
    connections
And the means to sexually combine

And to coddle countless progeny
But no longer have the guts
To digest for themselves or live free
Or know a meal from soup to nuts.

Figure 11-13

Schematic drawing of a longitudinal section through a sensory ending in the tegument of *Echinococcus granulosus*.
From Morseth, D.J. 1967. J. Parasitol. 53:492-500.

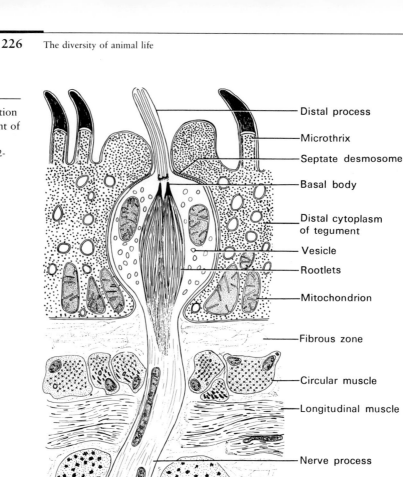

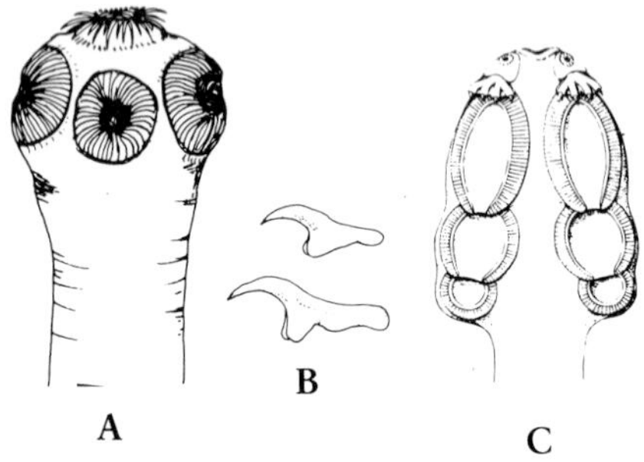

- Distal process
- Microthrix
- Septate desmosome
- Basal body
- Distal cytoplasm of tegument
- Vesicle
- Rootlets
- Mitochondrion
- Fibrous zone
- Circular muscle
- Longitudinal muscle
- Nerve process
- Glycogen

Figure 11-14

Two tapeworm scolices. **A**, Scolex of *Taenia solium* (pork tapeworm) with apical hooks and suckers. (Scolex of *Taeniarhynchus saginatus* is similar, but without hooks.) **B**, Hooks of *T. solium*. **C**, Scolex of *Acanthobothrium coronatum*, a tapeworm of sharks. This species has large leaflike sucker organs divided into chambers with apical suckers and hooks.

With rare exceptions, all cestodes require at least two hosts, and the adult is a parasite in the digestive tract of vertebrates. Often one of the intermediate hosts is an invertebrate.

The class Cestoda is divided into two subclasses: Cestodaria and Eucestoda. The Cestodaria is a small group found in a few lower vertebrates. Their bodies are not divided into separate proglottids and have only one set of reproductive organs (**monozoic**.) Their larvae have 10 hooks.

The Eucestoda contains the great majority of species in the class. With the exception of two small orders, the members of this subclass have the body divided into a series of proglottids and are thus referred to as **polyzoic**. Their larval forms all have six hooks, rather than 10 as in the Cestodaria. The main body of the worms, the chain of proglottids, is called the **strobila**. Typically, there is a **germinative zone** just behind the scolex where new proglottids are formed. As new proglottids are differentiated in front of it, each individual proglottid moves posteriorly in the strobila, and its gonads mature. The proglottid is usually fertilized by another proglottid in the same or a different strobila. The shelled embryos form in the uterus of the proglottid, and either they are expelled through a uterine pore or the entire proglottid is shed from the worm as it reaches the posterior end.

Some zoologists have maintained that the proglottid formation of cestodes represents "true" segmentation (metamerism), but we do not support this view. Segmentation of tapeworms is best considered a replication of sex organs to

Table 11-1   Common Cestodes of Humans

Common and scientific name	Means of infection; prevalence in humans
Beef tapeworm *(Taeniarhynchus saginatus)*	Eating rare beef; most common of all tapeworms in humans
Pork tapeworm *(Taenia solium)*	Eating rare pork; less common than *T. saginatus*
Fish tapeworm *(Diphyllobothrium latum)*	Eating rare or poorly cooked fish; fairly common in Great Lakes region of United States, and other areas of world where raw fish is eaten
Dog tapeworm *(Dipylidium caninum)*	Unhygienic habits of children (juveniles in flea and louse); moderate frequency
Dwarf tapeworm *(Vampirolepis nana)*	Juveniles in flour beetles; common
Unilocular hydatid *(Echinococcus granulosus)*	Cysts of juveniles in humans; infection by contact with dogs; common wherever humans are in close relationship with dogs and ruminants
Multilocular hydatid *(Echinococcus multilocularis)*	Cysts of juveniles in humans; less common than unilocular hydatid

increase reproductive capacity and is not related to the metamerism found in Annelida, Arthropoda, and Chordata (see pp. 113 and 313).

More than 1000 species of tapeworms are known to parasitologists. Almost all vertebrate species are infected. Normally, adult tapeworms do little harm to their hosts. The most common tapeworms found in humans are given in Table 11-1.

Taeniarhynchus saginatus: beef tapeworm

Structure

Taeniarhynchus saginatus (Gr. *tainia*, band, ribbon, + *rhynchos*, beak, snout) is referred to as the beef tapeworm, but it lives as an adult in the alimentary canal of humans. The juvenile form is found primarily in the intermuscular tissue of cattle. The mature adult may reach a length of 10 m or more. Its scolex has four suckers for attachment to the intestinal wall, but no hooks. A short neck connects the scolex to the strobila, which may be made up of as many as 2000 proglottids. The terminal gravid proglottids bear shelled, infective larvae (Figure 11-15) and are detached and shed in the feces.

The tapeworm shows some unity in its organization, for **excretory canals** in the scolex are also connected to the canals, two on each side, in the proglottids, and two longitudinal **nerve cords** from a **nerve ring** in the scolex run back into the proglottids also (Figure 11-16). Attached to the excretory ducts are the flame cells. Each mature proglottid also contains muscles and parenchyma as well as a complete set of male and female organs similar to those of a trematode.

In the order to which this species belongs, however, the vitellaria are typically a single, compact **vitelline gland** located just posterior to the ovaries. When the gravid proglottids break off and pass out with the feces, they usually crawl out of the fecal mass and onto vegetation nearby. There they may be picked up by grazing cattle. The proglottid ruptures as it dries up, further scattering the embryos on soil and grass. The embryos may remain viable on grass for as long as 5 months.

Figure 11-15

Life cycle of beef tapeworm, *Taeniarhynchus*. Ripe proglottids break off in human intestine, pass out in feces, crawl out of feces onto grass, and are ingested by cattle. Eggs hatch in cow's intestine, freeing oncospheres, which penetrate into muscles and encyst, developing into "bladder worms." Human eats infected rare beef, and cysticercus is freed in intestine where it attaches to the intestine wall, forms a strobila, and matures.

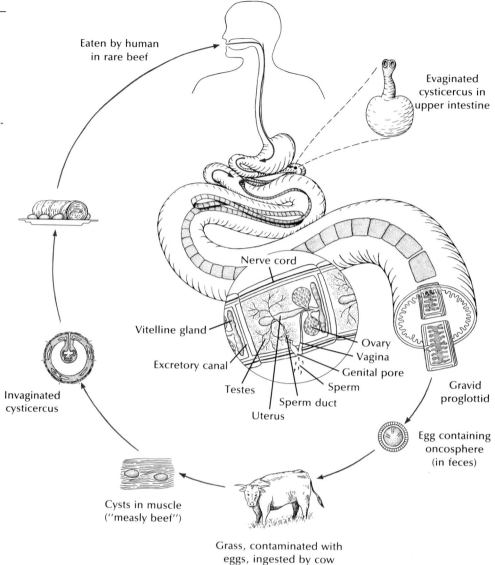

Eaten by human in rare beef

Evaginated cysticercus in upper intestine

Nerve cord

Vitelline gland

Excretory canal

Testes

Sperm duct

Uterus

Ovary

Vagina

Genital pore

Sperm

Gravid proglottid

Egg containing oncosphere (in feces)

Invaginated cysticercus

Cysts in muscle ("measly beef")

Grass, contaminated with eggs, ingested by cow

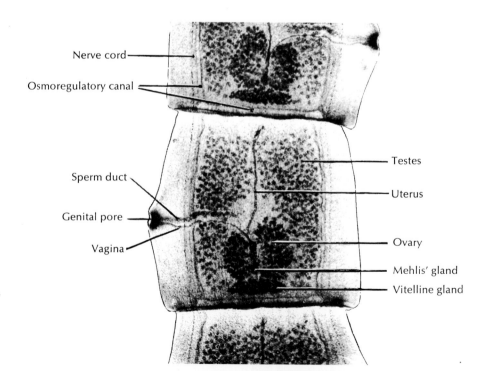

Figure 11-16

Photomicrograph of mature proglottid of *Taenia pisiformis*, dog tapeworm. Portions of two other proglottids also shown.
Courtesy General Biological Supply House, Inc., Chicago.

Nerve cord

Osmoregulatory canal

Sperm duct

Genital pore

Vagina

Testes

Uterus

Ovary

Mehlis' gland

Vitelline gland

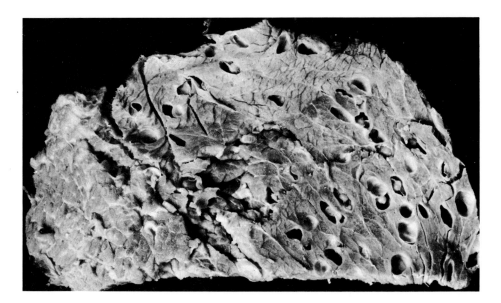

Figure 11-17

"Measly" pork showing cysts of bladder worms, *Taenia solium*. Cysticerci are up to 10 mm in largest diameter. Beef heavily infected with beef tapeworm has a similar appearance, but lighter infections may be much less obvious.
Photograph by F.M. Hickman.

Life cycle

When cattle swallow the eggs, the eggs hatch, and the six-hooked larvae (**oncospheres**) burrow through the intestinal wall into the blood or lymph vessels and finally reach voluntary muscle, where they encyst to become **bladder worms** (juveniles called **cysticerci**). There the juveniles develop an invaginated scolex but remain quiescent. When infected "measly" meat (Figure 11-17) is eaten by a suitable host, the cyst wall dissolves, the scolex evaginates and attaches to the intestinal mucosa, and new proglottids begin to develop. It takes 2 to 3 weeks for a mature worm to form. When a person is infected with one of these tapeworms, numerous gravid proglottids are expelled daily, sometimes crawling out the anus by themselves. Humans become infected by eating rare roast beef, steaks, and barbecues. Considering that about 1% of American cattle are infected, that 20% of all cattle slaughtered are not federally inspected, and that even in inspected meat one fourth of infections are missed, it is not surprising that tapeworm infection is fairly common. Infection is precluded when meat is thoroughly cooked.

Some other tapeworms

Taenia solium: pork tapeworm

The adult *Taenia solium* (Gr. *tainia,* band, ribbon) lives in the small intestine of humans, whereas the juveniles live in the muscles of pigs. The scolex has both suckers and hooks arranged on its tip (Figure 11-14) the **rostellum.** The life history of this worm is similar to that of the beef tapeworm, except that humans become infected by eating improperly cooked pork.

The pork tapeworm is much more dangerous than *T. saginatus* because the cysticerci, as well as the adults, can develop in humans. If eggs or proglottids are accidentally ingested by a human, the liberated embryos migrate to any of several organs and form cysticerci (Figure 11-17). The condition is called **cysticercosis.** Common sites are the eye or brain, and infection in such locations results in blindness or serious neurological symptoms.

Diphyllobothrium latum: fish tapeworm

The adult *Diphyllobothrium* (Gr. *dis,* double, + *phyllon,* leaf, + *bothrion,* hole, trench) is found in the intestine of humans, dogs, cats, and other mammals; the

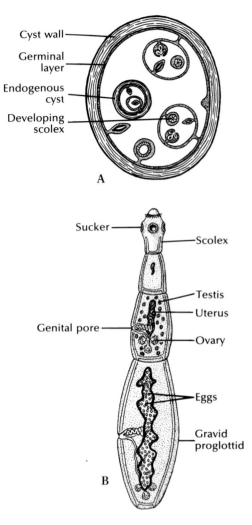

Cyst wall

Germinal layer

Endogenous cyst

Developing scolex

A

Sucker

Scolex

Testis

Uterus

Genital pore

Ovary

Eggs

Gravid proglottid

B

Figure 11-18

Echinococcus granulosus, dog tapeworm, which may be dangerous to humans. **A,** Early hydatid cyst or bladder-worm stage found in cattle, sheep, hogs, and sometimes humans produces hydatid disease. Humans acquire disease by unsanitary habits in association with dogs. When eggs are ingested, liberated larvae usually encyst in the liver, lungs, or other organs. Brood capsules containing scolices are formed from the inner layer of each cyst. The cyst enlarges, developing other cysts with brood pouches. It may grow for years, to the size of a basketball, necessitating surgery. **B,** Adult tapeworm lives in intestine of dog or other carnivore.

immature stages are in crustaceans and fish. With a length of up to 20 m, it is the largest of the cestodes that infect humans. Fish tapeworm infections are found all over the world; in the United States infections are most common in the Great Lakes region. In Finland, but apparently not other areas, the worm can cause a serious anemia.

Echinococcus granulosus: unilocular hydatid

The adult *E. granulosus* (Gr. *echinos*, hedgehog, + *kokkos*, kernel) (Figure 11-18, *B*) is found in dogs and other canines; the juveniles are found in more than 40 species of mammals, including humans, monkeys, sheep, reindeer, and cattle. Thus humans serve as an intermediate host in the case of this tapeworm. The juvenile stage is a special kind of cysticercus called a **hydatid cyst** (Gr. *hydatis*, watery vesicle). It grows slowly, but it can grow for a long time—up to 20 years—reaching the size of a basketball in an unrestricted site such as the liver. If the hydatid grows in a critical site, such as the heart or central nervous system, serious symptoms may appear in a much shorter time. The main cyst maintains a single or unilocular chamber, but within the main cyst, daughter cysts bud off, and each contains thousands of scolices. Each scolex will produce a worm when eaten by a canine. The only treatment is surgical removal of the hydatid.

PHYLUM NEMERTEA (RHYNCHOCOELA)

Nemerteans (nem-er′te-a) (Gr. *Nemertes*, one of the Nereids, unerring one) are often called the ribbon worms. Their name refers to the unerring aim of the proboscis, a long muscular tube that can be thrust out swiftly to grasp the prey. The phylum is also called Rhynchocoela (ring′ko-se′la) (Gr. *rhynchos*, beak, + *koilos*, hollow), which also refers to the proboscis. They are thread-shaped or ribbon-shaped worms; nearly all of them are marine. Some live in secreted gelatinous tubes. There are about 650 species in the group.

Nemertean worms are usually less than 20 cm long, although a few are several meters in length. *Lineus longissimus* (L. *linea*, line) is said to reach 30 m. Their colors are often bright, although most are dull or pallid. In the odd genus *Gorgonorhynchus* (Gr. *Gorgo*, name of a female monster of terrible aspect, + *rhynchos*, beak, snout) the proboscis is divided into many proboscides, which appear as a mass of wormlike structures when everted.

With a few exceptions, the general body plan of the nemerteans is similar to that of Turbellaria. Like the latter, their epidermis is ciliated and has many gland cells. Another striking similarity is the presence of flame cells in the excretory system. Recently rhabdites have been found in several nemerteans, including *Lineus*. However, nemerteans differ from flatworms in their reproductive system. They are mostly dioecious. In the marine forms there is a ciliated **pilidium larva** (Gr. *pilidion*, a small felt nightcap) (Figure 11-19). This helmet-shaped larva has a ventral mouth but no anus—another flatworm characteristic. It also has some resemblance to the trochophore larva that is found in annelids and molluscs. Other flatworm characteristics are the presence of bilateral symmetry and a mesoderm and the lack of a coelom. All in all, the present evidence seems to indicate that the nemerteans came from an ancestral form closely related to Platyhelminthes.

The nemerteans show some advances over the flatworms. One of these is the eversible **proboscis** and its sheath, for which there are no counterparts among Platyhelminthes. Another difference is the presence of an **anus** in the adult. These forms have a **complete digestive system,** the first to be found in the animal king-dom. They are also the simplest animals to have a **blood-vascular** system.

A few of the nemerteans are found in moist soil and fresh water, but by far the larger number are marine. At low tide they are often coiled up under stones. It seems probable that they are active at high tide and quiescent at low tide. Some nemerteans such as *Cerebratulus* (L. *cerebrum,* brain, + *ulus,* dim. suffix) often live in empty mollusc shells. The small species live among seaweed, or they may be found swimming near the surface of the water. Nemerteans are often secured by dredging at depths of 5 to 8 m or deeper. A few are commensals or parasites. *Prostoma rubrum* (Gr. *pro,* before, in front of, + *stoma,* month) which is 20 mm or less in length, is a well-known freshwater species.

CHARACTERISTICS

1. Bilateral symmetry; highly contractile body that is cylindrical anteriorly and flattened posteriorly
2. Three germ layers
3. Epidermis with cilia and gland cells; rhabdites in some
4. Body spaces with parenchyma, which is partly gelatinous
5. An **eversible proboscis,** which lies free in a cavity (rhynchocoel) above the alimentary canal
6. **Complete digestive system** (mouth to anus)
7. Body-wall musculature of outer circular and inner longitudinal layers with diagonal fibers between the two; sometimes another circular layer inside the longitudinal layer
8. **Blood-vascular system with two or three longitudinal trunks**
9. Acoelomate, although the rhynchocoel technically may be considered a true coelom
10. Nervous system usually a four-lobed brain connected to paired longitudinal nerve trunks or, in some, middorsal and midventral trunks
11. Excretory system of two coiled canals, which are branched with **flame cells**
12. Sexes separate with simple gonads; asexual reproduction by fragmentation; few hermaphrodites; **pilidium larvae** in some
13. No respiratory system
14. Sensory **ciliated pits** or **head slits** on each side of head, which communicate between the outside and the brain; tactile organs and ocelli (in some)
15. In contrast to Platyhelminthes, few parasitic nemerteans

CLASSIFICATION

Class Enopla (en′o-pla) (Gr. *enoplos,* armed). Proboscis usually armed with stylets; mouth opens in front of brain. Examples: *Amphiporus, Prostoma.*

Class Anopla (an′o-pla) (Gr. *anoplos,* unarmed). Proboscis lacks stylets; mouth opens below or posterior to brain. Examples: *Cerebratulus, Tubulanus, Lineus.*

Form and function

Nemerteans are slender worms and very fragile (Figure 11-20) with a great diver-sity in size. Longer ones are difficult to study in the laboratory. *Amphiporus* (Gr. *amphi,* both sides of, + *porus,* pore) (Figure 11-21), which is taken here as a representative type, is one of the smaller ones. It is from 20 to 80 mm long and about 2.5 mm wide. It is dorsoventrally flattened and has rounded ends. The body wall consists of an epidermis of ciliated columnar cells and layers of circular and longitudinal muscles (Figure 11-22, *A*). A partly gelatinous parenchyma fills the space around the visceral organs. Ocelli are located at the anterior end. The thick-lipped mouth is anteroventral, with the opening of the proboscis just above it.

The proboscis is not connected with the digestive tract but is an eversible organ that can be protruded from its cavity, the **rhynchocoel,** and used for defense

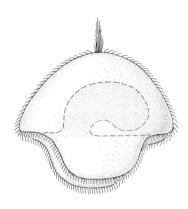

Figure 11-19

The pilidium larva, typical of most nemer-teans, is ciliated, free swimming, and has lateral lobes.

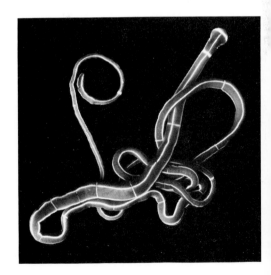

Figure 11-20

Ribbon worm *Tubulanus annulatus* (phylum Nemertea) may be several feet long. The anterior end has the enlarged, rounded cephalic lobe.

From *Adaptive radiation—the mollusks,* an Ency-clopaedia Brittanica film.

and catching prey (Figure 11-21). It lies within a sheath to which it is attached by muscles. The rhynchocoel is filled with fluid, and by muscular pressure on this fluid the anterior part of the tubular proboscis is everted, or turned inside out. The proboscis apparatus is an invagination of the anterior body wall, and its structure therefore duplicates that of the body wall. The retractor muscles attached at the end are used to retract the everted proboscis, much like inverting the tip of a finger of a glove by a string attached inside at its tip. The proboscis is armed with a sharp-pointed stylet. A frontal gland also opens at the anterior end by a pore.

Locomotion

Nemerteans can move with considerable rapidity by the combined action of their well-developed musculature and cilia. They glide mainly against a substratum; some species make use of muscular waves in crawling. Some nemerteans have the interesting method of protruding the proboscis, attaching themselves by means of the stylet, and then drawing the body up to the attached position.

Feeding and digestion

The nemerteans are carnivorous and voracious, eating either dead or living prey. In seizing their prey they thrust out the slime-covered proboscis, which quickly ensnares the prey by wrapping around it (Figure 11-21, A). The stylet also pierces and holds the prey. Then retracting the proboscis, the nemertean draws the prey near the mouth, where it is engulfed by the esophagus that is thrust out to meet it.

The digestive system is complete and extends straight through the length of the body to the terminal anus, lying ventral to the proboscis sheath. The esophagus is straight and opens into a dilated part of the tract, the stomach. The blind anterior end of the intestine as well as the main intestine is provided with paired **lateral ceca.** The alimentary tract is lined with ciliated epithelium, and in the wall of the esophagus there are glandular cells.

Digestion is largely extracellular in the intestinal tube, and when the food is ready for absorption, it passes through the cellular lining of the intestinal tract into the blood-vascular system. The indigestible material passes out the anus (Figure 11-21, B), in contrast to Platyhelminthes in which it leaves by the mouth.

Circulation

The blood-vascular system is simple and enclosed with a single dorsal vessel and two lateral vessels (Figure 11-22, B) connected by transverse vessels. All three longitudinal vessels join together anteriorly to form a type of collar. The blood is usually colorless, containing nucleated corpuscles. However, in some nemerteans the blood is red, green, yellow, or orange from the presence of pigments whose function in unknown. There is no heart, and the blood is propelled by the muscular walls of the blood vessels and by bodily movements.

Excretion and respiration

The excretory system contains a pair of lateral tubes with many branches and flame cells (Figure 11-22, B). Each lateral tube opens to the outside by one or more pores. Waste is picked up from the parenchymal spaces and blood by the flame cells and carried by the excretory ducts to the outside. Many of the protonephridia are so closely associated with the circulatory system that their function may be truly excretory, in contrast to their apparently osmoregulatory function in Platyhelminthes. Respiration occurs through the body surface.

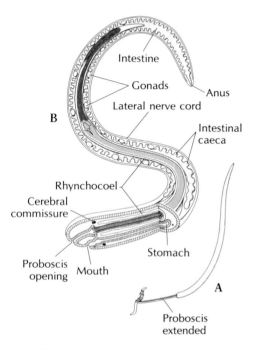

Figure 11-21

A, *Amphiporus,* with proboscis extended to catch prey. B, Structure of female nemertean worm *Amphiporus* (diagrammatic). Dorsal view to show proboscis.

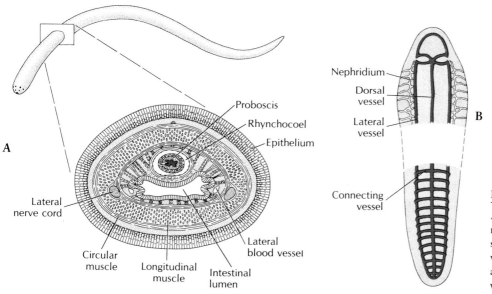

A

B

Figure 11-22

A, Diagrammatic cross section of female nemertean worm. **B**, Excretory and circulatory systems of anterior region of nemertean worm. Flame bulbs along nephridial canal are closely associated with lateral blood vessels.

Nervous system

The nervous system includes a brain composed of four fused ganglia, one pair dorsal and one pair ventral, united by commissures (connecting nerves). Five longitudinal nerves extend from the brain posteriorly—a large lateral trunk on each side of the body, paired dorsolateral trunks, and one middorsal trunk. These are connected by a network of nerve fibers. From the brain, nerves run to the proboscis, to the ocelli and other sense organs, and to the mouth and esophagus. In addition to the ocelli, there are other sense organs, such as tactile papillae, sensory pits and grooves, and probably auditory organs.

Reproduction and development

The reproductive system in *Amphiporus* is dioecious. The gonads in either sex lie between the intestinal ceca (Figure 11-21). From each gonad a short duct (gonopore) runs to the dorsolateral body surface. Eggs and sperm are discharged into the water, where fertilization occurs. Egg production in the females is usually accompanied by degeneration of the other visceral organs.

Nemerteans have a spiral, determinate cleavage. The mesoderm is derived partly from the endoderm and partly from the ectoderm. The rhyncocoel develops as a cavity in the mesoderm and is, therefore, technically a coelomic cavity, but it is not homologous to the coelom in higher forms.

A pilidium larva (Figure 11-19) develops, which bears a dorsal spike of fused cilia and a pair of lateral lobes. The entire larva is covered with cilia and has a mouth and alimentary canal but no anus. In some nemerteans the zygote develops directly without undergoing metamorphosis. The freshwater species, *Prostoma rubrum*, is hermaphroditic. A few nemerteans are viviparous.

Regeneration

Nemerteans have great powers of regeneration. At certain seasons some of them fragment by autotomy, and from each fragment a new individual develops. This is especially noteworthy in the genus *Lineus*. Fragments from the anterior region will produce a new individual more quickly than will one from the posterior part. Sometimes the proboscis is shot out with such force that it is broken off from the body. In such a case a new proboscis develops within a short time.

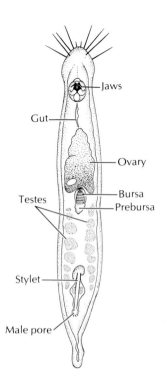

Figure 11-23

Gnathostomula jenneri (phylum Gnathosto-mulida) is a tiny member of the interstitial fauna between grains of sand or mud. Species in this family are among the most commonly encountered jaw worms, found in shallow water and down to depths of several hundred meters.
From Sterrer, W.E. 1972. Syst. Zool. 21:151.

PHYLUM GNATHOSTOMULIDA

The first species of the Gnathostomulida (nath′o-sto-myu′lid-a) (Gr. *gnathos*, jaw, + *stoma*, mouth, + L. *ulus*, dim. suffix) was observed in 1928 in the Baltic, but its description was not published until 1956. Since then jaw worms have been found in many parts of the world, including the Atlantic coast of the United States, and over 80 species in 18 genera have been described.

Gnathostomulids are delicate wormlike animals and are 0.5 to 1 mm long (Figure 11-23). They live in the interstitial spaces of very find sandy coastal sediments and silt and can endure conditions of very low oxygen. They are often found in large numbers and frequently in association with gastrotrichs, nematodes, ciliates, tardigrades, and other small forms.

Lacking a pseudocoel, a circulatory system, and an anus, the gnathostomulids show some similarities to the turbellarians and were at first included in that group. However, their parenchyma is poorly developed, and their pharynx is reminiscent of the rotifer mastax. The pharynx is armed with a pair of lateral jaws used to scrape fungi and bacteria off the substratum. And, although the epidermis is ciliated, each epidermal cell has but one cilium, a condition rarely found in the lower bilateral animals except in some gastrotrichs.

Gnathostomulids can glide, swim in loops and spirals, and bend the head from side to side. Sexual stages may include males, females, and hermaphrodites. Fertilization is internal.

The phyletic relationships of the Gnathostomulida are still obscure, but they seem to be intermediate between the Platyhelminthes and the pseudocoelomate phyla.

PHYLOGENY AND ADAPTIVE RADIATION
Phylogeny

Platyhelminthes and Nemertea are apparently closely related, with the flatworms being the more primitive. There can be little doubt that the bilaterally symmetrical flatworms were derived from a radial ancestor, perhaps one very similar to the planula larva of the cnidarians. Some investigators believe that this **planuloid ancestor** may have given rise to one branch of descendants that were sessile or free floating and radial, which became the Cnidaria, and another branch that acquired a creeping habit and bilateral symmetry, which became the Platyhelminthes. Bilateral symmetry is a selective advantage for creeping or swimming animals because sensory structures are concentrated on the anterior, the end that first encounters environmental stimuli (cephalization).

The transformation from a planuloid ancestor to an early platyhelminth involved a number of body modifications, such as an oral-aboral flattening, with the oral end becoming the ventral surface and the ventral surface adapting for locomotion with the aid of cilia and muscles. The small flatworms of the order Acoela seem to meet many of the requirements of an early ancestor of Platyhelminthes. They have several characteristics in common with the planula larva of the cnidarians, such as no epidermal basement membrane, no digestive system, a nerve plexus under the epidermis, and no distinct gonads. The acoeloid ancestor gave rise to the other orders of Turbellaria and the other classes in the phylum. It may be that the ancestral cestodes never had a digestive tract and therefore did not lose it in their adaptation to parasitism.

The Nemertea probably arose from flatworm stock: the body construction of ciliated epidermis, muscles, and mesenchyme-filled spaces is similar in both groups. The nemerteans are more advanced than the flatworms in having a complete digestive system, a vascular system, and a more highly organized nervous system.

The Gnathostomulida occupy a position intermediate between the Platyhelminthes and the pseudocoelomate phyla (Chapter 12). Monociliated cells are found in some gastrotrichs, and their protonephridial system is similar to that of gastrotrichs. Their jaws show similarities to the rotifer mastax. However, their complete ciliation, arrangement of body wall muscles, cleavage pattern, and reproductive system strongly suggest a relationship to the primitive Turbellaria.

Adaptive Radiation

The flatworm body plan, with its creeping adaptation, placed a selective advantage on bilateral symmetry and further development of cephalization, ventrodorsal regions, and caudal differentiation. Because of their body shape and metabolic requirements, early flatworms must have been well preadapted for parasitism and gave rise to symbiotic lines on numerous occasions. These lines produced descendants that were extremely successful as parasites, and many flatworms became very highly specialized for that mode of existence.

The ribbon worms have stressed the proboscis apparatus in their evolutionary diversity. Its use in capturing prey may have been secondarily evolved from its original function as a highly sensitive organ for exploring the environment. Although the ribbon worms have advanced beyond the flatworms in their complexity of organization, they have been dramatically less successful as a group. Perhaps the proboscis was so efficient as a predator tool that there was little selective pressure to explore parasitism, or perhaps some critical preadaptations were simply not present.

Likewise, the jaw worms have not radiated nor been nearly as successful as the flatworms. However, they have exploited the marine interstitial environment, particularly zones of very low oxygen concentration.

A *preadaptation* is an adaptation that was selected for in one environment or circumstance that *coincidentally* is adaptive in another environment. The term does not imply that the organism "prepared itself" for another environment.

SUMMARY

The Platyhelminthes, the Nemertea, and the Gnathostomulida are the most primitive phyla that are bilaterally symmetrical, a condition of adaptive value for actively crawling or swimming animals. They have neither a coelom nor a pseudocoel and are thus acoelomate. They are triploblastic and at the organ-system level of organization.

Of the classes of Platyhelminthes, the Turbellaria are mostly free living, and the Monogenea, Trematoda, and Cestoda are all parasitic. The more primitive orders of Turbellaria are endolecithal, but the more advanced Turbellaria and the other classes of flatworms are ectolecithal; that is, the yolk for embryonic development is not incorporated into the oocyte but is contributed by separate vitelline cells.

The outer surface of turbellarians is covered by a ciliated epidermis containing mucous cells and rod-shaped rhabdites, which function together in locomotion. Many have a dual-gland system that helps them adhere to the substratum. Planarians are mostly carnivorous and take in food through their eversible pharynx on their ventral side. Digestion is extracellular and intracellular. Osmoregulation is by flame cell protonephridia, and removal of metabolic wastes and respiration are across the body wall. Except for the most primitive turbellarians, flatworms have a ladder-type nervous system with motor, sensory, and association neurons. Various types of sensory receptors are found in the planarians, including photosensitive ocelli. Turbellarians are hermaphroditic, and many also reproduce asexually by fission.

Members of all the other classes of flatworms are covered by a nonciliated, syncytial tegument with a vesicular distal cytoplasm and cell bodies beneath superficial muscle layers. The Monogenea are important ectoparasites of fishes and have a direct life cycle (without intermediate hosts). These contrast with the digenetic trematodes, which have a mollusc intermediate host and almost always a vertebrate definitive host. The great amount of asexual reproduction that occurs in the intermediate host helps to increase the chances that some of the offspring will reach a definitive host.

The Digenea includes a number of important parasites of humans and domestic animals. *Clonorchis sinensis* is a liver fluke, and some species of *Schistosoma*, a trematode living in blood vessels, are important pathogens.

The cestodes, or tapeworms, generally have a scolex (holdfast) organ at their anterior end, followed by a long chain of proglottids. They live as adults in the digestive tract of vertebrates. Their tegument is similar in basic structure to that of trematodes, except that it is covered by minute projections called microtriches. The cestodes do not have a digestive tract. Tapeworms are almost all monoecious, and each proglottid contains a complete set of reproductive organs of both sexes. The shelled larva is passed in the feces, sometimes within a shed proglottid, and the juvenile develops in a vertebrate or invertebrate intermediate host. Some tapeworms are common in humans.

Members of the Nemertea have not been nearly as successful as a group, compared with the Platyhelminthes, but they are structurally more advanced than the flatworms. They have a complete digestive system with an anus and a true circulatory system. They are free living, mostly marine, and they capture their prey by ensnaring it with their long, eversible proboscis.

Members of the Gnathostomulida are tiny worms that live between sediment grains in the sea bottom in zones of low oxygen concentration. They have jaws similar to a rotifer mastax, an epidermis with a single cilium on each cell, and a digestive tract with no anus.

The flatworms and the cnidarians both probably evolved from a common ancestor (planuloid), of whose descendants some became sessile or free floating and radial, and some became creeping and bilateral. The more advanced flatworms were probably derived from an ancestor resembling the turbellarian order Acoela, whereas the nemerteans arose from platyhelminth stock.

Although they are acoelomate, the gnathostomulids show affinities with the pseudocoelomate phyla. They were probably derived from a primitive turbellarian.

Review questions

1. Why is bilateral symmetry of adaptive value for actively motile animals?
2. Match the terms in the right column with the classes in the left column:

    _____Turbellaria   a. Endoparasitic

    _____Monogenea   b. Free-living and commensal

    _____Trematoda   c. Ectoparasitic

    _____Cestoda

3. Give ten characteristics of the Platyhelminthes.
4. Distinguish two groups of turbellarians on the basis of yolk supply for the embryos and on structure of the pharynx.
5. Why is the Acoela considered the most primitive of the Turbellaria?
6. Briefly describe the body plan of turbellarians.
7. What do planarians eat, and how do they digest it?
8. Briefly describe the osmoregulatory system, the nervous system, and the sense organs of planaria.
9. Contrast asexual reproduction in Turbellaria, Trematoda, and Cestoda.
10. Contrast the typical life cycle of Monogenea with that of a digenetic trematode.
11. Describe and contrast the tegument of turbellarians, trematodes, and cestodes.
12. Answer the following questions with respect to both *Clonorchis* and *Schistosoma:* (a) how do humans become infected? (b) what is the general geographical distribution? (c) what are the main disease conditions produced?
13. Why is *Taenia solium* a more dangerous infection than *Taeniarhynchus saginatus?*
14. Name two cestodes for which humans can serve as intermediate hosts.
15. Define each of the following: scolex, microtriches, proglottids, strobila.
16. Give three differences between nemerteans and platyhelminths.
17. Where do gnathostomulids live?
18. Explain how a planuloid ancestor could have given rise to both the Cnidaria and the Platyhelminthes.
19. How would an acoeloid ancestor differ from a planuloid ancestor?

Selected references
See also general references for Part Two, p. 590.

Arme, C., and P.W. Pappas (eds.). 1983. Biology of the Eucestoda, 2 vols. New York, Academic Press, Inc. *The most up-to-date reference available on biology of tapeworms. Advanced; not an identification manual.*

Desowitz, R.S. 1981. New Guinea tapeworms and Jewish grandmothers. New York, W.W. Norton & Co. *Accounts of parasites and parasitic diseases of humans. Entertaining and instructive. Recommended for all students.*

Lumsden, R.D., and R. Specian. 1980. The morphology, histology, and fine structure of the adult stage of the cyclophyllidean tapeworm *Hymenolepis diminuta.* In H.P. Arai (ed.). Biology of the tapeworm *Hymenolepis diminuta.* New York, Academic Press, Inc. *Excellent presentation of the fine structure of the cestode most widely used as an experimental model.*

Schell, S.C. 1985. Handbook of trematodes of North America north of Mexico. Moscow, Idaho, University Press of Idaho. *Good for trematode identification.*

Schmidt, G.D. 1985. Handbook of tapeworm identification. Boca Raton, Florida, CRC Press. *Up-to-date manual for cestode identification.*

Strickland, G.T. 1984. Hunter's tropical medicine, ed. 6. Philadelphia, W.B. Saunders Co. *A valuable source of information on parasites of medical importance.*

Tyler, S. 1976. Comparative ultrastructure of adhesive systems in the Turbellaria. Zoomorphologie 84:1-76. *Describes the dual-gland system in turbellarians.*

CHAPTER 12

THE PSEUDOCOELOMATE ANIMALS

Phylum Rotifera

Phylum Gastrotricha

Phylum Kinorhyncha

Phylum Nematoda

Phylum Nematomorpha

Phylum Acanthocephala

Phylum Entoprocta

Position in Animal Kingdom

In the seven phyla covered in this chapter, the original blastocoel of the embryo persists as a space, or body cavity, between the enteron and body wall. Because this cavity lacks the peritoneal lining found in the true coelomates, it is called a **pseudocoel**, and the animals possessing it are called **pseudocoelomates.** Pseudocoelomates belong to the Protostomia division of the bilateral animals, but they are polyphyletic (not derived from a common ancestor).

Biological Contributions

The pseudocoel is a distinct advancement over the solid body structure of the acoelomates. It may be filled with fluid or may contain a gelatinous substance with some mesenchyme cells. In common with a true coelom, the pseudocoel presents certain adaptive potentials, although these are by no means realized in all members: (1) greater freedom of movement; (2) space for the development and differentiation of digestive, excretory, and reproductive systems; (3) a simple means of circulation or distribution of materials throughout the body; (4) a storage place for waste products to be discharged to the outside by excretory ducts; and (5) a hydrostatic organ. Since most pseudocoelomates are quite small, the most important functions of the pseudocoel are probably in circulation and as a means to maintain a high internal hydrostatic pressure.

The complete, mouth-to-anus digestive tract is found in these phyla and in all higher phyla.

THE PSEUDOCOELOMATES

Vertebrates and higher invertebrates have a true **coelom,** or peritoneal cavity, which is formed in the mesoderm during embryonic development and is therefore lined with a layer of mesodermal epithelium, the **peritoneum** (Figure 12-1). The pseudocoelomate phyla have a pseudocoel rather than a true coelom. It is derived from the embryonic blastocoel rather than from a secondary cavity within the mesoderm. It is a space not lined with peritoneum, between the gut and the mesodermal and ectodermal components of the body wall.

Seven distinct groups of animals belong to the pseudocoelomate category: Rotifera, Gastrotricha, Kinorhyncha, Nematoda, Nematomorpha, Acanthocephala, and Entoprocta. Since the first five of these groups have certain similarities, some authorities place them as classes in a phylum called Aschelminthes (as'kel-min'theez) (Gr. *askos,* bladder, + *helmins,* worm). However, they differ so much that any phyletic relationship is highly debatable, and other authorities consider them as separate phyla. Some group the five loosely as individual phyla under a superphylum Aschelminthes. The Entoprocta have sometimes been grouped with the Ectoprocta, together called the Bryozoa (moss animals). However, because the ectoprocts have a true coelom, they are usually considered a separate phylum, and the term "bryozoans" is currently taken to exclude the entoprocts.

However one classifies them, the pseudocoelomates are a heterogeneous assemblage of animals. Most of them are small; some are microscopic; some are fairly large. Some, such as the nematodes, are found in freshwater, marine, terrestrial, and parasitic habitats; others, such as the Acanthocephala, are strictly parasitic. Some have unique characteristics, such as the lacunar system of the acanthocephalans or the ciliated corona of the rotifers.

Even in such a diversified grouping, some characteristics are shared. All have a body wall of epidermis (often syncytial), a dermis, and muscles surrounding the pseudocoel. The digestive tract is complete (except in Acanthocephala), and it, along with the gonads and excretory organs, is within the pseudocoel and bathed in perivisceral fluid. The epidermis in many secretes a nonliving cuticle with some specializations such as bristles or spines.

A constant number of cells or nuclei in the individuals of a species, a condition known as **eutely,** is common to several of the groups, and in most of them there is an emphasis on the longitudinal muscle layer.

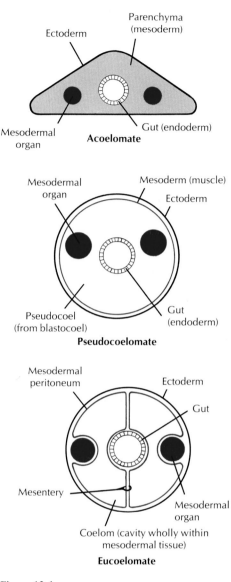

Figure 12-1

Acoelomate, pseudocoelomate, and eucoelomate body plans.

CHARACTERISTICS

1. Symmetry bilateral; unsegmented; triploblastic (three germ layers)
2. Body cavity a **pseudocoel**
3. Size mostly small; some microscopic; a few a meter or more in length
4. Body vermiform; body wall a **syncytial** or cellular epidermis with thickened cuticle, sometimes molted; muscular layers mostly of **longitudinal fibers;** cilia absent in several phyla
5. Digestive system (lacking in acanthocephalans) complete with mouth, enteron, and anus; pharynx muscular and well developed: **tube-within-a-tube arrangement;** digestive tract usually only an epithelial tube with **no definite muscle layer**
6. Circulatory and respiratory organs lacking
7. Excretory system of canals and protonephridia in some; cloaca that receives excretory, reproductive, and digestive products may be present
8. Nervous system of cerebral ganglia or of a circumenteric nerve ring connected to anterior and posterior nerves; sense organs of ciliated pits, papillae, bristles, and some eyespots
9. Reproductive system of gonads and ducts that may be single or double; sexes nearly always separate, with the male usually smaller than the female; eggs microscopic with shell often containing chitin
10. Development may be direct or within a complicated life history; cleavage mostly mosaic; **cell or nuclear constancy common**

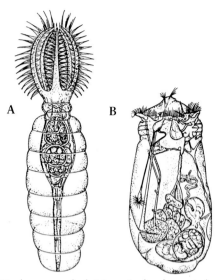

Stephanoceros fimbriatus   Asplanchna priodonta

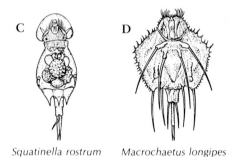

Squatinella rostrum   Macrochaetus longipes

Figure 12-2

Variety of body form in rotifers. **A,** *Stephanoceros* has five, long, fingerlike coronal lobes with whorls of short bristles. It catches its prey by closing its funnel when food organisms swim into it, and the bristly lobes prevent the prey from escaping **B,** *Asplanchna* is a pelagic, predatory genus with no foot. **C,** *Squatinella* has a semicircular, nonretractable, transparent hoodlike extension covering the head. **D,** *Macrochaetus* is dorsoventrally flattened.

A, From Voigt, M., and W. Koste. 1978. Rotaria, die Radertiere Mitteleuropas, ed. 2. Berlin, Borntraeger; B, from Edmonson, W.T. (ed.). 1959. Ward and Whipple's freshwater biology, ed. 2. New York, John Wiley & Sons, Inc.; C and D, from Ruttner-Kolisko, A. 1974. Die Binnengewasser 26(suppl.):1.

___ Phylum Rotifera

Rotifera (ro-tif′ e-ra) (L. *rota*, wheel, + *fera*, those that bear) derive their name from the characteristic ciliated crown, or **corona,** that, when beating, often gives the impression of rotating wheels. Rotifers range from 40 μm to 3 mm in length, but most are between 100 and 500 μm long. Some have beautiful colors, although most are transparent, and some have bizarre shapes (Figure 12-2). Their shapes are often correlated with their mode of life. The floaters are usually globular and saclike; the creepers and swimmers are somewhat elongated and wormlike; and the sessile types are commonly vaselike, with a cuticular envelope (lorica). Some are colonial. One of the best-known genera is *Philodina* (Gr. *philos*, fond of, + *dinos*, whirling) (Figure 12-3), which is often used for study.

Rotifers are a cosmopolitan group of about 1500 species, some of which are found throughout the world. Most of the species are freshwater inhabitants, a few are marine, some are terrestrial, and some are epizoic (live on the body of another animal) or parasitic.

Rotifers are adapted to many kinds of ecological conditions. Most species are benthic, living on the bottom or in vegetation of ponds or along the shores of freshwater lakes where they swim or creep about on the vegetation. A large proportion of the species that live in the water film between sand grains of beaches (meiofauna) are rotifers. Pelagic forms (Figure 12-2, *B*) are common in the surface waters of freshwater lakes and ponds, and they may exhibit cyclomorphosis, that is, seasonal variations in body form.

Many species of rotifers can endure long periods of desiccation, during which they resemble grains of sand. While in a desiccated condition, rotifers are very tolerant of temperature variations, especially those rotifers that dwell in mosses. True encystment occurs in only a few rotifers. On addition of water, desiccated rotifers resume their activity.

Strictly marine species are rather few in number. Some of the littoral (intertidal) species of the sea may be freshwater ones that are able to adapt to seawater.

CLASSIFICATION

Class Seisonidea (sy′son-id′e-a) (Gr. *seisōn*, earthen vessel, + *eidos*, form). Marine; elongated form; corona vestigial; sexes similar in size and form; females with pair of ovaries and no vitellaria; single genus (*Seison*) with two species; epizoic on gills of a crustacean (*Nebalia*).
Class Bdelloidea (del-oid′e-a) (Gr. *bdella*, leech, + *eidos*, form). Swimming or creeping forms; anterior end retractile; corona usually with pair of trochal discs; males unknown; parthenogenetic; two germovitellaria. Examples: *Philodina* (Figure 12-3), *Rotaria.*
Class Monogononta (mon′o-go-non′ta) (Gr. *monos*, one, + *gonos*, primary sex gland). Swimming or sessile forms; single germovitellarium; males reduced in size; eggs of three types (amictic, mictic, dormant). Examples: *Asplanchna* (Figure 12-2, *B*), *Epiphanes.*

Form and function

External features

The body of the rotifer is composed of a head bearing a ciliated corona, a trunk, and a posterior tail, or foot. It is covered with a cuticle and is nonciliated except for the corona.

The ciliated corona, or crown, surrounds a nonciliated central area of the head called the **apical field,** which may bear sensory bristles or papillae. The appearance of the head end depends on which of the several types of corona it has—usually a circlet of some sort, or a pair of trochal (coronal) discs (the term "trochal" comes from a Greek word meaning wheel). The cilia on the corona beat in succession, giving the appearance of a revolving wheel or pair of wheels. The mouth is located in the corona on the midventral side. The coronal cilia are used in both locomotion and feeding.

The trunk may be elongated, as in *Philodina* (Figure 12-3), or saccular in shape (Figure 12-2). It contains the visceral organs and often bears sensory antennae. It is covered by a transparent cuticle that in *Philodina* and others is superficially ringed so as to simulate segmentation but in many other forms is much thickened to form an outer case or **lorica,** often arranged in plates or rings.

The foot is narrower and usually bears one to four toes. Its cuticle may be ringed so that it is telescopically retractile. It is tapered gradually in some forms (Figure 12-3) and sharply set off in others (Figure 12-2). The foot is an attachment organ and contains pedal glands that secrete an adhesive material used by both sessile and creeping forms. In swimming pelagic forms, the foot is usually reduced. Rotifers move by creeping with leechlike movements aided by the foot, or by swimming with the coronal cilia, or both.

Internal features

Underneath the cuticle is the **syncytial epidermis,** which secretes the cuticle, and bands of **subepidermal muscles,** some circular, some longitudinal, and some running through the pseudocoel to the visceral organs. The **pseudocoel** is large, occupying the space between the body wall and the viscera. It is filled with fluid, some of the muscle bands, and a network of mesenchymal ameboid cells.

The digestive system is complete. Some rotifers feed by sweeping minute organic particles or algae toward the mouth by the beating of the coronal cilia. The cilia are able to sort out and dispose of the larger unsuitable particles. The pharynx (**mastax**) is fitted with a muscular portion that is equipped with hard jaws (**trophi**) for sucking in and grinding up the food particles. The constantly chewing pharynx is often a distinguishing feature of these tiny animals. Carnivorous species feed on protozoa and small metazoans, which they capture by trapping or grasping. The trappers have a funnel-shaped area around the mouth. When small prey swim into the funnel, the lobes fold inward to capture and hold them until they are drawn into the mouth and pharynx. The hunters have trophi that can be projected and used like forceps to seize the prey, bring it back into the pharynx, and then pierce it or break it up so that the edible parts can be sucked out and the rest discarded. The salivary and gastric glands are believed to secrete enzymes for extracellular digestion. Absorption occurs in the stomach.

The excretory system typically consists of a pair of **protonephridial tubules,** each with several **flame cells,** that empty into a common bladder. The bladder, by pulsating, empties into the **cloaca**—into which the intestine and oviducts also empty. The fairly rapid rate of pulsation of the protonephridia—one to four times per minute—would indicate that the protonephridia are important osmoregulatory organs. The water apparently enters through the mouth rather than across the cuticle; even marine species empty their bladder at frequent intervals.

The nervous system consists of a bilobed brain, dorsal to the mastax, that sends paired nerves to the sense organs, mastax, muscles, and viscera. Sensory organs include paired eyespots (in some species such as *Philodina*), sensory bristles and papillae, and ciliated pits and dorsal antennae.

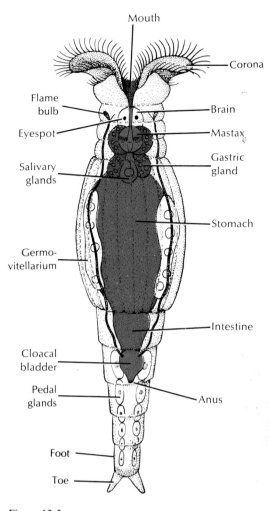

Figure 12-3

Structure of *Philodina* rotifer.

Reproduction

Rotifers are dioecious, but the males are usually smaller than the females. In the class Bdelloidea males are entirely unknown, and in Monogononta they seem to occur only for a few weeks of the year.

The female reproductive system in the Bdelloidea and Monogononta consists of combined ovaries and yolk glands (**germovitellaria**) and oviducts that open into the cloaca. Yolk is supplied to the developing ova by way of flow through cytoplasmic bridges, rather than as separate yolk cells as in many Platyhelminthes.

In the Bdelloidea (*Philodina*, for example), all females are parthenogenetic and produce diploid eggs that hatch into diploid females. These reach maturity in a few days. In the class Seisonidea the females produce haploid eggs that must be fertilized and that develop into either males or females. In the Monogononta,

Figure 12-4

The reproduction of some rotifers (class Monogononta) is parthenogenetic during the part of the year when environmental conditions are suitable. In response to certain stimuli, the females begin to produce haploid (N) eggs. If the haploid eggs are not fertilized, they hatch into haploid males. The males provide sperm to fertilize other haploid eggs, which then develop into diploid (2N), dormant eggs that can resist the rigors of winter. When suitable conditions return, the dormant eggs resume development, and a female hatches.

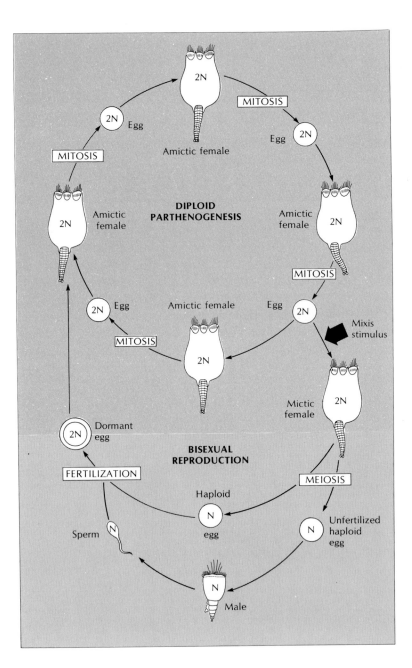

however, females produce two kinds of eggs (Figure 12-4). During most of the year diploid females produce thin-shelled, **diploid amictic eggs.** These develop parthenogenetically into diploid amictic females. However, such rotifers often live in temporary ponds or streams and are cyclic in their reproductive patterns. Any one of several environmental factors—for example, crowding, diet, or photoperiod (according to species)—may induce the amictic eggs to develop into diploid mictic females that will produce thin-shelled **haploid mictic eggs.** If these eggs are not fertilized, they develop into haploid males. But if fertilized, the eggs develop a thick, resistant shell and become dormant. They survive over winter ("winter eggs") or until environmental conditions are again suitable, at which time they hatch into diploid females. Dormant eggs are often dispersed by winds or birds, which may account for the peculiar distribution patterns of rotifers.

The male reproductive system includes a single testis and a ciliated sperm duct that runs to a genital pore (males usually lack a cloaca). The end of the sperm duct is specialized as a copulatory organ. Copulation is usually by hypodermic impregnation; that is, the penis can penetrate any part of the female body wall and inject the sperm directly into the pseudocoel.

Females hatch with adult features, needing only a few days' growth to reach maturity. Males often do not grow and are sexually mature at hatching.

Cell or nuclear constancy

Most structures in rotifers are syncytial, but the nuclei in the various organs are said to show a remarkable constancy in numbers in any given species (eutely). For example, E. Martini (1912) reported that in one species of rotifer he always found 183 nuclei in the brain, 39 in the stomach, 172 in the corona epithelium, and so on. Organisms with eutely show a precise genetic control of development. The cells are programmed to differentiate and divide an exact number of times, then halt when the appointed number is reached.

___ Phylum Gastrotricha

Gastrotricha (gas-tro-tri′ka) (N.L. fr. Gr. *gaster, gastros,* stomach or belly, + *thrix, trichos,* hair) includes small, ventrally flattened animals about 65 to 500 μm long, somewhat like rotifers but lacking the corona and mastax and having a characteristically bristly or scaly body. They are usually found gliding on the bottom, or on an aquatic plant or animal substrate, by means of their ventral cilia, or they compose part of the meiofauna in the interstitial spaces between bottom particles.

Gastrotrichs are found in both fresh and salt water. The 400 or so species are about equally divided between the two media. Many species are cosmopolitan, but only a few occur in both fresh water and the sea. Much is yet to be learned about their distribution.

Form and function

The gastrotrich (Figures 12-5 and 12-6) is usually elongated, with a convex dorsal surface bearing a pattern of bristles, spines, or scales, and a flattened ciliated ventral surface. Cells on the ventral surface may be monociliated or multiciliated. The head is often lobed and ciliated, and the tail end may be forked.

A syncytial epidermis is found beneath the cuticle. Longitudinal muscles are better developed than are circular ones, and in most cases they are unstriated.

Cells (other than eggs and sperm) in most animals contain two chromosomes that carry corresponding genes for a given trait. This condition is called *diploid,* or *2 N.* During maturation of eggs and sperm, the number of chromosomes is reduced to half *(haploid* or *N),* and each sex cell receives one of each pair of the chromosomes. *Parthenogenesis* is production of an individual from an egg without fertilization; therefore, such individuals are usually haploid, although there are some parthenogenetic individuals that are diploid. These concepts are explored further in Chapter 35.

Figure 12-5

Chaetonotus, a gastrotrich. **A,** Dorsal surface. **B,** Internal structure, ventral view.

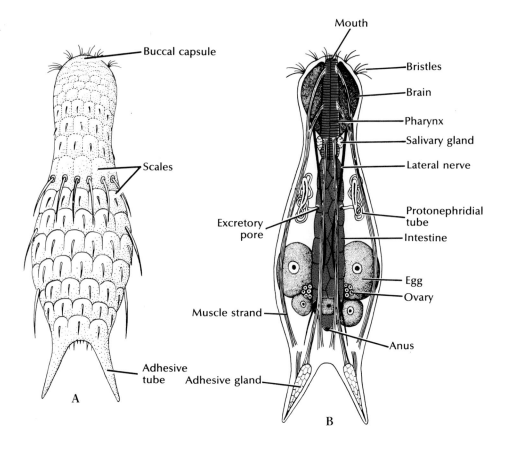

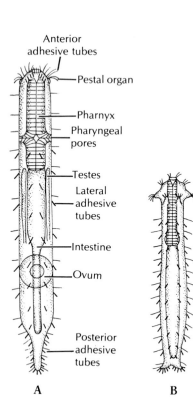

Anterior adhesive tubes

Pestal organ

Pharnyx

Pharyngeal pores

Testes

Lateral adhesive tubes

Intestine

Ovum

Posterior adhesive tubes

A   B

Figure 12-6

Gastrotrichs in the order Macrodasyida. **A,** *Macrodasys.* **B,** *Turbanella.*
From Hummon, W.D. 1982. In S.P. Parker (ed.). Synopsis and classification of living organisms, vol. 1. New York, McGraw-Hill Book Co.

Adhesive tubes secrete a substance for attachment. A dual-gland system for attachment and release is present, similar to that described for the Turbellaria (p. 217). The pseudocoel is somewhat reduced and contains no amebocytes.

The digestive system is complete and is made up of a mouth, a muscular pharynx, a stomach-intestine, and an anus (Figure 12-5, *B*). The food is largely algae, protozoa, and detritus, which are directed to the mouth by the head cilia. Digestion appears to be extracellular. The protonephridia are equipped with cyrtocytes rather than flame bulbs or flame cells. Cyrtocytes have a single flagellum enclosed in a cylinder of cytoplasmic rods.

The nervous system includes a brain near the pharynx and a pair of lateral nerve trunks. Sensory structures are similar to those in rotifers, except that the eyespots are generally lacking.

Gastrotrichs are hermaphroditic, although the male system of some is so poorly developed that they are functionally parthenogenetic females. The female reproductive system consists of one or two ovaries, a uterus, an oviduct, and a gonopore, which may open anteriorly to, or in common with, the anus.

Like the rotifers, chaetonotid gastrotrichs produce thin-walled, rapidly developing eggs and thick-shelled, dormant eggs. The thick-shelled eggs can withstand harsh environmental conditions and may survive dormancy for some years. Development is direct, and the juveniles have the same form as adults.

___ Phylum Kinorhyncha

Kinorhyncha (kin'o-ring'ka) (Gr. *kinein*, to move, + *rhynchos*, beak) are marine worms a little larger than rotifers and gastrotrichs but usually not more than 1 mm long. The phylum has also been called Echinodera, meaning spiny necked. About 75 species have been described.

Kinorhynchs are cosmopolitan, living from pole to pole, from intertidal areas to 6000 m in depth. Most live in mud or sandy mud, but some have been found in algal holdfasts, sponges, or other invertebrates. They feed mainly on diatoms. About 100 species have been reported. Among the best-known genera of the Kinorhyncha are *Echinoderes, Pycnophyes,* and *Kinorhynchus.*

Form and function

The body of the kinorhynch is divided into 13 segments, which bear spines but have no cilia (Figure 12-7). The retractile head has a circlet of spines with a small retractile proboscis. The body is flat underneath and arched above. The body wall is made up of a cuticle, a syncytial epidermis, and longitudinal epidermal cords, much like those of nematodes. The arrangement of the muscles is correlated with the segments, and circular, longitudinal, and diagonal muscle bands are all represented.

A kinorhynch cannot swim. In the silt and mud where it commonly lives, it burrows by extending the head into the mud and anchoring it with spines. It then draws the body forward until the head is retracted into the body. When disturbed, the kinorhynch draws in the head and protects it with a closing apparatus of cuticular plates (Figure 12-7).

The digestive system is complete, with a mouth at the tip of the proboscis, a pharynx, an esophagus, a stomach-intestine, and an anus. Kinorhynchs feed on diatoms or on organic material in the mud where they burrow.

The pseudocoel is filled with amebocytes containing fluid. The excretory system is made up of a multinucleated **solenocyte** protonephridium on each side of the tenth and eleventh segments. Each solenocyte has one long and one short flagellum.

The nervous system is in contact with the epidermis, with a multilobed brain encircling the pharynx, and with a ventral ganglionated nerve cord extending throughout the body. Sense organs are represented by eyespots in some and by the sensory bristles.

Sexes are separate, with paired gonads and gonoducts. There is a series of about six juvenile stages and a definitive, nonmolting adult.

___ Phylum Nematoda: Roundworms

It has been said that, if the earth were to disappear, leaving only the nematode worms, the general contour of the earth's surface would be outlined by the worms, because they are present in nearly every conceivable kind of ecological niche. Approximately 10,000 species of Nematoda (nem-a-to'da) (Gr., *nēmatos,* thread) have been named, but it has been estimated that if all species were known, the number would be nearer 500,000. They live in the sea, in fresh water, and in soil, from polar regions to the tropics, and from mountaintops to the depths of the sea. Good topsoil may contain billions of nematodes per acre. Nematodes also parasitize virtually every type of animal and many plants. The effects of nematode infestation on crops, domestic animals, and humans make this phylum one of the most important of all parasitic animal groups.

Free-living nematodes feed on bacteria, yeasts, fungal hyphae, and algae. They may be saprozoic or coprozoic (live in fecal material). Predatory species may eat rotifers, tardigrades, small annelids, and other nematodes. Many species feed on plant juices from higher plants, which they penetrate, sometimes causing agricultural damage of great proportions. Nematodes themselves may be prey for mites, insect larvae, and even nematode-capturing fungi.

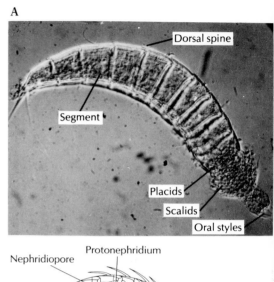

A

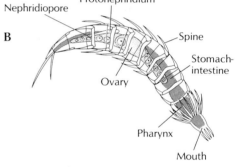

B

Figure 12-7

Echinoderes, a kinorhynch, is a minute marine worm. Segmentation is superficial. Head with its circle of spines is retractile. **A,** Photograph of unstained specimen. **B,** Diagram showing internal anatomy. **A,** Photograph by L.S. Roberts.

Virtually every species of vertebrate and many invertebrates serve as hosts for one or more types of parasitic nematodes. Nematode parasites in humans cause much discomfort, disease, and death, and in domestic animals they are the source of great economic loss.

CLASSIFICATION

Classification of the nematodes is somewhat more satisfactory at the order and superfamily level; the division into classes relies on characteristics that are not striking and that are difficult for the novice to distinguish. Two classes are usually recognized.

> **Class Phasmidia** (faz-mid′e-a) (Gr. *phasm*, phantom, + *idia*, dim. suffix) (**Secernentea**). Body with a pair of minute sensory pouches (phasmids) near posterior tip; similar pair of sense organs at anterior end (amphids) poorly developed; excretory system with one or two lateral canals, with or without associated glandular cells; both free-living and parasitic forms. Examples: *Rhabditis, Ascaris, Enterobius.*
>
> **Class Aphasmidia** (a′faz-mid′e-a) (Gr. *a*, without, + *phasm*, phantom, + *idia*, dim. suffix) (**Adenophorea**). Phasmids lacking, amphids usually well developed; excretory system of one or more renette cells (glandular); caudal and hypodermal glands common; mostly free living, but includes some parasites. Examples: *Dioctophyme, Trichinella, Trichuris.*

Form and function

Distinguishing characteristics of this large group of animals are their cylindrical shape; their flexible, nonliving cuticle; their lack of motile cilia or flagella (except in one species); the muscles of their body wall, which have several unusual features, such as the fact that the muscles run in a lontitudinal direction only, and eutely. Correlated with their lack of cilia, nematodes do not have protonephridia; their excretory system consists of one or more large gland cells opening by an excretory pore, or a canal system without gland cells, or both cells and canals together. Their pharynx is characteristically muscular with a triradiate lumen and resembles the pharynx of the gastrotrichs and of the kinorhynchs. Use of the pseudocoel as a hydrostatic organ is highly developed in the nematodes, and much of the functional morphology of the nematodes can be best understood in the context of the high **hydrostatic pressure** (turgor) in the pseudocoel.

Most nematode worms are less than 5 cm long, and many are microscopic, but some parasitic nematodes are more than 1 m in length.

The outer body covering is a relatively thick, noncellular **cuticle** secreted by the underlying epidermis (**hypodermis**). The hypodermis is syncytial, and its nuclei are located in four **hypodermal cords** that project inward (Figure 12-8). The dorsal and ventral hypodermal cords bear the longitudinal dorsal and ventral nerves, and the lateral cords bear excretory canals. The cuticle is of great functional importance to the worm, serving to contain the high hydrostatic pressure exerted by the fluid in the pseudocoel. The several layers of the cuticle are primarily of **collagen,** a structural protein also abundant in vertebrate connective tissue. Three of the layers are composed of crisscrossing fibers, which confer some longitudinal elasticity on the worm but severely limit its capacity for lateral expansion.

The body wall muscles of the nematodes are very unusual. They lie beneath the hypodermis and contract longitudinally only. There are no circular muscles in the body wall. The muscles are arranged in four bands, or quadrants, marked off by the four hypodermal cords (Figure 12-8). Each muscle cell has a contractile **fibrillar** portion (or **spindle**) and a noncontractile **sarcoplasmic** portion (cell

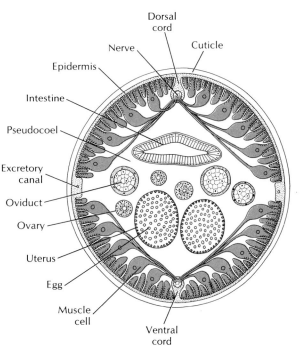

Dorsal
cord

Nerve

Cuticle

Epidermis

Intestine

Pseudocoel

Excretory
canal

Oviduct

Ovary

Uterus

Egg

Muscle
cell

Ventral
cord

Figure 12-8

Cross section of *Ascaris* (female).

body). The spindle is distal and abuts the hypodermis, and the cell body projects into the pseudocoel. The spindle is striated with bands of actin and myosin, reminiscent of vertebrate skeletal muscle (see Figure 6-14, p. 109 and p. 605). The cell bodies contain the nuclei and are a major depot for glycogen storage in the worm. From each cell body a process or **muscle arm** extends either to the ventral or the dorsal nerve. Though not unique to nematodes, this arrangement is very curious; in most animals nerve processes (axons, p. 681) extend to the muscle, rather than the other way around.

The fluid-filled pseudocoel, in which the internal organs lie, constitutes a hydrostatic skeleton. Hydrostatic skeletons, found in many invertebrates, lend support by transmitting the force of muscle contraction to the enclosed, noncompressible fluid. Normally, muscles are arranged antagonistically, so that movement is effected by contraction of one group of muscles and relaxation of the other. However, nematodes do not have circular body wall muscles to antagonize the longitudinal muscles; therefore, the cuticle must serve that function. It precludes radial expansion but permits some stretching and compression longitudinally. Thus comprehension of the cuticle on the side of muscular contraction and stretching on the opposite side are the forces that return the body to resting position when the muscles relax. This produces the characteristic thrashing motion seen in nematode movement. An increase in efficiency of this system can be achieved only by an increase in hydrostatic pressure. Consequently, the hydrostatic pressure in the nematode pseudocoel is much higher than is usually found in other kinds of animals that have hydrostatic skeletons but also have antagonistic muscle groups.

The alimentary canal of the nematode consists of a mouth (Figure 12-9), a muscular pharynx, a long nonmuscular intestine, a short rectum, and a terminal anus. Food material is sucked into the pharynx when the muscles in its anterior portion contract rapidly and open the lumen. Relaxation of the muscles anterior to the food mass closes the lumen of the pharynx, forcing the food posteriorly toward the intestine. The intestine is one cell layer thick. Food matter is moved

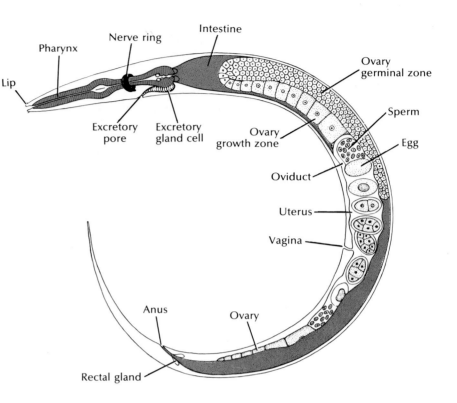

Figure 12-9

Rhabditis, a common free-living nematode that feeds on decaying plant and animal matter. Some species that feed on manure undergo a developmental arrest at the third-stage juvenile until they "hitchhike" to a new food supply on a dung beetle. In this drawing of a female, the intestine overlies and hides the germinal zone of the posterior ovary. As in most nematodes, the proximal end of the oviduct serves as a sperm storage area; the oocytes are penetrated by ameboid sperm as they pass through and then complete meiosis.

After Hirschmann, H., modified from Sasser, J.N., and W.R. Jenkins. 1960. Nematology. Chapel Hill, University of North Carolina Press.

The copulatory spicules of male nematodes are not true intromittent organs, since they do not conduct the sperm, but are another adaptation to cope with the high internal hydrostatic pressure. The spicules must hold the vulva of the female open while the ejaculatory muscles overcome the hydrostatic pressure in the female and rapidly inject sperm into her reproductive tract. Furthermore, nematode spermatozoa are unique among those studied in the animal kingdom in that they lack a flagellum and acrosome. Within the female reproductive tract, the sperm become ameboid and move by pseudopodial movement. Could this be another adaptation to the high hydrostatic pressure in the pseudocoel?

posteriorly by body movements and by additional food being passed into the intestine from the pharynx. Defecation is accomplished by muscles that simply pull the anus open, and the expulsive force is provided by the pseudocoelomic pressure.

The adults of many parasitic nematodes have an anaerobic energy metabolism; thus, a Krebs cycle and cytochrome system characteristic of aerobic metabolism are absent. Energy is derived through glycolysis and probably through some incompletely known electron transport sequences. Interestingly, some free-living nematodes and free-living stages of parasitic nematodes are obligate aerobes and have a Krebs cycle and cytochrome system.

A **ring of nerve tissue** and ganglia around the pharynx gives rise to small nerves to the anterior end and to two **nerve cords,** one dorsal and one ventral. **Sensory papillae** are concentrated around the head and tail. The **amphids** are a pair of somewhat more complex sensory organs that open on each side of the head at about the same level as the cephalic circle of papillae. The amphidial opening leads into a deep cuticular pit with sensory endings of modified cilia. The amphids are usually reduced in nematode parasites of animals, but most parasitic nematodes bear a bilateral pair of **phasmids** near the posterior end. They are similar in structure to the amphids.

Most nematodes are dioecious. The male is smaller than the female, and its posterior end usually bears a pair of **copulatory spicules.** Fertilization is internal, and eggs are usually stored in the uterus until deposition. After embryonation a juvenile worm hatches from the egg. The four juvenile stages are separated by a molt, or shedding, of the cuticle. Many parasitic nematodes have free-living juvenile stages. Others require an intermediate host to complete their life cycles.

Nematode parasites of humans

As mentioned previously, nearly all vertebrates and many invertebrates are parasitized by nematodes. A number of these are very important pathogens of humans and domestic animals. A few nematodes are common in humans in North Amer-

Table 12-1 Common Parasitic Nematodes of Humans in North America

Common and scientific names	Mode of infection; prevalence
Hookworm (*Ancylostoma duodenale* and *Necator americanus*)	Contact with juveniles in soil that burrow into skin; common in southern states
Pinworm (*Enterobius vermicularis*)	Inhalation of dust with ova and by contamination with fingers; most common worm parasite in United States
Intestinal roundworm (*Ascaris lumbricoides*)	Ingestion of embryonated ova in contaminated food; common in rural areas of Appalachia and southeastern states
Trichina worm (*Trichinella spiralis*)	Ingestion of infected pork muscle; occasional in humans throughout North America
Whipworm (*Trichuris trichiura*)	Ingestion of contaminated food or by unhygienic habits; usually common wherever *Ascaris* is found

ica (Table 12-1), but they and many others usually abound in tropical countries. Only a few will be mentioned in this discussion.

Ascaris lumbricoides: the large roundworm of humans

Because of its size and availability, *Ascaris* (Gr. *askaris,* intestinal worm) is usually selected as a type for study in zoology, as well as in experimental work. Thus, it is probable that more is known about the structure, physiology, and biochemistry of *Ascaris* than of any other nematode. This genus includes several species. One of the most common, *A. megalocephala,* is found in the intestine of horses. *A. lumbricoides* (Figure 12-10) is one of the most common parasites found in humans; recent surveys have shown a prevalence of up to 64% in some areas of the southeastern United States. The large roundworm of pigs, *A. suum,* is morphologically close to *A. lumbricoides,* and they were long considered the same species.

A female *Ascaris* may lay 200,000 eggs a day; they are passed out in the host's feces. Given suitable soil conditions, embryonation is complete within 2 weeks. Direct sunlight and high temperatures are rapidly lethal, but the eggs have

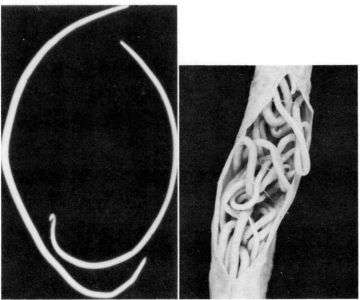

A   B

Figure 12-10

Ascaris suum of pigs, which is very similar morphologically to *Ascaris lumbricoides* of humans. **A,** Male and female. Male *(right)* has characteristic sharp kink in end of tail. **B,** Intestine of pig, nearly completely blocked by *A. suum* (threads were inserted to hold worms in place). Such heavy infections are also fairly common with *A. lumbricoides* in humans.

A, Photograph by F.M. Hickman; **B,** photograph by L.S. Roberts. From Schmidt, G.D., and L.S. Roberts. 1981. Foundations of parasitology, ed. 2. St. Louis, The C.V. Mosby Co.

Other ascarids are common in wild and domestic animals. Species of *Toxocara*, for example, are found in dogs and cats. Their life cycle is generally similar to that of *Ascaris*, but the juveniles often do not complete their tissue migration in adult dogs, remaining in the host's body in a stage of arrested development. Pregnancy in the female, however, stimulates the juveniles to wander, and they infect the embryos in the uterus. The puppies are then born with worms. These ascarids also survive in humans but do not complete their development, leading to an occasionally serious condition in children known as *visceral larva migrans*. This is a good argument for pet owners to practice hygienic disposal of canine wastes!

an amazing tolerance to other adverse conditions, such as desiccation or lack of oxygen. The eggs can remain viable for many months or even years in the soil. Infection usually occurs when eggs are ingested with uncooked vegetables or when children put soiled fingers or toys in their mouths. Unsanitary defecation habits "seed" the soil, and viable eggs remain long after all signs of the fecal matter have disappeared.

When embryonated eggs are swallowed by a host, the tiny juveniles hatch. They burrow through the intestinal wall into the veins or lymph vessels and are carried through the heart to the lungs. There they break out into the alveoli and are carried up the bronchi to the trachea. If the infection is large, they may cause a serious pneumonia at this stage. On reaching the pharynx, the juveniles are swallowed, passed through the stomach, and finally mature about 2 months after the eggs were ingested. In the intestine, where they feed on intestinal contents, the worms cause abdominal symptoms and allergic reactions, and in large numbers they may cause intestinal blockage. Perforation of the intestine with resultant peritonitis is not uncommon, and wandering worms may occasionally emerge from the anus or throat or may enter the trachea or eustachian tubes and middle ears.

Hookworms

Hookworms are so named because the anterior end curves dorsally, suggesting a hook. The most common species is *Necator americanus* (L. *necator,* killer), whose females are up to 11 mm long. The males can reach 9 mm in length. Large plates in their mouths (Figure 12-11) cut into the intestinal mucosa of the host where they suck blood and pump it through their intestines, partially digesting it and absorbing the nutrients. They suck much more blood than they need for food, and heavy infections cause anemia in the patient. Hookworm disease in children may result in retarded mental and physical growth and a general loss of energy.

Eggs are passed in the feces, and the juveniles hatch in the soil, where they live on bacteria. When human skin comes in contact with infested soil, the juveniles burrow through the skin to the blood, and reach the lungs and finally the intestine in a manner similar to that described for *Ascaris*.

Trichina worm

Trichinella spiralis (Gr. *trichinos*, of hair, + *-ella,* diminutive) is the tiny nematode responsible for the potentially lethal disease trichinosis. Adult worms burrow in

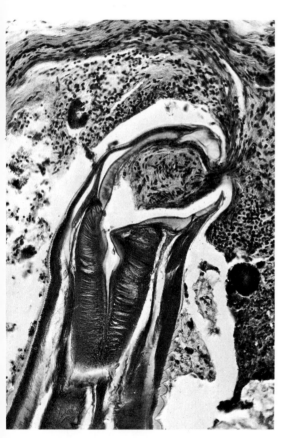

Figure 12-11

Section through anterior end of hookworm attached to human intestine. The cutting plates of the mouth pinch off a bit of mucosa from which the thick muscular pharynx sucks blood. Esophageal glands secrete an anticoagulant to prevent blood clotting.
AFIP No. 33810.

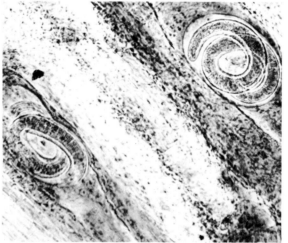

Figure 12-12

Muscle infected with trichina worm *Trichinella spiralis.* Juveniles may live 10 to 20 years in these cysts. If eaten in poorly cooked meat, the juveniles are liberated in the intestine. They quickly mature and release many juveniles into the blood of the host.

A

B

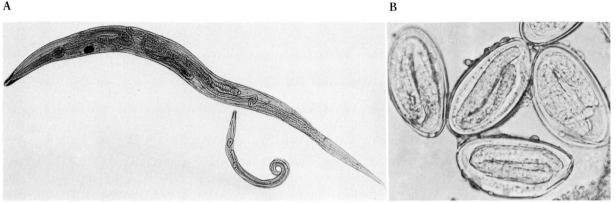

the mucosa of the small intestine where the female produces living young. The juveniles penetrate into blood vessels and are carried to the skeletal muscles where they coil up within a cyst that eventually becomes calcified (Figure 12-12). When meat containing cysts with live worms is swallowed, the juveniles are liberated into the intestine where they mature.

In addition to humans, *Trichinella spiralis* can infect many other mammals, including hogs, rats, cats, and dogs. Humans most often acquire the parasite by eating improperly cooked pork. Hogs become infected by eating garbage containing pork scraps with cysts or by eating infected rats.

Heavy infections may cause death, but lighter infections are much more common—about 2.4% of the population of the United States is infected.

Pinworms

The pinworm, *Enterobius vermicularis* (Gr. *enteron*, intestine, + *bios*, life), causes relatively little disease, but it is the most common helminth parasite in the United States, estimated at 30% in children and 16% in adults. The adults (Figure 12-13) live in the large intestine and cecum. The females, up to about 12 mm in length, migrate to the anal region at night to lay their eggs (Figure 12-13). Scratching the resultant itch effectively contaminates the hands and bedclothes. Eggs develop rapidly and become infective within 6 hours at body temperature. When they are swallowed, they hatch in the duodenum, and the worms mature in the large intestine.

Filarial worms

At least eight species of filarial nematodes infect humans, and some of these are major causes of diseases. Some 250 million people in tropical countries are infected with *Wuchereria bancrofti* (named for Otto Wucherer) or *Brugia malayi* (named for S.L. Brug), which places these species among the scourges of humanity. The worms live in the lymphatic system, and the females are as long as 100 mm. The disease symptoms are associated with inflammation and obstruction of the lymphatic system. The females release live young, the tiny microfilariae, into the blood and lymph. The microfilariae are picked up by mosquitoes as the insects feed, and they develop in the mosquitoes to the infective stage. They escape from the mosquito when it is feeding again on a human and penetrate the wound made by the mosquito bite.

The dramatic manifestations of elephantiasis are occasionally produced after long and repeated exposure to the worms. The condition is marked by an excessive growth of connective tissue and enormous swelling of affected parts, such as the scrotum, legs, arms, and more rarely, the vulva and breasts (Figure 12-14).

Figure 12-13

Pinworms, *Enterobius vermicularis*. **A,** Adult pinworms; the male is much smaller. **B,** Group of pinworm eggs, which are usually discharged at night around the anus of the host, who, by scratching during sleep, may contaminate fingernails and clothing. This may be the most common and widespread of all human helminth parasites. Courtesy Indiana University School of Medicine, Indianapolis.

Diagnosis of most intestinal roundworms is usually made by examination of a small bit of feces under the microscope and finding characteristic eggs. However, pinworm eggs are not often found in the feces because the female deposits them on the skin around the anus. The "Scotch tape method" is more effective. The sticky side of cellulose tape is applied around the anus to collect the eggs, then the tape is placed on a glass slide and examined under the microscope. Several drugs are effective against this parasite, but all members of a family should be treated at the same time, since the worm easily spreads through a household.

Another filarial worm causes river blindness and is carried by the blackfly. It infects more than 30 million people in parts of Africa, Arabia, Central America, and South America.

Phylum Nematomorpha

The popular name for the Nematomorpha (nem'a-to-mor'fa) (Gr. *nēma, nēmatos*, thread, + *morphe*, form) is "horsehair worms," based on an old superstition that the worms arise from horsehairs that happen to fall into the water, and they look something like hairs from a horse's tail. They were long included with the nematodes, with which they share the structure of the cuticle, presence of epidermal cords, longitudinal muscles only, and pattern of nervous system. However, since the early larval form of some species has a striking resemblance to the Priapulida, it is impossible to say to what group the nematomorphs are most closely related.

About 250 species of horsehair worms have been named. They are free living as adults and parasitic in arthropods as juveniles. As a group they have worldwide distribution and a variety of aquatic habitats and may be found in both running and standing water. Adults do not feed but will live almost anywhere in wet or moist surroundings if the oxygen is adequate. Juveniles do not emerge from the arthropod host unless there is water nearby. Adults are often seen wriggling slowly about in ponds or streams, with males being more active than females. The female discharges her eggs in water in long strings. Some juveniles, such as those of *Gordius* (named for an ancient king who tied an intricate knot), a cosmopolitan genus, are believed to encyst on vegetation that may later serve as food for a grasshopper or other arthropod. In the marine form *Nectonema* (Gr. *nēktos*, swimming, + *nēma*, thread), juveniles occur in hermit crabs and other crabs.

Form and function

Horsehair worms are extremely long and slender, with a cylindrical body. They may reach a length of 1.5 m with a diameter of only 3 mm, but most are smaller with a diameter of 1 to 2 mm. The anterior ends are usually rounded, and the posterior ends are rounded or have two or three caudal lobes (Figure 12-15).

The body wall is much like that of nematodes: a secreted cuticle, a hypodermis, and musculature of **longitudinal muscles** only. Ventral, or dorsal and ventral, but not lateral, hypodermal cords are present. In most nematomorphs the ventral nerve cord is connected to the ventral hypodermal cord by the **nervous lamella** (Figure 12-15).

The digestive system is vestigial. The pharynx is a solid cord of cells, and the intestine does not open to the cloaca. The larval forms absorb food from their arthropod hosts through the body wall, and the adults apparently live on stored nutrients.

Circulatory, respiratory, and excretory systems are lacking. There are a nerve ring around the pharynx and a midventral nerve cord. Each sex has a pair of gonads and a pair of gonoducts that empty into the cloaca. The young hatch from the eggs and somehow gain entry into the arthropod host. After several months in the hemocoel of the host, the matured worm emerges into the water. Curiously, if the host is a terrestrial insect, it is stimulated by an unknown mechanism to seek water.

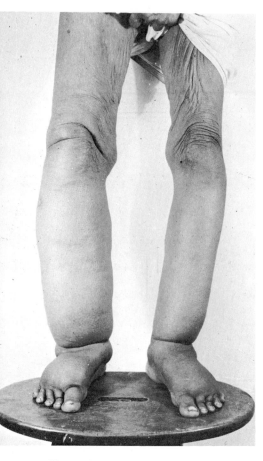

Figure 12-14

Elephantiasis of leg caused by adult filarial worms of *Wuchereria bancrofti*, which live in lymph passages and block the flow of lymph. Tiny juveniles, called microfilariae, are picked up with blood meal of mosquitoes, where they develop to infective stage and are transmitted to a new host. AFIP No. 44430-1.

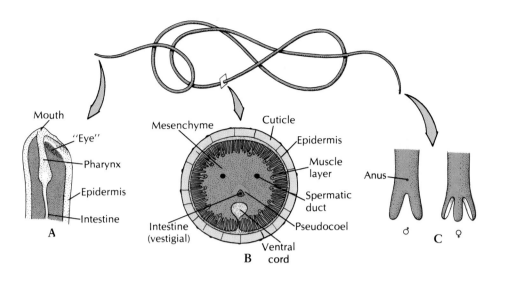

Figure 12-15

Structure of *Paragordius*, a nematomorph. **A**, Longitudinal section through the anterior end. **B**, Transverse section. **C**, Posterior end of male and female worms. Nematomorphs, or "horsehair worms," are very long and very thin. Their pharynx is usually a solid cord of cells and is nonfunctional. *Paragordius*, whose pharynx opens through to the intestine, is unusual in this respect and also in the possession of a photosensory organ ("eye").

___ Phylum Acanthocephala

The members of the phylum Acanthocephala (a-kan'tho-sef'a-la) (Gr. *akantha*, spine or thorn, + *kephalē*, head) are commonly known as "spiny-headed worms." The phylum derives its name from one of its most distinctive features, a cylindrical invaginable proboscis bearing rows of recurved spines, by which it attaches itself to the intestine of its host. All acanthocephalans are endoparasitic, living as adults in the intestines of vertebrates.

Various species range in size from less than 2 mm to more than 1 m in length, with the females of a species usually larger than the males. The body is usually bilaterally flattened, with numerous transverse wrinkles. The worms are usually cream color but may be yellowish or brown as a result of absorption of pigments from the intestinal contents.

Acanthocephalans inflict traumatic damage by penetrating the intestinal wall with the spiny proboscis. In many cases there is remarkably little inflammation, but in some species the inflammatory response of the host is intense. Great pain can be produced, particularly if the gut wall is completely perforated.

More than 500 species are known, most of which parasitize fish, birds, and mammals, and the phylum is worldwide in distribution. However, no species is normally a parasite of humans, although rarely humans are infected with species that usually occur in other hosts. *Macracanthorhynchus hirudinaceus* (Gr. *makros*, long, large, + *akantha*, spine, thorn, + *rhynchos*, beak) occurs throughout the world in the small intestine of pigs and occasionally in other mammals.

Larvae of spiny-headed worms develop in arthropods, either crustaceans or insects, depending on the species.

Form and function

In life the body is somewhat flattened, although it is usual for specimens to be treated with tap water before fixation so that fixed specimens are turgid and cylindrical (Figure 12-16, *C*).

The body wall is syncytial, and its surface is punctured by minute crypts 4 to 6 μm deep, which greatly increase the surface area of the tegument. About 80% of the thickness of the tegument is the radial fiber zone, which contains a **lacunar**

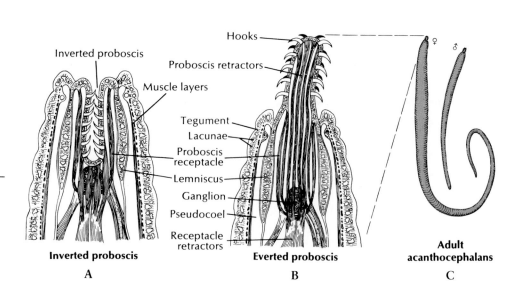

Figure 12-16

Structure of a spiny-headed worm (phylum Acanthocephala). **A** and **B**, Eversible spiny proboscis by which the parasite attaches to the intestine of the host, often doing great damage. Since they lack a digestive tract, food is absorbed through the tegument. **C**, Male is typically smaller than female.

system of ramifying fluid-filled canals (Figure 12-16, *A* and *B*). Curiously, the body-wall muscles are tubelike and filled with fluid. The lumina of the muscles are continuous with the lacunar system; therefore circulation of the lacunar fluid may well bring nutrients to and remove wastes from the muscles. There is no heart or other circulatory system, and contraction of the muscles would serve to move the lacunar fluid through the canals and muscles. Both longitudinal and circular body-wall muscles are present.

The proboscis, which bears rows of recurved hooks, is attached to the neck region (Figure 12-16) and can be inverted into a **proboscis receptacle** by retractor muscles. Attached to the neck region (but not within the proboscis) are two elongated **lemnisci** (extensions of the tegument and lacunar system) that may serve as reservoirs of the lacunar fluid from the proboscis when that organ is invaginated.

There is no respiratory system. When present, the excretory system consists of a pair of **protonephridia** with flame cells. These unite to form a common tube opening into the sperm duct or uterus.

The nervous system has a central ganglion within the proboscis receptacle and nerves to the proboscis and body. There are sensory endings on the proboscis and genital bursa.

Acanthocephalans have no digestive tract, and they must absorb all nutrients through their tegument. They can absorb various molecules by specific membrane transport mechanisms, and their tegument can carry out pinocytosis. The tegument bears some enzymes, such as peptidases, which can cleave several dipeptides, and the amino acids are then absorbed by the worm. Like cestodes, acanthocephalans require host dietary carbohydrate, but their mechanism for absorption of glucose is different. As glucose is absorbed, it is rapidly phosphorylated and compartmentalized, so that a metabolic "sink" is created into which glucose in the surrounding medium can flow. Glucose can diffuse down a concentration gradient into the worm because it is constantly removed as soon as it enters.

Acanthocephalans are dioecious. A pair of tubular **genital ligaments,** or **ligament sacs,** extends posteriorly from the end of the proboscis receptacle. The male has a pair of testes, each with a vas deferens, and a common ejaculatory duct that ends in a small penis. During copulation sperm are ejected into the vagina, travel up the genital duct, and escape into the pseudocoel.

In the female the ovarian tissue in the ligament sac breaks up into **ovarian balls** that rupture the ligament sacs and float free in the pseudocoel. One of the ligament sacs leads to a funnel-shaped uterine bell that receives the developing shelled embryos and passes them on to the uterus (Figure 12-17). An interesting and unique selective apparatus operates here. Fully developed embryos are slightly longer than immature ones, and they are passed on into the uterus, while immature eggs are retained for further maturation.

The shelled embryos, which are discharged in the feces of the vertebrate host, do not hatch until eaten by the intermediate host. For *M. hirudinaceus* this is any of several species of soil-inhabiting beetle larvae, especially scarabeids. Grubs of the June beetle (*Phyllophaga*) are frequent hosts. Here the larva (**acanthor**) burrows through the intestine and develops into the juvenile (**cystacanth**). Pigs are infected by eating the grubs. Multiple infections may do considerable damage to the pig's intestine, and perforations may occur.

Phylum Entoprocta

Entoprocta (en'to-prok'ta) (Gr. *entos,* within, + *proktos,* anus) is a small phylum of fewer than 100 species of tiny, sessile animals that superficially resemble hydroid cnidarians but have ciliated tentacles that tend to roll inward (Figure 12-18). Most entoprocts are microscopic, and none is more than 5 mm long. They are all stalked and sessile forms; some are colonial, and some are solitary. All are ciliary feeders.

With the exception of the genus *Urnatella* (L. *urna,* urn, + *ellus,* dim. suffix), all entoprocts are marine forms that have a wide distribution from the polar regions to the tropics. Most marine species are restricted to coastal and brackish waters and often grow on shells and algae. Some are commensals on marine annelid worms. Freshwater entoprocts occur on the underside of rocks in running water. *U. gracilis* is the only common freshwater species in North America (Figure 12-18, *B*).

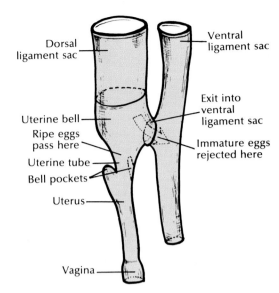

Figure 12-17

Scheme of the genital selective apparatus of a female acanthocephalan. It is a unique device for separating immature from mature fertilized eggs. Eggs containing larvae enter the uterine bell and pass on to the uterus and exterior. Immature eggs are shunted into the ventral ligament sac or into the pseudocoel to undergo further development.

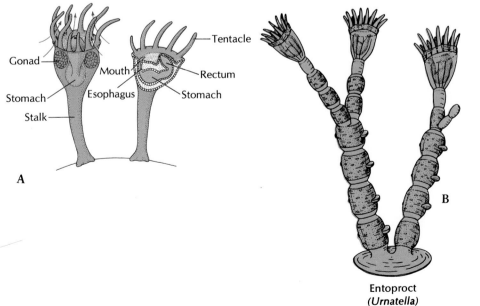

Entoproct
(*Urnatella*)

Figure 12-18

A, *Loxosoma,* a solitary entoproct. **B,** *Urnatella,* a freshwater entoproct, forms small colonies of two or three stalks from a basal plate. Both solitary and colonial entoprocts can reproduce asexually by budding, as well as sexually.

B, Modified from Con, C. 1936. Kamptozoa. In H.G. Bronn (ed.). Klassen und Ordnungen des Tier-Reichs, vol. 4, part 2. Leipzig, Akademische Verlagsgesselschaft.

Form and function

The body, or **calyx,** of the entoproct is cup shaped, bears a crown, or circle, of ciliated tentacles, and may be attached to a substratum by a single stalk and an attachment disc with adhesive glands, as in the solitary *Loxosoma* (Gr. *loxos,* crooked, + *soma,* body) (Figure 12-18, *A*), or by two or more stalks in colonial forms. Both tentacles and stalk are continuations of the body wall. The eight to 30 tentacles making up the crown are ciliated on their lateral and inner surfaces, and each can move individually. The tentacles can roll inward to cover and protect the mouth and anus but cannot be retracted into the calyx.

Movement is usually restricted in entoprocts, but *Loxosoma,* which lives in the tubes of marine annelids, is described as quite active, moving over the annelid and its tube freely.

The gut is U shaped and ciliated, and both the mouth and the anus open within the circle of tentacles. Entoprocts are **ciliary filter feeders.** Long cilia on the sides of the tentacles keep a current water containing protozoa, diatoms, and particles of detritus moving in between the tentacles. Short cilia on the inner surfaces of the tentacles capture the food and direct it downward toward the mouth.

The body wall consists of a cuticle, cellular epidermis, and longitudinal muscles. The pseudocoel is largely filled with a gelatinous parenchyma in which is embedded a pair of protonephridia and their ducts, which unite and empty near the mouth. There is a well-developed **nerve ganglion** on the ventral side of the stomach, and the body surface bears sensory bristles and pits. Circulatory and respiratory organs are absent. Exchange of gases occurs through the body surface, probably much of it through the tentacles.

Some species are monoecious, some dioecious, and some appear to be protandrous; that is, the gonad at first produces sperm and later eggs. The gonoducts open within the circle of tentacles.

Fertilized eggs develop in a depression, or brood pouch, between the gonopore and the anus. Entoprocts have a modified spiral cleavage pattern with mosaic blastomeres. The coeloblastula is formed by invagination. The trochophore-like larva is ciliated and free swimming. It has an apical tuft of cilia at the anterior end and a ciliated girdle around the ventral margin of the body. Eventually the larva settles to the substratum and inverts to form the adult.

——— Phylogeny and Adaptive Radiation

Phylogeny

Hyman (1951) grouped the Rotifera, Gastrotricha, Kinorhyncha, Nematoda, and Nematomorpha into a single phylum (Aschelminthes). All of these phyla share a certain combination of characteristics, including the fact that they are usually wormlike, have a cuticle secreted by an epidermis that is underlain by muscles not arranged in regular circular and longitudinal layers, have a brain that is a circumenteric nerve ring, have determinate cleavage and eutely, and lack a muscle layer in the intestine. Hyman contended that the evidences of relationships were so concrete and specific that they could not be disregarded. Nevertheless, most authors now consider that differences between the groups are sufficient to merit phylum status for each, although some accept the concept of the Aschelminthes as a superphylum. These phyla may well have been derived originally from the protostome line via an acoelomate common ancestor resembling the rhabdocoel flatworms. The possible relationship of the Gnathostomulida to the Gastrotricha and Rotifera was mentioned in Chapter 11.

Acanthocephalans are highly specialized parasites with a unique structure and have doubtless been so for millions of years. Any ancestral or other related group that would shed a clue to the phyletic relationships of the Acanthocephala is probably long since extinct. Like the cestodes, acanthocephalans have no digestive tract and must absorb all nutrients across the tegument, but the tegument of the two groups is quite different in structure. Also, acanthocephalans are pseudocoelomate and show eutely, as in the nematodes, although here, too, the structural and developmental differences are great. Thus, the Acanthocephala are an isolated phylum, not closely related to any known form.

The entoprocts were once included with the phylum Ectoprocta in a phylum called Bryozoa, but the ectoprocts are true coelomate animals, and many zoologists prefer to place them in a separate group. Ectoprocts are still often referred to as bryozoans. The Entoprocta may be distantly related to the Ectoprocta, but there is little evidence of close relationship. The entoprocts may have arisen as an early offshoot of the same line that led to the ectoprocts.

Adaptive radiation

Certainly the most impressive adaptive radiation in this group of phyla is shown by the nematodes. They are by far the most numerous in terms of both individuals and species, and they have been able to adapt to almost every habitat available to animal life. Their basic pseudocoelomate body plan, with the cuticle, hydrostatic skeleton, and longitudinal muscles, has proved generalized and plastic enough to adapt to an enormous variety of physical conditions. Free-living lines gave rise to parasitic forms on at least several occasions, and virtually all potential hosts have been exploited. All types of life cycle occur: from the simple and direct to the complex, with intermediate hosts; from normal dioecious reproduction to parthenogenesis, hermaphroditism, and alteration of free-living and parasitic generations. A major factor contributing to the evolutionary opportunism of the nematodes has been their extraordinary capacity to survive conditions suboptimal for viability, for example, the developmental arrests in many free-living and animal parasitic species and the ability to undergo cryptobiosis in many free-living and plant parasitic species.

SUMMARY

The phyla covered in this chapter possess a body cavity called a pseudocoel, which is derived from the embryonic blastocoel, rather than a secondary cavity in the mesoderm (coelom). Several of the groups exhibit eutely, a constant number of cells or nuclei in adult individuals of a given species.

The phylum Rotifera is composed of small, mostly freshwater organisms with a ciliated corona, which creates currents of water to draw planktonic food toward the mouth. The mouth opens into a muscular pharynx, or mastax, that is equipped with jaws.

Many rotifers can produce both amictic and mictic eggs. Amictic eggs are diploid and develop parthenogenetically into females. Mictic eggs are haploid and, if unfertilized, develop parthenogenetically into males. Fertilized mictic eggs develop into resistant, dormant eggs that can overwinter and give rise to females.

The Gastrotricha and Kinorhyncha are small phyla of tiny, aquatic pseudo-coelomates. Gastrotrichs move by cilia or adhesive glands, and kinorhynchs anchor and then pull themselves forward by the spines of their first segment.

By far the largest and most important of this group of phyla are the nematodes, of which there may be as many as 500,000 species in the world. They are more or less cylindrical, tapering at the ends, and covered with a tough, secreted cuticle. The body wall muscles are longitudinal only, and to function well in locomotion, such an arrangement must enclose a volume of fluid in the pseudo-coel at high hydrostatic pressure. This has a profound effect on most of nematodes' other physiological functions, such as ingestion of food, egestion of feces, excretion, and copulation. Most nematodes are dioecious, and there are four juvenile stages, each separated by a molt of the cuticle. Almost all animals and many plants have nematode parasites, and many other nematodes are free living in soil and aquatic habitats. Some nematode parasites of humans are the large round-worm (*Ascaris lumbricoides*), hookworms (for example, *Necator americanus*), trichina worm (*Trichinella spiralis*), pinworm (*Enterobius vermicularis*), and various filarial worms (for example, *Wuchereria bancrofti*). Some parasitic nematodes have part of their life cycle free living, some undergo a tissue migration in their host, and some have an intermediate host in their life cycle.

The Nematomorpha or horsehair worms are related to the nematodes and have parasitic juvenile stages in arthropods, followed by a free-living, aquatic, nonfeeding adult stage.

All Acanthocephalans are parasitic in the intestines of vertebrates as adults, and their juvenile stages develop in arthropods. They have an anterior, invaginable proboscis armed with spines, which they embed in the intestinal wall of their host. They do not have a digestive tract and so must absorb all nutrients across their tegument. Their tegument also bears a system of channels (lacunar system) that interconnects with the spaces in their tubelike muscles.

The Entoprocta are small, sessile, aquatic animals with a crown of ciliated tentacles encircling both the mouth and anus.

The Rotifera, Gastrotricha, Kinorhyncha, Nematoda, and Nematomorpha have been included by some workers in a phylum Aschelminthes, but most biologists believe that the groups are not sufficiently related to be encompassed by a single phylum. It is possible that they were derived from a common ancestor in the protostome line. Phylogenetic relationships of the Acanthocephala and Entoprocta are even more obscure. Of all these phyla, the Nematoda have achieved enormous evolutionary success and undergone great adaptive radiation.

Review questions

1. Give seven characteristics of pseudocoelomate animals.
2. Explain the difference between a true coelom and a pseudocoel.
3. What is the normal size of a rotifer; where is it found; and what are its major body features?
4. Explain the difference between mictic and amictic eggs of rotifers, and tell the adaptive value of each.
5. What is eutely?
6. What are the approximate lengths of gastrotrichs and kinorhynchs? Where are they found?
7. What is a hydrostatic skeleton?
8. Distinguish a solenocyte from a flame bulb protonephridium.
9. Explain two peculiar features of the body-wall muscles in nematodes.
10. What feature of body-wall muscles in nematodes requires a high hydrostatic pressure in the pseudocoelomic fluid for efficient function?

11. Explain the interaction of the cuticle, body wall muscles, and pseudocoelomic fluid in the locomotion of nematodes.
12. Explain how the high pseudocoelomic pressure affects feeding and defecation in nematodes.
13. Outline the life cycle of each of the following: *Ascaris lumbricoides,* hookworm, *Enterobius vermicularis, Trichinella spiralis, Wuchereria bancrofti.*
14. Where in the human body is each of the examples in question 13 found?
15. Where are juveniles and adults of nematomorphs found?
16. Give three differences and three similarities of nematodes and nematomorphs.
17. Describe the major features of the acanthocephalan body.
18. How do acanthocephalans get food?
19. Name four characteristics of entoprocts.
20. What are the "Aschelminth" phyla, and what do they have in common?
21. What is the most successful phylum covered in this chapter? Justify your answer.

Selected references

See also general references for Part Two, p. 590.

Croll, N.A., and B.E. Matthews. 1977. Biology of nematodes. New York, John Wiley & Sons, Inc. *Concise coverage of nematode biology.*

Lee, D.L., and H.J. Atkinson. 1977. The physiology of nematodes, ed. 2. New York, Columbia University Press. *Thorough review of nematode physiology.*

Poinar, G.O., Jr. 1983. The natural history of nematodes. Englewood Cliffs, N.J., Prentice-Hall, Inc. *Contains a great deal of information about these fascinating creatures, including free-living and plant and animal parasites.*

CHAPTER 13

THE MOLLUSCS

Phylum Mollusca

Position in Animal Kingdom

1. The molluscs are one of the major groups of true coelomate animals.
2. They belong to the protostome branch, or schizocoelous coelomates, and have spiral cleavage and determinate (mosaic) development.
3. All the organ systems are present and well developed.
4. Many molluscs have a trochophore larva similar to the trochophore larva of marine annelids and other marine protostomes.

Biological Contributions

1. In molluscs gaseous exchange occurs not only through the body surface as in lower invertebrates, but also in specialized respiratory organs in the form of gills or lungs.
2. Most classes have an open circulatory system with pumping heart, vessels, and blood sinuses. In most cephalopods the circulatory system is closed.
3. The efficiency of the respiratory and circulatory systems in the cephalopods has made greater body size possible. Invertebrates reach their largest size in some of the cephalopods.
4. They have a fleshy mantle that in most cases secretes a shell and is variously modified for a number of functions.
5. Features unique to the phylum are the radula and the muscular foot.
6. The highly developed direct eye of cephalopods is similar to the indirect eye of the vertebrates but arises as a skin derivative in contrast to the brain eye of vertebrates (convergent evolution, Principle no. 7, p. 9).

—— THE MOLLUSCS

The Mollusca (mol-lus′ka) (L. *molluscus,* soft) is one of the largest animal phyla after the Arthropoda. There are nearly 50,000 living species and some 35,000 fossil species. The name Mollusca indicates one of their distinctive characteristics, a soft body.

This very diverse group includes the chitons, tooth shells, snails, slugs, nudibranchs, sea butterflies, clams, mussels, oysters, squids, octopuses, and nautiluses. The group ranges from fairly simple organisms to some of the most complex of invertebrates, and in size from almost microscopic to the giant squid *Architeuthis.* These huge molluscs may grow to 18 m long, including their tentacles. They may weigh 450 kg (1000 pounds). The shells of some of the giant clams, *Tridacna gigas,* which inhabit the Indo-Pacific coral reefs, reach 1.5 m in length and weigh more than 225 kg. These are extremes, however, for probably 80% of all molluscs are less than 5 cm in maximum shell size. The phylum includes some of the most sluggish and some of the swiftest and most active of the invertebrates. It includes herbivorous grazers, predaceous carnivores, filter feeders, detritus feeders, and parasites.

Molluscs are found in a great range of habitats, from the tropics to polar seas, at altitudes exceeding 7000 m, in ponds, lakes, and streams, on mud flats, in pounding surf, and in open ocean from the surface to the abyssal depths. Most of them live in the sea, and they represent a variety of life-styles, including bottom feeders, burrowers, borers, and pelagic forms.

According to the fossil evidence, the molluscs originated in the sea, and most of them have remained there. Much of their evolution occurred along the shores, where food was abundant and habitats were varied. Only the bivalves and gastropods moved on to brackish and freshwater habitats. As filter feeders, the bivalves were unable to leave aquatic surroundings. Only the snails (gastropods) actually invaded the land. Terrestrial snails are limited in their range by their need for humidity, shelter, and the presence of calcium in the soil.

A wide variety of molluscs are used as food. Pearl buttons are obtained from shells of bivalves. The Mississippi and Missouri river basins have furnished material for most of this industry in the United States; however, supplies are becoming so depleted that attempts are being made to propagate bivalves artificially. Pearls, both natural and cultured, are produced in the shells of clams and oysters, most of them in a marine oyster, *Meleagrina*, found around eastern Asia.

Some molluscs are destructive. The burrowing shipworms, which are bivalves of several species (Figure 13-26), do great damage to wooden ships and wharves. To prevent the ravages of shipworms, wharves must be either creosoted or built of concrete (unfortunately, some ignore the creosote, and some bivalves bore into concrete). Snails and slugs frequently damage garden and other vegetation. In addition, snails often serve as intermediate hosts for serious parasites. The boring snail *Urosalpinx* rivals the sea star in destroying oysters.

In this chapter we explore the various major groups of molluscs, those that apparently met with little evolutionary success and those that have been enormously successful. All are based on the basic body plan of the Mollusca, which is described later in the chapter. Before proceeding, however, we must turn our attention briefly to a major anatomical feature that makes its appearance in the molluscs: the coelom.

___ Significance of the Coelom

Heretofore we have been studying animals that lack a true coelom. These follow several patterns. In the radiates with gelatinous mesoglea between the body surface and the enteron, diffusion of substances is a simple matter. In flatworms body spaces are filled with cellular parenchyma of endomesoderm; here the need for a better transport method is met by the extensive branching of the gastrovascular cavity and of the protonephridial system throughout the body. In the nemerteans this is aided by a system of blood vessels.

In the pseudocoelomates the parenchyma is replaced by spongy or open spaces: the pseudocoel. Fluid in the pseudocoel bathes the organs, providing a means of internal transport serviceable enough for small animals. Lacking mesenteries, the organs lie loose in the body cavity. The pseudocoelomates are all small; obviously such an arrangement would be unsuitable for larger forms.

In the coelomates the coelom develops as a **secondary cavity** within the mesoderm. The coelomic cavity is completely surrounded by mesodermal epithelium, called **parietal peritoneum** (Figure 12-1, p. 239). There is ample room in the coelom not only for organs but also for **mesenteries,** continuations of the peritoneum that hold the organs in place. The organs are themselves covered with **visceral peritoneum.** This ensures a more stable arrangement of organs with less

A

C

D

B

E

Figure 13-1

Molluscs: a diversity of life forms. The basic body plan of this ancient group has become variously adapted for different habitats. **A,** Marine snail *(Calliostoma annulata)*, class Gastropoda, subclass Prosobranchia. **B,** Nudibranch *(Hermissenda crassicornis)*, class Gastropoda, subclass Opisthobranchia. **C,** Chiton *(Tonicella lineata)*, class Polyplacophora. **D,** Octopus *(Octopus* sp.*)*, class Cephalopoda. **E,** Marine clam *(Donax denticulatus)* with siphons and foot extended, class Bivalvia.
A to D, Photographs by R. Harbo; E, photograph by K. Sandved.

crowding. The alimentary canal can become more muscular, more highly specialized, and more diversified without interfering with other organs, such as the heart, liver, or lungs.

The coelom performs other important functions. It is filled with **coelomic fluid,** and its lining is often ciliated to keep the fluid moving. Thus, it aids in the movement of materials, such as absorbed foods and metabolic wastes, from one place to another. In many smaller coelomates no other transport system is necessary. In animals with a vascular system, the mesenteries provide an ideal location for the network of blood vessels necessary to deliver blood to every organ.

The coelom can also serve as a hydrostatic skeleton. Circular and longitudinal body wall muscles, acting as antagonists, can contract or relax to vary the force exerted on the coelomic fluid and thus produce a variety of body movements.

Altogether the development of the coelom must be considered an important stepping stone in the evolution of larger and more complex forms. The three major phyla of coelomate protostomes are the molluscs, the annelids, and the arthropods. A number of smaller invertebrate phyla, the echinoderms, and the vertebrates are also coelomates. Ironically, despite the physiological advantages of a coelom, the molluscs do not seem to have exploited them. The coelom in molluscs is limited to a space around the heart, and perhaps around the gonads and part of the kidneys. Nevertheless, the coelom of molluscs develops embryonically in a manner similar to that of the annelids and represents in early form an important body plan to be exploited more fully in other groups.

CHARACTERISTICS
1. Body bilaterally symmetrical (bilateral asymmetry in some); unsegmented; usually with definite head
2. Ventral body wall specialized as a muscular **foot,** variously modified but used chiefly for locomotion
3. Dorsal body wall forms pair of folds called the **mantle,** which encloses the **mantle cavity,** is modified into **gills** or **lungs,** and secretes the **shell** (shell absent in some)
4. Surface epithelium usually ciliated and bearing mucous glands and sensory nerve endings
5. **Coelom** limited mainly to area around heart, and perhaps lumen of gonads and part of kidneys
6. Complex digestive system; rasping organ (**radula**) usually present; anus usually emptying into mantle cavity
7. **Open circulatory system** (mostly closed in cephalopods) of heart (usually three chambered), blood vessels, and sinuses; respiratory pigments in blood
8. Gaseous exchange by **gills, lungs, mantle,** or **body surface**
9. One or two kidneys (**metanephridia**) opening into the pericardial cavity and usually emptying into the mantle cavity
10. Nervous system of paired cerebral, pleural, pedal, and visceral ganglia, with nerve cords and subepidermal plexus; ganglia centralized in nerve ring in gastropods and cephalopods
11. Sensory organs of touch, smell, taste, equilibrium, and vision (in some); eyes highly developed in cephalopods
12. Internal and external **ciliary tracts** often of great functional importance

CLASSIFICATION
Useful characteristics for distinction of classes of molluscs are the type of foot and the type of shell. Several other characteristics are important in particular classes.

Class Caudofoveata (kaw'do-fo-ve-at'a) (L. *cauda*, tail, + *fovea*, small pit). Wormlike; shell, head, and excretory organs absent; radula usually present; mantle with chitinous cuticle and calcareous scales; oral pedal shield near anterior mouth; mantle cavity at posterior end with pair of gills; sexes separate; formerly united with solenogasters in class Aplacophora. Examples: *Chaetoderma, Limifossor.*

Class Solenogastres (so-len'o-gas'trez) (Gr. *solen*, pipe, + *gaster*, stomach): **solenogasters.** Wormlike; shell, head, and excretory organs absent; radula usually absent; mantle usually covered with scales or spicules; mantle cavity posterior, without true gills, but sometimes with secondary respiratory structures; foot represented by long, narrow, ventral pedal groove; hermaphroditic. Example: *Neomenia.*

Class Monoplacophora (mon'o-pla-kof'o-ra) (Gr. *monos*, one, + *plax*, plate, + *phora*, bearing). Body bilaterally symmetrical with a broad flat foot; a single limpetlike shell; mantle cavity with five or six pairs of gills; large coelomic cavities; radula present; six pairs of nephridia, two of which are gonoducts; separate sexes. Example: *Neopilina* (Fig. 13-7).

Class Polyplacophora (pol'y-pla-kof'o-ra) (Gr. *polys*, many, several, + *plax*, plate, + *phora*, bearing): **chitons.** Elongated, dorsoventrally flattened body with reduced head; bilaterally symmetrical; radula present; shell of eight dorsal plates; foot broad and flat; gills multiple, along sides of body between foot and mantle edge; sexes usually separate, with a trochophore but no veliger larva. Examples: *Mopalia* (Figure 13-8), *Tonicella* (Figure 13-1, C).

Class Scaphopoda (ska-fop'o-da) (Gr. *skaphe*, trough, boat, + *pous, podos*, foot): **tusk shells.** Body enclosed in a one-piece tubular shell open at both ends; conical foot; mouth with radula and tentacles; head absent; mantle for respiration; sexes separate; trochophore larva. Example: *Dentalium* (Figure 13-10).

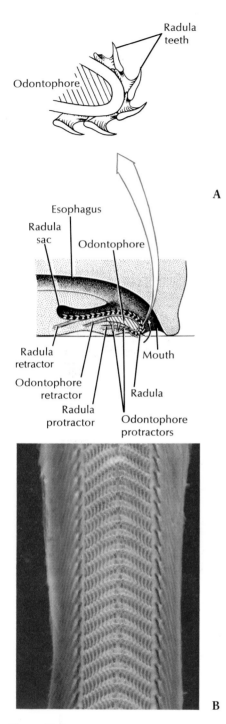

Figure 13-2

A, Diagrammatic longitudinal section of gastropod head showing the radula and radula sac. The radula moves back and forth over the odontophore cartilage. As the animal grazes, the mouth opens, the odontophore is thrust forward, the radula gives a strong scrape backward bringing food into the pharynx, and the mouth closes. The sequence is repeated rhythmically. As the radula ribbon wears out anteriorly, it is continually replaced posteriorly. **B,** Radula of a snail *(Cittarium pica)* prepared for microscopic examination.

B, Photograph by K. Sandved.

Class Gastropoda (gas-trop′o-da) (Gr. *gaster*, belly, + *pous, podos*, foot): **snails and others.** Body asymmetrical and shows effects of torsion; body usually in a coiled shell (shell uncoiled or absent in some); head well developed, with radula; foot large and flat; one or two gills, or with mantle modified into secondary gills or a lung; most with single auricle and single nephridium; nervous system with cerebral, pleural, pedal, and visceral ganglia; dioecious or monoecious, some with trochophore, typically with veliger, some without pelagic larva. Examples: *Busycon, Polinices* (Figure 13-14), *Physa, Helix, Aplysia* (Figure 13-20).

Class Bivalvia (bi-val′ve-a) (L. *bi*, two, + *valva*, folding door, valve) **(Pelecypoda): bivalves.** Body enclosed in a two-lobed mantle; shell of two lateral valves of variable size and form, with dorsal hinge; head greatly reduced, but mouth with labial palps; no radula; no cephalic eyes, a few with eyes on mantle margin; foot usually wedge shaped; gills platelike; sexes usually separate, typically with trochophore and veliger larvae. Examples: *Anodonta, Venus, Tagelus* (Figure 13-25), *Teredo* (Figure 13-26).

Class Cephalopoda (sef′a-lop′o-da) (Gr. *kephalē*, head, + *pous, podos*, foot): **squids and octopuses.** Shell often reduced or absent; head well developed with eyes and a radula; head with arms or tentacles; foot modified into siphon; nervous system of well-developed ganglia, centralized to form a brain; sexes separate, with direct development. Examples: *Loligo* (Figure 13-37), *Octopus* (Figure 13-1, *D*), *Sepia* (Figure 13-36).

____ Form and Function

The enormous variety, great beauty, and easy availability of the shells of molluscs have made shell collecting a popular pastime. However, many amateur shell collectors, even though able to name hundreds of the shells that grace our beaches, know very little about the living animals that created those shells and once lived in them. Reduced to its simplest dimensions, the mollusc body plan may be said to consist of a **head-foot** portion and a **visceral hump** portion. The head-foot is the more active area, containing the feeding, cephalic sensory, and locomotor organs. It depends primarily on muscular action for its function. The visceral hump is the portion containing digestive, circulatory, respiratory, and reproductive organs, and it depends primarily on ciliary tracts for its functioning. Two folds of skin, outgrowths of the dorsal body wall, make up a protective **mantle,** or **pallium,** which encloses a space between the mantle and body wall called the **mantle cavity (pallial cavity).** The mantle cavity houses the **gills (ctenidia)** or lung, and in some molluscs the mantle secretes a protective **shell** over the visceral hump. Modifications of the structures that make up the head-foot and the visceral hump produce the great confusion of different patterns making up this major group of animals. Greater emphasis on either the head-foot portion or the visceral hump portion can be observed in the various classes of molluscs.

Head-foot

Most molluscs have well-developed heads, which bear the mouth and some specialized sensory organs. Photosensory receptors range from fairly simple up to the complex eyes of the cephalopods. Tentacles are often present. Within the mouth is a structure unique to molluscs, the radula, and usually posterior to the mouth is the chief locomotor organ, or foot.

Radula

The radula is a rasping, protrusible, tonguelike organ found in all molluscs except the bivalves and most solenogasters. It is a ribbonlike membrane on which are mounted rows of tiny teeth that point backward (Figure 13-2). Complex muscles move the radula and its supporting cartilages (**odontophore**) in and out while the membrane is partly rotated over the tips of the cartilages. There may be a few or as many as 250,000 teeth, which, when protruded, can scrape, pierce, tear, or cut. The usual function of the radula is twofold: to rasp off fine particles of food material and to serve as a conveyor belt for carrying the particles in a continuous stream toward the digestive tract. As the radula wears away anteriorly, new rows of teeth are continuously replaced by secretion at its posterior end. The pattern and number of teeth in a row are specific for each species and are used in the classification of molluscs. Very interesting radular specializations, such as for boring through hard materials or for harpooning prey, are found in some forms.

Foot

The molluscan foot may be variously adapted for locomotion, for attachment to a substratum, or for a combination of functions. It is usually a ventral-solelike structure in which waves of muscular contraction effect a creeping locomotion (Figure 13-15). However, there are many modifications, such as the attachment disc of limpets, the laterally compressed "hatchet foot" of bivalves, or the siphon for jet propulsion in the squids and octopuses. Secreted mucus is often used as an aid to adhesion or as a slime tract by small molluscs that glide on cilia.

In snails and bivalves the foot is extended from the body hydraulically, by engorgement with blood. Burrowing forms can extend the foot into the mud or sand, enlarge it with blood pressure, then use the engorged foot as an anchor to draw the body forward. In pelagic (free-swimming) forms the foot may be modified into winglike parapodia, or thin, mobile fins for swimming.

Visceral hump

Mantle and mantle cavity

The mantle is a sheath of skin extending from the visceral hump that hangs down on each side of the body, protecting the soft parts and creating between itself and the visceral mass the space called the mantle cavity. The outer surface of the mantle secretes the shell (p. 266).

The mantle cavity plays an enormous role in the life of the mollusc. It usually houses the respiratory organs (gills or lung), which develop from the mantle, and the mantle's own exposed surface serves also for gaseous exchange. The products from the digestive, excretory, and reproductive systems are emptied into the mantle cavity. In aquatic molluscs a continuous current of water, kept moving by surface cilia or by muscular pumping, brings in oxygen and, in some forms, food; flushes out wastes; and carries reproductive products out to the environment. In aquatic forms the mantle is usually equipped with sensory receptors for sampling the environmental water. In cephalopods (squids and octopuses) the muscular mantle and its cavity create the jet propulsion used in locomotion. Many molluscs can withdraw the head or foot into the mantle cavity, which is surrounded by the shell, for protection.

In primitive form, the mollusc ctenidium (gill) consists of a long, flattened axis extending from the wall of the mantle cavity (Figure 13-3). Many leaflike gill filaments project from the central axis. Water is propelled by cilia between the gill filaments, and blood diffuses from an afferent vessel in the central axis through the filament to an efferent vessel. The direction of blood movement is opposite to the direction of water movement, thus establishing a countercurrent exchange mechanism (see p. 651). The two ctenidia are located on opposite sides of the mantle cavity and are arranged so that the cavity is functionally divided into an incurrent chamber and an excurrent chamber. Such gills are found in the more primitive gastropods, but they are variously modified in many molluscs.

Figure 13-3

Primitive condition of mollusc ctenidium. Circulation of water between the gill filaments is by cilia, and blood diffuses through the filament from the afferent vessel to the efferent vessel.

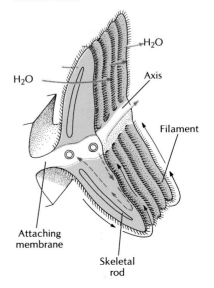

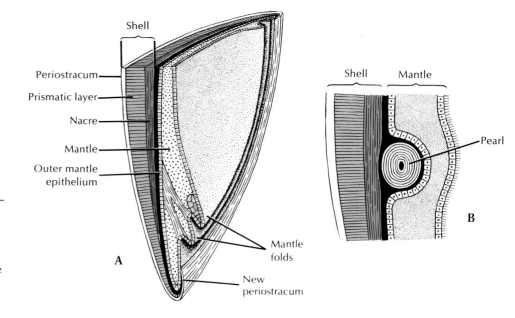

Figure 13-4

A, Diagrammatic vertical section of shell and mantle of a bivalve. The outer mantle epithelium secretes the shell; the inner epithelium is usually ciliated. **B,** Formation of pearl between mantle and shell as a parasite or bit of sand under the mantle becomes covered with nacre.

Shell

The shell of the mollusc, when present, is secreted by the mantle and is lined by it. Typically there are three layers (Figure 13-4). The **periostracum** is the outer horny layer, composed of an organic substance called conchiolin, which consists of quinone-tanned protein. It helps to protect the underlying calcareous layers from erosion by boring organisms. It is secreted by a fold of the mantle edge, and growth occurs only at the margin of the shell. On the older parts of the shell the periostracum often becomes worn away. The middle **prismatic layer** is composed of densely packed prisms of calcium carbonate laid down in a protein matrix. It is secreted by the glandular margin of the mantle, and increase in shell size occurs at the shell margin as the animal grows. The inner **nacreous layer** of the shell lies next to the mantle and is secreted continuously by the mantle surface, so that it increases in thickness during the life of the animal. The calcareous nacre is laid down in thin layers. Very thin and wavy layers produce the iridescent mother-of-pearl found in the abalones *(Haliotis),* the chambered nautilus *(Nautilus),* and many bivalves. Such shells may have 450 to 5000 fine parallel layers of crystalline calcium carbonate (aragonite) for each centimeter of thickness.

Freshwater molluscs usually have a thick periostracum that gives some protection against the acids produced in the water by the decay of leaf litter. In many marine molluscs the periostracum is relatively thin, and in some it is absent. There is a great variation in shell structure. Calcium for the shell comes from the environmental water or soil or from food. The first shell appears during the larval period and grows continuously throughout life.

Internal structure and function

Gaseous exchange occurs through the body surface, particularly the mantle, and in specialized respiratory organs such as ctenidia, secondary gills, and lungs. There is an **open circulatory system** with a pumping heart, blood vessels, and blood sinuses. Most cephalopods have a closed blood system with heart, vessels, and capillaries. The digestive tract is complex and highly specialized, according to the feeding habits of the various molluscs, and is usually provided with extensive ciliary tracts. Most molluscs have a pair of kidneys (**metanephridia,** a type of nephridium in which the inner end opens into the coelom by a **nephrostome**); the ducts of the kidneys in many forms also serve for the discharge of eggs and sperm.

The **nervous system** consists of several pairs of ganglia with connecting nerve cords, and is generally simpler than that of the annelids and arthropods. Neurosecretory cells have been identified in the nervous system that, at least in certain airbreathing snails, produce a growth hormone and function in osmoregulation. There are various types of highly specialized sense organs.

Reproduction and life history

Most molluscs are dioecious, although some are hermaphroditic. The free-swimming larva that emerges from the egg in primitive molluscs is the **trochophore**, which is also the primitive larval type of the annelids (Figure 13-5). In most primitive molluscs the trochophore metamorphoses directly into a small juvenile, but in more advanced forms, another free-swimming larval stage, the **veliger**, intervenes (Figure 13-6). The veliger has the beginnings of a foot, shell, and mantle. In many advanced molluscs the trochophore is passed in the egg, and a veliger hatches to become the only free-swimming stage. Cephalopods, freshwater and some marine snails, and some freshwater bivalves have no free-swimming larvae, and a juvenile hatches from the egg.

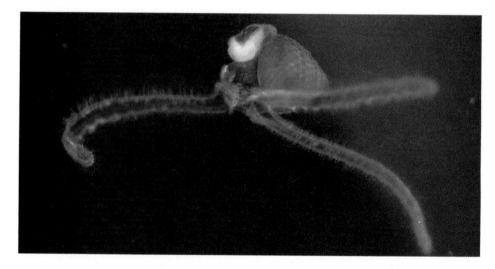

____ Classes of Molluscs

For more than 50 years five classes of living molluscs were recognized: Amphineura, Gastropoda, Scaphopoda, Bivalvia, and Cephalopoda. The discovery of *Neopilina* in the 1950s added another class (Monoplacophora), and Hyman (1967) contended that solenogasters and chitons make up separate classes (Polyplacophora and Aplacophora), lapsing the name Amphineura. Recognition of important differences between organisms such as *Chaetoderma* and the other solenogasters has led to the separation of the Aplacophora into the Caudofoveata and the Solenogastres (Boss, 1982).

Class Caudofoveata

Members of the class Caudofoveata are wormlike, marine organisms ranging from 2 to 140 mm in length (Figure 13-40). They are mostly burrowers and orient themselves vertically, with the terminal mantle cavity and gills at the entrance of the burrow. They feed on microorganisms and detritus. They have no shell, but their bodies are covered with calcareous scales. There are no spicules or scales on the oral pedal shield, an organ apparently associated with food selection and

Figure 13-5

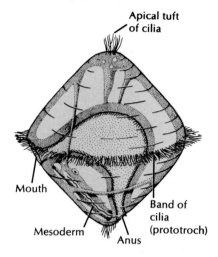

Generalized trochophore larva. Molluscs and annelids with primitive embryonic development have trochophore larvae, as do several other phyla.

Apical tuft of cilia

Mouth

Mesoderm

Anus

Band of cilia (prototroch)

Figure 13-6

Veliger of a snail, *Pedicularia*, swimming. The adults are parasitic on corals. The ciliated process (velum) develop from the prototroch of the trochophore (Fig. 13-5). Photograph by K. Sandved.

The trochophore larva (Figure 13-5) is minute, translucent, and more or less pear shaped and has a prominent circlet of cilia (prototroch) and sometimes one or two accessory circlets. It is found in molluscs and annelids with primitive embryonic development and is considered one of the evidences for common phylogenetic origin of the two phyla. Some form of trochophore-like larva is also found in marine turbellarians, nemertines, brachiopods, phoronids, sipunculids, and echiurids, and it probably reflects some phylogenetic relationship among all these phyla.

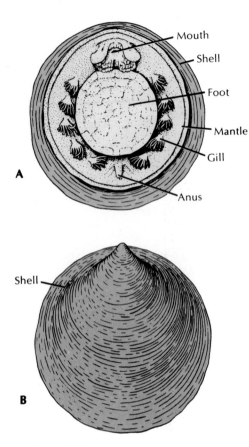

Figure 13-7

Neopilina, class Monoplacophora. Living specimens range from 3 mm to about 3 cm in length. **A,** Ventral view. **B,** Dorsal view.

intake. A radula is present, although reduced in some, and the sexes are separate. Since this little group has fewer than 70 species, it does not represent a successful variation on the mollusc body plan. However, its features may be closer to the ancestral mollusc than any other living group.

Class Solenogastres

The solenogasters (Figure 13-40) and the caudofoveates were formerly united in the class Aplacophora, and they are both marine, wormlike, shell-less, with calcareous scales or spicules in their integument, with reduced head, and without nephridia. The solenogasters, however, usually have no radula and no gills (although secondary respiratory structures may be present). Their foot is represented by a midventral, narrow furrow, the pedal groove. They are hermaphroditic. Rather than burrowing, solenogasters live free on the bottom, and they often live and feed on cnidarians. The solenogasters are also a small group, numbering about 180 species.

Class Monoplacophora

Until 1952 it was thought that the Monoplacophora were extinct; they were known only from Paleozoic shells. However, in that year living specimens of *Neopilina* (Gr. *neo,* new, + *pilos,* felt cap) were dredged up from the ocean bottom near the west coast of Costa Rica. Fewer than a dozen species of monoplacophorans are now known. These molluscs are small and have a low, rounded shell and a creeping foot (Figure 13-7). They have superficial resemblance to limpets, but unlike most other molluscs, a number of organs are serially repeated. Such serial repetition occurs to a more limited extent in the chitons. *Neopilina* has five pairs of gills, two pairs of auricles, six pairs of nephridia, one or two pairs of gonads, and a ladderlike nervous system with 10 pairs of pedal nerves. The mouth bears the characteristic radula.

Class Polyplacophora: chitons

The chitons (Gr. coat of mail, tunic) (Figures 13-8 and 13-9) represent a somewhat more successful mollusc group. They are rather flattened dorsoventrally and have a convex dorsal surface that bears eight articulating limy plates, or valves, hence their name Polyplacophora ("many plate bearers"). The plates overlap posteriorly and are usually dull colored to match the rocks to which the chitons cling. Their head and cephalic sensory organs are reduced, but photosensitive structures (esthetes), which have the form of eyes in some chitons, pierce the plates.

Most chitons are small (2 to 5 cm); the largest, *Cryptochiton* (Gr. *crypto,* hidden, + *chiton,* coat of mail), rarely exceeds 30 cm. They prefer rocky surfaces in intertidal regions, although some live at great depths. Most chitons are stay-at-home organisms, straying only very short distances for feeding. In feeding, the radula is projected from the mouth to scrape algae from the rocks. The radula is reinforced with the iron-containing mineral, magnetite. The chiton clings tenaciously to its rock with the broad, flat foot. If detached, it can roll up like an armadillo for protection.

The mantle forms a girdle around the margin of the plates, and in some species mantle folds cover part or all of the plates. Compared with the primitive condition, the mantle cavity has been extended along the side of the foot, and the gills have been increased in number. Thus, the gills are suspended from the roof of the mantle cavity along each side of the broad ventral foot. With the foot and the

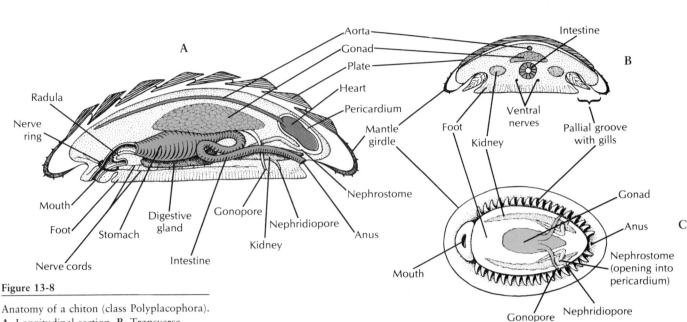

Figure 13-8

Anatomy of a chiton (class Polyplacophora). **A,** Longitudinal section. **B,** Transverse section. **C,** External ventral view.

mantle margin adhering tightly to the substrate, these grooves become closed chambers, open only at the ends. Water enters the grooves anteriorly, flows across the gills, and leaves posteriorly, bringing a continuous supply of oxygen to the gills. At low tide the margins of the mantle can be tightly pressed to the substratum to diminish water loss, but in some circumstances, the mantle margins can be held open for limited air breathing. A pair of **osphradia** (sense organs for sampling water) are found in the mantle grooves near the anus of many chitons.

Blood pumped by the three-chambered heart reaches the gills by way of an aorta and sinuses. A pair of kidneys (metanephridia) carries waste from the pericardial cavity to the exterior. Two pairs of longitudinal nerve cords are connected in the buccal region.

Sexes are separate in most chitons, and the trochophore larva metamorphoses directly into a juvenile, without an intervening veliger stage.

Class Scaphopoda

The Scaphopoda, commonly called the tusk shells or tooth shells, are benthic marine molluscs found from the subtidal zone to over 6000 m depth. They have a slender body covered with a mantle and a tubular shell open at both ends. In the scaphopods the molluscan body plan has taken a new direction, with the mantle wrapped around the viscera and fused to form a tube. Most scaphopods are 2.5 to 5 cm long, although they range from 4 mm to 25 cm long. *Dentalium* (L. *dentis*, tooth) is a common Atlantic genus.

The foot, which protrudes through the larger end of the shell, is used to burrow into mud or sand, always leaving the small end of the shell exposed to the water above (Figure 13-10). Respiratory water is circulated through the mantle cavity both by movements of the foot and ciliary action (Figure 13-10). Gaseous exchange occurs in the mantle, for gills are absent. Most of the food is detritus and protozoa from the substratum. It is caught on the cilia of the foot or on the mucus-covered, ciliated knobs of the long tentacles extending from the head, called captacula, and is conveyed to the nearby mouth. Eyes and tentacles are lacking. The radula carries the food to a crushing gizzard.

The sexes are separate, and the larva is a trochophore.

Figure 13-9

Mossy chiton, *Mopalia muscosa*. The upper surface of the mantle, or "girdle," is covered with hairs and bristles, an adaptation for defense.
Photograph by R. Harbo.

Figure 13-10

Scaphopoda. **A,** *Dentalium* (class Scaphopoda), with its tubular shell, burrows into soft mud or sand and feeds by means of its prehensile tentacles. Water enters and leaves by way of the posterior aperture. **B,** Internal anatomy of *Dentalium*.

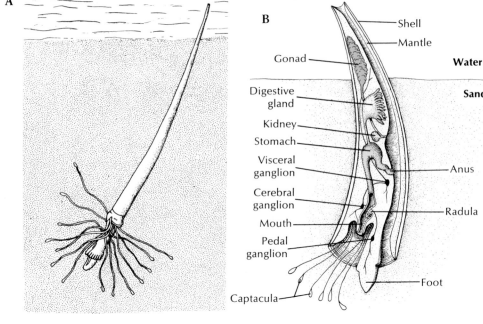

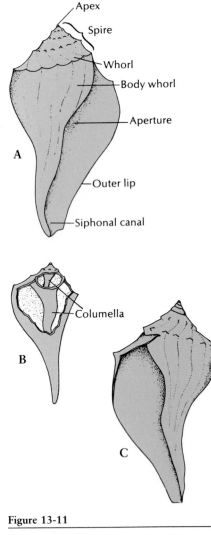

Figure 13-11

Shell of the whelk *Busycon*. **A** and **B,** *Busycon carica*, a dextral, or right-handed, shell. **C,** *B. contrarium*, a sinistral, or left-handed, shell.

Class Gastropoda

Among the molluscs the class Gastropoda is by far the largest and most successful, containing about 35,000 living and 15,000 fossil species. It is made up of members of such diversity that there is no single general term in our language that can apply to them as a group. They include snails, limpets, slugs, whelks, conchs, periwinkles, sea slugs, sea hares, and sea butterflies. They range from some of the most primitive of marine molluscs to the highly evolved terrestrial air-breathing snails and slugs. These animals are basically bilaterally symmetrical, but because of **torsion,** a twisting process that occurs in the veliger stage, the visceral mass has become asymmetrical.

The shell, when present, is always of one piece (**univalve**) and may be coiled or uncoiled. Starting at the apex, which contains the oldest and smallest whorl, the whorls become successively larger and spiral about the central axis, or **columella** (Figure 13-11). The shell may be right handed (**dextral**) or left handed (**sinistral**), depending on the direction of coiling. Dextral shells are far more common. The direction of coiling is genetically determined.

Gastropods range from microscopic forms to giant marine forms such as *Pleuroploca gigantea*, a snail with a shell up to 60 cm long, and the sea hare *Aplysia* (Figure 13-20), some species of which reach 1 m in length. Most of them, however, are between 1 and 8 cm in length. Some fossil gastropods were as much as 2 m long.

The range of gastropod habitats is large. In the sea, gastropods are common both in the littoral zones and at great depths, and some are even pelagic. Some are adapted to brackish water and others to fresh water. On land they are restricted by such factors as the mineral content of the soil and extremes of temperature, dryness, and acidity. Even so, they are widespread, and some have been found at great altitudes and some even in polar regions. Snails occupy all kinds of habitats: in small pools or large bodies of water, in woodlands, in pastures, under rocks, in mosses, on cliffs, in trees, underground, and on the bodies of other animals. They have successfully undertaken every mode of life except aerial locomotion.

Gastropods are usually sluggish, sedentary animals because most of them have heavy shells and slow locomotion. Some are specialized for climbing, swimming, or burrowing. Shells are their chief defense, although they are also protected by coloration and by secretive habits. Many snails have an **operculum,** a horny plate that covers the shell aperture when the body is withdrawn into the shell. Others lack shells altogether. Some are distasteful to other animals, and a few such as *Strombus* can deal an active blow with the foot, which bears a sharp operculum. Nevertheless, they are eaten by birds, beetles, small mammals, fish, and other predators. Serving as intermediate hosts for many kinds of parasites, especially trematodes, snails are often harmed by the larval stages of parasites.

Torsion

Of all the molluscs, only gastropods undergo torsion. Torsion is a peculiar phenomenon that moves the mantle cavity, which was originally (primitively) posterior, to the front of the body, thus twisting the visceral organs as well through a 180-degree rotation. It occurs during the veliger stage, and in some species it may take only a few minutes. Before torsion occurs, the embryo's mouth is anterior and the anus and mantle cavity are posterior (Figure 13-12). The change is brought about by an uneven growth of the right and left muscles that attach the shell to the head-foot.

After torsion, the anus and mantle cavity become anterior and open above the mouth and head. The left gill, kidney, and heart auricle are now on the right side, whereas the original right gill, kidney, and heart auricle are now on the left, and the nerve cords have been twisted into a figure eight. Because of the space available in the mantle cavity, the animal's sensitive head end can now be withdrawn into the protection of the shell, with the tougher foot forming a barrier to the outside.

Varying degrees of **detorsion** are seen in the opisthobranchs and pulmonates, and the anus opens to the right side or even to the posterior. However, both of these groups were derived from torted ancestors.

The curious arrangement that results from torsion poses a serious sanitation problem by creating the possibility of wastes being washed back over the gills (**fouling**) and causes us to wonder what strong evolutionary pressures selected for such a strange realignment of body structures. Several explanations have been

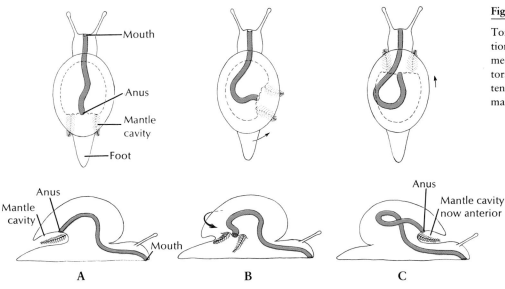

Figure 13-12

Torsion in gastropods. **A,** Ancestral condition before torsion. **B,** Hypothetical intermediate condition. **C,** Early gastropod, torsion complete; direction of crawling now tends to carry waste products back into mantle cavity, resulting in fouling.

proposed, none entirely satisfying. For example, sense organs of the mantle cavity (osphradia) would better sample water in the direction of travel. Certainly the consequences of torsion and the resulting need to avoid fouling have been very important in the subsequent evolution of gastropods. These consequences cannot be explored, however, until another unusual feature of gastropods—coiling—has been described.

Coiling

The coiling, or spiral winding, of the shell and visceral hump is not the same as torsion. Coiling may occur in the larval stage at the same time as torsion, but the fossil record shows that coiling was a separate evolutionary event and originated in gastropods earlier than torsion did. Nevertheless, all living gastropods have descended from coiled, torted ancestors, whether or not they now show these characteristics.

Early gastropods had a bilaterally symmetrical **planospiral** shell; that is, all the whorls lay in a single plane (Figure 13-13, *A*). Such a shell was not very compact, since each whorl had to lie completely outside the preceding one. Curiously, a few modern species had secondarily returned to the planospiral form. The compactness problem of the planospiral shell was solved by the **conispiral** shape, in which each succeeding whorl was at the side of the preceding one (Figure 13-13, *B*). However, this shape was clearly unbalanced, hanging as it was with much weight over to one side. Better weight distribution was achieved by shifting the shell upward and posteriorly, with the shell axis oblique to the longitudinal axis of the foot (Figure 13-22). The weight and bulk of the main body whorl, the largest whorl of the shell, pressed on the right side of the mantle cavity, however, and apparently interfered with the organs on that side. Accordingly, the gill, auricle, and kidney of the right side have been lost in all except primitive living gastropods, leading to a condition of *bilateral asymmetry*.

Although the loss of the right gill was probably an adaptation to the mechanics of carrying the coiled shell, that condition made possible a way to avoid the problem of torsion—fouling—that is displayed in most modern prosobranchs. Water is brought into the left side of the mantle cavity and out the right side, carrying with it the wastes from the anus and nephridiopore, which lie near the right side. Ways in which fouling is avoided in other gastropods are mentioned later in the chapter.

Feeding habits

Feeding habits of gastropods are as varied as their shapes and habitats, but all include the use of some adaptation of the radula. The majority of gastropods are herbivorous, rasping off particles of algae. Some herbivores are grazers, some are browsers, and some are planktonic feeders. *Haliotis,* the abalone (Figure 13-14, *A*), holds seaweed with the foot and breaks off pieces with the radula. Land snails forage at night for green vegetation.

Some snails, such as *Bullia* and *Buccinum,* are scavengers living on dead and decaying flesh; others are carnivores that tear their prey with the radular teeth. *Melongena* feeds on clams, especially *Tagelus,* the razor clam, thrusting its proboscis between the gaping shell valves. *Fasciolaria* and *Polinices* (Figure 13-14, *B*) feed on a variety of molluscs, preferably bivalves. *Urosalpinx cinerea,* the oyster borer, or tingle, drills holes through the shell of the oyster. Its radula, bearing three longitudinal rows of teeth, is used first to begin the drilling action, then the animal glides forward, everts an accessory boring organ through a pore in the anterior sole of its foot, and holds it against the shell, using a chemical agent to soften the

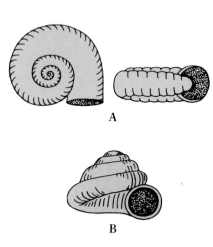

Figure 13-13

Coiling of the shell, which is independent of torsion. **A,** Two views of planospiral coiling, or coiling that occurs in a single plane. **B,** Conispiral coiling, which produces a cone-shaped shell.

A   B

Figure 13-14

A, Red abalone, *Haliotis rufescens.* This huge, limpetlike snail is prized as food and extensively marketed. Abalones are strict vegetarians, feeding especially on sea lettuce and kelp. **B,** Moon snail, *Polinices lewisii.* A common inhabitant of West Coast sand flats, the moon snail is a predator of clams and mussels. It uses its radula to drill neat holes through its victim's shell, through which the proboscis is then extended to eat the bivalve's fleshy body.
Photographs by D.W. Gotshall.

shell. Short periods of rasping alternate with long periods of chemical activity until a neat round hole is completed. With its proboscis inserted through the hole, the snail may feed continuously for hours or days, using the radula to tear away the soft flesh. *Urosalpinx* is attracted to its prey at some distance by sensing some chemical, probably one released in the metabolic wastes of the prey.

Cyphoma gibbosum and related species live and feed on gorgonians (phylum Cnidaria, Chapter 10) in shallow, tropical coral reefs. This snail is commonly known as the flamingo tongue. During normal activity the brightly colored mantle entirely envelops the shell, but it can be quickly withdrawn into the shell aperture when the animal is disturbed.

Members of the genus *Conus* (Figure 13-15) feed on worms, fish, and other gastropods. Their radula is highly modified for prey capture. A gland charges the radular teeth with a highly toxic venom. When *Conus* senses the presence of its prey, a single radular tooth is slid into position at the tip of the proboscis. When the proboscis strikes the prey, the tooth is expelled like a harpoon, and the poison quiets the prey at once. Some species of *Conus* can deliver very painful stings, and in several species the sting is lethal to humans. The venom consists of a mixture of toxic peptides that attack successive physiological targets in the neuromuscular system.

Some gastropods feed on organic deposits on the sand or mud. Others collect the same sort of organic debris but can digest only the microorganisms contained in it. Some sessile gastropods, such as some limpets, are ciliary feeders that use the gill cilia to draw in particulate matter that is rolled into a mucous ball and is carried to the mouth. Some sea butterflies secrete a mucous net to catch small planktonic forms; then they draw the web into the mouth.

After maceration by the radula or by some grinding device, such as the gizzard in the sea hare *Aplysia,* digestion is usually extracellular in the lumen of the stomach or digestive glands. In ciliary feeders the stomachs are sorting regions, and most of the digestion is intracellular in the digestive glands.

Internal form and function

Respiration in most gastropods is carried out by a **ctenidium** (two ctenidia in primitive prosobranchs) located in the mantle cavity, though some aquatic forms, lacking gills, depend on the mantle and skin. After the more advanced prosobranchs lost one of the gills, most of them lost half of the remaining one, and the

Figure 13-15

Cone shell, *Conus californicus.* Water is drawn into the mantle cavity through the extended siphon.
Photograph by D.W. Behrens.

Some species of *Conus* that feed on fish show amazing behavioral tactics to lure their prey. The snail lies concealed beneath the sand, and after it detects a suitable fish nearby (method of detection unknown), it extends its long, wormlike proboscis. When the fish attempts to consume this tasty morsel, the *Conus* stings it in the mouth and kills it. The snail emerges and engulfs the fish with its distensible stomach, then regurgitates the scales and bones some hours later.

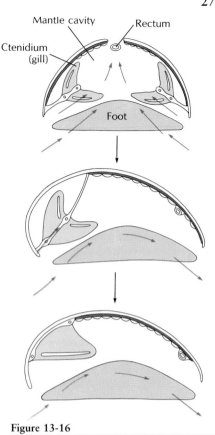

Figure 13-16

Evolution of the ctenidia in gastropods. **A,** Primitive prosobranchs with two ctenidia and excurrent water leaving the mantle cavity by a dorsal slit or hole. **B,** Condition after one ctenidium had been lost. **C,** Condition in advanced prosobranchs, in which filaments on one side of remaining gill are lost, and axis is attached to mantle wall.

central axis became attached to the wall of the mantle cavity (Figure 13-16). Thus they attained the most efficient gill arrangement for the way the water circulated through the mantle cavity (in one side and out the other).

The pulmonates have a highly vascular area in the mantle that serves as a **lung** (Figure 13-17). Most of the mantle margin seals to the back of the animal, and the lung opens to the outside by a small opening called the **pneumostome.** Many aquatic pulmonates must surface to expel a bubble of gas from the lung. To take in air, they curl the edge of the mantle around the pneumostome to form a siphon.

Most gastropods have a single nephridium (kidney). The circulatory and nervous sytems are well developed (Figure 13-17). The latter includes three pairs of ganglia connected by nerves. Sense organs include eyes or simple photoreceptors, statocysts, tactile organs, and chemoreceptors. The simplest type of gastropod eye is simply a cuplike indentation in the skin lined with pigmented photoreceptor cells. In many gastropods the eyecup contains a lens and is covered with a cornea. A sensory area called the **osphradium,** located at the base of the incurrent siphon of most gastropods, is known to be chemosensory in some forms, although its function may be mechanoreceptive in some and is still unknown in others.

There are both dioecious and monoecious gastropods. Many gastropods perform courtship ceremonies. During copulation in monoecious species there is an exchange of spermatozoa or spermatophores (bundles of sperm). Many terrestrial pulmonates eject a dart from a dart sac (Figure 13-17) into the partner's body to heighten excitement before copulation. After copulation each partner deposits its eggs in shallow burrows in the ground. The most primitive gastropods discharge ova and sperm into the seawater where fertilization occurs, and the embryos soon hatch as a free-swimming trochophore larvae. In most gastropods fertilization is internal.

Fertilized eggs encased in transparent shells may be emitted singly to float among the plankton or may be laid in gelatinous layers attached to the substratum. Some marine forms enclose the eggs, either in small groups or in large numbers, in tough egg capsules, in a wide variety of egg cases (Figure 13-18). The young generally emerge as veliger larvae (Figure 13-6), or they may spend the veliger stage in the case of capsule and emerge as young snails. Some species, including many freshwater snails, are ovoviviparous, brooding their eggs and young in the pallial oviduct.

Figure 13-17

Anatomy of a pulmonate snail.

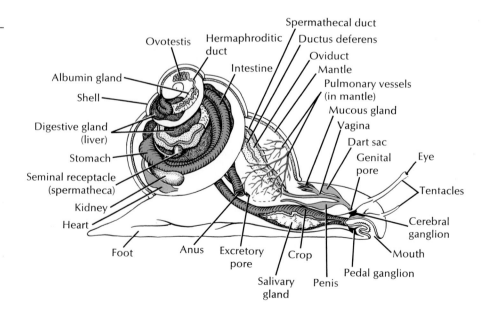

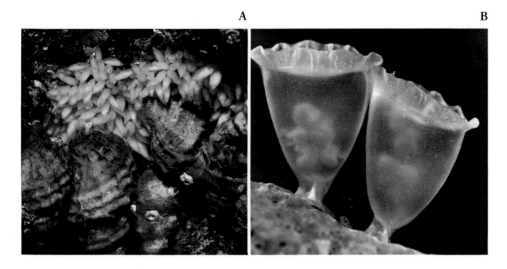

A   **B**

Figure 13-18

Eggs of marine gastropods. **A,** The wrinkled whelk, *Thais lamelosa,* lays egg cases resembling grains of wheat; each contains hundreds of eggs. **B,** Urnlike egg case of the common Caribbean drupe snail *(Drupa).*
A, Photograph by R. Harbo; **B,** photograph by K. Sandved.

Major groups of gastropods

There are three subclasses of gastropods: the Prosobranchia, Opisthobranchia, and Pulmonata.

Prosobranchia

The mantle cavity is anterior as a result of torsion, with the gill or gills lying in front of the heart. Water enters the left side and exits from the right side, and the edge of the mantle is often extended into a long siphon to further separate incurrent from excurrent flow (Figure 13-15). In the primitive prosobranchs with two gills (for example, the abalone *Haliotis* and the keyhole limpet *Diodora,* Figures 13-14, *A,* and 13-19, *B*), fouling is avoided by having the excurrent water go up and out through one or more holes in the shell above the mantle cavity.

Figure 13-19

A, Cowry, *Jenneria pustulata,* crawls over zoanthid cnidarians. The brightly patterned and polished shells of cowries have been used as ornaments for thousands of years. **B,** Keyhole limpet, *Diodora aspera,* a prosobranch gastropod with a hole in the apex through which the water leaves the mantle cavity.
A, Photograph by K. Sandved; **B,** photograph by R. Harbo.

A   **B**

A

Rhinophore   Oral tentacle   **B**

Figure 13-20

A, The sea hare, *Aplysia dactylomela,* crawls and swims across a tropical seagrass bed, assisted by large, winglike parapodia, here curled above the body. **B,** When attacked, the sea hare squirts a copious protective secretion from its "purple gland" in the mantle cavity.

Photographs by C.P. Hickman, Jr.

Figure 13-21

Aeolid nudibranchs crawl across a hydroid colony. Their long, dorsal cerata contain nematocysts, which the animals obtain from their cnidarian diet.

Photograph by K. Sandved.

Prosobranchs have one pair of tentacles. The sexes are usually separate. An operculum is often present.

This group contains most of the marine snails and some of the freshwater and terrestrial gastropods. They range in size from the periwinkles and small limpets *(Patella* and *Diodora)* (Figure 13-19, *B)* to the horse conch *(Pleuroploca),* the largest gastropod in America. Familiar examples of prosobranchs are the abalone *(Haliotis),* which has an ear-shaped shell; the whelk *(Busycon),* which lays its eggs in double-edged, disc-shaped capsules attached to a cord a meter long; the common periwinkle *(Littorina);* the moon snail *(Polinices,* Figure 13-14, *B);* the oyster borer *(Urosalpinx),* which bores into oysters and sucks out their juices; the rock shell *(Murex),* a European species that was used to make the royal purple of the ancient Romans; and the freshwater forms *(Goniobasis* and *Viviparus).*

Opisthobranchia

The opisthobranchs show partial or complete detorsion; thus, the anus and gill (if present) are displaced to the right side or rear of the body. Clearly, the fouling problem is obviated if the anus is moved away from the head toward the posterior. Two pairs of tentacles are usually found, and the second pair is often further modified (**rhinophores**), with platelike folds that apparently increase the area for chemoreception. The shell is typically reduced or absent. All are monoecious. The opisthobranchs are an odd assemblage of molluscs that include sea slugs, sea hares, sea butterflies, and canoe shells. They are nearly all marine; most of them are shallow-water forms, hiding under stones and seaweed; a few are pelagic. Currently nine or more orders of opisthobranchs are recognized, but for convenience they can be divided into two classical groups: tectibranchs, with gill and shell usually present, and nudibranchs, in which there is no shell or true gill, but in which secondary gills are present along the dorsal side or around the anus.

Among the tectibranchs are the large sea hare *Aplysia* (Figure 13-20), which has large, earlike anterior tentacles and a vestigal shell, and the pteropods, or sea butterflies *(Cavolina* and *Clione).* In pteropods the foot is modified into fins for swimming; thus, they are pelagic and form a part of the plankton fauna.

The nudibranchs are represented by the sea slugs, which are often brightly colored and carnivorous. The plumed sea slug *Aeolis,* which lives on sea anemones and hydroids, often draws the color of its prey into the elongated papillae (cerata) that cover its back. It also salvages the nematocysts of the hydroids for its own use. The frilled sea slug *Tridachia* is a lovely little green or blue and white form common in Florida waters. *Hermissenda* is one of the more common West Coast nudibranchs.

Pulmonata

The pulmonates show some detorsion and include the land and most freshwater snails and slugs (and a few brackish and saltwater forms). They have lost their ancestral ctenidia, but the vascularized mantle wall has become a lung, which fills with air by contraction of the mantle floor (some aquatic species have developed secondary gills in the mantle cavity). The anus and nephridiopore open near the pneumostome, and waste is expelled forcibly with air or water from the lung. They are monoecious. The aquatic species have one pair of nonretractile tentacles, at the base of which are the eyes; land forms have two pairs of tentacles, with the posterior pair bearing the eyes (Figure 13-22). Among the thousands of land species, some of the most familiar American forms are *Helix, Polygyra, Succinea, Anguispira, Zonitoides, Limax,* and *Agriolimax*. Aquatic forms are represented by *Helisoma, Lymnaea,* and *Physa. Physa* is a left-handed (sinistral) snail.

Class Bivalvia (Pelecypoda)

The Bivalvia are also known as Pelecypoda (pel-e-sip′o-da), or "hatchet-footed" animals, as their name implies (Gr. *pelekys,* hatchet, + *pous, podos,* foot). They are the bivalved molluscs that include the mussels, clams, scallops, oysters, and shipworms (Figures 13-23 to 13-26) and they range in size from tiny seed shells 1 to 2 mm in length to giant South Pacific clams *Tridacna,* which may reach more than 1 m in length and as much as 225 kg (500 pounds) in weight (Figure 13-34). Most bivalves are sedentary **filter feeders** that depend on ciliary currents produced by the gills to bring in food materials. Unlike the gastropods, they have no head, no radula, and very little cephalization.

Figure 13-22

A, Pulmonate land snail. Note two pairs of tentacles; the second, larger pair bears the eyes. B, Banana slug, *Ariolimax columbianus.* Note pneumostome.
Photographs by C.P. Hickman, Jr.

A

B

Figure 13-23

Bivalve molluscs. A, Mussels, *Mytilus edulis,* occur in northern oceans around the world; they form dense beds in the intertidal zone. A host of marine creatures lives protected beneath attached mussels. B, Scallops *(Chlamys opercularis)* swim to escape attack by starfish *(Asterias rubens).* When alarmed, these most agile of bivalves swim by clapping the two shell valves together.
A, Photograph by R. Harbo; B, photograph by D.P. Wilson.

A   B

Figure 13-24

Representing a group that has evolved from burrowing ancestors, the surface-dwelling bivalve *Pecten* sp. has developed sensory organs along their mantle edges (tentacles and a series of blue eyes).
Photograph by L.S. Roberts.

Most pelecypods are marine, but many live in brackish water and in streams, ponds, and lakes.

Form and function
Shell

Bivalves are laterally compressed, and their two shells (**valves**) are held together dorsally by a hinge ligament that causes the valves to gape ventrally. The valves are drawn together by adductor muscles that work in opposition to the hinge ligament (Figure 13-25, C and D). The umbo is the oldest part of the shell, and growth occurs in concentric lines around it (Figure 13-25, A).

Pearl production is the by-product of a protective device used by the animal when a foreign object (grain of sand, parasite, or other) becomes lodged between the shell and mantle. The mantle secretes many layers of nacre around the irritating object (Figure 13-4). Pearls are cultured by inserting particles of nacre,

Figure 13-25

Tagelus plebius, the stubby razor clam (class Bivalvia). **A,** External view of right valve. **B,** Inside of left shell showing scars where muscles were attached. The mantle was attached at the pallial line. **C** and **D,** Sections showing function of adductor muscles and hinge ligament. In **C** the adductor muscle is relaxed, allowing the hinge ligament to pull the valves apart. In **D** the adductor muscle is contracted, pulling the valves together.

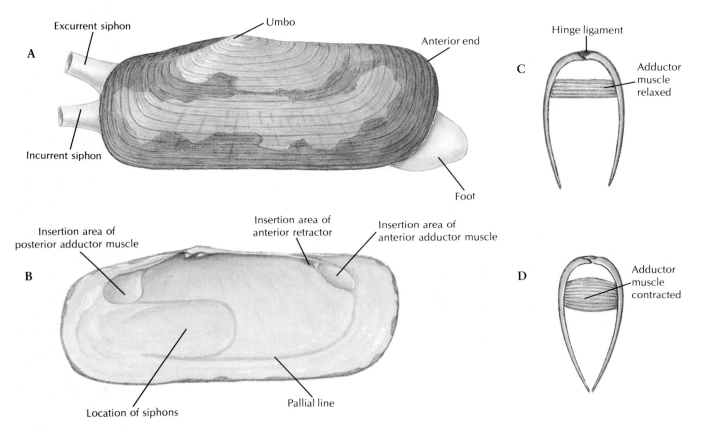

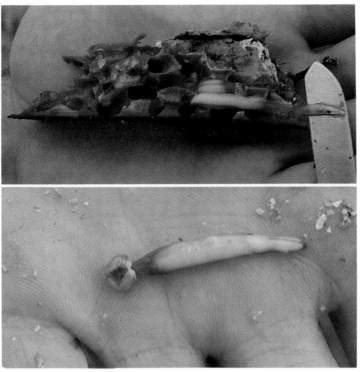

A

B

Figure 13-26

A, Shipworms are bivalves that burrow in wood, causing great damage to unprotected wooden hulls and piers. **B,** The two small, anterior valves, seen at left, are used as rasping organs to extend the burrow. Photographs by L.S. Roberts.

usually taken from the shells of freshwater clams, between the shell and mantle of a certain species of oyster and by keeping the oysters in enclosures for several years. *Meleagrina* is an oyster used extensively by the Japanese for pearl culture.

One might get the impression that a "cultured" pearl is somehow artificial or imitation. However, a cultured pearl is just as real or genuine as a "natural" pearl; the difference is that humans have stimulated its production rather than gathering it from a "wild" oyster.

Body and mantle

The **visceral mass** is suspended from the dorsal midline, and the muscular foot is attached to the visceral mass anteroventrally. The ctenidia hang down on each side, each covered by a fold of the mantle. The posterior edges of the mantle folds are modified to form dorsal excurrent and ventral incurrent apertures (Figure 13-27, *A*). In some marine bivalves the mantle is drawn out into long muscular siphons that allow the clam to burrow into the mud or sand and extend the siphons to the water above (Figure 13-27, *B* to *D*).

Figure 13-27

Adaptations of siphons in bivalves. **A,** In the northwest ugly clam, *Entodesma saxicola,* the incurrent and excurrent siphons are clearly visible. **B** to **D,** In many marine forms the mantle is drawn out into long siphons. In **A, B,** and **D** the incurrent siphon brings in both food and oxygen. In **C,** *Yoldia,* the siphons are respiratory; long ciliated palps feel about over the mud surface and convey food to the mouth. **A,** Photograph by R. Harbo.

A

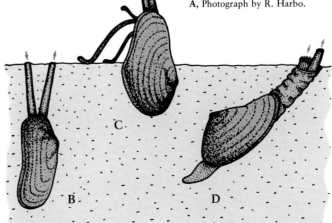

B

C

D

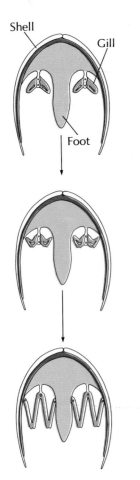

Figure 13-28

Evolution of bivalve ctenidia. By a great lengthening of individual filaments, the ctenidia became adapted for filter feeding and separated the incurrent chamber from the excurrent, suprabranchial chamber.

Figure 13-29

Section through heart region of a freshwater clam to show relation of circulatory and respiratory systems. Respiratory water currents: water is drawn in by cilia, enters gill pores, and then passes up water tubes to suprabranchial chambers and out excurrent aperture. Blood in gills exchanges carbon dioxide for oxygen. Blood circulation: ventricle pumps blood forward to sinuses of foot and viscera, and posteriorly to mantle sinuses. Blood returns from mantle to auricles; it returns from viscera to the kidney, and then goes to the gills, and finally to the auricles.

Locomotion

Pelecypods initiate movement by extending a slender muscular foot between the valves (Figure 13-27, *D*). Blood is pumped into the foot, causing it to swell and to act as an anchor in the mud or sand, then longitudinal muscles contract to shorten the foot and pull the animal forward.

Scallops and file shells are able to swim jerkily by clapping their valves together to create a sort of jet propulsion. The mantle edges can direct the stream of expelled water, so that the animals can swim in virtually any direction (Figures 13-23, *B*, and 13-24).

Gills

Gaseous exchange is carried on by both the mantle and the gills. The gills of most bivalves are highly modified for filter feeding; they are derived from the primitive ctenidia by a great lengthening of the filaments on each side of the central axis (Figure 13-28). As the ends of the long filaments became folded back toward the central axis, the ctenidial filaments took the shape of a long, slender W. The filaments lying beside each other became joined by ciliary junctions or tissue fusions, forming platelike **lamellae** with many vertical water tubes inside. Thus water enters the incurrent siphon, propelled by ciliary action, then enters the water tubes through pores between the filaments in the lamellae, proceeds dorsally into a common **suprabranchial chamber** (Figure 13-29), and then out the excurrent aperture.

Feeding

Most bivalves are filter feeders. The respiratory currents bring both oxygen and organic materials to the gills where ciliary tracts direct them to the tiny pores of the gills. Gland cells on the gills and labial palps secrete copious amounts of

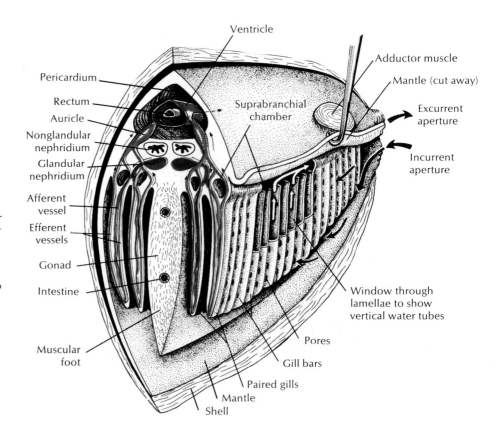

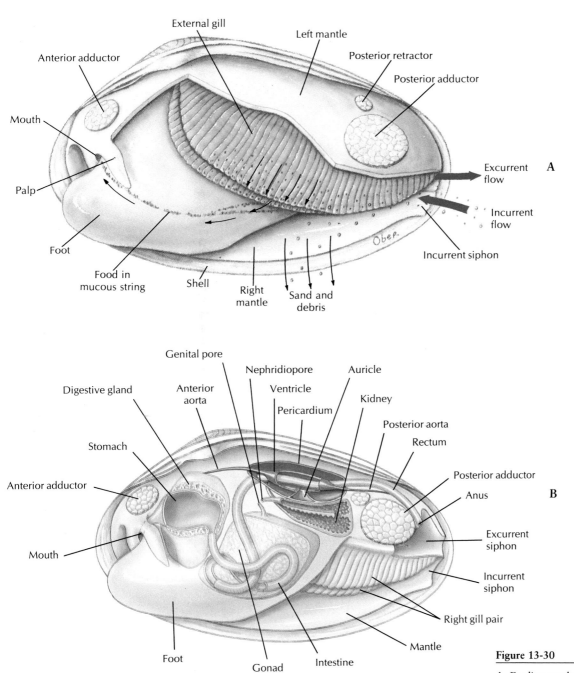

Figure 13-30

A, Feeding mechanism of freshwater clam. Left valve and mantle are removed. Water enters the mantle cavity posteriorly and is drawn forward by ciliary action to the gills and palps. As water enters the tiny openings of the gills, food particles are sieved out and caught up in strings of mucus that are carried by cilia to the palps and directed to the mouth. Sand and debris drop into the mantle cavity and are removed by cilia. **B,** Clam anatomy.

mucus, which entangles particles suspended in the water going through gill pores. These mucous masses slide down the outside of the gills toward food grooves at the lower edge of the gills (Figure 13-30). Heavier particles of sediment drop off the gills as a result of gravitational pull, but smaller particles travel along the food grooves toward the labial palps. The palps, being also grooved and ciliated, direct the mucous mass into the mouth.

Some bivalves, such as *Nucula* and *Yoldia,* are deposit feeders and have long proboscides attached to the labial palps (Figure 13-27, C). These can be protruded onto the sand or mud to collect food particles, in addition to the particles attracted by the gill currents.

Septibranchs, another group of bivalves, draw small crustaceans or bits of organic debris into the mantle cavity by sudden inflow of water created by the pumping action of a muscular septum in the mantle cavity.

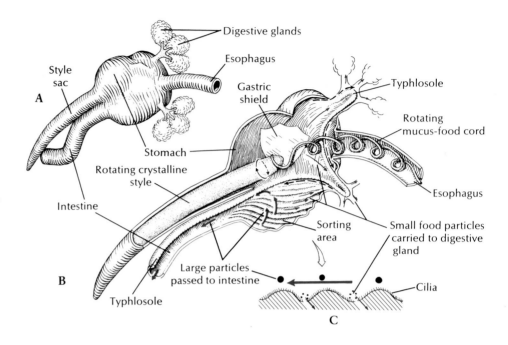

Figure 13-31

Stomach and crystalline style of ciliary-feeding clam. **A,** External view of stomach and style sac. **B,** Transverse section showing direction of food movements. Food particles in incoming water are caught in a cord of mucus that is kept rotating by the crystalline style. Ridged sorting areas direct large particles to the intestine and small food particles to digestive glands. **C,** Sorting action of cilia.

B, Modified from Morton, J.E. 1967. Mollusca, ed. 4. London, Hutchinson & Co.

Internal structure and function

The floor of the stomach of filter-feeding bivalves is folded into ciliary tracts for sorting the continuous stream of particles. A cylindrical style sac opening into the stomach secretes a gelatinous rod called the crystalline style, which projects into the stomach and is kept whirling by means of cilia in the style sac (Figure 13-31). Rotation of the style helps to dissolve its surface layers, freeing enzymes (such as amylase) that it contains, and to roll the mucous food mass. Dislodged particles are sorted, and suitable ones are directed to the digestive gland or picked up by amebocytes. Further digestion is intracellular.

The three-chambered heart, which lies in the pericardial cavity (Figure 13-30), is made up of two auricles and a ventricle and beats slowly, sometimes at the rate of about six times per minute. Part of the blood is oxygenated in the mantle and is returned to the ventricle through the auricles; the rest circulates through sinuses and passes in a vein to the kidneys, from there to the gills for oxygenation, and back to the auricles.

A pair of U-shaped kidneys (nephridial tubules) lies just ventral and posterior to the heart (Figure 13-30). The glandular portion of each tubule opens into the pericardium; the bladder portion empties into the suprabranchial chamber.

The nervous system consists of three pairs of widely separated ganglia connected by commissures and a system of nerves. Sense organs are poorly developed. They include a pair of statocysts in the foot, a pair of osphradia of uncertain function in the mantle cavity, tactile cells, and sometimes simple pigment cells on the mantle. Scallops *(Pecten, Chlamys)* have a row of small blue eyes along each mantle edge (Figures 13-23, *B*, and 13-24). Each is made up of cornea, lens, retina, and pigmented layer. Tentacles on the margin of the mantle of *Pecten* (Figure 13-24) and *Lima* have tactile and chemoreceptor cells.

Reproduction and development

Sexes are usually separate. Gametes are discharged into the suprabranchial chamber to be carried out with the excurrent flow. An oyster may produce 50 million eggs in a single season. In most bivalves fertilization is external. The embryo develops into trochophore, veliger, and spat stages (Figure 13-32).

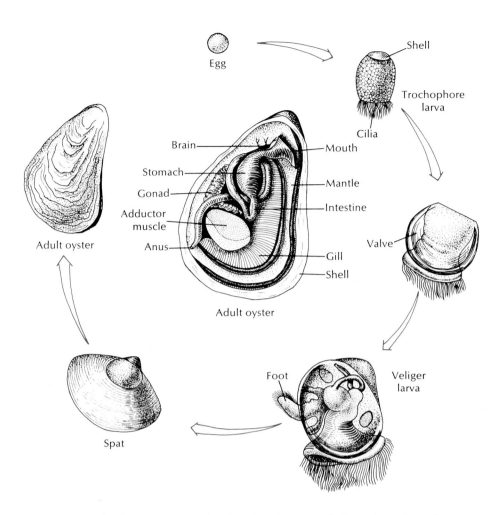

Adult oyster

Egg — Shell — Trochophore larva — Cilia

Brain — Mouth — Stomach — Mantle — Gonad — Intestine — Adductor muscle — Valve — Anus — Gill — Shell

Adult oyster

Foot — Veliger larva

Spat

Figure 13-32

Life cycle of the oyster. Oyster larvae swim about for approximately 2 weeks before settling down for attachment to become spats. Oysters take about 4 years to grow to commercial size.

In most freshwater clams, fertilization is internal. Eggs drop into the water tubes of the gills where they are fertilized by sperm entering with the incurrent flow. They develop there into a bivalved **glochidium larva** stage, which is a specialized veliger (Figure 13-33). When discharged, the glochidia are carried by water currents, and if they come in contact with a passing fish, they attach to its gills or skin and live as parasites for several weeks. Then they sink to the bottom to begin independent lives. Larval "hitchhiking" helps distribute a form whose locomotion is very limited.

Boring

Many pelecypods can burrow into mud or sand, but some have evolved a mechanism for burrowing into much harder substances, such as wood or stone.

Teredo, Bankia, and some other genera are called shipworms. They can be very destructive to wooden ships and wharves. These strange little clams have a long, wormlike appearance, with a pair of slender siphons on the posterior end that keep water flowing over the gills, and a pair of small globular valves on the anterior end with which they burrow (Figure 13-26). The valves have microscopic teeth so that they function as very effective wood rasps. The animals extend their burrows with an unceasing rasping motion of the valves. This sends a continuous flow of fine wood particles into the digestive tract where they are attacked intracellularly by the enzyme cellulase. Interestingly, cellulases are rarely produced by animals.

Some clams bore into rock. The piddock *(Pholas)* bores into limestone, shale, sandstone, and sometimes wood or peat. It has strong valves that bear

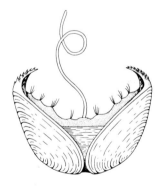

Figure 13-33

Glochidium, or larval form, of some freshwater clams. When the larva is released from brood pouch of mother, it may become attached to a fish's gill by clamping its valves closed. It remains as a parasite on the fish for several weeks. Its size is approximately 0.3 mm.

Figure 13-34

Clam *(Tridacna crocea)* lies buried in coral rock with greatly enlarged siphonal area visible. These tissues are richly colored and bear enormous numbers of symbiotic single-celled algae (zooxanthellae) that provide much of the clam's nutriment. Photograph by K. Sandved.

The enormous giant squid, *Architeuthis,* is very poorly known because no one has ever been able to study a living specimen. The anatomy has been studied from stranded animals, from those captured in the nets of fishermen, and from specimens found in the stomach of sperm whales. The mantle length is 5 to 6 m, and the head is up to one meter. They have the largest eyes in the animal kingdom: up to 25 cm (10 inches) in diameter. They apparently eat fish and other squids, and they are an important food item for sperm whales. They are thought to live on or near the sea bottom at a depth of 1000 m, but some have been observed swimming at the surface.

spines, which it uses to gradually cut away the rock while anchoring itself with its foot. *Pholas* may grow to 15 cm long and make rock burrows up to 30 cm long.

Class Cephalopoda

The Cephalopoda are the most advanced of the molluscs; in some respects they are the most advanced of all the invertebrates. They include the squids, octopuses, nautiluses, devilfish, and cuttlefish. All are marine, and all are active predators.

Cephalopods are the "head-footed" molluscs (Gr. *kephale,* head, + *pous, podos,* foot) in which the modified foot is concentrated in the head region. It is in the form of a funnel for expelling water from the mantle cavity. The anterior margin of the head is drawn out into a circle or crown of arms or tentacles.

Cephalopods range upward in size from 2 or 3 cm. The common squid of markets, *Loligo,* is about 30 cm long. The giant squid *Architeuthis* is the largest invertebrate known.

Fossil records of cephalopods go back to Cambrian times. The earliest shells were straight cones; others were curved or coiled, culminating in the coiled shell similar to that of the modern *Nautilus,* the only remaining member of the once flourishing nautiloids (Figure 13-35). Cephalopods without shells or with internal shells (such as octopuses and squids) are believed to have evolved from some early straight-shelled nautiloid.

The natural history of some cephalopods is fairly well known. They are marine animals and appear sensitive to the degree of salinity. Few are found in the Baltic Sea, where the water has a low salt content. Cephalopods are found at various depths. The octopus is often seen in the intertidal zone, lurking among rocks and crevices, but occasionally is found at great depths. The more active squids are rarely found in shallow water, and some have been taken at depths of 5000 m. *Nautilus* is usually found near the bottom in water 50 to 560 m deep, near islands in the southwestern Pacific.

Major groups of cephalopods

There are three subclasses of cephalopods: the Nautiloidea, which have two pairs of gills; the entirely extinct Ammonoidea; and the Coleoidea, which have one pair of gills. The Nautiloidea populated the Paleozoic and Mesozoic seas, but there survives only one genus, *Nautilus* (Figure 13-35), of which there are five or six species. The Ammonoidea were widely prevalent in the Mesozoic era but became extinct by the end of the Cretaceous period.

The subclass Coleoidea includes all living cephalopods except *Nautilus.* There are four orders of coleoids. Members of the order Sepioidea (cuttlefishes and their relatives) have a rounded or compressed, bulky body bearing fins (Figure 13-36). They have eight arms and two tentacles. Both the arms and the tentacles have suckers, but the tentacles bear suckers only at their ends (Figure 13-37). Members of the order Teuthoidea (squids, Figure 13-37) have a more cylindrical body but also have eight arms and two tentacles. The order Vampyromorpha (vampire squid) contains only a single, deepwater species. Members of the order Octopoda have eight arms and no tentacles (Figure 13-1, *D*). Their bodies are short and saclike, with no fins. The suckers in teuthoid squids are stalked (pedunculated), with horny rims bearing teeth; in octopuses the suckers are sessile and have no horny rims.

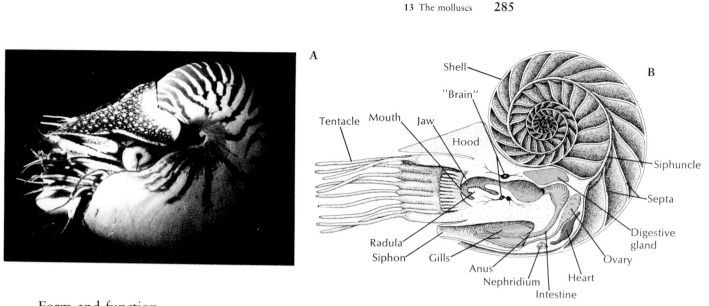

Figure 13-35

Nautilus, a cephalopod. **A**, Live *Nautilus*, feeding on a fish. **B**, Longitudinal section, showing gas-filled chambers of shell, and diagram of body structure.
A, Courtesy M. Butschler.

Form and function
Shell

Although early nautiloid and ammonoid shells were heavy, they were made buoyant by a series of **gas chambers,** as is that of *Nautilus* (Figure 13-35), enabling the animal to swim while carrying its shell. The shell of *Nautilus*, although coiled, is quite different from that of a gastropod. The shell is divided by transverse septa into internal chambers (Figure 13-35). The living animal inhabits only the last chamber. As it grows, it moves forward, secreting behind a new septum. The chambers are connected by a cord of living tissue called the **siphuncle,** which extends from the visceral mass. Cuttlefishes also have a small, curved shell, but it is entirely enclosed by the mantle. In the squids most of the shell has disappeared, leaving only a thin, horny strip called a pen, which is enclosed by the mantle. In *Octopus* (Gr. *oktos*, eight, + *pous, podos,* foot) the shell has disappeared entirely.

Locomotion

Cephalopods swim by forcefully expelling water from the mantle cavity through a ventral **funnel** (or **siphon**)—a sort of jet propulsion method. The funnel is mobile and can be pointed forward or backward to control direction; speed is controlled by the force with which water is expelled.

Figure 13-36

Sepia, the cuttlefish.
Photograph by D.P. Wilson.

It was formerly believed that the gas in the chambers of *Nautilus* was the product of secretion by the siphuncle, but recent investigation has shown that the function of the siphuncle is *not* to secrete gas but to *bail fluid* out of the unoccupied chambers. The shell and septa are composed of calcium carbonate and protein, and as the animal grows, the mantle covering the visceral mass secretes more septa and creates additional chambers. Each new chamber is initially filled with a fluid similar in ionic composition to that of the *Nautilus's* blood (and of seawater). The mechanism of fluid removal appears to involve the active secretion of ions into tiny intercellular spaces in the siphuncular epithelium, so that a very high local osmotic pressure is produced, and the water is drawn out of the chamber by osmosis. The gas in the chamber is only the respiratory gas from the siphuncle tissue that diffuses into the chamber as the fluid is removed. Thus, the gas pressure in the chamber is 1 atmosphere or less because it is in equilibrium with the gases dissolved in the seawater surrounding the *Nautilus*, which are in turn in equilibrium with air at the surface of the sea, despite the fact that the *Nautilus* may be swimming at 400 m beneath the surface. That the shell can withstand implosion by the surrounding 41 atmospheres (about 600 pounds per square inch), and that the siphuncle can remove water against this pressure are marvelous feats of natural engineering!

Squids and cuttlefishes are excellent swimmers. The squid body is streamlined and built for speed (Figure 13-37). Cuttlefishes swim more slowly. The lateral fins of squids and cuttlefishes serve as stabilizers, but they are held close to the body for rapid swimming.

Nautilus is active at night; its gas-filled chambers keep the shell upright. Although not as fast as the squid, it moves surprisingly well.

Octopus has a rather globular body and no fins (Figure 13-1, *D*). The octopus can swim backward by spurting jets of water from its funnel, but it is better adapted to crawling about over the rocks and coral, using the suction discs on its arms to pull or to anchor itself. Some deep-water octopods have the arms webbed like an umbrella and swim in a sort of medusa fashion.

External features

During the embryonic development of the cephalopod, the head and foot become indistinguishable. The ring around the mouth, which bears the arms, or tentacles, is considered to be derived from the anterior margin of the head.

The *Nautilus'* head, with its 60 to 90 or more tentacles, can be extruded from the opening of the body compartment of the shell (Figure 13-35). Its tentacles have no suckers but are made adhesive by secretions. They are used in searching for, sensing, and grasping food. Beneath the head is the funnel. The mantle, mantle cavity, and visceral mass are sheltered by the shell. Two pairs of gills are located in the mantle cavity.

A

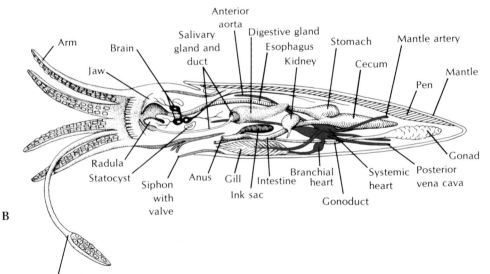

B

Figure 13-37

A, Squid *Loligo pealei* capturing a menhaden. **B,** Lateral view of squid anatomy, with the left half of the mantle removed.
A, Photograph by H.W. Pratt/BPS.

Internal features

The active life-style of cephalopods is reflected in their internal anatomy, particularly their respiratory, circulatory, and nervous systems.

Respiration and circulation. Except for the nautiloids, cephalopods have only one pair of gills. Since ciliary propulsion would not circulate enough water for their high oxygen requirements, there are no cilia on the gills. Instead, radial muscles in the mantle wall thin the wall and enlarge the mantle cavity, drawing water in. Strong circular muscles contract and expel the water forcibly through the funnel. A system of one-way valves prevents the water from being taken in through the funnel and expelled around the mantle margin.

Likewise, the open circulatory system found in other molluscs would be inadequate for cephalopods. Their circulatory system consists of a network of vessels, and the blood is conducted through the gill filaments in capillaries. Furthermore, the molluscan plan of circulation places the entire systemic circulation before the blood reaches the gills (in contrast to vertebrates, in which the blood leaves the heart and goes directly to the gills or lungs). This functional problem has been solved by the development of **accessory** or **branchial hearts** (Figure 13-37) at the base of each gill to increase the pressure of the blood going through the capillaries there.

Nervous and sensory systems. The nervous and sensory systems are more highly advanced in cephalopods than in other molluscs. The brain, the largest of any in the invertebrates, consists of several lobes with millions of nerve cells. Squids have giant nerve fibers (among the largest known in the animal kingdom), that are activated when the animal is alarmed and that initiate maximal contractions of the mantle muscles for a speedy escape.

Sense organs are well developed. Except in *Nautilus*, which has relatively simple eyes, cephalopods have highly advanced eyes with cornea, lens, chambers, and retina (Figure 13-38). Orientation of the eyes is controlled by the statocysts, which are larger and more complex than in other molluscs. The eyes are held in a constant relation to gravity, so that the slit-shaped pupils are always in a horizontal position. Octopods are apparently color-blind but can be taught to discriminate between shapes—for example, a square and a rectangle—and to remember such a discrimination for a considerable time. Experimenters find it easy to modify their behavior patterns by devices of reward and punishment. Octopods use their arms for tactile exploration and can discriminate between textures but apparently not between shapes. The arms are well supplied with both tactile and chemoreceptor cells. Cephalopods seem to lack a sense of hearing.

Communication

Little is known of the social behavior of nautiloids or deep-water coleoids, but inshore and littoral forms such as *Sepia, Sepioteuthis, Loligo,* and *Octopus* have been extensively studied. Although their tactile sense is well developed and they have some chemical sensitivity, visual signals are the predominant means of communication. These consist of a host of movements of the arms, fins, and body, as well as many color changes. The movements may range from minor body motions to exaggerated spreading, curling, raising, or lowering of some or all of the arms. Color changes are brought about by chromatophores, cells in the skin that contain pigment granules. Each elastic chromatophore is surrounded by tiny muscle cells whose contractions pull the cell boundary of the chromatophore outward, causing it to expand greatly. As the cell expands, the pigment becomes dispersed, changing the color pattern of the animal. When the muscles relax, the chromatophores return to their original size and the pigment becomes concentrated again. By

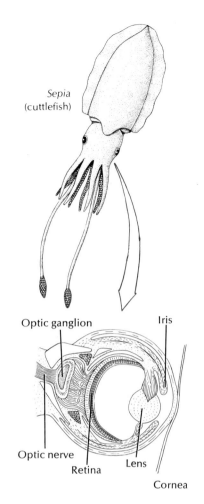

Sepia (cuttlefish)

Optic ganglion
Iris
Optic nerve
Retina
Lens
Cornea

Figure 13-38

Eye of a cuttlefish *(Sepia).* The structure of cephalopod eyes shows a high degree of convergent evolution with the eyes of vertebrates.

means of the chromatophores, which are under nervous and probably hormonal control, an elaborate system of changes in color and pattern is possible, including general darkening or lightening; flushes of pink, yellow, or lavender; and the formation of bars, stripes, spots, or irregular blotches. These may be used variously as danger signals, as protective coloring, in courtship rituals, and probably in other ways.

By assuming different color patterns of different parts of the body, a squid can transmit three or four different messages *simultaneously* to different individuals and in different directions, and it can instantaneously change any or all of the messages. Probably no other system of communication in invertebrates can convey so much information so rapidly.

Deep-water cephalopods may have to depend more on chemical or tactile senses than their littoral or surface cousins, but they also produce their own type of visual signals, for they have evolved many elaborate luminescent organs.

Most cephalopods other than nautiloids have another protective device. An ink sac that empties into the rectum contains an **ink gland** that secretes **sepia,** a dark fluid containing the pigment melanin, into the sac. When the animal is alarmed, it releases a cloud of ink, which may hang in the water as a blob or be contorted by water currents. The animal quickly departs from the scene, leaving the ink as a decoy to the predator.

Reproduction

Sexes are separate in cephalopods. In the male seminal vesicle the spermatozoa are encased in spermatophores and stored in a sac that opens into the mantle cavity. One arm of the adult male is modified as an intromittent organ, called a **hectocotylus,** which is used to pluck a spermatophore from his own mantle cavity and insert it into the mantle cavity of the female near the oviduct opening (Figure 13-39). Before copulation males often undergo color displays, apparently directed against rival males. Eggs are fertilized as they leave the oviduct and are then usually attached to stones or other objects. Some octopods tend their eggs. *Argonauta,* the paper nautilus, secretes a fluted "shell," or capsule, in which she broods her eggs.

Figure 13-39

Copulation in cephalopods. **A,** Mating cuttlefishes. **B,** Male octopus uses modified arm to deposit spermatophores in female mantle cavity to fertilize her eggs. Octopuses often tend their eggs during development.

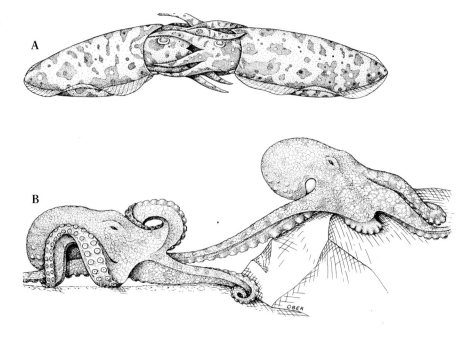

___ Phylogeny and Adaptive Radiation

The first molluscs probably arose during Precambrian times since fossils attributed to the Mollusca have been found in geological strata as old as the early Cambrian period. The primitive larva of the molluscs, which many still retain, and which is similar to that of marine annelids, is the trochophore. The type of embryonic cleavage (spiral) is also similar in molluscs and annelids. The ladderlike nervous system of some molluscs and the calcareous integumentary scales of the solenogasters and caudofoveates resemble those of some turbellarians. A flatworm type of ancestor may have given rise to the two main protostome groups: the nonsegmented molluscs and the segmentally arranged annelid-arthropod stem. The molluscs may have diverged somewhat later from a coelomate "protoannelid" before the annelid line developed metamerism. Some scientists have contended that the ancestral mollusc was metameric because they interpreted *Neopilina* (class Monoplacophora) as metameric and believed that metamerism was lost in more advanced forms. If this hypothesis were correct and metamerism is primitive in molluscs, the trait might be expected to show up in molluscan larvae, but there is no characteristic suggesting metamerism in the development of any known molluscan larvae. Most zoologists now suggest that the replication of body parts found in the monoplacophorans is pseudometamerism.

The primitive ancestral mollusc was probably a more or less wormlike, dorsoventrally flattened organism with a ventral gliding surface and a dorsal mantle with a chitinous cuticle and calcareous scales (Boss, 1982) (Figure 13-40). It had a posterior mantle cavity with two gills; a complete, straight gut; a radula; a ladderlike nervous system; and an open circulatory system with a heart. It crawled on the bottom in marine habitats. Among living molluscs the primitive condition is most nearly approached by the caudofoveates, although members of this class are burrowing, with the foot reduced to an oral shield. The solenogasters have lost the radula and gills, and the foot is represented by the ventral groove. Both these classes probably branched off from the primitive ancestor before the development of a solid shell, a distinct head with sensory organs, a ventral muscularized foot, and the dorsal visceral mass. The polyplacophorans probably also branched off early from the main lines of molluscan evolution, while the head was still fused with the mantle and before the veliger was established as the larva (Stasek, 1972). Possibly the polyplacophorans were closer to the monoplacophoran stem than shown in Figure 13-40 because they share a pseudometameric arrangement of organs.

Primitive monoplacophorans apparently gave rise to the higher molluscan classes, as well as the more advanced monoplacophorans, such as *Neopilina*. Fossil monoplacophoran shells are known that are very similar to early gastropods. These have an increased dorsoventral height (emphasis on the dorsal visceral mass), and they probably enabled the development of a "waist" between the visceral mass and the head-foot, freeing the head from close fusion with the mantle. Stasek views torsion in gastropods as arising from an ability to shift the mantle cavity to the side by muscular action, providing a space into which the head could be withdrawn to escape predators or desiccation. Some investigators believe that the Gastropoda are polyphyletic, perhaps being composed of several groups independently derived from monoplacophorans.

Some of the early monoplacophorans gave rise to groups that exploited burrowing habits rather than crawling on the bottom. Some of these had their mantle extended around to the ventral side to completely enclose the body and developed a tubular shell (Scaphopoda), and in others the mantle edges dropped

Fossils are remains of past life uncovered from the crust of the earth (Chapter 39). They can be actual parts or products of animals (teeth, bones, shells, and so on), petrified skeletal parts, molds, casts, impressions, footprints, and others. Soft and fleshy parts rarely leave recognizable fossils. Therefore we have no record of molluscs before they had shells, and there can be some doubt that certain early fossil shells are really remains of molluscs, particularly if the group they represent is now extinct. The issue of how to define a mollusc from hard parts alone was emphasized by Yochelson (1978), who said, "If scaphopods were extinct and soft parts were unknown, would they be called mollusks? I think not."

curtainlike over the sides of the body, while the shell became bivalved (Bivalvia). The lateral flattening of bivalves and the foot modifications were adaptations for burrowing. The most primitive bivalves were probably deposit feeders, but they later radiated with great success as filter feeders. Reduction of the head and loss of the radula were correlated with the development of burrowing and changes in feeding habits. Some of the bivalves, such as oysters and scallops, have returned secondarily to the surface, but their ancestors were burrowing filter feeders.

Some of the early monoplacophorans probably also gave rise to the cepha-

Figure 13-40

The classes of Mollusca, showing their derivations and relative abundance.
Derivations after K.J. Boss, personal communication.

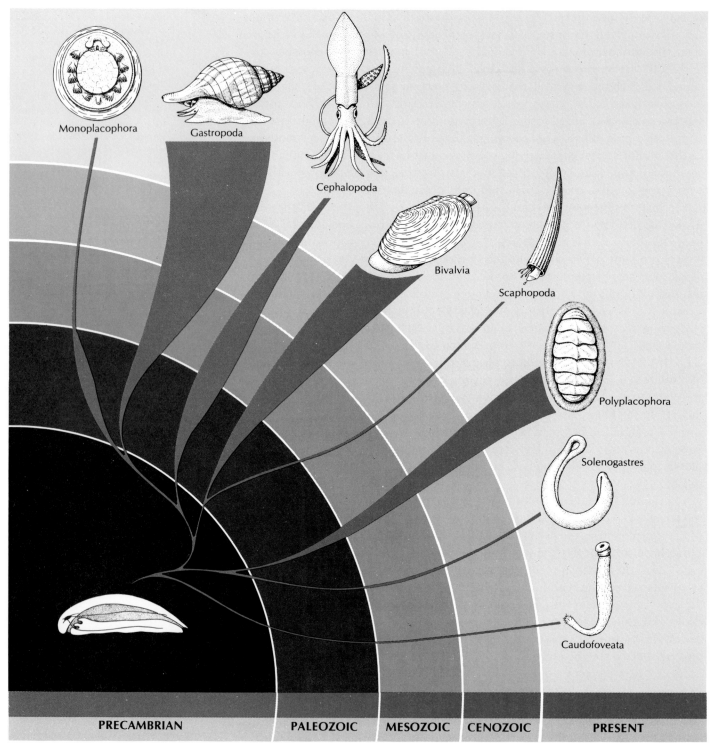

lopods. In these the mantle cavity was brought ventrally toward the head, with a concomitant increase in the length of the visceral mass. The foot became modified as a funnel, and the head tentacles developed into the arms. Stasek believes that the cephalopods were derived from the monoplacophorans before the other higher classes were delineated. The evolution of their chambered shell was a very important contribution to their freedom from the substratum and their ability to swim. The development of their respiratory, circulatory, and nervous systems is correlated with their predatory and swimming habits.

Most of the diversity among molluscs is related to their adaptation to different habitats and modes of life and to a wide variety of feeding methods, ranging from sedentary filter feeding to active predation. There are many adaptations for food gathering within the phylum and an enormous variety in radular structure and function, particularly among the gastropods.

The versatile glandular mantle has probably shown more plastic adaptive capacity than any other molluscan structure. Besides secreting the shell and forming the mantle cavity, it is variously modified into gills, lungs, siphons, and apertures, and it sometimes functions in locomotion, in the feeding processes, or in a sensory capacity. The shell, too, has undergone a variety of evolutionary adaptations.

SUMMARY

The Mollusca is one of the largest and most diverse phyla, its members ranging in size from very small organisms to the largest of invertebrates. Their basic body divisions are the head, the foot, and the visceral hump, which is usually covered by a shell. The majority are marine, but some are freshwater, and a few are terrestrial. They occupy a variety of niches. A number are economically important, and a few are medically important as hosts of parasites.

The molluscs are coelomate (have a coelom), although their coelom is limited to the area around the heart and gonads. The evolutionary development of a coelom was important because it enabled better organization of visceral organs and, in many of the animals that have it, an efficient hydrostatic skeleton. All of the most advanced invertebrates and vertebrates are coelomates.

The mantle and mantle cavity are important characteristics of molluscs. The mantle secretes the shell and overlies a part of the visceral hump to form a cavity housing the gills. The mantle cavity has been modified into a lung in some molluscs. The foot is usually a ventral, solelike, locomotor organ, but it may be variously modified. The radula, found in all molluscs except bivalves and solenogasters, is a protrusible, tonguelike organ with teeth, used in feeding. The shell has an outer layer composed of protein (periostracum), a middle layer of calcium carbonate and protein (prismatic), and an inner, calcareous, nacreous layer. Except for the closed circulatory system of cephalopods, the circulatory system of molluscs is open, with a heart and blood sinuses. Molluscs usually have a pair of nephridia connecting with the coelom, and a complex nervous system with a variety of sense organs. The primitive larva of molluscs is the trochophore, and most marine molluscs have a more advanced larva, the veliger.

The class Caudofoveata is a small, burrowing marine group that feeds with its anterior end down and its posterior mantle cavity and gills at the surface. These wormlike molluscs are the most primitive living members of the phylum. Another wormlike mollusc group is the class Solenogastres, which have lost the radula and feed on cnidarian polyps. Their foot is in the form of a long ventral groove.

The class Monoplacophora is a tiny, univalve marine group showing pseudometamerism. The Polyplacophora are more common, marine organisms with shells in the form of a series of eight plates. They are rather sedentary animals with a row of gills along each side of their foot. The Scaphopoda is a small marine group of burrowing animals with a tubular shell, open at both ends, and the mantle wrapped around the body.

The Gastropoda is the largest and most successful class of molluscs. Their interesting evolutionary history includes coiling, an elongation and spiraling of the visceral hump, and torsion, or the twisting of the posterior end to the anterior, so that the anus and head are at the same end. Coiling has led to a bilateral asymmetrical condition of the body, including loss of one nephridium, gill, and heart auricle. Torsion has led to the survival problem of fouling, which is the release of excreta over the head and in front of the gills, and this has been solved in various ways among different gastropods. Among the solutions to fouling are bringing water into one side of the mantle cavity and out the other (many prosobranchs), some degree of detorsion (opisthobranchs), and conversion of the mantle cavity into a lung (pulmonates). The Prosobranchia is the largest subclass, mostly marine, and they have a shell and usually an operculum. The Opisthobranchia are all marine, show partial or complete detorsion, and have a reduced shell or no shell. The Pulmonata are mostly freshwater or terrestrial and usually have a shell.

The Bivalvia are marine and freshwater, and their shell is divided into two valves joined by a dorsal ligament and held together by an adductor muscle. Most of them are filter feeders, drawing water through their gills by ciliary action. They are laterally compressed and usually burrow in soft substrates with their hatchet-shaped foot.

The members of the class Cephalopoda are the most advanced molluscs; they are all predators and many can swim rapidly. Their foot is modified into a funnel, through which water is forcefully expelled in swimming. They have arms and tentacles, which are used to capture prey with adhesive secretions or with suckers.

There is strong embryological evidence that the molluscs are related to the annelids, although the molluscs are not metameric. The enormous diversity of molluscs can be derived from a hypothetical ancestral mollusc that showed the basic body plan.

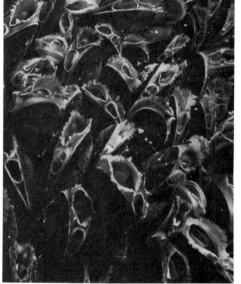

Blue mussels *Mytilus edulis*. Inhalant and exhalant siphons are clearly visible. Photograph by H.W. Pratt/BPS.

Review questions

1. Name five ways in which molluscs are important to humans.
2. How does the coelom develop embryologically? Why was the evolutionary development of the coelom important?
3. Give nine characteristics of molluscs.
4. Compare and contrast characteristics of the following classes of molluscs: Caudofoveata, Solenogastres, Monoplacophora, Polyplacophora, Scaphopoda, Gastropoda, Bivalvia, Cephalopoda.
5. Define the following: ctenidia, odontophore, periostracum, prismatic layer, nacreous layer, metanephridia, nephrostome, trochophore, veliger, glochidium, osphradium.
6. Briefly describe the habitat and habits of a typical chiton.
7. Define the following with respect to gastropods: operculum, columella, torsion, fouling, bilateral asymmetry, rhinophore, pneumostome.
8. Describe three ways to avoid fouling that have evolved in gastropods.
9. Describe four variations in feeding habits found in gastropods.
10. Distinguish among prosobranchs, opisthobranchs, and pulmonates.
11. Briefly describe how a typical bivalve feeds and how it burrows.

12. How is the ctenidium modified from the ancestral form in a typical bivalve?
13. What is the function of the siphuncle of cephalopods?
14. Describe how cephalopods swim and how they eat.
15. What cephalopod characteristics are particularly valuable for actively swimming, predaceous animals?
16. To what other major invertebrate groups are molluscs related, and what is the nature of the evidence for the relationship?
17. Briefly describe the characteristics of the primitive ancestral mollusc, and tell how each class of molluscs differs from the primitive condition with respect to each of the following: shell, radula, foot, mantle cavity and gills, circulatory system, and head.

Selected references

Abbott, R.T. 1974. American seashells, ed. 2. New York, Van Nostrand Reinhold Co., Inc. *Identification guide to 1500 Atlantic and Pacific species.*

Boss, K.J. 1982. Mollusca. In S.P. Parker (ed.). Synopsis and classification of living organisms, vol. 1. New York, McGraw-Hill Book Co. *Includes an account of the hypothetical ancestral mollusc and relation to Caudofoveata.*

Gosline, J.M., and M.D. DeMont. 1985. Jet-propelled swimming in squids. Sci. Am. **252**:96-103 (Jan.). *Mechanics of swimming in squid are analyzed; elasticity of collagen in mantle increases efficiency.*

Morris, P.A. (W.J. Clench [ed.]) 1973. A field guide to shells of the Atlantic and Gulf coasts and the West Indies, ed. 3. Boston, Houghton Mifflin Co. *An excellent revision of a popular handbook.*

Moynihan, M. 1985. Communication and noncommunication by cephalopods. Bloomington, Indiana University Press. *Readable summarization of our current understanding of communication in this remarkable group of molluscs.*

Roper, C.R.E., and K.J. Boss. 1982. The giant squid. Sci. Am. **246**:96-105 (April). *Many mysteries remain about the deep-sea squid, Architeuthis, because it has never been studied alive. It can reach a weight of 1000 pounds and a length of 18 m, and its eyes are as large as automobile headlights.*

Stasek, C.R. 1972. The molluscan framework. In M. Florkin and B.T. Scheer (eds.). Chemical zoology, vol. 8. New York, Academic Press, Inc. *Discusses phylogeny of molluscs and derivation of the classes from a functional viewpoint.*

Ward, P., L. Greenwald, and O.E. Greenwald. 1980. The buoyancy of the chambered nautilus. Sci. Am. **243**:190-203 (Oct.). *Reviews recent discoveries on how the nautilus removes the water from a chamber after secreting a new septum.*

Yochelson, E.L. 1978. An alternative approach to the interpretation of the phylogeny of ancient mollusks. Malacologia **17**:165-191. *Asserts that there are several classes of extinct molluscs.*

CHAPTER 14

THE SEGMENTED WORMS

Phylum Annelida

Position in Animal Kingdom

1. Annelids belong to the **protostome** branch of the animal kingdom and have spiral cleavage and determinate (mosaic) development.
2. Annelids have a true **coelom** (body cavity).
3. Annelids as a group show a primitive metamerism with comparatively few differences between the different somites.
4. All organ systems are present and well developed.

Biological Contributions

1. The introduction of **metamerism** by the group represents the greatest advancement of this phylum and lays the groundwork for the more highly specialized metamerism of the arthropods.
2. A true coelomic cavity reaches a high stage of development in this group.
3. Specialization of the head region into differentiated organs, such as the tentacles, palps, and eyespots of the polychaetes, is carried further in some annelids than in other invertebrates so far considered.

4. There are modifications of the **nervous system,** with cerebral ganglia (brain), two closely fused ventral nerve cords with giant fibers running the length of the body, and various ganglia with their lateral branches.
5. The circulatory system is much more complex than any we have so far considered. It is a closed system with muscular blood vessels and aortic arches ("hearts") for propelling the blood.
6. The appearance of the fleshy **parapodia,** with their respiratory and locomotor functions, introduces a suggestion of the paired appendages and specialized gills found in the more highly organized arthropods.
7. The well-developed **nephridia** in most of the somites have reached a differentiation that involves removal of waste from the blood as well as from the coelom.
8. Annelids are the most highly organized animals capable of complete regeneration. However, this ability varies greatly within the group.

—— PHYLUM ANNELIDA

Annelida (an-nel′i-da) (L. *annellus*, little ring, + *ida*, pl. suffix) consists of the segmented worms. It is a large phylum, numbering approximately 9000 species, the most familiar of which are the earthworms and freshwater worms (oligochaetes) and the leeches (hirudineans). However, approximately two thirds of the phylum is composed of the marine worms (polychaetes), which are less familiar to most people. Among the latter are many curious members; some are strange, even grotesque, whereas others are graceful and beautiful. They include the clamworms, plumed worms, parchment worms, scaleworms, lugworms, and many others. The annelids are true coelomates and belong to the protostome branch, with spiral and mosaic cleavage. They are a highly developed group in which the nervous system is more centralized and the circulatory system more complex than those of the phyla we have studied thus far.

The Annelida are worms whose bodies are divided into similar rings, or **segments,** arranged in linear series and externally marked by circular grooves called **annuli;** the name of the phylum is descriptive of this characteristic. Body segmentation, or **metamerism,** in the annelids is not merely an external feature but is also seen internally in the repetitive arrangement of organs and systems and in the partitioning off of segments (also called metameres or somites) by septa. Metamerism is not limited to annelids; it is shared by the arthropods (insects, crustaceans, and others), which are related to the annelids, and also by the vertebrates, in which it evolved independently.

Annelids are sometimes called "bristle worms" because, with the exception of the leeches, most annelids bear tiny chitinous bristles called **setae** (L. *seta,* hair or bristle). Short needlelike setae help anchor the somites during locomotion to prevent backward slipping; long, hairlike setae aid aquatic forms in swimming. Since many annelids either are burrowers or live in secreted tubes, the stiff setae also aid in preventing the worm from being pulled out or washed out of its home. Robins know from experience how effective the earthworms' setae are.

Annelids have a worldwide distribution and occur in marine and brackish waters, fresh water, and terrestrial soils. A few of the species may be called cosmopolitan.

Polychaetes are chiefly marine forms. Most are benthic, but some live free in the open sea. They are often divided for convenience into two groups (formerly the basis of subclasses): the sedentary polychaetes, and the errant or free-moving polychaetes. Sedentary polychaetes are mainly tubicolous; that is, they spend all or much of their time in tubes or permanent burrows. Many of them, especially those that live in tubes, have elaborate devices for feeding and respiration. In sabellids and serpulids the cirri or tentacles around the mouth give rise to great featherlike "branchial crowns" that are involved in both feeding and respiration (Figure 14-3). By means of ciliary currents water passes between the tentacles and carries food entangled in mucus to the mouth. The errant polychaetes have various habitats; some are strictly pelagic and others live in crevices or under rocks or shells, never straying far in open water.

Many polychaetes are euryhaline (can tolerate a wide range of environmental salinity) and occur in brackish water. Freshwater polychaete fauna is more diversified in warmer regions than in the temperate zones.

Clitellates (oligochaetes and leeches) occur predominantly in fresh water or terrestrial soils. Some freshwater species burrow in the bottom mud and sand and others among submerged vegetation. The swimming species usually have long setae, whereas the common earthworms, the most familiar example of the oligochaetes, have short setae.

Many of the leeches (class Hirudinea) are predators, and many are specialized for piercing their prey and feeding on blood or soft tissues. A few leeches are marine, but most of them live in fresh water or in damp regions. Suckers are typically found at both ends of the body for attachment to the substratum or to their prey. Some are adapted for forcing their pharynx or proboscis into soft tissues such as the gills of fish. The most specialized leeches, however, have sawlike chitinous jaws with which they can cut through tough skin. Many leeches live as carnivores on small invertebrates; some are temporary parasites; and some are permanent parasites, never leaving their host.

CHARACTERISTICS

1. Body **segmented;** symmetry bilateral
2. Body wall with outer circular and inner longitudinal muscle layers; outer transparent moist cuticle secreted by epithelium
3. **Chitinous setae** often present; setae absent in leeches
4. Coelom (schizocoel) well developed and divided by septa, except in leeches; coelomic fluid supplies turgidity and functions as hydrostatic skeleton
5. **Blood system closed** and segmentally arranged; respiratory pigments (hemoglobin, hemerythrin, or chlorocruorin) often present; amebocytes in blood plasma
6. Digestive system complete and not metamerically arranged
7. Respiratory gas exchange through skin, **gills,** or **parapodia**
8. Excretory system typically a **pair of nephridia for each metamere**
9. Nervous system with a double ventral nerve cord and a pair of ganglia with lateral nerves in each metamere; brain a pair of dorsal cerebral ganglia with connectives to cord
10. Sensory system of tactile organs, taste buds, statocysts (in some), photoreceptor cells, and eyes with lenses (in some)
11. Hermaphroditic or separate sexes; larvae, if present, are trochophore type; asexual reproduction by budding in some; spiral cleavage and mosaic development

CLASSIFICATION

The annelids are classified primarily on the basis of the presence or absence of parapodia, setae, metameres, and other morphological features. Because both the oligochaetes and the hirudineans (leeches) bear a saddlelike enlargement, called a **clitellum** (L. *clitellae,* packsaddle) (Figure 14-12, *B*), that is involved in reproduction, these two groups are often placed under the heading Clitellata (cli-tel-la′ta) and members are called clitellates. On the other hand, because both the Oligochaeta and the Polychaeta possess setae, some authorities place them together in a group called Chaetopoda (ke-top′o-da) (N.L. *chaeta,* bristle, from Gr. *chaitē,* long hair, + *pous, podos,* foot).

The Branchiobdellida, a group of small annelids that are parasitic or commensal on crayfish and show similarities to both oligochaetes and leeches, are here placed with the oligochaetes, but they are considered by some authorities to be a separate class. They have 14 or 15 segments and bear a head sucker.

One genus of leech, *Acanthobdella,* is a primitive type with some characteristics of leeches and some of oligochaetes; it is sometimes separated from the other leeches into a special class, Acanthobdellida, that characteristically has 27 somites, setae on the first five segments, and no anterior sucker.

Class Polychaeta (pol′e-ke′ta) (Gr. *polys,* many, + *chaitē,* long hair). Mostly marine; head distinct and bearing eyes and tentacles; most segments with parapodia (lateral appendages) bearing tufts of many setae; clitellum absent; sexes usually separate; gonads transitory; asexual budding in some; trochophore larva usually; mostly marine. Examples: *Nereis, Aphrodita, Glycera, Arenicola, Chaetopterus, Amphitrite.*
Class Oligochaeta (ol′i-go-ke′ta) (Gr. *oligos,* few, + *chaitē,* long hair). Body with conspicuous segmentation; number of segments variable; setae few per metamere; no parapodia; head absent; coelom spacious and usually divided by intersegmental septa; hermaphroditic; development direct, no larva; chiefly terrestrial and freshwater. Examples: *Lumbricus, Stylaria, Aeolosoma, Tubifex.*
Class Hirudinea (hir′u-din′e-a) (L. *hirudo,* leech, + *ea,* characterized by): **leeches.** Body with fixed number of segments (normally 34; 17 or 31 in some groups) with many annuli; body with oral and caudal suckers usually; clitellum present; no parapodia; setae absent (except *Acanthobdella*); coelom closely packed with connective tissue and muscle; development direct; hermaphroditic; terrestrial, freshwater, and marine. Examples: *Hirudo, Placobdella, Macrobdella.*

Body Plan

The annelid body typically has an anterior **prostomium,** a segmented body, and a terminal portion bearing the anus (**pygidium**). The prostomium and the pygidium are not considered metameres, but anterior segments often fuse with the prostomium to make up the head. New metameres form during development just in front of the pygidium; thus, the oldest segments are at the anterior end and the youngest segments are at the posterior.

The body wall is made up of strong circular and longitudinal muscles adapted for swimming, crawling, and burrowing and is covered with epidermis and a thin, outer layer of nonchitinous cuticle (Figure 14-1).

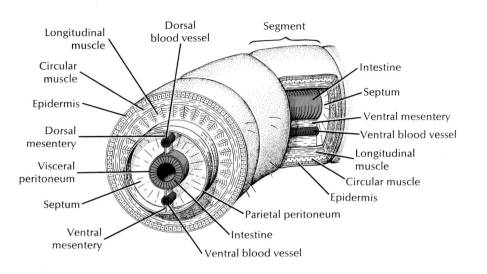

Figure 14-1

Annelid body plan.

In most annelids the coelom develops embryonically as a split in the mesoderm on each side of the gut (**schizocoel**), forming a pair of coelomic compartments in each segment. Each compartment is surrounded with **peritoneum** (a layer of mesodermal epithelium), which lines the body wall, forms dorsal and ventral **mesenteries,** and covers all the organs (Figure 14-1). Where the peritonea of adjacent segments meet, the **septa** are formed. These are perforated by the gut and longitudinal blood vessels. Not only is the coelom metamerically arranged, but practically every body system is affected in some way by this segmental arrangement.

Except in the leeches, the coelom is filled with fluid and serves as a **hydrostatic skeleton.** Because the volume of the fluid is essentially constant, contraction of the longitudinal body wall muscles causes the body to shorten and become larger in diameter, whereas contraction of the circular muscles causes it to lengthen and become thinner. Separation of the hydrostatic skeleton into a metameric series of coelomic cavities increases its efficiency greatly because the force of local muscle contraction is not transferred throughout the length of the worm. Widening and elongation can occur in restricted areas. Crawling motions are effected by alternating waves of contraction of longitudinal and circular muscles (peristaltic contraction) passing down the body. Segments in which longitudinal muscles are contracted widen and anchor themselves against burrow walls or other substratum while other segments, in which circular muscles are contracted, elongate and stretch forward. Forces powerful enough for burrowing as well as locomotion can thus be generated. Swimming forms use undulatory rather than peristaltic movements in locomotion.

Class Polychaeta

The polychaetes are the largest and most primitive class of annelids, with more than 5300 species, most of them marine. Although the majority of them are 5 to 10 cm long, some are less than 1 mm, and others may be as long as 3 m. Some are brightly colored in reds and greens; others are dull or iridescent. Some are picturesque, such as the "featherduster" worms (Figure 14-3).

Polychaetes live under rocks, in coral crevices, or in abandoned shells, or they burrow into mud or sand; some build their own tubes on submerged objects or in bottom material; some adopt the tubes or homes of other animals; some are pelagic, making up a part of the planktonic population. They are extremely abundant in some areas; for example, a square meter of mud flat may contain thousands of polychaetes. They play a significant part in marine food chains, since they are eaten by fish, crustaceans, hydroids, and many others.

Form and function

Polychaetes differ from other annelids in having a well-differentiated head with specialized sense organs; paired appendages, called parapodia, on most segments; and no clitellum (Figure 14-2). As their name implies, they have many setae, usually arranged in bundles on the parapodia. They show a pronounced differentiation of some body somites and a specialization of sensory organs practically unknown among clitellates.

In contrast to clitellates, polychaetes have no permanent sex organs, and they usually have separate sexes. Their development is indirect, for they undergo a form of metamorphosis that involves a trochophore larva.

Polychaetes were traditionally divided into two subclasses: Errantia and Sedentaria. These terms may still be used for descriptive purposes. Errant polychaetes (L. *errare*, to wander), include the free-moving pelagic forms, active burrowers, crawlers, and the tube worms that leave their tubes for feeding or breeding. Most of these, like the clam worm *Nereis* (Gr. name of a sea nymph) (Figure 14-2), are predatory forms equipped with jaws or teeth. They have a muscular eversible pharynx armed with teeth that can be thrust out with surprising speed and dexterity for capturing prey. Sedentary polychaetes rarely expose more than the head end from the tubes or burrows in which they live (Figure 14-3).

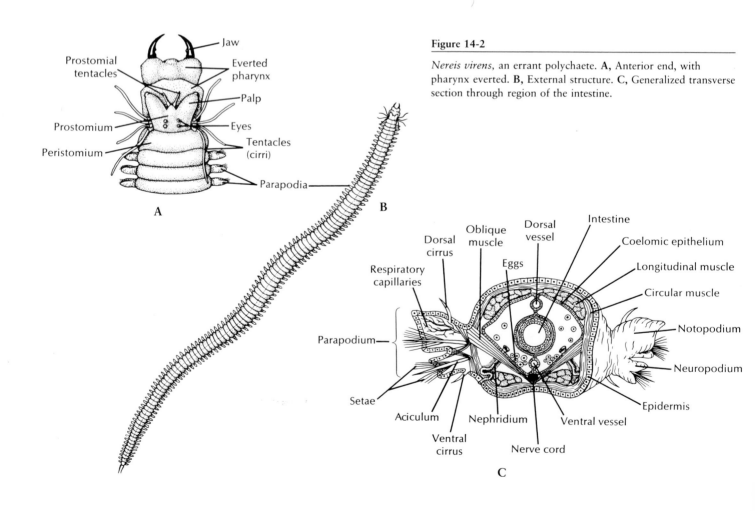

Figure 14-2

Nereis virens, an errant polychaete. **A,** Anterior end, with pharynx everted. **B,** External structure. **C,** Generalized transverse section through region of the intestine.

Figure 14-3

Polychaete tubeworms. **A,** *Protula* (family Serpulidae) builds a calcareous tube. **B,** *Spirographis spallanzani* (family Sabellidae) builds a noncalcareous tube. Photograph by C.P. Hickman, Jr.

Most sedentary tube and burrow dwellers are particle feeders, using ciliary or mucoid methods of obtaining food. The principal food source is plankton and detritus. Some, like *Amphitrite* (Gr. a mythical sea nymph) (Figure 14-4), with head peeping out of the mud, send out long extensible tentacles over the surface. Cilia and mucus on the tentacles entrap particles found on the sea bottom and move them toward the mouth.

The fanworms, or "featherduster" worms, are beautiful tubeworms, fascinating to watch as they emerge from their secreted tubes and unfurl their lovely tentacular crowns to feed. A slight disturbance, sometimes even a passing shadow, causes them to duck quickly into the safety of the homes they have built. Food attracted to the feathery arms, or **radioles,** by ciliary action is trapped in mucus and is carried down ciliated food grooves to the mouth (Figure 14-5). Particles too large for the food grooves are carried along the margins and dropped off. Further

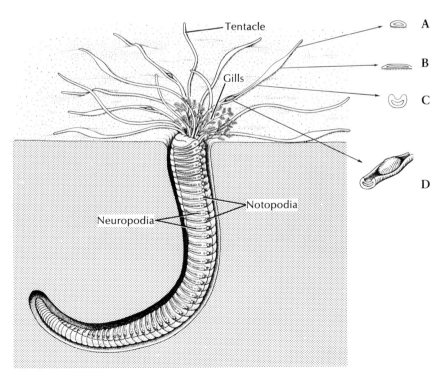

Figure 14-4

Amphitrite, which builds its tubes in mud or sand, extends long grooved tentacles out over the mud to pick up bits of organic matter. The smallest particles are moved along food grooves by cilia, larger particles by peristaltic movement. Its plumelike gills are blood red. **A,** Section through exploratory end of tentacle. **B,** Section through tentacle in area adhering to substratum. **C,** Section showing ciliary groove. **D,** Particle being carried toward mouth.

Figure 14-5

Sabella, a polychaete ciliary feeder, extends its crown of feeding radioles from its leathery secreted tube, reinforced with sand and debris. **A,** Anterior view of the crown. Cilia direct small food particles along grooved radioles to mouth and discard larger particles. Sand grains are directed to storage sacs and later are used in tube building. **B,** Distal portion of radiole showing ciliary tracts of pinnules and food grooves.

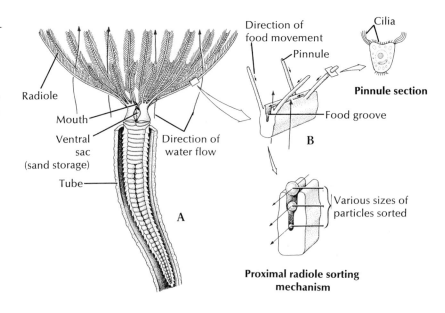

Figure 14-6

Chaetopterus, a sedentary polychaete (in U-tube) and *Phascolosoma,* a sipunculan worm (in center). *Chaetopterus* lives in a parchment tube through which it pumps water with its three pistonlike fans. The fans beat 60 times per minute to keep water currents moving. The winglike notopodia of the twelfth segment continuously secrete a mucous net that strains out food particles. As the net fills with food, the food cup rolls it into a ball and, when the ball is large enough (about 3 mm), the food cup bends forward and deposits the ball in a ciliated groove to be carried by cilia to the mouth and swallowed.

Courtesy The American Museum of Natural History, New York.

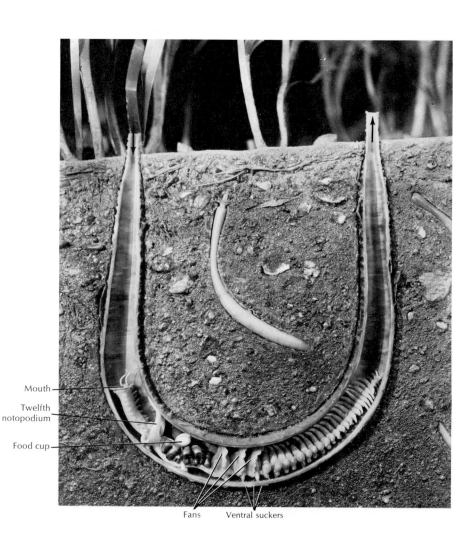

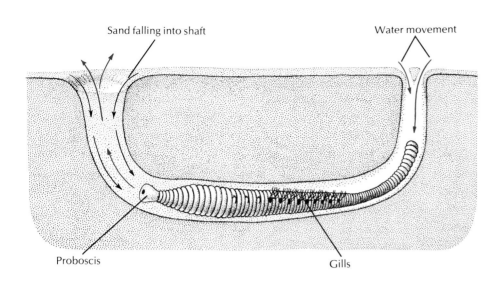

Sand falling into shaft

Water movement

Proboscis

Gills

Figure 14-7

Arenicola, the lugworm, lives in an L-shaped burrow in intertidal mud flats. It burrows by successive eversions and retractions of its proboscis. By peristaltic movements it keeps water filtering through the sand. The worm then ingests the food-laden sand.

sorting may occur near the mouth where only the small particles of food enter the mouth, and sand grains are stored in a sac to be used later in enlarging the tube.

Some worms, such as *Chaetopterus* (Gr. *chaitē,* long hair, + *pteron,* wing), secrete mucous filters through which they pump water to collect edible particles (Fig. 14-6). The lugworm *Arenicola* (L. *arena,* sand, + *colo,* inhabit) lives in an L-shaped burrow in which, by peristaltic movements, it causes water to flow. Food particles are filtered out by the sand at the front of its burrow, and it ingests the food-laden sand (Figure 14-7).

Tube dwellers secrete many types of tubes. Some are parchmentlike (Figure 14-5); some are firm, calcareous tubes attached to rocks or other surfaces (Figure 14-3, *A*); and some are simply grains of sand or bits of shell or seaweed cemented together with mucous secretions. Many burrowers in sand and mud flats simply line their burrows with mucus (Figure 14-7).

The polychaete typically has a head, or **prostomium,** which may or may not be retractile and which often bears eyes, antennae, and sensory palps (Figures 14-2 and 14-8). The first segment (**peristomium**) surrounds the mouth and may bear setae, palps, or, in predatory forms, chitinous jaws. Ciliary feeders may bear a tentacular crown that can be opened like a fan or withdrawn into the tube.

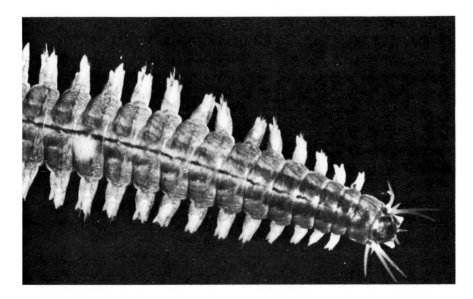

Figure 14-8

Nereis diversicolor, head and anterior segments. Note the well-defined segments, the lobed parapodia, and the prostomium with tentacles.
Photograph by D.P. Wilson.

Some polychaetes live most of the year as sexually immature animals called atokes, but during the breeding season a portion of the body develops into a sexually mature worm called an epitoke, which is swollen with gametes (Figure 14-9). One example is the palolo worm, which lives in burrows among the coral reefs of the South Seas. During the reproductive cycle, the posterior somites become swollen with gametes. During the swarming period, which occurs at the beginning of the last quarter of the October-November moon, these epitokes break off and swim to the surface. Just before sunrise, the sea is literally covered with them, and at sunrise they burst, freeing the eggs and sperm for fertilization. The anterior portions of the worms regenerate new posterior sections. A related form swarms in the Atlantic in the third quarter of the June-July moon. Swarming is of great adaptive value because the synchronous maturation of all the epitokes ensures the maximum number of fertilized eggs. However, it is very hazardous; many types of predators have a feast. In the meantime, the atoke remains safe in its burrow to produce another epitoke at the next cycle.

The trunk is segmented, and most segments bear appendages called **parapodia,** which may have lobes, cirri, setae, and other parts on them (Figure 14-8). The parapodia are used in crawling, swimming, or anchoring in tubes. They usually serve as the chief respiratory organs, although some polychaetes may also have gills. *Amphitrite,* for example, has three pairs of branched gills and long extensible tentacles (Figure 14-4). *Arenicola,* the lugworm (Figure 14-7), which burrows through the sand leaving characteristic castings at the entrance to its burrow, has paired gills on certain somites.

Sense organs are more highly developed in polychaetes than in oligochaetes and include eyes, nuchal organs (see below), and statocysts. Eyes, when present, may range from simple eyespots to well-developed organs. They are most conspicuous in errant worms. Usually the eyes are retinal cups, with rodlike photoreceptor cells lining the cup wall and directed toward the lumen of the cup. The highest degree of development is found in the family Alciopidae, which has large, image-resolving eyes similar in structure to those of some cephalopod molluscs (Figure 13-38), with cornea, lens, retina, and retinal pigment. The alciopid eye also has accessory retinas, a characteristic shared by deep-sea fishes and some deep-sea cephalopods. The accessory retinas are sensitive to different wavelengths of light, and their function may be to serve as a depth gauge for these pelagic animals because different wavelengths penetrate to different depths. Recent studies with electroencephalograms show that these eyes are especially well adapted to utilizing the dim light of the deep sea. Nuchal organs are ciliated sensory pits or slits that appear to be chemoreceptive, an important factor in food gathering. Some burrowing and tube-building polychaetes have statocysts that function in body orientation.

Reproductive systems are simple. Gonads appear as temporary swellings of the peritoneum and shed their gametes into the coelom. They are carried outside through gonoducts, through nephridia, or by rupture of the body wall. Fertilization is external, and the early larva is a trochophore.

Clam worms: Nereis

The clam worms, or sand worms as they are sometimes called, are errant polychaetes that live in mucus-lined burrows in or near low tide (Figure 14-8). Sometimes they are found in temporary hiding places, such as under stones, where they stay with their bodies covered and their heads protruding. They are most active at night, when they wiggle out of their hiding places and swim about or crawl over the sand in search of food.

The body, containing about 200 somites, may grow to 30 or 40 cm in length. The head is made up of a prostomium and a peristomium. The prostomium bears a pair of stubby palps, sensitive to touch and taste; a pair of short sensory tentacles; and two pairs of small dorsal eyes that are light sensitive. The peristomium bears the ventral mouth, a pair of chitinous jaws, and four pairs of sensory tentacles (Figure 14-2, *A*).

Each parapodium is formed of two lobes: a dorsal **notopodium** and a ventral **neuropodium** (Figure 14-2, *C*). Each lobe is supported by one or more chitinous spines (acicula). The parapodia bear setae and are abundantly supplied with blood vessels. The parapodia are used for both creeping and swimming and are manipulated by oblique muscles that run from the midventral line to the parapodia in each somite. The worm swims by lateral undulatory wriggling of the body—unlike the peristaltic movement of the earthworms. It can dart through the water with considerable speed. These undulatory movements can also be used to suck water into or pump it out of the burrow. The worm will usually adapt some kind of burrow if it can find one. When a worm is placed near a glass tube, it will wriggle in without hesitation.

The clam worm feeds on small animals, other worms, larval forms, and the like. It seizes them with its chitinous jaws, which are protruded through the mouth when the pharynx is everted. When the pharynx is withdrawn, the food is swallowed. Movement of the food through the alimentary canal is by peristalsis.

Other interesting polychaetes

Scale worms (Figure 14-10) are members of the family Polynoidae (Gr. *Polynoe*, the daughter of Nereus and Doris, a sea god and goddess), one of the most abundant and widespread of polychaete families. Their rather flattened bodies are covered with broad scales, modified from dorsal parts of the parapodia. Most are of modest size, but some are enormous (up to 190 mm long and 100 mm wide). They are carnivorous and feed on a wide variety of animals. Many are commensal, living in burrows of other polychaetes or in association with cnidarians, molluscs, or echinoderms.

Hermodice carunculata (Gr. *herma*, reef, + *dex*, a worm found in wood) (Figure 14-11) and related species are called fireworms. Their setae are hollow, brittle, and contain a poisonous secretion. When touched, the setae break off in the wound and cause skin irritation. They feed on corals, gorgonians, and other cnidarians.

The parchment worm *Chaetopterus* lives in a U-shaped, parchmentlike tube buried, except for the tapered ends, in sand or mud along the shore (Figure 14-6). The worm attaches to the side of the tube by ventral suckers. Fans (modified parapodia) on segments 14 to 16 pump water through the tube by rhythmical movements. A pair of enlarged parapodia in the twelfth segment secretes a long mucous bag that reaches back to a small food cup just in front of the fans. All the water passing through the tube is filtered through this mucous bag, the end of which is rolled up into a ball by cilia in the cup. When the ball is about the size of a BB shot, the fans stop beating and the ball of food and mucus is rolled forward by ciliary action to the mouth and is swallowed.

Figure 14-9

Eunice viridis, the Samoan palolo worm. The posterior segments make up the epitokal region, consisting of segments packed with gametes. Each segment has an eyespot on the ventral side. Once a year the worms swarm, and the epitokes detach, rise to the surface, and discharge their ripe gametes, leaving the water milky. By the next breeding season, the epitokes are regenerated. After W.M. Woodworth, 1907; from Fauvel, P. 1959. Annelides polychetes. Reproduction. In P.P. Grasse (ed.). Traite de Zoologie, vol. 5, part 1. Paris, Masson et Cie.

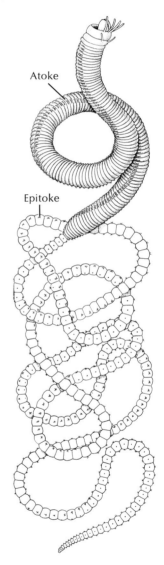

Atoke

Epitoke

Figure 14-10

The scale worm *Halosydna brevisetosa* often lives as a commensal in the tubes of certain other polychaetes.
Photograph by R. Harbo.

Class Oligochaeta

The more than 3000 species of oligochaetes are found in a great variety of sizes and habitats. They include the familiar earthworms and many species that live in fresh water. Most are terrestrial or freshwater forms, but some are parasitic, and a few live in marine or brackish water.

Oligochaetes, with few exceptions, bear setae, which may be long or short, straight or curved, blunt or needlelike, or arranged singly or in bundles. Whatever the type, they are less numerous in oligochaetes than in polychaetes, as is implied by the class name, which means "few long hairs." Aquatic forms usually have longer setae than do earthworms.

Figure 14-11

The fireworm *Hermodice carunculata* is shown here feeding on a gorgonian. Its setae are like tiny glass fibers and serve to ward off predators.
Photograph by L.S. Roberts.

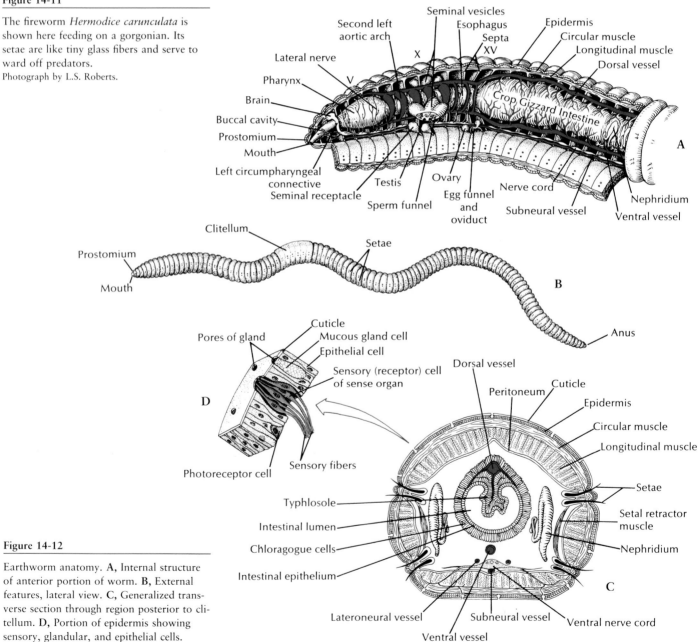

Figure 14-12

Earthworm anatomy. **A,** Internal structure of anterior portion of worm. **B,** External features, lateral view. **C,** Generalized transverse section through region posterior to clitellum. **D,** Portion of epidermis showing sensory, glandular, and epithelial cells.

Earthworms

The most familiar of the oligochaetes are the earthworms ("night crawlers"), which burrow in moist, rich soil, emerging at night to explore their surroundings. In damp, rainy weather they stay near the surface, often with mouth or anus protruding from the burrow. In very dry weather they may burrow several feet underground, coil up in a slime chamber, and become dormant. *Lumbricus terrestris* (L. *lumbricum,* earthworm), the form commonly studied in school laboratories, is approximately 12 to 30 cm long (Figure 14-12). Giant tropical earthworms may have from 150 to 250 or more segments and may grow to as much as 4 m in length. They usually live in branched and interconnected tunnels.

Form and function

In earthworms the mouth is overhung by a fleshy prostomium at the anterior end, and the anus is on the posterior end (Figure 14-12, *B*). In most earthworms each segment bears four pairs of chitinous setae (Figure 14-12, *C*), although in some oligochaetes each segment may have up to 100 or more. Each seta is a bristlelike rod set in a sac within the body wall and moved by tiny muscles (Figure 14-13). The setae project through small pores in the cuticle to the outside. In locomotion and burrowing, setae anchor parts of the body to prevent slipping. Earthworms move by peristaltic movement. Contractions of circular muscles in the anterior end lengthen the body, pushing the anterior end forward where it is anchored by setae; contractions of longitudinal muscles then shorten the body, pulling the posterior end forward. As these waves of contraction pass along the entire body, it is gradually moved forward.

Nutrition

Most oligochaetes are scavengers. Earthworms feed mainly on decayed organic matter, bits of leaves and vegetation, refuse, and animal matter. After being moistened by secretions from the mouth, food is drawn in by the sucking action of the muscular pharynx. The liplike prostomium aids in manipulating the food into position. The calcium from the soil swallowed with the food tends to produce a high blood calcium level. **Calciferous glands** along the esophagus secrete calcium ions into the gut and so reduce the calcium ion concentration of the blood. Calciferous glands are really ionoregulatory, rather than digestive, organs. They also function in regulating the acid-base balance of the body fluids, maintaining the pH at a fairly stable value.

Leaving the esophagus, food is stored temporarily in the thin-walled crop before being passed on into the gizzard, which grinds the food into small pieces. Digestion and absorption take place in the intestine. Along the dorsal side, the wall of the intestine is infolded to form a **typhlosole,** which greatly increases the absorptive and digestive surface (Figure 14-12, *C*). The digestive system secretes various enzymes to break down the food: pepsin, which acts on proteins; amylase, which acts on polysaccharides; cellulase, which acts on cellulose; and lipase, which acts on fats. The indigestible residue is discharged through the anus.

Surrounding the intestine and dorsal vessel and filling much of the typhlosole is a layer of yellowish **chlorogogue tissue** derived from the peritoneum. This tissue serves as a center for the synthesis of glycogen and fat, a function roughly equivalent to that of liver cells. The chlorogogue cells when ripe (full of fat) are released into the coelom where they float free as cells called **eleocytes** (Gr. *elaio,* oil, + *kytos,* hollow vessel [cell]), which transport materials to the body tissues. They apparently can pass from segment to segment and have been found to accu-

Aristotle called earthworms the "intestines of the soil." Some 22 centuries later Charles Darwin published his observations in his classic *The Formation of Vegetable Mould Through the Action of Worms.* He showed how worms enrich the soil by bringing subsoil to the surface and mixing it with the topsoil. An earthworm can ingest its own weight in soil every 24 hours, and Darwin estimated that from 10 to 18 tons of dry earth per acre pass through their intestines annually, thus bringing up potassium and phosphorus from the subsoil and also adding to the soil nitrogenous products from their own metabolism. They expose the mold to the air and sift it into small particles. They also drag leaves, twigs, and organic substances into their burrows closer to the roots of plants. Their activities are important in aerating the soil. Darwin's views were at odds with his contemporaries, who thought earthworms were harmful to plants. But recent research has amply confirmed Darwin's findings, and earthworm management is now practiced in many countries.

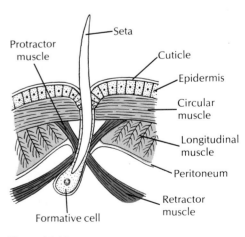

Figure 14-13

Seta with its muscle attachments showing relation to adjacent structures. Setae lost by wear and tear are replaced by new ones, which develop from formative cells.

mulate around wounds and regenerating areas, where they break down and release their contents into the coelom. Chlorogogue cells also function in excretion.

Circulation and respiration

Annelids have a double transport system: the coelomic fluid and the circulatory system. Food, wastes, and respiratory gases are carried by both coelomic fluid and blood in varying degrees. The blood is carried in a closed system of blood vessels, including capillary systems in the tissues. There are five main blood trunks, all running lengthwise through the body.

The **dorsal vessel** (single) runs above the alimentary canal from the pharynx to the anus. It is a pumping organ, provided with valves, and functions as the true heart. This vessel receives blood from vessels of the body wall and digestive tract and pumps it anteriorly into the five pairs of aortic arches. The function of the aortic arches is to maintain a steady pressure of blood into the ventral vessel.

The **ventral vessel** (single) serves as the aorta. It receives blood from the aortic arches and delivers it to the brain and rest of the body, giving off segmental vessels to the walls, nephridia, and digestive tract.

The blood contains colorless ameboid cells and a dissolved respiratory pigment, hemoglobin. The blood of some annelids may have respiratory pigments other than hemoglobin.

Earthworms have no special respiratory organs, but gaseous exchange is made in their moist skin, where oxygen is picked up and carbon dioxide is given off.

Excretion

A pair of **metanephridia,** the organs of excretion, is found in each somite except the first three and the last one. Each nephridium occupies parts of two successive

Figure 14-14

Nephridium of earthworm. Wastes are drawn into the ciliated nephrostome in one segment, then passed through the loops of the nephridium, and expelled through the nephridiopore of the next segment.

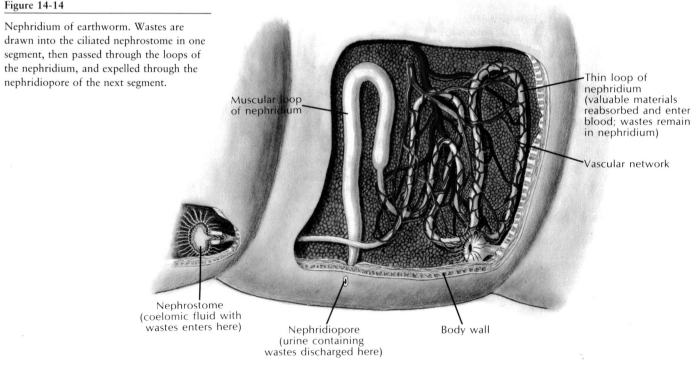

Muscular loop of nephridium

Thin loop of nephridium (valuable materials reabsorbed and enter blood; wastes remain in nephridium)

Vascular network

Nephrostome (coelomic fluid with wastes enters here)

Nephridiopore (urine containing wastes discharged here)

Body wall

somites (Figure 14-14). A ciliated funnel, the **nephrostome,** lies just anterior to an intersegmental septum and leads by a small ciliated tubule through the septum into the somite behind, where it connects with the main part of the nephridium. This part of the nephridium is made up of several complex loops of increasing size, which finally terminate in a bladderlike structure leading to an aperture, the **nephridiopore;** this opens to the outside near the ventral row of setae. By means of cilia, wastes from the coelom are drawn into the nephrostome and tubule, where they are joined by salts and organic wastes transported from blood capillaries in the glandular part of the nephridium. All the waste is discharged to the outside through the nephridiopore.

Aquatic oligochaetes excrete ammonia; terrestrial oligochaetes excrete the much less toxic urea. *Lumbricus* produces both, the level of urea depending somewhat on environmental conditions. Both urea and ammonia are produced by chlorogogue cells, which may break off and enter the nephridia directly, or their products may be carried by the blood. Some nitrogenous waste is also eliminated through the body surface.

Oligochaetes are largely freshwater animals, and even such terrestrial forms as earthworms must exist in a moist environment. Osmoregulation is a function of the body surface and the nephridia, as well as the gut and the dorsal pores. *Lumbricus* will gain weight when placed in tap water and lose it when returned to the soil. Salts as well as water can pass across the integument, salts apparently being carried by active transport.

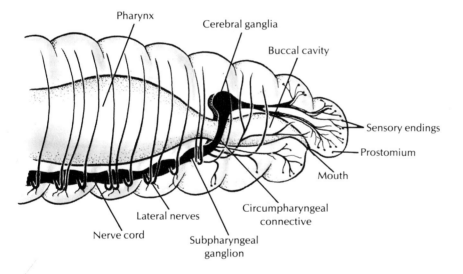

Pharynx

Cerebral ganglia

Buccal cavity

Sensory endings

Prostomium

Mouth

Circumpharyngeal connective

Subpharyngeal ganglion

Lateral nerves

Nerve cord

Figure 14-15

Anterior portion of earthworm and its nervous system. Note concentration of sensory endings in this region.

Nervous system and sense organs

The nervous system in earthworms (Figure 14-15) consists of a central system and peripheral nerves. The central system is made up of a pair of cerebral ganglia (the brain) above the pharynx and a pair of connectives passing around the pharynx connecting the brain with the first pair of ganglia in the nerve cord; a ventral nerve cord, really double, running along the floor of the coelom to the last somite; and a pair of fused ganglia on the nerve cord in each somite. Each pair of fused ganglia gives off nerves to the body structures, which contain both sensory and motor fibers.

Neurosecretory cells have been found in the brain and ganglia of annelids, both oligochaetes and polychaetes. They are endocrine in function and secrete neurohormones concerned with the regulation of reproduction, secondary sex characteristics, and regeneration.

Figure 14-16

Portion of nerve cord of earthworm showing arrangement of simple reflex arc *(in foreground)* and the three dorsal giant fibers that are adapted for rapid reflexes and escape movements. Ordinary crawling involves a succession of reflex acts, the stretching of one somite stimulating the next to stretch, and so on. Impulses are transmitted much faster in giant fibers than in regular nerves so that all segments can contract simultaneously when quick withdrawal into a burrow is necessary.

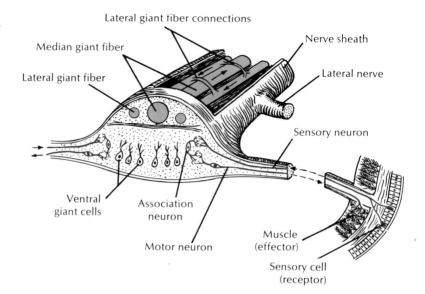

In the dorsal median giant fiber of *Lumbricus,* which is 90 to 160 μm in diameter, the speed of conduction has been estimated at 20 to 45 m/second, several times faster than in ordinary neurons of this species. This is also much faster than in polychaete giant fibers, probably because in the earthworms the giant fibers are enclosed in myelinated sheaths. The speed of conduction may be altered by changes in temperature.

For rapid escape movements most annelids are provided with from one to several very large axons commonly called **giant axons** (Figure 14-16), or giant fibers, located in the ventral nerve cord. Their large diameter increases the rate of conduction and makes possible simultaneous contractions of muscles in many segments.

Simple sense organs are distributed all over the body. Earthworms have no eyes but do have many lens-shaped photoreceptors in the epidermis. Most oligochaetes are negatively phototactic to strong light but positively phototactic to weak light. Many single-celled sense organs are widely distributed in the epidermis. What are presumably chemoreceptors are most numerous on the prostomium. There are many free nerve endings in the integument, which are probably tactile.

General behavior

Earthworms are among the most defenseless of creatures, yet their abundance and wide distribution indicate their ability to survive. Although they have no specialized sense organs, they are sensitive to many stimuli, such as mechanical, to which they react positively when it is moderate and negatively when it is a vibration (such as footfall near them), which causes them to retire quickly into their burrows; and light, which they avoid unless it is very weak. Chemical responses aid them in the choice of food.

Chemical as well as tactile responses are very important to the worm. It not only must be able to sample the organic content of the soil to find food, but also must sense its texture, acidity, and calcium content.

Experiments show that earthworms have some learning ability. They can be taught to avoid an electric shock, and thus an association reflex can be built up in them. Darwin credited earthworms with a great deal of intelligence in pulling leaves into their burrows, for they apparently seized the leaves by the narrow end, the easiest way for drawing a leaf-shaped object into a small hole. Darwin assumed that the seizure of the leaves by the worms did not result from random handling or from chance but was purposeful in its mechanism. However, investigations since Darwin's time have shown that the process is mainly one of trial and error, for they often seize a leaf several times before attaining the right position.

Reproduction and development

Earthworms are monoecious (hermaphroditic); that is, both male and female organs are found in the same animal (Figure 14-12, *A*). In *Lumbricus* the reproductive systems are found in somites 9 to 15. Two pairs of small testes and two pairs of sperm funnels are surrounded by three pairs of large seminal vesicles. Immature sperm from the testes mature in the seminal vesicles, then pass into the sperm funnels and down sperm ducts to the male genital pores in somite 15, where they are expelled during copulation. Eggs are discharged by a pair of small ovaries into the coelomic cavity, where they are picked up by ciliated funnels and are carried by oviducts to the outside through female genital pores on somite 14. Two pairs of seminal receptacles in somites 9 and 10 receive and store sperm from the mate during copulation.

Reproduction in earthworms may occur at any season, but copulation usually occurs at night during warm, moist weather (Figure 14-17). When mating, the worms extend their anterior ends from their burrows and bring their ventral surfaces together (Figure 14-18). They are held together by mucus secreted by the **clitellum** and by special ventral setae, which penetrate each other's bodies in the regions of contact. Sperm are discharged and travel to the seminal receptacles of the other worm in its seminal grooves. After copulation each worm secretes first a

Figure 14-17

Two earthworms in copulation. Their anterior ends point in opposite directions as their ventral surfaces are held together by mucous bands secreted by the clitella. Mutual insemination occurs during copulation. After separation each worm secretes a cocoon to receive its eggs and sperm.
Courtesy Carolina Biological Supply Co., Burlington, N.C.

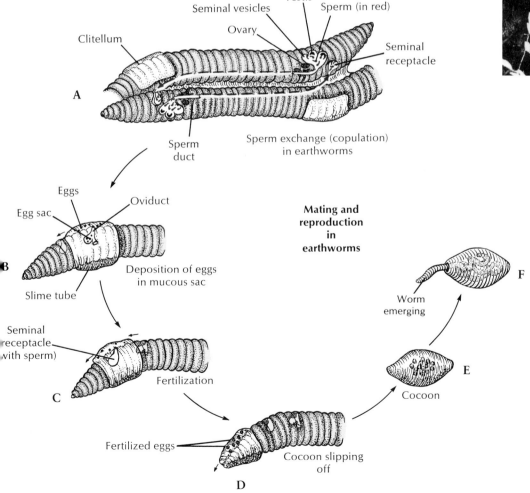

Sperm exchange (copulation) in earthworms

Mating and reproduction in earthworms

Clitellum

Seminal vesicles

Testis

Sperm (in red)

Ovary

Seminal receptacle

Sperm duct

A

Eggs

Oviduct

Egg sac

Deposition of eggs in mucous sac

Slime tube

B

Seminal receptacle (with sperm)

Fertilization

C

Fertilized eggs

Cocoon slipping off

D

Cocoon

E

Worm emerging

F

Figure 14-18

Earthworm copulation and formation of egg cocoons. **A,** Mutual insemination occurs during copulation; sperm from genital pore (somite 15) pass along seminal grooves to seminal receptacles (somites 9 and 10) of each mate. **B** and **C,** After worms separate, a slime tube formed over the clitellum passes forward to receive eggs from oviducts and sperm from seminal receptacles. **D,** As cocoon slips off over anterior end, its ends close and seal. **E,** Cocoon is deposited near burrow entrance. **F,** Young worms emerge in 2 to 3 weeks.

mucous tube and then a tough, chitinlike band that forms a **cocoon** around its clitellum. As the cocoon passes forward, eggs from the oviducts, albumin from the skin glands, and sperm from the mate (stored in the seminal receptacles) are poured into it. Fertilization of the eggs then takes place within the cocoon. When the cocoon leaves the worm, its ends close, producing a lemon-shaped body. Embryonation occurs within the cocoon, and the form that hatches from the egg is a young worm similar to the adult. It does not develop a clitellum until it is sexually mature.

Freshwater oligochaetes

Freshwater oligochaetes usually are smaller and have more conspicuous setae than do earthworms. They are more mobile than earthworms and tend to have better-developed sense organs. They are generally benthic forms that creep about on the bottom or burrow in the soft mud. Aquatic oligochaetes are an important food source for fishes. A few are ectoparasitic.

Some of the more common freshwater oligochaetes are the 1 mm long *Aeolosoma* (Gr. *aiolos,* quick-moving, + *soma,* body) (Figure 14-19, *B*), which contains red or green pigments, has bundles of setae, and is often found in hay cultures; the 2 to 4 mm long *Nais* (L. *nais,* water nymph), which is brownish and has two bundles of setae on anterior segments and four bundles of setae on each posterior segment; the 10 to 25 mm long *Stylaria* (Gr. *stylos,* pillar) (Figure 14-19, *A*), with setae arranged like those of *Nais,* a prostomium extended into a long process, and black eyespots; the 5 to 10 mm long *Dero,* (Gr. *dere,* neck or throat), which is reddish, lives in tubes, and usually has 3 to 4 pairs of tail gills; the 30 to 40 mm long *Tubifex* (L. *tubus,* tube, + *faciens,* to make or do) (Figure 14-19, *C*), which is reddish and lives with its head in mud at the bottom of ponds and its tail waving in the water; the 10 to 15 mm long *Chaetogaster* (N.L. *chaeta,* bristle, + *gastrula,* belly), which has only ventral bundles of setae; and *Enchytraeus* (Gr. *enchytraeus,* living in an earthen pot), small whitish worms that live both in moist soil and in water. Some oligochaetes, such as *Aeolosoma,* may form chains of zooids asexually by transverse fission (Figure 14-19, *B*).

____ Class Hirudinea: the Leeches

Leeches are found predominantly in freshwater habitats, but a few are marine, and some have even adapted to terrestrial life in warm, moist places. They are more abundant in tropical countries than in temperate zones. Some leeches attack human beings and are a nuisance.

Most leeches are between 2 and 6 cm in length, but some are smaller; some, including the "medicinal" leech, reach 20 cm, but the giant of all is the Amazonian *Haementeria* (Gr. *haimateros,* bloody) (Figure 14-20), which reaches 30 cm.

Leeches are found in a variety of patterns and colors: black, brown, red or olive green. They are usually flattened dorsoventrally.

Like the oligochaetes, leeches are hermaphroditic and have a clitellum, but this only appears during the breeding season. The clitellum secretes a cocoon for the reception of eggs. Leeches are more highly specialized than the oligochaetes. As fluid feeders and bloodsuckers, they have lost the setae used by the oligochaetes in locomotion and have developed suckers for attachment while sucking blood; their gut is specialized for storage of large quantities of blood.

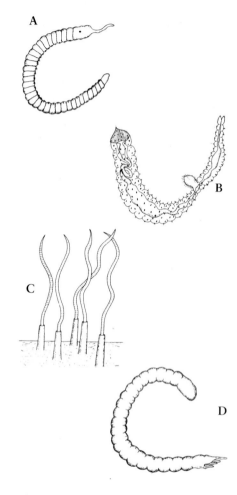

Figure 14-19

Some freshwater oligochaetes. **A,** *Stylaria* has the prostomium drawn out into a long snout. **B,** *Aeolosoma* uses cilia around the mouth to sweep in food particles, and it buds off new individuals asexually. **C,** *Tubifex* lives head down in long tubes. **D,** *Aulophorus* is provided with ciliated anal gills.

Form and function

Unlike other annelids, leeches have a fixed number of somites (usually 34; 17 or 31 in some groups), but they appear to have many more because each somite is marked by transverse grooves to form from two to 16 superficial rings (**annuli**) (Figure 14-21).

The coelom represents another difference between leeches and other annelids; leeches lack distinct coelomic compartments. In all but one species the septa have disappeared, and the coelomic cavity is filled with connective tissue and a system of spaces called **lacunae**. The coelomic lacunae form a regular system of channels filled with coelomic fluid, which in some leeches serves as an auxiliary circulatory system.

Most leeches creep with looping movements of the body, by attaching first one sucker and then the other and pulling up the body. Aquatic leeches can also swim with a graceful undulatory movement.

Figure 14-20

The world's largest leech, *Haementeria ghilianii*, on the arm of Dr. Roy K. Sawyer, who found it in French Guiana, South America.
Photograph by T. Branning.

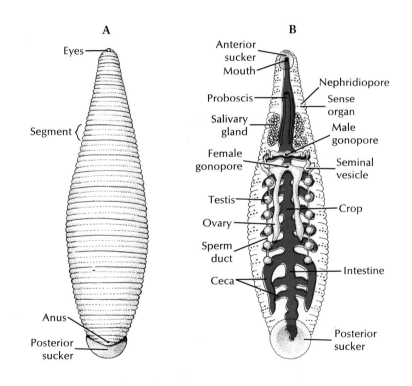

A

- Eyes
- Segment
- Anus
- Posterior sucker

B

- Anterior sucker
- Mouth
- Proboscis
- Salivary gland
- Female gonopore
- Testis
- Ovary
- Sperm duct
- Ceca
- Nephridiopore
- Sense organ
- Male gonopore
- Seminal vesicle
- Crop
- Intestine
- Posterior sucker

Figure 14-21

Structure of a leech, *Placobdella*. **A**, External appearance, dorsal view. **B**, Internal structure, ventral view.

Nutrition

Leeches are popularly considered parasitic, but it would be more accurate to call them predaceous. Even the true bloodsuckers rarely are host specific or remain on the host for a long period of time. Most freshwater leeches are active predators or scavengers equipped with a proboscis that can be extended to draw in small invertebrates or to take blood from cold-blooded vertebrates. Some freshwater leeches are true bloodsuckers, preying on cattle, horses, humans, and others. Some terrestrial leeches feed on insect larvae, earthworms, and slugs, which they hold by an oral sucker while using a strong sucking pharynx to draw in the food. Other terrestrial forms climb bushes or trees to reach warm-blooded vertebrates such as birds or mammals.

For centuries the "medicinal leech" *(Hirudo medicinalis)* was used for bloodletting because of the mistaken idea that bodily disorders and fevers were caused by an excess of blood. A 10 to 12 cm long leech can extend to a much greater length when distended with blood, and the amount of blood it can suck is considerable. Leech collecting and leech culture in ponds were practiced in Europe on a commercial scale during the nineteenth century. Wordsworth's poem "The Leech-Gatherer" was based on this use of the leech.

Leeches are once again being used medicinally. When fingers or toes are severed, microsurgeons can reconnect arteries but not the more delicate veins. Leeches are used to relieve congestion until the veins can grow back into the healing digit.

Leeches are highly sensitive to stimuli associated with the presence of a prey or host. They are attracted by and will attempt to attach to an object smeared with appropriate host substances, such as fish scales, oil secretions, or sweat. Those that feed on the blood of mammals are attracted by warmth, and the terrestrial haemadipsids of the tropics will converge on a person standing in one place.

Most leeches are fluid feeders. Many prefer to feed on tissue fluids and blood pumped from wounds already open. The true bloodsuckers, which include the so-called medicinal leech *Hirudo* (L. a leech) (Figure 14-22), have cutting plates, or "jaws," for cutting through tissues. Some species are true parasites, living permanently on their hosts.

Respiration and excretion

Gas exchange occurs only through the skin except in some of the fish leeches, which have gills. There are 10 to 17 pairs of nephridia, in addition to coelomocytes and certain other specialized cells that may also be involved in excretory functions.

Nervous and sensory systems

Leeches have two "brains"; one is in the head and is composed of six pairs of fused ganglia forming a ring around the pharynx, and one is in the tail and is composed of seven pairs of fused ganglia. The additional 21 pairs of ganglia are segmentally arranged along the double nerve cord. In addition to free sensory nerve endings and photoreceptor cells in the epidermis, there is a row of sense organs, called sensillae, in the central annulus of each segment; there are also a number of pigment cup ocelli.

Reproduction

Leeches are hermaphroditic but practice cross-fertilization during copulation. Sperm are transferred by a penis or by hypodermic impregnation (a spermatophore is expelled from one worm and penetrates the integument of the other). After copulation the clitellum secretes a cocoon that receives the eggs and sperm. Cocoons are buried in bottom mud, attached to submerged objects, or, in terrestrial species, placed in damp soil. Development is similar to that of oligochaetes.

Circulation

In leeches the coelom has been reduced, by the invasion of connective tissue and in some by a proliferation of chlorogogue tissue, to a system of coelomic sinuses and channels. Some orders of leeches retain the typical oligochaete circulatory system, and in these the coelomic sinuses act as an auxillary blood-vascular system. In other orders the traditional blood vessels have disappeared and the system of coelomic sinuses forms the only blood-vascular system. In those the blood (the equivalent of coelomic fluid) is propelled by contractions of certain longitudinal channels.

____ Evolutionary Significance of Metamerism

No truly satisfactory explanation has yet been given for the origins of metamerism and the coelom, although the subject has stimulated much speculation and debate over the years. All of the classical explanations of the origin of metamerism and the coelom have had important arguments leveled against them, and more than one may be correct, or none, as suggested by R.B. Clark (1964). The coelom and metamerism may have evolved independently in more than one group of animals, as, for example, in the chordates and in the protostome line. Clark stressed the functional and evolutionary significance of these features to the earliest animals that possessed them. He argued forcefully that the adaptive value of the coelom in

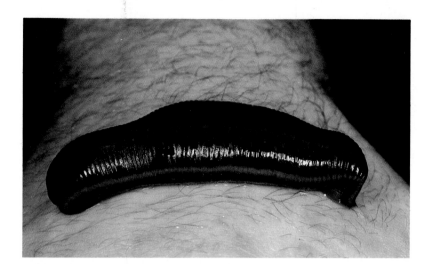

Figure 14-22

Hirudo medicinalis feeding on blood from human arm.
Photograph by C.P. Hickman, Jr.

the protostomes, at least, was as a **hydrostatic skeleton** in a burrowing animal. Thus, contraction of muscles in one part of the animal could act antagonistically on muscles in another part by transmission of the force of contraction through the enclosed constant volume of fluid in the coelom.

Although burrowing in the substrate may have selected for the coelom, certain other advantages accrued to its possessors. The coelomic fluid would have acted as a circulatory fluid for nutrients and wastes, making large numbers of flame cells distributed throughout the tissues unnecessary. Gametes could be stored in the spacious coelom for release simultaneously with other individuals in the population, thus enhancing chances of fertilization, and this would have selected for greater nervous and endocrine control. Finally, separation of the coelom into a series of compartments by septa, resulting in metamerism, would have greatly increased the capacity for powerful burrowing. The force exerted by muscle contraction on the fluid in one coelomic compartment would not be transmitted to the fluid in other compartments, making possible independent and separate movements by the separate metameres. Independent movements of metameres in different parts of the body would have placed selective value on a more sophisticated nervous system for control of the movements and led to elaboration of the central nervous system.

____ Phylogeny and Adaptive Radiation

Phylogeny

There are so many similarities in the early development of the molluscs, annelids, and primitive arthropods that there seems little doubt about their close relationship. It is thought that the common ancestor of the three phyla was some type of flatworm. Many marine annelids and molluscs have an early embryogenesis typical of protostomes, in common with some marine flatworms, suggesting a real, if remote, relationship. Annelids share with the arthropods an outer secreted cuticle and have a similar nervous system, and the lateral appendages (parapodia) of many marine annelids are similar to the appendages of certain primitive arthropods. The most important resemblance, however, probably lies in the segmented plan of the annelid and the arthropod body structure.

Which of the annelids came first? This has been the subject of a long and continuing debate. It was long assumed that the archiannelids, a small group of minute polychaetes, were the most primitive of the annelids. This view is no longer widely held. Rather than primitive, they appear to be polychaetes that have become secondarily simplified as a result of their small size and their adaptation to a psammolittoral life—that is, living in the interstitial spaces between the grains of sand or on the mud bottom. R.B. Clark (1964) suggested that the origin of annelids needs to be viewed in terms of the origin of the metameric arrangement of worms. He emphasizes that the only consistent correlation between the structure and function of a segmented coelom is a mechanical one. The division of the coelom, a hydrostatic organ, into units is a mechanical device limiting the transfer of hydrostatic pressure from one part of the coelom to another. There is an advantage in such a locomotor device to macroscopic animals, but not to microscopic ones, so there is little reason to assume it evolved first in the microscopic archiannelids. Most hypotheses of annelid origin have assumed that the polychaetes were the earliest segmented worms and that metamerism arose in connection with the development of lateral appendages (parapodia). However, the oligochaete body is adapted to vagrant burrowing in the substratum with a peristaltic movement that is highly benefited by a metameric coelom. On the other hand, polychaetes with well-developed parapodia are generally adapted to swimming and crawling in a medium too fluid for effective peristaltic locomotion. Although the parapodia do not prevent such locomotion, they do little to further it, and it seems unlikely that they evolved as an adaptation for swimming. Although the polychaetes are more primitive in some respects, such as in their reproductive system, some authorities have argued that the ancestral annelids were more similar to the oligochaetes in overall body plan and that these gave rise to the polychaetes and leeches. Others maintain that the oligochaetes developed from polychaete stock. The leeches are closely related to the oligochaetes and probably evolved from them in connection with a swimming existence and the abandonment of a burrowing mode of life.

Adaptive radiation

Annelids are an ancient group that has undergone extensive adaptive radiation. The basic body structure, particularly of the polychaetes, lends itself to almost endless modification. As marine worms, polychaetes have a wide range of habitats in an environment that is not physically or physiologically demanding. Unlike the earthworms, whose environment imposes strict physical and physiological selective pressures, the polychaetes have been free to experiment and thus have achieved a wide range of adaptive features.

A basic adaptive feature in evolution of annelids is their septal arrangement, resulting in fluid-filled coelomic compartments. Fluid pressure in these compartments is used as a hydrostatic skeleton in precise movements such as burrowing and swimming. Powerful circular and longitudinal muscles have been adapted for flexing, shortening, and lengthening the body.

There is a wide variation in feeding adaptations, from the sucking pharynx of the oligochaetes and the chitinous jaws of carnivorous polychaetes to the specialized tentacles and cirri of the ciliary feeders.

In polychaetes the parapodia have been adapted in many ways and for many functions, chiefly locomotion and respiration.

In leeches many adaptations, such as suckers, cutting jaws, pumping pharynx, distensible gut, and the production of hirudin, are related to their predatory and bloodsucking habits.

SUMMARY

The phylum Annelida is a large group including the marine polychaetes, earthworms, freshwater oligochaetes, and leeches. They are metameric, having their bodies divided into a linear series of segments, each of which contains representatives of most body systems. Annelids, with the exception of leeches, bear chitinous bristles called setae. They are cosmopolitan. Many are burrowing, and there are many filter feeders and predators.

The body is composed of many metameres. Because the coelom forms embryonically as a split in the mesoderm, it is lined with a mesodermal peritoneum, and the gut and other organs are suspended in mesenteries. The peritoneum between metameres forms septa. The fluid-filled coelomic cavities function as a hydrostatic skeleton except in the leeches.

Polychaetes are the largest class of annelids and are mostly marine. They have many setae on each somite, which are borne on paired parapodia. Parapodia show a wide variety of adaptations among polychaetes, including specialization for swimming, respiration, crawling, maintaining position in a burrow, pumping water through a burrow, and as accessory feeding organs. Errant polychaetes are mostly predaceous and have an eversible pharynx with jaws. Sedentary polychaetes rarely leave the burrows or tubes in which they live. Several styles of deposit and filter feeding are shown among the sedentary polychaetes. Polychaetes are dioecious, have a primitive reproductive system, have no clitellum, and practice external fertilization, and their larvae are trochophores.

The class Oligochaeta includes earthworms and many freshwater forms; they have a small number of setae per segment (compared to the Polychaeta) and no parapodia. Most earthworms have four pairs of setae per segment, which function to keep the body from slipping in the burrow. The circulatory system is closed, and the dorsal blood vessel is the main pumping organ. There is a pair of nephridia in most somites, and the nephrostome of each nephridium lies in the intersegmental septum of the somite immediately preceding it. Earthworms have the typical annelid nervous system: dorsal cerebral ganglia connected to a double, ventral nerve cord with segmental ganglia, running the length of the worm. Giant axons with a high speed of conduction are present. Oligochaetes are hermaphroditic and practice cross-fertilization. The clitellum plays an important role in reproduction, including secretion of mucus to surround the worms during copulation and secretion of a cocoon to receive the eggs and sperm and in which embryonation occurs. A small, juvenile worm hatches from the cocoon.

Freshwater oligochaetes have longer setae than earthworms, and some have gills.

The leeches (class Hirudinea) are mostly freshwater, although a few are marine and a few are terrestrial. They feed mostly on fluids; many are predators, some are temporary parasites, and a few are permanent parasites. The hermaphroditic leeches reproduce in a fashion similar to oligochaetes, with cross-fertilization and cocoon formation by the clitellum.

It is probable that the coelom and metamerism evolved as a hydrostatic skeleton in adaptation for strong and sustained burrowing, but its origin had several important evolutionary implications, leading to more efficient nervous and endocrine control systems.

Embryological evidence supports a phylogenetic relationship of the annelids with the molluscs and arthropods. The question of whether the oligochaetes arose from polychaete stock or vice versa is unsettled. Leeches probably evolved from an oligochaete ancestor.

Adaptive radiation in the annelids shows exploitation of the hydrostatic skeleton, setae, parapodia, and other organs for locomotion, respiration, and feeding. Leeches have become specialized for predation and bloodsucking habits.

Review questions

1. Name eight of the most important characteristics of the phylum Annelida.
2. Distinguish among the classes of the phylum Annelida.
3. Describe the annelid body plan, including the body wall, segments, coelom and its compartments, and coelomic lining.
4. Explain how the hydrostatic skeleton of the annelids helps them to burrow. How is the efficiency for burrowing increased by metamerism?
5. Describe at least three ways that various polychaetes obtain food.
6. Define each of the following: prostomium, peristomium, pygidium, radioles, parapodium, neuropodium, notopodium.
7. Explain the function of each of the following in earthworms: pharynx, calciferous glands, crop, gizzard, typhlosole, chlorogogue tissue.
8. Describe the main features of each of the following in earthworms: circulatory system, nervous system, excretory system.
9. Describe the function of the clitellum and the cocoon.
10. How are freshwater oligochaetes generally different from earthworms?
11. Name three freshwater oligochaetes and give a characteristic feature of each.
12. Describe the ways in which leeches obtain food.
13. What are the main differences in reproduction and development among the three classes of annelids?
14. What was the evolutionary significance of metamerism and the coelom to its earliest possessors?
15. What are the phylogenetic relationships between the molluscs, annelids, and arthropods? What is the evidence for these relationships?

Selected references

See also general references to Part Two, p. 590.

Clark, R.B. 1964. Dynamics in metazoan evolution. The origin of the coelom and segments. Oxford, England, Clarendon Press. *An important treatise giving the author's hypotheses on the subject.*

Dales, R.P. 1967. Annelids, ed. 2. London, the Hutchinson Publishing Group, Ltd. *A concise account of the annelids.*

Nicholls, J.C., and D. Van Essen. 1974. The nervous system of the leech. Sci. Am. 230:38-48 (Jan.). *Because the leech has large nerve cells and only a few neurons perform a given function, its nervous system is particularly appropriate for experimental studies.*

Hartman, O. 1968. Atlas of the errantiate polychaetous annelids from California. Los Angeles, University of Southern California, Allan Hancock Foundation. *This and the following reference have extensive keys for identification.*

Hartman, O. 1969. Atlas of the sedentariate polychaetous annelids from California. Los Angeles, University of Southern California, Allan Hancock Foundation.

C H A P T E R   1 5

T H E   A R T H R O P O D S

Phylum Arthropoda

Subphylum Trilobita

Subphylum Chelicerata

Position in Animal Kingdom

1. Arthropods belong to the **protostome** branch, or schizocoelous coelomates, of the animal kingdom, and in primitive forms they have spiral cleavage and mosaic development.
2. Arthropods have the characteristic structure of higher forms: bilateral symmetry, three germ layers, coelomic cavity, and organ systems.
3. Like the annelids, the arthropods have conspicuous segmentation, but their somites have greater variety and more grouping for specialized purposes; the somites have jointed appendages, with pronounced division of labor, resulting in greater variety of action.

Biological Contributions

1. **Cephalization** makes additional advancements, with centralization of fused ganglia and sensory organs in the head.
2. The **somites** have gone beyond the sameness of the annelid type and are now **specialized** for a variety of purposes, forming functional groups of somites (**tagmosis**).
3. The presence of paired **jointed appendages** diversified for numerous uses produces greater adaptability.

4. Locomotion is by extrinsic limb muscles, in contrast to the body musculature of annelids. **Striated muscles** are emphasized, thus ensuring rapidity of movement.
5. Although **chitin** is found in a few other forms below arthropods, its use is better developed in the arthropods. The **cuticular exoskeleton,** containing chitin, is a great advance over that of the annelids, making possible a wide range of adaptations.
6. The **tracheae** represent a breathing mechanism more efficient than that of most invertebrates.
7. The alimentary canal shows greater specialization by having chitinous teeth, compartments, and gastric ossicles.
8. Behavior patterns have advanced far beyond those of most invertebrates, with a higher development of **social** organization.
9. **Metamorphosis** is common in development.
10. Many arthropods have well-developed protective coloration and protective resemblances.

—PHYLUM ARTHROPODA

Phylum Arthropoda (ar-throp'o-da) (Gr. *arthron*, joint, + *pous, podos*, foot) is the most extensive phylum in the animal kingdom, composed of more than three fourths of all known species. Approximately 900,000 species of arthropods have been recorded, and probably as many more remain to be classified. Arthropods include the spiders, scorpions, ticks, mites, crustaceans, millipedes, centipedes, insects, and some others. In addition, there is a rich fossil record extending to the very late Precambrian period.

Arthropods are eucoelomate protostomes with well-developed organ systems, and they share with the annelids the property of conspicuous segmentation.

Arthropods have an exoskeleton containing chitin, and their primitive pattern is that of a linear series of similar somites, each with a pair of jointed appendages. However, the pattern of somites and appendages varies greatly in the phylum. There is a tendency for the somites to be combined or fused into functional groups, called **tagmata** (sing, **tagma**), for specialized purposes; the appendages are frequently differentiated and specialized for pronounced division of labor.

Few arthropods exceed 60 cm in length, and most are far below this size. The largest is the Japanese crab *Macrocheira* (Gr. *makros*, large + *cheir*, hand), which has approximately a 4 m span; the smallest is the parasitic mite *Demodex* (Gr. *dēmos*, fat, + *dex*, a wood worm), which is less than 0.1 mm long.

Arthropods are usually active, energetic animals. Judging by their great diversity and their wide ecological distribution, as well as by the vast number of species, their success is surpassed by no other group of animals.

Although arthropods compete with humans for food and spread serious diseases, they are essential in pollination of many food plants, and they also serve as food, yield drugs and dyes, and produce products such as silk, honey, and beeswax.

The arthropods are more widely and more densely distributed throughout all regions of the earth than are members of any other phylum. They are found in all types of environment from low ocean depths to very high altitudes, and from the tropics far into both north and south polar regions. Different species are adapted for life in the air; on land; in fresh, brackish, and marine waters; and in or on the bodies of plants and other animals. Some species live in places where no other form could survive.

Although all types—carnivorous, omnivorous, and symbiotic—occur in this vast group, the majority are herbivorous. Most aquatic arthropods depend on algae for their nourishment, and the majority of land forms live chiefly on plants. In diversity of ecological distribution, the arthropods have no rivals.

CHARACTERISTICS

1. Bilateral symmetry; **metameric body** divided into **tagmata** consisting of head and trunk; head, thorax, and abdomen; or cephalothorax and abdomen
2. **Jointed appendages;** primitively, one pair to each somite, but number often reduced; appendages often modified for specialized functions
3. **Exoskeleton of cuticle** containing protein, lipid, chitin, and often calcium carbonate secreted by underlying epidermis and shed (molted) at intervals
4. **Complex muscular system,** with exoskeleton for attachment, **striated muscles** for rapid actions, smooth muscles for visceral organs; no cilia
5. **Reduced coelom** in adult; most of body cavity consisting of hemocoel (sinuses, or spaces, in the tissues) filled with blood

6. Complete digestive system; mouthparts modified from appendages and adapted for different methods of feeding
7. Open circulatory system, with dorsal **contractile heart**, arteries, and hemocoel (blood sinuses)
8. Respiration by **body surface, gills, tracheae** (air tubes), or **book lungs**
9. Paired excretory glands called coxal, antennal, or maxillary glands present in some, homologous to metameric nephridial system of annelids; some with other excretory organs, called **malpighian tubules**
10. Nervous system of annelid plan, with dorsal brain connected by a ring around the gullet to a double nerve chain of ventral ganglia; fusion of ganglia in some species; well-developed sensory organs
11. Sexes usually separate, with paired reproductive organs and ducts; usually internal fertilization; oviparous or ovoviviparous; often with **metamorphosis**; parthenogenesis in a few forms

CLASSIFICATION

Subphylum Trilobita (tri'lo-bi'ta) (Gr. *tri*, three, + *lobos*, lobe): **trilobites.** All extinct forms; Cambrian to Carboniferous; body divided by two longitudinal furrows into three lobes; distinct head, thorax, and abdomen, biramous (two-branched) appendages.

Subphylum Chelicerata (ke-lis'-e-ra'ta) (Gr. *chēlē*, claw, + *keras*, horn, + *ata*, group suffix): **eurypterids, horseshoe crabs, spiders, ticks.** First pair of appendages modified to form chelicerae; pair of pedipalps and four pairs of legs; no antennae, no mandibles; cephalothorax and abdomen usually unsegmented.

    Class Merostomata (mer'o-sto'ma-ta) (Gr. *mēros*, thigh, + *stoma*, mouth, + *ata*, group suffix): **aquatic chelicerates.** Cephalothorax and abdomen; compound lateral eyes; appendages with gills; sharp telson; subclasses Eurypterida (all extinct) and Xiphosurida, the horseshoe crabs. Example: *Limulus.*

    Class Pycnogonida (pik'no-gon'i-da) (Gr. *pyknos*, compact, + *gony*, knee, angle): **sea spiders.** Small (3 to 4 mm), but some reach 500 mm; body chiefly cephalothorax; tiny abdomen; usually four pairs of long walking legs (some with five or six pairs); mouth on long proboscis; four simple eyes; no respiratory or excretory system. Example: *Pycnogonum.*

    Class Arachnida (ar-ack'ni-da) (Gr. *arachnē*, spider): **scorpions, spiders, mites, ticks, harvestmen.** Four pairs of legs; segmented or unsegmented abdomen with or without appendages and generally distinct from cephalothorax; respiration by gills, tracheae, or book lungs; excretion by malpighian tubules or coxal glands; dorsal bilobed brain connected to ventral ganglionic mass with nerves; simple eyes; sexes separate; chiefly oviparous; no true metamorphosis. Examples: *Argiope, Centruroides.*

Subphylum Crustacea (crus-ta' she-a) (L. *crusta*, shell, + *acea*, group suffix): **crustaceans.** Mostly aquatic, with gills; cephalothorax usually with dorsal carapace; biramous appendages, modified for various functions. Head appendages consisting of two pairs of antennae, one pair of mandibles, and two pairs of maxillae. Sexes usually separate. Development primitively with nauplius stage (see resume of classes, pp. 347 to 353).

Subphylum Uniramia (yu-ni-ra' me-a) (L. *unus*, one, + *ramus*, a branch): **insects and myriapods.** All appendages uniramous; head appendages consisting of one pair of antennae, one pair of mandibles, and one or two pairs of maxillae.

    Class Diplopoda (di-plop'o-da) (Gr. *diploos*, double, + *pous, podos*, foot): **millipedes.** Subcylindrical body; head with short antennae and simple eyes; body with variable number of somites; short legs, usually two pairs of legs to a somite; separate sexes; oviparous. Examples: *Julus, Spirobolus.*

    Class Chilopoda (ki-lop'-o-da) (Gr. *cheilos*, lip, + *pous, podos*, foot): **centipedes.** Dorsoventrally flattened body; variable number of somites, each with one pair of legs; one pair of long antennae; separate sexes; oviparous. Examples: *Cermatia, Lithobius, Geophilus.*

    Class Pauropoda (pau-rop'o-da) (Gr. *pauros*, small, + *pous, podos*, foot) **pauropods.** Minute (1 to 1.5 mm); cylindrical body consisting of double segments and bearing nine or 10 pairs of legs; no eyes. Example: *Pauropus.*

Class Symphyla (sym'fy-la) (Gr. *syn*, together, + *phylon*, tribe): **garden centipedes.** Slender (1 to 8 mm) with long, filiform antennae; body consisting of 15 to 22 segments with 10 to 12 pairs of legs; no eyes. Example: *Scutigerella.*
Class Insecta (in-sek'ta) (L. *insectus*, cut into): **insects.** Body with distinct head, thorax, and abdomen; pair of antennae; mouthparts modified for different food habits; head of six fused somites; thorax of three somites; abdomen with variable number, usually 11 somites; thorax with two pairs of wings (sometimes one pair or none) and three pairs of jointed legs; separate sexes; usually oviparous; gradual or abrupt metamorphosis. (A brief description of insect orders is given on pp. 382 to 385.)

COMPARISON OF ARTHROPODA WITH ANNELIDA
Similarities between Arthropoda and Annelida are as follows:

1. External segmentation
2. Segmental arrangement of muscles
3. Ventral nerve cord with metamerically arranged ganglia and dorsal cerebral ganglia
4. Spiral cleavage (found in some arthropods)

Arthropods differ from annelids in having the following:

1. Fixed number of segments (in adults)
2. Usually lack intersegmental septa
3. Pronounced tagmatization (compared with limited tagmatization in annelids)
4. Coelomic cavity reduced; main body cavity a hemocoel
5. Open (lacunar) circulatory system
6. Special mechanisms (gills, tracheae, book lungs) for respiration
7. Exoskeleton containing chitin
8. Jointed appendages
9. Compound eyes (also present in a few annelids) and other well-developed sense organs
10. Absence of cilia
11. Metamorphosis in many cases

____ Why Have Arthropods Been So Successful?

The success of the arthropods is attested to by their diversity, number of species, wide distribution, variety of habitats and feeding habits, and power of adaptation to changing conditions. Some of the structural and physiological patterns that have been helpful to them are briefly summarized in the following discussion.

1. A versatile exoskeleton. The arthropods possess an exoskeleton that is highly protective without sacrificing mobility. This skeleton is the **cuticle,** an outer covering secreted by the underlying epidermis. The cuticle is made up of an inner and usually thicker **endocuticle** and an outer, relatively thin **epicuticle.** The endocuticle contains **chitin** bound with protein. Chitin is a tough, resistant, nitrogenous polysaccharide that is insoluble in water, alkalis, and weak acids. Thus the endocuticle not only is flexible and lightweight but also affords protection, particularly against dehydration. In some crustaceans the chitin may make up as much as 60% to 80% of the endocuticle, but in insects it is probably not more than 40%. In most crustaceans the endocuticle in some areas is also impregnated with **calcium salts,** which reduce its flexibility. In the hard shells of lobsters and crabs, for instance, this calcification is extreme. The outer epicuticle is composed of protein and lipid. The protein is stabilized and hardened by tanning, adding further protection. Both the endocuticle and epicuticle are laminated, that is, composed of several layers each (see Figure 28-1, p. 593).

The cuticle may be soft and permeable or may form a veritable coat of armor. Between body segments and between the segments of appendages it is thin and flexible, creating movable joints and permitting free movements. In crustaceans and insects the cuticle forms ingrowths (apodemes) that serve for muscle attachment. It may also line the foregut and hindgut, line and support the trachea, and be adapted for biting mouthparts, sensory organs, copulatory organs, and ornamental purposes. It is indeed a versatile material.

The nonexpansible cuticular exoskeleton does, however, impose important restrictions on growth. To grow, an arthropod must shed its outer covering at intervals and grow a larger one—a process called **ecdysis,** or **molting.** Arthropods molt four to seven times before reaching adulthood, and some continue to molt after that. An exoskeleton is also relatively heavy and becomes proportionately heavier with increasing size. This tends to limit the ultimate body size.

2. Segmentation and appendages for more efficient locomotion. Typically each somite is provided with a pair of jointed appendages, but this arrangement is often modified, with both segments and appendages specialized for adaptive functions. The limb segments are essentially hollow levers that are moved by internal muscles, most of which are striated for rapid action. The jointed appendages are equipped with sensory hairs and have been modified and adapted for sensory functions, food handling, swift and efficient walking legs, and swimming appendages.

3. Air piped directly to cells. Most land arthropods have the highly efficient tracheal system of air tubes, which delivers oxygen directly to the tissues and cells and makes a high metabolic rate possible. This system also tends to limit body size. Aquatic arthropods breathe mainly by some form of gill that is quite efficient.

4. Highly developed sensory organs. Sensory organs are found in great variety, from the compound (mosaic) eye to those simpler senses that have to do with touch, smell, hearing, balancing, chemical reception, and so on. Arthropods are keenly alert to what goes on in their environment.

5. Complex behavior patterns. Arthropods exceed most other invertebrates in the complexity and organization of their activities. Innate (unlearned) behavior unquestionably controls much of what they do, but learning also plays an important part in the lives of many of them.

6. Reduced competition through metamorphosis. Many arthropods pass through metamorphic changes, including a larval form quite different from the adult in structure. The larval form is often adapted for eating a different kind of food from that of the adult, resulting in less competition within a species.

_____ Subphylum Trilobita

The trilobites probably had their beginnings before the Cambrian period, in which they flourished. They have been extinct some 200 million years, but were abundant during the Cambrian and Ordovician periods. Their name refers to the trilobed shape of the body, caused by a pair of longitudinal grooves. They were bottom dwellers, probably scavengers (Figure 15-1, _A_). Most of them could roll up like pill bugs, and they ranged from 2 to 67 cm in length.

The exoskeleton contained chitin, strengthened in some areas by calcium carbonate. The body was composed of three tagmata: head, thorax, and pygidium. The head was one piece but showed signs of former segmentation; the thorax had a variable number of somites; and the somites of the pygidium, at the posterior end, were fused into a plate. The head bore a pair of antennae, compound eyes, mouth, and four pairs of jointed appendages. Each body somite except the

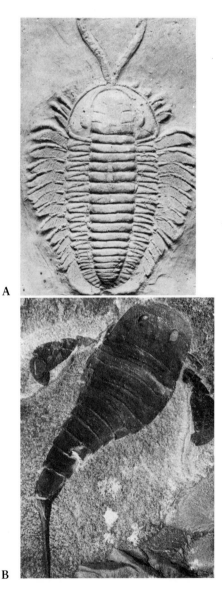

A

B

Figure 15-1

Fossils of early arthropods. **A,** Trilobite, dorsal view; plaster-cast impression. These animals were abundant in mid-Cambrian period. **B,** Eurypterid fossil; eurypterids flourished in Europe and North America from Ordovician to Permian periods. Photographs by F.M. Hickman.

last also bore a pair of biramous (two-branched) appendages. One of the branches had a fringe of filaments that may have served as gills.

Subphylum Chelicerata

The chelicerate arthropods are an ancient group that includes the eurypterids (extinct), horseshoe crabs, spiders, ticks and mites, scorpions, and sea spiders. They are characterized by having six pairs of appendages that include a pair of chelicerae, a pair of pedipalps, and four pairs of walking legs (a pair of chelicerae and five pairs of walking legs in horseshoe crabs). They have no mandibles and no antennae. Most chelicerates suck up liquid food from their prey.

Class Merostomata

Class Merostomata is represented by the eurypterids, all now extinct, and the xiphosurids, or horseshoe crabs, an ancient group sometimes referred to as "living fossils."

Eurypterids: subclass Eurypterida

The eurypterids, or giant water scorpions (Figure 15-1, *B*) were the largest of all fossil arthropods, some reaching a length of 3 m. Their fossils are found in rocks from the Ordovician to the Permian periods. They had many resemblances to the marine horseshoe crabs (Figures 15-2 and 15-3) and also to the scorpions, their land counterparts. The head had six fused segments and bore both simple and compound eyes and six pairs of appendages. The abdomen had 12 segments and a spikelike tail.

Ideas regarding their early habitats differ. Some believe the eurypterids evolved mainly in fresh water; others hold that they arose in brackish lagoons.

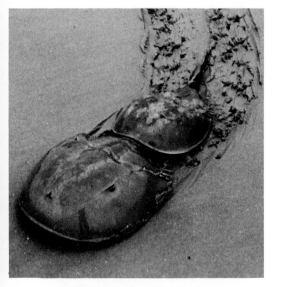

Figure 15-2

Mating of horseshoe crabs, *Limulus*. At high tide the female digs a depression in the sand for her eggs, which the male fertilizes externally before the hole fills up with sand. The male *(right)* is following the female *(left)* as she selects the spot for her eggs. Sometimes she is followed by several males. Photograph by L.L. Rue, III.

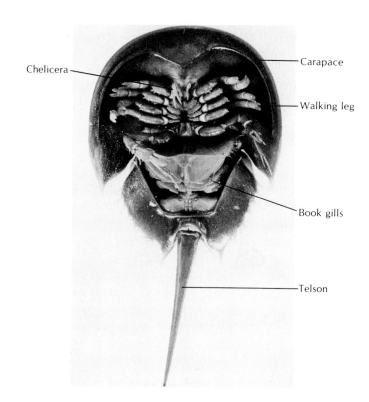

Figure 15-3

Ventral view of horseshoe crab *Limulus* (class Merostomata). They grow to 0.5 m in length.
Photograph by F.M. Hickman.

Horseshoe crabs: subclass Xiphosurida

The xiphosurids are an ancient marine group that dates from the Cambrian period. Our common horseshoe crab *Limulus* (L. *limus*, sidelong, askew) (Figure 15-2) goes back practically unchanged to the Triassic period. Only three genera (five species) survive today: *Limulus*, which lives in shallow water along the North American Atlantic coast; *Carcinoscorpius* (Gr. *karkinos*, crab, + *skorpiōn*, scorpion), along the southern shore of Japan; and *Tachypleus*, (Gr. *tachys*, swift, + *pleutēs*, sailor), in the East Indies and along the coast of southern Asia. They usually live in shallow water.

Xiphosurids have an unsegmented, horseshoe-shaped **carapace** (hard dorsal shield) and a broad abdomen, which has a long **telson,** or tailpiece. The cephalothorax bears five pairs of walking legs and a pair of chelicerae, whereas the abdomen has six pairs of broad, thin appendages that are fused in the median line (Figure 15-3). On some of the abdominal appendages, **book gills** (flat, leaflike gills) are exposed. There are two compound and two simple eyes on the carapace. The horseshoe crab swims by means of its abdominal plates and can walk with its walking legs. It feeds at night on worms and small molluscs, which it seizes with its chelicerae.

During the mating season the horseshoe crabs come to shore at high tide to mate (Figure 15-2). The female burrows into the sand where she lays her eggs, with one or more smaller males following her closely to add their sperm to the nest before she covers the eggs with sand. The eggs are warmed by the sun and protected from the waves until the young larvae hatch and return to the sea by another high tide. The larvae are segmented and are often called "trilobite larvae" because they resemble the trilobites, which are often considered to have been their ancestors.

Class Pycnogonida: sea spiders

Some sea spiders are only a few millimeters long, but others are much larger. They have small, thin bodies and usually four pairs of long, thin walking legs. In addition, they have a feature unique among arthropods: somites are reduplicated in some groups, so that they possess five or six pairs of legs instead of the "normal" four pairs. The males of many species bear a subsidiary pair of legs (**ovigers**) (Figure 15-4) on which they carry the developing eggs. The ovigers are often absent in females, and they are absent in males of at least one species. Many species also are equipped with chelicerae and palps.

The mouth is located at the tip of a long suctorial **proboscis** used to suck juices from cnidarians and soft-bodied animals. Most pycnogonids have four simple eyes. The circulatory system is limited to a simple dorsal heart, and excretory and respiratory systems are absent. The long, thin body and legs provide a large surface, in proportion to volume, that is evidently sufficient for diffusion of gases and wastes. Because of the small size of the body, the digestive system sends branches into the legs, and most of the gonads are also found there.

Sea spiders are found in all oceans, but they are most abundant in polar waters. *Pycnogonum* (Figure 15-5) is a common intertidal genus found on both the Atlantic and Pacific coasts of the United States; it has relatively short, heavy legs. *Nymphon* (Figure 15-4) is the largest genus of pycnogonids, with over 200 species, and it occurs from subtidal depths to 6800 m in all seas except the Black and Baltic seas.

Some authorities believe that the pycnogonids are more closely related to the crustaceans than to other arthropods (their larva is rather similar in appearance to

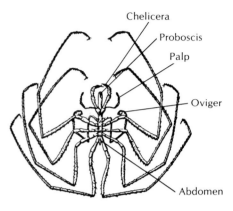

Figure 15-4

Pycnogonid, *Nymphon* sp. In this genus all the anterior appendages (chelicerae, palps, and ovigers) are present in both sexes, although ovigers are often not present in females of other genera.
From Hedgpeth, J.W. 1982. In S.P. Parker (ed.). Synopsis and classification of living organisms. New York, McGraw Hill Book Co.

Figure 15-5

Pycnogonum, a pycnogonid with relatively short legs. Females of this genus have none of the anterior appendages, and only males have ovigers.
From Hedgpeth, J.W. 1982. In S.P. Parker (ed.). Synopsis and classification of living organisms. New York, McGraw-Hill Book Co.

drain the fluid, or "urine," into the intestine. The rectal glands reabsorb most of the potassium and water, leaving behind such wastes as uric acid. By this cycling of water and potassium, species living in dry environments may conserve body fluids, producing a nearly dry mixture of urine and feces. Many spiders also have **coxal glands,** which are modified nephridia that open at the coxa, or base, of the first and third walking legs.

Spiders usually have eight **simple eyes,** each provided with a lens, optic rods, and a retina (Figure 15-7, *A*). They are used chiefly for perception of moving objects, but some, such as those of the hunting and jumping spiders, may form images. Since a spider's vision is usually poor, its awareness of its environment depends a great deal on its hairlike **sensory setae.** Every seta on its surface, whether or not it is actually connected to receptor cells, is useful in communicating some information about the surroundings, air currents, or changing tensions in the spider's web. By sensing the vibrations of its web, the spider can judge the size and activity of its entangled prey or can receive the message tapped out by a prospective mate.

Web-spinning habits

The ability to spin silk is an important factor in the lives of spiders, as it is in some other arachnids. Two or three pairs of spinnerets containing hundreds of microscopic tubes run to special abdominal **silk glands** (Figure 15-7, *B* and *C*). A scleroprotein secretion emitted as a liquid apparently hardens as a result of being pulled from the spinnerets and forms the silk thread. Spiders' silk threads are stronger than steel threads of the same diameter and are said to be second in strength only to fused quartz fibers. The threads will stretch one fifth of their length before breaking.

The web used for trapping insects is the use of silk familiar to most people. The kind of net varies with the species. Some are primitive and consist merely of a few strands of silk radiating out from a spider's burrow or place of retreat. Others

Figure 15-8

Grasshopper, snared and helpless in the web of a golden garden spider *(Argiope aurantia),* is wrapped in silk while still alive. If the spider is not hungry, the prize will be saved for a later meal.
Photograph by J.H. Gerard.

Figure 15-9

Fisher spider, *Dolomedes triton,* feeds on a minnow. This handsome spider feeds mostly on aquatic and terrestrial insects but occasionally captures small fishes and tadpoles. It pulls its paralyzed victim from the water, pumps in digestive enzymes, then sucks out the predigested contents.
Photograph by J.H. Gerard.

spin the beautiful, geometrical orb webs. However, spiders use silk threads for many purposes besides web making. They use them to line their nests; form sperm webs or egg sacs; build draglines; make bridge lines, warning threads, molting threads, attachment discs, or nursery webs; or wrap up their prey securely (Figure 15-8). Not all spiders spin webs for traps. Some, such as the wolf spiders (Figure 15-6), jumping spiders, and fisher spiders (Figure 15-9), simply chase and catch their prey.

Reproduction

Before mating, the male spins a small web, deposits a drop of sperm on it, and then picks the sperm up and stores it in the special cavities of his pedipalps. When he mates, he inserts the pedipalps into the female genital opening to store the sperm in his mate's seminal receptacles. Before mating, there is usually a courtship ritual. The female lays her eggs in a silken net, which she may carry about or may attach to a web or plant. A cocoon may contain hundreds of eggs, which hatch in approximately 2 weeks. The young usually remain in the egg sac for a few weeks and molt once before leaving it. Several molts occur before adulthood.

Are spiders really dangerous?

It is truly amazing that such small and helpless creatures as the spiders have generated so much unreasoning fear in the human heart. Spiders are timid creatures that, rather than being dangerous enemies to humans, are actually allies in the continuing battle with insects. The venom produced to kill the prey is usually harmless to humans. Even the most poisonous spiders bite only when threatened or when defending their eggs or young. The American tarantulas (Figure 15-10), despite their fearsome size, are *not* dangerous. They rarely bite, and their bite is not considered serious.

There are, however, two genera in the United States that can give severe or even fatal bites: *Latrodectus* (L. *latro*, robber, + *dektes*, biter), and *Loxosceles* (Gr. *loxos*, crooked, + *skelos*, leg). The most important species are *Latrodectus mactans*, the **black widow**, and *Loxosceles reclusa*, the **brown recluse**. The black widow is moderate to small in size and shiny black, with a bright orange or red "hourglass" on the underside of the abdomen (Figure 15-11, *A*) The venom is neurotoxic; that is, it acts on the nervous system. About four or five out of each 1000 bites reported have proved fatal.

The brown recluse is smaller than the black widow, is brown, and bears a

Figure 15-10

Tarantula, *Dugesiella hentzi.*
Photograph by J.H. Gerard.

A   **B**

Figure 15-11

A, Black widow spider, *Latrodectus mactans,* suspended on her web. Note the orange "hourglass" on the ventral side of her abdomen. **B,** Brown recluse spider, *Loxosceles reclusa,* is a small venomous spider. Note the small violin-shaped marking on its cephalothorax. The venom is hemolytic and dangerous.
Photograph by J.H. Gerard.

violin-shaped dorsal stripe on its back (Figure 15-11, *B*). Its venom is hemolytic rather than neurotoxic, producing death of the tissues and skin surrounding the bite. Its bite can be mild to serious and occasionally fatal.

Scorpions: order Scorpionida

Although scorpions are more common in tropical and subtropical regions, some also occur in temperate zones. Scorpions are generally secretive, hiding in burrows or under objects by day and feeding at night. They feed largely on insects and spiders, which they seize with the pedipalps and tear up with the chelicerae.

Sand-dwelling scorpions apparently locate their prey by sensing surface waves generated by movements of insects on or in the sand. These waves are picked up by compound slit sensilla located on the basitarsal segments of the legs. The scorpion can locate a burrowing cockroach 50 cm away and reach it in three or four quick orientation movements.

The scorpion tagmata are a rather short **cephalothorax,** which bears the appendages, a pair of large median eyes, and two to five pairs of small lateral eyes; a **preabdomen** of seven segments; and a long slender **postabdomen,** or tail, of five segments, which ends in a stinging apparatus (Figure 15-12, *A*). The chelicerae are small and three jointed; the pedipalps are large, chelate (pincerlike), and six jointed; and the four pairs of walking legs are eight jointed.

On the ventral side of the abdomen are curious comblike **pectines,** which are tactile organs used for exploring the ground and for sex recognition. The stinger on the last segment consists of a bulbous base and a curved barb that injects the venom. The venom of most species is not harmful to humans but may produce a painful swelling. However, the sting of certain species of *Androctonus* in Africa and *Centruroides* (Gr. *kenteō,* to prick, + *oura,* tail, + *oides,* form) in Mexico can be fatal unless antivenin is available.

Scorpions perform a complex mating dance, the male holding the female's chelae and stepping back and forth. He taps her genital area with his forelegs and stings her pedipalp. Finally, he deposits a spermatophore and pulls the female over it until the sperm mass is taken up in the female orifice. Scorpions are either ovoviviparous or truly viviparous; that is, the females brood their young within the female reproductive tract. After several months or a year of development, anywhere from six to 90 young are produced, depending on the species. The young, only a few millimeters long, crawl up on the mother's back until after the first molt. They mature in about a year.

Harvestmen: order Opiliones

Harvestmen, often known as "daddy longlegs," are common in the United States and other parts of the world (Figure 15-12, *B*). These curious creatures are easily distinguished from spiders by the fact that their abdomen and cephalothorax are broadly joined, without the constriction of the pedicel, and their abdomen shows external segmentation. They have four pairs of usually long, spindly legs, and they can cast off one or more of these without apparent ill effect if they are grasped by a predator (or human hand). The ends of their chelicerae are pincerlike, and they feed much more as scavengers than do spiders.

Ticks and mites: order Acari

Members of the order Acari are without doubt the most medically and economically important group of arachnids. They far exceed the other orders in number of individuals and species. Although about 30,000 species have been described, some authorities estimate that from 500,000 to 1 million species exist. Hundreds

Figure 15-12

A, Striped scorpion, *Centruroides vittatus* (order Scorpionida). This species is not considered dangerous to humans. **B,** A harvestman, *Leiobunum* sp. (order Opiliones). Harvestmen run rapidly on their stiltlike legs. They are especially noticeable during the harvesting season, hence the common name.

A, Photograph by J.H. Gerard; **B,** photograph by C.P. Hickman, Jr.

of individuals of several species of mites may be found in a small portion of leaf mold in forests. They are found throughout the world in both terrestrial and aquatic habitats, even extending into such inhospitable regions as deserts, polar areas, and hot springs. Many acarines are parasitic during one or more stages of their life cycle.

Most mites are 1 mm or less in length. Ticks, which make up only one suborder of the Acari, range from a few millimeters to occasionally 3 cm. A tick may become enormously distended with blood after feeding on its host.

Acarines differ from all other arachnids in having complete fusion of the cephalothorax and abdomen, with no sign of external division or segmentation (Figure 15-13). Their mouthparts are carried on a little anterior projection, the **capitulum.** The capitulum is mainly made up of the feeding appendages surrounding the mouth. On each side of the mouth is a chelicera, which functions in piercing, tearing, or gripping food. The form of the chelicerae varies greatly in different families. Lateral to the chelicerae is a pair of segmental pedipalps, which also vary greatly in form and function related to feeding. Ventrally the bases of the pedipalps are fused to form a **hypostome,** whereas a **rostrum,** or **tectum,** extends dorsally over the mouth. Adult mites and ticks usually have four pairs of legs, although there may be only one to three in some specialized forms.

Sperm are transferred directly by most acarines, but many species transfer sperm indirectly by means of a spermatophore. A larva with six legs hatches from the egg, and this is followed by one or more eight-legged nymphal stages before the adult stage is reached.

Many species of mites are entirely free living. Beetle mites (oribatid mites) (Figure 15-14) have a very convex dorsal surface and are abundant in humus and moss. There are some marine mites, but most aquatic species are found in fresh water. They have long, hairlike setae on their legs for swimming, and their larvae may be parasitic on aquatic invertebrates. Such abundant organisms must be important ecologically, but many acarines have more direct effects on our food supply and health. The spider mites (family Tetranychidae) are serious agricultural pests on fruit trees, cotton, clover, and many other plants. They suck out the contents of plant cells, causing a mottled appearance to the leaves, and construct a

A

B

Figure 15-13

A, Wood tick, *Dermacentor variabilis* (order Acari). Larvae, nymphs, and adults are all parasitic but drop off their hosts to molt to the next stage. **B,** Red velvet (harvest) mite, *Trombidium* sp. As with the chigger *(Trombicula),* only the larvae of *Trombidium* are parasitic. Nymphs and adults are free living and feed on insect eggs and small soil invertebrates.
Photographs by J.H. Gerard.

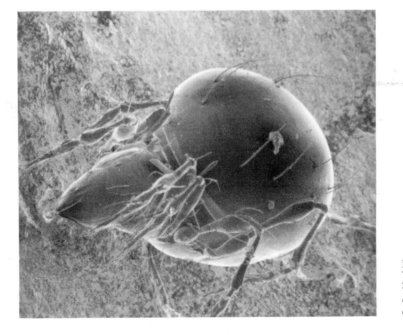

Figure 15-14

Scanning electron micrograph of *Oppia coloradensis,* a beetle mite.
Courtesy T. Woolley.

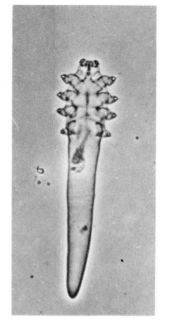

Figure 15-15

Demodex folliculorum, the human follicle mite.
From Desch, C., and W.B. Nutting. 1972. J. Parasitol. 58:169-178.

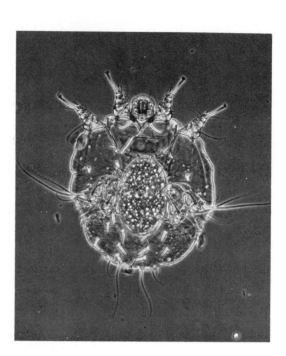

Figure 15-16

Sarcoptes scabiei, the itch mite.
Courtesy J. Georgi.

The inflamed welt and intense itching that follows a chigger bite is not the result of the chigger burrowing into the skin, as is popularly believed. Rather the chigger bites through the skin with its chelicerae and injects a salivary secretion containing powerful enzymes that liquefy skin cells. Human skin responds defensively by forming a hardened tube that the larva uses as a sort of drinking straw and through which it gorges itself with host cells and fluid. Scratching usually removes the chigger but leaves the tube, which is a source of irritation for several days.

protective web from silk glands opening near the base of the chelicerae. The larvae of the genus *Trombicula* are called chiggers or redbugs. They feed on the dermal tissues of terrestrial vertebrates, including humans, and may cause an irritating dermatitis; some species of chiggers transmit a disease called Asiatic scrub typhus. The hair follicle mite, *Demodex* (Figure 15-15), is apparently nonpathogenic in humans; it infects most of us although we are unaware of it. Other species of *Demodex* and other genera of mites cause mange in domestic animals. The human itch mite, *Sarcoptes scabiei* (Figure 15-16), causes intense itching as it burrows beneath the skin. Many species of ticks transmit diseases of humans and domestic animals. Species of *Dermacentor* (Figure 15-13, *A*) and other ticks transmit Rocky Mountain spotted fever, a poorly named disease because most cases occur in the

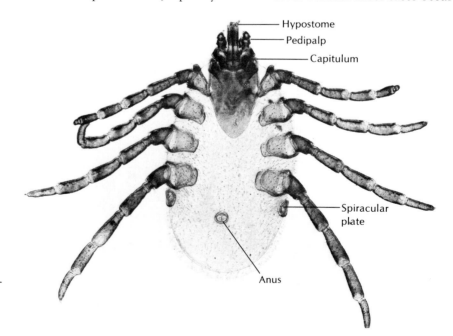

Figure 15-17

Boophilus annulatus, a tick that carries Texas cattle fever.
Courtesy J. Georgi.

eastern United States. *Dermacentor* also transmits tularemia and the agents of several other diseases. Texas cattle fever, also called red-water fever, is caused by a protozoan parasite transmitted by the cattle tick *Boophilus annulatus* (Figure 15-17). Many more examples could be cited.

Phylogeny and Adaptive Radiation

Phylogeny

The similarities between annelids and arthropods give strong support to the hypothesis that both phyla originated from a line of coelomate segmented protostomes, which in time diverged to form a protoannelid line with laterally located parapodia and one or more protoarthropod lines with more ventrally located parapodia. Some authors have contended that the Arthropoda is polyphyletic and that some or all of the present subphyla are derived from different annelid-like ancestors that underwent "arthropodization." The crucial development is the hardening of the annelid cuticle to form an arthropod exoskeleton, and most of the features that distinguish arthropods from annelids (p. 320) result from the stiffened exoskeleton. For example, the vital role of the coelomic compartments as a hydrostatic skeleton was gone; therefore, intersegmental septa were unnecessary, as was a closed circulatory system. Jointed appendages, of course, are required if the external surface is hard, and the body wall muscles of the annelid could be converted and inserted on the considerable inner surfaces of the cuticle for efficient movement of body parts. Compared to annelids, there was a great restriction in permeable surfaces for respiration and excretion. Thus arthropods *could* have evolved more than once. Separate origin of the chelicerates from the other arthropods seems probable, and some investigators feel that the crustaceans and uniramians, and probably the trilobites, were derived separately. The phylum Onychophora also may have come from one of the protoarthropod lines.

The morphological diversity of the Acari has suggested to acarologists that this order arose from at least two separate arachnid ancestors.

Adaptive radiation

Annelids show limited tagmatization and little differentiation of appendages. However, in arthropods the adaptive trend has been toward pronounced tagmatization by differentiation or fusion of somites, giving rise in more advanced groups to such tagmata as head and trunk; head, thorax, and abdomen; or cephalothorax (fused head and thorax) and abdomen. Primitive arthropods tend to have similar appendages, whereas the more advanced forms have appendages specialized for specific functions, or some appendages may be lost entirely.

Much of the amazing diversity in arthropods seems to have developed because of modification and specialization of their cuticular exoskeleton and their jointed appendages, resulting in a wide variety of locomotor and feeding adaptations.

SUMMARY

The Arthropoda is the largest, most successful phylum in the world. They are metameric, coelomate protostomes with well-developed organ systems. Most show marked tagmatization. They are extremely diverse and occur in all habitats capable of supporting life. Perhaps more than any other single factor, the success

W.S. Bristowe (1971) estimated that at certain seasons a Sussex field that had been undisturbed for several years had a population of 2 million spiders to the acre. He concluded that so many could not successfully compete except for the many specialized adaptations they had evolved. These include adaptations to cold and heat, wet and dry conditions, and light and darkness.

Some spiders capture large insects, some only small ones; web-builders snare mostly flying insects, whereas hunters seek those that live on the ground. Some lay eggs in the spring, others in the late summer. Some feed by day, others by night, and some have developed flavors that are distasteful to birds or to certain predatory insects. As it is with the spiders, so has it been with other arthropods; their adaptations are many and diverse and contribute in no small way to their long success.

of the arthropods is accounted for by adaptations made possible by their cuticular exoskeleton. Other important elements in their success are jointed appendages, tracheal respiration, efficient sensory organs, complex behavior, and metamorphosis.

The trilobites were a dominant Paleozoic subphylum, now extinct. Members of the subphylum Chelicerata have no antennae, and their main feeding appendages are chelicerae. In addition, they have a pair of pedipalps (which may be similar to the walking legs) and four pairs of walking legs. The class Merostomata includes the extinct eurypterids and the ancient, although still extant, horseshoe crabs. The class Pycnogonida contains the sea spiders, which are odd little animals with a large suctorial proboscis and vestigial abdomen. The great majority of living chelicerates are in the class Arachnida: spiders (order Araneae), scorpions (order Scorpionida), harvestmen (order Opiliones), ticks and mites (order Acari), and others.

The tagmata of spiders (cephalothorax and abdomen) show no external segmentation and are joined by a waistlike pedicel. Spiders are predaceous, and their chelicerae are provided with poison glands for paralyzing or killing their prey. They breathe by book lungs, tracheae, or both. Spiders can spin silk, which they use for a variety of purposes, including webs for trapping prey in some cases.

Scorpions are distinguished by their large, clawlike pedipalps and their clearly segmented abdomen, which bears a terminal stinging apparatus. Harvestmen have small, ovoid bodies with very long, slender legs. Their abdomen is segmented and broadly joined to the cephalothorax.

The cephalothorax and abdomen of ticks and mites are completely fused, and the mouthparts are borne on the anterior capitulum. They are the most numerous of any arachnids; some are important carriers of disease, and others are serious plant pests.

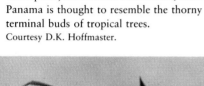

This spider, *Micrathena schreibersci,* from Panama is thought to resemble the thorny terminal buds of tropical trees.
Courtesy D.K. Hoffmaster.

Review questions

1. Give 10 characteristics of arthropods.
2. Name the subphyla of arthropods, and give a few examples of each.
3. Give four characteristics common to annelids and arthropods, and give five ways in which arthropods differ from annelids.
4. Briefly discuss the contribution of the cuticle to the success of arthropods, and name some other factors that have contributed to their success.
5. What is a trilobite?
6. What appendages are characteristic of chelicerates?
7. Briefly describe the appearance of each of the following: eurypterids, horseshoe crabs, pycnogonids.
8. What are the tagmata of arachnids, and which tagma bears all the appendages?
9. Describe the mechanism of each of the following with respect to spiders: feeding, excretion, sensory reception, web-spinning, reproduction.
10. What are the most important spiders in the United States that are dangerous to humans?
11. Distinguish each of the following orders from each other: Araneae, Scorpionida, Opiliones, Acarina.
12. What are some ways Acarina affect humans?
13. Some biologists suggest that the Arthropoda is polyphyletic. Explain why this could be so despite the characteristics shared by all arthropods.

Selected references

See also general references for Part Two, p. 590.

Brownell, P.H. 1984. Prey detection by the sand scorpion. Sci. Am. **251**:86-97 (Dec.). *The scorpion hunts at night, but it cannot see or hear its prey. Very sensitive mechanoreceptors on its legs detect vibrations in the sand.*

Burgess, J.W. 1976. Social spiders. Sci. Am. **234**:100-107 (Mar.). *Gregarious spiders are rare, but some form communal webs for catching prey.*

Foelix, R.F. 1982. Biology of spiders. Cambridge, Mass., Harvard University Press. *Attractive, comprehensive book with extensive references; of interest to both amateurs and professionals.*

Hadley, N.F. 1986. The arthropod cuticle. Sci. Am. **255**:104-112 (July). *Modern studies on the chemistry and structure of arthropod cuticle help to explain its remarkable properties.*

Harwood, R.F., and M.T. James. 1979. Entomology in human and animal health, ed. 7. New York, Macmillan, Inc. *Authoritative and readable account of medical entomology; coverage includes spiders, scorpions, ticks and mites.*

Jackson, R.R. 1985. A web-building jumping spider. Sci. Am. **253**:102-115 (Sept.). *This unusual jumping spider eats mostly other spiders, rather than insects. It often fastens its web to that of another species to invade the alien web and prey on the spider that built it.*

Kaston, B.J. 1978. How to know the spiders, ed. 3. Dubuque, Iowa, William C. Brown Co., Publishers. *Spiral-bound identification manual.*

McDaniel, B. 1979. How to know the ticks and mites. Dubuque, Iowa, William C. Brown Co., Publishers. *Useful, well-illustrated keys to genera and higher categories of ticks and mites in the United States.*

Shear, W.A. (ed.). 1986. Spiders: webs, behavior, and evolution. Stanford, CA, Stanford University Press. *Web building in all its many forms and patterns is detailed.*

CHAPTER 16

THE AQUATIC MANDIBULATES

Phylum Arthropoda

Subphylum Crustacea

Arthropods that possess mandibles (jawlike appendages) are known as mandibulates and have traditionally been united in the subphylum Mandibulata. As noted in the previous chapter, some authors think that the phylum Arthropoda is unnatural and polyphyletic and that arthropodization occurred more than once. In addition, many investigators now believe that there are sufficient differences between the crustaceans and the uniramians (insects, millipedes, centipedes, pauropods, and symphylans) to justify separation at least to subphylum level. Both the Crustacea and the Uniramia have, at least, a pair of **antennae,** a pair of **mandibles,** and a pair of **maxillae** on the head. These appendages perform sensory, masticatory, and food-handling functions, respectively. The body may consist of a head and trunk, but in the more advanced forms, a high degree of tagmosis (p. 318) has occurred so that there is a well-defined head, thorax, and abdomen. In most Crustacea one or more thoracic segments have become fused with the head to form a **cephalothorax.** The thoracic and abdominal appendages are mainly for walking or swimming, but in some groups they are highly specialized in function. The Crustacea are mainly marine; however, there are many freshwater and a few terrestrial species, whereas the uniramians are mainly terrestrial. There are numerous species of insects in freshwater habitats, and a few in marine.

——SUBPHYLUM CRUSTACEA

The subphylum Crustacea (L. *crusta,* shell) gets its name from the hard shells most of them bear. The 30,000 or more species in this class include lobsters, crayfish, shrimp, crabs, water fleas, copepods, barnacles, and some others. The majority are free living and free moving, but many are sessile, commensal, or parasitic. Although crustaceans differ from other arthropods in a variety of ways, the only truly distinguishing characteristic is that crustaceans are the only arthropods with **two pairs of antennae.**

General Nature of a Crustacean

In addition to the two pairs of antennae and a pair of mandibles, crustaceans have two pairs of maxillae on the head, followed by a pair of appendages on each body segment (in some crustaceans not all somites bear appendages). All appendages, except perhaps the first antennae, are primitively **biramous** (two main branches), and at least some of the appendages of present-day adults show that condition. Organs specialized for respiration, if present, are in the form of **gills.**

Most crustaceans have between 16 and 20 segments, but some primitive forms have 60 segments or more. The more advanced crustaceans tend to have fewer segments and increased tagmatization. The major tagmata are the head, thorax, and abdomen, but these are not homologous throughout the class (or even within some subclasses) because of varying degrees of fusion of segments (somites), for example, as in the cephalothorax.

The most advanced and by far the largest group of crustaceans is the class Malacostraca, which includes the lobsters, crabs, shrimps, beach fleas, sow bugs, and many others. These show a surprisingly constant arrangement of body segments and tagmata that is often referred to as the **caridoid facies*** and is considered the ancestral plan of the class (Figure 16-1). This typical body plan has a head of five (six embryonically) fused somites, a thorax of eight somites, and an abdomen of six somites (seven in a few species). At the anterior end is the nonsegmented **rostrum** and at the posterior end is the nonsegmented **telson,** which with the last abdominal somite and its **uropods** forms the tail fan in many forms.

In many crustaceans the dorsal cuticle of the head may extend posteriorly and around the sides of the animal to cover or be fused with some or all of the thoracic and abdominal somites. This covering is called the **carapace.** In some groups the carapace forms clamshell-like valves that cover most or all of the body.

---

*"Caridoid" derives from the scientific name of a group of crustaceans; "facies" means face or general appearance.

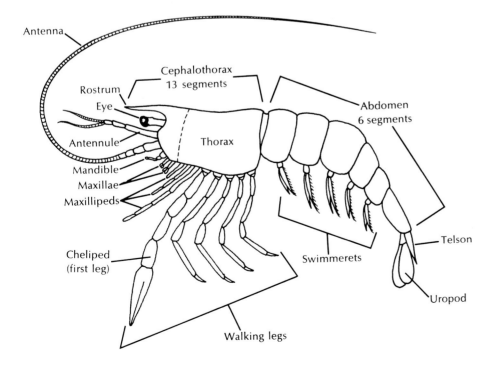

Figure 16-1

Archetypical plan of the Malacostraca. The maxillae and maxillipeds have been separated diagrammatically to illustrate the general plan. Typically in the living animal only the third maxilliped is visible externally.

In the decapods (including lobsters, shrimp, crabs, and others), the carapace covers the entire cephalothorax but not the abdomen.

___ Form and Function

Because of their size and easy availability, large crustaceans such as crayfish have been studied more than other groups. They are also commonly studied in introductory laboratory courses. Therefore many of the comments that follow apply specifically to crayfishes and their relatives.

External features

The bodies of crustaceans are covered with a secreted cuticle composed of chitin, protein, and limy, calcareous material. The harder, heavy plates of the larger crustaceans are particularly high in calcareous deposits. The hard protective cov-

Figure 16-2

Dorsal view of crayfish.
Photograph by F.M. Hickman.

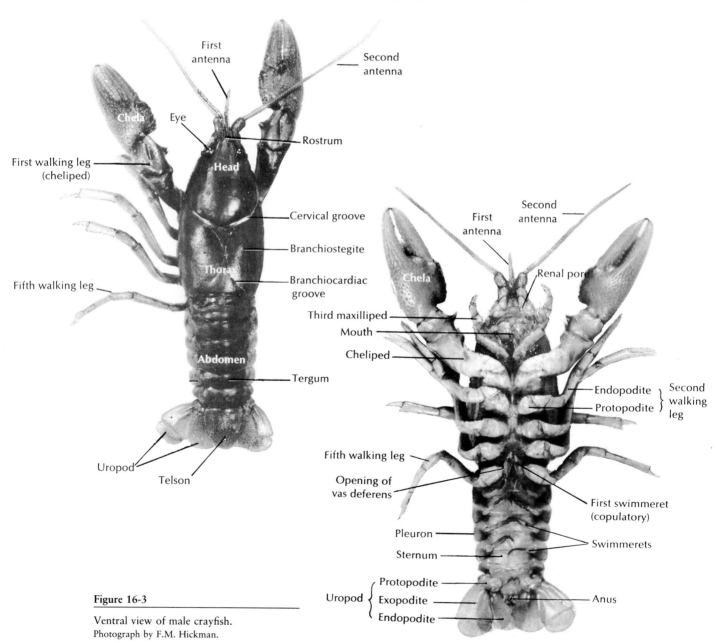

Figure 16-3

Ventral view of male crayfish.
Photograph by F.M. Hickman.

ering is soft and thin at the joints between the somites, allowing flexibility of movement. The carapace, if present, covers much or all of the cephalothorax; in decapods such as crayfish, all of the head and thoracic segments are enclosed dorsally by the carapace. Each somite not enclosed by the carapace is covered by a dorsal cuticular plate, or **tergum** (Figure 16-2), and a ventral transverse bar, the **sternum,** lies between the segmental appendages (Figure 16-3). The abdomen terminates in a telson, which is not considered a somite and bears the anus. (The telson may be homologous to the annelid pygidium.) In several classes the telson bears a pair of processes, forming the **caudal furca.**

The position of the **gonopores** varies according to sex and group of crustaceans. They may be on or at the base of a pair of appendages, at the terminal end of the body, or on somites without legs. In crayfish the openings of the vasa deferentia are on the median side at the base of the fifth pair of walking legs, and those of the oviducts are at the base of the third pair. In the female the opening to the seminal receptacle is located in the midventral line between the fourth and fifth pairs of walking legs.

Appendages

Members of the class Malacostraca (including crayfishes) and the primitive classes typically have a pair of jointed appendages on each somite (Figure 16-3), although the abdominal somites in most other classes do not bear appendages. Considerable specialization is evident in the appendages of the advanced crustaceans such as the crayfishes. However, all are variations of the basic, biramous plan, illustrated by a crayfish appendage such as a maxilliped (a thoracic limb modified to become a head appendage) (Figures 16-4 and 16-5). The basal portion, the **pro-**

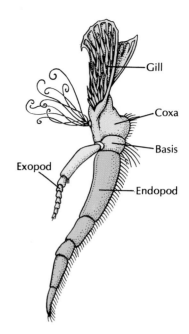

Figure 16-4

Parts of a biramous crustacean appendage (third maxilliped of a crayfish).

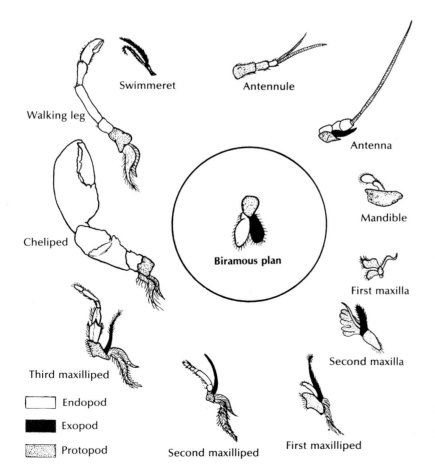

Figure 16-5

Appendages of the crayfish showing how they have become modified from the basic biramous plan, as found in a swimmeret.

Table 16-1 Crayfish appendages

Appendage	Protopod	Endopod	Exopod	Function
First antenna	3 segments, statocyst in base	Many-jointed feeler	Many-jointed feeler	Touch, taste, equilibrium
Second antenna	2 segments, excretory pore in base	Long, many-jointed feeler	Thin, pointed blade	Touch, taste
Mandible	2 segments, heavy jaw and base of palp	2 distal segments of palp	Absent	Crushing food
First maxilla	2 thin medial lamellae	Small unjointed lamella	Absent	Food handling
Second maxilla	2 bilateral lamellae, extra plate, epipod	1 small pointed segment	Dorsal plate, the scaphognathite (bailer)	Drawing currents of water into gills
First maxilliped	2 medial plates and epipod	2 small segments	1 basal segment, plus many-jointed filament	Touch, taste, food handling
Second maxilliped	2 segments plus gill	5 short segments	2 slender segments	Touch, taste, food handling
Third maxilliped	2 segments plus gill	5 larger segments	2 slender segments	Touch, taste, food handling
First walking leg (cheliped)	2 segments plus gill	5 segments with heavy pincer	Absent	Offense and defense
Second walking leg	2 segments plus gill	5 segments plus small pincer	Absent	Walking and prehension
Third walking leg	2 segments plus gill; genital pore in female	5 segments plus small pincer	Absent	Walking and prehension
Fourth walking leg	2 segments plus gill	5 segments, no pincer	Absent	Walking
Fifth walking leg	2 segments; genital pore in male; no gill	5 segments, no pincer	Absent	Walking
First swimmeret	In female reduced or absent; in male fused with endopod to form tube			In male, transferring sperm to female
Second swimmeret Male	Structure modified for transfer of sperm to female	Structure modified for transfer of sperm to female		
Female	2 segments	Jointed filament	Jointed filament	Creating water currents; carrying eggs and young
Third, fourth, and fifth swimmerets	2 short segments	Jointed filament	Jointed filament	Creating water currents; in female carrying eggs and young
Uropod	1 short, broad segment	Flat, oval plate	Flat, oval plate; divided into 2 parts with hinge	Swimming; egg protection in female

topod, bears a lateral **exopod** and a medial **endopod.** The protopod is made up of two joints (**basis** and **coxa**), whereas the exopod and endopod have from one to several segments each. Some appendages, such as the walking legs of crayfishes, have become secondarily uniramous. Medial or lateral processes may be found on crustacean limbs, called **endites** and **exites,** respectively, and an exite on the protopod is called an **epipod.** Epipods are often modified as gills. Table 16-1 shows how the various appendages have become modified from the biramous plan to fit specific functions.

Structures that have a similar basic plan and have descended from a common form are said to be homologous, whether they have the same function or not. Since the specialized walking legs, mouthparts, chelipeds, and swimmerets have all developed from a common biramous type but have become modified to perform different functions, they are all homologous to each other. In primitive forms they were all very similar to one another, a condition known as **serial homology.** During the evolution of this structural modification, some branches have been reduced, some lost, some greatly altered, and some new parts added. The crayfish and their allies possess the best examples of serial homology in the animal kingdom.

Internal features

The muscular and nervous systems and segmentation in the thorax and abdomen clearly show the metamerism inherited from the annelid ancestors, but there are marked modifications in other systems. Most of the changes involve concentration of parts in a particular region or else reduction or complete loss of parts, such as the intersepta.

Hemocoel

The major body space in the arthropods is not a true coelom but a blood-filled **hemocoel.** During the embryonic development of most arthropods, vestigial coelomic cavities open within the mesoderm of at least some somites. These are soon obliterated or may become continuous with the space between the developing mesodermal and ectodermal structures and the yolk. This space becomes the hemocoel and is thus not lined by a mesodermal peritoneum. In the crustaceans the only remaining coelomic compartments are the end sacs of the excretory organs and the space around the gonads.

Muscular system

Striated muscles make up a considerable part of the body of most Crustacea. The muscles are usually arranged in antagonistic groups: **flexors,** which draw a part toward the body, and **extensors,** which straighten it out. The abdomen of a crayfish has powerful flexors (Figure 16-6), which are used when the animal swims backward—its best means of escape. Strong muscles on either side of the stomach control the mandibles.

Respiratory system

Respiratory gas exchange in the smaller crustaceans takes place over thinner areas of the cuticle (for example, in the legs) or over the entire body, and specialized

The terminology applied by various workers to crustacean appendages has not been blessed with uniformity. At least two systems are in wide use. Alternative terms to those we have used, for example, are protopodite, exopodite, endopodite, basipodite, coxopodite, and epipodite. The first and second pairs of antennae may be referred to as the antennules and antennae, and the first and second maxillae are often called maxillules and maxillae. A rose by any other name . . .

Figure 16-6

Internal structure of male crayfish.

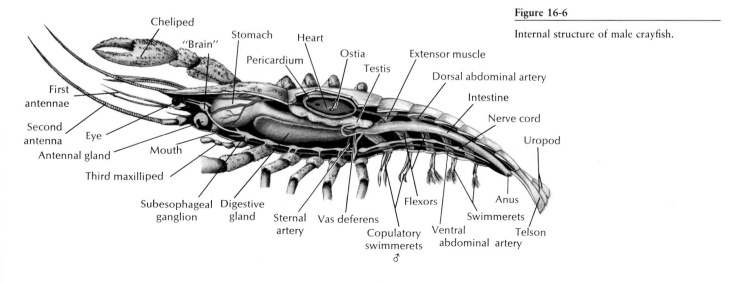

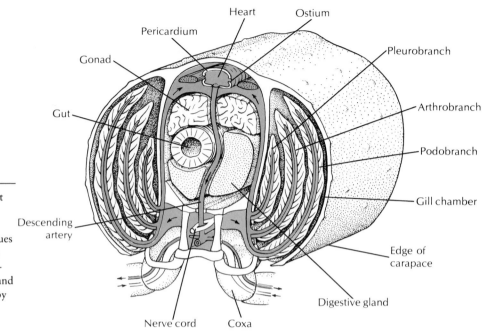

Figure 16-7

Diagrammatic cross section through heart region of crayfish showing direction of blood flow in this "open" blood system. Blood is pumped from heart to body tissues through arteries, which empty into tissue sinuses. Returning blood enters sternal sinus, is carried to gills for gas exchange, and then is carried back to pericardial sinus by branchiocardiac canals. Note absence of veins.

structures may be absent. The larger crustaceans have gills, which are delicate, featherlike projections with very thin cuticle. In the decapods the sides of the carapace enclose the gill cavity, which is open anteriorly and ventrally (Figure 16-7). Gills may project from the pleural wall into the gill cavity, from the articulation of the thoracic legs with the body, or from the thoracic coxopods. The latter two types are typical of crayfish. The "bailer," a part of the second maxilla, draws water over the gill filaments, into the gill cavity at the bases of the legs, and out of the gill cavity at the anterior.

Circulatory system

Crustaceans and other arthropods have an "open" or lacunar type of circulatory system. This means that there are no veins, but that the hemolymph (blood) leaves the heart by way of arteries, circulates through the hemocoel, and returns to venous sinuses, or spaces, instead of veins before it reenters the heart. The annelids have a closed system, as do the vertebrates.

A dorsal heart is the chief propulsive organ. It is a single-chambered sac of striated muscle. Hemolymph enters the heart from the surrounding pericardial sinus through paired ostia, with valves that prevent backflow into the sinus (Figure 16-6). From the heart the hemolymph enters one or more arteries. Valves in the arteries prevent a backflow of hemolymph. Small arteries empty into the tissue sinuses, which in turn often discharge into the large sternal sinus (Figure 16-7).

From there, afferent sinus channels carry hemolymph to the gills, if any, where oxygen and carbon dioxide are exchanged. The hemolymph is then returned to the pericardial sinus by efferent channels (Figure 16-6).

Hemolymph in arthropods is largely colorless. It includes ameboid cells of at least two types. Hemocyanin, a copper-containing respiratory pigment, or hemoglobin, an iron-containing pigment, may be carried in solution. Hemolymph has the property of clotting, which prevents its loss in minor injuries. Some ameboid cells release a thrombinlike coagulant that precipitates the clotting.

Excretory system

The excretory organs of adult crustaceans are a pair of tubular structures located in the ventral part of the head anterior to the esophagus (Figure 16-6). They are called **antennal glands** or **maxillary glands,** depending on whether they open at the base of the antennae or of the second maxillae. A few adult crustaceans have both. The excretory organs of the decapods are antennal glands, also called **green glands** in this group. Crustaceans do not have malpighian tubules.

The **end sac** of the antennal gland, which is derived from an embryonic coelomic compartment, consists of a small vesicle (**saccule**) and a spongy mass called a **labyrinth.** The labyrinth is connected by an **excretory tubule** to a dorsal **bladder,** which opens to the exterior by a pore on the ventral surface of the basal antennal segment (Figure 16-8). Hydrostatic pressure within the hemocoel provides the force for filtration of fluid into the end sac, and as the filtrate passes through the excretory tubule, it is modified by resorption of salts, amino acids, glucose, and some water and is finally excreted as urine.

Excretion of nitrogenous wastes (mostly ammonia) takes place by diffusion across thin areas of cuticle, especially the gills, and the so-called excretory organs function principally to regulate the ionic and osmotic composition of the body fluids. Freshwater crustaceans, such as the crayfish, are constantly threatened with overdilution of the blood by water, which diffuses across the gills and other water-permeable surfaces. The green glands, by forming a dilute, low-salt urine, act as an effective "flood-control" device. Some Na^+ and Cl^- are lost in the urine, however, but this is made up by active absorption of salt from the water by the gills. In marine crustaceans, such as lobsters and crabs, the kidney functions to adjust the salt composition of the hemolymph by selective modification of the salt content of the tubular urine. In these forms the urine remains isosmotic to the blood.

Nervous system

The nervous systems of crustaceans and annelids have much in common, although those of crustaceans have more fusion of ganglia (Figure 16-6). The brain is a pair of **supraesophageal ganglia** that supply nerves to the eyes and the two pairs of antennae. It is joined by connectives to the **subesophageal ganglion,** a fusion of at least five pairs of ganglia that supply nerves to the mouth, appendages, esophagus, and antennal glands. The double ventral nerve cord has a pair of ganglia for each somite and gives off nerves to the appendages, muscles, and other parts. In addition to this central system, there may be a sympathetic nervous system associated with the digestive tract.

Sensory system

Crustaceans have better-developed sense organs than do the annelids. The largest sense organs of the crayfish are the eyes and the statocysts. Tactile organs are widely distributed over the body in the form of **tactile hairs,** delicate projections of the cuticle that are especially abundant on the chelae, mouthparts, and telson. The chemical senses of taste and smell are found in hairs on the antennae, mouthparts, and other places.

A saclike **statocyst,** opening to the surface by a dorsal pore, is found on the basal segment of each first antenna of crayfishes. The statocyst contains a ridge that bears sensory hairs formed from the chitinous lining and grains of sand that serve as **statoliths.** Whenever the animal changes its position, corresponding

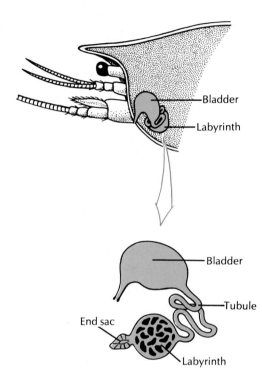

Figure 16-8

Scheme of antennal gland (green gland) of crayfish. (In natural position organ is much folded.) Most selective resorption takes place in the labyrinth, a complicated spongy mass. Some crustaceans lack a labyrinth, and the excretory tubule (nephridial canal) is a much-coiled tube.

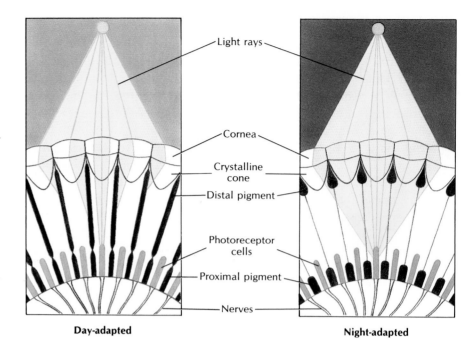

Day-adapted   **Night-adapted**

Figure 16-9

Portion of compound eye of arthropod showing migration of pigment in ommatidia for day and night vision. Five ommatidia represented in each diagram. In daytime each ommatidium is surrounded by a dark pigment collar so that each ommatidium is stimulated only by light rays that enter its own cornea (mosaic vision); in nighttime, pigment forms incomplete collars and light rays can spread to adjacent ommatidia (continuous, or superposition, image).
Redrawn from Moment, G.B. 1967. General zoology. Boston, Houghton Mifflin Co.

changes in the position of the grains on the sensory hairs are relayed as stimuli to the brain, and the animal can adjust itself accordingly. The cuticular lining of the statocyst is shed at each molting (ecdysis), and with it the sand grains are also lost. New grains are picked up through the dorsal pore after ecdysis.

The eyes in many crustaceans are compound, since they are made up of many photoreceptor units called **ommatidia** (Figure 16-9). Covering the rounded surface of each eye is a transparent area of the cuticle, the **cornea,** which is divided into many small squares or hexagons known as facets. These are the outer ends of the ommatidia. Each ommatidium behaves like a tiny eye and is composed of several kinds of cells arranged in a columnar fashion (Figure 16-9). Black pigment cells are found between adjacent ommatidia.

The movement of pigment in the arthropod compound eye permits it to adjust for different amounts of light. In each ommatidium are three sets of pigment cells: distal retinal, proximal retinal, and reflecting; these are so arranged that they can form a more or less complete collar or sleeve around each ommatidium. For strong light or day adaptation the distal retinal pigment moves inward and meets the outward-moving proximal retinal pigment so that a complete pigment sleeve is formed around the ommatidium (Figure 16-9). In this condition only rays that strike the cornea directly will reach the photoreceptor (retinular) cells, for each ommatidium is shielded from the others. Thus, each ommatidium will see only a limited area of the field of vision (a mosaic, or apposition, image). In dim light the distal and proximal pigments separate so that the light rays, with the aid of the reflecting pigment cells, have a chance to spread to adjacent ommatidia and to form a continuous, or superposition, image. This second type of vision is less precise but takes maximum advantage of the limited amount of light received.

Reproduction, life cycles, and endocrine function

Most crustaceans have separate sexes, and there are a variety of specializations for copulation among the different groups. The barnacles are monoecious but generally practice cross-fertilization. In some of the ostracods males are scarce, and reproduction is usually parthenogenetic. Most crustaceans brood their eggs in some manner: branchiopods and barnacles have special brood chambers, the copepods have egg sacs attached to the sides of the abdomen (Figure 16-19), and the malacostracans usually carry eggs and young attached to their appendages.

From the egg of a crayfish hatches a tiny juvenile with the same form as the adult and a complete set of appendages and somites. This type of development is described as direct: the juvenile that hatches from the egg resembles a miniature adult, and there is no larval form. In the majority of crustaceans, however, a larva quite unlike the adult in structure and appearance hatches from the egg, and the development is said to be indirect. Change from larva to an adult is **metamorphosis.** The primitive and most widely occurring larva in the Crustacea is the **nauplius** (Figure 16-10). The nauplius bears only three pairs of appendages: uniramous first antennae, biramous second antennae, and biramous mandibles. They all function as swimming appendages at this stage. Subsequent development may involve a gradual change to the adult body form, and appendages and somites are added through a series of molts, or assumption of the adult form may involve more abrupt changes. For example, the metamorphosis of a barnacle proceeds from a free-swimming nauplius to a larva with a bivalve carapace called a cyprid and finally to the sessile adult with calcareous plates.

Ecdysis

Ecdysis (ek′duh-sis) (Gr. *ekdyein,* to strip off), or molting, is necessary for the body to increase in size because the exoskeleton is nonliving and does not grow as the animal grows. Much of the crustacean's functioning, including its reproduction, behavior, and many metabolic processes, is directly affected by the physiology of the molting cycle.

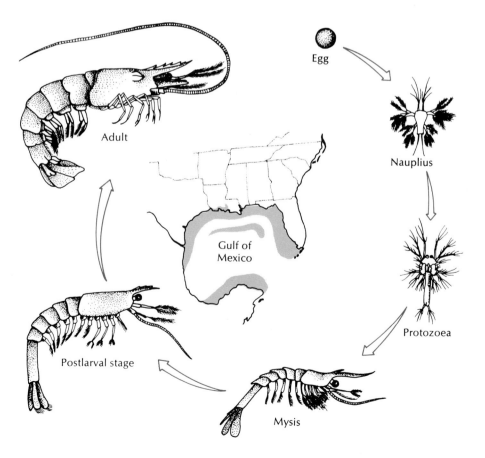

Egg

Nauplius

Adult

Gulf of Mexico

Protozoea

Postlarval stage

Mysis

Figure 16-10

Life cycle and distribution *(in blue)* of the Gulf shrimp *Pennaeus.* Pennaeids spawn at depths of 20 to 50 fathoms. The young larval forms make up part of the plankton fauna. Older shrimp spend their days hidden in the loose deposits on the bottom, coming up at night to feed.

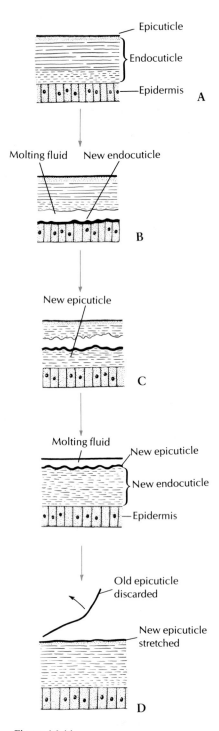

Figure 16-11

Cuticle secretion and resorption in preecdysis. **A,** Intermolt condition. **B,** Old endocuticle separates from epidermis, which secretes new epicuticle. **C,** As new endocuticle is secreted, molting fluid dissolves old endocuticle, and the solution products are resorbed. **D,** At ecdysis, little more than the old epicuticle is left to discard. In postecdysis, new cuticle is stretched and unfolded, and more endocuticle is secreted.
Modified from Schmidt, G.D., and L.S. Roberts. 1981. Foundations of parasitology, ed. 2. St. Louis, The C.V. Mosby Co.

The **cuticle,** which is secreted by an underlying cell layer, the epidermis, is composed of several layers. The outermost is the **epicuticle,** a very thin layer of lipid-impregnated protein. The bulk of the cuticle is made up of the several layers of the **endocuticle:** (1) the **pigmented layer,** which is just beneath the epicuticle and contains protein, calcium salts, and chitin; (2) the **calcified layer,** which contains more chitin and less protein and is heavily mineralized with calcium salts; and (3) the **uncalcified layer,** which is a relatively thin layer of chitin and protein.

Some time before the actual ecdysis, the epidermis separates from the uncalcified layer and secretes a new epicuticle (Figure 16-11). Enzymes are released through canals to the area above the new epicuticle. The enzymes begin to dissolve the old endocuticle, and the soluble products are resorbed and stored within the body of the crustacean. Some of the calcium salts are stored as gastroliths (mineral accretions) in the walls of the stomach. Finally, little more than the epicuticle is left. The animal swallows water, which it absorbs through its gut, and its blood volume increases greatly. The internal pressure causes the cuticle to split, and the animal pulls itself out of its old exoskeleton (Figure 16-12). Then follow a stretching of the still soft cuticle, redeposition of the salvaged inorganic salts and other constituents, hardening of the new cuticle, and deposition of more endocuticular material. During the period of molting, the animal is defenseless and remains hidden away.

When the crustacean is young, ecdysis must occur frequently to allow growth, and the molting cycle is relatively short. As the animal approaches maturity, intermolt periods become progressively longer, and in some species molting ceases altogether. During intermolt periods, increase in tissue mass occurs as water is replaced by living tissue.

Hormonal control of the ecdysis cycle

Although ecdysis is hormonally controlled, the cycle is often initiated by an environmental stimulus perceived by the central nervous system. Such stimuli may include temperature, day length, and humidity (in the case of land crabs). The action of the signal from the central nervous system is to decrease the production of a **molt-inhibiting hormone** by the **X-organ.** The X-organ is a group of neurosecretory cells in the medulla terminalis of the brain. In the crayfish and other decapods, the medulla terminalis is found in the eyestalk. The hormone is carried in the axons of the X-organ to the **sinus gland** (which is probably not glandular in function), also in the eyestalk, where it is released into the hemolymph.

When the level of molt-inhibiting hormone drops, release of a **molting hormone** from the **Y-organs** is promoted. The Y-organs are located beneath the adductor muscles of the mandibles and are homologus to the prothoracic glands of insects, which produce the hormone ecdysone. The action of the molting hormone is to initiate the processes leading to ecdysis (proecdysis). Once initiated, the cycle proceeds automatically without further action of hormones from either the X- or Y-organs.

Other endocrine functions

Not only does removal of eyestalks accelerate molting, it was also found over 100 years ago that crustaceans whose eyestalks have been removed can no longer adjust body coloration to background conditions. About 50 years later it was discovered that the defect was caused not by loss of vision but by loss of hormones in the eyestalks. Body color of crustaceans is largely a result of pigments in special branched cells (chromatophores) in the epidermis. Color change is achieved by concentration of the pigment granules in the center of the cells, which causes a

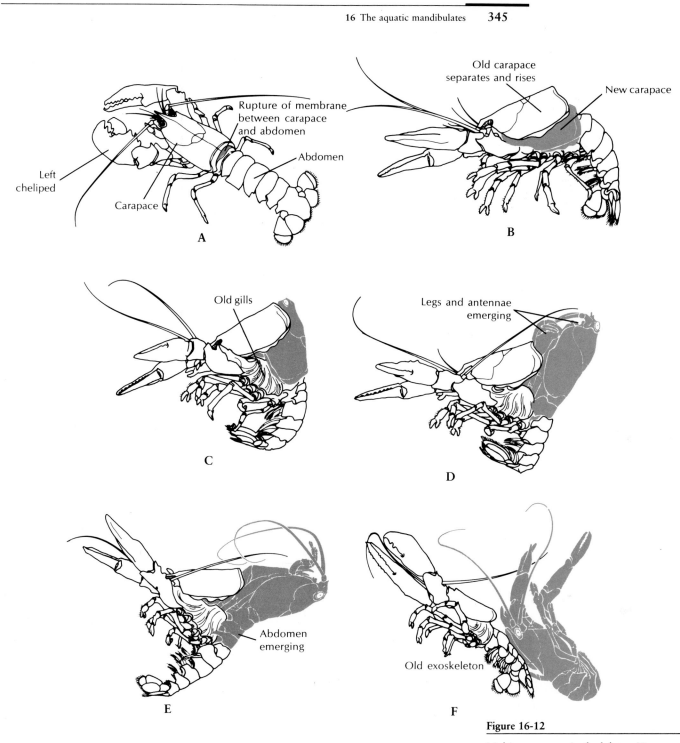

Figure 16-12

Molting sequence in the lobster *Homarus americanus*. **A,** Membrane between carapace and abdomen ruptures, and carapace begins slow elevation. This step may take up to 2 hours. **B** to **E,** Head, thorax, and finally abdomen are withdrawn. This process usually takes no more than 15 minutes. **F,** Immediately after ecdysis, chelipeds are desiccated and body is very soft. Lobster continues rapid absorption of water so that within 12 hours body increases about 20% in length and 50% in weight. Tissue water will be replaced by protein in succeeding weeks.

Drawn from photographs by D.E. Aiken.

lightening effect, or by dispersal of the pigment throughout the cells, which causes a darkening effect. The pigment behavior is controlled by hormones from neurosecretory cells in the eyestalk, as is migration of retinal pigment for light and dark adaptation in the eyes (Figure 16-9).

Release of neurosecretory material from the pericardial organs in the wall of the pericardium causes an increase in the rate and amplitude of the heartbeat.

Androgenic glands, which were first found in an amphipod (*Orchestia,* a common beach flea), occur in male malacostracans. These are not neurosecretory organs. Their secretion stimulates the expression of male sexual characteristics. Young malacostracans have rudimentary androgenic glands, but in females these glands fail to develop. If they are artificially implanted in a female, her ovaries transform to testes and begin to produce sperm, and her appendages begin to take

Neurosecretory cells are nerve cells that are modified for secretion of hormones. They are widespread in invertebrates and also occur in vertebrates. Cells in the vertebrate hypothalamus and in the posterior pituitary are good examples (see p. 717).

on male characteristics at the next molt. In isopods the androgenic glands are found in the testes; in all other malacostracans they are between the muscles of the coxopods of the last thoracic legs and partly attached near the ends of the vasa deferentia. Although females do not possess organs similar to androgenic glands, their ovaries produce one or two hormones that influence secondary sexual characteristics.

Hormones that influence other body processes in Crustacea may be present, and evidence suggests that a neurosecretory substance produced in the eyestalk regulates blood sugar level.

Feeding habits

Feeding habits and adaptations for feeding vary greatly among crustaceans. Many forms can shift from one type of feeding to another depending on environment and food availability, but fundamentally the same set of mouthparts is used by all. The mandibles and maxillae are involved in the actual ingestion; maxillipeds hold and crush food. In predators the walking legs, particularly the chelipeds, serve in food capture.

Many crustaceans, both large and small, are predatory, and some have interesting adaptations for killing their prey. One shrimplike form, *Lygiosquilla,* has on one of its walking legs a specialized digit that can be drawn into a groove and released suddenly to pierce a passing prey. The pistol shrimp *Alpheus* has one enormously enlarged chela that can be cocked like the hammer of a gun and snapped with a force that stuns its prey.

The food of crustaceans ranges from plankton, detritus, and bacteria, used by the **filter feeders,** to larvae, worms, crustaceans, snails, and fishes, used by **predators,** and dead animal and plant matter, used by **scavengers.** Filter feeders, such as the fairy shrimps, water fleas, and barnacles, use their legs, which bear a thick fringe of setae, to create water currents that sweep food particles through the setae. The mud shrimp *Upogebia* uses long setae on its first two pairs of thoracic appendages to strain food material from water circulated through its burrow by movements of its swimmerets.

Crayfishes have a two-part stomach (Figure 16-13). The first part contains a **gastric mill** in which food, already torn up by the mandibles, can be further ground up by three calcareous teeth into particles fine enough to pass through a setose filter in the second part; the food particles then pass into the intestine for chemical digestion.

Figure 16-13

Malacostracan stomach showing gastric "mill" and directions of food movements. Mill is provided with chitinous ridges, or teeth, for mastication, and setae for straining the food before it passes into the pyloric stomach.

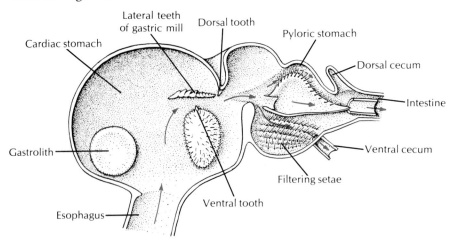

BRIEF RÉSUMÉ OF THE CRUSTACEANS

The crustaceans are an extensive group with many subdivisions. They have many patterns of structure, habitat, and mode of living. Some are much larger than the crayfish; others are smaller, even microscopic. Some are highly developed and specialized; others have simpler organization.

The reader should realize that the following summary of crustacean groups is misleadingly brief. Although all classes are mentioned, a complete presentation would require coverage of a surprisingly large number of taxa in the hierarchy below the class level. Among the Malacostraca, for example, some unusual taxa such as "infraorder" and "section" are used.

Class Cephalocarida

Cephalocarida (sef'a-lo-kar'i-da) (Gr. *kephalē*, head, + *karis*, shrimp, + *ida*, pl. suffix) (Figure 16-14) is a small group, with only four species described so far, that is found along both coasts of the United States, in the West Indies, and in Japan. They are 2 to 3 mm long and have been found in bottom sediments from the intertidal zone to a depth of 300 m. They are even more primitive than the oldest known fossil crustacean, dating from the Devonian period. For example, the thoracic limbs are very similar to each other, and the second maxillae are similar to the thoracic limbs. The second maxillae and the first seven thoracic legs may be considered triramous, bearing a pseudepipod in addition to the exopod and endopod. Cephalocarids do not have eyes, a carapace, or abdominal appendages. True hermaphrodites, they are unique among the Arthropoda in discharging both eggs and sperm through a common duct.

Class Remipedia

Remipedia (ri-mi-pee'dee-a) (L. *remipedes*, oar-footed) (Figure 16-15) is another very small group, described even more recently than the Cephalocarida. The single species was found in a marine cave in the Bahamas, and like the cephalocarids, it is very primitive. It has 31 to 32 trunk segments (thorax and abdomen), all bearing paired, biramous, swimming appendages that are all essentially alike. It has biramous antennules. Both pairs of maxillae and a pair of maxillipeds, however, are prehensile and apparently adapted for feeding. The shape of the swimming appendages is similar to that found in the Copepoda, but unlike copepods and cephalocarids, the swimming legs are directed laterally rather than ventrally.

Class Branchiopoda

Branchiopoda (bran'kee-op'o-da) (Gr. *branchia*, gills, + *pous, podos*, foot) also represents a primitive crustacean type. Four orders are recognized: **Anostraca** (fairy shrimp and brine shrimp, Figure 16-16, *B*), which lack a carapace; **Notostraca** (tadpole shrimp such as *Triops*, Figure 16-16, *A*), whose carapace forms a large dorsal shield covering most of the trunk somites; **Conchostraca** (clam shrimp such as *Lynceus*), whose carapace is bivalve and usually encloses the entire body; and **Cladocera** (water fleas such as *Daphnia*, Figure 16-16, *C*), with a carapace typically covering the entire body but not the head. Branchiopods have reduced first antennae and second maxillae. Their legs are flattened and leaflike (**phyllopodia**) and are the chief respiratory organs (hence the name branchiopods). The

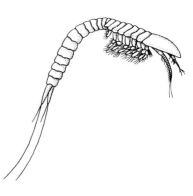

Figure 16-14

Cephalocarida.

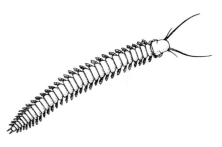

Figure 16-15

Remipedia.

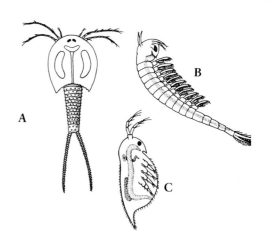

Figure 16-16

Branchiopoda. **A,** Tadpole shrimp, order Notostraca. **B,** Fairy shrimp, order Anostraca. **C,** *Daphnia,* order Cladocera.

legs also are used in filter feeding in most branchiopods, and in groups other than the cladocerans, they are used for locomotion as well.

Most branchiopods are freshwater forms. The most important and successful order is the Cladocera, which often forms a large segment of the freshwater zooplankton. Their reproduction is very interesting and is reminiscent of that occurring in some rotifers (Chapter 12). During the summer they often produce only females, by parthenogenesis, rapidly increasing the population. With the onset of unfavorable conditions, some males are produced, and eggs that must be fertilized are produced by normal meiosis. The fertilized eggs are highly resistant to cold and desiccation, and they are very important for survival of the species over the winter and for passive transfer to new habitats. Cladocera have mostly direct development, whereas other branchiopods have gradual metamorphosis.

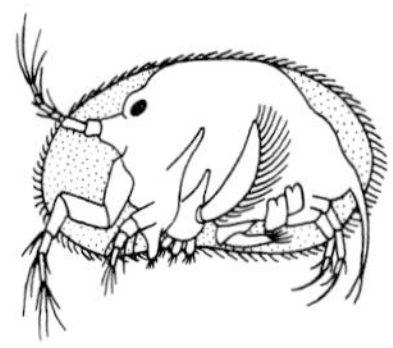

Class Ostracoda

Members of Ostracoda (os-trak′o-da) (Gr. *ostrakodes*, testaceous, that is, having a shell) are, like the conchostracans, enclosed in a bivalve carapace and resemble tiny clams, 0.25 to 8 mm long (Figure 16-17). Ostracods show considerable fusion of trunk somites, and numbers of thoracic appendages are reduced to two or none. Feeding and locomotion are principally by use of the head appendages. Most ostracods live on the bottom or climb on plants, but some are planktonic, burrowing, or parasitic. Feeding habits are diverse; there are particle, plant, and carrion feeders and predators. They are widespread in both marine and freshwater habitats. Development is gradual metamorphosis.

Class Mystacocarida

The Mystacocarida (mis-tak′o-kar′i-da) (Gr. *mystax*, mustache, + *karis*, shrimp, + *ida*, pl. suffix) is a class of tiny crustaceans (less than 0.5 mm long) that live in the interstitial water between sand grains of marine beaches (psammolittoral habitat) (Figure 16-18). Only 10 species have been described, but mystacocarids are widely distributed through many parts of the world. They are primitive in several characteristics and are believed to be related to the Copepoda.

Class Copepoda

The Copepoda (ko-pep′o-da) (Gr. *kōpē*, oar, + *pous, podos*, foot) is an important class of Crustacea, second only to the Malacostraca in numbers of species. The copepods are small (usually a few millimeters or less in length) and rather elongate, tapering toward the posterior. They lack a carapace and retain the simple, median, nauplius eye in the adult (Figure 16-19). They have a single pair of uniramous maxillipeds and four pairs of rather flattened, biramous, thoracic swimming appendages. The fifth pair of legs is reduced. The posterior part of the body is usually separated from the anterior, appendage-bearing portion by a major articulation. The first antennae are often longer than the other appendages. The Copepoda have been very successful and evolutionarily enterprising, with large numbers of symbiotic as well as free-living species. Many of the parasites are highly modified, and the adults may be so highly modified (and may depart so far from the description just given) that they can hardly be recognized as arthropods.

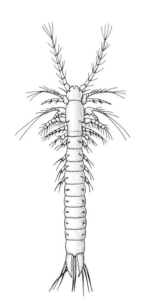

Figure 16-17

Ostracoda.

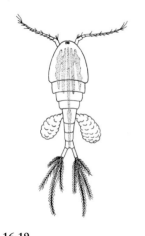

Figure 16-18

Mystacocarida.

Figure 16-19

Copepoda.

Ecologically, the free-living copepods are of extreme importance, often dominating the primary consumer level (p. 904) in aquatic communities. In many marine localities the copepod *Calanus* is the most numerous organism in the zooplankton and has the greatest proportion of the total biomass (p. 902). In other localities it may be surpassed in the biomass only by euphausids (p. 352). *Calanus* forms a major portion of the diet of such economically and ecologically important fish as herring, menhaden, sardines, and the larvae of larger fish and (along with euphausids) is an important food item for whales and sharks. Other genera commonly occur in the marine zooplankton, and some forms such as *Cyclops* and *Diaptomus* may form an important segment of the freshwater plankton. Many species of copepods are parasites of a wide variety of other marine invertebrates and marine and freshwater fish, and some of the latter are of economic importance. Some species of free-living copepods serve as intermediate hosts of parasites of humans, such as *Diphyllobothrium* (a tapeworm) and *Dracunculus* (a nematode), and of other animals.

Development in the copepods is indirect, and some of the highly modified parasites show striking metamorphoses.

Class Tantulocarida

The Tantulocarida (tan'tu-lo-kar'i-da) (L. *tantulus,* so small, + *caris,* shrimp) (Figure 16-20) is the most recently described class of crustaceans (1983). Only four species are known so far. They are tiny (0.15 to 0.2 mm) copepod-like ectoparasites of other deepsea benthic crustaceans. They have no recognizable head appendages and penetrate the cuticle of their hosts by a mouth tube. Their abdomen and all thoracic limbs are lost during metamorphosis to the adult.

Class Branchiura

Branchiura (brank-i-ur'a) (Gr. *branchia,* gills, + *ura,* tail) is a small group of primarily fish parasites, which, despite its name, has no gills (Figure 16-21). Members of this group are usually between 5 and 10 mm long and may be found on marine or freshwater fish. They typically have a broad, shieldlike carapace, compound eyes, four biramous thoracic appendages for swimming, and a short, unsegmented abdomen. The second maxillae have become modified as suction cups, enabling the parasites to move about on their fish host or even from fish to fish. Development is almost direct: there is no nauplius, and the young resemble the adults except in size and degree of development of the appendages.

Class Cirripedia

The Cirripedia (sir-ri-ped'i-a) (L. *cirrus,* curl of hair, + *pes, pedis,* foot) includes the barnacles (order Thoracica), which are usually enclosed in a shell of calcareous plates, as well as three smaller orders of burrowing or parasitic forms. Barnacles are sessile as adults and may be attached to the substrate by a stalk (goose barnacles) (Figure 16-22, *B*) or directly (acorn barnacles) (Figure 16-22, *A*). Typically the carapace (mantle) surrounds the body and secretes a shell of calcareous plates. The head is reduced, the abdomen is absent, and the thoracic legs are long, many-jointed cirri with hairlike setae. The cirri are extended through an opening between the calcareous plates to filter from the water the small particles on which the animal feeds (Figure 16-22). Although all barnacles are marine, they are often

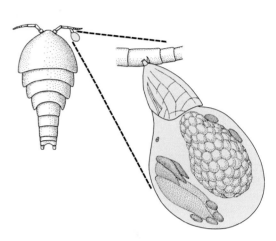

Figure 16-20

Tantulocarida. This curious little parasite is shown attached to the first antenna of its copepod host at left.
Modified from Boxshall and Lincoln. 1983. J. Crust. Biol. 3:1-16.

Figure 16-21

Fish louse (Branchiura).

Several authorities in recent years have adopted the concept of a class Maxillopoda to contain the Mystacocarida, Copepoda, Branchiura, and Cirripedia (see Bowman and Abele, 1982). Others strongly believe that this class should not be used (see Boxshall and Lincoln, 1983). We will retain the conventional separation of these groups, but point out this example of the continuing controversy over the higher taxa in Crustacea.

Figure 16-22

A, Giant barnacles, *Balanus nubilis* (class Cirripedia), of the Pacific coast may be 7.5 cm (3 inches) high and are the largest barnacles in the world. **B,** Common gooseneck barnacles, *Lepas anatifera.* Note the feeding legs, or cirri, on both *Lepas* and *Balanus.* Barnacles attach themselves to a variety of firm substrates, including rocks, pilings, and boat bottoms. A boat's speed may be decreased by as much as 35% because of increased friction caused by barnacles.

A, Photograph by R. Harbo; **B,** photograph by K. Sandved.

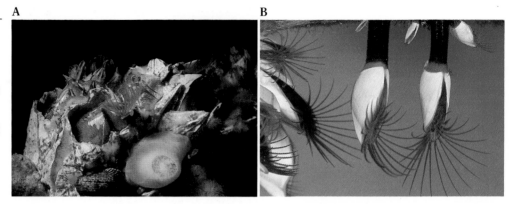

A   B

found in the intertidal zone and are therefore exposed to drying and sometimes fresh water for some periods of time. During these periods the aperture between the plates can be closed to a very narrow slit.

Barnacles are hermaphroditic and undergo a striking metamorphosis during development. Most hatch as nauplii, which soon become cyprid larvae, so called because of their resemblance to the ostracod genus *Cypris.* They have a bivalve carapace and compound eyes. The cyprid attaches to the substrate by means of its first antennae, which have adhesive glands associated with them, and begins its metamorphosis. This involves several dramatic changes, including secretion of the calcareous plates, loss of the eyes, and transformation of the swimming appendages to cirri.

Barnacles frequently foul ship bottoms by settling and growing there. So great may be their number that the speed of the ship may be reduced 30% to 40%, necessitating drydocking the ship to clean them off.

Figure 16-23

Life cycle of *Sacculina,* a rhizocephalan (class Cirripedia) parasite of crabs *(Carcinus).*

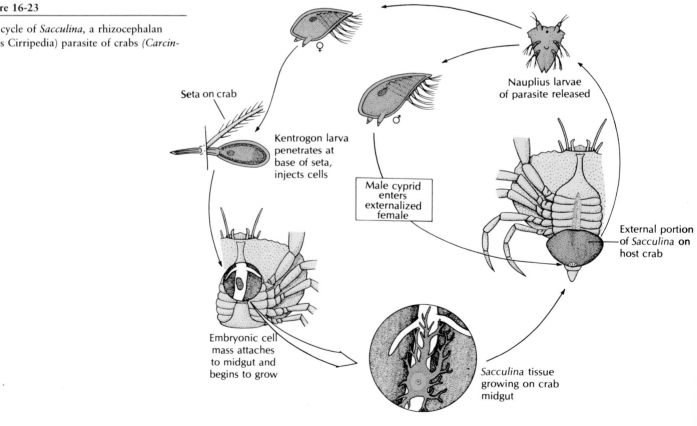

Seta on crab

Kentrogon larva penetrates at base of seta, injects cells

Male cyprid enters externalized female

Nauplius larvae of parasite released

External portion of *Sacculina* on host crab

Embryonic cell mass attaches to midgut and begins to grow

Sacculina tissue growing on crab midgut

Members of the cirripede order Rhizocephala, such as *Sacculina,* are highly modified parasites of crabs. They start life as cyprid larvae, just as other cirripedes, but when they find a host, most species metamorphose into a **kentrogon,** which injects cells of the parasite into the hemocoel of the crab (Figure 16-23). Eventually, rootlike absorptive processes grow throughout the crab's body, and the parasite's reproductive structures become externalized between the cephalothorax and the reflexed abdomen of the crab.

Class Malacostraca

The Malacostraca (mal-a-kos'tra-ka) (Gr. *malakos,* soft, + *ostrakon,* shell) is the largest class of Crustacea and shows great diversity. The diversity is indicated by the higher classification of the group, which includes three subclasses, 14 orders, and many suborders, infraorders, and superfamilies. We confine our coverage to mentioning a few of the most important orders. The characteristic caridoid facies of the malacostracans is described on p. 335.

Order Isopoda

The Isopoda (i-sop'o-da) (Gr. *isos,* equal, + *pous, podos,* foot) is one of the few crustacean groups to have successfully invaded terrestrial habitats in addition to freshwater and seawater habitats.

Isopods are commonly dorsoventrally flattened, lack a carapace, and have sessile compound eyes; their first pair of thoracic limbs are maxillipeds. The remaining thoracic limbs lack exopods and are similar, while the abdominal appendages bear the gills and, except the uropods, also are similar to each other (hence the name isopods).

Common land forms are the sow bugs, or pill bugs *(Porcellio* and *Armadillidium,* Figure 16-24, *A),* which live under stones and in damp places. Although they are terrestrial, they do not have the efficient cuticular covering and other adaptations possessed by insects to conserve water; therefore, they must live in moist conditions. *Asellus* (Figure 16-24, *B)* is a common freshwater form found

The exact position at which the reproductive structures become externalized from the crab's body is of great adaptive value for the rhizocephalan parasite. Because the crab's egg mass (if it had one) would be borne in this position, the crab treats the parasite as if it were a mass of eggs. It protects, ventilates, and grooms the parasite and actually assists in the parasite reproduction by performing spawning behavior at the appropriate time. It has been shown experimentally that the crab's grooming is necessary to the continued good health of the parasite. But what if the rhizocephalan's larva is so unlucky as to infect a male crab? No problem. During the parasite's internal growth in the male crab, the crab is castrated, and it becomes structurally and behaviorally like a female!

A   B

Figure 16-24

A, Four pill bugs, *Armadillidium vulgare* (order Isopoda), common terrestrial forms. **B,** Freshwater sow bug, *Asellus* sp., an aquatic isopod.

A, Photograph by C.P. Hickman, Jr.;

A

B

C

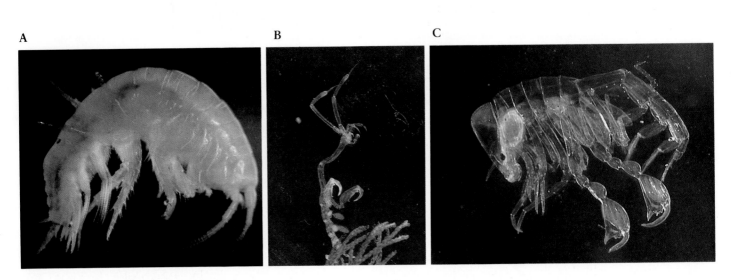

Figure 16-25

Marine amphipods. **A,** Free-swimming amphipod, *Anisogammarus* sp. **B,** Skeleton shrimp, *Caprella* sp., shown on a bryozoan colony, resemble praying mantids. **C,** *Phronima,* a marine pelagic amphipod, takes over the tunic of a salp (subphylum Urochordata, Chapter 22). Swimming by means of its abdominal swimmerets, which protrude from the opening of the barrel-shaped tunic, the amphipod maneuvers to catch its prey. The tunic is not seen in the photograph.
A, Photograph by R. Harbo; **B** and **C,** photographs by K. Sandved.

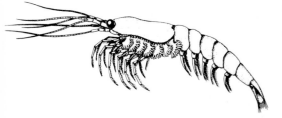

Figure 16-26

Meganyctiphanes (order Euphausiacea) "northern krill."

under rocks and among aquatic plants. *Ligia* is a common marine form that scurries about on the beach or rocky shore. Some isopods are parasites of fish or crustaceans and may be highly modified.

Development is essentially direct but may be highly metamorphic in the specialized parasites.

Order Amphipoda

Amphipoda (am-fip′o-da) (Gr. *amphis,* on both sides, + *pous, podos,* foot) resembles isopods in that the members lack a carapace and have sessile compound eyes and one pair of maxillipeds (Figure 16-25). However, they are usually compressed laterally, and their gills are in the typical thoracic position. Furthermore, their thoracic and abdominal limbs are each arranged in two or more groups that differ in form and function. For example, one group of abdominal legs may be for swimming and another for jumping. There are many marine amphipods, including some beach-dwelling forms (for example, *Orchestia,* one of the beach fleas), numerous freshwater species of the genera *Hyalella* and *Gammarus,* and a few parasites. Development is direct.

Order Euphausiacea

The Euphausiacea (u-faws-i-a′si-a) (Gr. *eu,* well, + *phausi,* shining bright, + L. *acea,* suffix, pertaining to) is a group of only about 90 species, but they are important as the oceanic plankton known as "krill." They are about 3 to 6 cm long, have a carapace that is fused with all the thoracic segments but does not entirely enclose the gills, have no maxillipeds, and have all thoracic limbs with exopods (Figure 16-26). Most are bioluminescent, with a light-producing substance in an organ called a **photophore.** Some species may occur in enormous swarms, up to 45 m² and extending 100 to 500 m in one direction. They form a major portion of the diet of baleen whales and many fishes. Eggs hatch as nauplii, and development is indirect.

Order Decapoda

Decapoda (de-cap′o-da) (Gr. *deka,* ten, + *pous, podos,* foot) have three pairs of maxillipeds and five pairs of walking legs, of which the first is modified to form pincers (chelae). They range in size from a few millimeters to the largest of all arthropods, the Japanese spider crab, whose chelae span 4 m. The crayfishes, lobsters, crabs, and "true" shrimp belong in this group (Figures 16-27 and 16-28). There are about 10,000 species of decapods, and the order is extremely diverse. They are very important ecologically and economically, and numerous species are items of food for humans.

The crabs, especially, exist in a great variety of forms. Although resembling the pattern of crayfish, they differ from the latter in having a broader cephalothorax and a reduced abdomen. Familiar examples along the seashore are the hermit crabs (Figure 16-27, *B*), which live in snail shells because their abdomens are not protected by the same heavy exoskeleton as the anterior parts are; the fiddler crabs, *Uca* (Figure 16-27, *C*), which burrow in the sand just below the high-tide level and come out to run about over the sand while the tide is out; and the spider crabs such as *Libinia* and the interesting decorator crabs *Oregonia* and others, which cover their carapaces with sponges and sea anemones for protective camouflage (Figure 16-28, *A*).

A

B

C

D

E

Figure 16-27

Decapod crustaceans. **A,** The bright orange tropical rock crab, *Grapsus grapsus,* is a conspicuous exception to the rule that most crabs bear cryptic coloration. **B,** The hermit crab, *Dardanus venosus,* which has a soft abdominal exoskeleton, lives in a snail shell that it carries about and into which it can withdraw for protection. **C,** The male fiddler crab, *Uca* sp., uses its enlarged cheliped to wave territorial displays and in threat and combat. **D,** The massive chelipeds of box crabs, *Calappa ocellata,* are well adapted for crushing the shells of their snail prey. Here the crab is "exhaling" water from its branchial chamber. **E,** Northern lobster *Homarus americanus.*
A and C, Photographs by C.P. Hickman, Jr.; **B** and **D,** photographs by K. Sandved; E, photograph by J.H. Gerard.

Figure 16-28

A, Decorator crab, *Oregonia gracilis,* covered with orange sponge. This species is one of several spider crab species that deliberately mask themselves with material from their immediate environment. This crab is so thoroughly masked with pieces of the sponge on which it sits that little of its carapace is visible. **B,** Painted shrimp, *Hymenocera picta. Hymenocera* is known to feed on sea stars (phylum Echinodermata, p. 415). **A,** Photograph by R. Harbo, **B,** photograph by K. Sandved.

PHYLOGENY AND ADAPTIVE RADIATION

The relationship of the crustaceans to other arthropods has long been a puzzle. According to a widely held hypothesis, the trilobites are the ancestors of the crustaceans, but some zoologists have proposed that both these groups evolved independently from different nonarthropod ancestors.

Some light was shed on the problem when the primitive crustacean *Hutchinsoniella macracantha* was discovered in 1954 in Long Island Sound. This form was assigned to a new subclass (now a class) of its own (Cephalocarida) because it has several very primitive characteristics (p. 347). Except for the eight thoracic legs, which lack the innermost branch, the thoracic limbs are all essentially three branched, or triramous. Some authorities see similarities between these appendages and those of the trilobites, and the many specialized limbs of other crustaceans could have been derived from the triramous cephalocarid appendages. The position of the Remipedia in crustacean phylogeny is yet to be established with confidence. The differentiation of both pairs of maxillae and a pair of maxillipeds for food handling is a more advanced character than is shown by cephalocarids. Remipedians show several similarities to fossil crustaceans from the Devonian and the Pennsylvanian periods.

The adaptive radiation demonstrated by the crustaceans is great, with all manner of aquatic niches exploited. They are unquestionably the dominant arthropod group in the marine environment, and they share dominance of freshwater habitats with insects. Invasion of terrestrial environments has been much more limited, with isopods being the only notable success. There are a few other terrestrial examples, such as land crabs. The most important class, in both numbers and diversity, is the Malacostraca, followed by the Copepoda. Both groups include planktonic filter feeders and numerous scavengers. Copepods have been particularly successful as parasites of both vertebrates and invertebrates, and it is clear that the present parasitic copepods are the products of numerous invasions of such niches.

—— SUMMARY

In addition to a pair of mandibles, the Crustacea and the Uniramia have a pair of antennae, a pair of maxillae, and often other head appendages. Their tagmata are a head and trunk or a head, thorax, and abdomen.

The Crustacea is a large, primarily aquatic subphylum. Crustaceans have two pairs of antennae, their appendages are primitively biramous, and many have a carapace.

The two branches of the crustacean leg are the exopod and the endopod, which are both attached to the basal protopod. Appendages are variously specialized, such as chelipeds, maxillipeds, swimmerets, and uropods.

All arthropods must periodically cast off their cuticle (ecdysis) and grow in dimensional size before the newly secreted cuticle hardens. Premolt and postmolt periods are hormonally controlled, as are several other processes, such as change in body color and expression of sexual characteristics.

Feeding habits vary greatly in Crustacea, and there are many predators, scavengers, filter feeders, and parasites. Respiration is through the body surface or by gills, and excretory organs take the form of maxillary or antennal glands. Circulation, as in other arthropods, is through an open system of sinuses (hemocoel), and a dorsal, tubular heart is the chief pumping organ. Most crustaceans have compound eyes composed of units called ommatidia. Sexes are usually separate.

The crustacean class Branchiopoda is characterized by the possession of phyllopodia and contains, among others, the order Cladocera, which is ecologically important as zooplankton. Members of the class Copepoda lack a carapace and abdominal appendages. They are abundant and are among the most important of the primary consumers in many freshwater and marine ecosystems. Most members of the class Cirripedia (barnacles) are sessile as adults, secrete a shell of calcareous plates, and filter feed by means of their thoracic appendages.

The Malacostraca is the largest crustacean class, and the most important orders are the Isopoda, Amphipoda, Euphausiacea, and Decapoda. All have both abdominal and thoracic appendages. Isopods lack a carapace and are usually dorsoventrally flattened. Amphipods also lack a carapace but are usually laterally flattened. Euphausiaceans are important oceanic plankton called krill. Decapods include crabs, shrimp, lobster, crayfish, and others; they have five pairs of walking legs (including the chelipeds) on their thorax.

Review questions

1. What are the tagmata and the appendages on the head of crustaceans? What are some other important characteristics of Crustacea?
2. Define each of the following: tergum, sternum, caudal furca, telson, protopod, exopod, endopod, epipod, endite.
3. What is meant by homologous structures? How do crustaceans show serial homology?
4. Distinguish a hemocoel from a coelom.
5. Briefly describe respiration and circulation in the crayfish.
6. Briefly describe the function of antennal and maxillary glands in Crustacea.
7. How does a crayfish detect changes in position?
8. What is the photoreceptor unit of a compound eye? How does this unit adjust to varying amounts of light?

9. What is a nauplius? What is the difference between direct and indirect development in the Crustacea?
10. Describe the molting process in Crustacea, including the action of the hormones.
11. Of the classes of Crustacea, the Branchiopoda, Ostracoda, Copepoda, Cirripedia, and Malacostraca are the most important. Distinguish them from each other.
12. Compare and contrast the Isopoda, Amphipoda, Euphausiacea, and Decapoda.
13. What is the significance of the Cephalocarida to hypotheses concerning the origin of crustaceans?

Selected references

See also general references for Part Two, p. 590.

Bliss, D.E. (ed. in chief). 1982-1985. The biology of Crustacea, vols. 1-9. New York, Academic Press, Inc. *Nine volumes of this series, planned as a 10-volume comprehensive publication, have been published. This series is a standard reference for all aspects of crustacean biology.*

Bowman, T.E., and L.D. Abele. 1982. Classification of the recent Crustacea. In L.G. Abele (ed.). The biology of Crustacea, vol. 1. Systematics, the fossil record, and biogeography. New York, Academic Press, Inc. *This classification is being followed by many authors. It recognizes the following classes: Cephalocarida, Branchiopoda, Remipedia, Maxillopoda (containing subclasses Mystacocarida, Cirripedia, Copepoda, and Branchiura), Ostracoda, and Malacostraca.*

Boxshall, G.A., and R.J. Lincoln. 1983. Tantulocarida, a new class of Crustacea ectoparasitic on other crustaceans. J. Crust. Biol. 3:1-16.

Cameron, J.N. 1985. Molting in the blue crab. Sci. Am. 252:102-109 (May). *The life cycle and development of the commercially valuable blue crab Callinectes sapidus is described. Studies on the chemistry of the molting process may have important economic benefits.*

Ritchie, L.E., and J.T. Høeg. 1981. The life history of *Lernaeodiscus porcellanae* (Cirripedia: Rhizocephala) and co-evolution with its porcellanid host. J. Crust. Biol. 1:334-347. *The life history and description of the maternal care given the parasite by its host. A fascinating story.*

Schmitt, W.L. 1965. Crustaceans. Ann Arbor, The University of Michigan Press. *A good little book by one of the most eminent students of the group; very interesting reading and easy to understand.*

Yager, J. 1981. Remipedia, a new class of Crustacea from a marine cave in the Bahamas. J. Crust. Biol. 1:328-333.

CHAPTER 17

THE TERRESTRIAL MANDIBULATES

Uniramians

Phylum Arthropoda

Classes Chilopoda, Diplopoda, Pauropoda, Symphyla, and Insecta

Along with the arachnids, the members of the subphylum Uniramia are primarily terrestrial arthropods. Only a few of them have returned to aquatic life, usually in fresh water.

The term "myriapod," meaning "many footed," is commonly used for a group of several classes of uniramians that have evolved a pattern of two tagmata—head and trunk—with paired appendages on most or all trunk somites. The myriapods include the Chilopoda (centipedes), Diplopoda (millipedes), Pauropoda (pauropods), and Symphyla (symphylans).

The insects have evolved a pattern of three tagmata—head, thorax, and abdomen—with appendages on the head and thorax but greatly reduced on or absent from the abdomen. Insects may have arisen from an early myriapod or protomyriapod form.

The uniramians have only one pair of antennae, and their appendages are always uniramous, never biramous like those of the crustaceans. Although some insect young are aquatic and have gills, the gills are not homologous to those of the crustaceans.

The insects and myriapods use tracheae to carry the respiratory gases directly to and from all body cells in a manner similar to the onychophorans and some of the arachnids.

Excretion is usually by malpighian tubules.

Figure 17-1

Centipede *Scolopendra* (class Chilopoda). Most segments have one pair of appendages each. First segment bears a pair of poison claws, which in some species can inflict serious wounds. Centipedes are carnivorous. Photograph by F.M. Hickman.

CLASS CHILOPODA

The Chilopoda (ki-lop'o-da) (Gr. *cheilos*, margin, lip, + *pous*, *podos*, foot), or centipedes, are land forms with somewhat flattened bodies that may contain from a few to 177 somites (Figure 17-1). Each somite, except the one behind the head and the last two in the body, bears a pair of jointed legs. The appendages of the first body segment are modified to form poison claws.

The head appendages are similar to those of an insect. There are a pair of antennae, a pair of mandibles, and one or two pairs of maxillae. A pair of eyes on the dorsal side of the head consists of groups of ocelli.

The digestive system is a straight tube into which salivery glands empty at the anterior end. Two pairs of malpighian tubules empty into the hind part of the intestine. There is an elongated heart with a pair of arteries to each somite. The heart has a series of ostia to provide for the return of the blood to the heart from the hemocoel. Respiration is by means of a tracheal system of branched air tubes that come from a pair of spiracles in each somite. The nervous system is typically arthropod, and there is also a visceral nervous system.

Sexes are separate, with unpaired gonads and paired ducts. Some centipedes lay eggs and others are viviparous. The young are similar to the adults.

Centipedes prefer moist places such as under logs, bark, and stones. They are very agile and are carnivorous in their eating habits, living on earthworms, cockroaches, and other insects. They kill their prey with their poison claws and then chew it with their mandibles. The common house centipede, *Scutigera* (L. *scutum*, shield, + *gera*, bearing), which has 15 pairs of legs, is often seen scurrying around bathrooms and damp cellars, where it catches insects. Most species are harmless to humans. Some of the tropical centipedes may reach a length of 30 cm.

CLASS DIPLOPODA

The Diplopoda (Gr. *diplos*, double, two + *pous*, *podos*, foot) are commonly called millipedes, which literally means "thousand feet" (Figure 17-2). Even though they do not have that many legs, they do have a large number of appendages, since each abdominal somite has two pairs of appendages, a condition that may have arisen from the fusion of pairs of somites. Their cylindrical bodies are made up of 25 to 100 somites. The short thorax consists of four somites, each bearing one pair of legs.

Figure 17-2

Seven-inch-long millipedes from Ecuador, mating. Note the typical doubling of appendages on most segments. Photograph by K. Sandved.

The heads bears two clumps of simple eyes and a pair each of antennae, mandibles, and maxillae. The general body structures are similar to those of centipedes, with a few variations here and there. Two pairs of spiracles on each abdominal somite open into air chambers that give off the tracheal air tubes. Two genital apertures are found toward the anterior end.

In most millipedes the appendages of the seventh somite are specialized for copulatory organs. After copulation the eggs are laid in a nest and are carefully guarded by the mother. The larval forms have only one pair of legs to each somite.

Millipedes are not as active as centipedes. They walk with a slow, graceful motion, not wriggling as the centipedes do. They prefer dark, moist places under logs or stones. They are herbivorous, feeding on decayed plant or animal matter, although sometimes they eat living plants. When disturbed, they often roll up into a coil. Common examples of this class are *Spirobolus* and *Julus*, both of which have wide distribution.

CLASS PAUROPODA

The Pauropoda (Gr. *pauros*, small, + *pous, podos*, foot) are a group of minute (2 mm or less), soft-bodied myriapods, numbering almost 500 species. They have a small head with branched antennae and no eyes, but they have a pair of sense organs that have the appearance of eyes (Figure 17-3, *A*). Their 12 trunk segments usually bear nine pairs of legs (none on the first or the last two segments). They have only one tergal plate covering each two segments.

Tracheae, spiracles, and circulatory system are lacking. Pauropods are probably most closely related to the diplopods but are considered to be more primitive.

Although widely distributed, the pauropods are the least well known of the myriapods. They live in moist soil, leaf litter, or decaying vegetation and under bark and debris. Representative genera are *Pauropus* and *Allopauropus*.

CLASS SYMPHYLA

The Symphyla (Gr. *sym*, together, + *phylon*, tribe) are small (2 to 10 mm) and have centipede-like bodies (Figure 17-3, *B*). They live in humus, leaf mold, and debris. *Scutigerella* (L. dim. of *Scutigera*) are often pests on vegetables and flowers, particularly in greenhouses. They are soft bodied, with 14 segments, 12 of which bear legs and one a pair of spinnerets. The antennae are long and unbranched. Only 160 species have been described.

Symphylans are eyeless but have sensory pits at the bases of the antennae. The tracheal system is limited to a pair of spiracles on the head and tracheal tubes to the anterior segments only.

CLASS INSECTA

The Insecta (L. *insectus*, cut into) are the most successful biologically of all the groups of arthropods. There are more species of insects than species in all the other classes of animals combined. The recorded number of insect species has been estimated to be close to one million, but this figure may represent only a fraction of the species that exist. There is also striking evidence that evolution is continuing

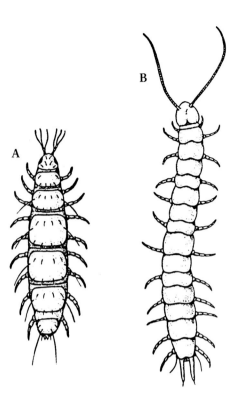

Figure 17-3

A, Pauropod. Pauropods are minute, whitish myriapods with three-branched antennae and nine pairs of legs. They live in leaf litter and under stones. They are eyeless but have sense organs that resemble eyes. **B,** *Scutigerella*, a symphylan, is a minute whitish myriapod that is sometimes a greenhouse pest.
B, Modified from Snodgrass, R.E. 1952. A textbook of arthropod anatomy. Ithaca, N.Y., Cornell University Press.

The mating behavior of *Scutigerella* is unusual. The male places a spermatophore at the end of a stalk. When the female finds it, she takes it into her mouth, storing the sperm in special buccal pouches. Then she removes the eggs from her gonopore with her mouth and attaches them to moss or lichen, or to the walls of crevices, smearing them during the handling with some of the semen and so fertilizing them. The young at first have only six or seven pairs of legs.

Figure 17-6

Hindleg of grasshopper. Muscles that operate the leg are found within a hollow cylinder of exoskeleton. Here they are attached to the internal wall, from which they manipulate segments of limb on the principle of a lever. Note pivot joint and attachment of tendons of extensor and flexor muscles, which act reciprocally to extend and flex the limb.

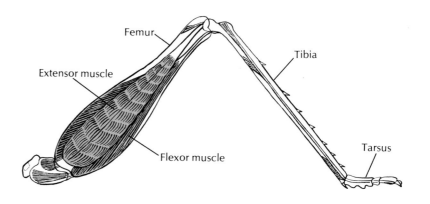

Figure 17-7

A, Praying mantis *Tenodera sinensis* (order Orthoptera), feeding on an insect. **B,** Praying mantis laying eggs. Egg mass is at lower right.
Photographs by J.H. Gerard.

A

B

Figure 17-8

Adaptive legs of worker honeybee. In the foreleg, toothed indentation covered with velum is used to comb out antennae. Spur on the middle leg removes wax from wax glands on the abdomen. Pollen picked up on body hairs is combed off by pollen brushes on the front and middle legs and deposited on pollen brushes of the hindlegs. Long hairs of pecten on the hindleg remove pollen from the brush of the opposite leg; then the auricle is used to press it into a pollen basket when the leg joint is flexed back. A bee carries her load in both baskets to the hive and pushes pollen into a cell, to be cared for by other workers.

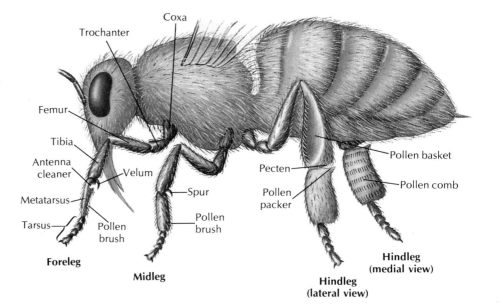

of the praying mantis are long and strong (Figure 17-7). The legs of honeybees show complex adaptations for collecting pollen (Figure 17-8).

The abdomen of insects is composed of nine to 11 segments; the eleventh, when present, is reduced to a pair of cerci (appendages at the posterior end). Larval or nymphal forms have a variety of abdominal appendages, but these are lacking in the adults. The end of the abdomen bears the external genitalia (Figure 17-4, *A*).

There are innumerable variations in body form among the insects. Beetles are usually thick and plump (Figure 17-9, *A*); damselflies, crane flies, and walking sticks are long and slender (Figure 17-9, *B*); aquatic beetles are streamlined; and cockroaches are flat, adapted to living in crevices. The ovipositor of the female ichneumon wasp is extremely long (Figure 17-10). The cerci form horny forceps in the earwigs and are long and many jointed in stoneflies and mayflies. Antennae are long in cockroaches and katydids, short in dragonflies and most beetles, knobbed in butterflies, and plumed in most moths.

Locomotion

Walking

When walking, most insects use a triangle of legs involving the first and last leg of one side together with the middle leg of the opposite side. In this way, insects keep three of their six legs on the ground, a tripod arrangement for stability. A slow-moving insect alternates its movement first on one side and then on the other. When it goes faster, the two phases tend to overlap and one side may begin before the other side has finished.

Some insects, such as the water strider *Gerris* (L. *gero*, to carry), are able to walk on the surface of water. The water strider has on its footpads nonwetting hairs that do not break the surface film of water but merely indent it. As it skates along, *Gerris* uses only the two posterior pairs of legs and steers with the anterior pair (Figure 17-11). The body of the marine water strider *Halobates* (Gr. *halos*, the sea, + *bātes*, one that treads), an excellent surfer on rough ocean waves, is further protected by a water-repellent coat of close-set hairs shaped like thick hooks.

A
B

Figure 17-9

A, A giant horned beetle *Diloboderus abderus* (order Coleoptera) from Uruguay. Though the ferocious-looking processes from the head and thorax might appear to be for pinching or stabbing an opponent, they actually are used to lift or pry up a rival of the same species away from resources. *B,* Walking sticks *Diapheromera femorata* (order Orthoptera), mating. This species is common in much of North America. It is wingless, and despite its camouflage as a twig, it is fed upon by numerous predators.
Photographs by K. Sandved.

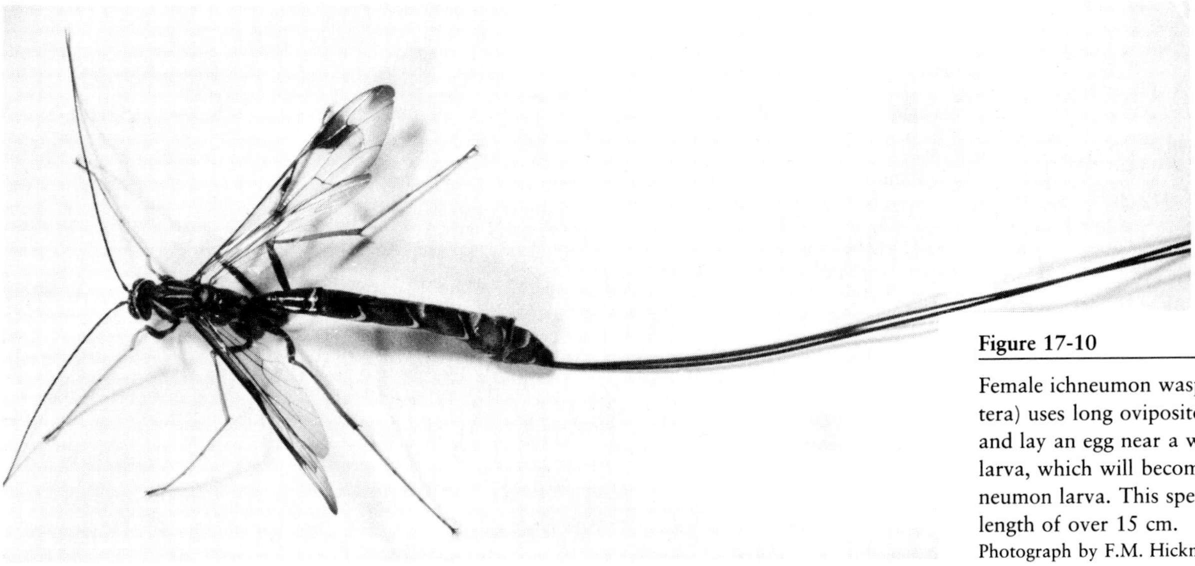

Figure 17-10

Female ichneumon wasp (order Hymenoptera) uses long ovipositor to bore into a tree and lay an egg near a wood-boring beetle larva, which will become food for the ichneumon larva. This specimen had an overall length of over 15 cm.
Photograph by F.M. Hickman.

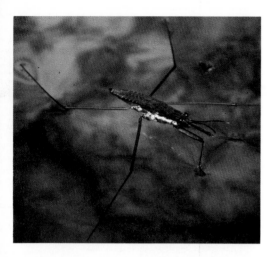

Figure 17-11

Water strider, *Gerris* sp. (order Hemiptera). The animal is supported on its long, slender legs by the water's surface tension. Photograph by J.H. Gerard.

Power of flight

Insects share the power of flight with birds and flying mammals. However, their wings have evolved in a different manner from that of the limb buds of birds and mammals and are not homologous to them. Insect wings are formed by outgrowths from the body wall of the mesothoracic and metathoracic segments and are composed of cuticle.

Most insects have two pairs of wings, but the Diptera (true flies) have only one pair, the hindwings being represented by a pair of tiny **halters** (balancers) that vibrate and are responsible for equilibrium during the flight. Males of the order Strepsiptera have only the hind pair of wings and an anterior pair of halteres. Males of the scale insects also have one pair of wings but no halteres. Some insects are wingless. Ants and termites, for example, have wings only on males, and on females during certain periods; workers are always wingless. Lice and fleas are always wingless.

Wings may be thin and membranous, as in flies and many others (Figure 17-10); thick and horny, such as the front wings of beetles (Figure 17-9, *A*); parchmentlike, such as the front wings of grasshoppers; covered with fine scales, as in butterflies and moths; or covered with hairs, as in caddis flies.

Wing movements are controlled by a complex of muscles in the thorax. Direct flight muscles are attached to a part of the wing itself; indirect flight muscles are not attached to the wing and cause wing movement by altering the shape of the thorax. The wing is hinged at the thoracic tergum and slightly laterally on a pleural process, which acts as a fulcrum (Figure 17-12). In all insects, the upstroke of the wing is effected by contraction of indirect muscles that pull the tergum down toward the sternum (Figure 17-12, *A*). In dragonflies and cockroaches the upstroke is accomplished by contraction of direct muscles attached to the wings lateral to the pleural fulcrum. In Hymenoptera and Diptera all the flight muscles are indirect. The upstroke occurs when the sternotergal muscles relax and longitudinal muscles in the thorax arch the tergum (Figure 17-12, *B*), pulling up the tergal articulations relative to the pleura. The upstroke in beetles and grasshoppers involves both direct and indirect muscles.

Flight muscle contraction has two basic types of neural control: **synchronous** and **asynchronous**. Synchronous muscles are found in the larger insects such as dragonflies and butterflies, in which a single volley of nerve impulses stimulates a muscle contraction and thus one wing stroke. Asynchronous muscles are found in the more specialized insects. Their mechanism of action is complex and depends on the storage of potential energy in resilient parts of the thoracic cuticle. As one

Figure 17-12

A, Flight muscles of insects such as cockroaches, in which upstroke is by indirect muscles and downstroke is by direct muscles. **B,** In insects such as flies and bees, both upstroke and downstroke are by indirect muscles. **C,** The figure-8 path followed by the wing of a flying insect during the upstroke and downstroke.

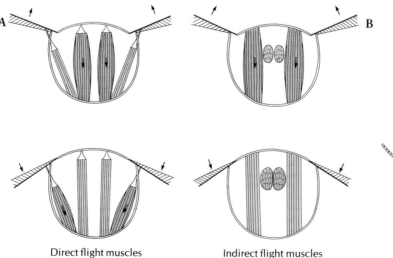

Direct flight muscles of locusts and dragonflies

Indirect flight muscles of flies and midges

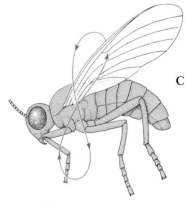

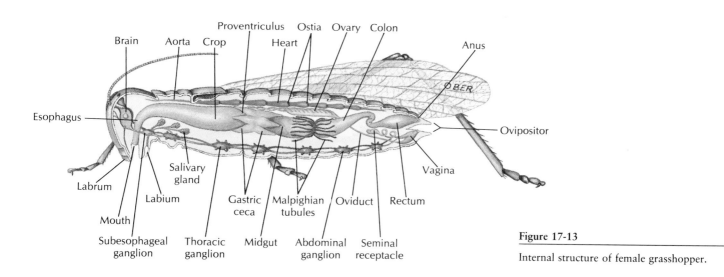

Figure 17-13

Internal structure of female grasshopper.

set of muscles contracts (moving the wing in one direction), they stretch the antagonistic set of muscles, causing them to contract (and move the wing in the other direction). Because the muscle contractions are not phase-related to nervous stimulation, only occasional nerve impulses are necessary to keep the muscles responsive to alternating stretch activation. Thus extremely rapid wing beats are possible. For example, butterflies (with synchronous muscles) may beat as few as four times per second. Insects with asynchronous muscles, such as flies and bees, may vibrate at 100 beats per second or more. The fruit fly *Drosophila* (Gr. *drosos*, dew, + *philos*, loving) can fly at 300 beats per second, and midges have been clocked at more than 1000 beats per second.

Obviously flying entails more than a simple flapping of wings; a forward thrust is necessary. As the indirect flight muscles alternate rhythmically to raise and lower the wings, the direct flight muscles alter the angle of the wings so that they act as lifting airfoils during both the upstroke and the downstroke, twisting the leading edge of the wings downward during the downstroke and upward during the upstroke. This produces a figure-eight movement (Figure 17-12, C) that aids in spilling air from the trailing edges of the wings. The quality of the forward thrust depends, of course, on several factors, such as variations in wing venation, how much the wings are tilted, and how they are feathered.

Flight speeds vary. The fastest flyers usually have narrow, fast-moving wings with a strong tilt and a strong figure-eight component. Sphinx moths and horseflies are said to achieve approximately 48 km (30 miles) per hour and dragonflies approximately 40 km (25 miles) per hour. Some insects are capable of long continuous flights. The migrating monarch butterfly *Danaus plexippus* (Gr. after Danaus, mythical king of Arabia) (Figure 17-22) travels south for hundreds of miles in the fall, flying at a speed of approximately 10 km (6 miles) per hour.

Internal Form and Function

Nutrition

The digestive system (Figure 17-13) consists of a foregut (mouth with salivary glands, esophagus, crop for storage, and gizzard for grinding); a midgut (stomach and gastric ceca); and a hindgut (intestine, rectum, and anus). Some digestion may take place in the crop as the food is mixed with enzymes from the saliva, but no absorption takes place there. The main site for digestion and absorption is the

Figure 17-14

Female *Pulex irritans*. The origin of the ocular bristle is beneath the eye.
Photograph by Jay Georgi. From Schmidt, G.D. and L.S. Roberts. 1985. Foundations of parasitology. St. Louis, The C.V. Mosby Co.

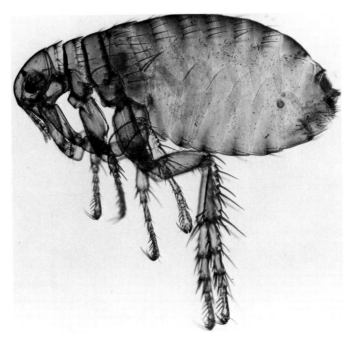

A

B

Figure 17-15

A, Hornworm, larval stage of a sphinx moth (order Lepidoptera). The more than 100 species of North American sphinx moths are strong fliers and mostly nocturnal feeders. Their larvae, called hornworms because of the large, fleshy posterior spine, are often pests of tomatoes, tobacco, and other plants. B, Hornworm parasitized by a tiny wasp *Apanteles,* which laid its egg inside the caterpillar. The wasp larvae have emerged, and their pupae are on the caterpillar's skin. Young wasps emerge in 5 to 10 days, but the caterpillar usually dies.
A, Photograph by C.P. Hickman, Jr.; B, photograph by J.H. Gerard.

midgut, and the ceca may increase the digestive and absorptive area. Little absorption of nutrients occurs in the hindgut (with certain exceptions, such as wood-eating termites), but this is a major area for resorption of water and some ions (see p. 369).

The majority of insects feed on plant juices and plant tissues. Such a food habit is called **phytophagous.** Some insects feed on specific plants; others, such as grasshoppers, will eat almost any plant. The caterpillars of many moths and butterflies eat the foliage of only certain plants. Certain species of ants and termites cultivate fungus gardens as a source of food.

Many beetles and the larvae of many insects live on dead animals (**saprophagous**). A number of insects are **predaceous,** catching and eating other insects as well as other types of animals (Figure 17-7). However, the so-called predaceous diving beetle *Cybister fimbriolatus* (Gr. *kybistētēr,* diver) has been found not to be as predaceous as supposed, but is largely a scavenger.

Many insects, adults as well as larvae, are **parasitic.** For instance, fleas (Figure 17-14) live on the blood of mammals, and the larvae of many varieties of wasps live on spiders and caterpillars (Figure 17-15). In turn, many are parasitized by other insects. Some of the latter are beneficial to humans by controlling the numbers of injurious insects. Parasitism of parasitic insects by other insects, a condition known as **hyperparasitism,** often becomes quite involved.

For each type of feeding, the mouthparts are adapted in a specialized way. The **sucking mouthparts** are usually arranged in the form of a tube and can pierce the tissues of plants or animals. This arrangement is well shown in the water scorpion (*Ranatra fusca*), a member of the order Hemiptera. This elongated, stick-like insect with a slender caudal respiratory tube has a beak in which there are four piercing, needlelike stylets made up of two mandibles and two maxillae. These parts are fitted together to form two tubes, a salivary tube for injecting saliva into the prey and a food tube for drawing out the body fluid of the prey. The mosquito also combines piercing with needlelike stylets and sucking through a food channel (Figure 17-16, *B*). In honeybees the labium forms a flexible and contractile "tongue" covered with many hairs. When the bee plunges its proboscis into nectar, the tip of the tongue bends upward and moves back and forth rapidly. Liquid

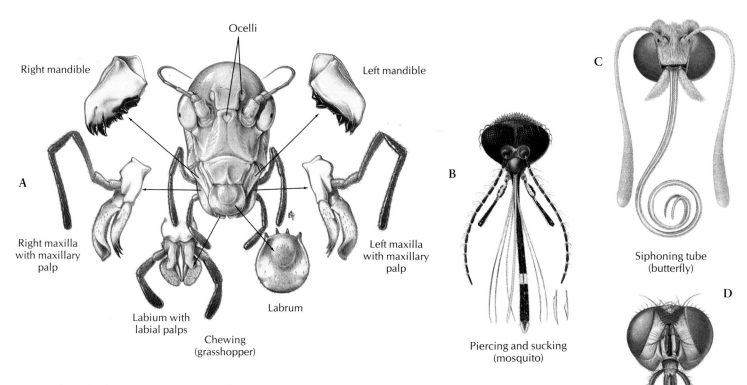

Ocelli

Right mandible

Left mandible

A

Right maxilla
with maxillary
palp

Left maxilla
with maxillary
palp

Labium with
labial palps

Labrum

Chewing
(grasshopper)

B

Piercing and sucking
(mosquito)

C

Siphoning tube
(butterfly)

D

Sponging
(housefly)

enters the tube by capillarity and is drawn up continuously by a pumping pharynx. In butterflies and moths, mandibles are usually absent, and the maxillae are modified into a long sucking proboscis (Figure 17-16, *C*) for drawing nectar from flowers. At rest the proboscis is coiled up into a flat spiral. In feeding it is extended, and fluid is pumped up by pharyngeal muscles.

Houseflies, blowflies, and fruit flies have **sponging** and **lapping mouthparts** (Figure 17-16, *D*). At the apex of the labium is a pair of large, soft lobes with grooves on the lower surface that serve as food channels. These flies lap up liquid food or liquefy food first with salivary secretions. Horseflies are fitted not only to sponge up surface liquids but to bite into the skin with slender, tapering mandibles and then sponge up blood.

Biting mouthparts such as those of the grasshopper and many other herbivorous insects are adapted for seizing and crushing food (Figure 17-16, *A*); those of most carnivorous insects are sharp and pointed for piercing their prey. The mandibles of chewing insects are strong, toothed plates whose edges can bite or tear while the maxillae hold the food and pass it toward the mouth. Enzymes secreted by the salivary glands add chemical action to the chewing process.

Circulation

A tubular heart in the pericardial cavity (Figure 17-13) moves hemolymph (blood) forward through the only blood vessel, a dorsal aorta. The heartbeat is a peristaltic wave. Accessory pulsatory organs help move the hemolymph into the wings and legs, and flow is also facilitated by various body movements. The hemolymph consists of plasma and amebocytes and apparently has little to do with oxygen transport.

Gas exchange

Terrestrial animals require efficient respiratory systems that permit rapid oxygen–carbon dioxide exchange but at the same time restrict water loss. In insects this is the function of the **tracheal system,** an extensive network of thin-walled tubes that

Figure 17-16

Four types of insect mouthparts. (See text for description of types and examples.) Illustration by George Venable.

Figure 17-17

A, Relationship of spiracle, tracheae, taenidia (chitinous bands that strengthen the tracheae), and tracheoles (diagrammatic). **B,** Generalized arrangement of insect tracheal system (diagrammatic). Air sacs and tracheoles not shown.

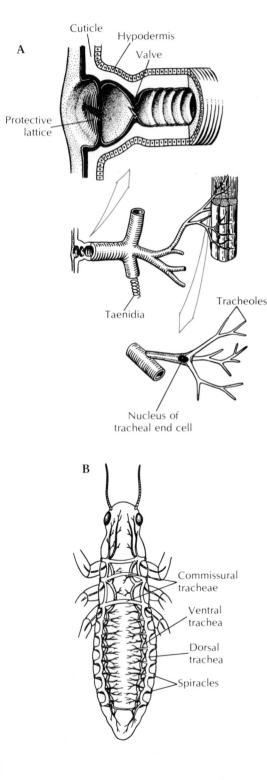

branch into every part of the body (Figure 17-17). The tracheal trunks open to the outside by paired **spiracles,** usually two on the thorax and seven or eight on the abdomen. A spiracle may be merely a hole in the integument, as in primary wingless insects, but it is usually provided with a valve or some sort of closing mechanism that cuts down water loss. The evolution of such a device must have been very important in enabling insects to move into drier habitats. The spiracle may also possess a filtering device such as a sieve plate or a set of interlocking bristles that may prevent entrance of water, parasites, or dust into the tracheae. The **tracheae** are composed of a single layer of cells and are lined with cuticle that is shed, along with the outer cuticle, during the molt. They are supported by spiral thickenings of the cuticle (called taenidia) that prevent their collapse. The tracheae branch out into smaller tubes, ending in very fine, fluid-filled tubules called **tracheoles** (lined with cuticle, but not shed at ecdysis), which branch into a fine network over the cells. In large insects the largest tracheae may be several millimeters in diameter but taper down to 1 to 2 μm. The tracheoles then taper to 0.5 to 0.1 μm in diameter. In one of the stages of the silkworm larva, it is estimated that there are 1.5 million tracheoles! Scarcely any living cell is located more than a few micrometers away from a tracheole. In fact, the ends of some tracheoles actually indent the membranes of the cells they supply, so that they terminate close to the mitochondria. The tracheal system affords an efficient system of transport without the use of oxygen-carrying pigments in the hemolymph.

The tracheal system may also include **air sacs,** which appear to be dilated tracheae without taenidia (Figure 17-19, *A*). They are thin walled and flexible and are located largely in the body cavity but also in appendages. In some insects the sacs may have functions other than respiratory. For example, they may allow internal organs to change in volume during instar stages without changing the shape of the larvae; they reduce the weight of large insects; or they may act as sound resonators or heat insulators.

In some very small insects, gas transport occurs entirely by diffusion along a concentration gradient. As oxygen is used, a reduced pressure develops in the tracheae and air is sucked in through the spiracles. Larger or more active insects employ some ventilation device for moving air in and out of the tubes. Usually muscular movements in the abdomen perform the pumping action that draws air in or expels it. In some insects—locusts, for example—additional pumping is provided by telescoping the abdomen, pumping with the prothorax, or thrusting the head forward and backward.

The tracheal system is primarily adapted for air breathing, but many insects (nymphs, larvae, and adults) live in water. In small, soft-bodied aquatic nymphs, the gaseous exchange may occur by diffusion through the body wall, usually into and out of a tracheal network just under the integument. The aquatic nymphs of stoneflies and mayflies are equipped with **tracheal gills,** which are thin extensions of the body wall containing a rich tracheal supply. The gills of dragonfly nymphs are ridges in the rectum (rectal gills) where gas exchange occurs as water moves in and out.

Excretion and water balance

Insects and spiders have a unique excretory system consisting of **malpighian tubules** that operate in conjunction with specialized glands in the wall of the rectum. The malpighian tubules, variable in number, are thin, elastic, blind tubules attached to the juncture between the midgut and hindgut (Figures 17-13

A   B

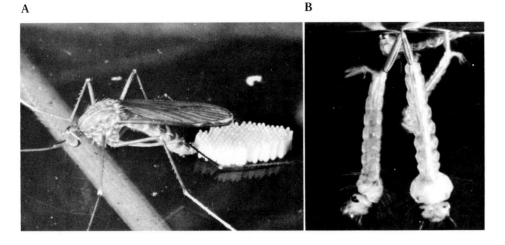

Figure 17-18

A, Mosquito *Culex* (order Diptera) lays her eggs in small packets or rafts on the surface of standing or slowly moving water. **B,** Mosquito larvae are the familiar wrigglers of ponds and ditches. To breathe, they hang head down, with respiratory tubes projecting through the surface film of water. Motion of vibratile tufts of fine hairs on the head brings a constant supply of food.
From *Mosquito*, an Encyclopaedia Britannica film.

and 17-19, *A*). The free ends of the tubules lie free in the hemocoel and are bathed in hemolymph.

The mechanism of urine formation in the malpighian tubules of herbivorous insects appears to depend on the active secretion of potassium into the tubules (Figure 17-19, *B*). This primary secretion of ions pulls water along with it by osmosis to produce a potassium-rich fluid. Other solutes and waste materials also are secreted or diffuse into the tubule. The predominant waste product of nitrogen metabolism in most insects is uric acid, which is virtually insoluble in water (see p. 645). This enters the upper end of the tubule, where the pH is slightly alkaline, as relatively soluble potassium acid urate. As the formative urine passes into the lower end of the tubule, the potassium combines with carbon dioxide, is reabsorbed as potassium bicarbonate, the pH changes to acidic (pH 6.6), and the insoluble uric acid precipitates out. As the urine drains into the intestine and

Although the diving beetle *Dytiscus* (Gr. *dytikos,* able to swim) can fly, it spends most of its life in the water as an excellent swimmer. It uses an "artificial gill" in the form of a bubble of air held under its wing covers. The bubble is kept stable by a layer of hairs on top of the abdomen and is in contact with the spiracles on the abdomen. Oxygen from the bubble diffuses into the tracheae and is replaced by diffusion of oxygen from the water. Thus, the bubble can last for several hours before the beetle must surface to replace it. Mosquito larvae are not good swimmers but live just below the surface, putting out short breathing tubes like snorkels to the surface for air (Figure 17-18). Spreading oil on the water, a favorite method of mosquito control, clogs the tracheae with oil and so suffocates the larvae. "Rattailed maggots" of the syrphid flies have an extensible tail that can stretch as much as 15 cm to the water surface.

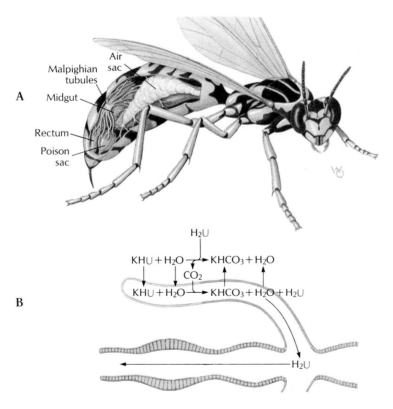

Malpighian tubules
Air sac

A   Midgut

Rectum

Poison sac

$$KHU + H_2O \rightarrow KHCO_3 + H_2O$$
$$CO_2$$
B $$KHU + H_2O \rightarrow KHCO_3 + H_2O + H_2U$$

H_2U

H_2U

Figure 17-19

Malpighian tubules of insect. **A,** Malpighian tubules are located at the juncture of the midgut and hindgut (rectum) as shown in the cutaway view of a wasp. **B,** Function of malpighian tubules. Solutes, especially potassium, are actively secreted into the tubules. Water and potassium acid urate follow. This fluid moves into the rectum where solutes and water are actively resorbed, leaving uric acid to be excreted.

passes through the hindgut, specialized rectal glands absorb chloride, sodium (and in some cases potassium), and water.

Since water requirements vary among different types of insects, this ability to cycle water and salts is very important. Insects living in dry environments may resorb nearly all water from the rectum, producing a nearly dry mixture of urine and feces. Leaf-feeding insects take in and excrete quantities of fluid. Freshwater larvae need to excrete water and conserve salts. Insects that feed on dry grains need to conserve water and excrete salt.

Nervous system

The nervous system in general resembles that of the larger crustaceans, with a similar tendency toward fusion of ganglia (Figure 17-13). A giant fiber system has been demonstrated in a number of insects. There is also a stomodeal nervous system that corresponds in function to the autonomic nervous system of vertebrates. Neurosecretory cells located in various parts of the brain have an endocrine function, but, except for their role in molting and metamorphosis, little is known of their activity.

Sense organs

Along with neuromuscular coordination, insects have unusually keen sensory perception. Their sense organs are mostly microscopic and are located chiefly in the body wall. Each type usually responds to a specific stimulus. The various organs are receptive to mechanical, auditory, chemical, visual, and other stimuli.

Mechanoreception

Mechanical stimuli, or those dealing with touch, pressure, vibration, and the like, are picked up by sensilla. A sensillum may be simply a seta, or hairlike process, connected with a nerve cell, a nerve ending just under the cuticle and lacking a seta, or a more complex organ (scolopophorous organ) consisting of sensory cells with their endings attached to the body wall. Such organs are widely distributed over the antennae, legs, and body.

Auditory reception

Airborne sounds may be detected by very sensitive setae (hair sensilla) or by tympanal organs. In tympanal organs a number of sensory cells (ranging from a few to hundreds) extend to a very thin tympanic membrane that encloses an air space in which vibrations can be detected. Tympanal organs are found in certain Orthoptera (Figure 17-4, *A*), Homoptera, and Lepidoptera. Some insects are fairly insensitive to airborne sounds but can detect vibrations reaching them through the substrate. Vibrations of the substrate are detected by organs usually on the legs.

Chemoreception

Chemoreceptors (for taste or smell) are usually bundles of sensory cell processes that are often located in sensory pits. These are usually on mouthparts, but in ants, bees, and wasps they are also found on the antennae, and butterflies, moths, and flies also have them on the legs. The chemical sense is generally keen, and some insects can detect certain odors for several kilometers. Many of the patterns of insect behavior such as feeding, mating, habitat selection, and host-parasite rela-

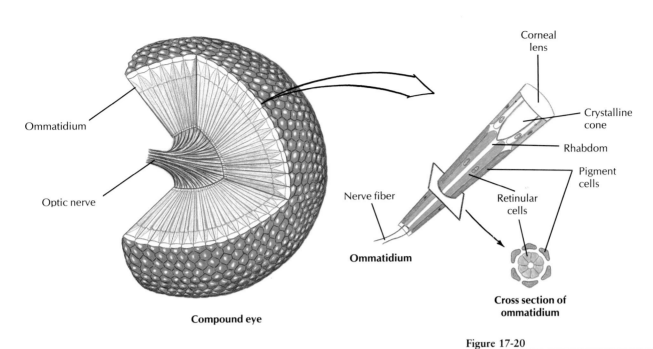

Compound eye

Corneal
lens

Crystalline
cone

Rhabdom

Pigment
cells

Retinular
cells

Nerve fiber

Ommatidium

**Cross section of
ommatidium**

Figure 17-20

Compound eye of an insect. A single ommatidium is shown enlarged to the right.

tions are mediated through the chemical senses. These senses are also involved in
the responses of insects to artificial repellents and attractants.

Visual reception

Insect eyes are of two types, simple and compound. Simple eyes are found in some
nymphs and larvae and in many adults. Most insects have three ocelli on the head.
Evidence indicates that the honeybee uses them to monitor light intensity but not
to form images.

Compound eyes are found in most adults and may cover most of the head.
They consist of thousands of ommatidia—6300 in the eye of a honeybee, for
example. The structure of the compound eye is similar to that of crustaceans
(Figure 17-20). An insect such as a honeybee can see simultaneously in almost all
directions around its body, but it is more myopic than the human, and images,
even of nearby objects, are fuzzy. However, most flying insects rate much higher
than humans in flicker-fusion tests. Flickers of light become fused in the human
eye at a frequency of 45 to 55 per second, but bees and blowflies can distinguish as
many as 200 to 300 separate flashes of light per second. This should be an advantage in analyzing a fast-changing landscape during flight.

A bee can distinguish colors, but its sensitivity begins in the ultraviolet range,
which the human eye cannot see, and extends into the orange; the honeybee
cannot distinguish shades of red from shades of gray.

Other senses

Insects also have well-developed senses for temperature, especially on the antennae and legs, and for humidity, proprioception, gravity, and others.

Neuromuscular coordination

Insects are active creatures with excellent neuromuscular coordination. Arthropod muscles are typically cross-striated, just as vertebrate skeletal muscles are. A
flea can leap a distance of 100 times its own length, and an ant can carry in its jaws
a load greater than its own weight. This sounds as though insect muscle were

In terms of proportionate body
length, the flea's jump would be
the equivalent of a 6-foot human
executing a standing high jump of
600 feet! Actually, the insect's
muscles are not entirely
responsible for its jump; they
cannot contract rapidly enough to
reach the required acceleration.
The flea depends on pads of
resilin, a protein with unusual
elastic properties, that is also
found in the wing-hinge ligaments
of many other insects. Resilin
releases 97% of its stored energy
on returning from a stretched
position, compared with only
85% in most commercial rubber.
When the flea prepares to jump,
it rotates its hind femurs and
compresses the resilin pads, then
engages a "catch" mechanism. In
effect, it has cocked itself. To
take off, the flea must exert the
relatively small muscular action
to unhook the catches, allowing
the resilin to expand.

stronger than that of other animals. Actually, however, the force a particular muscle can exert is related directly to its cross-sectional area, not its length. Based on maximum load moved per square centimeter of cross section, the strength of insect muscle is relatively the same as that of vertebrate muscle.

Reproduction

Sexes are separate in insects, and fertilization is usually internal. Insects have various means of attracting mates. The female moth gives off a powerful pheromone that can be detected for a great distance by the male. Fireflies use flashes of light; some insects find each other by means of sounds or color signals and by various kinds of courtship behavior.

Sperm are usually deposited in the vagina of the female at the time of copulation (Figure 17-21). In some orders the sperm are encased in spermatophores that may be transferred at copulation or deposited on the substratum to be picked up by the female. The male silverfish deposits a spermatophore on the ground, then spins signal threads to guide the female to it. During the evolutionary transition from aquatic to terrestrial life, spermatophores were widely used and copulation evolved much later.

Usually the sperm are stored in the spermatheca of the female in numbers sufficient to fertilize more than one batch of eggs. Many insects mate only once during their lifetime, and none mates more than a few times.

Insects usually lay a great many eggs. The queen honeybee, for example, may lay more than 1 million eggs during her lifetime. On the other hand, some flies are viviparous and bring forth only a single offspring at a time. Insect forms that make no provision for the care of their young may lay many more eggs than do insects that provide for their young or those that have a very short life cycle.

Most species normally lay their eggs in a particular type of place to which they are guided by visual, chemical, or other cues. Butterflies and moths lay their eggs on the specific kind of plant on which the caterpillar must feed. The tiger moth may look for a pigweed, the sphinx moth for a tomato or tobacco plant, and the monarch butterfly for a milkweed plant (Figure 17-22). Insects whose immature stages are aquatic lay their eggs in water (Figure 17-18, A). A tiny braconid wasp lays her eggs on the caterpillar of the sphinx moth where they will feed and pupate in tiny white cocoons (Figure 17-15, B). The ichneumon wasp, with unerring accuracy, seeks out a certain kind of larva in which her young will live as internal parasites. Her long ovipositors may have to penetrate 1 to 2 cm of wood to find the larva of a wood wasp or a wood-boring beetle in which she will deposit her eggs (Figure 17-10).

Metamorphosis and Growth

Early development occurs within the egg, and the hatching young escape from the egg in various ways. During the postembryonic development most insects change in form; that is, they undergo **metamorphosis** (Figure 17-22). During this period they must undergo a number of molts in order to grow, and each stage of the insect between molts is called an **instar**.

Although metamorphosis is not limited to insects, insects illustrate it more dramatically than any other group. The transformation, for instance, of the hickory horned devil caterpillar into the beautiful royal walnut moth represents an astonishing morphological change. In insects metamorphosis is associated with

Figure 17-21

Copulation in insects (see also Figure 17-9, B). **A,** Bluet damselflies *Enallagma* sp. (order Odonata) are common throughout North America. **B,** *Omura congrua* (order Orthoptera) are a kind of grasshopper found in Brazil.

Photographs by K. Sandved.

A

B

C

D

the evolution of wings, which are restricted to the reproductive stage where they can be of the most benefit.

Complete metamorphosis

Approximately 88% of insects undergo a complete metamorphosis, which separates the physiological processes of growth (larva) from those of differentiation (pupa) and reproduction (adult) (Figure 17-22). Each stage functions efficiently without competition with the other stages, for the larvae often live in entirely different surroundings and eat different foods from the adults. The wormlike larvae, which usually have chewing mouthparts, are known as caterpillars, maggots, bagworms, fuzzy worms, grubs, and so on. After a series of instars during which the wings are developing internally, the larva forms a case or cocoon about itself and becomes a pupa, or chrysalis, a nonfeeding stage in which many insects pass the winter. When the final molt occurs over winter, the full-grown adult emerges, pale and with wings wrinkled. In a short time the wings expand and harden, and the insect is on its way. The stages, then, are egg, larva (several instars), pupa, and adult. The adult undergoes no further molting. Insects that undergo complete metamorphosis are said to be **holometabolous** (Gr. *holo,* complete, + *metabolē,* change).

Gradual metamorphosis

Some insects undergo a type of gradual, or incomplete, metamorphosis. These include the grasshoppers, cicadas, and mantids, which have terrestrial young, and mayflies, stoneflies, and dragonflies, which lay their eggs in water. The young are called **nymphs** (or **naiads** if aquatic), and their wings develop externally as budlike

Figure 17-22

Complete metamorphosis. **A,** Monarch butterfly, *Danaus plexippus* (order Lepidoptera), lays eggs on a milkweed plant. **B,** Hatched larva feeds on milkweed leaves. **C,** When mature, the larva hangs on the milkweed plant and transforms into a chrysalis, or pupa, an inactive stage that does not feed and is covered by a cocoon or protective covering. **D,** Adult has emerged. The wings quickly expand and harden, pigmentation develops, and the butterfly goes on its way. **A,** Photograph by C.P. Hickman, Jr.; **B** to **D,** photographs by J.H. Gerard.

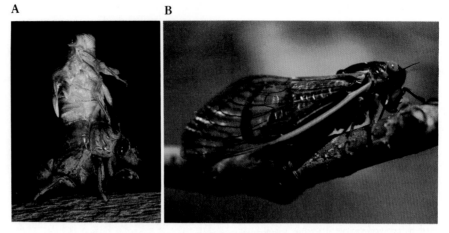

A   B

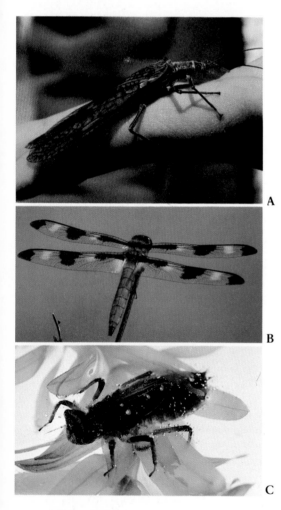

A

B

C

outgrowths in the early instars and increase in size as the animal grows by successive molts and becomes a winged adult (Figure 17-23). The aquatic naiads have tracheal gills or other modifications for aquatic life (Figure 17-24). The stages are egg, nymph (several instars), and adult. Insects that undergo gradual metamorphosis are called **hemimetabolous** (Gr. *hemi*, half, + *metabolē*, change).

Direct development

A few insects, such as silverfish and springtails, undergo direct development. The young, or juveniles, are similar to the adults except in size and sexual maturation. The stages are egg, juvenile, and adult. Such insects include the wingless insects (apterygote orders).

Physiology of metamorphosis

Metamorphosis in insects is controlled and regulated by hormones. Three major endocrine organs are involved in development through the larval stages to the pupa and eventually to the emergence of the adult. These organs are the **brain,** the **prothoracic (ecdysial) glands,** and the **corpora allata** (Figure 33-5, p. 715).

The intercerebral part of the brain and the ganglia of the nerve cord contain several groups of neurosecretory cells that produce an endocrine substance called the **brain hormone (ecdysiotropin).** These neurosecretory cells may send their axons to paired organs behind the brain, the **corpora allata,** which serve as a storage place for the brain hormone. The brain hormone is carried in the hemolymph to the prothoracic gland, a glandular organ in the head or the prothorax that is stimulated to produce the **molting hormone,** or **ecdysone** (ek′duh-sone). This hormone sets in motion certain processes that lead to the casting off of the old skin (ecdysis) by proliferation of the epidermal cells.

If the larval form is retained at the end of this process, it is called simple molting; if the insect undergoes changes into pupa or adult, it may be referred to as metamorphosis. Simple molting persists as long as a certain **juvenile hormone** (neotenine) is present in sufficient amounts, along with the molting hormone in the hemolymph, and each molting simply produces a larger larva. The juvenile hormone is produced by the corpora allata (Figure 33-5).

Even the kind of cuticle produced depends on the amount of juvenile hormone present. If only a small amount of this hormone is present, a pupal cuticle is the result. When the corpora allata cease to produce the juvenile hormone, the molting hormone alone is secreted into the hemolymph and the adult emerges (metamorphosis). Thus, the molting hormone is necessary for each molt but is modified by the juvenile hormone, whose action is to maintain larval characteristics in the young insect.

Experimental evidence shows that when the corpora allata (and thus the juvenile hormone) are removed surgically from the larva, the following molt will result in metamorphosis into the adult. Conversely, if the corpora allata from a young larva are transplanted into an old larva, the latter can be converted into a giant larva, because metamorphosis to the pupa or adult stage cannot occur. Many other experimental modifications on this theme have been performed. Progress has also been made in determining the chemical nature of the hormones (p. 716).

The mechanism of molting and metamorphosis just described is that found in holometabolous insects, but the same factors also apply in general to the molting nymphal stages of hemimetabolous insects, in which there are no pupal stages. A recent method of insect control involves the use of compounds that mimic the juvenile hormones. These prevent insects from becoming sexually competent when treated just before they become adults.

What factors initiate the sequence of secretion and the role of these three different hormones? How are they correlated with cyclical events in the life histories of insects? Experimentally it has been shown that in some insects low temperature activates the neurosecretory cells of the intercerebral gland of the brain, which then sets in motion the sequence of events already related. The chilling of the brain seems to be all important in the initiation of metamorphosis. Adults do not molt and grow, since the ecdysial glands degenerate after the last ecdysis. Many aspects of the control mechanism of these interesting processes have not yet been worked out.

Diapause

Many animals, including many types of insects, undergo a period of dormancy in their annual life cycle. In temperate zones there may be a period of winter dormancy, called hibernation, or a period of summer dormancy, called estivation, or both. It is well known that there are periods in the life cycle of many insects when eggs, larvae, pupae, or even adults remain dormant for a long time because external conditions of climate, moisture, and the like are too harsh or unfavorable for survival in states of normal activity. Thus, the life cycle is synchronized with periods of suitable environmental conditions and abundance of food. Most insects enter such a state when some factor of the environment, such as temperature, becomes unfavorable, and the state continues until conditions again become favorable.

However, some species have a prolonged arrest of growth that occurs irrespective of the environment, that is, whether or not favorable conditions prevail. This type of dormancy is called **diapause** (di′a-poz) (Gr. *dia,* through, dividing into two parts, + *pausis,* a stopping), and it is an important adaptation to survive adverse environmental conditions. Diapause is genetically determined in each species and sometimes varies between strains within a species, but it is usually set off by some certain signal. In the environment of the insect, such signals forecast adverse conditions to come, for example, the lengthening of shortening of the days. Thus photoperiod, or day length, is often the signal that initiates diapause. The arrival of a critical length of day starts the physiological machinery for establishing diapause, which continues until the proper day length or other signal is received.

A
B

Figure 17-25

Mimicry in butterflies. **A,** Monarch butterfly is distasteful to, and avoided by, birds because as a caterpillar it fed on the acrid milkweed. **B,** It is mimicked by the smaller viceroy butterfly, *Limenitis archippus,* which feeds on willows and is presumably tasteful to birds, but is not eaten because it so closely resembles the monarch in color and markings. This kind of mimicry is called ''Batesian'' mimicry.
Photographs by J.H. Gerard.

Figure 17-26

Camouflage in insects. **A,** *Estigena pardalis* (order Lepidoptera) in Java resembles a dead leaf. **B,** Bizarre processes from the thorax of a treehopper from Mexico, *Sphongophorus* sp. (order Homoptera), masquerade as parts of the twig where it feeds. **C,** Broken outlines and color of a katydid (*Dysonia* sp., order Orthoptera) in Costa Rica give it the appearance of the leaves on which it has been feeding.
Photographs by K. Sandved.

Diapause always occurs at the end of an active growth stage of the molting cycle so that, when the diapause period is over, the insect is ready for another molt. One species of the ant *Myrmica* reaches the third instar stage in late summer. Many of the larvae do not develop beyond this point until the following spring, even if temperatures are mild or if the larvae are kept in a warm laboratory.

Defense

Insects as a group display many colors. This is especially true of butterflies, moths, and beetles. Even in the same species the color pattern may vary in a seasonal way, and there also may be color differences between males and females. Some of the color patterns in insects are probably highly adaptive, such as those for **protective coloration, warning coloration,** and **mimicry** (Figures 17-25 and 17-26).

Besides color, insects have other methods of protecting themselves. The cuticular exoskeleton affords good protection for many of them; some, such as stinkbugs, have repulsive odors and tastes; others protect themselves by a good offense, for many are very aggressive and can put up a good fight (for example, bees and ants); and still others are swift in running for cover when danger threatens.

Many insects practice chemical warfare in a variety of ingenious ways. Some repel an assault by virtue of their bad taste, odor, or poisonous properties, others use chemical exudates that mechanically prevent a predator from attacking. The caterpillars of monarch butterflies (Figure 17-22, *A*) assimilate cardiac glycosides from certain species of milkweed (Asclepiadaceae); this substance confers unpal-

A
B
C

atability on the butterflies after metamorphosis and induces vomiting in their predators. The bombardier beetle, on the other hand, produces an irritating spray that it aims accurately at attacking ants or other enemies.

Behavior and Communication

The keen sensory perceptions of insects make them extremely responsive to many stimuli. The stimuli may be internal (physiological) or external (environmental), and the responses are governed by both the physiological state of the animal and the pattern of nerve pathways traveled by the impulses. Many of the responses are simple, such as orientation toward or away from the stimulus, for example, attraction of a moth to light, avoidance of light by a cockroach, or attraction of carrion flies to the odor of dead flesh.

Much of the behavior of insects, however, is not a simple matter of orientation but involves a complex series of responses. A pair of tumble bugs, or dung beetles, chews off a bit of dung, rolls it into a ball, and rolls the ball laboriously to where they intend to bury it, after laying their eggs in it (Figure 17-27). The cicada slits the bark of a twig and then lays an egg in each of the slits. The female potter wasp *Eumenes* scoops up clay into pellets, carries them one by one to her building site, and fashions them into dainty little narrow-necked clay pots, into each of which she lays an egg. Then she hunts and paralyzes a number of caterpillars, pokes them into the opening of a pot, and closes up the opening with clay. Each egg, in its own protective pot, hatches to find a well-stocked larder of food.

Much of such behavior is "innate"; that is, entire sequences of actions apparently have been genetically programmed. However, a great deal more learning is involved than was once believed. The potter wasp, for example, must learn where she has left her pots if she is to return to fill them with caterpillars one at a time. Social insects, which have been studied extensively, have been found capable of most of the basic forms of learning used by mammals. The exception is insight learning. Apparently insects, when faced with a new problem, cannot reorganize their memories to construct a new response.

Insects communicate with other members of their species by means of chemical, visual, auditory, and tactile signals. **Chemical signals** take the form of **pheromones,** which are substances secreted by one individual that affect the behavior or physiological processes of another individual. Pheromones include sex attractants, releasers of certain behavior patterns, trail markers, alarm signals, territorial markers, and the like. Like hormones, pheromones are effective in minute quantities. Social insects, such as bees, ants, wasps, and termites, can recognize a nestmate—or an alien in the nest—by means of identification pheromones. An intruder from another species is violently attacked at once. If the intruder is from the same species but another colony, there may be a variety of responses. It may be attacked and killed; it may be investigated but finally accepted; or it may be accepted but given less food than the others until it has had time to acquire the colony odor. Caste determination in termites, and to some extent in ants and bees, is determined by pheromones. In fact, pheromones are probably a primary integrating force in populations of social insects. Many insect pheromones have been extracted and chemically identified.

Sound production and **reception** (phonoproduction and phonoreception) in insects have been studied extensively, and although a sense of hearing is not present in all insects, this means of communication is meaningful to those insects that use it. Sounds serve as warning devices, advertisement of territorial claims, or courtship songs. The sounds of crickets and grasshoppers seem to be concerned with courtship and aggression. Male crickets scrape the modified edges of the

Figure 17-27

Tumble bugs, or dung beetles, *Canthon pilularis* (order Coleoptera), chew off a bit of dung, roll it into a ball, and then roll it to where they will bury it in soil. One beetle pushes while the other pulls. Eggs are laid in the ball, and the larvae feed on the dung. Tumble bugs are black, an inch or less in length, and common in pasture fields. Photograph by J.H. Gerard.

Some insects can memorize and perform in sequence tasks involving multiple signals in various sensory areas. Worker honeybees have been trained to walk through mazes that involved five turns in sequence, using such clues as the color of a marker, the distance between two spots, or the angle of a turn. The same is true of ants. Workers of one species of *Formica* learned a six-point maze at a rate only two or three times slower than that of laboratory rats. The foraging trips of ants and bees often wind and loop about in a circuitous route, but once the forager has found food, the return trip is relatively direct. One investigator suggests that the continuous series of calculations necessary to figure the angles, directions, distance, and speed of the trip and to convert it into a direct return could involve a stopwatch, a compass, and integral vector calculus. How the insect does it is unknown.

forewings together to produce their characteristic chirping. The long, drawn-out sound of the male cicada, a recruitment call, is produced by the vibrating membranes in a pair of organs on the ventral side of the basal abdominal segment.

There are many forms of **tactile communication,** such as tapping, stroking, grasping, and antennae touching, which evoke responses varying from recognition to recruitment and alarm. Certain kinds of flies, springtails, and beetles manufacture their own **visual signals** in the form of **bioluminescense.** The best known of the luminescent beetles are the fireflies, or lightning bugs (which are neither flies nor bugs, but beetles), in which the flash of light is used to locate a prospective mate. Each species has its own characteristic flashing rhythm produced on the ventral side of the last abdominal segments. The females flash an answer to the species-specific pattern to attract the males. This interesting "love call" has been adopted by species of *Photuris*, which prey on the male fireflies of a different species they have attracted (Figure 17-28).

Social behavior

Insects rank very high in the animal kingdom in their organization of social groups, and cooperation within the more complex groups depends heavily on chemical and tactile communication. Social communities are not all complex, however. Some community groups are temporary and uncoordinated, as are the hibernating associations of carpenter bees or the feeding gatherings of aphids. Some are coordinated for only brief periods, and some cooperate more fully, such as the tent caterpillars *Malacosoma,* that join in building a home web and feeding net. However, all of these are open communities with limited social behavior.

In the true societies of some orders, such as the Hymenoptera (honeybees and ants) and Isoptera (termites), a complex social life is necessary for the perpetuation of the species. Such societies are closed. In them all stages of the life cycle are involved, the communities are usually permanent, all activities are collective, and there is reciprocal communication and division of labor. The society is usually characterized by polymorphism, or **caste** differentiation.

The honeybees have one of the most complex organizations in the insect world. Instead of lasting one season, their organization continues for a more or less indefinite period. As many as 60,000 to 70,000 honeybees may be found in a single hive. Of these, there are three castes: a single sexually mature female, or **queen;** a few hundred **drones,** which are sexually mature males; and the **workers,** which are sexually inactive genetic females (Figure 17-29).

The workers take care of the young, secrete wax with which they build the six-sided cells of the honeycomb, gather the nectar from flowers, manufacture honey, collect pollen, and ventilate and guard the hive. One drone, sometimes more, fertilizes the queen during the mating flight, at which time enough sperm is stored in her spermatheca to last her a lifetime.

Castes are determined partly by fertilization and partly by what is fed to the larvae. Drones develop from unfertilized eggs (and consequently are haploid); queens and workers develop from fertilized eggs (and thus are diploid). Female larvae that are destined to become queens are fed royal jelly, a secretion from the salivary glands of the nurse workers. Royal jelly differs from the "worker jelly" fed to ordinary larvae, but the components in it that are essential for queen determination have not yet been identified. Honey and pollen are added to the worker diet about the third day of larval life. Female workers are prevented from maturing sexually by pheromones in the "queen substance," which is produced by the queen's mandibular glands. Royal jelly is produced by the workers only when the level of "queen substance" pheromone in the colony drops. This occurs when the

Figure 17-28

Firefly femme fatale, *Photuris versicolor,* eating a male *Photinus tanytoxus,* which she attracted with false mating signals. Photograph by J.E. Lloyd.

Figure 17-29

Queen bee surrounded by her court. The queen is the only egg layer in the colony. The attendants, attracted by her pheromones, constantly lick her body. As food is transferred from these bees to others, the queen's presence is communicated throughout the colony.
Photograph by K. Lorenzen © 1979. Educational Images, Lyons Falls, N.Y.

A   B

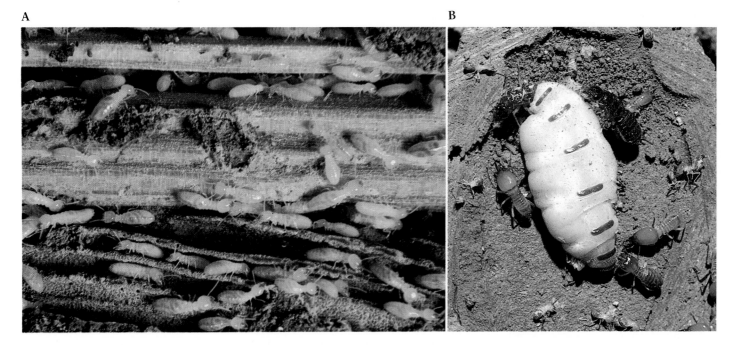

Figure 17-30

A, Termite workers, *Reticulitermes flavipes* (order Isoptera), eating yellow pine. Workers are wingless sterile adults that tend the nest, care for the young, and so forth. **B,** Termite queen (*Macrotermes bellicosus* from Ghana) becomes a distended egg-laying machine. The queen and several workers and soldiers are shown here.
A, Photograph by J.H. Gerard; **B,** photograph by K. Sandved.

queen becomes too old, dies, or is removed. Then the workers' ovaries develop, and they start enlarging a larval cell and feeding the larva the royal jelly that produces a new queen.

Honeybees have evolved an efficient system of communication by which, through certain body movements, their scouts inform the workers of the location and quantity of food sources (Figure 34-16, p. 745).

Termite colonies contain several castes, consisting of fertile individuals, both males and females, and sterile individuals (Figure 17-30). Some of the fertile individuals may have wings and may leave the colony, mate, lose their wings, and as **king** and **queen** start a new colony. Wingless fertile individuals may under certain conditions substitute for the king or queen. Sterile members are wingless and become **workers** and **soldiers.** Soldiers have large heads and mandibles and serve for the defense of the colony. As in bees and ants, caste differentiation is caused by extrinsic factors. Reproductive individuals and soldiers secrete inhibiting pheromones that are passed throughout the colony to the nymphs through a mutual feeding process, called **trophallaxis,** so that they become sterile workers. Workers also produce pheromones, and if the level of "worker substance" or "soldier substance" falls, as might happen after an attack by marauding predators, for example, compensating proportions of the appropriate caste are produced in the next generation.

Ants also have highly organized societies. Superficially, they resemble termites, but they are quite different (belong to a different order) and can be distinguished easily. In contrast to termites, ants are usually dark in color, are hard bodied, and have a constriction between the thorax and abdomen.

In ant colonies the males die soon after mating and the queen either starts her own new colony or joins some established colony and does the egg laying. The sterile females are wingless workers and soldiers that do the work of the colony: gather food, care for the young, and protect the colony. In many larger colonies there may be two or three types of individuals within each caste.

Ants have evolved some striking patterns of "economic" behavior, such as making slaves, farming fungi, herding "ant cows" (aphids or other homopterans) (Figure 17-31, *A*), sewing their nests together with silk (Figure 17-31, *B*), and using tools.

Figure 17-31

A, Ants attending treehopper nymphs on a jackfruit in Brazil. **B,** A weaver ant nest in Australia.

A, Photograph by K. Sandved; **B,** photograph by L.S. Roberts.

Insects and Human Welfare

Beneficial insects

Although most of us think of insects primarily as pests, humanity would have great difficulty in surviving if all insects were suddenly to disappear. Some of them produce useful materials: honey and beeswax from bees, silk from silkworms, and shellac from a wax secreted by the lac insects. More important, however, insects are necessary for the cross-fertilization of many fruits and other crops.

Very early in their evolution, insects and higher plants formed a relationship of mutual adaptations that have been to each other's advantage. Insects exploit flowers for food, and flowers exploit insects for pollination. Each floral development of petal and sepal arrangement is correlated with the sensory adjustment of certain pollinating insects. Among these mutual adaptations are amazing devices of allurements, traps, specialized structures, and precise timing.

Many predaceous insects, such as tiger beetles, aphid lions, ant lions, praying mantids, and lady bird beetles, destroy harmful insects. Some insects control harmful ones by parasitizing them or by laying their eggs where their young, when hatched, may devour the host. Dead animals are quickly consumed by maggots hatched from eggs laid in carcasses (Figure 17-32).

Insects and their larvae serve as an important source of food for many birds, fish, and other animals.

Harmful insects

Harmful insects include those that eat and destroy plants and fruits, such as grasshoppers, chinch bugs, corn borers, boll weevils, grain weevils, San Jose scale, and scores of others (Figure 17-33). Practically every cultivated crop has some insect pest. Lice, bloodsucking flies, warble flies, botflies, and many others attack humans or domestic animals or both. Malaria, carried by the *Anopheles* mosquito, is still one of the world's killers; yellow fever and filariasis are also transmitted by mosquitoes. Fleas carry plague, which at many times in history has almost wiped out whole human populations. The housefly is the vector of typhoid, as is the louse for typhus fever; the tsetse fly carries African sleeping sickness; and bloodsucking bug, *Rhodnius,* is a carrier of Chagas' disease. In addition there is tremendous destruction of food, clothing, and property by weevils, cockroaches,

Figure 17-32

Fly maggots (order Diptera) feeding on a deer carcass.
Photograph by L.L. Rue, III.

A   B   C

Figure 17-33

Insect pests. **A,** Japanese beetles, *Popillia japonica* (order Coleoptera), are serious pests of fruit trees and ornamental shrubs. They were introduced into the United States from Japan in 1917. **B,** Walnut caterpillars, *Datana ministra* (order Lepidoptera), defoliating a hickory tree. **C,** Corn ear worms, *Heliothis zea* (order Lepidoptera). An even more serious pest of corn is the infamous corn borer, an import from Europe in 1908 or 1909.

A, Photograph by L.L. Rue, III; **B** and **C,** photographs by J.H. Gerard.

ants, clothes moths, termites, and carpet beetles. Not the least of the insect pests is the bedbug, *Cimex,* a bloodsucking hemipterous insect that humans may have contracted, probably early in their evolution, from bats that shared their caves.

Control of insects

Because all insects are an integral part of the ecological communities to which they belong, their total destruction would probably do more harm than good. Food chains would be disturbed, some of our favorite birds would disappear, and the biological cycles by which dead animal and plant matter disintegrates and returns to enrich the soil would be seriously impeded. The beneficial role of insects in our environment has often been overlooked, and in our zeal to control the pests we have indiscriminately sprayed the landscape with extremely effective "broad-spectrum" insecticides that eradicate the good, as well as the harmful, insects. We have also found, to our chagrin, that many of the chemicals we have used persist in the environment and accumulate as residues in the bodies of animals higher up in the food chains. Also, many strains of insects have developed a resistance to the insecticides in common use.

In recent years an effort has been made to use pesticides that are specific in their targets. In addition to chemical control, other methods of control have been under intense investigation and experimentation.

The development of **insect-resistant crops** is one area of investigation. So many factors, such as yield and quality, are involved in developing resistant crops that the teamwork of specialists from many related fields is required. Some progress, however, has been made.

Several types of biological controls have been developed and are under investigation. All of these areas present problems but also show great possibilities. One is the use of bacterial and viral pathogens. A bacterium, *Bacillus thuringiensis,* has been used to control several lepidopteran pests (cabbage looper, imported cabbage worm, tomato worm), and several other bacteria show some potential. Many viruses that seem to have potential as insecticides have been isolated. However, specific viruses are difficult to rear and could be expensive to put into commercial production.

Introduction of natural predators or parasites of the insect pests has met with some success. In the United States the vedalia beetle was brought from Australia to counteract the work of the cottony-cushion scale on citrus plants, and numerous instances of control by use of insect parasites have been recorded.

Another approach to biological control is to interfere with the reproduction or behavior of insect pests with sterile males or with naturally occurring organic compounds that act as hormones or pheromones. Such research, although very

The sterile male approach has been used effectively in eradicating screwworm flies, a livestock pest. Large numbers of male insects, sterilized by irradiation, are introduced into the natural population; females that mate with the sterile flies lay infertile eggs.

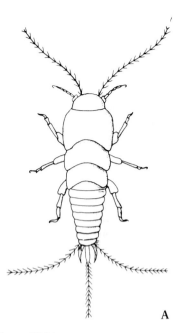

Figure 17-34

A, Silverfish *Lepisma* (order Thysanura) is often found in homes. **B** to **D,** Springtails (order Collembola). **B,** *Anurida.* **C** and **D,** *Orchesetta* in resting and leaping positions.

promising, is slow because of our limited understanding of insect behavior and the problems of isolating and identifying complex compounds that are produced in such minute amounts. Nevertheless, pheromones will probably play an important role in biological pest control in the future.

A systems approach referred to as **integrated pest management** is receiving increased attention. This involves integrated utilization of all possible, practical techniques to contain pest infestations at a tolerable level, for example, cultural techniques (resistant plant varieties, crop rotation, tillage techniques, timing of sowing, planting or havesting, and others), use of biological controls, and sparing use of insecticides.

CLASSIFICATION

Insects are divided into orders on the basis of wing structure, mouthparts, metamorphosis, and so on. Entomologists do not all agree on the names of the orders or on the limits of each order. Some choose to combine and others to divide the groups. However, the following synopsis of the orders is one that is rather widely accepted. (The most important orders are marked with an *.)

> **Subclass Apterygota** (ap-ter-y-go′ta) (Gr. *a,* not, + *pterygōtos,* winged) (**Ametabola**). Primitive **wingless** insects.
>> **Order Protura** (pro-tu′ra) (Gr. *protos,* first, + *oura,* tail). Minute (1 to 1.5 mm); no eyes or antennae; appendages on abdomen as well as thorax; live in soil and dark, humid places; slight, gradual metamorphosis.
>> **Order Diplura** (dip-lu′ra) (Gr. *diploos,* double, + *oura,* tail): **japygids.** Usually less than 10 mm; pale, eyeless; a pair of long terminal filaments or pair of caudal forceps; live in damp humus or rotting logs; development direct.
>> **Order Collembola** (col-lem′bo-la) (Gr. *kolla,* glue, + *embolon,* peg, wedge): **springtails** and **snow fleas.** Small (5 mm or less); no eyes; respiration by trachea or body surface; a springing organ folded under the abdomen for leaping (Figure 17-34, *B* to *D*); abundant in soil; sometimes swarm on pond surface film or on snowbanks in spring; development direct.
>> **Order Thysanura** (thy-sa-nu′ra) (Gr. *thysanos,* tassel, + *oura,* tail): **silverfish** and **bristletails** (Figure 17-34, *A*). Small to medium size; large eyes; long antennae; three long terminal cerci; live under stones and leaves and around human habitations; development direct.
> **Subclass Pterygota** (ter-y-go′ta) (Gr. *pterygōtos,* winged) (**Metabola**). **Winged insects** (some secondarily wingless) **with metamorphosis;** includes 97% of all insects.
>> **Superorder Exopterygota** (ek-sop-ter-i-go′ta) (Gr. *exo,* outside + *pterygōtos,* winged) (**Hemimetabola**). Metamorphosis hemimetabolous; wings develop externally on larvae; compound eyes present on larvae; larvae called **nymphs** (or **naiads** if aquatic).
>>> **Order Ephemeroptera** (e-fem-er-op′ter-a) (Gr. *ephēmeros,* lasting but a day, + *pteron,* wing): **mayflies** (Figure 17-35). Wings membranous; forewings larger than hindwings; adult mouthparts vestigial; nymphs aquatic, with lateral tracheal gills.
>>> **Order Odonata** (o-do-na′ta) (Gr. *odontos,* tooth, + *ata,* characterized by): **dragonflies, damselflies** (Figures 17-21, *A,* and 17-24, *B*). Large; membranous

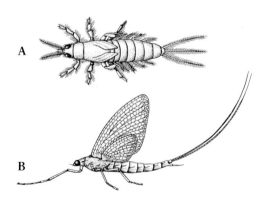

Figure 17-35

A and **B,** Mayfly (order Ephemeroptera) naiad and adult.

wings are long, narrow, net veined, and similar in size; long and slender body; aquatic nymphs with aquatic gills and prehensile labium for capture of prey.

*Order Orthoptera (or-thop'ter-a) (Gr. *orthos*, straight, + *pteron*, wing): grasshoppers (Fig. 17-4), locusts, crickets, cockroaches, walking sticks (Figure 17-9, *B*), praying mantids (Fig. 17-7). Wings, when present, with forewings thickened and hindwings folded like a fan under forewings; chewing mouthparts.

Order Dermaptera (der-map'ter-a) (Gr. *derma*, skin, + *pteron*, wing): earwigs (Figure 17-36). Very short forewings; large and membranous hindwings folded under forewings when at rest; biting mouthparts; forcepslike cerci.

Order Plecoptera (ple-kop'ter-a) (Gr. *plekein*, to twist, + *pteron*, wing): stoneflies (Figure 17-24, *A*). Membranous wings; larger and fanlike hindwings; aquatic nymph with tufts of tracheal gills.

*Order Isoptera (i-sop'ter-a) (Gr. *isos*, equal, + *pteron*, wing): termites (Figure 17-30). Small; membranous, narrow wings similar in size with few veins; wings shed at maturity; erroneously called "white ants"; distinguishable from true ants by broad union of thorax and abdomen; complex social organization.

Order Embioptera (em-bi-op'ter-a) (Gr. *embios*, lively, + *pteron*, wing): webspinners. Small; male wings membranous, narrow, and similar in size; wingless females; chewing mouthparts; colonial; make silk-lined channels in tropical soil.

Order Psocoptera (so-cop'ter-a) (Gr. *psoco*, rub small, + *pteron*, wing) (Corrodentia): psocids, "book lice," "bark lice." Body usually small, may be as large as 10 mm; membranous, narrow wings with few veins, usually held rooflike over abdomen when at rest; some wingless species; found in books, bark, bird nests, on foliage.

Order Zoraptera (zo-rap'ter-a) (Gr. *zōros*, pure, + *apterygos*, wingless): zorapterans. As large as 2.5 mm; membranous, narrow wings usually shed at maturity; colonial and termitelike.

Order Mallophaga (mal-lof'a-ga) (Gr. *mallos*, wool, + *phagein*, to eat): biting lice. As large as 6 mm; wingless; chewing mouthparts; legs adapted for clinging to host; live on birds and mammals.

*Order Anoplura (an-o-plu'ra) (Gr. *anoplos*, unarmed, + *oura*, tail): sucking lice (Figure 17-37). Depressed body; as large as 6 mm; wingless; mouthparts for

Figure 17-36

Earwig (order Dermaptera). Forcepslike cerci at posterior end are usually better developed in male and are used as organs for defense and offense. (Stained preparation, greatly enlarged.)
Photograph by F.M. Hickman.

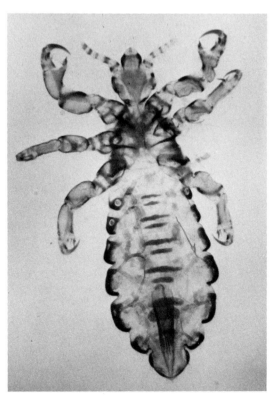

Figure 17-37

Pediculus humanus (order Anoplura), the head and body louse of humans.
Photograph by Warren Buss, from Schmidt, G.D., and L.S. Roberts. 1985. Foundations of parasitology, ed. 3. St. Louis, The C.V. Mosby Co.

Figure 17-38

A, *Papilio krishna* (order Lepidoptera) is a beautiful swallowtail butterfly from India. Members of the Papilionidae grace many areas of the world, both tropical and temperate, including North America. Compare the knobbed antennae with the plumed antennae in **B,** *Rothschildia jacobaea,* a saturniid moth from Brazil..*Hyalophora cecropia* is a common saturniid in North America. **C,** Paper wasp (order Hymenoptera) attending her pupae. **D,** *Curculio proboscideus,* the chestnut weevil, is a member of the largest family (Curculionidae) of the largest insect order (Coleoptera). This family includes many serious agricultural pests.

Photographs by K. Sandved.

piercing and sucking; adapted for clinging to warm-blooded host; includes the head louse, body louse, crab louse, others.

Order Thysanoptera (thy-sa-nop′ter-a) (Gr. *thysanos,* tassel, + *pteron,* wing): **thrips.** Length 0.5 to 5 mm (a few longer); wings, if present, long, very narrow, with few veins, and fringed with long hairs; sucking mouthparts; destructive plant-eaters, but some feed on insects.

*****Order Hemiptera** (he-mip′ter-a) (Gr. *hemi,* half, + *pteron,* wing) (Heteroptera): **true bugs.** Size 2 to 100 mm; wings present or absent; forewings with basal portion leathery, apical portion membranous; hindwings membranous; at rest, wings held flat over abdomen; piercing-sucking mouthparts; many with odorous scent glands; include water scorpions, water striders (Figure 17-11), bedbugs, squash bugs, assassin bugs, chinch bugs, stinkbugs, plant bugs, lace bugs, others.

*****Order Homoptera** (ho-mop′ter-a) (Gr. *homos,* same, + *pteron,* wing): **cicadas** (Figure 17-23), **aphids, scale insects, leafhoppers, treehoppers** (Figure 17-26, *B*). (Often included as suborder under Hemiptera.) If winged, either membranous or thickened front wings and membranous hindwings; wings held rooflike over body; piercing-sucking mouthparts; all plant-eaters; some destructive; a few serving as source of shellac, dyes, and so on; some with complex life histories.

Superorder Endopterygota (en-dop-ter-y-go′ta) (Gr. *endon,* inside, + *pterygotos,* winged) (Holometabola). Metamorphosis holometabolous; wings develop internally; larvae without compound eyes.

Order Neuroptera (neu-rop′ter-a) (Gr. *neuron,* nerve, + *pteron,* wing): **dobsonflies, ant lions, lacewings.** Medium to large size; similar, membranous wings with many cross veins; chewing mouthparts; dobsonflies with greatly enlarged mandibles in males, and with aquatic larvae; ant lion larvae (doodlebugs) make craters in sand to trap ants.

*****Order Coleoptera** (ko-le-op′ter-a) (Gr. *koleos,* sheath, + *pteron,* wing): **beetles** (Figures 17-9, *A,* 17-27, and 17-33, *A*), **fireflies** (Figure 17-28), **weevils** (Figure 17-38, *D*). The largest order of animals in the world; front wings (elytra) thick, hard, opaque; membranous hindwings folded under front wings at rest; mouthparts for biting and chewing; includes ground beetles, carrion beetles, whirligig beetles, darkling beetles, stag beetles, dung beetles, diving beetles, boll weevils, others.

Order Strepsiptera (strep-sip′ter-a) (Gr. *strepsis,* a turning, + *pteron,* wing): **stylops.** Females with no wings, eyes, or antennae; males with vestigial forewings and fan-shaped hindwings; females and larvae parasitic in bees, wasps, and other insects.

A

B

C

D

Order Mecoptera (me-kop′ter-a) (Gr. *mekos*, length, + *pteron*, wing): **scorpionflies** (Figure 17-39, *A* and *B*). Small to medium size; wings long, slender, with many veins; at rest, wings held rooflike over back; scorpion-like male clasping organ at end of abdomen; carnivorous; live in moist woodlands.

*Order Lepidoptera (lep-i-dop′ter-a) (Gr. *lepidos*, scale, + *pteron*, wing): **butterflies and moths.** Membranous wings covered with overlapping scales, wings coupled at base; mouthparts a sucking tube, coiled when not in use; larvae (caterpillars) with chewing mandibles for plant eating, stubby prolegs on the abdomen, and silk glands for spinning cocoons; antennae knobbed in butterflies and usually plumed in moths (Figure 17-38, *A* and *B*).

*Order Diptera (dip′ter-a) (Gr. *dis*, two, + *pteron*, wing): **true flies.** Single pair of wings, membranous and narrow; hindwings reduced to inconspicuous balancers (halteres); sucking mouthparts or adapted for sponging, lapping, or piercing; legless larvae called maggots or, when aquatic, wrigglers (Figure 17-18); includes crane flies, mosquitoes, moth flies, midges, fruit flies, flesh flies, houseflies, horseflies, botflies, blowflies, and many others.

Order Trichoptera (tri-kop′ter-a) (Gr. *trichos*, hair, + *pteron*, wing): **caddis flies** (Figure 17-39, *C* and *D*). Small, soft bodied; wings well veined and hairy, folded rooflike over hairy body; chewing mouthparts; aquatic larvae construct cases of leaves, sand, gravel, bits of shell, or plant matter, bound together with secreted silk or cement; some make silk feeding nets attached to rocks in stream.

*Order Siphonaptera (si-fon-ap′ter-a) (Gr. *siphon*, a siphon, + *apteros*, wingless); **fleas** (Figure 17-14). Small; wingless; bodies laterally compressed; legs adapted for leaping; no eyes; ectoparasitic on birds and mammals; larvae legless and scavengers.

*Order Hymenoptera (hi-men-op′ter-a) (Gr. *hymen*, membrane, + *pteron*, wing): **ants, bees, wasps** (Figure 17-38, *C*). Very small to large; membranous, narrow wings coupled distally; subordinate hindwings; mouthparts for biting and lapping up liquids; ovipositor sometimes modified into stinger, piercer, or saw (Figure 17-10); both social and solitary species, most larvae legless, blind, and maggotlike.

Phylogeny and Adaptive Radiation

Insect fossils, although not abundant, have been found in numbers sufficient to give a general idea of the evolutionary history of insects. Although several groups of marine arthropods, such as trilobites, crustaceans, and xiphosurans, were present in the Cambrian period, the first terrestrial arthropods—the scorpions and millipedes—did not appear until the Silurian period. The first insects, which were wingless, date from the Devonian period. By the Carboniferous period, several orders of winged insects, most of which are now extinct, had appeared.

Not all zoologists have put the same interpretation on the comparative data that are available, but certain general relationships are evident. It is believed that the insects arose from a myriapod ancestor that had paired leglike appendages. Both myriapods and insects have clearly defined heads with antennae and mandibles. However, the evolution of insects involved specialization of the next three segments to become the locomotor segments (thorax) and a loss or reduction of appendages on the rest of the body (abdomen). The primitively wingless apterygotes are undoubtedly the most primitive of the insects, and traits similar to those of the myriapods are found in them. Probably some ancestral form similar to the Protura or Thysanura gave rise to two major lines of winged insects, which differed in their ability to flex their wings. One of these led to the Odonata and Ephemeroptera, which have outspread wings that cannot be folded back over the abdomen. The other line branched into three groups, all of which were present by the Permian period. One group with hemimetabolous metamorphosis, chewing mouthparts, and cerci includes the Orthoptera, Dermaptera, Isoptera, and

Figure 17-39

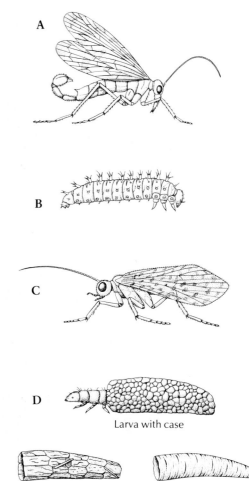

A, Male scorpionfly *Panorpa* (order Mecoptera) has recurved abdomen with scorpion-like claspers. **B,** Scorpionfly larva. **C,** Caddis fly (order Trichoptera). **D,** Several types of larval cases built by aquatic caddis fly larvae, often on the underside of stones.

Larva with case

Embioptera; another with hemimetabolous metamorphosis and a tendency toward sucking mouthparts includes the Thysanoptera, Hemiptera, and Homoptera and perhaps also the Psocoptera, Zoraptera, Mallophaga, and Anoplura, although there is some disagreement among authorities about the last group. Insects with holometabolous metamorphosis are the most specialized, and the Neuroptera, which were probably the earliest of these, may have given rise to the other endopterygote orders, with the social insects being the most advanced.

The adaptive nature of the insects has been stressed throughout this chapter. The direction and range of their adaptive radiation, both structurally and physiologically, have been amazingly varied. Whether it be in the area of habitat, feeding adaptations, means of locomotion, reproduction, or general mode of living, the adaptive achievements of the insects are truly remarkable.

SUMMARY

The uniramians include several small classes of myriapods and the largest class of animals, the Insecta. All have only one pair of antennae and have uniramous appendages.

The tagmata of the myriapods are head and trunk, and the most important groups are the centipedes (class Chilopoda) and millipedes (class Diplopoda). Centipedes are predators and bear one pair of legs on most somites. Millipedes are mostly scavengers and have two pairs of legs on most somites. All myriapods are terrestrial.

The Insecta are easily recognized by the combination of their tagmata (head, thorax, and abdomen) and the possession of three pairs of thoracic legs.

The evolutionary success of insects is largely explained by several features allowing them to exploit terrestrial habitats, such as waterproofing their cuticle and other mechanisms to minimize water loss and the ability to enter dormancy during adverse conditions.

Most insects bear two pairs of wings on their thorax, although some have one pair and some are wingless. Wing movements in some insects are controlled by synchronous, direct flight muscles, which insert directly on the base of the wings in the thorax, whereas others have asynchronous, indirect flight muscles, which move the wings by changing the shape of the thorax.

Feeding habits vary greatly among insects and can be described as phytophagous, saprophagous, predaceous, and parasitic. Within the general categories of biting, chewing, sucking, and piercing, there is an enormous variety of specialization reflecting the particular feeding habits of a given insect. Insects breathe by means of a tracheal system, which is a system of tubes that open by spiracles on the thorax and abdomen and conduct respiratory gases to and from internal tissues. Aquatic forms often have tracheal gills. Excretory organs are malpighian tubules. Insects possess efficient sense organs that can respond to mechanical, auditory, chemical, visual, and other stimuli.

Sexes are separate in insects, and fertilization is usually internal. Almost all insects undergo metamorphosis during development. In hemimetabolous (gradual) metamorphosis, the juvenile instars (nymphs) have externally developing wing buds (hence, exopterygote). The adult emerges at the last nymphal molt. In holometabolous (complete) metamorphosis, the juvenile instars (larvae) have internally developing wings (hence, endopterygote), and the last larval molt gives rise to a nonfeeding stage (pupa). A winged adult emerges at the final, pupal, molt. Hormonal control of metamorphosis is the same for both types: the ecdysiotropic hormone produced in the brain stimulates ecdysone production from the prothoracic glands, which initiates the processes of the premolt stage. The amount of juvenile hormone secreted by the corpora allata determines the degree of maturation at each molt. The ability to undergo a state of dormancy or diapause at some stage is of great value in surviving harsh conditions.

Insects can communicate with one another and affect one another's behavior or physiological state by various means, the most important of which is chemical signals (pheromones). Pheromones and some other stimuli coordinate the functioning of very complex insect societies, such as those of bees, ants, and termites. In each of these, functions in the society are divided among three or more castes.

Insects are important to human welfare, particularly because they pollinate food crop plants, control populations of other, harmful insects by predation and parasitism, and serve as food for other animals. Many insects are harmful to human interests because they feed on food and other crop plants, and many are carriers of important diseases of humans and domestic animals. There are grave difficulties with control of harmful insects by chemical insecticides, but these may be overcome by the development and more widespread use of natural and biological control methods.

Modern insects and myriapods show certain basic similarities, and insects are probably descended from one or more myriapod-like ancestors. The apterygote orders of insects are the most primitive, and the holometabolous orders are the most advanced.

Adaptive radiation and the evolutionary success of the arthropods have been enormous.

Review questions

1. Distinguish the following from each other: Diplopoda, Chilopoda, Insecta.
2. What characteristics of insects distinguish them from *all* other arthropods?
3. Explain why indirect flight muscles can beat much more rapidly than direct flight muscles.
4. How do insects walk?
5. What are the parts of the insect gut, and what are the functions of each?
6. Describe three different types of mouthparts found in insects, and tell how they are adapted for feeding on different foods.
7. Describe the tracheal system of a typical insect and explain why it is able to function efficiently without the use of oxygen-carrying pigments in the hemolymph. Why would a tracheal system not be suitable for humans?
8. Describe the unique excretory system of insects.
9. Describe the sensory receptors on insects for the various stimuli.
10. Explain the difference between holometabolous and hemimetabolous metamorphosis in insects, including the stages of each.
11. Describe the hormonal control of metamorphosis in insects, including the action of each hormone and where each is produced.
12. What is diapause, and what is its adaptive value?
13. Briefly describe three ways that insects have evolved to avoid predation.
14. Describe and give an example of each of four ways insects can communicate with each other.
15. What are the castes found in honeybees and in termites, and what is the function of each?
16. What are the mechanisms of caste determination in honeybees and termites?
17. What is trophallaxis?
18. Name several ways in which insects are beneficial to humans and several ways they are detrimental.
19. What are ways in which detrimental insects can be controlled? What is integrated pest management?
20. What is the most probable immediate ancestor of the insects, and to what lines did it apparently give rise?

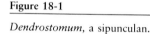

Figure 18-1

Dendrostomum, a sipunculan.

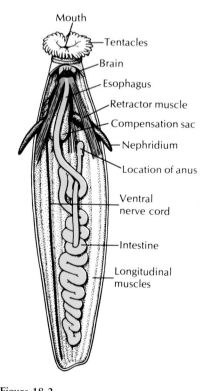

Mouth
Tentacles
Brain
Esophagus
Retractor muscle
Compensation sac
Nephridium
Location of anus
Ventral nerve cord
Intestine
Longitudinal muscles

Figure 18-2

Internal structure of *Sipunculus*.

Four of the phyla, Sipuncula, Echiura, Priapulida, and Pogonophora, are benthic (bottom-dwelling) marine worms that seem to have some affinity with the annelids. The first three have a variety of proboscis devices used in burrowing and food gathering. The pogonophores live in tubes, mostly in deep-sea mud, have long anterior tentacles, and lack a digestive tract. The Pentastomida, Onychophora, and Tardigrada have sometimes been grouped together and called the pararthropods because they have unjointed limbs with claws (at some stage) and a cuticle that undergoes molting and thus show a relationship with arthropods. The Pentastomida are entirely parasitic; the Onychophora are terrestrial but are limited to damp areas; the Tardigrada are found in marine, freshwater, and terrestrial habitats.

Phylum Sipuncula

The phylum Sipuncula (sigh-pun′kyu-la) (L. *sipunculus*, little siphon, + *ida*, pl. suffix) consists of benthic marine worms, predominantly littoral or sublittoral. They live sedentary lives in burrows in mud or sand (Figure 14-6, p. 300), occupy borrowed snail shells, or live in coral crevices or among vegetation. Some species construct their own rock burrows by chemical and perhaps mechanical means. More than half the species are restricted to tropical zones. Some are tiny, slender worms, but the majority range from 15 to 30 cm in length. Some of them are commonly known as "peanut worms" because, when disturbed, they can contract to a peanut shape (Figure 18-1).

Sipunculans have no segmentation or setae. They are most easily recognized by a slender retractile introvert, or proboscis, that is continually and rapidly being run in and out of the anterior end. The walls of the trunk are muscular. When the introvert is everted, the mouth can be seen at its tip surrounded by a crown of ciliated tentacles. Undisturbed sipunculans usually extend the anterior end from the burrow or hiding place and stretch out the tentacles to explore and feed. They are largely deposit feeders living on organic matter collected in mucus on the tentacles and moved to the mouth by ciliary action. The introvert is extended by hydrostatic pressure produced by contraction of the body wall muscles against the coelomic fluid. The lumen of the hollow tentacles is not connected to the coelom but rather to one or two blind, tubular compensatory sacs that lie along the esophagus (Figure 18-2). The sacs receive the fluid from the tentacles when the introvert is retracted. Retraction is effected by special retractor muscles. The surface of the introvert is often rough because of surface spines, hooks, or papillae.

There is a large, fluid-filled coelom traversed by muscle and connective tissue fibers. The digestive tract is a long tube that doubles back on itself to end in the anus near the base of the introvert (Figure 18-2). A pair of large nephridia open to the outside to expel waste-filled coelomic amebocytes; they also serve as gonoducts. Circulatory and respiratory systems are lacking, but the coelomic fluid contains red corpuscles that bear a respiratory pigment, hemerythrin, used in the transportation of oxygen. The nervous system has a bilobed cerebral ganglion just behind the tentacles and a ventral nerve cord extending the length of the body. The sexes are separate. Permanent gonads are lacking, and ovaries or testes develop seasonally in the connective tissue covering the origins of one or more of the retractor muscles. Sex cells are released through the nephridia. Asexual reproduction also occurs by transverse fission, the posterior one fifth of the parent constricting off to become the new individual. The larval form is usually a trochophore.

There are approximately 330 species and 16 genera, which are placed by some authorities into four families. The best-known genera are probably *Sipunculus*, *Phascolosoma* (Gr. *phaskōlos*, leather bag, pouch, + *sōma*, body), *Aspidosiphon* (Gr. *aspidos*, shield, + *siphōn*, siphon), and *Golfingia* (named by E.R. Lankester in honor of an afternoon of golfing at St. Andrews).

The early embryological development of sipunculans, echiurans, and annelids is almost identical, showing a very close relationship among the three. It is also similar to the molluscan development. The four phyla are grouped together by some authors into a supraphyletic assemblage called the "Trochozoa" because of the common possession of a trochophore larva. Other similarities, too, point to close relationship of the sipunculans to the echiurans and annelids, such as the nature of the nervous system and body wall. The sipunculans and echiurans are not metameric and thus are more primitive than annelids. They probably represent collateral evolutionary lines that branched from protoannelid stock before the origin of metamerism.

Phylum Echiura

The phylum Echiura (ek-ee-yur′a) (Gr. *echis*, viper, serpent, + *oura*, tail, + *ida*, pl. suffix) consists of marine worms that burrow into mud or sand or live in empty snail shells or sand dollar tests, rocky crevices, and so on. They are found in all oceans—most commonly in littoral zones of warm waters—but some have been found in polar waters and some have been dredged from depths of 2000 m. They vary in length from a few millimeters to 40 or 50 cm.

The echiurans have only about one third as many species (130) as the sipunculans, but they are much more diverse and are found in greater densities. There are two classes: Echiurida and Sactosomatida. Echiurida is much larger and includes two orders and five families.

The body of the echiuran is cylindrical and somewhat sausage shaped (Figure 18-3). Anterior to the mouth is a flattened, extensible proboscis which, unlike that of the sipunculids, cannot be retracted into the trunk. Echiurids are often called "spoonworms" because of the shape of the contracted proboscis in some worms. The proboscis, which contains the brain, is actually a cephalic lobe, probably homologous to the annelid prostomium. The proboscis has a ciliated groove leading to the mouth. While the animal lies buried, the proboscis can extend out over the mud for exploration and deposit feeding (Figure 18-4). *Bonellia viridis* picks up very small particles and moves them along the proboscis by cilia; larger particles are moved by a combination of cilia and muscular action or by muscular action alone. Unwanted particles can be rejected along the route to the mouth. The proboscis is short in some forms and long in others. *Bonellia*, which is only 8 cm long, can extend its proboscis to a meter in length.

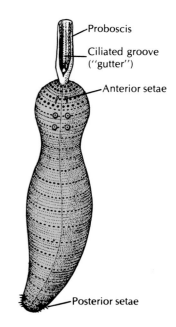

Figure 18-3

Echiurus, an echiurid common on both Atlantic and Pacific coasts. The shape of the proboscis lends them the common name of "spoon worms."

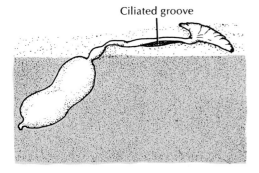

Ciliated groove

Figure 18-4

Tatjanellia (phylum Echiura) is a detritus feeder. Lying buried in the sand, it explores the surface with its long proboscis, which picks up organic particles and carries them along a ciliated groove to the mouth.
After Zenkevitch; modified from Dawydoff, C. 1959. Classe des Echiuriens. In P. Grassé (ed.). Traité de Zoologie, vol. 5. Paris, Masson et Cie.

In some species sexual dimorphism is pronounced, with the female being much the larger of the two. *Bonellia* has an extreme sexual dimorphism, and sex is determined in a very interesting way. At first the freeswimming larvae are sexually undifferentiated. Those that come into contact with the proboscis of a female become tiny males (1 to 3 mm long) that migrate to the female uterus. About 20 males are usually found in a single female. Larvae that do not contact a female proboscis metamorphose into females. It is not known whether the stimulus for male development is a chemical from the female proboscis, a matter of the chemical content of the environmental water, or a dimorphism in the eggs.

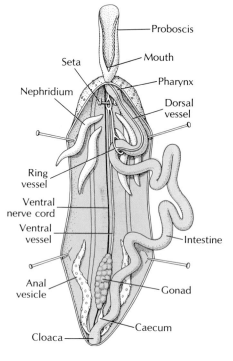

Figure 18-5

Internal anatomy of an echiuran.

One common form, *Urechis* (Gr. *oura*, tail, + *echis*, viper, serpent), lives in a U-shaped burrow in which it secretes a funnel-shaped mucous net. It pumps water through the net, capturing bacteria and fine particulate material in it. When loaded with food, the net is swallowed. *Lissomyena* (Gr. *lissos*, smooth, + *mys*, muscle) lives in empty gastropod shells in which it constructs galleries irrigated by rhythmical pumping of water and feeds on sand and mud drawn in by the irrigation process.

The muscular body wall is covered with cuticle and epithelium, which may be smooth or ornamented with papillae. There may be a pair of anterior setae or a row of bristles around the posterior end. The coelom is large. The digestive tract is long and coiled and terminates at the posterior end (Figure 18-5). A pair of anal sacs may have an excretory and osmoregulatory function. Most echiurans have a closed circulatory system with colorless blood but contain hemoglobin in coelomic corpuscles and certain body cells. Two to many nephridia serve mainly as gonoducts. A nerve ring runs around the pharynx and forward into the proboscis, and there is a ventral nerve cord. There are no specialized sense organs.

The sexes are separate, with a single gonad in each sex. The mature sex cells break loose from the gonads and leave the body cavity by way of the nephridia, and fertilization is usually external.

Early cleavage and trochophore stages are very similar to those of annelids and sipunculans. The trochophore stage, which may last from a few days to 3 months, according to the species, is followed by gradual metamorphosis to the wormlike adult.

Phylum Pogonophora

The phylum Pogonophora (po'go-nof'e-ra) (Gr. *pōgōn*, beard, + *phora*, bearing), or beardworms, was entirely unknown before the twentieth century. The first specimens to be described were collected from deep-sea dredgings off the coast of Indonesia in 1900. They have since been discovered in several seas, including the western Atlantic off the U.S. eastern coast. Some 80 species have been described so far; they have been divided into two orders: Athecanephria and Thecanephria.

These elongated tube-dwelling forms have left no known fossil record. Their closest affinity seems to be to the annelids.

Figure 18-6

Diagram of a typical pogonophoran. **A,** External features. The body, in life, is much more elongated than shown in this diagram. **B,** Position in tube.

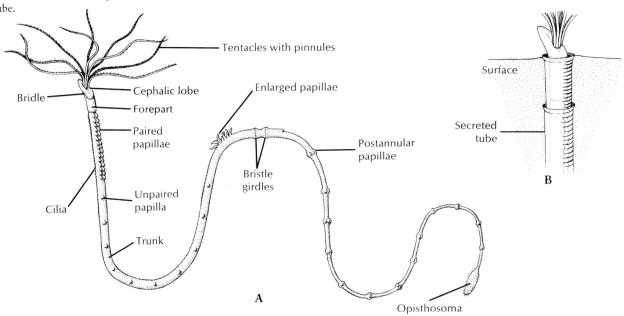

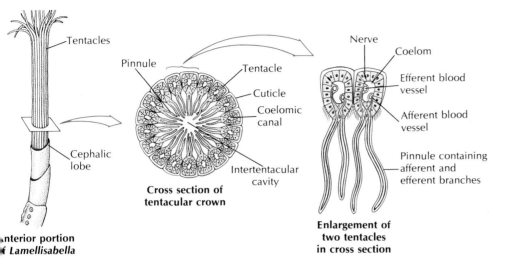

Tentacles

Pinnule

Cephalic lobe

Anterior portion of *Lamellisabella*

Tentacle

Cuticle

Coelomic canal

Intertentacular cavity

Cross section of tentacular crown

Nerve

Coelom

Efferent blood vessel

Afferent blood vessel

Pinnule containing afferent and efferent branches

Enlargement of two tentacles in cross section

Figure 18-7

Cross section of tentacular crown of pogonophore *Lamellisabella*. Tentacles arise from ventral side of forepart at base of cephalic lobe. Tentacles (which vary in number in different species) enclose a cylindrical space, with the pinnules forming a kind of food-catching network. Food may be digested in this pinnular meshwork and absorbed into the blood supply of tentacles and pinnules.

Most pogonophores live in the bottom ooze on the ocean floor, always below the intertidal zone and usually at depths of more than 200 m. This accounts for their delayed discovery, for they are obtained only by dredging. Their usual length is from 5 to 85 cm, with a diameter usually of a fraction of a millimeter. They are sessile animals that secrete very long chitinous tubes in which they live, probably extending the anterior end only for feeding. The tubes are generally oriented upright in the bottom ooze. The tube is usually about the same length as the animal, which can move up or down inside the tube but cannot turn around.

The beardworm has a long, cylindrical body covered with cuticle. The body is divided into a short anterior fore-part; a long, very slender trunk; and a small, segmented opisthosoma (Figure 18-6). At its anterior, the cephalic lobe bears from one to 260 long tentacles (the "beard" that gives the phylum its name), depending on the species. The tentacles are hollow extensions of the coelom and bear minute pinnules. For a part or all of their length the tentacles lie parallel with each other, enclosing a cylindrical intertentacular space into which the pinnules project (Figure 18-7).

The long trunk bears papillae and, about midway back, two rings of short toothed setae called girdles, which are used to grip the wall of the tube, allowing the two halves of the body to contract or extend independently in the tube. Posterior to the girdles, the trunk is very thin and easily broken when the animals are collected. In fact, the segmented tail end of the animal, or opisthosoma, was not found and described until after 1963! It is thicker than the trunk and is divided into five to 23 short segments that bear setae.

The body wall is composed of cuticle, epidermis, and circular and longitudinal muscles. The cuticle is similar in structure to that of annelids and sipunculans.

Pogonophores are remarkable in having no mouth or digestive tract, making their mode of nutrition a rather puzzling matter. Evidence suggests that they absorb nutrients dissolved in the seawater, such as glucose, amino acids, and fatty acids, through the pinnules and microvilli of their tentacles. This absorption is against a concentration gradient and is therefore an active transport process. Absorption of dissolved substances may be supplemented by phagocytosis and pinocytosis in some species. Pogonophorans are the *only nonparasitic metazoa* that lack any trace of a digestive system.

Sexes are separate, with a pair of gonads and a pair of gonoducts in the trunk section. The cleavage is unequal but atypical. It seems to be closer to radial than to

Among the most amazing animals found in the deep-water, Pacific rift communities (Chapter 41, p. 902) are the giant pogonophorans, *Riftia pachyptila* (p. 400). Much larger than any pogonophorans reported before this discovery, they measure up to 3m in length and 2 to 3 cm in diameter. Like other members of their phylum, they have no mouth or digestive tract. Their blood-red plume of tentacles provides a large surface area for absorption of nutrients from the seawater; the tentacles are partially fused into lamellae, with about 340 tentacles per lamella and 335 lamellae on each side, for a total of approximately 2.28×10^5 tentacles in the plume. Even so, it seems unlikely that sufficient dissolved nutrients would be present in seawater to support such a large animal. Evidence suggests that the bodies of the worms contain large numbers of symbiotic, chemoautotrophic bacteria in a highly vascularized organ in the trunk called the trophosome. The discovery of enzymes of sulfur metabolism in the trophosome indicates that the bacteria can oxidize the sulfide from the vent water and make use of the ATP and reducing power generated by sulfur oxidation to reduce and fix carbon dioxide. Thus, *Riftia* fixes enough carbon in the trophosome to nourish the rest of the worm.

spiral. The development of the apparent coelom is schizocoelic, not enterocoelic as was originally described. The worm-shaped embryo is ciliated but a poor swimmer. It is probably swept along by water currents until it settles.

Because the first specimens of Pogonophora that were dredged up lacked the segmented opisthosoma, Ivanov and other early workers, who believed they were working with whole specimens, described the coelom as trimeric (composed of three parts), like that of the hemichordates, and assumed that the organisms were deuterostomes. Ivanov also described the larval coelom as being trimeric. The later discovery of the segmented posterior end brought about some revision of the hypothesis. The adult coelom has proved to be polymeric, not trimeric. That fact and the schizocoelic development of the larva point toward an affinity with protostomes rather than deuterostomes. The pogonophore tubes were originally thought to resemble those of the hemichordate pterobranchs, but analysis of their amino acid and chitin content shows no relationship to the pterobranchs. Pogonophores have photoreceptor cells very similar to those of annelids (oligochaetes and leeches), and the structure of the cuticle, the makeup of the setae, and the segmentation of the opisthosoma all strongly suggest a relationship with the annelids. However, the phylogenetic position of the Pogonophora must be considered still unsettled until the embryology of more species of more than one family is studied.

Adaptive radiation has not been extensive. The chief areas of diversity are in the structure of the tentacular crown and the tube.

Phylum Priapulida

The Priapulida (pri'a-pyu'li-da) (Gr. *priapos*, phallus, + *ida*, pl. suffix) are a small group (only nine species) of marine worms found chiefly in the colder waters of both hemispheres. They have been reported along the Atlantic coast from Massachusetts to Greenland and along the Pacific coast from California to Alaska. They live in the mud and sand of the sea floor and range from intertidal zones to depths of several thousand meters.

Their cylindrical bodies are rarely more than 12 to 15 cm long. Most of them are burrowing predaceous animals that usually orient themselves upright in the mud with the mouth at the surface. They are adapted for burrowing by body contractions. *Tubiluchus* (L. *tubulus*, dim. of *tubus*, waterpipe) is a minute detritus feeder adapted to interstitial life in warm coralline sediments. *Maccabeus* (named for a Judean patriot who died in 160 BC) is a tiny tube-dweller discovered in muddy Mediterranean bottoms.

The body includes a proboscis, trunk, and usually one or two caudal appendages (Figure 18-8). The eversible proboscis is ornamented with papillae and ends with rows of curved spines that surround the mouth. The proboscis is used in sampling the surroundings as well as for the capture of small, soft-bodied prey. The genus *Maccabeus* has a crown of branchial tentacles around the mouth.

The trunk is not truly segmented but is superficially divided into 30 to 100 rings and is covered with tubercles and spines. The tubercles are probably sensory in function. The anus and urogenital pores are located at the posterior end of the trunk. The caudal appendages are hollow stems believed to be respiratory and probably chemoreceptive in function. The body is covered with a chitinous cuticle that is molted periodically throughout life.

The digestive system contains a muscular pharynx and a straight intestine and rectum (Figure 18-8). There is a nerve ring around the pharynx and a mid-

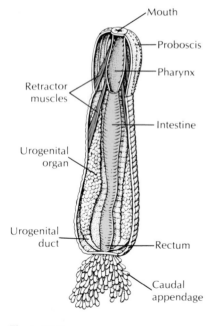

Figure 18-8

Major internal structures of *Priapulus*.

Labels on figure: Mouth, Proboscis, Pharynx, Retractor muscles, Intestine, Urogenital organ, Urogenital duct, Rectum, Caudal appendage ventral nerve cord. The coelom contains amebocytes and, at least in *Priapulus caudatus*, corpuscles bearing hemerythrin.

Sexes are separate. The paired urogenital organs are each made up of a gonad and clusters of solenocytes, both connected to a protonephridial tubule that carries both gametes and excretory products to the outside. The embryology is poorly known. In some the egg undergoes radial cleavage and develops into a stereogastrula. The larvae of *Priapulus* dig into the mud and become detritus feeders.

Long thought to be pseudocoelomate, priapulids were judged coelomate when nuclei were found in the membranes lining the body cavity, the membranes thus representing a peritoneum. However, a recent report maintains that the nuclei originate from amebocytes and that the membranes are acellular. Therefore the status of the priapulids is still unsettled, and their relationship to other groups is obscure.

Phylum Pentastomida

The Pentastomida (pen-ta-stom′i-da) (Gr. *pente*, five, + *stoma*, mouth), or tongue worms, are a phylum of about 60 to 70 species of wormlike parasites of the respiratory system of vertebrates. The adults live mostly in the lungs of reptiles, such as snakes, lizards, and crocodiles, but one species, *Reighardia sternae*, lives in the air sacs of terns and gulls, and another, *Linguatula serrata* (Gr. *lingua*, tongue), lives in the nasopharynx of canines and felines (and occasionally humans). Although more common in tropical areas, they are also found in North America, Europe, and Australia.

The adults range from 1 to 13 cm in length. Transverse rings give their bodies a segmented appearance (Figure 18-9). The body is covered with a chitinous cuticle that is molted periodically during larval stages. The anterior end may bear five short protuberances (hence the name Pentastomida). Four of these bear claws. The fifth bears the mouth and two pairs of sclerotized hooks for attachment to the host tissues (Figure 18-10). There is a simple straight digestive system, adapted for sucking. The nervous system, similar to that of annelids and arthropods, has paired ganglia along the ventral nerve cord. The only sense organs appear to be papillae. There are no circulatory, excretory, or respiratory organs.

Sexes are separate, and the females are usually larger than the males. A

Figure 18-9

Two pentastomids. **A,** *Pentastomum*, found in lungs of snakes and other vertebrates. Female is shown with some internal structures. **B,** Female *Armillifer*, a pentastomid with pronounced body rings. In parts of Africa and Asia, humans are parasitized by immature stages; adults (10 cm long or more) live in lungs of snakes. Human infection may occur from eating snakes or from contaminated food or water.

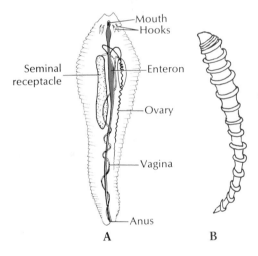

Figure 18-10

Anterior end of a pentastome. Note both the mouth *(arrow)* between the middle hooks and the apical sensory papillae. Coutesy J. Ubelaker. From Schmidt, G.D., and L.S. Roberts. 1981. Foundations of parasitology, ed. 2. St. Louis, The C.V. Mosby Co.

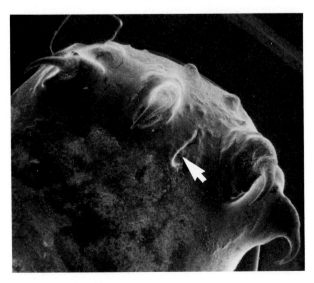

female may produce several million eggs, which pass up the trachea of the host, are swallowed, and pass out with the feces. The larvae hatch out as oval, tailed creatures with four stumpy legs. Most pentastomid life cycles require an intermediate vertebrate host such as a fish, a reptile, or, rarely, a mammal, that is eaten by the definitive vertebrate host. After ingestion by the intermediate host, the larva penetrates the intestine, migrates randomly in the body, and finally metamorphoses into a nymph. After growth and several molts, the nymph finally becomes encapsulated and dormant. When eaten by the final host, the juvenile finds its way to the lung, feeds on blood and tissue, and matures.

Several species have been found encysted in humans, the most common being *Armillifer armillatus* (L. *armilla*, ring, bracelet, + *fero*, to bear), but usually they cause few symptoms. *Linguatula serrata* is a cause of nasopharyngeal pentastomiasis, or "halzoun," a disease of humans in the Middle East and India.

The phylogenetic affinities of the Pentastomida are uncertain. They have some similarities to the Annelida, and some workers have believed that they are off-shoots from the polychaetes. Their larval appendages and molting cuticle, however, are arthropod characteristics. Their larvae resemble tardigrade larvae. Most modern taxonomists align them with the arthropods, but there is little agreement as to where they fit in that phylum. Because their modifications for parasitic life make it difficult to do more than guess at the free-living forms from which they arose, it seems best to keep them in a separate phylum.

Phylum Onychophora

Members of the phylum Onychophora (on-y-kof′o-ra) (Gr. *onyx*, claw, + *pherein*, to bear) are commonly called the "velvet worms," or "walking worms." They compose approximately 70 species of caterpillar-like animals, ranging from 1.4 to 15 cm in length. They live in rain forests and other moist, leafy habitats in tropical and subtropical regions and in some temperate regions of the Southern Hemisphere.

The fossil record of the onychophorans shows that they have changed little in their 500 million year history. A fossil form, *Aysheaia*, discovered in the Burgess shale deposit of British Columbia and dating back to mid-Cambrian times, is very much like the modern onychophorans. Onychophorans have been of unusual interest to zoologists because they share so many characteristics with both the annelids and the arthropods. They have been called, a bit too hopefully perhaps, the "missing link" between the two phyla. Onychophorans were probably far more common at one time than they are now. Today they are terrestrial and extremely retiring, coming out only at night or when the air is nearly saturated with moisture.

Form and function

External features

The onychophoran body is more or less cylindrical and shows no external segmentation except for the paired appendages (Figure 18-11). The skin is soft and velvety and is covered with a thin, flexible cuticle that contains protein and chitin. In structure and chemical composition it resembles arthropod cuticle; however, it never hardens like arthropod cuticle, and it is molted in patches rather than all at one time. The body is studded with tiny **tubercles,** some of which bear sensory bristles. The color may be green, blue, orange, dark gray, or black, and minute scales on the tubercles give the body an iridescent and velvety appearance. The

Antenna
Oral papilla
Oral lobes
First leg

Ventral view of head

In natural habitat

Figure 18-11

Peripatus, a caterpillar-like onychophoran that has both annelid and arthropod characteristics.

head bears a pair of large **antennae**, each with an annelid-like eye at the base (Figure 18-11). The ventral mouth has a pair of clawlike **mandibles** and is flanked by a pair of **oral papillae** from which a defensive secretion can be expelled.

The **unjointed legs** are short, stubby, and clawed. Locomotion is achieved by waves of contraction passing from anterior to posterior. When a segment is extended, the legs are lifted up and moved forward. The legs are more ventrally located than are the parapodia of annelids.

Internal features

The body wall is muscular like that of the annelids. The body cavity is a **hemocoel,** imperfectly divided into compartments, or sinuses, much like those of the arthropods. **Slime glands** on each side of the body cavity open on the oral papillae. When disturbed by a predator, the animal can eject from the slime glands two streams of a sticky substance that rapidly hardens.

The mouth, surrounded by lobes of skin, contains a dorsal tooth and a pair of lateral mandibles used for grasping and cutting prey. There is a muscular pharynx and a straight digestive tract (Figure 18-12). Most velvet worms are predaceous, feeding on caterpillars, insects, snails, worms, and the like. Some onychophorans live in termite nests and feed on termites.

Each segment contains a pair of **nephridia**, each nephridium with a vesicle, ciliated funnel and duct, and nephridiopore opening at the base of a leg. Absorptive cells in the midgut excrete crystalline uric acid, and certain pericardial cells function as nephrocytes, storing excretory products taken from the blood.

For respiration there is a **tracheal system** that ramifies to all parts of the body and communicates with the outside by many openings, or **spiracles,** scattered all over the body. The spiracles cannot be closed to prevent water loss, so although the tracheae are efficient, the animals are restricted to moist habitats. The tracheal system is somewhat different from that of arthropods and probably has evolved independently.

The open circulatory system has, in the pericardial sinus, a dorsal, tubular heart with a pair of ostia in each segment.

There are a pair of cerebral ganglia with connectives and a pair of widely separated nerve cords with connecting commissures. The brain gives off nerves to the antennae and head region, and the cords send nerves to the legs and body wall. Sense organs include the pigment cup ocelli, taste spines around the mouth, tactile papillae on the integument, and hygroscopic receptors that orient the animal toward water vapor.

Onychophorans are dioecious, with paired reproductive organs. The males usually deposit their sperm in spermatophores in the female seminal receptacle. The male deposits the spermatophores on the female's back, which may accumulate a number of them. White blood cells dissolve the skin beneath the spermato-

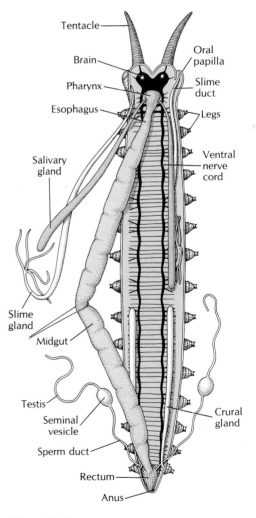

Tentacle
Brain
Pharynx
Esophagus
Oral papilla
Slime duct
Legs
Salivary gland
Ventral nerve cord
Slime gland
Midgut
Testis
Seminal vesicle
Sperm duct
Rectum
Anus
Crural gland

Figure 18-12

Internal anatomy of an onychophoran.

phores. The sperm can then enter the body cavity and migrate in the blood to the ovaries to fertilize the eggs. Onychophorans may be oviparous, ovoviparous, or viviparous. Only two Australian genera are oviparous, laying shell-covered eggs in moist places. In all other onychophorans the eggs develop in the uterus, and living young are produced. In some species there is a placental attachment between mother and young (viviparous); in others the young develop in the uterus without attachment (ovoviviparous).

Phylogeny

Onychophorans resemble the annelids with their soft body, nonjointed appendages, segmentally arranged nephridia, muscular body wall, pigment cup ocelli, and ciliated reproductive ducts. Arthropod characteristics are the cuticle, the tubular heart and open circulatory system, the presence of tracheae, a hemocoel for a body cavity, and the large size of the brain. They differ from either phylum in their scanty metamerism, structure of the mandibles, and the separate arrangement of the nerve cords. They are more primitive than insects and are somewhat like the centipedes in the arrangement of internal metamerism.

Some authors believe the onychophorans should be included with the arthropods, but that would involve redefining the phylum Arthropoda. Manton (1977) recommends placing the Onychophora with the myriapods and insects in the phylum Uniramia. Despite the onychophorans' obvious relationship to the myriapods and insects, however, most authors believe that the differences seem to warrant keeping them in a separate phylum.

—— Phylum Tardigrada

Tardigrada (tar-di-gray'da) (L. *tardus,* slow, + *gradus,* step), or "water bears," are minute forms usually less than a millimeter in length. Most of the 300 to 400 species are terrestrial forms that live in the water film surrounding mosses and lichens. Some live in freshwater algae or mosses of in the bottom debris, and a few are marine, inhabiting the interstitial spaces between sand grains, in both deep and shallow seawater. They share many characteristics with the arthropods.

Figure 18-13

Scanning electron micrograph of an aquatic tardigrade, *Pseudobiotus.*
Photographs by D.R. Nelson.

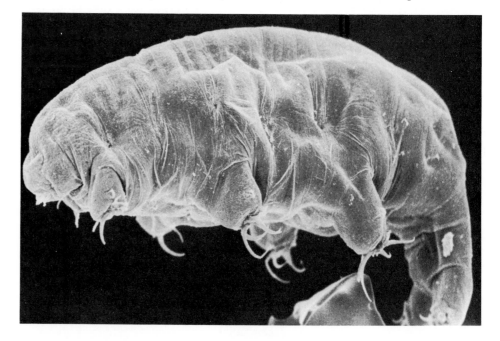

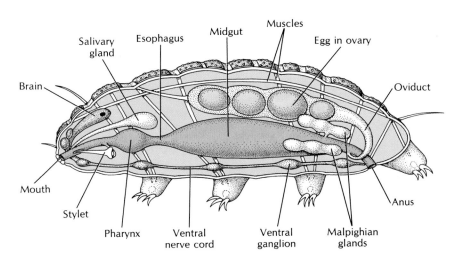

Figure 18-14

Internal anatomy of a tardigrade.

The body is elongated, cylindrical, or a long oval and is unsegmented. The head is merely the anterior part of the trunk. The trunk bears four pairs of short, stubby, unjointed legs, each armed with four to eight claws (Figure 18-13). The body is covered by a nonchitinous cuticle that is molted along with the claws and buccal apparatus four or more times in the life history. Cilia are absent. Common American genera are *Macrobiotus* (Gr. *makros,* large, + *biotos,* life), *Echiniscus* (Gr. *echinos,* hedgehog, + *iskos,* dim. suffix), and *Hypsibius* (Gr. *hypsos,* high height, + *bios,* life).

The mouth opens into a buccal tube that empties into a muscular pharynx that is adapted for sucking (Figure 18-14). Two needlelike stylets flanking the buccal tube can be protruded through the mouth. The stylets are used for piercing the cellulose walls of plant cells, and the liquid contents are then sucked in by the pharynx. Some tardigrades suck the body juices of nematodes, rotifers, and other small animals. Some, such as *Echiniscus,* expel feces when molting, leaving the feces in the discarded cuticle. At the junction of the stomach and rectum, three glands, thought to be excretory and often called Malpighian tubules, empty into the digestive system.

Most of the body cavity is a hemocoel, with the true coelom restricted to the gonadal cavity. There are no circulatory or respiratory systems, gaseous exchange occurring through the body surface.

The muscular system consists of a number of long muscle bands. Circular muscles are absent, but the hydrostatic pressure of the body fluid may act as a skeleton. Being unable to swim, the water bear creeps about awkwardly, clinging to the substrate with its claws.

The brain is large and covers most of the dorsal surface of the pharynx. Circumpharyngeal connectives link it to the subpharyngeal ganglion, from which the double ventral nerve cord extends posteriorly as a chain of four ganglia.

Sexes are separate in tardigrades. In some freshwater and moss-dwelling species, males are unknown and parthenogenesis seems to be the rule. In marine species, however, males and females occur with approximately equal frequency. Eggs of some species are highly ornate (Figure 18-15). Egg laying, like defecation, apparently occurs only at molting, when the volume of coelomic fluid is reduced. Females of some species deposit the eggs in the molted cuticle (Figure 18-16). Males gather around the old cuticle and shed sperm into it. Other species are fertilized internally but only at the time of molting.

Cleavage is holoblastic but atypical, and a stereogastrula is formed. Six pairs of coelomic pouches arise from the gut, but all except the last pair disaggregate to

Figure 18-15

Scanning electron micrograph of the highly ornate egg of the tardigrade, *Macrobiotus hufelandii.*
Photograph by D.R. Nelson.

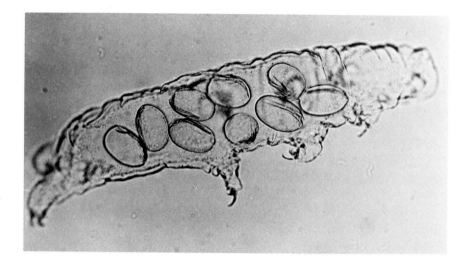

Figure 18-16

Molted cuticle of a tardigrade, containing a number of fertilized eggs.
From Sayre, R.M. 1969. Trans. Am. Microsc. Soc. 88:266-274.

form the buccal apparatus, pharynx, and body musculature. The last pair fuses to form the gonad. Thus, the gonocoel (which is enterocoelic) is the only true coelom left in the adult. Development is direct.

One of the most intriguing features of terrestrial tardigrades is their capacity to enter a state of suspended animation, called cryptobiosis (formerly called anabiosis), during which metabolism is virtually imperceptible; the organism can withstand harsh environmental conditions. Under gradual drying conditions, the water content of the body is reduced from 85% to only 3%, movement ceases, and the body becomes barrel shaped. In a cryptobiotic state tardigrades can resist temperature extremes, ionizing radiation, oxygen deficiency, and other adverse conditions and may survive for years. Activity resumes when moisture is again available.

The affinities of tardigrades are among the most puzzling of all animal groups. They have some similarities to the rotifers, particularly in their reproduction and their cryptobiotic tendencies, and some authors call them pseudocoelomates. Their embryology, however, would seem to put them among the coelomates. The nervous system indicates a relationship to the annelids and arthropods. Some authors place them close to the arthropods, particularly the mites. Their enterocoelic origin of the mesoderm is a deuterostome characteristic. For the present, at least, their status is quite uncertain.

A colony of giant beardworms at great depth near a hot sulphur vent along the Galapagos Trench, eastern Pacific Ocean. Photograph by Jack Donnelley, Woods Hole Oceanographic Institute. From Raven, P.H., and G.B. Johnson, 1986. Biology. St. Louis, Times Mirror/Mosby College Publishing.

SUMMARY

The seven small phyla covered in this chapter are grouped together here for convenience. The Sipuncula, Echiura, and probably the Pogonophora are related to the annelids; the Pentastomida, Onychophora, and Tardigrada seem closer to the arthropods. The relationship of the Priapulida to other coelomates is obscure.

Sipunculans are small, burrowing marine worms with an eversible introvert at their anterior end. The introvert bears tentacles with which they use for deposit-feeding. They are not metameric.

Echiurans are more diverse than sipunculans, but fewer in number of species. They are also burrowing marine worms, and most are deposit-feeders, with a proboscis anterior to their mouth. They also are not metameric.

Pogonophorans live in tubes on the deep ocean floor, and they are metameric. They have no mouth or digestive tract but apparently absorb nutrient by the crown of tentacles at their anterior end.

The Priapulida are a tiny group of burrowing, marine worms with an eversible proboscis. They are mostly predaceous, and in most species the mouth is surrounded by recurved spines for capturing prey. They are not metameric but show superficial segmentation.

The Pentastomida are wormlike parasites in the lungs and nasal passages of carnivorous vertebrates. They are clearly related to the arthropods.

The Onychophora are caterpillar-like animals found in humid, mostly tropical habitats. They are metameric and crawl by means of a series of unjointed, clawed appendages. They show both annelid and arthropod characteristics.

Tardigrades are minute animals, mostly terrestrial, living in the water film that surrounds mosses and lichens. They have eight unjointed legs and a nonchitinous cuticle. Their chief body cavity is a hemocoel, as in arthropods. They can undergo cryptobiosis, withstanding adverse conditions for long periods.

Review questions

1. Give three distinctive characteristics for each of the following, and describe each one's habitat: Sipuncula, Echiura, Pogonophora, Priapulida, Pentastomida, Onychophora, Tardigrada.
2. What do the members of each of the aforementioned groups feed upon?
3. What is the evidence that the Sipuncula and Echiura diverged from the protostome line before the origin of the annelids? Why are these phyla considered closely related?
4. What is the largest pogonophoran known? Where is it found, and how is it nourished?
5. Briefly describe the life cycle of a typical pentastomid.
6. Name three characteristics of onychophorans that are annelidlike, three that are arthropod-like, and three that differ from both groups.
7. What is the survival value of cryptobiosis in tardigrades?
8. How does the introvert of sipunculans differ from the proboscis of echiurans?

Selected references

See also general references for Part Two, p. 590.

Crowe, J.H., and A.F. Cooper, Jr. 1971. Cryptobiosis. Sci. Am. 225:30-36 (Dec.). *Cryptobiotic nematodes, rotifers, and tardigrades can withstand adverse conditions of astonishing rigor, yet perceptible metabolism continues in their state of suspended animation.*

Hackman, R.H., and M. Goldberg. 1975. Peripatus: its affinities and its cuticle. Science 190:582-583.

Jones, M.L. 1981. *Riftia pachyptila* Jones: observations on the vestimentiferan worm from the Galapagos Rift. Science 213:333-336. *This article concerns primarily the structure of the giant pogonophoran. Following it in the same issue of Science are five more papers on the worm's nutrition, blood, and bacterial symbionts.*

Manton, S.M. 1977. The Arthropoda: habits, functional morphology, and evolution. Oxford, England, Clarendon Press. *The Onychophora is considered part of the phylum Uniramia, along with the Myriapoda, the Hexapoda (Insecta) as subphyla.*

Peck, S.B. 1975. A review of the New World Onychophora with the description of a new cavernicolous genus and species from Jamaica. Psyche 82(3/4):341-358. *A key to the two families and eight genera of New World onychophorans.*

Rice, M.E., and M. Todorovic (eds.) 1975. Proceedings of the International Symposium on the biology of the Sipuncula and Echiura, 2 vols. Washington, D.C., National Museum of Natural History. *A series of technical articles, but much of interest for further reading on these two phyla.*

Southward, E.C. 1975. Fine structure and phylogeny of the Pogonophora. In E.J.W. Barrington and R.P.S. Jefferies (eds.). Protochordates. London, Zoological Society of London, no. 36.

Stephen, A.C., and S.J. Edmonds. 1972. The phyla Sipuncula and Echiura. London, British Museum (Natural History).

CHAPTER 19

THE LOPHOPHORATE ANIMALS

Phylum Phoronida

Phylum Ectoprocta (Bryozoa)

Phylum Brachiopoda

Position in Animal Kingdom

1. The lophophorate phyla possess a **true coelom,** that is, a body cavity lined with a layer of mesodermal epithelium called the peritoneum.
2. They belong to the **protostome** branch of the **bilateral** animals, but they have some of the characteristics of the deuterostomes.
3. The three phyla are usually grouped together because they all possess the crown of tentacles called a **lophophore,** which is specialized for sedentary filter feeding. The lophophore surrounds the mouth but not the anus, thus differing from the tentacular crown of Entoprocta.

Biological Contributions

1. The lophophore is a unique ridge that bears hollow, ciliated tentacles, and is an efficient, specialized filter-feeding device that forms a ciliated route, or trough, for trapping and directing food particles to the mouth.
2. The brachiopods and phoronids possess vascular systems for circulation of food nutrients and other materials.
3. The blood in phoronids possesses red blood corpuscles that contain hemoglobin for carrying oxygen.

——THE LOPHOPHORATES

The Phoronida are wormlike marine forms that live in secreted tubes in sand or mud or attached to rocks or shells. The Ectoprocta are minute forms, mostly colonial, whose protective cases often form encrusting masses on rocks, shells, or plants. The Brachiopoda are bottom-dwelling marine forms that superficially resemble molluscs because of their bivalved shells.

One might wonder why these three apparently different types of animals are lumped together in a group called lophophorates. Actually they have more in common than first appears. They are all coelomate; all have some protostome characteristics; all are sessile; and none has a distinct head. But these characteristics are also shared by other phyla. What really sets them apart from other phyla is the common possession of a ciliary feeding device called a **lophophore** (Gr. *lophos*, crest or tuft, + *phorein*, to bear).

A lophophore is a unique arrangement of ciliated tentacles borne on a ridge (a fold of the body wall), which surrounds the mouth but not the anus. The lophophore with its crown of tentacles contains within it an extension of the coelom, and the thin, ciliated walls of the tentacles are not only an efficient feeding device but also serve as a respiratory surface for exchange of gases between the environmental water and the coelomic fluid. The lophophore can usually be extended for feeding or withdrawn for protection.

In addition, all three phyla have a U-shaped alimentary canal, with the anus placed near the mouth but outside the lophophore. The coelom is divided into two compartments, the **mesocoel** and the **metacoel,** and the mesocoel extends into the hollow tentacles of the lophophore. The portion of the body that contains the mesocoel is known as the **mesosome,** and that containing the metacoel is the **metasome.** All three phyla have a free-swimming larval stage but are sessile as adults.

_____ Phylum Phoronida

The phylum Phoronida (fo-ron′i-da) (L. surname of Io, in mythology) is composed of approximately 10 species of small, wormlike animals that live on the bottom of shallow coastal waters, especially in temperate seas. They range from a few millimeters to 30 cm in length. Each worm secretes a leathery or chitinous tube in which it lies free, but which it never leaves. The tubes may be anchored singly or in a tangled mass on rocks, shells, or pilings or buried in the sand. The tentacles on the lophophore are thrust out for feeding, but if the animal is disturbed, it can withdraw completely into its tube.

The lophophore is made up of two parallel ridges curved in a horseshoe shape, the bend located ventrally and the mouth lying between the two ridges (Figure 19-1). The horns of the ridges are often coiled into twin spirals. Each ridge carries hollow ciliated tentacles, which, like the ridges themselves, are extensions of the body wall.

The cilia on the tentacles direct a water current toward a groove between the two ridges, which leads toward the mouth. Plankton and detritus caught in this current are carried by the cilia to the mouth. The anus lies dorsal to the mouth, outside the lophophore, flanked on each side by a nephridiopore (Figure 19-1). Water leaving the lophophore passes over the anus and nephridiopores, carrying away the wastes. Cilia in the stomach area of the U-shaped gut aid in food movement.

The body wall is made of cuticle, epidermis, and both longitudinal and circular muscles. The coelomic cavity is subdivided by mesenteric partitions into the mesocoel and metacoel. The phoronids have a closed system of contractile blood vessels but no heart; the red blood contains hemoglobin. There is a pair of metanephridia. A nerve ring sends nerves to the tentacles and body wall; a single giant motor fiber lies in the epidermis; and an epidermal nerve plexus supplies the body wall and epidermis.

There are both monoecious (the majority) and dioecious species of Phoronida, and at least one species reproduces asexually. Cleavage seems to be related to both the spiral and the radial types. The free-swimming, ciliated larva, called an actinotroch, metamorphoses into the adult, which sinks to the bottom, secretes a tube, and becomes sessile.

Phoronopsis californica is a large, orange form about 30 cm long found

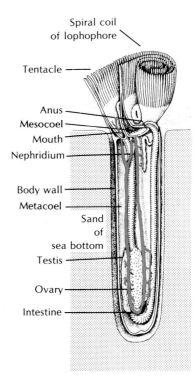

Figure 19-1

Internal structure of *Phoronis* (phylum Phoronida), in diagrammatic vertical section, showing one half of lophophore.

along the west coast of the United States. *Phoronis architecta* is a smaller (approximately 12 cm long) Atlantic coast species that has a very wide distribution.

___ Phylum Ectoprocta (Bryozoa)

The Ectoprocta (ek-to-prok'ta) (Gr. *ektos*, outside, + *proktos*, anus) have long been called bryozoans, or moss animals (Gr. *bryon*, moss, + (*zōon*, animal), a term that originally included the Entoprocta also. However, because the entoprocts are pseudocoelomates and have the anus located within the tentacular crown, they are no longer classed with the ectoprocts, which, like the other lophophorates, are eucoelomate and have the anus outside the circle of tentacles. Some authors continue to use the name "Bryozoa" but now exclude the entoprocts from the group.

Of the 4000 or so species of ectoprocts, few are more than 0.5 mm long. All are aquatic, both freshwater and marine, but are found largely in shallow waters. With very few exceptions they are colony builders. Ectoprocts have been very successful. They have left a rich fossil record since the Ordovician period. Marine forms today exploit all kinds of firm surfaces, such as shells, rocks, large brown algae, mangrove roots, and ship bottoms.

Each member of a colony lives in a tiny chamber, called a **zoecium,** which is secreted by its epidermis (Figure 19-2). Each individual, or **zooid,** consists of a feeding polypide and a case-forming cystid. The **polypide** includes the lophophore, digestive tract, muscles, and nerve centers. The **cystid** is the body wall of the animal, together with its secreted exoskeleton. The exoskeleton, or zoecium, may, according to the species, be gelatinous, chitinous, or stiffened with calcium and possibly also impregnated with sand. The shape may be boxlike, vaselike, oval, or tubular.

Some colonies form limy encrustations on seaweed, shells, and rocks; others form fuzzy or shrubby growths on erect, branching colonies that look like seaweed. Some ectoprocts might easily be mistaken for hydroids but can be distinguished under a microscope by the fact that their tentacles are ciliated (Figure 19-3). In some freshwater forms the individuals are borne on finely branching stolons that form delicate tracings on the underside of rocks or plants. Other freshwater ectoprocts are embedded in large masses of gelatinous material.

Figure 19-2

Small portion of freshwater colony of *Plumatella* (phylum Ectoprocta), which grows on the underside of rocks. These tiny individuals disappear into their chitinous zoecia when disturbed. Statoblasts are resistant capsules containing germinating cells.

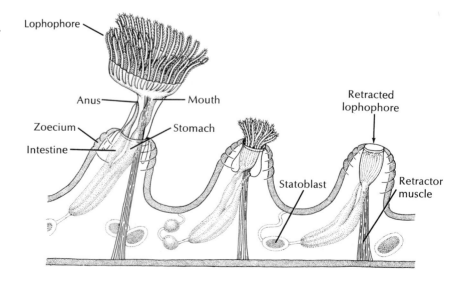

A

Although the zooids are minute, the colonies may be several centimeters in diameter, some encrusting colonies may be a meter or more in width (Figure 19-4), and erect forms may reach 30 cm or more in height. Freshwater ectoprocts may form mosslike colonies on the stems of plants or on rocks, usually in shallow ponds or pools. They may be able to slide along slowly on the object that supports them.

The polypide lives a type of jack-in-the-box existence, popping up to feed and then quickly withdrawing into its little chamber, which often has a tiny trapdoor (operculum) that shuts to conceal its inhabitant. To extend the tentacular crown, certain muscles contract, which increases the hydraulic pressure within the body cavity and pushes the lophophore out. Other muscles can contract to withdraw the crown to safety with great speed.

The lophophore ridge tends to be circular in marine ectoprocts (Figure 19-3, *A*) and U-shaped in freshwater species (Figure 19-3, *B*). When feeding, the animal extends the lophophore and spreads the tentacles out into a funnel. Cilia on the

Figure 19-3

A, Ciliated lophophore of *Flustrella,* a marine ectoproct. **B,** *Plumatella repens,* a freshwater bryozoan (phylum Ectoprocta). It grows in branching, threadlike colonies on the underside of rocks and vegetation in lakes, ponds, and streams.
A, Courtesy J.A. Cooke, Museum of Natural History; B, photograph by R. Vishniac.

Figure 19-4

Skeletal remains of a colony of *Membranipora,* a marine encrusting form of Ectoprocta. Each little oblong zoecium is the calcareous former home of a tiny ectoproct.
Photograph by B. Tallmark.

tentacles draw water into the funnel and out between the tentacles. Food particles caught by cilia in the funnel are drawn into the mouth, both by the pumping action of the muscular pharynx and by the action of cilia in the pharynx. Undesirable particles can be rejected by reversing the ciliary action, by drawing the tentacles close together, or by retracting the whole lophophore into the zoecium. Digestion in the ciliated, U-shaped digestive tract appears to be extracellular for protein and starches and intracellular for fats.

Respiratory, vascular, and excretory organs are absent. Gaseous exchange is through the body surface, and since the ectoprocts are small, coelomic fluid is adequate for internal transport. Coelomocytes engulf and store waste materials. There are a ganglionic mass and a nerve ring around the pharynx, but no sense organs are present. The coelom is divided by a septum into the anterior mesocoel in the lophophore and the larger posterior metacoel. Pores in the walls between adjoining zooids permit exchange of materials by way of the coelomic fluid.

Most colonies are made up of feeding individuals, but polymorphism also occurs. One type of modified zooid resembles a bird beak that snaps at small invading organisms that might foul a colony. Another type has a long bristle that sweeps away foreign particles.

Most ectoprocts are hermaphroditic. Some species shed eggs into the seawater, but most brood their eggs, some within the coelom and some externally in a special ovicell, which is a modified zoecium in which the embryo develops. Marine species have radial cleavage but a highly modified trochophore larva with a vibratile plume of sensory cilia, an adhesive sac, and a piriform organ. After swimming about for a time, the larva uses the vibratile plume to select a suitable site for settling, then attaches temporarily with a sticky acid mucopolysaccharide secreted from the piriform organ. Later, the adhesive sac produces acid mucopolysaccharide and protein secretions that effect a permanent attachment.

Freshwater species reproduce both sexually and asexually. Asexual reproduction is by budding or by means of **statoblasts,** which are hard, resistant capsules containing a mass of germinative cells that are formed during the summer and fall (Figure 19-2). When the colony dies in late autumn, the statoblasts are released, and in spring they give rise to new polypides and eventually to new colonies.

Brooding is often accompanied by degeneration of the lophophore and gut of the adults, the remains of which contract into minute dark balls, or **brown bodies.** Later, new internal organs may be regenerated in the old chambers. The brown bodies may remain passive or may be taken up and eliminated by the new digestive tract—an unusual kind of storage excretion.

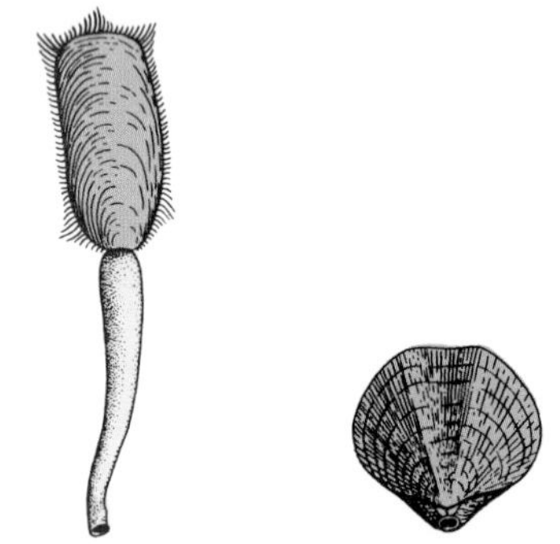

___ Phylum Brachiopoda

The Brachiopoda (brak-i-op'o-da) (Gr. *brachiōn*, arm, + *pous, podos,* foot), or lamp shells, are an ancient group. Although fewer than 300 species are now living, some 30,000 fossil species, which once flourished in the Paleozoic and Mesozoic seas, have been described. Modern forms have changed little from the early ones. *Lingula* (L. tongue) (Figure 19-5, *A*) is probably the most ancient of these "living fossils," having existed virtually unchanged since Ordovician times. Most modern brachiopod shells range between 5 to 80 mm in length, but some fossil forms reached 30 cm.

Brachiopods are attached, bottom-dwelling, marine forms that mostly prefer shallow water. Their name, which means "arm-footed," refers to the arms of the **lophophore.** Externally brachiopods resemble the bivalved molluscs in having

Lingula
(inarticulate)

Terebratella
(articulate)

A

B

Figure 19-5

Brachiopods. **A,** *Lingula,* an inarticulate brachiopod that normally occupies a burrow. The contractile pedicel can withdraw the body into the burrow. **B,** An articulate brachiopod, *Terebratella.* The valves have a tooth-and-socket articulation, and a short pedicel projects through the pedicel valve to attach to the substratum.

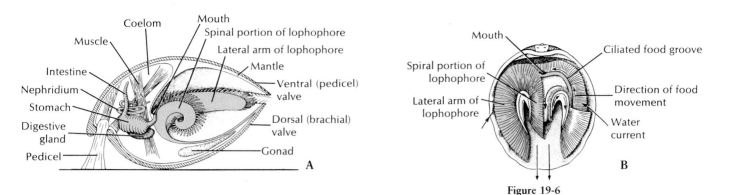

Figure 19-6

Phylum Brachiopoda. **A,** An articulate brachiopod (longitudinal section). **B,** Feeding
and respiratory currents. Large arrows show
water flow over lophophore; small arrows
indicate food movement toward mouth in
ciliated food groove.
B, Modified from Russell-Hunter, W.D. 1969. A
biology of higher invertebrates. New York, Macmillan, Inc.

two calcareous shell valves secreted by the mantle. They were, in fact, classed with
the molluscs until the middle of the nineteenth century. Brachiopods, however,
have dorsal and ventral valves instead of right and left lateral valves as do the
bivalve molluscs and, unlike the bivalves, most of them are attached to a substrate
either directly or by means of a fleshy stalk called a **pedicel** (or pedicle). Some,
such as *Lingula,* live in vertical burrows in sand or mud. Muscles open and close
the valves and provide movement for the stalk and tentacles.

In most brachiopods the ventral (pedicel) valve is slightly larger than the
dorsal (brachial) valve, and one end projects in the form of a short, pointed beak
that is perforated where the fleshy stalk passes through (Figure 19-5, *B*). In many
the shape of the pedicel valve is like that of the classic oil lamp of ancient Greece
and Rome, so that the brachiopods came to be known as the "lamp shells."

There are two classes of brachiopods, based on shell structure. The shell
valves of Articulata are connected by a hinge with an interlocking tooth-and-
socket arrangement, as in *Terebratella* (L. *terebratus,* a boring, + *ella,* dim. suffix); those of Inarticulata lack the hinge and are held together by muscles only, as
in *Lingula* and *Glottidia* (Gr. *glōttidos,* mouth of windpipe).

The body occupies only the posterior part of the space between the valves
(Figure 19-6), and extensions of the body wall form mantle lobes that line and
secrete the shell. The large horseshoe-shaped lophophore in the anterior mantle
cavity bears long, ciliated tentacles used in respiration and feeding. Ciliary water
currents carry food particles between the gaping valves and over the lophophore.
Food is caught on the tentacles and is carried in a ciliated food groove along the
arm of the lophophore to the mouth. Unwanted particles are carried down rejection tracts to the mantle lobe and carried out in ciliary currents. Organic detritus
and some algae are apparently the primary food sources. The brachiopod lophophore not only can create food currents, as do other lophophorates, but also seems
able to absorb dissolved nutrients directly from the environmental seawater.

The coelom, like that of other lophophorates, has an anterior mesocoel and
a posterior metacoel. One or two pairs of nephridia open into the coelom and
empty into the mantle cavity. Coelomocytes, which ingest particulate wastes, are
carried out by the nephridia. There is an open circulatory system with a contractile
heart. The lophophore and mantle are probably the chief site of gaseous exchange.
There is a nerve ring with a small dorsal and a larger ventral ganglion.

Sexes are separate, and paired gonads discharge gametes through the
nephridia. Most fertilizaton is external, but a few species brood their eggs and
young.

The development of brachiopods is similar in some ways to that of the
deuterostomes, with radial, mostly equal, holoblastic cleavage and the coelom
forming enterocoelically in the articulates. The free-swimming larva resembles the
trochophore. In the articulates, metamorphosis occurs after the larva has attached

by a pedicel. In the inarticulates, the juvenile resembles a minute brachiopod with a coiled pedicel in the mantle cavity. There is no metamorphosis. As the larva settles, the pedicel attaches to the substratum, and adult existence begins.

Phylogeny and Adaptive Radiation

The possession of a lophophore by all three phyla is considered evidence of their close relationship, but each phylum has specialized along is own lines and developed its own life-style. As a group they display characteristics intermediate between protostomes and deuterostomes. The coelom is divided into compartments as in many deuterostomes, although the protocoel is repressed. The absence of a protosome and lack of a head is probably correlated with the sessile habit and ciliary method of feeding. Their embryology shows both protostome and deuterostome characteristics, with radial cleavage and, in the brachiopods, enterocoelic development of the coelom. All three phyla have a trochophore type of larva. Since the blastopore becomes the mouth, they must be considered protostomes. Hyman suggested that the deuterostomes may have branched from the protostome line by way of the lophophorates.

All lophophorates are **filter feeders,** and most of their evolutionary diversification has been guided by this function. The tubes of phoronids vary according to their habitats. Various ectoprocts tend to build their protective exoskeletons of chitin or gelatin, which may or may not be impregnated with calcium and sand. Brachiopod variations occur largely in their shells and lophophores.

SUMMARY

The Phoronida, Ectoprocta, and Brachiopoda all bear a lophophore, which is a crown of ciliated tentacles surrounding the mouth but not the anus and containing an extension of the mesocoel. They are also sessile as adults, have a U-shaped digestive tract, and have a free-swimming larva. The lophophore functions as both a respiratory and a feeding structure, its cilia creating water currents from which food particles are filtered.

Phoronida are the least common of the lophophorates, living in tubes mostly in shallow coastal waters. The lophophore is thrust out of the tube for feeding.

Ectoprocts are abundant in marine habitats, living on a variety of submerged substrata, and a number of species are common in fresh water. Ectoprocts are colonial, and although each individual is quite small, the colonies are commonly several centimeters or more in width or height. Each individual lives in a chamber (zoecium), which is a secreted exoskeleton of chitinous, calcium carbonate, or gelatinous material.

Brachiopods were a very successful phylum in the Paleozoic era but have been declining since the early Mesozoic era. Their bodies and lophophores are covered by a mantle, which secretes a dorsal and a ventral valve (shell). They are usually attached to the substrate directly or by means of a pedicel.

The lophophorates have coelomic compartments that apparently correspond to two of the three compartments (mesocoel and metacoel) found in many deuterostomes. Their embryogenesis shows both protostome and deuterostome characteristics.

Review questions

1. What characteristics do the three lophophorate phyla have in common? What characteristics distinguish them from each other?
2. Define each of the following: lophophore, zoecium, zooid, polypide, cystid, brown bodies, statoblasts.
3. What is the evidence for placing the lophophorates in a phylogenetic position between the protostomes and the deuterostomes?
4. What are the coelomic compartments found in the lophophorates?
5. What is the difference in orientation of the valves of brachiopods compared to bivalve molluscs?
6. How is the lophophore of ectoprocts extended?

Selected references

See also general references for Part Two, p. 590.

American Society of Zoologists. 1977. Biology of lophophorates. Am. Zool. **17**(1):3-150. *A collection of 13 papers.*

Nielson, C. 1977. The relationships of the Entoprocta, Ectoprocta and Phoronida. Am. Zool. **17**:149-150. *On the basis of their development, the author considers phoronids deuterostomes and reunites Entoprocta with Ectoprocta.*

Richardson, J.R. 1986. Brachiopods. Sci. Am. **255**:100-106. (Sept.). *Reviews brachiopod biology and adaptations and contends that in the next few million years there may be an increase in the number of species, rather than a decline.*

Strathmann, R. 1973. Function of lateral cilia in suspension feeding of lophophorates (Brachiopoda, Phoronida, Ectoprocta). Mar. Biol. **23**:129-136. *In this and the following paper, the author contends that particle capture on the lophophore is the result of localized reversal of ciliary beat, and that mucus strands are not used.*

Strathmann, R.R. 1982. Cinefilms of particle capture by an induced local change of beat of lateral cilia of a bryozoan. J. Exp. Mar. Biol. Ecol. **62**:225-236.

Woollacott, R.M., and R.C. Zimmer (eds.). 1977. Biology of bryozoans. New York, Academic Press, Inc. *Contains 15 articles on ectoprocts. Advanced.*

CHAPTER 20

THE ECHINODERMS

Phylum Echinodermata

Position in Animal Kingdom

Phylum Echinodermata (e-ki′no-der′mata) (Gr. *echinos*, sea urchin, hedgehog, + *derma*, skin, + *ata*, characterized by) belongs to the **Deuterostomia** branch of the animal kingdom, the members of which are enterocoelous coelomates. The other phyla of this group are Chaetognatha, Hemichordata, and Chordata. Primitively, deuterostomes have the following embryological features in common: anus developing from or near the blastopore, and mouth developing elsewhere; coelom budded off from the archenteron (enterocoel); radial and regulative (indeterminate) cleavage; and endomesoderm (mesoderm derived from or with the endoderm) from enterocoelic pouches. Although only distantly related, the Echinodermata is the only major invertebrate group showing affinities with the vertebrates. The typical deuterostome embryogenesis is shown only by some of the lower vertebrates, but the similarity probably has real evolutionary meaning. Thus, the echinoderms, the chordates, and the lesser deuterostome phyla are presumably derived from a common ancestor. Nevertheless, their evolutionary history has taken the echinoderms to the point where they are very much unlike any other animal group.

Biological Contributions

There is one word that best describes the echinoderms: strange. They are a major group, but they occupy the end of a side branch of the phylogenetic tree, so none of their characteristics can be said to presage those of any more advanced group. They have a unique constellation of characteristics found in no other phylum. Although an enormous amount of research has been devoted to the echinoderms, we are still far from a satisfactory understanding of many aspects of their biology. As Libbie Hyman (1955) wrote, echinoderms are a "noble group especially designed to puzzle the zoologist."

Among the more striking of the features shown by the echinoderms are the system of coelomic channels composing the **water-vascular system**, the calcareous **dermal endoskeleton**, the **hemal system**, and the **metamorphosis** from bilateral larva to radial adult.

── THE ECHINODERMS

The echinoderms are marine forms and include the sea stars, brittle stars, sea urchins, sea cucumbers, and sea lilies. They represent a bizarre group sharply distinguished from all other members of the animal kingdom. Their name is derived from their external spines or protuberances. A calcareous endoskeleton is found in all members of the phylum, either in the form of plates or represented by scattered tiny ossicles.

The most noticeable characteristics of the echinoderms are (1) the spiny endoskeleton of plates, (2) the water-vascular system, (3) the pedicellariae, (4) the dermal branchiae, and (5) radial or biradial symmetry. Radial symmetry is not limited to echinoderms, but no other group with such complex organ systems has radial symmetry.

Echinoderms are an ancient group of animals extending back to the Cambrian period. Despite the excellent fossil record, the origin and early evolution of the echinoderms are still obscure. It seems clear that they descended from bilateral ancestors because their larvae are bilateral but become radially symmetrical later in their development. Many zoologists believe that early echinoderms were sessile and evolved radiality as an adaptation to the sessile existence. Bilaterality is of adaptive value to animals that travel through their environment, while radiality is of value to animals whose environment meets them on all sides equally. Hence, the body plan of present-day echinoderms seems to have been derived from one that was attached to the bottom by a stalk, had radial symmetry and radiating grooves (ambulacra) for food gathering, and had an upward-facing oral side. Attached forms were once plentiful, but only about 80 species, all in the class Crinoidea, still survive. Oddly, conditions have favored the survival of their free-moving descendants, although they are still quite radial, and among them are some of the most abundant marine animals. Nevertheless, in the exception that proves the rule (that bilaterality is adaptive for free-moving animals), at least three groups of echinoderms (two groups of echinoids and the holothuroids) seem to be evolving back toward bilaterality.

Echinoderms have no ability to osmoregulate and thus are rarely found in brackish waters. They are found in all oceans of the world and at all depths, from the intertidal to the abyssal regions. Often the most common animals in the deep ocean are echinoderms. The most abundant species found in the Philippine Trench (10,540 m) was a holothurian. Echinoderms are virtually all bottom dwellers, although there are a few pelagic species.

No parasitic echinoderms are known, but a few are commensals. On the other hand, a wide variety of other animals make their homes in or on echinoderms, including parasitic or commensal algae, protozoa, ctenophores, turbellarians, cirripedes, copepods, decapods, snails, clams, polychaetes, fish, and other echinoderms.

The asteroids, or sea stars (Figure 20-1), are commonly found in various types of bottom habitats, often on hard, rocky surfaces, but numerous species are at home on sandy or soft bottoms. Some species are particle feeders, but many are predators, feeding particularly on sedentary or sessile prey, since the sea stars themselves are relatively slow moving.

Ophiuroids—brittle stars, or serpent stars (Figure 20-11)—are by far the most active echinoderms, moving by their arms rather than by tube feet. A few species are reported to have swimming ability, and some burrow. They may be scavengers, browsers, or deposit or filter feeders. Some are commensal in large sponges, in whose water canals they may live in great numbers.

Holothurians, or sea cucumbers (Figure 20-21), are widely prevalent in all seas. Many are found on sandy or mucky bottoms, where they lie concealed. Compared with other echinoderms, holothurians are greatly extended in the oral-aboral axis. They are oriented with that axis more or less parallel to the substrate and lying on one side. Most are suspension or deposit feeders.

Echinoids, or sea urchins (Figure 20-16), are adapted for living on the ocean bottom and always keep their oral surface in contact with the substratum. The "regular" sea urchins prefer hard bottoms, but the sand dollars and heart urchins ("irregular" urchins) are usually found on sand. The regular urchins, which are radially symmetrical, feed chiefly on algae or detritus, while the irregulars, which are secondarily bilateral, feed on small particles.

Crinoids (Figure 20-26) stretch their arms out and up like a flower's petals and feed on plankton and suspended particles. Most living species become

Figure 20-1

Some sea stars (subclass Asteroidea) from the Pacific. **A,** Cushion star *Pteraster tessellatus* can secrete incredible quantities of mucus when disturbed, presumably as a defense reaction. **B,** Leather star *Dermasterias imbricata* lacks spines and feeds on sea anemones. **C,** *Pentagonaster duebeni* from the Great Barrier Reef is brilliant red and orange. **D,** *Crossaster papposus*, one of the sun stars, feeds on other sea stars. **E,** *Pisaster ochraceus* is the most abundant sea star on the Pacific coast of the United States, sometimes occurring in great numbers. This and many other species of sea stars show a diversity of individual coloration. A, B, D, and E, Photographs by R. Harbo; C, photograph by L.S. Roberts.

detached from their stems as adults, but they nevertheless spend most of their time on the substrate, holding on by means of aboral appendages called cirri.

The zoologist who admires the fascinating structure and function of echinoderms can share with the layperson an admiration of the beauty of their symmetry, often enhanced by bright colors. Many species are rather drab, but others may be orange, red, purple, blue, and often bicolor.

Because of the spiny nature of their structure, echinoderms are not often the prey of other animals—except other echinoderms (sea stars). Some fish have strong teeth and other adaptations that enable them to feed on echinoderms. A few mammals, such as sea otters, feed on sea urchins. In scattered parts of the world, humans relish sea urchin gonads, either raw or roasted on the half shell. Trepang, the cured body wall of certain large holothurians, is considered a delicacy, particularly in some Oriental countries. It is highly nutritious, since more than 50% is easily digestible protein, and it is said to impart a delicate flavor to soups.

Sea stars feed on a variety of molluscs, crustaceans, and other invertebrates. In some areas they may perform an important ecological role as a top carnivore in the habitat. Their chief economic impact is on clams and oysters. A single star may eat as many as a dozen oysters or clams in a day. To rid shellfish beds of these pests, quicklime is sometimes spread over areas where they abound. The quicklime damages the delicate epidermal membrane, destroying the dermal branchiae and ultimately the animal itself. Unfortunately, other soft bodied invertebrates are also damaged. However, the oysters remain with their shells tightly closed until the quicklime is degraded.

Echinoderms have been widely used in experimental embryology, for their gametes are usually abundant and easy to collect and handle in the laboratory. The investigator can follow the embryonic developmental stages with great accuracy. We know more about the molecular biology of sea urchin development than that of almost any other embryonic system. Artificial parthenogenesis was first discovered in sea urchin eggs, when it was found that, by treating the eggs with hypertonic seawater or subjecting them to a variety of other stimuli, development would proceed without the presence of sperm.

CHARACTERISTICS

1. Body unsegmented (nonmetameric) with **radial, pentamerous symmetry;** body rounded, cylindrical, or star shaped, with five or more radiating areas, or ambulacra, alternating with interambulacral areas
2. **No head or brain;** few specialized sensory organs; sensory system of tactile and chemoreceptors, podia, terminal tentacles, photoreceptors, and statocysts
3. Nervous system with circumoral ring and radial nerves; usually two or three systems of networks located at different levels in the body, varying in degree of development according to group
4. **Endoskeleton** of **dermal calcareous ossicles** with **spines** or of calcareous **spicules** in dermis; covered by an epidermis (ciliated in most); **pedicellariae** (in some)
5. A unique **water-vascular system** of coelomic origin that extends from the body surface as a series of tentacle-like projections (**podia,** or **tube feet**) that are protracted by increase of fluid pressure within them; an opening to the exterior (**madreporite** or **hydropore**) usually present
6. Locomotion by tube feet, which project from the ambulacral areas, by movement of spines, or by movement of arms, which project from central disc of body
7. Digestive system usually complete; axial or coiled; anus absent in ophiuroids
8. Coelom extensive, forming the perivisceral cavity and the cavity of the water-vascular system; coelom of enterocoelous type; coelomic fluid with amebocytes
9. Blood-vascular system (**hemal system**) much reduced, playing little if any role in circulation, and surrounded by extensions of coelom (**perihemal sinuses**); main circulation of body fluids (coelomic fluids) by peritoneal cilia
10. Respiration by **dermal branchiae, tube feet, respiratory tree** (holothuroids), and **bursae** (ophiuroids)
11. **Excretory organs absent**
12. Sexes separate (except a few hermaphroditic) with large gonads, single in holothuroids but multiple in most; simple ducts, with no elaborate copulatory apparatus or secondary sexual structures; fertilization usually external; eggs brooded in some
13. Development through **free-swimming, bilateral, larval stages** (some with direct development); metamorphosis to radial adult or subadult form
14. Autotomy and regeneration of lost parts conspicuous

CLASSIFICATION

There are about 6000 living and 20,000 extinct or fossil species of Echinodermata. The traditional classification of the echinoderms placed all the free-moving forms that were oriented with oral side down in the subphylum Eleutherozoa, containing most of the living species. The other subphylum, Pelmatozoa, contained mostly forms with stems and oral side up; most of the extinct classes and the living Crinoidea belong to this group. In recognition that the traditional subphyla were polyphyletic, the following classification proposed by Fell has been widely adopted. Difficulties yet remain, however (see phylogeny discussion, p. 431). Characteristics for strictly fossil classes are omitted here.

Subphylum Homalozoa (ho-mal'o-zo'a) (Gr. *homalos,* level, even, + *zōon,* animal). Ancient fossil echinoderms called carpoids, which were not radially symmetrical; fossil classes Homostelea, Homoiostelea, Stylophora (Figure 22-3, p. 449), and Ctenocystoidea.

Subphylum Crinozoa (krin'o-zo'a) (Gr. *krinon,* lily, + *zōon,* animal). Radially symmetrical, with rounded or cup-shaped theca and brachioles or arms. Attached by stem during part or all of life. Oral surface directed upward. Fossil classes: Eocrinoidea, Paracrinoidea, Cystoidea, and Blastoidea.

Class Crinoidea (krin-oi'de-a) (Gr. *krinon,* lily, + *eidos,* form, + *ea,* characterized by): sea lilies and feather stars. Aboral attachment stalk of dermal ossicles; mouth and anus on oral surface; five arms branching at base and bearing pinnules; ciliated ambulacral grooves on oral surface with tentacle-like tube feet for food gathering; spines, madreporite, and pedicellariae absent. Examples: *Antedon, Nemaster* (Figure 20-26).

Subphylum Asterozoa (as'ter-o-zo'a) (Gr. *aster,* star, + *zōon,* animal). Radially symmetrical, star-shaped echinoderms, unattached as adults.

Class Stelleroidea (stel'ler-oi'de-a) (L. *stella,* star, + Gr. *eidos,* form). With characteristics of subphylum, body a central disc with radially arranged rays or arms.

Subclass Somasteroidea (som'ast-er-oi'de-a) (Gr. *soma,* body, + *aster,* star, + *eidos,* form). Mostly extinct sea stars with primitive skeletal structure; single living species, *Platasterias latiradiata,* from deep water off western coast of Mexico.

Subclass Asteroidea (as'ter-oi'de-a) (Gr. *aster,* star, + *eidos,* form, + *ea,* characterized by): sea stars. Star-shaped echinoderms, with the arms not sharply marked off from the central disc; ambulacral grooves open, with tube feet on oral side; tube feet often with suckers; anus and madreporite aboral; pedicellariae present. Examples: *Asterias, Pisaster* (Figure 20-1).

Subclass Ophiuroidea (o'fe-u-roi'de-a) (Gr. *ophis,* snake, + *oura,* tail, + *eidos,* form): brittle stars and basket stars. Star shaped, with the arms sharply marked off from the central disc; ambulacral grooves closed, covered by ossicles; tube feet without suckers and not used for locomotion; pedicellariae absent. Examples: *Ophiura* (Figure 20-11, *A*), *Gorgonocephalus* (Figure 20-14, *B*).

Subphylum Echinozoa (ek'in-o-zo'a) (Gr. *echinos,* sea urchin, hedgehog, + *zōon,* animal). Unattached, globoid, discoid, or more or less cylindrical echinoderms without arms. Fossil classes: Helicoplacoidea, Edriosteroidea, Ophiocistioidea.

Class Echinoidea (ek'i-noi'de-a) (Gr. *echinos,* sea urchin, hedgehog, + *eidos,* form): sea urchins, sea biscuits, and sand dollars. More or less globular or disc-shaped echinoderms with no arms; compact skeleton or test with closely fitting plates; movable spines; ambulacral grooves covered by ossicles, closed; tube feet with suckers; pedicellariae present. Examples: *Arbacia, Strongylocentrotus* (Figure 20-16), *Lytechinus, Mellita.*

Class Holothuroidea (hol'o-thu-roi'de-a) (Gr. *holothourion,* sea cucumber + *eidos,* form): sea cucumbers. Cucumber-shaped echinoderms with no arms; spines absent; microscopic ossicles embedded in thick muscular wall; anus present; ambulacral grooves closed; tube feet with suckers; circumoral tentacles (modified tube feet); pedicellariae absent; madreporite plate internal. Examples: *Sclerodactyla, Parastichopus, Cucumaria* (Figure 20-21, *C*).

___ Class Stelleroidea

Sea stars: subclass Asteroidea

Although sea stars are not considered the most primitive living echinoderms, they demonstrate the basic features of echinoderm structure and function very well, and they are easily obtainable. Thus we will consider them first, then comment on the major differences shown by the other groups.

Sea stars are familiar along the shoreline where large numbers of them may aggregate on the rocks. Sometimes they cling so tenaciously that they are difficult to dislodge without tearing off some of the tube feet. They also live in muddy or sandy bottoms and among coral reefs. They are often brightly colored and range in size from a centimeter in greatest diameter to about a meter across. *Asterias* (Gr. *asteros,* a star) is one of the common genera of the east coast of the United States and is commonly studied in zoology laboratories. *Pisaster* (Gr. *pisos,* a pea, + *asteros,* a star) (Figure 20-1, *E*) is common on the west coast of the United States, as is *Dermasterias* (Gr. *dermatos,* skin, leather, + *asteros,* a star), the leather star (Figure 20-1, *B*).

Form and function
External features

Sea stars are composed of a central disc that merges gradually with the tapering arms (rays). They typically have five arms, but they may have more (Figure 20-1, *D*). The body is flattened, flexible, and covered with a ciliated pigmented epidermis. The mouth is centered on the under, or oral, side, surrounded by a soft peristomial membrane. An **ambulacral groove** along the oral side of each arm is bordered by movable **spines** that protect the rows of **tube feet** (podia) projecting from the groove (Figure 20-2). Viewed from the oral side, the large **radial nerve** can be seen in the center of each ambulacral groove (Figure 20-3, *C*), between the rows of tube feet. The nerve is very superficially located, covered only by thin epidermis. Under the nerve is an extension of the coelom and the radial canal of the water-vascular system, all of which are external to the underlying ossicles (Figure 20-3, *C*). In all other classes of living echinoderms except the crinoids,

Figure 20-2

External anatomy of asteroid. **A,** Aboral view. **B,** Oral view.

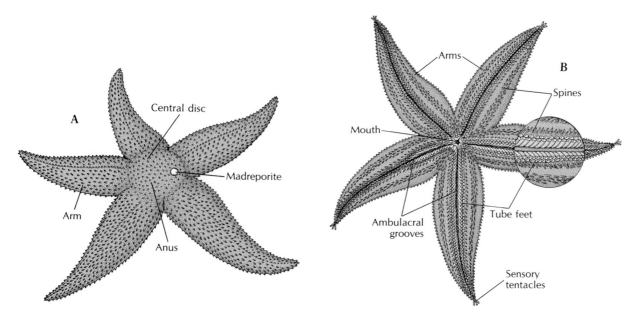

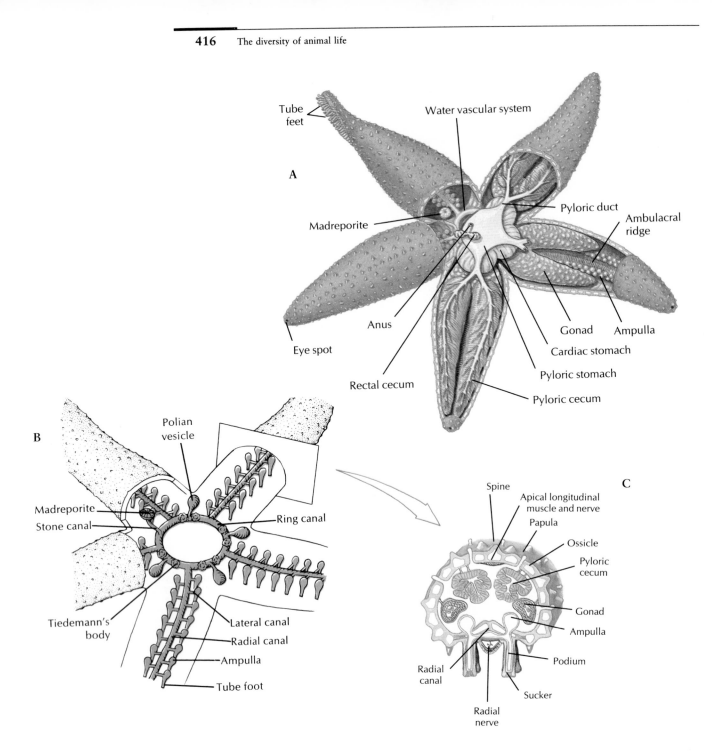

Figure 20-3

A, Internal anatomy of a sea star. B, Water-vascular system. Podia penetrate between ossicles. (Polian vesicles are not present in *Asterias.*) C, Cross section of arm at level of gonads, illustrating open ambulacral grooves.

these structures are covered over by ossicles or other dermal tissue; thus, the ambulacral grooves in asteroids and crinoids are said to be open, and those of the other groups are closed.

The aboral surface is usually rough and spiny, although the spines of many species are flattened, so that the surface appears smooth (Figure 20-1, *A*). Around the bases of the spine are groups of minute, pincerlike **pedicellariae**, bearing tiny jaws manipulated by muscles (Figure 20-4). These help keep the body surface free of debris, protect the papulae, and sometimes aid in food capture. The **papulae** (**dermal branchiae** or **skin gills**) are soft delicate projections of the coelomic cavity, covered only with epidermis and lined internally with peritoneum; they extend out through spaces between the ossicles and are concerned with respiration (Figures

20-3, *C*, and 20-4, *F*). Also on the aboral side are the inconspicuous anus and the circular **madreporite** (Figure 20-2, *A*), a calcareous sieve leading to the water-vascular system.

Endoskeleton

Beneath the epidermis of the sea star is a mesodermal endoskeleton of small calcareous plates, or **ossicles,** bound together with connective tissue. From these ossicles project the spines and tubercles that make up the spiny surface. Muscles in the body wall move the rays and can partially close the ambulacral grooves by drawing their margins together.

Coelom, excretion, and respiration

The coelomic compartments of the larval echinoderm give rise to several structures in the adult, one of which is a spacious body coelom filled with fluid. The fluid contains amebocytes (coelomocytes), bathes the internal organs, and projects into the papulae. The coelomic fluid is circulated around the body cavity and into the papulae by the ciliated peritoneal lining. Exchange of respiratory gases and excretion of nitrogenous waste, principally ammonia, take place by diffusion through the thin walls of the papulae and tube feet. Some wastes may be picked up by the coelomocytes, which migrate through the epithelium of the papulae or tube feet to the exterior, or the tips of papulae containing the waste-laden coelomocytes may be pinched off.

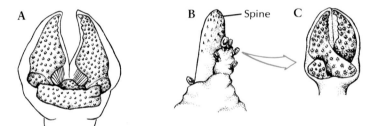

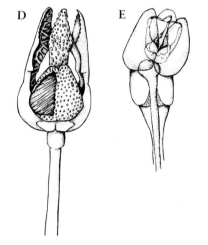

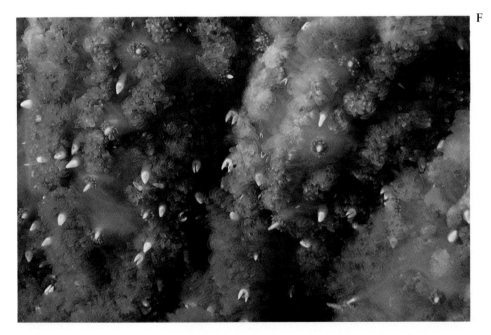

Figure 20-4

Pedicellariae of sea stars and sea urchins.
A, Forceps-type pedicellaria of *Asterias*.
B and **C,** Scissors-type pedicellariae of *Asterias;* size relative to spine is shown in **B.**
D, Tridactyl pedicellaria of *Strongylocentrotus*, cutaway showing muscle. **E,** Globiferous pedicellaria of *Strongylocentrotus*. **F,** Close-up view of the aboral surface of the sea star *Pycnopodia helianthoides*. Note the large pedicellariae, as well as the groups of small pedicellariae around the spines. Many thin-walled papulae can be seen.
A to **E,** Drawings by T. Doyle; **F,** photograph by R. Harbo.

Water-vascular system

The water-vascular system is another coelomic compartment and is unique to the echinoderms. Showing exploitation of hydraulic mechanisms to a greater degree than in any other animal group, it is a system of canals and specialized tube feet that, together with the dermal ossicles, has determined the evolutionary potential and limitations of this phylum. In sea stars the primary functions of the water-vascular system are locomotion and food gathering, in addition to respiration and excretion.

Structurally, the water-vascular system opens to the outside through small pores in the madreporite. The madreporite of asteroids is on the aboral surface (Figure 20-2, *A*) and leads into the **stone canal,** which descends toward the **ring canal** around the mouth (Figure 20-3, *B*). **Radial canals** diverge from the ring canal, one into the ambulacral groove of each ray. Also attached to the ring canal are four or five pairs of folded, pouchlike **Tiedemann's bodies** and from one to five **polian vesicles** (polian vesicles are absent in some sea stars, such as *Asterias*). The Tiedemann's bodies are thought to produce coelomocytes, and the polian vesicles are apparently for fluid storage.

A series of small **lateral canals,** each with a one-way valve, connects the radial canal to the cylindrical podia, or tube feet, along the sides of the ambulacral groove in each ray. Each podium is a hollow, muscular tube, the inner end of which is a muscular sac, the **ampulla,** that lies within the body coelom (Figure 20-3, *A* and *C*), and the outer end of which usually bears a **sucker.** Some species lack the suckers. The podia pass to the outside between the ossicles in the ambulacral groove.

The water-vascular system operates hydraulically and is an effective locomotor mechanism. The valves in the lateral canals prevent backflow of fluid into the radial canals. The tube foot has in its walls connective tissue that maintains the cylinder at a relatively constant diameter. On contraction of muscles in the ampulla, fluid is forced into the podium, extending it. Conversely, contraction of the longitudinal muscles in the tube foot retracts the podium, forcing fluid back into the ampulla. Contraction of muscles in one side of the podium bends the organ toward that side. Small muscles at the end of the tube foot can raise the middle of

The function of the madreporite is still obscure. One suggestion is that it allows rapid adjustment of hydrostatic pressure within the water-vascular system in response to changes in external hydrostatic pressure resulting from depth changes, as in tidal fluctuations and wave surges.

Figure 20-5

A, *Orthasterias koehleri* eating a clam. **B,** This *Pycnopodia helianthoides* has been overturned while eating a large sea urchin *Strongylocentrotus franciscanus.* This sea star has 20 to 24 arms and can range up to 1 m in diameter (arm tip to arm tip).
A, Photograph by R. Harbo; **B,** photograph by D.W. Gotshall.

A   B

A   B

the disclike end, creating a suction-cup effect when the end is applied to the substrate. It has been estimated that by combining mucous adhesion with suction, a single podium can exert a pull equal to 25 to 30 g. Coordinated action of all or many of the tube feet is sufficient to draw the animal up a vertical surface or over rocks. The ability to move while firmly adhering to the substrate is a clear advantage to an animal living in a sometimes wave-churned environment.

On a soft surface, such as muck or sand, the suckers are ineffective (numerous sand-dwelling species have no suckers), so the tube feet are employed as legs. Locomotion becomes mainly a stepping process. Most sea stars can move only a few centimeters per minute, but some very active ones can move 75 to 100 cm per minute; for example, *Pycnopodia* (Gr. *pyknos*, compact, dense, + *pous, podos,* foot) (Figure 20-5, *B*). When inverted, the sea star bends its rays until some of the tubes reach the substratum and attach as an anchor; then the sea star slowly rolls over.

Tube feet are innervated by the central nervous system (ectoneural and hyponeural systems, see below). Nervous coordination enables the tube feet to move in a single direction, although not in unison, so that the sea star may progress. If the radial nerve in an arm is cut, the podia in that arm lose coordination, although they can still function. If the circumoral nerve ring is cut, the podia in all arms become uncoordinated, and movement ceases.

Feeding and digestive system

The mouth on the oral side leads through a short esophagus to a large stomach in the central disc. The lower (cardiac) part of the stomach can be everted through the mouth during feeding (Figure 20-2, *B*), and excessive eversion is prevented by gastric ligaments. The upper (pyloric) part is smaller and connects by ducts to a pair of large **pyloric ceca (digestive glands)** in each arm (Figure 20-3, *A*). Digestion is mostly extracellular, although some intracellular digestion may occur in the ceca. A short intestine leads aborally from the pyloric stomach, and there is usually a few small, saclike **intestinal ceca** (Figure 20-3, *A*). The anus is inconspicuous, and some sea stars lack an intestine and anus.

Many sea stars are carnivorous and feed on molluscs, crustaceans, polychaetes, echinoderms, other invertebrates, and sometimes small fish. Sea stars consume a wide range of food items, but many show particular preferences (Figures 20-5 and 20-6). Some select brittle stars, sea urchins, or sand dollars, swallowing them whole and later regurgitating undigestible ossicles and spines (Figure

Figure 20-6

A, Crown-of-thorns star *Acanthaster planci* feeding on coral. **B,** Close-up view of arm of *A. planci*. Puncture wounds from these spines are painful; the spines are equipped with poison glands.

Photographs by J.L. Rotman.

Since 1963 there have been numerous reports of increasing numbers of the crown-of-thorns sea star (*Acanthaster planci* [Gr. *akantha*, thorn, + *asteros*, star]) (Figure 20-6) that were damaging large areas of coral reef in the Pacific Ocean. The crown-of-thorns star feeds on coral polyps, and it sometimes occurs in large aggregations, or "herds." Among the suggested reasons for its increase were human destruction of the giant triton, a gastropod that feeds on sea stars, or that dredging, blasting, and other activities had destroyed the creatures that feed on the sea star larvae. Various attempts have been made to correct the problem, although controversy continues as to whether a problem actually exists. Some have contended that we are observing a natural biological cycle and that, left alone, the situation will correct itself in time. Available data do not support this contention, however (Moran et al., 1986).

Figure 20-7

Hemal system of asteroids. The main peri-hemal channel is the thin-walled axial sinus, which encloses both the axial gland and the stone canal. Other features of the hemal system are shown.

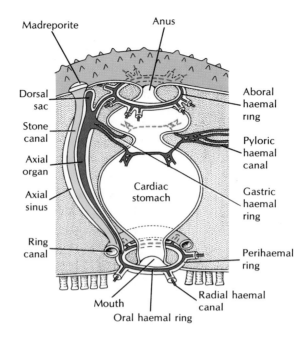

Figure 20-8

"Comet" form of the sea star *Linckia* in which one arm is regenerating the rest of the animal. Although asteroids characteristically can regenerate new arms as necessary if a part of the central disc is present, species of *Linckia* can accomplish this feat from an arm only.
Photograph by K. Sandved.

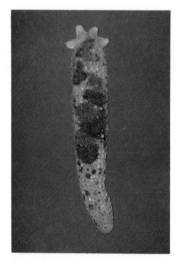

20-5). Some attack other sea stars, and if they are small compared to their prey, they may attack and begin eating at the end of one arm.

Some asteroids feed heavily on molluscs (Figure 20-5, *A*), and *Asterias* is a significant predator on commercially important clams and oysters. When feeding on a bivalve, a sea star will hump over its prey, attaching its podia to the valves, and then exert a steady pull, using its feet in relays. A force of some 1300 g can thus be exerted. In half an hour or so the adductor muscles of the bivalve fatigue and relax. With a very small gap available, the star inserts its soft everted stomach into the space between the valves and wraps it around the soft parts of the shellfish. After feeding, the sea star draws in its stomach by contraction of the stomach muscles and relaxation of body wall muscles.

Some sea stars feed on small particles, either entirely or in addition to carnivorous feeding. Plankton and other organic particles coming in contact with the animal's surface are carried by the epidermal cilia to the ambulacral grooves and then to the mouth.

Hemal system

The so-called hemal system is not very well developed in asteroids, and its function in all echinoderms is unclear. The hemal system has little or nothing to do with the circulation of body fluids. It is a system of tissue strands enclosing unlined sinuses and is itself enclosed in another coelomic compartment, the **perihemal channels** (Figure 20-7). The hemal system may be useful in distributing digested products, but its specific functions are not really known.

Nervous system

The nervous system consists of three units at different levels in the disc and arm. The chief of these systems is the **oral (ectoneural) system** composed of a **nerve ring** around the mouth and a main **radial nerve** into each arm. It appears to coordinate the tube feet. A **deep (hyponeural) system** lies aboral to the oral system, and an **aboral** system consists of a ring around the anus and radial nerves along the roof of the rays. An **epidermal nerve plexus** or nerve net freely connects these systems with the body wall and related structures. The epidermal plexus has been shown to coordinate the responses of the dermal branchiae to tactile stimulation—the only instance known in echinoderms in which coordination occurs through a nerve net.

The sense organs are not well developed. Tactile organs and other sensory cells are scattered over the surface, and an ocellus is at the tip of each arm. Their reactions are mainly to touch, temperature, chemicals, and differences in light intensity. Sea stars are usually more active at night.

Reproductive system, regeneration, and autotomy

Most sea stars have separate sexes. A pair of gonads lies in each interradial space (Figure 20-3, A). Fertilization is external and occurs in early summer when the eggs and sperm are shed into the water. The maturation and shedding of sea star eggs are stimulated by a secretion from neurosecretory cells located on the radial nerves.

Echinoderms can regenerate lost parts. Sea star arms can regenerate readily, even if all are lost. Stars also have the power of autotomy and can cast off an injured arm near the base. It may take months to regenerate a new arm.

Some species can regenerate a complete new sea star (Figure 20-8) from an arm that was broken off or removed if it contains a part (about one fifth) of the central disc. In former times fishermen used to dispatch sea stars they collected from their oyster beds by chopping them in half with a hatchet—a worse than futile activity. Some sea stars reproduce asexually under normal conditions by cleaving the central disc, each part regenerating the rest of the disc and missing arms.

Development

In some species the liberated eggs are brooded, either under the oral side of the animal or in specialized aboral structures, and development is direct, but in most species the embryonating eggs are free in the water and hatch to free-swimming larvae.

Early embryogenesis shows the typical primitive deuterostome pattern. Gastrulation is by invagination, and the anterior end of the archenteron pinches off to become the coelomic cavity, which expands in a U shape to fill the blastocoel. Each of the legs of the U, at the posterior, constricts to become a separate vesicle, and these eventually give rise to the main coelomic compartments of the body (**somatocoels**). The anterior portion of the U undergoes subdivision to form the **axo-** and **hydrocoels** (Figure 20-10). The left hydrocoel will become the water-vascular system, and the left axocoel will give rise to the stone canal and perihemal channels. The right axocoel and hydrocoel will disappear. The free-swimming larva has cilia arranged in bands and is called a **bipinnaria** (Figure 20-9, A). The ciliated tracts become extended into larval arms. Soon the larva grows three adhesive arms and a sucker at its anterior end and is then called a **brachiolaria**. At that time it attaches to the substratum, forms a temporary attachment stalk, and undergoes metamorphosis.

Metamorphosis involves a dramatic reorganization of a bilateral larva into a radial juvenile. The anteroposterior axis of the larva is lost, and *what was the left side becomes the oral surface, and the larval right side becomes the aboral surface* (Figure 20-10). Correspondingly, the larval mouth and anus disappear, and a new mouth and anus form on what was originally the left and right sides, respectively. The portion of the anterior coelomic compartment from the left side expands to form the ring canal of the water-vascular system around the mouth, and then it

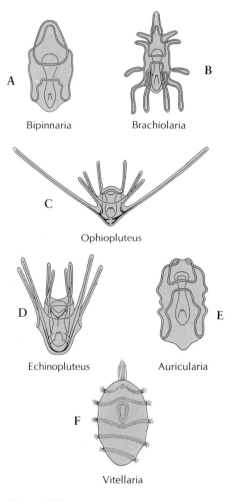

Figure 20-9

Larvae of echinoderms. **A,** Bipinnaria of asteroids. **B,** Brachiolaria of asteroids. **C,** Ophiopluteus of ophiuroids. **D,** Echinopluteus of echinoids. **E,** Auricularia of holothuroids. **F,** Vitellaria of crinoids.

The mesocoel and metacoel compartments of the coelom, as found in the lophophorates, are present in echinoderms. Echinoderms also have a more anterior compartment, the protocoel, although the protocoel often is not completely separated from the mesocoel. In echinoderms the metacoel, mesocoel, and protocoel are referred to as somatocoel, hydrocoel, and axocoel.

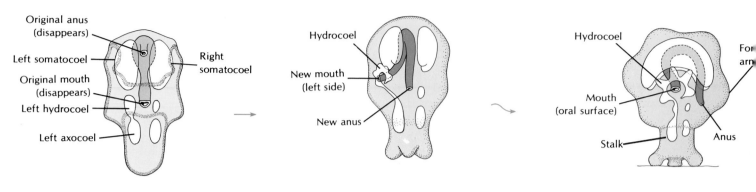

Left somatocoel becomes the oral coelom, and the right somatocoel becomes the aboral coelom. The left hydrocoel becomes the water-vascular system and the left axocoel the stone canal and perihemal channels. The right axo- and hydrocoel are lost.

Figure 20-10

Asteroid metamorphosis. The left somatocoel becomes the oral coelom, and the right somatocoel becomes the aboral coelom. The left hydrocoel becomes the water-vascular system and the left axocoel the stone canal and perihemal channels. The right axo- and hydrocoel are lost.

grows branches to form the radial canals. As the short, stubby arms and the first podia appear, the animal detaches from its stalk and begins life as a young sea star.

Brittle stars: subclass Ophiuroidea

The brittle stars are the largest of the major groups of echinoderms in numbers of species, and they are probably the most abundant also. They abound in all types of benthic marine habitats, even carpeting the abyssal sea bottom in many areas.

Form and function

Apart from the typical possession of five arms, the brittle stars are surprisingly different from the asteroids. The arms of brittle stars are slender and sharply set off from the central disc (Figure 20-11). They have no pedicellariae or papulae, and their ambulacral grooves are closed and covered with arm ossicles. The tube feet are without suckers; they aid in feeding but are of limited use in locomotion. In contrast to that in the asteroids, the madreporite of the ophiuroids is located on the oral surface, on one of the oral shield ossicles (Figure 20-12). Ampullae on the podia are absent, and force for protrusion of the podium is generated by a proximal muscular portion of the podium.

Each of the jointed arms consists of a column of articulated ossicles (the so-called **vertebrae**), connected by muscles and covered by plates. Locomotion is by arm movement. The arms are moved forward in pairs and are placed against

Figure 20-11

A, Brittle star *Ophiura lutkeni* (subclass Ophiuroidea). Brittle stars do not use their tube feet for locomotion but can move rapidly (for an echinoderm) by means of their arms. **B,** Basket star *Astrophyton muricatum* (subclass Ophiuroidea). Basket stars extend their many-branched arms to filter feed, usually at night.

A, Photograph by R. Harbo; **B,** photograph by D.W. Gotshall.

A

B

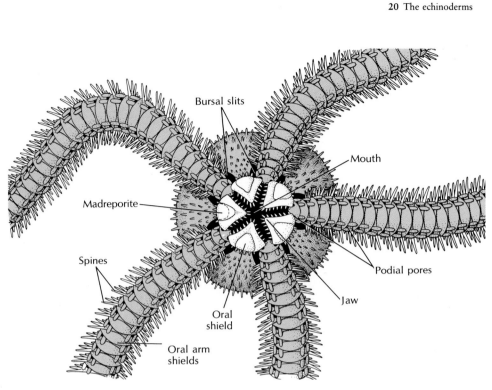

Figure 20-12

Oral view of spiny brittle star *Ophiothrix*.

the substratum, while one (any one) is extended forward or trailed behind, and the animal is pulled or pushed along in a jerky fashion.

The mouth is surrounded by five movable plates that serve as jaws (Figure 20-12). There is no anus. The skin is leathery, with dermal plates and spines arranged in characteristic patterns. Surface cilia are mostly lacking.

The visceral organs are confined to the central disc, since the rays are too slender to contain them (Figure 20-13). The stomach is saclike, and there is no intestine. Indigestible material is cast out of the mouth.

Five pairs of **bursae** (peculiar to ophiuroids) open toward the oral surface by genital slits at the bases of the arms. Water circulates in and out of these sacs for exchange of gases. On the coelomic wall of each bursa are small gonads that discharge into the bursa their ripe sex cells, which pass through the genital slits into the water for fertilization (Figure 20-14, *A*). Sexes are usually separate; a few ophiuroids are hermaphroditic. Some brood their young in the bursae; the young escape through the genital slits or by rupturing the aboral disc. The larva is called the ophiopluteus, and its ciliated bands extend onto delicate, beautiful larval arms (Figure 20-9, *C*). During the metamorphosis to the juvenile, there is no temporarily attached phase, as there is in asteroids.

Water-vascular, nervous, and hemal systems are similar to those of the sea stars. Each arm contains a small coelom, a radial nerve, and a radial canal of the water-vascular system.

Biology

Brittle stars tend to be secretive, living on hard bottoms where little or no light penetrates. They are generally negatively phototropic and insinuate themselves into small crevices between rocks, becoming more active at night. They are commonly fully exposed on the bottom in the permanent darkness of the deep sea. Ophiuroids feed on a variety of small particles, either browsing food from the bottom or filter feeding. Podia are important in transferring food to the mouth. Some brittle stars extend arms into the water and catch suspended particles in mucous strands between the arm spines.

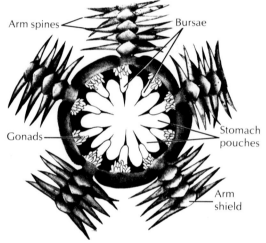

Figure 20-13

Ophiuroid with aboral disc wall cut away to show principal internal structures. The bursae are fluid-filled sacs in which water constantly circulates for respiration. They also serve as brood chambers. Only bases of arms are shown.

A

B

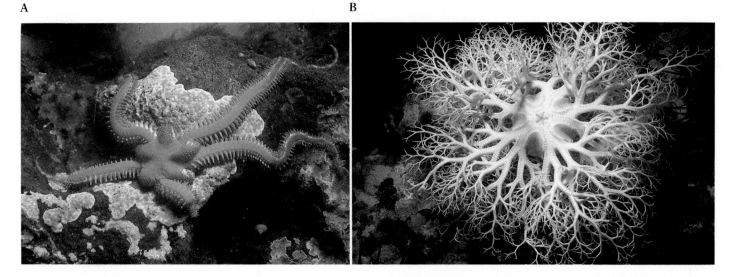

Figure 20-14

A, This brittle star *Ophiopholis aculeata* has its bursae swollen with eggs, which it is ready to expel. The arms have been broken and are regenerating. **B,** Oral view of a basket star *Gorgonocephalus eucnemis*, showing pentaradial symmetry.
Photographs by R. Harbo.

Regeneration and autotomy are even more pronounced in brittle stars than in sea stars. Many seem very fragile, releasing an arm or even part of the disc at the slightest provocation. Some can reproduce asexually by cleaving the disc; each progeny then regenerates the missing parts.

Some common ophiuroids along the coast of the United States are *Amphipholis* (Gr. *amphi*, both sides of, + *pholis*, horny scale) (viviparous and hermaphroditic), *Ophioderma* (Gr. *ophis*, snake, + *dermatos*, skin), *Ophiothrix* (Gr. *ophis*, snake, + *thrix*, hair), and *Ophiura* (Gr. *ophis*, snake, + *oura*, tail) (Figure 20-10, *A*). The basket stars *Gorgonocephalus* (Gr. *Gorgo*, name of a female monster of terrible aspect, + *kephalē*, a head) (Figure 20-14, *B*) and *Astrophyton* (Gr. *asteros*, star, + *phyton*, creature, animal) (Figure 20-11, *B*) have arms that branch repeatedly. Most ophiuroids are drab, but some are attractive, with variegated color patterns (p. 432).

____ Class Echinoidea

Sea urchins, sand dollars, and heart urchins

The echinoids have a compact body enclosed in an endoskeletal test, or shell. The dermal ossicles, which have become closely fitting plates, make up the test. Echinoids lack arms, but their tests reflect the typical pentamerous plan of the echi-

Figure 20-15

Diadema antillarum (class Echinoidea) is a common species in the West Indies and Florida and its spines are a hazard to swimmers.
Photograph by A. Kerstitch.

noderms in their five ambulacral areas. The most notable modification of the ancestral body plan is that the oral surface has expanded over the sides and top, so that the ambulacral areas extend up to the area around the anus (**periproct**). The majority of living species of sea urchins are referred to as "regular"; they have a hemispherical shape, radial symmetry, and medium to long spines (Figures 20-15 and 20-16). Sand dollars (Figure 20-17) and heart urchins (Figure 20-18) are "irregular" because the orders to which they belong have become secondarily bilateral; their spines are usually very short. Regular urchins move by means of their tube feet, with some assistance from their spines, and irregular urchins move chiefly by their spines (Figure 20-17). Some echinoids are quite colorful.

Echinoids have a wide distribution in all seas, from the intertidal regions to the deep oceans. Regular urchins often prefer rocky or hard bottoms, whereas sand dollars and heart urchins like to burrow into a sandy substrate. Distributed along one or both coasts of North America are common genera of regular urchins (*Arbacia* [Gr. Arbakēs, first king of Media], *Strongylocentrotus* [Gr. *strongylos*, round, compact, + *kentron*, point, spine] [Figure 20-16, *D*], and *Lytechinus* [Gr.

A
B
C
D
E

Figure 20-16

Diversity among regular sea urchins (class Echinoidea). **A,** Pencil urchin *Eucidaris tribuloides*. Members of this order are the most primitive living sea urchins, surviving since the Paleozoic era, and may be ancestral to all other extant echinoids. **B,** Slate-pencil urchin *Heterocentrotus mammilatus*. The large, triangular spines of this urchin were formerly used for writing on slates. **C,** Aboral spines of the intertidal urchin *Colobocentrotus atratus* are flattened and mushroom shaped, while the marginal spines are wedge shaped, giving the animal a streamlined form to withstand pounding surf. **D,** Purple sea urchin *Strongylocentrotus purpuratus* is common along the Pacific coast of North America where there is heavy wave action. **E,** *Astropyga magnifica* is one of the most spectacularly colored sea urchins, with bright-blue spots along its interambulacral areas.

A, Photograph by A. Kerstitch; B to D, photographs by R. Harbo; E, photograph by K. Sandved.

Figure 20-17

Two sand dollar species. **A,** *Encope grandis* as they are normally found burrowing near the surface on a sandy bottom. **B,** Removed from the sand. The short spines and petaloids on the aboral surface of this *Encope micropora* are easily seen.
Photographs by A. Kerstitch.

A

B

lytos, dissolvable, broken, + *echinos*, sea urchin]) and sand dollars (*Dendraster* [Gr. *dendron*, tree, stick, + *asteros*, star] and *Echinarachnius* [Gr. *echinos*, sea urchin, + *arachnē*, spider]). The West Indies–Florida region is rich in echinoderms, including echinoids, of which *Diadema* (Gr. *diadeō*, to bind around), with its long, needle-sharp spines, is a prominent example (Figure 20-15).

Form and function

The echinoid test is a compact skeleton of 10 double rows of plates that bear movable, stiff spines (Figure 20-19). The plates are firmly sutured. The five pairs of ambulacral rows are homologous to the five arms of the sea star and have pores (Figure 20-18, *B*) through which the long tube feet extend. The plates bear small tubercles on which the round ends of the spines articulate as ball-and-socket joints. The spines are moved by small muscles around the bases.

There are several kinds of pedicellariae, the most common of which are three jawed and are mounted on long stalks (Figure 20-4). Pedicellariae help keep the body clean and capture small organisms. The pedicellariae of many species bear poison glands, and the toxin paralyzes small prey.

The mouth of regular urchins is surrounded by five converging teeth. In some sea urchins branched gills (modified podia) encircle the peristome. The anus, genital openings, and madreporite are located aborally in the periproct region (Figure 20-19). The sand dollars also have teeth, and the mouth is located at about the center of the oral side, but the anus has shifted to the posterior margin or even the oral side of the disc, so that an anteroposterior axis and bilateral symmetry can be recognized. Bilateral symmetry is even more accentuated in the heart urchins, with the anus near the posterior on the oral side and the mouth moved away from the oral pole toward the anterior (Figure 20-18).

Inside the test (Figure 20-19) are the coiled digestive system and a complex chewing mechanism (in the regular urchins and in sand dollars), called **Aristotle's lantern** (Figure 20-20), to which the teeth are attached. A ciliated siphon connects the esophagus to the intestine and enables the water to bypass the stomach to concentrate the food for digestion in the intestine. Sea urchins eat algae and other organic material, which they graze with their teeth. Sand dollars have short club-shaped spines that move the sand and its organic contents over the aboral surface

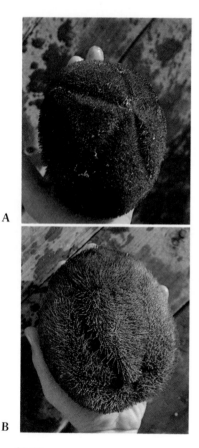

A

B

Figure 20-18

An irregular echinoid *Meoma*, one of the largest heart urchins (test up to 18 cm). *Meoma* occurs in the West Indies and from the Gulf of California to the Galápagos Islands. **A,** Aboral view. Anterior ambulacral area is not modified as a petaloid in the heart urchins, although it is in sand dollars. **B,** Oral view. Note curved mouth at anterior end and periproct at posterior end.
Photographs by L.S. Roberts.

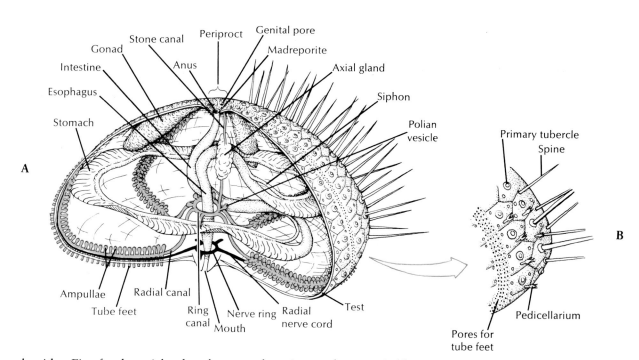

Figure 20-19

A, Internal structure of the sea urchin; water-vascular system in red. **B,** Detail of portion of endoskeleton.

and down the sides. Fine food particles drop between the spines and are carried by ciliated tracts on the oral side to the mouth.

The hemal and nervous systems are basically similar to those of the asteroids. The ambulacral grooves are closed, and the radial canals of the water-vascular system run just beneath the test, one in each of the ambulacral radii (Figure 20-19). The podia are supplied with ampullae within the test, each of which usually communicates with its podium by *two* canals through pores in the ambulacral plate; consequently, such pores in the plates are in pairs. The peristomial gills, where present, are of little or no importance in respiratory gas exchange, this function being carried out principally by the other podia. In the irregular urchins the respiratory podia are thin walled, flattened, or lobulate and are arranged in ambulacral fields called **petaloids** on the aboral surface. The irregular urchins also have short, suckered, single-pored podia in the ambulacral and sometimes interambulacral areas; these function in food handling.

Sexes are separate, and both eggs and sperm are shed into the sea for external fertilization. Some, such as certain of the pencil urchins, brood their young in depressions between the spines. The **echinopluteus larvae** (Figure 20-9, *D*) of nonbrooding echinoids may live a planktonic existence for several months and then metamorphose quickly into young urchins.

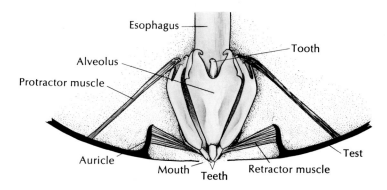

Figure 20-20

Aristotle's lantern, the complex mechanism used by the sea urchin for masticating its food. Five pairs of retractor muscles draw the lantern and teeth up into the test; five pairs of protractors push the lantern down and expose the teeth. Other muscles produce a variety of movements. Only major skeletal parts and muscles are shown in the diagram.

A

B

C

Figure 20-21

Sea cucumbers (class Holothuroidea). **A,** Common along the Pacific coast of North America, *Parastichopus californicus* grows to 50 cm in length. Its tube feet on the dorsal side are reduced to papillae and warts. **B,** In sharp contrast to most sea cucumbers, the surface ossicles of *Psolus chitonoides* are developed into a platelike armor. The ventral surface is a flat, soft, creeping sole, and the mouth (surrounded by tentacles) and anus are turned dorsally. **C,** Tube feet are found in all ambulacral areas of *Cucumaria miniata* but are better developed on its ventral side, shown here. Photographs by R. Harbo.

____ Class Holothuroidea

Sea cucumbers

In a phylum characterized by odd animals, class Holothuroidea contains members that both structurally and physiologically are among the strangest. These animals have a remarkable resemblance to the vegetable after which they are named (Figure 20-21). Compared with the other echinoderms, the holothurians are greatly elongated in the oral-aboral axis, and the ossicles are much reduced in most, so that the animals are soft bodied. Some species crawl on the surface of the sea bottom, others are found beneath rocks, and some are burrowers.

Common species along the east coast of North America are *Cucumaria frondosa* (L. *cucumis*, cucumber), *Sclerodactyla briareus* (Gr. *skleros*, hard, + *daktylos*, finger) (Figure 20-22), and the translucent, burrowing *Leptosynapta* (Gr. *leptos*, slender, + *synapsis*, joining together). Along the Pacific coast there are several species of *Cucumaria* (Figure 20-21, *C*) and the striking reddish brown *Parastichopus* (Gr. *para*, beside, + *stichos*, line or row, + *pous, podos*, foot) Figure 20-21, *A*), with very large papillae.

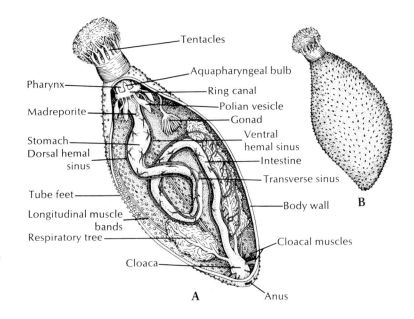

Figure 20-22

Anatomy of the sea cucumber *Sclerodactyla*. **A,** Internal. **B,** External. *Red,* Hemal system.

Form and function

The body wall is usually leathery, with tiny ossicles embedded in it (Figure 20-23), although a few species have large ossicles forming a dermal armor (Figure 20-21, *B*). Because of the elongate body form of the sea cucumbers, they characteristically lie on one side. In some species the locomotor tube feet are equally distributed to the five ambulacral areas (Figure 20-21, *C*) or all over the body, but most have tube feet well developed only in the ambulacra normally applied to the substratum (Figure 20-21, *A* and *B*). Thus, a secondary bilaterality is present, albeit of quite different origin from that of the irregular urchins. The side applied to the substratum has three ambulacra and is called the sole; the tube feet in the dorsal ambulacral areas, if present, are usually without suckers and may be modified as sensory papillae. All tube feet, except oral tentacles, may be absent in burrowing forms.

The oral tentacles are 10 to 30 retractile, modified tube feet around the mouth. The body wall contains circular and longitudinal muscles along the ambulacra.

The coelomic cavity is spacious and fluid filled and has many coelomocytes. Because of the reduction in the dermal ossicles, they no longer function as an endoskeleton, and the fluid-filled coelom has become a hydrostatic skeleton.

The digestive system empties posteriorly into a muscular **cloaca** (Figure 20-22). A **respiratory tree** composed of two long, many-branched tubes also empties into the cloaca, which pumps seawater into it. The respiratory tree serves for both respiration and excretion and is not found in any other group of living echinoderms. Gas exchange also occurs through the skin and tube feet.

The hemal system is more well developed in holothurians than in other echinoderms. The water-vascular system is peculiar in that the madreporite lies free in the coelom.

The sexes are separate, but some holothurians are hermaphroditic. Among the echinoderms, only the sea cucumbers have a single gonad, and this is considered a primitive characteristic. The gonad is usually in the form of one or two clusters of tubules that join at the gonoduct. Fertilization is external, and the free-swimming larva is called an **auricularia** (Figure 20-9, *E*). Some species brood the young either inside the body or somewhere on the body surface.

Biology

Sea cucumbers are sluggish, moving partly by means of their ventral tube feet and partly by waves of contraction in the muscular body wall. The more sedentary species trap suspended food particles in the mucus of their outstretched oral tentacles or pick up particles from the surrounding bottom. They then stuff the tentacles into the pharynx, one by one, sucking off the food material (Figure 20-24,

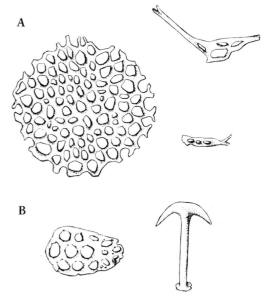

Figure 20-23

Ossicles of sea cucumbers are usually tiny and buried in the leathery dermis. They can be extracted from the tissue with caustic solutions and are important taxonomic characteristics. **A,** *Sclerodactyla.* **B,** *Leptosynapta.*
Drawings by T. Doyle.

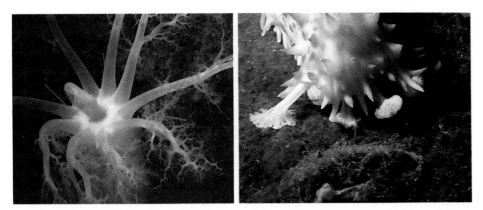

A   B

Figure 20-24

A, *Eupentacta quinquesemita* extends its tentacles to collect particulate matter in the water, then puts them one by one into its mouth and cleans the food from them. **B,** Moplike tentacles of *Parastichopus californicus* are used for deposit feeding on the bottom.
Photographs by R. Harbo.

A). Others crawl along, grazing the bottom with their tentacles (Figure 20-24, *B*).

Sea cucumbers have a peculiar power of what appears to be self-mutilation but is really a mode of defense. When irritated, some may cast out a part of their viscera by a strong muscular contraction that may either rupture the body wall or evert its contents through the anus. The lost parts are soon regenerated.

There is an interesting commensal relationship between some sea cucumbers and a small fish, *Carapus,* that uses the cloaca and respiratory tree of the sea cucumber as shelter.

____ Class Crinoidea

Sea lilies and feather stars

The crinoids are the most primitive of the living echinoderms. As fossil records reveal, crinoids were once far more numerous than now. They differ from other echinoderms by being attached during a substantial part of their lives. Sea lilies have a flower-shaped body that is placed at the tip of an attached stalk (Figure 20-25). The feather stars have long, many-branched arms, and the adults are free moving, though they may remain in the same spot for long periods (Figure 20-26). During metamorphosis feather stars become sessile and stalked, but after several months they detach and become free moving. Many crinoids are deep-water forms, but feather stars may inhabit shallow waters, especially in the Indo-Pacific and West-Indian–Caribbean regions, where the largest numbers of species are found.

Form and function

The body disc, or **calyx,** is covered with a leathery skin (**tegmen**) containing calcareous plates. The epidermis is poorly developed. Five flexible arms branch to form many more arms, each with many lateral **pinnules** arranged like barbs on a feather (Figure 20-25). The calyx and arms together are called the **crown.** Sessile

Figure 20-25

Crinoid structure. **A,** Feather star (stalked crinoid) with portion of stalk. Modern crinoid stalks rarely exceed 60 cm, but fossil forms were as much as 20 m long. **B,** Oral view of calyx of the crinoid *Antedon,* showing direction of ciliary food currents. Ambulacral grooves with podia extend from mouth along arms and branching pinnules. Food particles touching podia are tossed into ambulacral grooves and carried, tangled in mucus, by strong ciliary currents toward mouth. Particles falling on interambulacral areas are carried by cilia first toward mouth and then outward and finally dropped off the edge, thus keeping the oral disc clean.

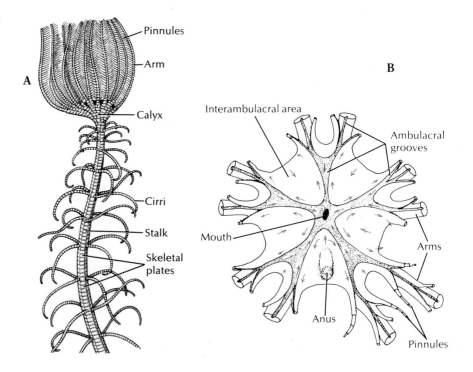

forms have a long, jointed **stalk** attached to the aboral side of the body. This stalk is made up of plates, appears jointed, and may bear **cirri.** Madreporite, spines, and pedicellariae are absent.

The upper (oral) surface bears the mouth, which opens into a short esophagus, from which the long intestine with diverticula proceeds aborally for a distance and then makes a complete turn to the **anus,** which may be on a raised cone (Figure 20-25, *B*). With the aid of tube feet and mucous nets, crinoids feed on small organisms that are caught in the ambulacral grooves. The **ambulacral grooves** are open and ciliated and serve to carry food to the mouth (Figure 20-25, *B*). Tube feet in the form of tentacles are also found in the grooves.

The water-vascular system has the basic echinoderm plan. The nervous system is made up of an oral ring and a radial nerve that runs to each arm. The aboral or entoneural system is more highly developed in crinoids than in most other echinoderms. Sense organs are scanty and primitive.

The sexes are separate. The gonads are simply masses of cells in the genital cavity of the arms and pinnules. The gametes escape without ducts through a rupture in the pinnule wall. Brooding is known in some forms. The **vitellaria** larvae (Figure 20-9, *F*) are free swimming for a time before they become attached and metamorphose. Most living crinoids are from 15 to 30 cm long, but some fossil species had stalks 25 m in length.

Figure 20-26

Nemaster spp. are crinoids found on Caribbean coral reefs. They extend their arms into the water to catch food particles while maintaining their body in a crevice. Photograph by D.W. Gotshall.

Phylogeny and Adaptive Radiation

Phylogeny

Based on the embryological evidence of the bilateral larvae of echinoderms, there can be little doubt that their ancestors were bilateral and that their radial symmetry is secondary. Some recent investigators believe that the radial symmetry arose in a free-moving echinoderm ancestor and that sessile groups were derived several times independently from the free-moving ancestors. However, this view does not account for the adaptive significance of the radial symmetry, that is, as an adaptation for the sessile existence. The more traditional view is that the first echinoderms were sessile, became radial as an adaptation to that existence, and then gave rise to the free-moving groups. Certainly, the most primitive living echinoderms are the crinoids, and the existence of a transitory stalked phase in the asteroids, which also have some primitive characteristics such as open ambulacra, supports the traditional sequence. It is also believed that the endoskeleton was an adaptation for a sessile existence and that the original function of the water-vascular system was in feeding.

Despite its wide usage, the current classification of the higher taxa in echinoderms may not reflect true relationships. For example, there are basic differences between the asteroids and ophiuroids, including the primitive characteristics of the asteroids mentioned above, yet these two groups are placed in the same class. The ophiopluteus and echinopluteus are quite similar, and larval similarities are often viewed as indicative of common ancestry, yet the ophiuroids and echinoids are separated into different subphyla.

The nature of the hypothetical preechinoderm has also been subject to debate. Some have held that it was a **dipleurula** ancestor, which was a creeping, soft-bodied animal with three, paired coelomic compartments. Hyman and others have suggested that it was a **pentactula** ancestor, more like a lophophorate animal, with tentacles around the mouth. According to the latter hypothesis, the water vascular system was derived from the five tentacles around one side, which contained extensions of the middle coelomic compartment of the ancestor.

Adaptive radiation

The radiation of the echinoderms has been determined by the limitations and potentials of their most important characteristics: radial symmetry, the water-vascular system, and their dermal endoskeleton. If their ancestors had a brain and specialized sense organs, these were lost in the adoption of radial symmetry. Thus, it is unsurprising that there are large numbers of creeping, benthic forms with filter-feeding, deposit-feeding, scavenging, and herbivorous habits, comparatively few predators, and very rare pelagic forms. In this light the relative success of the asteroids as predators is impressive and probably attributable to the extent to which they have exploited the hydraulic mechanism of the tube feet.

The basic body plan of echinoderms has severely limited their evolutionary opportunities to become parasites. Indeed, the most mobile of the echinoderms, the ophiuroids, which are also the ones most able to insert their bodies into small spaces, are the only group with significant numbers of commensal species.

—— SUMMARY

The phylum Echinodermata shows the characteristics of the Deuterostomia division of the animal kingdom. The echinoderms are an important marine group sharply distinguished from other phyla of animals. They have a radial symmetry but were derived from bilateral ancestors. They fill a variety of benthic niches, including particle feeders, browsers, scavengers, and predators.

The sea stars (class Stelleroidea, subclass Asteroidea) can be used to illustrate the echinoderms. They usually have five arms, which merge gradually with a central disc. Like other echinoderms, they have no head and few specialized sensory organs. The mouth is directed toward the substratum. They have dermal ossicles, respiratory papulae, and open ambulacral areas. Many sea stars have pedicellariae. Their water-vascular system is an elaborate hydraulic system derived from one of the coelomic cavities. Along the ambulacral areas, branches of the water-vascular system (tube feet) are important in locomotion, food gathering, respiration, and excretion. Many sea stars are predators, whereas others feed on small particles. As with other echinoderms, sea stars have a hemal system, enclosed by another coelomic compartment, which is of uncertain function. Sexes are separate, and reproductive systems are very simple. The bilateral, free-swimming larva becomes attached, transforms to a radial juvenile, then detaches and becomes a motile sea star.

Brittle stars (subclass Ophiuroidea) differ from asteroids in that their arms are slender and sharply set off from the central disc, they have no pedicellariae or ampullae, and their ambulacral grooves are closed. Their tube feet have no suckers, and their madreporite is on the oral side. They crawl by means of their arms, and they can move around more rapidly than other echinoderms. As in other echinoderms, their bilateral larva metamorphoses to a radial juvenile, but like sea urchins and sea cucumbers, there is no attached phase.

In sea urchins (class Echinoidea), the dermal ossicles have become closely fitting plates, the body is compact, and there are no arms. Their ambulacral areas are closed and extend up around their bodies. They move by means of tube feet or by their spines. Some urchins (sand dollars and heart urchins) are evolving a return to bilateral symmetry.

The dermal ossicles in sea cucumbers (class Holothuroidea) are very small; therefore the body wall is soft. Holothuroids are greatly elongated in the oral-

Brittle stars (subclass Ophiuroidea) are usually dull gray or brown, but some are quite showy, like these specimens from Puerto Rico.
Photograph by K. Sandved.

aboral axis and lie on their side. Because certain of the ambulacral areas are characteristically against the substratum, sea cucumbers have also undergone some return to bilateral symmetry. The tube feet around the mouth are modified into tentacles, with which they feed. They have an internal respiratory tree connected to the cloaca, and the madreporite hangs free in the coelom.

Sea lilies and feather stars (class Crinoidea) are the only group of living echinoderms, other than the asteroids, with open ambulacral areas. They are mucociliary particle feeders and lie with their oral side up. They become attached to the substratum by a stem during metamorphosis, although the feather stars later detach from the stem.

Although it is clear that the ancestor of the echinoderms was bilaterally symmetrical, a point still debated is whether radiality arose in a free-moving ancestor, which then gave rise to sessile groups, or whether a sessile ancestor evolved radiality and then gave rise to the free-moving forms. A hypothetical preechinoderm called the pentactula would relate the echinoderms to the lophophorates.

The basic body plan of echinoderms (including radial symmetry, water-vascular system, and dermal endoskeleton) has limited their radiation but has produced surprising evolutionary success.

Review questions

1. What is the constellation of characteristics possessed by echinoderms that is found in no other phylum?
2. How do we know that echinoderms were derived from an ancestor with bilateral symmetry?
3. Distinguish the following groups of echinoderms from each other: Crinoidea, Asteroidea, Ophiuroidea, Echinoidea, Holothuroidea.
4. What is an ambulacral groove, and what is the difference between open and closed ambulacral grooves?
5. Trace or make a rough copy of Figure 20-3, B, without the labels, then from memory label the parts of the water-vascular system of the sea star.
6. Briefly explain the mechanism of action of a sea star's tube foot.
7. Name the structures involved in the following functions in sea stars, and briefly describe the action of each: respiration, feeding and digestion, excretion, reproduction.
8. Compare the structures and functions in question 6 as they are found in brittle stars, sea urchins, sea cucumbers, and crinoids.
9. Briefly describe development in sea stars, including metamorphosis.
10. Match the groups in the left column with *all* correct answers in the right column.

_____ Crinoidea	a. Closed ambulacral grooves
_____ Asteroidea	b. Oral surface generally upward
_____ Ophiuroidea	c. With arms
_____ Echinoidea	d. Without arms
_____ Holothuroidea	e. Approximately globular or disc-shaped
	f. Elongated in oral-aboral axis
	g. With pedicellariae
	h. Madreporite internal
	i. Madreporite on oral plate

11. Define the following: pedicellariae, madreporite, respiratory tree, Aristotle's lantern.
12. What is some evidence that the ancestral echinoderm was sessile?
13. Give four examples of how echinoderms are important to humans.
14. What is a major difference in the function of the coelom in the holothurians compared with the other echinoderms?
15. How would the pentaradial symmetry of echinoderms be accounted for by the hypothesis of the pentactula ancestor?

Selected references

See also general references for Part Two, p. 590.

Benson, A.A., and R.F. Lee. 1975. The role of wax in oceanic food chains. Sci. Am. **232**:77-86 (Mar.). *The possession of enzymes to digest the wax in corals may account for the ability of* Acanthaster planci, *a sea star, to feed on coral polyps.*

Binyon, J. 1972. Physiology of echinoderms. Elmsford, N.Y., Pergamon Press, Inc.

Davidson, E.H., B.R. Hough-Evans, and R.J. Britten. 1982. Molecular biology of the sea urchin embryo. Science **217**:17-26. *Many fundamental insights into the process of embryogenesis have been revealed through studies of sea urchins.*

Dungan, M.L., T.E. Miller, and D.A. Thomson. 1982. Catastrophic decline of a top carnivore in the Gulf of California rocky intertidal zone. Science **216**:989-991. *A devastating disease led to a population crash of the predatory sea star,* Heliaster kubiniji. *Significant change in the intertidal community can be expected.*

Fell, H.B. 1982. Echinodermata. In S.P. Parker (ed.). Synopsis and classification of living organisms, vol. 2. New York, McGraw-Hill Book Co.

Moran, P.J., R.E. Reichelt, and R.H. Bradbury. 1986. An assessment of the geological evidence for previous *Acanthaster* outbreaks. Coral Reefs **4**:235-238. *The authors conclude that there is no geological evidence for prior outbreaks of the coral-eating sea star,* Acanthaster planci, *and thus we are premature in believing that outbreaks occur in natural cycles.*

CHAPTER 21

THE LESSER DEUTEROSTOMES

Phylum Chaetognatha

Phylum Hemichordata

The deuterostomes include, along with the Echinodermata, three other phyla: Chaetognatha, Hemichordata, and Chordata. Two of the chordate subphyla—Urochordata and Cephalochordata—are also invertebrate groups.

The term "lesser (or minor) deuterostomes" is usually used in reference to the hemichordates and chaetognaths, but only in the sense that they have a relatively small number of species. However, they are widespread and include some commonly found invertebrate forms. Often smaller groups deserve much more attention than is usually given them, for they contribute much to our understanding of evolutionary diversity and relationships.

These phyla have enterocoelous development of the coelom and some form of radial cleavage. The hemichordates were formerly included as a subphylum of the Chordata, but they are probably more closely related to the Echinodermata. The Chaetognatha apparently are not closely related to any other group.

——PHYLUM CHAETOGNATHA

A common name for the chaetognaths is arrowworms. They are all marine animals and are considered by some to be related to the nematodes and by others to be related to the annelids. However, they actually seem to be aberrant and show no distinct relationship to any other group. Only their embryology indicates their position as deuterostomes.

The name Chaetognatha (ke-tog′na-tha) (Gr. *chaitē,* long flowing hair, + *gnathos,* jaw) refers to the sickle-shaped bristles on each side of the mouth. This is not a large group, for only some 65 species are known. Their small, straight bodies resemble miniature torpedoes, or darts, ranging from 2.5 to 10 cm in length.

The arrowworms are all adapted for a planktonic existence, except for *Spadella* (Gr. *spadix,* palm frond, + *ella,* dim. suffix), a benthic genus. They usually swim to the surface at night and descend during the day. Much of the time they drift passively, but they can dart forward in swift spurts, using the caudal fin and

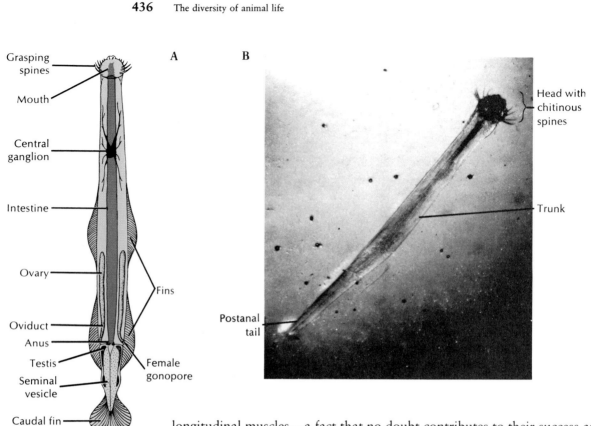

Grasping spines
Mouth
Central ganglion
Intestine
Ovary
Fins
Oviduct
Anus
Testis
Female gonopore
Seminal vesicle
Caudal fin

A

B

Head with chitinous spines
Trunk
Postanal tail

Figure 21-1

Arrowworm *Sagitta*. **A,** Internal structure. **B,** Head *(top)* is largely covered with hood formed from epidermis. When worm is engaged in catching its prey, hood is retracted to neck region. (Preserved specimen.) **B,** Photograph by L.S. Roberts.

longitudinal muscles—a fact that no doubt contributes to their success as planktonic predators. Horizontal fins bordering the trunk are used in flotation rather than in active swimming.

Form and Function

The body of the arrowworm is unsegmented and is made up of head, trunk, and postanal tail (Figure 21-1). On the underside of the head is a large vestibule leading to the mouth. The vestibule contains teeth and is flanked on both sides by curved chitinous spines used in seizing the prey. A pair of eyes is on the dorsal side. A peculiar hood formed from a fold of the neck can be drawn forward over the head and spines. When the animal captures prey, the hood is retracted, and the teeth and raptorial spines spread apart and then snap shut with startling speed. Arrowworms are voracious feeders, living on planktonic forms, especially copepods, and even small fish. When they are abundant, as they often are, they may have a substantial ecological impact. They are nearly transparent (Figure 21-1, *B*), a characteristic of adaptive value in their role as planktonic predators.

The body is covered with a thin cuticle and with epidermis that is single layered except along the sides of the body, where it is stratified in a thick layer. These are the only invertebrates with a many-layered epidermis.

Arrowworms are fairly advanced worms in that they have a complete digestive system, a well-developed coelom, and a nervous system with a nerve ring containing large dorsal and ventral ganglia and a number of lateral ganglia. Sense organs include the eyes, sensory bristles, and a U-shaped ciliary loop that extends over the neck from the back of the head and may detect water currents or may be chemosensory. However, vascular, respiratory, and excretory systems are entirely lacking.

Arrowworms are hermaphroditic with either cross- or self-fertilization. The eggs of *Sagitta* (L. arrow) are coated with jelly and are planktonic. Eggs of other arrowworms may be attached to the body and carried about for a time. The juveniles develop directly without metamorphosis. Chaetognath embryogenesis differs from that of other deuterostomes in that the coelom is formed by a backward extension from the archenteron rather than by pinched-off coelomic sacs. There is no true peritoneum lining the coelom. Cleavage is radial, complete, and equal.

The best-known genus is *Sagitta,* the common arrowworm (Figure 21-1).

PHYLUM HEMICHORDATA

Position in Animal Kingdom
1. Hemichordates belong to the deuterostome branch of the animal kingdom and are enterocoelous coelomates with radial cleavage.
2. Hemichordates show some of both echinoderm and chordate characteristics.
3. A chordate plan of structure is suggested by gill slits and a restricted dorsal tubular nerve cord.
4. Similarity to the echinoderms is shown in larval characteristics.

Biological Contributions
1. A **tubular dorsal nerve cord** in the collar zone may represent an early stage of the condition in chordates; a diffused net of nerve cells is similar to the uncentralized, subepithelial plexus of echinoderms.
2. The **gill slits** in the pharynx, which are also characteristic of chordates, are used primarily for filter feeding and only secondarily for breathing and are thus comparable to those in the protochordates.

The Hemichordata (hem'i-kor-da'ta) (Gr. *hemi,* half, + *chorda,* string, cord) are marine animals that were formerly considered a subphylum of the chordates, based on their possession of gill slits and a rudimentary notochord. However, it is now generally agreed that the so-called hemichordate notochord is really a stomochord and not homologous with the chordate notochord, so the hemichordates are given the rank of a separate phylum.

Hemichordates are vermiform bottom dwellers, living usually in shallow waters. Some are colonial and live in secreted tubes. Most are sedentary or sessile. Their distribution is fairly worldwide, but their secretive habits and fragile bodies make collecting them difficult.

Members of class Enteropneusta (Gr. *enteron,* intestine, + *pneustikos,* of, or for, breathing) (acorn worms) range from 20 mm to 2.5 m in length and 3 to 200 mm in breadth. Members of class Pterobranchia (Gr. *pteron,* wing, + *branchia,* gills) are smaller, usually 5 to 14 mm, not including the stalk. About 70 species of enteropneusts and three small genera of pterobranchs are recognized.

Hemichordates have the typical tricoelomate structure of deuterostomes.

CHARACTERISTICS
1. Soft bodied; wormlike or short and compact with stalk for attachment
2. Body divided into proboscis, collar, and trunk; coelomic pouch single in proboscis, but paired in other two; buccal diverticulum in posterior part of proboscis
3. Enteropneusta free moving and of burrowing habits; pterobranchs sessile, mostly colonial, living in secreted tubes
4. Circulatory system of dorsal and ventral vessels and dorsal heart
5. Respiratory system of gill slits (few or none in pterobranchs) connecting the pharynx with outside as in chordates
6. No nephridia; a single glomerulus connected to blood vessels may have excretory function
7. A subepidermal nerve plexus thickened to form dorsal and ventral nerve cords, with a ring connective in the collar; dorsal nerve cord of collar hollow in some
8. Sexes separate in Enteropneusta, with gonads projecting into body cavity; in pterobranchs reproduction may be sexual or asexual (in some) by budding; tornaria larva in some Enteropneusta

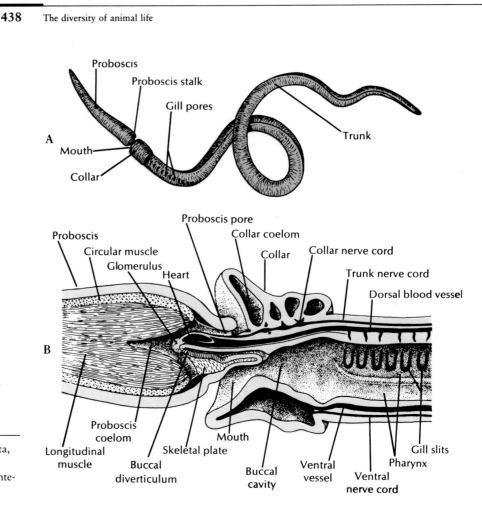

Figure 21-2

Acorn worm *Saccoglossus* (Hemichordata, class Enteropneusta). **A,** External lateral view. **B,** Longitudinal section through anterior end.

___ Class Enteropneusta

The enteropneusts, or acorn worms, are sluggish, wormlike animals that live in burrows or under stones, usually in mud or sand flats of intertidal zones. *Balanoglossus* (Gr. *balanos,* acorn, + *glōssa,* tongue) and *Saccoglossus* (Gr. *sakkos,* sac, strainer, + *glōssa,* tongue) (Figure 21-2) are common genera.

Form and function

The mucus-covered body is divided into a tonguelike proboscis, a short collar, and a long trunk (protosome, mesosome, and metasome).

Proboscis

The proboscis is the active part of the animal. It probes about in the mud, examining its surroundings and collecting food in mucous strands on its surface. These are carried by cilia to the groove at the edge of the collar, are directed to the mouth on the underside, and are swallowed. Large particles can be rejected by covering the mouth with the edge of the collar (Figure 21-3).

Burrow dwellers use the proboscis to excavate, thrusting it into the mud or sand and allowing cilia and mucus to move the sand backward. Or they may eat the sand and mud as they go, extracting its organic contents. They build U-shaped mucus-lined burrows, usually with two openings 10 to 30 cm apart and with the base of the U 50 to 75 cm below the surface. They can thrust the proboscis out the front opening for feeding. Defecation at the back opening builds characteristic spiral fecal mounds that leave a telltale clue to the location of the burrows.

In the posterior end of the proboscis is a small coelomic sac (protocoel) into which extends the **buccal diverticulum,** a slender, blindly ending pouch of the gut that reaches forward into the buccal region and was formerly believed to be a notochord. A slender canal connects the protocoel with a **proboscis pore** to the outside (Figure 21-2, *B*). The paired coelomic cavities in the collar also open by pores. By taking in water through the pores into the coelomic sacs, the proboscis and collar can be stiffened to aid in burrowing. Contraction of the body musculature then forces the excess water out through the gill slits, reducing the hydrostatic pressure and allowing the animal to move forward.

Branchial system

A row of **gill pores** is located dorsolaterally on each side of the trunk just behind the collar (Figure 21-3, *A*). These open from a series of gill chambers that in turn connect with a series of **gill slits** in the sides of the pharynx. No gills are attached to the gill slits, but some respiratory gaseous exchange occurs in the vascular branchial epithelium, as well as in the body surface. Ciliary currents keep a fresh supply of water moving from the mouth through the pharynx and out the gill slits and branchial chambers to the outside.

Feeding and the digestive system

Hemichordates are largely ciliary-mucus feeders. Behind the buccal cavity lies the large pharynx containing in its dorsal part the U-shaped gill slits (Figure 21-2, *B*). Since there are no gills, the primary function of the branchial mechanism of the pharynx is presumably food gathering. Food particles caught in mucus and brought to the mouth by ciliary action on the proboscis and collar are strained out of the branchial water that leaves through the gill slits and are directed along the ventral part of the pharynx and esophagus to the intestine, where digestion and absorption occur (Figure 21-3).

Circulatory and excretory systems

A middorsal vessel carries the colorless blood forward above the gut. In the collar the vessel expands into a sinus and a heart vesicle above the buccal diverticulum. Blood is then driven into a network of blood sinuses called the **glomerulus,** which partially surrounds these structures. The glomerulus is assumed to have an excretory function (Figure 21-2, *B*). Blood travels posteriorly through a ventral vessel below the gut, passing through extensive sinuses to the gut and body wall.

Nervous and sensory systems

The nervous system consists mostly of a subepithelial network, or plexus, of nerve cells and fibers to which processes of epithelial cells are attached. Thickenings of this net form dorsal and ventral nerve cords that are united posterior to the collar by a ring connective. The dorsal cord continues into the collar and furnishes many fibers to the plexus of the proboscis. The collar cord is hollow in some species and contains giant nerve cells with processes running to the nerve trunks. This primitive nerve plexus system is highly reminiscent of that of the cnidarians and echinoderms.

Sensory receptors include neurosensory cells throughout the epidermis (especially in the proboscis, a preoral ciliary organ that may be chemoreceptive) and photoreceptor cells.

Reproductive system and development

Sexes are separate in enteropneusts. Gonads are arranged in a dorsolateral row on each side of the anterior part of the trunk. Fertilization is external, and in some species a ciliated **tornaria** larva develops that at certain stages is so similar to the

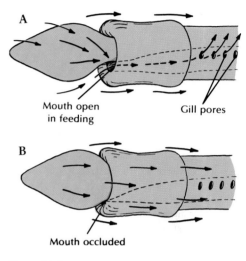

A Mouth open in feeding   Gill pores

B Mouth occluded

Figure 21-3

Food currents of enteropneust hemichordate. **A,** Side view of acorn worm with mouth open, showing direction of currents created by cilia on proboscis and collar. Food particles are directed toward mouth and digestive tract. Rejected particles move toward outside of collar. Water leaves through gill pores. **B,** When mouth is occluded, all particles are rejected and passed onto the collar. Nonburrowing and some burrowing hemichordates use this feeding method.

Modified from Russell-Hunter, W.D. 1969. A biology of the higher invertebrates. New York, Macmillan, Inc.

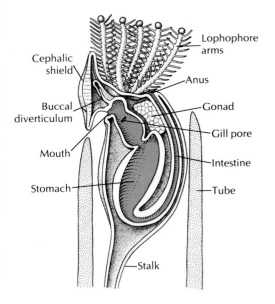

Figure 21-4

Cephalodiscus, a pterobranch hemichordate. These tiny (5 to 7 mm) forms live in coenecium tubes in which they can move about. Ciliated tentacles and arms direct currents of food and water toward mouth. These deep-sea animals may be close to the ancestral stock of echinoderms and chordates.

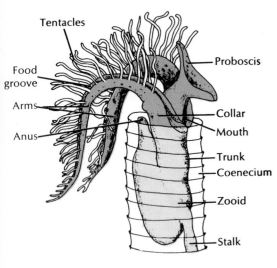

Figure 21-5

Rhabdopleura, a pterobranch hemichordate in its tube. Individuals live in branching tubes, connected by stolons, and protrude the tentacled lophophore for feeding.

echinoderm bipinnaria that it was once believed to be an echinoderm larva. The familiar *Saccoglossus* of American waters has direct development without a tornaria stage.

Class Pterobranchia

The basic plan of the class Pterobranchia is similar to that of the Enteropneusta, but certain structural differences are correlated with the sedentary life-style of pterobranchs. The first pterobranch ever reported was obtained by the famed *Challenger* expedition of 1872 to 1876. Although first placed among the Polyzoa (Entoprocta and Ectoprocta), its affinities to the hemichordates were later recognized. Only two genera (*Cephalodiscus* and *Rhabdopleura*) are known in any detail.

Pterobranchs are small animals, usually within the range of 1 to 7 mm in length, although the stalk may be longer. Many individuals of *Cephalodiscus* (Gr. *kephalē*, head, + *diskos*, disc) (Figure 21-4) live together in gelatinous tubes, which often form an anastomosing system. The zooids are not connected, however, and live independently in the tubes. Through apertures in these tubes, they extend their crown of tentacles. They are attached to the walls of the tubes by extensible stalks that can jerk the owners back into the tubes when necessary.

The body of *Cephalodiscus* is divided into the three regions—proboscis, collar, and trunk—characteristic of the hemichordates. There is only one pair of gill slits, and the alimentary canal is U-shaped, with the anus near the mouth. The proboscis is shield shaped. At the base of the proboscis are five to nine pairs of branching arms with tentacles containing an extension of the coelomic compartment of the mesosome, as in a lophophore. Ciliated grooves on the tentacles and arms collect food. Some species are dioecious, and others are monoecious. Asexual reproduction by budding may also occur.

In *Rhabdopleura* (Gr. *rhabdos*, rod, + *pleura*, a rib, the side), which is smaller than *Cephalodiscus*, the members remain together to form a colony of zooids connected by a stolon and enclosed in coenecium tubes (Figure 21-5). The collar in these forms bears two branching arms or lophophores. No gill clefts or glomeruli are present. New individuals are reproduced by budding from a creeping basal stolon, which branches on a substratum. None of the pterobranchs has a tubular nerve cord in the collar, but otherwise their nervous system is similar to that of the Enteropneusta.

Atubaria (Gr. *a*, prefix meaning absence of, + L. *tubus*, tube), a little-known genus, has no tube but attaches its stalk to colonial hydroids.

The fossil graptolites of the middle Paleozoic era are often placed as an extinct class under Hemichordata. Their tubular chitinous skeleton and colonial habits indicate an affinity with *Rhabdopleura*. They are considered important index fossils of the Ordovician and Silurian geological strata.

Phylogeny and Adaptive Radiation

Phylogeny

Hemichordate phylogeny has long been puzzling. Hemichordates share characteristics with both the echinoderms and the chordates. They share with the chordates the gill slits, which are used primarily for filter feeding and secondarily for breathing as in some of the protochordates. A short dorsal, somewhat hollow nerve cord in the collar zone foreshadows the nerve cord of the chordates. The buccal diverticulum in the hemichordate mouth cavity, which was long believed to be a rudimentary notochord homologous with the notochord of chordates, is now considered of no phylogenetic importance. The relationship to the echinoderms is striking. The early embryogenesis is remarkably like that of echinoderms, and the early tornaria larva is almost identical to the bipinnaria larva of asteroids. The similarity between the hydraulic action of the coelomic pouches and that of the watervascular system in echinoderms and the similarity in plan of the subepithelial nerve plexus of the two groups are further evidence of their relationship.

Within the phylum, the class Pterobranchia is considered more primitive than the class Enteropneusta and shows affinities with the Ectoprocta, Brachiopoda, and others because of its lophophore and sessile habits. Some believe that the pterobranchs may be similar to the common ancestors of both the hemichordates and the echinoderms.

Adaptive radiation

Because of their sessile lives and their habitat in secreted tubes in ocean bottoms, where conditions are fairly stable, the pterobranchs have undergone little adaptive divergence. They have retained a tentacular type of ciliary feeding. The enteropneusts, on the other hand, although sluggish, are more active than the pterobranchs. Having lost the tentaculated arms, they use a proboscis to trap small organisms in mucus, or they eat sand as they burrow and digest organic sediments from the sand. Their evolutionary divergence, although greater than that of the pterobranchs, is still modest.

—— SUMMARY

The arrowworms (phylum Chaetognatha) are a small group but are important as a component of marine plankton. They have a well-developed coelom and are effective predators, catching other planktonic organisms with the teeth and chitinous spines around their mouth.

Members of the phylum Hemichordata are marine worms that were formerly considered chordates because their buccal diverticulum was believed to be a notochord. However, like the chordates, some of them do have gill slits and a hollow, dorsal nerve cord. The divisions of their body (proboscis, collar, trunk) contain the typical deuterostome coelomic compartments (protocoel, mesocoel, metacoel). The hemichordate class Enteropneusta contains burrowing worms that feed on particles strained out of the water by their gill slits. Members of the class Pterobranchia are tube dwellers, filter feeding with tentacles. The hemichordates are important phylogenetically because they show affinities with the chordates, echinoderms, and lophophorates.

Review questions
 1. What are four morphological characteristics of Chaetognatha?
 2. What is the ecological importance of arrowworms?
 3. What characteristics do the Hemichordata have in common with the Chordata, and how do the two phyla differ?
 4. Distinguish the Enteropneusta from the Pterobranchia.
 5. What is the evidence that the Hemichordata are related both to the echinoderms and the lophophorate phyla?

Selected references

See also general references for Part Two, p. 590.

Alvarino, A. 1965. Chaetognaths. Oceanogr. Mar. Biol. Ann. Rev. 3:115-194.

Barrington, E.J.W. 1965. The biology of Hemichordata and Protochordata. San Francisco, W.H. Freeman & Co., Publishers. *Concise account of behavior, physiology, and reproduction of hemichordates, urochordates, and cephalochordates.*

Ghirardelli, E. 1968. Some aspects of the biology of the chaetognaths. Adv. Mar. Biol. 6:271-375.

T H E   C H O R D A T E S

Ancestry and Evolution, General Characteristics, Protochordates

Position in Animal Kingdom

Phylum Chordata (kor-da'ta) (L. *chorda*, cord) belongs to the Deuterostomia branch of the animal kingdom that includes the phyla Echinodermata, Chaetognatha, and Hemichordata. All of these phyla share many embryological features and are probably descended from an ancient common ancestor. Of all the deuterostomes, the Chordata are by far the most successful in an evolutionary sense. From humble beginnings, the chordates evolved a vertebrate body plan of unrivaled adaptability that always remains distinctive, while it provides almost unlimited scope for specialization in life-style, form, and function.

Biological Contributions

1. The **endoskeleton** of the vertebrates, evolved from the notochord of protochordates, permits continuous growth without molting, the attainment of large body size, and provides an efficient framework for muscle attachment.

2. The **perforated pharynx** of protochordates that originated as a filter-feeding device served as the framework for the subsequent evolution of true internal gills and jaws.

3. The adoption of a **predatory habit** by the early vertebrates and the accompanying evolution of a **highly differentiated brain** and **paired special sense organs** contributed in large measure to the successful adaptive radiation of the vertebrates.

4. The **paired appendages** that appeared in the aquatic vertebrates were successfully adapted later as jointed limbs for efficient locomotion on land or as wings for flight.

5. **Neoteny,** the attainment of sexual maturity in the larval body form, has occurred more than once among vertebrates and was probably important in the evolution of the earliest vertebrates.

—— THE CHORDATES

Animals most familiar to most people belong to the great phylum Chordata (kor-da'ta) (L. *chorda*, cord). Humans are members and share the characteristic from which the phylum derives its name—the **notochord** (Gr. *nōton*, back, + L. *chorda*, cord) (Figure 22-1). This structure is possessed by all members of the phylum, in either the larval or the embryonic stages or throughout life. The notochord is a rodlike, semirigid body of vacuolated cells, which extends, in most cases, the length of the body between the enteric canal and the central nervous system. Its primary purpose is to support and stiffen the body, that is, to act as a skeletal axis.

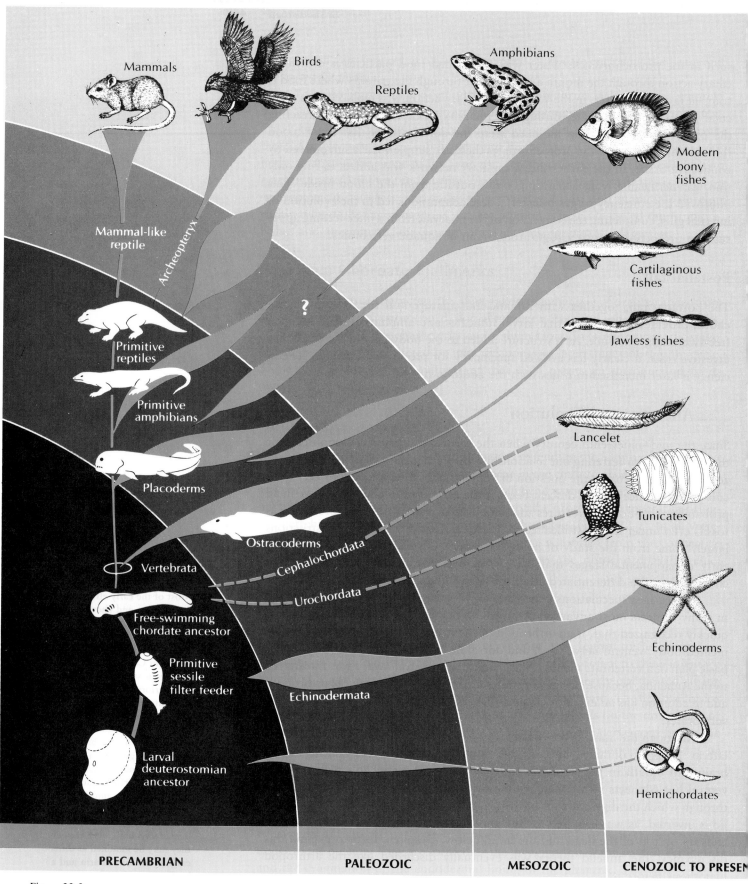

Figure 22-2

Family tree of the chordates, suggesting probable origin and relationships. Other schemes have been suggested and are possible. *White*, Extinct stem groups. *Black*, Living groups. *Red*, Lines of descent. The relative success in numbers of species of each group through geological time, as indicated by the fossil record, is suggested by the bulging and thinning of that group's line of descent. Dashed lines indicate a poor or nonexistent fossil record.

anus forms from the blastopore or at the end of the gastrula, and the mouth is formed as a secondary opening, usually on the opposite end. The coelom of both phyla is primitively enterocoelous: it is budded off from the archenteron of the embryo. (However, as we noted in the introduction, coelom development in the vertebrates is modified: the mesoderm arises as a solid mass of cells, rather than a pouch, which then *splits* to form a coelom. Thus the vertebrate coelom can be considered a schizocoel. Both echinoderm and chordate embryos show radial, regulative cleavage; that is, each of the early blastomeres has equivalent potentiality for supporting full development of a complete embryo. These characteristics are shared by brachiopods and pterobranchs (a hemichordate group), as well as by echinoderms, protochordates, amphioxus, and vertebrates. This is probably a natural grouping and almost certainly indicates interrelationships, although remote (Figure 22-2).

More recently, another piece of evidence linking chordates with the sea stars and their allies has come from the detailed study of a curious group of fossil echinoderms, the Stylophora. These small, nonsymmetrical forms have a head resembling a long-toed medieval boot, a series of branchial slits covered with flaps much like the gill openings of sharks, a postanal tail with a central rod resembling a notochord, and muscle blocks and a dorsal nerve cord (Figure 22-3). These creatures were apparently adapted to use their gill slits for filter feeding, as do the primitive chordates today. Are they the long-sought chordate ancestors? Unfortunately we are not yet, and perhaps will never be, in a position to know.

Subphylum Urochordata (Tunicata)

The urochordates ("tail-chordates") are the most successful of the invertebrate chordates. The approximately 2000 species are all marine and are widely distributed in all seas from near the shoreline to great depths. The urochordates are commonly called tunicates, a name suggested by the **tunic** that invests and protects the animal. The tunic, curiously composed largely of cellulose, may be thick and tough, or thin, delicate, and translucent. As adults, tunicates are highly specialized chordates, for in most species only the larval form, which resembles a microscopic tadpole, bears all the chordate hallmarks.

Urochordata is divided into three classes: **Ascidiacea** (Gr. *askiolion*, little bag, + *acea*, suffix), **Larvacea** (L. *larva*, ghost, + *acea*, suffix), and **Thaliacea** (Gr. *thalia*, luxuriance, + *acea*, suffix). Of these, the members of Ascidiacea, are by far the most common and best known. They are often called "sea squirts" because some species forcefully discharge a jet of water from the excurrent siphon when irritated. All but a few ascidian species are sessile animals, attached to rocks or other hard substrates such as pilings or the bottoms of ships (Figure 22-4). In many areas, they are among the most abundant of intertidal animals.

Ascidians may be solitary, colonial, or compound. Each of the solitary and colonial forms has its own test, but among the compound forms many individuals may share the same test (Figure 22-5). In some of these compound ascidians each member has its own incurrent siphon, but the excurrent opening is common to the group.

Solitary ascidians (Figures 22-4 and 22-6) are usually spherical or cylindrical forms bearing two projections: the **incurrent siphon,** which corresponds to the anterior end of the body, and the **excurrent siphon** that marks the dorsal side. When the sea squirt is expanded, water enters the incurrent siphon and passes into a capacious ciliated **pharynx** that is minutely subdivided by gill slits to form an elaborate basketwork. Water passes through the gill slits into an **atrial cavity** and out through the excurrent siphon.

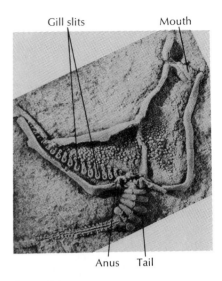

Figure 22-3

Fossil of a primitive echinoderm, *Cothurnocystis*, that lived during the Ordovician period (450 million years BP). It shows affinities with both echinoderms and chordates and may belong to a group that was ancestral to the chordates.
Courtesy R.P.S. Jeffries, The British Museum.

Figure 22-4

Typical solitary ascidian, the sea peach *Halocynthia pyriformis* of the northwestern Atlantic.
Photograph by C.P. Hickman, Jr.

Figure 22-5

Compound sea squirt *Botryllus schlosseri*, common in shallow coastal waters and rock tide pools. Each of the star-shaped patterns represents a colonial arrangement in which the arms of the star are individuals, each with its own incurrent siphon at the end of the arm. All are united centrally where they share a common test, forming a compound ascidian. (Approximately ×2.)
Photograph by D.P. Wilson.

Feeding depends on the formation of a mucous net that is secreted by a glandular groove, the **endostyle,** located along the midventral side of the pharynx. Cilia on gill bars of the pharynx pull the mucus into a sheet that spreads dorsally across the inner face of the pharynx. Food particles brought in the incurrent opening are trapped on the mucous net, which is then worked into a rope and carried posteriorly by cilia into the esophagus and stomach. Nutrients are absorbed in the midgut and indigestible wastes are discharged from the anus, located near the excurrent siphon.

The circulatory system consists of a ventral heart and two large vessels, one on either side of the heart; these vessels connect to a diffuse system of smaller vessels and spaces serving the pharyngeal basket (where respiratory exchange occurs), the digestive organs, gonads, and other structures. An odd feature found in no other chordate is that the heart drives the blood first in one direction for a few beats, then pauses, reverses its action, and drives the blood in the opposite direction for a few beats. Another remarkable feature is the presence of strikingly high amounts of the rare elements vanadium or niobium in the blood. The concentrations of vanadium in the sea squirt *Ciona* may reach 2 million times its concentration in sea water. The function of these rare metals in the blood is a mystery.

The nervous system is restricted to a nerve ganglion and a plexus of nerves that lie on the dorsal side of the pharynx. Beneath the nerve ganglion is located the **subneural gland,** which is connected by a duct to the pharynx. Apparently this gland samples the water coming into the pharynx, and may additionally carry on an endocrine function concerned with reproduction. A notochord is lacking in adult sea squirts.

Figure 22-6

Metamorphosis of a solitary ascidian from a free-swimming tadpole larva stage.

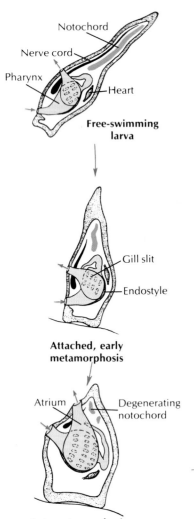

Notochord
Nerve cord
Pharynx
Heart
Free-swimming larva

Gill slit
Endostyle
Attached, early metamorphosis

Atrium
Degenerating notochord
Late metamorphosis

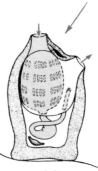

Adult

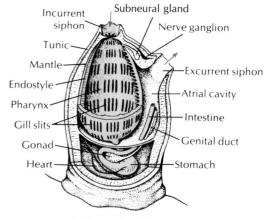

Incurrent siphon
Subneural gland
Tunic
Nerve ganglion
Mantle
Excurrent siphon
Endostyle
Atrial cavity
Pharynx
Intestine
Gill slits
Gonad
Genital duct
Heart
Stomach
Anatomy of adult

Sea squirts are hermaphroditic, with usually a single ovary and a single testis in the same animal. Germ cells are carried by ducts into the atrial cavity, and then into the surrounding water where fertilization occurs.

Of the four chief characteristics of chordates, adult sea squirts have only one: the pharyngeal gill slits. However, the larval form gives away the secret of their true relationship. The tadpole larva (Figure 22-6) is an elongate, transparent form with all four chordate characteristics: a notochord, a hollow dorsal nerve cord, a propulsive postanal tail, and a large pharynx with endostyle and gill slits. The larva does not feed but swims about for some hours before fastening itself vertically by its adhesive papillae to some solid object. It then undergoes a retrograde metamorphosis to become the sessile adult (Figure 22-6).

Tunicates of the class Thaliacea are barrel- or lemon-shaped pelagic forms with transparent, gelatinous bodies that, despite the considerable size that some species reach, are nearly invisible in sunlit surface waters. The cylindrical thaliacean body is typically surrounded by bands of circular muscle, with incurrent and excurrent siphons at opposite ends (Figure 22-7). Water pumped through the body by muscular contraction (rather than by cilia as in ascidians) is used for locomotion by a sort of jet propulsion, for respiration, and as a source of particulate food that is filtered out on mucous surfaces. Many are provided with luminous organs and give a brilliant light at night. Most of the body is hollow, with the viscera forming a compact mass on the ventral side.

The life histories of thaliaceans are often complex and are adapted to respond to sudden increases in their food supply. The appearance of a phytoplankton bloom, for example, is met by an explosive population increase leading to an extremely high density of thaliaceans. Common forms include *Doliolum* (Figure 22-7) and *Salpa,* both of which reproduce by an alternation of sexual and asexual generations. Thaliaceans are believed to have evolved from sessile ancestors as did the ascidians.

The third tunicate class, the Larvacea (=Appendicularia) are curious pelagic

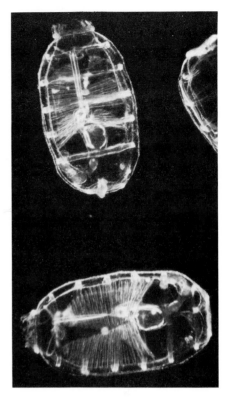

Figure 22-7

Thaliacean *Doliolum nationalis*—a member of the class Thaliacea. (Approximately ×25.)
Photograph by D.P. Wilson.

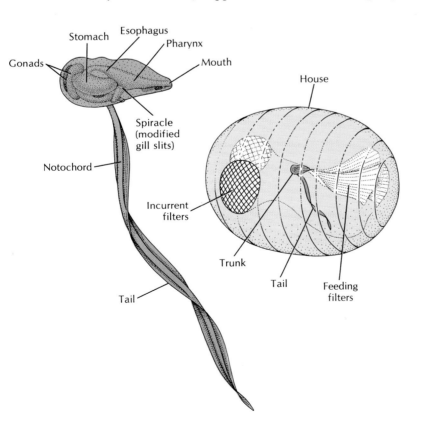

Figure 22-8

Larvacean adult *(left)* and as it appears within its delicate, transparent house *(right),* which is about the size of a walnut. When the feeding filters become clogged with food, the tunicate abandons its house and builds a new one.

creatures shaped like a bent tadpole. They feed by a method unique to the animal world. Each builds a delicate house, a delicate hollow sphere of mucus interlaced with filters and passages through which the water enters (Figure 22-8). Particulate food trapped on a feeding filter inside the house is drawn into the animal's mouth through a strawlike tube. When the filters become clogged with waste, which happens about every 4 hours, the larvacean abandons its house and builds a new house, a process that takes only a few minutes. Like the thaliaceans, the larvaceans can quickly build up dense populations when food is abundant. Scuba diving through the houses, which are about the size of walnuts, is likened to swimming through a snowstorm! Larvaceans are neotenous, that is, they are sexually mature animals that have retained the larval body form.

___ Subphylum Cephalochordata

The cephalochordates are the marine lancelets: slender, laterally compressed, translucent animals about 5 to 7 cm in length (Figure 22-9) that inhabit the sandy bottoms of coastal waters around the world. Lancelets originally bore the generic name *Amphioxus* (Gr. *amphi*, both ends, + *oxys*, sharp), but later surrendered by priority to *Branchiostoma (Gr. branchia*, gills, + *stoma*, mouth). This left amphioxus as a convenient trivial name for all of the some 25 species in this diminutive subphylum. Four species of amphioxus are found in North American coastal waters.

Amphioxus is especially interesting because it has the four distinctive characteristics of chordates in simple form, and in other ways it may be considered an early blueprint of the phylum. Water enters the mouth, driven by cilia in the buccal cavity, then passes through numerous gill slits in the pharynx where food is trapped in mucus, which is then moved into the intestine. Here the food particles are separated from the mucus and passed into a hepatic cecum where they are digested. As in the tunicates, the filtered water passes into an atrium, then leaves the body by an atriopore.

The closed circulatory system is complex for so simple a chordate (Figure 22-10). The flow pattern is remarkably similar to that of the primitive fishes, although there is no heart. Blood is pumped forward in the ventral aorta by peristaltic-like contractions of the vessel wall, then passes upward through the branchial arteries (aortic arches) in the gill bars to the dorsal aorta. From here the blood is distributed to the body tissues by microcirculation and then is collected in veins, which return it to the ventral aorta. The blood lacks erythrocytes and hemoglobin.

The nervous system is centered around a hollow nerve cord lying above the notochord. Pairs of spinal nerve roots emerge at each trunk myomeric (muscle)

Figure 22-9

Amphioxus. This interesting bottom-dwelling cephalochordate possesses the four distinctive chordate characteristics (notochord, dorsal nerve cord, pharyngeal gill slits, and postanal tail) that once made it the prime candidate for our vertebrate ancestor. However, because it also bears many specialized features, zoologists now consider it a divergent offshoot from the main line of chordate evolution.
Photograph by B. Tallmark.

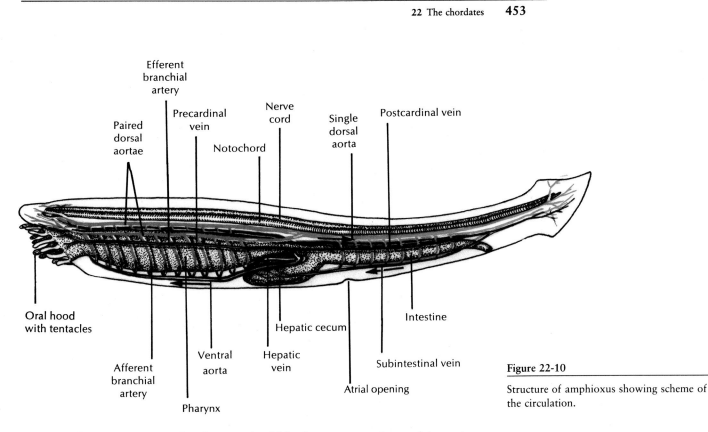

Efferent
branchial
artery

Precardinal
vein

Nerve
cord

Paired
dorsal
aortae

Notochord

Single
dorsal
aorta

Postcardinal vein

Oral hood
with tentacles

Hepatic cecum

Intestine

Afferent
branchial
artery

Ventral
aorta

Hepatic
vein

Subintestinal vein

Atrial opening

Pharynx

Figure 22-10

Structure of amphioxus showing scheme of
the circulation.

segment. Sense organs are simple, unpaired bipolar receptors located in various
parts of the body. The "brain" is a simple vesicle at the anterior end of the nerve
cord.

Sexes are separate. The sex cells are set free in the atrial cavity, then pass out
the atriopore to the outside where fertilization occurs. Cleavage is total (holoblastic) and a gastrula is formed by invagination. The larvae hatch soon after deposition and gradually assume the shape of adults.

No other chordate shows the basic diagnostic chordate characteristics as
clearly as the amphioxus. In addition to the four chordate anatomical hallmarks,
amphioxus possesses several structural features that foreshadow the vertebrate
plan. Among these are a liver diverticulum, a cecum that resembles the vertebrate
pancreas in secreting digestive enzymes, segmented trunk musculature, and the
basic circulatory plan of more advanced chordates.

Just where amphioxus belongs in the phylogeny of chordates is a much-
disputed point. Some regard amphioxus as a highly specialized or degenerative
member of the phylum, and certainly amphioxus lacks that most important of all
vertebrate distinguishing characteristics, a distinct head with its well-developed
special sense organs and the equipment for shifting to an active predatory mode of
life. Nevertheless the earliest vertebrate ancestor may well have emerged from a
form that closely resembled that of amphioxus.

Subphylum Vertebrata

The third subphylum of the chordates is the large and eminently successful Vertebrata, the subject of the next five chapters of this book. The subphylum Vertebrata shares the basic chordate characteristics with the other two subphyla, but in
addition it has a number of features that the others do not share. The characteristics that give the members of this group the name "Vertebrata" or "Craniata"
are a spinal column of vertebrae, which forms the chief skeletal axis of the body,
and a braincase, or cranium.

The once exalted position of
amphioxus in zoological circles
was put to rhyme by Philip Pope.
Sung to the tune of "Tipperary,"
the poem ends:

My notochord shall grow into
  a chain of vertebrae;
As fins my metapleural folds
  shall agitate the sea;
This tiny dorsal nervous tube
  shall form a mighty brain,
And the vertebrates shall
  dominate the animal domain.
Chorus

It's a long way from
  Amphioxus
It's a long way to us.
It's a long way from
  Amphioxus
To the meanest human cuss.
It's good-bye fins and gill slits,
Welcome skin and hair.
It's a long way from
  Amphioxus
But we came from there.

CHARACTERISTICS

1. Chief diagnostic features of chordates—**notochord, dorsal nerve cord, pharyngeal gill pouches,** and **postanal tail**—all present at some stage of the life cycle
2. **Integument** basically of two divisions, an outer **epidermis** of stratified epithelium from the ectoderm and an inner **dermis** of connective tissue derived from the mesoderm; many modifications of skin among the various classes, such as glands, scales, feathers, claws, horns, and hair
3. Notochord more or less replaced by the spinal column of vertebrae composed of cartilage or bone or both; distinctive **endoskeleton** consisting of vertebral column with the cranium, visceral arches, limb girdles, and two pairs of jointed appendages
4. **Many muscles** attached to the skeleton to provide for movement
5. Complete digestive system ventral to the spinal column and provided with large digestive glands, liver, and pancreas
6. Circulatory system consisting of the **ventral heart** of two to four chambers; a closed blood vessel system of arteries, veins, and capillaries; blood fluid containing red blood corpuscles with hemoglobin and white corpuscles; paired aortic arches connecting the ventral and dorsal aortas and giving off branches to the gills among the aquatic vertebrates; in the terrestrial types modification of the aortic arch plan into pulmonary and systemic systems
7. Well-developed **coelom** largely filled with the visceral systems
8. Excretory system consisting of **paired kidneys** (mesonephric or metanephric types in adults) provided with ducts to drain the waste to cloaca or anal region
9. Brain typically divided into five vesicles
10. Ten or 12 pairs of **cranial nerves** with both motor and sensory functions usually; a pair of spinal nerves for each primitive myotome; an **autonomic nervous system** in control of involuntary functions of internal organs
11. **Endocrine system** of ductless glands scattered through the body
12. Nearly always separate sexes; each sex containing paired gonads with ducts that discharge their products either into the cloaca or into special openings near the anus
13. **Body plan** consisting typically of **head, trunk,** and **postanal tail; neck** present in some, especially terrestrial forms; two pairs of appendages usually, although entirely absent in some; coelom divided into a pericardial space and a general body cavity; mammals with a thoracic cavity

___ Adaptations That Have Guided Vertebrate Evolution

From the earliest fishes to the most advanced mammals, the evolution of the vertebrates has been guided by the specialized basic adaptations of the living endoskeleton, pharynx and efficient respiration, advanced nervous system, and paired limbs.

Living Endoskeleton

The endoskeleton of vertebrates, as in the echinoderms, is an internal supportive structure and framework for the body. This is a departure in animal architecture, since invertebrate skeletons generally enfold the body. Exoskeletons and endoskeletons have their own particular set of advantages and limitations that are related to size (see marginal note on p. 596). For vertebrates, the living endoskeleton possesses an overriding advantage over the dead exoskeleton of arthropods: growing with the body as it does, the endoskeleton permits almost unlimited body size with much greater economy of building materials. Some vertebrates have become the most massive animals on earth. The endoskeleton forms an excellent jointed scaffolding for muscles and the muscles in turn protect the skeleton and cushion it from potentially damaging impact.

We should note that the vertebrates have not wholly lost the protective function of a firm external covering. The skull and the thoracic rib cage enclose and protect vulnerable organs. Most vertebrates are further protected with a tough integument, often bearing non-living structures such as scales, hair, and feathers that may provide insulation as well as physical security.

The endoskeleton was probably composed initially of cartilage that later gave way to bone. Cartilage forms a perfectly suitable endoskeleton, especially for an aquatic existence, and still is the first skeleton to appear in all larval and embryonic vertebrates. In the agnathans (hagfish and lampreys), the sharks and their kin, and even some primitive bony fishes such as sturgeons, the adult endoskeleton is composed strictly of cartilage. Bone appears in the endoskeleton of more advanced vertebrates, perhaps because it offers two clear advantages to cartilage. First, it serves as a reservoir for phosphate, an indispensable component of compounds with high-energy bonds, of membranes, and of DNA. Second, only bone could provide the structural strength required for life on land, where mechanical stresses on the endoskeleton are far greater than they are in water.

Pharynx and efficient respiration

The perforated pharynx (pharyngeal pouches) present in all chordates at some stage in their life cycle evolved as a filter-feeding apparatus. In primitive chordates (such as amphioxus), water with suspended food particles is drawn through the mouth by ciliary action and flows out through the gill slits where the food is trapped in mucus. As the protovertebrates shifted from filter-feeding to a predatory life-style, the pharynx became modified into a muscular feeding apparatus through which water could be pumped by expanding and contracting the pharyngeal cavity. Circulation to the internal gills was improved by the addition of capillary beds (lacking in protochordates) and the development of a ventral heart and muscular aortic arches. These changes supported the increased metabolic rate that would accompany the switch to an active life of selective predation.

Advanced nervous system

No single system in the body is more strongly associated with functional and structural advancement than is the nervous system. The protochordate nervous system consisted of a brainless nerve cord and rudimentary sense organs that were for the most part chemosensory in function. With the protovertebrate switch to a predatory life-style, a new set of selective pressures appeared. New sensory, motor, and integrative controls were now essential for the location and capture of larger prey items. Centralization of the nervous system, that is, the development of a true brain, probably developed as an anterior addition to the existing protochordate nerve cord. In short, the protovertebrates developed a new head, complete with a brain and external paired sense organs especially designed for distance reception. These included paired eyes with lenses and inverted retinas; pressure receptors, such as paired ears designed for equilibrium and later redesigned to include sound reception; and chemical receptors, including taste receptors and exquisitely sensitive olfactory organs.

Paired limbs

Pectoral and pelvic appendages are present in most vertebrates in the form of paired fins or jointed legs. These originated as swimming stabilizers and later became prominently developed into legs for locomotion on land. Jointed limbs are especially suited for life on land because they permit finely graded levering motions against a substrate.

ANCESTRY AND EVOLUTION OF THE VERTEBRATES
Candidates for the Vertebrate Ancestral Stock

As we have seen, many developmental similarities between the protochordates and echinoderms seem to place the echinoderms firmly in the position of the ancestral stock to the chordates. But what is the origin of the vertebrates, the most spectacularly successful group in the animal kingdom?

Position of amphioxus

There is agreement among students of evolution that the vertebrates arose gradually from a ciliated filter-feeding animal resembling protochordates. Without question, the most important development in the transition to the mobile vertebrate predatory life-style was the emergence of a distinct head. This new head appears as an addition to the existing protochordate body. With it came three structural advancements, all crucial to the subsequent success of the vertebrates: (1) an improved nervous system with complex paired sense organs for the detection of prey and enhanced environmental awareness, (2) a well-developed cranial skeleton and, later, jaws for the capture of prey, and (3) a muscular buccal pump and effective gill circulation for efficient gas exchange.

Although none of these structures is present in any protochordate, the cephalochordates have long been considered to be the most suitable structural ancestor of the vertebrates. As adults, amphioxus possesses all four chordate hallmarks plus several vertebrate hallmarks: segmented musculature, the beginning of optic and olfactory sense organs, a liver diverticulum, beginnings of a ventral heart, and separation of dorsal and ventral spinal roots in the vertebrate style. Little wonder that amphioxus once attained a pinnacle position among zoologists searching for their vertebrate ancestor.

But amphioxus' place in the sun was not to endure. When it was recognized that larval forms, not fully formed adults, represent past ancestral forms, amphioxus fell from favor. Nevertheless, although amphioxus is too specialized to provide a suitable framework for the vertebrate body plan, many zoologists today believe that the vertebrates arose gradually from a filter-feeding chordate that *resembled* present day cephalochordates.

Urochordata and the concept of recapitulation

After amphioxus, attention became focused on the alternative protochordate group, the Urochordata (tunicates). The urochordates are divided into three groups of which the ascidians (sea squirts) are the simplest and most common.

At first glance, more unlikely candidates for vertebrate ancestors could hardly be imagined. As adults, ascidians are virtually immobile forms surrounded by a tough, cellulose-containing tunic of variable color. Their adult life is spent in one spot attached to some submarine surface, filtering vast amounts of seawater from which they extract their planktonic food. As adults they lack a notochord, tubular nerve cord, postanal tail, sense organs, and segmented musculature. Superficially they resemble sponges far more than they resemble any known vertebrate. Yet the chordate nature of ascidians is abundantly evident in their larvae, which, because of their superficial resemblance to larval amphibians, are referred to as "tadpole larvae." These tiny, active, site-seeking forms have all the right qualifications for membership in the prevertebrate club: notochord, hollow dorsal nerve cord, gill

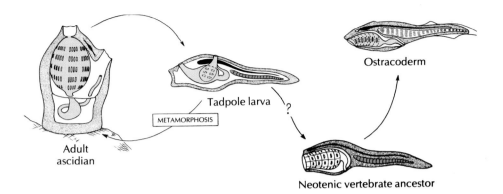

Figure 22-11

Garstang's hypothesis of vertebrate origins. Adult ascidians live on the sea floor but reproduce through a free-swimming tadpole larva. More than 500 million years ago, some larvae began to reproduce in the swimming stage. These evolved into the ostracoderms, the first known vertebrates.

slits, postanal tail, brain, and sense organs (an otolith balance organ and an eye complete with a lens).

The discovery of this form in 1869 not only placed the urochordates squarely in the vertebrate camp but greatly influenced the great German zoologist Ernst Haeckel in formulating his concept of recapitulation (biogenetic law, see Principle 8, p. 9). According to this hypothesis, adult stages of ancestors are repeated during the development of their descendants; in other words, the development of a living organism is an accurate record of past evolutionary history.

We recognize now that this record is very slurred and telescoped and must be interpreted with caution. But at the time that the nature of the ascidian tadpole larva was first understood, it was considered to be a relic of an ancient free-swimming chordate ancestor of the ascidians. Adult ascidians then came to be regarded as degenerate, sessile descendants of the ancient chordate form.

Garstang's hypothesis of chordate larval evolution

In 1928 W. Garstang in England introduced totally fresh thinking to the vertebrate ancestor debates. In effect, Garstang turned the sequence around: rather than the ancestral tadpole larva giving rise to a degenerative sessile ascidian adult, he suggested that sessile ascidian adults *were* the ancestral stock from which the tadpole larvae were evolved to seek out new habitats. Thus, the planktonic tadpole larva was visualized as an ascidian creation, evolved within the group to spread it far and wide. Garstang next suggested that at some point the tadpole larva became neotenous, becoming capable of maturing gonads and reproducing in the larval stage. With continued larval evolution, a new group of free-swimming animals would appear (Figure 22-11).

The best evidence for this hypothesis is found in the living tunicates today, especially among the two planktonic groups, the thaliaceans and the larvaceans (p. 451). In the latter group the basic larval form is retained throughout life; they are in effect neotenous tunicates, although extremely specialized.

Garstang departed from previous thinking by suggesting that evolution may occur in the larval stages of animals. Zoologists accepted this idea slowly because they were accustomed to thinking of developmental stages as being largely insulated from change, as embodied in the "biogenetic law."

_____ The Earliest Vertebrates: Jawless Ostracoderms

The earliest vertebrate fossils are fragments of bony armor discovered in Ordovician rock in Russia and in the United States. They were small, jawless creatures collectively called ostracoderms (os-trak′o-derm) (Gr. *ostrakon*, shell, + *derma*, skin), which belongs to the Agnatha division of the vertebrates. These earliest

Neoteny is well known among living amphibians. The larvae of the American axolotl, for example, may become sexually mature and reproduce without ever undergoing metamorphosis. In some lakes, generation after generation reproduce in the neotenous larval state, and true adults never occur. In other lakes having different environmental conditions, the axolotls metamorphose to reproducing adults. Other examples of neoteny are known; indeed, many believe the human species is neotenous! Certainly, neoteny *could* have occurred among the early chordates, and if it did, the sessile adult stage would be dropped, leaving a new free-swimming chordate.

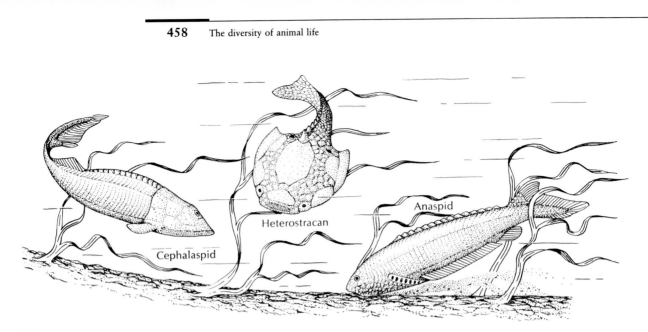

Figure 22-12

Three ostracoderms, jawless fishes of Silurian and Devonian times. Representatives of three of the best-known ostracoderm groups are illustrated as they might have appeared while searching for food on the floor of a Devonian sea or lake. All were filter feeders, drawing water and organic debris into the mouth, straining out the organic matter, and expelling the water through the gill openings and between the ventral head plates.

Erik Stensiö was the first paleozoologist to approach fossil anatomy with the same painstaking attention to minute detail that morphologists have long applied to the anatomical study of living fishes. He developed novel and exacting methods for gradually grinding away a fossil, a few micrometers at a time, to reveal internal features. He was able to reconstruct not only bony anatomy, but nerves, blood vessels, and muscles in numerous groups of Paleozoic and early Mesozoic fishes. His innovative methods are now widely used by paleozoologists.

ostracoderms, called **heterostracans** (Figure 22-12), lacked paired lateral fins that subsequent fishes found so important for stability. Their swimming movements must have been clumsy, although sufficient to propel them along the ocean bottom where they searched for food. They were probably filter-feeders, although their highly differentiated brain and relatively advanced sense organs suggest to some authorities that they may have been mobile predators that fed on soft-bodied animals. During the Devonian period, the heterostracans underwent a major radiation, resulting in the appearance of several peculiar-looking forms varying in shape and length of the snout, dorsal spines, and dermal plates. Without ever evolving fins or jaws, this group dominated the early Devonian period until eclipsed by another ostracoderm group, the **cephalaspids.**

The cephalaspids improved the efficiency of a benthic life by evolving paired fins. These fins, located just behind the head shield, provided control over pitch and yaw that ensured well-directed forward movement. The best-known genus in this group is *Cephalaspis* (Gr. *kephalē*, head, + *aspis*, shield) (Figure 22-12), the subject of a brilliant and classical series of studies by the Swedish paleozoologist E.A. Stensiö. A typical cephalaspid was a small animal, seldom exceeding 30 cm in length; it was covered by a well-developed armor—the head by a solid shield (rounded anteriorly) and the body by bony plates. It had no axial skeleton or vertebrae. The mouth was ventral and anterior, and it was jawless and toothless. Its paired eyes were located close to the middorsal line. At the lateroposterior corners of the head shield was a pair of flaplike fins. The trunk and tail appeared to be adapted for active swimming. Between the margin of the head shield and the ventral plates, there were 10 gill openings on each side. These fishes also had a lateral line system.

As a group, the ostracoderms were basically fitted for a simple, bottom-feeding life. Yet, despite their anatomical limitations, they enjoyed a respectable radiation in the Silurian and Devonian periods. Their overall contribution was enormous, because they provided a blueprint for subsequent vertebrate evolution. But they could not survive the competition of the more advanced jawed fishes that began to dominate the Devonian period, and in the end they disappeared.

Early Jawed Vertebrates

All jawed vertebrates, whether extinct or living, are collectively called gnathostomes ("jaw mouth") in contrast to jawless vertebrates, the agnathans ("without jaw"). The latter are also often referred to as cyclostomes ("circle mouth").

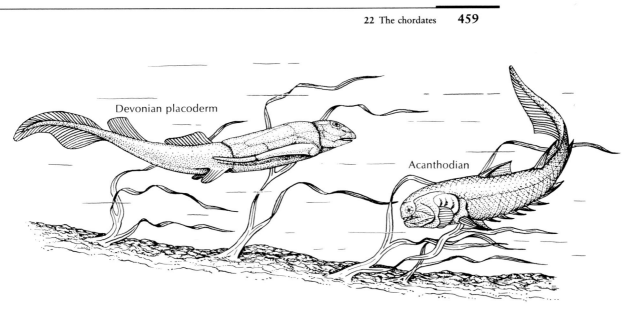

Figure 22-13

Early jawed fishes of the Devonian period, 400 million years ago. The placoderm *(left)* and a related acanthodian *(right)* were highly mobile and voracious forms. Although more successful than the less maneuverable ostracoderms, they eventually failed in competition with their successors, the bony and cartilaginous fishes.

The first jawed vertebrates to appear in the fossil record were the **placoderms** (plak′o-derm) (Gr. *plax*, plate, + *derma*, skin) (Figure 22-13). The advantages of jaws are obvious; they allow predation on large and active forms of food. Possessors of jaws would enjoy a great advantage over jawless vertebrates, which were restricted to a wormlike existence of sifting out organic debris and small organisms in the bottom mud.

Jaws arose through modifications of the first two of the serially repeated cartilaginous gill arches. The beginnings of this trend can, in fact, be seen in some of the jawless ostracoderms where the mouth became bordered by strong dermal plates that could be manipulated somewhat like jaws with the gill arch musculature. The more anterior arches were gradually modified to permit more efficient seizing, and the skin surrounding the mouth was modified into teeth. Eventually the anterior gill arches became bent into the characteristic position of vertebrate jaws, as seen in the placoderms (Figure 22-14).

Placoderms and their relatives evolved into a great variety of forms, some large and grotesque in appearance. They were armored fish covered with diamond-shaped scales or with large plates of bone. All became extinct by the end of the Paleozoic era.

Evolution of Modern Fishes and Tetrapods

Figure 22-14

How the vertebrates got their jaw. The mud-loving Silurian fish converted the first gill opening into a spiracle that opened on the upper head, allowing water to be drawn in without fouling the gills. The first gill bars, no longer required for gill support, became enlarged and armed with teeth. Relics of this transformation are seen during the development of sharks.

Reconstruction of the origins of the vast and varied assemblage of modern living vertebrates is, as we have seen, based largely on fossil evidence. Unfortunately the fossil evidence for the earliest vertebrates is often incomplete and tells us much less than we would like to know about subsequent trends in evolution. Affinities become much easier to establish as the fossil record improves. For instance, the descent of birds and mammals from reptilian ancestors has been worked out in a

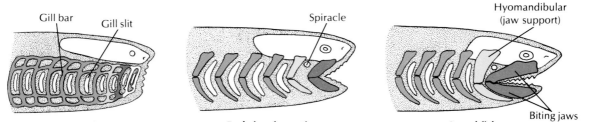

Gill bar   Gill slit   Spiracle   Hyomandibular (jaw support)

Jawless, filter-feeding ancestor   **Early jaw formation**   **Jawed fish**

Biting jaws

highly convincing manner from the relatively abundant fossil record available. By contrast, the ancestry of modern fishes is shrouded in uncertainty.

The Swedish paleontologist E. Jarvik has emphasized that the main vertebrate stem groups (such as cyclostomes, lungfishes, sharks, bony fishes, and stem tetrapods) became anatomically specialized some 400 to 500 million years ago and have changed relatively little since then. Thus main evolutionary lines, as seen in the fossil record, run back almost in parallel; if extended backwards to their illogical extreme, they would hardly ever meet. Obviously they must meet at some point in the distant past, but this exercise reveals that the crucial separations in vertebrate evolution occurred in the Cambrian period, perhaps even the Precambrian period, long before the fossil record became established for the convenience of paleozoologists.

Despite the difficulty of establishing early lines of descent for the vertebrates, they are clearly a natural, monophyletic group, distinguished by a great number of common characteristics. They have almost certainly descended from a common ancestor, the nature of which we have already discussed. Very early in their evolution, the vertebrates divided into two great stems, the agnathans and the gnathostomes. These two groups differ from each other in many fundamental ways, in addition to the obvious lack of jaws in the former group and their presence in the latter. Thus both groups are very old and of approximately the same age. On this basis we cannot say that agnathans are more "primitive" than gnathostomes, even though the latter have continued on a marvelous evolutionary advance that produced most of the modern fishes, all of the tetrapods, and the reader of this book. Although the agnathans are represented today only by the hagfishes and the lampreys, these creatures are also successful in their own way.

Ammocoete Larva of the Lamprey as a Chordate Archetype

The lampreys (jawless fishes of the class Cephalaspidomorphi, discussed in the next chapter) have a freshwater larval stage known as the ammocoete stage. In body form, appearance, life habit, and most anatomical details, the ammocoete larva resembles amphioxus. In fact, the lamprey larva was given the genus name *Ammocoetes* (Gr. *ammos*, sand, + *koite*, bed, referring to the preferred larval habitat) in the nineteenth century when it was erroneously thought to be an adult cephalochordate, closely allied to amphioxus. The ammocoete larva is so different from the adult lamprey that the mistake is understandable; not until it was shown to metamorphose into the adult lamprey was the exact relationship explained. This eel-like larva spends several years buried in the sand and mud of shallow

Figure 22-15

Ammocoete larva, freshwater larval stage of a sea lamprey. Although it resembles amphioxus in many ways, the ammocoete has a well-developed brain, median eyes, pronephric kidney, and other features lacking in amphioxus.

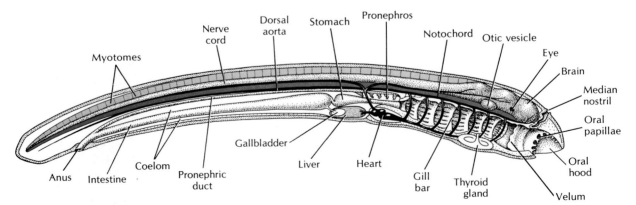

streams, until it finally emerges as an adult that may continue to live in fresh water (freshwater lampreys) or else may migrate to the sea (marine lampreys).

The lamprey larva has a long, slender body with an oral hood surrounding the mouth much like amphioxus (Figure 22-15). The ammocoete feeds like amphioxus, drawing water by ciliary action into the pharynx and pumping it out through gill slits. In the floor of the pharynx is an endostyle, as in amphioxus, that produces a food-ensnaring mucus that is passed directly to the intestine. The arrangement of body muscle into myotomes, the presence of a notochord serving as the chief skeletal axis, and the plan of the circulatory system all closely resemble these features in amphioxus.

The ammocoete does have several characteristics lacking in amphioxus that foreshadow the vertebrate body plan. These include a two-chambered heart (atrium and ventricle), two median eyes each consisting of a simple lens and receptor cells, a three-part brain (forebrain, midbrain, hindbrain), a thyroid gland, and a pituitary gland. The kidney is pronephric (p. 646) and conforms to the basic vertebrate plan. Instead of the numerous gill slits of amphioxus, there are only seven pairs of gill pouches and slits in ammocoetes (there are six pairs in shark embryos). The ammocoete also has a true liver replacing the hepatic cecum of amphioxus, a gallbladder, and pancreatic tissue (but no distinct pancreas gland).

Thus, the ammocoete larva clearly bears many vertebrate hallmarks and approaches the supposed blueprint of the ancestral chordate. Yet the ammocoete and amphioxus have much in common and share many true homologies, strengthening the conclusion that the protochordates and vertebrates share a common ancestry.

▬ SUMMARY

The phylum Chordata is named for the rodlike notochord that forms a stiffening body axis at some stage in the life cycle of every chordate. All chordates share four distinctive hallmarks that set them apart from all other phyla: notochord, dorsal tubular nerve cord, pharyngeal gill pouches, and postanal tail. Two of the three chordate subphyla are invertebrates and lack a well-developed head. They are the Urochordata (tunicates), most of which are sessile as adults, but all of which have a free-swimming larval stage; and the Cephalochordata (lancelets), fishlike forms that include the famous amphioxus.

The chordates are believed to have descended from echinoderm-like ancestors, probably in the Precambrian period, but the true origin of the chordates is not yet, and may never be, known with certainty. Taken as a whole, the chordates have a greater fundamental unity of organ systems and body plan than have many of the invertebrate phyla.

The subphylum Vertebrata, backboned members of the animal kingdom, are characterized as a group by having a well-developed head, and by their comparatively large size, high degree of motility, and distinctive body plan, which embodies several distinguishing features that have led to the eminent evolutionary success of the group. Most important of these are the living endoskeleton that allows continuous growth and provides a sturdy framework for efficient muscle attachment and action; a pharynx perforated with gill slits (lost or greatly modified in higher vertebrates) with vastly increased respiratory efficiency; advanced nervous system with clear separation of the brain and spinal cord; and paired limbs. The vertebrates are believed to have evolved by neoteny from a protochordate larval ancestor such as the ascidian tadpole larva or the larval amphioxus.

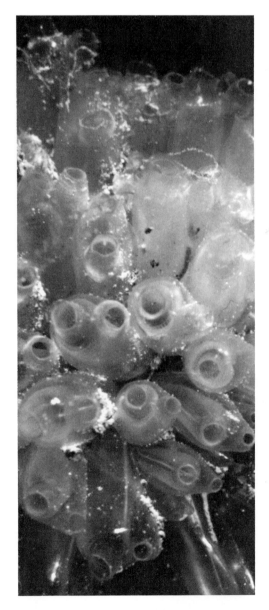

Sea squirts *(Clavelina sp.)*, sessile marine protochordates of the subphylum Urochordata.
Photograph by S.K. Webster, Monterey Bay Aquarium/BPS.

Review questions

1. What is the evidence that the four deuterostome phyla—Echinodermata, Chaetognatha, Hemichordata, and Chordata—should be considered a natural grouping?
2. Name four hallmarks shared by all chordates, and explain the function of each.
3. Name the two invertebrate chordate subphyla, and give three distinguishing characteristics of each.
4. Both sea squirts (urochordates) and lancelets (cephalochordates) are filter-feeders. Describe the feeding apparatus of a sea squirt and explain in what ways its mode of feeding is similar to, and different from, that of amphioxus.
5. Explain why it is necessary to know the life history of a sea squirt to understand why sea squirts are chordates.
6. What is distinctive about the way a larvacean tunicate feeds?
7. Explain why the body plan of amphioxus has been considered to represent a blueprint of the early vertebrate body plan.
8. Distinguish among the terms acrania, craniata, agnatha, and gnathostomata.
9. List four adaptations that guided vertebrate evolution, and explain how each has contributed to the success of vertebrates.
10. Discuss Garstang's hypothesis of chordate larval evolution.
11. Distinguish between ostracoderms and placoderms.

Selected references

See also general references for Part Two, p. 590.

Alexander, R.M. 1980. The chordates, ed. 2. Cambridge, Cambridge University Press. *Chapter 2 of the text treats the protochordates.*

Alldredge, A. 1976. Appendicularians. Sci. Am. **235**:94-102 (July). *Describes the biology of larvaceans, which build gossamer houses for trapping food.*

Barrington, E.J.W., and R.P.S. Jeffries (eds.). 1975. Protochordates. Symp. Zool. Soc. London, No. 36. New York, Academic Press, Inc. *Contributed chapters on aspects of amphioxus biology and on chordate evolution.*

Berrill, N.J. 1955. The origin of vertebrates. New York, Oxford University Press. *The author stresses the tunicates as the basic stock from which other protochordates and vertebrates arose. He believes that such a sessile filter-feeder was really the most primitive chordate and was not a mere degenerate side branch of chordate evolution.*

Bone, Q. 1972. The origin of chordates. Oxford Biology Readers, No. 18. New York, Oxford University Press. *Excellent synthesis of hypotheses and range of disagreements bearing on an unsolved riddle.*

Northcutt, R.G., and C. Gans. 1983. The genesis of neural crest and epidermal placodes: a reinterpretation of vertebrate origins. Quart. Rev. Biol. **58**:1-28. *Describes the protochrodate-vertebrate transition and the origin of the vertebrate head.*

CHAPTER 23

THE FISHES

Phylum Chordata

Classes Myxini, Cephalaspidomorphi, Chondrichthyes, and Osteichthyes

Fishes are the undisputed masters of the aquatic environment. Because they live in a habitat that is basically alien to humans, we have not always found it easy to appreciate the incredible success of these vertebrates. Plato considered fishes "senseless beings . . . which have received the most remote habitations as a punishment for their extreme ignorance." And average North Americans today are usually uninformed about fishes unless they happen to be fishermen or tropical fish enthusiasts.

Nevertheless, the world's fishes have enjoyed an adaptive radiation easily as spectacular as that of all the land vertebrates, with the possible exception of the mammals (the latter having succeeded in the air and in the sea as well as on land). Their numerous structural adaptations have produced a great variety of forms ranging from gracefully streamlined trout to grotesque creatures that dwell in the blackness of the ocean's abyssal depths. Considered either in numbers of species (some 21,700 named species) or in numbers of individuals (countless billions), the fishes at least equal, if not outnumber, the four terrestrial vertebrate classes combined.

Although fishes are the oldest vertebrate group, there is not the slightest evidence that, like their amphibian and reptile successors, they are declining from a period of earlier glory; certain groups of ancient fishes are extinct, but they have been replaced by successful modern fishes. There are indeed more bony fishes today than ever before, and no other group threatens their domination of the seas.

Their success can be attributed to perfect adaptation to their dense medium. A trout or pike can hang motionless in the water, varying its neutral buoyancy by adding or removing gas from the swim bladder, or it can dart forward or at angles, using its fins as brakes and tilting rudders. Fish have excellent olfactory and visual senses and a unique lateral line system, which with its exquisite sensitivity to water currents and vibrations provides a "distance touch" in water. Their gills are the most effective respiratory devices in the animal kingdom for extracting oxygen from water. With highly developed organs of salt and water exchange, bony fishes are excellent osmotic regulators, capable of fine-tuning their body fluid composition in their chosen freshwater or seawater environment. Fishes have evolved

Figure 23-1

Family tree of the fishes, showing their evolution through geological time. Widened areas in the lines of descent indicate periods of adaptive radiation and the relative success of each group. The lungfishes, for example, flourished in the Devonian period, but declined and are today represented by only three surviving genera. The sharks and rays radiated during the Carboniferous period and at that time were rulers of the sea (their successful contemporaries, the early chondrosteans, were freshwater fishes). The sharks came dangerously close to extinction during the Permian period but staged a recovery in the Mesozoic era and are a secure group today. Johnny-come-latelies in fish evolution are the spectacularly successful modern bony fishes, or teleosts, which make up 90% of all living fishes.

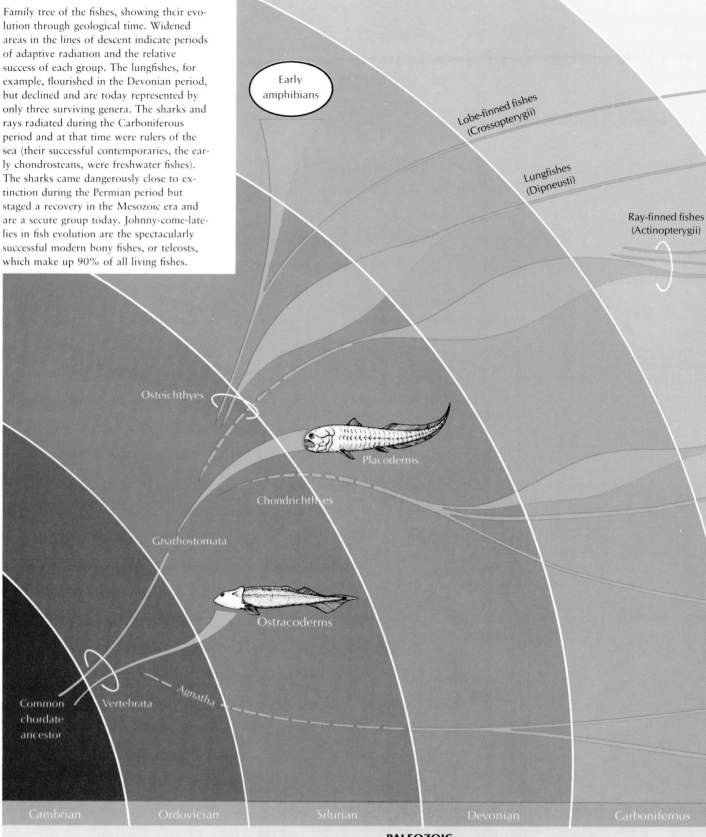

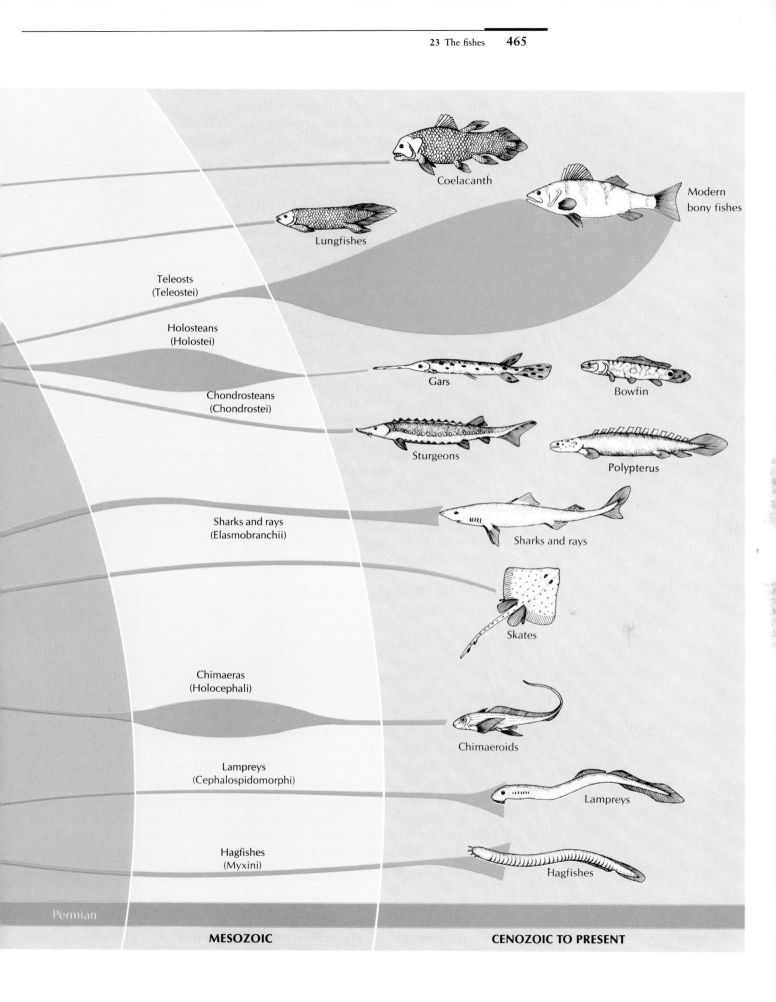

Coelacanth

Modern
bony fishes

Lungfishes

Teleosts
(Teleostei)

Holosteans
(Holostei)

Gars

Bowfin

Chondrosteans
(Chondrostei)

Sturgeons

Polypterus

Sharks and rays
(Elasmobranchii)

Sharks and rays

Skates

Chimaeras
(Holocephali)

Chimaeroids

Lampreys
(Cephalospidomorphi)

Lampreys

Hagfishes
(Myxini)

Hagfishes

Permian

MESOZOIC

CENOZOIC TO PRESENT

complex behavioral mechanisms for dealing with emergencies, and many have evolved elaborate reproductive behavior concerned with courtship, nest building, and care of the young. These are only a few examples of many such adaptations evident in this varied phylogenetic assemblage, which includes four of the eight vertebrate classes.

ANCESTRY AND RELATIONSHIPS OF MAJOR GROUPS OF FISHES

The fishes are of ancient ancestry, having descended from an unknown free-swimming protochordate ancestor (hypotheses of chordate and vertebrate origins were discussed in Chapter 22). Whatever their origin, during the Cambrian period, or perhaps even in the Precambrian, the earliest fishlike vertebrates branched into the jawless **agnathans** and the jawed **gnathostomes** (Figure 23-1). All vertebrates have descended from one or the other of these two ancestral groups.

The agnathans, the more primitive of the two groups, include the extinct ostracoderms and the living **hagfishes** and **lampreys,** fishes adapted as scavengers or parasites. The ancestry of hagfishes and lampreys is uncertain; they bear little resemblance to the extinct ostracoderms. Although hagfishes and lampreys superficially look much alike, they are in fact so different from each other that they have been assigned to separate classes by ichthyologists.

All remaining fishes have jaws and are called gnathostomes. They may have descended from the placoderms (p. 459) but, as with the agnathans, the fossil evidence is too fragmentary to assign ancestral groups with certainty. Whatever the early lines of descent, the gnathostomes branched into two major groups, the cartilaginous fishes (class Chondrichthyes) and the bony fishes (class Osteichthyes).

The **cartilaginous fishes**—sharks, skates, rays and chimaeras—have lost the heavy armor of the early placoderms and adopted cartilage rather than bone for the skeleton. Most are active predators with a sharklike body form that has undergone only minor changes over the ages. As a group, the sharks and their kin flourished during the Devonian and Carboniferous periods of the Paleozoic era but declined dangerously close to extinction at the end of the Paleozoic. They staged a recovery in the early Mesozoic and radiated into the modest but thoroughly successful assemblage of modern sharks.

The **bony fishes** (class Osteichthyes) are the dominant fishes today. We can recognize three great stems of descent. Of these, by far the most successful are the **ray-finned fishes** (actinopterygians) which radiated into the modern bony fishes. The other two lineages are the **lobe-finned fishes** (crossopterygians) from which the amphibians descended, and the **lungfishes** (dipneusts). Both are relic groups consisting of just a few living species—meager remnants of important stocks that flourished in the Devonian period of the Paleozoic. The distinction of being ancestors of the **tetrapods** (vertebrates with four legs, that is, all higher vertebrates) probably rests with the lobe-finned fishes, which are represented today by a single living species, the coelacanth.

CLASSIFICATION

The following broad classification mostly follows that of Nelson (1984). No one scheme is accepted by even a majority of ichthyologists. When we contemplate the incredible difficulty of ferreting out relationships among some 21,700 living species and a vast number of fossils of varying ages, we can appreciate why fish classification has undergone, and will continue to undergo, continuous change.

Subphylum Vertebrata

Superclass Agnatha (ag'na-tha) (Gr. *a*, not, + *gnathos*, jaw) (**Cyclostomata**). No jaws; cartilaginous skeleton; ventral fins absent; one or two semicircular canals; notochord persistent.

Class Myxini (mik-sy'ny) (Gr. *myxa*, slime): **hagfishes**. Mouth terminal with four pairs of tentacles; buccal funnel absent; nasal sac with duct to pharynx; gill pouches, five to 15 pairs; partially hermaphroditic. Examples: *Myxine*, *Bdellostoma*.

Class Cephalaspidomorphi (sef-a-lass'pe-do-morf'e) (Gr. *kephalē*, head, + *aspidos*, shield, + *morphē* form): **lampreys**. Mouth suctorial with horny teeth; nasal sac not connected to mouth; gill pouches, seven pairs. Examples: *Petromyzon*, *Lampetra*.

Superclass Gnathostomata (na'tho-sto'ma-ta) (Gr. *gnathos*, jaw, + *stoma*, mouth). Jaws present; usually paired limbs; three pairs of semicircular canals; notochord persistent or replaced by vertebral centra.

Class Chondrichthyes (kon-drik'thee-eez) (Gr. *chondros*, cartilage, + *ichthys*, fish): **cartilaginous fishes**. Cartilaginous skeleton; teeth not fused to jaws; no swim bladder; intestine with spiral valve.

Subclass Elasmobranchii (e-laz'mo-bran'kee-i) (Gr. *elasmos*, metal plate, + *branchia*, gills): **sharks, skates, rays**. Placoid scales or no scales; five to seven gill arches and gills in separate clefts along pharynx. Examples: *Squalus*, *Raja*.

Subclass Holocephali (hol'o-sef'a-li) (Gr. *holos*, entire, + *kephalē*, head): **chimaeras, or ghostfish**. Gill slits covered with operculum; jaws with tooth plates; single nasal opening; without scales; accessory clasping organs in male; lateral line an open groove. Examples: *Chimaera*, *Hydrolagus*.

Class Osteichthyes (os'te-ik'thee-eez) (Gr. *osteon*, bone, + *ichthys*, a fish) (**Teleostomi**): **bony fishes**. Body primitively fusiform but variously modified; skeleton mostly ossified; single gill opening on each side covered with operculum; usually swim bladder or lung.

Subclass Crossopterygii (cros-sop-te-rij'ee-i) (Gr. *krossoi*, fringe or tassels, + *pteryx*, fin, wing): **lobe-finned fishes**. Heavy bodied; paired fins lobed with internal skeleton of basic tetrapod type; premaxillae, maxillae present; scales large with tubercles and heavily overlapped; three-lobed diphycercal tail; skeleton with much cartilage; bony spines hollow; air bladder vestigial; gills hard with teeth; intestine with spiral valve; spiracle present. Example: *Latimeria*.

Subclass Dipneusti (dip-nyu'sti) (Gr. *di*, two, + *pneustikos*, of breathing): **lungfishes**. All median fins fused to form diphycercal tail; fins lobed or of filaments; scales of cycloid bony type; teeth of grinding plates; no premaxillae or maxillae; air bladder of single or paired lobes and specialized for breathing; intestine with spiral valve; spiracle absent. Examples: *Neoceratodus*, *Protopterus*, *Lepidosiren*.

Subclass Actinopterygii (ak'ti-nop-te-rij'ee-i) (Gr. *aktis*, ray, + *pteryx*, fin, wing): **ray-finned fishes**. Paired fins supported by dermal rays and without basal lobed portions; nasal sacs open only to outside. Examples: *Salmo*, *Perca*.

JAWLESS FISHES: SUPERCLASS AGNATHA

The living members of the Agnatha are represented by some 60 species almost equally divided between two classes: Myxini (hagfishes) and Cephalaspidomorphi (lampreys) (Figure 23-2). Members of both groups lack jaws, internal ossification, scales, and paired fins, and both share porelike gill openings and an eel-like body form. At the same time there are so many important differences, some of which are indicated in the following list, that they have been assigned to separate vertebrate classes.

Figure 23-2

Comparison of hagfish (class Myxini) and lamprey (class Cephalaspidomorphi), representatives of the superclass Agnatha.

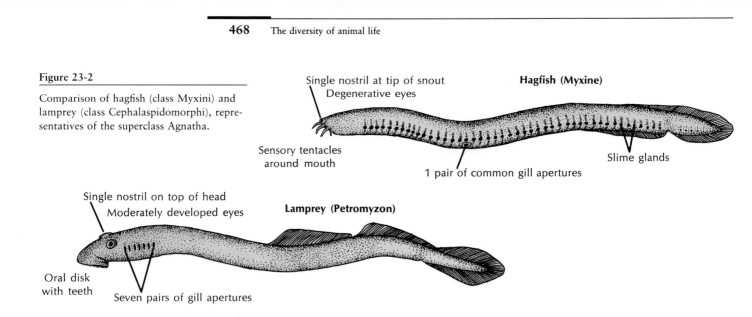

Single nostril at tip of snout
Degenerative eyes
Hagfish (Myxine)
Sensory tentacles around mouth
1 pair of common gill apertures
Slime glands

Single nostril on top of head
Moderately developed eyes
Lamprey (Petromyzon)
Oral disk with teeth
Seven pairs of gill apertures

CHARACTERISTICS

1. Body slender, **eel-like**, rounded, with **soft skin** containing **mucous glands** but **no scales**
2. Median fins with cartilaginous fin rays, but **no paired appendages**
3. **Fibrous** and **cartilaginous** skeleton; notochord persistent
4. Suckerlike oral disc with well-developed teeth in lampreys; biting mouth with two rows of eversible teeth in hagfishes
5. Heart with one atrium and one ventricle; aortic arches in gill region; blood with erythrocytes and leukocytes
6. Seven pairs of gills in lampreys; five to 16 pairs of gills in hagfishes
7. **Mesonephric kidney** in lampreys; **pronephric kidney** anteriorly and mesonephric kidney posteriorly in hagfishes
8. Dorsal nerve cord with differentiated brain; eight to 10 pairs of cranial nerves
9. Digestive system **without stomach;** intestine with spiral fold and cilia in lampreys; spiral fold and cilia absent in hagfishes
10. Sense organs of taste, smell, hearing; eyes moderately developed in lampreys but highly degenerate in hagfishes; one pair of **semicircular canals** (hagfishes) or two pairs (lampreys)
11. External fertilization; gonad single without duct; sexes separate; long larval stage (ammocoete) in lampreys; direct development with no larval stage in hagfishes

____ Hagfishes: Class Myxini

The hagfishes are an entirely marine group that feeds on dead or dying fishes, annelids, molluscs, and crustaceans. Thus, they are neither parasitic like lampreys nor predaceous, but are scavengers. There are only 32 described species of hagfishes, of which the best known in North America are the Atlantic hagfish *Myxine glutinosa* (Gr. *myxa*, slime) (Figure 23-2) and the Pacific hagfish *Eptatretus stouti* (NL *ept*, Gr. *hepta*, seven + *tretos*, perforated).

Hagfishes have long been of interest to comparative physiologists. Unlike any other vertebrate, they are isosmotic to seawater like marine invertebrates. They are the only vertebrates to have both pronephric and mesonephric kidneys in the adult, although only the latter forms urine. They have no fewer than four sets of hearts positioned at different places in the body to boost blood flow through their low-pressure circulatory system.

Hagfishes are renowned for their capacity to generate enormous quantities of slimy mucus from special glands positioned along the body. A single hagfish may produce half a liter of slime within minutes of capture. Slime production is

While the unique anatomical and physiological features of the strange hagfishes are of interest to biologists, hagfishes have not endeared themselves to either sports or commercial fishermen. In earlier days of commercial fishing mainly by gill nets and set lines, hagfish often bit into the bodies of captured fish and ate out the contents, leaving behind a useless sack of skin and bones. But as large and efficient otter trawls came into use, hagfishes ceased to be an important pest.

important to the animal for feeding and in defense, but it is a nuisance to the sportsman unfortunate enough to hook one.

The reproductive biology of hagfishes remains largely a mystery, despite a still unclaimed prize offered more than 100 years ago by the Copenhagen Academy of Science for information on the animal's breeding habits. It is known that although both male and female gonads are found in the same animal, only one gonad becomes functional. The females produce small numbers of surprisingly large, yolky eggs up to 3 cm in diameter. There is no larval stage and growth is direct.

Lampreys: Class Cephalaspidomorphi

All the lampreys of the Northern Hemisphere belong to the family Petromyzontidae (Gr. *petros,* stone, + *myzon,* sucking). The group name refers to the lamprey's habit of grasping a stone with its mouth to hold position in a current. The destructive marine lamprey *Petromyzon marinus* is found on both sides of the Atlantic Ocean (in America and Europe) and may attain a length of 1 m. *Lampetra* (L. *lambo,* to lick or lap up) also has a wide distribution in North America and Eurasia and ranges from 15 to 60 cm long. There are 20 species of lampreys in North America. About half of these belong to the nonparasitic brook type; the others are parasitic. The nonparasitic species have probably descended from the parasitic forms by degeneration of the teeth, alimentary canal, and so on. The genus *Ichthyomyzon* (Gr. *ichthyos,* fish, + *myzon,* sucking), which includes three parasitic and three nonparasitic species, is restricted to eastern North America. On the west coast of North America the chief marine form is *Lampetra tridentatus.*

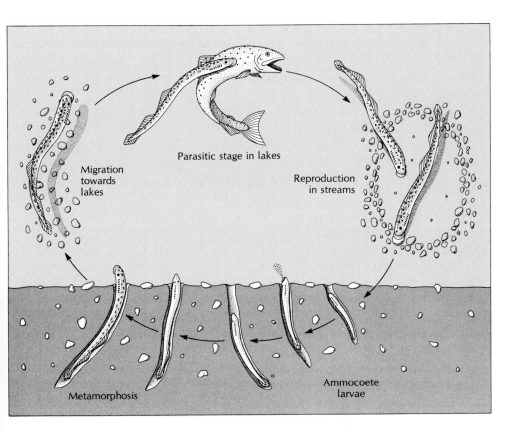

Figure 23-3

Life cycle of the landlocked form of the sea lamprey, *Petromyzon marinus.*

All lampreys ascend freshwater rivers or streams to breed. The marine forms are anadromous (Gr. *ana*, up, + *dramein*, to run); that is, they leave the sea where they spend their adult lives to run up rivers and streams to spawn. In North America all lampreys spawn in the spring. The males begin nest building and are joined later by females. Using their oral discs to lift stones and pebbles and vigorous body vibrations to sweep away light debris, they form an oval depression (Figure 23-3). At spawning, with the female attached to a rock to maintain her position over the nest, the male attaches to the dorsal side of her head. As the eggs are shed into the nest, they are fertilized by the male. The sticky eggs adhere to pebbles in the nest and quickly become covered with sand. The adults die soon after spawning.

The eggs hatch in about 2 weeks, releasing small larvae (ammocoetes), which are so unlike their parents that early biologists thought they were a separate species. The larva bears a remarkable resemblance to amphioxus and possesses the basic chordate characteristics in such simplified and easily visualized form that it has been considered a chordate archetype (p. 460). After absorbing the remainder of its yolk supply, the young ammocoete, now about 7 mm long, leaves the nest gravel and drifts downstream to burrow in some suitable sandy, low-current area. Here it remains for an extraordinarily long time, 3 to 7 years, before the larva rapidly metamorphoses into an adult. This change involves the development of larger eyes, the replacement of the hood by the oral disc with teeth, a shifting of the nostril to the top of the head, and the development of a rounder but shorter body.

Parasitic lampreys either migrate to the sea, if marine, or remain in fresh water, where they attach themselves by their suckerlike mouth to fish and, with their sharp horny teeth, rasp away the flesh and suck out the blood (Figure 23-3). To promote the flow of blood, the lamprey injects an anticoagulant into the wound. When the lamprey is gorged, it releases its hold but leaves the fish with a large, gaping wound that may prove fatal. The parasitic freshwater adults live a year or more before spawning and then die; the anadromous forms live longer.

The nonparasitic lampreys do not feed after emerging as adults, for their alimentary canal degenerates to a nonfunctional strand of tissue. Within a few months they also spawn and die.

The invasion of the Great Lakes by the landlocked sea lamprey *Petromyzon marinus* in this century has had a devastating effect on the fisheries. No lampreys were present in the Great Lakes west of Niagara Falls until the Welland Ship Canal was built in 1829. Even then nearly 100 years elapsed before sea lampreys were first seen in Lake Erie. After that the spread was rapid, and the sea lamprey was causing extraordinary damage in all the Great Lakes by the late 1940s. No fish species was immune from attack, but the lampreys preferred lake trout, and this multimillion dollar fishing industry was brought to total collapse in the early 1950s. Lampreys then turned to rainbow trout, whitefish, turbot, yellow perch, and lake herring, all important commercial species. These stocks were decimated in turn. The lampreys then began attacking chubs and suckers. Coincident with the decline in attacked species, the sea lampreys themselves began to decline after reaching a peak abundance in 1951 in Lakes Huron and Michigan and in 1961 in Lake Superior. The fall has been attributed both to depletion of food and to the effectiveness of control measures (mainly chemical larvicides in selected spawning streams). Lake trout, aided by a restocking program begun in 1967, are now recovering. Wounding rates are low in Lake Michigan but still high in Lake Huron. Fishery organizations are experimenting with the introduction into the

Great Lakes of species that appear to be more resistant to lamprey attack, such as kokanee salmon (landlocked Pacific sockeye salmon).

CARTILAGINOUS FISHES: CLASS CHONDRICHTHYES

There are nearly 800 living species in the class Chondrichthyes, an ancient, compact, and highly developed group. Although a much smaller and less diverse assemblage than the bony fishes, their impressive combination of well-developed sense organs, powerful jaws and swimming musculature, and predaceous habits ensures them a secure and lasting niche in the aquatic community. One of their distinctive features is their cartilaginous skeleton. Although there is some calcification here and there, bone is entirely absent throughout the class—a curious evolutionary feature, since the Chondrichthyes are derived from ancestors having well-developed bone.

With the exception of whales, sharks are the largest living vertebrates. The larger sharks may reach 15 m in length. The dogfish sharks so widely studied in zoological laboratories rarely exceed 1 m.

Sharks, Skates, and Rays: Subclass Elasmobranchii

The elasmobranchs are carnivores that track their prey using their lateral line system and large olfactory organs. Their vision is not well developed.

Fertilization is internal (a curiously advanced feature in an ancient group), and many sharks have evolved elaborate reproductive modes; some are live-bearers with gestation periods up to 2 years—the longest of any vertebrate.

There are five living orders of elasmobranchs, numbering about 760 species. All of the more notorious sharks belong to the order Lamniformes, including the requiem sharks, hammerhead sharks (Figure 23-4, *A*), whale sharks, mackerel sharks, nurse sharks, and sand sharks (Figure 23-4, *B*). The order Squaliformes includes the dogfish shark, familiar to generations of comparative anatomy students. The skates and several groups of rays (sawfish rays, electric rays, stingrays, eagle rays, manta rays, and devil rays) belong to the order Rajiformes.

Much has been written about the propensities of sharks to eat humans, both by those exaggerating their ferocious nature and by those seeking to write them off as harmless. It is true, as the latter group of writers argues, that sharks are by nature timid and cautious. But it also is a fact that certain of them are dangerous to humans and deserve respect. There are numerous authenticated cases of shark attacks by *Carcharodon* (Gr. *karcharos,* sharp, + *odous,* tooth), the great white shark (commonly reaching 6 m and often larger); the mako shark *Isurus* (Gr. *is,* equal, + *ouros,* tail); the tiger shark *Galeocerdo* (Gr. *galeos,* shark, + *kerdō,* fox); and the hammerhead shark *Sphyrna* (Gr. *sphyra,* hammer). More shark casualties have been reported from the tropical and temperate waters of the Australian region than from any other. During World War II there were several reports of mass shark attacks on the victims of ship sinkings in tropical waters.

Outside of North America, sharks and skates are commonly used for food. They make up about 1% of the present market for fish. There is a small market for shark leather, which has greater tensile strength and lasting qualities than mammalian leather.

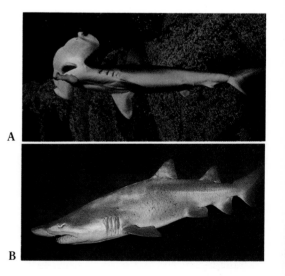

A

B

Figure 23-4

Sharks of the order Lamniformes (subclass Elasmobranchii). **A,** Hammerhead shark *Sphyrna tudes* may reach a length of 4.5 m. **B,** Sand tiger shark *Carcharias* sp. Large individuals of both species are considered highly dangerous.
Photographs by K. Sandved.

CHARACTERISTICS

1. **Body fusiform,** with a **heterocercal** caudal fin (Figure 23-14); paired pectoral and pelvic fins, two dorsal median fins; pelvic fins in male modified for "claspers"; fin rays present
2. **Mouth ventral; two olfactory sacs that do not open into the mouth cavity;** jaws present
3. Skin with **placoid** scales (Figure 23-17) and **mucous glands;** modified placoid scales for teeth
4. **Endoskeleton entirely cartilaginous;** notochord persistent; vertebrae complete and separate; appendicular, girdle, and visceral skeletons present
5. Digestive system with a J-shaped stomach and intestine with a spiral valve; liver, gallbladder, and pancreas present
6. Circulatory system of several pairs of aortic arches; dorsal and ventral aorta, capillary and venous systems, hepatic portal and renal portal systems; two-chambered heart; high concentrations of urea and trimethylamine oxide in blood
7. Respiration by means of five to seven pairs of gills with separate and exposed gill slits; **no operculum** except in chimaeras
8. No swim bladder or lung
9. Brain of two olfactory lobes, two cerebral hemispheres, two optic lobes, a cerebellum, and a medulla oblongata; 10 pairs of cranial nerves; three pairs of semicircular canals
10. Sexes separate; gonads paired; reproductive ducts open into cloaca; oviparous, ovoviviparous, or viviparous; direct development; fertilization internal
11. Kidneys of mesonephros (opisthonephros) type

Form and function

Although to most people sharks have a sinister appearance and fearsome reputation, they are at the same time among the most gracefully streamlined of all fishes. The body of a dogfish shark (Figure 23-5) is fusiform (spindle shaped). In front of the ventral mouth is a pointed **rostrum;** at the posterior end the vertebral column turns up to form the **heterocercal** tail. The fins consist of the paired **pectoral** and **pelvic** fins supported by appendicular skeletons, two median **dorsal** fins (each with a spine in *Squalus* [L. a kind of sea fish]), and a median **caudal** fin. A median **anal** fin is present in the smooth dogfish (*Mustelus* [L. *mustela,* weasel]). In the male, the medial part of the pelvic fin is modified to form a **clasper,** which is used in copulation. The paired **nostrils** (blind pouches) are ventral and anterior to the mouth (Figure 23-6). The lateral eyes are lidless, and behind each eye is a spiracle (remnant of the first gill slit). Five gill slits are found anterior to each pectoral fin. The tough, leathery skin is covered with **placoid scales** (Figure 23-17), which are modified anteriorly to form replaceable rows of teeth in both jaws (Figure 23-6). Placoid scales consist of dentin enclosed by an enamel-like substance, and they very much resemble the teeth of other vertebrates.

Sharks are well equipped for their predatory life. Their vision is less acute than that of most bony fishes, but this is more than compensated for by a keen

Figure 23-5

Dogfish shark *Squalus acanthias* (subclass Elasmobranchii).

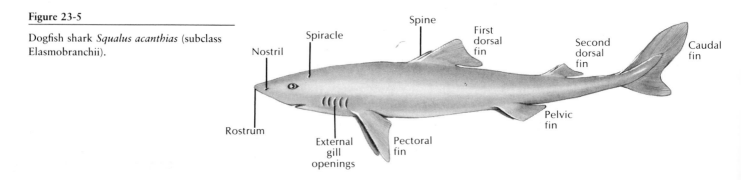

sense of smell used to guide them to food. A well-developed **lateral line system** serves as a "distance touch" in water for detecting and locating objects and moving animals (predators, prey, and social partners). It is composed of a canal system extending along the side of the body and over the head. The canal opens at intervals to the surface. Inside are special receptor organs (**neuromasts**) that are extremely sensitive to vibrations and currents in the water. Sharks also can detect and aim attacks at prey buried in the sand by sensing the bioelectric fields that surround all animals.

Internally the cartilaginous skeleton is made up of a **chondrocranium,** which houses the brain and auditory organs and partially surrounds the eyes and olfactory organs; a vertebral column; a visceral skeleton; and an appendicular skeleton. The jaws are suspended from the chondrocranium by ligaments and cartilages. Both the upper and the lower jaws are provided with many sharp, triangular teeth that, when lost, are replaced by other rows of teeth. Teeth serve to grasp the prey, which is usually swallowed whole. The muscles are segmentally arranged and are especially useful in the undulations of swimming.

The mouth cavity opens into the large **pharynx,** which contains openings to the separate gill slits and spiracles. A short, wide esophagus runs to the J-shaped stomach. A **liver** and **pancreas** open into the short, straight **intestine,** which contains the unique **spiral valve** that delays the passage of food and increases the absorptive surface (Figure 23-7). Attached to the short rectum is the **rectal gland,** which secretes a colorless fluid containing a high concentration of sodium chloride. It assists the kidney in regulating the salt concentration of the blood. The chambers of the **heart** are arranged in tandem formation (Figure 23-7), and the circulatory system is basically the same as that of the embryonic vertebrate and of the ammocoete.

The **mesonephric kidneys** are two long, slender organs above the coelom; they are drained by the **wolffian ducts,** which open into a single urogenital sinus at the **cloaca.** The wolffian ducts also carry the sperm from the testes of the male, which uses a clasper to deposit the sperm in the female oviduct. The müllerian

Figure 23-6

Head of sand tiger shark *Carcharias* sp. Note the series of successional teeth; each tooth generation grows toward the jaw margin and is eventually shed. Photograph by J.L. Rotman.

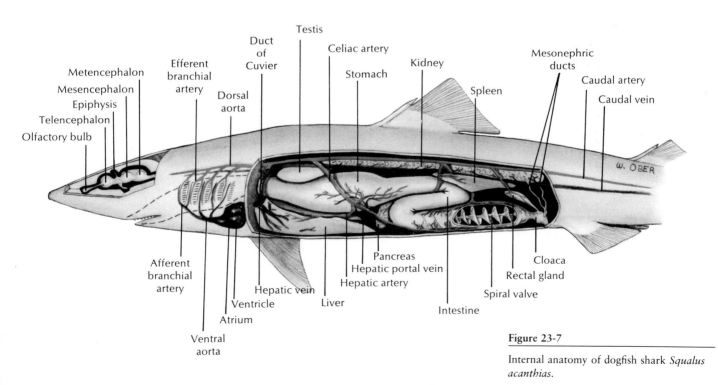

Figure 23-7

Internal anatomy of dogfish shark *Squalus acanthias.*

Figure 23-8

Egg case of a skate ("mermaid's purse"). The horny case protects the developing young. Skate egg cases have coiling tendrils on the corners that entangle seaweeds. Photograph by R. Harbo.

duct, or oviduct (paired), carries the eggs from the **ovary** and coelom and is modified into a **uterus** in which a primitive placenta may attach the embryo shark until it is born. Such a relationship is actually **viviparous reproduction**; other species simply retain the developing egg in the uterus without attachment to the mother's wall (**ovoviviparous reproduction**). Still others lay large, yolky eggs immediately after fertilization (**oviparous reproduction**). Some sharks and rays deposit their fertilized eggs in a horny capsule called the "mermaid's purse," which often is attached by tendrils to seaweed (Figure 23-8). Later the young shark emerges from this "cradle."

The embryonic brain is the basic tripartite vertebrate brain: forebrain, midbrain, and hindbrain. These three parts develop into five subdivisions or regions—telencephalon, diencephalon, mesencephalon, metencephalon, and myelencephalon—each with certain functions. There are 10 pairs of cranial nerves that are distributed largely to the head regions. Surrounding the spinal cord are the neural arches of the vertebrae. Along the spinal cord a pair of spinal nerves, with united dorsal and ventral roots, is distributed to each body segment.

Elasmobranchs have developed an interesting solution to the physiological problem of living in a hyperosmotic medium. Like the ancestors of the bony fishes, those of sharks and their kin lived in fresh water, only returning to the sea in the Triassic period (about 230 million years ago) after becoming thoroughly adapted to freshwater existence. The salt concentration of their body fluid was then much lower than the surrounding seawater; to prevent water from being drawn out of the body osmotically, the elasmobranchs have retained nitrogenous wastes, especially urea and trimethylamine oxide, in the blood. These solutes, combined with the blood salts, raised the blood solute concentration to slightly exceed that of seawater, eliminating an osmotic inequality between their bodies and the surrounding seawater.

Slightly more than half of all elasmobranchs are rays, a group that includes skates, electric rays, sawfishes, stingrays, eagle rays, and manta rays. All are specialized for bottom dwelling, with greatly enlarged pectoral fins that are fused to the head and used like wings in swimming. The gill openings are on the underside of the head, but the large spiracles are on top. Water for breathing is taken in through these spiracles to prevent clogging the gills, for the mouth is often buried

Figure 23-9

Yellow stingray *Urolophus jamaicensis* (subclass Elasmobranchii) of the Caribbean. Photograph by L.S. Roberts.

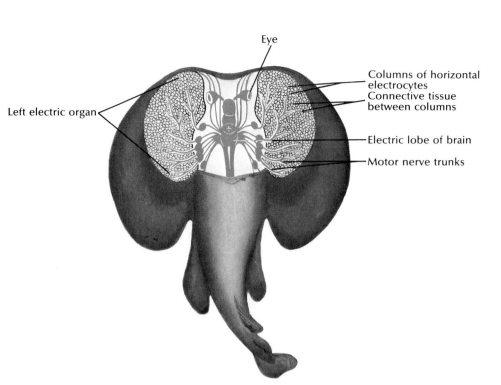

Eye

Columns of horizontal
electrocytes
Connective tissue
between columns

Left electric organ

Electric lobe of brain

Motor nerve trunks

Figure 23-10

Electric ray *Torpedo* with electric organs
uncovered from above. Organs are built up
of disclike, multinucleated cells called electrocytes. In addition to defensive and offensive purposes, electric organs may be used
for recognition among members of a species
when other methods of communication are
absent.

in sand. Their teeth are adapted for crushing their prey: molluscs, crustaceans, and
an occasional small fish.

The stingrays have a slender and whiplike tail that is armed with one or
more saw-edged spines with venom glands at the base (Figure 23-9). Wounds
from the spines are excruciatingly painful, and may heal slowly and with compli-
cations. Electric rays are sluggish fish with large electric organs on each side of the
head (Figure 23-10). Each organ is made up of numerous vertical stacks of disclike
cells connected in parallel so that when all the cells are discharged simultaneously,
a high-amperage current is produced that flows out into the surrounding water.
The voltage produced is relatively low but the power output may be several kilo-
watts—quite sufficient to stun prey or discourage predators. Electric rays were
used by the ancient Egyptians for a form of electrotherapy in the treatment of
afflictions such as arthritis and gout.

Chimaeras: Subclass Holocephali

The members of the small subclass Holocephali, distinguished by such suggestive
names as ratfish (Figure 23-11), rabbitfish, spookfish, and ghostfish, are remnants
of an aberrant line that diverged from the elasmobranchs at least 300 million years
ago (Carboniferous or Devonian period). Fossil chimaeras (ky-meer′uz) first
occurred in the Jurassic period, reached their zenith in the Cretaceous and early
Tertiary periods (120 million to 50 million years ago), and have declined ever
since.

Today there are only about 25 species extant. Anatomically they present an
odd mixture of sharklike and bony fish–like features. Instead of a toothed mouth,
their jaws bear large flat plates. The upper jaw is completely fused to the cranium,
a most unusual development in fishes. Their food is seaweed, molluscs, echinoderms, crustaceans, and fishes—a surprisingly mixed diet for such a specialized
grinding dentition. Chimaeras are not commercial species and are seldom caught.
Despite their grotesque shape, they are beautifully colored with a pearly irides-
cence.

Figure 23-11

Ratfish *Hydrolagus colliei,* a chimaera of
the Pacific coast of North America (subclass
Holocephali).
Photograph by R. Harbo.

BONY FISHES: CLASS OSTEICHTHYES (TELEOSTOMI)

In few other major animal groups do we see better examples of adaptive radiation than among the bony fishes. Their adaptations have fitted them for every aquatic habitat except the most completely inhospitable. Body form alone is indicative of this diversity. Some have fusiform, streamlined bodies and other adaptations for reducing friction. Predaceous, pelagic fish have trim, elongate bodies, powerful tail fins, and other mechanical advantages for swift pursuit. Sluggish bottom-feeding forms have flattened bodies for movement and concealment on the ocean floor. The elongate body of the eel is an adaptation for wriggling through mud and reeds and into holes and crevices. Some, such as pipefishes, are so whiplike that they are easily mistaken for filaments of marine algae waving in the current. Many other grotesque body forms are obviously cryptic or mimetic adaptations for concealment from predators or as predators. Such few examples cannot begin to express the amazing array of physiological and anatomical specializations for defense and offense, food gathering, navigation, and reproduction in the diverse aquatic habitats to which bony fishes have adapted themselves. More of these adaptations are described in the following pages.

CHARACTERISTICS

1. **Skeleton more or less bony,** representing the primitive skeleton; vertebrae numerous; notochord may persist in part; **tail usually homocercal**
2. Skin with mucous glands and with embedded dermal scales of three types: **ganoid, cycloid,** or **ctenoid;** some without scales; no placoid scales
3. Fins both median and paired, with **fin rays of cartilage or bone**
4. **Mouth terminal** with many teeth (some toothless); jaws present; olfactory sacs paired and may or may not open into mouth
5. Respiration by gills supported by bony gill arches and covered by a **common operculum**
6. **Swim bladder** often present with or without duct connected to pharynx
7. Circulation consisting of a two-chambered heart, arterial and venous systems, and characteristically four pairs of aortic arches; blood containing nucleated red cells
8. Nervous system of a brain with small olfactory lobes and cerebrum; large optic lobes and cerebellum; 10 pairs of cranial nerves; three pairs of semicircular canals
9. Sexes separate (sex reversal in some), gonads paired; fertilization usually external; larval forms may differ greatly from adults

CLASSIFICATION

Class Osteichthyes: bony fishes. Three subclasses, 42 living orders.
  Subclass Crossopterygii: lobe-finned fishes. Four extinct orders, one living order containing one species, *Latimeria chalumnae.*
  Subclass Dipneusti: lungfishes. Six extinct orders, two living orders containing three genera: *Neoceratodus, Lepidosiren,* and *Protopterus.*
  Subclass Actinopterygii: ray-finned fishes. Three superorders and 39 living orders.
    Superorder Chondrostei (kon-dros'tee-i) (Gr. *chondros,* cartilage, + *osteon,* bone): primitive ray-finned fishes. Ten extinct orders; two living orders containing the bichir *(Polypterus),* sturgeons, and paddlefish.
    Superorder Holostei (ho-los'tee-i) (Gr. *holos,* entire, + *osteon,* bone): **intermediate ray-finned fishes.** Four extinct orders; two living orders containing the bowfin *(Amia)* and gars *(Lepisosteus).*
    Superorder Teleostei (tel'e-os'tee-i) (Gr. *teleos,* perfect, + *osteon,* bone): **climax bony fishes.** Body covered with thin scales without bony layer (cycloid or ctenoid) or scaleless; dermal and chondral parts of skull closely united; caudal fin mostly homocercal; mouth terminal; notochord a mere vestige; swim bladder mainly a hydrostatic organ and usually not opened to the esophagus; endoskeleton mostly

Figure 23-12

Western suckers *Catastomus occidentalis* (order Cypriniformes, superorder Teleostei). These freshwater fishes are aptly named for their habit of grubbing along the bottom, drawing in food through their protrusible jaws with thick, fleshy lips. Photograph by D.W. Gotshall.

bony. According to Nelson (1984), there are 10 extinct orders and 35 living orders. The latter are comprised of 408 families, and approximately 21,000 living species, representing 96% of all living fishes. Seven of the larger orders are as follows:

Anguilliformes: 597 species; freshwater eels, moray eels, conger eels, snipe eels.

Salmoniformes: 320 species; pikes, whitefish, salmon, trout, smelts, deep-sea luminescent fishes.

Cypriniformes: about 2400 species; suckers (Figure 23-12), minnows, carp, electric eels.

Siluriformes: about 2200 species; catfish (Figure 23-13).

Atheriniformes: 235 species; flying fish, medakas, killifish, live-bearers.

Scorpaeniformes: about 1160 species; rockfish, searobins, greenlings, sculpins, poachers, scorpionfish.

Perciformes: about 7800 species; barracudas, mullets, perch, darters, sunfishes, grunters, croakers, Moorish idols, damsel fish, viviparous perch, wrasses, parrot fish, trumpeters, sand perch, stargazers, blennies, wolffish, eel pouts, mackerels, tunas, swordfish.

Origin, Evolution, and Diversity

The Osteichthyes are divided into three distinct groups: the **lobe-finned fishes** (Crossopterygii), the **lungfishes** (Dipneusti), and the **ray-finned fishes** (Actinopterygii). What are their origins? At present, it seems prudent to steer a middle course through current paleontological debate and assign equal rank to all three groups in the class Osteichthyes. In other words, it is impossible to decide which one of these three groups, if any, might have served as ancestral stock for the other two. It is apparent from fossil evidence that all three groups were distinct in the Devonian period, some 400 million years ago. They are believed to have descended from an acanthodian of the Silurian period (Figure 22-13, p. 459).

Lobe-Finned Fishes: Subclass Crossopterygii

The lobe-finned fishes occupy an important position in vertebrate evolution because the amphibians—indeed all tetrapod vertebrates—arose from one or more of their ancient members. The crossopterygians had nostrils (choanae) that opened into the mouth, lungs as well as gills, and paired **lobed fins.** They first appeared in Devonian times, a capricious period of alternating droughts and floods when their lungs would have been a decided asset, if not absolutely essential, for their survival. They used their strong lobed fins as four legs to scuttle from one disappearing swamp to another that offered more promise for a continuing aquatic existence.

The crossopterygians are divided broadly into two groups. The **rhipidistians** appeared in the Devonian period, flourished in the late Paleozoic era, and then disappeared. This is the stock from which the amphibians have descended (Figure 23-1). Among the primitive characteristics of the rhiphidistians were the fusiform shape, two dorsal fins, and a **heterocercal** tail (Figure 23-14). The paired fins bore sharp resemblances to a tetrapod limb, for they consisted of a basal arrangement of median or axial bones, with other bones radiating from these median ones. Some of the proximal bones seem to correspond to the three chief bones of the tetrapod limb.

The other group of crossopterygians was the **coelacanths.** These also arose in the Devonian period, radiated somewhat, and reached their evolutionary peak in the Mesozoic era. At the end of the Mesozoic era they nearly disappeared but left one remarkable surviving species, *Latimeria chalumnae* (named for M. Cour-

Figure 23-13

Yellow bullhead *Ictalurus natalis* (order Siluriformes, superorder Teleostei). The characteristic barbels are used to detect food. Photograph by J.H. Gerard.

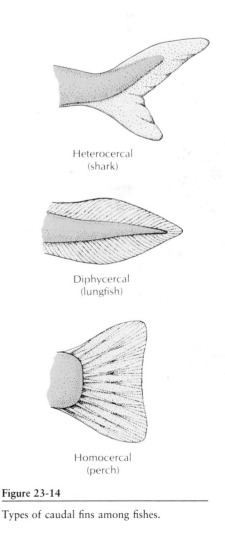

Heterocercal (shark)

Diphycercal (lungfish)

Homocercal (perch)

Figure 23-14

Types of caudal fins among fishes.

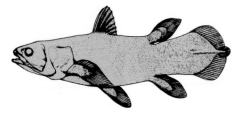

Figure 23-15

Coelacanth *Latimeria chalumnae*. This surviving marine relic of the crossopterygians that flourished some 350 million years ago has fleshy-based ("lobed") fins with which its ancestors used to pull themselves across land from pond to pond.

Courtesy Vancouver Public Aquarium, British Columbia.

tenay-Latimer, South African museum director) (Figure 23-15). Since the last coelacanths were believed to have become extinct 70 million years ago, the astonishment of the scientific world can be imagined when the remains of a coelacanth were found on a dredge off the coast of South Africa in 1938. An intensive search was begun in the Comoro Islands area near Madagascar where, it was learned, native Comoran fishermen occasionally caught them with hand lines at great depths. Numerous specimens have now been caught, many in excellent condition, although none has been kept alive beyond a few hours after capture.

The "modern" marine coelacanth is a descendant of the Devonian freshwater stock. The tail is of the **diphycercal** type (Figure 23-14) but possesses a small lobe between the upper and lower caudal lobes, producing a three-pronged structure (Figure 23-15). Coelacanths also show some degenerative features, such as more cartilaginous parts and a swim bladder that was either calcified or else persisted as a mere vestige. They also lack the internal nostril so characteristic of crossopterygians, but this is probably a secondary loss after the adoption of a deep-sea existence; obviously neither nostrils nor functional lungs have any relevance for such a life habit.

Lungfishes: Subclass Dipneusti

The lungfishes are another relic group of fishes, represented today by only three genera (Figure 23-16). These resemble the lobe-finned fishes in having lobe-shaped paired fins and lungs. However, all lungfishes, extinct or living, differ from the crossopterygians in several significant skeletal features, including the totally different origin of the internal nostrils.

Of the three surviving genera of lungfishes, the least specialized is *Neoceratodus* (Gr. *neos*, new, + *keratos*, horn, + *odes*, form), the living Australian lungfish, which may attain a length of 1.5 m. This lungfish is able to survive in stagnant, oxygen-poor water by coming to the surface and gulping air into its single lung, but it cannot live out of water. The South American lungfish *Lepidosiren* (L. *lepidus*, pretty, + *siren*, mythical mermaid) and the African lungfish *Protopterus* (Gr. *prōtos*, first, + *pteron*, wing) are evolutionary side branches of the Dipneusti, and they can live out of water for long periods of time. *Protopterus* lives in African streams and rivers that run completely dry during the dry season, with their mud beds baked hard by the hot tropical sun. The fish burrows down at the approach of the dry season and secretes a copious slime that is mixed with mud to form a hard cocoon in which it estivates until the rains return.

Neoceratodus (Australia)
Direct descendant of ancient lungfish
Cannot withstand complete drying up of water

Protopterus (Africa)
Side branch of lungfish evolution
Can burrow in mud when water dries up

Lepidosiren (South America)
Side branch of lungfish evoluton
Can burrow in mud when water dries up

Figure 23-16

The three surviving lungfishes of the subclass Dipneusti. The approximate range of each genus is shown on map insets.

Ray-Finned Fishes: Subclass Actinopterygii

The ray-finned fishes are an enormous assemblage containing all of our familiar bony fishes—some 21,000 species. The group had its beginnings in Devonian freshwater lakes and streams. They were small fish with large eyes, extended mouths, and a **heterocercal** tail (Figure 23-14). These earliest ray-finned fishes, known from the fossil record as palaeoniscids (pay'lee-o-nis'ids), were heavily armored with **ganoid scales** (Figure 23-17) and had functional lungs as well as gills. They looked distinctly different from the lungfishes and lobe-finned fishes with which they shared the Devonian swamps and rivers.

In their evolution the actinopterygians have passed through three stages. The most primitive are the chondrosteans (superorder Chondrostei), represented today by the freshwater and marine sturgeons, paddlefishes, and the bichir *Polypterus* (Gr. *poly*, many, + *pteros*, winged) of African rivers (Figure 23-18). *Polyp-*

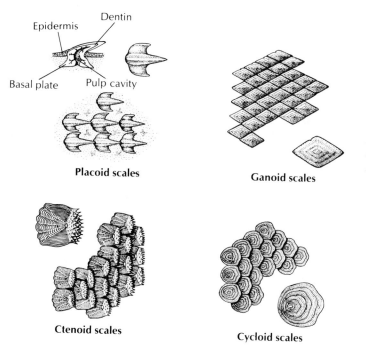

Figure 23-17

Types of fish scales. Placoid scales are small, conical toothlike structures characteristic of the Chondrichthyes. Diamond-shaped ganoid scales, present in primitive bony fishes such as the gar, are composed of layers of silvery enamel (ganoin) on the upper surface and bone on the lower. Advanced bony fishes have either cycloid or ctenoid scales. These are thin and flexible and are arranged in overlapping rows.

terus is an interesting relic with a lunglike swim bladder and many other primitive characteristics; it resembles an ancient palaeoniscid more than any other living descendant. There is no satisfactory explanation for the survival to the present of certain fishes such as this one and the coelacanth *Latimeria* when all of their kin perished millions of years ago.

The holosteans (superorder Holostei) is a second, less primitive group of ray-finned fishes. There were several lines of descent within this group, which flourished during the Triassic and Jurassic periods. They declined toward the end of the Mesozoic era as their successors, the teleosts, crowded them out. There are only two surviving genera, the bowfin *Amia* (Gr. tunalike fish) of shallow, weedy waters of the Great Lakes and Mississippi Valley, and the gars *Lepisosteus* (Gr.

Figure 23-18

Chondrostean fishes. **A,** Shovelnose sturgeon *Scaphirhynchus platorhyncus* of the Mississippi basin. **B,** Bichir *Polypterus bichir* of the African Congo. It is a nocturnal predator. **C,** Paddlefish *Polyodon spathula* of the Mississippi River reaches a length of 2 m and a weight of 90 kg. It is sought by commercial fishermen, but catches are now very low because of river pollution. **A** and **B,** Courtesy John G. Shedd Aquarium/ Patrice Ceisel; **C,** photograph by J.H. Gerard.

A

B

Figure 23-19

Holostean fishes. **A,** Bowfin *Amia calva.* **B,** Longnose gar *Lepisosteus osseus.* Both the bowfin and the gars live in the Great Lakes and Mississippi basin. The gar is a common fish of slow-moving streams, where it may hang motionless in the water, ready to snatch a passing fish.

Courtesy John G. Shedd Aquarium/Patrice Ceisel.

lepidos, scale, + *osteon,* bone) of eastern North America (Figure 23-19). The Holostei is now considered an artificial grouping because the gars appear to be more closely related to the teleosts (described below) than to the bowfin.

The teleosts (superorder Teleostei) are the third group of modern bony fishes (Figure 23-20). Diversity appeared early in teleost evolution, foreshadowing the truly incredible variety of body forms among teleosts today. The skeleton of primitive fish was largely ossified, but it returned to a partly cartilaginous condition among many of the chondrosteans and holosteans. Teleosts, however, have an internal skeleton almost completely ossified like the primitive fish.

There are a number of distinctive characteristics of the modern teleosts. The heavy armorlike scales of more primitive fish have been replaced by light, thin, and flexible **cycloid** and **ctenoid** scales. These look much alike (Figure 23-17) except that ctenoid scales have comblike ridges on the exposed edge that may be

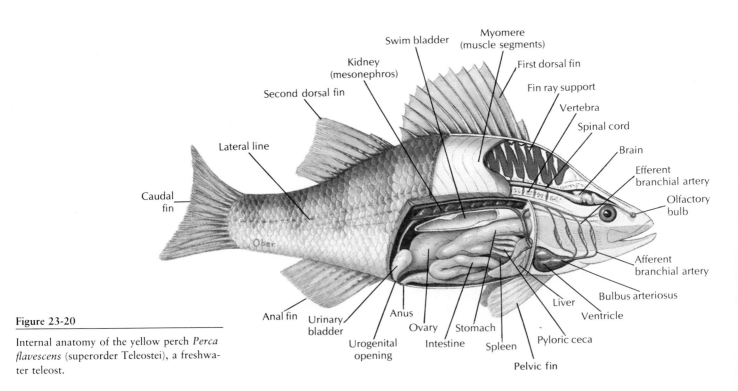

Figure 23-20

Internal anatomy of the yellow perch *Perca flavescens* (superorder Teleostei), a freshwater teleost.

an adaptation for improved swimming efficiency. Some teleosts, such as catfishes and sculpins, lack scales altogether. Nearly all teleosts have a **homocercal** tail, with the upper and lower lobes of about equal size (Figure 23-14). The lungs of primitive forms have been transformed in the teleosts to a **swim bladder** with a buoyancy function. Teleosts have highly maneuverable fins for control of body movement. In small teleosts the fins are provided with stout, sharp spines, thus making themselves prickly mouthfuls for would-be predators. With these adaptations (and many others), teleosts have evolved into the most successful of fishes.

STRUCTURAL AND FUNCTIONAL ADAPTATIONS OF FISHES
Locomotion in Water

To the human eye, some fishes appear capable of swimming at extremely high speeds. But our judgment is unconsciously tempered by our own experience that water is a highly resistant medium to move through. Most fishes, such as a trout or a minnow, can swim maximally about 10 body lengths per second, obviously an impressive performance by human standards. Yet when these speeds are translated into kilometers per hour it means that a 30 cm (1 foot) trout can swim only about 10.4 km (6.5 miles) per hour. As a general rule, the larger the fish the faster it can swim.

The propulsive mechanism of a fish is its trunk and tail musculature. The axial, locomotory musculature is composed of zigzag muscle bands (myotomes) that on the surface take the shape of a **W** lying on its side (Figure 23-21). Internally the muscle bands are deflected forward and backward in a complex fashion that apparently promotes efficiency of movement. The muscles are bound to broad sheets of tough connective tissue, which in turn tie to the highly flexible vertebral column.

Understanding how fishes swim can be approached by studying the motion of a very flexible fish such as an eel (Figure 23-22). The movement is serpentine, not unlike that of a snake, with waves of contraction moving backward along the body by alternate contraction of the myotomes on either side. The anterior end of the body bends less than the posterior end, so that each undulation increases in amplitude as it travels along the body. While undulations move backward, the bending of the body pushes laterally against the water, producing a **reactive force** that is directed forward, but at an angle. It can be analyzed as having two components: **thrust,** which is used to overcome drag and propels the fish forward, and **lateral force,** which tends to make the fish's head "yaw," or deviate from the

Measuring fish cruising speeds accurately is best done in a "fish wheel," a large ring-shaped channel filled with water that is turned at a speed equal and opposite to that of the fish. Much more difficult to measure are the sudden bursts of speed that most fish can make to capture prey or to avoid being captured. A hooked bluefin tuna was once "clocked" at 66 km per hour (41 mph); swordfish and marlin are thought to be capable of incredible bursts of speed approaching, or even exceeding, 110 km per hour (68 mph). Such high speeds can be sustained for no more than 1 to 5 seconds.

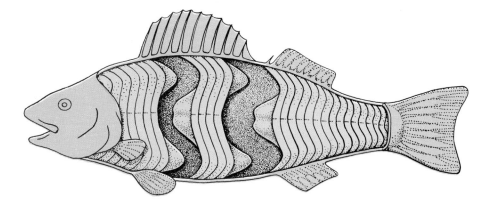

Figure 23-21

Trunk musculature of a teleost fish. Segmental myotomes are arranged as bands, each shaped like a W lying on its side. The musculature has been dissected away in two places to show internal anterior and posterior deflection of myotomes.

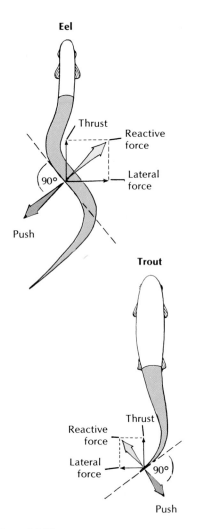

Figure 23-22

Movements of swimming fishes, showing
the forces developed by an eel-shaped and a
spindle-shaped fish.
From McFarland, W.N., and others. 1979. Vertebrate life. New York, Macmillan, Inc.

course in the same direction as the tail. This side-to-side head movement is very obvious in a swimming eel or shark, but many fishes have a large, rigid head with enough surface resistance to minimize yaw.

The movement of an eel is reasonably efficient at low speed, but its body shape generates too much frictional drag for rapid swimming. Fishes that swim rapidly, such as trout, are less flexible and limit the body undulations mostly to the caudal region (Figure 23-22). Muscle force generated in the large anterior muscle mass is transferred through tendons to the relatively nonmuscular caudal peduncle and tail where thrust is generated. This form of swimming reaches its highest development in the tunas, whose bodies do not flex at all. Virtually all the thrust is derived from powerful beats of the tail fin. Many fast oceanic fishes such as marlin, swordfish, amberjacks, and wahoo have swept-back tail fins shaped much like a sickle. Such fins are the aquatic counterpart of the high–aspect ratio wings of the swiftest birds (p. 550).

Swimming is the most economical form of animal locomotion, largely because aquatic animals are almost perfectly supported by their medium and need expend little energy to overcome the force of gravity. If we compare the energy cost per kilogram of body weight of traveling 1 km by different forms of locomotion, we find swimming to cost only 0.39 kcal (salmon) as compared to 1.45 kcal for flying (gull) and 5.43 for walking (ground squirrel). However, part of the unfinished business of biology is understanding how fish and aquatic mammals are able to move through the water while creating almost no turbulence. The secret lies in the way aquatic animals bend their bodies and fins (or flukes) to swim and in the friction-reducing properties of the body surface.

Neutral Buoyancy and the Swim Bladder

All fishes are slightly heavier than water because their skeletons and other tissues contain heavy elements that are present only in trace amounts in natural waters. To keep from sinking, sharks must always keep moving forward in the water. The asymmetrical (heterocercal) tail of a shark provides the necessary tail lift as it sweeps to and fro in the water, and the broad head and flat pectoral fins (Figure 23-5) act as angled planes to provide head lift. Sharks are also aided in buoyancy by having very large livers containing a special fatty hydrocarbon called **squalene** that has a density of only 0.86. The liver thus acts like a large sack of buoyant oil that helps to compensate for the shark's heavy body.

By far the most efficient flotation device is a gas-filled space. The **swim bladder** serves this purpose in the bony fishes. It arose from the paired lungs of the primitive Devonian bony fishes. Lungs were probably a ubiquitous feature of the Devonian freshwater bony fishes when, as we have seen, the alternating wet and dry climate would have made such an accessory respiratory structure essential for life. Swim bladders are present in most pelagic bony fishes but are absent in tunas, most abyssal fishes, and most bottom dwellers, such as flounders and sculpins.

By adjusting the volume of gas in the swim bladder, a fish can achieve neutral buoyancy and remain suspended indefinitely at any depth with no muscular effort. There are severe technical problems, however. If the fish descends to a greater depth, the swim bladder gas is compressed so that the fish becomes heavier and tends to sink. Gas must be added to the bladder to establish a new equilibrium buoyancy. If the fish swims up, the gas in the bladder expands, making the fish lighter. Unless gas is removed, the fish will rise with ever-increasing speed while the bladder continues to expand, until the fish pops helplessly out of the water. (This is a very real hazard for divers in helmeted diving suits, who must care-

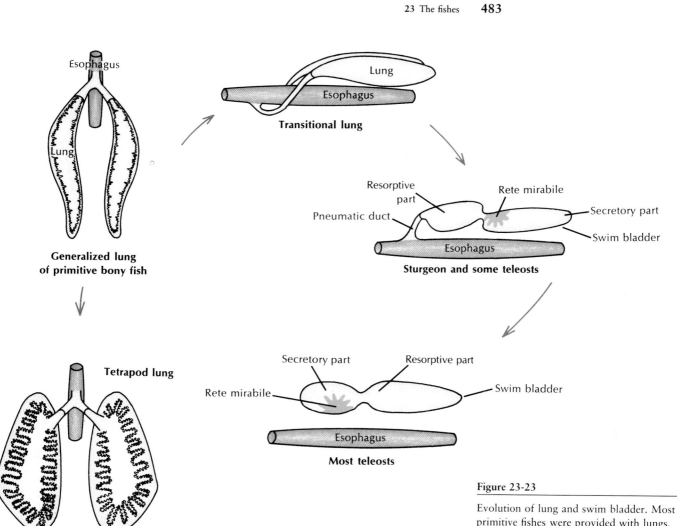

Esophagus

Lung

**Generalized lung
of primitive bony fish**

Lung

Esophagus

Transitional lung

Resorptive
part

Rete mirabile

Pneumatic duct

Secretory part

Esophagus

Swim bladder

Sturgeon and some teleosts

Tetrapod lung

Secretory part

Resorptive part

Rete mirabile

Swim bladder

Esophagus

Most teleosts

Figure 23-23

Evolution of lung and swim bladder. Most primitive fishes were provided with lungs, adaptations for the oxygen-depleted environments that existed during the evolution of the Osteichthyes. The lung originated as a diverticulum of the foregut. From this early, generalized lung two lines of evolution occurred. One led to the swim bladder of the modern teleost fish. Various transitional stages show that the swim bladder shifted to a dorsal position above the esophagus, becoming a buoyancy organ. The duct has been lost in most teleosts, and the swim bladder, with specialized gas secretion and reabsorption areas, is served by an independent blood supply. The second line of evolution has led to the tetrapod lung found in land forms. There has been extensive internal folding but no radical change in lung position.

fully adjust the air pressure to prevent overinflation or underinflation of their suits.)

Fishes adjust gas volume in the swim bladder in two ways. The less specialized fishes (trout, for example) have a **pneumatic duct** that connects the swim bladder to the esophagus (Figure 23-23); these forms must come to the surface and gulp air to charge the bladder and obviously are restricted to relatively shallow depths. More specialized teleosts have lost the pneumatic duct (upper diagram in Figure 23-23). In these fishes, the gas must originate in the blood and be secreted into the swim bladder. Gas exchange depends on two highly specialized areas: a **gas gland** that secretes gas into the bladder and a **resorptive area**, or "oval," that can remove gas from the bladder. The gas gland is supplied by a remarkable network of blood capillaries, called the **rete mirabile** ("marvelous net") that functions as a countercurrent exchange system to trap gases, especially oxygen, and prevent their loss to the circulation.

The amazing effectiveness of this device is exemplified by a fish living at a depth of 2400 m (8000 feet). To keep the bladder inflated at that depth, the gas inside (mostly oxygen, but also variable amounts of nitrogen, carbon dioxide, argon, and even some carbon monoxide) must have a pressure exceeding 240 atmospheres, which is much greater than the pressure in a fully charged steel gas cylinder. Yet the oxygen pressure in the fish's blood cannot exceed 0.2 atmosphere—equal to the oxygen pressure at the sea surface.

Physiologists who were at first baffled by the secretion mechanism now understand how it operates. In brief, the gas gland secretes lactic acid, which enters the blood, causing a localized high acidity in the rete mirabile that forces hemoglobin to release its load of oxygen. The capillaries in the rete are arranged so that the released oxygen accumulates in the rete, eventually reaching such a high pressure that the oxygen diffuses into the swim bladder. The final gas pressure attained in the swim bladder depends on the length of the rete capillaries; they are relatively short in fishes living near the surface, but are extremely long in deep-sea fishes.

Respiration

Fish gills are composed of thin filaments covered with a thin epidermal membrane that is folded repeatedly into platelike **lamellae** (Figure 23-24). These are richly supplied with blood vessels. The gills are located inside the pharyngeal cavity and are covered with a movable flap, the **operculum.** This arrangement provides excellent protection to the delicate gill filaments, streamlines the body, and makes possible a pumping system for moving water through the mouth, across the gills, and out the operculum. Instead of opercular flaps as in bony fishes, the elasmobranchs have a series of **gill slits** out of which the water flows. In both elasmobranchs and bony fishes the branchial mechanism is arranged to pump water continuously and smoothly over the gills, even though to an observer it appears that fish breathing is pulsatile. The flow of water is opposite to the direction of blood flow (countercurrent flow), the best arrangement for extracting the greatest possible amount of oxygen from the water. Some bony fishes can remove as much as 85% of the oxygen from the water passing over their gills. Very active fishes, such as herring and mackerel, can obtain sufficient water for their high oxygen demands only by continuously swimming forward to force water into the open mouth and across the gills. This is called ram ventilation. Such fish will be asphyxiated if placed in an aquarium that restricts free swimming movements, even though the water is saturated with oxygen.

A surprising number of fishes can live out of water for varying lengths of time by breathing air. Several devices are employed by different fishes. We have already described the lungs of the lungfishes, *Polypterus,* and the extinct crossopterygians. Freshwater eels often make overland excursions during rainy weather, using the skin as a major respiratory surface. The bowfin, *Amia* (Figure 23-19, *A*), has both gills and a lunglike swim bladder. At low temperatures it uses only its gills, but as the temperature and the fish's activity increase, it breathes mostly air with its swim bladder. The electric eel has degenerate gills and must supplement gill respiration by gulping air through its vascular mouth cavity. One of the best air breathers of all is the Indian climbing perch, which spends most of its time on land near the water's edge, breathing air through special air chambers above the much-reduced gills.

Osmotic Regulation

Fresh water is an extremely dilute medium with a salt concentration (0.001 to 0.005 gram moles per liter [M]) much below that of the blood of freshwater fishes (0.2 to 0.3 M). Water therefore tends to enter their bodies osmotically, and salt is lost by diffusion outward. Although the scaled and mucus-covered body surface is almost totally impermeable to water, water gain and salt loss do occur across the thin membranes of the gills. Freshwater fishes are **hyperosmotic regulators** that have several defenses against these problems (Figure 23-25). First, the excess water is pumped out by the **mesonephric** kidney (p. 646), which is capable of

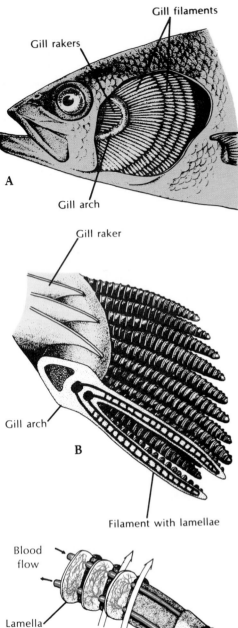

Figure 23-24

Gills of fish. Bony, protective flap covering the gills (operculum) has been removed, **A,** to reveal branchial chamber containing the gills. Four gill arches are on each side, each bearing numerous filaments. A portion of gill arch, **B,** shows gill rakers that project forward to strain out food and debris and gill filaments that project to the rear. A single gill filament, **C,** is dissected to show the blood capillaries within the platelike lamellae. Direction of water flow *(large arrows)* is opposite the direction of blood flow *(small arrows).*

forming very dilute urine. Second, special **salt-absorbing cells** located in the gill epithelium actively move salt ions, principally sodium and chloride, from the water to the blood. This, together with salt present in the fish's food, replaces diffusive salt loss. These mechanisms are so efficient that a freshwater fish devotes only a small part of its total energy expenditure to keeping itself in osmotic balance.

Marine bony fishes are **hypoosmotic regulators** that encounter a completely different set of problems. Having a much lower blood salt concentration (0.3 to 0.4 M) than the seawater around them (about 1 M), they tend to lose water and gain salt. The marine teleost fish quite literally risks drying out, much like a desert mammal deprived of water. Again, marine bony fishes, like their freshwater counterparts, have evolved an appropriate set of defenses (Figure 23-25). To compensate for water loss, the marine teleost drinks seawater. Although this behavior obviously brings needed water into the body, it is unfortunately accompanied by a great deal of unneeded salt. Unwanted salt is disposed of in two ways: (1) the

Figure 23-25

Osmotic regulation in freshwater and marine bony fishes. Freshwater fish maintains osmotic and ionic balance in its dilute environment by actively absorbing sodium chloride across gills (some salt enters with food). To flush out excess water that constantly enters body, glomerular kidney produces a dilute urine by reabsorbing sodium chloride. Marine fish must drink seawater to replace water lost osmotically to its salty environment. Sodium chloride and water are absorbed from stomach. Excess sodium chloride is secreted outward by gills. Divalent sea salts, mostly magnesium sulfate, are eliminated with feces and secreted by tubular kidney.
Modified from Webster, D., and M. Webster. 1974. Comparative vertebrate morphology. New York, Academic Press, Inc.

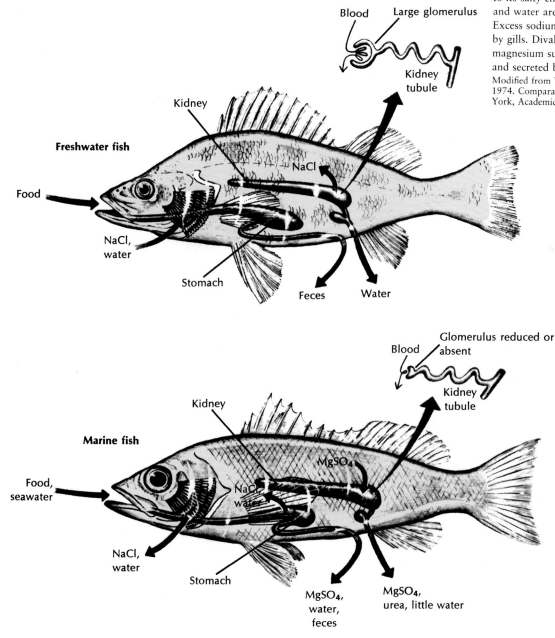

Perhaps 90% of all bony fishes are restricted to either a freshwater or a seawater habitat because they are incapable of osmotic regulation in the "wrong" habitat. Most freshwater fishes quickly die if placed in seawater, as will marine fishes placed in fresh water. However, some 10% of all teleosts can pass back and forth with ease between both habitats. These *euryhaline fishes* (Gr. *eurys,* broad, + *hals,* salt) are of two types: those such as many flounders, sculpins, and killifish that live in estuaries or certain intertidal areas where the salinity fluctuates throughout the day; or those such as salmon, shad, and eels, that spend part of their life cycle in fresh water and part in seawater.

major sea salt ions (sodium, chloride, and potassium) are carried by the blood to the gills where they are secreted outward by special **salt-secretory cells;** and (2) the remaining ions, mostly the divalent ions (magnesium, sulfate, and calcium), are left in the intestine and voided with the feces. However, a small but significant fraction of these residual divalent salts in the intestine, some 10% to 40% of the total, penetrates the intestinal mucosa and enters the bloodstream. These ions are excreted by the kidney. Unlike the freshwater fish kidney, which forms its urine by the usual filtration-resorption sequence typical of most vertebrate kidneys (pp. 648 to 650), the marine fish's kidney excretes divalent ions by tubular secretion. Since very little if any filtrate is formed, the glomeruli have lost their importance and disappeared altogether in some marine teleosts. The pipefishes and the goosefish, shown in Figure 23-27, are examples of "aglomerular" marine fishes.

___ Feeding Behavior

For any fish, feeding is one of the main concerns of day-to-day living. Although many a luckless angler would swear otherwise, the fact is that a fish devotes more time and energy to eating, or searching for food to eat, than to anything else. Throughout the long evolution of fishes, there has been unrelenting selective pressure for those adaptations that enable a fish to come out on the better end of the eat-or-be-eaten contest. Certainly the most far-reaching single event was the evolution of jaws. Their possessors were freed from a mud-grubbing or parasitic existence and could adopt a predatory mode of life. Improved means of capturing larger prey demanded stronger muscles, more agile movement, better balance, and improved special senses. More than any other aspect of its life habit, feeding behavior shapes the fish.

Most fishes are **carnivores** that prey on a myriad of animal foods from zooplankton and insect larvae to large vertebrates. Some deep-sea fishes are capable of eating victims nearly twice their own size—an adaptation for life in a world where meals are necessarily infrequent. Most advanced ray-finned fishes cannot masticate their food as we can because doing so would block the current of water across the gills. Some, however, such as the wolf eel (Figure 23-26), have molar-like teeth in the jaws for crushing their prey, which may include hard-bodied crustaceans. Others that do grind their food use powerful pharyngeal teeth in the throat. Most carnivores almost invariably swallow their prey whole, using sharp-pointed teeth in the jaws and on the roof of the mouth to seize their prey. The incompressibility of water makes the task even easier for many large-mouthed predators. When the mouth is opened, a negative pressure is created that sweeps the victim inside (Figure 23-27).

A second group of fishes are **herbivores** that eat flowering plants, algae, and grasses. Although the plant eaters are relatively few in number, they are crucial intermediates in the food chain, especially in freshwater rivers, lakes, and ponds that contain very little plankton.

The **filter-feeders** that crop the abundant microorganisms of the sea form a third and diverse group of fishes ranging from fish larvae to basking sharks. However, the most characteristic group of plankton feeders are the herringlike fishes (menhaden, herring, anchovies, capelin, pilchards, and others) that are for the most part **pelagic** (open-sea dwellers) and travel in large schools. Both phytoplankton and the smaller zooplankton are strained from the water with a sievelike device, the gill rakers (Figure 23-24). Because plankton feeders are the most abundant of all fishes, they are important food for numerous larger but less abundant carnivores. Many freshwater fishes also depend on plankton for food.

A fourth group of fishes are **omnivores** that feed on both plant and animal

Figure 23-26

Wolf eel, *Anarrhichthys ocellatus,* feeding on a sea cucumber *(Parastichopus* sp.) it has captured and pulled to the opening of its den.

Photograph by D.W. Gotshall.

food. Finally there are the **scavengers** that feed on organic debris (detritus) and the **parasites** that suck the body fluids of other fishes.

Digestion in most fishes follows the vertebrate plan. Except in several fishes that lack stomachs altogether, the food proceeds from stomach to tubular-intestine, which tends to be short in carnivores but may be extremely long and coiled in herbivorous forms. In the carp, for example, the intestine may be nine times the body length, an adaptation for the lengthy digestion required for plant carbohydrates. In carnivores, some protein digestion may be initiated in the acid medium of the stomach, but the principal function of the stomach is to store the often large and infrequent meals while awaiting their reception by the intestine.

Digestion and absorption proceed simultaneously in the intestine. A curious feature of ray-finned fishes, especially the teleosts, is the presence of numerous **pyloric ceca** (Figure 23-20) found in no other vertebrate group. Their primary function appears to be fat absorption, although all classes of enzymes (protein-, carbohydrate-, and fat-splitting) are secreted there. They number from two or three to several hundred in some advanced teleost species.

___ Migration

Eel

For centuries naturalists had been puzzled about the life history of the freshwater eel *Anguilla* (an-gwil′la) (L. eel), a common and commercially important species of coastal streams of the North Atlantic. Each fall, large numbers of eels were seen swimming down the rivers toward the sea, but no adults ever returned. And each spring countless numbers of young eels, called "elvers" (Figure 23-28), each about the size of a wooden matchstick, appeared in the coastal rivers and began swimming upstream. Beyond the assumption that eels must spawn somewhere at sea, the location of their breeding grounds was totally unknown.

The first clue was provided by two Italian scientists, Grassi and Calandruccio, who in 1896 reported that elvers were not larval eels but rather were relatively advanced juveniles. The true larval eels, they discovered, were tiny, leaf-shaped, completely transparent creatures that bore absolutely no resemblance to an eel. They had been called **leptocephali** (Gr. *leptos*, slender, + *kephalē*, head) by early naturalists, who never suspected their true identity. In 1905 Johann Schmidt, supported by the Danish government, began a systematic study of eel biology that he continued until his death in 1933. With the cooperation of captains of commercial vessels plying the Atlantic, thousands of the leptocephali were caught in different areas of the Atlantic with the plankton nets Schmidt supplied them. By noting where larvae in different stages of development were captured, Schmidt and his colleagues eventually reconstructed the spawning migrations.

When the adult eels leave the coastal rivers of Europe and North America, they swim steadily and apparently at great depth for 1 to 2 months until they reach the Sargasso Sea, a vast area of warm oceanic water southeast of Bermuda (Figure 23-28). Here, at depths of 300 m or more, the eels spawn and die. The minute larvae then begin an incredible journey back to the coastal rivers of Europe. Drifting with the Gulf Stream and preyed on constantly by numerous predators, they reach the middle of the Atlantic after 2 years. By the end of the third year they reach the coastal waters of Europe where the leptocephali metamorphose into elvers, with an unmistakable eel-like body form (Figure 23-28). Here the males and females part company; the males remain in the brackish waters of coastal rivers and estuaries while the females continue up the rivers, often traveling hundreds of miles upstream. After 8 to 15 years of growth, the females, now 1 m or

Figure 23-27

Goosefish *Lophius piscatorius* awaits its meal. Above its head swings a modified dorsal fin spine ending in a fleshy tentacle that contracts and expands in a convincing wormlike manner. When a fish approaches the alluring bait, the huge mouth opens suddenly, creating a strong current that sweeps the prey inside. In a split second all is over. Photograph by J.L. Rotman.

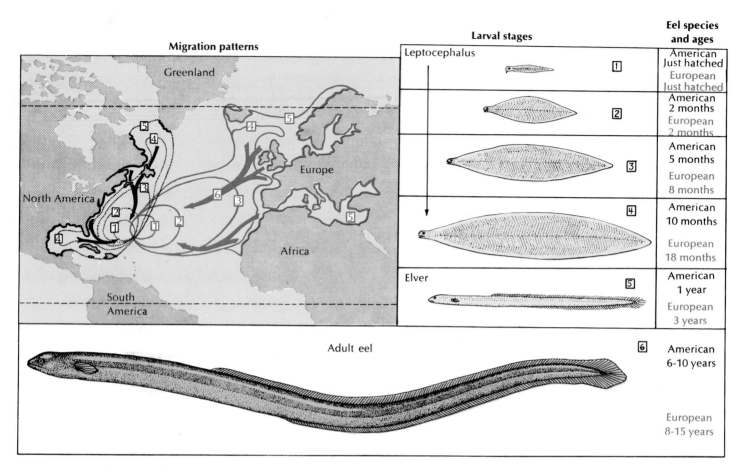

Migration patterns

Greenland

Europe

North America

Africa

South America

Adult eel

Larval stages

Eel species and ages

Leptocephalus

[1]	American Just hatched	
	European Just hatched	
[2]	American 2 months	
	European 2 months	
[3]	American 5 months	
	European 8 months	
[4]	American 10 months	
	European 18 months	
Elver [5]	American 1 year	
	European 3 years	
[6]	American 6-10 years	
	European 8-15 years	

Figure 23-28

Life histories of the European eel *(Anguilla anguilla)* and American eel *(A. rostrata).* *Red,* Migration patterns of European species. *Black,* Migration patterns of American species. Boxed numbers refer to stages of development. Note that the American eel completes its larval metamorphosis and sea journey in 1 year. It requires nearly 3 years for the European eel to complete its much longer journey.

more long, return to the sea to join the smaller males; both return to the ancestral breeding grounds thousands of miles away to complete the life cycle.

Schmidt found that the American eel *(Anguilla rostrata)* could be distinguished from the European eel *(A. vulgaris)* because it had fewer vertebrae—an average of 107 in the American eel as compared with an average 114 in the European species. Since the American eel is much closer to the North American coastline, it requires only about 8 months to make the journey.

Homing salmon

The life history of salmon is nearly as remarkable as that of the eel and certainly has received far more popular attention. Salmon are **anadromous;** that is, they spend their adult lives at sea but return to fresh water to spawn. The Atlantic salmon *(Salmo salar)* and the Pacific salmon (six species of the genus *Oncorhynchus* [on-ko-rink′us]) have this practice, but there are important differences among the seven species. The Atlantic salmon (as well as the closely related steelhead trout) make upstream spawning runs year after year. The six Pacific salmon species (king, sockeye, silver, humpback, chum, and Japanese masu) each make a single spawning run (Figure 23-29), after which they die.

The virtually infallible homing instinct of the Pacific species is legendary: after migrating downstream as a smolt, a sockeye salmon ranges many hundreds of miles over the Pacific for nearly 4 years, grows to 2 to 5 kg in weight, and then returns almost unerringly to spawn in the headwaters of its parent stream. Some straying does occur and is an important means of increasing gene flow and populating new streams.

Experiments by A.D. Hasler and others have shown that homing salmon are guided upstream by the characteristic odor of their parent stream. When the salmon finally reach the spawning beds of their parents (where they themselves were hatched), they spawn and die. The following spring, the newly hatched fry transform into smolts before and during the downstream migration. At this time they are imprinted (p. 736) with the distinctive odor of the stream, which is apparently a mosaic of compounds released by the characteristic vegetation and soil in the watershed of the parent stream. They also seem to imprint on the odors of other streams they pass while migrating downriver and use these odors in reverse sequence as a map during the upriver migration as returning adults.

How do salmon find their way to the mouth of the coastal river from the trackless miles of the open ocean? Salmon move hundreds of miles away from the coast, much too far to be able to detect their parent stream odor. There are experiments suggesting that some migrating fish, like birds, can navigate by orienting to the position of the sun. However, migrant salmon have been observed to navigate on cloudy days and at night, indicating that sun navigation, if used at all, cannot be the salmon's only navigational cue. Fish also (again, like birds) appear able to detect and navigate to the earth's magnetic field. Finally, fishery biologists concede that salmon may not require precise navigational abilities at all, but instead may use ocean currents, temperature gradients, and food availability to reach the general coastal area where "their" river is located. From this point, they would navigate by their imprinted odor map.

Reproduction and Growth

In a group as diverse as the fishes, it is no surprise to find extraordinary variations on the basic theme of sexual reproduction. Fortunately, most fishes favor a simple theme: they are **dioecious,** with **external fertilization** and **external development** of the eggs and embryos. This mode of reproduction is called **oviparous** (meaning "egg-producing"). However, as tropical fish enthusiasts are well aware, the ever-popular guppies and mollies of home aquaria bear their young alive after development in the ovarian cavity of the mother (Figure 23-30). These forms are said to

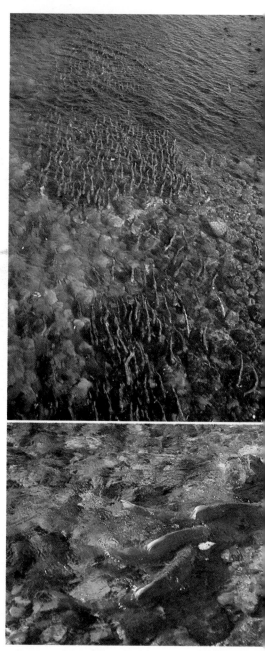

Figure 23-29

Migrating Pacific sockeye salmon. Photographs by M. Trim.

Figure 23-30

Surfperch *Hypsurus caryi* giving birth at the time it was captured. All of the West Coast surfperches (family Embiotocidae) are ovo-viviparous. Brood size ranges from 3 to 80 young depending on the size of the mother. Photograph by D.W. Gotshall.

Figure 23-31

Male banded jawfish *Opistognathus macrognathus* orally brooding its eggs. The male retrieves the female's spawn and incubates the eggs until they hatch. During brief periods when the jawfish is feeding, the eggs are left in the burrow.
Photograph by F. McConnaughey.

Figure 23-32

Lingcod *Ophiodon elongatus* guarding eggs, the bluish-white mass at left.
Photograph by R. Harbo.

be **ovoviviparous**, meaning "live egg–producing." There are even some sharks that develop some kind of placental attachment through which the young are nourished during gestation. These forms, like placental mammals, are **viviparous** ("alive-producing").

Let us return to the much more common oviparous mode of reproduction. Most fishes spawn at certain times or seasons within a restricted range of temperature. Temperature is critical both for successful spawning and for the survival of the sensitive eggs and young. Many, perhaps most, fish spawn in the spring or early summer when the water temperature is rising. Others, such as cod, many flatfishes, salmon, and certain trout species, spawn in fall or winter, often when water temperatures are at their lowest. In polar water, fishes may live out their existence at temperatures between 0° and −2° C.

Many marine fishes are extraordinarily profligate egg producers. Males and females come together in great schools and, without mating, release vast numbers of germ cells into the water to drift with the current. Large female cod may release 4 to 6 million eggs at a single spawning. Less than one in a million will survive the numerous perils of the ocean to reach reproductive maturity.

Unlike the minute, buoyant, transparent eggs of pelagic marine teleosts, those of near-shore species are larger, typically yolky, nonbuoyant, and adhesive. On the whole, fishes living in coastal waters where wave action and along-shore currents are prevalent dispose of their eggs in a more conservative manner. Some bury their eggs, many attach them to vegetation, some deposit them in nests, and some even incubate them in their mouths (Figure 23-31). Many coastal species guard their eggs (Figure 23-32). Intruders expecting an easy meal of eggs may be met with a vivid and often belligerent display by the guard, which is almost always the male.

Freshwater fishes almost invariably produce nonbuoyant eggs. Those, such as perch, that provide no parental care simply scatter their myriads of eggs among weeds or along the bottom. Freshwater fishes that do provide some form of egg care produce fewer, larger eggs that enjoy a better chance for survival.

Elaborate preliminaries to mating are the rule for freshwater fishes. The female Pacific salmon, for example, performs a ritualized mating "dance" with her breeding partner after arriving at the spawning bed in a fast-flowing, gravel-bottomed stream (Figure 23-33). She then turns on her side and scoops out a nest with her tail. As the eggs are laid by the female, they are fertilized by the male (Figure 23-33). After the female covers the eggs with gravel, the exhausted fish die and drift downstream.

Soon after the egg of an oviparous species is fertilized and laid, it takes up water and the outer layer hardens. Cleavage follows, and the blastoderm is formed, sitting astride a relatively enormous yolk mass. Soon the yolk mass is enclosed by the developing blastoderm, which then begins to assume a fishlike shape. The fish hatches as a larva that may be very different in appearance from the adult. The eyes and segmented muscles (myotomes) are well formed, but most conspicuous is the semitransparent, globular mass of yolk, so large that larval movement is nearly impossible (Figure 23-33). Not until the yolk is totally absorbed and the mouth and digestive tract are formed does the larva begin searching for its own food. After a period of growth the larva undergoes a metamorphosis, especially dramatic in many marine species such as the freshwater eel described previously (Figure 23-28). Body shape is refashioned, fin and color patterns change, and the animal becomes a juvenile bearing the unmistakable definitive body form of its species.

Female chum digging her nest.

Courting chum salmon. Male is nosing female's vent, possibly checking for spawning readiness.

Two coho salmon fertilizing eggs spawned from female between them.

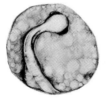

At fertilization sperm are guided to opening in egg (micropyle) by chemical attractant.

Developing embryo overlying large yolk sac.

Developing alvin. When yolk is absorbed, salmon emerges from gravel bed as fry and begins feeding.

Figure 23-33

Spawning of Pacific salmon and development of the egg and young.

Courtesy Joey Morgan/Hoot Productions, Ltd. From Childerhose, R.J., and M. Trim. 1979. Pacific salmon. Seattle, University of Washington Press. Reproduced by permission of the Minister of Supply and Services Canada.

Growth is temperature dependent. Consequently, fish living in temperate regions grow rapidly in summer when temperatures are high and food is abundant but nearly stop growing in winter. Seasonal growth is reflected as annual rings in the scales (Figure 23-34), a distinctive record of convenience to fishery biologists who wish to determine a fish's age. Unlike birds and mammals, which reach a definitive adult size, most fishes after attaining reproductive maturity continue to grow for as long as they live. This is possible because their dense medium is so buoyant that it offsets the pull of gravity. It also is probably a selective advantage for the species, since the larger the fish, the more germ cells it produces and the greater its contribution to future generations.

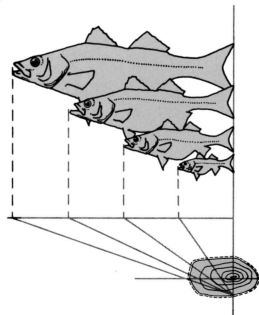

Figure 23-34

Scale growth. Fish scales disclose seasonal changes in growth rate. Growth is interrupted during winter, producing year marks (annuli). Each year's increment in scale growth is a ratio to the annual increase in body length. Otoliths (ear stones) and certain bones can also be used in some species to determine age and growth rate.

▬ SUMMARY

Fishes are poikilothermic, gill-breathing aquatic vertebrates with fins. They are the oldest vertebrates, having originated from an unknown chordate ancestor in the Cambrian period or possibly earlier. Four classes of fishes are recognized. Most primitive are the jawless hagfishes (class Myxini) and lampreys (class Cephalaspidomorphi), remnant groups having an eel-like body form without paired fins; a cartilaginous skeleton (although their ancestors, the ostracoderms, had bony skeletons); a notochord that persists throughout life; and a disclike mouth adapted for sucking or biting. All other vertebrates have jaws, a major development in vertebrate evolution. Members of the class Chondrichthyes (sharks, rays, skates, and chimaeras) are a successful group having a cartilaginous skeleton (a degenerative feature), paired fins, excellent sensory equipment, and an active, characteristically predaceous habit. The fourth class of fishes is the bony fishes (class Osteichthyes), which may be subdivided into three stems of descent. Two stems are relic groups, the lobe-finned fishes (subclass Crossopterygii, represented by one living species, the coelacanth) and the lungfishes (subclass Dipneusti). The third stem is the ray-finned fishes (subclass Actinopterygii), a huge and diverse modern assemblage containing nearly all of the familiar freshwater and marine fishes.

The modern bony fishes (teleost fishes) have radiated into nearly 21,000 species that reveal an enormous diversity of adaptations, body form, behavior, and habitat preference. Fishes swim by undulatory contractions of the body muscles, which generate thrust (propulsive force) and lateral force. Flexible fishes oscillate the whole body, but in more rapid swimmers the undulations are limited to the caudal region or tail fin alone.

Most pelagic bony fishes achieve neutral buoyancy in water using a gas-filled swim bladder, the most effective gas-secreting device known in the animal kingdom. The gills of fishes, having efficient countercurrent flow between water and blood, facilitate high rates of oxygen exchange. All fishes but the hagfishes show well-developed osmotic and ionic regulation, achieved principally by the kidneys and the gills.

With the exception of the jawless agnathans, all fishes have jaws that are variously modified for carnivorous, herbivorous, planktivorous, and omnivorous feeding modes.

Most fishes are migratory to some extent, and some, such as freshwater eels and anadromous salmon, make remarkable migrations of great length and precision. Fishes reveal an extraordinary range of sexual reproductive strategies. Most fishes are oviparous, but ovoviviparous and viviparous fishes are not uncommon. The reproductive investment may be in large numbers of germ cells with low survival (many marine fishes) or in fewer germ cells with greater parental care for better survival (freshwater fishes).

Review questions

1. Sketch a family tree for the fishes, placing the following groups in their proper relationship to each other: hagfishes; lampreys; gnathostomes; sharks, skates, and rays; lobe-finned fishes; lungfishes; ray-finned fishes. Refer to Figure 23-1.
2. Compare the morphology of hagfishes and lampreys, pointing out similarities and differences in the following features and systems: body shape, skeleton, notochord, mouth, circulatory system, digestive system, eyes, reproduction.

3. Describe the life cycle of the sea lamprey, *Petromyzon marinus,* and the history of its invasion of the Great Lakes.

4. In what ways are sharks well equipped for their predatory life style?

5. Give the common name(s) of the group or groups of fishes included within each of the following taxa: Agnatha, Osteichthyes, Cephalaspidomorphi, Chondrichthyes, Myxini, Gnathostomata, Holocephali, Crossopterygii, Dipneusti, Actinopterygii.

6. Explain how the bony fishes differ from the sharks and rays in the following systems or features: skeleton, tail shape, scales, buoyancy, respiration, position of mouth, reproduction.

7. Describe the discovery of a living lobe-finned fish, the coelacanth. What is the evolutionary significance of this group?

8. Give the geographic locations of the three surviving species of lungfishes.

9. Match the ray-finned fishes in the right column with the group to which each belongs in the left column:

_____chondrosteans	a. Perch
_____holosteans	b. Sturgeon
_____teleosts	c. Gar
	d. Salmon
	e. Paddlefish
	f. Bowfin

10. Give three characteristics of modern teleost fishes that distinguish them from the more primitive ray-finned fishes (chondrosteans and holosteans).

11. Compare the swimming movements of the eel with that of the trout, and explain why the latter is more efficient for rapid locomotion.

12. Explain the purpose and function of the swim bladder in teleost fishes. How is gas volume adjusted in the swim bladder?

13. What is meant by "countercurrent flow" as it applies to fish gills?

14. Compare the osmotic problem and the mechanism of osmotic regulation in freshwater and marine bony fishes.

15. Two principal groups of fishes, with respect to feeding behavior, are the carnivores and the filter-feeders. How are these two groups adapted for their feeding behavior?

16. Describe the life cycle of the European eel. How does the life cycle of the American eel differ from that of the European?

17. How do adult Pacific salmon find their way back to their parent stream to spawn?

18. What mode of reproduction in fishes is described by each of the following terms: oviparous, ovoviviparous, viviparous?

19. Reproduction in marine pelagic fishes and in freshwater fishes are distinctively different from each other. How and why do they differ?

Selected references

See also general references for Part Two, p. 590.

Bone, Q., and N.B. Marshall. 1982. Biology of fishes. New York, Chapman & Hall. *Concise, well-written, and well-illustrated primer on the functional processes of fishes.*

McKeown, B.A. 1984. Fish migration. Portland, Ore., Timber Press. *Comprehensive review of the subject.*

Moyle, P.B., and J.J. Cech, Jr. 1982. Fishes: an introduction to ichthyology. Englewood Cliffs, N.J., Prentice-Hall, Inc. *Textbook written in a lively style and stressing function and ecology rather than morphology; good treatment of the fish groups.*

Nelson, J.S. 1984. Fishes of the world, ed. 2. New York, John Wiley & Sons, Inc. *A modern and authoritative classification of all major groups of fishes.*

Pitcher, T.J. (ed.). 1986. The behavior of teleost fishes. Baltimore, Md., The Johns Hopkins University Press. *Contributed chapters on major functional topics in fish behavior.*

Potts, G.W., and R.J. Wootton (eds.). 1984. Fish reproduction: strategies and tactics. New York, Academic Press. *Contributed chapters on specific aspects of fish reproduction and development.*

Webb, P.W. 1984. Form and function in fish swimming. Sci. Amer. 251:72-82 (July). *Specializations of fish for swimming and analysis of thrust generation.*

CHAPTER 24

THE AMPHIBIANS

Phylum Chordata

Class Amphibia

The chorus of frogs beside a pond on a spring evening heralds one of nature's dramatic events. Masses of frog eggs soon hatch into limbless, gill-breathing, fishlike tadpole larvae. Warmed by the late spring sun, they feed and grow. Then, almost imperceptibly, a remarkable transformation takes place. Hindlegs appear and gradually lengthen. The tail shortens. The larval teeth are lost, and the gills are replaced by lungs. Eyelids develop. The forelegs emerge. In a matter of weeks the aquatic tadpole has completed its metamorphosis to an adult frog.

The early members of the class Amphibia (am-fib′e-a) (Gr. *amphi,* both or double, + *bios,* life), of which our chorusing frogs are among the more vociferous modern descendants, originated not in weeks but over millions of years by a lengthy series of almost imperceptible alterations that gradually fitted the vertebrate body plan for life on land. The origin of land vertebrates is no less a remarkable feat for this fact—a feat that incidentally would have a poor chance of succeeding today because well-established competitors make it impossible for a poorly adapted transitional form to gain a foothold.

Even now after some 350 million years of evolution, the amphibians are not completely land adapted; they are quasiterrestrial, hovering between aquatic and land environments. This double life is expressed in their name. Structurally they are between fishes and reptiles. Although adapted for a terrestrial existence, few can stray far from moist conditions. Many, however, have developed ways to keep their eggs out of open water where the larvae would be exposed to enemies.

The more than 3900 species of amphibians are grouped into three living orders: the salamanders (order Caudata), which are the least specialized and most aquatic of all amphibians; the frogs and toads (order Anura), which are the largest and most successful group of amphibians; and the highly specialized, secretive, earthwormlike tropical caecilians (order Gymnophiona).

▬ MOVEMENT ONTO LAND

The movement from water to land is perhaps the most dramatic event in animal evolution, since it involves the invasion of a habitat that in many respects is more hazardous for life. Life was conceived in water, animals are mostly water in composition, and all cellular activities occur in water. Nevertheless, organisms eventually invaded the land, carrying their watery composition with them. To survive and maintain this fluid matrix, it was necessary that various structural, functional, and behavioral changes evolve. Considering that almost every system in the body required some modification, it is remarkable that all vertebrates are basically alike in fundamental structural and functional pattern: whether aquatic or terrestrial, vertebrates are obviously descendants of the same evolutionary branch.

Amphibians were not the first living things to move onto land. Insects made the transition earlier and plants much earlier still. The pulmonate snails were experimenting with land as a suitable place to live about the same time the early amphibians were. Yet of all these, the amphibian story is of particular interest because their descendants became the most advanced animals on earth.

____ Physical Contrast between Aquatic and Land Habitats

Beyond the obvious difference in water content of aquatic and terrestrial habitats, there are several sharp differences between the two environments of significance to animals attempting to move from water to land.

Greater oxygen content of air

Air contains at least 20 times more oxygen than water. Air has approximately 210 ml of oxygen per liter; water contains 3 to 9 ml per liter, depending on temperature, presence of other solutes, and degree of saturation. Furthermore, the diffusion rate of oxygen is low in water. Consequently, aquatic animals must expend far more effort extracting oxygen from water than land animals expend removing oxygen from air.

Greater density of water

Water is approximately 1000 times denser than air and approximately 50 times more viscous. Although water is a much more resistant medium to move through, its high density, only a little less than that of animal protoplasm, buoys up the body. One of the major problems encountered by land animals was the need to develop strong limbs and remodel the skeleton to support their bodies in air.

Constancy of temperature in water

Natural bodies of water, containing a medium with tremendous thermal capacity, have little fluctuation in temperature. The temperature of the oceans remains almost constant day after day. In contrast, both the range and fluctuation in temperature are acute on land. Its harsh cycles of freezing, thawing, drying, and flooding, often in unpredictable sequence, presented severe thermal problems to terrestrial animals, which had to evolve behavioral and physiological strategies to protect themselves from thermal extremes. The most successful strategy, homeothermy (regulated constant body temperature), appeared in the birds and mammals.

Variety of land habitats

The variety of cover and shelter on land was a great inducement for its colonization. The rich offerings of terrestrial habitats include coniferous, temperate, and tropical forests, grasslands, deserts, mountains, oceanic islands, and polar regions. Even so, earth's hydrosphere (oceans, seas, lakes, rivers, and ice sheets), although offering a less diverse range of habitats, contains the greatest number and variety of living things on earth.

Opportunities for breeding on land

The provision of safe shelter for the protection of vulnerable eggs and young is much more readily accomplished on land than in water habitats.

ORIGIN AND RELATIONSHIPS OF AMPHIBIANS

The movement onto land required structural modifications of the fish body plan. Unlike a fish, which is supported and wetted by its medium and supplied with dissolved oxygen, a terrestrial animal must support its own weight, resist drying and rapid temperature change, and extract oxygen from air.

Appearance of Lungs

The Devonian period, beginning some 400 million years ago, was a time of mild temperatures and alternating droughts and floods. During dry periods, pools and streams began to dry up, water became foul, and the dissolved oxygen disappeared. Only those fishes able to use the abundance of atmospheric oxygen could survive such conditions. Gills were unsuitable because in air the filaments collapsed into clumps that soon dried out.

Virtually all freshwater fish surviving this period, including the lobe-finned fishes and the lungfishes, had a kind of lung developed as an outgrowth of the pharynx. It was relatively simple to enhance the efficiency of the air-filled cavity by improving its vascularity with a rich capillary network and by supplying it with arterial blood from the last (sixth) pair of aortic arches. Oxygenated blood was returned directly to the heart by a pulmonary vein to form a complete pulmonary circuit. Thus, the **double circulation** characteristic of all tetrapods originated: a systemic circulation, serving the body, and a pulmonary circulation, supplying the lungs.

Development of Limbs for Travel on Land

The evolution of limbs was also a product of difficult times during the Devonian period. When pools dried up, fishes were forced to move to another pool that still contained water. Only the lobe-finned fishes (crossopterygians) were preadapted for the task. They had strong lobed fins, used originally as swimming stabilizers, that could be adapted as paddles to lever their way across land in search of water. The pectoral fins were especially well developed, containing a series of skeletal elements in the fins and pectoral girdle that clearly foreshadowed the pentadactyl limb of tetrapods. We should note that the development first of strong fins and later of limbs did not happen so that fish could colonize land but rather as an adaptation to find water and continue living like fish. Land travel was simply and paradoxically a means for survival in water. But lungs and limbs were fortunate developments and essential specializations that preadapted vertebrates for life on land.

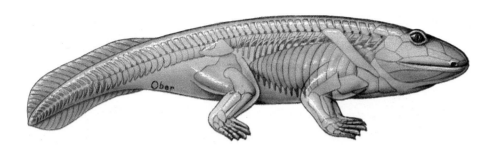

Figure 24-1

Reconstruction of the skeleton and body of the very early amphibian *Ichthyostega*. Modified from Jarvik, E. 1955. Sci. Monthly 80:141-154.

Earliest Amphibians

Evidence points to the lobe-finned fishes (crossopterygians) as ancestors of the modern amphibians. The lobe-fins, abundant and successful in the Devonian period, possessed lungs and strong, mobile fins. Their skull and tooth structure was similar to that of the earliest known amphibians, the labyrinthodonts, a distinct salamander-like group of the late Devonian period.

A representative of this group was a 350 million year old fossil called *Ichthyostega* (Gr. *ichthyos*, fish, + *stegos*, covering) (Figure 24-1). *Ichthyostega* possessed several new adaptations that equipped it for life on land. It had jointed, pentadactyl limbs for crawling on land, a more advanced ear structure for picking up airborne sounds, a foreshortening of the skull, and a lengthening of the snout that announced improved olfactory powers for detecting dilute airborne odors. Yet *Ichthyostega* was still fishlike in retaining a fish tail complete with fin rays and in having opercular (gill) bones.

The capricious Devonian period was followed by the Carboniferous period, characterized by a warm, wet climate during which mosses and large ferns grew in profusion on a swampy landscape. Conditions were ideal for the amphibians. They radiated quickly into a great variety of species, feeding on the abundance of insects, insect larvae, and aquatic invertebrates available: this was the age of amphibians.

With water everywhere, however, there was little selective pressure to encourage movement onto land and many amphibians actually improved their adaptations for living in water. Their bodies become flatter for moving about in shallow water. Many of the salamanders, which may have descended from the lepospondyls (Figure 24-2), developed weak limbs. The tail became better developed as a swimming organ. Even the anurans (frogs and toads), which are the most terrestrial of all amphibians, developed specialized hindlimbs with webbed feet better suited for swimming than for movement on land. All groups of amphibians use their porous skin as an accessory breathing organ. This specialization was encouraged by the swampy surroundings of the Carboniferous period but presented serious desiccation problems for life on land.

Amphibians' Contribution to Vertebrate Evolution

Amphibians have met the problems of independent life on land only halfway. To be sure, they made several important contributions to the transition that required the evolution of their descendants, the reptiles, to complete. Of crucial importance were the change from gill to lung breathing and the development of limbs for locomotion on land. Amphibians also show strengthening changes in the skeleton so that the body can be supported on land. A start was also made toward shifting special sense priorities from the lateral line system of fish to the senses of smell and

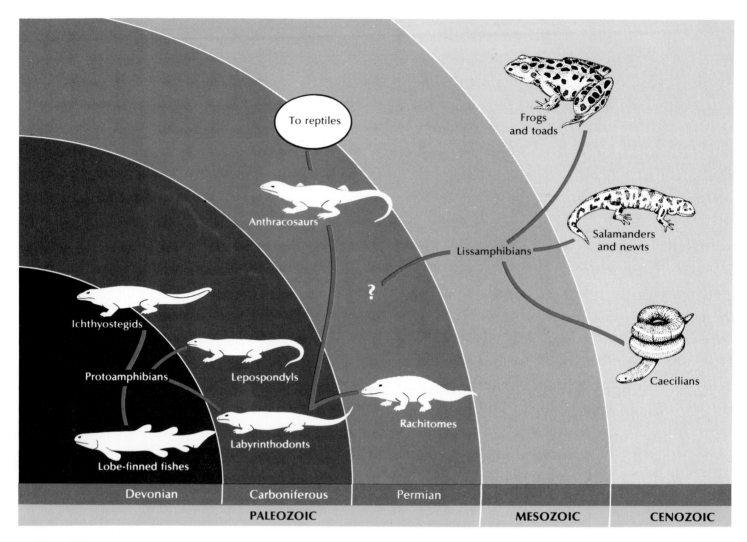

Devonian   Carboniferous   Permian

PALEOZOIC   **MESOZOIC**   **CENOZOIC**

Figure 24-2

Evolution of the amphibians. The early amphibians arose from the lobe-finned fishes and radiated successfully during the Carboniferous period (the age of amphibians). Modern amphibians have radiated from the Lissamphibians of the Mesozoic era, a group of uncertain connection to the Paleozoic amphibians.

hearing. For this, both the olfactory epithelium and the ear required redesigning to improve sensitivities to airborne odors and sounds.

Despite these modifications, the amphibians are basically aquatic animals. They are ectothermic; that is, their body temperature is determined by and varies with the environmental temperature. Their skin is thin, moist, and unprotected from desiccation in air. An intact frog loses water nearly as rapidly as a skinless frog. Most important, the amphibians remain chained to the aquatic environment by their mode of reproduction. Eggs are shed directly into the water or laid in moist surroundings and are externally fertilized (with very few exceptions). The larvae that hatch typically pass through an aquatic tadpole stage.

Many amphibians have developed ingenious devices for laying their eggs elsewhere to give their young protection and a better chance for life. They may lay eggs under logs or rocks, in the moist forest floor, in flooded tree holes, in pockets on the mother's back (Figure 24-3), or in folds of the body wall. One species of Australian frog even broods its young in its stomach.

However, it remained for the reptiles to complete the conquest of land with the development of a shelled (amniotic) egg, which finally freed the vertebrates from a reproductive attachment to the aquatic environment. With the appearance of reptiles at the end of the Paleozoic era, the halcyon era for the amphibians began to fade. The reptiles captured rule of both water and land and replaced most amphibians in both environments. From the survivors have descended the three modern orders of amphibians.

CHARACTERISTICS

1. Skeleton mostly bony, with varying number of vertebrae; ribs present in some, absent in others; notochord does not persist; **exoskeleton absent**
2. Body forms vary greatly from an elongated trunk with distinct head, neck, and tail to a compact, depressed body with fused head and trunk and no intervening neck
3. **Limbs, usually four (tetrapod),** although some are legless; forelimbs of some much smaller than hindlimbs, in others all limbs small and inadequate; **webbed feet often present;** no true nails or claws
4. **Skin smooth and moist with many glands,** some of which may be poison glands; pigment cells (chromatophores) common, of considerable variety; **no scales,** except concealed dermal ones in some
5. Mouth usually large with small teeth in upper or both jaws; **two nostrils open into anterior part of mouth cavity**
6. Respiration by lungs (absent in some salamanders), skin, and gills in some, either separately or in combination; external gills in the larval form and may persist throughout life in some
7. **Circulation with three-chambered heart,** two atria and one ventricle, and a double circulation through the heart; skin abundantly supplied with blood vessels
8. Ectothermal
9. Excretory system of paired mesonephric kidneys; urea main nitrogenous waste
10. Ten pairs of cranial nerves
11. Separate sexes; fertilization mostly internal in salamanders and caecilians, mostly external in frogs and toads; predominantly oviparous, some ovoviviparous or viviparous; metamorphosis usually present; **mesolecithal eggs with jellylike membrane coverings**

CLASSIFICATION

Order Gymnophiona (jim′no-fy′o-na) (Gr. *gymnos,* naked, + *ophineos,* of a snake) (**Apoda**): caecilians. Body wormlike; limbs and limb girdle absent; mesodermal scales present in skin of some; tail short or absent; 95-285 vertebrae; pantropical, 6 families, 34 genera, approximately 160 species.

Order Caudata (caw-dot′uh) (L. *caudatus,* having a tail) (**Urodela**): salamanders. Body with head, trunk, and tail; no scales; usually two pairs of equal limbs; 10-60 vertebrae; predominantly holarctic; 9 living families, 62 genera, approximately 350 species.

Order Anura (uh-nur′uh) (Gr. *an,* without, + *oura,* tail) (**Salientia**): frogs, toads. Head and trunk fused; no tail; no scales; two pairs of limbs; large mouth; lungs; 6-10 vertebrae including urostyle (coccyx); cosmopolitan, predominantly tropical; 21 living families; 301 genera; approximately 3400 species.

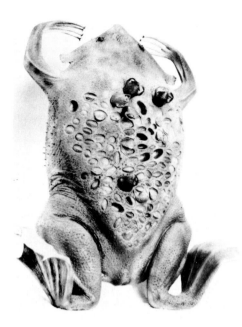

Figure 24-3

The female Surinam toad, carrying young on her back. As eggs are laid, the male assists in positioning them on the rough back skin of female. The skin swells, enclosing eggs. The approximately 60 young pass through the tadpole stage beneath the skin and emerge as small frogs. The Surinam "toad," actually a frog, is found mainly in Amazon and Orinoco river systems of equatorial South America.
Courtesy American Museum of Natural History.

STRUCTURE AND NATURAL HISTORY
Caecilians: Order Gymnophiona (Apoda)

The little-known order Gymnophiona (Gr. *gymnos,* naked, + *ophiona,* snake) contains some 160 species of burrowing, wormlike creatures commonly called **caecilians** (L. *caecilia,* kind of lizard, from *caecus,* blind) (Figure 24-2). They are distributed in tropical forests of South America (their principal home), Africa, and Southeast Asia. They are characterized by their long, slender body, small scales in the skin of some, many vertebrae, long ribs, no limbs, and terminal anus. The eyes are small, and most species are totally blind as adults (Figure 24-4). Replacing eyes, which would be useless for a subterranean existence, are special sensory tentacles on the snout. Because they are almost all burrowing forms, they are

Figure 24-4

A caecilian (order Gymnophiona [Apoda]). These legless and wormlike amphibians may reach lengths of 55 cm and diameters of 2 cm. Their body folds give them the appearance of a segmented worm. They have many sharply pointed teeth, a pair of tiny eyes mostly hidden beneath the skin, and a small tentacle between the eye and nostril; some forms have embedded mesodermal scales.

seldom seen by humans. Their food is mostly worms and small invertebrates, which they find underground. Fertilization is internal, and the male is provided with a protrusible copulatory organ. The eggs are usually deposited in moist ground near the water. The larvae may be aquatic, or the complete larval development may occur in the egg. In some species the eggs are carefully guarded in folds of the body during their development. Viviparity also is common among the more advanced caecilians, with the embryos obtaining nourishment probably by eating the wall of the oviduct.

Salamanders: Order Caudata (Urodela)

As its name suggests, the order Caudata consists of tailed amphibians: the salamanders and newts. This compact, natural group is the least specialized of all the amphibians. Although salamanders are found in almost all northern temperate and tropical regions of the world, most species occur in North America. Salamanders are typically small; most of the common North American salamanders are less than 15 cm long. Some aquatic forms are considerably longer, and the carnivorous Japanese giant salamander may exceed 1.5 m in length.

Salamanders have primitive limbs set at right angles to the body, with forelimbs and hindlimbs of approximately equal size. In some the limbs are rudimentary. One group, the sirens, with minute forelimbs and no hindlimbs at all, is so different from other salamanders that some authorities place them in a completely separate order, Trachystomata.

Salamanders are carnivorous, preying on worms, small arthropods, and small molluscs. Most eat only things that are moving. Since their food is rich in proteins, they do not usually store in their bodies great quantities of fat or glycogen. Like all amphibians, they are ectotherms and have a low metabolic rate.

Figure 24-5

Courtship and sperm transfer in the pigmy salamander *Desmognathus wrighti*. After juding the female's receptivity by the presence of her chin on his tail base, the male deposits a spermatophore on the ground, then moves forward a few paces. **A,** The white mass of the sperm atop a gelatinous base is visible at the level of the female's forelimb. The male moves ahead, the female following until the spermatophore is at the level of her vent. **B,** The female has recovered the sperm mass in her vent, while the male arches his tail, tilting the female upward and presumably facilitating recovery of the sperm mass.
Courtesy L. Houck.

Breeding behavior

Some salamanders are wholly aquatic throughout their life cycle, but most are terrestrial, living in moist places under stones and rotten logs, usually not far from water. They do not show as great a diversity of breeding habits as do frogs and toads. The eggs of most salamanders are fertilized internally, usually after the female picks up a packet of sperm (**spermatophore**) that previously has been deposited by the male on a leaf or stick (Figure 24-5). Aquatic species lay their eggs in clusters or stringy masses in the water. Terrestrial species deposit eggs in small, grapelike clusters under logs or in excavations in soft earth, and many species remain to guard the eggs (Figure 24-6). Unlike frogs and toads, which hatch into fishlike tadpole larvae, the embryos of salamanders hatch from their eggs resembling their parents. The larvae undergo metamorphosis in the course of development, but it is not nearly as revolutionary a change as is the metamorphosis of frog and toad tadpoles to the adult body form. American newts often have a terrestrial stage interposed between the aquatic larvae and the aquatic, breeding adults (Figure 24-7).

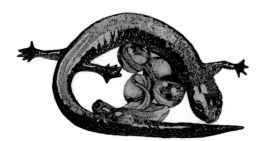

Figure 24-6

Red-backed salamander *Plethodon cinereus* encircles her hatching brood in their nest of forest humus. Females of many species of the large North American family Plethodontidae remain with their eggs during incubation and may even return to the same nest site year after year.
Drawn from a photograph by P.A. Zahl.

Respiration

All salamanders hatch with gills, but during development these are lost in all except the aquatic forms or in those that fail to undergo a complete metamorphosis. Gills would be useless for terrestrial salamanders, since the filaments would collapse and dry out to become functionless. Lungs, the characteristic respiratory organ of terrestrial vertebrates, replace the larval gills in most adult amphibians.

Yet many salamanders have dispensed with lungs altogether and thus bear the distinction of being the only vertebrates to have neither lungs nor gills. Members of the large family Plethodontidae, a group containing most of the familiar North American salamanders (Figure 24-6), are completely lungless, and some members of other caudate families exhibit reductions in lung development. In all amphibians the skin contains extensive vascular nets that serve in varying degrees for the respiratory exchange of oxygen and carbon dioxide. In lungless salamanders the efficiency of cutaneous respiration is increased by the penetration of a

Figure 24-7

Life history of the red-spotted newt, *Notophthalmus viridescens*. In many habitats the aquatic larva metamorphoses into a brightly colored "red eft" stage, which remains on land for 1 to 3 years before transforming into an aquatic adult.

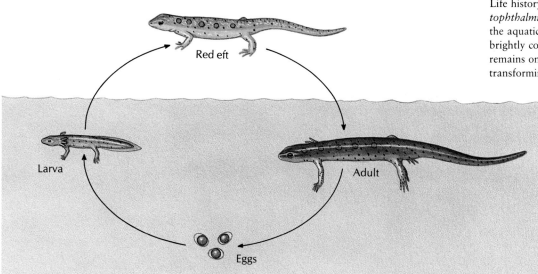

Red eft

Larva

Adult

Eggs

capillary network into the epidermis or by the thinning of the epidermis over superficial dermal capillaries. Cutaneous respiration is supplemented by the pumping of air in and out of the mouth where the respiratory gases are exchanged across the vascularized membranes of the buccal (mouth) cavity (buccopharyngeal breathing). Lungless plethodontid salamanders are believed to have originated in swift streams of the Appalachian mountains, where lungs would have been a disadvantage by providing too much buoyancy, and the water is so cool and well oxygenated that cutaneous respiration alone was sufficient for life.

Neoteny

Whereas most salamanders complete their development to the definitive adult body form by metamorphosis, there are some species that become sexually mature while retaining their gills and other larval characteristics. This condition is called **neoteny** (Gr. *neos*, young, + *teinen*, to extend). Some are permanent larvae, a genetically fixed condition in which the developing tissues fail to respond to the thyroid hormone that, in other amphibians, stimulates metamorphosis. This condition is called **obligatory neoteny.**

Examples of permanent larvae are mud puppies of the genus *Necturus* (Gr. *nekton*, swimming, + *oura*, tail) (Figure 24-8), which live on bottoms of ponds and lakes and keep their external gills throughout life, and the amphiuma (*Amphiuma means* [Gr. *amphi*, double, + *pneuma*, breath]) of the southeastern United States, which with its nearly useless, rudimentary legs superficially resembles an eel more than an amphibian.

There are other species of salamanders that become sexually mature and breed in the larval state, but unlike the permanent larvae, they may metamorphose to adults if environmental conditions change. This is called **facultative neoteny.**

Examples are species of the genus *Ambystoma* (Gr. *ambyx*, cup, + *stoma*, mouth) and of the genus *Triturus* (N.L. *Triton*, a sea god, + *oura*, tail). The American axolotl, *Ambystoma tigrinum*, widely distributed over Mexico and the southwestern United States, remains in the aquatic, gill-breathing, and fully reproductive larval form unless the water begins to dry up; then it metamorphoses to an adult, loses its gills, develops lungs, and assumes the appearance of an ordinary salamander.

Axolotls can be made to metamorphose by treating them with the thyroid hormone, thyroxine. Thyroxine is essential for normal metamorphosis in all amphibians. Recent research suggests that the pituitary gland fails to become fully active in neotenous forms and does not release thyrotropin, a pituitary hormone that is required to stimulate the production of thyroxine by the thyroid gland.

Although the British zoologist Walter Garstang is best remembered for his ideas on the origin of chordates and vertebrates, he also wrote numerous clever, unorthodox poems about larval forms and their development. Some of his verses were so complex as to require reference books to unravel, but the following charming poem about the axolotl can be enjoyed by all.

> *Ambystoma's* a giant newt who
>   rears in swampy waters,
> As other newts are wont to do,
>   a lot of fishy daughters:
> These axolotls, having gills,
>   pursue a life aquatic,
> But, when they should
>   transform to newts, are
>   naughty and erratic.
> They change upon compulsion,
>   if the water grows too foul,
> For then they have to use their
>   lungs, and go ashore to
>   prowl;
> But when a lake's attractive,
>   nicely aired, and full of food,
> They cling to youth perpetual,
>   and rear a tadpole brood.

Figure 24-8

Mud puppy *Necturus maculosus,* an example of a neotenic species. Gills are retained in the breeding adult.
Photograph by C.P. Hickman, Jr.

A   B

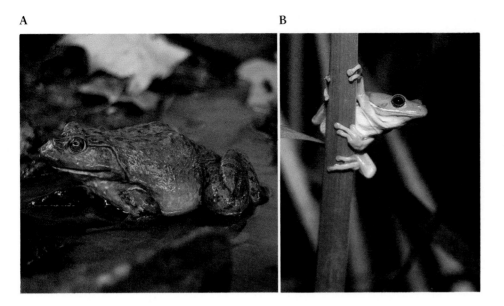

Figure 24-9

Two common North American frogs. **A,**
Bullfrog, *Rana catesbeiana*, largest American frog and mainstay of the frog leg epicurean market (family Ranidae). **B,** Green tree
frog, *Hyla cinerea*, a common inhabitant of
swamps of the southeastern United States
(family Hylidae). Note adhesive pads on the
feet.
Photographs by C.P. Hickman, Jr.

Frogs and Toads: Order Anura (Salientia)

The more than 3400 species of frogs and toads that make up the order Anura are
the most familiar and most successful of amphibians. The name of the order,
Anura, refers to an obvious group characteristic, the absence of tails in adults
(although all pass through a tailed larval stage during development). Frogs and
toads are specialized for a jumping mode of locomotion, as suggested by the
alternative order name, Salientia, meaning leaping.

The Anura are further distinguished from the Caudata by their larvae and a
dramatic metamorphosis during development. The eggs of most frogs hatch into a
tadpole ("polliwog") stage, with a long, finned tail, both internal and external
gills, no legs, specialized mouthparts for herbivorous feeding (salamander larvae,
in distinction, are carnivorous), and a highly specialized internal anatomy. They
look and act altogether differently from adult frogs. The metamorphosis of the
frog tadpole to the adult frog is thus a striking transformation. Neoteny is never
exhibited in frogs and toads as it is among salamanders.

The Anura are an old group—fossil frogs are known from the Jurassic period, 150 million years ago—and today they are a secure and successful group.
Frogs and toads occupy a great variety of habitats, despite their aquatic mode of
reproduction and water-permeable skin, which prevent them from wandering too
far afield from sources of water, and their ectothermy, which bars them from
polar and subarctic habitats.

Notwithstanding their success as a distinct group, the frogs and toads are
really a specialized side branch of amphibian evolution, and despite their popularity for education purposes—approximately 20 million are used each year in the
United States alone—they are not good representatives of the vertebrate body
plan. They lack a visible neck, the caudal vertebrae are fused into a urostyle
(coccyx), ribs are absent in most species, and the hindlegs are much enlarged for
leaping locomotion. The primitive and unspecialized salamander would be a
much superior choice for the zoology laboratory if frogs were not so readily available.

Frogs and toads are divided into 21 families. The best-known frog families in
North America are Ranidae, containing most of our familiar frogs, and Hylidae,
the tree frogs (Figure 24-9, *B*). True toads, belonging to the family Bufonidae (L.

In addition to their importance in
biomedical research and education, frogs have long served the
epicurean market. In 1979, for
example, nearly 6 million pounds
of frog legs, worth more than 9
million dollars, were consumed
by Americans alone. Most were
imported from overseas because
the domestic supply has been
hard hit by the draining and pollution of wetlands. Most vulnerable is the bullfrog, mainstay of
the frog leg market. Attempts to
raise them in farms have not been
successful, mainly because bullfrogs are voracious eating machines that prefer insects, crayfish, and other frogs and normally will accept only moving prey.

Figure 24-10

American toad *Bufo americanus* (family Bufonidae). This principally nocturnal yet familiar amphibian feeds on large numbers of insect pests as well as snails and earthworms. The warty skin contains numerous poison glands that produce a surprisingly poisonous milky fluid, providing the toad with excellent protection from a variety of potential predators.
Photograph by C.P. Hickman, Jr.

Figure 24-11

Gigantorana goliath (family Ranidae) of West Africa, the world's largest frog. This specimen weighed 3.3 kg (approximately 7½ pounds).
Courtesy American Museum of Natural History.

bufo, toad, + *idae,* suffix), have short legs, stout bodies, and thick skins usually with prominent warts (Figure 24-10). However, the term "toad" is used rather loosely to refer to more or less terrestrial members of several other families.

The largest anuran is the West African *Gigantorana (Conraua) goliath* (Gr. *gigantos,* giant, + *rana,* frog), which is more than 30 cm long from tip of nose to anus (Figure 24-11). This giant eats animals as big as rats and ducks. The smallest frog recorded is *Phyllobates limbatus* (Gr. *phyllon,* leaf, + *batēs,* climber), which is only approximately 1 cm long. This tiny frog, which can be more than covered by a dime, is found in Cuba. The largest American frog is the bullfrog, *Rana catesbeiana* (Figure 24-9, *A*), which reaches a head and body length of 20 cm.

Habitats and distribution

Probably the most abundant and successful of frogs are the approximately 260 species of the genus *Rana* (Gr. frog), found all over the temperate and tropical regions of the world except in New Zealand, the oceanic islands, and southern South America. They are usually found near water, although some, such as the wood frog *R. sylvatica,* spend most of their time on damp forest floors, often some distance from the nearest water. The wood frog probably returns to pools only for breeding in early spring. The larger bullfrogs *R. catesbeiana* and green frogs *R. clamitans* are nearly always found in or near permanent water or swampy regions. The leopard frog *R. pipiens* has a wider variety of habitats and, with all its subspecies and forms, is the most widespread of all the North American frogs. This is the species most commonly used in biology laboratories and for classical electrophysiological research. It has been found in some form in nearly every state, although sparingly represented along the extreme western part of the Pacific coast. It also extends far into northern Canada and as far south as Panama.

Within the range of any species of frogs, they are often restricted to certain habitats (for instance, to certain streams or pools) and may be absent or scarce in similar habitats of the range. The pickerel frog *(R. palustris)* is especially noteworthy in this respect because it is known to be abundant only in certain localized regions.

Most of the larger frogs are solitary in their habits except during the breeding season. During the breeding period most of them, especially the males, are very noisy. Each male usually takes possession of a particular perch, where he may remain for hours or even days, trying to attract a female to that spot. At times frogs are mainly silent, and their presence is not detected until they are disturbed. When they enter the water, they dart about swiftly and reach the bottom of the pool, where they kick up a cloud of muddy water. In swimming, they hold the

Figure 24-12

Figure 24-12

African clawed frog *Xenopus laevis* (family Pipidae). The claws, an unusual feature in frogs, are visible on the hindfeet. This is an alien species in California, where it is considered a serious pest.
Photograph by C.P. Hickman, Jr.

While native American amphibians continue to disappear as wetlands are drained, an exotic frog introduced into southern California has found the climate quite to its liking. The African clawed frog *Xenopus laevis* (Figure 24-12) is a voracious, aggressive, primarily aquatic frog that is rapidly displacing native frogs and fish from several waterways and is spreading rapidly. The species was introduced into North America in the 1940s when they were used extensively in human pregnancy tests. When more efficient tests appeared in the 1960s, some hospitals simply dumped surplus frogs into nearby streams, where the prolific breeders have become almost indestructible pests. As is so often the case with alien wildlife introductions, benign intentions frequently lead to serious problems.

forelimbs near the body and kick backward with the webbed hindlimbs, which propel them forward. When they come to the surface to breathe, only the head and foreparts are exposed, and, since they usually take advantage of any protective vegetation, they are difficult to see.

During the winter months most frogs hibernate in the soft mud of the bottom of pools and streams. Naturally their life processes are at a very low ebb during their hibernation period, and such energy as they need is derived from the glycogen and fat stored in their bodies during the spring and summer months. The more terrestrial frogs, such as tree frogs, bed down for the winter in the humus of the forest floor. They are tolerant of low temperatures, and many actually survive prolonged freezing of all the extracellular fluid, representing 35% of the body water. Such frost-tolerant frogs prepare for winter by accumulating glucose and glycerol in body fluids, which protects tissues from the normally damaging effects of ice crystal formation.

Adult frogs have numerous enemies, such as snakes, aquatic birds, turtles, raccoons, and humans; only a few tadpoles survive to maturity. Although usually defenseless, in the tropics and subtropics many frogs and toads are aggressive, jumping and biting at predators. Some defend themselves by feigning death. Most anurans can blow up their lungs so that they are difficult to swallow. When disturbed along the margin of a pond or brook, a frog often remains quite still; when it thinks it is detected, it jumps, not always into the water where enemies may be lurking but into grassy cover on the bank. When held in the hand, a frog may cease its struggles for an instant to put its captor off guard and then leap violently, at the same time voiding its urine. Their best protection is their ability to leap and their use of poison glands. Bullfrogs in captivity do not hesitate to snap at tormenters and are capable of inflicting painful bites.

Integument and coloration

The skin of the frog is thin and moist and is attached loosely to the body only at certain points. Histologically the skin is made up of two layers: an outer stratified **epidermis** and an inner spongy **dermis** (Figure 24-13). The outer layer of epidermal cells (which are shed periodically when a frog or toad "molts") contains deposits of **keratin,** a tough, fibrous protein that provides a certain measure of protection against abrasion and loss of water from the skin. The more terrestrial

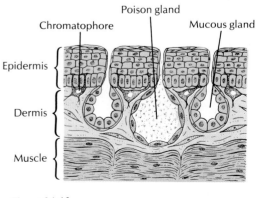

Figure 24-13

Section through frog skin.

Figure 24-14

Pigment cells (chromatophores). **A,** Pigment dispersed. **B,** Pigment concentrated. The pigment cell does not contract or expand; color effects are produced by streaming of cytoplasm, carrying pigment granules into cell branches for maximum color effect or to the center of the cell for minimum effect. Control over dispersal or concentration of pigment is mostly by light stimuli to eye acting through a pituitary hormone.

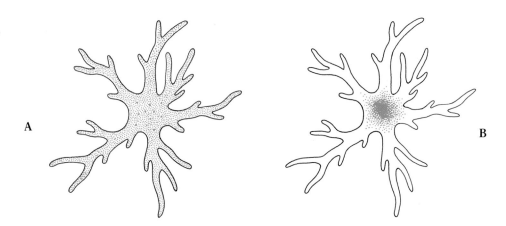

amphibians such as toads have especially heavy deposits of keratin. But amphibian keratin is soft, unlike the hard keratin that forms scales, claws, feathers, horns, and hair of the higher vertebrates.

The inner layer of the epidermis gives rise to two types of integumentary glands that grow down into the loose dermal tissues below. Small **mucous** glands secrete a protective mucous waterproofing onto the skin surface, and large **serous** glands produce a whitish, watery poison that is highly irritating to would-be predators. All amphibians produce a skin poison, but its effectiveness varies from species to species and with different predators. The poison of *Dendrobates,* a small South American frog, is used by Indian tribes to poison the points of their arrows. This and the poisons of other dendrobatid frogs are the most lethal animal secretions known, drop for drop more poisonous even than the venoms of sea snakes or any of the most poisonous arachnids.

Skin color in frogs is produced by special pigment cells, **chromatophores,** located mainly in the dermis. The chromatophores of amphibians, like those of many fishes and reptiles, are branched cells containing pigment that may be concentrated in a small area or dispersed throughout the branching processes to control skin coloration (Figure 24-14). Three types of chromatophores are recognized: uppermost are **xanthophores,** containing yellow, orange, or red pigments; beneath these lie **iridiophores,** containing a silvery, light-reflecting pigment; and lowermost are **melanophores,** containing black or brown melanin. The iridiophores act like tiny mirrors, reflecting light back through the xanthophores to produce the brightly conspicuous colors of many tropical frogs. Surprisingly perhaps, the green hues so common in North American frogs are produced not by green pigment but by an interaction of xanthophores containing a yellow pigment and underlying iridiophores that, by reflecting and scattering the light (Tyndall scattering), produce a blue color. The blue light is filtered by the overlying yellow pigment and thus appears green. Many frogs can adjust their color to blend with the background and thus camouflage themselves (Figure 24-15).

Skeletal and muscular systems

In amphibians, as in their fish ancestors, the well-developed **endoskeleton** of bone and cartilage provides a framework for the muscles in movement and protection for the viscera and nervous systems. But movement onto land and the necessity of transforming paddlelike fins into tetrapod legs capable of supporting the body's weight introduced a new set of stress and leverage problems. The changes are most noticeable in the anurans, whose entire musculoskeletal system is specialized for jumping and swimming by simultaneous extensor thrusts of the hindlimbs.

Figure 24-15

Cryptic coloration of the gray tree frog *Hyla versicolor.* Camouflage is so good that the presence of this frog is usually disclosed only at night by its resonant, flutelike trill. Photograph by C.P. Hickman, Jr.

Skeleton

In amphibians the vertebral column assumes a new role as a support from which the abdomen is slung and to which the limbs are attached. Since amphibians move with limbs instead of swimming with serial contractions of myotomic trunk musculature, the vertebral column has lost much of the original flexibility characteristic of fishes. Rather it has become a rigid frame for transmitting force from the hindlimbs to the body. The anurans are further specialized by an extreme shortening of the body. Typical frogs have only nine trunk vertebrae and a rodlike **urostyle,** which represents several fused caudal vertebrae (coccyx). The primitive, limbless caecilians, which obviously have not shared these specializations for tetrapod locomotion, may have as many as 285 vertebrae.

The frog skull is also vastly altered as compared with its crossopterygian ancestors; it is much lighter in weight and more flattened in profile and has fewer bones and less ossification. The front part of the skull, wherein are located the nose, eyes, and brain, is better developed, whereas the back of the skull, which contained the gill apparatus of fishes, is much reduced. Lightening of the skull was essential to mobility on land, and the other changes fitted the frog for its improved special senses and means for feeding and breathing.

The pattern of bones and muscles in the limbs is the typical tetrapod type. There are three main joints in each limb (hip, knee, and ankle; or shoulder, elbow, and wrist). The hand or foot is basically a five-rayed (pentadactyl) form with several joints in each of the digits. It is a repetitive system that can be plausibly derived from the bone structure of the crossopterygian fin, which is distinctly suggestive of the amphibian limb; it is not difficult to imagine how selective pressures through millions of years remodeled the former into the latter.

Musculature

The muscles of the limbs are presumably derived from the radial muscles that moved the fins of fishes up and down, but the muscular arrangement has become so complex in the tetrapod limb that it is no longer possible to see parallels between this and fin musculature. Despite its complexity, we can recognize two major groups of muscles on any limb: an anterior and ventral group that pulls the limb forward and toward the midline (protraction and adduction), and a second set of posterior and dorsal muscles that serves to draw the limb back and away from the body (retraction and abduction).

The trunk musculature, which in fishes is segmentally organized into powerful muscular bands (myotomes) for locomotion by lateral flexion, was much modified during amphibian evolution. The dorsal (epaxial) muscles are arranged to support the head and brace the vertebral column. The ventral (hypaxial) muscles of the belly are more developed in amphibians than in fishes, since they must support the viscera in air without the buoying assistance of water.

Respiration and vocalization

Amphibians use three respiratory surfaces for gas exchange: the skin (cutaneous breathing), the mouth (buccal breathing), and the lungs. These surfaces are used to varying degrees by different groups and under different environmental conditions. Frogs and toads show a much greater dependence on lung breathing than do salamanders, since cutaneous breathing, simple and direct as it obviously is, suffers from two disadvantages: (1) the skin must be kept thin and moist for gas exchange and consequently is too delicate for a wholly aerial life, and (2) the

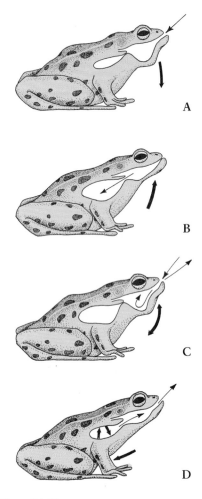

Figure 24-16

Breathing in frog. The frog, a positive-pressure breather, fills its lungs by forcing air into them. **A,** Floor of mouth is lowered, drawing air in through nostrils. **B,** With nostrils closed and glottis open, the frog forces air into its lungs by elevating floor of mouth. **C,** Mouth cavity rhythmically ventilates for a period. **D,** Lungs are emptied by contraction of body wall musculature and by elastic recoil of lungs.
Modified from Gordon, M.S., and others, 1968. Animal function: principles and adaptations. New York, Macmillan, Inc.

amount of gas exchange across the skin is mostly constant and cannot be varied to match changing demands of the body for more or less oxygen. Nevertheless, the skin continues to serve as an important supplementary avenue for gas exchange in anurans, especially during hibernation in winter. Even under normal conditions when lung breathing predominates, most of the carbon dioxide is lost across the skin while most of the oxygen is taken up across the lungs.

The lungs are supplied by pulmonary arteries (derived from the sixth aortic arches) and blood is returned directly to the left atrium by the pulmonary veins. Frog lungs are ovoid, elastic sacs with their inner surfaces divided into a network of septa that are in turn subdivided into small terminal air chambers called **alveoli.** The alveoli of the frog lung are much larger than those of more advanced vertebrates, and consequently the frog lung has a smaller relative surface available for gas exchange: the respiratory surface of the common *Rana pipiens* is about 20 cm^2 per cubic centimeter of air contained, compared with 300 cm^2 for humans. The problem in lung evolution was not the development of a good internal vascular surface, but rather the problem of moving air. A frog is a positive-pressure breather that fills its lungs by forcing air into them; this contrasts with the negative-pressure system of all the higher vertebrates. The sequence and explanation of breathing in a frog are shown in Figure 24-16. One can easily follow this sequence in a living frog at rest: rhythmical throat movements of mouth breathing may continue some time before flank movements indicate that the lungs are being emptied and refilled.

Both male and female frogs have **vocal cords,** but those of the male are much better developed. They are located in the **larynx,** or voice box. Sound is produced by passing air back and forth over the vocal cords between the lungs and a large pair of sacs (vocal pouches) in the floor of the mouth. The latter also serve as effective resonators in the male. The chief function of the voice is to attract mates. Most species utter characteristic sounds that identify them. Nearly everyone is familiar with the welcome springtime calls of the spring peeper, which produces a high-pitched sound surprisingly strident for such a tiny frog. Another sound familiar to residents of the more southern United States is the resonant "jug-o-rum" call of the bullfrog. The bass notes of the green frog are banjolike, and those of the leopard frog are long and guttural. Many frog sounds are now available on phonograph records.

Circulation

As in fishes, circulation in amphibians is a closed system of arteries and veins serving a vast peripheral network of capillaries through which blood is forced by the action of a single pressure pump, the heart.

The principal changes in circuitry involve the shift from gill to lung breathing. With the elimination of gills, a major obstacle to blood flow was removed from the arterial circuit. But two new problems arose. The first was to provide a blood circuit to the lungs. As we have seen, this was solved by converting the sixth aortic arch into pulmonary arteries to serve the lungs and by developing new pulmonary veins for returning oxygenated blood to the heart. The second and evidently more difficult evolutionary problem was to separate the new pulmonary circulation from the rest of the body's circulation in such a way that oxygenated blood from the lungs would be selectively sent to the body and deoxygenated venous return from the body would be selectively sent to the lungs. In effect this meant creating a double circulation consisting of separate pulmonary and systemic circuits. This was eventually solved by the development of a partition down the

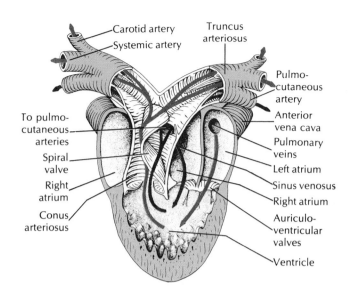

Carotid artery
Systemic artery
Truncus arteriosus
Pulmo-cutaneous artery
Anterior vena cava
Pulmonary veins
Left atrium
Sinus venosus
Right atrium
Auriculo-ventricular valves
Ventricle
To pulmo-cutaneous arteries
Spiral valve
Right atrium
Conus arteriosus

Figure 24-17

Structure of the frog heart. *Red arrows,* Oxygenated blood. *Black arrows,* Deoxygenated blood.

center of the heart, creating a double pump, one for each circuit. Amphibians and reptiles have made the separation to variable degrees, but the task was completed by the birds and mammals, which have a completely divided heart of two atria and two ventricles.

The frog heart (Figure 24-17) has two separate atria and a single undivided ventricle. Blood from the body (systemic circuit) first enters a large receiving chamber, the **sinus venosus,** which forces it into the **right atrium.** The **left atrium** receives freshly oxygenated blood from the lungs. Up to this point the deoxygenated blood from the body and oxygenated blood from the lungs are separated. But now both atria contract almost simultaneously, driving both right and left atrial blood into the single undivided **ventricle.** We should expect that complete admixture of the two circuits would happen here. In fact there is evidence that in at least some amphibians they remain mostly separated, so that when the ventricle contracts, oxygenated pulmonary blood is sent to the systemic circuit and deoxygenated systemic blood is sent to the pulmonary circuit. The **spiral valve** in the **conus arteriosus** (Figure 24-17) may play an important role in maintaining selective distribution. The matter is controversial and has defied complete analysis despite the application of advanced techniques using radiopaque media and high-speed cineradiography.

Feeding and digestion

Frogs are carnivorous like most other adult amphibians and feed on insects, spiders, worms, slugs, snails, millipedes, or nearly anything else that moves and is small enough to swallow whole. They snap at moving prey with their protrusible tongue, which is attached to the front of the mouth and is free behind. The free end of the tongue is highly glandular and produces a sticky secretion, which adheres to the prey. The teeth on the premaxillae, maxillae, and vomers are used to prevent escape of prey, not for biting or chewing. The digestive tract is relatively short in adult amphibians, a characteristic of most carnivores, and it produces a variety of enzymes for breaking down proteins, carbohydrates, and fats.

The larval stages of anurans (tadpoles) are usually herbivorous, feeding on pond algae and other vegetable matter; they have a relatively long digestive tract, since their bulky food must be submitted to time-consuming fermentation before useful products can be absorbed.

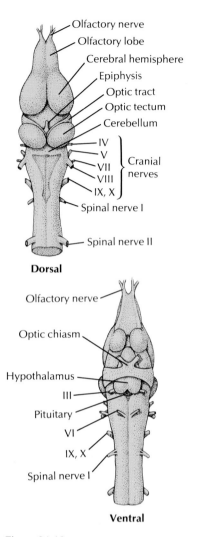

Olfactory nerve
Olfactory lobe
Cerebral hemisphere
Epiphysis
Optic tract
Optic tectum
Cerebellum
IV
V
VII   } Cranial
VIII   nerves
IX, X
Spinal nerve I

Spinal nerve II

Dorsal

Olfactory nerve

Optic chiasm

Hypothalamus

III

Pituitary

VI

IX, X

Spinal nerve I

Ventral

Figure 24-18

Brain of the frog, dorsal and ventral views.

Nervous system and special senses

The three fundamental parts of the brain—forebrain (telencephalon), concerned with the sense of smell; midbrain (mesencephalon), concerned with vision; and hindbrain (rhombencephalon), concerned with hearing and balance—have undergone dramatic developmental trends as the vertebrates moved onto land and improved their environmental awareness. In general there is increasing cephalization with emphasis on information processing by the brain and a corresponding loss of independence of the spinal ganglia, which are capable of only stereotyped reflexive behavior. Nonetheless, a headless frog preserves an amazing degree of purposive and highly coordinated behavior. With only the spinal cord intact, it maintains normal body posture and can with purposive accuracy raise its leg to wipe from its skin a piece of filter paper soaked in dilute acid. It will even use the opposite leg if the closer leg is held.

The forebrain (Figure 24-18) contains the olfactory center, which has assumed greatly increased importance for the detection of dilute airborne odors on land. The sense of smell is in fact one of the dominant special senses in frogs. The remainder of the forebrain, the cerebrum, is of little importance in amphibians and provides no hint of the magnificent development it is destined to attain in the higher mammals. Instead, complex integrative activities of the frog are located in the midbrain optic lobes.

The hindbrain is divided into an anterior cerebellum and a posterior medulla. The cerebellum (Figure 24-18) is concerned with equilibrium and movement coordination and is not well developed in amphibians, which stay close to the ground and are not noted for dexterity of movement. The cerebellum becomes vastly developed in the fast-moving birds and mammals. The medulla is really the enlarged anterior end of the spinal cord through which pass all sensory neurons except those of vision and smell. Here are located centers for auditory reflexes, respiration, swallowing, and vasomotor control.

The evolution of a semiterrestrial life for the amphibians has necessitated a reordering of sensory receptor priorities on land. The pressure-sensitive lateral line (acousticolateral) system of fishes remains only in the aquatic larvae of amphibians and in a few strictly aquatic adult amphibian species. This system of course can serve no useful purpose on land, since it was designed to detect and localize objects in water by reflected pressure waves. Instead the task of detecting airborne sounds devolved on the ear.

The **ear** of a frog is by higher vertebrate standards a primitive structure: a middle ear closed externally by a large **tympanic membrane** (eardrum) and containing a **stapes** (columella) that transmits vibrations to the inner ear. The latter contains the **utricle,** from which arise the semicircular canals, and a **saccule** bearing a diverticulum, the **lagena.** The lagena is partly covered with a **tectorial membrane** that in its fine structure is not unlike that of the much more advanced mammalian cochlea. In most frogs this structure is sensitive to low-frequency sound energy not greater than 4000 Hz (cycles per second); in the bullfrog the main frequency response is in the 100 to 200 Hz range, which matches the energy of the male frog's low-pitched call.

Vision is the dominant special sense in most amphibians (the mostly blind caecilians are obvious exceptions). Several modifications were required to adapt the fish eye for use on dry land. Lachrymal glands and eyelids evolved to keep the eye moist, wiped free of dust, and shielded from injury. Since the cornea is exposed to air, it is an important refractive surface, removing much of the burden from the lens of bending light rays and focusing the image on the retina. As in the fishes,

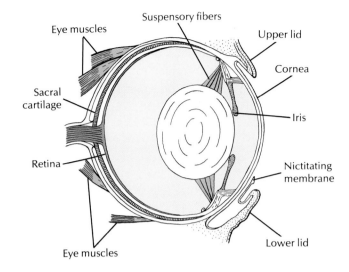

Figure 24-19

Amphibian eye.

accommodation (adjusting focus for near and distant objects) is accomplished by moving the lens. But unlike the eyes of most fishes, the amphibian eye at rest is adjusted for distant objects and is moved forward to focus on nearby objects.

The **retina** contains both **rods** and **cones,** the latter providing frogs with color vision. The **iris** contains well-developed circular and radial muscles and can rapidly expand or contract the aperture (pupil) to adjust to changing illumination. The upper lid of the eye is fixed, but the lower is folded into a transparent **nictitating membrane** capable of moving across the eye surface (Figure 24-19). In all, frogs and toads possess good vision, a fact of crucial importance to animals that rely on quick escape to avoid their numerous predators, and accurate movements to capture rapidly moving prey.

Other sensory receptors include tactile and chemical receptors in the skin, taste buds on the tongue and palate, and a well-developed olfactory epithelium lining the nasal cavity.

Reproduction

Because frogs and toads are ectothermic, they breed, feed, and grow only during the warmer seasons of the year. One of the first drives after the dormant period is breeding. In the spring males croak and call vociferously to attract females. When their eggs are mature, the females enter the water and are clasped by the males in a process called **amplexus** (L. *amplexus*, embrace). As the female lays the eggs, the male discharges seminal fluid containing sperm over the eggs to fertilize them. After fertilization, the jelly layers absorb water and swell (Figure 24-20). The eggs are laid in large masses, usually anchored to vegetation.

Development of the fertilized egg (zygote) begins almost immediately (Figure 24-21). By repeated division (cleavage) the egg is converted into a hollow ball of cells (blastula). This undergoes continued differentiation to form an embryo with a tail bud. At 6 to 9 days, depending on the temperature, a tadpole hatches from the protective jelly coats that had surrounded the original fertilized egg.

At the time of hatching, the tadpole has a distinct head and body with a compressed tail. The mouth is located on the ventral side of the head and is provided with horny jaws for scraping off vegetation from objects for food. Behind the mouth is a ventral adhesive disc for clinging to objects. In front of the mouth are two deep pits, which later develop into the nostrils. Swellings are found

Keeping a sharp image on the retina for approaching or receding objects requires accommodation, and this is accomplished in different ways by different vertebrates. The eye of the bony fishes and lampreys is adjusted for near vision; to focus on distant objects the lens must be moved backward. In amphibians, sharks, and snakes, and relaxed eye is focused on distant objects and the lens is moved *forward* to focus on nearby objects. In birds, mammals, and all reptiles except snakes, the lens accommodates by changing its *curvature* rather than by being moved forward or backward. The resting eye in these forms is adjusted for distant vision, and to focus on nearby objects the lens curvature is increased; that is, the lens is squeezed (or, in some, allowed to relax) into a rounder shape.

Figure 24-20

Eggs of American toad. Toads lay eggs in strings; frogs lay eggs in clusters.
Photograph by Priscilla Connell, Cincinnati Nature Center.

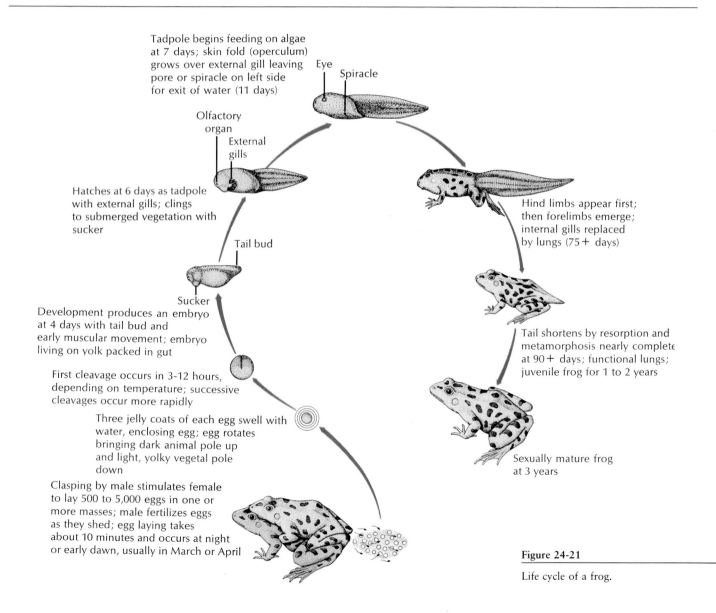

Tadpole begins feeding on algae at 7 days; skin fold (operculum) grows over external gill leaving pore or spiracle on left side for exit of water (11 days)

Eye
Spiracle

Olfactory organ
External gills

Hatches at 6 days as tadpole with external gills; clings to submerged vegetation with sucker

Tail bud

Sucker

Development produces an embryo at 4 days with tail bud and early muscular movement; embryo living on yolk packed in gut

First cleavage occurs in 3-12 hours, depending on temperature; successive cleavages occur more rapidly

Three jelly coats of each egg swell with water, enclosing egg; egg rotates bringing dark animal pole up and light, yolky vegetal pole down

Clasping by male stimulates female to lay 500 to 5,000 eggs in one or more masses; male fertilizes eggs as they shed; egg laying takes about 10 minutes and occurs at night or early dawn, usually in March or April

Hind limbs appear first; then forelimbs emerge; internal gills replaced by lungs (75 + days)

Tail shortens by resorption and metamorphosis nearly complete at 90 + days; functional lungs; juvenile frog for 1 to 2 years

Sexually mature frog at 3 years

Figure 24-21

Life cycle of a frog.

on each side of the head, and these later become external gills. There are finally three pairs of external gills, which are later replaced by three pairs of internal gills within the gill slits. On the left side of the neck region is an opening, the **spiracle** (L. *spiraculum*, air hole), through which water flows after entering the mouth and passing the internal gills. The hindlegs appear first, whereas the forelimbs are hidden for a time by the folds of the operculum. During metamorphosis the tail is resorbed, the intestine becomes much shorter, the mouth undergoes a transformation into the adult condition, lungs develop, and the gills are resorbed. The leopard frog usually completes its metamorphosis within 3 months; the bullfrog takes 2 or 3 years to complete the process.

Migration of frogs and toads is correlated with their breeding habits. Males usually return to a pond or stream in advance of the females, which they then attract by their calls. Some salamanders are also known to have a strong homing instinct, returning year after year to the same pool for reproduction guided by olfactory cues. The initial stimulus for migration in many cases is attributable to a seasonal cycle in the gonads plus hormonal changes that increase the frogs' sensitivity to temperature and humidity changes.

— SUMMARY

Amphibians are ectothermic quadrupedal vertebrates that have glandular skin and that breathe by lungs, gills, or skin. They are a transitional group, neither fully aquatic nor fully terrestrial. Although the group enjoyed a lengthy period of prominence in the late Paleozoic era, their aquatic mode of reproduction prevented a successful conquest of land. Most amphibians must return to water to lay their eggs, and their larvae are clearly fishlike. Direct development that omits the aquatic larval stage is not uncommon, however. Despite their insignificance as a group today, the amphibians made several important contributions of vertebrate evolution, which include a shift from gill to lung breathing, separation of pulmonary and systemic circulations, development of tetrapod limbs, restructuring of the skeleton to support the body in air, and redesigning of the senses of vision, smell, and hearing for use on land.

The modern amphibians consist of three evolutionary lines. The caecilians (order Gymnophiona) are a small tropical group of highly specialized, wormlike forms. The salamanders (order Caudata) are tailed amphibians that have retained the generalized four-legged body plan of their Paleozoic ancestors. Fertilization is internal following spermatophore transfer from male to female. Some salamanders are neotenic, becoming sexually mature while retaining gills and other larval characteristics.

The most successful of the modern amphibians are the frogs and toads (order Anura). All are specialized for a jumping mode of locomotion. Most inhabit wet habitats, but many toads and tree frogs are tolerant of arid environments. The integument is moist and contains numerous glands, some of which produce a poisonous irritant as protection against predators. Frogs are lung breathers, but the skin and mouth cavity are important supplementary avenues of gas exchange. A pulmonary circulation serves the lungs, and the heart is partly divided to keep pulmonary and systemic blood circuits mostly separated. All regions of the brain except the cerebellum are better developed in frogs and toads than in fishes. Most frogs have well-developed vocal cords, used principally by males to attract females. Fertilization is external. The eggs hatch into tadpole larvae that undergo a dramatic metamorphosis to the adult body form.

Longtail salamander *Eurycea longicauda*, a common plethodontid salamander.
Photograph by C.P. Hickman, Jr.

Review questions

1. As compared with the aquatic habitat, the terrestrial habitat offers both advantages and problems for an animal making the transition from water to land. Summarize how these differences might have influenced the early evolution of amphibians.
2. Describe the evolution of lungs and strong limbs in amphibians from fish ancestors and comment on the selective pressures responsible.
3. Give evidence to support the following statement: "Although amphibians made an important start toward movement onto land, they are still aquatic animals in many respects."
4. Give the literal meaning of the name Gymnophiona. What animals are included in this amphibian order, what do they look like, and where do they live?
5. What is the literal meaning of the order names Caudata and Anura? What features distinguish the members of these two orders from each other?
6. Describe the breeding behavior of a typical woodland salamander.
7. Explain the meaning of the term "neoteny" as it applies to the salamanders. What is the difference in the kind of neoteny that occurs in the mudpuppy *Necturus* and the kind that occurs in the American axolotl *Ambystoma?*
8. Describe the integument of a frog. What is responsible for skin color in frogs?
9. Describe respiration and circulation in amphibians.
10. Explain how the forebrain, midbrain, hindbrain, and the sensory structures with which each brain division is concerned have developed to meet the sensory requirements for amphibian life on land.
11. Briefly describe the reproductive behavior of frogs. In what important ways do frog and salamander reproduction differ?

Selected references

See also general references for Part Two, p. 590.

Duellman, W.E., and L. Trueb. 1986. Biology of amphibians. New York, McGraw-Hill Book Co. *Important comprehensive sourcebook of information on amphibians, extensively referenced and illustrated.*

Gibbons, W. 1983. Their blood runs cold: adventures with reptiles and amphibians. University, Ala., University of Alabama Press. *Delightful account of personal experiences of a herpetologist, filled with engaging stories and interesting facts.*

Goin, C.J., and O.B. Goin. 1978. Introduction to herpetology, ed. 3. San Francisco, W.H. Freeman & Co. Publishers. *Basic introductory text for the study of amphibians and reptiles.*

Halliday, T.R. and K. Adler (eds.). 1986. The encyclopedia of reptiles and amphibians. New York, Facts on File Inc. *Excellent authoritative reference work with high-quality illustrations.*

King, F.W., and J. Behler, 1979. The Audubon Society field guide to North American reptiles and amphibians. New York, Alfred A. Knopf, Inc.

Smith, H.M. 1978. Amphibians of North America. (Golden Field Guide Series.) Racine, Wisc., Western Publishing Co., Inc.

CHAPTER 25

THE REPTILES

Phylum Chordata

Class Reptilia

The class Reptilia (rep-til′e-a) (L. *repto*, to creep) were the first truly terrestrial vertebrates. With some 7000 species (approximately 300 species in the United States and Canada) occupying a great variety of aquatic and terrestrial habitats, they are clearly a successful group. Nevertheless, reptiles are perhaps remembered best for what they once were, rather than for what they are now. The age of reptiles, which lasted 160 million years, encompassing the Jurassic and Cretaceous periods of the Mesozoic era, saw the appearance of a great radiation of reptiles, many of huge stature and awesome appearance, that completely dominated life on land. Then they suddenly declined.

Of the dozen or so principal groups of reptiles that evolved, four remain today. The most successful of these are the lizards and snakes of the order Squamata. A second group is the crocodilians; having survived for 200 million years, they may finally be made extinct by humans. To a third group belong the turtles of the order Testudines, an ancient group that has somehow survived and remained mostly unchanged from its early reptile ancestors. The last group is a relic stock represented today by a sole surviving species, the tuatara of New Zealand.

Reptiles are easily distinguished from amphibians by several adaptations that permit them to live in arid regions and in the sea: habitats barred to amphibians by their reproductive requirements. Reptiles have a dry, scaly, virtually glandless skin that resists desiccation. Most important, reptiles lay their eggs on land; amphibians must lay their eggs in fresh water or in moist places. This seemingly simple difference was, in fact, a remarkable evolutionary achievement that was to have a profound impact on subsequent vertebrate evolution. To abandon totally an aquatic life, there evolved a sophisticated internally fertilized egg containing a complete set of life support systems. This **shelled egg** (known also as an **amniotic egg** because of the membranous amnion that encloses the embryo) could be laid on dry land (Figure 25-1). Within, the embryo floats and develops in an aquatic environment. It is provided with a yolk sac containing its food supply; another membrane, the allantois, serves as a surface for gas exchange through the calcareous or parchmentlike shell; the allantoic membrane also encloses a chamber for storing nitrogenous wastes that accumulate during development. The shelled egg allows a long developmental period and eliminates the vulnerable aquatic, feeding larval stage that is characteristic of amphibians.

Figure 25-1

Milk snake with its eggs. The shelled egg, provided with a self-contained aquatic environment, freed reptiles from having to find water in which to lay their eggs.
Photograph by L.L. Rue, III.

Figure 25-2

Evolution of the reptiles. The transition from certain labyrinthodont amphibians to reptiles occurred in the late Paleozoic to Mesozoic eras. This transition was effected by the development of an amniotic egg, which made land existence possible, although this egg may well have developed before the earliest reptiles had ventured far on land. Explosive radiation by the reptiles may have been due partly to the increased variety of ecological habitats into which they could move. The fossil record shows that lines arising from stem reptiles led to marine reptiles, dinosaurs, pterosaurs, crocodilians, turtles, lizards and snakes, and rhynchosaurs. Another radiation led to the mammals. Of this great assemblage, the only reptiles now in existence belong to four orders: Testudines, Crocodilia, Squamata, and Rhynchocephalia. How are the mighty fallen!

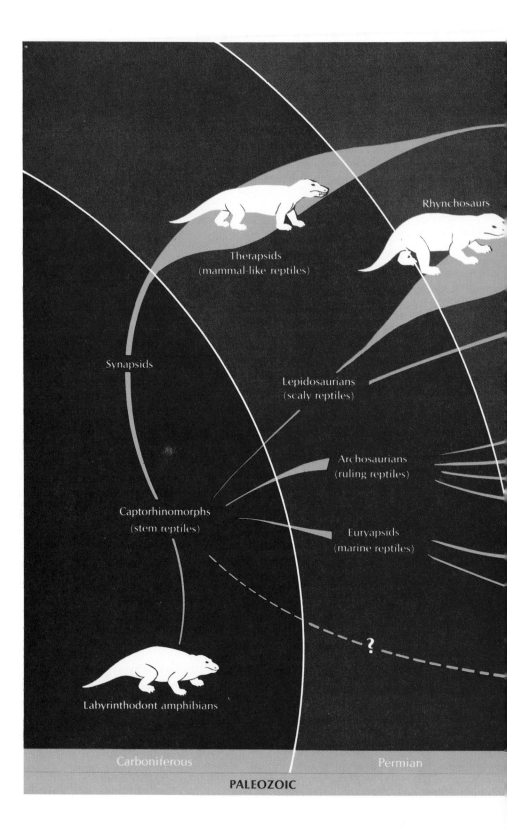

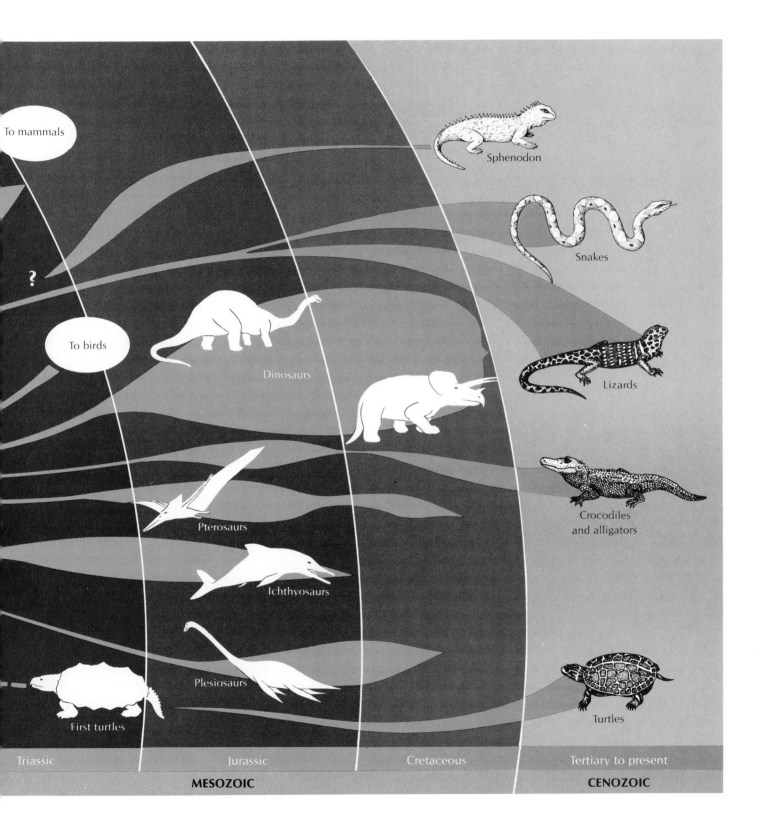

The early reptiles that developed this "land egg" must certainly have enjoyed an immediate advantage over the amphibians. They could hide their eggs in a protected location away from water—and away from the numerous aquatic creatures that fed freely on amphibian eggs each spring. With the evolution of this ultimate adaptation, conquest of land by the vertebrates was possible.

ORIGIN AND ADAPTIVE RADIATION OF REPTILES

Biologists generally agree that reptiles arose from the labyrinthodont amphibians sometime before the Permian period, which began approximately 280 million years ago. The oldest "stem reptiles" were the captorhinomorphs (Gr. *kaptō*, to gulp down, + *rhinos*, nose, + *morphē*, form) (Figure 25-2). These were small, lizardlike animals that probably fed mainly on insects, which were undergoing an adaptive radiation of their own at the same time and were already numerous. Early in their evolution, the captorhinomorphs rather quickly (in geological time sense) diverged into several specialized lineages and entered a long and spectacular evolutionary history that carried them to the present. Their descendants included the dinosaurs, marine reptiles, flying reptiles (pterosaurs [Gr. *pteron*, wing, + *sauros*, lizard]), and mammal-like reptiles (therapsids [Gr. *theraps*, attendant, + *ida*, suffix]), which produced the ancestral stock of the mammals. All were highly successful groups.

The adaptive radiation of reptiles, especially pronounced in the Triassic period (which followed the Permian period), corresponded to the appearance of new ecological habitats. These were provided by the climatic and geological changes that were taking place at that time, such as climatic variation from hot to cold, mountain building and terrain transformations, and a variety of plant life.

The Mesozoic era was the age of the great ruling reptiles. Then suddenly they disappeared near the close of the Cretaceous period approximately 65 to 80 million years ago. What caused their demise? The most exotic explanation is the recent hypothesis of L.W. Alvarez that the earth was struck with a huge asteroid several miles in diameter. The impact would have injected an enormous quantity of rock dust into the stratosphere where it would circle the globe for months, turning day into night. The absence of sunlight would have shut off photosynthesis, attacking food chains at their base. Both marine algae and land plants would have died, and all herbivorous and carnivorous animals would have become extinct except those feeding on insects or decaying vegetation. This hypothesis appears to be supported by the fossil record, which reveals that many groups of marine microorganisms that depended directly or indirectly on photosynthesis, as well as all animals larger than about 25 kg, became extinct at that time. But the Alvarez hypothesis, despite this and other supporting evidence, remains subject to debate among paleontologists.

Other changes were also taking place at that time. The ruling reptiles may not have been sufficiently adaptable to survive the combined effects of changing climate, the rapid spread of modern plants, and the appearance of aggressive and intelligent mammals. Competition from the mammals must have been particularly fierce. Yet some reptiles survived. Turtles had their protective shells; snakes and lizards evolved in habitats of dense forests and rocks where they could meet the competition of any tetrapod; and crocodiles, because of their size, stealth, and aggressiveness, had few enemies in their aquatic habitats.

Classical views of the dinosaurs and other Mesozoic reptiles as witless, cold-blooded beasts with inferior adaptability have recently been undergoing revision. Many biologists believe that the dinosaurs were alert creatures showing complex behavior equal to that of today's lizards and crocodilians. Some certainly traveled in organized family groups. Speculations drawn from ecological and paleontological data that the dinosaurs were warm blooded (endothermic) have generally been discounted, although because of their large size the dinosaurs probably enjoyed fairly stable body temperatures.

CHARACTERISTICS

1. Body varied in shape, compact in some, elongated in others; **body covered with an exoskeleton of horny epidermal scales** with the addition sometimes of bony dermal plates; **integument with few glands**
2. **Limbs paired, usually with five toes,** and adapted for climbing, running, or paddling; absent in snakes and some lizards
3. Skeleton well ossified; ribs with sternum forming a complete thoracic basket; **skull with one occipital condyle**
4. Respiration by lungs; **no gills;** cloaca used for respiration by some; branchial arches in embryonic life
5. **Three-chambered heart; crocodiles with four-chambered heart;** usually one pair of aortic arches
6. **Ectothermic;** some lizards and snakes behaviorally thermoregulate
7. **Metanephric kidney (paired); uric acid main nitrogenous waste**
8. Nervous system with the optic lobes on the dorsal side of brain; **12 pairs of cranial nerves** in addition to nervus terminalis
9. Sexes separate; **fertilization internal**
10. **Eggs covered with calcareous or leathery shells; extraembryonic membranes (amnion, chorion, yolk sac,** and **allantois)** present during embryonic life

CLASSIFICATION

Order Testudines (tes-tu´din-eez) (L. *testudo,* tortoise) (**Chelonia**): **turtles** (330 species). Body in a bony case of dermal plates with dorsal carapace and ventral plastron; jaws without teeth but with horny sheaths; quadrate immovable; vertebrae and ribs fused to shell; anus a longitudinal slit.

Order Squamata (squa-ma´ta) (L. *squamatus,* scaly, + *ata,* characterized by): snakes (2700 species), lizards (3000 species), amphisbaenids (130 species). Skin of horny epidermal scales or plates, which is shed; teeth attached to jaws; quadrate freely movable; vertebrae usually concave in front; anus a transverse slit.

    Suborder Sauria (sawr´e-a) (Gr. *sauros,* lizard) (**Lacertilia**): **lizards.** Body slender, usually with four limbs; rami of lower jaw fused; eyelids movable; copulatory organs paired.

    Suborder Serpentes (sur-pen´tes) (L. *serpere,* to creep) (**Ophidia**): **snakes.** Body elongate; limbs and ear openings absent; mandibles jointed anteriorly by ligaments; eyes lidless and immovable; tongue bifid and protrusible; teeth conical and on jaws and roof of mouth.

    Suborder Amphisbaenia (am´fis-bee´nee-a) (L. *amphis,* on both sides, + *baina,* to walk): **worm lizards.** Body elongate and of nearly uniform diameter; short tail; no legs (except one genus with short front legs); limb girdles vestigial; eyes hidden beneath skin; only one lung.

Order Crocodilia (croc´o-dil´e-a) (L. *crocodilus,* crocodile, + *ia,* pl. suffix) (**Loricata**): **crocodilians** (25 species). Four-chambered heart; vertebrae usually concave in front; forelimbs usually with five digits, hindlimbs with four digits; quadrate immovable; anus a longitudinal slit.

Order Rhynchocephalia (rin´ko-se-fay´le-a) (Gr. *rhynchos,* snout, + *kephalē,* head). Vertebrae biconcave; quadrate immovable; parietal eye fairly well developed and easily seen; anus a transverse slit. *Sphenodon* only species existing.

—— CHARACTERISTICS OF REPTILES THAT DISTINGUISH THEM FROM AMPHIBIANS

1. Reptiles have tough, dry, scaly skin offering protection against desiccation and physical injury. The skin consists of a thin **epidermis,** shed periodically, and a much thicker, well-developed **dermis.** The dermis is provided with **chromatophores,** the color-bearing cells that give many lizards and snakes their colorful hues. It is also the layer that, unfortunately for their bearers, is converted into alligator and snakeskin leather, so esteemed for expensive pocketbooks and shoes. The characteristic **scales** of reptiles are formed largely of keratin. They are derived mostly from the epidermis and thus are not homologous to fish scales, which are bony, dermal structures. In some reptiles, such as alligators, the scales remain throughout life, growing gradually to replace wear. In others, such as snakes and lizards, new scales grow beneath the old, which are then shed at intervals. Turtles add new layers of keratin under the old layers of the platelike scutes, which are modified scales. In snakes the old skin (epidermis and scales) is turned inside out when discarded; lizards split out of the old skin leaving it mostly intact and right side out, or it may slough off in pieces.

2. **The shelled egg of reptiles contains food and protective membranes for supporting embryonic development on dry land.** Reptiles lay their eggs in sheltered locations on land. The young hatch as lung-breathing juveniles rather than as aquatic larvae. The appearance of the shelled egg widened the division between the evolving amphibians and reptiles and, probably more than any other adaptation, contributed to the decline of amphibians and the ascendance of reptiles. The shelled egg is illustrated and described in some detail on pp. 785 and 786.

3. **The reptilian jaws are efficiently designed for applying crushing force to prey.** The jaws of fish and amphibians were designed for quick jaw closure, but once the prey was seized, little static force could be applied. In reptiles jaw muscles became larger, longer, and arranged for much better mechanical advantage.

4. **Reptiles have some form of copulatory organ, permitting internal fertilization.** Internal fertilization is obviously a requirement for a shelled egg, since the sperm must reach the egg before the egg is enclosed. Sperm from the paired testes are carried by the vasa deferentia to the copulatory organ, which is an evagination of the cloacal wall. The female system consists of paired ovaries and oviducts. The glandular walls of the oviducts secrete albumin and shells for the large eggs.

5. **Reptiles have a more efficient circulatory system and higher blood pressure than amphibians.** In all reptiles the right atrium, which receives unoxygenated blood from the body, is completely partitioned from the left atrium, which receives oxygenated blood from the lungs. In the crocodilians there are two com-

Figure 25-3

Internal structures of male crocodile.

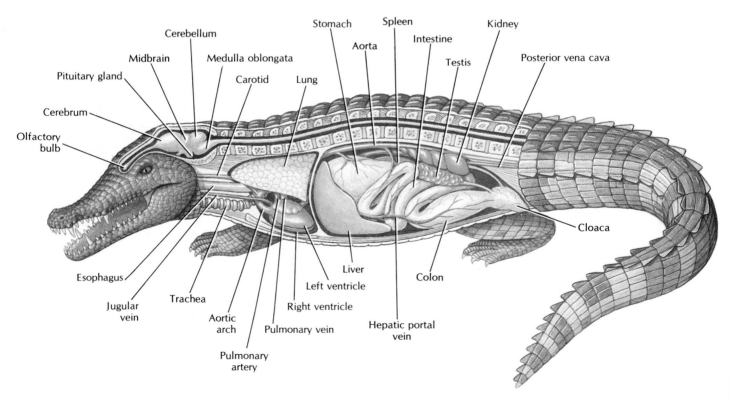

pletely separated ventricles as well (Figure 25-3); in other reptiles the ventricle is incompletely separated. The crocodilians are thus the first vertebrates with a four-chambered heart. Even in reptiles with incomplete separation of the ventricles, flow patterns within the heart prevent admixture of pulmonary (oxygenated) and systemic (unoxygenated) blood; all reptiles therefore have two functionally separate circulations.

6. **Reptile lungs are better developed than those of amphibians.** Reptiles depend almost exclusively on the lungs for gas exchange, supplemented by pharyngeal membrane respiration in some of the aquatic turtles. The lungs have a larger respiratory surface in reptiles than amphibians, and air is *sucked* into the lungs, as in higher vertebrates, rather than *forced* in by mouth muscles, as in the amphibians. Skin breathing, so important to most amphibians, has been completely abandoned by the reptiles.

7. **The reptilian kidneys are of the advanced metanephros type with their own passageways (ureters) to the exterior.** The kidneys are very efficient in producing small volumes of urine, thus conserving precious water. Nitrogenous wastes are excreted as uric acid, rather than urea or ammonia. Uric acid has a low solubility and precipitates out of solution readily; as a result the urine of many reptiles is a semisolid paste.

8. **All reptiles, except the limbless members, have better body support than the amphibians and more efficiently designed limbs for travel on land.** Many of the dinosaurs walked on powerful hindlimbs alone.

9. **The reptilian nervous system is considerably more advanced than the amphibian.** Although the reptile's brain is small, the **cerebrum** is increased in size relative to the rest of the brain. The crocodilians have the first true cerebral cortex (neopallium). Central nervous system connections are more advanced, permitting complex kinds of behavior unknown in the amphibians. Sense organs in general are well developed. All reptiles have a middle and an inner ear, but the sense of hearing is poorly developed in most. The lateral line system is entirely lost. Jacobson's organ, a unique sense organ, is a separate part of the nasal sac and communicates with the mouth; it is especially well developed in snakes and lizards. It is innervated by a branch of the olfactory nerve and is used in smelling the food in the mouth cavity. This organ in some form is found in other groups, including the amphibians.

CHARACTERISTICS AND NATURAL HISTORY OF REPTILIAN ORDERS
Turtles: Order Testudines (Chelonia)

The turtles are an ancient group that has plodded on from the Triassic period to the present with very little change in their early basic morphology. They are enclosed in shells consisting of a dorsal **carapace** (Sp. *carapacho*, covering) and a ventral **plastron** (Fr. breastplate). Clumsy and unlikely as they appear to be within their protective shells, they are nonetheless a varied and successful group that seems able to accommodate to human presence. The shell is so much a part of the animal that it is built in with the thoracic vertebrae and ribs. The shell is composed of two layers: an outer horny layer of keratin and an inner layer of bone. New layers of keratin are laid down beneath the old as the turtle grows and ages. Lacking teeth, the turtle jaw is provided with tough, horny plates for gripping and chewing food (Figure 25-4).

One consequence of living in a rigid shell with fused ribs is that a turtle

Figure 25-4

Snapping turtle, *Chelydra serpentina*, showing the absence of teeth. Instead, the jaw edges are covered with a horny plate. Photograph by C.P. Hickman, Jr.

The terms "turtle," "tortoise," and "terrapin" are applied variously to different members of the turtle order. In North American usage, they are all correctly called turtles. The term "tortoise" is frequently given to land turtles, especially the large forms. British usage of the terms is different: "tortoise" is the inclusive term, intertidal areas where the salinity fluctuates throughout the day; or those such as salmon, shad, and eels, that spend part of their life cycle in fresh water and part in seawater.

cannot expand its chest to breathe. Turtles solved this problem in their own way, by employing certain abdominal and pectoral muscles as a "diaphragm." Air is drawn in by contracting limb flank muscles to make the body cavity larger. Exhalation is also active and is accomplished by drawing the shoulder girdle back into the shell, thus compressing the viscera and forcing air out of the lungs. Many water turtles gain enough oxygen by just pumping water in and out of the mouth cavity; this enables them to remain submerged for long periods when inactive. When active they must lung-breathe more frequently.

A turtle's brain, like that of other reptiles, is small, never exceeding 1% of the body weight. The cerebrum, however, is larger than that of an amphibian, and turtles are able to learn a maze about as quickly as a rat. Turtles have both a middle and an inner ear, but sound perception is poor. Not unexpectedly, therefore, turtles are virtually mute (the biblical "voice of the turtle" refers to the turtledove), although many tortoises utter grunting or roaring sounds during mating (Figure 25-5). Compensating for poor hearing are a good sense of smell, acute vision, and color perception evidently as good as that of humans.

Turtles are oviparous. Fertilization is internal and all turtles, even the marine forms, bury their shelled, amniotic eggs in the ground. Usually considerable care is exercised in constructing the nest, but once the eggs are deposited and covered, the female deserts them.

The great marine turtles, buoyed by their aquatic environment, may reach 2 m in length and 725 kg in weight. One is the leatherback. The green turtle, so named because of its greenish body fat, may exceed 360 kg, although most individuals of this economically valuable and heavily exploited species seldom live long enough to reach anything approaching this size. Some land tortoises may weigh several hundred kilograms, such as the giant tortoises of the Galápagos Islands that so intrigued Darwin during his visit there in 1835. Most tortoises are rather slow moving; an hour of determined trudging carries a large Galápagos tortoise approximately 300 m. Their low metabolism probably explains their longevity, for some are believed to live more than 150 years.

Figure 25-5

Mating Galápagos tortoises. During copulation, males make loud grunting roars that may be heard for long distances.
Photograph by C.P. Hickman, Jr.

Turtles are the Methuselahs of vertebrates, and individuals of five species are known to have lived a century or more. However, except for longevity records based on specimens in captivity, reports of vast life spans are of dubious value; a box tortoise captured some years ago with "1850" carved on its plastron is an example of evidence that should be accepted with a healthy measure of skepticism.

The shell, like a medieval coat of armor, offers obvious advantages. The head and appendages can be drawn in for protection. The familiar box tortoise *(Terrapene carolina)* has a plastron that is hinged, forming two movable parts that can be pulled up against the carapace so tightly that one cannot force a knife blade between the shells. Some turtles, such as the large eastern snapping turtle *(Chelydra serpentina)*, have reduced shells, making complete withdrawal for protection quite impossible. Snappers, however, have another formidable defense, as their name implies (Figure 25-4). Ferocious and short tempered, they are often referred to as the "tigers of the pond." They are entirely carnivorous, living on fish, frogs, waterfowl, or almost anything that comes within reach of their powerful jaws. The alligator snapper lures unwary fish into its mouth with a "bait" (Figure 25-6). Snappers are wholly aquatic and come ashore only to lay their eggs.

Lizards, Snakes, and Worm Lizards: Order Squamata

The lizards and snakes of the order Squamata are the most recent products of reptile evolution and by all odds currently the most successful, comprising approximately 95% of all known living reptiles. The modern lizards began their adaptive radiation during the Cretaceous period of the Mesozoic era when the great dinosaurs were at the climax of their dominance of land. Lizards were successful because of their adaptability, which allowed them to occupy a variety of habitats and adopt diverse body forms and functional specializations.

Snakes appeared during the late Cretaceous period, probably from a group of lizards whose descendants include the Gila monster and monitor lizards, although the fossil record of snakes is poor. Legless species have evolved in other reptile groups that took up a burrowing existence. But none has refined this specialized life-style as have the snakes, which have radiated into terrestrial, aquatic, and arboreal niches. Two adaptations in particular characterize all snakes. One is the extreme elongation of the body and accompanying displacement and rearrangement of the internal organs. The other is the highly mobile jaw apparatus that enables snakes to swallow prey much larger than the snake's own diameter.

The order Squamata is divided into suborders Sauria (Lacertilia), the lizards; Serpentes (Ophidia), the snakes; and Amphisbaenia, the worm lizards. Structural features are even more reduced in the amphisbaenians than they are in snakes. Most amphisbaenians are limbless, and have both eyes and ears completely hidden beneath the skin.

Lizards: suborder Sauria

The lizards are an extremely diversified group, including terrestrial, burrowing, aquatic, arboreal, and aerial members. Among the more familiar groups in this varied suborder are the **geckos** (Figure 25-7), small, agile, mostly nocturnal forms with adhesive toe pads that enable them to walk upside down and on vertical surfaces; the **iguanas,** often brightly colored New World lizards with ornamental

Figure 25-6

Alligator snapping turtle *Macroclemys temmincki*, although larger than the eastern snapper, is less active—and for good reason. It lies on the bottom, mouth agape, luring its prey by undulating a pink, wormlike protrusion from its tongue. Photograph by P.C.H. Pritchard.

Figure 25-7

Tokay *(Gekko gecko)* of Southeast Asia may reach 35 cm (14 inches) in length and is the most aggressive of the geckos. This species has a true voice and is named after the strident, repeated *to-kay, to-kay* call. Photograph by J.H. Gerard.

Figure 25-8

Marine iguana *(Amblyrhynchus cristatus)* of the Galápagos Islands. This is the only marine lizard in the world.
Photograph by C.P. Hickman, Jr.

Figure 25-9

A chameleon snares a dragonfly. After cautiously edging close to its target, the chameleon suddenly lunges forward, anchoring its tail and feet to a branch. A split second later, it launches its sticky-tipped, foot-long tongue to trap the prey. The eyes of this common European chameleon *(Chamaeleo chamaeleon)* are swiveled forward to provide binocular vision and excellent depth perception.
Photograph by J. Andrada.

crests, frills, and throat fans, and a group that includes the remarkable marine iguana of the Galápagos Islands (Figure 25-8); the **skinks,** with elongate bodies and reduced limbs; and the **chameleons,** a group of arboreal lizards, mostly of Africa and Madagascar. The chameleons are entertaining creatures that catch insects with the sticky-tipped tongue that can be flicked accurately and rapidly to a distance greater than their own body length (Figure 25-9).

Unlike turtles, snakes, and crocodilians, which have distinctive body forms and ways of life, the lizards have radiated extensively into a variety of habitats and reveal a bewildering array of functional and behavioral specializations. The great majority of lizards have four limbs and relatively short bodies, but in many the limbs are degenerate, and a few such as the glass lizards (Figure 25-10) are completely limbless.

Despite their diversity of form, lizards as a group bear a number of characteristics that distinguish them from snakes. One is the lower jaw. Even in limbless forms the two halves are firmly united at the mandibular symphysis. This makes it impossible for lizards to swallow oversize meals as can snakes, which have "floating" jaw elements. Lizards have teeth, but in no species are they developed into

Figure 25-10

Glass lizard *(Ophisaurus* sp.) of the south-eastern United States. This legless lizard feels stiff and brittle to the touch and has an extremely long, fragile tail that readily fractures when the animal is struck or seized. Most specimens, such as this one, have only a partly regenerated tip to replace a much longer tail previously lost. Glass lizards can be readily distinguished from snakes by the deep, flexible groove running along each side of the body. They feed on worms, insects, spiders, birds' eggs, and small reptiles.
Photograph by L.L. Rue, III.

fangs, and only two species, the Gila monster and Mexican beaded lizard, are known to be venomous.

Most lizards have movable eyelids, whereas a snake's eyes are permanently covered with a transparent cap. Lizards have keen vision for daylight, although one group, the nocturnal geckos, has pure rod retinas. Most lizards have an external ear that snakes lack. The inner ear of lizards is variable in structure, but as with other reptiles, hearing does not play an important role in the lives of most lizards. Geckos are exceptions because the males are strongly vocal (to announce territory and discourage other males from approaching), and it is reasonable to assume that they can hear their own vocalizations. Other species of lizards vocalize in defensive behavior.

Many lizards live in the world's hot and arid regions, aided by several adaptations for desert life. Since their skin lacks glands, water loss by this avenue is much reduced. They produce a semisolid urine with a high content of crystalline uric acid. This is an excellent adaptation for conserving water and is found in other groups living successfully in arid habitats (birds, insects, and pulmonate snails). Some, such as the Gila monster of the southwestern United States deserts, store fat in their tails, which they draw on during drought to provide energy and metabolic water (Figure 25-11). The way many lizards keep their body temperature relatively constant by behavioral thermoregulation is described in Chapter 30.

Figure 25-11

Gila monster *(Heloderma suspectum)* of southwestern United States desert regions. This lizard and the congeneric Mexican beaded lizard are the only venomous lizards known. These brightly colored, clumsy-looking lizards feed principally on birds' eggs, nestling birds and mammals, and insects. Unlike poisonous snakes, the gila monster secretes venom from glands in its lower jaw. The bite is painful to humans but seldom fatal.
Photograph by L.L. Rue, III.

Figure 25-12

Black rat snake *(Elaphe obsoleta obsoleta)* swallowing a chipmunk.
Photograph by L.L. Rue, III.

Snakes: suborder Serpentes

Snakes are entirely limbless and lack both the pectoral and pelvic girdles (the latter persists as a vestige in pythons and boas). The numerous vertebrae of snakes, shorter and wider than those of tetrapods, permit quick lateral undulations through grass and over rough terrain. The ribs increase rigidity of the vertebral column, providing more resistance to lateral stresses. The elevation of the neural spine gives the numerous muscles more leverage.

The highly specialized skull and feeding apparatus of snakes, which enable them to eat prey several times their own diameter, are perhaps their most remarkable specialization. The two halves of the lower jaw (mandibles) are joined only by muscles and skin, allowing them to spread widely apart. Furthermore, many of the skull bones are so loosely articulated that the entire skull can flex asymmetrically to accommodate oversized prey (Figure 25-12). Since the snake must keep breathing during the slow process of swallowing, the tracheal opening is thrust forward between the two mandibles.

The cornea of the snake's eye is permanently protected with a transparent membrane, which, together with a lack of eyeball mobility, gives snakes the cold, unblinking stare that most people find so unnerving. Most snakes have relatively poor vision, with the tree-living snakes of the tropical forest being a conspicuous exception (Figure 25-13). Some arboreal snakes possess excellent binocular vision that helps them track prey through branches where scent trails would be impossible to follow.

It was mentioned earlier that snakes lack any superficial indication of an ear. This, together with the absence of any obvious response to aerial sounds, led to the widespread opinion that snakes are totally deaf. But snakes do have internal ears, and recent work has shown quite clearly that within a limited range of low frequencies (100 to 700 Hz), hearing in snakes compares favorably with that of most lizards. Snakes are also quite sensitive to vibrations carried in the ground.

Nevertheless, for most snakes it is the chemical senses and not vision and hearing that are employed to hunt their prey. In addition to the usual olfactory areas in the nose, which are not well developed, there are **Jacobson's organs,** a pair of pitlike organs in the roof of the mouth. These are lined with an olfactory epithelium and are richly innervated. The forked tongue, flicking through the air,

Figure 25-13

Parrot snake *(Leptophis ahaetulla).* The slender body of this Central American tree snake is an adaptation for sliding along branches without weighing them down.
Photograph by C.P. Hickman, Jr.

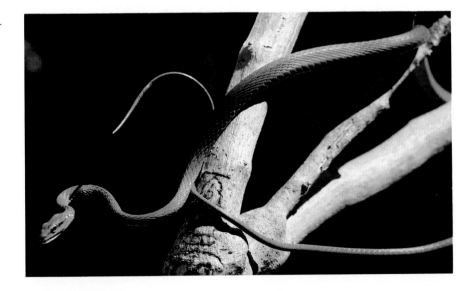

Figure 25-14

Timber rattlesnake *(Crotalus horridus)* flicks its tongue to smell its surroundings. Scent particles trapped on the tongue's surface are transferred to Jacobson's organs, olfactory organs in the roof of the mouth. Note the heat-sensitive pit organ between the nostril and eye.
Photograph by L.L. Rue, III.

picks up the scent particles and conveys them to the mouth; the tongue is then drawn past Jacobson's organs or the tips of the forked tongue are inserted directly into the organs (Figure 25-14). Information is then transmitted to the brain where scents are identified.

As with lizards, the snake's body is entirely covered with a tough, impervious skin. The skin is not elastic, and when required to stretch, as it must after the snake has enjoyed a hefty meal, it does so in a novel manner. The hard scales are set together, sometimes overlapping like shingles on a roof, with the skin folded inward between the scales. When the snake swallows a large object, the skin folds are pulled out straight, leaving the scales separated like islands on the skin.

Locomotion is an obvious problem for the legless animal, and snakes have discovered several solutions. The most typical pattern of movement is **lateral undulation** (Figure 25-15, *A*). Movement follows an S-shaped path, with the snake propelling itself by exerting lateral force against surface irregularities. The snake seems to "flow," since the moving loops appear stationary with respect to the ground. Lateral undulatory movement is fast and efficient under most but not all circumstances. **Concertina movement** (Figure 25-15, *B*) enables a snake to move in a narrow passage, as when climbing a tree by using the irregular channels in the bark. The snake extends forward while bracing S-shaped loops against the sides of the channel. To advance in a straight line as when stalking prey, many snakes use **rectilinear movement** (Figure 25-15, *C*), a form of locomotion that utilizes the large abdominal scutes. Two or three zones of scutes make frictional contact with the ground, while the rest of the body is moved forward. This effective but slow movement requires very loose, flexible skin and special muscle and

Figure 25-15

Snake locomotion. **A,** Lateral undulation. **B,** Concertina motion. **C,** Rectilinear motion.

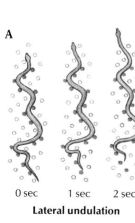
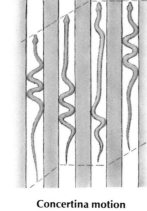

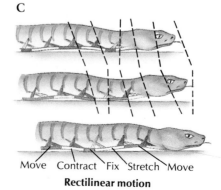

A

0 sec   1 sec   2 sec

Lateral undulation

B

Concertina motion

C

Move Contract Fix Stretch Move

Rectilinear motion

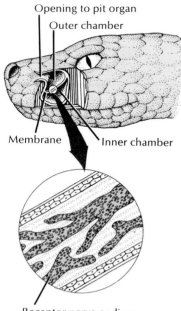

Figure 25-16

Pit organ of rattlesnake, a pit viper. Cutaway shows location of a deep membrane that divides the pit into inner and outer chambers. Heat-sensitive nerve endings are concentrated in the membrane.

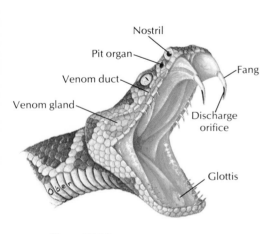

Figure 25-17

Head of rattlesnake showing the venom apparatus.

bone structure. **Sidewinding** is a fourth form of movement that enables desert vipers to move with surprising speed across loose, sandy surfaces with minimum surface contact. The sidewinder rattlesnake moves by throwing its body forward in loops with its body lying at an angle of about 60 degrees to its direction of travel.

Snakes of the subfamily Crotalinae within the family Viperidae are called **pit vipers** because of special heat-sensitive pits on their heads, between the nostrils and the eyes (Figures 25-14 and 25-16). All of the best-known North American poisonous snakes are pit vipers, such as the several species of rattlesnakes, the water moccasin and the copperhead. The pits are supplied with a dense packing of free nerve endings from the fifth cranial nerve. They respond to radiant energy in the long-wave infrared (5000 to 15,000 nm) and are especially sensitive to the heat emitted by the warm-bodied birds and mammals that comprise their food (infrared wavelengths of about 10,000 nm). Some measurements suggest that the pit organs can distinguish temperature differences of only 0.003° C from a radiating surface. Pit vipers use the pit organs to track warm-blooded prey and to aim strikes with great accuracy, as effectively in total darkness as in daylight. Boa constrictors and pythons also have heat receptors, but the anatomy is quite different from that of pit vipers, suggesting that they probably evolved independently.

All vipers have a pair of teeth on the maxillary bones modified as fangs. These lie in a membrane sheath when the mouth is closed. When the viper strikes, a special muscle and bone lever system erects the fangs when the mouth opens (Figure 25-17). The fangs are driven into the prey by the thrust and venom is injected into the wound through a channel in the fangs. A viper immediately releases its prey after the bite and waits until it is paralyzed or dead. Then the snake swallows the prey whole.

Approximately 8000 bites and 12 deaths from pit vipers are reported each year in the United States.

The tropical and subtropical countries are the homes of most species of snakes, both of the venomous and nonvenomous varieties. Even there, less than one third of snakes are venomous. The nonvenomous snakes kill their prey by constriction (Figure 25-18) or by biting and swallowing. Their diet tends to be restricted, many feeding principally on rodents, whereas others feed on fishes, frogs, and insects. Some African, Indian, and Neotropical snakes are egg eaters.

Poisonous snakes are usually divided into four groups based on their type of fangs. The vipers (family Viperidae) have highly developed tubular fangs at the front of the mouth; the group includes the American pit vipers previously mentioned and the Old World true vipers, which lack facial heat-sensing pits. Among the latter are the common European adder and the African puff adder. A second family of poisonous snakes (family Elapidae) has short, permanently erect fangs so that the venom must be injected by chewing. In this group are the cobras (Figure 25-19), mambas, coral snakes, and kraits. The highly poisonous sea snakes are usually placed in a third family (Hydrophiidae). The very large family Colubridae, which contains most of the familiar (and nonvenomous) snakes, does include at least two poisonous snakes that have been responsible for human fatalities: the African boomslang and the African twig snake. Both are rear-fanged snakes that normally use their venom to quiet struggling prey.

The saliva of all harmless snakes possesses limited toxic qualities, and it is logical that evolution should have stressed this toxic tendency. Snake venoms have

Figure 25-18

Nonvenomous African house snake
(Boaedon fuliginosus) constricting a mouse
before swallowing it.
Photograph by C.P. Hickman, Jr.

traditionally been divided into two types. The neurotoxic type acts mainly on the nervous system, affecting the optic nerves (causing blindness) or the phrenic nerve of the diaphragm (causing paralysis of respiration). The hemolytic type breaks down the red blood corpuscles and blood vessels and produces extensive extravasation of blood into the tissue spaces. In fact, most snake venoms are complex mixtures of various fractions that attack different organs in specific ways; they seldom can be assigned categorically to one or the other of the traditional types.

The toxicity of a venom is determined by the median lethal dose on laboratory animals (LD_{50}). By this standard the venoms of the Australian tiger snake and some of the sea snakes appear to be the most deadly of poisons drop for drop. However, several larger snakes are more dangerous. The aggressive king cobra, which may exceed 5.5 m in length, is the largest and probably the most dangerous of all poisonous snakes. In India, where snakes come in constant contact with people, some 200,000 snakebites cause more than 9000 deaths each year.

Most snakes are **oviparous** (L. *ovum*, egg, + *parere*, to bring forth) species that lay their shelled, elliptical eggs beneath rotten logs, under rocks, or in holes dug in the ground (Figure 25-1). Most of the remainder, including all the American pit vipers, except the tropical bushmaster, are **ovoviviparous** (L. *ovum*, egg, + *vivus*, living, + *parere*, to bring forth), giving birth to well-formed young. Very few snakes are **viviparous** (L. *vivus*, living, + *parere*, to bring forth); a primitive placenta forms, permitting the exchange of materials between the embryonic and maternal bloodstreams. Snakes are able to store sperm and can lay several clutches of fertile eggs at long intervals after one mating.

Worm lizards: suborder Amphisbaenia

The somewhat inappropriate common name "worm lizards" describes a group of highly specialized, burrowing forms that are neither worms nor true lizards but certainly are related to the latter. The name of the suborder literally means "walk on both sides" (or both ends), in reference to their peculiar ability to move backward nearly as effectively as forward. They have elongate, cylindrical bodies of nearly uniform diameter, and most lack any trace of external limbs (Figure 25-20). The soft skin is divided into numerous rings, and these rings, combined with the absence of visible eyes and ears (both are hidden under the skin), make the amphisbaenians look like earthworms. The resemblance, although superficial, is the kind of structural convergence that often occurs when two totally unrelated

Figure 25-19

Yellow cobra *(Naja flava)* of Africa. Cobras
erect the front of the body and flatten the
neck as a threat display before attacking.
All cobras are extremely poisonous.
Photograph by L.L. Rue, III.

The LD_{50} (median lethal dose) is a standardized procedure originally developed by pharmacologists for assaying the toxicity of drugs. In practice, small samples of laboratory animals, usually mice, are exposed to a graded series of doses of the drug or toxin. The dose that kills 50% of the animals in the test period is recorded as the LD_{50}.

Figure 25-20

A worm lizard of the suborder Amphisbaenia. Amphisbaenians are burrowing forms having a curious mixture of lizardlike and snakelike characteristics. The species pictured, *Amphisbaena alba,* is widely distributed in South America.

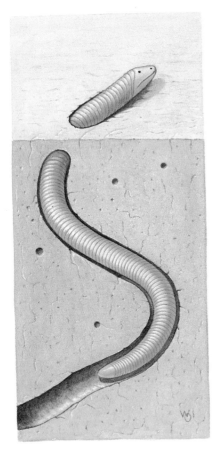

groups come to occupy similar habitats. The amphisbaenians have an extensive distribution in South America and tropical Africa. In the United States, one species, *Rhineura florida,* is found in Florida where it is known as the "graveyard snake."

___ Crocodiles and Alligators: Order Crocodilia

The modern crocodiles are the largest living reptiles. They are what remain of a group once abundant in the Jurassic and Cretaceous periods. Having managed to survive virtually unchanged for some 160 million years, the modern crocodilians face a forbidding, and perhaps short, future in a world dominated by humans.

Most crocodiles have relatively long slender snouts; alligators (Sp. *el lagarto,* lizard) have shorter and broader snouts. With their powerful jaws and sharp teeth, they are formidable antagonists. The "man-eating" members of the group are found mainly in Africa and Asia. The estuarine crocodile (*Crocodylus porosus*), found in southern Asia, and the Nile crocodile (*C. niloticus*) (Figure 25-21) grow to great size (adults weighing 1000 kg have been reported) and are much feared. Both species are swift and aggressive, eating any bird or mammal they can drag from the shore to water where, by rapidly turning over and over, they violently tear the prey apart. They then gulp down the pieces. Crocodiles are known to attack animals as large as cattle, deer, and people.

Alligators (Figure 25-22) are less aggressive than crocodiles and certainly far less dangerous to humans. Large alligators are powerful animals nevertheless, and adults have almost no enemies but humans. The chink in their formidable armor is the developmental stages. Nests left unguarded by the mother are almost certain to be discovered and raided by any of several mammals that relish eggs, and the young hatchlings may be devoured by large fish.

Alligators and crocodiles are oviparous. Usually 20 to 50 eggs are laid in a mass of vegetation and guarded by the mother. The mother hears vocalizations from the hatching young and responds by opening the nest to allow the hatchlings to escape. An odd feature about alligator reproduction is that the incubation temperature of the eggs determines the sex ratio of the offspring. Low nest temperatures produce only females, whereas high nest temperatures produce only males. This results in highly unbalanced sex ratios in some areas. For example, in one Louisiana study area, female hatchlings outnumbered males five to one.

Figure 25-21

Nile crocodile *(Crocodylus niloticus)* basking. Note the slender snout and the lower jaw tooth that fits *outside* the upper jaw. Alligators lack this feature.
Photograph by C.P. Hickman, Jr.

Figure 25-22

American alligators were nearing extinction in the southern United States, but now, under protection, they are increasing to the point of nuisance (Figure 25-22).

The Tuatara: Order Rhynchocephalia

The order Rhynchocephalia is represented by a single living species, the tuatara (*Sphenodon punctatum* [Gr. *sphenos*, wedge, + *odontos*, tooth]) of New Zealand (Figure 25-23). This animal is the sole survivor of a group of primitive reptiles that otherwise became extinct 100 million years ago. The tuatara was once widespread on the North Island of New Zealand but is now restricted to islets of Cook Strait and off the northern coast of North Island where, under protection from the New Zealand government, it may recover.

The tuatara is a lizardlike form 66 cm long or less that lives in burrows often shared with petrels. They are slow-growing animals with a long life; one is recorded to have lived 77 years.

The tuatara has captured the interest of biologists because of its numerous primitive features that are almost identical to those of Mesozoic fossils 200 million years old. These features include a primitive skull structure found in early Permian reptiles that were ancestors to the modern lizards. Tuataras also bear a well-developed parietal eye, complete with evidences of a retina (Figure 25-23), and a complete palate. They lack a copulatory organ. A specialized feature is the teeth, which are fused wedgelike to the edge of the jaws rather than being set in sockets. *Sphenodon* represents one of the slowest rates of evolution known among the vertebrates.

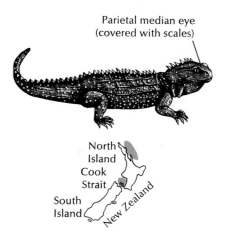

Parietal median eye (covered with scales)

North Island
Cook Strait
South Island
New Zealand

Figure 25-23

Tuatara *(Sphenodon punctatum)*, the only living representative of order Rhynchocephalia. This "living fossil" reptile has a well-developed parietal "eye" with retina and lens on top of the head. The eye is covered with scales and is considered nonfunctional but may have been an important sense organ in early reptiles. The tuatara is found only in New Zealand.

SUMMARY

The reptiles evolved from labyrinthodont amphibians during the late Paleozoic era, some 300 million years ago. Their success as the first truly terrestrial vertebrates is attributed to the evolution of the shelled, amniotic egg, whose internal membranes provide the developing embryo with its own independent water and food supplies. Reptiles are also distinguished from amphibians by their dry, scaly skin that prevents water loss; more powerful jaws; internal fertilization; and more advanced circulatory, respiratory, excretory, and nervous systems. Like amphibians, reptiles are ectotherms, but most exercise considerable behavioral control over their body temperature.

The modern reptiles are descendants of survivors of the Mesozoic age of reptiles when a burst of reptile evolution produced a worldwide fauna of great diversity. The turtles (order Testudines) with their distinctive shells have changed little in design since the Triassic period. Turtles are a small group of long-lived terrestrial, semiaquatic, aquatic, and marine species. They lack teeth. All are oviparous and all, including the marine forms, bury their eggs.

The lizards, snakes, and worm lizards (order Squamata) comprise 95% of all living reptiles. Lizards are a diversified and successful group adapted for walking, running, climbing, swimming, and burrowing. They are distinguished from snakes by typically having two pairs of legs (some species are legless), united lower jaw halves, movable eyelids, external ears, and absence of fangs. Many lizards are well adapted for survival under hot and arid desert conditions.

Snakes, in addition to being entirely limbless, are characterized by their elongate bodies and an elastic connection between the two halves of the lower jaw that permits the halves of the jaw to spread widely during swallowing. Most snakes rely on the chemical senses, especially Jacobson's organs, to hunt prey, rather than on the weakly developed visual and auditory senses. Two groups of snakes (pit vipers and boids) have unique infrared-sensing organs for tracking warm-bodied prey. Many snakes are venomous. The worm lizards (amphisbaenians) are a small tropical group of legless, burrowing forms with both eyes and ears hidden beneath the skin.

The crocodiles and alligators (order Crocodilia) are the largest living reptiles and have the most complex social behavior. Most are endangered species today.

The tuatara of New Zealand (order Rhynchocephalia) is a relict species and sole survivor of a group that otherwise disappeared 100 million years ago. It bears several primitive features that are almost identical to those of Mesozoic fossil reptiles.

Review questions

1. Name the major groups of reptiles, both extinct and living, that descended from the captorhinomorphs (stem reptiles). (Refer to both Figure 25-2 and the text.)
2. At the end of the Mesozoic era the ruling reptiles disappeared along with several invertebrate groups and several lineages of plants. What may have been responsible for these extinctions?
3. Describe at least six ways in which the reptiles are more advanced functionally or structurally than the amphibians.
4. Many zoologists believe that the evolution of the shelled (amniotic) egg was the single most important adaptation contributing to the great success of the reptiles. Describe this egg and explain its importance.
5. Describe the principal characteristics of turtles (order Testudines) that would distinguish them from any other reptile order.
6. In what ways are lizards and snakes (order Squamata) better adapted to hot and arid desert conditions than amphibians?
7. Name three anatomical characteristics of snakes that distinguish them from any lizard (remember that some lizards are legless).
8. What is the function of Jacobson's organ of snakes?
9. What is the function of the "pit" of pit vipers?
10. What distinguishes the three forms of snake movement known as lateral undulation, concertina, and rectilinear?

11. What is the difference in the structure or location of the fangs of a rattlesnake, a cobra, and an African boomslang?
12. What is meant by the abbreviation LD_{50}?
13. Most snakes are oviparous, but some are ovoviviparous or viviparous. What do these terms mean and what would you have to know to be able to assign a particular snake to one of these reproductive modes?
14. How do crocodiles and alligators differ from each other?
15. Why is the tuatara (order Rhynchocephalia) of special interest to biologists? Where would you have to go to see one in its natural habitat?

Selected references

See also general references for Part Two, p. 590.

Bellairs, A., and J. Attridge. 1975. Reptiles. London, Hutchinson University Library. *Paperback, well-written review of evolution, structure, and function of reptiles.*

Gibbons, W. 1983. Their blood runs cold: adventures with reptiles and amphibians. University, Ala., University of Alabama Press. *Lots of interesting reptile lore in this engaging book.*

Goin, C.J., O.B. Goin, and G.R. Zug. 1978. Introduction to herpetology, ed. 3. San Francisco, W.H. Freeman & Co. Publishers. *Basic college-level textbook.*

Halliday, T.R., and K. Adler (eds.). 1986. The encyclopedia of reptiles and amphibians. New York, Facts on File Inc. *Excellent authoritative reference work with high-quality illustrations. Highly recommended.*

Russel, D.A. 1982. The mass extinctions of the late Mesozoic. Sci. Am. **246**:58-65 (Jan.). *Details the evidence that the catastrophic disruption of the biosphere and the extinction of the dinosaurs at the close of the Mesozoic era were caused by the impact on earth of a large asteroid.*

CHAPTER 26

THE BIRDS

Phylum Chordata

Class Aves

Of the vertebrates, birds of the class Aves (ay'veez) (L. pl. of *avis,* bird) are the most studied, the most observable, the most melodious, and many think the most beautiful. With 8600 species distributed over nearly the entire earth, birds far outnumber all other vertebrates except the fishes. Birds are found in forests and deserts, in mountains and prairies, and on all oceans. Four species are known to have visited the North Pole, and one, a skua, was seen at the South Pole. Some birds live in total blackness in caves, finding their way about by echolocation, and others dive to depths greater than 45 m to prey on aquatic life. The "bee" hummingbird of Cuba, weighing in at only 1.8 g, is the smallest vertebrate endotherm.

The single unique feature that distinguishes birds from other animals is their feathers. If an animal has feathers, it is a bird; if it lacks feathers, it is not a bird. No other vertebrate group bears such an easily recognizable and foolproof identification tag.

There is great uniformity of structure among birds. Despite approximately 130 million years of evolution, during which they proliferated and adapted themselves to specialized ways of life, we have no difficulty recognizing a bird as a bird. In addition to feathers, all birds have forelimbs modified into wings (although they may not be used for flight); all have hindlimbs adapted for walking, swimming, or perching; all have horny beaks; and all lay eggs. Probably the reason for this great structural and functional uniformity is that birds evolved into flying machines. This fact greatly restricts diversity, so much more evident in other vertebrate classes. For example, birds do not begin to approach the diversity seen in their warm-blooded evolutionary peers, the mammals, a group that includes forms as unlike as a whale, porcupine, bat, and giraffe.

Birds share with mammals the highest organ system development in the animal kingdom. But a bird's entire anatomy is designed around flight and its perfection. An airborne life for a large vertebrate is a highly demanding evolutionary challenge. A bird must, of course, have wings for support and propulsion. Bones must be light and hollow yet serve as a rigid airframe. The respiratory system must be highly efficient to meet the intense metabolic demands of flight and serve also as a thermoregulatory device to maintain a constant body temperature. A bird must have a rapid and efficient digestive system to process an energy-rich diet; it must have a high metabolic rate; and it must have a high-

A

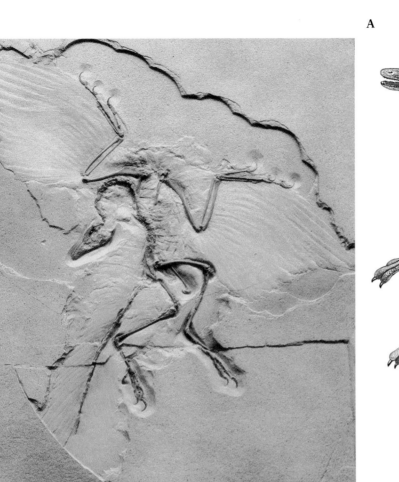

B

Figure 26-1

Archaeopteryx, the 150 million year old ancestor of modern birds. **A,** Cast of the second and most nearly perfect fossil of *Archaeopteryx,* which was discovered in a Bavarian stone quarry. **B,** Reconstruction of *Archaeopteryx.*

A, Courtesy American Museum of Natural History.

pressure circulatory system. Above all, birds must have a finely tuned nervous system and acute senses, especially superb vision, to handle the complex problems of headfirst, high-velocity flight.

ORIGIN AND RELATIONSHIPS

Approximately 150 million years ago, a flying animal drowned and settled to the bottom of a tropical lake in what is now Bavaria, Germany. It was rapidly covered with a fine silt and eventually fossilized. There it remained until discovered in 1861 by a workman splitting slate in a limestone quarry. The fossil was approximately the size of a crow, with a skull not unlike that of modern birds except that the beaklike jaws bore small bony teeth set in sockets like those of reptiles (Figure 26-1). The skeleton was decidedly reptilian with a long bony tail, clawed fingers, and abdominal ribs. It might have been classified as a reptile except that it carried the unmistakable imprint of **feathers,** those marvels of biological engineering that only birds possess. The finding was dramatic because it proved beyond reasonable doubt that birds had evolved from reptiles.

Archaeopteryx (ar-kee-op′ter-ix, meaning "ancient wing"), as the fossil was named, was an especially fortunate discovery because the fossil record of birds is disappointingly meager. The bones of birds are lightweight and quickly disintegrate, so that only under the most favorable conditions will they fossilize. Nev-

Figure 26-2

One of the strangest birds in a strange land, the flightless cormorant of the Galápagos Islands (*Nannopterum harrisi*) dries its wings after a fishing forage. It is a superb swimmer, propelling itself through the water with its feet to catch fishes and octopuses. The flightless cormorant is an example of a carinate bird (having a keeled sternum) that has lost the keel and the ability to fly.

Photograph by C.P. Hickman, Jr.

The bodies of flightless birds are dramatically redesigned. All of the restrictions of flight are removed. The keel of the sternum is lost, and the heavy flight muscles (as much as 17% of the body weight of flying birds), as well as other specialized flight apparatus, disappear. Since body weight is no longer a restriction, flightless birds tend to become large. Several extinct flightless birds were enormous: the giant moas of New Zealand weighed more than 225 kg (500 pounds) and the elephantbird of Madagascar, the largest bird that ever lived, probably weighed nearly 450 kg (about 1000 pounds) and stood nearly 2 m tall.

ertheless, there are certain localities where bird fossils are relatively abundant. One of these is the famous Rancho La Brea tarpits in Los Angeles where in one pit alone were found 30,000 fossil birds representing 81 species. By 1952 over 780 different fossil species had been recorded. Although most of these are relatively recent fossils, enough intermediate forms are known to provide a reasonable picture of bird evolution from the Jurassic period, when *Archaeopteryx* lived, to recent times (Figure 26-3). Two well-known fossil birds in particular deserve mention. One was *Ichthyornis* (ik-thee-or'nis), a small, ternlike sea bird that lived along the shores of North America's inland sea during the Cretaceous period about 100 million years ago, 50 million years after *Archaeopteryx*. The other was *Hesperornis*, a flightless, loonlike diving bird (Figure 26-3). Both were essentially modern birds but still retained certain reptilian features, including toothed jaws and a reptilelike jaw articulation. Nevertheless, by the close of the Cretaceous period, about 63 million years ago, the characteristics of modern birds had been thoroughly molded. There remained only the emergence and proliferation of the modern orders of birds. Hundreds of thousands of bird species have appeared and nearly as many have disappeared, following *Archaeopteryx* to extinction. Only a minute fraction of these nameless species have been discovered as fossils.

Most paleontologists agree that the ancestors of both birds and dinosaurs were derived from a stem group of archosaurian reptiles called thecodonts (Figure 25-2, pp. 516 and 517). Birds probably have a monophyletic origin (evolved from a single ancestor). However, existing birds are divided into two broad groups: (1) **ratite** (rat'ite) (L. *ratitus*, marked like a raft, from *ratis*, raft), the large flightless birds (ostriches, emus, rheas, cassowaries) and the kiwis, which have a flat sternum with poorly developed pectoral muscles, and (2) carinate (L. *carina*, keel), the flying birds that have a keeled sternum on which the powerful flight muscles insert. This division originated from the view that the flightless birds (ostrich, emu, kiwi, rhea) represented a separate line of descent that never attained flight. This idea is now completely rejected. The ostrichlike ratites clearly have descended from flying ancestors. Furthermore, not all carinate, or keeled, birds can fly and many of them even lack keels (Figure 26-2). Flightlessness has appeared independently among many groups of birds; the fossil record reveals flightless wrens, pigeons, parrots, cranes, ducks, auks, and even a flightless owl. Penguins are flightless although they use their wings to "fly" through the water. Flightlessness has almost always evolved on islands where few terrestrial predators are found. The flightless birds living on continents today are the large ratites (ostrich, rhea, cassowary, emu), which can run fast enough to escape predators. The ostrich can run 70 km (42 miles) per hour, and claims of speeds of 96 km (60 miles) per hour have been made.

It may seem paradoxical that birds with their agile, warm-blooded, colorful, and melodious way of life should have descended from lethargic, cold-blooded, and silent reptiles. Yet the numerous anatomical affinities of the two groups are abundant evidence of close kinship and led the great English zoologist Thomas Henry Huxley (p. 854) to call birds merely "glorified reptiles." This unflattering description causes bird lovers to answer, "But how wondrously glorified."

Ancient and modern birds are divided into two subclasses. (A complete classification of the orders of living birds appears on pp. 559 to 561.)

Subclass Archaeornithes (ar'ke-or'ni-theez) (Gr. *archaios*, ancient, + ornis, ornithos, bird). Fossil birds. This included *Archaeopteryx* and possibly one or two other genera.

Subclass Neornithes (ne-or'ni-theez) (Gr. *neos*, new, + ornis, bird). Modern birds are placed in this group. Some extinct species with teeth are also included here because of their likeness to modern forms.

CHARACTERISTICS

1. Body usually spindle shaped, with four divisions: head, neck, trunk, and tail; **neck disproportionately long** for balancing and food gathering
2. Limbs paired with the **forelimbs usually adapted for flying;** posterior pair variously adapted for perching, walking, and swimming; foot with four toes (chiefly)
3. Epidermal **covering of feathers** and **leg scales;** thin integument of epidermis and dermis; no sweat glands; oil or preen gland at root of tail; **pinna of ear rudimentary**
4. **Skeleton fully ossified with air cavities;** skull bones fused with **one occipital condyle;** jaws covered with **horny beaks;** small ribs; vertebrae tend to fuse, especially the terminal ones; sternum well developed with keel or reduced with no keel; **no teeth**
5. Nervous system well developed, with brain and 12 pairs of cranial nerves
6. Circulatory system of **four-chambered heart,** with the **right aortic arch persisting;** reduced renal portal system; nucleated red blood cells
7. Endothermic
8. Respiration by slightly expansible lungs, with thin **air sacs** among the visceral organs and skeleton; **syrinx** (voice box) near junction of trachea and bronchi
9. Excretory system of metanephric kidney; ureters open into cloaca; **no bladder;** semisolid urine; uric acid main nitrogenous waste
10. Sexes separate; testes paired, with the vas deferens opening into the cloaca; **females with left ovary and oviduct only;** copulatory organ in ducks, geese, ratites, and a few others
11. Fertilization internal; **amniotic eggs with much yolk and hard calcareous shells;** embryonic membranes in egg during development; **incubation external;** young active at hatching (**precocial**) or helpless and naked (**altricial**); sex determination by females (heterogametic)

FORM AND FUNCTION

Just as an airplane must be designed and built according to rigid aerodynamic specifications if it is to fly, so too must birds meet stringent structural requirements if they are to stay airborne. All the special adaptations found in flying birds contribute to two things: more power and less weight. Flight by humans became possible when they developed an internal combustion engine and learned how to reduce the weight-to-power ratio to a critical point. Birds did this millions of years ago. But birds must do much more than fly. They must feed themselves and convert food into high-energy fuel; they must escape predators; they must be able to repair their own injuries; they must be able to air-condition themselves when overheated and heat themselves when too cool; and, most important of all, they must reproduce themselves.

Feathers

A feather is very lightweight, yet possesses remarkable toughness and tensile strength. A typical **contour feather** consists of a hollow **quill,** or calamus, thrust into the skin and a **shaft,** or rachis, which is a continuation of the quill and bears numerous **barbs** (Figure 26-4). The barbs are arranged in a closely parallel fashion and spread diagonally outward from both sides of the central shaft to form a flat, expansive, webbed surface, the **vane.** There may be several hundred barbs in the vane.

If the feather is examined with a microscope, each barb appears to be a miniature replica of the feather with numerous parallel filaments called **barbules** set in each side of the barb and spreading laterally from it. There may be 600 barbules on each side of a barb, adding up to more than 1 million barbules for the feather. The barbules of one barb overlap the barbules of a neighboring barb in a herringbone pattern and are held together with great tenacity by tiny hooks. Should two adjoining barbs become separated—and considerable force is needed

The vivid color of feathers is of two kinds: pigmentary and structural. Red, orange, and yellow feathers are colored by pigments, called lipochromes, deposited in the feather barbules as they are formed. Black, brown, red-brown, and gray colors are from a different pigment, melanin. The blue feathers of blue jays, indigo buntings, and bluebirds depend not on pigment but on the scattering of shorter wavelengths of light by particles within the feather; these are structural colors. Blue feathers are usually underlain by melanin, which absorbs certain wavelengths, thus intensifying the blue. Such feathers look the same from any angle of view. Green colors are almost always a combination of yellow pigment and blue feather structure. Another kind of structural color is the beautiful iridescent color of many birds, which ranges from red, orange, copper, and gold to green, blue, and violet. Iridescent color is based on interference that causes light waves to reinforce, weaken, or eliminate each other. Iridescent colors may change with the angle of view; the quetzal, for example, looks blue from one angle and green from another. In the animal kingdom, only tropical reef fishes can vie with birds for intensity and vividness of color.

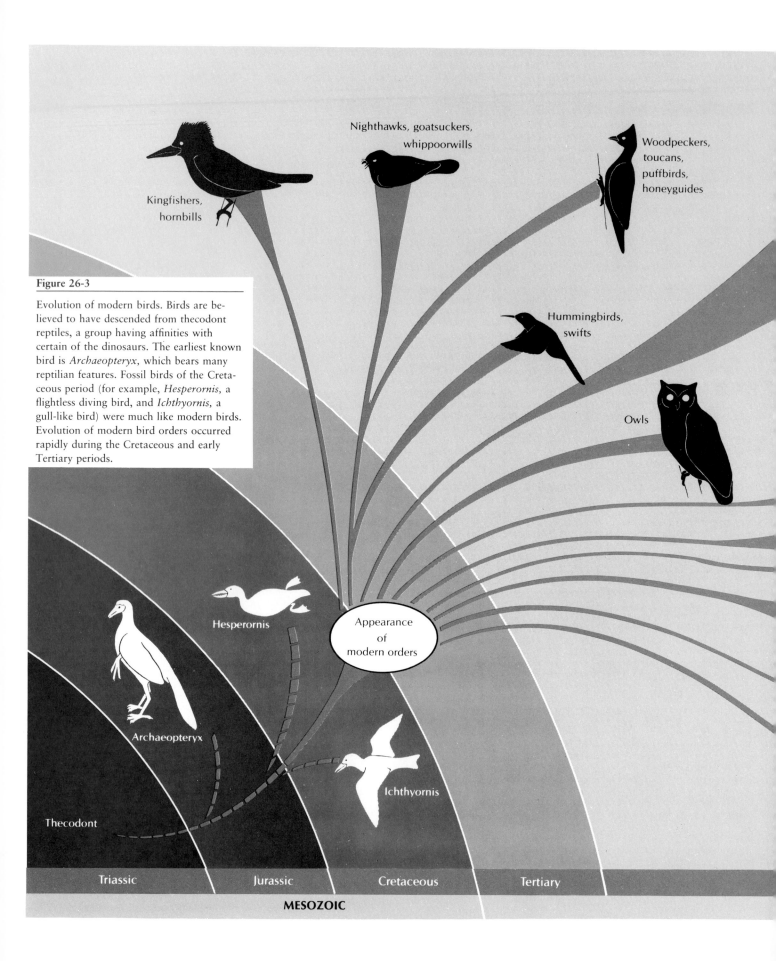

Figure 26-3

Evolution of modern birds. Birds are believed to have descended from thecodont reptiles, a group having affinities with certain of the dinosaurs. The earliest known bird is *Archaeopteryx*, which bears many reptilian features. Fossil birds of the Cretaceous period (for example, *Hesperornis*, a flightless diving bird, and *Ichthyornis*, a gull-like bird) were much like modern birds. Evolution of modern bird orders occurred rapidly during the Cretaceous and early Tertiary periods.

Kingfishers, hornbills

Nighthawks, goatsuckers, whippoorwills

Woodpeckers, toucans, puffbirds, honeyguides

Hummingbirds, swifts

Owls

Hesperornis

Appearance of modern orders

Archaeopteryx

Ichthyornis

Thecodont

Triassic   Jurassic   Cretaceous   Tertiary

MESOZOIC

Perching songbirds

Cuckoos, roadrunners

Cranes, rails, coots, bustards

Pheasants, quails, turkeys, domestic fowl

Pigeons, doves

Parrots, parakeets

Loons

Gulls, terns, snipe, sandpipers, plovers, auks, puffins

Hawks, eagles, falcons, condors, buzzards

Pelicans, gannets, boobies, frigates, cormorants

Herons, storks, ibises, flamingos

Swans, geese, ducks

Albatrosses, petrels, fulmars, shearwaters

Flightless birds

Penguins

Quaternary

CENOZOIC

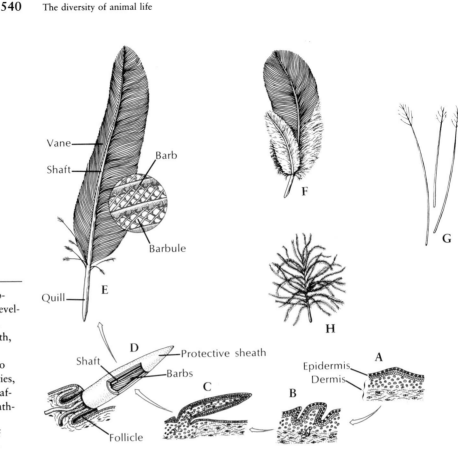

Figure 26-4

Types of bird feathers and their development. **A** to **E,** Successive stages in the development of a vaned, or contour, feather. Growth occurs within a protective sheath, **D,** that splits open when growth is complete, allowing the mature feather to spread flat. **F** to **H,** Other feather varieties, including a pheasant vane feather with aftershaft, **F,** filoplumes, **G,** and down feather, **H.**

Principally after Welty, J.C. 1982. The life of birds, ed. 3. Philadelphia, W.B. Saunders Co.

to pull the vane apart—they are instantly zipped together again by drawing the feather through the fingertips. The bird, of course, does this with its bill, and much of a bird's time is occupied with preening to keep its feathers in perfect condition.

Types of feathers

There are different types of bird feathers, serving different functions. **Contour feathers** (Figure 26-4, *E*) give the bird its outward form and are the type we have already described. Contour feathers that extend beyond the body and are used in flight are called **flight feathers. Down feathers** (Figure 26-4, *H*) are soft tufts hidden beneath the contour feathers. They are soft because their barbules lack hooks. They are especially abundant on the breast and abdomen of water birds and on the young of game birds and function principally to conserve heat. **Filoplume feathers** (Figure 26-4, *G*) are hairlike, degenerate feathers; each is a weak shaft with a tuft of short barbs at the tip. They are the "hairs" of a plucked fowl. They have no known function. The bristles around the mouths of flycatchers and whippoorwills are probably modified filoplumes. A fourth type of highly modified feather, called **powder-down feathers,** is found on herons, bitterns, hawks, and parrots. Their tips disintegrate as they grow, releasing a talclike powder that helps to waterproof the feathers and give them metallic luster.

Origin and development

Feathers are epidermal structures that evolved from the reptilian scale; indeed, a developing feather closely resembles a reptile scale when growth is just beginning. We can imagine that in its evolution, the scale elongated and its edges frayed outward until it became the complex feather of birds. Strangely enough, although modern birds possess both scales (especially on their feet) and feathers, no intermediate stage between the two has been discovered on either fossil or living forms.

Like a reptile's scale, a feather develops from an epidermal elevation overlying a nourishing dermal core (Figure 26-4). However, instead of flattening like a scale, the feather rolls into a cylinder or feather bud and is covered with epidermis. This feather bud sinks in slightly at its base and comes to lie in a feather follicle from which the feather will protrude. A layer of keratin is produced around the cylinder or bud and encloses the pulp cavity of blood vessels. This surface layer of keratin splits away from the deeper layer to form a sheath. The deeper layer becomes frayed distally to form parallel ridges, the median one grows large to form the shaft (contour feathers), and the others, the barbs. The sheath bursts, and the barbs spread flat to form the vane (Figure 26-4 *A* to *E*).

The pulp cavity of the quill dries up when growth is finished, leaving it hollow, with openings (umbilici) at its two ends. If the feather is to be a down feather, the sheath bursts and releases the barbs without the formation of a shaft or vane. Pigments (lipochromes and melanins) are added to the epidermal cells during growth in the follicle.

Molting

When fully grown, a feather, like mammalian hair, is a dead structure. The shedding, or molting, of feathers is a highly orderly process. Except in penguins, which molt all at once, feathers are discarded gradually to avoid the appearance of bare spots. Flight and tail feathers are lost in exact pairs, one from each side, so that balance is maintained (Figure 26-5). Replacements emerge before the next pair is lost, and most birds can continue to fly unimpaired during the molting period; only ducks and geese are completely grounded during the molt. Nearly all birds molt at least once a year, usually in late summer after the nesting season. Many birds also undergo a second partial or complete molt just before breeding season, to equip them with their breeding finery, so important for courtship display.

Figure 26-5

Osprey, *Pandion haliaetus* (order Falconiformes), landing on nest. Note alulas *(top arrows)* and new primary feathers *(side arrows)*. Feathers are molted in sequence in exact pairs so that balance is maintained during flight.
Photograph by B. Tallmark.

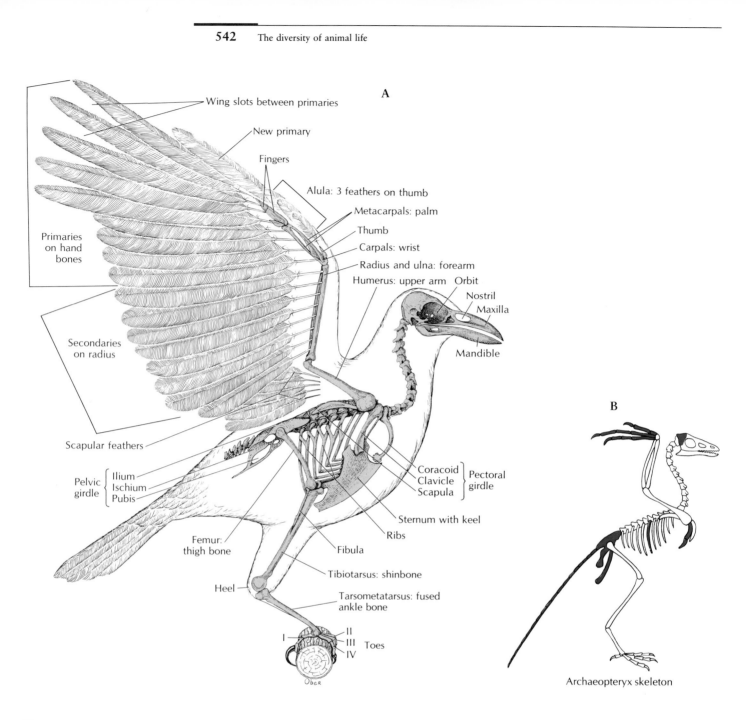

A

Wing slots between primaries

New primary

Fingers

Alula: 3 feathers on thumb

Metacarpals: palm

Thumb

Carpals: wrist

Radius and ulna: forearm

Humerus: upper arm   Orbit

Nostril

Maxilla

Mandible

Primaries on hand bones

Secondaries on radius

Scapular feathers

Pelvic girdle { Ilium
Ischium
Pubis

Coracoid
Clavicle   } Pectoral girdle
Scapula

Sternum with keel

Ribs

Femur: thigh bone

Fibula

Tibiotarsus: shinbone

Heel

Tarsometatarsus: fused ankle bone

I
II
III
IV   Toes

Ober

B

Archaeopteryx skeleton

Figure 26-6

A, Skeleton of a crow showing portions of the flight feathers. **B,** Skeleton of *Archaeopteryx* showing bones (color) that have been lost or greatly modified in modern birds.

Movement and Integration

Skeleton

One of the major adaptations that allows a bird to fly is its light skeleton (Figure 26-6). Bones are phenomenally light, delicate, and laced with air cavities (Figure 26-7), yet they are strong. The skeleton of a frigate bird with a 2.1 m (7 foot) wingspan weighs only 114 grams (4 ounces), less than the weight of all its feathers. A pigeon skull weighs only 0.21% of its body weight; the skull of a rat by comparison weighs 1.25%. The bird skull is mostly fused into one piece. The braincase and orbits are large to accommodate a bulging brain and the large eyes needed for quick motor coordination and superior vision.

The anterior skull bones are elongated to form a beak. The lower mandible is a complex of several bones that hinge on two small movable bones, the quadrates. This provides a double-jointed action that permits the mouth to open wide. The upper mandible, consisting of premaxillae, maxillae, and other bones, is usually fused to the forehead, but in many birds—parrots, for instance—the upper jaw is hinged also. This adaptation allows greater flexibility of the beak in food manipulation and provides insect-catching species with a wider gap for successful feeding on the wing.

The vertebral column of birds is highly specialized for flight. Its most distinctive feature is its rigidity. Most of the vertebrae except those of the neck are fused together and with the pelvic girdle to form a stiff but light framework to support the legs and provide rigidity for flight. To assist in this rigidity, the ribs are mostly fused with the vertebrae, pectoral girdle, and sternum. Except in the flightless birds, the sternum bears a large, thin keel that provides for the attachment of the powerful flight muscles. Of the body box, only the 8 to 24 (according to the species) vertebrae of the neck remain fully flexible.

The bones of the forelimbs have become highly modified for flight. They are hollow (for lightness) and reduced in number, and several are fused together. Despite these alterations, the bird wing is clearly a rearrangement of the basic vertebrate tetrapod limb from which it arose, and all the elements—upper arm, forearm, wrist, and fingers—are represented in modified form (Figure 26-6). The birds' legs have undergone less pronounced modification than the wings, since they are still designed principally for walking, as well as for perching and occasionally for swimming, as were those of their reptilian ancestors.

Muscular system

The locomotor muscles of the wings are relatively massive to meet the demands of flight. The largest of these is the **pectoralis,** which depresses the wings in flight. Its antagonist is the **supracoracoideus** muscle, which raises the wing (Figure 26-8). Surprisingly, perhaps, this latter muscle is not located on the backbone (anyone who has been served the back of a chicken knows that it offers little meat) but is positioned under the pectoralis on the breast. It is attached by a tendon to the upper side of the humerus of the wing so that it pulls from below by an ingenious "rope-and-pulley" arrangement. Both of these muscles are anchored to the keel. Thus, with the main muscle mass low in the body, aerodynamic stability is improved.

The main leg muscle mass is located in the thigh, surrounding the femur, and a smaller mass lies over the tibiotarsus (shank or "drumstick"). Strong but thin tendons extend downward through sleevelike sheaths to the toes. Consequently the feet are nearly devoid of muscles, explaining the thin, delicate appearance of the bird leg. This arrangement places the main muscle mass near the bird's center of gravity and at the same time allows great agility to the slender, lightweight feet. Since the feet are made up mostly of bone, tendon, and tough, scaly skin, they are highly resistant to damage from freezing. When a bird perches on a branch, an ingenious toe-locking mechanism (Figure 26-9) is activated, which prevents the bird from falling off its perch when asleep. The same mechanism causes the talons of a hawk or owl to automatically sink deeply into its victim as the legs bend under the impact of the strike. The powerful grip of a bird of prey was described by L. Brown*:

> When an eagle grips in earnest, one's hand becomes numb, and it is quite impossible to tear it free, or to loosen the grip of the eagle's toes with the other hand. One just has to wait until the bird relents, and while waiting one has ample time to realize that an animal such as a rabbit would be quickly paralyzed, unable to draw breath, and perhaps pierced through and through by the talons in such a clutch.

*Brown, L. 1970. Eagles. New York, Arco Publishing Co., Inc.

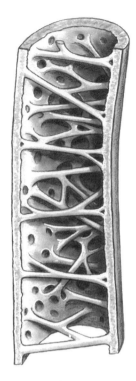

Figure 26-7

Hollow wing bone of a songbird showing the stiffening struts and air spaces that replace bone marrow. Such "pneumatized" bones are remarkably light and strong.

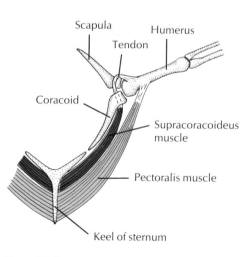

Figure 26-8

Flight muscles of a bird are arranged to keep the center of gravity low in the body. Both major flight muscles are anchored on the sternum keel. Contraction of the pectoralis pulls the wing downward. Then, as the pectoralis relaxes, the supracoracoideus contracts and, acting as a pulley system, pulls the wing upward.

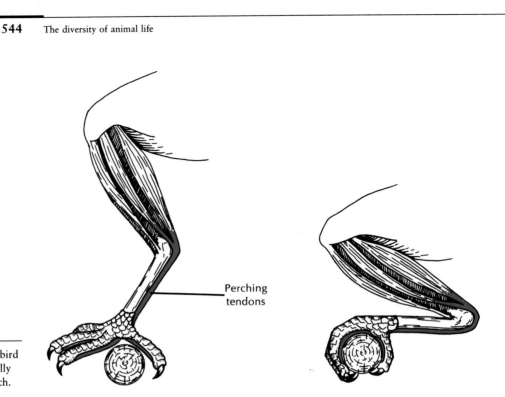

Figure 26-9

Perching mechanism of a bird. When a bird settles on a branch, tendons automatically tighten, closing the toes around the perch.

Birds have lost the long reptilian tail, still fully evident in *Archaeopteryx,* and have substituted a pincushion-like muscle mound into which the tail feathers are rooted. It contains a bewildering array of tiny muscles, as many as 1000 in some species, which control the crucial tail feathers. But the most complex muscular system of all is found in the neck of birds; the thin and stringy muscles, elaborately interwoven and subdivided, provide the bird's neck with the ultimate in vertebrate flexibility.

Nervous and sensory system

A bird's nervous and sensory system accurately reflects the complex problems of flight and a highly visible existence, in which it must gather food, mate, defend territory, incubate and rear young, and correctly distinguish friend from foe. The brain of a bird has well-developed **cerebral hemispheres, cerebellum,** and **midbrain tectum** (optic lobes) (Figure 26-10). The **cerebral cortex**—the portion in mammals that becomes the chief coordinating center—is thin, unfissured, and poorly developed in birds. But the core of the cerebrum, the **corpus striatum,** has enlarged into the principal integrative center of the brain, controlling such activities as eating, singing, flying, and all the complex instinctive reproductive activities. Relatively intelligent birds, such as crows and parrots, have larger cerebral hemispheres than do less intelligent birds such as chickens and pigeons. The **cerebellum** is a crucial coordinating center where muscle-position sense, equilibrium sense, and visual cues are all assembled and used to coordinate movement and balance. The **optic lobes,** laterally bulging structures of the midbrain, form a visual association apparatus comparable to the visual cortex of mammals.

Except in flightless birds and in ducks, the senses of smell and taste are poorly developed in birds. This deficiency, however, is more than compensated by good hearing and superb vision, the keenest in the animal kingdom. As in mammals, the bird ear consists of three regions: (1) the **external ear,** a sound-conducting canal extending to the **eardrum,** (2) the **middle ear,** containing a rodlike **columella** that transmits vibrations, and (3) the **inner ear,** where the organ of hearing, the **cochlea,** is located. The bird cochlea is much shorter than the coiled mamma-

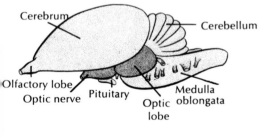

Figure 26-10

Bird brain showing principal divisions.

lian cochlea, yet birds can hear roughly the same range of sound frequencies as humans. Actually, the bird ear far surpasses that of humans in capacity to distinguish differences in intensities and to respond to rapid fluctuations in pitch.

The bird eye resembles that of other vertebrates in gross structure but is relatively larger, less spherical, and almost immobile; instead of turning their eyes, birds turn their heads with their long and flexible necks to scan the visual field. The light-sensitive **retina** (Figure 26-11) is elaborately equipped with rods (for dim night vision) and cones (for color vision). Cones predominate in day birds, and rods are more numerous in nocturnal birds. A distinctive feature of the bird eye is the **pecten,** a highly vascularized organ attached to the retina near the optic nerve and jutting out into the vitreous humor (Figure 26-11). The pecten is thought to provide nutrients to the eye. It may do more, but its function remains largely a mystery.

The position of a bird's eyes in its head is correlated with its life habits. Vegetarians that must avoid predators have eyes placed laterally to give a wide view of the world; predaceous birds such as hawks and owls have eyes directed to the front. In birds of prey and some others, the **fovea,** or region of keenest vision on the retina, is placed in a deep pit, which makes it necessary for the bird to focus exactly on the source. Many birds, moreover, have two foveae on the retina (Figure 26-11): the central one for sharp monocular views and the posterior one for binocular vision. Woodcocks can probably see binocularly both forward and backward. Bitterns, in their freezing stance of bill pointing up, can also see binocularly. The visual acuity of a hawk is believed to be eight times that of a human (enabling it to see clearly a crouching rabbit more than a mile away), and an owl's ability to see in dim light is more than 10 times that of a human. Birds have good color vision, especially toward the red end of the spectrum.

___ Food, Feeding, and Digestion

Birds have evolved along with food resources in nearly every environment on earth. In their early evolution, most birds were carnivorous, feeding principally on insects. Insects were well established on the earth's surface in both variety and numbers long before birds made their appearance, and they presented an enormously valuable food resource only partly exploited by amphibians and reptiles. With the advantage of flight, birds could hunt insects on the wing and carry their assault to insect refuges mostly inaccessible to their earthbound tetrapod peers. Today, there is a bird to hunt nearly every insect; they probe the soil, search the bark, scrutinize every leaf and twig, and drill into insect galleries hidden in tree trunks.

Other animal foods (worms, molluscs, crustaceans, fish, frogs, reptiles, mammals, as well as other birds) all found their way into the diet of birds. A very large group, nearly one fifth of all birds, feeds on nectar. Some birds are omnivores (often termed **euryphagous,** or "wide-eating" species) that will eat whatever is seasonally abundant. However, omnivorous birds must compete with numerous other omnivores for the same broad spectrum of food. Others are specialists (called **stenophagous,** or "narrow-eating" species) that have the pantry to themselves—but at a price. Should the food specialty be reduced or destroyed for some reason (disease, adverse climate, and the like), their very survival may be jeopardized.

The beaks of birds are strongly adapted to specialized food habits—from generalized types such as the strong, pointed beaks of crows, to grotesque, highly specialized ones in flamingoes, hornbills, and toucans (Figure 26-12). The beak of

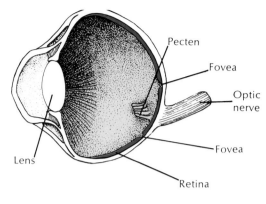

Figure 26-11

Hawk eye has all the structural components of the mammalian eye, plus a peculiar pleated structure, the pecten, believed to provide nourishment to the retina. The extraordinarily keen vision of the hawk is attributed to the extreme density of cone cells in the foveae: 1.5 million per fovea compared with 0.2 millions for humans. Each hawk eye has two foveae as opposed to one in humans, meaning that each hawk eye focuses on two objects simultaneously—the better to select its next meal!

That the vision of at least some birds extends into the near ultraviolet range may surprise people accustomed to assuming that human color vision approaches evolutionary perfection. In humans, ultraviolet light below 400 nm is filtered out by pigments in the lens and the macula, an adaptation that reduces chromatic aberration, which becomes severe at short wavelengths. Birds also possess ultraviolet-filtering pigments in their eyes, but these are located within the retinal cones rather than the lens. Recently it was shown that several species of hummingbirds can see in the near ultraviolet range down to 370 nm. This may help attract them to flowers having "nectar guides," which are striking patterns, visible only in the ultraviolet part of the spectrum, that have evolved to guide pollinating insects to these flowers.

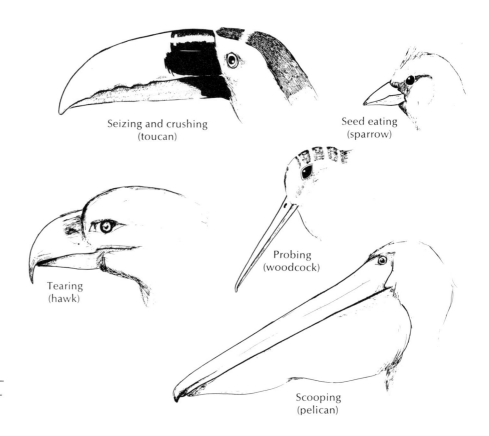

Seizing and crushing
(toucan)

Seed eating
(sparrow)

Tearing
(hawk)

Probing
(woodcock)

Scooping
(pelican)

Figure 26-12

Some bills of birds showing variety of adaptations.

a woodpecker is a straight, hard, chisel-like device. Anchored to a tree trunk with its tail serving as a brace, the woodpecker delivers powerful, rapid blows to build nests or expose the burrows of wood-boring insects. It then uses its long, flexible, barbed tongue to seek out insects in their galleries. The woodpecker's skull is especially thick to absorb shock.

How much do birds eat? By a peculiar twist of reality, the commonplace "to eat like a bird" is supposed to signify a diminutive appetite. Yet birds, because of their intense metabolism, are voracious feeders. Small birds eat relatively more than large birds because their metabolic rate is greater. This happens because the oxygen consumption increases only about three fourths as rapidly as body weight. For example, the resting metabolic rate (oxygen consumed per gram of body weight) of a hummingbird is 12 times that of a pigeon and 25 times that of a chicken. A 3 g hummingbird may eat 100% of its body weight in food each day, an 11 g blue tit about 30%, and a 1880 g domestic chicken, 3.4%. Obviously the weight of food consumed also depends on water content of the food, since water has no nutritive value. A 57 g Bohemian waxwing was estimated to eat 170 g of watery *Cotoneaster* berries in one day—three times its body weight! Seed-eaters of equivalent size might eat only 8 g of dry seeds per day.

Birds rapidly process their food with efficient digestive equipment. A shrike can digest a mouse in 3 hours, and berries will pass completely through the digestive tract of a thrush in just 30 minutes. Furthermore, birds utilize a very high percentage of the food they eat. There are no teeth in the mouth, and the poorly developed salivary glands mainly secrete mucus for lubricating the food and the slender, horn-covered tongue. There are few taste buds, although all birds can taste to some extent. Hummingbirds and some others have sticky tongues, and woodpeckers have tongues that are barbed at the end. From the short **pharynx** a relatively long, muscular, elastic **esophagus** extends to the **stomach.** Many birds have an enlargement (**crop**) at the lower end of the esophagus that serves as a storage chamber.

In pigeons, doves, and some parrots the crop not only stores food but also produces milk by the breakdown of epithelial cells of the lining. This "bird milk" is regurgitated by both male and female into the mouth of the young squabs. It has a much higher fat content than cow's milk.

The stomach proper consists of a **proventriculus,** which secretes gastric juice, and the muscular **gizzard,** which is lined with horny plates that serve as millstones for grinding the food. To assist in the grinding process, birds swallow coarse, gritty objects or pebbles, which lodge in the gizzard. Certain birds of prey such as owls form pellets of indigestible materials, for example, bones and fur, in the proventriculus and eject them through the mouth. At the junction of the intestine with the rectum are paired **ceca,** which may be well developed in some birds. Two **bile ducts** from the **gallbladder** or liver and two or three **pancreatic ducts** empty into the duodenum, or first part of the intestine. The **liver** is relatively large and bilobed. The terminal part of the digestive system is the **cloaca,** which also receives the genital ducts and ureters; in young birds the dorsal wall of the cloaca bears the **bursa of Fabricius,** which processes the B lymphocytes that are important in the immune response (p. 618).

—— Circulation, Respiration, and Excretion

Circulatory system

The general plan of bird circulation is not greatly different from that of mammals. The four-chambered heart is large, with strong ventricular walls; thus, birds share with mammals a complete separation of the respiratory and systemic circulations. However, the right aortic arch, instead of the left as in the mammals, leads to the dorsal aorta. The two jugular veins in the neck are connected by a cross vein, an adaptation for shunting the blood from one jugular to the other as the head is turned around. The brachial and pectoral arteries to the wings and breast are unusually large.

The heartbeat is extremely fast, and, as in mammals, there is an inverse relationship between heart rate and body weight. For example, a turkey has a heart rate at rest of about 93 beats per minute, a chicken has 250 beats per minute, and a black-capped chickadee has 500 beats per minute when asleep, which may increase to a phenomenal 1000 beats per minute during exercise. Blood pressure in birds is roughly equivalent to that in mammals of similar size.

Bird's blood contains nucleated, biconvex red corpuscles that are somewhat larger than those of mammals. The phagocytes, or mobile ameboid cells, of the blood are unusually active and efficient in birds in the repair of wounds and in destroying microbes.

Respiratory system

The respiratory system of birds differs radically from the lungs of reptiles and mammals and is marvelously adapted for meeting the high metabolic demands of flight. In birds the finest branches of the bronchi, rather than ending in saclike alveoli as in mammals, are developed as tube-like **parabronchi** through which the air flows continuously. Also unique is the extensive system of nine interconnecting **air sacs** that are located in pairs in the thorax and abdomen and are even extended by tiny tubes into the centers of the long bones (Figure 26-13). The air sacs are

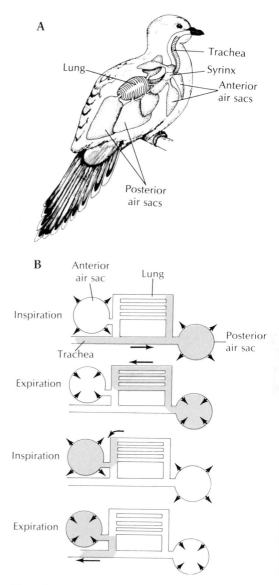

Figure 26-13

Respiratory system of a bird. **A,** Lungs and air sacs. One side of the bilateral air sac system is shown. **B,** Movement of a single volume of air through the bird's respiratory system. Two full respiratory cycles are required to move the air through the system. **B,** From Schmidt-Nielsen, K. 1979. Animal physiology: adaptation and environment. New York, Cambridge University Press.

connected to the lungs in such a way that perhaps 75% of the inspired air bypasses the lungs and flows directly into the posterior air sacs, which serve as reservoirs for fresh air. On expiration, this oxygenated air is shunted through the lung and collected in the anterior air sacs. From there it flows directly to the outside. The air flow sequence is shown in Figure 26-13. The advantage of such a system is obvious: the lungs receive fresh air during both inspiration and expiration. An almost continuous stream of oxygenated air is passed through a system of richly vascularized parabronchi. Although many details of the bird's respiratory system are not yet understood, it is clearly the most efficient of any vertebrate system.

In addition to performing its principal respiratory function, the air sac system helps cool the bird during vigorous exercise. A pigeon, for example, produces about 27 times more heat when flying than when at rest. The air sacs have numerous diverticula that extend inside the larger pneumatic bones of the pectoral and pelvic girdles, wings, and legs. Because they contain warmed air, they provide considerable buoyancy to the bird.

Excretory system

The relatively large paired metanephric kidneys are attached to the dorsal wall in a depression against the sacral vertebrae and pelvis. Urine passes by way of **ureters** to the **cloaca.** There is no urinary bladder. The kidney is composed of many thousands of **nephrons,** each consisting of a renal corpuscle and a nephric tubule. Urine is formed in the usual way by glomerular filtration followed by selective modification of the filtrate in the tubule (the details of this sequence are described on pp. 648 to 650).

Birds, like the reptiles from which they evolved, excrete their nitrogenous wastes as uric acid, rather than urea, an adaptation that originated with the evolution of the shelled egg. In the shelled egg, all excretory products must remain in the eggshell with the growing embryo. If urea were produced, it would quickly accumulate in solution to toxic levels. Uric acid, however, crystallizes *out* of solution and can be stored harmlessly within the egg shell. Thus, from an embryonic necessity was born an adult virtue. Because of uric acid's low solubility, a bird can excrete 1 g of uric acid in only 1.5 to 3 ml of water, whereas a mammal may require 60 ml of water to excrete 1 g of urea. The concentration of uric acid occurs almost entirely in the cloaca, where it is combined with fecal material, and the water is reabsorbed to form a white paste. Thus, despite having kidneys that are much less effective in true concentrative ability than mammalian kidneys, birds can excrete uric acid nearly 3000 times more concentrated than that in the blood. Even the most effective mammalian kidneys—those of certain desert rodents—can excrete urea only about 25 times the plasma concentration.

Marine birds (also marine turtles) have evolved a unique solution for excreting the large loads of salt eaten with their food and in the seawater they drink. Seawater contains about 3% salt and is three times saltier than a bird's body fluids. Yet the bird kidney cannot concentrate salt in urine above about 0.3%. The problem is solved by special **salt glands,** one located above each eye (Figure 26-14). These glands are capable of excreting a highly concentrated solution of sodium chloride—up to twice the concentration of seawater. The salt solution runs out the internal or external nostrils, giving gulls, petrels, and other sea birds a perpetual runny nose. The development of the salt gland in some birds depends on how much salt the bird takes in its diet. For example, a race of mallard ducks living a semimarine life in Greenland has salt glands 10 times larger than those of ordinary freshwater mallards.

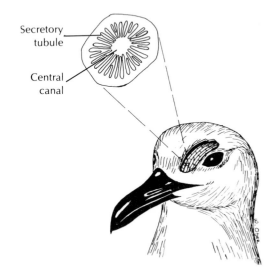

Secretory tubule

Central canal

Figure 26-14

Salt glands of a marine bird (gull). One salt gland is located above each eye. Each gland consists of several lobes arranged parallel to each other. One lobe is shown in cross section, much enlarged. Salt is secreted into many radially arranged tubules, then flows into a central canal that leads into the nose.

___ Flight

What prompted the evolution of flight in birds, the ability to rise free of earth-bound concerns, as almost every human has dreamed of doing? Much as we may envy birds their conquest of the air, we also recognize that evolution of flight was the pragmatic result of complex selective pressures; certainly the forerunners of birds did not take up flight just to enjoy a new experience. The air was a relatively unexploited habitat stocked with flying insects for food. Flight also offered escape from terrestrial predators and opportunity to travel rapidly and widely to establish new breeding areas and to benefit from year-round favorable climate by migrating north and south with the seasons.

The fossil evidence is too meager to provide us with a recorded history of the origin of bird flight, but it must have happened in one of two ways: birds began to fly by climbing to a high place and gliding down, or by flapping their way into the air from the ground. The "ground-up" hypothesis holds that birds were ground-dwelling runners with primitive wings used to snare insects. With continued enlargement the protowings eventually enabled the running animal to flap its way into the air. The more widely favored "trees-down" hypothesis suggests that birds passed through an arboreal apprenticeship of tree climbing, leaping through the trees, parachuting, gliding, and finally fully powered flight. One thing seems certain: feathers were an absolute requirement for flight and the evolution of flight and of feathers must have progressed together. There is absolutely no support for the idea that bird ancestors were originally membrane-winged flyers, like bats, that later developed feathers.

Bird wing as a lift device

Bird flight, especially the familiar flapping flight of birds, is complex. Despite careful analysis by conventional aerodynamic techniques and high-speed photography, it is not well understood. Nevertheless, we know that the bird wing is an airfoil that is subject to recognized laws of aerodynamics. It is adapted for high lift at low speeds, and, not surprisingly perhaps, it resembles the wings of early low-speed aircraft. The bird wing is streamlined in cross section, with a slightly concave (cambered) lower surface and with small, tight-fitting feathers where the leading edge meets the air (Figure 26-15). Air slips efficiently over the wing, creating lift with minimum drag. Some lift is produced by positive pressure against the undersurface of the wing. But on the upper side, where the airstream must travel farther and faster over the convex surface, a negative pressure is created that provides more than two thirds of the total lift.

The lift-to-drag ratio of an airfoil is determined by the angle of tilt (angle of attack) and the airspeed (Figure 26-15). A wing carrying a given load can pass through the air at high speed and small angle of attack or at low speed and larger angle of attack. But as speed decreases, a point is reached at which the angle of attack becomes too steep; turbulence appears on the upper surface, lift is destroyed, and stalling occurs. Stalling can be delayed or prevented by placing a **wing slot** along the leading edge so that a layer of rapidly moving air is directed across the upper wing surface. Wing slots were and still are used in aircraft traveling at a low speed. In birds, two kinds of wing slots have developed: (1) the **alula,** or group of small feathers on the thumb (Figures 25-5 and 26-6), which provides a midwing slot, and (2) **slotting between the primary feathers,** which provides a wing-tip slot. In a number of songbirds, these together provide stall-preventing slots for nearly the entire outer (and aerodynamically more important) half of the wing.

Figure 26-15

Air patterns formed by the airfoil, or wing, moving from right to left. **A,** Normal flight with a low angle of attack. As air moves smoothly over the wing, areas of negative pressure on the upper wing surface and high pressure on the lower wing surface create lift. **B,** Appearance of lift-destroying turbulence on the upper wing surface when the angle of attack becomes too great. Stalling occurs. **C,** Prevention of stalling by directing a layer of rapidly moving air over upper surface with a wing slot.
From Welty, J.C. 1982. The life of birds, ed. 3. Philadelphia, W.B. Saunders Co.

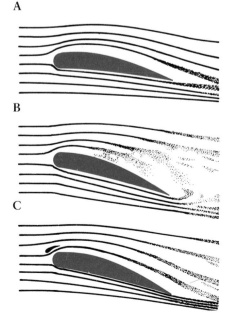

Basic forms of bird wings

Bird wings vary in size and form because the successful exploitation of different habitats has imposed special aerodynamic requirements. Four types of bird wings are easily recognized.*

Elliptical wings

Birds that must maneuver in forested habitats, such as sparrows, warblers, doves, woodpeckers, and magpies (Figure 26-16, *A*), have elliptical wings. This type has a **low aspect ratio** (ratio of length to width). The outline of a sparrow wing is almost identical to that of the British Spitfire fighter plane of World War II fame— also a highly maneuverable flyer. Elliptical wings are highly slotted between the primary feathers; this helps prevent stalling during sharp turns, low-speed flight, and frequent landing and takeoff. Each separated primary feather behaves as a narrow wing with a high angle of attack, providing high lift at low speed. The high maneuverability of the elliptical wing is exemplified by the tiny chickadee, which, if frightened, can change course within 0.03 second.

High-speed wings

Birds that feed on the wing, such as swallows, hummingbirds, and swifts, or that make long migrations, such as plovers, sandpipers, terns and gulls, (Figure 26-16, *B*), have wings that sweep back and taper to a slender tip. They are rather flat in section, have a moderately high aspect ratio, and lack the wing-tip slotting characteristic of the preceding group. Sweepback and wide separation of the wing tips reduce "tip vortex," a drag-creating turbulence that tends to develop at wing tips. The fastest birds, such as sandpipers, clocked at 175 km (109 miles) per hour, belong to this group.

Soaring wings

The oceanic soaring birds have **high-aspect ratio** wings resembling those of sailplanes. This group includes albatrosses frigate birds, and gannets (Figure 26-16, *C*). Such long, narrow wings lack wing slots and are adapted for high speed, high lift, and dynamic soaring. They have the highest aerodynamic efficiency of all

---

*Savile, D.B.O. 1957. Adaptive evolution in the avian wing. Evolution 11:212-224.

Figure 26-16

Four basic forms of bird wings.

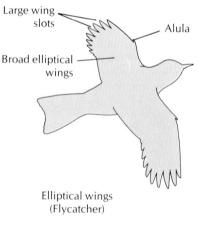

A

Large wing slots

Alula

Broad elliptical wings

Elliptical wings
(Flycatcher)

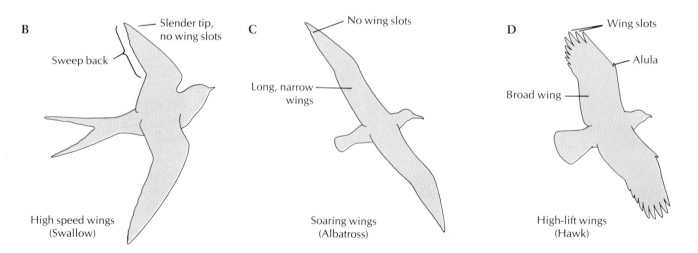

B

Slender tip, no wing slots

Sweep back

High speed wings
(Swallow)

C

No wing slots

Long, narrow wings

Soaring wings
(Albatross)

D

Wing slots

Alula

Broad wing

High-lift wings
(Hawk)

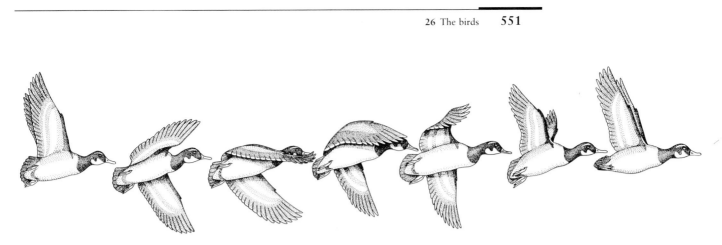

Figure 26-17

In normal flapping flight of strong fliers like ducks, the wings sweep downward and forward fully extended. Thrust is provided by the primary feathers at the wing tips. To begin the upbeat the wing is bent, bringing it upward and backward. The wing then extends, ready for the next downbeat.

wings but are less maneuverable than the wide, slotted wings of land soarers. Dynamic soarers have learned how to exploit the highly reliable sea winds, using adjacent air currents of different velocities.

High-lift wings

Vultures, hawks, eagles, owls, and ospreys (Figure 26-16, *D*)—predators that carry heavy loads—have wings with slotting, alulas, and pronounced camber, all of which promote high lift at low speed. Many of these birds are land soarers, with broad, slotted wings that provide the sensitive response and maneuverability required for static soaring in the capricious air currents over land.

Flapping Flight

This basic form of flight is so complex that complete analysis is still not possible—yet young birds fly almost perfectly on their maiden flight. More than a century ago an English zoologist reared swallow fledglings in a space so confining that they could not fully extend their wings. Yet when released at the age when swallows normally fly, they flew immediately and without practice.

In flapping flight, the primary feathers at the wing tips generate thrust, while the secondary feathers of the inner wing, which do not move so far or so fast, act as an airfoil, providing lift. Greatest power is applied on the downstroke. The primary feathers are bent upward and twist to a steep angle of attack, biting into the air like a propeller (Figure 26-17). The entire wing (and the bird's body) is pulled forward. On the upstroke, the primary feathers bend in the opposite direction so that their upper surfaces twist into a positive angle of attack to produce thrust, just as the lower surfaces did on the downstroke. A powered upstroke is essential for hovering flight, as in hummingbirds, and is important for fast, steep takeoffs by small birds with elliptical wings.

MIGRATION AND NAVIGATION

Perhaps it was inevitable that birds, having mastered the art of flight, would use this ability to make the long, arduous seasonal migrations that have captured human wonder and curiosity. The term **migration** refers to the regular, extensive, seasonal movements that birds make between their summer breeding regions and their wintering regions. The chief advantage seems obvious: it enables birds to live in an optimal climate all the time, where abundant and unfailing sources of food are available to sustain their intense metabolism. Migrations also provide optimal conditions for rearing young when demands for food are especially great. Broods are largest in the far north where the long summer days and the abundance of insects combine to provide parents with ample food-gathering opportunity. Predators are relatively rare in the north, and the brief, once-a-year appearance of vulnerable young birds does not encourage the buildup of predator populations. Migration also vastly increases the amount of space available for breeding and reduces aggressive territorial behavior. Furthermore, migration favors homeostasis by allowing birds to avoid climatic extremes.

Migration Routes

Most migratory birds have well-established routes trending north and south. Since most birds (and other animals) live in the Northern Hemisphere, where most of the earth's land mass is concentrated, most birds are south-in-winter and north-in-summer migrants. Of the 4000 or more species of migrant birds (a little less than half the total bird species), most breed in the more northern latitudes of the hemisphere; the percentage of migrants in Canada is far higher than the percentage of migrants in Mexico, for example. Some use different routes in the fall and spring (Figure 26-18). Some, especially certain aquatic species, complete their migratory routes in a very short time. Others, however, make the trip in a leisurely manner, often stopping along the way to feed. Some of the warblers are known to take 50 to 60 days to migrate from their winter quarters in Central America to their summer breeding areas in Canada.

Not all members of a species migrate at the same time; there is a great deal of straggling so that some members do not reach the summer breeding grounds until others are well along with their nesting. Many of the smaller species migrate at night and feed by day; others migrate chiefly in the daytime; and many swimming and wading birds migrate by either day or night. The height at which they fly varies greatly. Migrants tend to fly higher over water than over land and higher at night than during the day.

Many birds are known to follow landmarks, such as rivers and coastlines, but others do not hesitate to fly directly over large bodies of water in their routes. Some birds have very wide migration lanes, whereas others, such as certain sandpipers, are restricted to very narrow ones, keeping well to the coastlines because of their food requirements.

Some species are known for their long-distance migrations. The arctic tern, greatest globe spanner of all, breeds north of the Arctic Circle and in winter is found in the Antarctic regions. This species is also known to take a circuitous route in migrations from North America, passing over to the coastlines of Europe and Africa and then to their winter quarters, a trip that may exceed 18,000 km (11,200 miles). Other birds that breed in Alaska follow a more direct line down the Pacific coast of North and South America.

Many small songbirds also make great migration treks (Figure 26-18). Africa is a favorite wintering ground for European birds, and many fly there from Central Asia as well.

Stimulus for Migration

Humans have known for centuries that the onset of the reproductive cycle of birds is closely related to season. Only within the last 50 years, however, has it been proved that the lengthening days of late winter and early spring stimulate the development of the gonads and accumulation of fat—both important internal changes that predispose birds to migrate northward. There is evidence that increasing day length stimulates the anterior lobe of the pituitary into activity. The

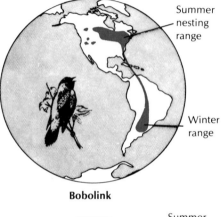

Bobolink

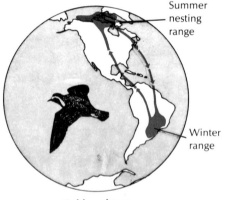

Golden plover

Figure 26-18

Migrations of the bobolink and golden plover. The bobolink commutes 22,500 km (14,000 miles) each year between nesting sites in North America and its wintering range in Argentina, a phenomenal feat for such a small bird. Although the breeding range has extended to colonies in western areas, these birds take no shortcuts but adhere to the ancestral eastern seaboard route. The golden plover flies a loop migration, striking out across the Atlantic in its southward autumnal migration but returning in the spring via Central America and the Mississippi Valley because ecological conditions are more favorable in these areas at that time.

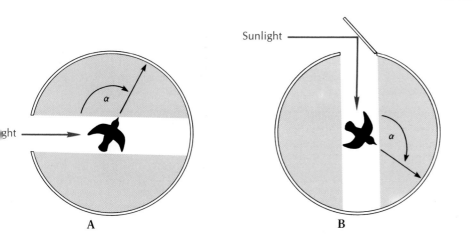

Sunlight

ght

α

α

A

B

Figure 26-19

Gustav Kramer's experiments with suncompass navigation in starlings. **A,** In a windowed, circular cage, the bird fluttered to align itself in the direction it would normally follow if it were free. **B,** When the true angle of the sun is deflected with a mirror, the bird maintains the same relative position to the sun. This shows that these birds use the sun as a compass. The bird navigates correctly throughout the day, changing its orientation to the sun as the sun moves across the sky.

release of pituitary gonadotropic hormone in turn sets in motion a complex series of physiological and behavioral changes, resulting in gonadal growth, fat deposition, migration, courtship and mating behavior, and care of the young.

Direction Finding in Migration

Numerous experiments suggest that most birds navigate chiefly by sight. Birds recognize topographical landmarks and follow familiar migratory routes—a behavior assisted by flock migration, during which navigational resources and experience of older birds can be pooled. But in addition to visual navigation, birds make use of a variety of orientation cues at their disposal. Birds have an innate time sense, a built-in clock of great accuracy; they have an innate sense of direction; and very recent work adds credence to an old, much debated hypothesis that birds can detect and navigate by the earth's magnetic field. All of these resources are inborn and instinctive, although a bird's navigational abilities may improve with experience.

Recent experiments by German ornithologists G. Kramer and E. Sauer and American ornithologist S. Emlen have demonstrated convincingly that birds can navigate by celestial cues: the sun by day and the stars by night. Using special circular cages, Kramer concluded that birds possess a built-in time sense that enables them to maintain compass direction by referring to the sun, regardless of the time of day (Figure 26-19). This is called **sun-azimuth orientation** (*azimuth*, compass bearing of the sun). Sauer's and Emlen's ingenious planetarium experiments strongly suggest that some birds, probably many, are able to detect and navigate by the North Star axis around which the constellations appear to rotate.

Some of the remarkable feats of bird navigation still defy rational explanation. Most birds undoubtedly use a combination of environmental and innate cues to migrate. Migration is a rigorous undertaking; the target is often small, and natural selection relentlessly prunes off individuals making errors in migration, leaving only the best navigators to propagate the species.

SOCIAL BEHAVIOR AND REPRODUCTION

The adage says "birds of a feather flock together," and many birds are indeed highly social creatures. Especially during the breeding season, sea birds gather, often in enormous colonies, to nest and rear young (Figure 26-20). Land birds, with some conspicuous exceptions, such as starlings and rooks, tend to be less

In the early 1970s W.T. Keeton showed that the flight bearings of homing pigeons were significantly disturbed by magnets attached to the birds' heads, or by minor fluctuations in the geomagnetic field. But until recently the nature and position of a magnetic receptor in pigeons remained a mystery. Deposits of a magnetic substance called magnetite (Fe_3O_4) have now been discovered in the neck musculature of pigeons and migratory white-crowned sparrows. If this material were coupled to sensitive muscle receptors, as has been proposed, the structure could serve as a magnetic compass that would enable birds to detect and orient their migrations to the earth's magnetic field.

gregarious than sea birds during breeding and to seek isolation for rearing their brood. But these same species that covet separation from their kind during breeding may aggregate for migration or feeding. Togetherness offers advantages: mutual protection from enemies, greater ease in finding mates, less opportunity for individual straying during migration, and mass huddling for protection against low night temperatures during migration. Certain species, such as pelicans (Figure 26-21), may use highly organized cooperative behavior to feed. At no time are the highly organized social interactions of birds more evident than during the breeding season, as they stake out territorial claims, select mates, build nests, incubate and hatch their eggs, and rear their young.

Reproductive System

In the male the paired **testes** and accessory ducts are similar to those in many other vertebrates. From the **testes** the **vasa deferentia** run to the cloaca. Before being discharged, the sperm are stored in the **seminal vesicle,** the enlarged distal end of the vas deferens. This seminal vesicle may become so large with stored sperm during the breeding season that it causes a cloacal protuberance. The high body temperature, which tends to inhibit spermatogenesis in the testes, is probably counteracted by the cooling effect of the abdominal air sacs. The testes of birds undergo a great enlargement at the breeding season, as much as 300 fold, and then shrink to tiny bodies afterward. Some birds, including ducks and geese, have a large, well-developed **copulatory organ** (penis), provided with a groove on its dorsal side for the transfer of sperm. However, in the more advanced birds, copulation is a matter of bringing the cloacal surfaces into contact, usually while the male stands on the female's back (Figure 26-22). Some swifts copulate in flight.

In the female of most birds, only the left ovary and oviduct develop; those on the right dwindle to vestigial structures. Eggs discharged from the ovary are

Figure 26-20

Part of a colony of Australian gannets *Sula serrator*. Order Pelecaniformes.
Photograph by C.P. Hickman, Jr.

A

B

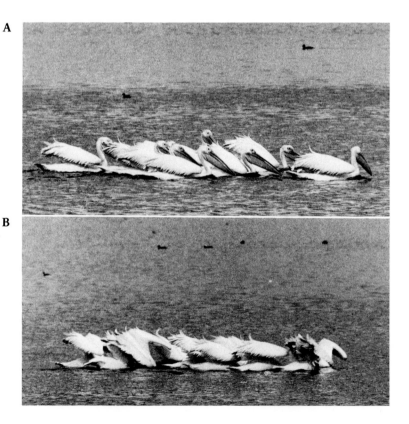

Figure 26-21

Cooperative feeding behavior of white pelicans *Pelecanus onocrotalus*. **A,** Pelicans on Lake Nakuru, East Africa, form a horseshoe to drive fish together. **B,** Then they plunge simultaneously to scoop up fishes in their huge bills. Pelicans were attracted to the lake in mid-1960s to feed on fish *(Tilapia grahami)* introduced to control malaria by eating mosquito larva. (The pictures were taken 2 seconds apart.) Order Pelecaniformes.
Photographs by B. Tallmark.

Figure 26-22

Copulation in birds. In advanced bird species, the male lacks a penis. The male copulates by standing on the back of the female, pressing his cloaca against that of the female, and passing sperm to the female.

picked up by the expanded end of the oviduct, the **ostium** (Figure 26-23). The oviduct runs posteriorly to the cloaca. While the eggs are passing down the oviduct, **albumin,** or egg white, from special glands is added to them; farther down the oviduct, the shell membrane, shell, and shell pigments are also secreted about the egg. Fertilization takes place in the upper oviduct several hours before the layers of albumin, shell membranes, and shell are added. Sperm remain alive in the female oviduct for many days after a single mating. Hen eggs show good fertility for 5 or 6 days after mating, but then fertility drops rapidly. However, the occasional egg will be fertile as long as 30 days after separation of the hen from the rooster.

Mating Systems

The two most common types of mating systems in animals are **monogamy,** in which an individual mates with only one partner each breeding season, and **polygamy,** in which an individual mates with two or more partners each breeding period. Monogamy is rare in most animal groups, but in birds it is the general rule: more than 90% of the birds are monogamous. In a few bird species such as swans and geese, partners are chosen for life and often remain together throughout the year. Seasonal monogamy is more common, however, in the great majority of migrant birds, which pair up during the breeding season but lead independent lives the rest of the year.

One reason that monogamy is much more common among birds than among mammals is that female birds are not equipped, as mammals are, with a built-in food supply for the young. Thus, the ability of the two sexes to provide parental care, especially food for the young, is more equal in birds than in mammals. A female bird will choose a male whose parental investment in their young is apt to be high and avoid a male that has mated with another female. If the male had mated with another female, he could at best divide his time between his two mates and might even devote most of his attention to the alternate mate. Thus, females enforce monogamy.

Monogamy in birds is also encouraged by the need for the male to secure and defend a territory before he can attract a mate. The male may sing a great deal to announce his presence to females and to discourage rival males from entering his territory. The female wanders from one territory to another, seeking a male with foraging territory that offers the best chances for reproductive success. Usually a male is able to defend an area that provides just enough resources for one nesting female.

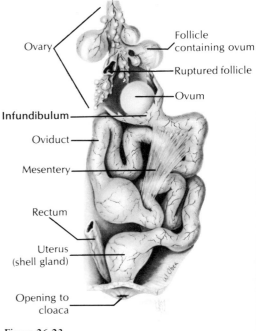

Ovary

Follicle containing ovum

Ruptured follicle

Ovum

Infundibulum

Oviduct

Mesentery

Rectum

Uterus (shell gland)

Opening to cloaca

Figure 26-23

Reproductive system of female bird.

The term *"polygamy"* ("many marriages") is used when the sex of the individual possessing a plurality of mates is not specified. The most common form of polygamy is polygyny ("many females"), in which a male mates with more than one female. Much less common is polyandry ("many males"), in which a female mates with more than one male per breeding season.

Figure 26-24

Dominant male sage grouse surrounded by several hens that have been attracted by his "booming" display. Order Galliformes. Photograph by L.L. Rue, III.

The most common form of polygamy in birds, when it occurs, is **polygyny,** in which a male mates with more than one female. In many species of grouse, the males gather in a collective display ground, the **lek,** which is divided into individual territories, each vigorously defended by a displaying male (Figure 26-24). There is nothing of value in the lek to the female except the male, and all he can offer are his genes, for only the females care for the young. Usually there are a dominant male and several subordinate males in the lek. Competition among males for females is intense, but the females appear to choose the dominant male for mating because, presumably, social rank correlates with genetic quality.

Nesting and Care of Young

To produce offspring, all birds lay eggs that must be incubated by one or both parents. The eggs of most songbirds require approximately 14 days for hatching; those of ducks and geese require at least twice that long. Most of the duties of incubation fall on the female, although in many instances both parents share the task, and occasionally only the male performs this work.

Most birds build some form of nest in which to rear their young. Some birds simply lay their eggs on the bare ground or rocks and make no pretense of nest building. Others build elaborate nests such as the pendant nests constructed by orioles, the delicate lichen-covered nests of hummingbirds (Figure 26-25) and flycatchers, the chimney-shaped mud nests of cliff swallows, the floating nests of rednecked grebes, and the huge brush pile nests of Australian brush turkeys. Most birds take considerable pains to conceal their nests from enemies. Woodpeckers, chickadees, bluebirds, and many others place their nests in tree hollows or other cavities; kingfishers excavate tunnels in the banks of streams for their nests; and birds of prey build high in lofty trees or on inaccessible cliffs. A few birds such as the American cowbird and the European cuckoo build no nests at all but simply lay their eggs in the nests of birds smaller than themselves. When the eggs hatch, the foster parents care for the young.

Nesting success is very low with many birds, especially in altricial species (see the following paragraph). One investigation of 170 altricial bird nests reported that only 21% produced at least one young. Of the many causes of nesting failures, predation by snakes, skunks, chipmunks, blue jays, crows, and others is by far the chief factor. Birds of prey probably have a much higher percentage of reproductive success.

Newly hatched birds are of two types: **precocial** and **altricial.** The precocial

Figure 26-25

Ruby-throated hummingbird *Archilochus colubris* in nest built of plant down and spider webs and decorated on the outside with lichens. The female builds the nest, incubates the two pea-sized eggs, and rears the young with no assistance from the male. These frail-looking but pugnacious little birds make arduous seasonal migrations between Canada and Mexico. Order Apodiformes.

Photograph by L.L. Rue, III.

Altricial
One-day-old meadow lark

Precocial
One-day-old ruffed grouse

Figure 26-26

Comparison of 1-day-old altricial and precocial young. The altricial meadowlark *(top)* is born nearly naked, blind, and helpless. The precocial ruffed grouse *(bottom)* is covered with down and is alert, strong legged, and able to feed itself.

young, such as quail, fowl, ducks, and most water birds, are covered with down when hatched and can run or swim as soon as their plumage is dry (Figure 26-26). The altricial ones, on the other hand, are naked and helpless at birth and remain in the nest for a week or more. The young of both types require care from parents for some time after hatching. They must be fed, guarded, and protected against rain and sun. The parents of altricial species must carry food to their young almost constantly, for most young birds will eat more than their weight each day. This enormous food consumption explains the rapid growth of the young and their quick exit from the nest. The food of the young, depending on the species, includes worms, insects, seeds, and fruit. Pigeons and doves are peculiar in feeding their young "pigeon milk."

BIRD POPULATIONS

Bird populations, like those of other animal groups, vary in size from year to year. Snowy owls, for example, are subject to population cycles that closely follow cycles in their food supply, mainly rodents. Voles, mice, and lemmings in the north have a fairly regular 4-year cycle of abundance; at population peaks, predator populations of foxes, weasels, and buzzards, as well as snowy owls, increase because there is abundant food for rearing their young. After a crash in the rodent

Figure 26-27

A, Starling with insect larva. Starlings are omnivorous. They eat mostly insects in spring and summer and shift to wild fruits in fall. **B,** Colonization of North America by starlings *Sturnus vulgaris* (order Passeriformes) after the introduction of 120 birds into Central Park in New York City in 1890. There are now perhaps 100 million starlings in the United States alone, testimony to the great reproductive potential of birds.

A, Photograph by L.L. Rue, III; **B,** modified from Fisher, J., and R.T. Peterson. 1971. Birds. London, Aldus Books Ltd.

A

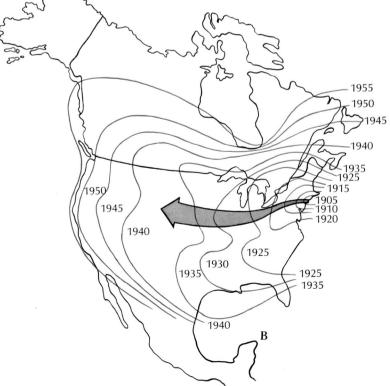

B

Figure 26-28

"Martha," the last passenger pigeon *Ecto-pistes migratorius*, died in the Cincinnati Zoo in 1914. Passenger pigeons traveled and roosted in huge masses with nesting areas exceeding 80 sq km (about 30 square miles). In 1869 hunters shot 7.5 million pigeons from a single nesting site.
Courtesy Smithsonian Institution, Washington, D.C.

population, snowy owls move south, seeking alternative food supplies. They occasionally appear in large numbers in southern Canada and the northern United States, where their total absence of fear of humans makes them easy targets for thoughtless hunters.

Occasionally the activities of people bring about spectacular changes in bird distribution. Both starlings (Figure 26-27) and house sparrows have been accidentally or deliberately introduced into numerous countries, and they have become the two most abundant bird species on earth, with the exception of domestic fowl.

Humans also are responsible for the extinction of many bird species. More than 80 species of birds have, since 1695, followed the last dodo to extinction. Many died naturally, victims of changes in their habitat or competition with better-adapted species. But several have been hunted to extinction, among them the passenger pigeon (Figure 26-28), which only a century ago darkened the skies over North America in incredible numbers estimated in the billions (Figure 26-29). Today, game bird hunting is a well-managed industry in the United States and Canada, and while hunters kill millions of game birds each year, none of the 74 bird species legally hunted is endangered. Unfortunately, many game birds die indirectly as the result of eating lead pellets, which they mistake for seeds, or from the crippling effect of embedded pellets. One survey revealed that 44% of the Canada geese reaching the southern Mississippi Valley contained embedded lead shot. Nevertheless, hunting interests, by acquiring large areas of wetlands for migratory bird refuges and sanctuaries, have probably overall benefited nongame as well as game birds.

Our most destructive effects on birds are usually unintentional. The draining of marshes—more than 99% of Iowa's once extensive marshland is now farmland—has destroyed waterfowl nesting. Deforestation has likewise had great impact on tree-nesting species. The vertical appendages of civilization, such as television towers, monuments, tall buildings, and electrical transmission towers and lines, take a fearful toll during bird migration in bad weather. Most birds, through their impressive reproductive potential, can replace in numbers those that become victims of human activities. Someone has calculated that a single pair of

Figure 26-29

Shooting passenger pigeons for sport in Louisiana during the nineteenth century. Relentless sport and market hunting eventually dropped the population too low to sustain colonial breeding.
Courtesy Culver Pictures.

robins, producing two broods of four young a season, would leave 19,500,000 descendants in 10 years, should all survive that long. But, although many birds have accommodated to the heavy-handed influence of humans on their environment, and some such as robins, house sparrows, and starlings even thrive on it, most birds find it adverse, and to some species it is lethal.

CLASSIFICATION

Class Aves (birds) is made up of about 27 orders of living birds and a few fossil orders. More than 8600 species and many subspecies have been described. Probably only relatively few species remain to be discovered and named, but many subspecies are added yearly. With their powers of flight and wide distribution, most species of birds are more easily detected than are many animals. Only those that are solitary, shy, and restricted to remote regions have a chance of remaining undiscovered for any length of time. Altogether, the species are grouped into 170 families. Of the 27 recognized orders, 20 are represented by North American species.

The first four orders in this classification are the **ratite,** or flightless, birds, although as noted earlier, flightlessness is not restricted to these orders. The remaining orders are the **carinate** birds (birds with a keeled sternum).

> **Order Struthioniformes** (stroo´thi-on-i-for´meez) (L. *struthio,* ostrich, + *forma,* form): ostriches (Figure 26-30). The flightless ostrich of Africa *(Struthio camelus)* is the largest of living birds, with some specimens being 2.4 m tall and weighing 135 kg. The feet are provided with only two toes of unequal size covered with pads, which enable the birds to travel rapidly through sandy country.
> **Order Rheiformes** (re´i-for´meez) (Gr. mythology, *Rhea,* mother of Zeus, + form): rheas. These flightless birds are restricted to South America and are often called the American ostrich.
> **Order Casuariiformes** (kazh´u-ar´ee-i-for´meez) (N.L. *Casuarius,* type genus, + form): cassowaries, emus. This is a group of flightless birds found in Australia, New Guinea, and a few other islands. Some specimens may reach a height of 1.5 m.
> **Order Apterygiformes** (ap´te-rij´i-for´meez) (Gr. *a,* not, + *pteryx,* wing, + form): kiwis. Kiwis are flightless birds about the size of the domestic fowl, found only in New Zealand. The three species belong to the genus *Apteryx.* Only the merest vestige of a wing is present. The egg is extremely large for the size of the bird.
> **Order Tinamiformes** (tin-am´i-for´meez) (N.L. *Tinamus,* type genus, + form): tinamous. These are flying birds found in South America and Mexico. They resemble the ruffed grouse and are classed as game birds. There are more than 60 species in this order.
> **Order Sphenisciformes** (sfe-nis´i-for´meez) (Gr. *sphēniskos,* dim. of *sphen,* wedge, from the shortness of the wings, + form): penguins. Penguins are web-footed, marine swimmers of the southern seas, from Antarctica to the Galápagos Islands. Although carinate birds, they use their wings as paddles rather than for flight. The largest penguin is the emperor penguin *(Aptenodytes forsteri)* of the Antarctic which breeds in enormous rookeries on the shores of that region.
> **Order Gaviiformes** (gay´vee-i-for´meez) (L. *gavia,* bird, probably sea mew, + form): loons. Loons are remarkable swimmers and divers with short legs and heavy bodies. They live exclusively on fish and small aquatic forms. The familiar great northern diver *(Gavia immer)* is found mainly in northern waters of North America and Eurasia.
> **Order Podicipediformes** (pod´i-si-ped´i-for´meez) (L. *podex,* rump, + *pes, pedis,* foot): grebes. These are short-legged divers with lobate-webbed toes. The pied-billed grebe *(Podilymbus podiceps)* is a familiar example of this order. Grebes are most common in old ponds where they build their raftlike floating nests. Worldwide distribution.
> **Order Procellariiformes** (pro-sel-lar´ee-i-for´meez) (L. *procella,* tempest, + form): albatrosses, petrels, fulmars, shearwaters. All are marine birds with tubular nostrils (Figure 26-31). In wingspan (more than 3.6 m in some), albatrosses are the largest of flying birds. *Diomedea* is a common genus of albatrosses. Worldwide distribution.

Figure 26-30

Ostrich *Struthio camelus* of Africa. The largest of all living birds. Order Struthioniformes.
Photograph by C.P. Hickman, Jr.

Figure 26-31

The waved albatross *Diomedea irrorata* of the Galápagos Islands, like others of the order Procellariiformes, has a hooked bill with distinctive tubular nostrils.
Photograph by C.P. Hickman, Jr.

Figure 26-32

Brown pelican *Pelecanus occidentalis*. Pelicans catch fish by diving from the air.
Order Pelecaniformes.
Photograph by C.P. Hickman, Jr.

Figure 26-33

Ruffed grouse *Bonasa umbellus* "drumming" to attract a female. Order Galliformes.
Photograph by N. Bolen.

Order Pelecaniformes (pel′e-can-i-for′meez) (Gr. *pelekan*, pelican, + form): **pelicans** (Figure 26-32), **cormorants, gannets, boobies, and so on.** These are fish-eaters with throat pouch and all four toes of each foot included within the web. Mostly colonial nesters. Worldwide distribution, especially in the tropics.

Order Ciconiiformes (si-ko′nee-i-for′meez) (L. *ciconia*, stork, + form): **herons, bitterns, storks, ibises, spoonbills, flamingoes.** These are long-necked, long-legged, mostly colonial waders. A familiar eastern North American representative is the great blue heron *(Ardea herodias)*, which frequents marshes and ponds. Worldwide distribution.

Order Anseriformes (an′ser-i-for′meez) (L. *anser*, goose, + form): **swans, geese, ducks.** The members of this order have broad bills with filtering ridges at their margins, a foot web restricted to the front toes, and a long breastbone with a low keel. The common domestic mallard duck is *Anas platyrhynchos*. Worldwide distribution.

Order Falconiformes (fal′ko-ni-for′meez) (L. *falco*, falcon, + form): **eagles, hawks, vultures, falcons, condors, buzzards.** These are the great diurnal birds of prey. All are strong fliers with keen vision. Worldwide distribution.

Order Galliformes (gal′li-for′meez) (L. *gallus*, cock, + form): **quail, grouse** (Figure 26-33), **pheasants, ptarmigan, turkeys, domestic fowl.** These are chickenlike vegetarians with strong beaks and heavy feet. Some of the most desirable game birds are in this order. The bobwhite quail *(Colinus virginianus)* is found all over the eastern half of the United States. The ruffed grouse *(Bonasa umbellus)*, or partridge, is found in about the same region, but in the woods instead of the open pastures and grain fields, which the bobwhite frequents. Worldwide distribution.

Order Gruiformes (groo′i-for′meez) (L. *grus*, crane, + form): **cranes, rails, coots, gallinules.** These are prairie and marsh breeders. Worldwide distribution.

Order Charadriiformes (ka-rad′ree-i-for′meez) (N.L. *Charadrius*, genus of plovers, + form): **gulls, oyster catchers, plovers, sandpipers, terns, woodcocks, turnstones, lapwings, snipe, avocets, phalaropes, skuas, skimmers, auks, puffins.** All are shorebirds. They are strong fliers and are usually colonial. Worldwide distribution.

Order Columbiformes (co-lum′bi-for′meez) (L. *columba*, dove, + form): **pigeons, doves.** All have short necks, short legs, and a short, slender bill. Worldwide distribution.

Order Psittaciformes (sit′ta-si-for′meez) (L. *psittacus*, parrot, + form): **parrots, parakeets.** These have a hinged and movable upper mandible. Pantropical distribution.

Order Cuculiformes (ku-koo′li-for′meez) (L. *cuculus*, cuckoo, + form): **cuckoos, roadrunners.** The common cuckoo *(Cuculus canorus)* of Europe lays its eggs in the nests of smaller birds, which rear the young cuckoos. The American cuckoos, black billed and yellow billed, usually rear their own young. Worldwide distribution.

Order Strigiformes (strij′i-for′meez) (L. *strix*, screech owl, + form): **owls.** Owls are nocturnal predators with large eyes, powerful beaks and feet, and silent flight. Worldwide distribution.

Order Caprimulgiformes (kap′ri-mul′ji-for′meez) (L. *caprimulgus*, goatsucker, + form): **goatsuckers, nighthawks** (Figure 26-34), **whippoorwills.** The birds of this group are night and twilight feeders with small, weak legs and wide mouths fringed with bristles. The whippoorwills *(Antrostomus vociferus)* are common in the woods of the eastern states, and the nighthawk *(Chordeiles minor)* is often seen and heard in the evening flying around city buildings. Worldwide distribution.

Order Apodiformes (up-pod′i-for′meez) (Gr. *apous*, sandmartin; footless, + form): **swifts, hummingbirds.** These are small birds with short legs and rapid wingbeat. The familiar chimney swift *(Chaetura pelagica)* fastens its nest in chimneys by means of saliva. A swift found in China *(Collocalia)* builds a nest of saliva that is used by the Chinese for soup making. Most species of hummingbirds are found in the tropics, but there are 14 species in the United States, of which only one, the ruby-throated hummingbird, is found in the eastern part of the country. Most hummingbirds live on nectar, although some also catch insects. Worldwide distribution.

Order Coliiformes (ka-ly′i-for′meez) (Gr. *kolios*, green woodpecker, + form): **mousebirds.** These are small birds of uncertain relationship. Restricted to southern Africa.

Figure 26-34

Female nighthawk *Chordeiles minor* broods two 1-day-old young, their down barely visible under her breast. Nighthawks originally nested on beaches and in old stumps, but with the spread of civilization they discovered building roofs to be ideal nesting sites. Their nasal "peent" is a familiar summer evening sound above city streets. Order Caprimulgiformes.
Photograph by C.P. Hickman, Jr.

Order Trogoniformes (tro-gon´i-for´meez) (Gr. *trōgon*, gnawing, + form): **trogons.** Richly colored birds. Pantropical distribution.

Order Coraciiformes (ka-ray´see-i-for´meez or kor´uh-sigh´uh-for´meez) (N.L. *coracii* from Gr. *korakias*, a kind of chough [akin to *korax*, raven or crow], + form): **kingfishers** (Figure 26-35), **hornbills, and so on.** These birds have strong, prominent bills and colorful plumage. In the eastern half of the United States, the belted kingfisher *(Megaceryle alcyon)* is common along most waterways of any size. It makes a nest in a burrow in a high bank or cliff along a water course. Worldwide distribution.

Order Piciformes (pis´i-for´meez) (L. *picus*, woodpecker, + form): **woodpeckers, toucans, puffbirds, honeyguides.** Birds of this order have two of the toes extending forward and two backward and a highly specialized bill (chisel-like in woodpeckers). All nest in cavities. There are many species of woodpeckers in North America, the most common of which are the flickers and the downy, hairy, red-bellied, redheaded, and yellow-bellied woodpeckers. The largest is the pileated woodpecker, which is usually found in deep and remote woods. Worldwide distribution.

Order Passeriformes (pas´er-i-for´meez) (L. *passer*, sparrow, + form): **perching songbirds.** This is the largest order of birds, containing 56 families and 60% of all birds. Most have a highly developed syrinx. Their feet are adapted for perching on thin stems and twigs. The young are altricial. To this order belong many birds with beautiful songs such as the skylark, nightingale, hermit thrush, mockingbird, meadowlark, robin, and hosts of others. Others of this order, such as the swallow, magpie, starling, crow, raven, jay, nuthatch, and creeper, have no songs worthy of the name. Worldwide distribution.

— SUMMARY

The 8600 species of living birds are egg-laying, endothermic vertebrates covered with feathers and having the forelimbs modified as wings. Birds have evolved from archosaurian reptiles during the Mesozoic era and bear numerous reptilian morphological characters. The oldest known fossil bird, *Archaeopteryx* from the Jurassic period of the Mesozoic era, was even more reptilelike than modern birds.

The adaptations of birds for flight are of two basic kinds: those reducing body weight and those promoting more power for flight. Feathers, the hallmark of birds, are complex derivatives of reptilian scales and combine lightness with

Figure 26-35

Kookaburra *Dacelo novaeguineae*, the wood kingfisher of Australia, famous for its cry that resembles demented laughter. Order Coraciiformes.
Photograph by C.P. Hickman, Jr.

strength, water repellency, and high insulative value. Body weight is further reduced by elimination of some bones, fusion of others (to provide rigidity for flight), and the presence in many bones of hollow, air-filled spaces. The light, horny bill, replacing the heavy jaws and teeth of reptiles, serves as both hand and mouth for all birds and is variously adapted for different feeding habits.

Adaptations that provide power for flight include a high metabolic rate and body temperature coupled with an energy-rich diet; a highly efficient respiratory system consisting of a system of air sacs arranged to pass air through the lungs during both inspiration and expiration; powerful flight and leg muscles arranged to place muscle weight near the bird's center of gravity; and an efficient, high-pressure circulation.

Birds have keen eyesight, good hearing, poorly developed sense of smell, and superb coordination for flight. The metanephric kidneys produce uric acid as the principal nitrogenous waste.

Birds fly by applying the same aerodynamic principles as an airplane and using similar equipment: wings for lift and support, a tail for steering and landing control, and wing slots for control at low flight speed. Flightlessness in birds is unusual but has evolved independently in several bird orders, usually on islands where terrestrial predators are absent; all are derived from flying ancestors.

Bird migration refers to regular movements between summer nesting places and wintering regions. Spring migration to the north where more food is available for nestlings enhances reproductive success. Many cues are used for direction finding in migration, including innate sense of direction and ability to navigate by the sun, the stars, or the earth's magnetic field.

The highly developed social behavior of birds is manifested in vivid courtship displays, mate selection, territorial behavior, and incubation of eggs and care of the young.

Review questions

1. Explain the significance of the discovery of *Archaeopteryx*. Why did this fossil prove beyond reasonable doubt that birds evolved from reptiles?
2. Birds are broadly divided into two groups—ratite and carinate. Explain what these terms mean and briefly discuss the appearance of flightlessness in birds.
3. The special adaptations of birds all contribute to two essentials for flight—more power and less weight. Explain how each of the following contributes to one or the other (or both) of these two essentials: feathers, skeleton, muscle distribution, digestive system, circulatory system, respiratory system, excretory system, reproductive system.
4. How do marine birds rid themselves of excess salt?
5. In what ways are the bird's ears and eyes specialized for their life-style?
6. Explain how the bird wing is designed to provide lift. What design features help to prevent stalling at low flight speeds?
7. Describe the four basic forms of bird wings.
8. What are the advantages of seasonal migration for birds?
9. Describe the different navigational resources birds may use in long-distance migration.
10. What are some of the advantages of social aggregation among birds?
11. More than 90% of all bird species are monogamous. Explain why monogamy is so much more common among birds than among mammals.
12. Briefly describe an example of polygyny among birds.
13. Define the terms precocial and altricial as they relate to birds.
14. Offer some examples of how human activities have been harmful to birds.

Selected references

See also general references for Part Two, p. 590.

Baker, R. (ed.). 1981. The mystery of migration. New York, The Viking Press. *Handsomely illustrated, popularized account.*

Burton, R. 1985. Bird behavior. New York, Alfred A. Knopf, Inc. *Well-written and well-illustrated summary of bird behavior.*

Emlen, S.T. 1975. The stellar-orientation system of a migratory bird. Sci. Am. **233**:102-111. (Aug.). *Describes fascinating research with indigo buntings, revealing their ability to navigate by the center of celestial rotation at night.*

Feduccia, A. 1980. The age of birds. Cambridge, Mass., Harvard University Press. *Semipopular but authoritative account of bird evolution. Excellent text and illustrations.*

Phillips, J.G., P.J. Butler, and P.J. Sharp. 1985. Physiological strategies in avian biology. London, Chapman & Hall. *Functional topics include locomotion, migration, thermoregulation, osmoregulation, and reproduction.*

Terres, J.K. 1980. The Audubon Society encyclopedia of North American birds. New York, Alfred A. Knopf, Inc. *Comprehensive, authoritative, and richly illustrated.*

Welty, J.C. 1982. The life of birds, ed. 3. Philadelphia, W.B. Saunders Co. *Among the best of the ornithology texts; lucid style and excellent illustrations.*

CHAPTER 27

THE MAMMALS

Phylum Chordata

Class Mammalia

Mammals, with their highly developed nervous system and numerous ingenious adaptations, occupy almost every environment on earth that supports life. Although not a large group (about 4000 species as compared with 8600 species of birds, approximately 21,700 species of fishes, and 800,000 species of insects), the class Mammalia (mam-may′lee-a) (L. *mamma*, breast) is overall the most biologically successful group in the animal kingdom, with the possible exception of the insects. Many potentialities that dwell more or less latently in other vertebrates are highly developed in mammals. Mammals are exceedingly diverse in size, shape, form, and function. They range in size from the recently discovered Kitti's hog-nosed bat, weighing only 1.5 g, to the whales, some of which exceed 100 tons.

Yet, despite their adaptability and in some instances because of it, mammals have been influenced by the presence of humans more than any other group of animals. We have domesticated numerous mammals for food and clothing, as beasts of burden, and as pets. We use millions of mammals each year in biomedical research. We have introduced alien mammals into new habitats, occasionally with benign results but more frequently with unexpected disaster. Although history provides us with numerous warnings, we continue to overcrop valuable wild stocks of mammals. The whale industry has threatened itself with total collapse by exterminating its own resource—a classic example of self-destruction in the modern world, in which competing segments of an industry are intent only on reaping all they can today as though tomorrow's supply were of no concern whatever. In some cases destruction of a valuable mammalian resource has been deliberate, such as the officially sanctioned (and tragically successful) policy during the Indian wars of exterminating the bison to drive the Plains Indians into starvation. Although commercial hunting has declined, the ever-increasing human population with the accompanying destruction of wild habitats has harassed and disfigured the mammalian fauna. Approximately 300 species and subspecies of mammals are considered endangered by the International Union for the Conservation of Nature and Natural Resources (IUCN), including all cetaceans, cats, otters, and primates (except humans).

We are becoming increasingly aware that our presence on this planet as the

most powerful product of organic evolution makes us responsible for the character of our natural environment. Since our welfare has been and continues to be
closely related to that of the other mammals, it is clearly in our interest to preserve
the natural environment of which all mammals, ourselves included, are a part. We
need to remember that nature can do without us but we cannot exist without
nature.

ORIGIN AND RELATIONSHIPS

In the early Mesozoic era, long before the great dinosaurs had reached the peak of
their evolutionary success, a group of reptiles with mammal-like characteristics
appeared. These were **therapsids** (Figure 27-1). The evolution of the therapsids
and their descendants was accompanied by several structural changes that
brought them ever closer to full mammalian status. The clumsy limbs of the reptile
that stuck out laterally were replaced by straight legs held close to the body, which
provided speed and efficiency for hunting. Since reptilian stability was sacrificed
by raising the animal from the ground, the muscular coordination center of the
brain, the cerebellum, took on a greatly expanded role. Among the many changes
in the bony structure of the head was the separation of air and food passages in the
mouth. This enabled the animal to breathe while holding prey in its mouth. A
stronger jaw joint made possible prolonged chewing and some predigestion of
food. And at some point, probably in the late Permian period, the premammals
acquired those two most characteristic of all mammalian identification tags: hair
and mammary glands.

Most of the living mammals belong to the subclass Theria and have
descended from a common ancestor of the Jurassic period some 150 million years
ago. However, monotremes (subclass Prototheria), the egg-laying mammals of
Australia, Tasmania, and New Guinea, are so different from the others and possess so many reptilian characteristics that they are believed to have descended
from an entirely different mammal-like reptile. The separation of Prototheria and
Theria probably occurred approximately 50 million years earlier in the Triassic
period. The geological record during the Jurassic and Cretaceous periods is fragmentary, in large part because the mammals of these periods were creatures the
size of a rat or smaller, with fragile bones that fossilized only under the most ideal
circumstances.

When the dinosaurs vanished near the beginning of the Cenozoic era, the
mammals suddenly expanded. This point, about 70 million years ago, marks the
beginning of the age of mammals. This is partly attributed to the numerous ecological niches vacated by the reptiles, into which the mammals could move as their
divergent adaptations fitted them. There were other reasons for their success.
Mammals were agile, warm blooded, and insulated with hair; they had developed
placental reproduction and suckled their young, thus dispensing with vulnerable
eggs and nests; and they were more intelligent than any other animal alive. During
the Eocene and Oligocene epochs of the Tertiary period (55 to 30 million years
ago), the mammals flourished and reached their peak. In terms of number of
species, this was the golden age of mammals. They have declined somewhat in
number since then, especially within the last million years, because humans
became formidable adversaries of wildlife. Nevertheless, mammals as a whole are
a secure group, dominating the land environment now as they did 50 million years
ago.

Figure 27-1

Evolution of the mammals. Mammals take their origin from the therapsid reptiles, which were probably warm blooded. The most primitive mammals are the monotremes, which lay reptilelike eggs. They may have arisen from a group of therapsid reptiles distinct from those that gave rise to the viviparous therians, or "true" mammals. The pouched marsupials probably arose in the Jurassic period, although some authorities believe they separated from the placental eutherians in the Cretaceous period. The great radiation of modern mammal orders occurred during the Cretaceous and Tertiary periods.

CHARACTERISTICS

1. **Body covered with hair,** but reduced in some
2. **Integument with sweat, scent, sebaceous, and mammary glands**
3. Skeletal features: skull with **two occipital condyles, seven cervical vertebrae** (usually), ribs attached only to thoracic vertebrae, and often an elongated tail
4. Mouth with **heterodont teeth** on both jaws; lower jaw of single enlarged bone, the **dentary**
5. **Movable eyelids** and fleshy external ears
6. Four limbs (reduced or absent in some) adapted for many forms of locomotion
7. Circulatory system of a four-chambered heart, **persistent left aorta,** and **nonnucleated, biconcave red blood corpuscles**
8. Respiratory system of lungs and voice box; **secondary (false) palate** separates air and food passages; **muscular diaphragm** separates thoracic and abdominal cavities
9. Excretory system of metanephros kidneys and ureters that usually open into a bladder
10. Brain highly developed, especially **neocerebrum;** 12 pairs of cranial nerves
11. Endothermic and homeothermic
12. Cloaca present only in monotremes
13. Separate sexes; reproductive organs of a penis, testes (usually in a scrotum), ovaries, oviducts, and vagina
14. Internal fertilization; **eggs develop in a uterus** with **placental attachment** (except in monotremes); **fetal membranes (amnion, chorion, allantois);** sex determination by males (heterogametic)
15. Young nourished by **milk from mammary glands**

Since mammals and birds both evolved from reptiles, there are many structural similarities among the three groups. It is, in fact, much easier to point to numerous resemblances between the mammals and the reptiles than to point to characteristics that are unique to mammals. **Hair** is the most obvious mammalian characteristic, although it is vastly reduced in some (such as whales) and although reptilian scales, from which hair is derived, may persist (such as on tails of rats and beavers). A second unique characteristic of mammals is the method of nourishing their young with **milk-secreting glands;** reptiles have nothing remotely similar. Although less obvious, several important differences are present in cranial and jaw structure and jaw articulation. Placental mammals have **diphyodont teeth** (milk teeth replaced by a permanent set of teeth) rather than reptilian **polyphyodont teeth** (successive sets of teeth). But the single most important factor contributing to the success of mammals is the remarkable development of the **neocerebrum** permitting a level of adaptive behavior, learning, curiosity, and intellectual activity far beyond the capacity of any reptile.

Class Mammalia is divided into two subclasses: Prototheria and Theria. Subclass **Prototheria** includes the monotremes, or egg-laying mammals. Subclass **Theria** includes two infraclasses, the **Metatheria,** with one order (marsupials), and the **Eutheria,** with 19 orders, all of which are placental mammals. A complete classification is found on pp. 584 to 587.

STRUCTURAL AND FUNCTIONAL ADAPTATIONS OF MAMMALS
Integument and Its Derivatives

The mammalian skin and its modifications especially distinguish mammals as a group. As the interface between the animal and its environment, the skin is strongly molded by the animal's way of life. In general the skin is thicker in mammals than in other classes of vertebrates, although as in all vertebrates it is made up of

A hair is more than a strand of keratin. It consists of three layers: the medulla or pith in the center of the hair, the cortex with pigment granules next to the medulla, and the outer cuticle composed of imbricated scales. The hair of different mammals shows a considerable range of structure. It may be deficient in cortex, such as the brittle hair of deer, or it may be deficient in medulla, such as the hollow, air-filled hairs of the wolverine, so favored by northerners for trimming the hoods of parkas because it resists frost accumulation. The hairs of rabbits and some others are scaled to interlock when pressed together. Curly hair, such as that of sheep, grows from curved follicles.

Figure 27-2

American beaver *Castor canadensis* (order Rodentia, family Castoridae) cutting up a trembling aspen tree. This second largest rodent has a heavy, waterproof pelage consisting of long, tough guard hairs overlying the thick, silky underhair so valued in the fur trade. Other adaptations for an aquatic life are nostrils and ear canals provided with valves, webbed hindfeet, flexible forefeet for gripping branches, a flat and broad paddlelike tail for swimming, diving, and signaling, and a mouth that closes behind the incisors to permit gnawing underwater. Photograph by L.L. Rue, III.

epidermis and **dermis** (see Figure 28-1, p. 593). Among the mammals the dermis becomes much thicker than the epidermis. The epidermis is relatively thin where it is well protected by hair, but in places that are subject to much contact and use, such as the palms or soles, its outer layers become thick and cornified with keratin.

Hair

Hair is especially characteristic of mammals, although humans are not very hairy creatures, and in whales hair is reduced to only a few sensory bristles on the snout. A hair grows out of a hair follicle that, although an epidermal structure, is sunk into the dermis of the skin (see Figure 28-1, p. 593). The hair grows continuously by rapid proliferation of cells in the follicle. As the hair shaft is pushed upward, new cells are carried away from their source of nourishment and die, turning into the same dense type of keratin that constitutes nails, claws, hooves, and feathers.

Mammals characteristically have two kinds of hair forming the **pelage** (fur coat): (1) dense and soft **underhair** for insulation and (2) coarse and longer **guard hair** for protection against wear and to provide coloration. The underhair traps a layer of insulating air; in aquatic animals, such as the fur seal, otter, and beaver, it is so dense that it is almost impossible to wet it. In water the guard hairs become wet and mat down over the underhair forming a protective blanket (Figure 27-2).

When a hair reaches a certain length, it stops growing. Normally it remains in the follicle until a new growth starts, whereupon it falls out. In humans, hair is shed and replaced throughout life. But in most mammals, there are periodic molts of the entire coat.

In the simplest cases, such as foxes and seals, the coat is shed once each year during the summer months. Most mammals have two annual molts, one in the spring and one in the fall. The summer coat is always much thinner than the winter coat and is usually a different color. Several of the northern mustelid carnivores, such as the weasel, have white winter coats and colored summer coats. It was once believed that the white inner pelage of arctic animals conserves body heat by reducing radiation loss, but recent research has shown that dark and white pelages radiate heat equally well. The winter white of arctic animals is simply camouflage in a land of snow. The varying hare of North America has three annual molts: the white winter coat is replaced by a brownish gray summer coat, and this is replaced in autumn by a grayer coat, which is soon shed to reveal the winter white coat beneath (Figure 27-3). The white fur of arctic mammals in winter (leukemism) is not to be confused with albinism, caused by a recessive gene that blocks pigment formation. Albinos have red eyes and pinkish skin, whereas arctic animals in their winter coats have dark eyes and often dark-colored ear tips, noses, and tail tips.

Outside the Arctic, most mammals wear somber colors that are protective. Often the species is marked with "salt-and-pepper" coloration or a disruptive pattern that helps make it inconspicuous in its natural surroundings. Examples are the spots of leopards and fawns and the stripes of tigers. Other mammals, such as skunks, advertise their presence with conspicuous warning coloration.

The hair of mammals has become modified to serve many purposes. The bristles of hogs, vibrissae on the snouts of most mammals, and spines of porcupines and their kin are examples.

A   B

Figure 27-3

Snowshoe, or varying, hare *Lepus americanus* (order Lagomorpha) in brown summer coat, **A**, and white winter coat, **B**. In winter, extra hair growth on the hindfeet broadens the animal's support in snow. Snowshoe hares are common residents of the taiga (northern coniferous forests) and are an important food for lynxes, foxes, and other carnivores. Population fluctuations of hares and their predators are closely related. Photographs by L.L. Rue, III.

Vibrissae, commonly called "whiskers," are really sensory hairs that provide a sensitive tactile sense to many mammals. The slightest movement of a vibrissa generates impulses in sensory nerve endings that travel to special sensory areas in the brain. The vibrissae are especially long in nocturnal and burrowing animals.

Porcupines, hedgehogs, echidnas, and a few other mammals have developed an effective and dangerous spiny armor; the spines of the common North American porcupine break off at the bases when struck and, aided by backward-pointing hooks on the tips, work deeply into their victims. To assist slow learners, such as dogs, in understanding what they are dealing with, porcupines rattle the spines and prominently display the white markings on the quills toward their tormentors (Figure 27-4).

Horns and antlers

Three kinds of horns or hornlike substances are found in mammals. **True horns,** found in ruminants such as sheep and cattle, are hollow sheaths of keratinized epidermis that embrace a core of bone arising from the skull. Horns are not normally shed, are not branched (although they may be greatly curved), and are found in both sexes.

Antlers of the deer family are entirely bone when mature. During their annual growth, antlers develop beneath a covering of highly vascular soft skin called "velvet" (Figure 27-5). When growth of the antlers is complete just before the breeding season, the blood vessels constrict and the stag tears off the velvet by rubbing the antlers against trees. The antlers are dropped after the breeding season. New buds appear a few months later to herald the next set of antlers. For several years each new pair of antlers is larger and more elaborate than the previous set. The annual growth of antlers places a strain on the mineral metabolism, since during the growing season a large moose or elk must accumulate 50 or more pounds of calcium salts from its vegetable diet.

The **rhinoceros horn** is the third kind of horn. Hairlike horny fibers arise from the dermal papillae and are cemented together to form a single horn.

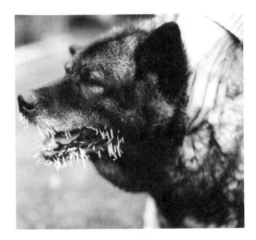

Figure 27-4

Dogs are frequent victims of the porcupine's impressive armor. Unless removed (usually by a veterinarian) the quills will continue to work their way deeper into the flesh causing great distress and may even lead to the victim's death. Photograph by R.E. Treat.

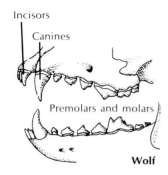

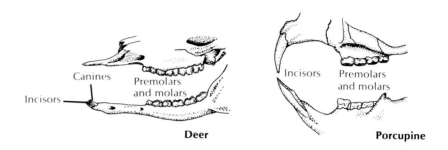

Wolf　　　　**Deer**　　　　**Porcupine**

Figure 27-8

Adaptations of mammal tooth patterns for different diets. Sharp canines of the wolf are adapted for stabbing, and premolars and molars are for cutting rather than grinding. Browsing deer has predominantly grinding teeth; lower incisors and canines bite against a horny pad in the upper jaw. Porcupine has no canines; self-sharpening incisors are used for gnawing.
Modified from Carrington, R. 1968. The mammals. New York, Life Nature Library, Time-Life Books, Inc.

Figure 27-9

Eastern chipmunk *Tamias striatus* with cheek pouches stuffed with seeds to be carried to an underground cache. It will try to store at least a half bushel of food for the winter. It hibernates but awakens periodically to eat some of its cached food. Order Rodentia, family Sciuridae.
Photograph by L.L. Rue, III.

Food and Feeding

Mammals have exploited an enormous variety of food sources; some mammals require highly specialized diets, whereas others are opportunistic feeders that thrive on diversified diets. In all, food habits and physical structure are inextricably linked. A mammal's adaptations for attack and defense and its specializations for finding, capturing, reducing, swallowing, and digesting food all determine a mammal's shape and habits.

Teeth, perhaps more than any other single physical characteristic, reveal the life-style of a mammal (Figure 27-8). It has been claimed that, if all mammals except humans were extinct and represented only by fossil teeth, we could still construct a classification nearly as correct as the one we have now, which is based on all anatomical features. All mammals have teeth, except certain whales, monotremes, and anteaters, and their modifications are correlated with what the mammal eats.

Typically, mammals have a **diphyodont** dentition, that is, two sets of teeth: a set of deciduous, or milk, teeth that is replaced by a set of permanent teeth. In any given species, mammalian teeth are modified to perform specialized tasks such as cutting, nipping, gnawing, seizing, tearing, grinding, or chewing. Teeth differentiated in this manner in the individual are called **heterodont,** in contrast to the uniform, **homodont** dentition characteristic of lower vertebrates.

Usually four types of teeth are recognized. **Incisors,** with simple crowns and slightly sharp edges, are mainly for snipping or biting; **canines,** with long conical crowns, are specialized for piercing; **premolars,** with compressed crowns and one or two cusps, are suited for shearing and slicing; and **molars,** with large bodies and variable cusp arrangement, are for crushing and mastication. Molars always belong to the permanent set.

Feeding specializations

On the basis of food habits, animals may be divided into herbivores, carnivores, omnivores, and insectivores.

Herbivorous animals that feed on grasses and other vegetation form two main groups: **browsers** or **grazers,** such as the ungulates (horses, swine, deer, antelope, cattle, sheep, and goats), and the **gnawers** and **nibblers,** such as the rodents (Figure 27-9) and rabbits. In herbivores the canines are suppressed, whereas the molars are broad and high crowned and bear enamel ridges for grinding. Rodents have chisel-shaped incisors that grow throughout life and must be worn away to keep pace with their continual growth.

Herbivorous mammals have a number of interesting adaptations for dealing with their massive diet of plant food. **Cellulose,** the structural carbohydrate of plants, is a potentially nutritious foodstuff, composed of long chains of glucose units. However, the glucose molecules in cellulose are linked by a type of chemical bond that few enzymes can attack. No vertebrates synthesize cellulose-splitting enzymes. Instead, the herbivorous vertebrates harbor a microflora of anaerobic bacteria in huge fermentation chambers in the gut. These bacteria break down and metabolize the cellulose, releasing a variety of fatty acids, sugars, and starches that the host animal can absorb and use.

In some herbivores, such as horses and rabbits, the gut has a capacious sidepocket, or diverticulum, called a **cecum,** which serves as a fermentation chamber and absorptive area. Hares, rabbits, and some rodents often eat their fecal pellets (**coprophagy**), giving the food a second pass through the fermenting action of the intestinal bacteria. Coprophagy may also provide an opportunity for the animal to obtain vitamins produced by the cecal bacteria.

The **ruminants** (cattle, bison, buffalo, goats, antelopes, sheep, deer, giraffes, and okapis) have a huge **four-chambered stomach** (Figure 27-10). When a ruminant feeds, grass passes down the esophagus to the **rumen,** where it is broken down by the rich microflora and then formed into small balls of cud. At its leisure the ruminant returns the cud to its mouth where the cud is deliberately chewed at length to crush the fiber. Swallowed again, the food returns to the rumen where it is digested by the cellulolytic bacteria. The pulp passes to the **reticulum,** then to the **omasum,** and finally to the **abomasum** ("true" stomach), where proteolytic enzymes are secreted and normal digestion takes place.

Herbivores in general have large, long digestive tracts and must eat a considerable amount of plant food to survive. A large African elephant weighing 6 tons must consume 135 to 150 kg (300 to 400 pounds) of rough fodder each day to obtain sufficient nourishment for life.

Carnivorous mammals feed mainly on herbivores. This group includes foxes, weasels, cats, dogs, wolverines, fishers, lions, and tigers. Carnivores are well-equipped with biting and piercing teeth and powerful clawed limbs for killing their prey. Since their protein diet is much more easily digested than is the woody food of herbivores, their digestive tract is shorter and the cecum small or absent. Carnivores eat separate meals and have much more leisure time for play and exploration.

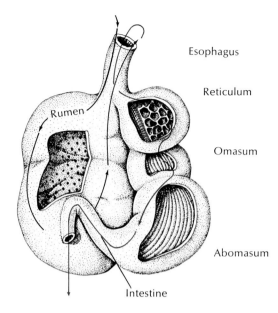

Figure 27-10

Ruminant's stomach. Food passes first to the rumen (sometimes through the reticulum) and then is returned to the mouth for chewing (chewing the cud, or rumination) *(black arrow).* After reswallowing, food returns to the rumen or passes directly to the reticulum, omasum, and abomasum for final digestion *(red arrow).*

Figure 27-11

Lioness *Panthera leo* with Thompson gazelle she killed. Lions stalk prey, then charge suddenly to surprise the victim. They lack stamina for a long chase. Lions will gorge themselves with the kill, then sleep and rest for periods as long as a week before eating again. Order Carnivora, family Felidae.
Photograph by L.L. Rue, III.

The renowned fecundity of meadow mice, and the effect of removing the natural predators from rodent populations, is felicitously expressed in this excerpt from Thornton Burgess's "Portrait of a Meadow Mouse":

He's fecund to the nth degree
In fact this really seeems to be
His one and only honest claim
To anything approaching fame.
In just twelve months, should all survive,
A million mice would be alive—
His progeny. And this, 'tis clear,
Is quite a record for a year.
Quite unsuspected, night and day
They eat the grass that would be hay.
On any meadow, in a year,
The loss is several tons, I fear.
Yet man, with prejudice for guide,
The checks that nature doth provide
Destroys. The meadow mouse survives
And on stupidity he thrives.

Figure 27-12

Alaskan brown bear *Ursus arctos* (order Carnivora) with a large salmon captured during the spawning run. All bears are opportunistic omnivores that take advantage of any seasonally abundant resource. They are highly dependent on vegetable food but also kill small mammals, scavenge, and rob all bees' nests they can find and enter. Bears are the least carnivorous of the carnivores. Photograph by L.L. Rue, III.

In general, carnivores lead more active—and by human standards more interesting—lives than do the herbivores. Since a carnivore must find and catch its prey, there is a premium on intelligence; many carnivores, such as the cats, are noted for their stealth and cunning in hunting prey (Figure 27-11). Although evolution seems to have favored the carnivores, their success has led to a selection of herbivores capable either of defending themselves or of detecting and escaping carnivores. Thus, for the herbivores, there has been a premium on keen senses and agility. Some herbivores, however, survive by virtue of their sheer size (for example, elephants) or by defensive group behavior (for example, muskoxen).

Humans have changed the rules in the carnivore-herbivore contest. Carnivores, despite their intelligence, have suffered much from human presence and have been virtually exterminated in some areas. Herbivores, on the other hand, especially the rodents with their potent reproductive ability, have consistently defeated our most ingenious efforts to banish them from our environment. The problem of rodent pests in agriculture has intensified; we have removed carnivores, which served as the herbivores' natural population control, but have not been able to devise a suitable substitute.

Omnivorous mammals live on both plants and animals for food. Examples are pigs, raccoons, rats, bears (Figure 27-12), humans, and most other primates. Many carnivorous forms also eat fruits, berries, and grasses when hard pressed. The fox, which usually feeds on mice, small rodents, and birds, eats frozen apples, beechnuts, and corn when its normal food sources are scarce.

Insectivorous mammals subsist chiefly on insects and grubs. Examples are moles, shrews, anteaters, and most bats. The insectivorous category is not a sharply distinguished one, however, because many omnivores, carnivores, and even some herbivores eat insects on occasion (sometimes by accident).

For most mammals, searching for food and eating occupy most of their active life. Seasonal changes in food supplies are considerable in temperate zones. Living may be easy in the summer when food is abundant, but in winter many carnivores must range far and wide to eke out a narrow existence. Some migrate to regions where food is more abundant. Others hibernate and sleep the winter months away.

But there are many provident mammals that build up food stores during periods of plenty. This habit is most pronounced in rodents, such as squirrels, chipmunks, gophers, and certain mice. All tree squirrels—red, fox, and gray—collect nuts, conifer seeds, and fungi and bury these in caches for winter use. Often each item is hidden in a different place (scatter hoarding) and marked by a scent to assist relocation in the future. The chipmunk (Figure 27-9) is one of the greatest providers because it spends the autumn months collecting nuts and seeds. Some of its caches may exceed a bushel in size.

Body weight and food consumption

The relationship between body size and metabolic rate is discussed in relation to food consumption of birds (p. 546) and is treated again in Chapter 36 (p. 796). The smaller the animal, the greater is its metabolic rate and the more it must eat relative to its body size. This happens because the metabolic rate of an animal—and therefore the amount of food it must eat to sustain this metabolic rate—varies in rough proportion to the surface area rather than to the body weight. Surface area is proportional to a 0.7 power of body weight, and the amount of food a mammal (or bird) eats is also roughly proportional to a 0.7 power of its body weight. This means that as the size of animals gets smaller, their metabolic rate (usually measured as oxygen consumption per gram of body weight) becomes more intense. A 3 g mouse will consume *per gram body weight* five times more food than does a 10 kg dog and about 30 times more food than does a 50,000 kg elephant. One can easily see why small mammals (shrews, bats, and mice) must spend much more time hunting and eating food than do large mammals. The smallest shrews weighing only 2 g may eat more than their body weight each day and will starve to death in a few hours if deprived of food (Figure 27-13). In contrast, a large carnivore can remain fat and healthy with only one meal every few days. The mountain lion is known to kill an average of one deer a week, although it will kill more frequently when game is abundant.

___ Migration

Migration is a much more difficult undertaking for mammals than for birds; not surprisingly, few mammals make regular seasonal migrations, preferring instead to center their activities in a defined and limited home range. Nevertheless, there are some striking examples of mammalian migrations. More migrators are found in North America than on any other continent.

An example is the barren-ground caribou of Canada and Alaska, which undertakes direct and purposeful mass migrations spanning 160 to 1100 km (100 to 700 miles) twice annually (Figure 27-14). From winter ranges in the boreal forests (taiga), they migrate rapidly in late winter and spring to calving ranges on the barren grounds (tundra). The calves are born in mid-June. Harassed by warble and nostril flies that bore into their flesh and by mosquitoes and wolves, the caribou move southward in July and August, feeding little along the way. In September they reach the forest, feeding there almost continuously on low ground vegetation. Mating (rut) occurs in October.

The drastic decline in caribou populations in recent years has been attributed to several factors, including habitat deterioration resulting from exploration and development in the North, but especially to excessive hunting. For example, the Western Arctic herd in Alaska exceeded 250,000 caribou in 1970. Following 5 years of heavy unregulated hunting, a 1976 census revealed only about 65,000 animals left. Hunting by people was then restricted and aerial hunting of wolves (a major predator of the calves) was opened. These changes produced a remarkable recovery. By 1980 the herd had increased to 140,000 and was expected to reach its original population of 250,000 by 1985.

The plains bison, before its deliberate near extinction by humans, made huge circular migrations to separate summer and winter ranges.

The longest mammal migrations are made by the oceanic seals and whales. One of the most remarkable migrations is that of the fur seal, which breeds on the Pribilof Islands approximately 300 km (190 miles) off the coast of Alaska and north of the Aleutian Islands. From wintering grounds off southern California the females journey as much as 2800 km (1700 miles) across open ocean, arriving in the spring at the Pribilofs where they congregate in enormous numbers (Figure 27-15). The young are born within a few hours or days after arrival of the cows. Then the bulls, having already arrived and established territories, collect harems of cows, which they guard with vigilance. After the calves have been nursed for approximately 3 months, cows and juveniles leave for their long migration southward. The bulls do not follow but remain in the Gulf of Alaska during the winter.

Figure 27-13

The masked (common) shrew *Sorex cincereus* (order Insectivora) feeding on a deer mouse it has killed. Note small size of shrew relative to much larger mouse. This tiny but fierce mammal, with a prodigious appetite for insects, mice, snails, and worms, spends most of its time underground and so is seldom seen by humans. Shrews are a primitive group believed to closely resemble the insectivorous ancestors of placental mammals.
Photograph by C.G. Hampson.

Figure 27-14

Barren-ground caribou, *Rangifer tarandus groenlandicus*, of Canada and Alaska. **A,** Adult male caribou in autumn pelage and antlers in velvet. **B,** Summer and winter ranges of some major caribou herds in Canada and Alaska (other herds not shown occur on Baffin Island and in western and central Alaska). The principal spring migration routes are indicated by arrows; routes vary considerably from year to year. Photograph by C.P. Hickman, Jr.

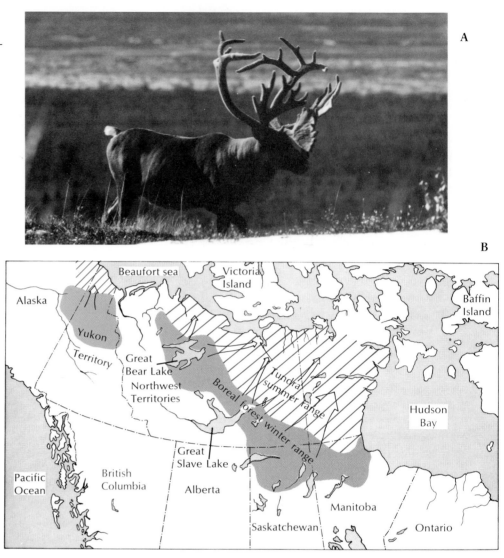

Figure 27-15

Annual migrations of the fur seal, showing the separate wintering grounds of males and females. Both males and females of the larger Pribilof population migrate in early summer to the Pribilof Islands, where the females give birth to their pups and then mate with the males.

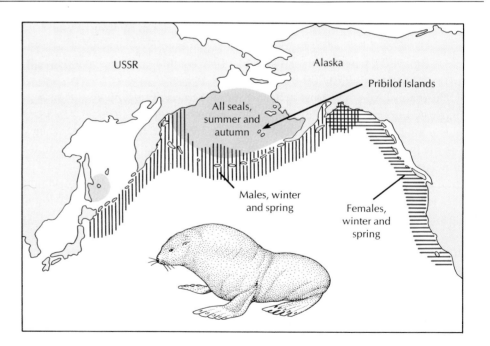

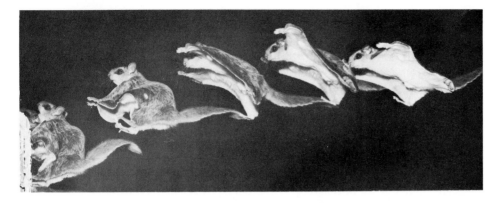

Figure 27-16

Flying squirrel *Glaucomys sabrinus* coming in for a landing, photographed with stroboscopic flash. Area of undersurface is nearly trebled when gliding skin is spread. Glides of 40 to 50 m are possible; good maneuverability during flight is achieved by adjusting the position of the gliding skin with special muscles. Flying squirrels are nocturnal and have superb night vision. Order Rodentia, family Sciuridae.
Photograph by C.G. Hampson.

Although we might expect bats, the only winged mammals, to use their gift of flight to migrate, few of them do. Most spend the winter in hibernation. The four species of American bats that do migrate, the red bat (Figure 27-17), silver-haired bat, hoary bat, and Brazilian free-tailed bat, spend their summers in the northern or western states and their winters in the southern United States or Mexico.

Flight and Echolocation

Mammals have not exploited the skies to the same extent that they have the terrestrial and aquatic environments. However, many mammals scamper about in trees with amazing agility; some can glide from tree to tree, and one group, the bats, is capable of full flight. Gliding and flying evolved independently in several groups of mammals, including the marsupials, rodents, flying lemurs, and bats. Anyone who has watched a gibbon perform in a zoo realizes there is something akin to flight in this primate, too. Among the arboreal squirrels, all of which are nimble acrobats, by far the most efficient is the flying squirrel (Figure 27-16). These forms actually glide rather than fly, using the gliding skin that extends from the sides of the body.

Bats, the only group of flying mammals, are nocturnal insectivores and thus occupy a niche left vacant by birds (Figure 27-17). Their outstanding success is

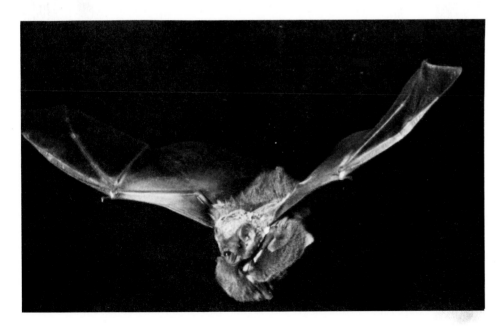

Figure 27-17

Red bat *Lasiurus borealis* (order Chiroptera) in flight with four young. This species is unusual in giving birth to three or four young; most bats have one or two. The mother carries the young until their combined weight may exceed her own weight; they then are left in the roost, usually a tree. Mortality among the young is rather high. These medium-sized bats are distributed over all the eastern and southern United States. They are strong fliers, migrating northward in spring and southward in autumn.
Photograph by L.L. Rue, III.

attributed to two things: flight and the capacity to navigate by echolocation. Together these adaptations enable bats to fly and avoid obstacles in absolute darkness, to locate and catch insects with precision, and to find their way deep into caves (another habitat largely ignored by both mammals and birds) where they sleep away the daytime hours.

Research has been concentrated on members of the family Vespertilionidae, to which most of the common North American bats belong. When in flight, these bats emit short pulses 5 to 10 msec in duration in a narrow directed beam from the mouth. Each pulse is frequency modulated; that is, it is highest at the beginning, up to 100,000 hertz (Hz, cycles per second), and sweeps down to perhaps 30,000 Hz at the end. Sounds of this frequency are ultrasonic to the human ear, which has an upper limit of about 20,000 Hz. When the bat is searching for prey, it produces about 10 pulses per second. If a prey is detected, the rate increases rapidly up to 200 pulses per second in the final phase of approach and capture. The pulses are spaced so that the echo of each is received before the next pulse is emitted, an adaptation that prevents jamming. Since the transmission-to-reception time decreases as the bat approaches an object, it can increase the pulse frequency to obtain more information about the object. The pulse length is also shortened as the bat nears the object.

The external ears of bats are large, like hearing trumpets, and shaped variously in different species. Less is known about the bat's inner ear, but it obviously is capable of receiving the ultrasonic sounds emitted. Biologists believe that bat navigation is so refined that the bat builds up a mental image of its surroundings from echo scanning that is virtually as complete as the visual image from eyes of diurnal animals.

Bats have undergone some adaptive radiation, yet for reasons not fully understood, all are nocturnal, even the fruit-eating bats that use vision and olfaction to find their food instead of sonar. The tropics and subtropics have many nectar-feeding bats that are important pollinators for a wide variety of chiropterophilous ("bat-loving") plants. The flowers of these plants open at night, are white or light in color, and emit a musky, batlike odor that the nectar-feeding bats find attractive.

The famed tropical vampire bat has razor-sharp incisors that it uses to shave away the epidermis of its prey, exposing underlying capillaries. After infusing an anticoagulant to keep the blood flowing, it laps up its meal and stores it in a specially modified stomach. It is said that dogs can hear an approaching vampire's sonar and thus awaken and escape.

Reproduction

Most mammals have definite mating seasons, usually in the winter or spring and timed to coincide with the most favorable time of the year for rearing the young after birth. Many male mammals are capable of fertile copulation at any time, but the female mating function is restricted to a time during a periodic cycle, known as the **estrous cycle.** The female receives the male only during a relatively brief period known as **estrus,** or heat (Figure 27-18).

The estrous cycle is divided into stages marked by characteristic changes in the ovary, uterus, and vagina. **Proestrus,** or period of preparation, when new ovarian follicles grow, is followed by **estrus,** when mating occurs. Almost simultaneously the ovarian follicles burst, releasing the eggs (**ovulation**), which are fertilized. In all placental mammals the fertilized egg then implants itself in the

Echolocation is believed to be used by many insectivores (e.g., shrews and tenrecs) but is crudely developed as compared with bats. The toothed whales, however, have a highly developed capacity to locate objects by echolocation. Totally blind sperm whales that are in perfect health have been captured with food in their stomachs. Although the mechanism of sound production and reception remains imperfectly understood, it is thought that low- and high-frequency clicks produced in the sinus passages are focused into a narrow beam by a lens-shaped body in the forehead (the "melon"). Returning echos are channeled through oil-filled sinuses in the lower jaw to the inner ear. Toothed whales can apparently determine the size, shape, speed, distance, direction, and density of objects in the water and know the position of every whale in the pod.

Figure 27-18

African lions *Panthera leo* (order Carnivora, family Felidae) mating. Lions breed at any season, although predominantly in spring and summer. During the short period a female is receptive, she may mate repeatedly. Three or four cubs are born after gestation of 100 days. Once the mother introduces the cubs into the pride, they are treated with affection by both adult males and females. Cubs go through an 18- to 24-month apprenticeship learning how to hunt and then are frequently driven from the pride to manage for themselves.
Photograph by L.L. Rue, III.

uterine wall and pregnancy follows. However, should mating and fertilization not occur, estrus is followed by **metestrus,** a period of repair. This stage is followed by **diestrus,** during which the uterus becomes small and anemic. The cycle then repeats itself, beginning with proestrus.

How often females are in heat varies greatly among the different mammals. Animals that have only a single estrus during the breeding season are called **monestrous;** those that have a recurrence of estrus during the breeding season are called **polyestrous.** Dogs, foxes, and bats belong to the first group; field mice and squirrels are all polyestrous as are many mammals living in the more tropical regions of the earth. The Old World monkeys and humans have a somewhat different cycle in which the postovulation period is terminated by **menstruation,** during which the lining of the uterus (endometrium) collapses and is discharged with some blood. This is called a **menstrual cycle** and is described in Chapter 35.

There are three different patterns of reproduction in mammals. One pattern is represented by the egg-laying mammals, the monotremes. The duck-billed platypus has one breeding season each year. The ovulated eggs, usually two, are fertilized in the oviduct. As they continue down the oviduct, various glands add albumin and then a thin, leathery shell to each egg. When laid, the eggs are about the size of a robin's egg. The platypus lays its eggs in a burrow nest where they are incubated for about 12 days. After hatching, the young are fed milk (which they obtain by licking, not suckling) for a prolonged period. Thus, in monotremes there is no gestation (period of pregnancy) and the developing embryo draws on nutrients stored in the egg, much as do the embryos of reptiles and birds. But in common with all other mammals, the monotremes rear their young on milk.

The pouched mammals (marsupials) (Figure 27-19) exhibit a second pattern of reproduction. The physiology of gestation and lactation may be complicated in members of this group, which have a brief gestation period. In red kangaroos the first pregnancy of the season is followed by a 33-day gestation, after which the young ("joey") is born, crawls to the pouch without assistance from the mother, and attaches to a nipple. The mother immediately becomes pregnant again, but the presence of a suckling young in the pouch arrests development of the new embryo in the uterus at about the 100-cell stage. This period of arrest, called embryonic diapause, lasts approximately 235 days during which time the first joey

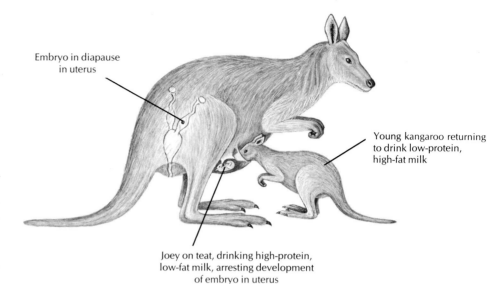

Embryo in diapause
in uterus

Young kangaroo returning
to drink low-protein,
high-fat milk

Joey on teat, drinking high-protein,
low-fat milk, arresting development
of embryo in uterus

Figure 27-19

Kangaroos have a complicated reproductive pattern in which the mother may have three young in different stages of development dependent on her at once.

Adapted from Austin, C.R., and R.V. Short (eds). 1972. Reproductive patterns, vol. 4. In Reproduction in mammals. New York, Cambridge University Press.

is growing in the pouch. When the joey leaves the pouch, the uterine embryo resumes development and is born about a month later. The mother again becomes pregnant, but because the second joey is suckling, once again development of the new embryo is arrested. Meanwhile, the first joey returns to the pouch from time to time to suckle. At this point the mother has three young of different ages dependent on her: a joey on foot, a joey in the pouch, and a diapause embryo in the uterus. There are variations on this remarkable sequence—not all marsupials have developmental delays like kangaroos, and some do not even have pouches—but in all, the young are born at an extremely early stage of development and undergo prolonged development while dependent on a teat.

The third pattern of reproduction is that of the placental mammals, the eutherians. In this most successful of the mammal groups, the reproductive investment is in gestation. The embryo remains in the mother's uterus, nourished by food supplied through the placenta, an intimate connection between mother and young. The length of gestation varies greatly. In general, the larger the mammal, the longer the gestation time. For example, mice have a gestation period of 21 days; rabbits and hares, 30 to 36 days; cats and dogs, 60 days; cattle, 280 days; and elephants, 22 months. But there are important exceptions (nature seldom offers perfect correlations). Baleen whales, the largest mammals, carry their young for only 12 months, while bats, no larger than mice, have gestation periods of 4 to 5 months. The condition of the young at birth also varies. An antelope bears its young well furred, eyes open, and able to run about. Newborn mice, however, are blind, naked, and helpless. We all know how long it takes a human baby to gain its footing. Human growth is in fact slower than that of any other mammal, and this is one of the distinctive attributes that sets us apart from other mammals (p. 796).

The number of young produced by mammals in a season depends on many factors. Usually the larger the animal, the smaller the number of young in a litter. One of the greatest factors involved is the number of enemies a species has. Small rodents, which serve as prey for many carnivores, usually produce more than one litter of several young each season. Meadow mice are known to produce as many as 17 litters of four to nine young in a year. Most carnivores have but one litter of three to five young per year. Large mammals, such as elephants and horses, give birth to a single young with each pregnancy. An elephant produces an average of 4 calves during her reproductive life of perhaps 50 years.

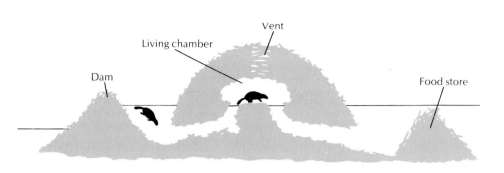

Figure 27-20

Each beaver colony constructs its own lodge in a pond it creates by damming a stream. Each year the mother bears four or five young; when the third litter arrives, the 2-year-olds are driven out of the colony. They will establish new colonies elsewhere.

Territory and Home Range

Many mammals have territories—areas from which individuals of the *same* species are excluded. In fact, many wild mammals, like many people, are basically unfriendly to their own kind, especially so to their own sex during the breeding season. If the mammal dwells in a burrow or den, this area forms the center of its territory. If it has no fixed address, the territory is marked out, usually with the highly developed scent glands described earlier in this chapter. Territories vary greatly in size, of course, depending on the size of the animal and its feeding habits. The grizzly bear has a territory of several square miles, which it guards zealously against all other grizzlies.

Mammals usually use natural features of their surroundings in staking their claims. These are marked with secretions from the scent glands or by urinating or defecating. When an intruder knowingly enters another's marked territory, it is immediately placed at a psychological disadvantage. Should a challenge follow, the intruder almost invariably breaks off the encounter in a submissive display characteristic for the species.

A beaver colony is a family unit, and beavers are among several mammalian species in which the male and female form a strong monogamous bond that lasts a lifetime. Because beavers invest considerable time and energy in constructing a lodge and dam and storing food for winter (Figure 27-20), the family, especially the adult male, vigorously defends its real estate against intruding beavers. Most of the work of building dams and lodges is undertaken by male beavers, but the females help when not occupied with their young.

An interesting exception to the strong territorial nature of most mammals is the prairie dog, which lives in large, friendly communities called prairie dog "towns" (Figure 27-21). When a new litter has been reared, the adults relinquish

Figure 27-21

Family of prairie dogs *Cynomys ludovicianus* (order Rodentia). These highly social prairie dwellers are plant eaters that provide an important source of food to many animals. They live in elaborate tunnel systems so closely interwoven that they form "towns" of as many as 1000 individuals. Towns are subdivided into wards, in turn divided into coteries, the basic family unit, containing one or two adult males, several females, and their litters. Although prairie dogs display ownership of burrows with territorial calls, they are friendly with inhabitants of adjacent burrows. The name "prairie dogs" derives from the sharp, doglike bark they make when danger threatens. Photograph by L.L. Rue, III.

the old home to the young and move to the edge of the community to establish a new home. Such a practice is totally antithetical to the behavior of most mammals, which drive off the young when they are self-sufficient.

The **home range** of a mammal is a much larger foraging area surrounding a defended territory. Home ranges are not defended in the same way as is a territory; home ranges may, in fact, overlap, producing a neutral zone used by the owners of several territories for seeking food.

Mammal Populations

A population of animals includes all the members of a species that interbreed and share a particular space (Chapter 41). All mammals (like other organisms) live in communities, each composed of numerous populations of different animal and plant species. Each species is affected by the activities of other species and by other changes, especially climatic, that occur. Thus, populations are always changing in size. Populations of small mammals are lowest before the breeding season and greatest just after the addition of the new members. Beyond these expected changes in population size, animal populations may fluctuate from other causes.

Irregular fluctuations are commonly produced by variations in climate, such as unusually cold, hot, or dry weather, or by natural catastrophes, such as fires, hailstorms, and hurricanes. These are density-independent causes because they affect a population whether it is crowded or dispersed. However, the most spectacular fluctuations are density dependent; that is, they are correlated with population crowding (p. 914). Cycles of abundance are common among many rodent species.

The population peaks and mass migrations of the Scandinavian and arctic North American lemmings are well known. Lemmings breed all year round, although more in the summer than in the winter. The gestation period is only 21 days; young born at the beginning of the summer are weaned in 14 days and are capable of reproducing by the end of the summer. At the peak of their population density, having devastated the vegetation by tunneling and grazing, they begin long, mass migrations to find new undamaged habitats for food and space. They swim across streams and small lakes as they go but cannot distinguish these from large lakes, rivers, and the sea, in which they drown. Since lemmings are the main diet of many carnivorous mammals and birds, any change in lemming population density affects all their predators as well.

The varying hare (snowshoe rabbit) of North America shows 10-year cycles

In his book *The Arctic* (1974. Montreal, Infacor, Ltd.), Canadian naturalist Fred Bruemmer describes the growth of lemming populations in arctic Canada:

"After a population crash one sees few signs of lemmings; there may be only one to every 10 acres. The next year, they are evidently numerous; their runways snake beneath the tundra vegetation, and frequent piles of rice-sized droppings indicate the lemmings fare well. The third year one sees them everywhere. The fourth year, usually the peak year of their cycle, the populations explode. Now more than 150 lemmings may inhabit each acre of land and they honeycomb it with as many as 4000 burrows. Males meet frequently and fight instantly. Males pursue females and mate after a brief but ardent courtship. Everywhere one hears the squeak and chitter of the excited, irritable, crowded animals. At such times they may spill over the land in manic migrations."

Figure 27-22

Changes in population of varying hare and lynx in Canada as indicated by pelts received by the Hudson's Bay Company. The abundance of lynx (predator) follows that of the hare (prey).

From Fundamentals of ecology, ed. 3, by Eugene P. Odum. Copyright © 1971 by W.B. Saunders Company. Reprinted by permission of Holt, Rinehart and Winston, CBS College Publishing.

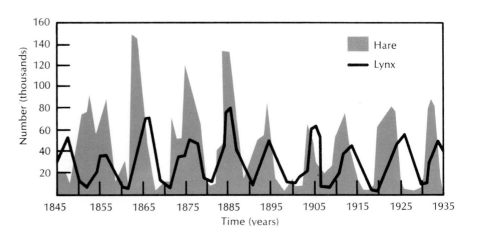

in abundance. The well-known fecundity of rabbits enables them to produce litters of three or four young as many as five times per year. The density may increase to 4000 hares competing for food in each square mile of northern forests. Predators (owls, minks, foxes, and especially lynxes) also increase (Figure 27-22). Then the population crashes precipitously for reasons that have long been a puzzle to scientists. Rabbits die in great numbers, not from lack of food or from an epidemic disease (as was once believed) but evidently from some density-dependent psychogenic cause. As crowding increases, hares become more aggressive, show signs of fear and defense, and stop breeding. The entire population reveals symptoms of pituitary–adrenal gland exhaustion, an endocrine imbalance called "shock disease," which results in death. These dramatic crashes are not well understood. Whatever the causes, population crashes that follow superabundance, although harsh, are clearly advantageous to the species because the vegetation is allowed to recover, providing the survivors with a much better chance for successful breeding.

HUMANS AND MAMMALS

Some 10,000 years ago, at the time people developed agricultural methods, they also began the domestication of mammals. Dogs were certainly among the first to be domesticated, probably entering voluntarily into their human dependence. The dog is an extremely adaptable and genetically plastic species derived from wolves. Much less genetically variable and certainly less social than dogs is the domestic cat, probably derived from an African race of wildcat. Wildcats look like oversized domestic cats and are still widespread in Africa and Eurasia. The domestication of cattle, buffaloes, sheep, and pigs probably came much later. It is believed that the beasts of burden—horses, camels, oxen, and llamas—probably were subdued by early nomadic peoples. Certain domestic species no longer exist as wild animals—for example, the one-humped Arabian camel, the llama, and the alpaca of South America. All of the truly domestic animals breed in captivity and have become totally dependent on humans; many have been molded by selective breeding to yield characteristics that are desirable for our purposes.

Some mammals hold special positions as "domestic" animals. The elephant has never been truly domesticated because it will not breed in captivity. In Asia, adults are captured and submit to a life of toil with astonishing docility. The reindeer of northern Scandinavia are domesticated only in the sense that they are "owned" by nomadic peoples who continue to follow them in their seasonal migrations. The eland of Africa is undergoing experimental domestication in several places. It is placid, gentle, and immune to native diseases and produces excellent meat.

Mammals can of course be enemies of humans. Rodents and rabbits are capable of inflicting staggering damage to growing crops and stored food (Figure 27-23). We have provided an inviting forage for rodents with our agriculture and convenienced them further by removing their natural predators. Rodents also carry various diseases. Bubonic plague and typhus are carried by house rats. Tularemia, or rabbit fever, is transmitted to humans by the wood tick carried by rabbits, woodchucks, muskrats, and other rodents. Rocky Mountain spotted fever is carried to humans by ticks from ground squirrels and dogs. Trichina worms and tapeworms are acquired by humans who eat the meat of infected hogs, cattle, and other mammals.

In the introduction to this chapter, we alluded to the discouraging exploi-

Figure 27-23

Brown rat *Rattus norvegicus* (order Rodentia, family Muridae). Living all too successfully beside human habitations, the brown rat not only causes great damage to food stores but also spreads disease, including bubonic plague (a disease carried by infected fleas that greatly influenced human history in medieval Europe), typhus, infectious jaundice, *Salmonella* food poisoning, and rabies.
Photograph by L.L. Rue, III.

tation of the whales as one example of our inability to reconcile human needs with the preservation of wildlife. This dilemma is explored in the final chapter of this book. The extermination of a species for commercial gain is so totally indefensible that no debate is required. Once a species is extinct, no amount of scientific or technical ingenuity will bring it back. What has taken millions of years to evolve can be destroyed in a decade of thoughtless exploitation. Many people are concerned with the awesome impact we have on wildlife, and there is more determination today to reverse a regrettable trend than ever before. If given a chance, mammals will usually make spectacular recoveries from human depredations, as have the sea otter and the saiga antelope, both once in danger of extinction and now numerous. Paradoxically, in Africa where conservationists wage a seesaw battle with opposing interests, it appears that the commercial gain of tourism and game cropping will do more to save the fauna than the outraged concern of preservationists.

CLASSIFICATION

The classification of living mammalian orders given here follows that of Corbet and Hill (1980) and Eisenberg (1981).

Subclass Prototheria (pro'to-thir'e-a) (Gr. *prōtos*, first + *thēr*, wild animal). The egg-laying mammals.

Order Monotremata (mon'o-tre'ma-tah) (Gr. *monos*, single, + *trēma*, hole): **egg-laying mammals: duck-billed platypus, spiny anteater.** The three species in this order are from Australia, Tasmania, and New Guinea. The most noted member of the order is the duck-billed platypus *(Ornithorhynchus anatinus)*. The spiny anteater, or echidna *(Tachyglossus)*, has a long, narrow snout adapted for feeding on ants, its chief food. Monotremes represent the only order that is oviparous and there is no known group of extinct mammals from which they can be derived. Their fossils date from the Pleistocene epoch.

Subclass Theria (thir'e-a) (Gr. *thēr*, wild animal).

Infraclass Metatheria (met'a-thir'e-a) (Gr. *meta*, after, + *thēr*, wild animal). The marsupial mammals.

Order Marsupialia (mar-su'pe-ay'le-a) (Gr. *marsypion*, little pouch): **pouched mammals: oppossums, kangaroos, koalas, Tasmanian wolves, wombats, bandicoots, numbats, and others** (Figure 27-24). These are primitive mammals characterized by an abdominal pouch, the **marsupium,** in which they rear their young. Although the young are nourished in the uterus for a short time, there is rarely a placenta present. This order has 254 species; only the opossum is found in the Americas, but the order is the dominant group of mammals in Australia.

Infraclass Eutheria (yu-thir'e-a) (Gr. **eu,** true, + *thēr,* wild animals). The placental mammals.

Order Edentata (ee-den-ta'ta) (L. *edentatus,* toothless): **anteaters, armadillos, sloths.** The 29 species of this order are toothless (anteaters) or have simple, rootless molars that grow throughout life (sloths and armadillos). Most live in South and Central America, although the nine-banded armadillo *(Dasypus novemcinctus)* is common in the southern United States.

Order Pholidota (fol'i-do'ta) (Gr. *pholis,* horny scale): **pangolins.** In this order there is one genus *(Manis)* with seven species. They are an odd group of animals whose bodies are covered with overlapping horny scales that have arisen from fused bundles of hair. Their home is in tropical Asia and Africa.

Order Macroscelidea (mak-ro-sa-lid'e-a) (Gr. *makros,* large, + *skelos,* leg): **elephant shrews.** The 15 species of this order are secretive mammals with long legs, a snout-like nose adapted for foraging for insects, large eyes, and are widespread in Africa.

Order Lagomorpha (lag'o-mor'fa) (Gr. *lagos,* hare, + *morphē,* form): **rabbits, hares, pikas.** This order has 54 species. Lagomorphs have long, constantly growing incisors, like rodents, but unlike rodents, they have an additional pair of peglike incisors growing behind the first pair. All lagomorphs are herbivores with cosmopolitan distribution.

Figure 27-24

Long-nosed potoroo, a nocturnal rat kangaroo. Order Marsupialia.
Photograph by C.P. Hickman, Jr.

Order Rodentia (ro-den′che-a) (L. *rodere*, to gnaw): **gnawing mammals: squirrels, rats, woodchucks.** The rodents, comprising nearly 40% of all mammalian species, are characterized by two pairs of razor-sharp incisors used for gnawing through the toughest pods and shells for food. With their impressive reproductive powers, adaptability, and capacity to invade all terrestrial habitats, they are of great ecological significance. Important families of this order, consisting of 1591 species, are **Sciuridae** (squirrels (Figure 27-25) and woodchucks), **Muridae** (rats and house mice), **Castoridae** (beavers), **Erethizontidae** (porcupines), **Geomyidae** (pocket gophers), and **Cricetidae** (hamsters, deer mice, gerbils, voles, lemmings).

Order Insectivora (in-sec-tiv′o-ra) (L. *insectum*, an insect, + *vorare*, to devour): **insect-eating mammals: shrews, (Figure 27-13), hedgehogs, tenrecs, moles.** The principal food of animals in this order of 343 species is insects. The most primitive of placental mammals, they are widely distributed over the world except Australia and New Zealand. Insectivores are small, sharp-snouted animals that spend a great part of their lives underground. The shrews are among the smallest mammals known.

Order Carnivora (car-niv′o-ra) (L. *caro*, flesh, + *vorare*, to devour): **flesh-eating mammals: dogs, wolves, cats, bears, weasels.** To this extensive order of 240 species belong some of the stongest and most intelligent of animals. They all have predatory habits, and their teeth are especially adapted for tearing flesh. They are distributed all over the world except in the Australian and Antarctic regions where there are no native forms. Among the more familiar families are **Canidae** (the dog family), consisting of dogs, wolves, foxes, and coyotes; **Felidae** (the cat family), whose members include the domestic cats, tigers, lions, cougars, and lynxes; **Ursidae** (the bear family), made up of bears; and **Mustelidae** (the fur-bearing family), containing the martens, skunks, weasels, otters, badgers, minks, and wolverines.

Order Pinnipedia (pi-ni-peed′e-a) (L. *pinna*, feather, + *ped*, foot): **sea lions, seals, and walruses.** This order consists of 34 species. The limbs of these aquatic carnivores have been modified as flippers for swimming. They are all saltwater forms, and their food consists mostly of fish.

Order Scandentia (skan-dent′e-a) (L. *scandentis*, climbing): **tree shrews.** There are 16 species in this order. Tree shrews are small, squirrel-like mammals of the tropical rain forests of southern and southeastern Asia. Despite their name, many are not especially well-adapted for life in trees, and some are almost completely terrestrial.

Order Dermoptera (der-mop′ter-a) (Gr. *derma*, skin, + *pteron*, wing): **flying lemurs.** There are two species in this order. These are related to the true bats and consist of the single genus *Galeopithecus*. They are found in the Malay peninsula in the East Indies. They are not lemurs (which are primates) and cannot fly in the strict sense of the word, but glide like flying squirrels.

Order Chiroptera (ky-rop′ter-a) (Gr. *cheir*, hand, + *pteron*, wing): **bats.** The wings of bats, the only true flying mammals, are modified forelimbs in which the second to fifth digits are elongated to support a thin integumental membrane for flying. The first digit (thumb) is short with a claw. There are many families and 950 species of bats the world over. The common North American forms are the little brown bat *(Myotis)*, the free-tailed bat *(Tadarida)*, which lives in the Carlsbad Caverns, and the large brown bat *(Eptesicus)*. In the Old World tropics the fruit bats, or "flying foxes," *(Pteropus)* are the largest of all bats, with a wingspread of 1.2 to 1.5 m; they live chiefly on fruits.

Order Primates (pry-may′teez) (L. *prima*, first): **highest mammals: lemurs, monkeys, apes, humans, and others.** This order stands first in the animal kingdom in brain development, possessing especially large cerebral hemispheres. Of the 179 species most are arboreal, apparently derived from tree-dwelling insectivores. The primates represent the end product of a line that branched off early from other mammals and have retained many primitive characteristics. It is believed that their tree-dwelling habits of agility in capturing food or avoiding enemies were largely responsible for their advances in brain structure. As a group they are generalized with five digits (usually provided with flat nails) on both forelimbs and hindlimbs. All have their bodies covered with hair except humans. Forelimbs are often adapted for grasping, as are the hindlimbs sometimes. The group is

Figure 27-25

Arctic ground squirrel, *Spermophilus parryi*, one of the most northern in distribution of the large squirrel family.
Photograph by C.P. Hickman, Jr.

singularly lacking in claws, scales, horns, and hoofs. There are two suborders.

Suborder Prosimii (pro-sim′ee-i) (Gr. *pro*, before, + *simia*, ape): **lemurs, tree shrews, tarsiers, lorises, pottos.** These are primitive arboreal primates, with their second toe provided with a claw and a long nonprehensile tail. They look like a cross between squirrels and monkeys. They are found in the forests of Madagascar, Africa, the Malay peninsula, and the Philippines. Their food consists of both plants and small animals.

Suborder Anthropoidea (an′thro-poi′de-a) (Gr. *anthropos*, man): **monkeys, gibbons, apes, humans.** There are three superfamilies.

Superfamily Ceboidea (se-boi′de-a) (Gr. *kebos*, long-tailed monkey): **Platyrhinii.** These are New World monkeys, characterized by the broad flat nasal septum, nonopposable thumb, prehensile tail, and the absence of ischial callosities and cheek pouches. Familiar members of this superfamily are the capuchin monkey *(Cebus)* of the organ grinder, the spider monkey *(Ateles)*, and the howler monkey *(Alouatta)*.

Superfamily Cercopithecoidea (sur′ko-pith′e-koi′de-a) (Gr. *kerkos*, tail, + *pithekos*, monkey): **Catarrhini.** These Old World monkeys have the external nares close together, and many have internal cheek pouches. They never have prehensile tails, there are callused ischial tuberosities on their buttocks, and their thumbs are opposable. Examples are the savage mandrill *(Cynocephalus)*, the rhesus monkey *(Macaca)* (Figure 27-26), widely used in biological investigation, and the proboscis monkey *(Nasalis)*.

Superfamily Hominoidea (hom′i-noi′de-a) (L. *homo, hominis,* man). The higher (anthropoid) apes and humans make up this superfamily. Their chief characteristics are lack of a tail and lack of cheek pouches. There are two families: Pongidae and Hominidae. The Pongidae family includes the higher apes, gibbon *(Hylobates)*, orangutan *(Simia)*, chimpanzee *(Pan)*, and gorilla *(Gorilla)*. The other family, Hominidae, is represented by a single living species *(Homo sapiens)*, modern humans. Humans differ from the members of family Pongidae in being more erect, in having shorter arms and larger thumbs, and in having lighter jaws with smaller front teeth. Most of the apes also have much more prominent ridges over the eyes. Many differences between the human and the anthropoid apes are associated with higher human intelligence, human speech centers in the brain, and the absence of the arboreal habit.

Order Cetacea (see-tay′she-a) (L. *cetus,* whale): **fishlike mammals: whales, dolphins, porpoises.** This order of 76 species is well adapted for aquatic life. Their anterior limbs are modified into broad flippers; the posterior limbs are absent. Some have a fleshy dorsal fin and the tail is divided into transverse fleshy flukes. The nostrils are represented by a single or double blowhole on top of the head. They have no hair except a few hairs on the muzzle, no skin glands except the mammary and those of the eye, no external ear, and small eyes. The order is divided into the **toothed whales,** represented by dolphins, porpoises, and sperm whales; and the **baleen whales** represented by the rorquals, right whales, and gray whales. The baleen whales are generally larger than toothed whales. The blue whale, a rorqual, is the largest animal that has ever lived. Rather than teeth, baleen whales have a peculiar straining device of whalebone (baleen) attached to the palate, used to filter plankton from the water.

Order Sirenia (sy-re′ne-a) (Gr. *seiren*, sea nymph): **sea cows and manatees.** The sirenians are large, clumsy, aquatic mammals with large head, no hindlimbs, and forelimbs modified into flippers. The sea cow (dugong) of tropical coastlines of east Africa, Asia, and Australia and three species of manatees of the Caribbean area, Amazon River, and west Africa comprise the only living species. A fifth species, the large Steller's sea cow, was hunted to extinction by humans in the mid eighteenth century.

Order Proboscidea (pro′ba-sid′e-a) (Gr. *proboskis*, elephant's trunk, from *pro*, before, + *boskein*, to feed): **proboscis mammals: elephants.** These are the largest of living land animals. The two upper incisors are elongated as tusks, and the molar teeth are well developed. There are two species of elephants: the Indian *(Elephas maximus)*, with relatively small ears, and the African *(Loxodonta africana)*, with

Figure 27-26

Rhesus monkey *Macaca mulatta.* Order Primates, superfamily Cercopithecoidea. Photograph by Irene Vandermolen, Leonard Rue Enterprises.

large ears. The Asiatic or Indian elephant has long been domesticated and is trained to do heavy work. The taming of the African elephant is more difficult but was done extensively by the ancient Carthaginians and Romans, who employed them in their armies.

Order Hyracoidea (hy'ra-coi'de-a) (Gr. *hyrax*, shrew): **hyraxes (coneys)** (Figure 27-27). There are five species in this order. Coneys are herbivores that are restricted to Africa and Syria. They have some resemblance to short-eared rabbits but have teeth like rhinoceroses, with hooves on their toes and pads on their feet. They have four toes on the front feet and three toes on the back.

Order Perissodactyla (pe-ris'so-dak'ti-la) (Gr. *perissos*, odd, + *dactylos*, toe): **odd-toed hoofed mammals: horses, asses, zebras, tapirs, rhinoceroses.** The odd-toed hoofed mammals have an odd number (one or three) of toes, each with a cornified hoof (Figure 27-28). Both the Perissodactyla, of which there are 17 species, and the Artiodactyla are often referred to as **ungulates** (L. *ungula*, hoof), or hoofed mammals, with teeth adapted for chewing. The horse family (Equidae), which also includes asses and zebras, has only one functional toe. Tapirs have a short proboscis formed from the upper lip and nose. The rhinoceros *(Rhinoceros)* includes several species found in Africa and Southeast Asia. All are herbivorous.

Order Artiodactyla (ar'te-o-dak'ti-la) (Gr. *artios*, even, + *daktylos*, toe): **even-toed hoofed mammals: swine, camels, deer, hippopotamuses, antelopes, cattle, sheep, goats.** Most of these ungulates have two toes, although the hippopotamus and some others have four (Figure 27-28). Each toe is sheathed in a cornified hoof. Many, such as the cow, deer, and sheep have horns. Many are ruminants, that is, animals that chew the cud. Like Perissodactyla, they are strictly herbivorous. The group is divided into nine living families and many extinct ones and includes some of the most valuable domestic animals. This extensive order, consisting of 184 species, is commonly divided into three suborders: the **Suina** (pigs, peccaries, and hippopotamuses), the **Tylopoda** (camels), and the **Ruminantia** (deer, giraffes, sheep, cattle, and so on).

Order Tubulidentata (tu'byu-li-den-ta'ta) (L. *tubulus*, tube, + *dens*, tooth): **aardvark.** The name "aardvark" is Dutch for earth pig, a peculiar animal with a piglike body found in Africa. The order is represented by one species.

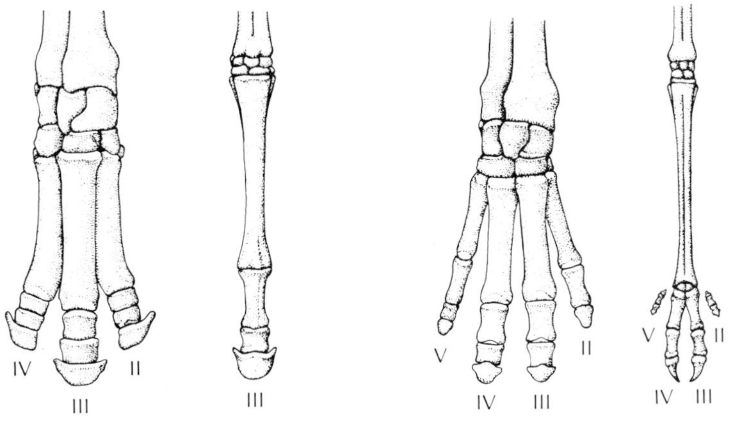

Figure 27-27

Rock hyrax, a colonial mammal widely distributed in Africa. Order Hyracoidea. Photograph by B. Tallmark.

Figure 27-28

Odd-toed and even-toed ungulates. The rhinoceros and horse (order Perissodactyla) are odd-toed; the hippopotamus and deer (order Artiodactyla) are even-toed. The lighter, faster animals run on only one or two toes.

IV   II   III   V   II   V   II
III   III   IV   III   IV   III
Rhinoceros   **Horse**   **Hippopotamus**   **Deer**

SUMMARY

Mammals are endothermic and homeothermic vertebrates whose bodies are insulated by hair and who nurse their young with milk. The approximately 4000 species of mammals are descended from a mammal-like reptile (therapsid) of the Jurassic period of the Mesozoic era. They diversified rapidly during the Tertiary period of the Cenozoic era to become the most intelligent and advanced animals on earth.

American marten.
Photograph by Irene Vandermolen, Leonard Rue Enterprises.

Mammals are named for the glandular milk-secreting organs of the female (rudimentary in the male), a unique adaptation which, combined with prolonged parental care, buffers the infants from the demands of foraging for themselves and eases the transition to adulthood. Hair, the integumentary outgrowth that covers most mammals, serves variously for mechanical protection, thermal insulation, protective coloration, and waterproofing. Mammalian skin is rich in glands: sweat glands that function in evaporative cooling, scent glands used in social interactions, and sebaceous glands that secrete lubricating skin oil. All placental mammals have deciduous teeth that are replaced by permanent teeth (diphyodont dentition). The four groups of teeth—incisors, canines, premolars, and molars—may be highly modified in different mammals for specialized feeding tasks, or they may be absent.

The food habits of mammals strongly influence their body form and physiology. Herbivorous mammals have special adaptations for harboring the intestinal microflora that break down cellulose of the woody diet, and they have developed adaptations for detecting and escaping predators. Carnivorous mammals feed mainly on herbivores, have a simple digestive tract, and have developed adaptations for a predatory life. Omnivores feed on both plant and animal foods. Insectivores feed mainly on insects.

Many marine, terrestrial, and aerial mammals migrate; some migrations, such as those of fur seals and caribou, are extensive. Migrations are usually made toward favorable climatic and optimal food and calving conditions, or to bring the sexes together for mating.

Mammals with true flight, the bats, are nocturnal and thus avoid direct competition with birds. Most employ ultrasonic echolocation to navigate and feed in darkness.

The most primitive living mammals are the egg-laying monotremes, a remnant group today. All other mammals are viviparous. Embryos of marsupials are born underdeveloped and complete their early growth in the mother's pouch, nourished by milk. The largest and most successful mammals are the placentals in which the embryo undergoes an extensive development in the uterus, nourished by the placenta, a specialization of the embryonic membranes. The unqualified success of mammals as a group cannot be attributed to greater organ system perfection, but rather to their impressive overall adaptability—the capacity to fit more perfectly in total organization to environmental conditions and thus exploit virtually every habitat on earth.

Review questions

1. In the early Mesozoic a group of mammal-like reptiles, the therapsids, appeared that later gave rise to the true mammals. What new structural adaptations appeared in this group that made them more like mammals than reptiles?

2. The Age of Mammals began about 70 million years ago and was marked by a great radiation of mammal groups. Give some reasons why mammals were so successful.

3. Hair is believed to have evolved in the therapsids in response to the need for insulation, but modern mammals have adapted hair for several other purposes. Describe these.

4. What is distinctive about each of the following: horns of the ruminants, antlers of the deer family, and the horn of the rhinoceros? Briefly describe the growth cycle of antlers.

5. Describe the location and principal function(s) of each of the following skin glands: sweat glands (of two kinds, eccrine and apocrine), scent glands, sebaceous glands, and mammary glands.

6. Define the terms "diphyodont" and "heterodont" and explain how both terms apply to mammalian dentition.

7. Describe the food habits of each of the following groups: herbivores, carnivores, omnivores, and insectivores. Give the common names of several mammals belonging to each group.
8. Most herbivorous mammals depend on cellulose as their main energy source, yet no mammal synthesizes cellulose-splitting enzymes. How are the digestive tracts of mammals specialized for symbiotic digestion of cellulose?
9. Describe the annual migrations of barren-ground caribou and fur seals.
10. Explain what is distinctive about the life-style and mode of navigation in bats.
11. Describe and distinguish the patterns of reproduction in monotremes, marsupials, and placental mammals. What aspects of mammalian reproduction are present in *all* mammals but in no other class of vertebrates?
12. Distinguish between territory and home range in mammals.
13. What is the difference between density-dependent and density-independent causes of population fluctuations in mammals?
14. Describe the hare-lynx population cycle, considered a classic example of a prey-predator relationship (Figure 27-22). From your examination of the cycle, can you formulate a hypothesis for the explanation of the oscillations?
15. What do the terms Prototheria, Theria, Metatheria, Eutheria, Monotremata, and Marsupalia literally mean, and what mammals are grouped under each taxon?

Selected references

See also general references for Part Two, p. 590.

Corbet, G.B., and J.E. Hill. 1980. A world list of mammalian species. Ithaca, Cornell University Press.

Eisenberg, J.F. 1981. The mammalian radiations: an analysis of trends in evolution, adaptation, and behavior. Chicago, University of Chicago Press. *Wide-ranging, authoritative synthesis of mammalian evolution and behavior.*

Hall, E.R., and K.R. Kelson. 1959. The mammals of North America, 2 vols. New York, The Ronald Press Co. *Full descriptions of species and subspecies with distribution maps. Taxonomic keys, records, and revealing line drawings of skull characteristics are included in this authoritative work.*

Kanwisher, J.W., and S.H. Ridgeway. 1983. The physiological ecology of whales and porpoises. Sci. Am. **248**:110-120 (June). *Their adaptations for deep dives are described.*

Kemp, T.S. 1982. Mammal-like reptiles and the origin of mammals. New York, Academic Press, Inc. *Comprehensive synthesis. The final chapter summarizes the early chapters and offers a model of evolutionary history of the mammal-like reptiles and primitive mammals.*

Macdonald, D. (ed.). 1984. The encyclopedia of mammals. New York, Facts on File, Inc. *Lucidly written and uncompromisingly thorough treatment of all mammal groups, enhanced by well-chosen photographs and color artwork.*

Myers, J.H., and C.J. Krebs. 1974. Population cycles in rodents. Sci. Am. **230**:38-46 (June). *Population fluctuations appear to be associated with periodic changes in genetic makeup.*

References to Part Two

The references below pertain to groups covered in more than one chapter of Part Two. They include a number of valuable field manuals that aid in identification, as well as general texts.

Alexander, R.M. 1981. The chordates. Cambridge, Cambridge University Press. *Comparative anatomy and physiology of chordates.*

Barnes, R.D. 1987. Invertebrate zoology, ed. 5. Philadelphia, Saunders College/Holt, Rinehart & Winston. *Authoritative, detailed coverage of the invertebrate phyla.*

Barrington, E.J.W. 1979. Invertebrate structure and function, ed. 2. New York, John Wiley & Sons, Inc. *Excellent account of function in major invertebrate groups.*

Fotheringham, N. 1980. Beachcomber's guide to Gulf Coast marine life. Houston, Texas, Gulf Publishing Co. *Coverage arranged by habitats. No keys, but most common forms that occur near shore can be identified.*

Gosner, K.L. 1979. A field guide to the Atlantic seashore: invertebrates and seaweeds of the Atlantic coast from the Bay of Fundy to Cape Hatteras. The Peterson Field Guide Series. Boston, Houghton Mifflin Co. *A helpful aid for students of the invertebrates found along the northeastern coast of the United States.*

Grzimek, H.C.B. (ed.). 1984. Grzimek's animal life encyclopedia. 13 vols. New York, Van Nostrand Reinhold Company. *Comprehensive overview of the animal kingdom, beautifully illustrated with numerous full-color drawings and photographs, with emphasis on animal behavior. Although heavily weighted toward the higher vertebrates, the set is nevertheless a goldmine of information for zoologists of all persuasions.*

Hyman, L.H. 1940-1967. The invertebrates, 6 vols. New York, McGraw-Hill Book Co. *Informative discussions on the phylogenies of most of the invertebrates are treated in this outstanding series of monographs. Vol. 1 contains a discussion of the colonial hypothesis of the origin of metazoa, and vol. 2 contains a discussion of the origin of bilateral animals, body cavities, and metamerism.*

Jameson, E.W., Jr. 1981. Patterns of vertebrate biology. New York, Springer-Verlag. *Synoptic topical approach to vertebrate natural history.*

Kaplan, E.H. 1982. A field guide to coral reefs of the Caribbean and Florida. The Peterson Field Guide Series, Boston, Houghton Mifflin Co. *More than just a field guide, this little book has much information on biology of coral reefs and the animals found there.*

Morris, R.H., D.P. Abbott, and E.C. Haderlie. 1980. Intertidal invertebrates of California. Stanford, Stanford University Press. *An essential reference on the most important invertebrates of the intertidal zone in California. Contains 900 color photographs.*

New Larousse Encyclopedia of Animal Life. 1980. New York, Bonanza Books. *Comprehensive illustrated survey of the animal world, treating all major groupings with general principles of form and function, and with well-written accounts of representative species.*

Parker, S.P. (ed.). 1982. Synopsis and classification of living organisms, 2 vols. New York, McGraw-Hill Book Co. *A comprehensive reference to the classification of living organisms with descriptions of taxa above the generic level. Contains 8200 synoptic articles on the biology of many of the groups.*

Pearse, V., J. Pearse, M. Buchsbaum, and R. Buchsbaum. 1987. Living invertebrates. Palo Alto, Ca., Blackwell Scientific Publications, and Pacific Grove, Ca., Boxwood Press. *Readable account of invertebrates, many photographs.*

Pennak, R.W. 1978. Freshwater invertebrates of the United States, ed. 2. New York, John Wiley & Sons, Inc. *Contains keys for identification of freshwater invertebrates, with brief account of each group. Indispensable for freshwater biologists.*

Ricketts, E.F., J. Calvin, and J.W. Hedgpeth. (revised by D.W. Phillips). 1985. Between Pacific tides, ed. 5. Stanford, Stanford University Press. *A revision of a classic work in marine biology. It stresses the habits and habitats of the Pacific coast invertebrates, and the illustrations are revealing. It includes an excellent, annotated systematic index and bibliography.*

Russell-Hunter, W.D. 1979. A life of invertebrates. New York, Macmillan, Inc. *Rather than emphasizing classification and anatomy, this text helps the reader understand the "life-styles" of the various groups.*

Schmidt, G.D., and L.S. Roberts. 1985. Foundations of parasitology, ed. 3. St. Louis, The C.V. Mosby Co. *Highly readable and up-to-date account of parasitic protozoa, worms, and arthropods.*

Smith, D.L. 1977. A guide to marine coastal plankton and marine invertebrate larvae. Dubuque, Iowa, Kendall/Hunt Publishing Co. *Valuable manual for identification of marine plankton, which is usually not covered in most field guides.*

Smith, R.I., and J.T. Carlton (eds.). 1975. Light's manual: Intertidal invertebrates of the central California coast, ed. 3. Berkeley, University of California Press. *Has keys for identification of intertidal invertebrates from central California.*

Young, J.Z. 1981. The life of vertebrates, ed. 3. Oxford, Clarendon Press. *This is the updated, comprehensive, classic of vertebrate biology.*

PART THREE

ACTIVITY OF LIFE

C.P. Hickman, Jr.

CHAPTER 28

SUPPORT, PROTECTION, AND MOVEMENT

INTEGUMENT AMONG VARIOUS GROUPS OF ANIMALS

The integument is the outer covering of the body, a protective wrapping that includes the skin and all structures derived from or associated with the skin, such as hair, setae, scales, feathers, and horns. In most animals it is tough and pliable, providing mechanical protection against abrasion and puncture and forming an effective barrier against the invasion of bacteria. It may provide moisture proofing against fluid loss or gain. The skin helps protect the underlying cells against the damaging action of the ultraviolet rays of the sun. In addition to being a protective cover, the skin serves a variety of important regulatory functions. For example, in warm-blooded (homeothermic) animals, it is vitally concerned with temperature regulation, since most of the body's heat is lost through the skin; it contains mechanisms that cool the body when it is too hot and slow heat loss when the body is too cold. The skin contains sensory receptors that provide essential information about the immediate environment. It has excretory functions and in some forms respiratory functions as well. Through skin pigmentation the organism can make itself more or less conspicuous. Skin secretions can make the animal sexually attractive or repugnant or provide olfactory cues that influence behavioral interactions between individuals.

Invertebrate Integument

Many protozoa have only the delicate cell or plasma membranes for external coverings; others, such as *Paramecium*, have developed a protective pellicle. Most multicellular invertebrates, however, have more complex tissue coverings. The principal covering is a single-layered **epidermis.** Some invertebrates have added a secreted noncellular **cuticle** over the epidermis for additional protection.

The molluscan epidermis is delicate and soft and contains mucous glands, some of which secrete the calcium carbonate of the shell. Cephalopod molluscs (squids and octopuses) have developed a more complex integument, consisting of cuticle, simple epidermis, layer of connective tissue, layer of reflecting cells (iridocytes), and thicker layer of connective tissue.

Arthropods have the most complex of invertebrate integuments, providing not only protection but also skeletal support. The development of a firm exoskeleton and jointed appendages suitable for the attachment of muscles has been a key

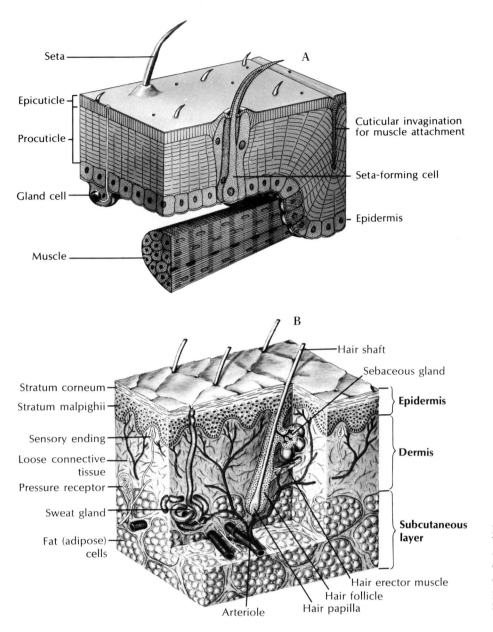

Figure 28-1

A, Structure of insect integument. This reconstruction shows block of integument drawn at point where cuticle invaginates to provide exoskeletal muscle attachment. **B,** Structure of human skin.

feature in the extraordinary evolutionary success of this phylum, the largest of animal groups. The arthropod integument consists of a single-layered **epidermis** (also called more precisely **hypodermis**), which secretes a complex cuticle of two zones (Figure 28-1, *A*). The inner zone, the **procuticle,** is composed of protein and chitin (a polysaccharide) laid down in layers (lamellae) much like the veneers of plywood. The outer zone of cuticle, lying on the external surface above the procuticle, is the thin **epicuticle.** The epicuticle is a nonchitinous complex of proteins and lipids that provides a protective moisture-proofing barrier to the integument.

The arthropod cuticle may remain as a tough but soft and flexible layer, as it is in many microcrustaceans and insect larvae. However, it may be hardened by either of two ways. In the decapod crustaceans, for example, crabs and lobsters, the cuticle is stiffened by **calcification,** the deposition of calcium carbonate. In insects hardening occurs when the protein molecules bond together with stabilizing cross-linkages within and between the adjacent lamellae of the procuticle. The result of the process, called **sclerotization,** is the formation of a highly resistant and insoluble protein, **sclerotin.** Arthropod cuticle is one of the toughest materials synthesized by animals; it is strongly resistant to pressure and tearing and can withstand boiling in concentrated alkali, yet it is light, having a specific mass of only 1.3 (1.3 times the weight of water).

When arthropods molt, the epidermal cells first divide by mitosis. Enzymes secreted by the epidermis dissolve most of the procuticle; the digested materials are then absorbed and consequently are not lost to the body. Then in the space beneath the old cuticle a new epicuticle and procuticle are formed. After the old cuticle is shed, the new cuticle is thickened and calcified or sclerotized.

Vertebrate Integument

The basic plan of the vertebrate integument, as exemplified by human skin (Figure 28-1, *B*), includes a thin, outer stratified epithelial layer, the **epidermis,** derived from ectoderm and an inner, thicker layer, the **dermis,** or true skin, which is of mesodermal origin.

Although the epidermis is thin and appears simple in structure, it gives rise to most derivatives of the integument, such as hair, feathers, claws, and hooves. The dermis, containing blood vessels, collagenous fibers, nerves, pigment cells, fat cells, and fibroblasts, supports, cushions, and nourishes its overlying partner, which is devoid of blood vessels.

The epidermis consists usually of several layers of cells. The basal part is made up of columnar cells that undergo frequent mitosis to renew the layers that lie above. As the outer layers of cells are displaced upward by new generations of cells beneath, an exceedingly tough, fibrous protein called **keratin** accumulates in the interior of the cells. Gradually, keratin replaces all metabolically active cytoplasm. The cell dies and is eventually shed, lifeless and scalelike. Such is the origin of dandruff as well as a significant fraction of household dust. This process is called **keratinization,** and the cell, thus transformed, is said to be "**cornified.**" Cornified cells, highly resistant to abrasion and water diffusion, comprise the stratum corneum. This epidermal layer becomes especially thick in areas exposed to persistent pressure or friction, such as calluses and the human palms and soles.

The **dermis,** as already mentioned, mainly serves a supportive role for the epidermis. Nevertheless, true bony structures, where they occur in the integument, are always dermal derivatives. Heavy bony plates were common in primitive ostracoderms and placoderms of the Paleozoic era but occur in few living fishes. Scales of contemporary fishes are bony dermal structures that have evolved from the bony armor of the Paleozoic fishes but are much smaller and more flexible. Although of dermal origin, fish scales are intimately associated with the thin, overlying epidermis; in some species the scales protrude through the epidermis, but typically the epidermis forms a continuous sheath that is reflected under the overlapping scales (Figure 28-2). Dermal bone also forms the flat bones of the skull and gives rise to antlers, which are outgrowths of dermal frontal bone.

Animal coloration

The colors of animals may be vivid and dramatic when serving as important recognition marks or as warning coloration, or they may be subdued or cryptic when used for camouflage. Integumentary color is usually produced by pigments, but in many insects and in some vertebrates, especially birds, certain colors are produced by the physical structure of the surface tissue, which reflects certain light

---

The reptiles were the first to exploit the adaptive possibilities of the remarkably tough protein keratin. The reptilian epidermal scale that develops from keratin is a much lighter and more flexible structure than the bony, dermal scale of fishes, yet it provides excellent protection from abrasion and desiccation. Scales may be overlapping structures, as in snakes and some lizards, or develop into plates, as in turtles and crocodilians. In birds, keratin found new uses. Feathers, beaks, and claws, as well as scales, are all epidermal structures composed of dense keratin. Mammals continued to capitalize on keratin's virtues by turning it into hair, hooves, claws, and nails. As a result of its keratin content, hair is by far the strongest material in the body. It has a tensile strength comparable to that of rolled aluminum and is nearly twice as strong, weight for weight, as the strongest bone.

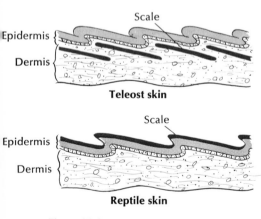

Figure 28-2

Integument of bony fishes and reptiles. Bony (teleost) fishes have bony scales from dermis, and reptiles have horny scales from epidermis. Dermal scales of fishes are retained throughout life. Epidermal scales of reptiles are shed periodically.

wavelengths and eliminates others. Colors produced this way are called **structural color,** and they are responsible for the most beautifully iridescent and metallic hues to be found in the animal kingdom. Many butterflies and beetles and a few fishes thus share with birds the distinction of being the earth's most-resplendent animals. Certain structural colors of feathers are caused by minute, air-filled spaces or pores that reflect white light (white feathers) or some portion of the spectrum (for example, Tyndall blue coloration produced by scattering of light [see marginal note, p. 539]). Iridescent colors that change hue as the animal's angle shifts with respect to the observer are produced when light is reflected from several layers of thin, transparent film. By phase interference, light waves reinforce, weaken, or eliminate each other to produce some of the purest and most brilliant colors we know.

More common than structural colors in animals are **pigments** (biochromes), an extremely varied group of large molecules that reflect light rays. In crustaceans and ectothermic vertebrates these pigments are contained in large cells with branching processes, called **chromatophores** (Figure 28-3, *A*). The pigment may concentrate in the center of the cell in an aggregate too small to be visible, or it may disperse throughout the cell and its processes, providing maximum display. The chromatophores of the cephalopod molluscs are entirely different (Figure 28-3, *B*). Each is a small saclike cell filled with pigment granules and surrounded by muscle cells that, when contracted, stretch the whole cell out into a highly pigmented sheet. When the muscles relax, the elastic chromatophore quickly shrinks to a small sphere. With such pigment cells the squids and octopuses can alter their color more rapidly than any other animal.

The most widespread of animal pigments are the **melanins,** a group of black or brown polymers that are responsible for the various earth-colored shades that most animals wear. Yellow and red colors are often caused by **carotenoid** pigments, which are frequently contained within special pigment cells called **xanthophores.** Most vertebrates are incapable of synthesizing their own carotenoid pigments but must obtain them directly or indirectly from plants. Two entirely different classes of pigments called ommochromes and pteridines are usually responsible for the yellow pigments of molluscs and arthropods. Green colors are rare; when they occur, they are usually produced by yellow pigment overlying blue structural color. **Iridophores,** a third type of chromatophore, contain crystals of guanine or some other purine, rather than pigment. They produce a silvery or metallic effect by reflecting light.

By vertebrate standards, the mammals are a somber-colored group. Most mammals are more or less color blind, a deficiency that is doubtless connected with the lack of bright colors in the group. Exceptions are the brilliantly colored skin patches of some baboons and mandrills. Significantly, the primates have color vision and thus can appreciate such eye-catching ornaments. The muted colors of mammals are caused by melanin, which is deposited in growing hair by dermal melanophores.

Injurious effects of sunlight

The familiar vulnerability of the human skin to sunburn reminds us of the potentially damaging effects of ultraviolet radiation on protoplasm. Many animals, such as protozoa and flatworms, if exposed to the sun in shallow water are damaged or killed by ultraviolet radiation. Most land animals are protected from such damage by the screening action of special body coverings, for example, the cuticle of arthropods, the scales of reptiles, and the feathers and fur of birds and mam-

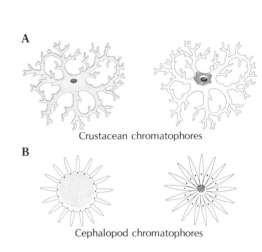

A

Crustacean chromatophores

B

Cephalopod chromatophores

Figure 28-3

Chromatophores. **A,** The crustacean chromatophore showing the pigment dispersed and concentrated. Vertebrate chromatophores are similar. **B,** The cephalopod chromatophore is an elastic capsule surrounded by muscle fibers that, when contracted (left), stretch out the capsule to expose the pigment.

mals. Humans, however, lack such special coverings and must depend on thickening of the epidermis (**stratum corneum**) and on epidermal pigmentation for protection. Most ultraviolet radiation is absorbed in the epidermis, but about 10% penetrates the dermis. Damaged cells in both the epidermis and dermis release histamine and other vasodilator substances that cause blood vessel enlargement in the dermis and the characteristic red coloration of sunburn. Light skins suntan through the formation of the pigment **melanin** in the deeper epidermis and by "pigment darkening," that is, the photooxidative blackening of bleached pigment already present in the epidermis. Regrettably, white Americans pay dearly for their sun worship with approximately 400,000 cases of skin cancer each year.

SKELETAL SYSTEMS

Skeletons are supportive systems that provide rigidity to the body, surfaces for muscle attachment, and protection for vulnerable body organs. The familiar bone of the vertebrate skeleton is only one of several kinds of supportive and connective tissues serving various binding and weight-bearing functions, which are described in this discussion.

Hydrostatic Skeletons

Not all skeletons are rigid; many invertebrate groups use their body fluids as an internal hydrostatic skeleton. The muscles in the body wall of the earthworm, for example, have no firm base for attachment but develop muscular force by contracting against the coelomic fluids, which are enclosed within a limited space and are incompressible, much like the hydraulic brake system of an automobile.

Alternate contractions of the circular and longitudinal muscles of the body wall enable the worm to thin and thicken, setting up backward-moving waves of motion that propel the animal forward. Earthworms and other annelids are helped by the septa that separate the body into more or less independent compartments. An obvious advantage is that if a worm is punctured or even cut into pieces, each part can still develop pressure and move. Worms that lack internal compartments, for example, the lugworm *Arenicola,* are rendered helpless if the body fluid is lost through a wound.

Rigid Skeletons

Rigid skeletons differ from hydrostatic skeletons in one fundamental way: rigid skeletons consist of rigid elements, usually jointed, to which muscles can attach. Muscles can only contract; to be lengthened they must be extended by the pull of an antagonistic set of muscles. Rigid skeletons provide the anchor points required by opposing sets of muscles, such as flexors and extensors.

There are two principal types of rigid skeletons: the **exoskeleton,** typical of molluscs and arthropods, and the **endoskeleton,** characteristic of echinoderms and vertebrates. The invertebrate exoskeleton may be mainly protective, but it may also perform a vital role in locomotion. An exoskeleton may take the form of a shell, a spicule, or a calcareous, proteinaceous, or chitinous plate. It may be rigid, as in molluscs, or jointed and movable, as in arthropods. Unlike the endoskeleton, which grows with the animal, the exoskeleton is often a limiting coat of armor that must be periodically molted to make way for an enlarged replacement. Some

From the viewpoint of structural mechanics, the arthropod-type exoskeleton is perhaps a better arrangement for small animals than a vertebrate-type endoskeleton because a hollow cylindrical tube can support much more weight without collapsing than can a solid cylindrical rod of the same material and weight. Arthropods can thus enjoy both protection and structural support from their exoskeleton. But for larger animals the hollow tube loses its advantage, because the rigidity of a cylinder decreases rapidly as the radius is increased. For a very large animal the hollow cylinder would be completely impractical. If made thick enough to support the body weight, it would be too heavy to lift; but if kept thin and light, it would be extremely sensitive to buckling or shattering on impact. Finally, can you imagine the sad plight of a large animal when it shed its exoskeleton to molt?

invertebrate exoskeletons, such as the shells of snails and bivalves, grow with the animal.

The vertebrate endoskeleton is formed inside the body and is composed of bone and cartilage, which are forms of dense connective tissue. Bone not only supports and protects but is also the major body reservoir for calcium and phosphorus. In higher vertebrates the red blood cells and certain white blood cells are formed in the bone marrow.

Cartilage

Cartilage and bone are the characteristic vertebrate supportive tissues. The **notochord,** the semirigid axial rod of protochordates, vertebrate larvae, and embryos, is also a primitive vertebrate supportive tissue. Except in the most primitive chordates, for example, amphioxus and the cyclostomes, the notochord is surrounded or replaced by the backbone during embryonic development. The notochord is composed of large, vacuolated cells and is surrounded by layers of elastic and fibrous sheaths. It is a stiffening device, preserving body shape during locomotion (p. 444).

Cartilage is the major skeletal element of primitive vertebrates. The jawless fishes (agnathans) and elasmobranchs have purely cartilaginous skeletons, which strangely enough is a degenerative feature, since their Paleozoic ancestors had bony skeletons. Higher vertebrates as adults have principally bony skeletons with some cartilage interspersed. Cartilage is a soft, pliable, characteristically deeplying tissue. Unlike most connective tissues, which are quite variable in form, cartilage is basically the same wherever it is found. The basic form, **hyaline cartilage,** has a clear, glassy appearance (see Figure 6-11, *B*, p. 108). It is composed of cartilage cells (**chondrocytes**) surrounded by firm complex protein gel interlaced with a meshwork of collagenous fibers. Blood vessels are virtually absent. In addition to forming the cartilaginous skeleton of the primitive vertebrates and that of all vertebrate embryos, hyaline cartilage makes up the articulating surfaces of many bone joints of higher adult vertebrates and the supporting tracheal, laryngeal, and bronchial rings.

Cartilage similar to hyaline cartilage occurs in some invertebrates, for example in the radula of gastropod molluscs and in the lophophore of brachiopods. The cartilage of cephalopod molluscs is of a special type with long, branching processes that resemble the cells of vertebrate bone.

Bone

Bone is a living tissue that differs from other connective and supportive tissues by having significant deposits of inorganic calcium salts laid down in an extracellular matrix. Its structural organization is such that bone has nearly the tensile strength of cast iron, yet is only one third as heavy.

Bone is never formed in vacant space but is always laid down by replacement of areas occupied by some form of connective tissue. Most bones develop from cartilage (**endochondral** ["within-cartilage"] **bone**). The embryonic cartilage is gradually eroded, leaving it extensively honeycombed; bone-forming cells then invade these areas and begin depositing calcium salts around the strandlike remnants of the cartilage. A second type of bone is **membranous bone,** which develops directly from sheets of embryonic cells. In higher vertebrates membranous bone is restricted to bones of the face and cranium; the remainder of the skeleton is endochondral bone. But once bone is fully formed, there is no difference in the

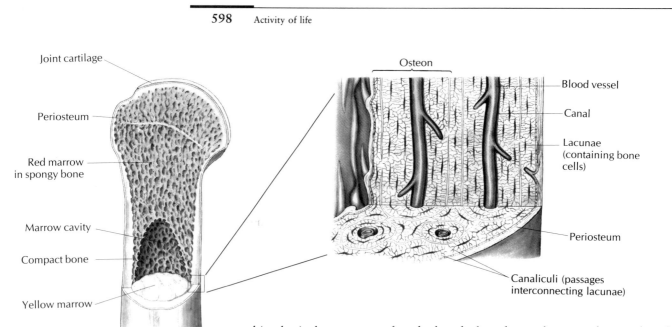

Figure 28-4

Structure of bone, showing the appearance of spongy and compact bone. The enlarged section shows how bone cells and the dense calcified matrix are arranged into units called osteons. Bone cells are entrapped within cell-like lacunae but receive nutrients from the circulatory system via tiny canaliculi that interlace the calcified matrix. Bone cells are known as osteoblasts when they are building bone, but in mature bone shown here, they become resting osteocytes. Bone is covered with compact connective tissue called periosteum.

histological structure of endochondral and membranous bone; they look the same.

Fully formed bone, however, may vary in density. **Spongy** (or cancellous) **bone** consists of an open, interlacing framework of bony tissue, oriented to give maximum strength under the normal stresses and strains that the bone receives. All bone develops first as spongy bone, but some bones, through further deposition of bone salts, become **compact.** Compact bone is dense, appearing absolutely solid to the unaided eye. Both spongy and compact bone are found in the typical long bones of the body (Figure 28-4).

Microscopic structure of bone

Compact bone is composed of a calcified bone matrix arranged in concentric rings. The rings contain cavities (**lacunae**) filled with bone cells (**osteocytes**) that are interconnected by many minute passages (**canaliculi**). These serve to distribute nutrients throughout the bone. This entire organization of lacunae and canaliculi is arranged into an elongated cylinder called an **osteon** (**haversian system**) (Figure 28-4). Bone consists of bundles of osteons cemented together and interconnected with blood vessels.

Bone growth is a complex restructuring process, involving both its destruction internally by bone-resorbing cells (**osteoclasts**) and its deposition externally by bone-building cells (**osteoblasts**). Both processes occur simultaneously so that the marrow cavity inside grows larger by bone resorption while new bone is laid down outside by bone deposition. Bone growth responds to several hormones, particularly the **parathyroid hormone,** which stimulates bone resorption, and **calcitonin,** which inhibits bone resorption. These two hormones, together with a derivative of vitamin D, are responsible for maintaining a constant level of calcium in the blood (p. 723).

Plan of the vertebrate skeleton

The vertebrate skeleton is composed of two main divisions: the **axial skeleton,** which includes the skull, vertebral column, sternum, and ribs, and the **appendicular skeleton,** which includes the limbs (or fins or wings) and the pectoral and pelvic girdles (Figure 28-5). Not surprisingly, the skeleton has undergone extensive remodeling in the course of vertebrate evolution. The move from water to land forced dramatic changes in body form. With cephalization, that is, the concentration of brain, sense organs, and food-gathering and respiratory apparatus in the head, the skull became the most intricate portion of the skeleton. The lower vertebrates have a larger number of skull bones than the more advanced vertebrates. Some fish have 180 skull bones (a source of frustration to paleontologists); amphibians and reptiles, 50 to 95; and mammals, 35 or fewer. Humans have 29 skull bones. There is a basic plan of homology in the skull elements of vertebrates from fish to human beings; evolution has meant reduction in numbers of bones through loss and fusion in accordance with size and functional changes.

The vertebral column is the main stiffening axis of the postcranial skeleton. In fishes it serves much the same function as the notochord from which it is derived; that is, it provides points for muscle attachment and prevents telescoping of the body during muscle contraction. Since fish musculature is similar throughout the trunk and tail, fish vertebrae are differentiated only into trunk and caudal vertebrae.

With the evolution of tetrapods, the vertebral column became structurally adapted to withstand new regional stresses transmitted to the column by the two pairs of appendages. In the higher tetrapods, the vertebrae are differentiated into **cervical** (neck), **thoracic** (chest), **lumbar** (back), **sacral** (pelvic), and **caudal** (tail) vertebrae. In birds and also in humans the caudal vertebrae are reduced in number

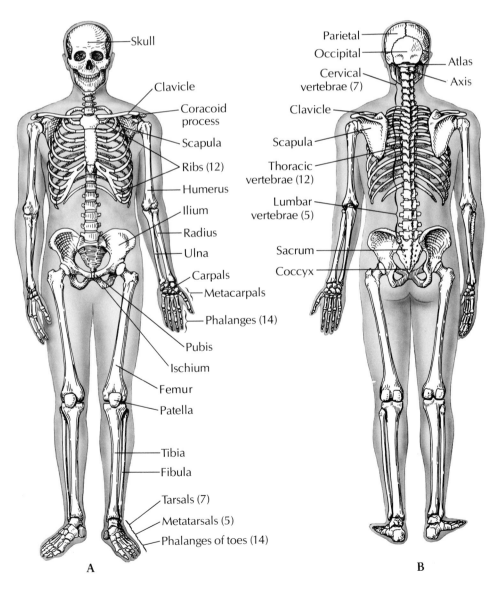

Figure 28-5

Human skeleton. **A,** Ventral view. **B,** Dorsal view. Numbers in parentheses indicate number of bones in that unit. In comparison with other mammals, the human skeleton is a patchwork of primitive and specialized parts. Erect posture, brought about by specialized changes in legs and pelvis, enabled the primitive arrangement of arms and hands (arboreal adaptation of human ancestors) to be used for manipulation of tools. Development of the skull and brain followed as a consequence of the premium natural selection put on dexterity, better senses, and ability to appraise environment.

and size, and the sacral vertebrae are fused. The number of vertebrae varies among the different animals. The python seems to lead the list with 435. In humans (Figure 28-5) there are 33 in the child, but in the adult 5 are fused to form the **sacrum** and 4 to form the **coccyx.** Besides the sacrum and coccyx, humans have 7 cervical, 12 thoracic, and 5 lumbar vertebrae. The number of cervical vertebrae (7) is constant in nearly all mammals.

The first two cervical vertebrae, the **atlas** and the **axis,** are modified to support the skull and permit pivotal movements. The atlas bears the globe of the head much as the mythological Atlas bore the earth on his shoulders. The axis, the second vertebra, permits the head to turn from side to side.

Ribs are long or short skeletal structures that articulate medially with vertebrae and extend into the body wall. Primitive forms have a pair of ribs for every vertebra; they serve as stiffening elements in the connective tissue septa that separate the muscle segments and thus improve the effectiveness of the muscle contractions. Many fishes have both dorsal and ventral ribs, and some have numerous riblike intermuscular bones as well—all of which increase the difficulty and reduce the pleasure of eating certain kinds of fish. Higher vertebrates have a reduced number of ribs, and some, such as the familiar leopard frog, have no ribs at all. Others, such as elasmobranchs and some amphibians, have very short ribs. Humans have 12 pairs of ribs, but approximately one person in 20 has a thirteenth pair. In mammals the ribs together form the thoracic basket, which supports the chest wall and prevents collapse of the lungs.

Most vertebrates, fishes included, have paired appendages. All fishes except the agnathans have thin pectoral and pelvic fins that are supported by the pectoral and pelvic girdles, respectively. Forms above the fishes (except caecilians, snakes, and limbless lizards) have two pairs of **pentadactyl** (five-toed) limbs, also supported by girdles. The pentadactyl limb is similar in all tetrapods, alive and extinct; even when highly modified for various modes of life, the elements are rather easily homologized.

Modifications of the basic pentadactyl limb for life in different environments involve the distal elements much more frequently than the proximal, and it is far more common for bones to be lost or fused than for new ones to be added. Horses and their relatives developed a foot structure for fleetness by elongation of the third toe. In effect, a horse stands on its third fingernail (hoof), much like a ballet dancer standing on the tips of the toes (see Figure 27-28, p. 587). The bird wing is a good example of distal modification. The bird embryo bears 13 distinct wrist and hand bones (carpals and metacarpals), which are reduced to three in the adult.

Figure 28-6

The chief difference between male and female skeletons is the structure of the pelvis. The female pelvis has less depth with broader, less sloping ilia, a more circular bony ring (pelvic canal), a wider and more rounded pubic arch, and a shorter and wider sacrum. Most structures of the female pelvis are correlated with childbearing functions. In the evolution of the human skeleton, the pelvis has changed more than any other part because it has to support the weight of the erect body. (Anterior view.)

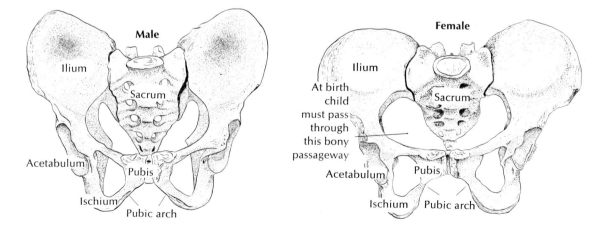

Most of the finger bones (phalanges) are lost, leaving four bones in three digits (see p. 542). The proximal bones (humerus, radius, and ulna), however, are slightly modified in the bird wing.

In nearly all tetrapods the pelvic girdle (Figure 28-6) is firmly attached to the axial skeleton, since the greatest locomotory forces transmitted to the body come from the hindlimbs. The pectoral girdle, however, is much more loosely attached to the axial skeleton, providing the forelimbs with greater freedom for manipulative movements.

In humans the pectoral girdle is made up of two scapulae and two clavicles; the arm is made up of the humerus, ulna, radius, eight carpals, five metacarpals, and 14 phalanges. The pelvic girdle consists of three fused bones—ilium, ischium, and pubis; the leg is made up of the femur, patella, tibia, fibula, seven tarsals, five metatarsals, and 14 phalanges. Each bone of the leg has its counterpart in the arm with the exception of the patella. This correspondence between anterior and posterior parts is an example of **serial homology.**

ANIMAL MOVEMENT

Movement is a distinctive characteristic of animals. Plants may show movement, but this usually results from changes in turgor pressure or growth rather than from specialized contractile proteins as in animals. Animal movement occurs in many forms in animal tissues, ranging from barely discernible streaming of cytoplasm to frank movements of powerful striated muscles of vertebrates. It has become evident that most animal movement depends on a single fundamental mechanism: **contractile proteins,** which can change their form to elongate or contract. This contractile machinery is always composed of ultrafine fibrils—fine filaments, striated fibrils, or tubular fibrils (microtubules)—arranged to contract when powered by **ATP.** By far the most important protein contractile system is the **actomyosin system,** composed of two proteins, **actin** and **myosin.** This is an almost universal biomechanical system found from protozoa to vertebrates; it performs a long list of diverse functional roles. In this discussion we examine the three principal kinds of animal movement: ameboid, ciliary, and muscular.

Ameboid Movement

Ameboid movement is a form of movement especially characteristic of amebas and other protozoa; it is also found in many wandering cells of higher animals, such as white blood cells, embryonic mesenchyme, and numerous other mobile cells that move through the tissue spaces. Ameboid cells change their shape by sending out and withdrawing **pseudopodia** (false feet) from any point on the cell surface. Beneath the plasmalemma lies a nongranular layer, the gel-like **ectoplasm,** which encloses the more liquid **endoplasm** (see Figure 8-3, p. 138).

Optical studies of an ameba in movement suggest the outer layer of ectoplasm surrounds a rather fluid core of endoplasm. As the pseudopod extends, the inner endoplasm fountains out to the periphery, changing from the sol to the gel state. The newly formed ectoplasm then slips posteriorly under the plasmalemma and is changed to endoplasm at the rear to begin another cycle. Although no completely satisfactory analysis exists, it seems certain that ameboid movement is based on the same fundamental contractile system that powers vertebrate muscles: an actomyosin machinery driven by ATP.

___ Ciliary Movement

Cilia are minute, hairlike, motile processes that extend from the surfaces of the cells of many animals. They are a particularly distinctive feature of ciliate protozoa, but except for the nematodes in which motile cilia are absent and the arthropods in which they are rare, cilia are found in all major groups of animals. Cilia perform many roles either in moving small animals such as protozoa through their aquatic environment or in propelling fluids and materials across the epithelial surfaces of larger animals.

Cilia are of remarkably uniform diameter (0.2 to 0.5 μm) wherever they are found. The electron microscope has shown that each cilium contains a peripheral circle of nine double microtubules and an additional two microtubules in the center (see Figure 8-2, p. 137). (Exceptions to the 9 + 2 arrangement have been noted; sperm tails of flatworms have but one central microtubule.) A **flagellum** is a whiplike structure longer than a cilium and usually present singly or in small numbers at one end of a cell. They are found in members of flagellate protozoa, in animal spermatozoa, and in sponges. The main difference between a cilium and a flagellum is in their beating pattern rather than in their structure, since both look alike internally. A flagellum beats symmetrically with snakelike undulations so that the water is propelled parallel to the long axis of the flagellum. A cilium, in contrast, beats asymmetrically with a fast power stroke in one direction followed by a slow recovery during which the cilium bends as it returns to its original position. The water is propelled parallel to the ciliated surface (Figure 28-7).

According to the currently favored theory of ciliary movement, the microtubules behave as "sliding filaments" that move past one another much like the sliding filaments of vertebrate skeletal muscle that is described in the next discussion. During contraction, microtubules on the concave side slide outward past microtubules on the convex side to increase curvature of the cilium; during the recovery stroke, microtubules on the opposite side slide outward to bring the cilium back to its starting position. For such a system to work, the microtubules must be interconnected by molecular bridges, which in fact, can be seen with the electron microscope.

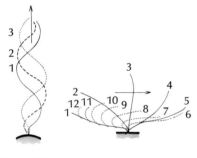

Figure 28-7

Flagellum beats in wavelike undulation, propelling water parallel to the main axis of the flagellum. Cilium propels water in direction parallel to the cell surface.
With permission from Sleigh, M.A. 1974. Cilia and flagella. Copyright by Academic Press, Inc. (London), Ltd.

___ Muscular Movement

Contractile tissue is most highly developed in muscle cells called **fibers.** Although muscle fibers themselves can do work only by contraction and cannot actively lengthen, they can be arranged in so many different configurations and combinations that almost any movement is possible.

Types of vertebrate muscle

Vertebrate muscle is broadly classified on the basis of the appearance of fibers when viewed with a light microscope. **Striated muscle** appears transversely striped (striated), with alternating dark and light bands (Figure 28-8, *A*). We can recognize two types of striated muscle: **skeletal** and **cardiac muscle.** A third kind of vertebrate muscle is **smooth** (or visceral) **muscle,** which lacks the characteristic alternating bands of the striated type.

Skeletal muscle is typically organized into sturdy, compact bundles or bands (Figure 28-8, *A*). It is called skeletal muscle because it is attached to skeletal elements and is responsible for movements of the trunk, appendages, respiratory organs, eyes, mouthparts, and so on. Skeletal muscle **fibers** are extremely long, cylindrical, multinucleate cells that may reach from one end of the muscle to the other. They are packed into bundles called **fascicles** (L. *fasciculus*, small bundle), which are enclosed by tough connective tissue. The fascicles are in turn grouped into a discrete **muscle** surrounded by a thin connective tissue layer. Most skeletal muscles taper at their ends, where they connect to bones by tendons. Other muscles, such as the ventral abdominal muscles, are flattened sheets.

In most fishes, amphibians, and to some extent reptiles, there is a segmented organization of muscles alternating with the vertebrae. The skeletal muscles of higher vertebrates, by splitting, fusion, and shifting, have developed into specialized muscles best suited for manipulating the jointed appendages that have evolved for locomotion on land. Skeletal muscle contracts powerfully and quickly but fatigues more rapidly than does smooth muscle. Skeletal muscle is sometimes called **voluntary muscle** because it is stimulated by motor fibers and is under conscious cerebral control.

Smooth muscle lacks the striations typical of skeletal muscle (Figure 28-8, *B*). The cells are long, tapering strands, each containing a single nucleus. Smooth muscle cells are organized into sheets of muscle circling the walls of the alimentary canal, blood vessels, respiratory passages, and urinary and genital ducts. Smooth muscle is typically slow acting and can maintain prolonged contractions with very little energy expenditure. It is under the control of the autonomic nervous system; thus, unlike skeletal muscle, its contractions are involuntary and unconscious. The principal functions of smooth muscles are to push the material in a tube, such as the intestine, along its way by active contractions or to regulate the diameter of a tube, such as a blood vessel, by sustained contraction.

Cardiac muscle, the seemingly tireless muscle of the vertebrate heart, combines certain characteristics of both skeletal and smooth muscle (Figure 28-8, *C*). It is fast acting and striated like skeletal muscle, but contraction is under involuntary autonomic control like smooth muscle. Actually the autonomic nerves serving the heart can only speed up or slow down the rate of contraction; the heartbeat originates within specialized cardiac muscle, and the heart continues to beat even after all autonomic nerves are severed. Until recently, cardiac muscle was believed to be **syncytial** (a tissue with many nuclei not separated into discrete cells by cell membranes) with branching, interconnected fibers. Histologists, their understanding vastly increased by the electron microscope, now consider cardiac muscle to be comprised of closely opposed, but separate, uninucleate cell fibers.

Types of invertebrate muscle

Smooth and striated muscles are also characteristic of invertebrate animals, but there are many variations of both types and even instances in which the structural and functional features of vertebrate smooth and striated muscle are combined in the invertebrates. Striated muscle appears in invertebrate groups as diverse as the primitive cnidarians and the advanced arthropods. The thickest muscle fibers known, approximately 3 mm in diameter and 6 cm long, are those of giant barnacles and of Alaska king crabs living along the Pacific coast of North America. These cells are so large that they can be readily cannulated for physiological studies and are understandably popular with muscle physiologists.

It is not possible in this short space to describe adequately the tremendous diversity of muscle structure and function in the vast assemblage of invertebrates. We will mention only two functional extremes.

Bivalve molluscan muscles contain fibers of two types. One kind is striated

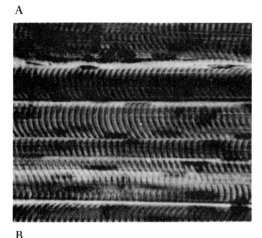

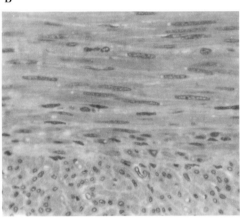

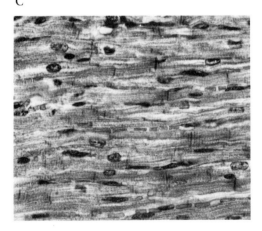

Figure 28-8

Photomicrographs of types of vertebrate muscle. **A,** Skeletal muscle (human) showing several striated fibers (cells) lying side by side. **B,** Smooth muscle (human) showing absence of striations. Note elongate nuclei in the long fibers. **C,** Cardiac muscle (monkey). Note the vertical bars, called intercalated discs, joining separate fibers end to end.

A, Courtesy J.W. Bamberger; B and C, courtesy Carolina Biological Supply Co.

Figure 28-10

Sliding filament model, showing how thick and thin filaments interact during contraction. **A,** Muscle relaxed. **B,** Muscle contracted.

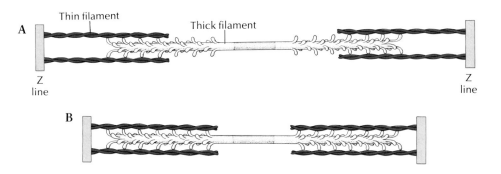

action. As contraction continues, the **Z** lines are pulled closer together (Figure 28-10).

At the time the sliding filament model was proposed, the nature of the mechanism that pulls the thin filaments past the thick was largely a mystery. In the intervening years, however, there have been important improvements in biochemical techniques and in the technology of electron microscopy and x-ray diffraction; these advances have permitted much better visualization of the complex architecture of the contractile machinery.

Each myosin molecule, shown in Figure 28-11, *A*, is composed of two polypeptide chains, each having a club-shaped "head." Lined up as they are in a bundle to form a thick filament (Figure 28-11, *B*), the double heads of each myosin molecule face outward from the center of the filament. These heads act as the molecular cross bridges that interact with the thin filaments during contraction.

The thin filaments are more complex because they are composed of three different proteins. The backbone of the thin filament is a double strand of the protein actin, twisted into a double helix. Surrounding the actin filament are two thin strands of another protein, **tropomyosin,** that lie near the grooves between

Figure 28-11

Molecular structure of thick and thin filaments of skeletal muscle. **A,** The myosin molecule is composed of two peptides coiled together and expanded at their ends into a globular head. **B,** The thick filament is composed of a bundle of myosin molecules with the globular heads extended outward. **C,** The thin filament consists of a double strand of actin surrounded by two tropomyosin strands. A globular protein complex, troponin, occurs in pairs at every seventh actin unit. Troponin is a calcium-dependent switch that controls the interaction between actin and myosin.

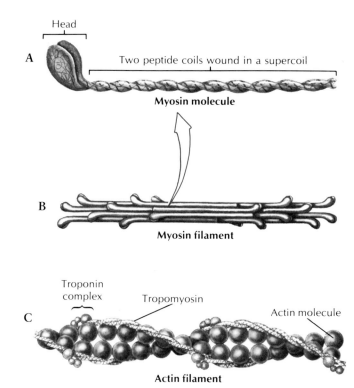

the actin strands. Each tropomyosin strand is itself a double helix as shown in Figure 28-11, C.

The third protein of the thin filament is **troponin,** a complex of three globular proteins located at intervals along the filament. Troponin is a calcium-dependent switch that acts as the control point in the contraction process.

For shortening to occur, the cross bridges must attach, swivel, detach, and reattach at a point farther along the thin filament. Biochemical studies suggest that the attach-pull-release cycle occurs in a series of steps (Figure 28-12). First, myosin binds and then splits a molecule of ATP. The release of bond energy from ATP activates the myosin head, which attaches to the adjacent actin strand and swings 45 degrees, at the same time releasing the ADP molecule. This is the power stroke that pulls the actin filament a distance of about 10 nm, and it comes to an end when another ATP molecule binds to the myosin head, inactivating the site. Thus, each cycle requires the expenditure of energy in the form of ATP.

As long as the muscle is stimulated and new ATP becomes available, the attach-pull-release cycle can repeat again and again, 50 to 100 times per second, pulling the thick and thin filaments past each other. While the distance each sarcomere can shorten is very small, this distance is multiplied by the thousands of sarcomeres lying end to end in a muscle fiber. Thus, a strongly contracted muscle may shorten as much as one-third its resting length.

Stimulation of contraction

To contract skeletal muscle, it must of course be stimulated. If the nerve supply to a muscle is severed, the muscle **atrophies,** or wastes away. Skeletal muscle fibers are grouped into **motor units,** each of which is a small number of muscle fibers innervated by a single axon. As the nerve fiber approaches the muscle fibers, it splays out into many terminal branches. Each branch attaches to a muscle fiber by a special structure, called a **synapse,** or **myoneural junction** (Figure 28-13). At the synapse is a tiny gap, or cleft, that thinly separates nerve fiber and muscle fiber. In the nerve fiber terminus is stored a chemical, **acetylcholine,** which is released when

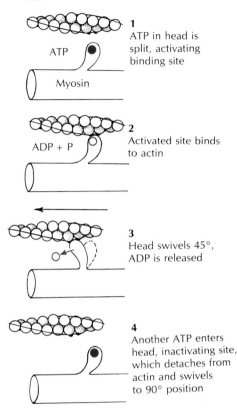

Figure 28-12

Successive steps in the attach-pull-release cycle that pulls actin past myosin and shortens the filament length.

Actin

1 ATP in head is split, activating binding site

ATP

Myosin

2 Activated site binds to actin

ADP + P

3 Head swivels 45°, ADP is released

4 Another ATP enters head, inactivating site, which detaches from actin and swivels to 90° position

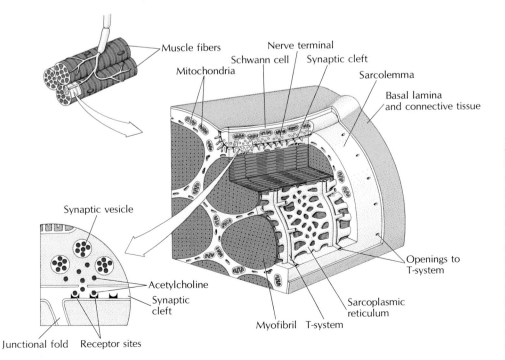

Muscle fibers
Nerve terminal
Schwann cell
Synaptic cleft
Mitochondria
Sarcolemma
Basal lamina and connective tissue
Synaptic vesicle
Acetylcholine
Synaptic cleft
Junctional fold   Receptor sites
Myofibril   T-system
Sarcoplasmic reticulum
Openings to T-system

Figure 28-13

Section of vertebrate skeletal muscle showing nerve-muscle synapse (myoneural junction), sarcoplasmic reticulum, and connecting transverse tubules (T system). Arrival of a nerve impulse at the synapse triggers the release of acetylcholine into synaptic cleft (*inset* at left). The binding of transmitter molecules to receptors leads to the generation of a membrane depolarization. This spreads across the sarcolemma, into the T system, and to the sarcoplasmic reticulum where the sudden release of calcium sets in motion the contractile machinery of the myofibril.

a nerve impulse reaches the synapse. This substance is a chemical mediator that diffuses across the narrow junction and acts on the muscle fiber membrane to generate an electrical depolarization. The depolarization spreads rapidly through the muscle fiber, causing it to contract. Thus the synapse is a special chemical bridge that couples together the electrical activities of nerve and muscle fibers.

For a long time physiologists were puzzled as to how the depolarization at the myoneural junction could spread quickly enough through the fiber to cause simultaneous contraction of all the densely packed filaments within. Then it was discovered that vertebrate skeletal muscle contains an elaborate communication system that performs just this function. This is the endoplasmic reticulum (called the **sarcoplasmic reticulum** in muscle), a system of fluid-filled channels running parallel to the myofilaments and communicating with the sarcolemma that surrounds the fiber (Figure 28-13). The system is ideally arranged for speeding the electrical depolarization from the myoneural junction to the myofilament within.

How does the electrical depolarization activate the contraction process? In the resting, unstimulated muscle, shortening does not occur because the thin tropomyosin strands lie in a position on the actin filament that prevents the myosin heads from attaching to actin. When the muscle is stimulated and the electrical depolarization arrives at the sarcoplasmic reticulum surrounding the fibrils, calcium ions are released. Some of the calcium binds to the control protein troponin. Troponin immediately undergoes conformational changes that allow tropomyosin to move out of its blocking position, and the attach-pull-release cycle is set in motion. Shortening will continue as long as nerve impulses arrive at the neuromuscular junction and free calcium remains available around the microfilament. But when stimulation stops, the calcium is quickly pumped back into the sarcoplasmic reticulum. Troponin resumes its original shape, tropomyosin moves back into its blocking position on actin, and the muscle relaxes.

Energy for contraction

ATP is the immediate source of energy for muscle, but the amount present will sustain contraction for only a fraction of a second. However, vertebrate muscle contains a much larger reservoir of high-energy phosphate, creatine phosphate. This compound contains even more free bond energy than ATP (p. 78) and thus can readily transfer its bond energy to ADP to form ATP.

$$\text{Creatine phosphate} + \text{ADP} \rightleftharpoons \text{ATP} + \text{Creatine}$$

The reserves of creatine phosphate are soon depleted in rapidly contracting muscle and must be restored by the oxidation of carbohydrate. The major store of carbohydrate in muscle is glycogen. In fact, about three fourths of all the glycogen in the body is stored in muscle (most of the rest is stored in the liver). Glycogen can be readily converted into glucose-6-phosphate, the first stage of glycolysis that leads into mitochondrial respiration and the generation of ATP (p. 82).

If muscular contraction is not too vigorous or too prolonged, glucose can be completely oxidized to carbon dioxide and water by **aerobic metabolism.** During prolonged or heavy exercise, however, the blood flow to the muscles, although greatly increased above the resting level, is insufficient to supply oxygen as rapidly as required for the complete oxidation of glucose. When this happens, the contractile machinery receives its energy largely by **anaerobic glycolysis,** a process that does not require oxygen (p. 89). The ability to take advantage of this anaerobic pathway, although not nearly as efficient as the aerobic one, is of great importance; without it, all forms of heavy muscular exertion would be impossible.

Human muscle tissue develops before birth, and a newborn child's complement of skeletal muscle fibers is all that he or she will ever have. But while a male weight-lifter and a young girl have a similar number of muscle fibers, he may be several times her strength because repeated high-intensity, short-duration exercise has induced the synthesis of additional actin and myosin filaments. Each fiber has hypertrophied, becoming larger and stronger. Endurance exercise such as long-distance running produces a very different response. Fibers do not become greatly stronger but develop more mitochondria and myoglobin and become adapted for a high rate of oxidative phosphorylation. These changes, together with the development of more capillaries serving the fibers, lead to increased capacity for long-duration activity.

During anaerobic glycolysis, glucose is degraded to lactic acid with the release of energy. This is used to resynthesize creatine phosphate, which in turn passes the energy to ADP for the resynthesis of ATP. Lactic acid accumulates in the muscle and diffuses rapidly into the general circulation. If the muscular exertion continues, the buildup of lactic acid causes enzyme inhibition and fatigue. Thus, the anaerobic pathway is a self-limiting one, since continued heavy exertion leads to exhaustion. The muscles incur an **oxygen debt** because the accumulated lactic acid must be oxidized by extra oxygen. After the period of exertion, oxygen consumption remains elevated until all of the lactic acid has been oxidized or resynthesized to glycogen.

—SUMMARY

An animal is wrapped in a protective covering, the integument, which may be as simple as the delicate plasma membrane of a protozoan or as complex as the skin of a mammal. The arthropod exoskeleton is the most complex of invertebrate integuments, consisting of a two-layered cuticle secreted by a single-layered epidermis. It may be hardened by calcification or sclerotization. Vertebrate integument consists of two layers: the epidermis, which gives rise to various derivatives such as hair, feathers, and claws; and the dermis, which supports and nourishes the epidermis. It also is the origin of bony derivatives such as fish scales and deer antlers.

Integument color is of two kinds: structural color, produced by refraction or scattering of light by particles in the integument, and pigmentary color, produced by pigments that are usually confined to special pigment cells (chromatophores).

Skeletons are supportive systems that may be hydrostatic or rigid. The hydrostatic skeletons of several soft-walled invertebrate groups depend on body wall muscles that contract against a noncompressible internal fluid of constant volume. Rigid skeletons have evolved with attached muscles that act with the supportive skeleton to produce movement. In higher animals, two forms of skeleton have appeared. Arthropods have an external skeleton, which must be periodically shed to make way for an enlarged replacement. The vertebrates developed an internal skeleton, a framework formed of cartilage or bone, that can grow with the animal, while, in the case of bone, additionally serving as a reservoir of calcium and phosphate.

Animal movement, whether in the form of cytoplasmic streaming, ameboid movement, or the contraction of an organized muscle mass, depends on specialized contractile proteins. The most important of these is the actomyosin system, which in higher animals is usually organized into elongate thick and thin filaments that slide past one another during contraction. When a muscle is stimulated, an electrical depolarization is conducted into the muscle fibers through the sarcoplasmic reticulum, causing the release of calcium. Calcium binds to a protein troponin complex associated with the thin actin filament. This causes tropomyosin to shift out of its blocking position and allows the myosin heads to cross-bridge with the actin filament. Powered by ATP, the myosin heads swivel back and forth to pull the thick and thin filaments past each other. Phosphate bond energy for contraction is supplied by carbohydrate fuels through a storage intermediate, creatine phosphate.

Salmon leaping falls.
Courtesy G.B. Kelez, U.S. Fish and Wildlife Service.

Review questions

1. Describe the structure of the arthropod integument, and explain the difference in the way the cuticle is hardened in crustaceans and in insects.
2. Distinguish between epidermis and dermis in vertebrate integument, and describe the structural derivatives of these two layers.
3. What is the difference between structural colors and colors based on pigments? How do the chromatophores of vertebrates and cephalopod molluscs differ in structure and function?
4. Explain how human skin develops protection against the damaging effects of the ultraviolet rays of the sun.
5. Explain what a hydrostatic skeleton is and how it helps in movement, and name one invertebrate group in which hydrostatic skeletons are important in locomotion.
6. What is hyaline cartilage? Compare its distribution and function in lower and higher vertebrates.
7. What is the difference between endochondral and membranous bone? Between spongy and compact bone?
8. Discuss the role of osteoclasts, osteoblasts, parathyroid hormone, and calcitonin in bone growth.
9. Name the major skeletal components included in the axial and appendicular skeleton.
10. Describe the interaction of endoplasm and ectoplasm in ameboid movement.
11. Compare the structure and function of a cilium with those of a flagellum.
12. Describe the structural and functional features that distinguish each of the three types of vertebrate muscle.
13. What functional features set molluscan smooth muscle and insect fibrillar muscle apart from any known vertebrate muscle?
14. Explain how skeletal muscle shortens according to the sliding filament hypothesis.
15. Describe in sequence the events in muscle stimulation, explaining the role of each of the following: motor unit, myoneural junction, acetylcholine, sarcoplasmic reticulum, calcium, troponin, and tropomyosin.
16. Describe the immediate and reserve sources of energy for muscle contraction. Under what circumstances is an oxygen debt incurred during muscle contraction?

Selected references

See also general references for Part Three, p. 752.

Alexander, R.M. 1982. Locomotion in animals. New York, Chapman and Hall. *Concise, fully comparative treatment. Introduced with a discussion of "sources of power" followed by treatment of mechanisms and energetics of locomotion on land, in water, and in the air. Undergraduate level.*

Caplan, A.J. 1984. Cartilage. Sci. Am. **251**:84-94 (Oct.). *Structure, aging, and development of vertebrate cartilage.*

Fogden, M., and P. Fogden. 1979. Animals and their colors: camouflage, warning coloration, courtship and territorial display, mimicry. New York, Crown Publishers, Inc. *Semipopular treatment, lavishly illustrated.*

Hadley, N.F. 1986. The arthropod cuticle. Sci. Am. **255**: 104-112 (July). *Describes properties of this complex covering that accounts for much of the adaptive success of arthropods.*

McMahon, T.A. 1984. Muscles, reflexes, and locomotion. Princeton, Princeton University Press. *Comprehensive, ranging from basic muscle mechanics to coordinated motion. Though sprinkled with mathematical models, the text is lucid throughout.*

Spearman, R.I.C. 1973. The integument: a textbook of skin biology. Cambridge, Cambridge University Press. *Comparative treatment, embracing both invertebrates and vertebrates.*

Spearman, R.I.C. (ed.). 1977. Comparative biology of skin. Symposia of the Zoological Society of London, no. 39, New York, Academic Press. *Contributed chapters by specialists, especially strong treatment of invertebrate integument. Advanced.*

I N T E R N A L   F L U I D S

Circulation, Immunity,

and Gas Exchange

Single-celled organisms live a contact existence with their environment. Nutrients and oxygen are obtained and wastes are released directly across the cell surface. These animals are so small that no special internal transport system other than the normal streaming movements of the cytoplasm is required. Even some primitive multicellular forms such as sponges, cnidarians, and flatworms have such a simple internal organization and low rate of metabolism that no circulatory system is needed. However, diffusion is an inadequate mechanism for gaseous exchange and nutrient transport over long distances. Therefore most of the more advanced multicellular organisms, because of their size and complexity, require a specialized circulatory (vascular) system.

In addition to serving these primary transport needs, circulatory systems have acquired additional functions. Hormones are moved about, finding their way to target organs where they assist the nervous system to integrate body function. Water, electrolytes, and the many other constituents of the body fluids are distributed and exchanged between different organs and tissues. The body's defenses against microbial invasion are centered in the vascular system. The warm-blooded birds and mammals depend heavily on the blood circulation to conserve or dissipate heat as required for the maintenance of constant body temperature.

—— INTERNAL FLUID ENVIRONMENT

The body fluid of a single-celled animal is in the cellular cytoplasm, in which the various membrane systems and organelles of the cell are suspended. In multicellular animals the body fluids are divided into two main phases, the **intracellular** and the **extracellular**. The intracellular phase (also called intracellular fluid) is the fluid inside each of the body's cells. The extracellular phase (or fluid) is the fluid outside and surrounding the cells which buffers these sites of the body's crucial metabolic activities from the potentially deleterious physical and chemical changes occurring outside them (Figure 29-1, *A*). The importance of the extracellular fluid was first emphasized by the great French physiologist Claude Bernard (Figure 29-2).

In animals that have closed circulatory systems (vertebrates, annelids, and a few other invertebrate groups), the extracellular fluid is further subdivided into

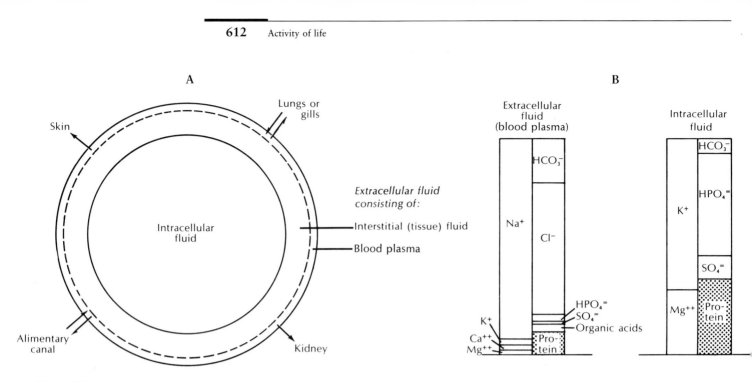

A

B

Figure 29-1

Fluid compartments of the body. **A,** All body cells can be represented as belonging to a single large fluid compartment that is completely surrounded and protected by extracellular fluid *(milieu intérieur)*. This fluid is further subdivided into plasma and interstitial fluid. All exchanges with the environment occur across the plasma compartment. **B,** Electrolyte composition of extracellular and intracellular fluids. Total equivalent concentration of each major constituent is shown. Equal amounts of anions (negatively charged ions) and cations (positively charged ions) are in each fluid compartment. Sodium and chloride, the major plasma electrolytes, are virtually absent from intracellular fluid (actually they are present in low concentration). Note the much higher concentration of protein inside the cells.

blood plasma and **interstitial fluid** (Figure 29-1, *A*). The blood plasma is contained within the blood vessels, whereas the interstitial fluid, or tissue fluid as it is sometimes called, occupies the space immediately around the cells. Nutrients and gases passing between the vascular plasma and the cells must traverse this narrow fluid separation. The interstitial fluid is constantly formed from the plasma by filtration through the capillary walls.

Composition of the Body Fluids

All these fluids—plasma, interstitial, and intracellular—differ from one another in solute composition, but all have one feature in common: they are mostly water. Despite their firm appearance, animals are 70% to 90% water. Humans, for example, are about 70% water by weight: of this, 50% is intracellular water, 15% is interstitial fluid water, and the remaining 5% is in the blood plasma. Figure 29-1, *A* illustrates how the plasma space serves as the pathway of exchange between the cells of the body and the outside world. This exchange of respiratory gases, nutrients, and wastes is accomplished by specialized organs (kidney, lung, gill, alimentary canal), as well as by the integument.

The body fluids contain many inorganic and organic substances in solution. Principal among these are the inorganic electrolytes and proteins. Figure 29-1, *B*, shows that **sodium, chloride,** and **bicarbonate** are the chief extracellular electrolytes, whereas **potassium, magnesium, phosphate,** and **proteins** are the major intracellular electrolytes. These differences are prominent; they are always maintained despite the continuous flow of materials into and out of the cells of the body. The two subdivisions of the extracellular fluid—plasma and interstitial fluid—have similar compositions except that the plasma has more proteins, which are too large to filter through the capillary wall into the interstitial fluid.

Composition of blood

Among the lower invertebrates that lack a circulatory system (such as flatworms and cnidarians), it is not possible to distinguish a true "blood." These forms possess a clear, watery tissue fluid containing some primitive phagocytic cells, some protein, and a mixture of salts similar to seawater. The "blood," or **hemo-**

lymph, of higher invertebrates is more complex. Invertebrates with closed circulatory systems maintain a clear separation between blood contained within blood vessels and tissue (interstitial) fluid surrounding the vessels.

In vertebrates, blood is a complex liquid tissue composed of plasma and formed elements, mostly corpuscles, suspended in the plasma. When the red blood corpuscles and other formed elements are separated from the fluid components by centrifugation, the blood is found to be approximately 55% plasma and 45% formed elements.

The composition of mammalian blood is as follows:

Plasma
1. Water (90%)
2. Dissolved solids, consisting of the plasma proteins (albumin, globulins, fibrinogen), glucose, amino acids, electrolytes, various enzymes, antibodies, hormones, metabolic wastes, and traces of many other organic and inorganic materials
3. Dissolved gases, especially oxygen, carbon dioxide, and nitrogen

Formed elements (Figure 29-4)
1. Red blood cells (erythrocytes), containing hemoglobin for the transport of oxygen and carbon dioxide
2. White blood cells (leukocytes), serving as scavengers and as defensive cells
3. Platelets (thrombocytes), functioning in blood coagulation

The plasma proteins are a diverse group of large and small proteins that perform numerous functions. The major protein groups are (1) **albumin,** the most abundant plasma protein, which constitutes 60% of the total; (2) the **globulins** (α_1, α_2, β, and γ), a diverse group of high–molecular weight proteins (35% of the total) that includes immunoglobulins and various metal-binding proteins; and (3) **fibrinogen,** a very large protein that functions in blood coagulation.

Red blood cells, or **erythrocytes,** are present in enormous numbers in the blood, approximately 5.4 billion per milliliter of blood in adult men and 4.8 billion in women. They are formed continuously from large nucleated **erythroblasts** in the red bone marrow. Here hemoglobin is synthesized and the cells divide several times. In mammals the nucleus shrinks during development to a small remnant and eventually disappears altogether. Many other characteristics of a typical cell also are lost: ribosomes, mitochondria, and most enzyme systems.

Figure 29-2

French physiologist Claude Bernard (1813-1878), one of the most influential of nineteenth-century physiologists. Bernard believed in the constancy of the *milieu intérieur* ("internal environment"), which is the extracellular fluid bathing the cells. He pointed out that it is through the *milieu intérieur* that foods and wastes and gases are exchanged and chemical messengers are distributed.
From Fulton, J.F., and L.G. Wilson. 1966. Selected readings in the history of physiology. Springfield, Ill., Charles C Thomas, Publisher.

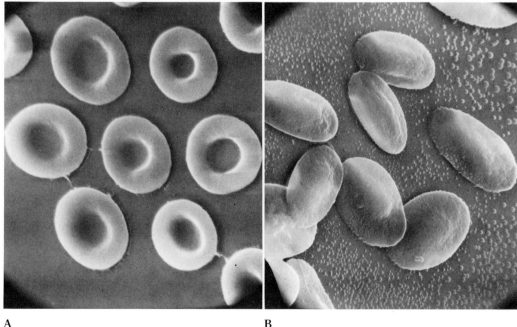

A   **B**

Figure 29-3

Mammalian and amphibian red blood corpuscles. **A,** The erythrocytes of a gerbil are biconcave discs containing hemoglobin and surrounded by a tough stroma. **B,** The frog erythrocytes are convex discs, each containing a nucleus, which is plainly visible in the scanning electron micrograph as a bulge in the center of each cell. (Magnifications: mammalian erythrocytes, $\times 6300$; frog erythrocytes, $\times 2400$.)
Courtesy P.P.C. Graziadei.

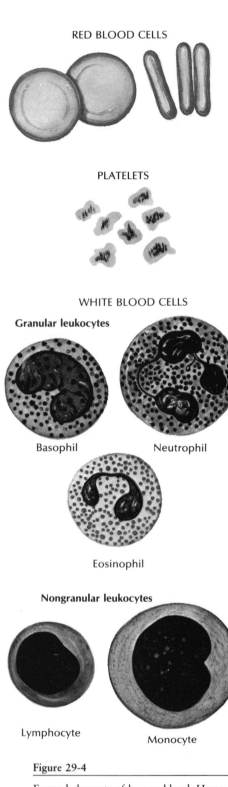

RED BLOOD CELLS

PLATELETS

WHITE BLOOD CELLS

Granular leukocytes

Basophil

Neutrophil

Eosinophil

Nongranular leukocytes

Lymphocyte

Monocyte

Figure 29-4

Formed elements of human blood. Hemoglobin-containing red blood cells of humans and other mammals lack nuclei, but those of all lower vertebrates have nuclei. Various leukocytes provide a wandering system of protection for the body. Platelets participate in the body's clotting mechanism.

From Anthony, C.P., and G.A. Thibodeau. 1983. Textbook of anatomy and physiology, ed. 11. St. Louis, The C.V. Mosby Co.

What is left is a biconcave disc consisting of a baglike membrane, the **stroma,** packed with about 280 million molecules of the blood-transporting pigment **hemoglobin** (several respiratory pigments, in addition to hemoglobin, are found in various invertebrates; see marginal note, p. 634). Approximately 33% of the erythrocyte's weight is hemoglobin. The biconcave shape (Figure 29-3, *A*) is a mammalian innovation that provides a larger surface for gas diffusion than would a flat or spherical shape. All other vertebrates have nucleated erythrocytes that are usually ellipsoidal rather than round discs (Figure 29-3, *B*).

The erythrocyte has an average life span of approximately 4 months. During this time it may journey 700 miles, squeezing repeatedly through the capillaries, which are sometimes so narrow that the erythrocyte must bend to get through. When at last it fragments, it is quickly engulfed by large scavenger cells called **macrophages,** located in the liver, bone marrow, and spleen. The iron from the hemoglobin is salvaged to be used again, while the rest of the heme is converted to **bilirubin,** a bile pigment. It is estimated that 10 million erythrocytes are produced and another 10 million destroyed every second in the human body.

The white blood cells, or **leukocytes,** form a wandering system of protection for the body. In adults they number only approximately 7.5 million per milliliter of blood, a ratio of one white cell to 700 red cells. There are several kinds of white blood cells: **granulocytes** (subdivided into neutrophils, basophils, and eosinophils), **lymphocytes,** and **monocytes** (Figure 29-4). The role of the leukocytes in the body's defense mechanisms will be discussed later.

___ Hemostasis: Prevention of Blood Loss

Animals must have ways of preventing the rapid loss of body fluids after an injury. Since blood is flowing and is under considerable hydrostatic pressure, it is especially vulnerable to hemorrhagic loss.

When a vessel is damaged, its smooth muscle in the wall contracts, sometimes completely closing the lumen and preventing any blood flow through the injured vessel. This primitive but highly effective means of preventing hemorrhage is used by invertebrates and vertebrates alike. Beyond this first defense against blood loss, all vertebrates, as well as some of the larger, active invertebrates with high blood pressures, have special cellular elements and proteins in the blood that are capable of forming plugs, or clots, at the injury site.

In higher vertebrates **blood coagulation** is the dominant hemostatic defense. Blood clots form as a tangled network of fibers from one of the plasma proteins, **fibrinogen.** The transformation of fibrinogen into a **fibrin** meshwork (Figure 29-5) that entangles blood cells to form a gel-like clot is catalyzed by the enzyme thrombin. Thrombin is normally present in the blood in an inactive form called **prothrombin,** which must be activated for coagulation to occur.

The blood platelets (Figure 29-4) play a vital role in this process. Platelets—minute, colorless, incomplete cells lacking nuclei—are present in large numbers in the blood. When the normally smooth inner surface of a blood vessel is disrupted, either by a break or by deposits of a cholesterol-lipid material, the platelets rapidly adhere to the surface and release a variety of substances. These factors, which were released from damaged tissue, and calcium ions, initiate the conversion of prothrombin to the active thrombin. Thrombin, in turn, converts soluble fibrinogen to the gel form of the protein, fibrin. The stages in the formation of fibrin are summarized in Figure 29-6.

The catalytic sequence in this scheme is unexpectedly complex, involving a series of plasma protein factors, each normally inactive until activated by a pre-

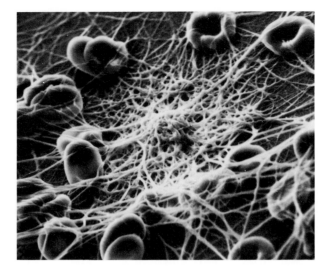

Figure 29-5

Human red blood cells entrapped in fibrin clot. Clotting is initiated after tissue damage by the disintegration of platelets in the blood, resulting in a complex series of intravascular reactions that end with the conversion of a plasma protein, fibrinogen, into long, tough, insoluble polymers of fibrin. Fibrin and entangled erythrocytes form the blood clot, which arrests bleeding. An aggregation of platelets probably underlies the raised mass of fibrin in center.

Scanning electron micrograph courtesy N.F. Rothman.

vious factor in the sequence. The sequence behaves like a "cascade" with each reactant in the sequence leading to a large increase in the amount of the next reactant. At least 13 different plasma coagulation factors have been recognized. A deficiency of a single factor can delay or prevent the clotting process. Why has such a complex clotting mechanism evolved? Probably it is necessary to allow for a number of initial clotting responses to a variety of internal or external hemorrhage stimuli. At the same time, it cannot be fooled by ambiguous stimuli into forming dangerous intravascular clots when no injury has occurred.

Several kinds of clotting abnormalities in humans are known. One of these, hemophilia, is a condition characterized by the failure of the blood to clot, so that even insignificant wounds can cause continuous severe bleeding. It is caused by a rare mutation (about one in 10,000) on the X sex chromosome, resulting in an inherited lack of one of the platelet factors in males and in homozygous females. Called the "disease of kings," it once ran through several interrelated royal families of Europe, apparently having originated from a mutation in one of Queen Victoria's parents.

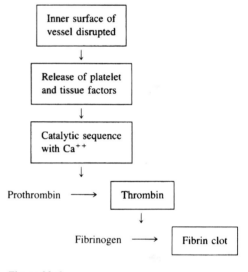

Figure 29-6

Stages in the formation of fibrin.

DEFENSE MECHANISMS OF THE BODY
Phagocytosis

Most animals have one or more mechanisms to protect themselves against the invasion of a foreign body or infectious agent. These may be coincidental attributes of certain structures (for example, a tough skin or high stomach acidity), or they may be characteristics evolved as adaptations for defense. To initiate defenses, an animal's cells must "know" when a substance is foreign; they must recognize "nonself." **Phagocytosis,** a feeding mechanism in protozoa, is a nonself recognition strategy in almost all metazoa. A cell that has this ability is a **phagocyte.** In phagocytosis the invading particle is engulfed within an invagination of the phagocyte's cell membrane (Figure 29-7). The invagination becomes pinched off, and the particle is thereby enclosed in an intracellular vacuole. Lysosomes empty digestive enzymes into the vacuole to destroy the particle. In metazoan invertebrates the cells performing this function are known as **amebocytes** (sometimes another name, depending on the group of animals). If the particle is too large for phagocytosis, the amebocytes may gather around it and wall it off. In humans

Hemophilia is one of the best-known cases of sex-linked inheritance in humans. Actually two different loci on the X chromosome are involved. Classical hemophilia (hemophilia A) accounts for about 80% of persons with the condition, and the remainder have Christmas disease (hemophilia B). The allele at each locus results in the deficiency of a different platelet factor.

A

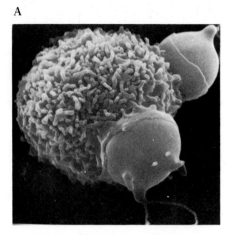

B

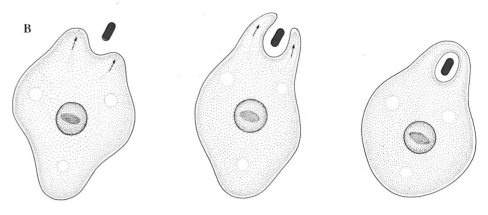

Figure 29-7

Phagocytosis. **A,** Scanning electron micrograph of macrophage ingesting two worn-out erythrocytes. Thin cytoplasmic extensions of the macrophage are surrounding the erythrocyte surface. **B,** Diagrammatic representation.
Photograph by J.P. Revel, California Institute of Technology.

there are **fixed** and **mobile** phagocytes. The fixed phagocytes taken together form the **reticuloendothelial system** (RE system) and are found in the liver, spleen, lymph nodes, and other tissues. As the blood circulates through the RE system, it filters out and destroys particles and spent red blood cells. The mobile phagocytes circulate in the blood and include the granulocytic leukocytes, especially neutrophils, and the monocytes, which become phagocytes (Figure 29-4). When monocytes move into the tissue from the blood, they differentiate into active phagocytes called **macrophages.**

___ Acquired Immune Response in Vertebrates

Vertebrates have a specialized system of nonself recognition that results in increased resistance to a *specific* foreign substance or invader after repeated exposures. A substance that stimulates an immune response is called an **antigen** (Gr. *anti*, against, + *genos*, birth). Antigens may be any of a variety of substances that have a molecular weight greater than 3000, most commonly proteins, and are usually (but not always) foreign to the host. There are two arms of the immune

Figure 29-8

Stimulation of a humoral immune response by an antigen. *1,* Macrophage consumes antigen, partially digests it, and displays portions on its surface, along with class II MHC protein. *2,* T helper cell recognizes antigen and class II protein on macrophage and is activated. *3,* T helper then activates B cell, which carries same antigen and class II protein on its surface. *4,* Activated B cell multiplies, finally producing many plasma cells which secrete antibody. *5,* Some of B cell progeny become memory cells. *6,* Antibody produced by plasma cells binds to antigen and stimulates macrophages to consume antigen (opsonization).

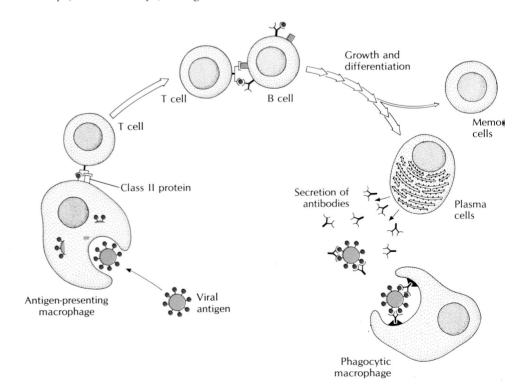

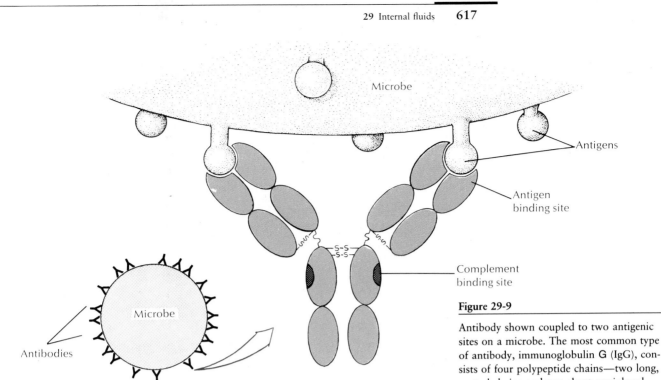

Figure 29-9

Antibody shown coupled to two antigenic sites on a microbe. The most common type of antibody, immunoglobulin **G** (IgG), consists of four polypeptide chains—two long, central chains and two short, peripheral chains—arranged in a **Y** configuration. The two long chains are held together by disulfide bonds (—**S**—**S**—). Each antibody couples through two receptors, which are antigen-binding sites having geometrical and chemical specificity for the antigens.

response, known as **humoral** (Figure 29-8) and **cellular.** Humoral immunity is based on **antibodies** (Figure 29-9), which are dissolved in and circulate in the blood, whereas cellular immunity is associated with cell surfaces. Although the two types interact, humoral immunity seems to be more important in a variety of bacterial infections, whereas the cellular response is of particular importance in tissue rejection reactions and a variety of viral, fungal, and parasitic infections.

Basis of self and nonself recognition

Nonself recognition is very specific; if tissue from one individual is transplanted into another individual in the same species, the graft will grow for a time and then die as immunity against it arises. Without immunosuppression, tissue grafts can only grow successfully if they are between identical twins or between individuals of highly inbred strains of animals. It has been found in recent years that the molecular basis for this nonself recognition depends on certain proteins imbedded in the cell surface. These proteins are coded by certain genes, now known as the **major histocompatibility complex** (MHC), because they were discovered in tissue graft experiments. The MHC proteins are among the most variable known, and unrelated individuals almost always have different genes. There are two types of MHC proteins: class I and class II. Class I proteins are found on the surface of virtually all cells, whereas class II MHC proteins are found only on certain cells participating in the immune response, such as certain lymphocytes and macrophages.

In 1980 the Nobel Prize was awarded to Baruj Benacerraf, Jean Dausset, and George Snell for their many years of research on the identification and action of the histocompatibility antigens.

Antibodies

Antibodies are proteins called **immunoglobulins.** The basic antibody molecule consists of four polypeptide strands: two identical light chains and two identical heavy chains, held together in a **Y**-shape by disulfide bonds and hydrogen bonds (Figure 29-9). The amino acid sequence toward the ends of the **Y** varies in both the

A major problem of immunology is understanding how the mammalian genome could contain the information needed to produce at least a million different antibodies. The recently discovered answer seems to be that antibody genes occur in pieces, rather than as continuous stretches of DNA, and that the antigen-recognizing sites (variable regions) of the heavy and light chains of the antibody molecules are pieced together from information supplied by separate DNA sequences, which can be shuffled to vastly increase the diversity of the gene products. The immense repertoire of antibodies is achieved in part by complex gene rearrangements (the exact mechanism of DNA switching is not yet clear) and in part by frequent somatic mutations that produce additional variation in protein structure of the variable regions of the heavy and light antibody chains.

Many aspects of immunology have been greatly assisted by the recent discovery of a method for producing stable clones of cells that will produce only one kind of antibody. Such monoclonal antibodies will bind to only *one kind* of antigenic determinant (most proteins bear many different antigenic determinants and thus stimulate the body to produce complex mixtures of antibodies). Monoclonal antibodies are made by fusing normal antibody-producing plasma cells with a continuously growing plasma cell line, producing a hybrid of the normal cell with one that can divide indefinitely in culture. This is called a hybridoma. Clones are grown from the hybrids, becoming "factories" that produce almost unlimited quantities of one specific antibody. Although clinical trials with monoclonal antibodies in cancer detection and therapy are only beginning, the hybridoma techniques discovered in 1975 have already become one of the most important research tools for the immunologist.

heavy and light chains, according to the specific antibody molecule (the **variable region**), and this determines which antigen the antibody can bind with. Each of the ends of the Y forms a cleft that acts as the antigen-binding site (Figure 29-9), and the specificity of the molecule depends on the shape of the cleft and the properties of the chemical groups that line its walls. The rest of the antibody is known as the **constant region,** but it also varies to some extent. The constant region determines the role of the antibody in the immune response but not the antigen it recognizes, for example, whether the antibody is secreted or held on a cell surface.

Generation of a humoral response

The immune response is due primarily to the type of leukocytes called **lymphocytes.** There are two broad categories: **T lymphocytes (T cells)** and **B lymphocytes (B cells).** B cells have antibody molecules on their surface and give rise to cells that actively secrete antibodies into the blood. T cells have surface receptors that bind antigens, but the receptors are somewhat different in structure from antibodies. There are a vast array of B cells, each bearing on its surface molecules of antibody that will bind with one particular antigen, even though that antigen may have never been present in the body previously. There is probably an equal number of different T cells with receptors for specific antigens.

When an antigen is introduced into the body, it binds to a specific antibody on the surface of the appropriate B cell, but this is usually insufficient to activate the B cell to multiply. Some of the antigen is taken up by macrophages that partially digest it and then incorporate portions of the antigen into their own cell surface (Figure 29-8). The macrophages also secrete a substance known as **interleukin-1.** Fragments of the antigen on the surfaces of the macrophages are recognized by a subset of T cells called **T helper cells** (T_H), in conjunction with the class II MHC protein on the macrophage surface. Both the class II protein and the antigen must be present; neither is effective alone. The T_H cells then activate the B cell that has the same processed antigen and a class II MHC protein on its surface; interleukin-1 is also necessary for this activation. The B cell multiplies rapidly and produces many **plasma cells,** which secrete large quantities of antibody for a period of time, then die. Thus if the amount of the antibody (**titer**) is measured soon after the antigen is injected, little or none can be detected. The titer rises rapidly as the plasma cells secrete antibody, then it may decrease somewhat as they die and the antibody is degraded (Figure 29-10). However, if another dose of antigen (the **challenge**) is given, there is no lag, and the antibody titer rises quickly to a higher level than after the first dose. This is the **secondary response,** and it occurs because some of the activated B cells gave rise to long-lived **memory cells.** There are many more memory cells present in the body than the original B cell with the appropriate antibody on its surface, and they rapidly multiply to produce additional plasma cells.

Actions of antibodies

Antibodies can destroy an invader (antigen) in a number of ways. A foreign particle, for example, becomes coated with antibody molecules as their variable regions become bound to it. Phagocytic macrophages recognize the projecting constant regions and are stimulated to engulf the particle. This process is called **opsonization.** Another important process, particularly in the destruction of bacterial cells, is the interaction with **complement.** Complement is a series of 12 enzymes activated by bound antibodies, which actually punch holes in the bacterial cell surface.

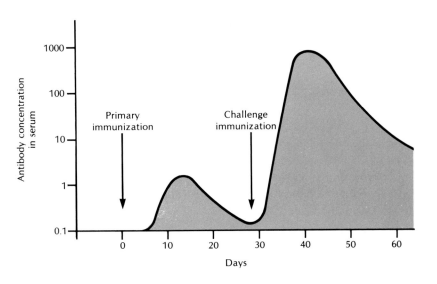

Figure 29-10

Typical immunoglobulin response after primary and challenge immunizations. The secondary response is a result of the large numbers of memory cells produced after the primary B-cell activation.

AIDS (acquired immune deficiency syndrome) is an extremely serious disease in which the ability to mount an immune response is crippled. The first case was recognized in 1981, and by the end of 1986 over 28,000 cases had been reported in the United States alone, of whom almost 25,000 had died. AIDS patients are continuously plagued with infections by agents that cause insignificant problems in persons with normal immune responses. Although a variety of cell types are infected, the disease virus preferentially invades and destroys T4 lymphocytes. Normally, T4 cells make up 60% to 80% of the T-cell population; in AIDS they can become too rare to be detected.

Types of T cells

In addition to the T helper cells, three other T cell types are known. **T inducer cells** trigger the maturation of T lymphocytes from precursors into functionally distinct cells. **Cytotoxic T cells** help defend against viruses by killing cells that have been infected with a virus. Cytotoxic T cells recognize class I MHC protein on the cell surface of infected cells in conjunction with viral protein that was left on the surface when the virus invaded the cell. The action of cytotoxic T cells is mediated by T_H cells. **T suppressor cells,** interact with T_H cells in complex ways, but in general their effect on cytotoxic cells is the opposite of T_H cells. They turn off the T_H cells and dampen the immune response. T suppressor cells are very important in the regulation of the immune response.

Certain biochemical markers on the surface of the T cells allow classification of the four types into two: T inducer and T helper cells are known as T4, and T suppressor and cytotoxic T cells are known as T8. T4 cells recognize class II MHC proteins in association with antigenic proteins on other cell surfaces, and T8 cells recognize class I MHC proteins along with foreign proteins.

Only a few years ago transplantation of organs from one person to another seemed impossible. Then physicians began to transplant kidneys and depress the immune response to the recipient. It was very difficult to immunosuppress the recipient enough that the new organ would not be rejected and at the same time not leave the patient defenseless against infection. Cyclosporine, a new drug derived from a fungus, was discovered and now not only kidneys, but hearts, lungs, and livers can be transplanted. Cyclosporine inhibits the T cell growth factor and affects cytotoxic T cells more than T helpers. It has no effect on other white cells or on healing mechanisms, so that a patient can still mount an immune response but not reject the transplant. However, the patient must continue to take cyclosporine, because if the drug is stopped, the body will recognize the transplanted organ as foreign and reject it.

The cell-mediated response

Some immune responses involve little (if any) antibody, but depend on the action of cells only. In cell-mediated immunity the antigen is also processed by macrophages, but only T cells respond. The T cells with the specific receptors for the particular antigen are activated and proliferate. They produce soluble effector molecules called **lymphokines.** Examples of lymphokines are the **interferons** and **interleukin-2.** Interferons are a family of 20 to 25 proteins that act on other cells to increase their resistance to viral infection. Interleukin-2 causes proliferation of enlarged populations of mature cytotoxic, suppressor, and helper T cells. It also bolsters the **natural killer cells,** which are lymphocyte-like cells that can kill virus-infected and tumor cells in the absence of antibody. They do not directly interact with lymphocytes or recognize antibody.

Like humoral immunity, cell-mediated immunity shows a secondary response as a result of large numbers of memory T cells that were produced from the original activation. For example, a second tissue graft (challenge) between the same donor and host will be rejected much more quickly than the first.

Mast cells are found in the dermis and other tissues. They have one of the classes of antibody bound to their surface by the constant region (the variable region of the antibody is exposed). Binding of antigen to the antibody in their cell membrane causes them to secrete the histamine and other substances quickly. Some people become sensitized to certain antigens, and when they are exposed to these antigens, excessive amounts of histamine are secreted, producing the symptoms of *allergy*. In some instances, the conditions may be so severe that life-threatening consequences result. People with allergies take medicine to counter the effects of excessive histamine (antihistamines).

Inflammation

Inflammation is a vital process in the mobilization of the body's defense against an invading organism and in the repair of the damage thereafter. Although inflammation is a nonspecific process, it is greatly influenced by the body's prior immunizing experience with the invader. Also, a more noxious foreign substance will produce more intense inflammation. Tissue damage initiates release of pharmacologically active substances (such as histamine) from **mast cells** in the area. These substances increase the diameter of nearby small blood vessels and also increase their permeability. Thus redness and warmth result from the increased amount of blood in the area (hence "inflammation"), and more proteins and fluid escape into the tissue, causing swelling. The first phagocytic line of defense is the neutrophils, which may last only a few days, then macrophages (either fixed or differentiated from monocytes) become predominant. "Pus" of an infection is formed principally from exuded tissue fluid and spent phagocytes. Activation of complement by fixed antibody stimulates opsonization by macrophages and increased release of substances from mast cells. As appropriate T cells arrive at the site, binding of antigen to their surface stimulates them to release lymphokines, one of which attracts macrophages and inhibits their migration away from the area of inflammation.

We are unaware of the majority of minute invasions that are always occurring. Most are efficiently disposed of and heal leaving little or no trace. However, if tissue damage has been more severe, fibrous connective tissue (scar) will be deposited in the area.

___ Blood Group Antigens

ABO blood types

Blood differs chemically from person to person, and when two different (incompatible) blood types are mixed, **agglutination** (clumping together) of erythrocytes results. Naturally occurring antigens on the membranes of red blood cells are the basis of these chemical differences. The best known of these inherited immune systems is the ABO blood group. The antigens A and B are inherited as dominant genes. Thus, as shown in Table 29-1, an individual with, for example, genes *A/A* or *A/O* develops A antigen (blood type A). The presence of a B gene produces B antigens (blood type B), and for the genotype *A/B* both A and B antigens develop on the erythrocytes (blood type AB).

There is an odd feature about the ABO system. Normally we would expect that a type A individual would develop antibodies against type B blood only if B cells were introduced into the body. In fact, type A persons always have anti-B antibodies in their blood, even without the prior exposure to type B blood. Similarly, type B individuals carry anti-A antibodies. Type AB blood has neither anti-A nor anti-B antibodies (since if it did, it would destroy its own blood cells), and type O blood has both anti-A and anti-B antibodies.

We see then that the blood group names identify their *antigen* content. Persons with type O blood are called universal donors because, lacking antigens, their blood can be infused into a person with any blood type. Even though it contains anti-A and anti-B antibodies, these are so diluted during transfusion that they do not react with A or B antigens in a recipient's blood. In practice, however, clinicians insist on matching blood types to prevent any possibility of incompatibility.

Table 29-1 Major Blood Groups

Blood type	Genotype	Antigens on red blood cells	Antibodies in serum	Can give blood to	Can receive blood from	Frequency in United States (%)		
						Whites	Blacks	Asians
O	O/O	None	Anti-A and anti-B	All	O	45	48	31
A	A/A, A/O	A	Anti-B	A, AB	O, A	41	27	25
B	B/B, B/O	B	Anti-A	B, AB	O, B	10	21	34
AB	A/B	AB	None	AB	All	4	4	10

Rh factor

Karl Landsteiner, an Austrian—later American—physician, discovered the ABO blood groups in 1900. In 1940, 10 years after receiving the Nobel Prize, he made still another famous discovery. This was a blood group called the Rh factor, named after the Rhesus monkey in which it was first found. Approximately 85% of white individuals in the United States have the factor (positive) and the other 15% do not (negative). Rh-positive and Rh-negative bloods are incompatible; shock and even death may follow their mixing when Rh-positive blood is introduced into an Rh-negative person who has been sensitized by an earlier transfusion of Rh-positive blood. Rh incompatibility accounts for a peculiar and often fatal form of anemia of newborn infants called **erythroblastosis fetalis.** If an Rh-negative mother has an Rh-positive baby (father is Rh-positive), she can become immunized by the fetal blood during the birth process. Anti-Rh antibodies can cross the placenta during a subsequent pregnancy and agglutinate the fetal blood.

Erythroblastosis fetalis can now be prevented by giving an Rh-negative mother anti-Rh antibodies just after the birth of each Rh-positive child. These antibodies remain long enough to neutralize any Rh-positive fetal blood cells that may have entered her circulation, thus preventing her own antibody machinery from being stimulated to produce the Rh-positive antibodies. Active, permanent immunity is blocked. If the mother has already developed an immunity, however, the baby may be saved by an immediate, massive transfusion of blood free of antibodies.

CIRCULATION

The circulatory system of vertebrates is made up of a system of tubes, the **blood vessels,** and a propulsive organ, the **heart.** This is a **closed circulation** because the circulating medium, the **blood,** is confined to vessels throughout its journey from the heart to the tissues and back again. Many invertebrates have an **open circulation;** the blood is pumped from the heart into blood vessels that open into tissue spaces. The blood circulates in direct contact with the tissues and then reenters open blood vessels to be propelled forward again. In invertebrates having open circulatory systems, there is no clear separation of the extracellular fluid into plasma and interstitial fluids, as there must be in closed systems. Closed systems are more suitable for large and active animals because the blood can be moved rapidly to the tissues needing it. In addition, flow to various organs can be readjusted to meet changing needs by varying the diameters of the blood vessels.

The closed circulatory system of vertebrates works with the **lymphatic system.** This is a fluid "pickup" system. It re-collects tissue fluid (lymph) that has been squeezed out through the walls of the capillaries and returns it to the blood circulation. In a sense "closed" circulatory systems are not absolutely closed because fluid is constantly leaking out into the tissue spaces. However, this leakage is but a small fraction of the total blood flow.

Although it seems obvious to us today that blood flows in a circuit, the first correct description of blood flow by the English physician William Harvey initially received vigorous opposition when published in 1628. Centuries before, Galen had taught that air enters the heart from the windpipe and that blood is able to pass from one ventricle to the other through "pores" in the interventricular septum. He also believed that blood first flowed out of the heart into all vessels, arteries, and veins alike and then returned to the heart by these same vessels—an idea of ebb and flow of the blood.

Even though almost nothing about this concept was right, it was still doggedly trusted at the time of Harvey's publication. Harvey's conclusions were based on sound experimental evidence. He used a variety of animals for his experiments, including a little snake found in English meadows. By tying ligatures on arteries, he noticed that the region between the heart and ligature swelled up. When veins were tied off, the swelling occurred beyond the ligature. When blood vessels were severed, blood flowed in arteries from the cut end nearest the heart; the reverse happened in veins. By means of such experiments, Harvey worked out a correct scheme of blood circulation, even though he could not see the capillaries that connected the arterial and venous flows.

___ Plan of the Circulatory System

All vertebrate vascular systems have certain features in common. A **heart** pumps the blood into **arteries** that branch and narrow into **arterioles** and then into a vast system of **capillaries.** Blood leaving the capillaries enters **venules** and then **veins** that return the blood to the heart. Figure 29-11 compares the circulatory systems of fish and mammals. The principal differences in circulation involve the role of heart in the transformation from gill to lung breathing.

The fish heart contains two main chambers in series: the **atrium** (or **auricle**) and the **ventricle.** There are also two subsidiary chambers (not shown in Figure 29-11): the **sinus venosus,** which precedes the atrium, and the **conus arteriosus,**

Figure 29-11

Plan of circulatory system of fish *(left)* and mammal *(right)*. *Red,* Oxygenated blood. *Dark red,* Deoxygenated blood.

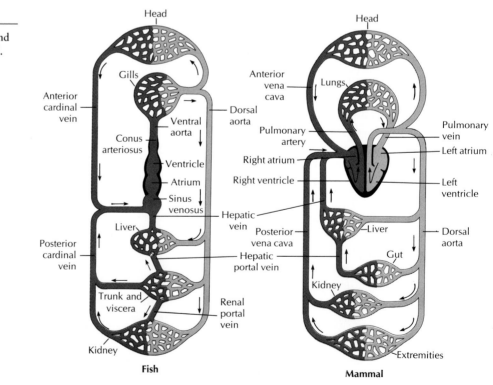

which follows the heart and contains valves that prevent backward flow of the blood when the heart is relaxed. Blood makes a single circuit through the fish's vascular system; it is pumped from the heart to the gills where it is oxygenated, and then flows into the dorsal aorta to be distributed to the body organs. After passing through the capillaries of the body organs and musculature, it returns by veins to the heart. In this circuit the heart must provide sufficient pressure to push the blood through two sequential capillary systems, one in the gills and the other in the organ tissues. The principal disadvantage of the single-circuit system is that the gill capillaries offer so much resistance to blood flow that the pressure drops considerably before entering the dorsal aorta. This system can never provide high and continuous blood pressure to the body organs.

The evolution of land forms with lungs and their need for highly efficient blood delivery resulted in the introduction of a **double** circulation. One **systemic** circuit with its own pump provides oxygenated blood to the capillary beds of the body organs; another **pulmonary** circuit with its own pump sends deoxygenated blood to the lungs. Rather than actually developing two separate hearts, the existing two-chambered heart was divided down the center into four chambers—really two two-chambered hearts lying side-by-side.

Such a great change in the vertebrate circulatory plan, involving not only the heart but the attendant plumbing as well, took many millions of years to evolve. The partial division of the atrium and ventricle began with the ancestors of present-day lungfishes. The course of the blood through this double circuit is shown in the right diagram of Figure 29-11.

Heart

The mammalian heart (Figure 29-12) is a muscular organ located in the thorax and covered by a tough, fibrous sac, the **pericardium.** As we have seen, the higher vertebrates have a four-chambered heart. Each half consists of a thinner-walled atrium and a thicker-walled ventricle. Heart (cardiac) muscle is a unique type of muscle found nowhere else in the body. It resembles striated muscle, but the cells are branched, and the dense end-to-end attachments between the cells are called intercalated discs (Figure 28-8, p. 603).

There are two sets of valves. **Atrioventricular (A-V) valves** separate the cavities of the atrium and ventricle in each half of the heart. These permit blood to flow from atrium to ventricle while preventing backflow. Where the great arteries, the **pulmonary** from the right ventricle and the **aorta** from the left ventricle, leave the heart, **semilunar valves** prevent backflow.

The contraction of the heart is called **systole** (sis´to-lee), and the relaxation, **diastole** (dy-as´to-lee) (Figure 29-13). The rate of the heartbeat depends on age, sex, and especially exercise. Exercise may increase the **cardiac output** (volume of blood forced from either ventricle each minute) more than fivefold. Both the heart **rate** and the **stroke volume** increase. Heart rates among vertebrates vary with the general level of metabolism and the body size. The cold-blooded codfish has a heart rate of approximately 30 beats per minute; a warm-blooded rabbit of about the same weight has a rate of 200 beats per minute. Small animals have higher heart rates than do large animals. The heart rate in an elephant is 25 beats per minute, in a human 70 per minute, in a cat 125 per minute, and in a mouse 400 per minute. In the tiny 4 g shrew, the smallest mammal, the heart rate approaches a prodigious 800 beats per minute. We must marvel that the shrew's heart can sustain such a frantic pace throughout this animal's life, brief as it is.

The heart rests only during the short interval between contractions. The

Figure 29-12

Human heart. Deoxygenated blood *(black arrows)* enters right side of heart and is pumped to the lungs. Oxygenated blood *(red arrows)* returning from the lungs enters left side of the heart and is pumped to the body.

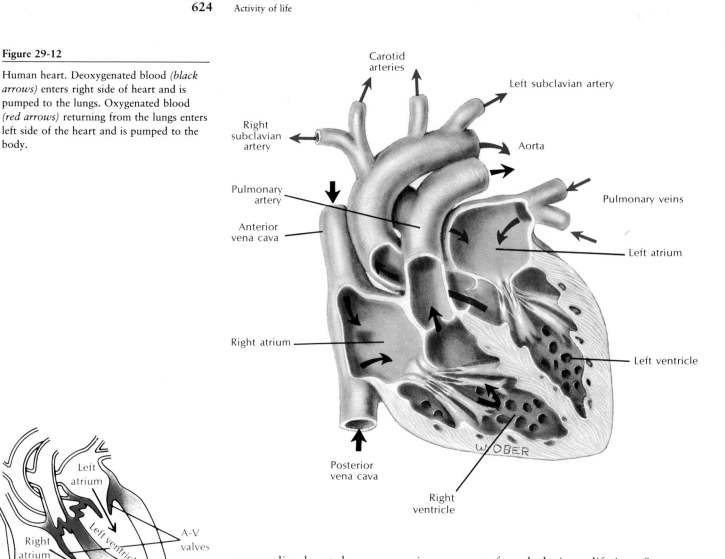

**Heart at rest
(diastole)**

**Heart during contraction
(systole)**

Figure 29-13

Human heart in systole and diastole.

mammalian heart does an amazing amount of work during a lifetime. Someone has calculated that the heart of a human approaching the end of a normal lifetime has beat some 2.5 billion times and pumped 300,000 tons of blood!

Excitation of the heart

The heartbeat originates in a specialized muscle tissue, the **sinus node**, located in the right atrium near the entrance of the caval veins (Figure 29-14). This tissue serves as the **pacemaker** of the heart. The contraction spreads across the two atria to the **atrioventricular (A-V) node.** At this point the electrical activity is conducted very rapidly to the apex of the ventricle through specialized fibers (atrioventricular bundle) and then spreads more slowly up the walls of the ventricles. This arrangement allows the contraction to begin at the apex or "tip" of the ventricles and spread upward to squeeze out the blood in the most efficient way; it also ensures that both ventricles contract simultaneously.

Although the vertebrate heart can beat spontaneously—and the excised fish or amphibian heart does beat for hours in a balanced salt solution—the heart rate is normally under nervous control. The control (cardiac) center is located in the medulla and sends out two sets of motor nerves. Impulses sent along one set, the **vagus** (parasympathetic) nerves, apply a brake action to the heart rate, and impulses sent along the other set, the accelerator (sympathetic) nerves, speed it up. Both sets of nerves terminate in the sinus node, thus guiding the activity of the pacemaker.

The cardiac center in turn receives sensory information about a variety of stimuli. Pressure receptors (sensitive to blood pressure) and chemical receptors (sensitive to carbon dioxide and pH) are located at strategic points in the vascular system. This information is used by the cardiac center to increase or reduce the heart rate and cardiac output in response to activity or changes in body position. The heart is thus controlled by a series of feedback mechanisms that keep its activity constantly attuned to body needs.

Coronary circulation

It is no surprise that an organ as active as the heart needs a very good blood supply of its own. The heart muscle of the frog and other amphibians is so thoroughly channeled with spaces between the muscle fibers that sufficient oxygenated blood is squeezed through by the heart's own pumping action. In birds and mammals, however, the heart muscle is very thick and has such a high rate of metabolism that it must have its own vascular (**coronary**) circulation. The coronary arteries break up into an extensive capillary network surrounding the muscle fibers and provide them with oxygen and nutrients. Heart muscle has an extremely high oxygen demand, removing 80% of the oxygen from the blood, in contrast to most other body tissues, which remove only approximately 30%.

Arteries

All vessels leaving the heart are called arteries whether they carry oxygenated blood (aorta) or deoxygenated blood (pulmonary artery). To withstand high, pounding pressures, arteries are invested with layers of both elastic and tough inelastic connective tissue fibers (Figure 29-15). The elasticity of the arteries allows them to yield to the surge of blood leaving the heart during systole and then to squeeze down on the fluid column during diastole. This smooths out the blood pressure. Thus the normal arterial pressure in humans varies only between a high of 120 mm Hg (systole) and a low of 80 mm Hg (diastole) (usually expressed as 120/80 or 120 over 80), rather than dropping to zero during diastole as we might expect in a fluid system with an intermittent pump.

As the arteries branch and narrow into **arterioles,** the walls become mostly smooth muscle. Contraction of this muscle narrows the arterioles and reduces the flow of blood. The arterioles thus control the blood flow to body organs, diverting it to where it is needed most. The blood must be given a hydrostatic pressure sufficient to overcome the resistance of the narrow passages through which the blood must flow. Consequently, large animals tend to have higher blood pressure than do small animals.

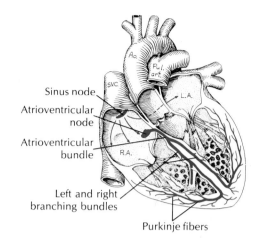

Figure 29-14

Neuromuscular mechanisms controlling heartbeat. Arrows indicate spread of excitation from the sinus node, across the atria, to the atrioventricular (A-V) node. Wave of excitation is then conducted very rapidly to ventricular muscle over the specialized atrioventricular bundle and Purkinje fibers.

If one of the coronary arteries becomes blocked by a small blood clot, the person has a heart attack (a "coronary"). The blockage is usually the final event after a progressive narrowing of the diameter of the artery by fatty depositions of cholesterol in its walls. The portion of the heart muscle served by the branch of the coronary artery that is blocked is starved for oxygen. It may be replaced by scar tissue if the person survives.

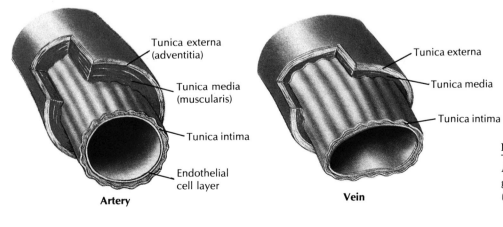

Figure 29-15

Artery and vein, showing layers. Note greater thickness of the muscularis layer (tunica media) in the artery.

Blood pressure was first measured in 1733 by Stephen Hales, an English clergyman with unusual inventiveness and curiosity. He tied his mare, which was "to have been killed as unfit for service," on her back and exposed the femoral artery. This he cannulated with a brass tube, connecting it to a tall glass tube with the windpipe of a goose. The use of the windpipe was both imaginative and practical; it gave the apparatus flexibility "to avoid inconveniences that might arise if the mare struggled." The blood rose 8 feet in the glass tube and bobbed up and down with the systolic and diastolic beats of the heart. The weight of the 8-foot column of blood was equal to the blood pressure. We now express this as the height of a column of mercury, which is 13.6 times heavier than water. Hales' figures, expressed in millimeters of mercury, indicate that he measured a blood pressure of 180 to 200 mm Hg, about normal for a horse.

Today, blood pressure in humans is most commonly and easily measured with an instrument called a **sphygmomanometer.** Air is used to inflate a cuff on the upper arm to a pressure sufficient to close the arteries in the arm. As air is slowly released from the cuff, a person with a stethoscope held over the brachial artery (in the crook of the elbow) can hear the first spurts of blood through the artery when the pressure in the cuff allows the artery to open slightly. This is equivalent to the systolic pressure. As the pressure in the cuff is decreased, the sound heard with the stethoscope finally disappears when the blood is running smoothly through the artery. The pressure at which the sound disappears is the diastolic pressure.

Capillaries

The Italian Marcello Malpighi was the first to describe the capillaries in 1661, thus confirming the existence of the minute links between the arterial and venous systems that Harvey knew must be there but could not see. Malpighi studied the capillaries of the living frog's lung, which incidentally is still one of the simplest and most vivid preparations for demonstrating capillary blood flow.

The capillaries are present in enormous numbers, forming extensive networks in nearly all tissues (Figure 29-16). In muscle there are more than 2000 per square millimeter (1,250,000 per square inch), but not all are open at once. Indeed, perhaps less than 1% are open in resting skeletal muscle. But when the muscle is active, all the capillaries may open to bring oxygen and nutrients to the working muscle fibers and to carry away metabolic wastes.

Capillaries are extremely narrow, averaging less than 10 μm in diameter in mammals, which is hardly any wider than the red blood cells that must pass through them. Their walls are formed in a single layer of thin **endothelial** cells, held together by a delicate basement membrane and connective tissue fibers. Capillaries have a built-in leakiness that allows water and most dissolved substances except proteins to filter through into the interstitial space.

Because of the low permeability of the capillaries to protein, the solute concentration in the blood is higher than in the interstitial fluid; therefore, there is a net osmotic pressure difference, due to the protein, of about 25 mm Hg in mammals. The hydrostatic pressure of the blood at the arteriole end of the capillaries is about 40 mm Hg in humans. Thus there is a **net filtration pressure** (the hydrostatic pressure, which tends to force fluid out, less the osmotic pressure, which tends to draw water back in) of about 15 mm Hg at the arteriole end of the capillaries (Figure 29-17). Water and dissolved materials are forced out of the capillaries and circulate through the tissue space. As the blood proceeds through the narrow capillary, the blood pressure decreases steadily to perhaps 15 mm Hg. At this point the hydrostatic pressure is less than the osmotic pressure because of the

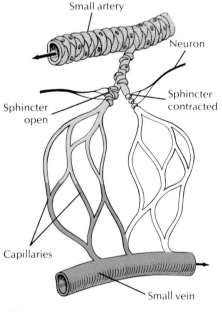

Figure 29-16

Capillary bed. Precapillary sphincters control blood flow through the capillaries.

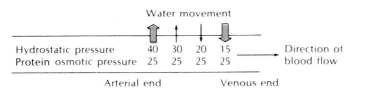

Figure 29-17

Fluid movement across the wall of a capillary. At the arterial end of the capillary, hydrostatic (blood) pressure exceeds protein osmotic pressure contributed by plasma proteins, and a plasma filtrate (shown as "water movement") is forced out. At the venous end, protein osmotic pressure exceeds the hydrostatic pressure, and fluid is drawn back in. In this way plasma nutrients are carried out into the interstitial space where they can enter cells, and metabolic end products from the cell are drawn back into the plasma and carried away.

plasma proteins, still approximately 25 mm Hg, and water is drawn back into the capillaries.

Thus it is the balance between the hydrostatic pressure and protein osmotic pressure that determines the direction of capillary fluid shift. Normally, water is forced out of the capillary at the arteriole end, where hydrostatic pressure exceeds osmotic pressure, and drawn back into the capillary at the venule end, where osmotic pressure exceeds hydrostatic pressure. Any fluid left behind is picked up and removed by the **lymph capillaries.**

Veins

The venules and veins into which the capillary blood drains for its return journey to the heart are thinner walled, less elastic, and of considerably larger diameter than their corresponding arteries and arterioles (Figure 29-15). Blood pressure in the venous system is low, from approximately 10 mm Hg, where capillaries drain into venules, to approximately zero in the right atrium. Because pressure is so low, the venous return gets assists from valves in the veins, muscles surrounding the veins, and the rhythmical pumping action of the lungs. If it were not for these mechanisms, the blood might pool in the lower extremities of a standing animal— a very real problem for people who must stand for long periods. Veins that lift blood from the extremities to the heart contain valves that divide the long column of blood into segments. When the muscles around the veins contract, as in even slight activity, the blood column is squeezed upward and cannot slip back because of the valves. The well-known risk of fainting while standing at stiff attention in hot weather can usually be prevented by deliberately pumping the leg muscles. The negative pressure created in the thorax by the inspiratory movement of the lungs also speeds the venous return by sucking the blood up the large vena cava into the heart.

The actual situation is a bit more complicated because there is a small hydrostatic pressure in the interstitial fluid, and a small amount of protein does leak through the capillary wall. The protein tends to accumulate at the venule end of the capillary, building up a small osmotic pressure there. Although actual calculation of the pressure differences must take into account the interstitial fluid hydrostatic and osmotic pressures, the principle of the capillary fluid shift is as we have presented it.

Lymphatic system

The lymphatic system is an extensive network of thin-walled vessels that is separate from the circulatory system. The system arises as blind-ended lymph capillaries in most tissues of the body. These unite to form larger and larger lymph vessels, which finally drain into veins in the lower neck (Figure 29-18).

The lymphatic system is an accessory drainage system for the body. As we have seen, the blood pressure in the arteriole end of the capillaries forces a plasma filtrate through the capillary walls and into the interstitial space. This interstitial (tissue) fluid bathing the cells is a clear, nearly colorless liquid. Interstitial fluid and plasma are nearly identical except that interstitial fluid contains very little protein, which was screened out as the plasma was squeezed through the capillary walls. Most of the interstitial fluid returns to the vascular system at the venous end of the capillaries by the capillary fluid-shift mechanism described earlier. Usually, however, outflow from the capillaries slightly exceeds backflow. This difference is gathered up and returned to the circulatory system by lymphatic vessels. Intersti-

Figure 29-18

Human lymphatic system, showing major vessels, **A,** and a detail of the blood and lymphatic capillaries, **B.**

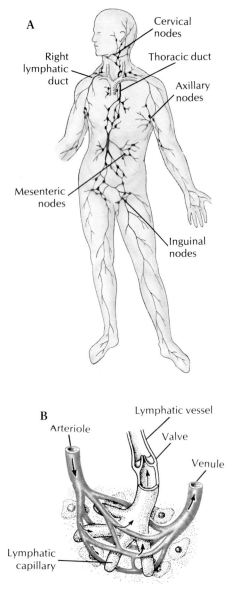

tial fluid is referred to as **lymph** as soon as it enters the lymph vessels. The rate of lymph flow is very low, a minute fraction of the blood flow.

Lymph nodes (Figure 29-18) are located at strategic intervals along the lymph vessels. Their role in the body's defense against disease is important because they contain large numbers of macrophages and both T and B lymphocytes. The lymph nodes and the spleen are major locations in which antigens are taken up by macrophages, and lymphocytes are activated and proliferate to generate the immune response.

RESPIRATION

The energy bound up in food is released by oxidative processes, usually with oxygen as the terminal electron acceptor. As oxygen is used by the body cells, carbon dioxide is produced; this process is called **respiration.** Small aquatic animals such as the one-celled protozoa obtain what oxygen they need by direct diffusion from the environment. Carbon dioxide is also lost by diffusion to the environment. Such a simple solution to the problem of gas exchange is really only possible for very small animals (less than 1 mm in diameter) that have a large surface relative to their volume or those having very low rates of metabolism.

As animals became larger and evolved a waterproof covering, specialized devices such as lungs and gills developed that greatly increased the effective surface for gas exchange. But, because gases diffuse so slowly through protoplasm, a circulatory system was necessary to distribute the gases to and from the deep tissues of the body. Even these adaptations were inadequate for advanced animals with their high rates of cellular respiration. The solubility of oxygen in the blood plasma is so low that plasma alone could not carry enough to support metabolic demands. With the evolution of special oxygen-transporting blood proteins such as hemoglobin, the oxygen-carrying capacity of the blood was greatly increased. Thus what began as a simple and easily satisfied requirement resulted in the evolution of several complex and essential respiratory and circulatory adaptations.

Problems of Aquatic and Aerial Breathing

How an animal respires is largely determined by the nature of its environment. The two great arenas of animal evolution—water and land—are vastly different in their physical characteristics. The most obvious difference is that air contains far more oxygen—at least 20 times more–than does water.

Water at 5° C fully saturated with air contains approximately 9 ml of oxygen per liter, compared with air, which contains 210 ml of oxygen per liter (21%). The solubility of oxygen in water decreases as the temperature rises. For example, water at 15° C contains approximately 7 ml of oxygen per liter, and at 35° C, only 5 ml of oxygen per liter. The relatively low concentration of oxygen dissolved in water is the greatest respiratory problem facing aquatic animals. Unfortunately, it is not the only one. Oxygen diffuses much more slowly in water than in air, and water is much denser and more viscous than air. All of this means that successful aquatic animals must have evolved very efficient ways of removing oxygen from water. Yet even the most advanced fishes with highly efficient gills and pumping mechanisms may use as much as 20% of their energy just extracting oxygen from water. By comparison, a mammal uses only 1% to 2% of its resting metabolism to breathe.

It is essential that respiratory surfaces be thin and always kept wet to allow diffusion of gases between the environment and the underlying circulation. This is hardly a problem for aquatic animals, immersed as they are in water, but it is a very real problem for air breathers. To keep the respiratory membranes moist and protected from injury, air breathers have in general developed invaginations of the body surface and then added pumping mechanisms to move air in and out. The lung is the best example of a successful solution to breathing on land. In general, **evaginations** of the body surface, such as gills, are most suitable for aquatic respiration; **invaginations**, such as lungs, are best for air breathing. We can now consider the specific kinds of respiratory organs employed by animals.

Cutaneous respiration

Protozoa, sponges, cnidarians, and many worms respire by direct diffusion of gases between the organism and the environment. We have noted that this kind of **integumentary respiration** is not adequate when the gases must diffuse further than 1 mm across living tissue. However, the body plans of many multicellular animals greatly increase the surface of the body relative to the mass (for example, the canal systems of sponges and the body shape of flatworms) thus making possible this mode of gas exchange. Integumentary respiration frequently supplements gill or lung breathing in larger animals such as amphibians and fishes. An eel can exchange 60% of its oxygen and carbon dioxide through its highly vascular skin. During their winter hibernation, frogs exchange all their respiratory gases through the skin while they remain submerged in ponds or springs.

Gills

Gills of various types are more effective respiratory devices for life in water. Gills may be simple **external** extensions of the body surface, such as the **dermal papulae** of sea stars (p. 416) or the **branchial tufts** of marine worms (p. 299) and aquatic amphibians. Most efficient are the **internal gills** of fishes (p. 484) and arthropods. Fish gills are thin filamentous structures, richly supplied with blood vessels arranged so that blood flow is opposite to the flow of water across the gills. This arrangement, called **countercurrent flow,** provides for the greatest possible extraction of oxygen from water. Water flows over the gills in a steady stream, pushed and pulled by an efficient branchial pump, and often assisted by the fish's forward movement through the water. Mollusc ctenidia also take advantage of countercurrent flow.

Lungs

Gills are unsuitable for life in air because, when removed from the buoying water medium, the gill filaments collapse and stick together; a fish out of water rapidly asphyxiates despite the abundance of oxygen around it. Consequently air-breathing vertebrates possess lungs, highly vascularized internal cavities. Lungs of a sort are found in certain invertebrates (pulmonate snails, scorpions, some spiders, some small crustaceans), but these structures cannot be very efficiently ventilated.

Lungs that can be ventilated more efficiently are characteristic of the terrestrial vertebrates. The most primitive vertebrate lungs are those of lungfishes (Dipneusti), which use them to supplement, or even replace, gill respiration during

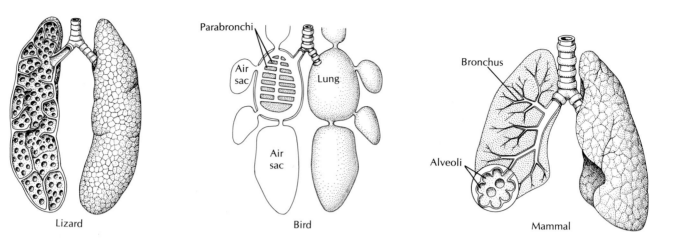

Lizard

Parabronchi

Air sac

Lung

Air sac

Bird

Bronchus

Alveoli

Mammal

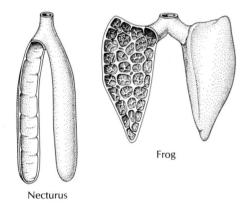

Frog

Necturus

Figure 29-19

Internal structures of lungs among vertebrate groups. Generally, the evolutionary trend has been from simple sacs with little exchange surface between blood and air spaces to complex, lobulated structures, each with complex divisions and extensive exchange surfaces.

The student will notice that the same word (trachea, pl., tracheae) is used for certain respiratory structures in vertebrates as in some arthropods. However, they are entirely different in evolutionary origin and morphology. About the only thing they have in common is that they are tubes through which air passes.

periods of drought. Although of simple construction, the lungfish lung is supplied with a capillary network in its largely unfurrowed walls, a tubelike connection to the pharynx, and a primitive ventilating system for moving air in and out of the lung.

Amphibians also have simple baglike lungs, whereas in higher forms the inner surface area is vastly increased by numerous lobulations and folds (Figure 29-19). This increase is greatest in the mammalian lung, which is complexly divided into many millions of small sacs (**alveoli**), each veiled by a rich vascular network. It has been estimated that human lungs have a total surface area of from 50 to 90 m²—50 times the area of the skin surface—and contain 1000 miles of capillaries.

Moving air into and out of lungs was an evolutionary design problem that has been, of course, solved. However, we wonder whether an imaginative biological engineer, if given the proper resources, couldn't come up with a better design. Unlike the efficient one-way flow of water across fish gills, air must enter and exit a lung through the same channel. Furthermore, a tube of some length—the bronchi, trachea, and mouth cavity—connects the lungs to the outside. This is a "dead-air space" containing a volume of air that shuttles back and forth with each breath, adding to the difficulty of properly ventilating the lungs. In fact, lung ventilation is so inefficient that in normal breathing only approximately one sixth of the air in the lungs is replenished with each inspiration. Birds, however, have vastly improved the efficiency of lung ventilation with their system of air sacs (see Figure 26-13, p. 547 and Figure 29-19).

Tracheae

Insects and certain other terrestrial arthropods (centipedes, millipedes, and some spiders) have a highly specialized type of respiratory system; in many respects it is the simplest, most direct, and most efficient respiratory system found in active animals. It consists of a system of tubes (**tracheae**) that branch repeatedly and extend to all parts of the body (p. 368). The smallest end channels (**tracheoles**), less than 1 μm in diameter, sink into the plasma membranes of the body cells. Air enters the tracheal system through valvelike openings (**spiracles**) on each side of the body, and oxygen diffuses directly to all cells of the body. Carbon dioxide diffuses out in the opposite direction. Some insects can ventilate the tracheal system with body movements; the familiar telescoping movement of the bee abdomen is an example. The tracheal system is simple because blood is not needed to transport the respiratory gases; the cells have a direct pipeline to the outside.

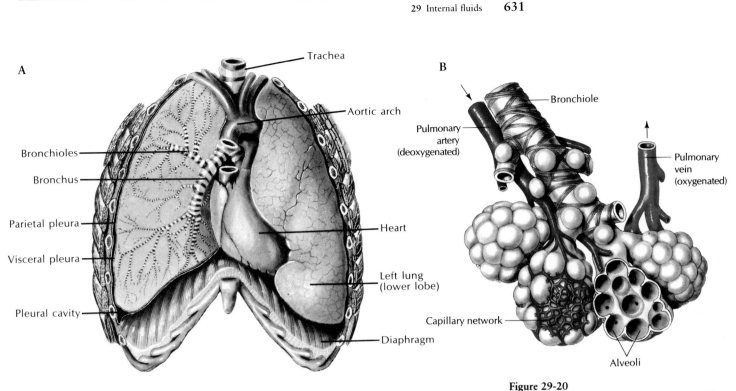

Figure 29-20

A, Lungs of human with right lung shown in section. **B,** Terminal portion of bronchiole showing air sacs with their blood supply. Arrows show direction of blood flow.

Respiration in Humans

In mammals the respiratory system is made up of the following: the nostrils (external nares); the **nasal chamber,** lined with mucus-secreting epithelium; the **posterior nares,** which connect to the **pharynx** where the pathways of digestion and respiration cross; the **epiglottis,** a flap that folds over the **glottis** (the opening to the larynx) to prevent food from going the wrong way in swallowing; the **larynx,** or voice box; the **trachea,** or windpipe; and two **bronchi,** one to each lung (Figure 29-20). Within the lungs each bronchus divides and subdivides into small tubes (**bronchioles**) that lead to the air sacs (**alveoli**) (Figure 29-21). The walls of the alveoli are thin and moist to facilitate the exchange of gases between the air sacs and the adjacent blood capillaries. Air passageways are lined with mucus-secreting ciliated epithelium, which plays an important role in conditioning the air before it reaches the alveoli. There are partial cartilage rings in the walls of the

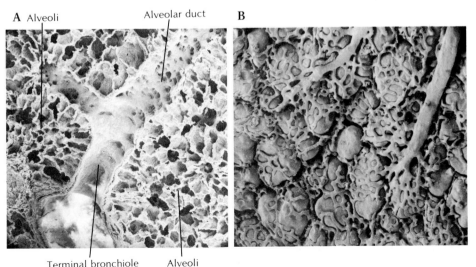

Figure 29-21

A, Scanning electron micrograph of mammalian lung. (×35.) **B,** Resin cast of lung, showing the extensive capillary network surrounding each alveolus. (×150.)

A, Courtesy E.E. Morrison; **B,** from Tissues and organs: a text-atlas of scanning electron microscopy, by Richard G. Kessel and Randy H. Kardon. W.H. Freeman and Co., Publishers. Copyright © 1979.

tracheae, bronchi, and even some of the bronchioles to prevent those structures from collapsing.

In its passage to the air sacs the air undergoes three important changes: (1) it is filtered free from most dust and other foreign substances, (2) it is warmed to body temperature, and (3) it is saturated with moisture.

The lungs consist of a great deal of elastic connective tissue and some muscle. They are covered by a thin layer of tough epithelium known as the **visceral pleura.** A similar layer, the **parietal pleura,** lines the inner surface of the walls of the chest (Figure 29-20). The two layers of the pleura are in contact and slide over one another as the lungs expand and contract. The "space" between the pleura, called the **pleural cavity,** contains a partial vacuum, which helps keep the lungs expanded to fill the pleural cavity. Therefore no real pleural space exists; the two pleura rub together, lubricated by interstitial fluid (lymph). The chest cavity is bounded by the spine, ribs, and breastbone, and floored by the **diaphragm,** a dome-shaped, muscular partition between the chest cavity and abdomen.

Mechanism of breathing

The chest cavity is an air-tight chamber. In **inspiration** the ribs are pulled upward, the diaphragm is contracted and flattened, and the chest cavity is enlarged. The resultant increase in volume of chest cavity and lungs causes the air pressure in the lungs to fall below atmospheric pressure: air rushes in through the air passageways to equalize the pressure. **Expiration** is a less active process than inspiration. When the muscles relax, the ribs and diaphragm return to their original position, and the chest cavity size decreases. The elastic lungs then deflate and force the air out.

Coordination of breathing

Respiration must adjust itself to changing requirements of the body for oxygen. Breathing is normally involuntary and automatic but can come under voluntary control. Normal, quiet breathing is regulated by neurons centered in the medulla oblongata of the brain. These neurons spontaneously produce rhythmical bursts that lead to contraction of the diaphragm and the intercostal muscles between the ribs. The rhythm and depth of breathing are precisely regulated by the amount of carbon dioxide in the blood. Exercise raises the carbon dioxide level, and the breathing rate increases. Actually, the effects of carbon dioxide are caused by the increase in blood hydrogen ion concentration (p. 635). The increase in hydrogen ions stimulates the respiratory center in the medulla oblongata, leading to increased rate and depth of respiration.

Composition of inspired, expired, and alveolar airs

The composition of expired and alveolar airs is not identical. Air in the alveoli contains less oxygen and more carbon dioxide than does the air that leaves the lungs. Inspired air has the composition of atmospheric air. Expired air is really a mixture of alveolar and inspired airs. The composition of the three kinds of air is shown in Table 29-2.

The water given off in expired air depends on the relative humidity of the external air and the activity of the person. At ordinary room temperature and with a relative humidity of approximately 50%, an individual in performing light work loses approximately 350 ml of water from the lungs each day.

It is well known that swimmers can remain submerged much longer if they vigorously hyperventilate first to blow off carbon dioxide from the lungs. This delays the overpowering urge to surface and breathe. The practice is dangerous because blood oxygen is depleted just as rapidly as without prior hyperventilation, and the swimmer may lose consciousness when the oxygen supply to the brain drops below a critical point. This practice has caused several documented drownings among swimmers attempting long underwater swimming records.

Table 29-2 Composition of Respired Air

	Inspired air (vol %)	Expired air (vol %)	Alveolar air (vol %)
Oxygen	20.96	16	14.0
Carbon dioxide	0.04	4	5.5
Nitrogen	79.00	80	80.5

Gaseous exchange in lungs

The diffusion of gases takes place in accordance with the laws of physical diffusion; that is, the gases pass from regions of higher concentration to those of lower concentration. The partial pressure of a gas refers to that pressure which the gas exerts in a mixture of gases, in other words, the part of the total pressure because of that component. If the atmospheric pressure at sea level is equivalent to 760 mm Hg, the partial pressure of oxygen is 21% (percentage of oxygen in air) of 760, or 159 mm Hg in dry air. The partial pressure of oxygen in the lung alveoli is greater (100 mm Hg pressure) than it is in venous blood of lung capillaries (40 mm Hg pressure) (Figure 29-22). Oxygen then naturally diffuses into the capillaries. In a similar manner the carbon dioxide in the blood of the lung capillaries has a higher concentration (46 mm Hg) than has this same gas in the lung alveoli (40 mm Hg), so that carbon dioxide diffuses from the blood in the alveoli.

In the tissues respiratory gases also move according to their concentration gradients (Figure 29-22). The concentration of oxygen in the blood (100 mm Hg pressure) is greater than in the tissues (0 to 30 mm Hg pressure), and the carbon

Because of the weight of water, the hydrostatic pressure increases the equivalent of 1 atmosphere for every 10 m of depth in seawater, and the pressure of the air supplied to a diver must be increased correspondingly so that it can be drawn into the lungs. Under the increased pressure, additional air dissolves in the blood, the amount depending on the depth and time at depth of the dive. If the diver ascends slowly, the gas comes out of solution imperceptibly and is breathed out from the lungs. However, if the ascent is too rapid, the air comes out of solution and forms bubbles in the blood and other tissues, a condition known as *decompression sickness* or *the bends*. The result is painful and, if severe, can cause paralysis or death.

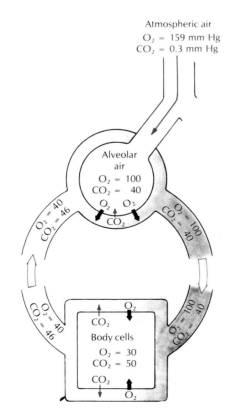

Atmospheric air
O_2 = 159 mm Hg
CO_2 = 0.3 mm Hg

Alveolar air
O_2 = 100
CO_2 = 40

O_2 = 40
CO_2 = 46

O_2 = 100
CO_2 = 40

O_2 = 100
CO_2 = 40

O_2 = 40
CO_2 = 46

CO_2
Body cells
O_2 = 30
CO_2 = 50

Figure 29-22

Exchange of respiratory gases in lungs and tissue cells. Numbers represent partial pressures in millimeters of mercury (mm Hg).

dioxide concentration in the tissues (45 to 68 mm Hg pressure) is greater than that in blood (40 mm Hg pressure). The gases in each case diffuse from a high to a low concentration.

Transport of gases in blood

In some invertebrates the respiratory gases are simply carried dissolved in the body fluids. However, the solubility of oxygen is so low in water that it is adequate only for animals with low rates of metabolism. For example, only approximately 1% of a human's oxygen requirement can be transported in this way. Consequently in many invertebrates and in the vertebrates, nearly all of the oxygen and a significant amount of the carbon dioxide are transported by special colored proteins, or **respiratory pigments,** in the blood. In most animals (all vertebrates) these respiratory pigments are packaged into blood cells. This is necessary because, if this amount of respiratory pigment were free in blood, the blood would have the viscosity of syrup and would hardly flow through the blood vessels at all.

The most widespread respiratory pigment in the animal kingdom is **hemoglobin,** a red, iron-containing protein present in all vertebrates and many invertebrates. Each molecule of hemoglobin is made up of 5% **heme,** an iron-containing compound giving the red color to blood, and 95% **globin,** a colorless protein. The heme portion of the hemoglobin has a great affinity for oxygen; each gram of hemoglobin (there are approximately 15 g of hemoglobin in each 100 ml of human blood) can carry a maximum of approximately 1.3 ml of oxygen; each 100 ml of fully oxygenated blood contains approximately 20 ml of oxygen. Of course, for hemoglobin to be of value to the body it must hold oxygen in a loose, reversible chemical combination so that it can be released to the tissues. The actual amount of oxygen with which hemoglobin can combine depends on small structural changes in the hemoglobin molecule. These structural changes result from variation in the oxygen partial pressure surrounding the blood cells. When the oxygen concentration is high, as it is in the capillaries of the lung alveoli, the hemoglobin can actually combine with more oxygen than it can when the oxygen concentration is low, as it is in the systemic capillaries in the body tissues. Thus when the blood enters regions of low oxygen partial pressure, it is forced to give up or unload its oxygen because of its structural change under these conditions. The relationship of carrying capacity to surrounding oxygen concentration is shown by oxygen dissociation curves (Figure 29-23). As these curves show, the

Although hemoglobin is the only vertebrate respiratory pigment, several other respiratory pigments are known among the invertebrates. *Hemocyanin,* a blue, copper-containing protein, is present in the crustaceans and most molluscs. Among other pigments is *chlorocruorin* (klora-croo'o-rin), a green-colored, iron-containing pigment found in four families of polychaete tube worms. Its structure and oxygen-carrying capacity are very similar to those of hemoglobin, but it is carried free in the plasma rather than being enclosed in blood corpuscles. *Hemerythrin* is a red pigment found in some polychaete worms. Although it contains iron, this metal is not present in a heme group (despite the name of the pigment!), and its oxygen-carrying capacity is poor.

Figure 29-23

Oxygen dissociation curves. Curves show how amount of oxygen that can bind to hemoglobin is related to oxygen partial pressure. **A,** Small animals have blood that gives up oxygen more readily than does the blood of large animals. **B,** Hemoglobin is also sensitive to carbon dioxide partial pressure. As carbon dioxide enters blood from the tissues, it shifts the curve to the right, decreasing affinity of hemoglobin for oxygen. Thus the hemoglobin unloads more oxygen in the tissues where carbon dioxide concentration is higher.

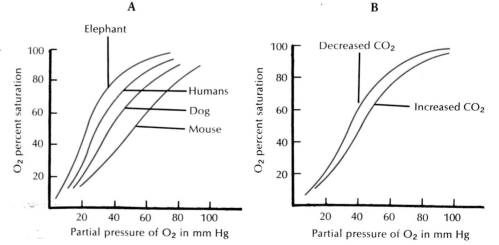

lower the surrounding oxygen tension, the greater the quantity of oxygen released. This is an important characteristic because it allows more oxygen to be released to those tissues which need it most (have the lowest partial pressure of oxygen). More oxygen is loaded in the lungs and more unloaded in the tissues than if the combination and release depended on concentration gradients alone.

Another characteristic facilitating the release of oxygen to the tissues is the sensitivity of oxyhemoglobin to carbon dioxide. Carbon dioxide shifts the oxygen dissociation curve to the right (Figure 29-23, *B*), a phenomenon that has been called the **Bohr effect** after the Danish scientist who first described it. Therefore as carbon dioxide enters the blood from the respiring tissues, it encourages the release of additional oxygen from the hemoglobin. The opposite event occurs in the lungs; as carbon dioxide diffuses from the venous blood into the alveolar space, the oxygen dissociation curve shifts back to the left, allowing more oxygen to be loaded onto the hemoglobin.

Transport of carbon dioxide by the blood

The same blood that transports oxygen to the tissues from the lungs must carry carbon dioxide back to the lungs on its return trip. However, unlike oxygen that is transported almost exclusively in combination with hemoglobin, carbon dioxide is transported in three major forms.

1. Most of the carbon dioxide, approximately 67%, is converted in the red blood cells into bicarbonate and hydrogen ions by undergoing the following series of reactions:

$$CO_2 + H_2O \rightleftharpoons H_2CO_3$$
Carbonic
acid

This reaction would normally proceed very slowly, but an enzyme in the red blood cells, **carbonic anhydrase,** catalyzes the reaction, enabling it to proceed almost instantly. As soon as carbonic acid forms, it instantly and almost completely ionizes as follows:

$$H_2CO_3 \rightleftharpoons HCO_3^- + H^+$$
Carbonic   Bicarbonate   Hydrogen
acid   ion   ion

The hydrogen ion is buffered by several buffer systems in the blood, thus preventing a severe decrease in blood pH. The bicarbonate ion remains in solution in the plasma and red blood cell water since, unlike carbon dioxide, bicarbonate is extremely soluble.

2. Another fraction of the carbon dioxide, approximately 25%, combines reversibly with hemoglobin. It is carried to the lungs where the hemoglobin releases it in exchange for oxygen.

3. A third small fraction of the carbon dioxide, approximately 8%, is carried as the physically dissolved gas in the plasma and red blood cells.

Unfortunately for humans and other higher animals, hemoglobin has an affinity for carbon monoxide that is about 200 times greater than its affinity for oxygen. Consequently, even when carbon monoxide is present in the atmosphere at lower concentrations than oxygen, it tends to displace oxygen from hemoglobin to form a stable compound called carboxyhemoglobin. Air containing only 0.2% carbon monoxide may be fatal. Because of their higher respiratory rate, children and small animals are poisoned more rapidly than adults. Carbon monoxide is becoming an atmospheric contaminant of ever-increasing proportions as the world's population and industrialization continue to increase rapidly.

SUMMARY

The fluid in the body, whether intracellular, plasma, or interstitial, is mostly water, but it has many substances dissolved in it, including electrolytes and proteins. Mammalian blood consists of the fluid plasma and the formed elements, including red and white blood cells and platelets. The plasma has many dissolved solids, as well as dissolved gases. Mammalian red blood cells lose their nucleus during their development and contain the oxygen-carrying pigment, hemoglobin. White blood cells are important defensive elements. Platelets are vital in the process of clotting, which is necessary to prevent excess blood loss when the blood vessel is damaged. They release a series of factors that activate prothrombin to thrombin, an enzyme that causes fibrinogen to be changed to the gel form, fibrin.

Phagocytosis is one of the most important defense mechansims in most animals. Specific immune responses of vertebrates are stimulated by antigens. The two arms of the immune response are humoral (mediated by antibodies) and cellular (associated with cell surfaces). The basis of self–nonself recognition lies in the genes of the major histocompatibility complex (MHC), which control production of certain proteins in cell surfaces. Antibodies are immunoglobulins, and each molecule is made up of two longer chains of amino acids (heavy chains) and two shorter chains (light chains) that have variable regions that bind with specific antigens. A humoral response is stimulated when an antigen is taken up by a macrophage, the macrophage presents parts of the antigen to T lymphocytes (T helper cells), the T helpers stimulate specific B lymphocytes to proliferate, and finally plasma cells are produced that manufacture antibody to the antigen. Some of the activated B cells become memory cells, which mediate the secondary response. Among other actions, antibodies help destroy invading substances by opsonization and by lysis via complement. Other types of T cells are T inducers, cytotoxic T cells, and T suppressors. Cellular immunity depends on T cells that secrete lymphokines. Inflammation is a nonspecific process that is greatly influenced by the body's prior immunizing experience to a particular antigen. People have genetically determined antigens in the surfaces of their red blood cells (ABO blood groups and others); blood types must be compatible in transfusions, or the transfused blood will be agglutinated by antibodies in the recipient.

In a closed circulatory system, the heart pumps blood into arteries, then into arterioles of smaller diameter, through the bed of fine capillaries, through venules, and finally through the veins, which lead back to the heart. In fishes, which have a two-chambered heart with a single atrium and a single ventricle, the blood is pumped to the gills and then directly to the systemic capillaries throughout the body without first returning to the heart. The four-chambered heart of birds and mammals is more efficient than the two-chambered heart of fishes because the blood is pumped through the capillary bed of the lungs by one ventricle, then returned to the heart and pumped through the systemic circulation by another ventricle. One-way flow of blood during the heart's contraction (systole) and relaxation (diastole) is assured by valves between the atria and the ventricles and between the ventricles and the pulmonary arteries and the aorta. Although the heart can beat spontaneously, its rate is controlled by parasympathetic and sympathetic nerves from the central nervous system. The heart muscle uses a great deal of oxygen and has a well-developed coronary blood circulation. The walls of arteries are thicker than those of veins, and the connective tissue in the walls of arteries allows them to expand during systole and contract during diastole. Normal arterial blood pressure (hydrostatic) of humans in systole is 120 mm Hg and in diastole, 80 mm Hg. It decreases to about 40 mm Hg at the arteriolar end of the capillaries, about 15 mm Hg at the venule end of the capillaries, 10 mm Hg in the veins, and finally to near zero at the right atrium. Because the capillary walls are permeable to water, and there is a net osmotic pressure because of proteins in the plasma, water enters the surrounding tissue at the arteriole end and reenters the blood at the venule end of the capillaries. Not all the fluid reenters the blood at this point, but the remainder is collected by the lymphatic system and is returned to the blood through the thoracic duct.

Very small animals can depend on diffusion between the external environment and their tissues or cytoplasm for transport of respiratory gases, but larger animals require specialized organs, such as gills, tracheae, or lungs, for this function. Gills and lungs provide an increased surface area for exchange of respiratory gases between the blood and the environment. Since simple solution of gases in the blood may be insufficient for respiratory needs, many animals have special respiratory pigments and other mechanisms to help transport oxygen and carbon dioxide. The pigment in vertebrates, hemoglobin, undergoes small structural changes that enable it to combine with more oxygen at higher oxygen concentrations than it can at lower concentrations. This makes it possible for hemoglobin to load up more oxygen in the lungs and unload more in the tissues than if the process depended on concentration gradients alone. Carbon dioxide is carried from the tissues to the lungs in the blood as the bicarbonate ion, in combination with hemoglobin, and as the dissolved gas.

Review questions

1. Name the chief intracellular electrolytes and the chief extracellular electrolytes.
2. What is the fate of spent erythocytes in the body?
3. Outline or briefly describe the sequence of events that leads to blood coagulation.
4. Phagocytosis is an important defense mechanism in most animals. How are phagocytes classified? Name two kinds of cells that are phagocytic.
5. What is the molecular basis of self and nonself recognition?
6. What is the difference between humoral and cellular immunity?
7. Outline the sequence of events in a humoral immune response from the introduction of antigen to the production of antibody.
8. Define the following: plasma cell, secondary response, memory cell, complement, opsonization, titer, challenge, lymphokine, natural killer cell.
9. Name and describe the function of the four types of T cells.
10. Distinguish between class I and class II MHC proteins.
11. Describe a typical inflammatory response.
12. Give the genotypes of each of the following blood types: A, B, O, AB. What happens when a person with type A gives blood to a person with type B? With type AB? With type O?
13. Trace the flow of blood through the heart and pulmonary circulation, naming each of the chambers, valves, arteries, and veins through which the blood passes.
14. Explain why there is a net filtration pressure at the arteriole end of capillaries in humans, and explain the movement of fluid through the walls of the capillaries.
15. What is an advantage of a fish's gills for breathing in water and a disadvantage for gill breathing on land?
16. Trace the route of inspired air in humans from the nostrils to the smallest chamber of the lungs.
17. Explain the mechanism of breathing and how breathing is controlled.
18. Explain how oxygen is carried in the blood, including specifically the role of hemoglobin. Answer the same question but with regard to carbon dioxide transport.

Selected references

See also general references for Part Three, p. 752.

Bellanti, J.A. 1985. Immunology III. Philadelphia, W.B. Saunders Co. *A good introductory immunology textbook.*

Kennedy, R.C., J.L. Melnick, and G.R. Dreesman. 1986. Anti-idiotypes and immunity. Sci. Am. **255**:48-56 (July). *Certain portions of antibodies (idiotypes) stimulated by an antigen themselves stimulate production of antibodies against the idiotypes (anti-idiotypes). These ramify into a intricate network of antibodies against antibodies that help regulate the immune system.*

Laurence, J. 1985. The immune system in AIDS. Sci. Am. **253**:84-93 (Dec.). *Good description of the effects of the AIDS virus on the immune response, especially T4 lymphocytes.*

Lawn, R.M., and G.A. Vehar. 1986. The molecular genetics of hemophilia. Sci. Am. **254**:48-54 (Mar.). *The gene coding for the factor in the clotting cascade in the most common form of hemophilia has been isolated and cloned with recombinant DNA techniques. Prospects are good for a safe (virus-free), abundant source of the factor for treating hemophiliacs.*

Marrack, P., and J. Kappler. 1986. The T cell and its receptor. Sci. Am. **254**:36-45 (Feb.). *The molecule on the surface of T cells that recognizes antigen along with MHC protein has been identified.*

Milstein, C. 1980. Monoclonal antibodies. Sci. Am. **243**:66-74 (Oct.). *What monoclonal antibodies are and how they are produced. Together with Georges J.F. Köhler, Milstein received the 1984 Nobel Prize for developing the technique of monoclonal antibody production. They shared the prize with Niels K. Jerne for his anti-idiotype hypotheses (see Kennedy et al., 1986).*

Perutz, M.F. 1978. Hemoglobin structure and respiratory transport. Sci. Am. **240**:92-125 (Dec.). *Hemoglobin transports oxygen and carbon dioxide between the lungs and tissues by clicking back and forth between two structures. Perutz and J.C. Kendrew won the Nobel Prize in 1962 for discovering the structure of hemoglobin.*

Playfair, J.H.L. 1982. Immunology at a glance. Oxford, Blackwell Scientific Publications. *Good diagrams with an excellent summary of modern immunology, but it takes more than a "glance."*

Randall, D.J., W.W. Burggren, A.P. Farrell, and M.S. Haswell. 1981. The evolution of air breathing in vertebrates. Cambridge, Eng., Cambridge University Press. *Traces the physiology of air breathing from aquatic ancestors.*

Robinson, T.F., S.M. Factor, and E.H. Sonnenblick. 1986. The heart as a suction pump. Sci. Am. **254**:84-91 (June). *Suggest that filling of heart in diastole is aided by elastic recoil of energy from systole.*

Tonegawa, S. 1985. The molecules of the immune system. Sci. Am. **253**:122-131 (Oct.). *Excellent review of structure and function of antibody molecules, T-cell surface receptors, and the genetics of antibody diversity.*

Zucker, M.B. 1980. The functioning of the blood platelets. Sci. Am. **242**:86-103 (June). *The small blood elements that act to stop blood flow from a wound also perform complex roles in health and disease.*

CHAPTER 30

HOMEOSTASIS

Osmotic Regulation, Excretion, and Temperature Regulation

At the beginning of the preceding chapter we described the double-layered environment of the body's cells: the extracellular fluid, which immediately surrounds the cells, and the external environment of the outside world.

The life-supporting metabolic activities that occur within the body's cells can proceed only as long as they are bathed by a protective extracellular fluid of relatively constant composition and are protected from extremes in environmental temperature. Yet an animal's world is seldom constant, varying not only in temperature but in the nutrients and other materials necessary for life. The animal itself requires a steady supply of these materials for its metabolic activities, which in turn produce heat and a continuous flow of products and wastes. Thus, there are many elements within and without the animal that threaten to throw the protected cellular system out of balance.

Obviously body composition and stability is not a static thing; it operates as a **dynamic steady state.** This means that a fixed composition is maintained despite the continuous shifting of components within the system. This kind of internal regulation was termed **homeostasis** (meaning "same state") by W.B. Cannon (Figure 30-1). The internal environment is not kept absolutely constant, however, but rather is held within limits of fluctuations that the body can tolerate without disruption of function.

Homeostasis is maintained by the coordinated activities of numerous body systems, such as the circulatory system, nervous system, endocrine system, and especially the organs that serve as sites of exchange with the external environment, which include the kidneys, lungs or gills, alimentary canal, and skin. Through these organs oxygen, foodstuffs, minerals, and other constituents of the body fluids enter, water is exchanged, heat is lost, and metabolic wastes are eliminated.

We will look first at the problems of controlling the internal fluid environment in aquatic animals. Next we will briefly examine how these problems are solved by terrestrial animals and consider the function of the organs that regulate the internal state. Finally we will look at the different ways animals solve the problem of living in a world of changing temperatures.

Figure 30-1

Walter Bradford Cannon (1871-1945), Harvard professor of physiology who coined the term "homeostasis" and developed the concept originated by French physiologist Claude Bernard.
From Fulton, J.F., and L.G. Wilson. 1966. Selected readings in the history of physiology. Springfield, Ill., Charles C Thomas, Publisher.

WATER AND OSMOTIC REGULATION
How Marine Invertebrates Meet Problems of Salt and Water Balance

Most marine invertebrates are in osmotic equilibrium with their seawater environment. They have body surfaces that are permeable to salts and water so that their body fluid concentration rises or falls in conformity with changes in concentrations of seawater. Because such animals are incapable of regulating their body fluid osmotic pressure, they are referred to as **osmotic conformers.** Invertebrates living in the open sea are seldom exposed to osmotic fluctuations because the ocean is a highly stable environment. Oceanic invertebrates have, in fact, very limited abilities to withstand osmotic change. If they should be exposed to dilute seawater, they die quickly because their body cells cannot tolerate dilution and are helpless to prevent it. These animals are restricted to living in a narrow salinity range and are said to be **stenohaline** (Gr. *stenos,* narrow, + *hals,* salt). An example is the marine spider crab, represented in Figure 30-2.

Conditions along the coasts and in estuaries and river mouths are much less constant than those of the open ocean. Here animals must be able to withstand large and often abrupt salinity changes as the tides move in and out and mix with fresh water draining from rivers. These animals are referred to as **euryhaline** (Gr. *eurys,* broad, + *hals,* salt), meaning that they can survive a wide range of salinity change. Most coastal invertebrates also show varying powers of **osmotic regulation.** For example, the brackish-water shore crab can resist body fluid dilution by dilute (brackish) seawater (Figure 30-2). Although the body fluid concentration falls, it does so less rapidly than the fall in seawater concentration. This crab is a **hyperosmotic regulator** because in a dilute environment it can maintain the salt concentration of its blood above that of the surrounding water.

What is the advantage of hyperosmotic regulation over osmotic conformity, and how is this regulation accomplished? The advantage is that by regulating against excessive dilution, thus protecting the body cells from extreme changes, these crabs can successfully live in the physically unstable but biologically rich coastal environment. Their powers of regulation are limited, however, because if the water is highly diluted, their regulation fails and they die.

To understand how the brackish-water shore crab and other coastal invertebrates achieve hyperosmotic regulation, let us examine the problems they face. First, the salt concentration of the internal fluids is greater than in the dilute seawater outside. This causes a steady osmotic influx of water. As with the membrane osmometer placed in a sugar solution (p. 61), water diffuses inward because it is more concentrated outside than inside. The shore crab is not nearly as permeable as a membrane osmometer—most of its shelled body surface is, in fact, almost impermeable to water—but the thin respiratory surfaces of the gills are highly permeable. Obviously the crab cannot insulate its gills with an impermeable hide and still breathe. The problem is solved by removing the excess water through the action of the kidney (the antennal gland located in the crab's head).

The second problem is salt loss. Again, because the animal is saltier than its environment, it cannot avoid loss of ions by outward diffusion across the gills. Salt is also lost in the urine. This problem is solved by special salt-secreting cells in the gills that can actively remove ions from the dilute seawater and move them into the blood, thus maintaining the internal osmotic concentration. This is an **active transport** process that requires energy because ions must be transported against a

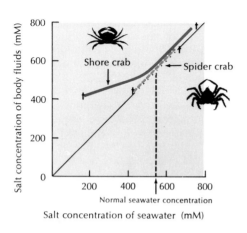

Figure 30-2

Salt concentration of body fluids of two crabs as affected by variations in the seawater concentration. The 45-degree line represents equal concentration between body fluids and seawater. Since the spider crab cannot regulate its body-fluid salt concentration, it conforms to whatever changes happen in the external seawater environment. The shore crab, however, can regulate osmotic concentration of its body fluids to some degree because in dilute seawater the shore crab can hold its body-fluid concentration above the seawater concentration. For example, when seawater is 200 mM, the shore crab's body fluid concentration is approximately 430 mM. Crosses at ends of lines indicate tolerance limits of each species.

concentration gradient, that is, from a lower salt concentration (in the dilute seawater) to an already higher one (in the blood).

Invasion of Fresh Water

Some 400 million years ago, during the Silurian and Lower Devonian periods, the major groups of jawed fishes began to penetrate brackish-water estuaries and then gradually freshwater rivers. Before them lay a new, unexploited habitat already stocked with food in the form of insects and other invertebrates, which had preceded them into fresh water. However, the advantages of this new habitat were traded off for a tough physiological challenge: the necessity of developing effective osmotic regulation.

Freshwater animals must keep the salt concentration of their body fluids higher than that of the water. Water enters their bodies osmotically, and salt is lost by diffusion outward. Their problems are similar to those of the brackish-water crab, but more severe and unremitting. Fresh water is much more dilute than are coastal estuaries, and there is no retreat, no salty sanctuary into which the freshwater animal can retire for osmotic relief. It must and has become a permanent and highly efficient hyperosmotic regulator.

The scaled and mucus-covered body surface of a fish is about as waterproof as any flexible surface can be. In addition, freshwater fishes have several defenses against the problems of water gain and salt loss. First, water that inevitably enters by osmosis across the gills is pumped out by the kidney, which is capable of forming a very dilute urine. Second, special salt-absorbing cells located in the gills move salt ions, principally sodium and chloride, from the water to the blood. This, together with salt present in the fish's food, replaces diffusive salt loss. These mechanisms are so efficient that a freshwater fish devotes only a small part of its total energy expenditure to keeping itself in osmotic balance. Osmotic regulation in fishes is described on p. 484 and illustrated in Figure 23-25, p. 485.

Crayfishes, aquatic insect larvae, mussels, and other freshwater animals are also hyperosmotic regulators and face the same hazards as freshwater fishes; they tend to gain too much water and lose too much salt. Not surprisingly, all of these forms solved these problems in the same direct way that fishes did. They excrete the excess water as urine and they actively absorb salt from the water by some salt-transporting mechanism on the body surface.

Amphibians living in water also must compensate for salt loss by actively absorbing salt from the water (Figure 30-3). They use their skin for this purpose. Physiologists learned some years ago that pieces of frog skin continue to transport sodium and chloride actively for hours when removed and placed in a specially balanced salt solution. Fortunately for biologists, but unfortunately for frogs, these animals were so easily collected and maintained in the laboratory that frog skin became a favorite membrane system for studies of ion-transport phenomena.

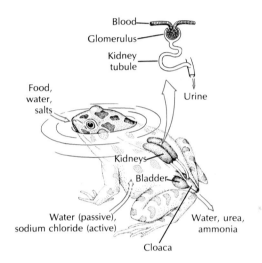

Figure 30-3

Water and solute exchange in a frog. Water enters the highly permeable skin and is excreted by the kidney. The skin also actively transports ions (sodium chloride) from the environment. The kidney forms a dilute urine by reabsorbing sodium chloride. Urine flows into the urinary bladder, where, during temporary storage, most of the remaining sodium chloride is removed and returned to the blood.
Modified from Webster, D., and H. Webster. 1974. Comparative vertebrate morphology. New York, Academic Press, Inc.

Return of Fishes to the Sea

The marine bony fishes maintain the salt concentration of their body fluids at approximately one third that of seawater (body fluids = 0.3 to 0.4 gram moles per liter [M]; seawater = 1M). They are **hypoosmotic regulators** because their body fluids are substantially more dilute than their seawater environment. Bony fishes living in the oceans today are descendants of earlier freshwater bony fishes that moved back into the sea during the Triassic period approximately 200 million

When we express seawater or body fluid concentration in molarity, we are saying that the osmotic strength is equivalent to the molar concentration of an ideal solute having the same osmotic strength. In fact, seawater and animal body fluids are not ideal solutions because they contain electrolytes that dissociate in solution. A 1 M solution of sodium chloride (which dissociates in solution) has a much greater osmotic strength than a 1 M solution of glucose, an ideal solute that does not dissociate in solution. Consequently, biologists usually express the osmotic strength of a biological solution in osmolarity rather than in molarity. A 1 osmolar solution exerts the same osmotic pressure as a 1 M solution of a nonelectrolyte.

The high concentration of urea in the blood of sharks and their kin—more than 100 times as high as in mammals—could not be tolerated by most other vertebrates. In the latter, such high concentrations of urea disrupt the peptide bonds of proteins, altering protein configuration. Sharks have adapted biochemically to the presence of the urea that permeates all their body fluids, even penetrating freely into the cells. So accommodated are the elasmobranchs to urea that their tissues cannot function without it, and the heart will stop beating in its absence.

years ago. The return to their ancestral sea was probably prompted by unfavorable climatic conditions on land and the deterioration of freshwater habitats, but we can only guess at the reasons. During the many millions of years that the freshwater fishes were adapting themselves so well to their environment, they established a body fluid concentration equivalent to approximately one-third that of seawater, thus setting the pattern for all the vertebrates that were to evolve later, whether aquatic, terrestrial, or aerial. The ionic composition of vertebrate body fluid is remarkably similar to that of dilute seawater too, a fact that is undoubtedly related to their marine heritage.

When some of the freshwater bony fishes of the Triassic period ventured back to the sea, they encountered a new set of problems. Having a much lower internal osmotic concentration than the seawater around them, they lost water and gained salt. Paradoxically, the marine bony fish literally risks drying out, much like a desert mammal deprived of water. Osmotic regulation of the marine fishes is described on p. 485.

In brief, to compensate for water loss, the marine teleost drinks seawater. This is absorbed from the intestine, and the major sea salt, sodium chloride, is carried by the blood to the gills, where specialized salt-secreting cells transport it back into the surrounding sea. The ions remaining in the intestinal residue, especially magnesium, sulfate, and calcium, are voided with the feces or excreted by the kidney. In this roundabout way, marine fishes rid themselves of the excess sea salts they have drunk, resulting in a net gain of water, which replaces the water lost by osmosis. Samuel Taylor Coleridge's ancient mariner, surrounded by "water, water, everywhere, nor any drop to drink" would undoubtedly have been tormented even more had he known of the marine fishes' simple solution for thirst. A marine fish carefully regulates the amount of seawater it drinks, consuming only enough to replace water loss and no more.

The cartilaginous sharks and rays (elasmobranchs) solve their water balance problems in a completely different way. This primitive group is almost totally marine. The salt composition of shark's blood is similar to that of the bony fishes, but the blood also carries a large content of organic compounds, especially urea and trimethylamine oxide. Urea is, of course, a metabolic waste that most animals quickly excrete in the urine. The shark kidney, however, conserves urea, causing it to accumulate in the blood. The blood urea, added to the usual blood electrolytes, raises the blood osmotic pressure to exceed slightly that of seawater. In this way the sharks and their kin turn an otherwise useless waste material into an asset, eliminating the osmotic problem encountered by the marine bony fishes.

How Terrestrial Animals Maintain Salt and Water Balance

The problems of living in an aquatic environment seem small indeed compared with the problems of life on land. Since animal bodies are mostly water, all metabolic activities proceed in water, and life itself originated in water, it would seem that animals were meant to stay in water. Yet many animals, like the plants preceding them, moved onto land, carrying their watery composition with them. Once on land, the terrestrial animals continued their adaptive radiation, undaunted by the threat of desiccation, until they became abundant even in some of the most arid parts of the earth.

Terrestrial animals lose water by evaporation from respiratory and body surfaces, excretion in the urine, and elimination in the feces. Such losses are

Table 30-1   Water Balance in the Human and the Kangaroo Rat, a Desert Rodent

	Human (%)	Kangaroo rat (%)
Gains		
Drinking	48	0
Free water in food	40	10
Metabolic water	12	90
Losses		
Urine	60	25
Evaporation (lungs and skin)	34	70
Feces	6	5

Data in part from Schmidt-Nielsen, K. 1972. How animals work. New York, Cambridge University Press.

replaced by water in the food, drinking water if it is available, and formation of **metabolic water** in the cells by oxidation of foodstuffs, especially carbohydrates. Certain insects, such as desert roaches, certain ticks and mites, and the mealworm, are able to absorb water vapor directly from atmospheric air. In some desert rodents, the metabolic water gain may constitute most of the animals' water intake.

Particularly revealing is a comparison of water balance in the human being, a nondesert mammal that drinks water, with that of the kangaroo rat, a desert rodent that may drink no water at all (Table 30-1).

The kangaroo rat gains all of its water from its food (90% as metabolic water derived from the oxidation of foodstuffs, 10% as free moisture in the food). Even though humans eat foods with a much higher water content than the dry seeds that make up much of the kangaroo rat's diet, people must still drink half their total water requirement.

The excretion of wastes presents a special problem in water conservation. The primary end product of protein breakdown is ammonia, a highly toxic material. Fishes can easily excrete ammonia across their gills, since there is an abundance of water to wash it away. The terrestrial insects, reptiles, and birds have no convenient way to rid themselves of toxic ammonia; instead, they convert it into uric acid, a nontoxic, almost insoluble compound. This enables them to excrete a semisolid urine with little water loss. The use of uric acid has another important benefit. Reptiles and birds lay amniotic eggs enclosing the embryos, their stores of food and water, and whatever wastes that accumulate during development. By converting ammonia to uric acid, the developing embryo's waste can be precipitated into solid crystals, which are stored harmlessly within the egg until hatching.

Marine birds and turtles have evolved a unique solution for excreting the large loads of salt eaten with their food. Located above each eye is a special **salt gland** capable of excreting a highly concentrated solution of sodium chloride—up to twice the concentration of seawater. In birds the salt solution runs out the nares (see p. 548). Marine lizards and turtles, like Alice in Wonderland's Mock Turtle, shed their salt gland secretion as salty tears. Salt glands are important accessory organs of salt excretion in these animals because their kidney cannot produce a concentrated urine, as can the mammalian kidney.

Given ample water to drink, humans can tolerate extremely high temperatures while preventing a rise in body temperature. Our ability to keep cool by evaporation was impressively demonstrated more than 200 years ago by a British scientist who remained for 45 minutes in a room heated to 260° F (126° C). A steak he carried in with him was thoroughly cooked, but he remained uninjured and his body temperature did not rise. Sweating rates may exceed 3 liters of water per hour under such conditions and cannot be long tolerated unless the lost water is replaced by drinking. Without water, a human continues to sweat unabatedly until the water deficit exceeds 10% of the body weight, when collapse occurs. With a water deficit of 12% a human is unable to swallow even if offered water, and death occurs when the water deficit reaches about 15% to 20%. Few people can survive more than a day or two in a desert without water. Thus, people are not physiologically well adapted for desert climates but prosper there nonetheless by virtue of their technological culture.

INVERTEBRATE EXCRETORY STRUCTURES

In such a variety of groups as make up the invertebrates, it is hardly surprising that there is a great variety of morphological structures serving as excretory organs. Many protozoa and some freshwater sponges have special excretory organelles called contractile vacuoles. The more advanced invertebrates have excretory organs that are basically tubular structures, forming urine by first producing an ultrafiltrate or fluid secretion of the blood. This enters the proximal end of the tubule and is modified continuously as it flows down the tubule. The final product is urine.

Contractile Vacuole

The tiny, spherical, intracellular vacuole of protozoa and freshwater sponges is not a true excretory organ, since ammonia and other nitrogenous wastes of metabolism readily leave the cell by direct diffusion across the cell membrane into the surrounding water. The contractile vacuole is really an organ of water balance. Because the cytoplasm of freshwater protozoa is considerably saltier than their freshwater environment, they tend to draw water into themselves by osmosis. In *Amoeba proteus* this excess water collects in numerous tiny vesicles surrounding the single thin membrane of the contractile vacuole (Figure 30-4). These vesicles then fuse with the vacuolar membrane, emptying their contents (a weak salt solution into the contractile vacuole. This grows larger as water accumulates within it. Finally the vacuole is emptied through a pore on the surface, and the cycle is rhythmically repeated. Although the mechanism for filling is not fully understood, it is noteworthy that a layer of mitochondria surrounds the contractile vacuole. These are believed to provide energy for reabsorbing salts from the tiny vesicles as they are formed. This mechanism would create a hyposmotic solution for excretion, while conserving valuable salts.

Nephridium

The most common type of invertebrate excretory organ is the nephridium, a tubular structure designed to rid the body of wastes and excess water. One of the simplest arrangements is the flame cell system (or **protonephridium**) of the acoelomate flatworms and some pseudocoelomates.

In planaria and other flatworms the protonephridial system takes the form of two highly branched duct systems distributed throughout the body (Figure 30-5). Fluid enters the system through specialized "flame cells," moves slowly into and down the tubules, and is excreted through pores that open at intervals on the body surface. The rhythmical beat of the ciliary tuft, which resembles a tiny flickering flame, creates a negative pressure that draws fluid through delicate interdigitations between the flame cell and the tubule cell and drives it into the tubular portion of the system. In the tubule, certain molecules and ions are recovered by reabsorption, leaving wastes behind to be expelled.

Note that the flame cell system with its extensive branching throughout the body is very unlike the condensed kidneys of vertebrates and the more advanced invertebrates. This is necessary in primitive invertebrates lacking a circulatory system which, in more advanced animals, carries the wastes to tubules which are gathered together into a compact excretory organ.

The protonephridium just described is a **closed** system. The tubules are closed on the inner end and urine is formed from a fluid that must first enter the

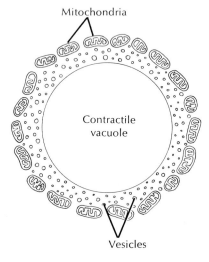

Figure 30-4

Contractile vacuole of *Amoeba proteus*, surrounded by a layer of tiny vesicles that fill with fluid, then empty into the vacuole. The layer of vesicles is bounded by mitochondria that probably provide energy for adjusting the salt content of the tiny vesicles.

Mitochondria

Contractile vacuole

Vesicles

tubules by being transported across flame cells. A more advanced type of nephridium is the **open,** or "true," nephridium (**metanephridium**) that is found in several of the eucoelomate phyla such as the annelids, the molluscs, and several smaller phyla (the earthworm nephridium is shown in Figure 14-14, p. 306). The true nephridium is more advanced than the protonephridium in two important ways. First, the tubule is open at *both* ends, allowing fluid to be swept into the tubule through a ciliated funnel-like opening (**nephrostome**). Second, the true nephridium is surrounded by a network of blood vessels that assists in urine formation by reabsorbing salts and other valuable materials from the fluid in the tube. Despite these improvements, the basic process of urine formation is the same in protonephridia and true nephridia: fluid flows continuously through a tubule while materials are added here and taken away there, until urine is formed. We will see that the advanced kidneys of vertebrates operate in basically the same way.

___ Arthropod Kidney

The **antennal glands** of crustaceans form a single, paired tubular structure located in the ventral part of the head. Their structure and function are described on p. 341. These excretory devices are an advanced design of the basic nephridial organ. However, they lack open nephrostomes. Instead, a protein-free filtrate of the blood (ultrafiltrate) is formed in the end sac by the hydrostatic pressure of the blood. In the tubular portion of the gland, the filtrate is modified by the selective reabsorption of certain salts and the active secretion of others. Thus, crustaceans have excretory organs that are basically vertebrate-like in the functional sequence of urine formation.

Insects and spiders have a unique excretory system consisting of **malpighian tubules** that operate in conjunction with specialized glands in the wall of the rectum (Figure 17-19, p. 369). These thin, elastic, blind malpighian tubules are closed and lack an arterial supply. Consequently, urine formation cannot be initiated by blood ultrafiltration as in the crustaceans and vertebrates. Instead, salts, largely potassium, are actively secreted into the tubules. This primary secretion of ions creates an osmotic drag that pulls water, solutes, and nitrogenous wastes, especially uric acid, into the tubule. Uric acid enters the upper end of the tubule as soluble potassium urate, which precipitates out as insoluble uric acid in the proximal end of the tubule. Once the formative urine drains into the rectum, most of the water and potassium are reabsorbed by specialized rectal glands, leaving behind uric acid and other wastes that are disposed of in the feces. This unique excretory system is ideally suited for life in dry environments and has contributed to the great success of insects on land.

___ VERTEBRATE KIDNEY
___ Ancestry and Embryology

Reconstructing the evolution of the vertebrate kidney is admittedly imperfect because soft organs such as kidneys are seldom preserved in fossils. Fortunately, a reasonably accurate record has persisted in the embryological development of the kidneys of living forms. (Embryology is highly conservative; developmental sequences, once fixed in the genes, are not easily modified; see Principle 8, p. 9.) From this record it is believed that the kidney of the earliest vertebrates extended the length of the coelomic cavity and was made up of segmentally arranged tubules, each resembling an invertebrate nephridium. Each tubule opened at one

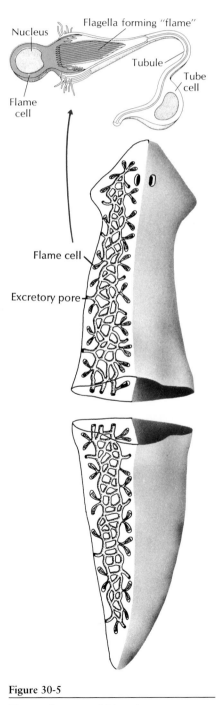

Figure 30-5

Flame cell system of *Planaria.*

Figure 30-6

Evolution of male vertebrate kidney from archinephric prototype. *Red,* Functional structures. *Light red,* Degenerative or undeveloped parts.

Archinephros: Ancestral prototype of vertebrate kidney and kidney found in embryo of hagfish; from prototype, three successive kidneys evolved during vertebrate evolution

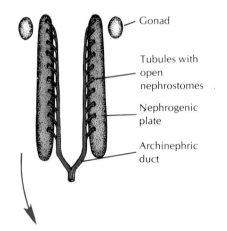

Gonad

Tubules with open nephrostomes

Nephrogenic plate

Archinephric duct

Pronephros: Functional kidney in adult hagfish and embryonic fishes and amphibians; fleeting existence in embryonic amniotes

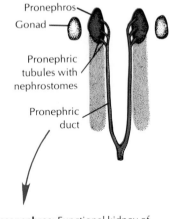

Pronephros
Gonad

Pronephric tubules with nephrostomes

Pronephric duct

Mesonephros: Functional kidney of embryonic amniotes; contributes to opisthonephros of anamniotes

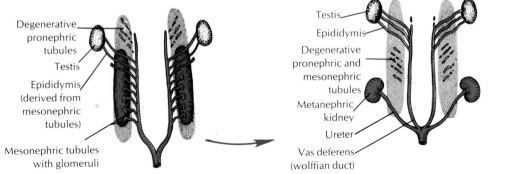

Degenerative pronephric tubules

Testis

Epididymis (derived from mesonephric tubules)

Mesonephric tubules with glomeruli

Testis

Epididymis

Degenerative pronephric and mesonephric tubules

Metanephric kidney

Ureter

Vas deferens (wolffian duct)

Metanephros: Functional kidney of adult amniotes

end into the coelom by a nephrostome and at the other end into a common **archinephric duct.** This ancestral kidney has been called the **archinephros** ("first kidney"), and a segmental kidney very similar to an archinephros is found in the embryos of hagfishes and caecilians (Figure 30-6). Almost from the beginning, the reproductive system, which develops beside the excretory system from the same segmental blocks of trunk mesoderm, made use of the nephric ducts as a convenient conducting system for reproductive products. Thus, even though the two systems have nothing functionally in common, they are closely associated in their use of common ducts.

Kidneys of living vertebrates developed from this primitive plan. During embryonic development of higher vertebrates, there is a succession of three developmental stages of kidneys: **pronephros, mesonephros,** and **metanephros.** In all vertebrate embryos, the pronephros is the first and most primitive kidney to appear. It is located anteriorly in the body and becomes part of the persistent kidney only in adult hagfishes. In all other vertebrates it degenerates during development and is replaced by a more centrally located mesonephros. The mesonephros is the functional kidney of embryonic amniotes (reptiles, birds, and mammals), and contributes to the adult kidney (called an opisthonephros) of the anamniotes (fishes and amphibians).

The most advanced vertebrate kidney, the metanephros of the amniotes, is distinguished in several ways from the pronephros and mesonephros. It is more caudally located and it is a much larger, more compact structure containing a very large number of nephric tubules. It is drained by a new duct, the **ureter,** which developed when the old archinephric duct was relinquished to the reproductive system of the male for sperm transport. Thus the three successive kidney types—pronephros, mesonephros, metanephros—succeed each other embryologically, and to some extent phylogenetically as well. Each type gives rise to a more functionally advanced adult kidney, each more compact and located more caudally than its predecessor.

Vertebrate Kidney Function

The kidneys of humans and other vertebrates play a critical role in the body's economy. Their failure means death. In this respect they are neither more nor less important than are the heart, lungs, and liver. The kidney is a part of many interlocking mechanisms that maintain **homeostasis**—constancy of the internal environment. However, the kidney's share in this regulatory council is an especially large one. It must individually monitor and regulate most of the major constituents of the blood and several minor constituents as well. In addition, it labors to remove a variety of potentially harmful substances that animals deliberately or unconsciously eat, drink, or inhale.

Perhaps even more remarkable is the way in which the kidney does its job.

These small organs, which make up less than 0.5% of the body weight in humans, receive nearly 25% of the total cardiac output, amounting to approximately 2000 liters of blood per day! This vast blood flow is channeled to approximately 2 million **nephrons,** which make up the bulk of the two human kidneys. Each nephron is a tiny excretory unit consisting of a pressure filter (**glomerulus**) and a long **nephric tubule.** Urine formation begins in the glomerulus where an ultrafiltrate of the blood is squeezed into the nephric tubule by the hydrostatic blood pressure. The ultrafiltrate then flows steadily down the twisted tubule. During its travel some substances are added to the ultrafiltrate and others are subtracted from it. The final product of this process is urine.

All mammalian kidneys are paired structures that lie embedded in fat, anchored against the dorsal abdominal wall. The two **ureters,** 25 to 30 cm (10 to 12 inches) long in humans, extend from the **renal pelvis** to the dorsal surface of the **urinary bladder.** Urine is discharged from the bladder by way of the single urethra (Figure 30-7). In the male the urethra is the terminal portion of the reproductive system as well as of the excretory system. In the female the urethra is solely excretory in function, opening to the outside just anterior to the vagina.

Since each of the thousands of nephrons in the kidney forms urine independently, each is in a way a tiny, self-contained kidney that produces a minuscule amount of urine—perhaps only a few nanoliters per hour. This amount, multiplied by the number of nephrons in the kidney, produces the total urine flow. The kidney is an "in parallel" system of independent units. However, these "independent" nephrons actually work together to create large osmotic gradients in the kidney medulla. This makes it possible for the mammalian kidney to concentrate urine whose salt concentration is well above that of the blood.

As previously indicated, the nephron, with its pressure filter and tubule, is intimately associated with the blood circulation (Figure 30-8). Blood from the

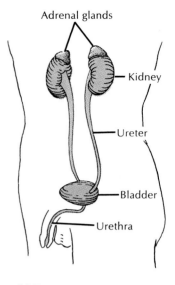

Figure 30-7

Urinary system of the human male.

Figure 30-8

Structure of a nephron and collecting duct of human kidney.

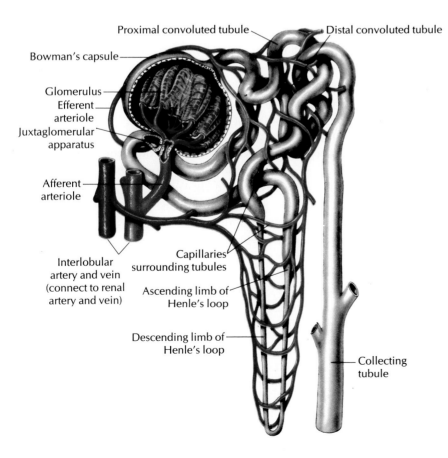

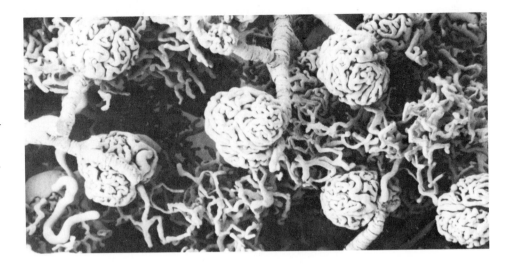

Figure 30-9

Scanning electron micrograph of a cast of the microcirculation of the mammalian kidney, showing several glomeruli and associated blood vessels. Bowman's capsule, which normally surrounds each glomerulus, has been digested away in preparing the cast.

From Tissue and Organs: a text-atlas of scanning electron microscopy, by Richard G. Kessel and Randy H. Kardon. W.H. Freeman and Company. Copyright 1979.

aorta is delivered to the kidney by way of the large **renal artery,** which breaks up into a branching system of smaller arteries. The arterial blood flows to each nephron through an **afferent arteriole** to the **glomerulus** (glo-mer′ yoo-lus), which is a tuft of blood capillaries enclosed within a thin, cuplike **Bowman's capsule** (Figure 30-9). Blood leaves the glomerulus via the **efferent arteriole.** This vessel immediately branches into an extensive system of capillaries, the **peritubular capillaries,** which completely surround the nephric tubules. Finally the blood from these many capillaries is collected by veins that unite to form the **renal vein.** This vein returns the blood to the vena cava.

Glomerular filtration

Let us now return to the glomerulus, where the process of urine formation begins. The glomerulus acts as a specialized mechanical filter in which a protein-free filtrate resembling plasma is driven by the blood pressure across the capillary walls and into the fluid-filled space of Bowman's capsule. As shown in Figure 30-10, the net filtration pressure is the difference between the blood pressure in the glomerular capillaries, believed to be approximately 45 mm Hg, and the opposing colloid osmotic and hydrostatic back pressures. Most important of these negative pressures is the colloid (protein) osmotic pressure, which is created because the proteins are too large to pass the glomerular membrane. The unequal distribution of protein causes the water concentration of the plasma to be less than the water concentration of the ultrafiltrate in Bowman's capsule. The osmotic gradient created, approximately 25 mm Hg, opposes filtration. Although small, a net filtration pressure of 10 mm Hg is sufficient to force the ultrafiltrate that is formed down the nephric tubule.

The nephric tubule consists of several segments. The first segment, the **proximal convoluted tubule,** leads into a long, thin-walled, hairpin loop called the **loop of Henle** (Figure 30-8). This loop drops deep into the medulla of the kidney and then returns to the cortex to join the third segment, the **distal convoluted tubule.** The collecting duct empties into the kidney **pelvis,** a cavity that collects the urine before it passes into the **ureter,** on its way to the **urinary bladder** (Figure 30-7).

Tubular modification of the formative urine

The ultrafiltrate that enters this complex tubular system must undergo extensive modification before it becomes urine. Approximately 200 liters of filtrate is formed each day by the average person's kidneys. Obviously the loss of this vol-

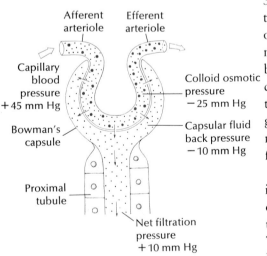

Figure 30-10

Pressures determining the net filtration pressure in a glomerulus. The net filtration pressure is the blood hydrostatic pressure, less the blood osmotic pressure resulting from plasma proteins, less the capsular hydrostatic pressure.

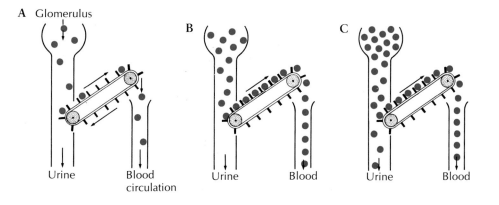

A Glomerulus

Urine   Blood
circulation

B

Urine   Blood

C

Urine   Blood

Figure 30-11

The mechanism for the tubular reabsorption of glucose can be likened to a conveyor belt running at constant speed. **A,** When the concentration of glucose in the filtrate is low, all is reabsorbed. **B,** When the glucose concentration in the filtrate has reached the transport maximum, all carrier sites for glucose are occupied. If the glucose rises further, **C,** as in the disease diabetes mellitus, some glucose escapes the carriers and appears in the urine.
Adapted from Pitts, R.F. 1963. Physiology of the kidney and body fluids. Chicago, Year Book Medical Publishers.

ume of body water, not to mention the many other valuable materials present in the filtrate, cannot be tolerated. How does tubular action convert the plasma filtrate into urine?

Two general processes are involved: **tubular reabsorption** and **tubular secretion.** Since the nephric tubules are at all points in close contact with the peritubular capillaries, materials can be transferred from the tubular lumen to the capillary blood plasma (tubular reabsorption) or from the blood plasma to the tubular lumen (tubular secretion).

Tubular reabsorption

The plasma contains a great variety of ions and molecules. With the exception of the plasma proteins, which are too large to pass the glomerular filter, all the plasma components are filtered and most are reabsorbed. Some vital materials, such as glucose and amino acids, are completely reabsorbed. Others, such as sodium, chloride, and most other minerals, undergo variable reabsorption. That is, some are strongly reabsorbed and others weakly reabsorbed, depending on the body's need to conserve each mineral. Much of this reabsorption is by **active transport,** in which cellular energy is used to transport materials from the tubular fluid, across the cell, and into the peritubular blood that returns them to the general circulation.

For most substances there is an upper limit to the amount of substance that can be reabsorbed. This upper limit is termed the **transport maximum** for that substance. For example, glucose is normally completely reabsorbed by the kidney because the transport maximum for the glucose reabsorptive mechanism is well above the amount of glucose normally present in the plasma filtrate (Figure 30-11).

Unlike glucose, most of the mineral ions are excreted in the urine in variable amounts. Their excretion is regulated. The reabsorption of sodium, the dominant cation in the plasma, illustrates the flexibility of the reabsorption process. Approximately 600 g of sodium is filtered by the human kidneys every 24 hours. Nearly all of this is reabsorbed, but the exact amount is precisely matched to sodium intake. With a sodium chloride intake of 4 g a day, the kidney excretes 4 g and reabsorbs 596 g each day. A person on a low-salt diet of 0.3 g of sodium chloride a day still maintains salt balance because only 0.3 escapes reabsorption. But with a very high salt intake, much above 10 g a day, the kidney cannot excrete sodium as fast as it enters. The unexcreted sodium chloride holds additional water in the body fluids, and the person begins to gain weight. (The salt intake of the average American is about 10 g a day, approximately 20 times more than the body needs, and three times more than is considered acceptable for those predisposed to high blood pressure.)

It may seem odd that the kidney can excrete only 10 g of sodium chloride per

In the disease diabetes mellitus ("sweet running through"), glucose rises to abnormally high concentrations in the blood plasma (hyperglycemia) because the hormone insulin, which enables the body cells to take up glucose, is deficient. As the blood glucose rises above a normal level of about 100 mg/100 ml of plasma, the concentration of glucose in the filtrate also rises, and more glucose must be reabsorbed by the proximal tubule. Eventually a point is reached (about 300 mg/100 ml of plasma) at which the reabsorptive capacity of the tubular cells is saturated. This is the transport maximum for glucose. Should the plasma glucose continue to rise, glucose spills over into the urine. In untreated diabetes the victim's urine tastes sweet, thirst is unrelenting, and the body wastes away despite a large food intake. In England the disease for centuries was appropriately called the "pissing evil."

day when approximately 600 g is filtered. It would appear that more sodium could be excreted by simply allowing more to escape tubular reabsorption. Unfortunately for salt lovers, however, the reabsorption of salt is not completely flexible. Some 80% to 85% of the salt and water filtered is reabsorbed in the proximal tubule; this is an **obligatory reabsorption** because it is governed entirely by a physical process (the osmotic pressure of the solutes) and cannot be controlled physiologically. In the distal tubule, however, sodium reabsorption is controlled by **aldosterone,** a steroid hormone of the adrenal cortex. This is called **facultative reabsorption,** meaning that the reabsorption can be adjusted physiologically according to need. We can say that proximal reabsorption is involuntary and distal reabsorption is voluntary, although, of course, we are not aware of the adjustments the kidney is performing on our behalf. The flexibility of distal reabsorption varies considerably in different animals: it is restricted in humans but very broad in many rodents. These differences have appeared because selective pressures during evolution have resulted in rodents adapted for dry environments. They must conserve water and at the same time excrete considerable sodium. Humans, however, were not designed to accommodate the large salt appetites many have. Our closest relatives, the great apes, are vegetarians with an average salt intake of less than 0.5 g a day.

As mentioned previously, salt reabsorption is controlled by the steroid hormone aldosterone. But what regulates the release of aldosterone from the adrenal cortex? Experiments performed to answer this question have revealed a remarkable system of delicate homeostatic control. The secretion of aldosterone is regulated mainly by the enzyme **renin,** produced by the **juxtaglomerular apparatus,** a complex of cells located in the afferent arteriole at its junction with the glomerulus (Figure 30-8). Renin is released in response to a low blood sodium level or to low blood pressure (which can occur if the blood volume drops too low). Renin then initiates a series of events that culminate in the production of **angiotensin,** a blood protein that has several related effects. First, it stimlates the release of aldosterone, which acts in turn to increase sodium reabsorption by the distal tubule. Second, it increases the secretion of ADH (vasopressin, discussed later in the chapter), a hormone that promotes water conservation by the kidney. Third, it increases the blood pressure. Finally, it stimulates thirst. These actions of angiotensin tend to reverse the circumstances (low blood sodium and low blood pressure and/or blood volume) that triggered the secretion of renin. Sodium and water are conserved, and blood volume and blood pressure are restored to normal.

The medical profession is in agreement that too much dietary salt causes high blood pressure (hypertension) in humans, but why it does so has been a mystery. Since angiotensin is a potent vasoconstrictor, it has long been suspected that a derailment of the renin-angiotensin system is responsible for some forms of hypertension. However, according to new findings, excess salt in the diet may indirectly produce hypertension by causing the release of a natriuretic (salt-releasing) hormone. This hormone has two effects: it increases sodium excretion and it constricts arterioles, leading to increased blood pressure.

Tubular secretion

In addition to reabsorbing large amounts of materials from the plasma filtrate, the kidney tubules are able to secrete certain substances into the tubular fluid. This process, which is the reverse of tubular reabsorption, enables the kidney to build up the urine concentrations of materials to be excreted, such as hydrogen and potassium ions, drugs, and various foreign organic materials. The distal tubule is the site of most tubular secretion.

In the kidney of bony marine fishes, reptiles, and birds, tubular secretion is a much more highly developed process than it is in mammalian kidneys. Marine bony fishes actively secrete large amounts of magnesium and sulfate, which are by-products of their mode of osmotic regulation (p. 485). Reptiles and birds excrete uric acid instead of urea as their major nitrogenous waste. This material is actively secreted by the tubular epithelium. Since uric acid is nearly insoluble, it forms crystals in the urine and requires little water for excretion. Thus, the excretion of uric acid is an important adaptation for water conservation.

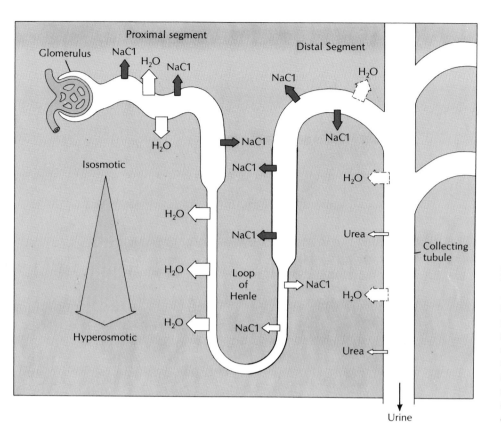

Figure 30-12

Mechanism of urine concentration in mammals. Sodium and chloride are pumped from the ascending limb of loop of Henle, and water is withdrawn passively from descending limb. Sodium chloride, together with urea reabsorbed from collecting duct, raise osmotic concentration in kidney medulla, creating osmotic gradient for the controlled reabsorption of water from the collecting duct.

Water excretion

The osmotic pressure of the blood is closely regulated by the kidney. When fluid intake is high, the kidney excretes a dilute urine, saving salts and excreting water. When fluid intake is low, the kidney conserves water by forming a concentrated urine. A dehydrated person can concentrate urine to approximately four times blood osmotic concentration.

The capacity of the kidney of mammals and some birds to produce a concentrated urine involves the loop of Henle, the long hairpin loop between the proximal and distal tubules that extends into the renal medulla. The loops constitute a **countercurrent multiplier system.** Flow is in opposite directions in the two limbs, hence the name "countercurrent."

The functional characteristics of this system are as follows. Sodium chloride is actively transported out of the thick portion of the ascending limb and into the surrounding tissue fluid (Figure 30-12). As the interstitium surrounding the loop becomes more concentrated with solute, water is passively withdrawn from the descending limb (which, unlike the ascending limb, is permeable to water but impermeable to sodium and chloride). The tubular fluid in the hairpn, now more concentrated, moves up the ascending limb where still more sodium chloride is pumped out. In this way the effect of active ion transport in the ascending limb is multiplied as more water is withdrawn from the descending limb and more concentrated fluid is presented to the ascending limb ion pump. Also contributing significantly to tissue fluid concentration at the bottom of the hairpin loop is urea, which is reabsorbed from the collecting duct that lies parallel to the loop (Figures 30-12 and 30-13).

The final adjustment of urine concentration does not occur in the loops of Henle but in the collecting ducts. Formative urine that enters the distal tubule

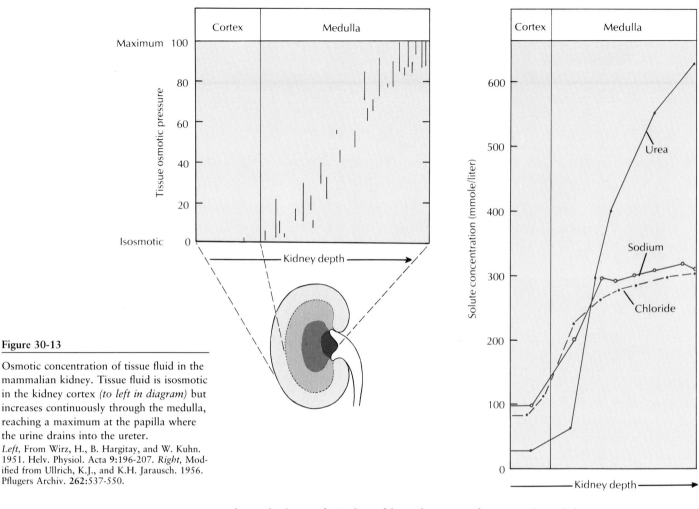

Figure 30-13

Osmotic concentration of tissue fluid in the mammalian kidney. Tissue fluid is isosmotic in the kidney cortex *(to left in diagram)* but increases continuously through the medulla, reaching a maximum at the papilla where the urine drains into the ureter.

Left, From Wirz, H., B. Hargitay, and W. Kuhn. 1951. Helv. Physiol. Acta **9**:196-207. *Right,* Modified from Ullrich, K.J., and K.H. Jarausch. 1956. Pflugers Archiv. **262**:537-550.

from the loop of Henle is dilute (because of active salt withdrawal) and is diluted still more by the active reabsorption of more sodium chloride in the distal tubule. The formative urine, low in solutes but carrying urea, now flows down into the collecting duct. Because of the high concentration of solutes surrounding the collecting duct, water is withdrawn from the urine. As the urine becomes more concentrated, urea also diffuses out and adds to the high osmotic pressure in the kidney medulla (Figure 30-13). The amount of water saved and the final concentration of the urine depend on the permeability of the walls of the collecting duct. This is controlled by the **antidiuretic hormone** (ADH, or vasopressin), which is released by the posterior pituitary gland (neurohypophysis). The release of this hormone is governed in turn by special receptors in the brain that constantly sense the osmotic pressure of the blood. When the blood osmotic pressure increases, as during dehydration, more ADH is released from the pituitary gland. ADH increases the permeability of the collecting duct, probably by expanding the size of pores in the walls of the duct. Then, as the fluid in the collecting duct passes through the hyperosmotic region of the kidney medulla, water diffuses through the pores into the surrounding interstitial fluid and is carried away by the blood circulation (Figure 30-8). The urine loses water and becomes more concentrated. Given this sequence of events for dehydration, it is not difficult to guess how the system responds to overhydration: the pituitary stops releasing ADH, the pores in the collecting duct walls close, and a large volume of dilute urine is excreted.

The varying ability of different mammals to form a concentrated urine is closely correlated with the length of the loops of Henle. The beaver, which has no need to conserve water in its aquatic environment, has short loops and can concentrate its urine to only approximately twice that of the blood plasma. Humans, with relatively longer loops, can concentrate urine 4.2 times that of the blood. As we would anticipate, desert mammals have much greater urine-concentrating powers. The camel can produce a urine eight times the plasma concentration, the gerbil 14 times, and the Australian hopping mouse 22 times. In this creature, the greatest urine concentrator of all, the loops of Henle extend to the tip of a long renal papilla that pushes out into the mouth of the ureter.

TEMPERATURE REGULATION

We have seen that a fundamental problem facing an animal is keeping its internal environment in a state that permits normal cell function. Biochemical activities are sensitive to the chemical environment and our discussion thus far has examined how the chemical environment is stabilized. Biochemical reactions are also extremely sensitive to temperature. Most enzyme reactions double—sometimes even triple—their rate with every 10° C rise in temperature. Temperature, therefore, is a severe constraint for animals, all of which seek biochemical stability. When the body temperature drops too low, metabolic processes are slowed and the amount of energy the animal can muster for activity and reproduction is reduced. If the body temperature rises too high, metabolic reactions become imbalanced and enzymatic activity is impaired or even destroyed. Thus animals can successfully function only in a restricted range of temperature, usually between 0° to 40° C. Animals must either find a habitat where they do not have to contend with temperature extremes, or they must develop the means of regulating their body temperature independent of temperature extremes.

Ectothermy and Endothermy

The terms "cold-blooded" and "warm-blooded" have long been used to divide animals into two groups: invertebrates and lower vertebrates that feel cold to the touch, and those, such as humans, other mammals, and birds, that do not. It is true that the body temperature of mammals and birds is usually (though not always) warmer than the air temperature, but a "cold-blooded" animal is not necessarily cold. Tropical fishes, and insects and reptiles basking in the sun, may have body temperatures equaling or surpassing those of mammals. Moreover, many "warm-blooded" mammals hibernate, allowing their body temperature to approach the freezing point of water. Thus the terms "warm-blooded" and "cold-blooded" are hopelessly subjective and nonspecific but are so firmly entrenched in our vocabulary that most biologists find it easier to accept the usage than to try to change people.

The terms **poikilothermic** (variable body temperature) and **homeothermic** (constant body temperature) are frequently used by zoologists as alternatives to "cold-blooded" and "warm-blooded," respectively. These terms are more precise and more informative, but they still offer difficulties. For example, deep sea fishes live in an environment that has no perceptible temperature change. Even though their body temperature is absolutely stable, day in and day out, few would argue that deep sea fishes are homeotherms. Furthermore, among the homeothermic birds and mammals there are many that allow their body temperature to change between day and night, or, as with hibernators, between seasons.

Physiologists prefer yet another way to describe body temperatures, one that reflects the fact that an animal's body temperature is a balance between heat gain and heat loss. All animals produce heat from cellular metabolism, but in most the heat is conducted away as fast as it is produced. In these animals, the **ectotherms**— and the overwhelming majority of all animals belong to this group—the body temperature is determined solely by the environment. Alternatively, some animals conserve enough of the body heat they produce to elevate their own body temperature. Since the source of their body heat is internal, they are called **endotherms.** These favored few in the animal kingdom are the birds and mammals, as well as a few reptiles and fast-swimming fishes, and certain insects that are at least partially endothermic. Endothermy allows birds and mammals to stabilize their internal temperature so that biochemical processes and nervous system functions can proceed at steady high levels of activity. Endotherms can thus remain active in winter and exploit habitats denied to ectotherms.

___ How Ectotherms Achieve Temperature Independence

Behavioral adjustments

Although ectotherms cannot control their body temperature, they are not total slaves to temperature. First, ectotherms often have the option of seeking out areas in the environment where the temperature is favorable to their activities. Some ectotherms, such as desert lizards, exploit hour-to-hour changes in solar radiation to keep their body temperature relatively constant (Figure 30-14). In the early morning they emerge from their burrows and bask in the sun with their bodies flattened to absorb heat. As the day warms, they turn to face the sun to reduce the body area exposed, and raise their bodies from the hot substrate. In the hottest part of the day they may retreat to their burrows. Later they emerge to bask as the sun sinks lower and the air temperature drops.

These behavioral patterns help to maintain a relatively steady body temperature of 36° to 39° C while the air temperature varies between 29° and 44° C. Some lizards can tolerate intense midday heat without shelter. The desert iguana of the southwestern United States prefers a body temperature of 42° C when active and can tolerate a rise to 47° C, a temperature that is lethal to all birds and

Figure 30-14

How a lizard regulates its body temperature by its behavior. In the morning, the lizard absorbs the sun's heat through its head while keeping the rest of its body protected from the cool morning air. At noon, with its body temperature high, it seeks shade from the hot sun. Later, it emerges and lies parallel to the sun's rays.

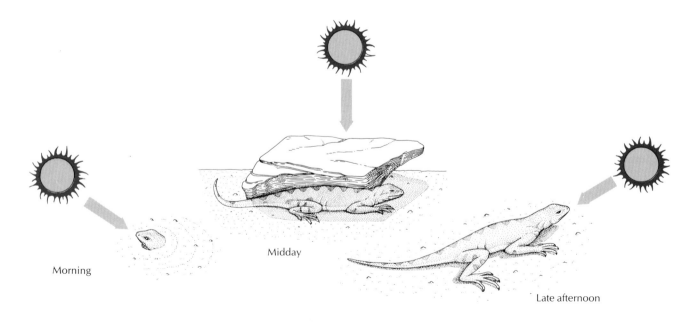

Morning

Midday

Late afternoon

mammals and most other lizards. The term "cold-blooded" clearly does not apply to these animals.

Metabolic adjustments

Even under initially unfavorable temperature conditions and without the help of the behavioral adjustments just described, most ectotherms can adjust their metabolic rates to the prevailing temperature so that the intensity of metabolism remains mostly unchanged. This is called **temperature compensation** and it involves complex biochemical and cellular adjustments. These adjustments enable a fish or a salamander, for example, to benefit from the same level of activity in both warm and cold environments. Thus, whereas endotherms achieve metabolic homeostasis by regulating their body temperature, ectotherms accomplish much the same by directly regulating their metabolism. This also is a form of homeostasis.

Temperature Regulation in Endotherms

Most mammals have body temperatures between 36° and 38° C, somewhat lower than those of birds, which range between 40° and 42° C. This constant temperature is maintained by a delicate balance between heat production and heat loss— not a simple matter when these animals are constantly alternating between periods of rest and bursts of activity.

Heat is produced by the animal's metabolism, which includes the oxidation of foodstuffs, basal cellular metabolism, and muscular contraction. Heat is lost by radiation and conduction to a cooler environment and by the evaporation of water. A bird or mammal can control both processes of heat production and heat loss within rather wide limits. If the animal becomes too cool, it can generate heat by increasing muscular activity (exercise or shivering) or decrease heat loss by increasing its insulation. If it becomes too warm, it can decrease heat production and increase heat loss. We will examine these processes in the examples that follow.

Adaptations for hot environments

Despite the harsh conditions of deserts—intense heat during the day, cold at night, and scarcity of water, vegetation, and cover—many kinds of animals live there successfully. The smaller desert mammals are mostly fossorial (fitted for digging burrows) and nocturnal. The lower temperature and higher humidity of burrows help to reduce water loss by evaporation. As explained earlier in this chapter (p. 643), desert animals such as the kangaroo rat and the American desert ground squirrels can, if necessary, derive all the water they need from their dry food, thus drinking no water at all. Such animals produce a highly concentrated urine and form almost completely dry feces.

The large desert ungulates obviously cannot escape the desert heat by living in burrows. Animals such as camels and desert antelopes (gazelle, oryx, and eland) possess a number of adaptations for coping with heat and dehydration. Those of the eland are shown in Figure 30-15. The mechanisms for controlling water loss and preventing overheating are closely linked. The glossy, pallid color of the fur reflects direct sunlight, and the fur itself is an excellent insulation that works to keep heat out. Heat is lost by convection and conduction from the underside of the eland where the pelage is very thin. Fat tissue of the eland, an essential food

Figure 30-15

Physiological and behavioral adaptations of the common eland for maintaining heat balance in the hot, arid savannah of central Africa.

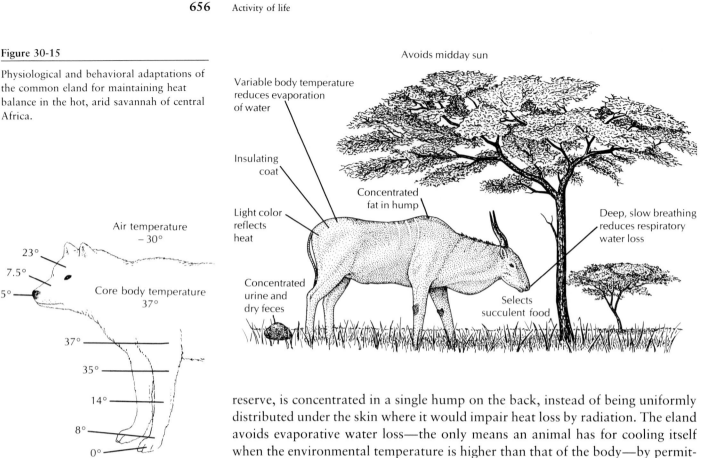

Figure 30-15

Physiological and behavioral adaptations of the common eland for maintaining heat balance in the hot, arid savannah of central Africa.

Avoids midday sun

Variable body temperature reduces evaporation of water

Insulating coat

Light color reflects heat

Concentrated fat in hump

Deep, slow breathing reduces respiratory water loss

Concentrated urine and dry feces

Selects succulent food

Air temperature – 30°

23°
7.5°
5°

Core body temperature 37°

37°
35°
14°
8°
0°

37°   36°

37°   36°

14°   13°

Figure 30-16

Countercurrent heat exchange in the leg of an arctic wolf. The upper diagram shows how the extremities cool when the animal is exposed to low air temperatures. The lower diagram depicts a portion of the front leg artery and vein, showing how heat is exchanged between outflowing arterial and inflowing venous blood.

reserve, is concentrated in a single hump on the back, instead of being uniformly distributed under the skin where it would impair heat loss by radiation. The eland avoids evaporative water loss—the only means an animal has for cooling itself when the environmental temperature is higher than that of the body—by permitting its body temperature to decrease during the cool night and then increase slowly during the day as the body stores heat. Only when the body temperature reaches 41° C must the eland prevent further rise through **evaporative cooling** by sweating and panting. Water is also conserved by means of concentrated urine and dry feces. All of these adaptations are also found developed to a similar or even greater degree in camels, the most perfectly adapted of all large desert mammals.

Adaptations for cold environments

In cold environments mammals use two major mechanisms to maintain homeothermy: (1) **decreased conductance,** that is, reduction of heat loss by increasing the effectiveness of the insulation, and (2) **increased heat production.**

In all mammals living in the cold regions of the earth, fur thickness (especially the dense and soft underhair) increases in winter, sometimes by as much as 50%. But the body extremities (legs, tail, ears, nose) of arctic mammals cannot be insulated as well as can the trunk. To prevent these parts from becoming major avenues of heat loss, they are allowed to cool to low temperatures, often approaching the freezing point. As warm arterial blood passes into a leg, for example, heat is shunted direct from artery to vein and carried back to the core of the body (Figure 30-16). This device prevents the loss of valuable body heat through the poorly insulated distal regions of the leg. A consequence of this **peripheral (countercurrent) heat exchange system** is that the legs and feet must operate at low temperatures. The temperatures of the feet of the arctic fox and barren-ground caribou are just above the freezing point; in fact, the temperature may be below 0° C in the footpads and hooves. To keep feet supple and flexible at such low temperatures, fats in the extremities have very low melting points, perhaps 30° C lower than ordinary body fats.

In severely cold conditions all mammals can produce more heat by **augmented muscular activity** through exercise or shivering. We are all familiar with the effectiveness of both activities. A person can increase heat production as much as 18-fold by violent shivering when maximally stressed by cold. Another source of heat is the increased oxidation of foodstuffs, especially brown fat stores. This mechanism is called **nonshivering thermogenesis.**

Small mammals the size of lemmings, voles, and mice meet the challenge of cold environments in a different way. Small animals are not as well insulated as large mammals because there is an obvious practical limit to how much pelage a mouse, for example, can carry before it becomes an immobile bundle of fur. Consequently these forms have successfully exploited the excellent insulating qualities of snow by living under it in runways on the forest floor, where incidentally their food is also located. In this **subnivean environment** the temperature seldom drops below $-5°$ C even though the air temperature above may fall to $-50°$ C. The snow insulation decreases thermal conductance from small mammals in the same way that pelage does for large mammals. Living beneath the snow is really a type of avoidance response to cold.

Adaptive Hypothermia in Birds and Mammals

Endothermy is energetically expensive. Whereas an ectotherm can survive for weeks in a cold environment without eating, an endotherm must always have energy resources to supply its high metabolic rate. The problem is especially acute for small birds and mammals which, because of their intense metabolism, may have to consume food each day equal to their own body weight to maintain homeothermy. It is not surprising then that a few small birds and mammals have evolved ways to abandon homeothermy for periods ranging from a few hours to several months, and allow their body temperature to fall until it equals or remains just above the surrounding air temperature.

Some very small birds and mammals, such as hummingbirds (Figure 30-17) and bats, maintain normal high body temperatures when active but allow their

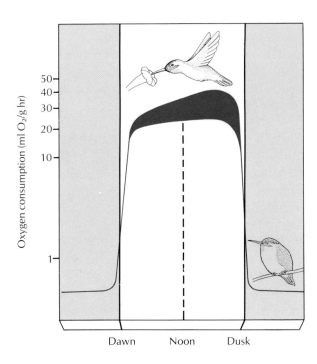

Figure 30-17

Daily torpor in hummingbirds. Body temperature and oxygen consumption are high when hummingbirds are active during the day but drop to ¹⁄₂₀ these levels when they enter torpor in the evening. Torpor vastly lowers demands on the bird's limited energy reserves.
Adapted from Lasiewski, R.C. 1963. Physiol. Zool. 36:122-140.

body temperature to drop profoundly when inactive and asleep. This is called **daily torpor,** an adaptive hypothermia that provides enormous energy saving to small endotherms that are never more than a few hours away from starvation at normal body temperatures.

Many small and medium-sized mammals in northern temperate regions solve the problem of winter scarcity of food and low temperature by entering a prolonged and controlled state of dormancy: **hibernation.** True hibernators, such as ground squirrels, jumping mice, marmots, and woodchucks (Figure 30-18), prepare for hibernation by building up large amounts of body fat. Entry into hibernation is gradual. After a series of "test drops" during which body temperature decreases a few degrees and then returns to normal, the animal cools to within a degree or less of the ambient temperature. Metabolism decreases to a fraction of normal. In the ground squirrel, for example, the respiratory rate decreases from a normal rate of 200 per minute to four or five per minute, and the heart rate from 150 to five beats per minute. During arousal the hibernator both shivers violently and employs nonshivering thermogenesis to produce heat.

Some mammals, such as bears, badgers, raccoons, and opossums, enter a state of prolonged sleep in winter with little or no decrease in body temperature. This is not true hibernation. Bears of the northern forest den-up for several months. A bear's heart rate may decrease from 40 to 10 beats per minute, but body temperature remains normal and the bear is awakened if sufficiently disturbed. One intrepid but reckless biologist learned how lightly a bear sleeps when he crawled into a den and attempted to measure the bear's rectal temperature with a thermometer!

SUMMARY

Throughout life, matter and energy pass through the body, producing perturbations of the internal physiological state. Homeostasis, the ability of an organism to maintain internal stability despite such perturbations, is a characteristic of all living systems. Homeostasis involves the coordinated activity of several physiological and biochemical mechanisms, and it is possible to relate major advances in animal evolution to increasing internal independence from the consequences of environmental change. In this chapter we have examined two aspects of homeostasis: (1) the varying ability of animals to stabilize the osmotic and chemical composition of the blood, and (2) the capacity of animals to free themselves from the constraints of temperature change.

Most marine invertebrates must either depend on the stability of the ocean to which they conform osmotically, or they must be able to tolerate wide fluctuations in environmental salinity. Some of the latter show limited powers of osmotic regulation, that is, the capacity to resist internal osmotic change, through the evolution of specialized regulatory organs. All animals living in fresh water are hyperosmotic to their environment and have developed mechanisms for recovering salt from the environment and pumping out excess water that enters the body osmotically.

All vertebrate animals, except the primitive hagfishes, show excellent osmotic homeostasis. The marine bony fishes maintain their body fluids distinctly hypoosmotic to their environment by drinking seawater and physiologically distilling it. The elasmobranchs (sharks and their kin) have adopted a strategy of near-osmotic conformity by retaining urea in the blood.

The kidney is the most important organ for regulating the chemical and osmotic composition of the blood. In all metazoa the kidney is some variation on a

Figure 30-18

Hibernating woodchuck *Marmota monax* (order Rodentia) in den exposed by road-building work sleeps on, unaware of the intrusion. Woodchucks begin hibernating in late September while the weather is still warm and may sleep 6 months. The animal is rigid and decidedly cold to the touch. Breathing is imperceptible, as slow as one breath every 5 minutes. Although it appears to be dead, it will awaken if the den temperature drops dangerously low.
Photograph by L.L. Rue, III.

basic theme: a tubular structure that forms urine by introducing a fluid secretion or ultrafiltrate of the blood into a tubule in which it is selectively modified to form urine. The terrestrial vertebrates have especially sophisticated kidneys, since they must be able to regulate closely the water content of the blood by balancing off gains and expenditures. The basic excretory unit is the nephron, composed of a glomerulus in which an ultrafiltrate of the blood is formed, and a long nephric tubule in which the formative urine is selectively modified by the tubular epithelium. Water, salts, and other valuable materials are passed by reabsorption to the peritubular circulation, and certain wastes are passed by secretion from the circulation to the tubular urine. All mammals and some birds can produce urine more concentrated than blood by means of a countercurrent multiplier system localized in the loops of Henle, a specialization not found in lower vertebrates.

Temperature has a profound effect on the rate of biochemical reactions and, consequently, on the metabolism and activity of all animals. Animals may be classified according to whether body temperature is variable (poikilothermic) or stable (homeothermic), or by the source of body heat, whether external (ectothermic) or internal (endothermic).

Ectotherms partially free themselves from thermal constraints by seeking out habitats with favorable temperatures, by behavioral thermoregulation, or by adjusting their metabolism to the prevailing temperature through biochemical alterations.

The endothermic birds and mammals differ from ectotherms in having a much higher rate of metabolic heat production and a much lower rate of heat conductance from the body. They maintain constant body temperature by balancing heat production with heat loss.

Small mammals in hot environments for the most part escape intense heat and reduce evaporative water loss by burrowing. Large mammals employ several strategies for dealing with direct exposure to heat, including reflective insulation, heat storage by the body, and evaporative cooling.

Endotherms in cold environments maintain their body temperature by decreasing heat loss with thickened pelage or plumage and peripheral cooling, and by increasing heat production through shivering or nonshivering thermogenesis.

Adaptive hypothermia is a strategy used by small mammals and birds to blunt energy demands during periods of inactivity (daily torpor) or periods of prolonged cold and minimal food availability (hibernation).

Review questions

1. Define homeostasis, and explain why body fluid stability is considered a *dynamic* steady state.
2. Distinguish between the following pairs of terms: osmotic conformity and osmotic regulation; stenohaline and euryhaline; hyperosmotic and hypoosmotic.
3. Explain why marine bony fish, but not freshwater bony fish, must drink sea water to maintain osmotic balance.
4. Most marine invertebrates are osmotic conformers. How does their body fluid composition differ from that of the cartilaginous sharks and rays, which are also in near-osmotic equilibrium with their environment?
5. What strategy does the kangaroo rat use that allows it to exist in the desert without drinking any water?
6. In what animals would you expect to find a salt gland? What is its function?
7. Relate the function of the contractile vacuole to the following experimental observations: to expel an amount of fluid equal in volume to the volume of the animal required 4 to 53 minutes for some freshwater protozoa, and between 2 and 5 hours for some marine species.

8. How does a protonephridium differ structurally and functionally from a true nephridium (metanephridium)? In what ways are they similar?

9. Explain why it is possible to reconstruct the evolution of the vertebrate kidney with the expectation of some accuracy even though we have no fossil record of kidney evolution.

10. In what ways does the nephridium of an earthworm parallel the human nephron in structure and function?

11. Describe what happens during the following stages in urine formation in the mammalian nephron: filtration, tubular reabsorption, tubular secretion.

12. Explain how the cycling of sodium chloride between the descending and ascending limbs of the loop of Henle in the mammalian kidney, and the special permeability characteristics of these tubules, produces high interstitial fluid osmotic concentrations in the kidney medulla.

13. Explain how the antidiuretic hormone (ADH) controls the excretion of water in the mammalian kidney.

14. Define the following terms and comment on the limitations (if any) of each in describing the thermal relationships of animals: poikilothermy, homeothermy, ectothermy, endothermy.

15. Defend the statement: "Both ectotherms and endotherms achieve metabolic homeostasis in unstable temperature environments, but they do so by employing completely different physiological strategies."

16. Large mammals live successfully in deserts and in the arctic. Describe the different adaptations mammals use to maintain homeothermy in each environment.

17. Explain why it is advantageous for certain small birds and mammals to abandon homeothermy during brief or extended periods of their lives.

Selected references
See also general references for Part Three, p. 572.

Beeuwkes, R. 1982. Renal countercurrent mechanisms, or how to get something for (almost) nothing. In C.R. Taylor, et al. (eds.), A companion to animal physiology. New York, Cambridge University Press. *Clear explanation of the countercurrent multiplier mechanism of the mammalian kidney.*

Hardy, R.N. 1983. Homeostasis, ed. 2. The Institute of Biology's Studies in Biology, no. 63, London, Edward Arnold. *Introduces the history of the homeostasis concept; temperature and osmotic regulation are treated in the final chapter.*

Rankin, J.C. and J. Davenport. 1981. Animal osmoregulation. New York, John Wiley and Sons, Inc. *Concise and selective treatment.*

Riegel, J.A. 1972. Comparative physiology of renal excretion. New York, Hafner Publishing Co. *Excellent survey of both vertebrate and invertebrate excretory systems.*

Schmidt-Nielsen, K. 1972. How animals work. Cambridge, Cambridge University Press. *Mechanisms of temperature regulation among vertebrates is engagingly discussed.*

Schmidt-Nielsen, K. 1981. Countercurrent systems in animals. Sci. Am. **244**:118-128 (May). *Explains how countercurrent systems transfer heat between fluids moving in opposite directions.*

Smith, H.W. 1953. From fish to philosopher. Boston, Little, Brown & Co. *Classic account of vertebrate kidney evolution.*

CHAPTER 31

DIGESTION AND NUTRITION

All organisms require energy to maintain their highly ordered and complex structure. This energy is chemical bond energy that is released by transforming complex compounds acquired from the organism's environment into simpler ones.

The ultimate source of energy for life on earth is the sun. Sunlight is captured by chlorophyll molecules in green plants, which transform a portion of this energy into chemical bond energy (food energy). Green plants are **autotrophic** organisms; they require only inorganic compounds absorbed from their surroundings to provide the raw materials for synthesis and growth. Most autotrophic organisms are the chlorophyll-bearing **phototrophs**, although some, the chemosynthetic bacteria, are **chemotrophs**; they gain energy from inorganic chemical reactions.

Almost all animals are **heterotrophic organisms** that depend on already synthesized organic compounds of plants and other animals to obtain the materials they will use for growth, maintenance, and reproduction of their kind. Since the food of animals, normally the complex tissues of other organisms, is usually too bulky to be absorbed directly by the body cells, it must be broken down, or digested, into soluble molecules that are small enough to be used.

Animals may be divided into a number of categories on the basis of dietary habits. **Herbivorous** animals feed mainly on plant life. **Carnivorous** animals feed mainly on herbivores and other carnivores. **Omnivorous** forms eat both plants and animals.

The ingestion of foods and their simplification by digestion are only initial steps in nutrition. Foods reduced by digestion to soluble, molecular form are **absorbed** into the circulatory system and are **transported** to the tissues of the body. There they are **assimilated** into the protoplasm of the cells. Oxygen is also transported by the blood to the tissues, where food products are **oxidized**, or burned to yield energy and heat. Much food is not immediately used but is **stored** for future use. Then the wastes produced by oxidation must be **excreted**. Food products unsuitable for digestion are eliminated. The sum total of all these processes is called **metabolism.**

In this chapter we will first examine the feeding adaptations of animals. Next we will discuss digestion and absorption of foodstuffs. We will close with a consideration of nutritional requirements of animals.

FEEDING MECHANISMS

Only a few animals can absorb nutrients directly from their external environment. Some blood parasites and intestinal parasites may derive all their nourishment as primary organic molecules by surface absorption. Some aquatic invertebrates may soak up part of their nutritional needs directly from the water. For most animals,

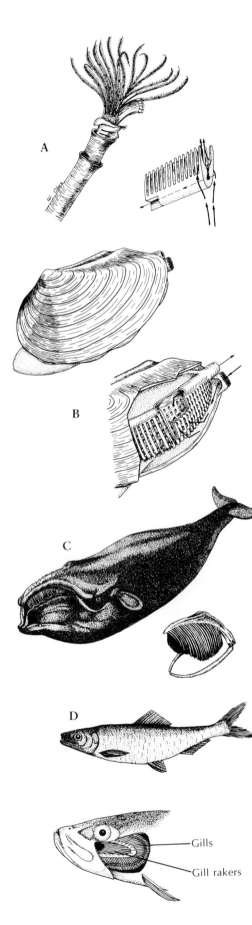

however, working for their meals is the main business of living, and the specializations that have evolved for food procurement are almost as numerous as the species of animals. In this brief discussion we consider some of the major food-gathering devices.

Feeding on Particulate Matter

Drifting microscopic particles are found in the upper hundred meters of the ocean. Most of this multitude is **plankton,** plant and animal microorganisms too small to do anything but drift with the ocean's currents. The rest is organic debris, the disintegrating remains of dead plants and animals. Although this oceanic swarm of plankton forms a rich life domain, it is unevenly distributed. The heaviest plankton growth occurs in estuaries and areas of upwelling, where there is an abundant nutrient supply. It is preyed on by numerous larger animals, invertebrates and vertebrates, using a variety of feeding mechanisms. One of the most important, successful, and widely employed methods for feeding to have evolved is **filter feeding** (Figure 31-1). The majority of filter feeders use ciliated surfaces to produce currents that draw drifting food particles into their mouths. Most filter-feeding invertebrates, such as the tube-dwelling polychaete worms, bivalve molluscs, hemichordates, and most of the protochordates, entrap the particulate food on mucous sheets that convey the food into the digestive tract. Others, such as the fairy shrimps, water fleas, and barnacles, use sweeping movements of their setae-fringed legs to create water currents and entrap food, which is then transferred to the mouth. In the freshwater developmental stages of certain insect orders, the organisms use fanlike arrangements of setae or spin silk nets to entrap food.

Filter feeding includes both sessile and free-swimming animals. Sessile filter feeders must in general accept whatever is present in the filtered water, although many have mechanisms to permit some selectivity of particle size and the rejection of inedible or harmful materials. Free-swimming filter feeders have the advantage of being able to swim through their food and thus can be much more selective in

---

Figure 31-1

Some filter feeders and their feeding mechanisms. **A,** The marine fan worm (class Polychaeta, phylum Annelida) has a crown of tentacles. Numerous cilia on the edges of the tentacles draw water *(solid arrows)* between pinnules where food particles are entrapped in mucus; the particles are then carried down a "gutter" in the center of the tentacle to the mouth *(broken arrows).* **B,** Bivalve molluscs (class Bivalvia, phylum Mollusca) use their gills as feeding devices, as well as for respiration. Water currents created by cilia on the gills carry food particles into the inhalant siphon and between slits in the gills where they are entangled in a mucous sheet covering the gill surface. The particles are then transported by ciliated food grooves to the mouth (not shown). Arrows indicate direction of water movement. **C,** Whalebone whales (class Mammalia, phylum Chordata) filter out plankton, principally large crustaceans called "krill," with whalebone, or baleen. Water enters the swimming whale's open mouth by the force of the animal's forward motion and is strained out through the more than 300 horny baleen plates that hang down like a curtain from the roof of the mouth. Krill and other plankters caught in the baleen are periodically wiped off with the huge tongue and swallowed. **D,** Herring and other filter-feeding fishes (class Osteichthyes, phylum Chordata) use gill rakers, which project forward from the gill bars into the pharyngeal cavity to strain out plankters. Herring swim almost constantly, forcing water and suspended food into the mouth; food is strained out by the gill rakers, and the water passes out the gill openings.

their feeding. Filter feeding has frequently evolved as a secondary modification among representatives of groups that are primarily selective feeders. Examples are many of the microcrustaceans; fishes such as herring, menhaden, and basking sharks; certain birds such as the flamingo; and the largest of all animals, the baleen (whalebone) whales. The vital importance of one component of the plankton, the diatoms, in supporting a great pyramid of filter-feeding animals is stressed by N.J. Berrill*:

> A humpback whale . . . needs a ton of herring in its stomach to feel comfortably full—as many as five thousand individual fish. Each herring, in turn, may well have 6000 or 7000 small crustaceans in its own stomach, each of which contains as many as 130,000 diatoms. In other words, some 400 billion yellow-green diatoms sustain a single medium-sized whale for a few hours at most.

Another type of particulate feeding exploits the deposits of organic detritus that accumulates on and in the substratum; this is called **deposit feeding.** Some deposit feeders, such as many annelids and some of the hemichordates, simply pass the substrate through their bodies, taking from it whatever provides nourishment. Others, such as the scaphopod molluscs, certain primitive bivalve molluscs, and some of the sedentary and tube-dwelling polychaete worms, use appendages to gather up organic deposits some distance from the body and move them toward the mouth.

Feeding on Food Masses

Some of the most interesting animal adaptations are those that have evolved for procuring and manipulating solid food. Such adaptations and the animals bearing them are partly shaped by what the animal eats.

Predators must be able to locate, capture, hold, and swallow prey. Most carnivorous animals simply seize the food and swallow it intact, although some employ toxins that paralyze or kill the prey at the time of capture. Although no true teeth appear among the invertebrates, many have beaks or toothlike structures for biting and holding. A familiar example is the carnivorous polychaete *Nereis,* which possesses a muscular pharynx armed with chitinous jaws that can be everted with great speed to seize prey (Figure 14-2, p. 298). Once a capture is made, the pharynx is retracted and the prey is swallowed. The teeth of lower vertebrates—those of fish, amphibians, and reptiles—are used principally to grip the prey and prevent its escape until it can be swallowed whole. Snakes and some fishes can swallow enormous meals. This, together with the absence of limbs, is associated with some striking feeding adaptations in these groups: recurved teeth for seizing and holding prey and distensible jaws and stomachs to accommodate their large and infrequent meals. Birds lack teeth, but the bills are often provided with serrated edges or the upper bill is hooked for seizing and tearing apart prey.

Many invertebrates are able to reduce food size by shredding devices (such as the shredding mouthparts of many crustaceans) or by tearing (such as the beaklike jaws of the cephalopod molluscs). Insects have three pairs of appendages on their heads that serve variously as jaws, chitinous teeth, chisels, tongues, or sucking tubes. Usually the first pair serves as crushing teeth; the second as grasping jaws; and the third, as a probing and tasting tongue.

*Berrill, N.J. 1958. You and the universe. New York, Dodd, Mead & Co.

Figure 31-2

Structure of human molar tooth. The tooth is built of three layers of calcified tissue covering: enamel, which is 98% mineral and the hardest material in the body; dentin, which composes the mass of the tooth and is approximately 75% mineral; and cementum, which forms a thin covering over the dentin in the root of the tooth and is very similar to dense bone in composition. The pulp cavity contains loose connective tissue, blood vessels, nerves, and tooth-building cells. The roots of the tooth are anchored to the wall of the socket by a fibrous connective tissue layer called the periodontal membrane.
Modified from Netter, F.H. 1959. The CIBA collection of medical illustrations, vol. 3. Summit, N.J., CIBA Pharmaceutical Products, Inc.

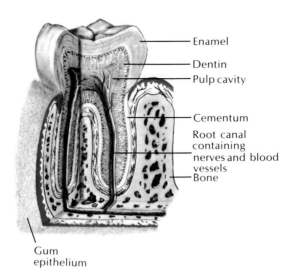

True mastication, that is, the chewing of food as opposed to tearing or crushing, is found only among the mammals. Mammals characteristically have four different types of teeth, each adapted for specific functions. **Incisors** are designed for biting, cutting, and stripping; **canines** are for seizing, piercing, and tearing; **premolars** and **molars,** at the back of the jaw, are for grinding and crushing. This basic pattern, well illustrated in human dentition (Figures 31-2 and 31-3), is often greatly modified in animals having specialized food habits (Figure 27-8, p. 572). Herbivores have suppressed canines but well-developed molars with enamel ridges for grinding. The well-developed, self-sharpening incisors of rodents grow throughout life and must be worn away by gnawing to keep pace with growth. Some teeth have become so highly modified that they are no longer useful for biting or chewing food. An elephant's tusk (Figure 31-4) is a modified upper incisor used for defense, attack, and rooting, and the male wild boar has modified canines that are used as weapons.

Herbivorous, or plant-eating, animals, whether vertebrate or invertebrate, have evolved special devices for crushing and cutting plant material. Despite its abundance on earth, the woody cellulose that encloses plant cells is an indigestible and useless material to many animals; some herbivores, however, make use of

Figure 31-3

Human deciduous and permanent teeth. Partly dissected skull of a 5-year-old child, showing milk (deciduous) teeth and permanent teeth. Milk teeth begin to erupt at 6 months and are gradually replaced by the permanent teeth beginning at approximately 6 years of age. There are 20 deciduous teeth, 5 on each side of each jaw, and 32 permanent teeth, 8 on each side of each jaw. These 8 are arranged as follows: 2 incisors, 1 canine (also called cuspid), 2 premolars (bicuspids), 3 molars. The last molar, known as the wisdom tooth, erupts between ages of 17 and 25 or not at all. Upper permanent molars are not seen in this frontal view.
Modified from Arey, L.B., 1965. Developmental anatomy. Philadelphia, W.B. Saunders Co.

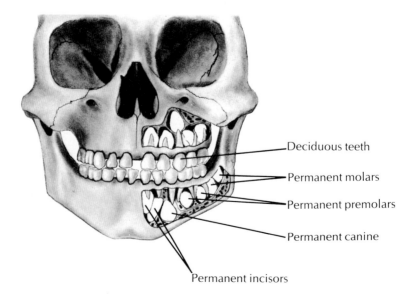

intestinal microorganisms to digest cellulose, once it is ground up. Thus, herbivores are able to digest food that the carnivores cannot, and in doing so, convert plant material into first-grade protein for carnivores and omnivores. One highly specialized group, the cud-chewing ruminants, is described on p. 573. Certain invertebrates have scraping mouthparts, such as the radula of snails. Insects such as locusts have grinding and cutting mandibles; herbivorous mammals such as horses and cattle use wide, corrugated molars for grinding. All these mechanisms disrupt the tough cellulose cell wall to accelerate its digestion by intestinal microorganisms, as well as to release the cell contents for direct enzymatic breakdown.

_____ Feeding on Fluids

Fluid feeding is especially characteristic of parasites, but it is certainly practiced among free-living forms as well. Some internal parasites (endoparasites) simply absorb the nutrient surrounding them, unwittingly provided by the host, whereas others bite and rasp off host tissue, suck blood, and feed on the contents of the host's intestine. External parasites (ectoparasites) such as leeches, lampreys, parasitic crustaceans, and insects use a variety of efficient piercing and sucking mouthparts to feed on blood or other body fluid. Unfortunately for humans and other warm-blooded animals, the ubiquitous mosquito excels in its bloodsucking habit. Alighting gently, the mosquito sets about puncturing its prey with an array of six needlelike mouthparts (Figure 17-16, p. 367). One of these is used to inject an anticoagulant saliva (responsible for the irritating itch that follows the "bite" and serving as a vector for microorganisms causing malaria, yellow fever, encephalitis, and other diseases); another mouthpart is a channel through which the blood is sucked. It is of little comfort that only the female of the species dines on blood. Far less annoying to people are the free-living butterflies, moths, and aphids that suck up plant fluids with long, tubelike mouthparts.

_____ DIGESTION

In the process of digestion, which means literally "carrying asunder," organic foods are mechanically and chemically broken down into small units for absorption. Even though food solids consist principally of carbohydrates, proteins, and fats, the very components that make up the body of the consumer, these components must nevertheless be reduced to their simplest molecular units before they can be used. Each animal reassembles some of these digested and absorbed units into organic compounds of the animal's own unique pattern. Cannibals enjoy no special metabolic benefit from eating their own kind; they digest their victims just as thoroughly as they do food of another species.

In the protozoans and sponges digestion is entirely **intracellular** (Figure 31-5). The food particle is enclosed within a food vacuole by phagocytosis (see pp. 64 and 615), and digestive enzymes are added. The products of digestion, the simple sugars, amino acids, and other molecules, are absorbed into the cell cytoplasm where they may be used directly or, in the case of multicellular animals, may be transferred to other cells. Food wastes are simply extruded from the cell.

There are important limitations to intracellular digestion, however. Only particles small enough to be phagocytized can be accepted, and every cell must be capable of secreting all of the necessary enzymes, and of absorbing the products into the cytoplasm. These limitations probably led to the evolution of an **alimen-**

Figure 31-4

An African elephant loosening soil from a salt lick with its tusk. Elephants use their powerful modified incisors in many ways in the search for food and water: plowing up the ground for roots, prying apart branches to reach the edible cambium, and drilling into dry riverbeds for water.
Photograph by C.P. Hickman, Jr.

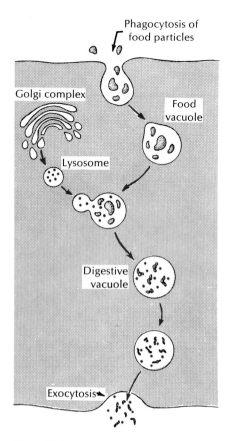

Phagocytosis of food particles

Golgi complex

Food vacuole

Lysosome

Digestive vacuole

Exocytosis

Figure 31-5

Intracellular digestion. Lysosomes containing digestive enzymes (lysozymes) are produced within the cell, possibly by the Golgi complex. Lysosomes fuse with food vacuoles and release enzymes that digest the enclosed food. Usable products of digestion are absorbed into the cytoplasm, and indigestible wastes are expelled.

tary system in which **extracellular** digestion of large food masses could take place. In extracellular digestion certain cells lining the **lumen** (cavity) of the alimentary canal specialize in forming various digestive secretions, whereas others function largely, or entirely, in absorption. Many of the lower metazoans, such as the radiates, turbellarian flatworms, and ribbon worms (nemerteans), practice both intracellular and extracellular digestion. With the evolution of greater body complexity and the appearance of complete mouth-to-anus alimentary systems, extracellular digestion became emphasized, together with increasing regional specialization of the digestive tract. In the most advanced animals, for example arthropods and vertebrates, digestion is almost entirely extracellular. The ingested food is exposed to various mechanical, chemical, and bacterial treatments, to different acidic and alkaline phases, and to digestive juices that are added at appropriate stages as the food passes through the alimentary canal.

Action of Digestive Enzymes

Mechanical processes of cutting and grinding by teeth and muscular mixing by the intestinal tract are important in digestion. However, the reduction of foods to small, absorbable units relies principally on chemical breakdown by **enzymes.** Enzymes are the highly specific organic catalysts essential to the orderly progression of virtually all life processes.

It must be stated that, although digestive enzymes are probably the best known and most studied of all enzymes, they represent but a small fraction of the numerous enzymes, perhaps thousands, that ultimately regulate all processes in the body. The digestive enzymes are **hydrolytic** enzymes, or **hydrolases,** so called because food molecules are split by the process of **hydrolysis;** that is, the breaking of a chemical bond by adding the components of water across it:

$$R—R + H_2O \xrightarrow[\text{enzyme}]{\text{Digestive}} R—OH + H—R$$

In this general enzymatic reaction, R—R represents a food molecule that is split into two products, R—OH and R—H. Usually these reaction products must in turn be split repeatedly before the original molecule is reduced to its numerous subunits. Proteins, for example, are composed of hundreds, or even thousands, of interlinked amino acids, which must be completely separated before the individual amino acids can be absorbed. Similarly, carbohydrates must be reduced to simple sugars. Fats (lipids) are reduced to molecules of glycerol and fatty acids, although some fats, unlike proteins and carbohydrates, may be absorbed without being completely hydrolyzed first. There are specific enzymes for each class of organic compounds. These enzymes are located in various regions of the alimentary canal in a sort of "enzyme chain," in which one enzyme may complete what another has started; the product moves along posteriorly for still further hydrolysis.

Motility in the Alimentary Canal

Food is moved through the digestive tract by **cilia** or by specialized **musculature,** and often by both. Movement is usually by cilia in the acoelomate and pseudocoelomate metazoa that lack the mesodermally derived gut musculature of the true coelomates, as well as in the eucoelomate bivalve molluscs in which the coelom is weakly developed. In animals with well-developed coeloms, the gut is usually lined with two opposing layers of muscle: a longitudinal layer, in which

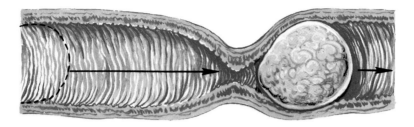

Figure 31-6

Peristalsis. Food is pushed along before a wave of circular muscle contraction.
From Schottelius, B.A., and D.D. Schottelius, 1978. A textbook of physiology, ed. 18. St. Louis, The C.V. Mosby Co.

the smooth muscle fibers run parallel with the length of the gut, and a circular layer, in which the muscle fibers embrace the circumference of the gut. This arrangement is ideal for mixing and propelling foods. The most characteristic gut movement is **peristalsis** (Figure 31-6). In this movement a wave of circular muscle contraction sweeps down the gut for some distance, pushing the food along before it. The peristaltic waves may start at any point and move for variable distances. Also characteristic of the gut are **segmentation** movements that divide and mix the food.

ORGANIZATION OF THE ALIMENTARY CANAL
Receiving Region

The first region of the alimentary canal consists of devices for feeding and swallowing. These include the parts of the mouth (for example, mandibles, jaws, teeth, radula, bills), the **buccal cavity,** and muscular **pharynx.** Most metazoans other than the filter feeders have **salivary glands** (buccal glands) that produce lubricating secretions containing mucus to assist in swallowing. Salivary glands often have other specialized functions, such as the secretion of toxic enzymes for quieting struggling prey and the secretion of salivary enzymes to begin digestion. The salivary secretion of the leech, for example, is a complex mixture containing an anesthetic (making its bite nearly painless) and several enzymes that dissolve tissues surrounding the bite and prevent blood coagulation.

Salivary **amylase** is a carbohydrate-splitting enzyme that begins the hydrolysis of plant and animal starches and is found only in certain herbivorous molluscs, some insects, and in the primate mammals, including humans. Starches are long polymers of glucose. Salivary amylase does not completely hydrolyze starch, but breaks it down mostly into two glucose fragments called **maltose.** Some free glucose, as well as longer fragments of starch, is also produced. When the food mass (bolus) is swallowed, salivary amylase continues to act for some time, digesting perhaps half of the starch before the enzyme is inactivated by the acidic environment of the stomach. Further starch digestion resumes beyond the stomach in the intestine.

The tongue is a vertebrate innovation that assists in food manipulation and swallowing. In humans, swallowing begins with the tongue pushing the moistened food bolus toward the pharynx. The nasal cavity is reflexively closed by raising the soft palate. As the food slides into the pharynx, the epiglottis is tipped down over the windpipe, nearly closing it. Some particles of food may enter the opening of the windpipe but are prevented from going farther by contraction of laryngeal muscles. Once in the esophagus, the bolus is forced smoothly toward the stomach by peristaltic contraction of the esophageal muscles.

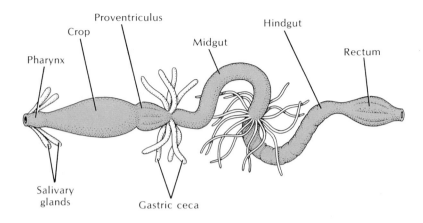

Insect digestive system (cockroach). The proventriculus is a gizzard containing chitinous teeth for grinding food.

Conduction and Storage Region

The **esophagus** of vertebrates and many invertebrates serves to transfer food to the digestive region. In many animals the esophagus is expanded into a **crop** (Figure 31-7), used for food storage before digestion. Among vertebrates, only birds have a crop; this serves to store and soften food (grain, for example) before it passes to the stomach, or to allow mild fermentation of the food before it is regurgitated to nestlings.

Region of Grinding and Early Digestion

In most vertebrates, and in some invertebrates, the **stomach** provides for initial digestion as well as for storage and mixing of the food with digestive juices. Mechanical breakdown of food, especially plant food with its tough cellulose cell walls, is often continued in herbivorous animals by grinding and crushing devices in the stomach. The muscular **gizzard** of terrestrial oligochaete worms, many arthropods, and birds, is assisted by stones and grit swallowed along with the food (annelids and birds) or by hardened linings (for example, the chitinous teeth of the insect proventriculus (Figure 31-7), and the calcareous teeth of the gastric mill of crustaceans).

Digestive diverticula—blind tubules or pouches arising from the main passage—often supplement the stomach of many invertebrates. They are usually lined with a multipurpose epithelium having cells specialized for secreting mucus or digestive enzymes, or having absorption or storage functions. Examples include the ceca of polychaete annelids, digestive glands of bivalve molluscs, the hepatopancreas of crustaceans, and the pyloric ceca of sea stars.

The stomach of carnivorous and omnivorous vertebrates is typically a U-shaped muscular tube provided with glands that produce a proteolytic enzyme and a strong acid, the latter an adaptation that probably arose for killing prey and checking bacterial activity. When food reaches the stomach, the **cardiac sphincter** opens reflexively to allow the food to enter, then closes to prevent regurgitation back into the esophagus. In humans, gentle peristaltic waves pass over the filled stomach at the rate of approximately three each minute; churning is most vigorous at the intestinal end where food is steadily released into the duodenum, the first region of the intestine. Approximately 2 liters of **gastric juice** is secreted each day by deep, tubular glands in the stomach wall (Figure 31-8). Two types of cells

The stomach contains both a strong acid, having a concentration some 4 million times that found in the blood, and a powerful proteolytic enzyme. Thus, it seems remarkable that the stomach mucosa is not digested by its own secretions. That it is not is a result of another gastric secretion, mucin, a highly viscous organic compound that coats and protects the mucosa from both chemical and mechanical injury. We should note that, despite the popular misconception of an "acid stomach" being unhealthy, a notion carefully nourished in media advertising, stomach acidity is normal and essential. Sometimes, however, the protective mucous coating fails, allowing the gastric juices to begin digesting the stomach. The result is a peptic ulcer.

line these glands: **chief cells,** which secrete **pepsin,** and **parietal cells,** which secrete **hydrochloric acid.** Pepsin is a **protease** (protein-splitting enzyme) that acts only in an acid medium—pH 1.6 to 2.4. It is a highly specific enzyme that splits large proteins by preferentially breaking down certain peptide bonds scattered along the peptide chain of the protein molecule. Although pepsin, because of its specificity, cannot completely degrade proteins, it effectively breaks them up into a number of small polypeptides. Protein digestion is completed in the intestine by other proteases that can together split all peptide bonds.

Rennin is a milk-curdling enzyme with only weak proteolytic activity found in the stomach of ruminant mammals (distinct from renin, an enzyme produced in the kidney, see p. 650). By clotting and precipitating the milk proteins, it apparently slows the movement of milk through the stomach. Rennin extracted from the stomachs of calves is used to make cheese. Human infants lack rennin and digest milk proteins with acidic pepsin, the same way adults do.

The secretion of the gastric juices is intermittent. Although a small volume of gastric juice is secreted continuously, even during prolonged periods of starvation, secretion is normally increased by the sight and the smell of food, the presence of food in the stomach, and emotional states such as anxiety and anger.

Perhaps the most unique and classic investigation in the field of digestion was made by U.S. Army surgeon William Beaumont during the years 1825 to 1833. His subject was a young, hard-living French Canadian voyageur, named Alexis St. Martin, who in 1822 had accidentally shot himself in the abdomen with a musket, the blast "blowing off integuments and muscles of the size of a man's hand, fracturing and carrying away the anterior half of the sixth rib, fracturing the fifth, lacerating the lower portion of the left lobe of the lungs, the diaphragm, and perforating the stomach." Miraculously the wound healed, but a permanent opening, or fistula, was formed that permitted Beaumont to see directly into the stomach (Figure 31-9). St. Martin became a permanent, although temperamental, patient in Beaumont's care, which included food and housing. Over a period of 8

Figure 31-8

Scanning electron micrograph of the folded epithelium of the human stomach. Epithelial cells, resembling cobblestone paving, cover the stomach surface and line gastric glands, which secrete hydrochloric acid and pepsinogen. The arrow indicates the opening of one gland.
From Pfeiffer, C.J. 1970. J. Ultrastruct. Res. 33:252.

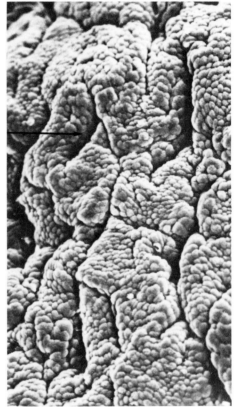

Figure 31-9

Dr. William Beaumont at Fort Mackinac, Michigan Territory, collecting gastric juice from Alexis St. Martin.
From Myer, J.S. 1919. Life and letters of Dr. William Beaumont. St. Louis, The C.V. Mosby Co.

years, Beaumont was able to observe and record how the lining of the stomach changed under different psychological and physiological conditions, how foods changed during digestion, the effect of emotional states on stomach motility, and many other facts about the digestive processes of his famous patient.

____ Region of Terminal Digestion and Absorption: the Intestine

Cells of the intestinal mucosa, like those of the stomach mucosa, are subjected to considerable wear and tear and undergo constant replacement. Cells deep in the crypt between adjacent villi divide rapidly and migrate up the villus. In mammals the cells reach the tip of the villus in about 2 days. There they are shed, along with their membrane enzymes, into the lumen at the rate of some 17 billion a day along the length of the human intestine. Before they are shed, however, these cells differentiate into absorptive cells that transport nutrients into the network of blood and lymph vessels, once digestion is complete.

The importance of the intestine varies widely among animal groups. In invertebrates that have extensive digestive diverticula in which food is broken down and phagocytized, the intestine may serve only as a pathway for conducting wastes out of the body. In other invertebrates with simple stomachs and in all vertebrates, the intestine is equipped for both digestion and absorption.

Devices for increasing the internal surface area of the intestine are highly developed in vertebrates, but are generally absent among invertebrates. Perhaps the most direct way to increase the absorptive surface of the gut is to increase its length. Coiling of the intestine is common among all vertebrate groups and reaches its highest development in mammals, in which the length of the intestine may exceed eight times the total length of the body. Although a coiled intestine is rare among invertebrates, other strategies for increasing surface are sometimes found. For example, the **typhlosole** of terrestrial oligochaete worms (Figure 14-12, C, p. 304), an inward folding of the dorsal intestinal wall that runs the full length of the intestine, effectively increases the internal surface area of the gut in a narrow body lacking space for a coiled intestine.

The more primitive fishes (lampreys and sharks) have longitudinal or spiral folds in their intestines. Higher vertebrates have developed elaborate folds and minute fingerlike projections called **villi**, some 10 to 40/mm², that give fresh intestinal tissue the appearance of velvet (Figure 31-10). Also the electron microscope reveals that each cell lining the intestinal cavity is bordered by hundreds of short, delicate processes called **microvilli** (Figure 31-10). These processes, together with

Figure 31-10

Scanning electron micrograph of a rat intestine showing the numerous fingerlike villi that project into the lumen and vastly increase the effective absorptive and secretory surface of the intestine. (×21.)

From Tissues and organs: a text-atlas of scanning electron microscopy, by Richard G. Kessel and Randy H. Kardon. W.H. Freeman and Company. Copyright © 1979.

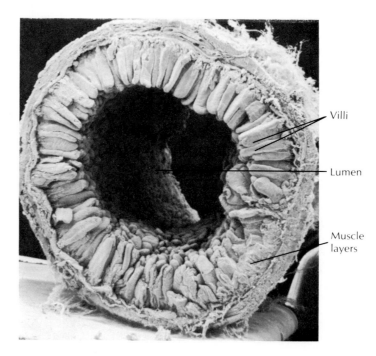

Villi

Lumen

Muscle layers

larger villi and intestinal folds, may make the internal surface of the intestine more than a million times greater than a smooth cylinder of the same diameter. The total surface area of the human intestine is approximately 300 m². The absorption of food molecules is enormously facilitated as a result.

Digestion in the vertebrate small intestine

Food is released into the small intestine through the **pyloric sphincter,** which relaxes at intervals to allow entry of acidic stomach contents into the initial segment of the small intestine, the **duodenum.** Two secretions are poured into this region: **pancreatic juice** and **bile.** Both of these secretions have a high bicarbonate content, especially the pancreatic juice, which effectively neutralizes the gastric acid, raising the pH of the liquefied food mass, now called **chyme,** from 1.5 to 7 as it enters the duodenum. This change in pH is essential because all the intestinal enzymes are effective only in a neutral or slightly alkaline medium.

Pancreatic enzymes

Approximately 2 liters of pancreatic juice is secreted each day. The pancreatic juice contains several enzymes of major importance in digestion. Two powerful proteases, **trypsin** and **chymotrypsin,** continue the enzymatic digestion of proteins begun by pepsin, which is now inactivated by the alkalinity of the intestine. Trypsin and chymotrypsin, like pepsin, are highly specific proteases that split apart peptide bonds deep inside the protein molecule. The hydrolysis of the peptide linkage may be shown as:

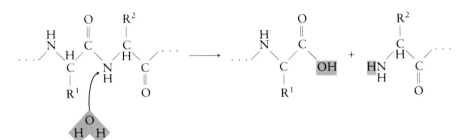

Pancreatic juice also contains **carboxypeptidase,** which splits amino acids off the carboxyl ends of polypeptides; **pancreatic lipase,** which hydrolyzes fats into fatty acids and glycerol; **pancreatic amylase,** which is a starch-splitting enzyme identical to salivary amylase in its action; and **nucleases,** which degrade RNA and DNA to nucleotides.

Membrane enzymes

The cells lining the intestine have digestive enzymes embedded in their surface membrane that continue the digestion of carbohydrates, proteins, and phosphate compounds. Until recently it was thought that these enzymes were secreted into the intestinal lumen as an intestinal "juice." Now it is known that the enzymes are actually a part of the microvillus membrane (Figure 31-11, *D*) where, as attached glycoproteins, they accomplish much of the work of digestion. Among the membrane digestive enzymes is **aminopeptidase.** It splits terminal amino acids from the amino end of short peptides in a manner similar to that of the pancreatic enzyme carboxypeptidase. Several **disaccharidases** (enzymes that split 12-carbon sugar molecules into 6-carbon units) are also part of the microvillus membrane. These include **maltase,** which splits maltose into two molecules of glucose (Figure 31-

Although milk is the universal food of newborn mammals and one of the most complete human foods, most adult humans cannot digest milk because they are deficient in lactase, the enzyme that hydrolyzes lactose (milk sugar). Lactose intolerance is genetically determined. It is characterized by abdominal bloating, cramps, flatulence, and watery diarrhea, all appearing within 30 to 90 minutes after ingesting milk or its unfermented by-products. (Fermented dairy products, such as yogurt and cheese, create no intolerance problems.)

Northern Europeans and their descendants, which include the majority of North American whites, are most tolerant of milk. Many other ethnic groups are generally intolerant to lactose, including the Japanese, Chinese, Jews in Israel, Eskimos, South American Indians, and most African blacks. Only about 30% of North American blacks are tolerant; those who are tolerant are mostly descendants of slaves brought from east and central Africa where dairying is traditional and tolerance to lactose is high. The widespread intolerance to lactose is a matter of concern to international agencies that are planning food distribution programs in developing countries, as well as in school lunch programs in developed countries.

12); **sucrase,** which splits sucrose to fructose and glucose; and **lactase,** which breaks down lactose (milk sugar) into glucose and galactose. Also present is **alkaline phosphatase,** which attacks a variety of phosphate compounds.

Bile

Bile is secreted by the cells of the liver into the **bile duct,** which drains into the upper intestine (duodenum). Between meals the bile is collected in the **gallbladder,** an expansible storage sac that releases the bile when stimulated by the presence of fatty food in the duodenum. Bile contains no enzymes. It is made up of water, bile salts, and pigments. The bile salts (mainly sodium taurocholate and sodium glycocholate) are essential for the complete absorption of fats, which, because of their tendency to remain in large, water-resistant globules, are especially resistant to enzymatic digestion. **Bile salts** reduce the surface tension of fats, so that they are broken up into small droplets by the churning movements of the intestine. This greatly increases the total surface exposure of fat particles, giving the fat-splitting lipases the opportunity to hydrolyze them. The golden yellow color of bile is produced by the **bile pigments,** which are breakdown products of hemoglobin from worn-out red blood cells. The bile pigments also give the feces its characteristic color.

The great versatility of the liver should be emphasized. Bile production is only one of the liver's many functions; it is a storehouse for glycogen, production center for the plasma proteins, site of protein synthesis and detoxification of protein wastes, destruction of worn-out red blood cells, center for metabolism of fat, amino acids, and carbohydrates, and many others.

Absorption

Most digested foodstuffs are absorbed from the small intestine, where the numerous finger-shaped **villi** provide an enormous surface area through which materials can pass from the intestinal lumen into the circulation (Figure 31-11). Little food is absorbed in the stomach because digestion is still incomplete and because of the

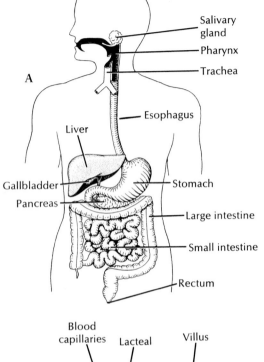

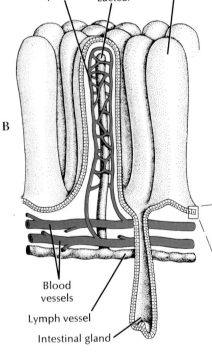

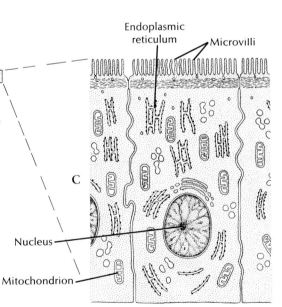

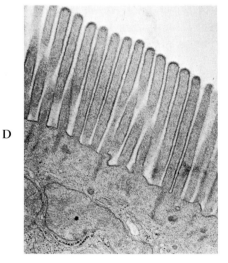

Figure 31-11

A, Human digestive system. **B,** Portion of mucosa lining of intestine, showing fingerlike villi. **C,** Section of single mucosal lining cell. **D,** Microvilli on surface of mucosal cell, rat intestine. (×16,400.)

D, Electron micrograph courtesy J.D. Berlin.

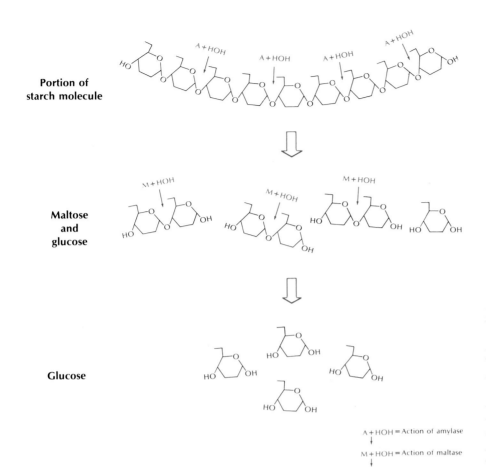

Portion of starch molecule

Maltose and glucose

Glucose

A + HOH = Action of amylase

M + HOH = Action of maltase

Figure 31-12

Digestion (hydrolysis) of starch. Starch is composed of long chains of glucose units. They are first cleaved into disaccharide residues (maltose) by the salivary enzyme, amylase. Some glucose may also be split off at the ends of starch chains. The intestinal enzyme maltase then completes the hydrolysis by cleaving the maltose molecules into glucose. A molecule of water is inserted into each enzymatically split bond.

limited surface exposure. Some materials, however, such as drugs and alcohol, are absorbed in part there, which explains their rapid action.

Carbohydrates are absorbed almost exclusively as simple sugars (for example, glucose, fructose, and galactose) because the intestine is virtually impermeable to polysaccharides. Proteins, too, are absorbed principally as their amino acid subunits, although it is believed that small amounts of small proteins or peptide fragments may sometimes be absorbed. Simple sugars and amino acids are transferred across the intestinal epithelium by both passive and active processes.

Immediately after a meal these materials are in such high concentration in the gut that they readily diffuse into the blood, where their concentration is initially lower. However, if absorption were passive only, we would expect transfer to cease as soon as the concentrations of a substance became equal on both sides of the intestinal epithelium. This would permit much of the valuable foodstuff to be lost in the feces. In fact, very little is lost because passive transfer is supplemented by an **active transport** mechanism located in the epithelial cells that picks up the food molecules and transfers them into the blood. Materials are thus moved *against* their concentration gradient, a process that requires the expenditure of energy. Although not all food products are actively transported, those that are, such as glucose, galactose, and most of the amino acids, are handled by transport mechanisms that are specific for each kind of molecule.

As already described, fat droplets are emulsified by bile salts and then digested by pancreatic lipase. Triglycerides are thus broken down into fatty acids and monoglycerides, which are absorbed by simple diffusion. However, free fatty acids never enter the blood. Instead, during their passage through the intestinal epithelial cells, the fatty acids are resynthesized into triglycerides that pass out of the cells and into the lacteals. From the lacteals, the fat droplets enter the lymph system and eventually get into the blood by way of the thoracic duct. After a fatty meal, even a peanut butter sandwich, the presence of numerous fat droplets in the blood imparts a milky appearance to the blood plasma.

___ Region of Water Absorption and Concentration of Solids

In the large intestine the indigestible remnants of digestion are consolidated by the reabsorption of water to form solid or semi-solid feces for removal from the body by **defecation.** The reabsorption of water is of special significance in insects, especially those living in dry environments which must—and do—conserve nearly all water entering the rectum. Specialized **rectal glands** absorb water and ions as needed, leaving behind fecal pellets that are almost completely dry. In reptiles and birds, which also produce nearly dry feces, most of the water is reabsorbed in the cloaca. A white paste-like feces is formed containing both indigestible food wastes and uric acid.

In humans, the colon contains enormous numbers of bacteria that enter the sterile colon of the newborn infant. In the adult approximately one third of the dry weight of feces is bacteria; these include harmless bacilli as well as cocci that can cause serious illness if they should escape into the abdomen or bloodstream. Normally the body's defenses prevent invasion of such bacteria. The bacteria break down organic wastes in the feces and provide some nutritional benefit by synthesizing certain vitamins (vitamin K and small quantities of some of the B vitamins), which are absorbed by the body.

___ REGULATION OF FOOD INTAKE

Most animals unconsciously adjust food intake to balance energy expenditure. If energy expenditure is increased by, for example, increased physical activity, food intake is increased accordingly. Most vertebrates, from fish to mammals, eat for calories rather than bulk because, if the diet is diluted with fiber, they respond by eating more. Similarly, intake is adjusted downward following a period of several days when caloric intake is too high.

Food intake is regulated in large part by a "hunger" center located in the hypothalamus region of the brain. Blood sugar level has an important influence on this center because hunger pangs coincide with decreasing levels of blood glucose. While most animals seem able to stabilize their weight at normal levels with ease, many humans cannot. It is becoming clear that many obese people do not in fact eat more food than thin people. Rather they have a reduced capacity to burn off excess calories by "nonshivering thermogenesis." The defect has been traced to **brown fat,** a diffuse tissue located in adults in the chest, upper back, and near the kidneys (newborn mammals, including human infants, have much more brown fat than adults). In normal people an increased caloric intake induces brown fat tissue to dissipate the excess energy as heat. This is referred to as "diet-induced thermogenesis." The capacity is diminished in people tending toward obesity because they have less brown fat or because their brown fat does not respond to hypothalamic signals as it should. There are other reasons for obesity in addition to the fact that many people simply eat too much. Fat stores are supervised by the hypothalamus, which is set at a point that may be higher or lower than the population norm. A high setpoint can be lowered somewhat by exercise, but as dieters are painfully aware, the body defends its fat stores with remarkable tenacity.

Brown fat, the tissue that mediates nonshivering thermogenesis, is brown because it is packed with mitochondria containing large quantities of iron-bearing cytochrome molecules. In ordinary body cells, ATP is generated by a gradient of hydrogen ions produced during the flow of electrons down the respiratory chain (chemiosmotic hypothesis of oxidative phosphorylation, p. 86). This ATP is then used to power various cellular processes. In brown fat cells, the hydrogen ions bypass the coupling of oxidation with ATP production and generate heat instead. Thermogenesis is activated by the sympathetic nervous system, which responds to signals from the hypothalamus.

NUTRITIONAL REQUIREMENTS

The food of animals must include **carbohydrates, proteins, fats, water, mineral salts,** and **vitamins.** Carbohydrates and fats are required as fuels for energy demands of the body and for the synthesis of various substances and structures. Proteins (actually the amino acids of which they are composed) are needed for the synthesis of the body's specific proteins and other nitrogen-containing compounds. Water is the solvent for the body's chemistry and a major component of all the body fluids. The inorganic salts are required as the anions and cations of body fluids and tissues and form important structural and physiological components throughout the body. The vitamins are accessory food factors that are often built into the structure of many of the enzymes of the body.

All animals require these broad classes of nutrients, although there are differences in the amounts and kinds of food required. Of these basic food classes some nutrients are used principally as fuels (carbohydrates and lipids), whereas others are required principally as structural and functional components (proteins, minerals, and vitamins). Any of the basic foods (proteins, carbohydrates, and fats) can serve as fuel to supply energy requirements, but no animal can thrive on fuels alone. A **balanced diet** must satisfy all metabolic requirements of the body—requirements for energy, growth, maintenance, reproduction, and physiological regulation.

The recognition years ago that many diseases of humans and domesticated animals were caused by or associated with dietary deficiencies led biologists to search for specific nutrients that would prevent such diseases. These studies eventually yielded a list of **"essential" nutrients** for humans and other animal species studied. The essential nutrients are those that are needed for normal growth and maintenance and that *must* be supplied in the diet. In other words, it is "essential" that these nutrients be in the diet because the animal cannot synthesize them from other dietary constituents. Nearly 30 organic compounds (amino acids and vitamins) and 21 elements have been established as essential. If we consider that the body contains thousands of different organic compounds, the list given in the margin is remarkably short. Animal cells have marvelous powers of synthesis, enabling them to build compounds of enormous variety and complexity from a small, select group of raw materials.

In the average diet of Americans and Canadians, approximately 50% of the total calories (energy content) comes from carbohydrates and 40% comes from lipids. Proteins, essential as they are for structural needs, supply only a little more than 10% of the total calories of the average North American's diet. Carbohydrates are widely consumed because they are more abundant and cheaper than proteins or lipids. Actually humans and many other animals can subsist on diets devoid of carbohydrates, provided sufficient total calories and essential nutrients are present. Eskimos, before the decline of their native culture, lived on a diet that was high in fat and protein and very low in carbohydrate.

Lipids are needed principally to provide energy. However, at least three fatty acids are essential for humans because they cannot be synthesized. Much interest and research have been devoted to lipids in our diets because of the association between fatty diets and the disease **atherosclerosis** (narrowing of the arteries). The matter is complex, but there is evidence that atherosclerosis may occur when the diet is high in saturated lipids (lipids with no double bonds in the carbon chains of the fatty acids) but low in polyunsaturated lipids (two or more double bonds in the carbon chains.) Eskimos who live along the western shores of Greenland and whose diet consists largely of fish have a remarkably low incidence of heart

Table 31-1 Human nutrient requirements*

Amino acids
Phenylalanine	Methionine
Lysine	Cystine
Isoleucine	Tryptophan
Leucine	Threonine
Valine	

Polyunsaturated fatty acids
  Arachidonic
  Linoleic
  Linolenic

Water-soluble vitamins
  Thiamine (B₁)
  Riboflavin (B₂)
  Niacin
  Pyridoxine (B₆)
  Pantothenic acid
  Folic acid
  Vitamin B₁₂
  Biotin
  Choline
  Ascorbic acid (C)

Fat-soluble vitamins
  A, D, E, and K

Minerals
Calcium	Silicon
Phosphorus	Vanadium
Sulfur	Tin
Potassium	Nickel
Chlorine	Selenium
Sodium	Manganese
Magnesium	Iodine
Iron	Molybdenum
Fluorine	Chromium
Zinc	Cobalt
Copper	

*Modified from Scrimshaw, N.S., and V.R. Young. 1976. Sci. Am. 235:50-64 (Sept.).

By genetic manipulation, plant breeders have been able to produce cereal proteins having much higher nutritional quality than the parent cereals. An example is triticale, an intergeneric cross between wheat and rye, which has more protein and a much higher lysine content than either of its parents.

attacks, cerebrovascular accidents, rheumatoid arthritis and other inflammatory diseases. Fish oils contain "omega minus 3" (or simply omega−3) fatty acids in which the last double bond is on the third to last carbon of the chain. Recent research suggests that diets containing omega−3 fatty acids reduce the incidence of heart disease. Most North Americans, however, prefer a diet high in saturated fats. Such diets promote elevated serum cholesterol and low-density lipoproteins (LDL), which may deposit in platelike formations in the lining of major arteries.

Proteins are expensive foods and restricted in the diet. Proteins, of course, are not themselves the essential nutrients, but rather contain essential amino acids. Of the 20 amino acids commonly found in proteins, nine and possibly 11 are essential to humans (Table 31-1). The rest can be synthesized. Generally, animal proteins have more of the essential amino acids than do proteins of plant origin. All nine of the essential amino acids must be present simultaneously in the diet for protein synthesis. If one or more are missing, the use of the other amino acids will be reduced proportionately; they cannot be stored and are broken down for energy. Thus, heavy reliance on a single plant source will inevitably lead to protein deficiency. This problem can be corrected if two kinds of plant proteins having complementary strengths in essential amino acids are ingested together. For example, a balanced protein diet can be prepared by mixing wheat flour, which is poor only in lysine, with a legume (peas or beans), which is a good source of lysine but deficient in methionine and cystine. Each plant complements the other by having adequate amounts of those amino acids that are deficient in the other.

Because animal proteins are so nutritious, they are in great demand in all countries. North Americans eat far more animal proteins than do Asians and Africans; on the average a North American eats 66 g of animal protein a day, supplemented by milk, eggs, cereals, and legumes. In the Middle East the individual consumption of protein is 14 g, in Africa 11 g, and in Asia 8 g.

Undernourishment and malnourishment are the world's major health problems today. The United Nations Food and Agricultural Organization estimates that 450 million people are chronically undernourished. The World Bank, using different calculations, has placed the figure at 1 billion—nearly one fourth of the human population. Growing children and pregnant and lactating women are especially vulnerable to the devastating effects of malnutrition. Cell proliferation and growth in the human brain are most rapid in the final months of pregnancy and the first year after birth. Adequate protein for neuron development is a requirement during this critical time to prevent neurological dysfunction. The brains of children who die of protein malnutrition during the first year of life have 15% to 20% fewer brain cells than those of normal children (Figure 31-13).

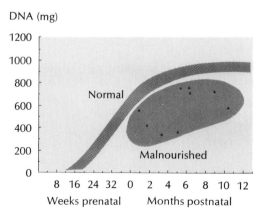

Figure 31-13

Effect of early malnutrition on cell number (measured as total DNA content) in the human brain. This graph shows that malnourished infants (colored oval) have far fewer brain cells than do normal infants (gray growth curve).

From Winick, M. 1976. Malnutrition and brain development. New York, Oxford University Press.

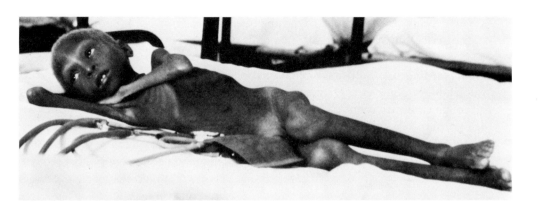

Figure 31-14

Biafran refugee child suffering severe malnutrition.
From Hosp. Tribune 8:1, 1974.

Malnourished children who survive this period suffer permanent brain damage and cannot be helped by later corrective treatment (Figure 31-14).

The major cause of the world's precarious food situation is recent rapid population growth. The world population was 2 billion in 1930, reached 3 billion in 1960, passed 5 billion during 1986, and probably will be 6 billion by the turn of the century. Eighty million people are added each year. The equivalent of the total U.S. population is added to the world every 30 months. Thus, the search for new ways to increase food production and distribution takes on a desperate urgency.

It was fitting that the 1970 Nobel Prize for Peace should go to Dr. Norman Borlaug, who developed several of the new wheat varieties that ushered in the Green Revolution of the 1960s. The Green Revolution has enabled developing countries to avoid catastrophic hunger and starvation by introducing improved plant varieties, chemical fertilizers, irrigation, and mechanization. Advances have been so successful that 25 countries, including India and China, are now net exporters of grain. As of 1987, there was a substantial world surplus of food. Nevertheless, all of the elements contributing to the Green Revolution depend mainly on the availability of fossil fuels. Research is now focusing on less energy-intensive technology, such as improved biological nitrogen fixation, greater photosynthetic efficiency, more efficient nutrient and water uptake, and the development of genetic resistance to pests. To be successful, this promising approach will require collaborative genetic research by Western scientists and the developing countries; this seems essential to the building of self-reliant food systems in the energy-poor developing world. For the next decade at least, the world capability to produce food promises to increase faster than the human population. In the long run, however, food shortages seem inevitable, and mass famine probable, unless the human population is stabilized.

Two different types of severe food deficiency are recognized: marasmus, general undernourishment from a diet low in both calories and protein, and kwashiorkor, protein malnourishment from a diet adequate in calories but deficient in protein. Marasmus (Gr. *marasmos,* to waste away) is common in infants weaned too early and placed on low-calorie–low-protein diets; these children are listless, and their bodies waste away. Kwashiorkor is a West African word describing a disease a child gets when displaced from the breast by a newborn sibling. This disease is characterized by retarded growth, anemia, weak muscles, a bloated body with typical pot belly, acute diarrhea, susceptibility to infection, and high mortality. Ten million children the world over are seriously undernourished or malnourished, and 1 million children die of hunger each year in India alone.

Vitamins

A vitamin is a relatively simple organic compound that is not a carbohydrate, fat, protein, or mineral and that is required in very small amounts in the diet for some specific cellular function. Vitamins are not sources of energy but are often associated with the activity of important enzymes that have vital metabolic roles. Plants and many microorganisms synthesize all the organic compounds they need; animals, however, have lost certain synthetic abilities during their long evolution and depend ultimately on plants to supply these compounds. Vitamins therefore represent synthetic gaps in the metabolic machinery of animals.

Vitamins are usually classified as fat soluble (soluble in fat solvents such as ether) or water soluble. The water-soluble ones include the B complex and vitamin

Vitamin A deficiency afflicts millions of people in the Western Pacific, Far East, semiarid zones of Africa, and parts of Latin America. Night blindness is an early sign of deficiency. Severe deficiencies lead to xerophthalmia, a disease of the eye in which the cornea and conjunctiva become dry and infected. If the disease is untreated, the cornea softens and perforates, resulting in total blindness (keratomalacia). Worldwide, 50,000 to 100,000 children become blind each year from this disease.

C. The family of B vitamins, so grouped because the original B vitamin was subsequently found to consist of several distinct compounds, tends to be found together in nature. Almost all animals, vertebrate and invertebrate, require the B vitamins; they are "universal" vitamins. The dietary need for vitamin C and the fat-soluble vitamins A, D, E, and K tends to be restricted to the vertebrates, although some are required by certain invertebrates. Even within groups of close relationship, vitamin requirements are relative, not absolute. A rabbit does not require vitamin C, but guinea pigs and humans do. Some songbirds require vitamin A, but others do not.

SUMMARY

Autotrophic organisms (mostly green plants), using inorganic compounds as raw materials, capture the energy of sunlight through photosynthesis and produce complex organic molecules. Heterotrophic organisms (bacteria, fungi, and animals) use the organic compounds synthesized by plants, and the chemical bond energy stored therein, for their own nutritional and energy needs.

A large group of animals of very different levels of complexity feed by filtering out minute organisms and other particulate matter from the water. Others, known as deposit feeders, feed on organic detritus deposited in the substrate. Selective feeders, on the other hand, have evolved mechanisms for manipulating larger food masses, including various devices for seizing, scraping, boring, tearing, biting, and chewing. Fluid feeding is characteristic of endoparasites, which may absorb food across the general body surface, and of ectoparasites, herbivores, and predators that have developed specialized mouthparts for piercing and sucking.

Digestion is the process of breaking down food mechanically and chemically into molecular subunits for absorption. Digestion is intracellular in the protozoa and sponges. In more complex metazoans it is supplemented, and finally replaced entirely, by extracellular digestion, which takes place in sequential stages in a tubular cavity, the alimentary canal. Food is received in the mouth and mixed with lubricating saliva, then passed down the esophagus to regions where the food may be stored (crop), or ground (gizzard), or acidified and subjected to early digestion (vertebrate stomach). Among vertebrates, most digestion occurs in the small intestine. Enzymes from the pancreas and membrane enzymes embedded in the intestinal mucosal cells hydrolyze proteins, carbohydrates, fats, nucleic acids, and various phosphate compounds. The liver secretes bile, which contains salts that emulsify fats. Once foodstuffs are digested, they are absorbed as molecular subunits (monosaccharides, amino acids, and fatty acids) into the blood or lymph vessels of the villi of the small intestine. The large intestine (colon) serves mainly to absorb water and minerals from the food wastes as they pass through. It also contains symbiotic bacteria that produce certain vitamins.

Most animals balance food intake with energy expenditure. Food intake is regulated primarily by a hunger center located in the hypothalamus. In mammals, should caloric intake exceed energy requirements, the excess calories normally are dissipated as heat in specialized brown fat tissue. A deficiency in this response is one cause of human obesity.

All animals require a balanced diet containing both fuels (mainly carbohydrates and lipids) and structural and functional components (proteins, minerals, and vitamins). For every multicellular animal, certain amino acids, lipids, vitamins, and minerals are "essential" dietary factors that cannot be produced by the animal's own synthetic machinery. Animal proteins are better-balanced sources of amino acids than are plant proteins, which tend to lack one or more essential amino acids. Undernourishment (marasmus) and protein malnourishment (kwashiorkor) are among the world's major health problems, afflicting millions of people.

Vitamins are simple organic compounds required for specific cellular functions. Most are associated with metabolic enzyme systems.

Review questions

1. Distinguish between the following pairs of terms: autotrophic and heterotrophic; phototrophic and chemotrophic; herbivores and carnivores; omnivores and insectivores; anabolic and catabolic.
2. Filter feeding is one of the most important methods of feeding among animals. Explain what its characteristics, advantages, and limitations are, and name three different groups of animals that are filter feeders.
3. An animal's feeding adaptations are an integral part of an animal's behavior and usually shape the appearance of the animal itself. Discuss the contrasting feeding adaptations of carnivores and herbivores.
4. Explain how food is propelled through the digestive tract.
5. Compare intracellular with extracellular digestion and explain the advantages of the latter over intracellular digestion.
6. What structural modifications vastly increase the internal surface area of the intestine, and why is this large surface area important?
7. Trace the digestion and final absorption of a carbohydrate (starch) in the vertebrate gut, naming the carbohydrate-splitting enzymes, where they are found, the breakdown products of starch digestion, and in what form they are finally absorbed.
8. As in question 7, trace the digestion and final absorption of a protein.
9. Explain how fats are emulsified, digested, and absorbed in the vertebrate gut.
10. Explain the phrase "diet-induced thermogenesis" and relate it to the problem of obesity in some people.
11. Name the basic classes of foods that serve mainly as (1) fuels and as (2) structural and functional components.
12. Explain what is meant by the term "essential nutrients."
13. Explain the difference between saturated and unsaturated lipids and comment on the current interest in these compounds as they relate to human health.
14. What is meant by "protein complementarity" among plant foods?
15. What is the Green Revolution? Comment on its promise and its weakness.
16. Define a vitamin. What are the water-soluble and the fat-soluble vitamins?

Selected references

See also general references for Part Three, p. 752.

Carr, D.E. 1971. The deadly feast of life. Garden City, N.Y., Doubleday & Co. *What and how animals eat, told with insight and wit.*

Doyle, J. 1985. Altered harvest: agriculture, genetics, and the fate of the world's food supply. New York, Viking Penguin Inc. *Examines the politics of the agricultural revolution and the environmental and biological costs of the American food production system.*

Jennings, J.B. 1973. Feeding, digestion and assimilation in animals, ed. 2. New York, St. Martin's Press, Inc. *A general, comparative approach. Excellent account of feeding mechanisms in animals.*

Kretchmer, N., and W. van B. Robertson (eds.). 1978. Human nutrition. Articles from Scientific American. New York, Scientific American, Inc.

Lloyd, L.W., B.E. McDonald, and E.B. Crampton. 1978. Fundamentals of nutrition, ed. 2. San Francisco, W.H. Freeman & Co. *Integrated approach to what nutrients are and how they are metabolized.*

Logue, A.W. 1986. The psychology of eating and drinking. New York, W.H. Freeman and Co. *Experimental basis for human eating and drinking behavior, written especially for those with eating and drinking problems.*

Magee, D.F. and A.F. Dalley, II. 1986. Digestion and the structure and function of the gut. Basel (Switzerland), S. Karger AG. *Comprehensive treatment of mammalian (mostly human) digestion.*

Moog, F. 1981. The lining of the small intestine. Sci. Am. **245:**154-176 (Nov.). *Describes how mucosal cells actively process foods.*

CHAPTER 32

NERVOUS COORDINATION

Nervous System and Sense Organs

The nervous system originated in a fundamental property of protoplasm: irritability. Each cell responds to stimulation in a manner characteristic of that type of cell. But certain cells have become highly specialized for receiving stimuli and for conducting impulses to various parts of the body. Through evolutionary changes, these cells have become organized into a vast communications network, and the most complex of all body systems. Indeed, the human brain is unparalleled in complexity among structures known to humans. The endocrine system is also part of the body's communication network and interacts continuously with the nervous system in the control of body function. Functionally, the nervous system differs from the endocrine system in its capacity to monitor external as well as internal changes and to respond immediately to such changes. Nervous responses are measured in milliseconds, whereas the fastest endocrine responses are measured in seconds.

The evolution of the nervous system has been correlated with the development of bilateral symmetry and cephalization. Along with this development, animals acquired exteroceptors and associated ganglia. The basic plan of the nervous system is to code sensory information and transmit it to regions of the central nervous system, where it is processed into appropriate action. This action may be any of several types, such as simple reflexes, automatic behavior patterns, conscious perception, or learning processes.

In this chapter we will examine neurons in some detail since they are the central functional elements of nervous systems. Then we will explore the evolution of nervous systems, focusing at some length on nervous integration in the vertebrates. The final section of the chapter is devoted to the senses.

——THE NEURON: FUNCTIONAL UNIT OF THE NERVOUS SYSTEM

The neuron is a cell body with all its processes. Neurons assume many shapes, depending on their function and location; a typical kind is shown diagrammatically in Figure 32-1. From the nucleated body extend **processes** of two types. All but the simplest nerve cells have one or more cytoplasmic **dendrites.** As the name dendrite suggests (Gr. *dendron,* tree), these are often profusely branched. They are the nerve cell's receptive apparatus, often receiving information from several different sources at once. Some of these inputs are excitatory, others are inhibitory.

From the nucleated cell body extends a single **axon** (Gr. *axon,* axle), often a long fiber (they can be meters in length in a large mammal), relatively uniform in diameter, that typically carries impulses away from the cell body. In vertebrates and some higher invertebrates, the axon is usually covered with an insulating sheath.

Neurons are commonly classified as **afferent,** or sensory; **efferent,** or motor; and **interneurons,** which are neither sensory or motor, but connect neurons with other neurons. Afferent and efferent neurons lie mostly outside the central nervous system (brain and nerve cord), whereas interneurons, which make up 99% of all nerve cells in the human body, lie entirely within the central nervous system. Afferent neurons are connected to **receptors,** which convert some environmental stimulus into nerve impulses. The impulses are carried by the afferent neurons into the central nervous system. Here the impulses may be perceived as conscious sensation. Impulses also move to efferent neurons, which carry them out by the peripheral system to **effectors,** such as muscles or glands.

In higher vertebrates, nerve processes (usually axons) are usually bundled together in a well-formed wrapping of connective tissue to form a **nerve** (Figure 32-2). The cell bodies of these nerve processes are located either somewhere in the central nervous system or in ganglia, which are discrete bundles of nerve cell bodies located outside the central nervous system.

Surrounding the neurons are nonnervous neuroglial cells (or simply "glial" cells) that have a special relationship to the nerve cells. Glial cells are extremely numerous in the vertebrate brain, where they outnumber nerve cells 10 to 1 and may make up almost half the volume of the brain. Some glial cells form intimate insulating sheaths of lipid-containing **myelin** around nerve fibers. Vertebrate peripheral nerves are often enclosed by myelin laid down in concentric rings by special glial cells called **Schwann cells** (Figure 32-3). These form a **myelin sheath** that insulates the axon and greatly facilitates the transmission of impulses in these fibers, as will be discussed later. The functional roles of other glial cells remain to be clarified. Since many glial cells lie between blood vessels and nerve cells, they may nourish the nerve cells metabolically. It is certain that glial cells serve in the regenerative processes that follow injury and disease.

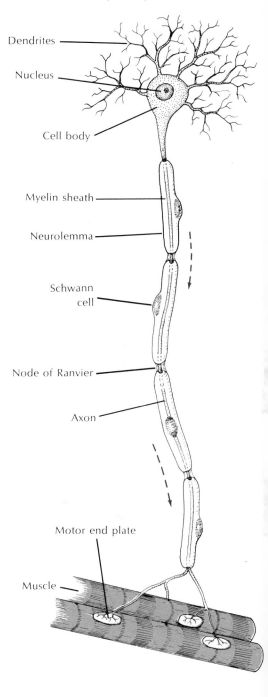

Figure 32-1

Structure of a motor (efferent) neuron.

Dendrites

Nucleus

Cell body

Myelin sheath

Neurolemma

Schwann cell

Node of Ranvier

Axon

Motor end plate

Muscle

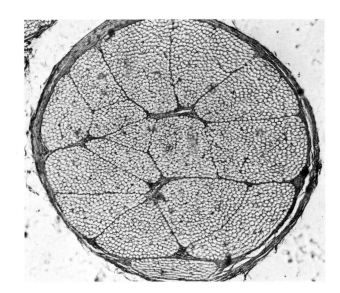

Figure 32-2

Cross section of nerve showing cut ends of nerve processes (*small white circles*). Such a trunk may contain thousands of both afferent and efferent fibers.

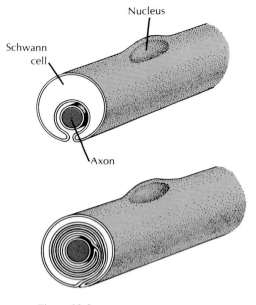

Figure 32-3

Development of the myelin sheath. The Schwann cell grows around the axon, then rotates around it, enclosing the axon in a tight, multilayered, insulating myelin sheath.

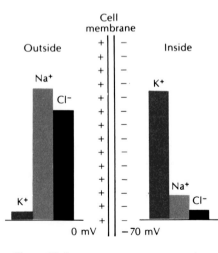

Figure 32-4

Ionic composition inside and outside a resting nerve cell. An active sodium pump located in the cell membrane drives sodium to the outside, keeping its concentration low inside. Potassium concentration is high inside, and although the membrane is "leaky" to potassium, this ion is held inside by the repelling positive charge outside the membrane.

Nature of the Nerve Impulse

The nerve impulse is the chemical-electrical message of nerves, the common functional denominator of all nervous system activity. Despite the incredible complexity of the nervous system of advanced animals, nerve impulses are basically alike in all nerves and in all animals. It is an "all-or-none" phenomenon; either the fiber is conducting an impulse, or it is not. Because all impulses are alike, the only way a nerve fiber can vary its effect on the tissue it innervates is by changing the frequency of impulse conduction. Frequency change is the language of a nerve fiber. A fiber may conduct no impulses at all or very few per second up to a maximum approaching 1000 per second. The higher the frequency (or rate) of conduction, the greater is the level of excitation.

The resting potential

To understand what happens when an impulse is conducted down a fiber, we need to know something about the resting, undisturbed fiber. Nerve cell membranes, like all cell membranes, have special permeability properties that create ionic imbalances. The interstitial fluid surrounding nerve cells contains relatively high concentrations of sodium (Na^+) and chloride (Cl^-) ions, but a low concentration of potassium ions (K^+). Inside the neuron, the ratio is reversed: the K^+ concentration is high but the Na^+ and Cl^- concentrations are low (Figure 32-4; see also Figure 29-1, *B*, p. 612). These differences are pronounced; there is approximately 10 times more Na^+ outside than in and 25 to 30 times more K^+ inside than out.

When at rest, the nerve cell membrane is selectively permeable to K^+, which can pass through the membrane by way of passive ion-specific channels. The permeability to Na^+ and Cl^- is nearly zero because these channels are closed in the resting membrane. K^+ tends to diffuse outward through the membrane, following the concentration gradient. But since Cl^- cannot follow, each K^+ that leaves the axon gives a positive charge to the outside of the membrane. The continued diffusion outward creates a charged membrane that is positive outside and negative inside. The positive charge outside quickly reaches a level that prevents any more K^+ from diffusing out of the axon. Now the resting membrane is at equilibrium, with a **resting membrane potential** that exactly balances the concentration gradient that forces K^+ out. The resting potential is usually -70 mv (millivolts), with the inside of the membrane negative to the outside.

The action potential

The nerve impulse is a rapidly moving change in electrical potential called the **action potential** (Figure 32-5). It is a very rapid and brief depolarization of the nerve fiber membrane; in fact, not only is the resting potential abolished, but in most nerve fibers the potential actually reverses for an instant so that the outside becomes negative as compared to the inside. Then, as the action potential moves ahead, the membrane returns to its normal resting potential ready to conduct another impulse. The entire event occupies approximately a millisecond. Perhaps the most significant property of the nerve impulse is that it is self-propagating; once started, the impulse moves ahead automatically, much like the burning of a fuse.

What causes the reversal of polarity in the cell membrane during passage of an action potential? We have seen that the resting potential depends on the high membrane permeability (leakiness) to K^+, some 50 to 70 times greater than the permeability to Na^+. When the action potential arrives at a given point, Na^+ channels suddenly open, permitting a flood of Na^+ to diffuse into the axon from the outside. Actually only a very minute amount of Na^+ moves across the membrane—less than one millionth of the Na^+ outside—but this sudden rush of positive ions wipes out the local membrane resting potential. The membrane is **depolarized** and an electrical "hole" is created. Potassium ions, finding their electrical barrier gone, begin to move out. Then, as the action potential passes on, the membrane quickly regains its resting properties. It becomes once again practically impermeable to Na^+, and the outward movement of K^+ is checked.

The rising phase of the action potential is associated with the rapid influx (inward movement) of Na^+ (Figure 32-5). When the action potential reaches its peak, the Na^+ permeability is restored to normal, and K^+ permeability briefly increases above the resting level. This causes the action potential to decrease rapidly toward the resting membrane level.

The sodium pump

The resting cell membrane has a very low permeability to sodium ions. Nevertheless some Na^+ leaks across, even in the resting condition. When the axon is active, Na^+ flows inward with each passing impulse, and, although the amount is very small, it is obvious that the ionic gradient would eventually disappear if the Na^+ was not moved back out again. This is accomplished by the sodium pump, a complex of protein subunits embedded in the plasma membrane of the axon. Each Na^+ pump uses the energy stored in ATP to transport Na^+ from the inside to the outside of the membrane. The name of the pump is something of a misnomer, since, as in several other cell membranes, it carries K^+ in as it pumps Na^+ out. The exchange of Na^+ for K^+ is not equal because three sodium ions on the inside are pumped out for every two potassium ions pumped in. An average neuron contains 100 to 200 sodium ion pumps per square micrometer of membrane surface. When working near capacity, each pump can transport about 200 sodium ions and 130 potassium ions per second (the actual rate is adjusted to meet the needs of the cell). Since more sodium ions are moved out of the axon than potassium ions are moved in, the sodium ion pump produces a net outward movement of positive charges. This of course helps to maintain the transmembrane polarity, which is positive outside.

_____ High-Speed Conduction

Although the ionic and electrical events associated with action potentials are much the same throughout the animal kingdom, this is not the case for the speed at which action potentials move down nerve axons. Conduction velocities vary enormously from nerve to nerve and from animal to animal—from as slow as 0.1 m/sec in sea anemones to as fast as 120 m/sec in some mammalian motor axons. In most invertebrates, speed of conduction is closely related to the diameter of the axon. Small axons conduct slowly because internal resistance to current flow is high. Where fast conduction velocities are important for a quick response, such as in locomotion to capture prey or to avoid capture, axon diameters are larger. The giant axon of the squid is nearly 1 mm in diameter and carries impulses 10 times faster than ordinary fibers in the same animal. The squid's giant axon innervates the animal's mantle musculature and is used for powerful mantle contractions when the animal swims by jet propulsion. Similar giant axons enable earthworms,

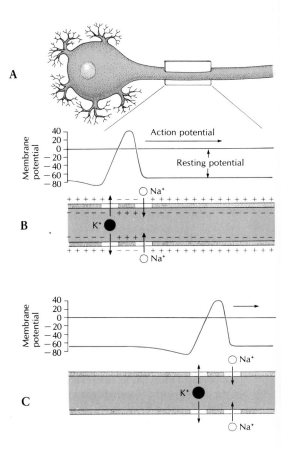

Figure 32-5

Conduction of the action potential of a nerve impulse. The impulse originates in cell body of the neuron (**A**) and moves toward the right. **B** and **C** show the electrical event and associated changes in localized membrane permeability to sodium and potassium. The position of the action potential in **C** is shown about 4 milliseconds after **B**. When the impulse arrives at a point, sodium gates are opened, allowing sodium ions to rush in. Sodium inflow reverses the membrane polarity, making the inner surface of the axon positive and the outside negative. Sodium gates then close and potassium gates open. Potassium ions can now penetrate the membrane and restore the normal resting potential. (Although the impulse is moving from left to right, it may be easier to follow the time course of events by reading the graphs from right to left.)

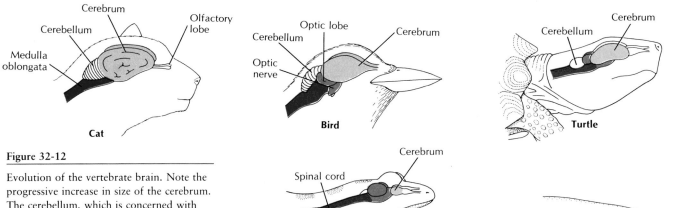

Figure 32-12

Evolution of the vertebrate brain. Note the progressive increase in size of the cerebrum. The cerebellum, which is concerned with equilibrium and motor coordination, is largest in animals whose balance and precise motor movements are well developed (fishes, birds, and mammals).

the best. This "great ravelled knot," as the British physiologist Sir Charles Sherrington called the human brain, in fact may be so complex that it will never be able to understand its own function.

The primitive three-part brain is made up of prosencephalon, mesencephalon, and rhombencephalon (forebrain, midbrain, and hindbrain) (Figure 32-13; Table 32-1). The prosencephalon and rhombencephalon each divide again to form the five-part brain characteristic of the adults of all vertebrates. The five-part brain includes the telencephalon, diencephalon, mesencephalon, metencephalon, and myelencephalon. From these divisions the different functional brain structures arise.

The impressive evolutionary improvement of the vertebrate brain has accompanied the increased powers of locomotion and greater environmental awareness of the more advanced vertebrates. In the primitive vertebrate brain, each of the three parts was concerned with one or more special senses: the prosencephalon with the sense of smell, the mesencephalon with vision, and the rhombencephalon with hearing and balance. These primitive but very fundamental concerns of the brain have been in some instances amplified and in others reduced or overshadowed during continued evolution as sensory priorities were shaped by the animal's habitat and way of life.

Figure 32-13

Brain of a fish, showing the basic organization of the vertebrate brain.

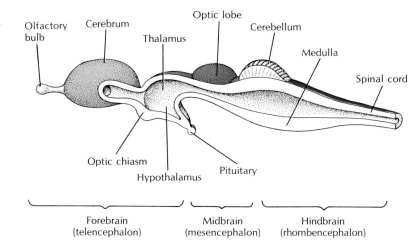

Table 32-1 Divisions of the Vertebrate Brain

Embryonic vesicle	Main component in adults
Prosencephalon (forebrain)	
Telencephalon	Olfactory bulb and tracts
	Cerebral cortex
	Basal ganglia
	Limbic system
Diencephalon	Thalamus
	Hypothalamus
	Pituitary gland
Mesencephalon (midbrain)	
Mesencephalon	Tectum
	Tegmentum
Rhombencephalon (hindbrain)	
Metencephalon	Cerebellum
	Pons
Myelencephalon	Medulla

The brain is made up of both white and gray matter, with the gray matter on the outside (in contrast to the spinal cord in which the gray matter is inside). The gray matter of the brain is mostly in the convoluted **cortex.** In the deeper white matter of the brain, myelinated bundles of nerve fibers connect the cortex with lower centers of the brain and spinal cord or connect one part of the cortex with another. Also in deeper portions of the brain are clusters of nerve cell bodies (gray matter) that provide synaptic junctions between the neurons of higher (cortical) centers and those of lower centers.

The **medulla,** the most posterior division of the brain, is really a conical continuation of the spinal cord. The medulla, together with the more anterior midbrain, constitutes the "brain stem," an area in which numerous vital and largely subconscious activities are controlled, such as heartbeat, respiration, vasomotor tone, and swallowing. The brain stem contains the nuclei of all the cranial nerves except the first two (olfactory and optic) and is traversed by many sensory and motor fiber tracts. Although it is small in size and largely hidden by the much enlarged "higher" centers, it is actually the most vital brain area. Damage to higher centers may result in severely debilitating loss of sensory or motor function or of higher mental processes (e.g., learning, thinking, memory), but damage to the brain stem is usually fatal.

The **pons,** between the medulla and the midbrain, is made up of a thick bundle of fibers that carry impulses from one side of the cerebellum to the other.

The **cerebellum,** lying above the medulla, is concerned with equilibrium, posture, and movement (Figure 32-14). Its development is directly correlated with the animal's mode of locomotion, agility of limb movement, and balance. It is usually weakly developed in amphibians and reptiles, which are relatively clumsy forms that stay close to the ground, and is well developed in the more agile bony fishes. It reaches its apogee in birds and mammals, in which it is greatly expanded and folded. The cerebellum does not initiate movements but operates as a precision error-control center, or servomechanism, that programs a movement initiated somewhere else, such as in the motor cortex. Primates and especially humans,

In neurophysiological usage a *nucleus* is a small aggregation of nerve cell bodies within the central nervous system.

Fron
lob

A

Te

Hemispheric specialization has long been considered a unique human trait, but was recently discovered in the brains of songbirds in which one side of the brain is specialized for song production.

Hemispheric specialization

The cerebral cortex is incompletely divided into two hemispheres by a deep longitudinal fissure. The right and left hemispheres are bridged through the **corpus callosum,** a fiber tract lying between and connecting the hemispheres. Through the corpus callosum the two hemispheres are able to transfer information and coordinate mental activites.

Until recently it was believed that one hemisphere, almost always the left, becomes functionally dominant over the other during childhood. This concept is now recognized as misleading. It is now known that the left and right brain hemispheres are specialized for entirely different functions: the left brain hemisphere (controlling the right side of the body) for language development, mathematical and learning capabilities, and sequential thought processes; and the right brain hemisphere (controlling the left side of the body) for spatial, musical, artistic, intuitive, and perceptual activities. It has long been known that even extensive damage to the right hemisphere may cause varying degrees of left-sided paralysis but has little effect on intellect. Conversely, damage to the left hemisphere usually has disastrous effects on intellect. Since these differences in brain symmetry and function exist at birth, they appear to be inborn rather than the result of developmental or environmental effects as previously believed.

Fluid circulation in the brain

The nerve and glial cells of the brain and spinal cord are surrounded by a clear, colorless, **cerebrospinal fluid.** The brain actually floats in the cerebrospinal fluid that circulates around it (Figure 32-16) and cushions it from shocks. The cerebrospinal fluid is not a filtrate of the blood plasma that bathes all the other cells of the body, but it is a special fluid that is *secreted* by capillary networks in the brain. This means that many native and foreign substances in the blood cannot enter the cerebrospinal fluid and therefore cannot reach the nerve and glial cells, even though these same substances can diffuse easily from the blood into other body tissues. There is, in effect, a **blood-brain barrier** between cerebral blood vessels and the surrounding cerebrospinal fluid that discriminates against certain materials in the blood. In this way the brain keeps its environment more closely controlled and protected than do other body organs.

The peripheral nervous system

The peripheral nervous system consists of nerve cells or extensions of nerve cells that lie *outside* the central nervous system. It is a communications system for the conduction of sensory and motor information between the brain and all parts of the body. The peripheral nervous system can be broadly subdivided into afferent and efferent components. As shown in the following outline, the efferent system is much more complex, consisting of a somatic nervous system and an autonomic nervous system.

    A. Afferent system (sensory)
    B. Efferent system (motor)
       1. Somatic nervous system
       2. Autonomic nervous system
          a. Sympathetic nervous system
          b. Parasympathetic nervous system

Afferent system

Afferent (sensory) neurons carry signals from receptors in the periphery of the body to the central nervous system. The afferent neuron of a reflex arc (Figure

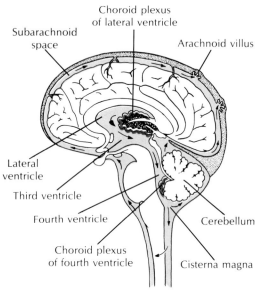

Figure 32-16

Circulation of cerebrospinal fluid (CSF). CSF is secreted by the choroid plexus and by capillaries in the brain. It circulates through all spaces and is drained into the venous system through the arachnoid villi.

32-11) is representative of all afferent pathways in the peripheral nervous system. One long nerve process extends from the cell body in the dorsal root ganglion just outside the spinal cord to innervate receptors; another process passes from the cell body into the central nervous system where it connects with other neurons.

Efferent system
Somatic nervous system

Nerve fibers of the somatic division of the peripheral nervous system pass from the brain or the spinal cord to skeletal muscle fibers. These are called motor neurons because they control muscle movement. They release acetylcholine as the transmitter substance at the nerve endings. Motor neurons in the reflex arc (Figure 32-11) are representative of the functional position of the somatic nervous system. The somatic nervous system is a "voluntary" system because the movement of skeletal muscles is normally under cerebral control.

Autonomic nervous system

The autonomic nerves govern the involuntary functions of the body that are not consciously affected. The cerebrum has no direct control over autonomic nerves as it has over the somatic nervous system. Thus, we cannot by volition stimulate or inhibit their action, although some people (for example, yoga adepts) manage to control such involuntary processes as heart rate, and others may be conditioned to control involuntary responses by biofeedback techniques (instrumental conditioning). Autonomic nerves control the movements of the alimentary canal and heart, the contraction of the smooth muscle of the blood vessels, urinary bladder, iris of eye, and others, and the secretions of various glands.

Autonomic nerves originate in the brain or spinal cord as do the nerves of the somatic nervous system, but unlike the latter, the autonomic fibers synapse once after leaving the cord and before arriving at the effector organ. These synapses are located outside the spinal cord in clusters of cell bodies called ganglia. Fibers passing from the cord to the ganglia are called preganglionic autonomic fibers; those passing from the ganglia to the effector organs are called postganglionic fibers. Thus, the autonomic is a *two-neuron* efferent system.

Subdivisions of the autonomic system are the **parasympathetic** and the **sympathetic** systems. Most organs in the body are innervated by both sympathetic and parasympathetic fibers, and their actions are antagonistic (Figure 32-17). If one fiber speeds up an activity, the other slows it down. However, neither kind of nerve is exclusively excitatory or inhibitory. For example, parasympathetic fibers inhibit heartbeat but excite peristaltic movements of the intestine; sympathetic fibers increase heartbeat but slow down peristaltic movement.

The **parasympathetic** system consists of motor nerves, some that emerge from the brain stem by certain cranial nerves and others that emerge from the sacral (pelvic) region of the spinal cord by certain spinal nerves (Figure 32-17). Parasympathetic fibers *excite* the stomach and intestine, muscles of the urinary bladder wall, bronchi, constrictor of the iris, salivary glands, and coronary arteries. They *inhibit* the heart, and sphincters of the intestine and the urinary bladder (allowing them to relax).

In the **sympathetic** division the nerve cell bodies of the preganglionic fibers are located in the thoracic and upper lumbar areas of the spinal cord. The fibers pass out through the ventral roots of the spinal nerves, separate from these, and go to the sympathetic ganglia, which are paired and form a chain on each side of the spinal column. From these ganglia some of the postganglionic fibers run through spinal nerves to the limbs and body wall, where they innervate the blood vessels of the skin, the smooth muscles of the hair follicles, the sweat glands, and so on.

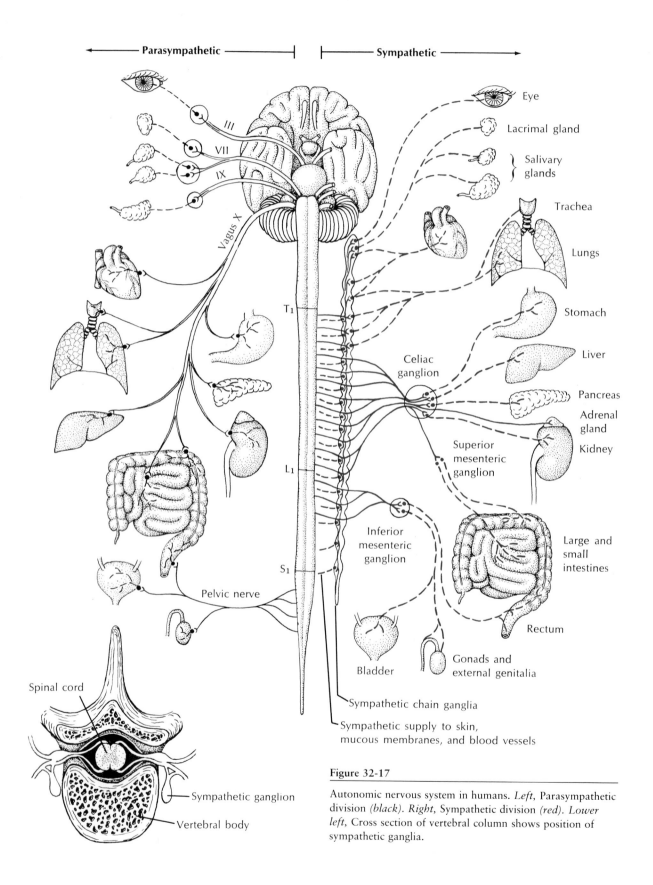

<- Parasympathetic -| |- Sympathetic ->

III

VII

IX

Vagus X

Eye

Lacrimal gland

Salivary glands

Trachea

Lungs

Stomach

Liver

Celiac ganglion

Pancreas

Adrenal gland

Kidney

Superior mesenteric ganglion

T₁

L₁

S₁

Inferior mesenteric ganglion

Large and small intestines

Pelvic nerve

Rectum

Bladder

Gonads and external genitalia

Sympathetic chain ganglia

Sympathetic supply to skin, mucous membranes, and blood vessels

Spinal cord

Sympathetic ganglion

Vertebral body

Figure 32-17

Autonomic nervous system in humans. *Left*, Parasympathetic division *(black)*. *Right*, Sympathetic division *(red)*. *Lower left*, Cross section of vertebral column shows position of sympathetic ganglia.

Other fibers run to the abdominal organs as the splanchnic nerves. Sympathetic fibers *excite* the heart, blood vessels, sphincters of the intestines and the urinary bladder, dilator muscles of the iris, and others. They *inhibit* the stomach, intestine, bronchial muscles, and coronary arterioles. The importance of these responses in emergency reactions (fight, flight, fear, and rage) will be described in the next chapter (p. 726).

All preganglionic fibers, whether sympathetic or parasympathetic, release **acetylcholine** at the synapse with the postganglionic cells. The terminations of the parasympathetic and sympathetic nervous systems release different types of chemical transmitter substances. The parasympathetic postganglionic fibers release **acetylcholine** at their endings, whereas the sympathetic fibers release **norepinephrine** (also called noradrenaline). This difference is another important characteristic distinguishing the two parts of the autonomic nervous system.

SENSE ORGANS

Animals require a constant inflow of information from the environment to regulate their lives. Sense organs are specialized receptors designed for detecting environmental status and change. An animal's sense organs are its first level of environmental perception; they are data input channels for the brain.

A **stimulus** is some form of energy—electrical, mechanical, chemical, or radiant. The sense organ transforms the energy form of the stimulus it receives into nerve impulses, the common language of the nervous system. In a very real way, sense organs are biological transducers. A microphone, for example, is a transducer that converts mechanical (sound) energy into electrical energy. Like the microphone that is sensitive only to sound, sense organs are, as a rule, specific for one kind, or **modality,** of stimulus energy. Eyes respond only to light, ears to sound, pressure receptors to pressure, and chemoreceptors to chemical molecules. All of these different forms of energy are converted into nerve impulses.

Since all nerve impulses are qualitatively alike, how do animals perceive and distinguish the different **sensations** of varying stimuli? The answer is that the real perception of sensation is done in localized regions of the brain, where each sense organ has its own hookup. Impulses arriving at a particular sensory area of the brain can be interpreted in only one way. This is why pressure on the eyeball causes us to see "stars" or other visual patterns; the mechanical distortion of the eye initiates impulses in the optic nerve fibers that are perceived as light sensations. Although such an operation probably could never be done, a deliberate surgical switching of optic and auditory nerves would cause the recipient to literally see thunder and hear lightning!

Classification of Receptors

Receptors are traditionally classified on the basis of their location. Those near the external surface, called **exteroceptors,** keep the animal informed about the external environment. Internal parts of the body are provided with **interoceptors,** which pick up stimuli from the internal organs. Muscles, tendons, and joints have **proprioceptors,** which are sensitive to changes in the tension of muscles and provide the organism with a sense of body position.

Another way of classifying receptors is based on the form of energy to which the receptors respond, such as **chemical, mechanical, photo,** or **thermal.**

Figure 32-18

Pheromone-producing glands of an ant. After E.O. Wilson and W.M. Bossert, Recent Progress in Hormone Research, 19:673-716, 1963.

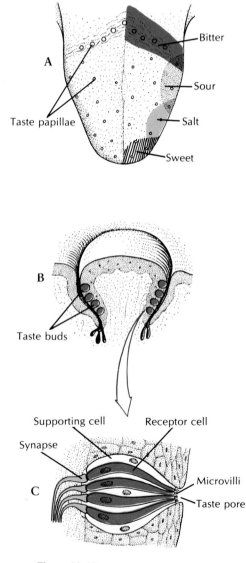

Figure 32-19

Taste receptors. **A,** Surface of human tongue showing regions of maximum sensitivity to the four primary taste sensations. **B,** Position of taste buds on a taste papilla. **C,** Structure of a taste bud.

Chemoreception

Chemoreception is the most primitive and most universal sense in the animal kingdom. It probably guides the behavior of animals more than any other sense. The most primitive animals, protozoa, use **contact chemical receptors** to locate food and adequately oxygenated water and to avoid harmful substances. These receptors elicit an orientation behavior toward or away from the chemical source called **chemotaxis.** More advanced animals have specialized **distance chemical receptors.** These are often developed to a truly amazing degree of sensitivity. Distance chemoreception, usually referred to as the sense of smell or olfactory sense, guides feeding behavior, location and selection of sexual mates, territorial and trail marking, and alarm reactions of numerous animals.

The social insects produce species-specific compounds, called **pheromones,** which constitute a highly developed chemical language. Pheromones are organic compounds released by an animal that affect the physiology or behavior of another individual. Ants, for example, are walking batteries of glands (Figure 32-18) that produce numerous chemical signals. These include releaser pheromones, such as alarm and trail pheromones, and primer pheromones that alter the endocrine and reproductive systems of different castes in the colony. Insects have a variety of chemoreceptors on the body surface for sensing specific pheromones, as well as other, nonspecific odors.

In all vertebrates and insects the senses of **taste** and **smell** are clearly distinguishable. Although there are similarities between taste and smell receptors, in general the sense of taste is more restricted in response and is less sensitive than the sense of smell. Taste and smell centers are also located in different parts of the brain.

In higher forms, **taste buds** are found in the mouth cavity and especially on the tongue (Figure 32-19) where they provide a means for judging foods before they are swallowed. A taste bud consists of a cluster of several receptor cells surrounded by supporting cells and is provided with a small external pore through which the slender tips of the sensory cells project. Because they are subject to the wear and tear of abrasive and spicy foods, taste buds have a short life of only about 5 days and are continually being replaced.

The four basic taste sensations possessed by humans—sour, salt, bitter, and sweet—are each attributable to a different kind of taste bud. The tastes for salt and sweet are found mainly at the tip of the tongue, bitter at the base of the tongue, and sour along the sides of the tongue. Of these, the bitter taste is by far the most sensitive, since it serves as an early warning system against potentially dangerous foods, many of which are bitter.

The sense of smell is more complex than taste, and the basic nature of odor reception is still unknown. The olfactory endings are located in a special epithelium covered with a sheet of mucus, positioned deep in the nasal cavity (Figure

32-20). There are perhaps 20 million olfactory receptors in the human nose, each ending in several projecting, cilia-like filaments. It is believed that the terminal cilia are the receptor sites, but they are so small that it has been difficult to obtain electrophysiological recordings from single cells.

Even humans, a species not renowned for detecting smells, can discriminate perhaps 10,000 different odors. The human nose can detect one 25 millionth of 1 mg of mercaptan, the odoriferous principle of the skunk. This averages out to approximately 1 molecule per sensory ending.

Although our olfactory abilities have been overshadowed by other sense organs, vision especially, many animals rely on olfaction for their very survival. A dog explores new surroundings with its nose in the same way we do with our eyes. A dog's sense of smell is justifiably renowned; with some odorous sources a dog's nose is at least a million times as sensitive as ours.

Since flavor of food depends a great deal on odors reaching the olfactory epithelium through the throat passage, taste and smell are easily confused. All the various "tastes" other than the four basic ones (sweet, sour, bitter, salt) are really the result of the flavors reaching the sense of smell in this manner. Food loses its appeal during a common cold because a stuffy nose blocks off odors rising from the mouth cavity.

Smells are distinctive, identifiable, and memorable, and yet nearly impossible to describe to another person except in vague, nonspecific terms. This difficulty of defining odor qualities objectively is another reason why olfactory research has lagged far behind other areas of sensory physiology. Of the many hypotheses that have been advanced to explain olfaction, a favored one proposes that a scent molecule fits into a corresponding receptor site, much like a key into a lock. This interaction somehow alters membrane permeability and depolarizes the receptor cell, which triggers a nerve impulse.

Mechanoreception

Mechanoreceptors are sensitive to quantitative forces such as touch, pressure, stretching, sound, vibration, and gravity—in short, they respond to movement. Animals require a steady flow of information from mechanoreceptors to interact with their environment, feed themselves, maintain normal posture, and to walk, swim, or fly.

Touch and pain

The **pacinian corpuscle,** a relatively large mechanoreceptor that registers deep touch and pressure in mammalian skin, illustrates the general properties of mechanoreceptors. They are common in the deep layers of skin (Figure 28-1, *B,* p. 593), connective tissue surrounding muscles and tendons, and the abdominal mesenteries. Each corpuscle consists of a nerve terminus surrounded by a capsule of numerous, concentric, onionlike layers of connective tissue (Figure 32-21). Pressure at any point on the capsule distorts the nerve ending, producing a graded **receptor potential.** This is a local flow of electric current. Progressively stronger stimuli lead to correspondingly stronger **receptor** potentials until a **threshold current** is produced; this initiates an action potential in the sensory nerve fiber. Stronger stimuli will produce a burst of action potentials. However, if the pressure applied is sustained, the corpuscle quickly adjusts to the new shape and no longer responds. This is referred to as **adaptation** and is characteristic of many kinds of touch receptors, which are admirably suited to detecting a sudden mechanical

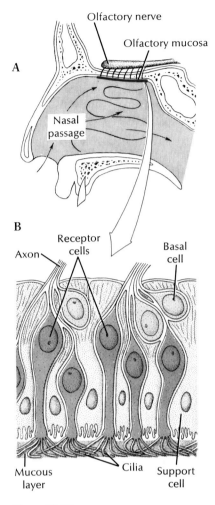

Figure 32-20

Human olfactory epithelium. **A,** The epithelium is a patch of tissue positioned in the roof of the nasal cavity. **B,** It is composed of supporting cells, basal cells, and olfactory receptor cells with cilia protruding from their free ends.

14. Give the meaning of the statement, "The idea that all sense organs behave as biological transducers is a uniting concept in sensory physiology."
15. Chemoreception in vertebrates and insects is mediated through the clearly distinguishable senses of taste and smell. Contrast these two senses in humans in terms of anatomical location and nature of the receptors, and sensitivity to chemical molecules.
16. Explain how the ultrasonic detectors of certain nocturnal moths are adapted to help them escape an approaching bat.
17. Outline the place theory of pitch discrimination as an explanation of the human ear's ability to distinguish between sounds of different frequencies.
18. Explain how the semicircular canals of the ear are designed to detect rotation of the head in any directional plane.
19. Prepare an unlabeled copy of Figure 32-31 and then, without reference to this figure, correctly label the following structures: sclera, choroid coat, retina, cornea, iris, pupil, lens, aqueous humor, vitreous humor, fovea, optic nerve, eye muscle.
20. Explain what happens when light strikes a dark-adapted rod that leads to the generation of a nerve impulse. What is the difference between rods and cones in their sensitivity to light?

Selected references

See also general references for Part Three, p. 752.

Bullock, T.H., R. Orkand, and A. Grinnell. 1977. Introduction to nervous systems. San Francisco, W.H. Freeman & Co. *Excellent comparative treatment.*

Dunant, Y. and M. Israel. 1985. The release of acetylcholine. Sci. Am. 252:58-66 (April). *Recent studies have altered prevailing views of the events at a synapse during impulse transmission.*

Hudspeth, A.J. 1983. The hair cells of the inner ear. Sci. Am. 248:54-64 (Jan.). *How these biological transducers work.*

Milne, L., and M. Milne. 1972. The senses of animals and men. New York, Antheneum Publishers. *Well written account of animal senses. Undergraduate level.*

Nathan, P. 1982. The nervous system, ed. 2. Oxford, Eng., Oxford University Press. *One of the best of several semipopular accounts of the nervous system.*

Parker, D.E. 1980. The vestibular apparatus. Sci. Am. 243: 118-135 (Nov.). *Functioning of the inner ear and other sense organs of balance and orientation.*

Snyder, S.H. 1985. The molecular basis of communication between cells. Sci. Am. 253:132-141 (Oct.). *Describes the different actions of neurotransmitters.*

Stevens, C.F. 1979. The neuron. Sci. Am. 241:54-65 (Sept.). *Its function is detailed.*

Stryker, L. 1987. The molecules of visual excitation. Sci. Am., 257:42-50 (July). *Describes the cascade of reactions following the absorption of light by a rod cell and leading to the generation of a nerve impulse.*

Werblin, F.S. 1973. The control of sensitivity in the retina. Sci. Am. 228:70-79 (Jan.). *Studies of neuron interactions in the retina help to explain its versatility over wide-ranging light conditions.*

CHAPTER   33

CHEMICAL COORDINATION

Endocrine System

─── HORMONAL INTEGRATION

The endocrine system is the second great integrative system controlling the body's activities. Endocrine glands secrete **hormones** (Gr. *hormon,* to excite), chemical compounds that are transported by the blood to some part of the body where they initiate definite physiological responses.

Endocrine glands are small, well-vascularized ductless glands composed of groups of cells arranged in cords or plates. Since the endocrine glands have no ducts, their only connection with the rest of the body is by the bloodstream; they must capture their raw materials from the blood and secrete their finished hormonal products into it. Consequently, it is not surprising that the endocrine glands receive enormous blood flows. The thyroid gland is said to have the highest blood flow per unit of tissue weight of any organ in the body. **Exocrine glands,** in contrast, are provided with ducts for discharging their secretions onto a free surface. Examples of exocrine glands are sweat glands and sebaceous glands of skin, salivary glands, and the various enzyme-secreting glands lining the walls of the stomach and intestine.

The classical definition of a hormone given in the first paragraph, like so many other generalizations in biology, may have to be altered as new information appears. For one thing, some hormones, such as certain neurosecretions, may never enter the general circulation at all. Furthermore, there is good evidence that many of the traditional hormones, such as insulin, are synthesized in minute amounts in a variety of nonendocrine tissues (nerve cells, for example) where they may function as local **tissue factors:** substances that stimulate cell growth or some biochemical process. Most hormones, however, are blood borne and therefore diffuse into every tissue space in the body. This is quite unlike the discrete action of the nervous system with its network of cablelike nerve fibers that selectively send messages to specific points.

The ubiquitous distribution of hormones makes it possible for certain hormones, such as the growth hormone of the pituitary gland, to affect most, if not all cells during specific stages of cellular differentiation. Other hormones, however, produce highly specific responses only in certain target cells and at certain times. But how is this specificity accomplished? Since hormones are distributed everywhere by the circulation, there must be some mechanism that limits hormone action to specific target cells. This is provided by **receptor molecules** located on or in the responding cells. A hormone will engage only those cells that display the receptor that, by virtue of its specific molecular shape, will bind with the hormone

Figure 33-1

Founders of endocrinology. **A,** Sir William M. Bayliss (1860-1924). **B,** Ernest H. Starling (1866-1927).

From Fulton, J.F., and L.G. Wilson. 1966. Selected readings in the history of physiology. Springfield, Ill., Charles C Thomas, Publisher.

molecule. Other cells are insensitive to the hormone's presence because they lack the specific receptors.

Compared with the nervous system, the endocrine is slow acting because of the time required for a hormone to reach the appropriate tissue, cross the capillary endothelium, and diffuse through tissue fluid to, and sometimes into, cells. Thus, the minimum response time is a matter of seconds and may be much longer. Furthermore, hormone responses in general are much longer lasting than those under nervous control. Where a sustained effect is required, as in many metabolic and growth processes, or where some concentration or secretion rate must be maintained at a particular level, we expect to find endocrine control.

However, the nervous and endocrine systems really function as a single, united system. There is no sharp separation between the two. As we shall see, the nervous system is itself an endocrine organ that controls most endocrine function. Conversely, several hormones act on the nervous system and may significantly affect many kinds of animal behavior.

Endocrinology is a comparatively young division of animal physiology. Its birthdate is usually given as 1902, the year two English physiologists, W.M. Bayliss and E.H. Starling, demonstrated the action of an internal secretion (Figure 33-1). They were interested in determining how the pancreas secreted its digestive juice into the small intestine at the proper time of the digestive process. In an anesthetized dog they tied off a section of the small intestine beyond the duodenum (the part of the intestine next to the stomach) and removed all nerves leading to this tied-off loop, but left its blood vessels intact. Bayliss and Starling found that the injection of hydrochloric acid into the blood serving this intestinal loop had no effect on the secretion of pancreatic juice, but when they introduced 0.4% hydrochloric acid directly inside the intestinal loop, a pronounced flow of pancreatic juice into the duodenum occurred through the pancreatic duct. When they scraped off some of the mucous membrane lining of the intestine and mixed it with acid, they found the injection of this extract into the blood caused an abundant flow of pancreatic juice.

They concluded that when the partly digested and slightly acid food from the stomach arrives in the small intestine, the hydrochloric acid reacts with something in the mucous lining to produce an internal secretion, or chemical messenger, which is conveyed by the bloodstream to the pancreas, causing it to secrete pancreatic digestive juices. They called this messenger **secretin.** In a 1905 Croonian lecture at the Royal College of Physicians, Starling first used the word "hormone," a general term to describe all such chemical messengers, since he correctly surmised that secretin was only the first of many hormones that remained to be described.

Mechanisms of Hormone Action

How do hormones exert their effects? Obviously, it is much easier to observe the physiological effect of a hormone than to determine what the hormone does to produce the effect. Recently, however, progress has been made toward understanding the specificity of hormones: it seems to depend on specific receptor sites on or in target cells.

Cell surface receptors and the second messenger concept

In many hormone actions, when the hormone arrives at its target cell, it binds to a receptor site on the membrane. The combination of hormone and receptor causes activation of an enzyme, adenylate cyclase, which is coupled to the membrane receptor (Figure 33-2). Adenylate cyclase converts ATP in the cytoplasm to cyclic AMP. This compound is formed when two of three phosphate groups are split from ATP and the remaining phosphate combines with a carbon atom of the adjacent ribose molecule of ATP to form a ring, making it "cyclic" (Figure 33-3). The cyclic AMP thus generated acts as a "second messenger" that relays the hormone's message to the cell's biochemical machinery, where it alters (usually stimulates) some cellular process (Figure 33-2). Since many molecules of cyclic AMP may be manufactured after a single hormone molecule has been bound, the message is amplified, perhaps many thousands of times.

Cyclic AMP mediates the actions of many hormones, including glucagon, epinephrine, adrenocorticotropic hormone (ACTH), thyrotropic hormone (TSH), melanophore-stimulating hormone (MSH), and vasopressin. With the exception of epinephrine, these are all peptides—small proteins but much too large to penetrate the cell membrane. All act *indirectly* through an immobile receptor on the cell surface.

Cytoplasmic receptors

Several hormones, including all of the steroids (for example, estrogen, testosterone, and aldosterone), diffuse into cells where they bind selectively to cytoplasmic receptor molecules found only in the target cells. The hormone-receptor complex can then diffuse into the nucleus where it binds directly to certain proteins (non-

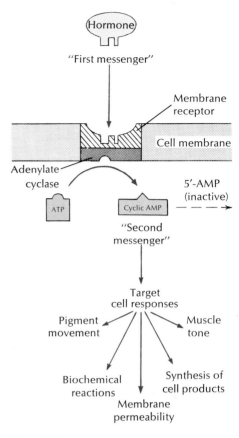

Figure 33-2

Second-messenger concept of hormone action. Many hormones act through cyclic AMP. The hormone is carried by the bloodstream from the endocrine gland to the target cell where it selectively combines with an immobile receptor in the plasma membrane. This interaction stimulates the enzyme adenylate cyclase to catalyze the formation of cyclic AMP (second messenger). Cyclic AMP acts intracellularly to initiate any of several changes, depending on the kind of cell.

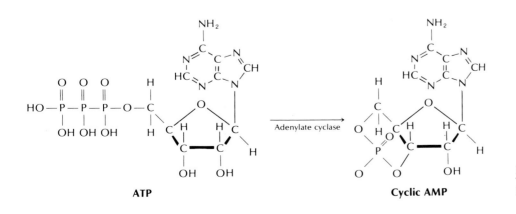

Figure 33-3

Formation of cyclic AMP.

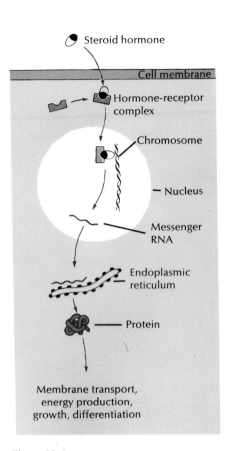

Figure 33-4

Concept of gene regulation by steroid hormones and thyroxine. The hormone penetrates the cell membrane to combine with a mobile cytoplasmic receptor. The complex enters the nucleus where it stimulates the transcription of messenger RNA. This is translated to specific proteins in the cytoplasm.

histones) in the chromosomes. As a result, gene transcription is increased, and messenger RNA molecules are synthesized on specific sequences of DNA. Moving from the nucleus into the cytoplasm, the newly formed messenger RNA initiates the formation of new proteins, thus setting in motion the hormone's observed effect (Figure 33-4). Thyroxine and the insect-molting hormone ecdysone are also believed to act through this mechanism.

As compared with hormones acting through the cyclic AMP mechanism, steroids have a more *direct* effect on protein synthesis because they combine with a mobile receptor and move with it into the nucleus to couple with chromosomal proteins.

All hormones are low-level signals. Even when an endocrine gland is secreting maximally, the hormone is so greatly diluted by the large volume of blood it enters that its plasma concentration seldom exceeds 10^{-9} M (or one billionth of a 1 M concentration). Since hormones have far-reaching and often powerful influences on cells, it is evident that their effects are vastly amplified at the cellular level.

How Hormone Secretion Rates Are Controlled

Hormones influence cell functions by altering the rates of a large range of biochemical processes. Some affect enzyme activity and thus alter cellular metabolism, some change membrane permeability, some regulate the synthesis of cellular proteins, and some stimulate the release of hormones from other endocrine glands. Since these are all dynamic processes that must adapt to changing metabolic demands, they must be regulated, not merely activated, by the appropriate hormones. This is achieved by varying hormone output from the gland. However, the concentration of hormone in the plasma depends on two factors: its rate of secretion and the rate at which it is inactivated and removed from the circulation. Consequently, an endocrine gland requires information about the level of its own hormone(s) in the plasma to control its secretion.

Many hormones, especially those of the pituitary gland, are controlled by negative feedback systems that operate between the glands secreting the hormones and the target cells. A feedback pattern is one in which the output is constantly compared with a set point. For example, ACTH, secreted by the pituitary, stimulates the adrenal gland (the target cells) to secrete cortisol. As the cortisol level in the plasma rises, it acts on the pituitary gland to inhibit the release of ACTH. Thus, any deviation from the set point (a specific plasma level of cortisol) leads to corrective action in the opposite direction. Such a system is highly effective in preventing extreme oscillations in hormone output. However, hormonal feedback systems are not identical to a rigid "closed-loop" system like the thermostat that controls the central heating system in a house, because they may be altered by input from the nervous system or by metabolites or other hormones.

INVERTEBRATE HORMONES

During the last half century physiologists have shown that many invertebrates have endocrine integrative systems that approach the complexity of the vertebrate endocrine system. Not surprisingly, however, there are few if any homologies between invertebrate and vertebrate hormones. The invertebrate phyla have different functional systems, growth patterns, and reproductive processes from vertebrates and have been separated from them phylogenetically for a vast span of time.

In many invertebrate phyla, the principal source of hormones is **neurosecretory cells,** specialized nerve cells capable of synthesizing and secreting hormones. Their products, called neurosecretions, or neurosecretory hormones, are discharged directly into the circulation. Neurosecretion is in fact important in vertebrates as well. It is an ancient physiological activity, and because it serves as a crucial link between the nervous and endocrine systems, we believe that hormones first evolved as nerve cell secretions. Later, nonnervous endocrine glands appeared, especially among the vertebrates, but remained chemically linked to the nervous system by the neurosecretory hormones.

Neurosecretory hormones have been found in all the larger invertebrate phyla, including the cnidarians, flatworms, nematodes, molluscs, annelids, arthropods, and echinoderms. Most extensively studied, however, have been the insects, and we will limit our discussion to that group.

In insects, as in other arthropods, growth is a series of steps in which the rigid, nonexpansible exoskeleton is periodically discarded and replaced with a new, larger one. Almost all insects undergo a process of metamorphosis (p. 372), in which there is a series of juvenile stages, each requiring the formation of a new exoskeleton, and each ending with a molt. In some orders the change to the adult form is gradual. In others the adult is separated from the larval stages by a quiescent form, the pupa, and the change to the adult is abrupt. Hormonal control of both types is the same.

Insect physiologists have discovered that molting and metamorphosis are controlled by the interaction of two hormones, one favoring growth and the differentiation of adult structures and the other favoring the retention of juvenile structures. These two hormones are the **molting hormone** (also referred to as **ecdysone** [ek′duh-sone]), produced by the prothoracic gland, and the **juvenile hormone,** produced by the corpora allata (Figure 33-5). The structure of both

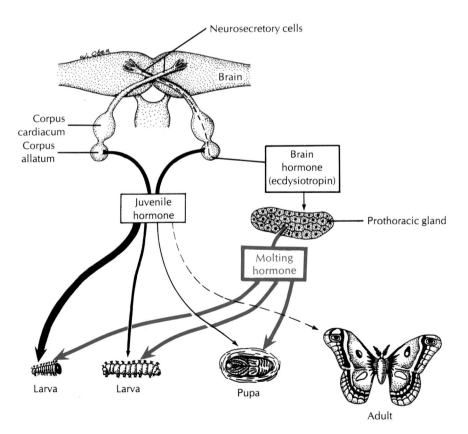

Larva   Larva   Pupa   Adult

Figure 33-5

Endocrine control of molting in a moth. Moths mate in the spring or summer, and eggs soon hatch into the first of several larval stages (called instars). After the final larval molt, the last and largest larva (caterpillar) spins a cocoon in which it pupates. The pupa overwinters, and an adult emerges in the spring to start a new generation. Two hormones interact to control molting and pupation. The molting hormone, produced by the prothoracic gland and stimulated by a separate brain hormone (ecdysiotropin), favors molting and the formation of adult structures. These effects are inhibited, however, by the juvenile hormone, produced by the corpora allata. Juvenile hormone output declines with successive molts, and the larva undergoes adult differentiation.

hormones has recently been determined. It required extraction from 1000 kg (about 1 ton) of silkworm pupae to show that the molting hormone is steroid. The juvenile hormone has an entirely different structure.

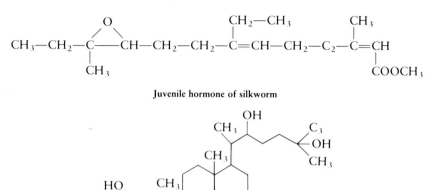

Juvenile hormone of silkworm

Molting hormone (α-ecdysone) of silkworm

The molting hormone is under the control of **ecdysiotropin.** This hormone is a polypeptide (molecular weight about 5000) that is produced by neurosecretory cells of the brain, transported down axons, and stored in the corpora allata, from where it is released into the blood. At intervals during juvenile growth, ecdysiotropin is released into the blood and stimulates the release of molting hormone. Molting hormone appears to act directly on the chromosomes to set in motion the changes resulting in a molt. The molting hormone favors the development of adult structures. It is held in check, however, by the juvenile hormone, which favors the development of juvenile characteristics. During juvenile life the juvenile hormone predominates and each molt yields another larger juvenile. Finally the output of juvenile hormone decreases and the final juvenile molt occurs.

Chemists have synthesized several potent analogs of the juvenile hormone, which hold great promise as insecticides. Minute quantities of these synthetic analogs induce abnormal final molts or prolong or block development. Unlike the usual chemical insecticides, they are highly specific and do not contaminate the environment.

___ VERTEBRATE ENDOCRINE GLANDS AND HORMONES

In the remainder of this chapter we describe some of the best understood and most important of the vertebrate hormones. The hormones of reproduction are discussed in the next chapter. Space does not permit us to deal with all the hormones and hormonelike substances that have been discovered. The mammalian hormonal mechanisms are the best understood, since laboratory mammals and humans have always been the objects of the most intensive research. Research with the lower vertebrates has shown that all vertebrates share similar endocrine organs. All vertebrates have a pituitary gland, for example, and all have thyroid glands, adrenal glands (or the special cells of which they are composed), and gonads. Nevertheless, there are some important differences in the functional roles that the hormones of these glands have among the different vertebrates.

The precise location of ecdysiotropin in the brain of pupal tobacco hornworms was revealed by N. Agui by delicate microdissection. Using a human eyebrow hair as a dissecting instrument, he was able to isolate the single cell in each brain hemisphere that contained ecdysiotropin activity. Thus, only two cells, each about 20 μm in diameter, produce this insect's total supply of ecdysiotropin. Agui also showed that the corpora allata are the storage-release sites for ecdysiotropin, not the corpora cardiaca as previously believed. In an age when sophisticated instrumentation has removed much of the tedium (and some of the creativity) from research, it is refreshing to learn that certain biological mysteries succumb only to skillful use of the human hand.

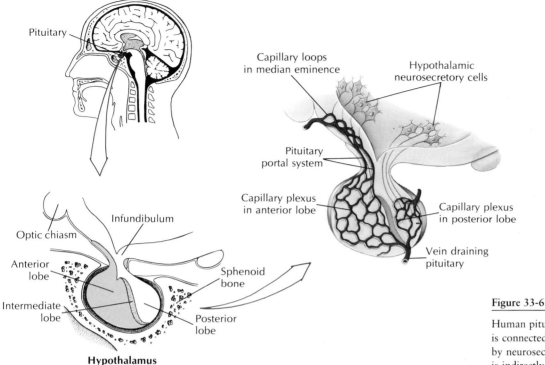

Figure 33-6

Human pituitary gland. The posterior lobe is connected directly to the hypothalamus by neurosecretory fibers. The anterior lobe is indirectly connected to the hypothalamus by a portal circulation beginning in the median eminence and ending in the anterior pituitary.

Hormones of the Pituitary Gland and Hypothalamus

The pituitary gland, or **hypophysis,** is a small gland (0.5 g in humans) lying in a well-protected position between the roof of the mouth and the floor of the brain (Figure 33-6). It is a two-part gland having a double embryological origin. The **anterior pituitary** (adenohypophysis) is derived embryologically from the roof of the mouth. The **posterior pituitary** (neurohypophysis) arises from a ventral portion of the brain, the **hypothalamus,** and is connected to it by a stalk, the **infundibulum.** Although the anterior pituitary lacks any *anatomical* connection to the brain, it is nonetheless *functionally* connected to it by a special portal circulatory system. A portal circulation is one that delivers blood from one capillary bed to another (Figure 33-6).

Anterior pituitary

The anterior pituitary consists of an **anterior lobe** (pars distalis) and an **intermediate lobe** (pars intermedia) as shown in Figure 33-6. The anterior lobe, despite its minute dimensions, produces at least six protein hormones. All but one of these six are **tropic hormones** that regulate other endocrine glands (Table 33-1).

The **thyrotropic hormone** (TSH) regulates the production of thyroid hormones by the thyroid gland. The **adrenocorticotropic hormone** (ACTH) stimulates the adrenal cortex. Two of the tropic hormones are commonly called **gonadotropins** because they act on the gonads (ovary of the female, testis of the male). These are the **follicle-stimulating hormone** (FSH) and the **luteinizing hormone** (LH) (in the male the luteinizing hormone goes by a different name, interstitial cell stimulating hormone [ICSH], but it is the same hormone chemically). A fifth tropic hormone is **prolactin,** which stimulates milk production in the female mammary glands and has a variety of other effects in the lower vertebrates. The functions of the two gonadotropins and prolactin are discussed in Chapter 35 in connection with the hormonal control of reproduction.

When pituitary deficiency is recognized early in life, the child can be treated with injections of human growth hormone and achieve normal stature. However, the hormone has been very expensive because it must be extracted from pituitaries of cadavers. In addition, growth hormone obtained in this manner occasionally contains the agent of Creutzfeldt-Jakob disease, which eventually causes dementia and death. Recently, the gene for human growth hormone has been inserted into bacteria by recombinant DNA techniques (p. 834), and the bacteria produce the hormone. By this means a cheaper source of growth hormone has become available, and patients are free of the danger of infection with a fearful disease agent.

Table 33-1 Hormones of the Vertebrate Pituitary and Hypothalamus: Chemical Nature and Actions

	Hormone	Chemical nature	Principal action
Adenohypophysis			
Anterior lobe	Thyrotropin (TSH)	Glycoprotein	Stimulates thyroid to secrete thyroid hormones
	Adrenocorticotropin (ACTH)	Polypeptide	Stimulates adrenal cortex to secrete steroid hormones
	Gonadotropins		
	1. Follicle-stimulating hormone (FSH)	Glycoprotein	Stimulates gamete production and secretion of sex hormones
	2. Luteinizing hormone (LH, ICSH)	Glycoprotein	Stimulates sex hormone secretion and ovulation
	Prolactin (LTH)	Protein	Stimulates mammary gland growth and secretion in mammals; various reproductive and nonreproductive functions in lower vertebrates
	Growth hormone (GH)	Protein	Stimulates growth
Intermediate lobe	Melanophore-stimulating hormone (MSH)	Polypeptide	Pigment dispersion in melanophores of ectotherms; function unclear in endotherms
Neurohypophysis			
Posterior lobe	Vasopressin (ADH)	Octapeptide	Antidiuretic effect on kidney
	Oxytocin	Octapeptide	Stimulates milk ejection and uterine contraction
	Vasotocin	Octapeptide	Antidiuretic activity
	Isotocin and three others in lower vertebrates	Octapeptides	Functions uncertain
Hypothalamus	Thyrotropin-releasing hormone (TRH)		
	Corticotropin-releasing hormone (CRH)		
	Follicle-stimulating hormone–releasing hormone (FSH-RH)		
	Luteinizing hormone–releasing hormone (LH-RH)		
	Prolactin release–inhibiting factor (PIF)	All polypeptides	Control release of anterior and intermediate lobe hormones
	Prolactin–releasing factor (PRF)		
	Melanophore-stimulating hormone–releasing factor (MRF)		
	Melanophore-stimulating hormone–release-inhibiting factor (MIF)		
	Growth hormone–releasing factor (GH-RF)		
	Growth hormone release–inhibiting hormone (GHR-IF)		

The sixth hormone of the anterior lobe is the **growth hormone** (also called somatotropic hormone). This hormone performs a vital role in governing body growth through its stimulatory effect on cellular mitosis and protein synthesis, especially in new tissue of young animals. If produced in excess, the growth hormone causes giantism. A deficiency of this hormone in the human child results in a midget.

In lower vertebrates, the intermediate lobe (Figure 33-6) produces **melano-phore-stimulating hormone** (MSH), which controls the dispersion of the pigment melanin within the melanophores of amphibians, enabling them to better match their background. In birds and mammals, MSH is produced by cells in the anterior pituitary rather than the intermediate lobe (birds and some mammals lack an intermediate lobe altogether), but its physiological function remains unclear. MSH appears to have little to do with pigmentation in the endotherms, even though it will cause darkening of the skin in humans if injected into the circulation. Until recently, many endocrinologists thought MSH to be a vestigial hormone, but interest has been rekindled by studies that show it to have an effect on memory enhancement and on growth of the mammalian fetus.

Hypothalamus and neurosecretion

Because of the strategic importance of the pituitary in influencing most of the hormonal activities in the body, the pituitary was once called the body's "master gland." This description is not appropriate, however, since the tropic hormones are regulated by a higher council, the neurosecretory centers of the hypothalamus. The hypothalamus is itself under the ultimate control of the brain. The hypothalamus contains groups of neurosecretory cells, which are specialized giant nerve cells (Figure 33-7). These cells manufacture polypeptide hormones, called releasing hormones (or "factors"), which then travel down the nerve fibers to their endings in the median eminence. Here they enter a capillary network to complete their journey to the anterior pituitary by way of a short pituitary portal system. The hypothalamic hormones then stimulate or inhibit the release of the various anterior pituitary hormones. Ten hypothalamic releasing hormones in all have been discovered since the demonstration in 1955 of a corticotropin-releasing hormone. There appear to be one or more releasing hormones regulating each of the six pituitary tropic hormones (Table 33-1). Several of the releasing hormones have now been isolated in pure state and characterized chemically. All are peptides.

Posterior pituitary

The hypothalamus is also the source of two hormones of the posterior lobe of the pituitary. They are formed in neurosecretory cells in the hypothalamus, then transported down the infundibular stalk and into the posterior lobe, ending in proximity to blood capillaries, which the hormones enter when released (Figure 33-6). In a sense the posterior lobe is not a true endocrine gland, but a storage and release center for hormones manufactured entirely in the hypothalamus. The two posterior lobe hormones of mammals, oxytocin and vasopressin, are chemically very much alike; both are polypeptides consisting of eight amino acids and are referred to as octapeptides (Figure 33-8). These hormones are among the fastest-acting hormones in the body, since they are capable of producing a response within seconds of their release from the posterior lobe.

Oxytocin has two important specialized reproductive functions in adult female mammals. It stimulates contraction of the uterine smooth muscles during parturition (birth of the young). In clinical practice, oxytocin is used to induce delivery during a difficult labor and to prevent uterine hemorrhage after birth. The second action of oxytocin is that of milk ejection by the mammary glands in response to suckling. Although present, oxytocin has no known function in the male.

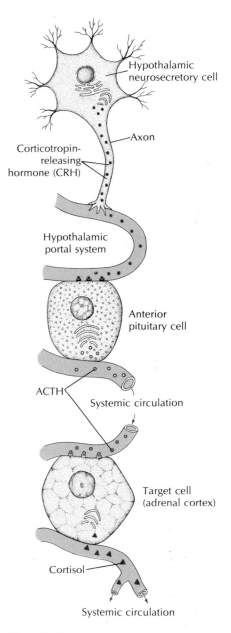

Figure 33-7

Relationship of hypothalamic, pituitary, and target-gland hormones. The hormone sequence controlling the release of cortisol from the adrenal cortex is used as an example.

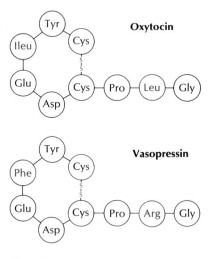

Figure 33-8

Posterior lobe hormones of humans. Both oxytocin and vasopressin consist of eight amino acids (the two sulfur-linked cysteine molecules are considered to be a single amino acid, cystine). Oxytocin and vasopressin are identical except for amino acid substitutions in the red positions.

The radioimmunoassay technique developed by Solomon Berson and Rosalyn Yalow about 1960 after a decade of intensive study has revolutionized endocrinology and neurochemistry. First, antibodies to the hormone of interest (insulin, for example) are prepared by injecting guinea pigs or rabbits with the hormone. Then, a fixed amount of radioactively labeled insulin and unlabeled insulin antibodies is mixed with the sample of blood plasma to be measured. The native insulin in the blood plasma and the radioactive insulin compete for antibodies. The more insulin there is in the sample, the less radioactive insulin will bind to the antibodies. Bound and unbound insulin are then separated, and their radioactivities are measured together with those of appropriate standards to determine the amount of insulin present in the blood sample. The method is so incredibly sensitive that it can measure the equivalent of a cube of sugar dissolved in one of the Great Lakes. Yalow was awarded the Nobel Prize in 1977. Berson died in 1972; he did not receive the prize because it is never awarded posthumously.

Vasopressin, the second posterior lobe hormone, acts on the kidney to restrict urine flow, as already described on p. 652. It is therefore often referred to as the **antidiuretic hormone** (ADH). Vasopressin has a second, weaker effect of increasing the blood pressure through its generalized constrictor effect on the smooth muscles of the arterioles. Although the name "vasopressin" unfortunately suggests that the vasoconstrictor action is the hormone's major effect, it is probably of little physiological importance, except perhaps to help sustain the blood pressure during a severe hemorrhage.

All jawed vertebrates secrete two posterior lobe hormones that are quite similar to those of mammals. All are octapeptides, but there is some variation in structure because of amino acid substitutions in three of eight amino acid positions in the molecule.

Of all the posterior lobe hormones, **vasotocin** has the widest phylogenetic distribution and is believed to be the parent hormone from which the other octapeptides have evolved. It is found in all vertebrate classes except mammals. It is a water-balance hormone in amphibians, especially toads, in which it acts to conserve water by (1) increasing permeability of the skin (to promote water absorption from the environment), (2) stimulating water reabsorption from the urinary bladder, and (3) decreasing urine flow. The action of vasotocin is best understood in amphibians, but it appears to play some water-conserving role in birds and reptiles as well.

Brain neuropeptides

The blurred distinction between the endocrine and nervous systems is nowhere more evident than in the brain, where numerous hormonelike neuropeptides recently have been discovered. More than a dozen neuropeptides (short chains of amino acids) have been identified, and many are known to lead double lives. They are capable of behaving both as hormones, carrying signals from gland cells to their targets, and as neurotransmitters, relaying signals between nerve cells. For example, both oxytocin and vasopressin have been discovered at widespread sites in the brain by radioimmunochemical methods. Apparently related to this is the fascinating observation that people and experimental animals injected with minute quantities of vasopressin experience enhanced learning and improved memory. As far as we can tell, this effect of vasopressin in brain tissue has nothing to do with its well-known antidiuretic function in the kidney (p. 652).

Just as amazing was the discovery of several hormones in the cerebral cortex and hippocampus, such as gastrin and cholecystokinin (p. 728), which long had been supposed to function only in the gut. We have a good idea of what these hormones do in the gastrointestinal tract, but what functional roles do they play in the brain?

Among the dramatic developments in this field was the discovery in 1975 of the endorphins and enkephalins, substances that bind with opiate receptors and are important in pain perception (see marginal note on p. 700). The endorphins and enkephalins are also found in brain circuits that modulate several other functions unrelated to pain, such as control of blood pressure, body temperature, and body movement. Even more intriguing, the endorphins are derived from the same chemical precursor that gives rise to the anterior pituitary hormones ACTH and MSH. It is clear that we have discovered in the brain a complex family of compounds whose functions and interrelationships are not yet clear. This is currently an active area of biomedical research, and we seem to be on the threshold of some exciting biological discoveries.

Hormones of Metabolism

Another important group of hormones adjusts the delicate balance of metabolic activities in the body. The rates of chemical reactions within cells are often regulated by long sequences of enzymes. Although such sequences are complex, each step in a pathway is mostly self-regulating as long as the equilibrium between substrate, enzyme, and product remains stable. However, hormones may alter the activity of crucial enzymes in a metabolic process, thus accelerating or inhibiting the entire process. It should be emphasized that hormones never initiate enzymatic processes. They simply alter their rate, speeding them up or slowing them down. The most important hormones of metabolism are those of the thyroid, parathyroid, adrenal glands, and pancreas.

Thyroid hormones

The two thyroid hormones, **thyroxine** and **triiodothyronine,** are secreted by the thyroid gland. This large endocrine gland is located in the neck region of all vertebrates. The thyroid is made up of thousands of tiny spherelike units, called follicles, where thyroid hormone is synthesized, stored, and released into the bloodstream as needed. The size of the follicles, and the amount of stored thyroxine they contain, depends on the activity of the gland (Figure 33-9).

One of the unique characteristics of the thyroid is its high concentration of **iodine;** in most animals this single gland contains well over half the body store of iodine. The epithelial cells of the thyroid follicles actively trap iodine from the blood and combine it with the amino acid tyrosine, creating the two thyroid hormones. Each molecule of thyroxine contains four atoms of iodine. Triiodothyronine is identical to thyroxine, except that it has three instead of four iodine atoms. Thyroxine is formed in much greater amounts than triiodothyronine, but both hormones have two important actions. One is to promote normal growth and development of the nervous system of growing animals. The other is to stimulate the metabolic rate.

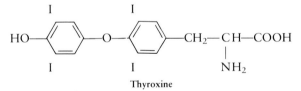

Thyroxine

Undersecretion of thyroid hormone dramatically impairs growth, especially of the nervous system. The human **cretin,** a mentally retarded dwarf, is the tragic result of thyroid malfunction from a very early age. Conversely, the oversecretion of thyroid hormones causes precocious development, particularly in lower vertebrates. Frogs and toads undergo a dramatic metamorphosis from aquatic tadpole without lungs or legs to semiterrestrial or terrestrial adult with lungs, four legs, and a completely remodeled alimentary canal. This transformation occurs when the thyroid gland becomes active at the end of larval premetamorphosis. Stimulated by a rise in the blood thyroxine level, metamorphosis and climax occur (Figure 33-10).

The control of oxygen consumption and heat production in birds and mammals is the best-known action of the thyroid hormones. The thyroid maintains metabolic activity of homeotherms (birds and mammals) at a normal level. Too much thyroid hormone will speed up body processes as much as 50%, resulting in irritability, nervousness, fast heart rate, intolerance of warm environments, and

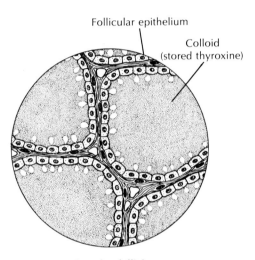

Inactive follicles

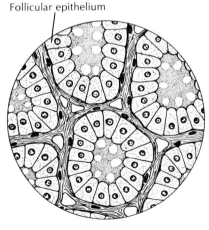

Active follicles

Figure 33-9

Appearance of thyroid gland follicles viewed through the microscope. (About ×350.) When inactive, the follicles are distended with colloid, the storage form of thyroxine, and the epithelial cells are flattened. When active, the colloid disappears as thyroxine is secreted into the circulation, and the epithelial cells become greatly enlarged.

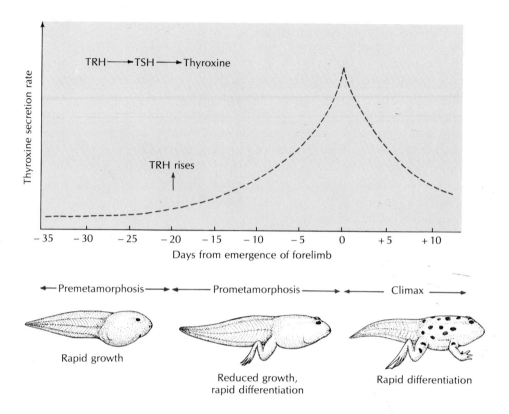

Figure 33-10

Effect of thyroxine on frog growth and metamorphosis. The release of TRH from the hypothalamus at the end of premetamorphosis sets in motion the hormonal changes (increased TSH and thyroxine) leading to metamorphosis. Thyroxine levels are maximal at the time the forelimbs emerge.
Modified from Bentley, P.J. 1976. Comparative vertebrate endocrinology. Cambridge, Eng., Cambridge University Press.

loss of body weight despite increased appetite. Too little thyroid hormone slows metabolic activities, which can result in loss of mental alertness, slowing of the heart rate, muscular weakness, increased sensitivity to cold, and weight gain. One important function of the thyroid gland is to help animals adapt to cold environments by increasing their heat production. Thyroxine in some way causes cells to produce more heat and store less chemical energy (ATP); in other words, thyroxine *reduces* the efficiency of the cellular oxidative phosphorylation system (pp. 85 and 87). This is why many cold-adapted mammals have heartier appetites and eat more food in winter than in summer even though their activity is about the same in both seasons. In winter, a larger portion of the food is being converted directly into body-warming heat.

The synthesis and release of thyroxine and triiodothyronine are governed by **thyrotropic hormone** (TSH) from the anterior pituitary gland (Table 33-1). TSH controls the thyroid through a beautiful example of negative feedback. If the thyroxine level in the blood decreases, more TSH is released, which returns the thyroxine level to normal. Should the thyroxine level rise too high, it acts on the anterior pituitary to inhibit TSH release. With declining TSH output, the thyroid is less stimulated and the blood thyroxine level returns to normal. Such a system is obviously very effective in damping out oscillations in hormone output by the target gland. It can be overridden, however, by neural stimuli, such as exposure to cold, which can directly stimulate increased release of TSH.

The control of thyroid activity involves another component, the thyrotropin-releasing hormone (TRH) of the hypothalamus. As noted earlier, TRH is part of a higher regulatory council that controls the tropic hormones of the anterior pituitary. But, if the hypothalamus releasing hormone controls the anterior pituitary, what controls release of the releasing hormone? At present, there is no general agreement on the answer, although there is evidence that the thyroid hormones have a negative feedback effect on the hypothalamus as well as on the anterior pituitary.

Some years ago, a condition called **goiter** was common among people living in the Great Lakes region of the United States and Canada, as well as other parts of the world, such as the Swiss Alps. This type of goiter is an enlargement of the thyroid gland caused by a deficiency of iodine in the food and water. By striving to produce thyroid hormone with not enough iodine available, the gland hypertrophies, sometimes so much that the entire neck region becomes swollen (Figure 33-11). Iodine deficiency goiter is seldom seen in North America because of the widespread use of iodized salt. However, it is estimated that even today 200 million people suffer from goiter worldwide, mostly in high mountain areas of Latin America, Europe, and Asia.

Hormonal regulation of calcium metabolism

Closely associated with the thyroid gland and often buried within it are the parathyroid glands. These tiny glands occur as two pairs in humans but vary in number and position in other vertebrates. They were discovered at the end of the nineteenth century when the fatal effects of "thyroidectomy" were traced to the unknowing removal of the parathyroid glands as well as the thyroid gland.

In many animals, including humans, removal of the parathyroid glands causes the blood calcium to decrease rapidly. This results in a serious increase in nervous system excitability, severe muscular spasms and tetany, and finally death.

The parathyroid glands are vitally concerned with the maintenance of the normal level of calcium in the blood. Actually, three hormones are involved in the stabilization of both calcium and phosphorus in the blood. They are **parathyroid hormone** (PTH), produced by the parathyroid gland; **calcitonin,** produced by specialized cells (C cells) in the thyroid gland; and a hormonal metabolite of vitamin D called **1,25-dihydroxyvitamin D** ($1,25\text{-}[OH]_2D$). Before considering how these factors interact, it will be helpful to summarize mineral metabolism in bone, a densely packed storehouse of both calcium and phosphorus.

Bone contains approximately 98% of the body calcium and 80% of the phosphorus. Although bone is second only to teeth as the most durable material in the body, as evidenced by the survival of fossil bones for millions of years, it is in a state of constant turnover in the living body. Bone-building cells (**osteoblasts**) withdraw calcium and phosphorus (as phosphate) from the blood and deposit them in a complex crystalline form around previously formed organic fibers. Bone-resorbing cells (**osteoclasts**), present in the same bone, tear down bone by engulfing it and releasing the calcium and phosphate into the blood. These opposing activities allow bone to constantly remodel itself, especially in the growing animal, for structural improvements to counter new mechanical stresses on the body. They additionally provide a vast and accessible reservoir of minerals that can be withdrawn as the body needs them for its general cellular requirements.

If the blood calcium should decrease slightly, the parathyroid gland increases its output of parathyroid hormone. This stimulates the osteoclasts to destroy bone adjacent to these cells, thus releasing calcium and phosphate into the bloodstream and returning the blood calcium level to normal. Should the calcium in the blood rise above normal, the parathyroid gland decreases its output of parathyroid hormone. The parathyroid hormone level varies inversely with blood calcium level, as shown in Figure 33-12.

The second calcium-regulating hormone, calcitonin, is secreted when the blood level of calcium begins to rise too high (Figure 33-12). Calcitonin lowers the blood calcium level by inhibiting bone resorption by the osteoclasts. It thus pro-

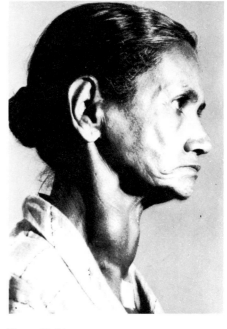

Figure 33-11

Thyroid goiter in a woman from western Colombia, an endemic goiter area of South America.
Courtesy Dr. Eduardo Gaitan, University of Mississippi Medical Center.

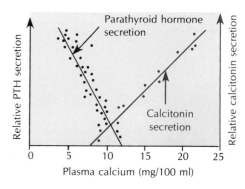

Figure 33-12

How parathyroid hormone (PTH) and calcitonin secretion rates respond to changes in blood calcium level in a mammal.
After Copp, D.H. 1969. J. Endocrinol. **43:**137-161.

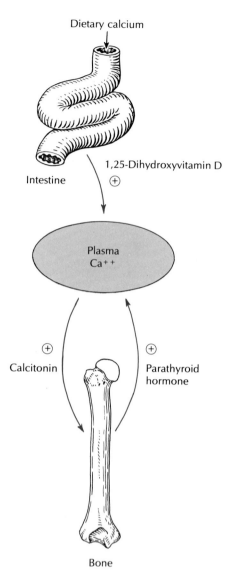

Dietary calcium

Intestine

1,25-Dihydroxyvitamin D
(+)

Plasma
Ca++

(+)
Calcitonin

(+)
Parathyroid
hormone

Bone

Figure 33-13

Regulation of blood calcium in birds and mammals.

tects the body against a dangerous increase in the blood calcium level, just as parathyroid hormone protects it from a dangerous decrease in blood calcium. The two act together to smooth out oscillations in blood calcium (Figure 33-13).

The third factor involved in calcium metabolism, 1,25-dihydroxyvitamin D, is an active hormonal form of vitamin D. Vitamin D, like all vitamins, is a dietary requirement. But unlike other vitamins, vitamin D may also be synthesized in the skin from a precursor by irradiation with ultraviolet light from the sun. Vitamin D is then converted in a two-step oxidation to 1,25-dihydroxyvitamin D. This steroid hormone is essential for active calcium absorption by the gut (Figure 33-13). It also promotes the synthesis of a protein that transports calcium in the blood.

A deficiency of vitamin D causes rickets, a disease characterized by low blood calcium and weak, poorly calcified bones that tend to bend under postural and gravitational stresses. Rickets has been called a disease of northern winters, when sunlight is minimal. It was once common in the smoke-darkened cities of England and Europe.

Hormones of the adrenal cortex

The vertebrate adrenal gland is a double gland consisting of two very different kinds of tissue: **interrenal tissue,** called **cortex** in mammals, and **chromaffin** tissue, called **medulla** in mammals (Figure 33-14). The mammalian terminology of cortex (meaning "bark") and medulla (meaning "core") arose because in this group of vertebrates the interrenal tissue completely surrounds the chromaffin tissue like a cover. Although in the lower vertebrates the interrenal and chromaffin tissues are usually separated, the mammalian terms "cortex" and "medulla" are so firmly fixed in our vocabulary that we commonly use them for all vertebrates instead of the more correct terms "interrenal" and "chromaffin."

Biochemists have found that the adrenal cortex contains at least 30 different compounds, all of them closely related lipoid compounds known as steroids. Only a few of these compounds, however, are true steroid *hormones;* most are various intermediates in the synthesis of steroid hormones from **cholesterol** (Figure 33-15). The corticosteroid hormones are commonly classified into three groups, according to their function:

1. **Glucocorticoids,** such as **cortisol** (Figure 33-15) and **corticosterone,** are concerned with food metabolism, inflammation, and stress. They cause the conversion of nonglucose compounds, particularly amino acids and fats, into glucose. This process is called **gluconeogenesis.** Cortisol, cortisone, and corticosterone are also **antiinflammatory.** Because several diseases of humans are inflammatory diseases (for example, allergies, hypersensitivity, and arthritis, p. 620), these corticosteroids have important medical applications. They must be used with great

Figure 33-14

Paired adrenal glands of a human, showing gross structure and position on the upper poles of the kidneys. Steroid hormones are produced by the outer cortex. The sympathetic hormones epinephrine and norepinephrine are produced by the inner medulla.

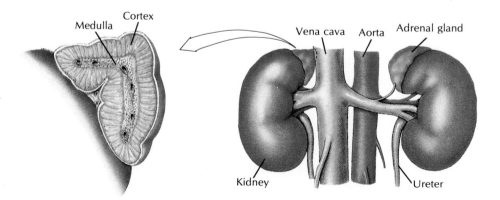

Medulla   Cortex

Vena cava   Aorta   Adrenal gland

Kidney   Ureter

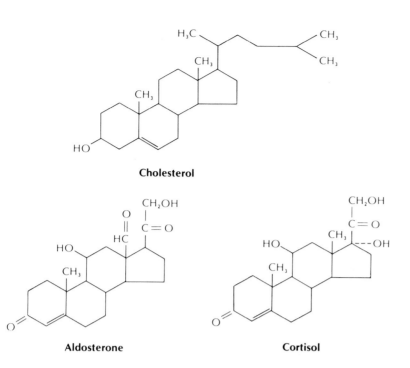

Cholesterol

Aldosterone

Cortisol

Figure 33-15

Hormones of the adrenal cortex. Cortisol (a glucocorticoid) and aldosterone (a mineralocorticoid) are two of several steroid hormones synthesized from cholesterol in the adrenal cortex.

care, however, since, if administered in excess, they may suppress the body's normal repair processes and lower resistance to infectious agents.

2. **Mineralocorticoids,** the second group of corticosteroids, are those that regulate salt balance. **Aldosterone** (Figure 33-15) and **deoxycorticosterone** are the most important steroids of this group. They promote the tubular reabsorption of sodium and chloride and the tubular excretion of potassium by the kidney. Since sodium usually is in short supply in the diet of many animals and potassium is in excess, it is obvious that the mineralocorticoids play vital roles in preserving the correct balance of blood electrolytes. We may also note that the mineralocorticoids *oppose* the antiinflammatory effect of cortisol and cortisone. In other words, they promote the *inflammatory* defense of the body to various noxious stimuli. Although these opposing actions of the corticosteroids seem self-defeating, they actually are not. They are necessary to maintain readiness of the body's defenses for any stress or disease threat, yet prevent these defenses from becoming so powerful that they turn against the body's own tissues.

3. **Sex hormones,** such as testosterone, estrogen, and progesterone, are produced primarily by the ovaries and testes (p. 769). The adrenal cortex is also a minor source of certain steroids that mimic the action of testosterone. These sex hormone–like secretions are of little physiological significance, except in certain disease states of humans.

The synthesis and secretion of the corticosteroids are controlled principally by ACTH of the anterior pituitary (Figure 33-7). As with pituitary control of the thyroid, a negative feedback relationship exists between ACTH and the adrenal cortex: an increase in the level of corticosteroids suppresses the output of ACTH; a decrease in the blood steroid level increases ACTH output. ACTH is also controlled by the corticotropin-releasing hormone (CRH) of the hypothalamus.

Hormones of the adrenal medulla

The adrenal medulla secretes two structurally similar hormones: **epinephrine** (adrenaline) and **norepinephrine** (noradrenaline). Norepinephrine is also released at the endings of sympathetic nerve fibers throughout the body, where it serves as

The adrenal steroid hormones, especially the glucocorticoids, are remarkably effective in relieving the *symptoms* of rheumatoid arthritis, allergies, and various connective tissue, skin, and blood disorders. Following the report in 1948 of Dr. P.S. Hench and his colleagues at the Mayo Clinic that cortisone dramatically relieved the pain and crippling effects of advanced arthritis, the steroid hormones were hailed by the media as "wonder drugs." Optimism was soon dimmed, however, when it became apparent that severe side effects (salt retention, high blood pressure, peptic ulcers, thinning of bones, and decreased resistance to infection) always attended long-term administration of the antiinflammatory steroids. Although more recently developed synthetic steroids cause less harmful side effects, they lull the adrenal cortex into inactivity and may permanently impair the body's capacity to produce its own steroids. Today steroid therapy is applied with caution, since it is realized that the inflammatory response is a necessary part of the body's defenses.

a neurotransmitter (p. 697) to carry neural signals across the gap that separates the fiber and the organ it innervates. The adrenal medulla has the same embryological origin as sympathetic nerves; in many respects the adrenal medulla is nothing more than an overgrown sympathetic nerve ending.

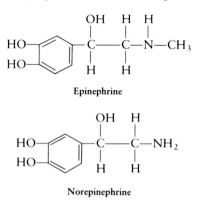

Epinephrine

Norepinephrine

It is not surprising then that the adrenal medulla hormones have the same general effects on the body that the sympathetic nervous system has. These effects center on emergency functions of the body, such as fear, rage, fight, and flight, although they have important integrative functions in more peaceful times as well. We are all familiar with the increased heart rate, tightening of the stomach, dry mouth, trembling muscles, general feeling of anxiety, and increased awareness that attends sudden fright or other strong emotional states. These effects are attributable to increased activity of the sympathetic nervous system and to the rapid release into the blood of epinephrine from the adrenal medulla.

Epinephrine and norepinephrine have many other effects of which we are not as aware, including constriction of the arterioles (which, together with the increased heart rate, increases the blood pressure), mobilization of liver glycogen and fat stores to release glucose and fatty acids for energy, increased oxygen consumption and heat production, hastening of blood coagulation, and inhibition of the gastrointestinal tract. All of these changes in one way or another tune up the body for emergencies.

Insulin from the islet cells of the pancreas

The pancreas is both an exocrine and an endocrine organ. The *exocrine* portion produces pancreatic juice, a mixture of digestive enzymes that is conveyed by ducts to the digestive tract. Scattered within the extensive exocrine portion of the

Figure 33-16

Insulin is synthesized in the beta cells of the pancreas as inactive proinsulin and is then converted to the hormone insulin by removal of the peptide that connects the A and B chains. **A,** The amino acid sequence of proinsulin. **B,** Chain folding as derived from x-ray analysis.

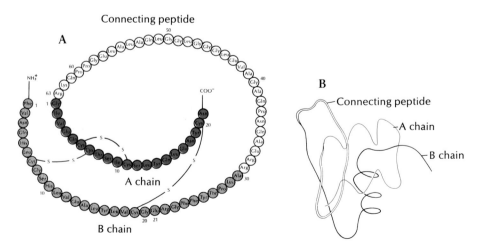

pancreas are numerous small islets of tissue, called **islets of Langerhans.** This is the *endocrine* portion of the gland. The islets are without ducts and secrete their hormones directly into blood vessels that extend throughout the pancreas.

Two polypeptide hormones are secreted by different cell types within the islets: **insulin** (Figure 33-16), produced by the **beta cells,** and **glucagon,** produced by the **alpha cells.** Insulin and glucagon have antagonistic actions of great importance in the metabolism of carbohydrates and fats. Insulin is essential for the use of blood glucose by cells, especially skeletal muscle cells. Insulin somehow allows glucose in the blood to be transported into body cells. Without insulin, the cells cannot use glucose, even if there is an abnormally high blood glucose level (hyperglycemia), and sugar appears in the urine. Lack of insulin also inhibits the uptake of amino acids by skeletal muscle, and fats and muscle are broken down to provide energy. The body cells actually starve while the urine abounds in the very substance the body craves. The disease, called diabetes mellitus, afflicts nearly 5% of the human population in varying degrees of severity. If left untreated, it leads inexorably to emaciation, coma, and death.

The first extraction of insulin in 1921 by two Canadians, Frederick Banting and Charles Best, was one of the most dramatic and important events in the history of medicine. Many years earlier two German scientists, J. Von Mering and O. Minkowski, discovered that surgical removal of the pancreas of dogs invariably caused severe symptoms of diabetes, resulting in the animal's death within a few weeks. Many attempts were made to isolate the diabetes preventive factor, but all failed because powerful protein-splitting digestive enzymes in the exocrine portion of the pancreas destroyed the hormone during extraction procedures. Following a hunch, Banting, in collaboration with Best and his physiology professor J.J.R. Macleod, tied off the pancreatic ducts of several dogs. This caused the exocrine portion of the gland with its hormone-destroying enzyme to degenerate but left the islets' tissue healthy, since they were independently served by their own blood supply. Banting and Best then successfully extracted insulin from these glands. Injected into another dog, the insulin immediately lowered the blood sugar level (Figure 33-17). Their experiment paved the way for the commercial extraction of insulin from slaughterhouse animals. It meant that millions of people with diabetes, previously doomed to invalidism or death, could look forward to more normal lives.

Glucagon, the second hormone of the pancreas, has several effects on carbohydrate and fat metabolism that are opposite to the effects of insulin. For example, glucagon raises the blood glucose level, whereas insulin lowers it. Glucagon and insulin do not have the same effects in all vertebrates, and in some, glucagon is lacking altogether. Glucagon is an example of a hormone that operates through the cyclic AMP second-messenger system.

Figure 33-17

Charles H. Best and Sir Frederick Banting in 1921 with the first dog to be kept alive by insulin.

From Fulton, J.F., and L.G. Wilson. 1966. Selected readings in the history of physiology. Springfield, Ill., Charles C Thomas, Publisher.

___ Hormones of Digestion

Gastrointestinal function is coordinated by a family of hormones produced by endocrine cells scattered throughout the gut. Although together they constitute the largest endocrine organ in the body, they have long been neglected by endocrinologists because it has been impossible to study them by applying the classical method of surgical removal of a gland, followed by examination of the effect of its absence. However, by the mid-1970s seven gut hormones had been chemically purified or defined. Recently several of these hormones were discovered in the nervous system, where they may serve as neurotransmitters. If this dual role of the gut hormones is confirmed by current research, it would be another example of

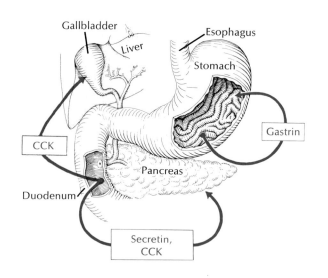

Figure 33-18

Three hormones of digestion. Arrows show source and target of three gastrointestinal hormones.

nature's conservative capacity to put the same cellular products to completely different uses in unrelated systems.

The three best-understood gut hormones are gastrin, cholecystokinin (CCK), and secretin (Figure 33-18). **Gastrin** is a small polypeptide hormone produced in the mucosa of the pyloric portion of the stomach. When food enters the stomach, gastrin stimulates the secretion of hydrochloric acid by the stomach wall. Gastrin is an unusual hormone in that it exerts its action on the same organ from which it is secreted. **CCK** is also a polypeptide hormone and has a striking structural resemblance to gastrin, suggesting that the two arose by duplication of ancestral genes. CCK has two distinct functions. It stimulates gallbladder contraction and thus increases the flow of bile salts into the intestine; it also stimulates an enzyme-rich secretion from the pancreas. The third gut hormone is **secretin,** the first hormone to be discovered (p. 712). Like CCK, it stimulates pancreatic secretion, but instead of being rich in enzymes, the secretion contains bicarbonate, which rapidly neutralizes stomach acid.

All of the gastrointestinal hormones are peptides that bind to surface receptors on target tissues and act through the second messenger, cyclic AMP.

SUMMARY

Hormones are chemical messengers synthesized by special endocrine glands and transported by the bloodstream to target cells where they affect cell function by altering specific biochemical processes. Specificity of response is ensured by the presence of protein receptors on or in the target cells that bind only selected hormones. Hormone effects are vastly amplified in the target cells by acting through one or the other of two basic mechanisms. Many hormones, including epinephrine, glucagon, vasopressin, and some anterior pituitary hormones, cause production of cyclic AMP, a "second messenger" that relays the hormone's message from a surface receptor to the cell's biochemical machinery. The alternative mechanism relays the action of steroid hormones by operating through cytoplasmic receptors. A hormone-receptor complex is formed that moves into the cell's nucleus to induce protein synthesis by setting gene transcription in motion.

Most invertebrate hormones are products of neurosecretory cells. The best understood invertebrate endocrine system is that controlling molting and metamorphosis in insects. An insect juvenile grows by passing through a series of molts under the control of two hormones, one (the molting hormone) favoring molting to an adult and the other (the juvenile hormone) favoring retention of juvenile characteristics. The molting hormone is under the control of a brain neurosecretory hormone called ecdysiotropin.

The vertebrate endocrine system is orchestrated by the pituitary gland. The anterior lobe of the pituitary produces at least six hormones. Five of these are tropic hormones that regulate subservient endocrine glands: adrenocorticotropic hormone (ACTH), which stimulates the adrenal cortex; thyrotropic hormone (TSH); follicle-stimulating hormone (FSH) and luteinizing hormone (LH), which act on the ovaries and testes; and prolactin, which plays several diverse roles, including the stimulation of milk production. A sixth anterior pituitary hormone is the growth hormone that governs body growth. The intermediate lobe of the pituitary produces melanophore-stimulating hormone (MSH), which controls melanophore dispersion in amphibians. The release of all of the anterior and intermediate lobe hormones is regulated in part by hypothalamic neurosecretory products called releasing hormones. The hypothalamus also produces two neurosecretory hormones, oxytocin and vasopressin, which are stored and released from the posterior lobe of the pituitary.

The recent application of ultrasensitive radioimmunochemical techniques has revealed many neuropeptides in the brain, several of which behave as neurotransmitters in the brain but as hormones elsewhere in the body.

Several hormones play important roles in regulating cellular metabolic activities. The two thyroid hormones, thyroxine and triiodothyronine, promote normal growth and nervous system development, and they control the rate of cellular metabolism. Calcium metabolism is regulated principally by two antagonistic hormones: parathyroid hormone from the parathyroid glands and calcitonin from the thyroid gland. A hormonal derivative of vitamin D, 1,25-dihydroxyvitamin D, is essential for calcium absorption from the gut.

The steroid hormones of the adrenal cortex are glucocorticoids, which stimulate glucose formation from nonglucose sources; mineralocorticoids, which regulate blood electrolyte balance; and certain of the sex hormones. The adrenal medulla is the source of epinephrine and norepinephrine, which have many effects, including assisting the sympathetic nervous system in emergency responses.

Review questions

1. Provide definitions for the following: hormone, endocrine gland, exocrine gland, hormone receptor molecule.
2. Outline the famous experiment of Bayliss and Starling that marks the birth of endocrinology. What might their *hypothesis* have been?
3. Hormone receptor molecules are the key to understanding the specificity of hormone action on target cells. Describe and distinguish between receptors located on the cell surface and those located in the cytoplasm of target cells. Name two hormones whose action is mediated through each receptor type.
4. What is the importance of feedback systems in the control of hormone output? Offer an example of a hormonal feedback pattern.
5. Explain how the three hormones involved in insect growth—molting hormone, juvenile hormone, and ecdysiotropin—interact in molting and metamorphosis.
6. Name six hormones produced by the anterior lobe of the pituitary gland. Explain how the secretion of these hormones is controlled by neurosecretory cells in the hypothalamus.
7. Describe the chemical nature and function of two posterior lobe hormones, oxytocin and vasopressin. What is distinctive about the way these neurosecretory hormones are secreted as compared with the neurosecretory release hormones that control the anterior pituitary hormones?

8. What are endorphins and enkephalins?
9. What are the two most important functions of the thyroid hormones?
10. Explain how you would interpret the graph in Figure 33-12 to show that PTH and calcitonin act in a complementary way to control the blood calcium level.
11. Describe the principal functions of the two major groups of adrenal cortical steroids, the glucocorticoids and the mineralocorticoids. To what extent do these names provide clues to their function?
12. Where are the hormones epinephrine and norepinephrine produced and what is their relationship to the sympathetic nervous system and its response to emergencies?
13. Explain the actions of the hormones of the islets of Langerhans on the blood glucose level. What is the consequence of insulin insufficiency as in the disease diabetes mellitus?
14. Name three hormones of the gastrointestinal tract and explain how they assist in the coordination of gastrointestinal function.

Selected references

See also general references for Part Three, p. 752.

Bloom, F.E. 1981. Neuropeptides. Sci. Am. **245**:148-168 (Oct.). *Recent research on the brain peptides is described.*

Goldsworthy, G.J., J. Robinson, and W. Mordue. 1981. Endocrinology. New York, John Wiley & Sons, Inc. *Concise comparative approach.*

Gorbman, A., W.W. Dickhoff, S.R. Vigna, N.B. Clark, and C. Ralph. 1983. Comparative endocrinology. New York, John Wiley & Sons, Inc. *Authoritative and up-to-date text, although the comparative treatment is limited to vertebrates.*

Hadley, M.E. 1984. Endocrinology. Englewood Cliffs, N.J., Prentice-Hall, Inc. *Undergraduate-level textbook in vertebrate endocrinology.*

Highnam, K.C., and L. Hill. 1977. The comparative endocrinology of the invertebrates, ed. 2. New York, American Elsevier Publishers, Co., Inc. *Clearly written, well-illustrated comparative treatment.*

Snyder, S.H. 1985. The molecular basis of communication between cells. Sci. Am. **253**:132-141 (Oct.). *Review of vertebrate hormonal communication.*

C H A P T E R 3 4

A N I M A L B E H A V I O R

For as long as people have walked the earth, their lives have been touched by, indeed interwoven with, the lives of other animals. They hunted and fished for them, domesticated them, ate them and were eaten by them, made pets of them, revered them, hated and feared them, immortalized them in art, song, and verse, fought them, and loved them. The very survival of ancient peoples depended on knowledge of wild animals. To stalk them, people had to know the ways of the quarry. As the hunting society of primitive people gave way to agricultural civilizations, an awareness was retained of the interrelationship with other animals.

This is still evident today. Zoos attract more visitors than ever before; wildlife television shows are increasingly popular; game-watching safaris to Africa constitute a thriving enterprise; and millions of pet animals share the cities with us—more than a half million pet dogs in New York City alone. Although people have always been interested in the behavior of animals, it has been interpreted in a scientific fashion only during the past century and a half. Several different aspects of behavior have been the focus of various scientists. Two of the more important topics of concentration have been the diversity of species—typical behavior under natural conditions, a subject known as **ethology,** and the study of social behavior, or **sociobiology.**

—— DEVELOPMENT OF ANIMAL BEHAVIOR

In 1973 the Nobel Prize in physiology and medicine was awarded to three pioneering zoologists, Karl von Frisch, Konrad Lorenz, and Niko Tinbergen (Figure 34-1). The citation stated that these three were the principal architects of the new science of ethology. It was the first time any contributor to the behavioral sciences was so honored, and it meant that the discipline of animal behavior, which is rooted in the work of Charles Darwin, had arrived.

The term "ethology," meaning literally "character study," was first used in the late eighteenth century to signify the interpretation of character through the study of gesture. It was an appropriate term for the new field of animal behavior, which had as its objective the study of motor patterns, that is, actions of animals, with the anticipation that such study would reveal the true characters of animals just as the interpretation of human gestures might reveal the true characters of people. This decidedly restricted interpretation of ethology as a purely descriptive study of the "habits" of animals was considerably modified during the period in which ethology was established as a science, 1935 to 1950. Today we may define ethology as a discipline that involves the study of the total repertoire of behavior,

A both simple and complex, that animals employ in their natural environment as they resolve the problems of survival and reproduction.

The aim of ethologists has been to describe the behavior of an animal in its *natural habitat.* Most ethologists have been naturalists. Their laboratory has been the out-of-doors, and early ethologists gathered their data by field observation. They also conducted experiments, often with nature providing the variables, but increasingly ethologists have manipulated the variables for their own purposes by using animal models, playing recordings of animal vocalizations, altering the habitat, and so on. Modern ethologists also conduct many experiments in the laboratory where they can test their predictions under closely controlled conditions. However, ethologists usually take pains to compare laboratory observations with observations of free-ranging animals in undisturbed natural environments. They recognize that it makes no more sense to try to study the natural behavior of an animal divorced from its natural surroundings than it does to try to interpret a structural adaptation of an animal apart from the function it serves.

With infinite patience Lorenz, Tinbergen, and their colleagues watched and catalogued the activities and vocalizations of animals during feeding, courtship, and nest building, as well as seemingly insignificant behavioral movements such as head scratching, stretching postures, and turning and shaking movements. These studies concentrated largely on innate motor patterns used for communication within a species.

One of the great contributions of Lorenz and Tinbergen was to demonstrate that behavioral traits are measurable entities like anatomical or physiological traits. This was to become the central theme of ethology: behavioral traits can be isolated and measured, and they have evolutionary histories. Lorenz and Tinbergen showed that behavior is not the wavering, transient, unpredictable phenomenon often depicted by earlier writers. In short, behavior is genetically mediated. It is apparent that, if behavior is determined by genes in the same way that genes determine morphological and physiological characters (ethologists have found abundant evidence that it is), then behavior evolves and is adaptive. Thus, modern behavioral study is founded on the recognition that the Darwinian view of evolution holds for behavioral traits, as well as for anatomical and functional characteristics.

——PRINCIPLES OF ETHOLOGY

The ethologists, through step-by-step analysis of the behavior of animals in nature, focused on the relatively invariant components of behavior. From such studies emerged several concepts that were first popularized in Tinbergen's influential book, *The Study of Instinct* (1951).

The basic concepts of ethology can be approached by considering the egg-rolling response of the greylag goose (Figure 34-2), described by Lorenz and Tinbergen in a famous paper published in 1938. If Lorenz and Tinbergen presented a female greylag goose with an egg a short distance from her nest, she would rise, extend her neck until the bill was just over the egg, then contract her neck, pulling the egg carefully into the nest.

Figure 34-1

A, Konrad Lorenz. **B,** Karl von Frisch. **C,** Niko Tinbergen.
A, Photograph by Thomas McAvoy, Life Magazine, © 1955 Time Inc.,
B, photograph courtesy W.S. Hoar; **C,** Photograph by Larry Shaffer.

Although this behavior appeared to be intelligent, Tinbergen and Lorenz noticed that if they removed the egg once the goose had begun her retrieval, or if the egg being retrieved slipped away and rolled down the outer slope of the nest, the goose would continue the retrieval movement without the egg until she was again settled comfortably on her nest. Then, seeing that the egg had not been retrieved, she would begin the egg-rolling pattern all over again.

Thus the bird performed the egg-rolling behavior as if it were a program which, once initiated, had to run to completion. Lorenz and Tinbergen termed this stereotyped behavior a **fixed-action pattern** (FAP): a motor pattern that is mostly invariable in its performance. The goose did not have to learn the movement; it was a "prewired" or innate skill.

Further experiments by Tinbergen disclosed that the greylag goose was not terribly discriminating about what she retrieved. Almost any smooth and rounded object placed outside the nest would trigger the egg-rolling behavior; even a small toy dog and a large yellow balloon were dutifully retrieved. But once the goose settled down on such objects, they obviously did not feel right and she discarded them.

Lorenz and Tinbergen realized that the presence of the egg outside the nest must act as a stimulus, or trigger, that released the fixed action pattern. Lorenz termed the triggering stimulus a **releaser**: a simple feature in the environment that would trigger a certain innate behavior. Or, because the animal usually responded to some specific aspect of the releaser (sound, shape, or color, for example) the effective stimulus was called a **sign stimulus.** Ethologists have described hundreds of examples of sign stimuli. In every case the response is highly predictable. For example, the alarm call of adult herring gulls always releases a crouching freeze response in the chicks. Or, to cite an example given in an earlier chapter (p. 701), certain nocturnal moths take evasive maneuvers or drop to the ground when they hear the ultrasonic cries of bats that feed on them; most other sounds do not release this response.

Sometimes exaggerated sign stimuli release an exaggerated response. Tinbergen found that an oyster catcher, offered a choice between its own egg and a giant egg four times as large, would attempt to incubate the giant egg and ignore its own (Figure 34-3). Similarly, a greylag goose allowed to choose between a goose egg and volleyball near its nest, would invariably recover the volleyball. These examples of exceptionally effective signals are called **supernormal stimuli.** Obviously a goose has no need for a volleyball in its nest, but these experiments do give us some insight into the nature of sign stimuli: they are usually simple cues, a fact that reduces the chance that the signal might be misunderstood. Natural selection will favor those combinations of genes that boost the signal value (in this case, large size of the object), even though this may occasionally lead to inappropriate behavior.

Lorenz and Tinbergen proposed that the nervous system of animals must have special filtering units, or "centers," capable of releasing a fixed-action pattern when stimulated by the approximate sign stimulus. They called this filter-trigger complex an **innate releasing mechanism** (IRM). The most important characteristic of the IRM is that it behaves like a programmed motor message that, once begun, produces a coordinated muscle performance without requiring any further sensory input. And the response is completely functional the first time it is performed (the animal "knows" how with no learning), once the animal is the right age and in the proper motivational state (only an adult *female* greylag goose *in a nest* will respond to an egg near the nest).

The IRM dramatically illustrates the stereotyped, predictable, and pro-

Figure 34-2

Egg-rolling movement of the greylag goose *Anser anser.*
From Lorenz, K., and N. Tinbergen. 1938. Zeit. Tierpsychol. 2:1-29.

Figure 34-3

Oyster catcher *Haematopus ostralegus* attempts to roll a giant egg model into its nest while ignoring its own egg.
From Tinbergen, N. 1951. The study of instinct. Oxford, Eng., Oxford University Press.

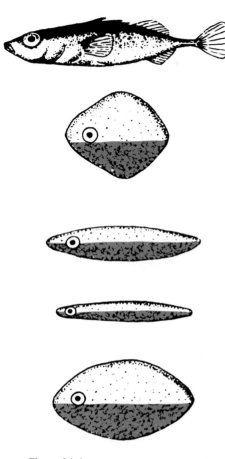

Figure 34-4

Stickleback models. The carefully made model of the stickleback *(top)* without a red belly is attacked much less frequently than the four simple red-bellied models.
From Tinbergen, N. 1951. The study of instinct. Oxford, Eng., Oxford University Press.

grammed nature of much animal behavior. This is even more evident when an FAP is released inappropriately. In the spring the male three-spined stickleback selects a territory, which it defends vigorously against other males. The underside of the male becomes bright red and the approach of another redbellied male will release a threat posture or even an aggressive attack. Tinbergen's suspicion that the red belly of the male served as a releaser for aggression was reinforced when a passing red postal truck evoked attacks from the males in his aquarium. Tinbergen then carried out experiments using a series of models which he presented to the males. He found that they vigorously attacked any model bearing a red stripe, even a plump lump of wax with a red underside. Yet a carefully made model that closely resembled a male stickleback but lacked the red belly was ignored (Figure 34-4). Tinbergen discovered other examples of FAPs released by simple sign stimuli. Male English robins furiously attacked a bundle of red feathers placed in their territory but ignored a stuffed juvenile robin without the red feathers (Figure 34-5).

We have seen in the examples above that there are costs to the releaser-IRM system because they may lead to improper responses. Fortunately for red-bellied sticklebacks and redbreasted English robins, their aggressive response toward red works appropriately most of the time because other red objects are uncommon in the worlds of these animals. But why don't these and other animals simply *reason* out the correct response rather than relying on stereotyped responses? Ethologists suggest that under conditons that are relatively consistent and predictable, automatic preprogrammed responses may be the most efficient. Thinking about or learning the correct response may take too much time. Releasers have the advantage of focusing the animal's attention on the relevant signal, and IRMs enable an animal to respond rapidly where speed may be essential for survival.

The concepts of releasers, IRMs, and FAPs, however, have been vigorously criticized by some neurophysiologists who were reluctant to believe that the IRM was a specific center of nerve cells (as Lorenz believed) designed to perform just one motor program (the FAP) in response to a specific sign stimulus. Instead, physiologists argued that the central nervous system operates as an integrated whole and that other neural processes also play important roles. Although the IRM concept has been modified somewhat in recent years, there is increasing evidence that IRMs do exist in a sense and that they generate FAPs. For example, in the large nudibranch *Tritonia* (a sluglike mollusc), a giant nerve cell has been

Figure 34-5

Two models of the English robin. The bundle of red feathers is attacked by male robins, whereas the stuffed juvenile bird *(right)* without a red breast is ignored.
From Tinbergen, N. 1951. The study of instinct. Oxford, Eng., Oxford University Press; after Lack, D. 1943. The life of the robin. Cambridge, Eng., Cambridge University Press.

discovered that, when stimulated, plays out a complete stereotyped escape-swimming response that is normally evoked when the animal comes in contact with its enemy, a sea star. This cell looks very much like the IRM envisioned by early ethologists. In vertebrates, executive neurons or brain "centers" for IRMs *may* not exist, but the brain works as though they do.

Instinct and Learning

From the beginning, the mostly invariable and predictable (i.e., stereotyped) nature of IRMs suggested to ethologists that they were dealing with inherited, or **innate** behavior. Many kinds of stereotyped behavior appear suddenly in animals and are indistinguishable from similar behavior performed by older, experienced individuals. Orb-weaving spiders "know" how to build their webs without practice, and male crickets "know" how to court females without lessons from more experienced crickets or by learning from trial and error. To such behaviors the term **instinct**, meaning "driven from within," is applied. It is easy to understand why instinctive behavior is an important adaptation for survival, especially in lower forms that never know their parents. They must be equipped to respond to the world immediately and correctly as soon as they emerge into it. It is also evident that more advanced animals with longer lives and with parental care or other opportunities for social interactions may improve or change their behavior by learning.

Learning and the diversity of behavior

Learning can be defined as the modification of behavior through experience. We should add that, like innate behavior, behavioral modifications by learning are usually adaptive. But unlike innate patterns, which emerge completely functional the first time the animal performs them, learning requires a change in previously existing activity. For example, newly hatched gull chicks crouch down in response to moving objects overhead, an innate and clearly adaptive response to the danger from overflying predators. As the chick grows, it becomes more discriminating and begins to lose its general fear of overflying birds. Its loss of sensitivity to this particular stimulus is a simple kind of learning called **habituation.** For gull chicks, habituation to overflying objects is appropriate, since running and hiding from every passing shadow would consume time and energy better applied to more productive activities.

However, gull chicks do not come to ignore *all* birds flying overhead. Absolute habituation in this instance would be even less appropriate than no habituation at all, since birds of prey are genuine predators of gull chicks. Chicks that fail to hide from an overflying hawk are less likely to survive than chicks that have retained a healthy fear of this predator.

Lorenz and Tinbergen discovered that chicks discriminate predatory birds from harmless songbirds and ducks on the basis of shape, especially the length of the head and neck. When chicks were presented with a silhouette having a hawk-like shape and a short neck such that the head protruded only slightly in front of the wings, they crouched in alarm. Long-necked silhouettes were ignored or aroused only mild interest (Figure 34-6). In another experiment, a model was built having symmetrical wings, with the head and tail shaped so that either the front or the rear of the dummy could be regarded as the head or as the tail (Figure 34-7). When sailed in the direction of the long-necked head, the model resembled a goose and caused no alarm; when sailed in the direction of the short-necked model,

Figure 34-6

Models used by Lorenz and Tinbergen for the study of predator reactions in young fowl. Young gull chicks crouch in alarm when hawk silhouettes *(red)* pass overhead but ignore shapes of harmless birds.
From Tinbergen, N. 1951. The study of instinct. Oxford, Eng., Oxford University Press.

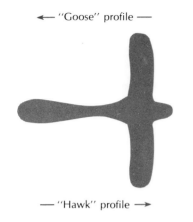

← "Goose" profile —

— "Hawk" profile →

Figure 34-7

Model that drew positive responses from gull chicks when sailed to the right (simulating a hawk) but none when sailed to the left (simulating a goose).

however, it resembled a hawk and did cause alarm. Obviously both the shape of the bird and the direction of its motion are important in recognition.

Sometimes learning is demonstrated in surprising and amusing ways. Tinbergen describes one such incident.*

> In order to sail our models, which crossed a meadow where the birds were feeding or resting, at a height of about 10 yards along a wire, running from one tree to another 50 yards away, either Lorenz or I had to climb a tree and mount the dummy we wanted to test out. One family of geese (which also reacted to some of our dummies) very soon associated tree-climbing humans with something dreadful to come, and promptly called the alarm and walked off when one of us went up.

These observations with models might suggest incorrectly that gull chicks (as well as pheasant and turkey chicks that react similarly) have an innate ability to distinguish short-necked predators from harmless birds having longer necks. But, in fact, subsequent experiments demonstrated that newly hatched chicks, which are at first alarmed by anything that passes overhead—even a falling leaf—gradually become habituated to familiar objects. The chick learns that songbirds and shorebirds are common and harmless features of its world. They never become accustomed to short-necked predators because these are seldom seen. It is the unfamiliar that arouses fear.

The alarm response of herring gull chicks is an example of a simple behavior that becomes altered as the result of maturation and experience. As they grow, the chicks store information about their world and become increasingly selective in their alarm response.

What can we say about the role of genes and the role of the environment in shaping instinctive and learned behavior? We defined an instinctive behavior as one that emerges in complete form the first time the animal reacts to the appropriate stimulus. It must have a genetic foundation because without genes there can be no behavior. Specifically, the animal's genotype contains instructions that result in the construction of a specific neural organization, which permits certain types of behavior. We are not saying that an instinctive behavioral attribute is determined solely by information contained in specific chromosomal loci. The genetic code is an information-generating device that depends on an environment that supplies materials and provides order for embryonic development. A genotype remains just a genotype if the developing organism cannot obtain the substances required to form tissues, organs, and a nervous system.

Learning depends on experience encountered by the organism as it interacts with its environment. It also depends on internal programming because the things an organism learns to do best are determined by the genetic blueprint. The nervous system must be designed to facilitate the acquisition of learning at specific stages in the organism's development. In other words, through its genetically determined development, the brain possesses properties that prepare for its eventual use in the modification of behavior. Learned behavior, like instinctive behavior, contains both genetic and environmental components.

One kind of learned behavior that clearly illustrates the interaction of heredity and environment is **imprinting.** As soon as a newly hatched gosling or duckling is strong enough to walk, it follows its mother away from the nest. After it has followed the mother for some time, it follows no other animal. But if the eggs are

*Tinbergen, N. 1961. The herring gull's world: a study of the social behavior of birds. New York, Basic Books, Inc. Publishers.

hatched in an incubator or if the mother is separated from the eggs as they hatch, the goslings follow the first large, moving object they see. As they grow, the young geese prefer the artificial "mother" to anything else, including their true mother. The goslings are said to be imprinted on the artificial mother.

Imprinting was observed at least as early as the first century AD when the Roman naturalist Pliny the Elder wrote of "a goose which followed Lacydes as faithfully as a dog." Konrad Lorenz was the first to study the imprinting phenomenon objectively and systematically. When Lorenz hand-reared goslings, they formed an immediate and permanent attachment to him and waddled after him wherever he went (Figure 34-1, *A*). They could no longer be induced to follow their own mother or another human being. Lorenz found that the imprinting period is confined to a brief *sensitive* period in the individual's early life and that once established the imprinted bond is retained for life.

What imprinting shows is that the goose or duck brain (or the brain of numerous other birds and mammals that show imprinting-like behavior) is designed to accommodate the imprinting experience. The animal's genotype is provided with an internal template that permits the animal to recognize its mother soon after hatching. Natural selection favors the evolution of animals having a brain structure that imprints in this way because following the mother and obeying her commands are important for survival. The fact that a gosling can be made to imprint to a mechanical toy duck or a human under artificial conditions is a cost to the system that can be tolerated; the disadvantages of the system's simplicity are outweighed by the advantages of its reliability.

Let us cite one final example to complete our consideration of instinct and learning. The males of many species of birds have characteristic territorial songs that identify the singers to other birds and announce territorial rights to other males of that species. Like many other songbirds, the male white-crowned sparrow must learn the song of its species by hearing the song of its father. If the sparrow is hand reared in acoustic isolation in the laboratory, it develops an abnormal song (Figure 34-8). But if the isolated bird is allowed to hear recordings of normal white-crowned sparrow songs during a critical period of 10 to 50 days after hatching, it learns to sing normally. It even imitates the local dialect it hears.

It might appear from this that song characteristic is determined by learning alone. However, if during the critical learning period the isolated male white-crowned sparrow is played a recording of another sparrow species, even a closely related one, it does not learn the song. It learns only the song appropriate to its own species. Thus, although the song must be learned, the brain has been pro-

Figure 34-8

Sound spectrograms of songs of white-crowned sparrows *Zonotrichia leucophrys*. *Above,* Natural song of wild bird; *below,* abnormal song of isolated bird.

Sound spectrograms by M. Konishi, from Alcock, J. 1979. Animal behavior: an evolutionary approach, ed. 2. Sunderland, Mass., Sinauer Associates, Inc.

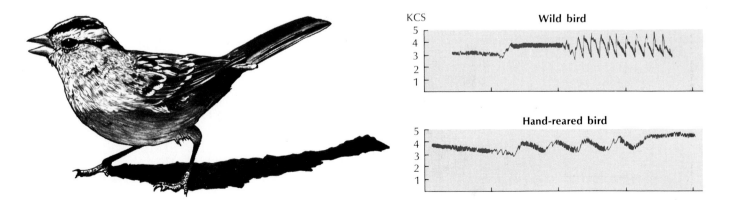

grammed in its development to recognize and learn vocalizations produced by males of its species alone. The sparrow *learns* by example, but its attention has been *innately* narrowed to focus on the appropriate example. Learning the wrong song would result in behavioral chaos, and natural selection quickly eliminates those genotypes that permit such mistakes to occur.

— SOCIAL BEHAVIOR

When we think of "social" animals we are likely to think of highly structured honeybee colonies, herds of antelope grazing on the African plains (Figure 34-9), schools of herring, or flocks of starlings. But social behavior is by no means limited to such obvious examples of animals *of the same species* living together in which individuals influence one another.

In the broad sense, any kind of interaction resulting from the response of one animal to another of the same species represents social behavior. Even a pair of rival males squaring off for a fight over the possession of a female is a social interaction, although our perceptual bias as people might encourage us to label it antisocial. Thus social aggregations are only one kind of social behavior, and indeed not all animal aggregations are social.

Clouds of moths attracted to a light at night, larval barnacles attracted to a common float, or trout gathering in the coolest pool of a stream are animal groupings responding to *environmental* signals. Social aggregations, on the other hand, depend on *animal* signals. They remain together and do things together by influencing one another.

Of course, not all animals showing sociality are social to the same degree. All sexually reproducing species must at least cooperate enough to achieve fertilization; but among most mammals, breeding is about the only adult sociality to occur. However, swans, geese, albatrosses, and beavers, to name just a few, form strong monogamous bonds that last a lifetime. Whether adult sociality is strongly or weakly developed, the most persistent social bonds usually form between mothers and their young, and these bonds for birds and mammals usually terminate at fledging or weaning.

Figure 34-9

Mixed herd of topi and common zebra grazing on the savannah of tropical Africa. Photograph by C.P. Hickman, Jr.

Figure 34-10

Yellow baboons resting and grooming. It is virtually impossible for a predator to approach such a group without being detected.

Courtesy Stuart and Jeanne Altmann.

___ Advantages of Sociality

Living together may be beneficial in many ways. Each species profits in its own particular way; what confers adaptive value to one species may not for another. One obvious benefit for social aggregations is defense, both passive and active, from predators. Musk-oxen that form a passive defensive circle when threatened by a wolf pack are much less vulnerable than an individual facing the wolves alone.

As an example of active defense, a breeding colony of gulls, alerted by the alarm calls of a few, attacks predators *en masse;* this is certain to discourage a predator more effectively than individual attacks. The members of a prairie dog town, although divided into social units called coteries, cooperate by warning each other with a special bark when danger threatens. Thus, every individual in a social organization benefits from the eyes, ears, and noses of all other members of the group (Figure 34-10).

Predators may also be distracted by the confusion effect created by large numbers of prey grouped together. A fish that can chase down and capture a lone crustacean may be unable to concentrate on a single individual in a large aggregation. Predators may even be frightened by a large aggregate of prey, which they would eat if encountered singly.

Sociality offers several benefits to animal reproduction. It facilitates encounters between males and females which, for solitary animals, may consume much time and energy. Sociality also helps synchronize reproductive behavior through the mutual stimulation that individuals have on one another. Among colonial birds the sounds and displays of courting individuals set in motion prereproductive endocrine changes in other individuals. Because there is more social stimulation, large colonies of gulls produce more young per nest than do small colonies. Furthermore, the parental care that social animals provide their offspring increases survival of the brood. Social living provides opportunities for individuals to give aid and to share food with young other than their own. Such interactions

On the Galápagos Islands, hunting by humans in the last century had so greatly thinned the giant tortoise population on one island that the few surviving males and females seldom if ever met. Lichens grew on the females' backs because there were no males to scrub them off during mating! Research personnel saved the tortoise from inevitable extinction by collecting them in a pen, where they began to reproduce.

within a social network have resulted in some intricate cooperative behavior among parents, their young, and their kin.

Of the many other advantages of social organization noted by ethologists, we will mention only a few in this brief treatment: cooperation in hunting for food; huddling for mutual protection from severe weather; opportunities for division of labor, which is especially well developed in the social insects with their caste systems; and the potential for learning and transmitting useful information through the society.

Observers of a seminatural colony of macaques *(Macaca fuscata)* in Japan recount an interesting example of passage of tradition in a society. The macaques were provisioned with sweet potatoes and wheat at a feeding station on the beach of an island colony. One day a young female named Imo was observed washing the sand off a sweet potato in seawater. Still later, when the young members of the troop became mothers, they waded into the sea to wash their potatoes; their offspring imitated them without hesitation. The tradition was firmly established in the troop.

Some years later Imo, an adult, discovered that she could separate wheat from sand by tossing a handful of sandy wheat in the water; allowing the sand to sink, she would scoop up the floating wheat to eat. Again, within a few years, wheat sifting became a tradition in the troop.

Imo's peers and social inferiors copied her innovations most readily. The adult males, her superiors in the social hierarchy, would not adopt the practice but continued laboriously to pick wet sand grains off their sweet potatoes and scour the beach for single grains of wheat.

If social living offers so many benefits, why have not all animals become social through natural selection? The answer is that a solitary existence offers its own set of advantages. In the diverse array of ecological situations in nature, species extract their own optimal ways of life. Species that survive by camouflage from potential predators profit by being well spaced out. Large predators benefit from a solitary existence for a different reason, their requirement for a large supply of prey. Thus, there is no overriding adaptive advantage to sociality that inevitably selects against the solitary way of life. It depends on the ecological situation.

___ Aggression and Dominance

Many animal species are social because of the numerous benefits that sociality offers. This requires cooperation. At the same time animals, like governments, tend to look out for their own interests. In short, they are in competition with one another because of limitations in the common resources that all require for life. Animals may compete for food, water, sexual mates, or shelter, when such requirements are limited in quantity and are therefore worth fighting over.

Much of what animals do to resolve competition is called **aggression,** which we may define as an offensive physical action, or threat, to force others to abandon something they own or might attain. Many ethologists consider aggression to be part of a somewhat more inclusive interaction called **agonistic** (Gr. contest) **behavior,** referring to any activity related to fighting, whether it be aggression, defense, submission, or retreat.

Contrary to the widely held notion that aggressive behavior aims at the destruction or at least defeat of an opponent, most aggressive encounters are ritualized duels that lack the atmosphere of violence that we usually associate with fighting. Many species possess specialized weapons such as teeth, beaks, claws, or horns that are used for protection from, or predation on, other species. Although potentially dangerous, such weapons are seldom used in any effective way against members *of their own species.*

Animal aggression within the species seldom results in injury or death because animals have evolved many symbolic aggressive displays that carry mutually understood meanings. Fights over mates, food, or territory become ritualized jousts rather than bloody, no-holds-barred battles. When fiddler crabs spar for territory, their large claws usually are only slightly opened. Even in the most intense fighting when the claws are used, the crabs grasp each other in a way that prevents reciprocal injury. Rival male poisonous snakes engage in stylized bouts by winding themselves together; each attempts to butt the other's head with its own until one becomes so fatigued that it retreats. The rivals never bite each other. Many species of fish contest territorial boundaries with lateral display threats, the males puffing themselves up to look as threatening as possible. The encounter is usually settled when either animal perceives itself to be obviously inferior, folds up its fins, and swims off. Rival giraffes engage in largely symbolic "necking" matches in which two males standing side by side wrap and unwrap their necks around each other (Figure 34-11). Neither uses its potentially lethal hooves on the other, and neither is injured.

Thus animals fight as though programmed by rules that prevent serious injury. Fights between rival bighorn rams are spectacular to watch, and the sound of clashing horns may be heard for hundreds of meters (Figure 34-12). But the skull is so well protected by the massive horns that injury occurs only by accident. Nevertheless, despite these constraints, aggressive encounters can on occasion be true fights to the death. If African male elephants are unable to resolve dominance conflicts painlessly with ritual postures, they may resort to incredibly violent battles, with each trying to plunge its tusks into the most vulnerable parts of the opponent's body.

More commonly, however, the loser of ritualized encounter may simply run away, or signal defeat by a specialized subordination ritual. If it becomes evident

Figure 34-11

Male Masai giraffes *Giraffa camelopardalis* fight for social dominance. Such fights are largely symbolic, seldom resulting in injury. Photograph by L.L. Rue, III.

Figure 34-12

Male bighorn sheep *Ovis canadensis* fight for social dominance during the breeding season. Photograph by L.L. Rue, III.

A, A dog approaches

B

Figure 34-13

Darwin's principle of antithesis as exemplified by the postures of dogs. **A,** A dog approaches another dog with hostile, aggressive intentions. **B,** The same dog is in a humble and conciliatory state of mind. The signals of aggressive display have been reversed.

From Darwin, C. 1873. Expression of the emotions in man and animals. New York, D. Appleton & Co.

to him that he is going to lose anyway, he is better off communicating his submission as quickly as possible and avoiding the cost of a real thrashing. Such submissive displays that signal the end of a fight may be almost the opposite of threat displays (Figure 34-13). In his book *Expression of the Emotions in Man and Animals* (1873), Charles Darwin described the seemingly opposite nature of threat and appeasement displays as the "principle of antithesis." The principle remains accepted by ethologists today.

Why doesn't the victor of an aggressive contest kill its opponent? A defeated wolf or dog presents its vulnerable neck to the victor as a sign of complete submission. Although the dominant wolf could easily kill the defeated foe and thus remove a competitor, it never does so. The display of submission has effectively inhibited further aggression by the winner. The best explanation for aggressive restraint is that the winner has little to gain by continuing the fight. His superiority is already assured. By continuing the aggression he merely endangers himself, since a defeated opponent fighting for his life might inflict a wound. It is not difficult to see how natural selection would favor genes that induce aggressive restraint. Aggression that is inappropriate runs counter to the maximization of individual fitness. It is maladaptive and consequently is selected against.

The winner of an aggressive competition is dominant to the loser, the subordinate. For the victor, dominance means enhanced access to all the contested resources that contribute to reproductive success: food, mates, territory, and so on. In a social species, dominance interactions often take the form of a dominance hierarchy. One animal at the top wins encounters with all other members in the social group; the second in rank wins all but those with the top-ranking individual.

Such a simple, linear hierarchy was first observed in chicken societies by Schjelderup-Ebbe, who called the hierarchy a "peck-order." Once social ranking is established, actual pecking diminishes and is replaced by threats, bluffs, and bows. Top hens and cocks get unquestioned access to feed and water, dusting areas, and the roost. The system works because it reduces the social tensions that would constantly surface if animals had to fight all the time over social position.

Still, life for the subordinates in any social order is likely to be hard. They are the expendables of the social group. They almost never get a chance to reproduce, and when times get difficult they are the first to die. During times of food scarcity, the death of the weaker members helps to protect the resource for the stronger

Figure 34-14

Two yellow baboons fighting. Fights over social status and females frequently result in injuries, but seldom is either combatant killed in the encounter.

Courtesy Stuart and Jeanne Altmann.

members. Rather than sharing food, the population excess is sacrificed. This is not viewed by contemporary behaviorists as resulting from some direct, purposeful "good for the species" process, however; rather it results as a *consequence* of the individual advantage that the stronger, dominant individuals possess during such circumstances. It is still a matter of survival of the fittest.

___ Territoriality

Territorial ownership is another facet of sociality in animal populations. A **territory** is a fixed area from which intruders of the same species are excluded. This involves defending the area from intruders and spending long periods of time on the site being conspicuous. Territorial defense has been observed in numerous animals: insects, crustaceans, other invertebrates, fish, amphibians, lizards, birds, and mammals, including humans.

Territoriality is generally an alternative to dominance behavior, although both systems may be observed operating in the same species. A territorial system may work well when the population is low, but break down with increasing population density to be replaced with dominance hierarchies with all animals occupying the same space.

Like every other competitive endeavor, territoriality carries both costs and advantages. It is beneficial when it ensures access to limited resources, *unless* the territory boundaries cannot be maintained with little effort. The presumed benefits of a territory are, in fact, numerous: uncontested access to a foraging area; enhanced attractiveness to females, thus reducing the problems of pair-bonding, mating, and rearing the young; reduced disease transmission; reduced vulnerability to predators. But the advantages of holding a territory begin to wane if the individual must spend most of the time in boundary disputes with neighbors.

Most of the time and energy required for territoriality are expended when the territory is first established. Once the boundaries are located they tend to be respected, and aggressive behavior diminishes as territorial neighbors come to recognize each other. Indeed, neighbors may look so peaceful that an observer who was not present when the territories were established may conclude (incorrectly) that the animals are not territorial. A "beachmaster" sea lion (that is, a dominant male with a harem) seldom quarrels with his neighbors who have their own territories to defend. However, he must be on constant vigilance against bachelor bulls who challenge the beachmaster for harem privileges.

Of all vertebrate classes, birds are the most conspicuously territorial. Most male songbirds establish territories in the early spring and defend these vigorously against all males of the same species during spring and summer when mating and nesting are at their height. A male song sparrow, for example, has a territory of approximately three fourths of an acre. In any given area, the number of song sparrows remains approximately the same year after year. The population remains stable because the young occupy territories of adults that die or are killed. Any surplus in the song sparrow population is excluded from territories and thus not able to mate or nest.

Sea birds such as gulls, gannets, boobies, and albatrosses occupy colonies that are divided into very small territories just large enough for nesting (Figure 34-15). These birds' territories cannot include their fishing grounds, since they all forage in the sea where the food is always shifting in location and shared by all.

Territorial behavior is not as prominent with mammals as it is with birds. Mammals are less mobile than birds, and this makes it more difficult for them to

Sometimes the space defended moves with the individual. This individual distance, as it is called, can be observed as the spacing between swallows or pigeons on a wire, in gulls lined up on the beach, or in people queued up for a bus.

Figure 34-15

Gannet nesting colony. Note precise spacing of nests, with each occupant just beyond the pecking distance of its neighbors.
Photograph by C.P. Hickman, Jr.

patrol a territory for trespassers. Instead, many mammals have **home ranges.** A home range is the total area an individual traverses in its activities. It is not an exclusive defended preserve but overlaps with the home ranges of other individuals of the same species.

For example, the home ranges of baboon troops overlap extensively, although a small part of each range becomes the recognized territory of each troop for its exclusive use. Home ranges may shift considerably with the seasons. A baboon troop may have to shift to a new range during the dry season to obtain water and better grass. Elephants, before their movements were restricted by humans, made long seasonal migrations across the African savannah to new feeding ranges. However, the home ranges established for each season were remarkably consistent in size.

Not surprisingly, the size of the home range increases with the body size of mammals, since larger animals require more foraging area to satisfy their energy requirements. Accordingly, the home range may be 1000 square meters for a field mouse, 40 hectares (about 100 acres) for a deer, 150 square kilometers for a grizzly bear, and 4000 square kilometers (about 1500 square miles) for an African hunting dog. In general, carnivores require more foraging space than herbivores, because any given area supports more plant food than animal food.

Animal Communication

Social animals, including people, must be able to communicate with each other. Only through communication can one animal influence the behavior of another. Compared to the enormous communicative potential of human speech, however, nonhuman communication is severely restricted. Whereas human communication is based mainly, although by no means exclusively, on sounds, animals may communicate by sounds, scents, touch, and movement. Indeed any sensory channel may be used, and in this sense animal communication has richness and variety.

Unlike our language, which is composed of words with definite meanings that may be rearranged to generate an almost infinite array of new meanings and images, communication of other animals consists of a limited repertoire of signals. Typically, each signal conveys one and only one message. These messages cannot be divided or rearranged to construct *new kinds* of information. A single message from the sender may, however, contain several bits of relevant information for the receiver.

The song of a cricket announces to an unfertilized female the species of the sender (males of different species have different songs), his sex (only males sing), his location (source of the song), and his social status (only a male able to defend the area around his burrow sings from one location). This is all crucial information to the female and accomplishes a biological function. But there is no way for the male to alter his song to provide additional information concerning food, predators, or habitat, which might improve his mate's chances of survival and thus enhance his own fitness.

The limitations of communication are especially evident in the invertebrates and lower vertebrates. Signals are characteristically stereotyped, and the responses highly predictable and constant throughout the species. This does not mean that such communication is always lacking in intensity and versatility, however. Of the two contrasting examples that follow, mate attraction in silkworm moths illustrates an extreme case of stereotyped, single-message communication that has evolved to serve a single biological function: mating. Yet, in the same group, the insects, we find one of the most sophisticated and complex of all nonhuman communication systems, the symbolic language of bees.

Chemical sex attraction in moths

Virgin female silkworm moths have special glands that produce a chemical sex attractant to which the males are sensitive. Adult males smell with their large bushy antennae, covered with thousands of sensory hairs that function as receptors. Most of these receptors are sensitive to the chemical attractant (a complex alcohol called bombykol, from the name of the silkworm *Bombyx mori*) and to nothing else.

To attract the male, the female merely sits quietly and emits a minute amount of bombykol, which is carried downwind. When a few molecules reach the male's antennae, he is stimulated to fly upwind in search of the female. His search is at first random, but, when by chance he approaches within a few hundred yards of the female, he encounters a concentration gradient of the attractant. Guided by the gradient, he flies toward the female, finds her, and copulates with her.

In this example of chemical communication, the attractant bombykol, which is really a pheromone, serves as a signal to bring the sexes together. Its effectiveness is ensured because natural selection favors the evolution of males with antennal receptors sensitive enough to detect the attractant at great distances (several miles). Males with a genotype that produces a less sensitive sensory system fail to locate a female and thus are reproductively eliminated from the population.

Language of bees

Honeybees are able to communicate the location of food resources when these sources are too distant to be located easily by individual bees. Communication is done by dances, which are mainly of two forms. The form having the most communicative richness is the **waggle dance** (Figure 34-16). Bees most commonly execute these dances when a forager has returned from a rich source, carrying

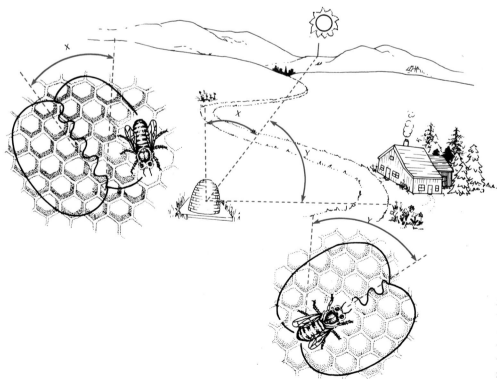

Figure 34-16

Waggle dance of the honeybee, used to communicate both the direction and the distance of a food source. The straight run of the waggle dance indicates direction according to the position of the sun.

The significance of the bee dances was discovered in the 1920s by the German zoologist Karl von Frisch, one of the recipients of the 1973 Nobel Prize. Despite detailed and extensive experiments by von Frisch that supported his original interpretations of the bee dances, the experiments have been criticized, especially by the American biologist A. Wenner, who suggested that the correlation between dance symbolism and food location is accidental. He argued that foraging bees bring back odors characteristic of the food source, and that recruits are stimulated by the dances to search for flowers bearing those odors. Few biologists were prepared to accept this interpretation, but Wenner's beneficial skepticism stimulated more rigorously controlled experiments (Gould, 1976), which established more conclusively than ever before that the bee dances communicate both distance and direction information and that the bees use this information in searching for food. The dances of bees are among the true wonders of the natural world.

either nectar in her stomach or pollen grains packed in basketlike spaces formed by hairs on her legs (p. 362). The waggle dance is roughly in the pattern of a figure-eight made against the vertical surface of the comb. One cycle of the dance consists of three components: (1) a circle with a diameter about three times the length of the bee, (2) a straight run while waggling the abdomen from side to side 13 to 15 times per second, and (3) another circle, turning in the opposite direction from the first. This dance is repeated many times with the circling alternating clockwise and counterclockwise.

The straight, waggle run is the important information component of the dance. Waggle dances are performed almost always in clear weather, and the direction of the straight run is related to the position of the sun. If the forager has located food directly toward the sun, she will make her waggle run straight upward over the vertical surface of the comb. If food was located 60 degrees to the right of the sun, her waggle run is 60 degrees to the right of vertical. We see then that the waggle run points at the same angle relative to the vertical as the food is located relative to the sun.

Distance information about the food source is also coded into bee dances. If the food is close to the hive (less than 50 m), the forager employs a simpler dance called the **round dance.** The forager simply turns a complete clockwise circle, then turns, and completes a counterclockwise circle, a performance that is repeated many times. Other workers cluster around the scout and become stimulated by the dance as well as by the odor of nectar and pollen grains from flowers she has visited. The recruits then fly out and search in all directions but do not stray far. The round dance carries the message that food is to be found in the vicinity of the hive.

If the food source is farther away, the round dances become waggle dances, which provide both distance and direction information. The tempo of the waggle dance is inversely related to the food distance. If the food is about 100 m away, each figure-eight cycle lasts about 1.25 seconds; if 1000 m away, it lasts about 3 seconds; and if about 8 km away (5 miles), is lasts 8 seconds. When food is plentiful, the bees may not dance at all. But when food is scarce, the dancing becomes intense, and the other workers cluster about the returning scouts and follow them through the dance patterns.

Communication by displays

A display is a kind of behavior or series of behaviors that serves a communicative purpose. The release of sex attractant by the female moth and the dances of bees just described are examples of displays; so are the alarm calls of herring gulls, song of the white-crowned sparrow, courtship dance of the sage grouse, and "eyespots" of the hindwings of certain moths, which are quickly exposed to startle potential predators. Of course, just about anything an animal does communicates *something* to other animals that see, hear, or smell it. A true display, on the other hand, is a behavior pattern that has been modified through evolution to make it increasingly effective in serving a communicative function. This process is called **ritualization.** Through ritualization, simple movements or traits become more intensive, conspicuous, or precise, and their original undifferentiated function acquires signal value. The result of such intensification is to reduce the possibility of misunderstanding.

The elaborate pair-bonding behavior of the blue-footed boobies (Figure 34-17) exemplifies this point. These displays are performed with maximum intensity when the birds come together after a period of separation. The male at right in the

Figure 34-17

Pair of Galápagos blue-footed boobies *Sula nebouxii* displays to each other. The male *(right)* is sky pointing; the female *(left)* is parading. Such vivid, stereotyped, communicative displays serve to maintain reciprocal stimulation and cooperative behavior during courtship, mating, nesting, and care of the young.

illustration is sky pointing: the head and tail are pointed skyward and the wings are swiveled forward in a seemingly impossible position to display their glossy upper surfaces to the female. This is accompanied by a high, piping whistle. The female at left, for her part, is parading. She goose steps with exaggerated slow deliberation, lifting each brilliant blue foot in turn, as if holding it aloft momentarily for the male to admire. Such highly personalized displays, performed with droll solemnity, appear comical, even inane to the observer. Indeed the boobies, whose name is derived from the Spanish word "bobo" meaning clown, presumably were so designated for their amusing antics.

Needless to say, for the birds, amusement plays no part in the ceremonies. The exaggerated nature of the displays ensures that the message is not missed or misunderstood. Such displays are essential to establish and maintain a strong pair bond between male and female. This requirement also explains the repetitious nature of the displays that follow one another throughout courtship and until egg laying. Redundancy of displays maintains a state of mutual stimulation between male and female, ensuring the degree of cooperation necessary for copulation and subsequent incubation and care of the young. A sexually aroused male has little success with an uninterested female.

Do Animals Have Mental Experiences?

Most people, although aware of the enormous capabilities of the human brain that separate us from other animal species, take for granted that animals have feelings and awareness analogous to our own. As we watch animals interacting in behavior that resembles human behavior, we may conclude more or less intuitively that animals have humanlike mental experiences. However, most behavioral scientists have deep-seated objections to even considering such a possibility, believing that because mental experiences in animals cannot be observed, measured, or verified, it is consequently meaningless to suggest that they may exist. Strict behaviorists, who argue that even human mental experiences are unique qualities that cannot be objectively defined, rebel against the notion of animal awareness as anthropomorphic, that is, ascribing human characteristics to other species. Yet, as D.R. Griffin argues in his perceptive book *The Question of Animal Awareness,* "it is actually no more anthropomorphic . . . to postulate mental

Whether apes are able to combine signals (that is, "words") to produce sentences—an ability that is crucial to true language—cannot be answered conclusively yet. Furthermore, although apes use signals for requests, commands, and pleas, they do not appear able to make statements aimed at changing the behavior or beliefs of the other party. Critics of the first ape-language studies charged that most of the signed utterances of apes were prompted by the teacher. They argued that the investigators had fallen victim to the "Clever Hans" error, after a horse prodigy of the turn of the century that appeared to do sums in his head and tap out answers with his hoof, but was in fact responding to subtle and inadvertent cues from the trainer. More recent studies appear to have dealt satisfactorily with this criticism by isolating the ape from the teacher during the training sessions.

experiences in another species than to compare its bony structure, nervous system, or antibodies with our own." Griffin points to a curious duality in behavioral studies: rats, pigeons, and monkeys are used as models in behavioral investigations on the implicit assumption that principles revealed from such studies are applicable to human behavior, yet when the question of mental experiences arises, behaviorists reject evolutionary continuity between humans and other animals.

Communication Between Animals and Humans

The reluctance of many behaviorists to acknowledge the existence of even simple mental experiences among animals stems in large part from the enormous difference in versatility and complexity that separates human language from any other known animal communication system. One criterion that we mentioned earlier as supposedly unique for human language is that words with discrete meanings and syllables, many of which are meaningless in themselves, can be arranged into almost endless combinations to generate a vast array of new meanings.

Until recently, it was accepted that this characteristic was absent from all nonhuman communication. However, in the early 1970s a female chimpanzee named Washoe was taught to employ *and combine* gestures, using words from the American Sign Language for the deaf, much as people use spoken words. The discovery that manual gestures and expressive motions were much more appropriate than vocalizations in communicating with apes was considered a major breakthrough in behavioral research. Since Washoe, sign-language studies have been extended to other chimpanzees and to gorillas; several have acquired "vocabularies" of several hundred reliable signs, some invented by the apes themselves.

Clearly, one problem in assessing the versatility of animal communication is understanding what sensory channel an animal is using. The signals may be visual displays (Figure 34-18), odors, vocalizations, tactile vibrations, or electrical currents (as, for example, among certain fishes). Even more difficult is establishing two-way communication between animals and humans, since the investigator must translate meanings into symbols the animal can understand. Furthermore, people are poor social partners for most animals. Nevertheless, when the appropriate communication channel is used, limited communication is possible.

The animal behaviorist Irven De Vore has reported how choosing the proper channel for dialogue can have more than academic interest*:

> One day on the savanna I was away from my truck watching a baboon troop when a young juvenile came and picked up my binoculars. I knew if the glasses disappeared into the troop they'd be lost, so I grabbed them back. The juvenile screamed. Immediately every adult male in the troop rushed at me—I realized what a cornered leopard must feel like. The truck was 30 or 40 feet away. I had to face the males. I started smacking my lips very loudly, a gesture that says as strongly as a baboon can, "I mean you no harm." The males came charging up, growling, snarling, showing their teeth. Right in front of me they halted, cocked their heads to one side—and started lip-smacking back to me. They lip-smacked. I lip-smacked, "I mean you no harm." "I mean *you* no harm." It was, in retrospect, a marvelous conversation. But while my lips talked baboon, my feet edged me toward the truck until I could leap inside and close the door.

We alluded earlier in this chapter to the stereotyped, mechanical nature of instinctive behavior. Yet in watching the interactions between mates during courtship, nest building, and care of the young, it is difficult for the observer to avoid

Figure 34-18

Male baboon shows his white eyelids as an aggressive display. Although subtle, this display communicates a well-understood meaning to other baboons. Photograph by B. Tallmark.

*DeVore, I. 1972. In The marvels of human behavior. Washington, D.C., National Geographic Society.

anthropomorphic interpretations of behavior that resembles human behavior. It is natural and easy to describe animal behavior in terms of human behavior by using words such as "love," "deceit," "happiness," and "gentleness." This is not necessarily false, especially for the higher primates, but the ethologist must always take care to interpret every animal response by the simplest mechanism that is known to work.

Animals are capable of highly organized behavior in the absence of any intelligent appreciation of its purpose. Sometimes instinctive behavior misfires, and such incidents often emphasize its stereotyped nature. The following excerpt contains a perfect example of the automatic release of inappropriate behavior in a gannet colony*:

> A male of an old pair flew into his nest. Normally he would bite his mate on the head with some violence and then go through a long and complicated meeting ceremony, an ecstatic display confined to members of a pair. Unfortunately, the female had caught her lower mandible in a loop of fish netting that was firmly anchored in the structure of the nest. Every time she tried to raise her head to perform the meeting ceremony with her mate she merely succeeded in opening her upper mandible whilst the lower remained fixed in the netting. So she apparently threatened the male with widely gaping beak and he immediately responded by attacking her. With each attack she lowered and turned away her bill (the way in which a female gannet appeases an aggressive male). At once the male stopped biting her and she again turned to greet him but simply repeated the beak-opening and drew another attack. And so it went on despite the fact that these two birds had been mated for years and that the netting, the cause of all the trouble, was clearly visible.

Despite such limitations to the adaptiveness of instinctive behavior, it obviously functions beautifully most of the time. In recognizing that reasoning and insight are not required for effective, highly organized behavior, we should not conclude that lower animals are, as Descartes proclaimed in the seventeenth century, nothing more than machines. Although the gannet in this example lacked the "intelligence" to free his mate by purposefully disentangling her beak from the netting, he was capable of appropriately analyzing the thousands of strategic choices he must make during his lifetime: how to find and hold a mate, where to build a nest and how to defend it, how to locate and catch evasive marine food, and what to do when the environment changes. All this and more requires endless behavioral adjustments to new situations. Conceivably this might be accomplished by a machine, but only by one of staggering complexity.

▬ SUMMARY

Ethology is the study of the behavior, both innate and learned, of animals in their natural habitat. Ethologists have shown that behavioral traits are genetically determined and thus are adaptive and evolve by natural selection.

Much innate (instinctive or inherited) behavior of animals is stereotyped; that is, it is highly predictable and invariable in performance. Ethologists have observed and cataloged numerous stereotyped motor acts, called fixed action patterns, which are triggered, or "released," by specific, and usually simple, environmental stimuli, called sign stimuli. Stereotyped behavior is triggered and coordinated through a "prewired" neural circuit classically called an innate releasing mechanism.

Innate behavior may be modified by learning through experience. A simple

---

*Nelson, B. 1968. Galápagos: islands of birds. London, Longmans, Green & Co., Ltd.

kind of learning behavior is habituation, which is the reduction or elimination of a behavioral response in the absence of any reward or punishment. The modification of the alarm response of herring gull chicks is described as an example of habituation. Another form of learning is imprinting, the lasting recognition bond that forms early in life between the young of many social animals and their mothers.

Social behavior is the behavior of a species when the members interact with one another. In social organizations, animals tend to remain together, communicate with each other, and usually resist intrusions by "outsiders." The advantages of sociality include cooperative defense from predators, improved reproductive performance and parental care of the young, cooperative searching for food, and transmission of useful information through the society. Because social animals compete with one another for common resources (such as food, sexual mates, and shelter), conflicts are often resolved by a form of overt hostility called aggression. Most aggressive encounters between conspecifics are stylized bouts involving more bluff than intent to injure or kill. Dominance hierarchies, in which a priority of access to common resources is established by aggression, is common in social organizations. Territoriality is a related alternative to dominance. A territory is a defended area from which intruders of the same species are excluded.

Communication, often considered the essence of social organization, is the means by which animals influence the behavior of other animals, using sounds, scents, visual displays, touch, or other sensory signals. As compared to the richness of human language, animals communicate with a very limited repertoire of signals. One of the most famous examples of animal communication is that of the symbolic dances of honeybees. Birds communicate by calls and songs and, especially, by visual displays. By ritualization, simple movements have evolved into conspicuous signals having definite meanings.

Review questions

1. Define the term ethology as it is used today, and comment on the aims of, and methods employed by, ethologists.
2. The egg-rolling behavior of greylag geese is an excellent example of stereotyped behavior. Interpret this behavior within the framework of classical ethology, using the terms releaser, sign stimulus, innate releasing mechanism, and fixed-action pattern. Interpret the territorial defense behavior of male three-spined sticklebacks in the same context.
3. Using silhouettes of birds, Lorenz and Tinbergen found that the perception of "hawkness" by gull chicks was mostly a matter of relative neck and tail length. Short necked silhouettes evoked alarm whereas long-necked silhouettes were ignored. These findings could be interpreted to mean that gull chicks enter the world with an innate fear of hawks. What is the evidence against this interpretation?
4. Two kinds of simple learning are habituation and imprinting. Distinguish between these two types of learning and offer an example of each from the living world.
5. The idea that behavior must be *either* innate or learned has been called the "nature versus nurture" controversy. Cite evidence that such a strict dichotomy does not exist. Comment on the role of an animal's genetic instructions in determining innate and learned behavior.
6. Discuss the advantages of sociality for animals. If social living has so many advantages, why do many animals live alone successfully?
7. Suggest why aggression, which might seem to be a counterproductive form of behavior, exists among social animals. Do you think aggression might be more characteristic of K-selected or r-selected species (refer to Chapter 41)?
8. What is the selective advantage, to the probable winner as well as the loser, of ritualized aggression over unrestrained fighting?

9. Of what use is a territory to an animal and how is a territory established and kept? What is the difference between territory and home range?
10. Comment on the limitations of animal communication as compared to that of humans.
11. The dance language used by returning forager honey bees to specify the location of food is a remarkable example of complex communication among "simple" animals. How is direction and distance information coded into the waggle dance of the bees?
12. What is meant by "ritualization" in display communication? What is the adaptive significance of ritualization?
13. Early efforts by humans to communicate vocally with chimpanzees were almost total failures. Recently, however, researchers have learned how to communicate successfully with apes. How was this done?

Selected references

Alcock, J. 1984. Animal behavior: an evolutionary approach, ed. 3. Sunderland, Mass., Sinauer Associates, Inc. *Clearly written and well-illustrated discussion of the genetics, physiology, ecology, and history of behavior in an evolutionary perspective.*

Eaton, G.G. 1976. The social order of Japanese macaques. Sci. Am. **235**:96-106 (Oct.). *Long-term observations of a troop of macaques.*

Gould, J.L. 1976. The dance-language controversy. Q. Rev. Biol. **51**:211-244.

Gould, J.L. 1982. Ethology: the mechanisms and evolution of behavior. New York, W.W. Norton & Co. *This is an excellent introduction to classical and modern ethology.*

Grier, J.W. 1984. Biology of animal behavior. St. Louis, The C.V. Mosby Co. *Lucid, broad perspective of animal behavior with strong emphasis on ecological and evolutionary aspects.*

Griffin, D.R. 1981. The question of animal awareness, ed. 2. New York, The Rockefeller University Press. *An important and provocative book about a controversial question.*

Ghiglieri, M.P. 1985. The social ecology of chimpanzees. Sci. Am. **252**:102-113 (June). *Social structure of wild chimpanzees.*

Huber, F. and J. Thorson. 1985. Cricket auditory communication. Sci. Am. **253**:60-68 (Dec.). *How female crickets respond to the mating song of the male, and the brain neuronal machinery underlying the response.*

Lorenz, K.Z. 1952. King Solomon's ring. New York, Thomas Y. Crowell Co., Inc. *One of the most delightful books ever written about the behavior of animals.*

Savage-Rumbaugh, E.S. 1986. Ape language: from conditioned response to symbol. New York, Columbia University Press. *Details the author's studies as well as the general area of ape language.*

References to Part Three

The books listed in the following selection are mainly textbooks covering wide areas of physiology. They vary considerably in depth and in the level of background in biology and chemistry required of the reader for a full understanding, as indicated in the annotations.

Annual Review of Physiology. 1939-present. Palo Alto, CA, Annual Reviews, Inc. *Contributed review articles selected from all major disciplines of physiology: published annually.*

Davson, H., and M.B. Segal. 1975-1980. Introduction to physiology, 5 vols. London, Academic Press, Ltd. *Advanced, mostly human physiology.*

Eckert, R., and D. Randall. 1982. Animal physiology, ed. 2. *Special emphasis on general principles and on membrane, neural, and sensory physiology.*

Gordon, M.S. (ed.) 1982. Animal function: principles and adaptations, ed. 4. New York, Macmillan, Inc. *Graduate-level vertebrate physiology.*

Guyton, A.C. 1981. Textbook of medical physiology, ed. 6. Philadelphia, W.B. Saunders Co. *A detailed but readable treatment of medical physiology.*

Hoar, W.S. 1983. General and comparative physiology, ed. 3. Englewood Cliffs, N.J., Prentice-Hall, Inc. *Perhaps the best-balanced synopsis of comparative physiology.*

Marshall, P.T., and G.M. Hughes. 1980. Physiology of mammals and other vertebrates, ed. 2. Cambridge, Eng., Cambridge University Press. *Undergraduate comparative physiology; clearly presented and illustrated organ-system approach.*

Miller, J. 1978. The body in question. New York, Random House, Inc. *Selective, historically grounded view of human body function, based on a television series.*

Prosser, C.L. (ed.). 1973. Comparative animal physiology, ed. 3. Philadelphia, W.B. Saunders Co. *Advanced treatise.*

Prosser, C.L. 1986. Adaptational biology: molecules to organisms. New York, John Wiley & Sons. *A wide-ranging synthesis of general biological principles and the interrelationships between disciplines, from molecular to whole-organism biology. Advanced.*

Schmidt-Nielsen, K. 1983. Animal physiology: adaptation and environment, ed. 3. New York, Cambridge University Press. *Clearly written, selective treatment of comparative physiology, emphasizing physiological adaptations to the environment.*

Smith, A. 1986. The body. New York, Viking Penguin, Inc. *A fascinating store of facts about the human body.*

Vander, A.J., J.H. Sherman, and D.S. Luciano. 1985. Human physiology: the mechanisms of the body function, ed. 4. New York, McGraw-Hill Book Co. *Excellent intermediate-level human physiology text.*

Young, J.Z. 1981. The life of vertebrates, ed. 3. Oxford, Eng., Oxford University Press. *The comprehensive treatment and trenchant writing style of this classic have been retained in this updating, which has increased emphasis on physiology, ecology, and behavior.*

Witherspoon, J.D. 1984. Human physiology. New York, Harper & Row, Publishers. *Engagingly written text, stressing principles.*

PART FOUR

CONTINUITY AND EVOLUTION OF ANIMAL LIFE

C.P. Hickman, Jr.

CHAPTER 35

THE REPRODUCTIVE PROCESS

All living organisms are capable of giving rise to new organisms similar to themselves (Principle 11, p. 10). If we admit that all living things are mortal, that every organism is endowed with a life span that must eventually end, we must acknowledge the indispensability of reproduction. Like Samuel Butler who concluded that a chicken is just an egg's way of making another egg, many biologists consider the ability to reproduce to be the ultimate objective of all life processes. Indeed, evolution on earth depends on reproduction, for it is only through *differential reproduction*—the reproduction of some organisms more than others—that changes in populations through time can occur.

The word "reproduction" implies replication, and it is true that biological reproduction almost always yields a reasonable facsimile of the parent unit. However, sexual reproduction, practiced by the majority of animals, promotes the *diversity* needed for survival in a world of constant change. At least for multicellular animals sexual reproduction offers enormous advantages over asexual reproduction, as we shall explain later. The reproductive process, whether sexual or asexual, embodies a basic pattern: (1) the conversion of raw materials from the environment into the offspring or sex cells that develop into offspring of a similar constitution and (2) the transmission of a hereditary pattern or code (DNA) from the parents (Principle 18, p. 12).

▬ NATURE OF THE REPRODUCTIVE PROCESS

The two fundamental modes of reproduction are asexual and sexual. In **asexual** reproduction there is only one parent and there are no special reproductive organs or cells. Each organism is capable of producing genetically identical copies of itself as soon as it becomes an adult. The production of copies is marvelously simple and direct and typically rapid. **Sexual** reproduction (Figure 35-1) as a rule involves two parents, each of which contributes special **sex cells,** or **gametes,** that in union develop into a new individual. The **zygote** formed from this union receives genetic material from *both* parents and accordingly is different from both. The combination of genes produces a genetically unique individual, still bearing the characteristics of the species but also bearing traits that make it different from its parents.

Sexual reproduction, by recombining the parental characters, tends to multiply variations and makes possible a richer and more diversified evolution. Mechanisms for interchange of genes between individuals are more limited in organisms with only asexual reproduction. This would seem to explain why forms that can reproduce only asexually are limited mostly to unicellular forms, which can mul-

A   B   C

Figure 35-1

Sexual reproduction usually involves two parents. Fertilization—union of the egg and sperm—may be external as in, **A**, amphibians (American toad), or internal as in, **B**, insects (gypsy moths) and, **C**, mammals (elk).
A and **B**, Photographs by L.L. Rue, III; **C**, photograph by L. Rue, Jr.

tiply rapidly enough to offset the disadvantages of repeated replication of identical products.

Of course, in asexual organisms such as fungi and bacteria that are haploid (bear only one set of genes), mutations are immediately expressed and evolution can proceed quickly. In sexual animals, on the other hand, a gene mutation is often not expressed immediately, since it may be masked by its normal partner on the homologous chromosome. (Homologous chromosomes are those that pair during meiosis and have genes controlling the same characteristics.) There is only a remote chance that both members of a gene pair will mutate in the same way at the same moment.

Asexual Reproduction

Asexual reproduction is found as a rule only among the simpler forms of life, such as bacteria, protozoa, cnidarians, bryozoans, and a few others. Even in those animal phyla where it occurs, most members employ sexual reproduction as well. In these groups, asexual reproduction ensures rapid increase in numbers when the differentiation of the organism has not advanced to the point of forming highly specialized gametes. Asexual reproduction is rare among the higher invertebrates and the vertebrates; where it appears in the vertebrates it is always by parthenogenesis (see below).

The forms of asexual reproduction in invertebrates are binary fission, budding (both internal and external), fragmentation, and multiple fission. **Binary fission** is common among bacteria and protozoa and to a limited extent among metazoa. In this method the body of the parent is divided into two approximately equal parts, each of which grows into an individual similar to the parent. Fission may be either transverse or longitudinal. **Budding** is an unequal division of the organism. The new individual arises as an outgrowth (bud) from the parent. This bud develops organs like those of the parent and then usually detaches itself. If the bud is formed on the surface of the parent, it is an external bud, but in some cases internal buds, or **gemmules,** are produced. Gemmules are collections of many cells surrounded by a dense covering in the body wall. When the body of the parent disintegrates, each gemmule gives rise to a new individual. External budding is common in the cnidarians, and internal budding in the freshwater sponges. Bryozoa also have a form on internal bud called statoblast. **Fragmentation** is a method in which an organism breaks into two or more parts, each capable of becoming a

It would be a mistake to conclude that asexual reproduction is in any way a "defective" form of reproduction relegated to the minute forms of life that have not yet discovered the joys of sex. Given the facts of their abundance, that they have persisted on earth for 3.5 billion years, and that they form the roots of the food chain on which all higher forms depend, the single-celled asexual organisms are both resoundingly successful and supremely important. For these forms the advantages of asexual reproduction are its rapidity (many bacteria divide every half hour) and simplicity (no sex cells to produce and no time and energy expended in finding a mate).

complete animal. This method is found among the Platyhelminthes, Rhynchocoela, and Echinodermata. In **multiple fission** the nucleus divides repeatedly before division of the cytoplasm, giving rise to many daughter cells almost simultaneously. Multiple fission occurs in a number of protozoan forms.

____ Sexual Reproduction

The essential feature of sexual reproduction is the involvement of *two genetically different parents that combine their genetic material to produce a cell having a new genotype.* The individuals sharing parenthood are characteristically of different **sexes,** male and female (there are exceptions among sexually reproducing bacteria and protozoa in which sexes are lacking). The distinction between male and female is based, not on any differences in parental size or appearance, but on the size and mobility of the sex cells they produce. The **ovum** (egg) is produced by the female. Ova are large (because of stored yolk to sustain early development), nonmotile, and produced in relatively small numbers. The **spermatozoon** (sperm) is produced by the male. Sperm are small, motile, and produced in enormous numbers. Each is a stripped-down package of highly-condensed genetic material designed for the single mission of finding and fertilizing the egg.

There is another crucial event that distinguishes sexual from asexual reproduction: **meiosis,** a distinctive type of gamete-producing nuclear division. As will be described later (p. 762), meiosis differs from ordinary cell division (mitosis) in being a double division. The chromosomes split once, but the cell divides *twice,* producing four cells, each with half the original number of chromosomes (the haploid number). Meiosis is followed by fertilization in which two haploid gametes are combined to restore the normal (diploid) chromosomal number of the species.

The new cell (zygote), which now begins to divide by mitosis, has equal numbers of chromosomes from each parent and accordingly is different from each. It is a unique individual bearing a random assortment of parental characteristics. This is the great strength of sexual reproduction, the "master adaptation" that keeps feeding new varieties into the population.

Many protozoans reproduce by both sexual and asexual modes of reproduction. When sexual reproduction does occur, it may or may not involve male and female gametes. Sometimes two mature sexual parents merely join together to exchange nuclear material or merge cytoplasm. It is not possible in these cases to distinguish sexes.

The male-female distinction is more clearly evident in the metazoa. Organs that produce the germ cells are known as **gonads.** The gonad that produces the sperm is called the **testis** (Figure 35-2) and that which forms the egg, the **ovary** (Figure 35-3). The gonads represent the **primary sex organs,** the only sex organs found in certain groups of animals. Most metazoa, however, have various **accessory sex organs** that transfer and receive sex cells (such as penis, vagina, oviducts, and uterus). In the primary sex organs the sex cells undergo many complicated changes during their development, the details of which are described in a later discussion. In our present discussion we will distinguish biparental reproduction from two alternatives: parthenogenesis and hermaphroditism.

Biparental reproduction

Biparental reproduction is the common and familiar method of sexual reproduction involving separate and distinct male and female individuals. Each has its own reproductive system and produces only one kind of sex cell, spermatozoon or ovum, but never both. Nearly all vertebrates and many invertebrates have separate sexes, and such a condition is called **dioecious** (Gr. *di-*, two, + *oikos*, house).

Parthenogenesis

Parthenogenesis ("virgin origin") is the development of an embryo from an unfertilized egg. Spontaneous, or natural parthenogenesis is known to occur in rotifers, some nematodes, crustaceans, and insects, and in several species of fish, amphibians and desert lizards. Often several generations of asexual parthenogenetic reproduction alternate with biparental sexual reproduction in which the egg is fertilized. In many species of freshwater rotifers, for example, diploid females produce meiotically reduced eggs at some time of the year (p. 242). If not fertilized, these eggs will develop into haploid males. If fertilized by a male, the egg develops into a dormant embryo that hatches, usually the following season, into a diploid female.

In some cases parthenogenesis appears to be the only form of reproduction. Certain populations of whiptail lizards in the American southwest are clones consisting solely of females. The females are diploid because, during meiosis, the chromosomes duplicate themselves to become tetraploid before undergoing the meiotic reductional division to form eggs. The eggs, now diploid, begin development without fertilization.

Hermaphroditism

Animals that have both male and female organs in the same individual are called hermaphrodites, and the condition is called hermaphroditism (from a combination of the names of the Greek god Hermes and goddess Aphrodite). In contrast to the dioecious state of separate sexes, hermaphrodites are **monoecious** (Gr. *monos*, single + *oikos*, house), meaning that both male and female organs are in the same organism. Many invertebrate animals (most flatworms, some hydroids, annelids, and crustaceans) and a few vertebrates (some fishes) are hermaphroditic. Most avoid self-fertilization by exchanging germ cells with each other. For example, although the earthworm bears both male and female organs, its eggs are fertilized by the copulating mate and vice versa (p. 309). Another way of preventing self-fertilization is by developing the eggs and sperm at different times.

___ What Good Is Sex?

The question "What good is sex?" appears to have an easy answer: it serves the purpose of reproduction. But if we rephrase the question to ask, "Why do so many animals reproduce sexually rather than asexually?" the answer is not so apparent. Because sexual reproduction is so nearly universal among animals, it might be inferred that it must be highly advantageous. Yet it is easier to list disadvantages to sex than advantages. Sexual reproduction is complicated, requires more time, and uses much more energy than asexual reproduction. Mating partners must come together, or at least come within proximity of each other, and coordinate their activities to produce young. Many biologists believe that an even more troublesome problem is the "cost of meiosis." A female that reproduces asexually passes all of her genes to her offspring. But when she reproduces sexually the genome is divided during meiosis and only half her genes flow to the next generation. Anoth-

From time to time claims arise that spontaneous parthenogenetic development to term has occurred in humans. A British investigation of about 100 cases in which the mother denied having had intercourse revealed that in nearly every case the child possessed characteristics not present in the mother, and consequently must have had a father. Nevertheless, mammalian eggs very rarely will spontaneously start developing into embryos without fertilization. In certain strains of mice, such embryos will develop into fetuses and then die. The most remarkable instance of parthenogenetic development among the higher vertebrates has been found in turkeys in which certain strains, selected for their ability to develop without sperm, grow to reproducing adults.

Variety may make sexual reproduction a winning strategy for the unstable environment, but some biologists believe that for higher animals sexual reproduction is unnecessary and may even be maladaptive. In animals (humans, for example) in which most of the young survive to reproductive age, there is no demand for novel recombinations to cope with changing habitats. One offspring appears as successful as the next in each habitat. Significantly, asexual reproduction (by parthenogenesis) has evolved in several species of fish and in a few amphibians and reptiles. Such species are exclusively asexual, suggesting that where it has been possible to overcome the numerous constraints to making the transition, asexuality wins out. Would all vertebrates do better to live sexlessly, avoiding the costs of sex? Possibly, but despite the costs, it is unlikely that higher animals if given the choice would opt for sexless lives—especially the males, which would necessarily become extinct. Woody Allen has pointed out that bisexuality, if nothing else, doubles the chance of getting a date on Saturday night.

Most aquatic vertebrates have no need for a penis, since sperm and eggs are liberated into the water in close proximity to each other. However, in terrestrial (and some aquatic) vertebrates that bear their young alive or enclose the egg within a shell, sperm must be transferred to the female. In most birds, this is a rather haphazard process of simply presenting cloaca to cloaca. Only reptiles and mammals have a true penis. In mammals the normally flaccid organ is erected when engorged with blood. Many mammals, although not humans, possess a bone in the penis (baculum), which presumably helps with rigidity. The baculum has a highly variable shape in different species and is a favored object for bewildering comparative anatomy students during practical examinations.

er cost is wastage in the production of males, many of whom fail to reproduce and thus consume resources that could be applied to the reproduction of asexual forms.

Clearly, the costs of sexual reproduction are substantial. How are they offset? Biologists have disputed this question for years without producing an answer that satisfies everyone. Many biologists believe that sexual reproduction, with its breakup and recombination of genomes, keeps producing novel genotypes that *in times of environmental change* may survive and reproduce whereas most others die. Variability, advocates of this viewpoint argue, is sexual reproduction's trump card.

But is variability worth the biological costs of sexual reproduction? The underlying problem keeps coming back: asexual organisms, because they can have more offspring in a given time, appear to be more fit in Darwinian terms. And yet sexuality is determinedly maintained in animals. There is considerable evidence that asexual reproduction is most successful in colonizing new environments. When habitats are empty what matters most is rapid reproduction; variability matters little. But as habitats become more crowded, competition between species for resources increases. Selection becomes more intense and genetic variability— new genotypes produced by recombination in sexual reproduction—furnishes the diversity on which natural selection can act. There are many invertebrates that use both sexual and asexual reproduction, thus enjoying the advantages each has to offer. Whatever the explanation for sex, it seems here to stay.

PLAN OF REPRODUCTIVE SYSTEMS

The basic plan of the reproductive systems is similar in all animals, although differences in reproductive habits, methods of fertilization, and so on have produced many variations. In vertebrate animals the reproductive and excretory systems are often referred to as the **urinogenital system** because of their close anatomical connection. This association is very striking during embryonic development.

The reproductive and excretory systems of the male are usually more intimately connected than they are in the female. For example, in male fishes and amphibians the duct that drains the kidney (**wolffian duct**) also serves as the sperm duct. In male reptiles, birds, and mammals in which the kidney develops its own independent duct (**ureter**) to carry away waste, the old wolffian duct becomes exclusively a sperm duct (**vas deferens**). In all these forms, with the exception of mammals, the ducts open into a **cloaca** (derived, appropriately, from the Latin meaning "sewer"), a common chamber into which the intestinal, reproductive, and excretory canals empty. Higher mammals have no cloaca; instead the urinogenital system has its own opening separate from the anal opening. The **oviduct** of the female is an independent duct that does, however, open into the cloaca in forms that have a cloaca.

The plan of the reproductive system in vertebrates includes (1) **gonads** that produce the sperm and eggs, (2) **ducts** to transport the gametes, (3) **accessory organs** for transferring and receiving gametes, (4) **accessory glands** (exocrine and endocrine) to provide secretions necessary to facilitate and synchronize the reproductive process, and (5) **organs** for storage before and after fertilization. This plan is modified among the various vertebrates, and some of the structures may be lacking altogether.

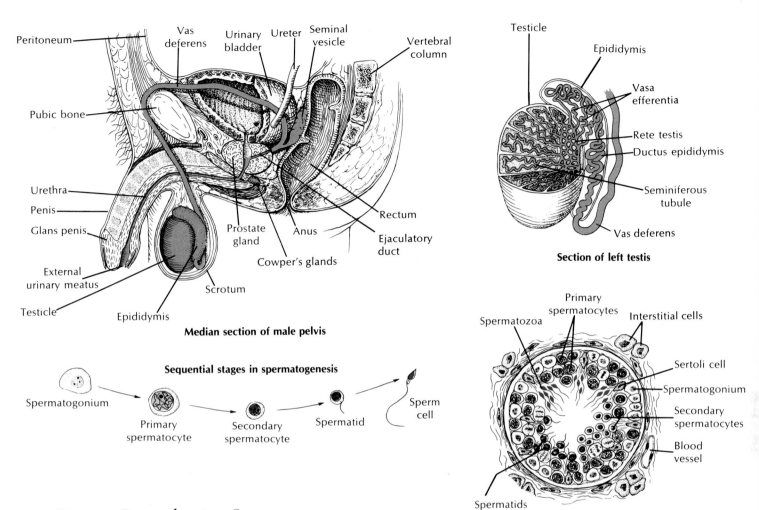

Median section of male pelvis

Sequential stages in spermatogenesis

Section of left testis

Cross section of one seminiferous tubule, showing different stages in spermatogenesis

Figure 35-2

Human male reproductive system.

___ Human Reproductive System

Male reproductive system

The human male reproductive system (Figure 35-2) includes testes, vasa efferentia, vasa deferentia, penis, and glands.

The paired **testes** are the locus of sperm production. Each testis is made up of numerous **seminiferous tubules,** in which the sperm develop (Figure 35-2), and the **interstitial tissue** lying along the tubules, which produces the male sex hormone (testosterone). The two testes are housed in the scrotal sac, which in many mammals hangs down as an appendage of the body. This strange and seemingly insecure arrangement provides an environment of slightly lower temperature, since in at least some forms (including humans) sperm apparently do not form at temperatures maintained within the body.

The sperm are conveyed from the seminiferous tubules to the **vasa efferentia,** small tubes passing to a coiled **vas epididymis** (one for each testis). The epididymis is connected by a **vas deferens** to the **urethra.** From this point the urethra serves to carry both sperm and urinary products through the penis, or external intromittent organ.

Three pairs of glands open into the reproductive channels: **seminal vesicles, prostate glands,** and **Cowper's glands.** Fluid secreted by these glands furnishes food to the sperm, lubricates the passageways for the sperm, and counteracts the acidity of the urine so that the sperm are not harmed.

Female reproductive system

The female reproductive system (Figure 35-3) contains ovaries, oviduct, uterus, vagina, and vulva.

The paired ovaries, slightly smaller than the male testes, contain many thousands of eggs (ova). Each egg develops within a **graafian follicle** that enlarges and finally ruptures to release the mature egg (Figure 35-3). During the fertile period of the woman, approximately 13 eggs mature each year, and usually the ovaries alternate in releasing an egg. Since the female is fertile for only some 30 years, only 300 to 400 eggs have a chance to reach maturity; the others degenerate and are absorbed.

The **oviducts,** or fallopian tubes, are egg-carrying tubes with funnel-shaped openings for receiving the eggs when they emerge from the ovary. The oviduct is

Figure 35-3

Human female reproductive system.

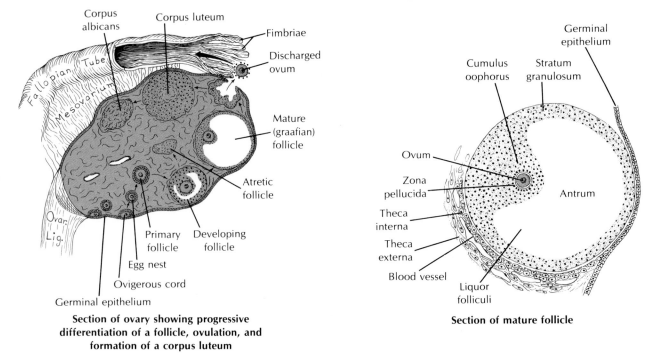

Median section through female pelvis

**Section of ovary showing progressive
differentiation of a follicle, ovulation, and
formation of a corpus luteum**

Section of mature follicle

lined with cilia for propelling the egg in its course. The two ducts open into the upper corners of the **uterus,** or womb, which is specialized for housing the embryo during the 9 months of its intrauterine existence. It is provided with thick muscular walls, many blood vessels, and a specialized lining: the **endometrium.** The uterus varies with different mammals. It was originally paired but tends to fuse in higher forms.

The **vagina** is a muscular tube adapted for receiving the male's penis and for serving as the birth canal during expulsion of the fetus from the uterus. Where the vagina and the uterus meet, the uterus projects down into the vagina to form the **cervix.**

The external genitalia of the female, or vulva, include folds of skin, the **labia majora** and **labia minora,** and a small erectile organ, the **clitoris.** The opening into the vagina is normally reduced in size in the virgin state by a membrane, the **hymen.**

FORMATION OF REPRODUCTIVE CELLS

The animal body has two basically different types of cells: the somatic cells, which are differentiated for specialized functions and die with the individual, and the germinal cells, some of which may contribute to the formation of a zygote and thereby to a new generation. The germinal cells are set aside at the beginning of embryonic development, usually in the endoderm, and migrate to the gonads. The germinal cells, or primordial germ cells, develop into eggs and sperm—nothing else. The other cells of the gonads are somatic cells. They cannot form eggs or sperm, but they are necessary aids in the development of the germinal cells (gametogenesis).

Origin and Migration of Germ Cells

The actual tissue from which the gonads arise appears in early development as a pair of ridges, or pouches, growing into the coelom from the dorsal coelomic lining on each side of the gut near the anterior end of the mesonephros.

Surprisingly perhaps, the primordial ancestors of the germ cells do not arise in the developing gonad but in the yolk-sac endoderm. From studies with frogs and toads, it has been possible to trace the germ cell line back to the fertilized egg, in which a localized area of "germinal cytoplasm" can be identified in the vegetal pole of the uncleaved egg mass. This material can be followed through subsequent cell divisions of the embryo until it becomes situated in primitive sex cells located deep in the endoderm of the yolk sac. From here they migrate to the developing gonads. The primordial sex cells are the future stock of gametes for the animal. Once in the gonad they begin to divide by mitosis, increasing their numbers from a few dozen to several thousand.

At first the gonad is sexually indifferent. In mammals the indifferent gonad has in inherent tendency to become an ovary. In rabbits, for example, the removal of the fetal gonads before they have differentiated will invariably produce a female, even if the rabbit is a genetic male. In the normal male, however, male-determining genes on the Y chromosome regulate the formation of a testis-determining substance that organizes the developing gonad into a testis instead of an ovary. Once formed, the testis secretes the steroid **testosterone.** This hormone masculinizes the fetus, causing the differentiation of penis, scrotum, and the male ducts and glands. It also destroys the incipient breast primordia, but leaves behind

For every structure in the reproductive system of the male or female, there is a homologous structure in the other. This happens because during early development male and female characteristics begin to differentiate from the embryonic genital ridge and two duct systems that at first are identical in both sexes. Under the influence of the sex hormones, the genital ridge develops into the testes of the male and the ovaries of the female. One duct system (wolffian) becomes ducts of the testes in the male and a vestigial structure adjacent to the ovaries in the female. The other duct (müllerian) develops into the oviducts, uterus, and vagina of the female and into the small, vestigial appendix of the testes in the male. Similarly, the clitoris and labia of the female are homologous to the penis and scrotum of the male, since they develop from the same embryonic structures.

the nipples as a reminder of the indifferent ground plan from which both sexes develop.

If the animal is a genetic female, no special stimulus is required for female development. The *absence* of male-determining genes allows the gonad to follow its inherent tendency to become an ovary. This arrangement makes sense for mammals, in which the fetus develops within a female's body (the mother) and is constantly exposed to the mother's sex hormones. If female hormones fostered female development, male embryos would not develop normally unless specially protected from the mother's hormones.

The genetics of sex determination are treated in Chapter 37 (p. 810).

Meiosis: Maturation Division of Germ Cells

Every body cell contains *two* chromosomes bearing genes for the same set of characteristics, and the two members of each pair usually, but not always, have the same size and shape. The members of such a pair are called **homologous** chromosomes. Thus, each cell normally has two genes coding for a given trait, one on each of the homologues. These may be alternative forms of the same gene, and if so they are **allelic genes,** or **alleles.** Sometimes only one of the alleles has an effect on the organism, although both are present in each cell, and either may be passed on to the progeny as a result of meiosis and subsequent fertilization.

During an individual's growth, all the chromosomes of the mitotically dividing cells are replicated during the S period of each cell cycle (Figure 4-22, p. 65), so that each new cell contains the double set of chromosomes. In the reproductive organs the germ cells are formed by a kind of maturation division, called meiosis, which *separates* the double sets of chromosomes. If it were not for this reductional division, the union of egg and sperm would produce an individual with twice as many chromosomes as the parents. Continuation of this process in just a few generations could yield body cells with astronomical numbers of chromosomes.

Meiosis consists of *two* nuclear divisions in which the chromosomes divide only once (Figure 35-4). The result is that mature gametes (eggs and sperm) have only *one* member of each homologous chromosome pair, or a haploid (n) number of chromosomes. In humans the zygotes and all body cells normally have the diploid number (2n), or 46 chromosomes; the gametes have the haploid number (n), or 23.

Most of the unique features of meiosis occur during the prophase of the first meiotic division (Figure 35-4). The two members of each pair of homologous chromosomes come into side-by-side contact (**synapsis**) to form a **bivalent.** Each chromosome of the bivalent has already replicated to form two chromatids, each of which will become a new chromosome. The two chromatids are joined at one point, the centromere, so that each bivalent is made up of two pairs of chromatids, or *four* future chromosomes, and is thus called a **tetrad.** The position or location of any gene on a chromosome is the gene **locus** (pl., **loci**), and in synapsis all gene loci on a chromatid normally lie exactly opposite the corresponding loci on the homologous chromatid. Toward the end of prophase, the chromosomes shorten and thicken and are ready to enter into the first meiotic division. In contrast to mitosis, the centromeres holding the chromatids together *do not divide* at the beginning of anaphase. As a result, one of each pair of double-stranded chromosomes (**dyads**) is pulled toward each pole by the microtubules of the division spindle. Therefore, at the end of the first meiotic division, the daughter cells contain *one of each* of the homologous chromosomes, so the total chromosome number has been reduced to the haploid. However, because the chromatids are still joined by the centromeres, each cell contains the 2n amount of DNA.

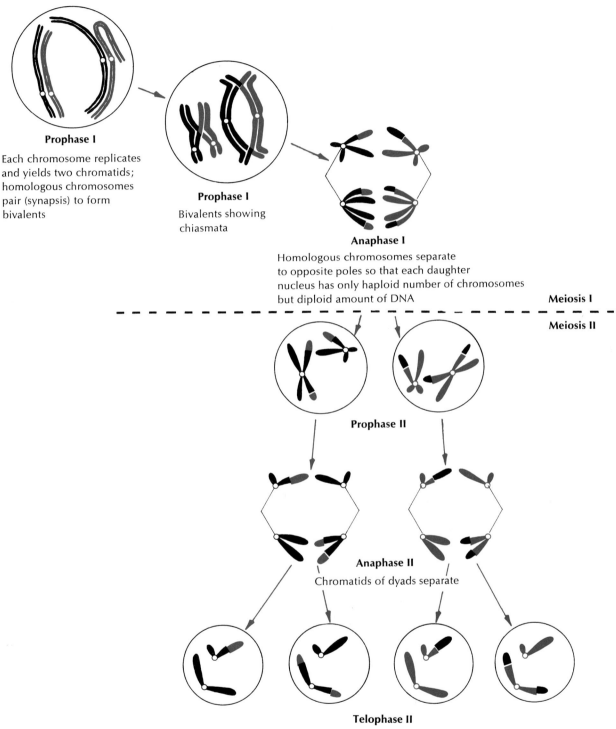

Prophase I

Each chromosome replicates and yields two chromatids; homologous chromosomes pair (synapsis) to form bivalents

Prophase I

Bivalents showing chiasmata

Anaphase I

Homologous chromosomes separate to opposite poles so that each daughter nucleus has only haploid number of chromosomes but diploid amount of DNA

Meiosis I

Meiosis II

Prophase II

Anaphase II

Chromatids of dyads separate

Telophase II

Four haploid cells (gametes) formed, each with haploid amount of DNA

Figure 35-4

Meiosis in a sex cell with two pairs of chromosomes. Compare this figure with Figure 4-23, p. 67, showing mitosis.

Figure 35-5

Section of a seminiferous tubule containing male germ cells. More than 200 long, highly coiled seminiferous tubules are packed in each human testis. As the germ cells differentiate into fertile sperm, they move inward from the periphery of the tubule toward its center. This scanning electron micrograph reveals numerous long tails of mature spermatozoa in the tubule's central cavity (lumen). The differentiation of sperm cells is the result of both mitotic and meiotic divisions. (×525.)

From Tissues and organs: a text-atlas of scanning electron microscopy, by Richard G. Kessel and Randy H. Kardon. W.H. Freeman and Company. Copyright © 1979.

The second meiotic division more closely resembles the events in mitosis. The dyads are split at the beginning of anaphase by division of the centromeres, and single-stranded chromosomes move toward each pole. Thus, by the end of the second meiotic division, the cells have the haploid number of chromosomes and n amount of DNA. Each chromatid of the original tetrad exists in a separate nucleus. Four cells (gametes) are formed, each containing one complete haploid set of chromosomes and only one allele of each gene.

Crossing over

An important event occurs when the chromatids of a bivalent exchange parts with the adjacent homologous (nonsister) chromatid. This phenomenon, called **crossing over,** is clearly shown in Figure 35-4. Crossing over is important because the hereditary material is redistributed between homologous chromatids in one bivalent. The chromosomes exchange equivalent sections bearing allelic genes for the same traits, so that each chromatid contains a full set of genes. But the genes are in new combinations.

While the chromosomes are in synapsis, a strand of one chromatid becomes joined with the homologous chromatid. As prophase continues, the homologues begin to move apart, revealing **chiasmata,** the connection points where crossing over has occurred. There may be one or more chiasmata present in each bivalent, depending on the number of times the adjacent homologues have joined. When the chiasmata pull apart, the exchange is complete. The resulting four gametes at the end of meiosis are all genetically different (Figure 35-4).

Figure 35-6

Mammalian seminiferous tubule. The large primary spermatocyte will give rise to four spermatids by meiosis. Each spermatid will differentiate into a mature spermatozoon. Sertoli cells provide support and protection and probably nutrition for the developing germ cells. (×1100.)

Light micrograph courtesy A. Wayne Vogl.

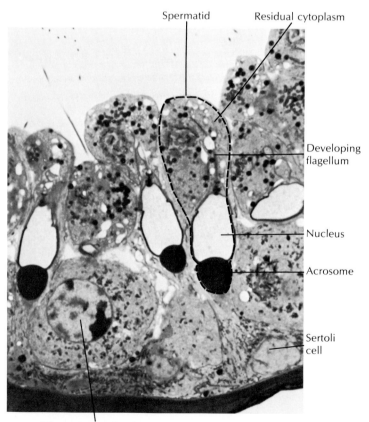

Spermatid

Residual cytoplasm

Developing flagellum

Nucleus

Acrosome

Sertoli cell

Primary spermatocyte

Gametogenesis

The series of transformations that results in the formation of mature gametes (germ cells) is called gametogenesis. Although the same essential processes are involved in the maturation of both sperm and eggs, there are some important differences. Gametogenesis in the testis is called **spermatogenesis,** and in the ovary it is called **oogenesis.**

Spermatogenesis

The walls of the seminiferous tubules contain the differentiating sex cells arranged in a stratified layer five to eight cells deep (Figure 35-5). The outermost layers contain **spermatogonia** (Figure 35-2), which have increased in number by ordinary mitosis. Each spermatogonium increases in size and becomes a **primary spermatocyte.** Each primary spermatocyte then undergoes the first meiotic division, as described previously, to become two **secondary spermatocytes.**

Each secondary spermatocyte enters the second meiotic division without the intervention of a resting period. The resulting cells are called **spermatids** (Figure 35-6), and each contains the haploid number (23 in humans) of chromosomes. A spermatid may have all maternal, all paternal, or both maternal and paternal chromosomes in varying proportions. Without further divisions the spermatids are transformed into mature sperm by losing a great deal of cytoplasm, condensing the nucleus into a head, and forming a whiplike, flagellar tail (Figure 36-7).

From following the divisions of meiosis, it can be seen that each primary spermatocyte gives rise to four functional sperm, each with the haploid number of chromosomes (Figure 35-8).

Figure 35-7

Types of sperm. *Left,* A semidiagrammatic enlargement of the anterior end of the human spermatozoon.

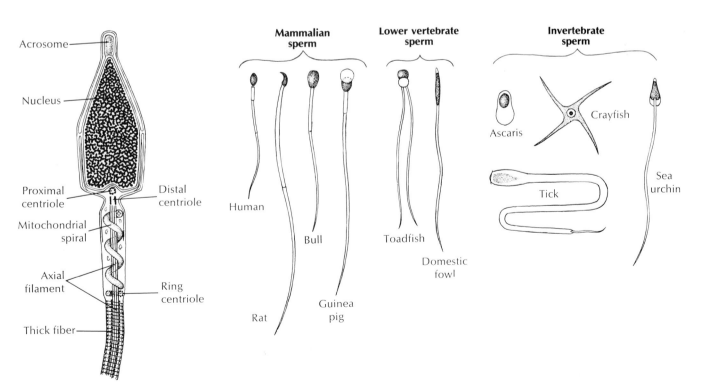

Acrosome

Nucleus

Proximal centriole

Distal centriole

Mitochondrial spiral

Axial filament

Ring centriole

Thick fiber

Mammalian sperm

Human

Bull

Rat

Guinea pig

Lower vertebrate sperm

Toadfish

Domestic fowl

Invertebrate sperm

Ascaris

Crayfish

Tick

Sea urchin

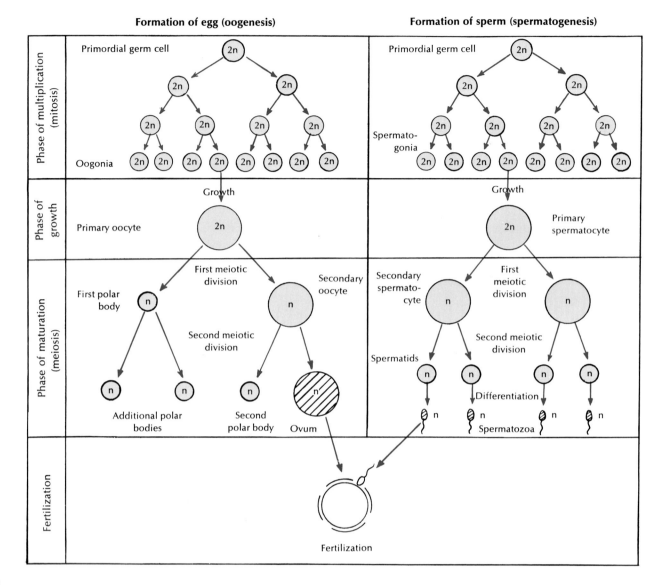

Figure 35-8

Gametogenesis compared in eggs and sperm. *n,* Haploid chromosome number.

Oogenesis

The early sperm cells in the ovary, called **oogonia,** increase in number by ordinary mitosis (Figure 35-8). Each oogonium contains the diploid number of chromosomes. In females after puberty, one of these oogonia typically develops each menstrual month into a functional egg. After the oogonia cease to increase in number, they grow in size and become **primary oocytes.** Before the first meiotic division, the chromosomes in each primary oocyte meet in pairs, paternal and maternal homologues, just as in spermatogenesis. When the first maturation (reduction) division occurs, the cytoplasm is divided unequally. One of the two daughter cells, the **secondary oocyte,** is large and receives most of the cytoplasm; the other is very small and is called the **first polar body** (Figure 35-8). Each of these daughter cells, however, has received half the nuclear material or chromosomes.

In the second meiotic division, the secondary oocyte divides into a large **ootid** and a small polar body. If the first polar body also divides in this division, which sometimes happens, there are three polar bodies and one ootid. The ootid grows into a functional **ovum;** the polar bodies are nonfunctional and disintegrate. The formation of the nonfunctional polar bodies is necessary to enable the egg to get rid of excess chromosomes, and the unequal cytoplasmic division makes possible a large cell with sufficient yolk for the development of the young. Thus, the mature ovum has the haploid number of chromosomes, the same as the sperm. However, each primary oocyte gives rise to only *one* functional gamete instead of four as in spermatogenesis.

_____ Gametes and Their Specializations

Eggs

During **oogenesis** the egg becomes a highly specialized, very large cell containing condensed food reserves for subsequent growth. It is of interest that the prolonged growth and enormous accumulation of food reserves in the oocyte occur **before** the meiotic, or maturation, divisions begin. (This contrasts with spermatogenesis, in which differentiation of the mature sperm occurs only **after** the meiotic divisions.) When the maturation divisions do occur, once the growth phase of the oocyte is complete, they are, as already described, highly unequal: the two meiotic divisions produce one very large mature ovum and two or three polar bodies.

In most vertebrates the egg does not actually complete all the meiotic divisions before fertilization occurs. The general rule is that the egg completes the first meiotic division and proceeds to the metaphase stage of the second meiotic division, at which point progress stops. The second meiotic division is completed and the second polar body extruded only if the egg is activated by fertilization.

The most obvious feature of egg maturation is the deposition of yolk. Yolk, usually stored as granules or more organized platelets, is not a definite chemical substance but may be lipid or protein or both. In insects and vertebrates, all having more or less yolky eggs, the yolk may be synthesized within the egg from raw materials supplied by the surrounding follicle cells, or preformed lipid or protein yolk may be transferred by pinocytosis from follicle cells to the oocyte.

The result of the enormous accumulation of yolk granules and other nutrients (glycogen and lipid droplets) is that an egg grows well beyond the normal limits that force ordinary body (somatic) cells to divide. A young frog oocyte 50 μm in diameter, for example, grows to 1500 μm in diameter when mature after 3 years of growth in the ovary, and the volume has increased by a factor of 27,000. Bird eggs attain even greater absolute size; a hen egg will increase 200 times in volume in only the last 6 to 14 days of rapid growth preceding ovulation.

Thus, eggs are remarkable exceptions to the otherwise universal rule that organisms are composed of relatively minute cellular units. This creates a surface area–to–cell volume ratio problem, since everything that enters and leaves the ovum (nutrients, respiratory gases, wastes, and so on) must pass through the cell membrane. As the egg becomes larger, the available surface per unit of cytoplasmic volume (mass) becomes smaller. As we would anticipate, the metabolic rate of the egg gradually diminishes until, when mature, the ovum is in a sort of suspended animation awaiting fertilization. However, large size is not the only factor leading to quiescence. There is increasing evidence that enzyme and nucleic acid inhibitors, which directly repress metabolic and synthetic activity in the egg, appear toward the end of maturation.

Figure 35-9

These three sex hormones all show the basic four-ring steroid structure. The female sex hormone estradiol-17β (an estrogen) is a C_{18} (18-carbon) steroid with an aromatic A ring *(first ring to left)*. The male sex hormone testosterone is a C_{19} steroid with a carbonyl group (C=O) on the A ring. The female pregnancy hormone progesterone is a C_{21} steroid, also bearing a carbonyl group on the A ring.

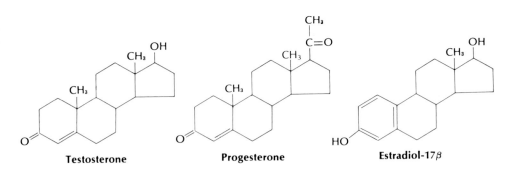

Testosterone   Progesterone   Estradiol-17β

The widely heralded first "test-tube baby," born in England in July 1978, was a medical and technological triumph, the result of years of collaborative research by R.G. Edwards and P. Steptoe. More properly termed external human fertilization (EHF), the procedure is used to overcome infertility caused usually by a malfunction or malformation of the oviduct that prevents fertilization. First, an egg must be collected from the ovary during the final stages of maturation just before ovulation occurs. A special instrument that combines a fiberoptic light and syringe is introduced through a small incision in the abdominal wall, and the egg is sucked out of the ovarian follicle. The egg is placed in a specially compounded solution in a laboratory dish and fertilized with the husband's sperm. The egg is allowed to develop to the eight-cell stage (about 2 days) until the uterus has reached a receptive condition. The embryo is then introduced into the uterus via a tube passed through the cervix and is allowed to implant. The results of this technically simple step are unpredictable, and only about 20% of replanted embryos develop to advanced stages of gestation. Despite the low rate of success, several hundred babies conceived by EHF have been born in the United States, England, and Australia. This, together with recent improvements in the technique, is encouraging news to childless couples who are able to produce normal eggs and sperm.

turn governed by neurosecretory centers in the hypothalamus of the brain. These cells produce neurohormones (polypeptide releasing hormones, pp. 718 and 719) that control synthesis and release of FSH and LH. Through this control system environmental factors such as light, nutrition, and stress may influence reproductive cycles. In many mammals, two separate releasing hormones have been identified: follicle-stimulating hormone–releasing hormone (FSH-RH) and luteinizing hormone–releasing hormone (LH-RH). In humans, however, it appears that both FSH and LH are controlled by a single releasing hormone, LH-RH.

The menstrual cycle (L. *mensis,* month) consists of three distinct phases: **menstrual phase, follicular phase,** and **luteal phase.** Menstruation (the "period") signals the menstrual phase, when part of the lining of the uterus (endometrium) degenerates and sloughs off, producing the menstrual discharge. By day 3 of the cycle the blood levels of FSH and LH begin to rise slowly, prompting some of the ovarian follicles to begin growing. As the follicles grow, they begin to secrete estrogen. As estrogen levels in the blood increase, the uterine endometrium begins to thicken and uterine glands within the endometrium enlarge. By day 10 most of the ovarian follicles that began to develop at day 3 now degenerate, leaving only one (sometimes two or three) to continue ripening until it appears like a blister on the surface of the ovary.

At day 13 or 14 in the cycle, a surge of LH from the pituitary induces the largest follicle to rupture (**ovulation**) releasing the egg onto the ovarian surface. During this critical period, the mature egg must be fertilized within a few hours or it will die. During the luteal phase, a **corpus luteum** ("yellow body") forms from the wall of the follicle that ovulated (Figure 35-11). The corpus luteum, responding to the continued stimulation of LH, secretes progesterone in addition to estrogen. Progesterone ("before bearing [gestation]"), as its name implies, stimulates the uterus to undergo the final maturational changes that prepare it for gestation.

The uterus is thus fully ready to house and nourish the embryo by the time the latter settles out onto the uterine surface, usually about 7 days after ovulation. If fertilization has *not* occurred, the corpus luteum disappears, and its hormones are no longer secreted. Since the uterine lining (endometrium) depends on progesterone and estrogen for its maintenance, their disappearance causes the endometrial lining to deteriorate, leading to the menstrual discharge. However, if the egg has been fertilized and has implanted, the corpus luteum continues to supply the essential sex hormones needed to maintain the mature uterine endometrium. During the first few weeks of pregnancy the developing placenta itself begins to produce the sex hormones progesterone and estrogen and soon replaces the corpus luteum in this function.

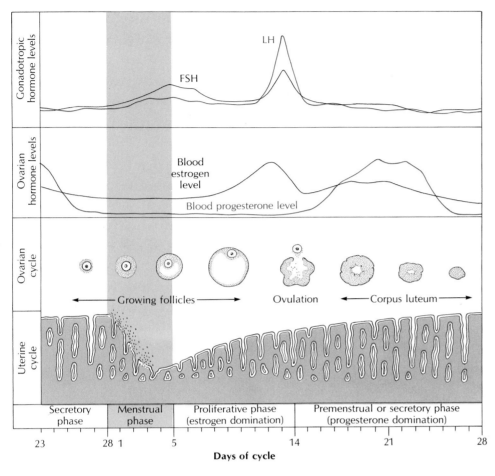

Growing follicles → Ovulation ← Corpus luteum →

| Secretory phase | Menstrual phase | Proliferative phase (estrogen domination) | Premenstrual or secretory phase (progesterone domination) |

23   28 1   5   14   21   28

Days of cycle

Figure 35-10

Human menstrual cycle, showing changes in blood hormone levels and uterine endometrium during the 28-day cycle. FSH promotes maturation of the ovarian egg follicles, which secrete estrogen. Estrogen prepares the uterine endometrium and causes a surge in LH, which in turn stimulates the corpus luteum to secrete progesterone. Progesterone production will persist only if the egg is fertilized; without pregnancy progesterone level declines and menstruation follows.

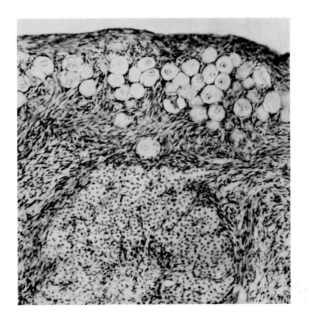

Figure 35-11

Section of cat ovary showing corpus luteum *(large body in lower half of photograph)* and primordial follicles *(above).*
Courtesy J.W. Bamberger.

Is mother's milk superior to artificial formulas? The question is important because only about one baby in three is breast fed, even to 1 month of age, although the trend among new mothers is to breast feed. (By comparison, in 1900 50% of all mothers breast fed their babies until *at least* 12 months of age.) Human milk is different from cow's milk and more complex than a listing of its macronutrient composition (protein, carbohydrate, lipid) suggests. Milk contains several factors known to play vital roles in growth and development, although the function of others remains to be understood. For example, antibodies in human milk help the newborn resist infections. Among the several hormones present in milk is a recently identified class of hormones known as growth factors. The most intensively studied of these is the epidermal growth factor (EGF), which accelerates differentiation of the gastrointestinal tract, stimulates production of digestive enzymes, accelerates maturation of lung tissue (especially important for premature babies), and hastens epidermal growth throughout the body. The numerous effects of EGF are only now being catalogued by researchers, but it seems abundantly clear that no commercial formula can duplicate mother's milk.

As pregnancy advances, progesterone and estrogen prepare the breasts for milk production. The actual secretion and release of milk after birth (lactation) are the result of two other hormones, **prolactin** and **oxytocin.** Milk is not secreted during pregnancy because the placental sex hormones inhibit the release of prolactin by the pituitary. The placenta, like the corpus luteum that preceded it, thus becomes a special endocrine gland of pregnancy. After delivery, many mammals eat the placenta (afterbirth), a behavior that serves to remove telltale evidence of a birth from potential predators.

The male sex hormone **testosterone** (Figure 35-9) is manufactured by the **interstitial cells** of the testes. Testosterone is necessary for the growth and development of the male accessory sex structures (penis, sperm ducts, glands), for development of secondary male sex characters (hair distribution, voice quality, bone and muscle growth), and for male sexual behavior. The same pituitary hormones that regulate the female reproductive cycle, FSH and LH, are also produced in the male, where they guide the growth of the testes and their testosterone secretion.

___ SUMMARY

Reproduction is a universal property of all living organisms. Asexual reproduction, characteristic of simple life forms, is a rapid and direct process by which a single organism produces genetically identical copies of itself. It may occur by fission, budding, fragmentation, or sporulation. Sexual reproduction involves the production of sex cells (gametes), usually by two parents, which combine by fertilization to form a zygote that develops into a new individual. The sex cells are formed by meiosis, reducing the number of chromosomes to haploid, and the diploid chromosome number is restored at fertilization. Sexual reproduction recombines parental characters and thus reshuffles and amplifies genetic diversity; this is important for evolution. Two alternatives to typical biparental reproduction are parthenogenesis, the development of an unfertilized egg, and hermaphroditism, the presence of both male and female organs in the same individual.

Sexual reproduction exacts heavy costs in time and energy, requires cooperative investments in mating, and results in a 50% loss of genetic representation of each parent in the offspring. Sex may have evolved because it produces variable offspring with superior fitness for environmental change.

The male reproductive system of humans includes the testes, composed of seminiferous tubules in which millions of sperm develop, a duct system (vasa efferentia and vas deferens) that joins the urethra, glands (seminal vesicles, prostate, Cowper's), and the penis. The human female system includes the ovaries, containing thousands of eggs within follicles; egg-carrying oviducts; uterus; and vagina.

In vertebrates the primordial germ cells arise in the yolk sac endoderm, then migrate to the gonad. In mammals the gonad will become a testis in response to masculinizing signals from the Y chromosome of the male, and an ovary in the absence of such signals in the female.

In bisexual animals the genetic material is distributed to the offspring in the gametes (eggs and sperm), produced in the process of meiosis. Each somatic cell in an organism has two chromosomes of each kind (homologous chromosomes) and is thus diploid. Meiosis separates the homologous chromosomes, so that each gamete has half the somatic chromosome number (haploid). In the first meiotic division the centromeres do not divide, and each daughter cell receives one of each of the replicated homologous chromosomes with the chromatids still attached to the centromere. At the beginning of the first meiotic division, the replicated homologous chromosomes come to lie alongside each other (synapsis), forming a bivalent. The gene loci on one set of chromatids lie opposite the corresponding loci on the homologous chromatids. Portions of the adjacent chromatids can exchange with the nonsister chromatids (crossing over) to produce new genetic combinations. At the second meiotic division, the centromeres divide, completing the reduction in chromosome number and amount of DNA. The diploid number is restored when the male and female gametes fuse to form the zygote.

Germ cells mature in the gonads by a process called gametogenesis (spermatogenesis in the male and oogenesis in the female), involving both mitosis and meiosis. In spermatogenesis, each primary spermatocyte gives rise by meiosis and growth to four motile sperm, each bearing the haploid number of chromosomes. In oogenesis, each primary oocyte gives rise to only one mature, nonmotile, haploid ovum. The remaining nuclear material is disposed of as polar bodies. During oogenesis the egg accumulates large food reserves.

Reproductive cycles are physiologically controlled by hormones to optimize conditions for development of the young. The menstrual cycle of primates and the estrous cycle of all other mammals are orchestrated by pituitary hormones, follicle-stimulating hormone (FSH) and luteinizing hormone (LH), that control the production of steroid sex hormones by the gonads. Estrogens in the female and testosterone in the male control the growth of the accessory sex structures and the secondary sex characteristics.

Review questions

1. Define asexual reproduction, and describe four forms of asexual reproduction in invertebrates.
2. Define sexual reproduction and explain why meiosis contributes to one of its great strengths.
3. Explain why gene mutations in asexual organisms lead to much more rapid evolutionary change than do gene mutations in sexual forms.
4. Define two alternatives to biparental reproduction—parthenogenesis and hermaphroditism—and offer a specific example of each from the animal kingdom.
5. Define the terms dioecious and monoecious.
6. Name the general location and give the function of the following reproductive structures: seminiferous tubules, vas deferens, urethra, seminal vesicles, graafian follicle, oviducts, endometrium.
7. Name the principal phases of meiosis. What events in meiosis clearly distinguish meiosis from mitosis? Which of the two meiotic divisions most closely resembles mitosis?
8. Draw or trace Figure 35-8, putting in only the circles and arrows. Then completely label the events in oogenesis and spermatogenesis without reference to Figure 35-8. How do oogenesis and spermatogenesis differ?
9. Define, and distinguish between, the terms oviparous, ovoviviparous, and viviparous.
10. How do the two kinds of mammalian reproductive cycles—estrous and menstrual—differ from each other?
11. Explain how the female hormones FSH, LH, and estrogen interact during the menstrual cycle to bring about ovulation and, subsequently, formation of the corpus luteum.
12. Explain the function of the corpus luteum in the menstrual cycle. Compare what happens to the corpus luteum when the ovulated egg is fertilized, and when it is not fertilized.

Selected references

Bell, G. 1982. The masterpiece of nature: the evolution and genetics of sexuality. Berkeley, University of California Press. *Scholarly synthesis of ideas on the meaning of sex. Advanced treatment.*

Daly, M., and M. Wilson. 1978. Sex, evolution, and behavior: adaptations for reproduction. Belmont, Calif., Wadsworth Publishing Co., Inc. *Wide-ranging, well-argued account of sexual strategies.*

Grobstein, C. 1979. External human fertilization. Sci. Am. **240**:57-67 (June). *The procedure and the issues it raises are explored.*

Halliday, T. 1982. Sexual strategy: survival in the wild. Chicago, University of Chicago Press. *Semi-popular treatment of sexual strategies, especially vertebrate mating systems, rested in a framework of natural selection. Well-chosen illustrations.*

Jones, R.E. 1984. Human reproduction and sexual behavior. Englewood Cliffs, N.J., Prentice-Hall, Inc. *Attractive, well-written, and well-illustrated undergraduate textbook.*

Margulis, L., and D. Sagan. 1986. Origins of sex: three billion years of genetic recombination. New Haven, Yale University Press. *Authors trace the origins of sex back to the trading of genes in bacteria and argue that sex, a process that produces a genetically new individual, became accidentally linked to reproduction, the creation of new organisms.*

Segal, S.J. 1974. The physiology of human reproduction. Sci. Am. **231**:52-62 (Sept.). *Explains how understanding the hormonal control of reproduction makes possible the development of modern contraceptive methods.*

C H A P T E R   3 6

PRINCIPLES OF

DEVELOPMENT

Growth and differentiation are fundamental characteristics of life (see Principles 22 and 23, p. 13). In Chapter 6 a brief survey of early embryonic development was given: types of eggs, fertilization and formation of the zygote, cleavage, and early morphogenesis. In this chapter we will first consider another dimension of development: the coordinating and regulating processes that guide the destiny of a growing embryo. The latter part of the chapter is devoted to the embryology of amniotes, especially humans, and the special adaptations that support development.

The phenomenon of development is a remarkable, and in many ways awesome, process. How is it possible that a tiny, spherical fertilized human egg, scarcely visible to the naked eye, can unfold into a fully formed, unique person, consisting of thousands of billions of cells, each cell performing a predestined functional or structural role? How is this marvelous unfolding controlled? Obviously all the information needed is contained within the egg, principally in the genes of the egg's nucleus. Thus, all development originates from the structure of the nuclear DNA molecules and in the egg cytoplasm surrounding the nucleus. But knowing where the blueprint for development resides is very different from understanding how this control system guides the conversion of a fertilized egg into a fully differentiated animal. This remains a major—many consider it *the* major—unsolved problem of biology. It has stimulated a vast amount of research on the processes and phenomena involved; some early, and in many cases tentative, answers have emerged.

——EARLY CONCEPTS: PREFORMATION VERSUS EPIGENESIS

Early scientists and lay people alike speculated at length about the mystery of development long before the process was submitted to modern techniques of biochemistry, molecular biology, tissue culture, and electron microscopy. An early and persistent idea was that the young animal was preformed in the egg and that development was simply a matter of unfolding what was already there. Some claimed they could actually see a miniature of the adult in the egg or the sperm (Figure 36-1). Even the more cautious argued that all the parts of the embryo were in the egg, needing only to unfold, but so small and transparent they could not be

Figure 36-1

Preformed human infant in sperm as imagined by seventeenth-century Dutch histologist Niklass Hartsoeker, one of the first to observe sperm with a microscope of his own construction. Other remarkable pictures published during this period depicted the figure sometimes wearing a nightcap! From N. Hartsoeker, 1694, Essai de Dioprique.

seen. The concept of **preformation** was strongly advocated by most seventeenth- and eighteenth-century naturalist-philosophers.

William Harvey, the great physiologist of blood circulation fame, published a book on embryology in his old age (1651) that contained the conclusion "Omne vivum ex ova" ("Every living thing from the egg"). Harvey could not accept the naive doctrine of preformation but was unable to offer experimental evidence to refute it.

In 1759 German embryologist Kaspar Friedrich Wolff clearly showed that in the earliest developmental stages of the chick, there was no embryo, only an undifferentiated granular material that became arranged into layers. These continued to thicken in some areas, to become thinner in others, to fold, and to segment, until the body of the embryo appeared. Wolff called this **epigenesis** ("origin upon or after"), an idea that the fertilized egg contains building material only, somehow assembled by an unknown directing force. Current ideas of development are essentially epigenetic in concept, although we know far more about what directs growth and differentiation.

Development describes the progressive changes in an individual from its beginning to maturity. Development in sexual multicellular organisms usually begins with a fertilized egg that divides mitotically to produce all of the many kinds of cells in the body. This generation of cellular diversity is called **differentiation.** Differentiation is not simply an unfolding of tissues of organs, but rather is the appearance of structures not present at an earlier stage. At each stage of development new structures arise from conditions created in preceding stages. But once a particular structure embarks on a course of final differentiation, it becomes irrevocably committed. It no longer depends on the stages that preceded it, nor does it have the option of forming anything different. Once a structure becomes committed, it is said to be **determined.** Thus, differentiation is **progressive,** it is **irreversible,** and the final differentiated tissue is relatively **stable.**

DIRECT AND INDIRECT DEVELOPMENT

Even before fertilization, the egg is programmed to follow a special developmental course. The eggs of most marine invertebrates and those of many freshwater vertebrates typically develop into feeding larval forms that look and behave in a totally different manner from the mature adult. In such **indirect** development, the presence of a larval stage enables the animal to exploit microplankton or other food supplies and thus enhance its chances for a successful **metamorphosis** to the juvenile-adult body form. The eggs of animals having indirect development often contain rather limited food reserves and are programmed for rapid differentiation into tiny, free-swimming larvae that must find their own food to sustain further growth (Figure 36-2).

The eggs of mammals, birds, and reptiles show **direct** development; that is, the egg develops directly into a juvenile without passing through a highly specialized larval phase. In direct development, either the egg contains a large amount of stored food, as do the huge, yolky bird and reptile eggs, or the early embryo develops a nourishing, parasitic relationship with the maternal body, as do the small, nearly yolkless eggs of mammals. Direct development is more leisurely than indirect. The eggs are programmed to develop first into extensive sheets of unspecialized cells and then to differentiate into the tissues and organs of a juvenile.

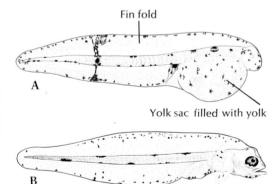

Fin fold

Yolk sac filled with yolk

Figure 36-2

Fish embryos showing yolk sac. **A,** The 1-day-old larva of a marine flounder has a large yolk sac. **B,** By the time the 10-day-old larva has developed a mouth and primitive digestive tract, its yolk supply has been exhausted. It must now catch its own food to survive and continue growing.

From Hickman, C.P., Jr.: Washington State Fisheries Research Papers 2:38-47, 1959.

___ FERTILIZATION

Fertilization is the union of male and female gametes to form a **zygote**. This process accomplishes two things: it provides for the recombination of paternal and maternal genes, thus restoring the original diploid number of chromosomes characteristic of the species; and it activates the egg to begin development.

For a species to survive, it must ensure that fertilization will occur and that enough progeny will survive to continue the population. Most marine invertebrates and many marine fishes simply set their eggs and sperm adrift in the ocean and rely on the random swimming movements of sperm to make chance encounters with eggs. Even though an egg is a large target for a sperm, the enormous dispersing effect of the ocean, the short life span of the gametes (often just a few minutes), and the limited range of the tiny sperm all conspire against an egg and a sperm coming together. Accordingly, each male releases countless millions of sperm at spawning. The odds against fertilization are further reduced by coordinating the time and place of spawning of both parents.

Ensuring that some eggs are fertilized, however, is not enough. The ocean is a perilous environment for a developing animal, and most never make it to maturity. Most marine invertebrates and fishes are strong r-strategists (p. 915), with high reproductive rates and high mortality. Thus, the females produce huge numbers of eggs. The common gray cod of the North American east coast regularly spawns 4 to 6 million eggs, of which only two or three eggs on the average will reach maturity. A single oyster may lay 25 to 60 million eggs at a time. Animals that provide more protection to their young produce fewer eggs. Internal fertilization, which is characteristic of flatworms, gastropod molluscs, and some other invertebrates, as well as sharks, reptiles, birds, and mammals, avoids dispersion of the gametes and protects them. However, even with internal fertilization, vast numbers of sperm must be released by the male into the female tract. Furthermore, the events of ovulation and insemination must be closely synchronized, and the gametes must remain viable for several hours to accomplish fertilization. Sperm may have to travel a considerable distance to reach the egg in the female genital tract, many parts of which may be hostile to sperm. Experiments with rabbits have shown that of the approximately 10 million sperm released into the female vagina, only about 100 reach the site of fertilization.

___ Oocyte Maturation

During oogenesis, described in the preceding chapter, the egg accumulates yolk reserves in preparation for future growth. The nucleus also grows rapidly in size during egg maturation, although not as much as the cell as a whole. It becomes bloated with nuclear sap and so changed in appearance that it was often given another name in older literature: the **germinal vesicle.**

Large amounts of both DNA and RNA accumulate during oogenesis. Early in the oogenesis of large amphibian eggs, the chromosomes become vastly expanded by thin loops thrown out laterally from the chromosomal axis; because of their fuzzy appearance, the chromosomes at this time are called **lampbrush chromosomes** (Figure 38-13, p. 829). Ribosomal RNA is rapidly synthesized on DNA templates as each chromomere puffs out in lampbrush loops, exposing the double helix of DNA. Messenger RNA is also undergoing intense synthesis by the nucleus during oocyte growth. The messenger RNA subsequently controls the synthesis of protein in the early embryo. In addition, a considerable quantity of DNA appears in the cytoplasm where it becomes bound to mitochondria and yolk platelets. Its function is unknown. Toward the end of oocyte maturation, this intense nucleic acid and protein synthesis gradually winds down. The lampbrush chromosomes contract and the ribosomal RNA and proteins mix with the egg cytoplasm.

The oocyte is now poised for the maturation (meiotic) divisions, which rid it of excess chromosomal material (in the polar bodies, described in Chapter 35, p. 766) and ready it for fertilization. A vast amount of synthetic activity has preceded this stage, all of which has packed the egg with reserve materials required for subsequent development.

In many vertebrates the primary oocyte proceeds with the maturation divisions only after it has approached its ultimate size. Even then, however, the divisions typically are interrupted by one or two periods of arrested development before maturation is completed. Human eggs, for example, begin their first meiotic divisions in the ovaries of the prenatal fetus. At birth, all the eggs of the human female infant are developmentally arrested in the diplotene stage of first meiotic prophase. This has been referred to as the "dictyate" stage; the egg is still a primary oocyte, and thus it remains until it is ovulated, a delay of perhaps 12 to 45 years! The first meiotic division is finally completed just before ovulation, and the second meiotic division occurs after the egg is fertilized.

Fertilization and Activation

We began this section with the statement that fertilization accomplishes two ends: (1) the fusion of sperm and egg to create a new individual, and (2) activation of the quiescent egg to restore metabolic activity and start cleavage. Normally it is sperm contact that activates the egg. However, we have already seen that sperm—and thus males—are not always required for development. Some animals, such as certain species of rotifers, crustaceans, insects, fishes, and desert lizards, are naturally parthenogenetic (p. 757), and the eggs of many species can be artificially induced to develop without sperm fertilization. Thus neither sperm contact nor the paternal genome is *always* essential for egg activation, although, in the great majority of species, artificial parthenogenesis will not take an embryo very far down the developmental path.

Activation is a dramatic event. In sea urchin eggs, the contact of the spermatozoon with the egg surface sets off almost instantaneous changes in the egg cortex. At the point of contact a **fertilization cone** appears into which the sperm head is later drawn (Figure 36-3). As soon as the first sperm enters the egg, important changes occur in the egg membrane that block the entrance of additional sperm which, in marine eggs especially, may quickly surround the egg in swarming numbers and bind to its surface (Figure 36-4). The entrance of more than one sperm (**polyspermy**) must be prevented because the union of more than two haploid nuclei would be ruinous for normal development. In the sea urchin egg, entrance of the first sperm is instantly followed by an electrical potential change in the egg membrane that prevents additional sperm from fusing with the membrane. This is followed by the elevation of a **fertilization membrane** (Figure 36-5), a permanent barrier that completes the egg's protection against polyspermy. In mammals relatively few sperm even reach the egg, and a rapid block to polyspermy is not important and seems not to occur.

Once the successful sperm binds to the egg surface, **cortical granules,** which form a layer beneath the plasma membrane, rupture and release materials that fuse together to build up a new egg surface, the **hyaline layer.** The hyaline layer lies between the inner plasma membrane and the outer fertilization membrane. Addi-

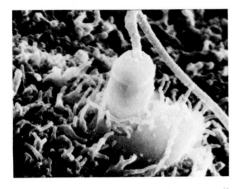

Figure 36-3

The dramatic moment of fertilization of a sea urchin egg. This scanning electron micrograph shows the sperm head attached to the egg surface. A fertilization cone, shown lifting from the surface, and the rapidly elongating microvilli, soon engulf the sperm. (×29,000.)
Courtesy Gerald Schatten, Florida State University.

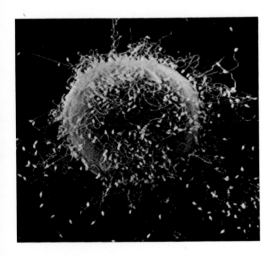

Figure 36-4

Binding of sperm to the surface of a sea urchin egg. Only one sperm penetrates the egg surface, the others being blocked from entrance by rapid changes in the egg membrane. Unsuccessful sperm soon fall away from the egg. (×1425.)
Courtesy Gerald Schatten, Florida State University.

Figure 36-5

Fertilization of an egg. **A,** Many sperm swim to the egg. **B,** The first sperm to penetrate the protective jelly envelope and contact the egg membrane causes the fertilization cone to rise and engulf the sperm head. **C,** The fertilization membrane begins to form at the site of penetration and spreads around the entire egg, preventing the entrance of additional sperm. **D,** Male and female pronuclei approach one another, lose their nuclear membranes, swell, and fuse. **E,** The mitotic spindle forms, signaling creation of a zygote and heralding the first cleavage of a new embryo.

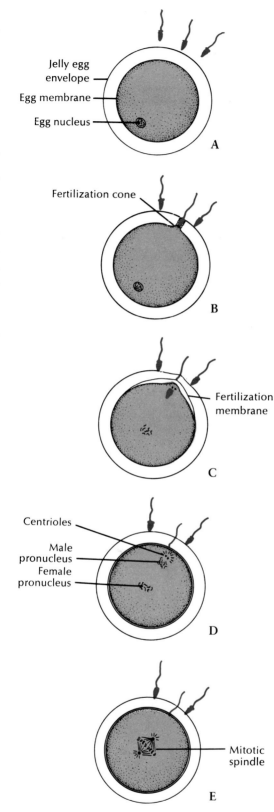

tional sperm continue to bind to the egg for a few seconds, but as the hyaline layer forms, all but the successful sperm detach.

The **cortical reaction,** as these changes are called, is a crucial event in development. It seems to produce a complete molecular reorganization of the egg cortex. It also serves to remove one or more inhibitors that have blocked the energy-yielding systems and protein synthesis in the egg and have kept the egg in its quiescent, suspended-animation state. Almost immediately after the cortical reaction, polyribosomes form from the enormous supply of monoribosomes stored in the cytoplasm and begin producing protein. Normal metabolic activity is restored. The reduction divisions, if not completed, are brought to completion. The male and female pronuclei then fuse, and the zygote enters into cleavage.

CLEAVAGE AND EARLY DEVELOPMENT

During cleavage the zygote divides repeatedly to convert the large, unwieldly cytoplasmic mass into a large number of small, maneuverable cells (called **blastomeres**) clustered together like a mass of soap bubbles. There is no growth during this period, only subdivision of mass, which continues until normal somatic cell size and nucleocytoplasmic ratios are attained. At the end of cleavage the zygote has been divided into many hundreds of thousands of cells (about 1000 in polychaete worms, 9000 in amphioxus, and 700,000 in frogs). There is a rapid increase in DNA content during cleavage as the number of nuclei and the amount of DNA are doubled with each division. Apart from this, there is little change in chemical composition or displacement of constituent parts of the egg cytoplasm during cleavage. **Polarity,** that is, a polar axis, is present in the egg, and this establishes the direction of cleavage and subsequent differentiaton of the embryo. Usually cleavage is very regular although enormously affected by the quantity of yolk present and whether cleavage is radial or spiral, as described previously (p. 101).

Regulative and Mosaic Development

In many species belonging to the Deuterostomia division of the animal kingdom (for example, echinoderms and vertebrates), the blastomeres of the early embryo are capable of regulating, or adjusting, their development to imposed damage. If, for example, a sea urchin embryo at the four-cell stage is placed in calcium-free seawater and gently shaken, the blastomeres will separate. Replaced in normal seawater, each separated blastomere will develop into a complete, although small, larva (Figure 36-6, *A*). Early blastomeres of frog and rabbit embryos also can be separated and yield complete embryos. Such embryos are classified as **regulative** because separated blastomeres can regulate themselves into well-proportioned embryos. One can even fuse two sea urchin embryos together to produce one giant, but otherwise normal, larva. This is another form of regulation, because

Figure 36-6

Regulative and mosaic cleavage. **A,** Regulative (indeterminate) cleavage. Each of the early blastomeres (such as that of the sea urchin) when separated from the others develops into a small pluteus larva. **B,** Mosaic (determinate) cleavage. In the mollusc, when the blastomeres are separated, each gives rise to only a part of an embryo. The larger size of one of the defective larvae is the result of the formation of a polar lobe *(P)* composed of clear cytoplasm of the vegetal pole, which this blastomere alone receives.

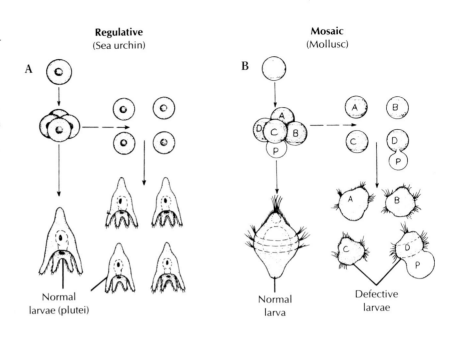

Regulative
(Sea urchin)

Mosaic
(Mollusc)

Normal larvae (plutei)

Normal larva

Defective larvae

Mammalian eggs are highly regulative. Human identical twins are monozygotic because they develop from a single zygote. Since most monozygotic twins share the same placenta, the separation of the embryos must have occurred after implantation when the inner cell mass, from which the embryo will form, consists of dozens of cells (a small percentage of monozygotic twins separate fully at cleavage and each twin has its own placenta). It is possible to fuse two mammalian embryos together to form one giant embryo that develops normally. Recently, mice have been reared from three fused embryos representing six parents!

each of the original embryos is now contributing to only one-half the fused embryo.

The eggs of many invertebrates having spiral cleavage and belonging to the Protostomia division of the animal kingdom (for example, flatworms, nemerteans, molluscs, and annelids) lack this early pliancy. If the early blastomeres of the embryo of one of these animals are separated, each will continue to develop for a period as though it were still part of the whole; each forms a defective, partial embryo (Figure 36-6, *B*). Such eggs are classified as **mosaic** and are said to have mosaic (**determinate**) development.

These two types of development point to two separate mechanisms of embryonic determination. In mosaic eggs, localization of organ-forming regions occurs by segregation of cytoplasm within the egg even before cleavage begins. After cleavage begins, a cell's fate is determined by the portion of egg cytoplasm that it acquires during cleavage. With such mosaic eggs it is often possible to map out the fates of specific areas on the uncleaved egg surface that are known to be presumptive for specific structures (Figure 36-7). In regulative eggs, localization happens later in cleavage. Once localization is established in either type of egg, blastomeres cannot be damaged or removed from the embryo without corresponding defects in development.

Figure 36-7

Segregation of organ-forming regions in, **A,** an ascidian (tunicate) egg and, **B,** an eight-cell embryo. Although the tunicates are deuterostomes, they have mosaic eggs like the protostomes. Mapping of prospective regions in this way would be impossible in a highly regulative mammalian egg.

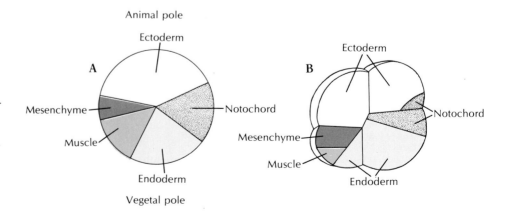

Animal pole

Ectoderm

Mesenchyme

Muscle

Notochord

Endoderm

Vegetal pole

Ectoderm

Mesenchyme

Muscle

Notochord

Endoderm

Significance of the Cortex

Early cleavage proceeds independent of nuclear genetic information, guided instead by information deposited in the egg during its maturation. It was once believed that the visible particulate material in the cytoplasm had determinative properties. However, it was soon discovered that if the egg's cytoplasm is thoroughly displaced by centrifugation, the embryo still develops normally. The only thing not affected by strong centrifugation of the egg is the plasma membrane and the **cortex,** a gelatinous layer just beneath the plasma membrane. The embryo will also develop perfectly when up to 50% of the egg's cytoplasm is removed with a micropipette.

These experiments show conclusively that the cortical layer of the egg, and not the cytoplasm, contains an invisible but dynamic organization that determines the pattern of cleavage. Cortical organization is at first labile (especially so in regulative eggs) but soon becomes regionally fixed and irreversible. Thus, as cleavage progresses, the cortex and the cytoplasm enclosed become segregated into territories having specific determinative properties. This explains why different blastomeres bear different cytodifferentiation properties.

Delayed Nucleation Experiments

Another kind of experiment that demonstrates the importance of specific cortical regions of the egg was first carried out in 1928 by Hans Spemann, a German embryologist. Spemann put ligatures of human hair around newt eggs (amphibian eggs similar to frog eggs) just as they were about to divide, constricting them until they were almost, but not quite, separated into two halves (Figure 36-8). The nucleus lay in one half of the partially divided egg; the other side was anucleate, containing only cytoplasm. The egg then completed its first cleavage division on the side containing the nucleus; the anucleate side remained undivided. Eventually, when the nucleated side had divided into about 16 cells, one of the cleavage nuclei would wander across the narrow cytoplasmic bridge to the anucleate side. Immediately this side began to divide.

With both halves of the embryo containing nuclei, Spemann drew the ligature tight, separating the two halves of the embryo. He then watched their devel-

Figure 36-8

Spemann's delayed nucleation experiments. Two kinds of experiments were performed. **A,** Hair ligature was used to partly divide an uncleaved fertilized newt egg. Both sides contained part of the gray crescent. The nucleated side alone cleaved until a descendant nucleus crossed over the cytoplasmic bridge. Then both sides completed cleavage and formed two complete embryos. **B,** Hair ligature was placed so that the nucleus and gray crescent were completely separated. The side lacking the gray crescent became an unorganized piece of belly tissue; the other side developed normally.

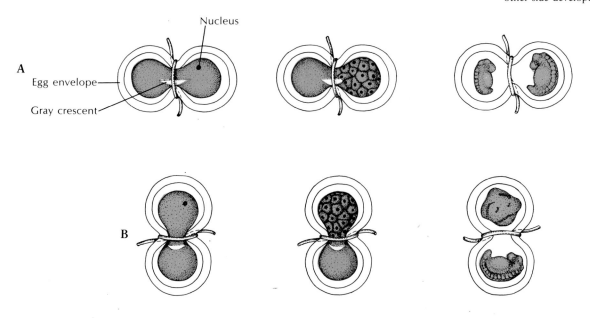

Nucleus

A

Egg envelope

Gray crescent

B

opment. Usually two complete embryos resulted. Although the one embryo possessed only one-sixteenth the original nuclear material, and the other contained fifteen sixteenths, they both developed normally. The one-sixteenth embryo was initially smaller, but it caught up in size in about 140 days. This showed that every nucleus of the 16-cell embryo contained a complete set of genes; all were equivalent.

Sometimes, however, Spemann observed that the nucleated half of the embryo developed only into an abnormal ball of "belly" tissue, although the half that received the delayed nucleus developed normally. Why should the more generously endowed fifteen-sixteenths embryo fail to develop and the small one-sixteenth embryo live? The explanation, Spemann discovered, depended on the position of the **gray crescent**, a pigment-free area on the egg surface. In amphibian eggs the gray crescent forms at the moment of fertilization and determines the plane of bilateral symmetry of the future animal. If one half of the constricted embryo lacked a part of the gray crescent, it would not develop.

Obviously, then, there must be cytoplasmic inequalities involved. The egg cortex in the area of the gray crescent contains substances that are essential for normal development. Since all the nuclei of the 16-cell embryo are equivalent, each capable of supporting full development, it is clear that the cytoplasmic environment is crucial to nuclear expression. The nuclei are all alike, but the cytoplasm (or cortex) throughout the embryo is not all alike. In some way chemically different regions of the egg, created during the early growth of the egg (oogenesis), are segregated out into specific cells during early cleavage. Thus, although all nuclei have the same information content, cytoplasmic substances surrounding the nucleus determine what part of the genome will be expressed and when (see Principle 25, p. 14).

PROBLEM OF DIFFERENTIATION

The egg cortex contains the primary guiding force for early development (cleavage and blastula formation), and synthetic activities during this period are supported by nucleic acids and reserves laid down in the egg cytoplasm *before* fertilization. However, continued differentiation past cleavage or the blastula stage requires the action of genetic information in the nuclear chromosomes. There is considerable evidence to show that this is so. If the maternal nucleus is removed from an activated but unfertilized frog egg, the anucleate egg will cleave more or less normally but arrest at the outset of gastrulation, when nuclear information is required.

Hybridization experiments have also been carried out in which the maternal nucleus is removed from the egg by microsurgery, just after the egg has been fertilized, but before the pronuclei have time to fuse. Such a zygote, called an **andromerogone**, contains *maternal* cytoplasm and a *paternal* (sperm) nucleus. The egg will cleave to the bastula stage normally; whether it will gastrulate depends on the compatibility of the sperm with the maternal cytoplasm. (Blastula and gastrula formation are described in Chapter 6, pp. 102-104.) If the egg was fertilized with a sperm of the same species, it develops into a haploid (but otherwise normal) tadpole. If it was fertilized with a sperm of a closely related species, it may develop to the neurula stage before development stops. If fertilized with a distantly related species, its development will arrest in the blastula or early gastrula stage. These differences are explained as varying degrees of immunological incompatibilities that appear when the paternal (sperm) genes begin coding for proteins in the embryo. The more foreign the sperm, the greater the incompatibility and the earlier the stage of arrest.

Finally, the most direct evidence that nuclear genes become active at or shortly before gastrulation is that the production of messenger RNA increases sharply at this time, followed by the production of transfer and ribosomal RNA. This has been demonstrated with the use of radioactively labeled RNA precursors.

____ Gene Expression

Gastrulation is a critical time of orderly and integrated cell movements. By the spreading of epithelial sheets of cells (epiboly), by folding processes (invagination and evagination) that segregate single epithelial layers and by splitting processes (delamination) that segregate multiple layers, the three prospective germ layers—ectoderm, mesoderm, and endoderm—are formed. This is followed by the rapid differentiation of germ layers into rudimentary, and later functional, tissues and organs.

As cells differentiate, they obviously use only a part of the instructions their nuclei contain. Cells that are differentiating into a thyroid gland are not concerned with that part of the genome that codes for a striated muscle, for example. The unneeded genes are in some way switched off. They are not destroyed, however, because a nucleus from a cell at an advanced stage of development transplanted into an activated egg whose nucleus has been removed can support development of a complete organism (see marginal note).

The basic problem is **gene expression.** What determines that a particular blastomere of, say, a 100-cell embryo will differentiate into muscle or skin or thyroid gland? If genes are the same in all nuclei of the early embryo, the only way differences can develop is through interaction between these nuclei and the surrounding cytoplasm. We have already seen that the basic polarity of the egg and the organizing qualities of the egg cortex provide an early opportunity for such interaction.

Let us briefly summarize what is now known of the transmission of genetic information. Genetic information is coded in the sequence of nucleotides in DNA molecules (see Principle 19, p. 12 and Chapter 38). DNA serves as a template for the synthesis of messenger RNA in the nucleus. Messenger RNA then migrates out through nuclear pores into the cytoplasm, where it attaches to a ribosome. Here the messenger RNA serves as a template for the synthesis of specific proteins (see Principle 20, p. 12). In this way cytoplasmic proteins are formed that may be specific for that cell.

Evidence to date suggests that at the beginning of development most nuclear genes are inactive. Only small amounts of messenger RNA are being produced on the DNA templates. As development proceeds, new cytoplasmic proteins appear, indicating that more genic DNA is producing messenger RNA. Evidently different genes are activated (or "derepressed") in different parts of the young embryo, and this differential gene activity is responsible for embryonic differentiation.

What is the mechanism by which genes are repressed and then derepressed at specific times during development? We do not know at present, although several possible genetic control mechanisms have been identified (p. 833). Whatever the mechanism is, it seems certain that the kinds of cytoplasm present in different cells determine what genes come into action. Nucleocytoplasmic interactions form the basis of the organized differentiation of tissues that characterizes animal development.

Two Oxford scientists, J.B. Gurdon and R. Laskey, were able to grow a normal, reproductive adult frog from an unfertilized egg containing the nucleus of a differentiated intestinal cell of a frog tadpole. The technique they used had been developed by R. Briggs and T.J. King at Indiana University in 1952. Using minute glass needles, they were able to pluck out the nucleus from an egg and replace it with a nucleus from a fully differentiated cell. These experiments are of great significance because they demonstrate that a cell, or actually the nucleus from a cell, can be forced backward from its specialized state and once again make available all of its genetic information.

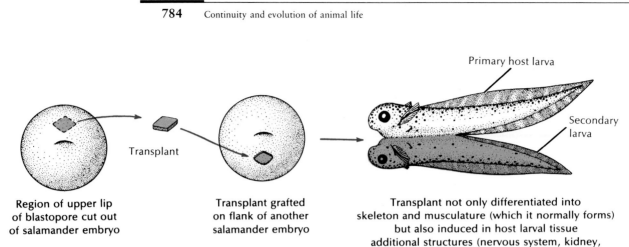

Primary host larva

Secondary larva

Transplant

Region of upper lip of blastopore cut out of salamander embryo

Transplant grafted on flank of another salamander embryo

Transplant not only differentiated into skeleton and musculature (which it normally forms) but also induced in host larval tissue additional structures (nervous system, kidney, enteron, etc.) necessary to form a whole second embryo

Figure 36-9

Spemann's famous organizer experiment.

Embryonic Inductors and Organizers

Embryonic induction is a widespread phenomenon in development. It is the capacity of one tissue to evoke a specific developmental response in another. The classic experiments were reported in 1924 by Hans Spemann and Hilde Mangold. When a piece of dorsal blastopore lip from a salamander gastrula is transplanted into a ventral or lateral position of another salamander gastrula, it invaginates and develops a notochord and muscle somites. It also induces the *host* ectoderm to form a neural tube. Eventually a whole system of organs develops where the graft was placed, and this grows into a nearly complete secondary embryo (Figure 36-9). This creature is composed partly of grafted tissue and partly of induced host tissue.

It was soon found that *only* grafts from the dorsal lip of the blastopore were capable of inducing the formation of a complete or nearly complete secondary embryo. This area corresponds to the presumptive areas of notochord, muscle somites, and prechordal plate. It was also found that only ectoderm of the host would develop a nervous system in the graft and that the reactive ability is greatest at the early gastrula stage and declines as the recipient embryo gets older.

Spemann called the dorsal lip area the **primary organizer** because it was the only region capable of inducing the development of a complete embryo in the host. This region can be traced back to the gray crescent of the undivided fertilized egg, a cortical area that, as we have already seen, is crucial in directing early development of amphibian embryos (Figure 36-8). Many other examples of embryonic induction have been discovered, both in amphibians, a favorite for study, and in other vertebrate and invertebrate species.

Efforts to discover the chemical nature of the inductor have met with much less success. Embryologists were dismayed to find that a great variety of denatured animal materials could cause inductive responses. Obviously the dorsal lip graft did not "organize" a particular differentiation; rather it evoked a response that was already part of the induced tissue's total developmental capacity and repertoire. The inductor appears to play a permissive rather than an instructive role by providing a favorable environment in which the stimulated tissue releases a preprogrammed response.

In normal development, induction operates by cell-to-cell contact. However, it may not always require cell-to-cell contact because, when small pieces of inductors and responding tissues are cultured together but separated from one another by a porous nitrocellulose filter, induction still occurs. The inducing substance must be a chemical signal that diffuses through the pores of the filter to reach the responding tissue.

Usually a tissue that has differentiated acts as an inductor for an adjacent undifferentiated tissue. Timing is important. Once a primary inductor sets in motion a specific developmental pattern in one tissue, numerous secondary inductions follow. What emerges is a sequential pattern of development involving not only inductors but cell movement, changes in adhesive properties of cells, and cell proliferation. This is why developmental patterns are so difficult to unravel. There is no master control panel that directs development, but rather a sequence of local patterns in which one step in development is a subunit of another.

The chemical nature of inductors is still unknown. Despite the enormous research effort that has been devoted to induction, the phenomenon has remained so resistant to experimental unveiling that embryologists find themselves with little more insight into the question than they had half a century ago. Nevertheless, in showing that one step in development is a necessary requisite for the next, Hans Spemann's induction experiments were among the most significant events in experimental embryology.

___AMNIOTES AND THE AMNIOTIC EGG

Reptiles, birds, and mammals form a natural grouping of vertebrates distinguished by having an amniotic egg. They are called **amniotes,** meaning that they develop an amnion, one of the extraembryonic membranes that make the development of these forms unique among animals.

As rapidly growing living organisms, embryos have the same basic animal requirements as adults—nutrients, fuels, oxygen, and disposal of wastes. For the embryos of the many invertebrates that develop in the external environment, gas exchange is a simple matter of direct diffusion. The egg may contain only enough yolk to sustain the embryo through hatching, then it must begin to feed itself. After hatching, the embryo, now a free-swimming larva, is on its own.

Yolk enclosed in a membranous **yolk sac** is a conspicuous feature of all fish embryos (Figure 36-2). The yolk is gradually used up as the embryo grows; the yolk sac shrinks and finally is enclosed within the body of the embryo. The mass of yolk is an **extraembryonic** structure, since it is not really a part of the embryo proper, and the yolk sac is an **extraembryonic membrane.** Bird and reptile eggs are also provided with large amounts of yolk to support early development. In birds (direct development), the yolk reaches relatively massive proportions, since it must nourish a baby bird to a much more advanced stage of development at hatching than a larval fish.

In abandoning an aquatic life for a land existence, the first terrestrial animals had to evolve a sophisticated egg containing a complete life-support system. Thus appeared the **amniotic egg,** equipped to protect and support the growth of embryos on dry land. In addition to the **yolk sac** containing the nourishing yolk are three other membranous sacs: **amnion, chorion,** and **allantois.** All are referred to as extraembryonic membranes because, again, they are accessory structures that develop beyond the embryonic body and are discarded when the embryo hatches. (We should note that since the amnion is only one of four extraembryonic membranes, the "amniotic" egg could just as well be called a "chorionic" or an "allantoic" egg.)

The **amnion** is a fluid-filled bag that encloses the embryo and provides a

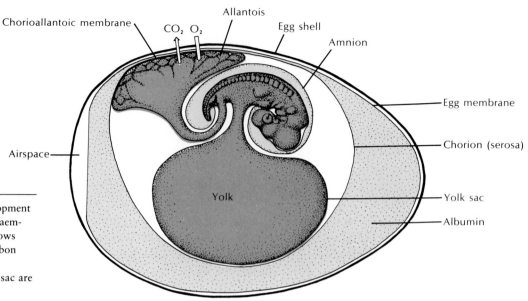

Figure 36-10

Amniotic egg at early stage of development showing a chick embryo and its extraembryonic membranes. Porous shell allows gaseous exchange of oxygen and carbon dioxide. Circulatory channels from embryo's body to allantois and yolk sac are not shown.

private aquarium for development (Figure 36-10). Floating freely in this aquatic environment, the embryo is fully protected from shocks and adhesions. The evolution of this structure, from which the amniotic egg takes its name, was crucial to the successful habitation of land.

The **allantois,** another component in the support system for embryos of land animals, is a bag that grows out of the hindgut of the embryo (Figure 36-10). It collects the wastes of metabolism (mostly uric acid). At hatching, the young animal breaks its connection with the allantois and leaves it and its refuse behind in the shell.

The **chorion** (also called **serosa**) is an outermost extraembryonic membrane that completely encloses the rest of the embryonic system. It lies just beneath the shell (Figure 36-10). As the embryo grows and its need for oxygen increases, the allantois and chorion fuse to form a **chorioallantoic membrane.** This double membrane is provided with a rich vascular network, connected to the embryonic circulation. Lying just beneath the porous shell, the vascular chorioallantoic membrane serves as a provisional lung across which oxygen and carbon dioxide can freely exchange. Although nature did not plan for it, the chorioallantoic membrane of the chicken egg has been used extensively by generations of experimental embryologists as a place to culture chick and mammalian tissues.

The great importance of the amniotic egg to the establishment of a land existence cannot be overemphasized. Amphibians must return to water to lay their eggs. But reptiles, even before they took to land, developed the amniotic egg with its self-contained aquatic environment enclosed by a tough outer shell. Protected from drying out and provided with yolk for nourishment, such eggs could be laid on dry land, far from water. Reptiles were thus freed from aquatic life and could become the first true terrestrial tetrapods.

Incidentally, the sexual act itself comes from the requirement that the egg be fertilized *before* the eggshell is wrapped around it, if it is to develop. Thus, the male must introduce the sperm into the female tract so that the sperm can reach the egg before it passes to that part of the oviduct where the shell is secreted. Hence, as one biologist put it, it is the eggshell, and not the devil, that deserves the blame for the happy event we know as sex.

__HUMAN DEVELOPMENT

The amniotic egg, for all its virtues, has one basic flaw: placed neatly in a nest, it makes fine food for other animals. It was left for the mammals to evolve an alternative for early development: allow the embryo to grow within the protective confines of the mother's body. This has resulted in important modifications in the development of mammals as compared with other vertebrates. The earliest mammals, descended from early reptiles, were egg layers. Even today the most primitive mammals, the monotremes (for example, duck-billed platypus, spiny anteater), lay large yolky eggs that closely resemble bird eggs. In the marsupials (pouched mammals such as the opossum and kangaroo), the embryos develop for a time within the mother's uterus. But the embryo does not "take root" in the uterine wall, as do the embryos of the more advanced **placental mammals,** and consequently it receives little nourishment within the mother. The young of marsupials are therefore born immature and are sheltered and nourished in a pouch of the abdominal wall.

All other mammals, the placentalians, nourish their young in the uterus by means of a **placenta.**

__Early Development of the Human Embryo

The eggs of all placental mammals, although relatively enormous on a cellular scale, are small by egg standards. The human egg is approximately 0.1 mm in diameter and barely visible to the unaided eye. It contains very little yolk. After fertilization in the oviduct, the dividing egg begins a 5-day journey down the oviduct toward the uterus, propelled by a combination of ciliary action and muscular peristalsis. Cleavage is very slow: 24 hours for the first cleavage and 10 to 12 hours for each subsequent cleavage. By comparison, frog eggs cleave once every hour. Cleavage produces a small ball of 20 to 30 cells (called the morula) within which appears a fluid-filled cavity, the **segmentation** cavity (Figure 36-11). This is comparable to the blastocoel of a frog's egg (Figure 6-6, p. 103). The embryo is now called a **blastocyst.**

At this point, development of the mammalian embryo departs radically from that of lower vertebrates. A mass of cells, called the **inner cell mass,** develops on one side of the peripheral cell, or **trophoblast,** layer (Figure 36-11). The inner cell mass forms the embryo, whereas the surrounding trophoblast forms the placenta. When the blastocyst is approximately 6 days old and composed of approximately 100 cells, it contacts and implants into the uterine endometrium (Figure 36-12). Very little is known of the forces involved in implantation, or why, incidentally,

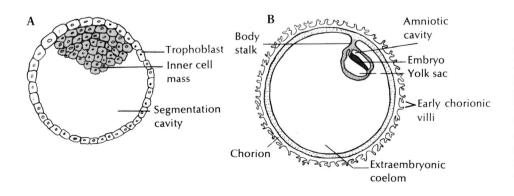

Figure 36-11

Early development of human embryo. **A,** Blastocyst at about 7 days. The trophoblast gives rise to the chorion, which later becomes part of the placenta; the inner cell mass becomes the embryo. **B,** Blastocyst at about 15 days. Note the amniotic cavity and yolk sac; primary villi are developing from the chorion to begin establishment of the placenta.

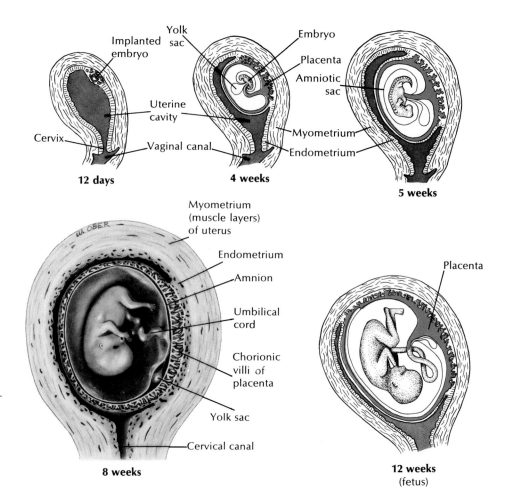

Figure 36-12

Development of a human embryo from 12 days to 12 weeks. The 8-week-old embryo and uterus are drawn to actual size. As the embryo grows, the placenta develops into a disclike structure attached to one side of the uterus. Note the vestigial yolk sac.

the intrauterine birth control devices (coil and loop) are so effective in preventing successful implantation. On contact, the trophoblast cells proliferate rapidly and produce enzymes that break down the epithelium of the uterine endometrium. This allows the blastocyst to sink into the endometrium. By the eleventh or twelfth day the blastocyst, totally buried, has eroded through the walls of capillaries and small arterioles; this releases a pool of blood that bathes the embryo. At first the minute embryo derives what nourishment it requires by direct diffusion from the surrounding blood. But very soon a remarkable fetal-maternal structure, the **placenta,** develops to assume these exchange tasks.

___ Placenta

The placenta is a marvel of biological engineering. Serving as a provisional gas-exchange organ, intestine, and kidney for the embryo, it performs elaborate selective activities without ever allowing the maternal and fetal blood to intermix (very small amounts of fetal blood may regularly escape into the maternal system; this causes no harm *unless* an Rh-*negative* mother who is sensitized to the Rh antigen is carrying an Rh-positive fetus [p. 621]). The placenta permits the entry of foodstuffs, hormones, vitamins, and oxygen and the exit of carbon dioxide and metabolic wastes. Its action is highly selective; it allows some materials to enter that are chemically quite similar to others that are rejected.

The two circulations are physically separated at the placenta by an exceedingly thin membrane (in places only 2 μm thick), across which materials are transferred by diffusive interchange. The transfer occurs across thousands of tiny, fingerlike projections, called **chorionic villi,** which develop from the original chorion membrane (Figures 36-11 and 36-12). These projections sink like roots into the uterine endometrium after the embryo implants. As development proceeds and embryonic demands for food and gas exchange increase, the great proliferation of villi in the placenta vastly increases its total surface area. Although the human placenta at term measures only 18 cm (7 inches) across, its total absorbing surface is approximately 13 square meters—50 times the surface area of the skin of the newborn infant.

Since the mammalian embryo is protected and nourished by the placenta, what becomes of the various embryonic membranes of the amniotic egg whose functions are no longer required? Surprisingly perhaps, all of these special membranes are still present, although they may be serving a new function. The yolk sac is retained, and during early development it is the source of lymphocytes that later migrate to the spleen and liver (later still they migrate again to the bone marrow) to originate the body's immune system. The amnion remains unchanged, a protective water jacket in which the embryo floats. The remaining two extraembryonic membranes, the allantois and chorion, have been totally redesigned. The allantois is no longer needed as a urinary bladder. Instead it becomes the stalk, or **umbilical cord,** that links the embryo physically and functionally with the placenta. The chorion, the outermost membrane, forms most of the placenta itself.

One of the most intriguing questions the placenta presents is why it is not rejected by the mother's tissues. The placenta is a uniquely successful foreign transplant, or **allograft.** It (as well as the embryo) is genetically foreign in the sense that it consists of tissue derived not only from the mother but also from the father. We should expect it to be rejected by the uterine tissues, just as a piece of a child's skin is rejected by the child's mother when a surgeon attempts a grafting transplant (p. 617). The placenta in some way circumvents the normal rejection phenomenon, a matter of greatest interest to immunologists seeking ways to transplant tissues and organs successfully.

____ Development of Systems and Organs

Pregnancy may be divided into four phases: the first phase, lasting 6 or 7 days, is the period of cleavage and blastocyst formation and ends when the blastocyst implants in the uterus. The second phase, about 2 weeks, is the period of gastrulation and formation of the neural plate. The third phase, called the **embryonic period,** is a crucial and sensitive period of primary organ system differentiation. This phase ends at about the eighth week of pregnancy. The last phase, known as the **fetal period,** is characterized by rapid growth, proportional changes in body parts, and final preparation for birth.

During gastrulation the three germ layers are formed. These differentiate, as we have seen, first into primordial cell masses and then into specific organs and tissues. During this process, cells become increasingly committed to specific directions of differentiation.

Derivatives of ectoderm: nervous system and nerve growth

The brain, spinal cord, and nearly all the outer epithelial structures of the body develop from the primitive ectoderm. They are among the earliest organs to appear. Just above the notochord, the ectoderm thickens to form a **neural plate** (Figure 36-13). The edges of this plate rise up, fold, and join together at the top to create an elongated, hollow **neural tube.** The neural tube gives rise to most of the

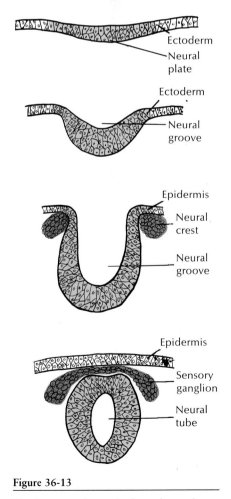

Figure 36-13

Development of neural tube and neural crest from neural plate ectoderm (cross section).

nervous system: anteriorly it enlarges and differentiates into the brain, cranial nerves, and eyes; posteriorly it forms the spinal cord and spinal motor nerves. Sensory nerves arise from special **neural crest** cells pinched off from the neural tube before it closes.

How are the billions of nerve axons in the body formed? What directs their growth? Biologists were intrigued with these questions that seemed to have no easy solutions. Since a single nerve axon may be more than a meter in length (for example, motor nerves running from the spinal cord to the toes), it seemed impossible that a single cell could spin out so far. It was suggested that a nerve fiber grew from a series of preformed protoplasmic bridges along its route. The answer had to await the development of one of the most powerful tools available to biologists, the cell culture technique.

In 1907 embryologist Ross G. Harrison discovered that he could culture living neuroblasts (embryonic nerve cells) for weeks outside the body by placing them in a drop of frog lymph hung from the underside of a cover slip. Watching nerves grow for periods of days, he saw that each nerve fiber was the outgrowth of a single cell. As the fibers extended outward, materials for growth flowed down the axon center to the growing tip where they were incorporated into new protoplasm.

The second question—what directs nerve growth—has taken longer to unravel. An idea held well into the 1940s was that nerve growth is a random, diffuse process. It was believed that the nervous system developed as an equipotential network, or blank slate, that later would be shaped by usage into a functional system. The nervous system just seemed too incredibly complex for us to imagine that nerve fibers could find their way selectively to so many predetermined destinations. Yet it appears that this is exactly what they do! Recent work on invertebrate nervous systems indicates that each of the billions of nerve cell axons acquires a distinct identity that in some way directs it along a specific pathway to its destination. Many years ago Harrison observed that a growing nerve axon terminated in a "growth cone," from which extend numerous tiny threadlike pseudopodial processes (filopodia) (Figure 36-14). These are constantly reaching out and searching the environment in all directions for the pathway having the molecular labels that only that axon's growth cone recognizes. This chemical guidepost system, which must, of course, be genetically directed, is just one example of the amazing precision that characterizes the entire process of differentiation.

Derivatives of endoderm: digestive tube and survival of gill arches

In the frog embryo the primitive gut makes its appearance during gastrulation with the formation of an internal cavity, the **archenteron.** From this simple endodermal cavity develop the lining of the digestive tract, the lining of the pharynx and lungs, most of the liver and pancreas, the thyroid and parathyroid glands, and the thymus.

The **alimentary canal** develops from the primitive gut and is folded off from the yolk sac by the growth and folding of the body wall (Figure 36-15). The ends of the tube open to the exterior and are lined with ectoderm, whereas the rest of the tube is lined with endoderm. The **lungs, liver,** and **pancreas** arise from the foregut.

Among the most intriguing derivatives of the digestive tract are the pharyngeal (gill) arches and pouches, which make their appearance in the early embryonic stages of all vertebrates (Figure 36-16). In fishes, the gill arches develop into gills and supportive structures and serve as respiratory organs. When the early

The tissue culture technique developed by Ross G. Harrison is now used extensively by scientists in all fields of active biomedical research, not just by embryologists. The great impact of the technique has been felt only in recent years. Harrison was twice considered for the Nobel Prize (1917 and 1933), but he failed ever to receive the award because, ironically, the tissue culture method was then believed to be "of rather limited value."

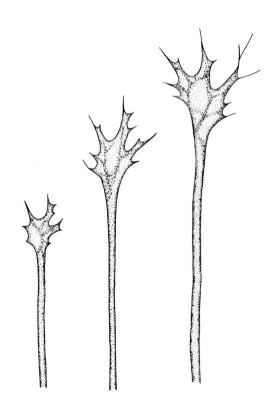

Figure 36-14

Growth cone at the growing tip of a nerve axon. Materials for growth flow up the axon to the growth cone from which numerous threadlike filopodia extend. These serve as a pioneering guidance system for the developing axon.

vertebrates moved onto land, gills were unsuitable for aerial respiration and were replaced by lungs.

   Why then do gill arches persist in the embryos of terrestrial vertebrates? Certainly not for the convenience of biologists who use these and other embryonic structures to reconstruct lines of vertebrate descent. Even though the gill arches serve no respiratory function in either the embryos or adults of terrestrial vertebrates, they remain as necessary primordia for a great variety of other structures. For example, the first arch and its endoderm-lined pouch (the space between adjacent arches) form the upper and lower jaws and inner ear of higher vertebrates. The second, third, and fourth gill pouches contribute to the tonsils, parathyroid gland, and thymus. We can understand then why gill arches and other fishlike structures appear in early mammalian embryos. Their original function has been abandoned, but the structures are retained for new purposes. It is the great conservatism of early embryonic development that has so conveniently provided us with a telescoped evolutionary history.

Derivatives of mesoderm: support, movement, and beating heart

The intermediate germ layer, the mesoderm, forms the vertebrate skeletal, muscular, and circulatory structures and the kidney. As vertebrates have increased in size and complexity, the mesodermally derived supportive, movement, and transport structures make up an even greater proportion of the body bulk.

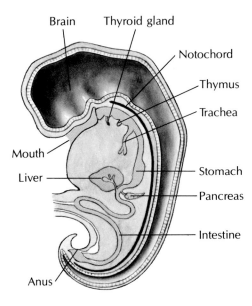

Figure 36-15

Derivatives of the alimentary canal of a human embryo.

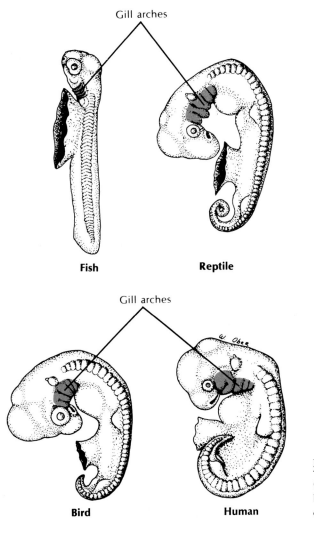

Figure 36-16

Comparison of gill arches of different vertebrate embryos at equivalent stages of development.

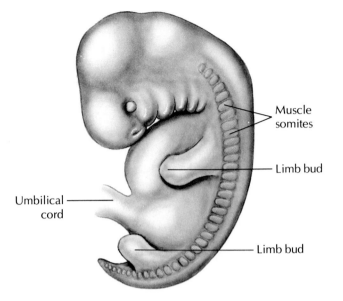

Muscle somites

Limb bud

Umbilical cord

Limb bud

Figure 36-17

Human embryo showing muscle somites.

Most **muscles** arise from the mesoderm along each side of the spinal cord (Figure 36-17). This mesoderm divides into a linear series of somites (38 in humans), which by splitting, fusion, and migration become the muscles of the body and axial parts of the skeleton. The **limbs** begin as buds from the side of the body. Projections of the limb buds develop into fingers and toes.

Although the primitive mesoderm appears after the ectoderm and endoderm, it gives rise to the first functional organ, the embryonic heart. Guided by the underlying endoderm, clusters of precardiac mesodermal cells move ameba-like into a central position between the underlying primitive gut and the overlying neural tube. Here the heart is established, first as a single, thin tube.

Even while the cells group together, the first twitchings are evident. In the chick embryo, a favorite and nearly ideal animal for experimental embryology studies, the primitive heart begins to beat on the second day of the 21-day incubation period; it begins beating before any true blood vessels have formed and before there is any blood to pump. As the ventricle primordium develops, the spontaneous cellular twitchings become coordinated into a feeble but rhythmical beat. Then, as the atrium develops behind the ventricle, followed by the development of the sinus venosus behind the atrium, the heart rate quickens. Each new heart chamber has an intrinsic beat that is faster than its predecessor.

Finally a specialized area of heart muscle called the **sinoatrial** node develops in the sinus venosus and takes command of the entire heartbeat. This becomes the heart's **pacemaker.** As the heart builds up a strong and efficient beat, vascular channels open within the embryo and across the yolk. Within the vessels are the first primitive blood cells suspended in plasma.

The early development of the heart and circulation is crucial to continued embryonic development because without a circulation the embryo could not obtain materials for growth. Food is absorbed from the yolk and carried to the embryonic body; oxygen is delivered to all the tissues, and carbon dioxide and other wastes are carried away. The embryo is totally dependent on these extraembryonic support systems, and the circulation is the vital link between them.

Fetal Circulation

During fetal development the placenta must function for the lungs because they are, of course, nonfunctional in the fetus, which has no access to air. Thus, the blood must be diverted to the placenta from the fetal lungs, and done in such a

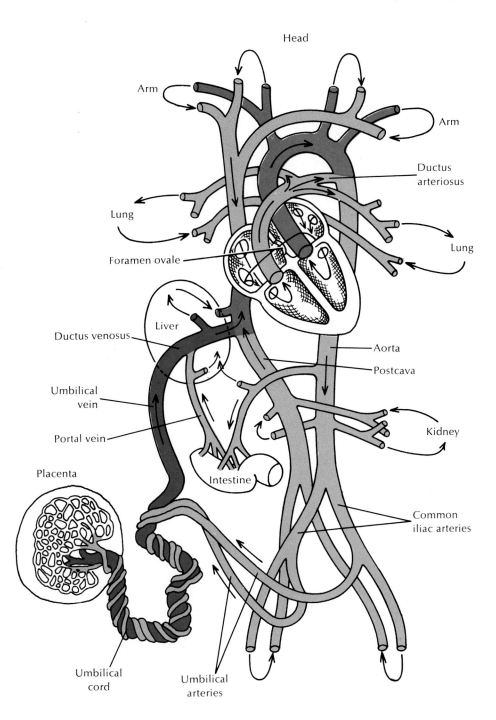

Head

Arm

Arm

Ductus
arteriosus

Lung

Lung

Foramen ovale

Liver

Ductus venosus

Aorta

Postcava

Umbilical
vein

Portal vein

Kidney

Placenta

Intestine

Common
iliac arteries

Umbilical
cord

Umbilical
arteries

Figure 36-18

Fetal circulation. Blood carrying food and oxygen from the placenta is brought by the umbilical vein to the liver, is shunted through the ductus venosus to the vena cava where it mixes with deoxygenated blood returning from fetal tissues, and then is carried to the right atrium. Two fetal shortcuts bypass the nonfunctioning lungs. Some blood goes directly from the right atrium to the left atrium by an opening, the foramen ovale, and so enters the systemic circulation. Some blood passes from the pulmonary trunk into the aorta by a temporary duct, the ductus arteriosus. At birth, placental exchange ceases; the ductus venosus, foramen ovale, umbilical arteries, and ductus arteriosus all close; and regular postnatal pulmonary and systemic circuits take over.

way that instantaneous changeover to the adult pattern of circulation is allowed when the newborn infant takes its first breath. The fetal circulation with its shunts are shown in Figure 36-18, and explained in the legend. At birth, the placental bloodstream is cut off, and the pulmonary circulation opens as the infant draws its first lung-expanding breath.

___ Birth (Parturition)

Hormonal changes preceding birth

During pregnancy the placenta gradually takes over most of the functions of regulating growth and development of the uterus and the fetus. As an endocrine gland, it secretes estradiol and progesterone, hormones that are secreted by the ovaries and corpus luteum in the early periods of pregnancy (p. 770). The placenta also produces **chorionic gonadotropin,** a hormone that now assumes the role of the LH and FSH pituitary hormones. These cease their secretions at about the second month of pregnancy. Chorionic gonadotropin maintains the corpus luteum so that the latter may continue secreting progesterone and estradiol necessary for the integrity of the placenta. In the later stages of pregnancy the placenta becomes a totally independent endocrine organ, not requiring support from either the corpus luteum or the pituitary gland.

What stimulates birth? Why does pregnancy not continue indefinitely? What factors produce the onset of labor (the rhythmical contractions of the uterus)? Thus far we have only tentative answers to these questions. It has long been known that estrogens stimulate uterine contractions, whereas progesterone, which is also secreted by the placenta, inhibits uterine activity. Just before birth there is a very sharp increase in the plasma level of estrogens and a decrease in progesterone. This appears to remove the "progesterone block" that keeps the uterus in a quiescent state throughout pregnancy. Thus, labor contractions can proceed.

Recently much more information on the control of birth has emerged, involving **prostaglandins,** a group of hormonelike long-chain fatty acids. Prostaglandins have complex actions on virtually all body tissues and organs. They are especially powerful in stimulating contractions and, when used pharmacologically, are very effective in inducing labor or abortion. There is increasing evidence from studies on sheep (which, like human females, have an endocrine placenta during pregnancy) that the fetus itself controls the onset of labor. In some way not yet understood, the fetus, when it reaches normal term weight, releases an endocrine signal that stimulates the placenta to produce prostaglandins and thus initiates labor.

Stages in childbirth

The first major signal that birth is imminent is the labor pains, caused by the rhythmical contractions of the uterine musculature. These are usually slight at first and occur at intervals of 15 to 30 minutes. They gradually become more intense, longer in duration, and more frequent. Some women compare the contractions to

Figure 36-19

Birth of a baby.
Courtesy American Museum of Natural History.

menstrual cramps. To some they are painful; to others they are not. Labor contractions may last anywhere from 4 to 24 hours, usually longer with the first child.

Childbirth occurs in three stages. In the first stage the neck (cervix), or opening of the uterus into the vagina, is enlarged by the pressure of the baby in its bag of amniotic fluid, which may be ruptured at this time. In the second stage the baby is forced out of the uterus and through the vagina to the outside (Figure 36-19). In the third stage the placenta, or afterbirth, is expelled from the mother's body, usually within 10 minutes after the baby is born.

Multiple births

Many mammals give birth to more than one offspring at a time or to a litter, each member of which has come from a separate egg. Most higher mammals, however, have only one offspring at a time, although occasionally they may have plural young. The armadillo *(Dasypus)* is almost unique among mammals in giving birth to four young at one time—all of the same sex, either male or female, and all derived from one zygote.

Human twins may come from one zygote (identical twins) or two zygotes (nonidentical, or fraternal, twins). Triplets, quadruplets, and quintuplets may include a pair of identical twins. The other babies in such multiple births usually come from separate zygotes. Fraternal twins do not resemble each other more than other children born separately in the same family, but identical twins are, of course, strikingly alike and always of the same sex. Embryologically, each member of fraternal twins has its own placenta, chorion, and amnion. Usually (but not always) identical twins share the same chorion and the same placenta, but each has its own amnion. Sometimes identical twins fail to separate completely and form Siamese twins, in which the organs of one may be a mirror image of the organs of the other.

The frequency of twin births in comparison to single births is approximately 1 in 86, that of triplets 1 in 86^2, and that of quadruplets approximately 1 in 86^3. The frequency of identical twin births to all births is about the same the world over, whereas the frequency of fraternal births varies with race and country. In the United States, three-fourths of all twin births are dizygotic (fraternal), whereas in Japan only a little more than one-fourth are dizygotic. The tendency for fraternal twinning (but apparently not identical twinning) seems to run in family lines; fraternal twinning (but not identical twinning) also increases in frequency as mothers get older.

LONGEVITY, AGING, AND DEATH

We will end our consideration of development with one aspect that some might wish to ignore: the inevitability of death. Each animal has a characteristic life span. The allotted lifetime of a human being would appear to be the biblical 3 score and 10 years. Even today, despite the achievements of modern medicine, the average life span is still about 70 years. Life spans among other animals vary enormously, and it is a common understanding that large animals live longer than small ones. The Indian elephant may reach 70 years; a hippopotamus or rhinoceros, 50 years; a horse, 45 years; a black bear or gorilla, 25 years; a beaver, 20 years; dogs and cats, 15 years; a gray squirrel, 10 years; a chipmunk, 5 years; a house mouse, 2 years; and a shrew, 1 year. Among birds, some parrots and owls may exceed 60 years (the record appears to be a 68-year-old eagle owl). Cranes and ostriches may live 40 years. However, most of our familiar songbirds are fortunate to reach 10 years, even under the benign conditions of captivity.

Why Do Humans Live So Long?

Reptiles are noted for longevity, and centenarians are common among giant Galápagos tortoises. However, they are probably the only animals that can routinely outlive people. We live far longer—perhaps three times longer—than other mammals of our body size. Our exceptional longevity has been attributed to **neoteny:** retention of larval or juvenile characteristics in the adult form. "The life course of man runs slowly," wrote Louis Bolk, Dutch advocate of this hypothesis. Compared with the development of any other primate, ours is severely retarded. We are born as helpless embryos and have a protracted infancy, childhood, and juvenility. Nearly 30% of our life span is devoted to growing. Our bones ossify late. Birth occurs when the brain is just one-fourth its ultimate size and the brain continues to grow at rapid fetal rates for the first year after birth.

The nineteenth-century physiologist Max Rubner had a hypothesis of aging that has recently received renewed support. He concluded that all mammals in their lifetimes use up about the same number of calories per unit of body weight. A mouse that lives for 2 years lives its life far more intensely than an elephant that lives for 70 years. This relationship is explained by **scaling theory,** an interpretation of the structural and functional consequences of a change in size (or in scale) among similar groups of animals. For example, the rate of oxygen consumption decreases consistently with increasing body size. This means that among mammals—the group that reveals scaling relationships most clearly—small mammals consume more oxygen and generate more heat in proportion to their body weight than do large mammals. When we plot the logarithm of body weight against the logarithm of oxygen consumption (Figure 36-20), we obtain a straight line with a negative slope of -0.25. Thus, the rate of oxygen consumption of a 10 kg animal is only three-fourths that of a 1 kg animal. Certain physiological properties such as heart rate and breathing rate are scaled, like oxygen consumption, to body weight. In other words, heart rates and breathing rates slow down as animals increase in size. If the life span in years of various mammals is multiplied by their heart rates, we arrive at the extraordinary conclusion that the hearts of all mammals, regardless of size, beat about the same number of times—800 million—in their allotted life spans. Because the breathing rate is about one-fourth the heart rate, a mammal will breathe about 200 million times. Lifetime is scaled to life's pace, and all mammals, should they live out their "natural" lives, will survive about the same length of *biological* time. But we humans deviate dramatically from our mammalian kin: our hearts beat three times as often, and we take three times as many breaths, as any other mammal of comparable body size.

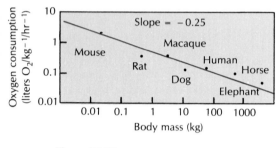

Figure 36-20

Rate of oxygen consumption ($\log_{10}$) of various mammals plotted against body weight ($\log_{10}$). Points tend to fall along a straight line on logarithmic coordinates, with a slope of -0.25.

The Aging Process

The outward signs of aging are obvious, but the basic process remains obscure. Until recently, aging was attributed to progressively defective cellular function that followed an orderly program of senescence encoded in the cell's genes. However, rather than animals having "suicide instructions" built into their genetic repertoire, the weight of evidence supports the general hypothesis that aging is a mixed variety of side effects from genes that confer reproductive advantages early in life.

Current hypotheses view aging as the result of random accumulation of molecular damage. Errors in somatic cell DNA gradually accumulate as animals age. Cells have mechanisms for mending DNA damage, but these repair mechanisms become less effective in old animals. There is much more visible damage in chromosomal structure in cells from aged animals than from young animals. As cells age, defective proteins, especially inactive enzymes, begin to accumulate in the cells. It is possible, too, that the hormonal control systems that regulate enzyme activity may deteriorate with senescence, leading to a breakdown in intercellular communication.

There is ample evidence that degenerative molecular cross-linking between side groups of several kinds of macromolecules occurs with aging. Cross-linked "age pigment" appears in aging cells, and even nuclear chromatin becomes progressively cross-linked with aging. The extracellular fibrous protein **collagen,** which accounts for 40% of all body protein, has long been known to undergo cross-linking as it ages. Because collagen is a major component of tendons, ligaments, bone, and the lining of blood vessels, its deteriorative hardening causes senescent alterations in these structures.

Why do animals age at all? Although the cause of senescence is poorly understood, its value as a deteriorative process that inevitably leads to death is obvious: it prevents disastrous overpopulation. All organisms have impressive reproductive potential. For example, a pair of houseflies in a single summer could produce 19×10^{19} offspring and descendants, if all reproduce and live at least until autumn. Under these circumstances the earth would disappear beneath a thick blanket of flies. Another, more fundamental reason is that death is essential to natural selection. Natural selection is equated with reproductive success (see p. 860). But individuals cannot remain as reproductive or postreproductive adults indefinitely because they would compete with their own offspring for limited resources. No offspring could survive in such a world. Senescence is only one of several agencies that lead to death—disease, starvation, predation, and adverse climate all take their toll—and few animals, other than most humans, live out their full potential life span.

Parenthetically, we must admit that unicellular organisms may have what is in one sense "eternal life": the parent cell does not die; it divides to produce two or more progeny and continues to do so as long as conditions compatible with its survival persist. Nevertheless, unicellular organisms are as subject to natural selection as are multicellular ones. As environmental conditions change, only organisms within the range of variation in the population that have the characteristics allowing survival will transmit those characteristics to their progeny.

SUMMARY

Developmental biology is concerned with the emergence of order and complexity during the development of a new individual from a fertilized egg, and with the control of this process. The early preformation concept of development gave way in the eighteenth century to the theory of epigenesis, which holds that development is the progressive appearance of new structures that arise as the products of antecedent development.

Development of reptiles, birds, and mammals is direct, proceeding by a more or less prolonged period of growth toward the adult body form. Fish, amphibians, and many members of all major invertebrate groups develop indirectly, with one or more larval stages interposed in the life cycle.

When the egg and sperm combine at fertilization, metabolic activity of the previously quiescent egg is suddenly renewed, and important changes occur in the egg cortex that prepare the egg for cleavage. During cleavage the zygote divides repeatedly, producing numerous small cells (blastomeres).

The eggs of most invertebrate groups develop during cleavage as rigid mosaics of blastomeres having predestined developmental fates. Echinoderm and vertebrate eggs show varying degrees of regulative development, in which the blastomeres do not at first have readily identifiable developmental fates.

Despite the different developmental fates of embryonic cells, all cells have the same number and kinds of genes, as has been demonstrated by nuclear transplantation experiments. During development, certain parts of each cell's genome are expressed while the remainder are switched off (repressed) but not destroyed. Differential gene expression in embryonic cells is controlled by cytoplasmic substances that are unequally distributed in the egg during egg maturation. Early development depends on information supplied by nucleic acids and reserves deposited in the egg cytoplasm before fertilization; with the approach of gastrulation, a process of cellular reorganization when germ layers (ectoderm, mesoderm, and endoderm) are formed, nuclear genes become active.

The harmonious differentiation of tissues depends in large part on induction, the ability of one tissue to react to the development response of another. Induction guides a sequence of permissive events along a specific developmental pathway. Inductors are chemical signals produced by responding cells. Their chemical nature is still unknown.

Reptiles, birds, and mammals are amniotes: terrestrial animals that develop extraembryonic membranes during embryonic life. The four membranes are the amnion, allantois, chorion, and yolk sac, each serving a specific life-support function for the embryo that develops within a self-contained egg (reptiles and birds) or within the maternal uterus (mammals).

The mammalian embryo is nourished by the placenta, a complex fetal-maternal structure that develops in the uterine wall. During pregnancy the placenta becomes an independent nutritive, endocrine, and regulatory organ for the embryo.

The germ layers formed at gastrulation differentiate into tissues and organs. The ectoderm gives rise to the skin and nervous system; the endoderm gives rise to the alimentary canal, pharynx, lungs, and certain glands; and the mesoderm forms the muscular, skeletal, circulatory, and excretory systems.

Before birth the human baby is nourished from the placenta by a special fetal circulation. At birth the umbilical blood supply is suddenly interrupted, and the circulatory shunts that exist during fetal development are closed; the pulmonary circuit opens and the infant's circulation becomes independent.

Childbirth is initiated by strong uterine contractions (labor), followed by expulsion of the baby into the birth canal. After delivery the placenta detaches from the uterine wall and is expelled.

The life span of animals has been correlated with the metabolic pace of life, as predicted by scaling theory. Among mammals, small species have more intense respiratory rates and faster heart rates than large species and age more rapidly. Humans, however, live much longer than any other mammal. Our exceptional longevity has been attributed to neoteny, the retention of juvenile characteristics in the adult, and to protracted growth extending nearly one-third the life expectancy.

Aging and death are inevitable for all multicellular animals. Aging results from the gradual accumulation of molecular damage, although the specific causes remain obscure.

Review questions

1. How did Kasper Friedrich Wolff's concept of epigenesis differ from the early notion of preformation?
2. What is the difference between direct and indirect development?
3. Describe the events that follow contact of a spermatozoon with an egg and explain what is meant by the term "activation."
4. Compare and explain the results of removing one blastomere from a four-cell stage embryo showing regulative development, with a four-cell stage embryo showing mosaic development.
5. What is the evidence that the egg's cortex has a dynamic organization that guides the pattern of cleavage?
6. Refer to Figure 36-8 and explain why the results of Spemann's experiments A and B differed.
7. What is the evidence for the statement, "Nucleocytoplasmic interactions form the basis of the organized differentiation of tissues"?
8. Describe the famous "organizer" experiment of Hans Spemann and Hilde Mangold and explain its significance.
9. What are the four extraembryonic membranes of the amniotic egg of a bird or reptile and what is the function of each membrane?
10. What is the fate of the four extraembryonic membranes of the amniotic egg in mammals that develop a placenta?
11. Explain what the "growth cone" that Ross Harrison observed at the ends of growing nerve fibers has to do with the direction of nerve growth.
12. Name two organ system derivatives of each of the three primary germ layers (ectoderm, mesoderm, and endoderm).
13. What endocrine changes precede birth of a human baby?
14. What is the evidence that the exceptional longevity of humans is the consequence of neoteny?
15. While the causes of aging are poorly understood, can you suggest why aging and death are essential for the evolution of life on earth?

Selected references

Epel, D. 1977. The program of fertilization. Sci. Am. **237**:128-138 (Nov.). *Describes the fusion of egg and sperm and the initial events of development.*

Gilbert, S.F. 1985. Developmental biology. Sunderland, Mass., Sinauer Associates. *Combines descriptive and mechanistic aspects; good selection of examples from many animal groups.*

Goodman, C.S., and M.J. Bastiani. 1984. How embryonic nerve cells recognize one another. Sci. Am. **251**:58-66 (Dec.). *Research with insect larvae shows that developing neurons follow pathways having specific molecular labels.*

Karp, G., and N.J. Berrill. 1981. Development, ed. 2. New York, McGraw-Hill Book Co. *College-level textbook, emphasizing mechanisms.*

Saunders, J.W. 1982. Developmental biology: patterns, problems, principles. New York, Macmillan, Inc. *Textbook emphasizing mechanisms of development with helpful comprehensive chapter summaries.*

Slack, J.M.W. 1984. From egg to embryo: determinative events in early development. New York, Cambridge University Press. *Emphasis on pattern formation; comparative approach.*

Stewart, A.D., and D.M. Hunt. 1982. The genetic basis of development. Tertiary Level Biology, New York, John Wiley & Sons. *Emphasis on genetic systems, exemplified with well-chosen experiments.*

CHAPTER 37

PRINCIPLES OF INHERITANCE

Heredity establishes the continuity of life forms. Although offspring and parents in a particular generation may look different, there is nonetheless a basic sameness that runs from generation to generation for any species of plant or animal. In other words, "like begets like." Yet children are not precise replicas of their parents. Some of their characteristics show resemblances to one or both parents, but they also demonstrate many traits not found in either parent. What is actually inherited by an offspring from its parents is a certain type of germinal organization (**genes**) that, under the influence of environmental factors, guides the orderly sequence of differentiation of the fertilized egg into a human being, bearing the unique physical characteristics as we see them. Each generation hands on to the next the instructions required for maintaining continuity of life (see Principle 14, p. 10).

The gene is the unit entity of inheritance, the germinal basis for every characteristic that appears in an organism (see Principle 15, p. 11). In a sense the primary function of an organism is to reproduce genes. The organism is a device, a vehicle for the transfer of genes from one generation to its descendants. It is a repository in which a portion of the gene "pool" of the population has been temporarily entrusted.

The study of what genes are and how they work is the science of genetics. It is a science that deals with the underlying causes of *resemblance,* as seen in the remarkable fidelity of reproduction, and of *variation,* which is the working material for organic evolution. Genetics has shown that all living forms use the same information storage, transfer, and translation system, and thus it has provided an explanation for both the stability of all life and its probable descent from a common ancestral form. This is one of the most important unifying concepts of biology.

▬ MENDEL'S INVESTIGATIONS

The first man to formulate the cardinal principles of heredity was Gregor Johann Mendel (1822-1884) (Figure 37-1), who lived with the Augustinian monks at Brünn (Brno), Moravia. At that time Brünn was a part of Austria, but now it is in the central part of Czechoslovakia. While conducting breeding experiments in a small monastery garden from 1856 to 1864, Mendel examined with great care the progeny of many thousands of plants. He worked out in elegant simplicity the laws governing the transmission of characters from parent to offspring. His dis-

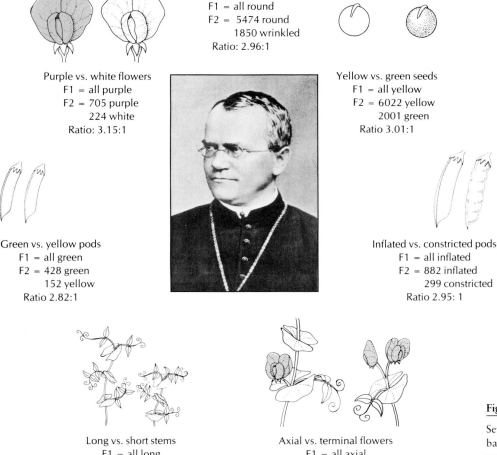

Round vs. wrinkled seeds
F1 = all round
F2 = 5474 round
    1850 wrinkled
Ratio: 2.96:1

Purple vs. white flowers
F1 = all purple
F2 = 705 purple
    224 white
Ratio: 3.15:1

Yellow vs. green seeds
F1 = all yellow
F2 = 6022 yellow
    2001 green
Ratio 3.01:1

Green vs. yellow pods
F1 = all green
F2 = 428 green
    152 yellow
Ratio 2.82:1

Inflated vs. constricted pods
F1 = all inflated
F2 = 882 inflated
    299 constricted
Ratio 2.95: 1

Long vs. short stems
F1 = all long
F2 = 787 long
    277 short
Ratio 2.84:1

Axial vs. terminal flowers
F1 = all axial
F2 = 651 axial
    207 terminal
Ratio 3.14:1

Figure 37-1

Seven experiments on which Gregor Mendel based his postulates. These are the results of monohybrid crosses for first and second generations.
Photograph from Historical Pictures Service, Chicago.

coveries, published in 1866, were of great significance, coming just after Darwin's publication of *The Origin of Species.* Yet these discoveries remained unappreciated and forgotten until 1900—some 35 years after the completion of the work and 16 years after Mendel's death.

Mendel's classic observations were based on the garden pea because it had been produced in pure strains by gardeners over a long period of time by careful selection. For example, some varieties were definitely dwarf and others were tall. A second reason for selecting peas was that they were self-fertilizing but also capable of cross-fertilization. To simplify his problem Mendel chose single characteristics and characteristics that were sharply contrasted. He carefully avoided mere quantitative and intermediate characteristics. He selected pairs of contrasting characteristics, such as tall plants, dwarf plants, smooth seeds, and wrinkled seeds (Figure 37-1).

Mendel crossed plants having one of these characteristics with others having the contrasting characteristic. He did this by removing the stamens from a flower to prevent self-fertilization and then placing, on the stigma of this flower, pollen

In not reporting conflicting findings, which must surely have arisen as they do in any original research, Mendel has been accused of "cooking" his results. The chances are, however, that he carefully avoided ambiguous material to strengthen his central message, which we still regard as an exemplary achievement in experimental analysis.

from the flower of the plant that had the contrasting characteristic. He also prevented the experimental flowers from being pollinated from other sources such as wind and insects. When the cross-fertilized flower bore seeds, he noted the kind of plants (hybrids) that were produced from the planted seeds. Subsequently he crossed these hybrids among themselves to see what would happen.

CHROMOSOMAL BASIS OF INHERITANCE

In bisexual animals special sex cells, or gametes (eggs and sperm, p. 756), are responsible for providing the genetic information to the offspring. Scientific explanation for genetic principles required a study of germ cells and their behavior, which meant working backward from certain visible results of inheritance to the mechanism responsible for such results. The nuclei of sex cells were early suspected of furnishing the real answer to the mechanism. This applied especially to the **chromosomes** (p. 66), since they appeared to be the only entities passed on in equal quantities from both parents to offspring.

When Mendel's laws were rediscovered in 1900, their parallelism with the cytological behavior of the chromosomes was obvious. Later experiments showed that the mechanism of heredity could be definitively assigned to the chromosomes. The next problem was to find out how chromosomes affected the hereditary pattern.

Mendel knew nothing of the cytological basis of heredity, since chromosomes and genes were unknown to him. Although we can admire Mendel's power of intellect in his discovery of the principles of inheritance without knowledge of chromosomes, the principles are certainly easier to understand if we first examine chromosomal behavior, especially in meiosis (p. 762).

MENDELIAN LAWS OF INHERITANCE
Mendel's First Law

In one of Mendel's original experiments, he pollinated pure-line tall plants with the pollen of pure-line dwarf plants. He found that all the progeny, the first generation (F_1), were tall, just as tall as the tall parents of the cross. The reciprocal cross—dwarf plants pollinated with tall plants—gave the same result. This always happened no matter which way the cross was made. Obviously, this kind of inheritance was not a blending of two characteristics, since none of the progeny was of intermediate size.

Next Mendel selfed (self-fertilized) the tall F_1 plants and raised several hundred progeny, the second (F_2) generation. This time, *both* tall and dwarf plants appeared. Again, there was no blending (no plants of intermediate size), but the appearance of dwarf plants from all tall parental plants was surprising. The dwarf characteristic, present in the grandparents but not in the parents, had reappeared. When he counted the actual number of tall and dwarf plants in the F_2 generation, he discovered that there were almost exactly three times more tall plants than dwarf ones.

Mendel then repeated this experiment for the six other contrasting characteristics that he had chosen, and in every case he obtained ratios very close to 3:1 (Figure 37-1). At this point it must have been clear to Mendel that he was dealing with hereditary determinants for the contrasting characteristics that did not blend when brought together. Even though the dwarf characteristic disappeared in the

F_1 generation, it reappeared fully expressed in the F_2 generation. He realized that the F_1 generation plants carried determinants (which he called "factors") of both tall and dwarf parents, even though only the tall characteristic was expressed in the F_1 generation.

Mendel called the tall factor **dominant** and the short **recessive.** Similarly, the other pairs of characters that he studied showed dominance and recessiveness. Whenever a dominant factor (gene) is present, the recessive one cannot produce an effect. The recessive factor will show up *only* when both factors are recessive, or in other words, in a pure condition.

In representing his crosses, Mendel used letters as symbols; dominant characteristics were represented by capital letters, and for recessive characteristics he used the corresponding lowercase letters. Modern geneticists still follow this custom. Thus the factors for pure tall plants might be represented by T/T, the pure recessive by t/t, and the mix, or hybrid, of the two plants by T/t. The slash mark is to indicate that the genes are on different chromosomes. When the gametes unite in any fertilization, a zygote is formed. The zygote bears the complete genetic constitution of the organism. All the gametes produced by T/T must necessarily be T, whereas those produced by t/t must be t. Therefore a zygote produced by union of the two must be T/t, or a **heterozygote.** On the other hand, the pure tall plants *(T/T)* and pure dwarf plants *(t/t)* are **homozygotes,** meaning that the factors (genes) are alike on the homologous chromosomes. A cross involving only one pair of contrasting characteristics is called a monohybrid cross.

In the cross between tall and dwarf plants there are two types of *visible* characteristics: tall and dwarf. These are called **phenotypes.** On the basis of genetic formulas there are three *hereditary* types: T/T, T/t, and t/t. These are called **genotypes.** A genotype is a gene combination *(T/T, T/t, or t/t),* and the phenotype is the appearance of the organism (tall or dwarf).

In diagram form, one of Mendel's original crosses (tall plant and dwarf plant) could be represented as follows:

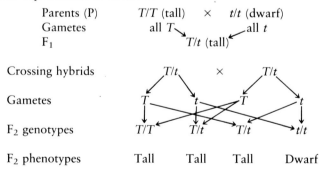

Parents (P)	T/T (tall) × t/t (dwarf)	
Gametes	all T ⟍ ⟋ all t	
F_1	T/t (tall)	

Crossing hybrids	T/t × T/t
Gametes	T t T t
F_2 genotypes	T/T T/t T/t t/t
F_2 phenotypes	Tall Tall Tall Dwarf

In other words, all possible combinations of F_1 gametes in the F_2 zygotes will yield a $3:1$ phenotype ratio and a $1:2:1$ genotype ratio. It is convenient in such crosses to use the checkerboard method devised by Punnett for representing the various combinations resulting from a cross. In the F_2 cross the following scheme would apply:

Sperm \ Eggs	½ T	½ t
½ T	¼ T/T (homozygous tall)	¼ T/t (hybrid tall)
½ t	¼ T/t (hybrid tall)	¼ t/t (homozygous dwarf)

Ratio: 3 tall to 1 dwarf.

The next step was an important one because it enabled Mendel to test his hypothesis that every plant contained nonblending factors from both parents. He self-fertilized the plants in the F_2 generation; that is, the stigma of a flower was fertilized by the pollen of the same flower. The results showed that self-pollinated F_2 dwarf plants produced only dwarf plants, whereas one third of the F_2 tall plants produced tall and the other two thirds produced both tall and dwarf in the ratio of $3:1$, just as the F_1 plants had done. Genotypes and phenotypes were as follows:

$$F_2 \text{ plants:} \quad \text{Tall} \begin{cases} \frac{1}{4}\ T/T \xrightarrow{\text{Selfed}} \text{all } T/T \text{ (homozygous tall)} \\ \frac{1}{2}\ T/t \xrightarrow{\text{Selfed}} 1\ T/T : 2\ T/t : 1\ t/t \text{ (3 tall : 1 dwarf)} \end{cases}$$

$$\text{Dwarf } \tfrac{1}{4}\ t/t \xrightarrow{\text{Selfed}} \text{all } t/t \text{ (homozygous dwarf)}$$

This experiment showed that the dwarf plants were pure because they at all times gave rise to short plants when self-pollinated; the tall plants contained both pure tall and hybrid tall. It also demonstrated that, although the dwarf characteristic disappeared in the F_1 plants, which were all tall, the characteristic for dwarfness appeared in the F_2 plants.

Mendel reasoned that the factors for tallness and dwarfness were units that did not blend when they were together. The F_1 generation (the first generation of hybrids, or first filial generation) contained both of these units or factors, but when these plants formed their germ cells, the factors separated so that each germ cell had only one factor. In a pure plant both factors were alike; in a hybrid they were different. He concluded that individual germ cells were always pure with respect to a pair of contrasting factors, even though the germ cells were formed from hybrids in which the contrasting characteristics were mixed.

This idea formed the basis for his first principle, the **law of segregation,** which states that whenever two factors are brought together in a hybrid, they segregate into separate gametes that are produced by the hybrid. Either one of the pair of genes of the parent passes with equal frequency to the gametes. We now understand that the factors segregate because there are two alleles for the trait, one on each chromosome of a homologous pair, but the gametes receive only one of each in meiosis. (Alleles are alternative forms of genes for the same trait; Chapter 35, p. 762.)

Mendel's great contribution was his quantitative approach to inheritance. This really marks the birth of genetics, since, before Mendel, people believed that traits were blended like mixing together two colors of paint, a notion that unfortunately still lingers in the minds of many. If this were true, variability would be lost in hybridization between individuals. With particulate inheritance, on the other hand, different variations are retained and can be shuffled about and resorted like blocks.

Testcross

When one of the alleles is dominant, the offspring of a cross are all of the same phenotypes whether they are homozygous or heterozygous. For instance, in Mendel's experiment of tall and dwarf characteristics, it is impossible to determine the genetic constitution of the tall plants of the F_2 generation by mere inspection of the tall plants. Three fourths of this generation are tall, but which of them are heterozygotes?

As Mendel reasoned, the test is to cross the questionable individuals with pure recessives. If the tall plant is homozygous, all the plants in such a testcross are tall, thus:

> Parents T/T(tall) $\times$ t/t(dwarf)
> Offspring T/t (hybrid tall)

If, on the other hand, the tall plant is heterozygous, half of the offspring are tall and half dwarf, thus:

> Parents $T/T \times t/t$
> Offspring T/t (tall) or t/t (dwarf)

The testcross is often used in modern genetics for the analysis of the genetic constitution of the offspring, as well as for a quick way to make desirable homozygous stocks of animals and plants.

Intermediate inheritance

In some cases neither allele is completely dominant over the other, and the heterozygote phenotype shows characteristics intermediate between or even quite distinct from those of the parents. In the four-o'clock flower *(Mirabilis)* (Figure 37-2), homozygotes are red or white flowered, but heterozygotes have pink flowers. In a certain strain of chickens, a cross between black and splashed white produces offspring that are not gray, but a distinctive color called blue Andalusian. In each case, if the F_1s are crossed, the F_2s have a ratio of 1:2:1 in colors, or 1 red:2 pink:1 white and 1 black:2 blue:1 white, respectively. This can be illustrated for the four-o'clocks as follows:

	(red flower)		(white flower)	
Parents	R/R	$\times$	R'/R'	
Gametes	all R		all R'	
F_1		R/R'		
		(all pink)		
Crossing hybrids	R/R'	$\times$	R/R'	
Gametes	R,R'		R,R'	
F_2 genotypes	R/R	R/R'	R/R'	R'/R'
F_2 phenotypes	Red	Pink	Pink	White

In this kind of a cross, the heterozygote is indeed a blending of both red and white characters. It is easy to see how observations of this type would encourage the notion of the blending theory of inheritance. However, in the cross of red and white four-o'clock flowers, *only* the hybrid is a phenotypic blend; the homozygous strains breed true to the parental phenotypes (red or white).

—— Mendel's Second Law

Thus far we have considered crosses involving alleles of a single gene (monohybrid cross). Mendel also carried out experiments on peas that differed from each other by two or more genes, that is, experiments involving two or more phenotypic characters.

Mendel had already established that tall plants were dominant to dwarf. He also noted that crosses between plants bearing yellow cotyledons and plants bearing green cotyledons produced plants with yellow cotyledons in the F_1 generation; therefore yellow was dominant to green. The next step was to make a cross between plants differing in these two characteristics. When a tall plant with yellow

When neither of the alleles is recessive, it is customary to represent both by capital letters and to distinguish them by the addition of a "prime" sign (R') or by superscript letters, for example, R^r (equals red flowers) and R^w (equals white flowers).

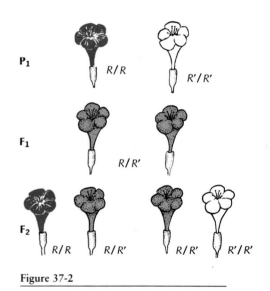

Figure 37-2

Cross between red and white four-o'clock flowers. Red and white are homozygous; pink is heterozygous.

cotyledons *(A/A B/B)* was crossed with a dwarf plant with green cotyledons *(a/a b/b)*, the F_1 plants were tall and yellow as expected *(A/a B/b)*.

Parents	*A/A B/B* (tall, yellow)	×	*a/a b/b* (dwarf, green)
Gametes	all *AB*		all *ab*
F_1		*A/a B/b* (tall, yellow)	

When the F_1 hybrids were crossed with each other, the result was the following four different phenotypes of plants in a ratio of 9:3:3:1:

9 tall with yellow cotyledons
3 tall with green cotyledons
3 dwarf with yellow cotyledons
1 dwarf with green cotyledons

Mendel already knew that a cross between two plants bearing a single pair of alleles of the genotype *A/a* would yield a 3:1 ratio. Similarly, a cross between two plants with the genotypes *B/b* would yield the same 3:1 ratio. If we examine *only* the tall and dwarf phenotypes in the outcome of the dihybrid experiment, they total up to 12 tall and 4 dwarf, giving a ratio of 3:1. Likewise, a total of 12 plants have yellow cotyledons and 4 plants have green—again a 3:1 ratio. Thus the monohybrid ratio prevails for both traits when they are considered independently. The 9:3:3:1 ratio is nothing more than a combination of the two 3:1 ratios.

$$3:1 \times 3:1 = 9:3:3:1$$

When one of the alleles is unknown, it can be designated by a dash *A/—)*. This designation can also be used when it is immaterial whether the genotype is heterozygote or homozygote, as when we total all of a certain phenotype. The dash could be either *A* or *a*.

The F_2 genotypes and phenotypes are as follows:

1 *A/A B/B* 2 *A/a B/B* 2 *A/A B/b* 4 *A/a B/b*	9 *A/—B/—*	9 tall yellow
1 *A/A b/b* 2 *A/a b/b*	3 *A/—b/b*	3 tall green
1 *a/a B/B* 2 *a/a B/b*	3 *a/a B/—*	3 dwarf yellow
1 *a/a b/b*	1 *a/a b/b*	1 dwarf green

The results of this experiment show that the segregation of alleles for plant height is entirely independent of the segregation of alleles for cotyledon color. Neither has any influence on the other. This is the basis of Mendel's second law, the **law of independent assortment,** which states: whenever two or more pairs of contrasting characteristics are brought together in a hybrid, the alleles of different pairs segregate independently of one another. The reason is that during meiosis the member of any pair of homologous chromosomes received by a gamete is independent of the other chromosomes it receives. Of course, independent assortment assumes that the genes are on different chromosomes. If they were on the same chromosome, they would assort together unless crossing over occurred (p. 814).

The dihybrid experiment is shown in the Punnett square below:

Parents	A/A B/B	×	a/a b/b
	(tall yellow)		(dwarf green)
Gametes	all AB		all ab
F₁		A/a B/b	
		(hybrid tall, hybrid yellow)	
Crossing hybrids	A/a B/b	×	A/a B/b
Gametes	AB, Ab, aB, ab		AB, Ab, aB, ab
F₂		(see checkerboard)	

	AB	Ab	aB	ab
AB	A/A B/B Pure tall Pure yellow	A/A B/b Pure tall Hybrid yellow	A/a B/B Hybrid tall Pure yellow	A/a B/b Hybrid tall Hybrid yellow
Ab	A/A B/b Pure tall Hybrid yellow	A/A b/b Pure tall Pure green	A/a B/b Hybrid tall Hybrid yellow	A/a b/b Hybrid tall Pure green
aB	A/a B/B Hybrid tall Pure yellow	A/a B/b Hybrid tall Hybrid yellow	a/a B/B Pure dwarf Pure yellow	a/a B/b Pure dwarf Hybrid yellow
ab	A/a B/b Hybrid tall Hybrid yellow	A/a b/b Hybrid tall Pure green	a/a B/b Pure dwarf Hybrid yellow	a/a b/b Pure dwarf Pure green

Ratio: 9 tall yellow to 3 tall green; 3 dwarf yellow to 1 dwarf green.

One way to estimate numbers of progeny from a cross with a given genotype or phenotype is to construct a Punnett square and count them up. With a monohybrid cross, this is easy; with a dihybrid cross, a Punnett square is rather laborious; and with a trihybrid cross, it is tedious. We can make such estimates much more easily by taking advantage of simple probability calculations. The basic assumption is that all the genotypes of gametes of one sex have an equal chance of uniting with all the genotypes of gametes of the other sex, in proportion to the numbers of each present. This is generally true when the sample size is large enough, and the actual numbers observed come close to those predicted by the laws of probability.

We may define probability as follows:

$$\text{Probability (p)} = \frac{\text{Number of times an event happens}}{\text{Total number of trials or possibilities for the event to happen}}$$

For example, the probability (p) of a coin falling heads when tossed is ½ because the coin has two sides. The probability of rolling a three on a die is ⅙ because the die has six sides.

The probability of independent events occurring together (ordered events) involves the **product rule,** which is simply the product of their individual probabilities. When two coins are tossed together, the probability of getting two heads is ½ × ½ = ¼, or 1 chance in 4. The probability of rolling two threes simultaneously with two dice is as follows:

$$\text{Probability of two threes} = ⅙ × ⅙ = \frac{1}{36}$$

Note, however, that a small sample size may give a result quite different from that predicted. Thus if we tossed the coin three times and it fell heads each time, we would not be much surprised. But if we tossed the coin 1000 times, and the number of times it fell heads diverged very much from 500, we would strongly suspect that there was something wrong with the coin or with the way we were tossing it.

We can use the product rule to predict the ratios of inheritance in monohybrid or dihybrid (or larger) crosses if the genes sort independently in the gametes (as they did in all of Mendel's experiments). In other words, the mechanism of placing A into a gamete is independent of the mechanism of putting a into a gamete. Therefore in a monohybrid cross the probability that a sperm carries the dominant is $\frac{1}{2}$, and the same applies to an egg. In a dihybrid cross involving A/a and B/b, the same thing applies; the probability of any gene appearing in a gamete is $\frac{1}{2}$. Now we can apply the product rule to an F_1 plant $A/a\ B/b$ to determine the frequency of each kind of gamete:

Probability of gamete being $AB = \frac{1}{2} \times \frac{1}{2} = \frac{1}{4}$
Probability of gamete being $Ab = \frac{1}{2} \times \frac{1}{2} = \frac{1}{4}$
Probability of gamete being $aB = \frac{1}{2} \times \frac{1}{2} = \frac{1}{4}$
Probability of gamete being $ab = \frac{1}{2} \times \frac{1}{2} = \frac{1}{4}$

From this point the probabilities for the genotype in each box of the Punnett square on p. 807 can be easily derived as follows:

Probability of plant being $A/A\ B/B = \frac{1}{4} \times \frac{1}{4} = \frac{1}{16}$
Probability of plant being $A/A\ B/b = \frac{1}{4} \times \frac{1}{4} = \frac{1}{16}$

and so on for all 16 boxes.

Collecting all similar phenotypes together we get:

Tall, yellow cotyledon	$\frac{9}{16} = 9$	
Tall, green cotyledon	$\frac{3}{16} = 3$	
Dwarf, yellow cotyledon	$\frac{3}{16} = 3$	
Dwarf, green cotyledon	$\frac{1}{16} = 1$	

Thus we have the $9:3:3:1$ ratio, derived by the product rule. In fact, one quickly learns by experience how to determine the ratios of phenotypes without using either the Punnett squares or the product rule. In a dihybrid ($9:3:3:1$ ratio), for instance, those phenotypes that make up the dominants of each gene are $\frac{9}{16}$ of the whole F_2 generation; each of the $\frac{3}{16}$ phenotypes consists of one dominant and one recessive; and the $\frac{1}{16}$ phenotype consists of two recessives.

The F_2 ratios in any cross involving more than one pair of contrasting pairs can be found by combining the ratios in the cross of one pair of alleles. Thus the number of genotypes is $(3)^n$ and the proportion of phenotypes $(3:1)^n$ when one allele is dominant and the other recessive. For example, let us suppose that in a cross of two pairs of alleles the phenotypes are in the ratio of $(3:1)^2$, or $9:3:3:1$. The genotypes in such a cross are $(3)^2$, or 9. If three pairs of characteristics are involved (trihybrid cross), the proportions, or ratios, of the phenotypes are then $(3:1)^3$, or $27:9:9:9:3:3:3:1$. The genotypes are $(3)^3$, or 27. Obviously, as the number of gene pairs increases, the number of phenotypes and genotypes rises steeply.

Multiple Alleles

Previously we defined alleles as the alternate forms of a gene. Many dissimilar alleles may occupy the same gene locus on a chromosome but not of course all at one time. Thus more than two alternative genes may affect the same character. An example is the set of multiple alleles that affects coat color in rabbits. The different alleles are C (normal color), c^{ch} (chinchilla color), c^h (Himalayan color), and c (albino). The four alleles fall into a dominance series with C dominant over everything. The dominant allele is always written to the left and the recessive to the right:

$$C/c^h = \text{Normal color}$$
$$c^{ch}/c^h = \text{Chinchilla color}$$
$$c^h/c = \text{Himalayan color}$$

Multiple alleles arise through mutations at the gene locus over long periods of time. Any gene may mutate in many different places (p. 836) if given time and thus can give rise to slightly different genes or alleles at the same locus. Examples of multiple alleles in humans are the ABO blood types and Rh factors (Chapter 29, p. 620).

____ Gene Interaction

The types of crosses previously described are simple in that the characters involved result from the action of a single gene, but many cases are known in which the characters are the result of two or more genes. Mendel probably did not appreciate the real significance of the genotype, as contrasted with the visible character—the phenotype. We now know that many different genotypes may be expressed as a single phenotype.

Also, many genes have more than a single effect. A gene for eye color, for instance, may be the ultimate cause of eye color, yet at the same time it may be responsible for influencing the development of other characters as well. A gene at one locus may mask or prevent the expression of a gene at another locus acting on the same trait. Another case of gene interaction is that in which several sets of alleles may produce a cumulative effect on the same character, or **polygenic inheritance.**

Polygenic inheritance

Several characteristics in humans are influenced by multiple genes. In such cases the characters, instead of being sharply marked, show continuous variation between two extremes. This is sometimes called **blending,** or **quantitative inheritance.** In this kind of inheritance the children are often more or less intermediate between the two parents.

One illustration of such a type is the degree of pigmentation in crosses between the black and white human races. The cumulative genes in such crosses have a quantitative expression. Many genes are believed to be involved in skin pigmentation, but we will simplify our explanation by assuming that there are only two pairs of independently assorting genes. Thus a person with very dark pigment has two genes for pigmentation on separate chromosomes (A/A B/B). Each dominant allele contributes one unit of pigment. A person with very light pigment has alleles (a/a b/b) that contribute no color. (Freckles that commonly appear in the skin of very light people represent pigment contributed by entirely separate genes.) The offspring of very dark and very light parents would have an intermediate skin color (A/a B/b).

The children of parents having intermediate skin color show a range of skin color, depending on the number of genes for pigmentation they inherit. Their skin color ranges from very dark (A/A B/B), to dark (A/A B/b or A/a B/B), intermediate

The description in Chapter 29 assumed that inheritance of the Rh factor was as a single dominant gene. Actually, the genetics of the Rh factor are much more complicated than was believed when the factor was discovered. Some authorities believe that three genes located close together on the same chromosome are involved, whereas others think the system is one gene with many alleles. In 1968 a revision of the single-gene concept listed 37 alleles necessary to account for the phenotypes then known. Furthermore, the frequency of the various alleles varies greatly between whites, Orientals, and blacks.

The inheritance of eye color in humans is another example of gene interaction. One allele (B) determines whether pigment is present in the front layer of the iris. This allele is dominant over the allele for the absence of pigment (b). The genotypes B/B and B/b pigment generally produce brown eyes, and b/b produces blue eyes. However, these phenotypes are greatly affected by many modifier genes influencing, for example, the amount of pigment present, the tone of the pigment, and its distribution. Thus a person with B/b may even have blue eyes if modifier genes determine a lack of pigment, thus explaining the rare instances of a brown-eyed child of blue-eyed parents.

(*A/A b/b* or *A/a B/b* or *a/a B/B*), light (*A/a b/b* or *a/a B/b*), to very light (*a/a b/b*). It is thus possible for parents heterozygous for skin color to produce children with darker or lighter colors than themselves.

The relationships can be seen in the following diagram:

Parents	*A/A B/B*	×	*a/a b/b*
	(very dark)		(very light)
Gametes	*AB*		*ab*
F₁	*A/a B/b*	×	*A/a B/b*
	(intermediate)		(intermediate)
Gametes	*AB, Ab, aB, ab*		*AB, Ab, aB, ab*

The F₂ generation shows this ratio:

1 *A/A B/B*	Very dark
2 *A/a B/B*	Dark
2 *A/A B/b*	Dark
4 *A/a B/b*	Intermediate
1 *A/A b/b*	Intermediate
2 *A/a b/b*	Light
1 *a/a B/B*	Intermediate
2 *a/a B/b*	Light
1 *a/a b/b*	Very light

Collecting the phenotypes, the ratio becomes: 1 very dark: 4 dark: 6 intermediate: 4 light: 1 very light.

When there are many genes involved in the production of traits, this situation may be represented in graphs as distribution curves, usually bell shaped. A graph of frequency distribution for height indicates that the height of most people is near the average, indicated by the high point of such a bell-shaped curve, and that few people are much shorter or taller than average. If one locus is much more important than others, the alleles at this locus might split the graph into two or three bell-shaped curves representing different genotypes. Because we never find this kind of graph for the height of human populations, we know that humans do not have a principle for tall and dwarf lines like that of garden peas.

Sex Determination and Sex-Linked Inheritance

Before the importance of the chromosomes in heredity was realized in the early 1900s, how gender was determined was totally unknown. The first really scientific clue to the determination of sex came in 1902 when C. McClung observed that bugs (Hemiptera) produced two kinds of sperm in approximately equal numbers. One kind contained among its regular set of chromosomes a so-called accessory chromosome that was lacking in the other kind of sperm. Since all the eggs of these species had the same number of haploid chromosomes, half the sperm would have the same number of chromosomes as the eggs, and half of them would have one chromosome less. When an egg was fertilized by a spermatozoon carrying the accessory (sex) chromosome, the resulting offspring was a female; when fertilized by the spermatozoon without an accessory chromosome, the offspring was a male. There are therefore two kinds of chromosomes: X chromosomes, which determine sex (and sex-linked traits), and **autosomes,** which determine the other body traits. The particular type of sex determination just described is often called the XX-XO type, which indicates that the females have two X chromosomes and the males only one X chromosome (the O indicates absence of the chromosome). The XX-XO method of sex determination is depicted in Figure 37-3.

Later, other types of sex determination were discovered. In humans and

Speculation on how sex was determined in animals produced several incredible beliefs, for example, that the two testicles of the male contain different types of semen, one begetting males, the other females. It is not difficult to imagine the abuse and mutilation of domestic animals that occurred when attempts were made to alter the sex ratios of herds. Another idea asserted that sex of the offspring was determined by the more heavily sexed parent. An especially masculine father should produce sons, an effeminate father only daughters. Such ideas were not testable and have lingered until recently.

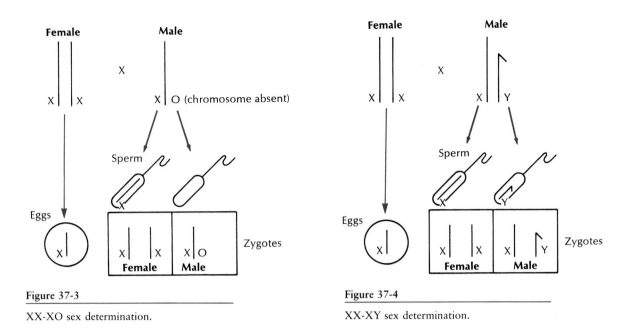

Figure 37-3

XX-XO sex determination.

Figure 37-4

XX-XY sex determination.

many other forms each sex has the same number of chromosomes; however, the sex chromosomes (XX) are alike in the female but unlike (XY) in the male. Hence the human egg contains 22 autosomes + 1 X chromosome; the sperm are of two kinds; half carry 22 autosomes + 1 X and half bear 22 autosomes + 1 Y. The Y chromosome is much smaller than the X. At fertilization, when two X chromosomes come together, the offspring are female; when X and Y chromosomes come together, the offspring are male. The XX-XY kind of determination is shown in Figure 37-4.

A third type of sex determination is found in birds, moths, and butterflies in which the male has two X (or sometimes called ZZ) chromosomes and the female an X and a Y (or ZW).

Since the X chromosome carries several genes that are essential for normal cellular function and the Y chromosome carries none, every individual must have *at least* one X chromosome. And of course the XX-XY system ensures that they will: the female has two, the male one. To keep the balance between X-chromosome genes and autosome genes the same in the female as in the male, one of the female's X chromosomes is always inactivated. This happens early in development when the embryo consists of only a few cells, according to the widely accepted Lyon hypothesis (after Mary Lyon, who discovered this and other aspects of human sex determination). At this time it is strictly a matter of chance whether the maternal or paternal X chromosome is the one inactivated in each cell. All descendants of that cell, however, retain the same inactivated chromosome. Consequently, tissues of adult females are a patchwork of cells (mosaic) expressing the genes of one or the other X chromosome, never both. Interestingly, the inactivated X chromosome condenses in the interphase nucleus of most somatic cells into a tiny, darkly staining object called a Barr body (Figure 37-5). This furnishes a simple method for determining whether an individual is a genetic male or female, as has been done, for example, in examination of entrants in women's athletic events.

Genetic sex is just one aspect of sex determination. Having an XX or an XY genetic constitution does not in itself produce a female or male. The embryo is at first totally indifferent sexually, despite its genetic sex. Even the gonad is sexually indifferent and bipotential, since it can differentiate into either a testis or an ovary.

We now know that not all animals with dioecious reproduction have their sex determined chromosomally. Several invertebrate examples are known (see marginal note, p. 391), and the sex of alligators is determined by the temperature at which the eggs are incubated (p. 530). Some coral reef fishes actually change sex during their lifetimes, starting off as females, then becoming males.

Normally this odd feature is of no consequence in human females, but it has a dramatic effect in cats. Tortoise-shell (calico, tabby) cats, which are spotted black and yellow, are *always* females; the males are either all black or all yellow. This happens because the genes for fur color are located on the X chromosome; that is, they are sex linked. A male, having only one X chromosome, will be either black or yellow. But a female has two X chromosomes and, if she is heterozygous for fur color, her fur will be a mosaic of yellow and black patches. Each patch represents the descendants of an early embryonic skin cell having one operating and one inactive X chromosome.

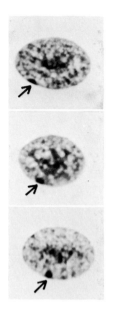

Figure 37-5

Nuclei from three epithelial cells taken from the mouth of a normal human female. The darkly staining Barr bodies *(arrows)* are the heterochromatinized (inactive) X chromosome.

Courtesy M.L. Barr.

Figure 37-6

Sex-linked inheritance of red-green color blindness in humans. **A,** Carrier mother and normal father produce color blindness in one half of their sons but in none of their daughters. **B,** Half of both sons and daughters of carrier mother and color-blind father are color blind.

Gonad differentiation was discussed in Chapter 35 (p. 761), and we pointed out there that the primordial gonad has an inherent tendency to become an ovary, and the rest of the body to become female, *unless* a male-determining gene on the Y chromosomes redirects gonadal differentiation into a testis. As the presumptive gonad differentiates into an ovary or testis, it begins to secrete sex hormones (androgens by the testes, estrogens by the ovaries). These hormones influence the other developing organs to become male or female.

Sex-linked inheritance

It has long been known that the inheritance of some characteristics depends on the sex of the parent carrying the gene and the sex of the offspring. One of the best-known sex-linked traits of humans is hemophilia (Chapter 29, p. 615). Another example is red-green color blindness in which red and green colors are indistinguishable to varying degrees. Color-blind men greatly outnumber color-blind women. When color blindness does appear in women, their fathers are color blind. Furthermore, if a woman with normal vision who is a carrier of color blindness bears sons, half of them are likely to be color blind, regardless of whether the father had normal or affected vision. How are these observations explained?

The color-blindness and hemophilia defects are recessive traits carried on the X chromosome that are visibly expressed either when both genes are defective in the female or when only one defective gene is present in the male. The inheritance pattern of these defects is shown for color-blindness in Figure 37-6. When the mother is a carrier and the father is normal, half of the sons but none of the daughters are color blind. However, if the father is color blind and the mother is a carrier, half of the sons *and* half of the daughters are color blind (on the average and in a large sample). It is easy to understand then why such defects are much more prevalent in males: a single sex-linked recessive gene in the male has a visible effect. What would be the outcome of a mating between a homozygous normal woman and a color-blind man?

Another example of a sex-linked character was discovered by Thomas Hunt Morgan (1910) in *Drosophila*. The normal eye color of this fly is red, but mutations for white eyes do occur. The genes for eye color are known to be carried on the X chromosome. If a white-eyed male and a red-eyed female are crossed, all the F_1 offspring are red eyed because this trait is dominant (Figure 37-7). If these F_1 offspring are interbred, all the females of F_2 have red eyes, half the males have red eyes, and the other half have white eyes. No white-eyed females are found in this generation; only the males have the recessive character (white eyes). The gene for

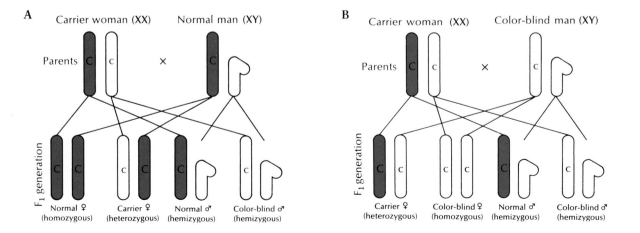

Figure 37-7

Sex-linked inheritance of eye color in fruit fly *Drosophila*. Genes for eye color are carried on X chromosome; Y carries no genes for eye color. Normal red is dominant to white. Homozygous red-eyed female mated with white-eyed male gives all red-eyed in F₁. F₂ ratios from F₁ cross are one homozygous red-eyed female and one heterozygous red-eyed female to one red-eyed male and one white-eyed male.

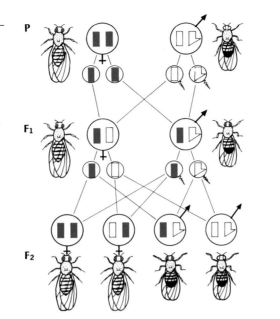

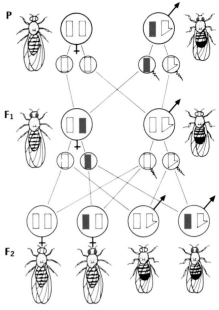

Figure 37-8

Reciprocal cross of Figure 37-7 (homozygous white-eyed female with red-eyed male) gives white-eyed males and red-eyed females in F₁. F₂ shows equal numbers of red-eyed and white-eyed females and red-eyed and white-eyed males.

being white eyed is recessive and should appear in a homozygous condition. However, since the male has only one X chromosome (the Y does not carry a gene for eye color), white eyes appear whenever the X chromosome carries the gene for this trait. If the reciprocal cross is made in which the females are white eyed and the males red eyed, all the F₁ females are red eyed and all the males are white eyed (Figure 37-8). If these F₁ offspring are interbred, the F₂ generation shows equal numbers of red-eyed and white-eyed males and females.

If the allele for red eyes is represented by W, white eyes by w, the female sex chromosome by X, and the male sex chromosome lacking a gene for eye color by Y, the following diagrams show the inheritance of this eye color:

Parents	X^W/X^W (red ♀)		×	X^w/Y (white ♂)
F₁	X^W/X^w (red ♀)		×	X^W/Y (red ♂)
Gametes	X^W, X^w			X^W, Y
F₂ genotypes	X^W/X^W	X^W/Y	X^w/Y	X^w/X^W
F₂ phenotypes	Red ♀	Red ♂	White ♂	Red ♀

Reciprocal cross

Parents	X^w/X^w (white ♀)		×	X^W/Y (red ♂)
F₁	X^w/X^W (red ♀)		×	X^w/Y (white ♂)
Gametes	X^w, X^W			X^w, Y
F₂ genotypes	X^w/X^w	X^w/Y	X^W/X^w	X^W/Y
F₂ phenotypes	White ♀	White ♂	Red ♀	Red ♂

___ Autosomal Linkage and Crossing Over

Linkage

Since Mendel's laws were rediscovered in 1900, it has become apparent that, contrary to Mendel's second law, not all factors segregate independently. Indeed, many traits are inherited together. Since the number of chromosomes in any organism is relatively small compared with the number of traits, each chromosome must contain many genes. All genes present on a chromosome are said to be **linked.** Linkage simply means that the genes are on the same chromosome, and all

We might wonder how these observations on sex-linked defects are reconciled with the inactivation of one X chromosome in females. If the normal X chromosome was the one inactivated in cells of heterozygous females, would the person not show the recessive trait because the active chromosome was the one with the gene? The answer is that which X chromosome is inactivated is entirely random in cells of the embryo, and there are almost always enough cells with normal X chromosomes that the recessive trait is not expressed.

Geneticists commonly use the word "linkage" in two somewhat different senses. Sex linkage refers to inheritance of a trait on the sex chromosomes, and thus its phenotypic expression depends on the sex of the organism and the factors already discussed. Autosomal linkage, or simply, linkage, refers to inheritance of the genes on a given autosomal chromosome. Letters used to represent such genes are normally written without a slash mark between them, indicating that they are on the same chromosome. For example, *AB/ab* shows that genes *A* and *B* are on the same chromosome, and *a* and *b* are on the homologous chromosome. Interestingly, Mendel studied seven characteristics of garden peas, which assorted independently because they were on seven different chromosomes. If he had studied eight characteristics, he would not have found independent assortment in two of the traits because garden peas have only seven pairs of homologous chromosomes.

genes present on homologous chromosomes belong to the same linkage groups. Therefore there should be as many linkage groups as there are chromosome pairs.

In *Drosophila*, in which this principle has been worked out most extensively, there are four linkage groups that correspond to the four pairs of chromosomes found in these fruit flies. Usually, small chromosomes have small linkage groups, and large chromosomes have large groups.

Crossing over

Linkage, however, is usually not complete. If we perform an experiment in which animals such as *Drosophila* are crossed, we find that linked traits separate in some percentage (usually small) of the offspring. Separation of genes located on the same chromosome occurs because of **crossing over.**

As described in Chapter 35 (p. 762), during the protracted prophase of the first meiotic division, homologous chromosomes break and exchange equivalent portions; genes "cross over" from one chromosome to its homologue, and vice versa (Figure 37-9). Each chromosome consists of two sister chromatids held together by means of a synaptonemal complex. Breaks and exchanges occur at corresponding points on nonsister chromatids. (Breaks and exchanges also occur between sister chromatids but usually have no genetic significance because sister chromatids are identical.) Crossing over is a means for exchanging genes between homologous chromosomes and as such greatly increases the amount of genetic recombination. The frequency of crossing over varies depending on the species,

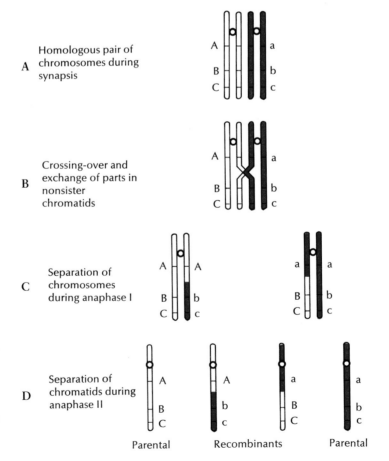

Figure 37-9

Crossing over during meiosis. Nonsister chromatids exchange portions, so that none of the resulting gametes is genetically the same as any other. Gene A is farther from gene B than B is from C; therefore gene A is more frequently separated from B in crossing over than B is from C.

A   Homologous pair of chromosomes during synapsis

B   Crossing-over and exchange of parts in nonsister chromatids

C   Separation of chromosomes during anaphase I

D   Separation of chromatids during anaphase II

Parental   Recombinants   Parental

but usually at least one and often several crossovers occur each time chromosomes pair.

Because the frequency of recombination is proportional to the distance between loci, the comparative linear position of each locus can be determined. Laborious genetic experiments over many years have produced gene maps that indicate the position of more than 500 genes distributed on the four *Drosophila* chromosomes.

___ Chromosomal Aberrations

Structural and numerical deviations from the norm that affect many genes at once are called chromosomal aberrations. They are sometimes called chromosomal mutations, but most cytogeneticists prefer to use the term "mutation" to refer to qualitative changes within a gene; gene mutations will be discussed on p. 836.

Despite the incredible precision of meiosis, chromosomal aberrations do occur, and they are more common than one might think. They have great economic benefit in agriculture. Unfortunately, they are also responsible for many human genetic malformations. It is estimated that five out of every 1000 humans are born with *serious* genetic defects attributable to chromosomal anomalies. An even greater number of embryos with chromosomal defects are aborted spontaneously, far more than ever reach term.

Changes in chromosome numbers are called **euploidy** when there is the addition or deletion of whole chromosome sets and **aneuploidy** when a single chromosome is added to or subtracted from a diploid set. The most common kind of euploidy is **polyploidy**, the carrying of one or more additional sets of chromosomes. Such aberrations are much more common in plants than in animals. Animals are much less tolerant of chromosomal aberrations because sex determination requires a delicate balance between the numbers of sex chromosomes and autosomes. Many domestic plant species are polyploid (cotton, wheat, apples, oats, tobacco, and others), and perhaps 40% of flowering plants are believed to have originated in this manner. Horticulturists favor polyploids and often try to develop them because they have more intensely colored flowers and more vigorous vegetative growth.

Aneuploidy is usually caused by nondisjunctional separation of chromosomes during meiosis. If a pair of chromosomes fails to separate during the first or second meiotic divisions, both members go to one pole and none to the other. This results in one gamete having n − 1 number of chromosomes and another having n + 1 number of chromosomes. If the n − 1 gamete is fertilized by a normal n gamete, the result is a **monosomic** animal. Survival is rare because the lack of one chromosome gives an uneven balance of genetic instructions. **Trisomy,** the result of the fusion of a normal n gamete and an n + 1 gamete, is much more common, and several kinds of trisomic conditions are known in humans. Perhaps the most familiar is **trisomy 21,** or **Down syndrome.** As the name indicates, it involves an extra chromosome 21 combined with the chromosome pair 21, and it is caused by nondisjunction of that pair during meiosis. It occurs spontaneously, and there is seldom any family history of the abnormality. However, the risk of its appearance rises dramatically with increasing age of the mother; it occurs 40 times as often in women over 40 years old as among women between the ages of 20 and 30.

Structural aberrations involve whole sets of genes within a chromosome. A portion of a chromosome may be reversed, placing the linear arrangement of genes in reverse order (inversion); nonhomologous chromosomes may exchange

A syndrome is a group of symptoms associated with a particular disease or abnormality, although not every symptom is necessarily shown by every patient with the condition. In 1866 an English physician, John Langdon Down, described the syndrome that we know is caused by trisomy 21. Because of Down's belief that the facial features of affected individuals were mongoloid in appearance, the condition has been known as mongolism. The resemblances are superficial, however, and the currently accepted names are trisomy 21 and Down syndrome. Among the numerous characteristics of the condition, the most disabling is severe mental retardation. This, as well as other conditions caused by chromosomal aberrations and several other birth defects, can be diagnosed *prenatally* by a procedure involving amniocentesis. The physician inserts a hypodermic needle through the abdominal wall of the mother and into the fluids surrounding the fetus (*not into* the fetus) and withdraws some of the fluid, which contains some fetal cells. The cells are grown in culture, their chromosomes are examined, and other tests are done. If a severe birth defect is found, the mother has the option of having an abortion performed. As an extra "bonus," the sex of the fetus is learned after amniocentesis. How?

sections (translocation); entire blocks of genes may be lost (deletion); or an extra section of chromosome may attach to a normal chromosome (duplication). These are all structural changes that usually do not produce phenotypic changes. Duplications, although rare, are important for evolution because they supply additional genetic information that may enable new functions.

── SUMMARY

Genes are the unit entities that determine all the characteristics of an organism and are inherited by offspring from their parents. Genes may be dominant, recessive, or intermediate; the recessive gene in the heterozygous genotype will not be expressed in the phenotype but requires the homozygous condition for overt expression. In a monohybrid cross involving a dominant gene and its recessive allele (both parents homozygous), the F_1 generation will all be heterozygous, whereas the F_2 genotypes will occur in a $1:2:1$ ratio, and the phenotypes in a $3:1$ ratio. This demonstrates Mendel's law of segregation. Heterozygotes in intermediate inheritance show phenotypes intermediate between the homozygous phenotypes, or sometimes they show a different phenotype altogether, with corresponding alterations in the phenotypic ratios. Dihybrid crosses (in which the alleles for the two traits are carried on separate pairs of homologous chromosomes) demonstrate Mendel's law of independent assortment, and the phenotypic ratios will be $9:3:3:1$ with dominant and recessive characters. The ratios for monohybrid and dihybrid crosses can be determined by construction of a Punnett square, but the laws of probability allow calculation of the ratios in crosses of two or more characters much more easily.

Genes can have more than two alleles, and different combinations of alleles can produce different phenotypic effects. Nonallelic genes can interact, as in polygenic inheritance, in which one gene affects the expression of another gene.

Gender is determined in most animals by the sex chromosomes; in humans, fruit flies, and many other animals, females have two X chromosomes, and males have an X and a Y. A gene on the X chromosome shows sex-linked inheritance and will produce an effect in the male, even if it is recessive, because the Y chromosome does not carry a corresponding allele. All genes on a given autosomal chromosome are linked and do not assort independently unless crossing over occurs. Crossing over increases the amount of genetic recombination in a population. Since the frequency of crossing over between two genes increases with the distance between the loci of the genes on the chromosome, gene maps of each chromosome can be constructed.

Occasionally, nondisjunction of one of the chromosomes in meiosis occurs, and one of the gametes ends up with one chromosome too many and the other with n − 1 chromosomes. Resulting zygotes usually do not survive; humans with 2n + 1 chromosomes may live, but they are born with serious defects, such as Down syndrome.

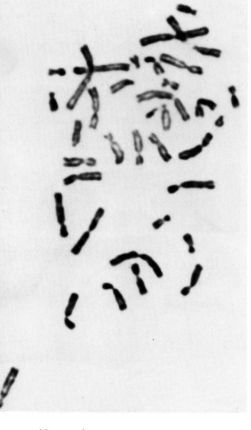

Human chromosomes.
Courtesy L.R. Emmons.

Review questions

1. What is the relationship between homologous chromosomes and alleles?
2. Define the following: dominant, recessive, zygote, heterozygote, homozygote, phenotype, genotype, monohybrid cross, dihybrid cross.
3. Diagram by Punnett square a cross between individuals with the following genotypes: $A/a \times A/a$; $A/a\ B/b \times A/a\ B/b$.
4. Concisely state Mendel's law of segregation and his law of independent assortment.
5. Assuming brown eyes (B) are dominant over blue eyes (b), and neglecting the effect of any modifier genes, determine the genotypes of all the following individuals. The blue-eyed son of two brown-eyed parents marries a brown-eyed daughter whose mother was brown eyed and whose father was blue eyed. Their child is blue eyed.
6. Recall that red color (R) in four-o'clock flowers is incompletely dominant over white (R'). In the following crosses, give the genotypes of the gametes produced by each parent and the flower color of the offspring: $R/R' \times R/R'$, $R'/R' \times R/R'$, $R/R \times R/R'$, $R/R \times R'/R'$.
7. A brown mouse is mated with two female black mice. When each female has produced several litters of young, the first female has had 48 black and the second female has had 14 black and 11 brown young. Can you deduce the pattern of inheritance of coat color and the genotypes of the parents?
8. Rough coat (R) is dominant over smooth coat (r) in guinea pigs, and black coat (B) is dominant over white (b). If a homozygous rough black is mated with a homozygous smooth white, give the appearance of each of the following: F_1; F_2; offspring of F_1 mated with smooth, white parent; offspring of F_1 mated with rough, black parent.
9. Assume right-handedness (R) dominates over left-handedness (r) in humans, and that brown eyes (B) are dominant over blue (b). A right-handed, blue-eyed man marries a right-handed, brown-eyed woman. Their two children are right handed/blue eyed and left handed/brown eyed. The man marries again, and this time the woman is right handed and brown eyed. They have 10 children, all right handed and brown eyed. What are the genotypes of the man and the two wives?
10. In *Drosophila*, red eyes are dominant to white and the recessive characteristic is on the X chromosome. Vestigial wings (v) are recessive to normal wings (V). What will be the appearance of the following crosses: $X^W/X^w\ V/v \times X^w/Y\ v/v$, $X^w/X^w\ V/v \times X^W/Y\ V/v$?
11. Assume that color blindness is a recessive character on the X chromosome. A man and woman with normal vision have the following offspring: daughter with normal vision who has one color-blind and one normal son; daughter with normal vision who has six normal sons; and a color-blind son who has a daughter with normal vision. What are the probable genotypes of the individuals?
12. Distinguish the following: euploidy, aneuploidy, and polyploidy; monosomy and trisomy.

Selected references

Goodenough, U. 1984. Genetics, ed. 2. Philadelphia, Saunders College. *Good molecular treatment.*

Jenkins, J.B. 1979. Genetics, ed. 2. Boston, Houghton Mifflin Co. *Comprehensive introductory test.*

Klug, W., and M. Cummings. 1983. Concepts of genetics. Columbus, Ohio, Charles E. Merrill Publishing Co. *A shorter text.*

Mange, A.P., and E.J. Mange. 1980. Genetics: human aspects. Philadelphia, Saunders College. *A readable, introductory text concentrating on the genetics of the animal species of greatest concern to most of us.*

Suzuki, D., A. Griffiths, and R. Lewontin. 1981. An introduction to genetic analysis, ed. 2. San Francisco, W.H. Freeman & Co. *Good introductory text.*

C H A P T E R   3 8

M O L E C U L A R   G E N E T I C S

Genetic transmission depends on the replication of molecules of deoxyribonucleic acid (DNA) (see Principle 18, p. 12). That nucleic acids and not proteins were the basic material of heredity was first demonstrated a little over 40 years ago, and it has been just over 30 years since it was proposed that the base sequence in DNA could code for the amino acids in a protein (see Principle 19, p. 12). As a result of the discovery and characterization of the restriction endonucleases (bacterial enzymes that cleave DNA at particular base sequences) between 1962 and 1972, there has been an explosion of knowledge in molecular genetics. This chapter explores the characteristics of nucleic acids, the physical nature of the genetic apparatus, and the replication and translation of the information in nucleic acids into gene products.

CHEMICAL STRUCTURE OF NUCLEIC ACIDS

All cells contain two kinds of nucleic acids: deoxyribonucleic acid (DNA), which is the genetic material of the chromosomes of cells, and ribonucleic acid (RNA), which functions in protein synthesis. Both DNA and RNA are polymers built of repeated units called **nucleotides.** Each nucleotide contains three parts: a **sugar,** a **nitrogenous base,** and a **phosphate group.** The sugar is a pentose (5-carbon) sugar; in DNA it is **deoxyribose** and in RNA it is **ribose** (Figure 38-1).

The nitrogenous bases of nucleotides are of two types: pyrimidines, which consist of a single, 6-membered ring, and purines, which are composed of two fused rings. Both of these types of compounds contain nitrogen as well as carbon in their rings, which is why they are called "nitrogenous" bases. Five bases in all occur in DNA and RNA (Figure 38-2). The pyrimidines are thymine, cytosine, and uracil, and the purines are adenine and guanine (Table 38-1). (Other bases are known to occur rarely in DNA and RNA.)

The sugar, phosphate group, and nitrogenous base are linked as shown in the generalized scheme for a nucleotide:

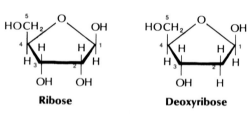

Ribose   **Deoxyribose**

Figure 38-1

Ribose and deoxyribose, the pentose sugars of nucleic acids. A carbon atom lies in each of the four corners of the pentagon (labeled 1 to 4). Ribose has a hydroxyl group (−OH) and a hydrogen on the number 2 carbon; deoxyribose has 2 hydrogens at this position.

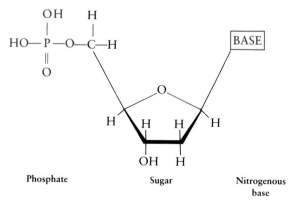

Phosphate   Sugar   Nitrogenous base

818

Table 38-1 Chemical Components of DNA and RNA

	DNA	RNA
Purines	Adenine	Adenine
	Guanine	Guanine
Pyrimidines	Cytosine	Cytosine
	Thymine	Uracil
Sugar	2-Deoxyribose	Ribose
Phosphate	Phosphoric acid	Phosphoric acid

In DNA the backbone of the molecule is built of phosphoric acid and deoxyribose sugar; to this backbone are attached the nitrogenous bases (Figure 38-3). However, one of the most interesting and important discoveries about the nucleic acids is that DNA is not a single polynucleotide chain but rather consists of *two* complementary chains that are precisely cross-linked by specific hydrogen bonding of purine and pyrimidine bases. It was found that the number of adenines is equal to the number of thymines, and the number of guanines equals the number of cytosines. This fact suggested a pairing of bases: adenine with thymine (AT) and guanine with cytosine (GC) (p. 826).

PURINES

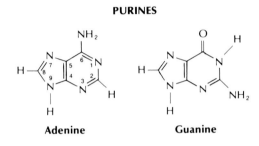

Adenine   Guanine

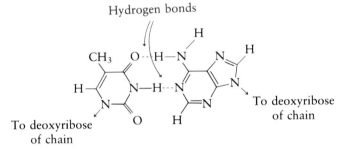

Thymine—adenine

PYRIMIDINES

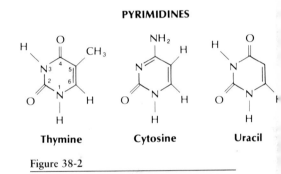

Thymine   Cytosine   Uracil

Figure 38-2

Purines and pyrimidines of DNA and RNA.

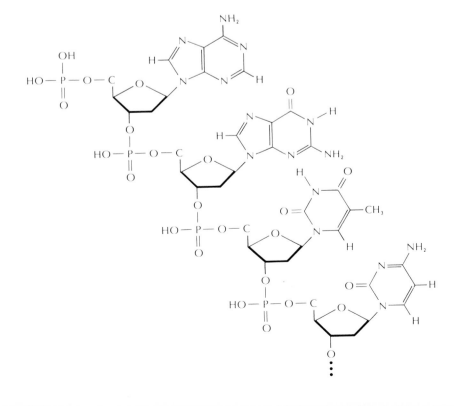

Figure 38-3

Section of DNA. Polynucleotide chain is built of a backbone of phosphoric acid and deoxyribose sugar molecules. Each sugar holds a nitrogenous base side arm. Shown from top to bottom are adenine, guanine, thymine, and cytosine.

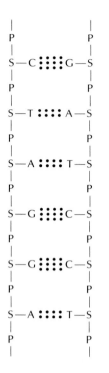

Figure 38-4

Double chain structure of DNA. The uprights of the "ladder" are the sugar-phosphate backbones of each DNA strand, and the connecting rungs are the paired nitrogenous bases, AT or GC.

It was originally believed that the asexual prokaryotes could only stamp out carbon copies of the parental cells, genetic change occurring solely when a mutation intervened. However, several processes of genetic recombination are now known in bacteria. For example, when certain viruses infect bacterial cells, the viruses may pick up a fragment of the bacterial DNA and transfer it to new cells during the course of subsequent infection (transduction). Sometimes DNA in the environment (released by cell death or other processes) can somehow penetrate the cell wall and membrane of the bacteria and be incorporated into their genetic complement (transformation). Under certain circumstances, bacteria of differing "mating types" can actually form cytoplasmic bridges between each other and transfer segments of DNA (conjugation). During conjugation, plasmids are exchanged.

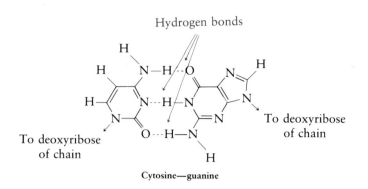

Cytosine—guanine

The result is a ladder structure (Figure 38-4). However, the ladder is twisted into a **double helix,** with about 10 base pairs for each complete turn of the helix (Figure 38-5).

RNA is similar to DNA in structure except that it consists of a *single* polynucleotide chain. In RNA one of the four bases, thymine (T), is replaced by uracil (U), and the sugar deoxyribose is replaced by ribose. In other respects the single RNA chain is joined together like each of the two DNA chains. RNA functions in the transcription and translation of genetic information (Principle 20, p. 12).

PHYSICAL STRUCTURE OF GENETIC MATERIAL
Prokaryotes, Plasmids, and Viruses

Prokaryotes (bacteria and cyanobacteria, p. 45) have a single, circular chromosome comprised of a helical, double strand of DNA. It has little protein associated with it and is not enclosed in a nucleus. Although it is visible with the electron microscope, the bacterial chromosome never condenses into the bodies we identify as chromosomes during eukaryotic mitosis and meiosis. Virtually all bacterial species and most individual cells carry an additional source of genetic information: the **plasmid.** Plasmids are not part of the bacterial chromosome, although in some instances they may be incorporated into it. They are normally present as small circles of double-stranded DNA, comprising from a fraction of 1% to 3% of the cell's genome. Nevertheless, they may carry important genetic information, such as ability to withstand a variety of toxic agents, including antibiotics; virulence in many disease-causing bacteria of plants and animals; and information for conjugation. Plasmids are regarded by some workers as separate organisms, even simpler than viruses, which live as endosymbionts in bacteria. Among the reasons for considering plasmids organisms is that they can be exchanged by conjugation between bacteria of different species and even genera. They cannot, however, replicate outside a host cell. Plasmids have been an important tool in recombinant DNA technology, described later in this chapter. Although best known in bacteria, plasmids also occur in yeast, and there is some evidence that they are present in vertebrate cells.

Another genetic system unable to replicate by itself is the **virus.** Viruses are essentially molecules of nucleic acid (DNA or RNA) encased in a protein coat. After penetrating the host cells, which may be prokaryotes or eukaryotes, they induce the host to replicate viral nucleic acid and protein, thus manufacturing more virus particles. For RNA viruses, the RNA must first be copied into DNA by an enzyme called **reverse transcriptase.**

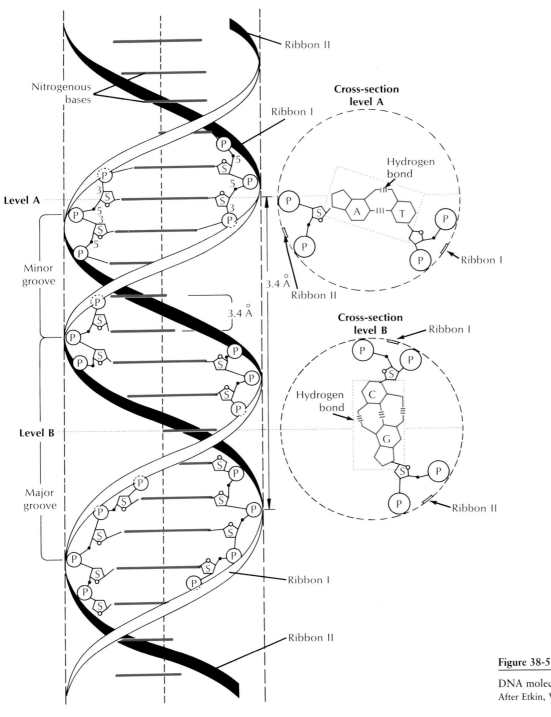

Figure 38-5

DNA molecule.
After Etkin, W. 1973. BioScience **23**:653.

_____ Eukaryotes

Mitochondria and plastids

Plastids in plant cells (for example, chloroplasts) and mitochondria, found in most eukaryotic cells, are self-replicating and have their own complement of DNA. The DNA is in the form of small circles with little protein, reminiscent of plasmids. Although mitochondrial and plastid DNA carries genetic information for some of their proteins and their ribosomes, some of the proteins of these organelles are specified by nuclear genes.

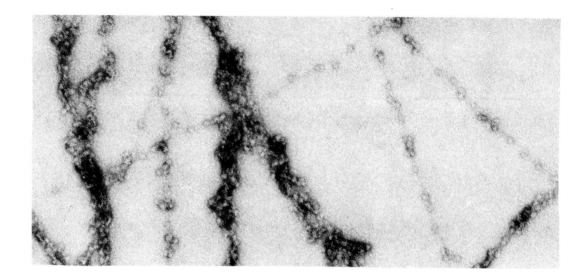

Figure 38-6

Chromatin prepared from chicken red blood cells. Some strands are stretched and demonstrate individual nucleosomes. (Original magnification ×100,000.)
Electron micrograph courtesy C.L.F. Woodcock.

Chromosomes

The eukaryotic nuclear chromosome carries much more DNA than any of the foregoing, and its organization and regulation are much more complex. The nucleus of a typical human somatic cell contains 2000 times as much DNA as a typical bacterial cell. If the DNA were pulled out of the nucleus as a continuous string, it would measure 2 m, compared with a 1.5 mm length of DNA from one bacterial cell. Clearly, eukaryote chromosomes must be exquisitely organized to direct the synthesis of proteins, be capable of exact duplication before each cell division, and avoid becoming hopelessly entangled during cell division.

During interphase the chromosomes are not condensed into bodies as in mitosis and meiosis but are present in a state referred to as **chromatin,** whose structure cannot be distinguished with the light microscope. In special preparations, the chromatin can be extracted from nuclei and visualized under the electron microscope (Figure 38-6). Its appearance is that of beads on a string. The "beads" are called **nucleosomes,** and it is believed that they are the structural quantum in chromosomes. They are formed by a molecule of DNA winding around molecules of proteins. The currently accepted model of nucleosomes is shown in Figure 38-7. The proteins, called **histones,** have long been known to comprise a large proportion of the protein associated with eukaryotic DNA. There are five kinds of histones. The "inner histones," around which the DNA winds, are in the form of octomers: two molecules each of four different histone types. The fifth type is found on the DNA in the "spacer" regions between the nucleosomes. The histones are highly basic; 25% of the amino acids in the inner histones and 37% of those in the spacer histone consist of lysine, arginine, and histidine. This property is important in holding them in close association with the acidic

Figure 38-7

Current model of chromatin showing nucleosome organization. The DNA molecule winds one and three-fourths turns around a group of eight histone molecules, two each of four different kinds. A fifth class of histones (H_1) is found in the spacer region.

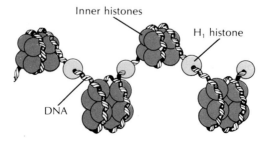

DNA. The nature of higher-order packaging (coiling, folding, and so on) of chromatin is still unclear, but the spacer histones apparently are important factors. Nonhistone proteins may also be important. Whatever the structure, it must be modified to permit genetic expression. Part of the thread must be unraveled and the nucleosome structure altered as RNA is synthesized.

Finally, the threads of chromatin must be condensed into the rodlike mitotic chromosome at each cell division (see Figure 4-24, p. 68). Evidence indicates that this compactness is achieved by assembly of the chromatin into a large series of radial loops (Figure 38-8), probably on a scaffold of nonhistone protein. Stabilization of the scaffold involves either calcium or copper ions. There is some evidence that a scaffold exists during interphase, and its reorganization may bring about the condensation of the mitotic chromosome.

By prophase in mitosis, the chromosomes have already replicated (during interphase); therefore the chromatids can be seen joined together by their **centromere** (point of spindle fiber attachment). The position of the centromere is characteristic for a particular chromosome and can be used to help identify it. Centromere positions may be described as **metacentric** (arms of the chromatids equal on each side), **acrocentric** (centromere closer to one end), and **telocentric** (centromere at the very end) (Figure 38-9). The staining reactions of different parts of the chromosomes vary. Some parts stain densely (**heterochromatin**), and some parts take up little stain (**euchromatin**). It is believed that heterochromatic regions of the chromosomes represent areas where the genes are inactive. The Barr body (p. 811) is an entire chromosome of heterochromatin. The particular regions of euchromatin and heterochromatin change during development as different sets of genes become active or inactive. Some authors use these terms to designate active and inactive (dispersed and condensed) interphase chromatin; genetic expression does not occur from condensed chromatin.

Normally manifestations of gene activity on chromosomes, that is, transcription of DNA into RNA, cannot be seen with the light microscope. However, in a few instances they can. In the salivary glands of *Drosophila* and some other insects, the chromatin strands replicate repeatedly without undergoing mitosis and are known as **polytene chromosomes** (Figure 38-10). These giant chromo-

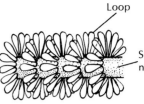

Figure 38-8

Probable arrangement of chromatin in condensed chromosome. The loops extend radially from a central scaffolding of nonhistone protein.

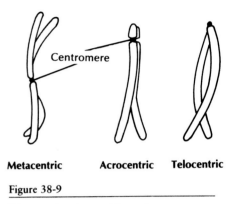

Metacentric   Acrocentric   Telocentric

Figure 38-9

Diagram showing positions of centromere on different metaphase chromosomes.

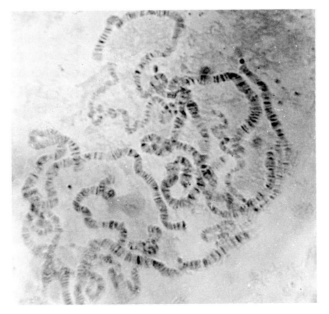

Figure 38-10

Chromosomes from a salivary gland cell of larval fruit fly *Drosophila*. These are among the largest chromosomes found in animal cells. Such chromosomes are called polytene because they are made up of many replicated chromatin strands. These chromosomes are not confined to salivary glands but are also known to occur in other organs, such as gut and malpighian tubules of most dipteran insects. The technique for their study is simply to crush salivary glands in a drop of acetocarmine, between a cover glass and a slide, so that the chromosomes are set free from the nuclei and are spread out as shown in the photograph.
Courtesy General Biological Supply House, Inc., Chicago.

somes are easily distinguishable with the light microscope. The chromatin strands are too close for efficient synthesis of RNA along them, but as genes in a particular region become active, they separate to form an observable "puff" (Figure 38-11).

GENE THEORY
Gene Concept

The term "gene" (Gr. *genos*, descent) was coined by W. Johannsen in 1909 to refer to the hereditary factors of Mendel (1865). Both cytological and genetic studies showed that genes, although of unknown chemical nature, were the fundamental units of inheritance. They were regarded as indivisible units of the chromosomes on which they were located. Studies with multiple mutant alleles demonstrated that alleles are in fact divisible by recombination; that is, *portions* of a gene are separable, and they have a fine structure. Furthermore, parts of many genes in eukaryotes are separated by sections of DNA that do not specify a part of the finished product (introns, described later).

As the chief functional unit of genetic material, genes determine the basic architecture of every cell, the nature and life of the cell, the specific protein syntheses, the enzyme formation, the self-reproduction of the cell, and, directly or indirectly, the entire metabolic function of the cell. Because of their ability to mutate, to be assorted and shuffled around in different combinations, genes have become the basis for our modern interpretation of evolution. Genes are molecular patterns that can maintain their identities for many generations, can be self-duplicated in each generation, and can control cell processes by allowing their specificities to be copied.

One Gene—One Enzyme Hypothesis

Since genes act to produce different phenotypes, we may infer that their action follows the scheme: gene → gene product → phenotypic expression. Furthermore, we may suspect that the gene product is usually a protein, because proteins, acting as enzymes, antibodies, hormones, and structural elements throughout the body, are the single most important group of biomolecules.

The first clear, well-documented study to link genes and enzymes was carried out on the common bread mold *Neurospora* by Beadle and Tatum in the early 1940s. This organism was ideally suited to a study of gene function for several reasons: these molds are much simpler to handle than fruit flies, they grow readily in well-defined chemical media, and they are haploid organisms that are consequently unencumbered with dominance relationships. Furthermore, mutations were readily induced by irradiation with ultraviolet light. Ultraviolet light–induced mutants, grown and tested in specific nutrient media, were found to have single-gene mutations that were inherited in accord with Mendelian principles of segregation. Each mutant strain was shown to be defective in one enzyme, which prevented that strain from synthesizing one or more complex molecules. Putting it another way, the ability to synthesize a particular molecule was controlled by a single gene.

From these experiments Beadle and Tatum set forth an important and exciting formulation: **one gene produces one enzyme** (see Principle 21, p. 13). For this work they were awarded the Nobel Prize in 1958. The new hypothesis was soon

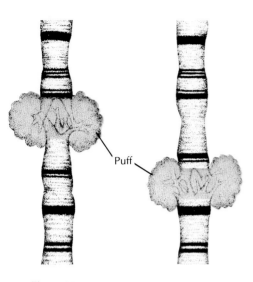

Figure 38-11

Puffing in one of the bands of a salivary gland chromosome of a midge larva *Chironomus*. Swelling, or puff, indicates activity in a region where RNA (and perhaps some DNA) is being produced, and it may include single bands or a group of adjacent ones. Puffs always include the same bands that occur in a definite sequence during development of the larva.

Puff

validated by the research of others who studied other biosynthetic pathways. Hundreds of inherited disorders, including dozens of human hereditary diseases, are caused by single mutant genes that result in the loss of a specific essential enzyme. We now know that a particular protein may be made of several chains of amino acids (polypeptides), each of which may be specified by a different gene, and not all proteins specified by genes are enzymes (for example, antibodies and hormones). Furthermore, genes direct the synthesis of various kinds of RNA. Therefore a gene now may be defined as **a nucleic acid sequence (usually DNA) that encodes a functional polypeptide or RNA sequence.**

___STORAGE AND TRANSFER OF GENETIC INFORMATION

The positive identification of the nucleic acids as the repository of genetic information came in the early 1950s after many decades of research. Although DNA was first isolated by Miescher (who called it nuclein) in 1869, just 4 years after Mendel published his findings, it was not until 1944 that O.T. Avery and his associates suggested that DNA is the genetic material in bacteria. However, many geneticists were reluctant to accept these conclusions, arguing that Avery's findings were a special case having little relevance to inheritance in other organisms. Then, in 1952, Hershey and Chase established beyond all reasonable doubt that DNA is the heritable material. This launched intensive activity in laboratories all over the world, culminating in the momentous discovery of the structure of DNA by James Watson and Francis Crick at Cambridge University in 1953. A lively and absorbing personal account of the discovery is given in Watson's book *The Double Helix.*

The Watson-Crick model of DNA structure had to suggest plausible answers to such problems as (1) how specific directions are transmitted from one generation to another, (2) how DNA could control protein synthesis, and (3) how the DNA molecule could duplicate itself. Classical genetics and cytology had shown how a cell divides to form two cells and how each cell receives a set of chromosomes with their genes identical in structure to the preexisting set. But nothing was known about how a chemical substance could carry out the specifications required by the genetic substance of a gene. The elegance of the Watson-Crick hypothesis lies in the perfect manner in which it fits the data and in the way that it can be tested.

Maurice Wilkins succeeded in getting sharp x-ray diffraction patterns that revealed three major periodic spacings in crystalline DNA. These periodicities were interpreted by Watson and Crick as the space distance between successive nucleotides in the DNA chains, the width of the chain, and the distance between successive turns of the helix, respectively. The x-ray diffraction photograph, with certain limitations, also gave indications of the spatial arrangement of some of the atoms within the large molecule. Watson and Crick came up with the idea that the molecule is bipartite, with an overall helical configuration. Accordingly, their model showed that the DNA molecule looks like two interlocked coils. Each "coil" consists of a sugar-phosphate backbone held together by phosphodiester bonds. The two interlocked coils are held together by hydrogen bonds between the purine and pyrimidine bases that form the core of the double helix (Figure 38-5).

Base pairing

The two DNA strands run in opposite directions, that is, they are **antiparallel.** This is evident from an examination of Figure 38-5. The two strands are also **complementary**—the sequence of bases along one strand specifies the sequence of bases along the other strand. Each strand could thus serve as a **template** for the production of its complement.

From examination of their original wire model of DNA (Figure 1-9, p. 12), Watson and Crick proposed a base-pairing rule. Adenine at one point on a strand must bond with thymine at the corresponding point on the other strand. Similarly, guanine must bond with cytosine. This happens because only these pairs are of proper size and arrangement of hydrogens and acceptor atoms for efficient bonding.

The purine bases adenine and guanine (A and G) are large, whereas the pyrimidine bases thymine and cytosine (T and C) are small. To fit into the structure of the DNA molecule, each pair must consist of one large and one small base; thus the small bases T and C, T and T, or C and C could not bond in pairs because they would not meet in the middle. The big bases (A and G) could not bond in any combination because they would produce a bulge in the helix. Neither can A and C nor G and T (although in these pairs one is large and one is small) form bond mates because their hydrogen bonds would not pair off properly. Thus there is only one correct way for the bases to bond: A with T and G with C. This is shown in the circled cross sections to the right of the DNA molecule in Figure 38-5.

Although all DNA molecules have the same general pattern, each one is unique because of the varying sequence of bases. This sequence spells out the genetic instructions that determine the sequence of amino acids in a protein organized by the gene (see Principle 18, p. 12). This genetic message, or "code," is written in a language of only four symbols (adenine, guanine, thymine, and cytosine) and is translated into protein amino acid sequences (see Principle 19, p. 12). The coding will be discussed subsequently.

____ DNA Replication

Every time the cell divides, the structure of the nuclear DNA must be precisely copied in the daughter cells. The wonder of the DNA double helix model was that its structure suggested how it could be replicated: each existing strand of the DNA molecule serves as a template for the synthesis of a new strand. This is the essential requirement of a genetic molecule, which was not satisfied by other models (such as single or triple helix) proposed before the Watson-Crick model. The two strands of the helix could unwind to expose the nucleotide bases. The double helix then separates, perhaps like a zipper, and new nucleotides are incorporated to synthesize new strands. Each of the four kinds of nucleotides consists of one of the DNA bases, one sugar unit, and three phosphate groups. Only one of the three phosphate groups is used in making the backbone of the new strand; the other two are necessary to provide energy for synthesis.

Because of the strict complementarity (base-pairing) rule, each exposed base will pair only with its complementary base. Thus each of the two strands will form a double helix identical with the parental molecule. Note that each daughter molecule will contain one parental strand and one new strand (Figure 38-12).

An enzyme, **DNA polymerase,** is essential for replication. The enzyme does not determine which of the four bases will be added to the new chain (base selection is determined by complementation with the base sequence in the parent

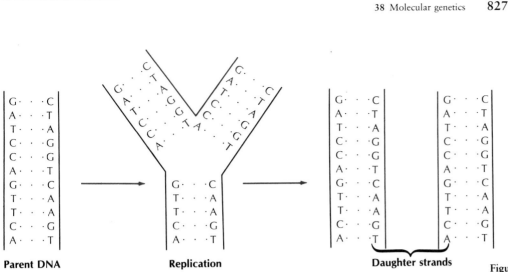

Parent DNA   **Replication**   **Daughter strands**

Figure 38-12

Replication of DNA. The parent strands of DNA part, and DNA polymerase synthesizes daughter strands using the base sequence of parent strands as a template.

strand), but it does hold the reacting molecules in position and vastly accelerates the reaction.

If replication proceeds as predicted, a replication "fork" should be visible where the double helix is "unzipping" during replication. Forks have in fact been seen in the circular bacterial chromosomes and plasmids. How the fork is formed is still unclear. Nor is it yet known how the two intertwined strands of DNA untwist during the replication. They may rotate during the process, but it is inconceivable that the entire chromosomal DNA molecule, millions of base pairs in length, flips about within the chromosome. When the hydrogen bonds holding base pairs together are broken during replication, the two strands move apart from one another, even though in the double helical form they appear to be interlocked.

DNA coding by base sequence

Since DNA is the genetic material and is composed of a linear sequence of base pairs, an obvious extension of the Watson-Crick model is that the sequence of base pairs in DNA codes for, and is colinear with, the sequence of amino acids in a protein (see Principle 19, p. 12). The coding hypothesis had to account for the way a string of four different bases—a four-letter alphabet—could dictate the sequence of 20 different amino acids.

In the coding procedure, obviously there cannot be a 1:1 correlation between four bases and 20 amino acids. If the coding unit (often called a word, or **codon**) consists of two bases, only 16 words (4^2) can be formed, which cannot account for 20 amino acids. Therefore the protein code must consist of at least three bases or three letters because 64 possible words (4^3) can be formed by four bases when taken as triplets. This means that there could be a considerable redundancy of triplets (codons), since DNA codes for just 20 amino acids. Later work by Crick confirmed that nearly all of the amino acids are specified by more than one code triplet (Table 38-2).

DNA shows a surprising stability, both in prokaryotes and in eukaryotes. Interestingly, it is susceptible to damage by harmful chemicals in the environment and radiation. Such damage is usually not permanent because cells have an efficient repair system. Various types of damage and repair are known, one of which is called **excision repair.** Ultraviolet irradiation often causes adjacent pyrimidines to link together by covalent bonds (dimerize), preventing transcription and replication. A series of several enzymes "recognizes" the area of the damaged strand

Table 38-2 Genetic Code: Codons (Code Triplets) between Messenger RNA and Specific Amino Acids

Codons	Amino acid
GCU, GCC, GCA, GCG	Alanine
CGU, CGC, CGA, CGG, AGA, AGG	Arginine
AAU, AAC	Asparagine
GAU, GAC	Aspartic acid
UGU, UGC	Cysteine
GAA, GAG	Glutamic acid
CAA, CAG	Glutamine
GGU, GGC, GGA, GGG	Glycine
CAU, CAC	Histidine
AUU, AUC, AUA	Isoleucine
CUU, CUC, CUA, CUG, UUA, UUG	Leucine
AAA, AAG	Lysine
AUG	Methionine; initiation of message
UUU, UUC	Phenylalanine
CCU, CCC, CCA, CCG	Proline
AGU, AGC, UCU, UCC, UCA, UCG	Serine
ACU, ACC, ACA, ACG	Threonine
UGG	Tryptophan
UAU, UAC	Tyrosine
GUU, GUC, GUA, GUG	Valine
UAA, UAG, UGA	Termination of message of one gene

and excises the pair of dimerized pyrimidines and several bases following them. DNA polymerase then synthesizes the missing strand along the remaining one, according to the base-pairing rules, and the enzyme **DNA ligase** joins the end of the new strand to the old one.

_____ Transcription and the Role of Messenger RNA

Information is coded in DNA, but DNA does not participate directly in protein synthesis. It is obvious that an intermediary is required. This intermediary is another nucleic acid called **messenger RNA (mRNA)** (see Principle 20, p. 12). Recall that RNA differs from DNA in three important ways: (1) it is _single_ stranded and not a double helix, (2) it has ribose sugar in its nucleotides instead of deoxyribose, and (3) it has the pyrimidine uracil (U) instead of thymine (T).

Ribosomal, transfer, and messenger RNAs are **transcribed** directly from DNA (Figure 38-13), with each of the many mRNAs being determined by a gene. In this process of making a complementary copy of one strand or gene of DNA in the formation of mRNA, an enzyme, **RNA polymerase,** is needed. (In fact, each type of RNA [ribosomal, transfer, and messenger] is transcribed by a specific type of RNA polymerase.) The mRNA contains a sequence of bases that complements the bases in one of the two DNA strands just as the DNA strands complement each other. Thus A in the coding DNA strand is replaced by U in mRNA; C is replaced by G; G is replaced by C; and T is replaced by A. Only one of the two chains is used as the template for RNA synthesis because only one bears the AUG codon that initiates a message (Table 38-2). The reason why only one strand of the double-stranded DNA is a "coding strand" is that mRNA otherwise would always be formed in complementary pairs, and enzymes also would be synthesized in

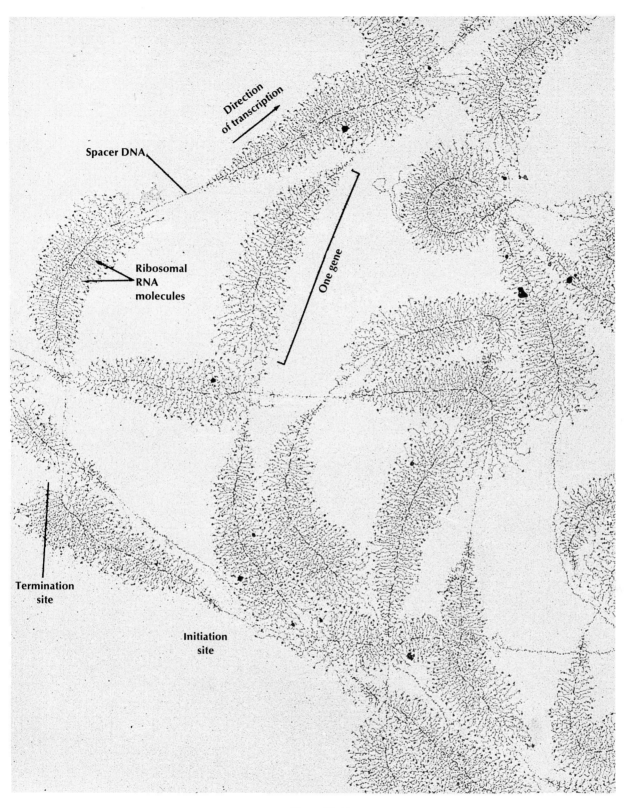

Figure 38-13

Genes in action. This remarkable electron micrograph shows DNA transcribing ribosomal RNA molecules in the oocyte nucleoli of the spotted newt *Diemictylus viridescens*. These are tandemly repeated genes separated by spacer (nontranscribing) segments of DNA. The newest and shortest ribosomal RNA transcripts are close to the initiation site on the gene. When completed molecules reach the termination site, they are freed to move out of the nucleus into the cytoplasm, where they form ribosomes. Enormous quantities of ribosomes are synthesized in maturing amphibian eggs. (×25,000.)

From Miller, O.L., Jr., and D.L. Beatty. Science **164** (cover). 1969. Copyright 1969 by the American Association for the Advancement of Science.

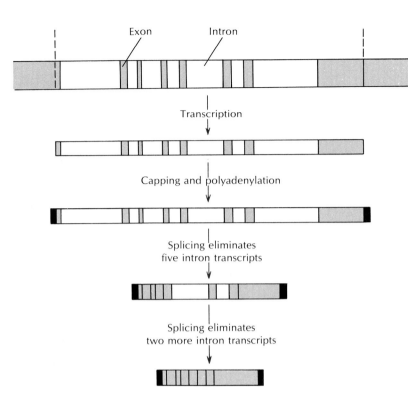

Figure 38-14

Transcription and maturation of ovalbumin gene of chicken. The entire gene of 7700 base pairs is transcribed to form the primary mRNA, then the cap of methyl guanine and the polyadenylate tail are added. After the introns are spliced out, the mature mRNA is transferred to the cytoplasm.

Redrawn from Chambon, P. 1981. Sci. Am. **244**:60-71 (May).

complementary pairs. In other words, two different enzymes would be produced for every DNA coding sequence instead of one. This would certainly lead to metabolic chaos.

Genes on the DNA of prokaryotes are coded on a continuous stretch of DNA, which is transcribed into mRNA and then translated (see the following section). It was assumed that this was also the case for eukaryote genes until the surprising discovery a few years ago that stretches of DNA are transcribed in the nucleus into mRNA that does not encode the finished product. In other words, the mRNA that is translated into protein in the cytoplasm is not the same mRNA that is transcribed in the nucleus; pieces of the nuclear mRNA have been spliced out (Figure 38-14). It was thus discovered that many genes are split, interrupted by sequences of bases that do not code for the final product, and the mRNA transcribed from them must be edited or "matured" before translation in the cytoplasm. The intervening segments of DNA are now known as **introns,** while those that code for part of the mature RNA and are translated into gene product are called **exons.** Before the mRNA leaves the nucleus, the introns are spliced out and a methylated guanine "cap" is added at one end, while a tail of adenine nucleotides (poly-*A*) is added at the other (Figure 38-14). The cap and the poly-*A* tail are characteristic of mRNA molecules.

It has now been found that the genes coding for many (but not all) proteins are split; in mammals the genes coding for the histones and for interferon are on continuous stretches of DNA. In lymphocyte differentiation the parts of the split genes coding for immunoglobulins are actually *rearranged* during development, so that different proteins result from subsequent transcription and translation. This partially accounts for the enormous diversity of antibodies manufactured by the descendants of the lymphocytes (p. 618).

Even more recently it has been found that a number of introns can *self-catalyze* their own cutting and splicing. They have base sequences that are complementary to other base sequences in the introns, and this mediates folding of the molecules so that complementary base sequences pair. Like protein enzymes, particular shapes of the molecules (brought about by the folding) are apparently necessary for the RNAs to have catalytic activity. Some RNAs have now been found that can catalyze the assembly of other RNAs, and this fully qualifies them as enzymes.

___ Translation: Final Stage in Information Transfer

The **translation** process takes place on **ribosomes,** granular structures composed of protein and **ribosomal RNA (rRNA)** (see Principle 20, p. 12). The mRNA molecules fix themselves to the ribosomes to form a messenger RNA–ribosome complex. Since only a short section of an mRNA molecule is in contact with a single ribosome, the mRNA usually fixes itself to several ribosomes at once. The entire complex, called a **polyribosome** or **polysome,** allows several proteins of the same kind to be synthesized at once, one on each ribosome of the polyribosome (Figure 38-15).

The assembly of proteins on the mRNA–ribosome complex requires the action of another kind of RNA called **transfer RNA (tRNA).** The tRNA molecules collect the free amino acids from the cytoplasm and deliver them to the polysome, where they are assembled into a protein. There is a special tRNA molecule for every amino acid. Furthermore each tRNA is accompanied by a special **activating enzyme.** These enzymes are necessary to sort out and attach the correct amino acid to each tRNA by a process called **loading.**

The tRNAs are surprisingly large molecules that are folded in a complicated way in the form of a cloverleaf (Figure 38-16). On this cloverleaf a special sequence of three bases (the **anticodon**) is exposed in just the right way to form base pairs with complementary bases (the codon) in the mRNA. The anticodon of the tRNA is the key to the correct sequencing of amino acids in the protein being assembled.

For example, alanine is assembled into a protein when it is signaled by the codon GCG in an mRNA. This translation is accomplished by alanine tRNA in which the anticodon is CGC. The alanine tRNA is first loaded with alanine by its activating enzyme. The loaded alanine tRNA enters the polysome where it fits precisely into the right place on the mRNA strand. Then the next loaded tRNA specified by the mRNA code (glycine tRNA, for example) enters the polysome and attaches itself beside the alanine tRNA. The two amino acids are united with a peptide bond (with the energy from a molecule of guanosine triphosphate), and the alanine tRNA falls off. The process continues stepwise as the protein chain is

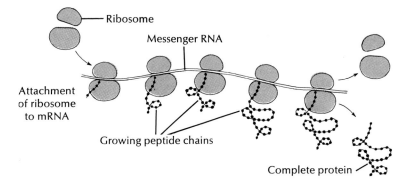

Figure 38-15

How the protein chain is formed. As ribosomes move along messenger RNA, the amino acids are added stepwise to form the polypeptide chain.

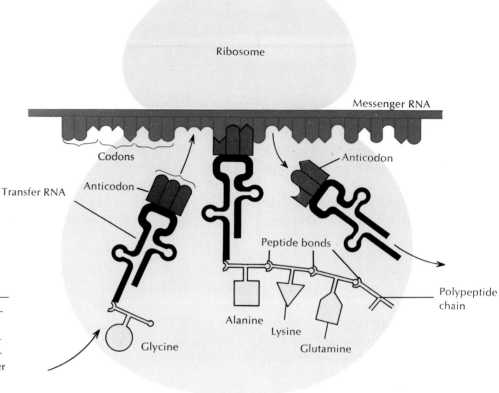

Figure 38-16

Formation of polypeptide chain on messenger RNA. As ribosome moves down messenger RNA molecule, transfer RNA molecules with attached amino acids enter ribosome *(left)*. Amino acids are joined together into polypeptide chain, and transfer RNA molecules leave ribosome *(right)*.

built (Figure 38-15). A protein of 500 amino acids can be assembled in less than 30 seconds.

_____ Regulation of Gene Function

In Chapter 36 we saw how the orderly differentiation of an organism from fertilized egg to adult requires the involvement of genetic material at every stage of development. Developmental biologists have provided convincing evidence that every cell in a developing embryo is genetically equivalent (see Principle 25, p. 14). This was first demonstrated by Spemann's delayed nucleation experiments (p. 781) and more recently confirmed with the elegant nuclear transplantation experiments of Gurdon and his associates in England (p. 783). Thus it is clear that as tissues differentiate, they use only a part of the genetic instruction present in every cell (unless chromatin diminution occurs [p. 833]). Certain genes express themselves only at certain times and not at other times. Indeed, there is reason to believe that in a particular cell or tissue, most of the genes are inactive at any given moment. The problem in development is to explain how, if every cell has a full gene complement, certain genes are "turned on" and produce proteins that are required for a particular developmental stage while the other genes remain silent.

Actually, although the developmental process brings the question of gene activation clearly into focus, gene regulation is necessary throughout an organism's existence. The cellular enzyme systems that control all functional processes obviously require genetic regulation because enzymes have powerful effects even in minute amounts. Enzyme synthesis must be responsive to the influences of supply and demand.

The discoveries of two French scientists, J. Monod and F. Jacob, contributed greatly to our understanding of gene regulation. Using bacteria, they described the **operon model.** Knowledge of gene regulation in prokaryotes is important in recombinant DNA technology, but we now know that the operon concept has little if any applicability to gene regulation in eukaryotes.

Gene regulation in eukaryotes

There are a number of different phenomena in eukaryotic cells that could serve as control points. Brown (1981) categorized possible genetic mechanisms for control as those in which the genes are altered (diminution, amplification, rearrangement, and modification), those in which gene expression is modulated (transcriptional, posttranscriptional, and translational control), and those that result from the organization of multigene families. Although thorough discussion of all of these mechanisms is beyond the scope of this book, a few examples are discussed.

Chromatin diminution

During differentiation of somatic cells or nuclei of some nematodes, protozoa, crustaceans, and insects, some chromatin or even entire chromosomes are discarded, so that only the germ cells (or micronuclei in protozoa) retain the entire genetic complement. Clearly, the products coded by the lost genes cannot be expressed by the cells. How widely applicable this mechanism may be has yet to be determined.

Gene amplification

Polytenization of the chromosomes in some insects has already been mentioned. This is an example of a case in which large amounts of a particular product, for example, rRNA, may be required at a given stage. The macronucleus of ciliates may contain multiple copies of some genes (amplification), while other genes are lost (diminution), and the micronucleus retains the full genetic complement.

Gene arrangement

The rearrangement of genes for immunoglobulins during differentiation of lymphocytes was mentioned previously. Such rearrangement may also be responsible for a switch in mating types in yeast.

DNA modification

An important mechanism for turning genes off appears to be methylation of cytosine residues in the 5' position, that is, adding a methyl group (CH_3-) to the carbon of the 5' position in the cytosine ring (Figure 38-17, A). This usually happens when the cytosine is next to a guanine residue; thus the bases in the complementary DNA strand would also be a cytosine and a guanine (Figure 38-17, B). When the DNA is replicated, an enzyme recognizes the CG sequence and quickly methylates the daughter strand, maintaining the gene in an inactive state.

Transcriptional control

Some hormones act by entering the cell and affecting gene transcription directly. Progesterone, for example, enters the cells of the chicken oviduct and binds with a receptor protein, and this complex binds with the chromatin in the nucleus. The effect is to turn on genes that code for specific cellular products; that is, molecules of mRNA are produced that result in synthesis of egg albumin and other substances.

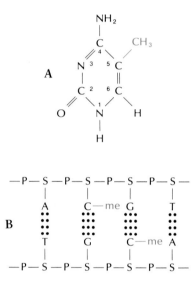

Figure 38-17

Some genes in eukaryotes are apparently turned off by the methylation of some cytosine residues in the chain. **A,** Structure of 5-methyl cytosine. **B,** Cytosine residues next to guanine are those that are methylated in a strand, thus allowing both strands to be symmetrically methylated.

In a recent report, the gene for fetal hemoglobin was turned on in an adult by treatment with a drug that causes demethylation of DNA. Because the fetal hemoglobin is functionally adequate, treatment with the drug results in a dramatic improvement in the clinical condition of persons with sickle cell anemia and beta thalassemia, which are caused by a defect in the gene for one of the adult hemoglobin chains.

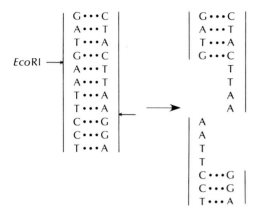

EcoRI →

Figure 38-18

Action of restriction endonuclease, *Eco*R1. Such enzymes recognize specific base sequences that are palindromic (a palindrome is a word spelled the same backward and forward). *Eco*R1 leaves "sticky ends," which are easily matched to other DNA fragments cleaved by the same enzyme and linked by DNA ligase.

Genetic Engineering

Progress in our understanding of genetic mechanisms on the molecular level, as discussed in the last few pages, has been almost breathtaking in the last few years. We can expect many more discoveries in the near future. This progress has largely been the result of the effectiveness of many biochemical techniques now used in molecular biology. We have space to describe only a few briefly.

One of the most important tools in this technology is a series of enzymes called **restriction endonucleases.** Each of these enzymes, derived from bacteria, cleaves double-stranded DNA at particular sites determined by the particular base sequences at that point. Many of these endonucleases cut the DNA strands so that one has several bases projecting farther than the other strand (Figure 38-18), leaving what are called "sticky ends." When these DNA fragments are mixed with others that have been cleaved by the same endonuclease, and in the presence of DNA ligase, they tend to anneal (join) by the rules of complementary base pairing.

If the DNA annealed after cleavage by the endonuclease is from two different sources, for example, a plasmid and a mammal, the product is **recombinant DNA.** To make use of the recombinant DNA, the modified plasmid must be cloned in bacteria. The bacteria are treated with dilute calcium chloride to make them more susceptible to taking up the recombinant DNA, but the plasmids do not enter most of the cells present. The bacterial cells that have taken up the recombinant DNA can be identified if the plasmid has a marker, for example, resistance to an antibiotic. Then, only the bacteria that will grow in the presence of the antibiotic are those that have absorbed the recombinant DNA. Some viruses have also been used as carriers for recombinant DNA. Plasmids and viruses that carry recombinant DNA are called **vectors.** The vectors retain the ability to replicate in the bacterial cells; therefore the mammalian gene is amplified.

Clearly, for this technique to be useful, the cloned bacteria must manufacture (and the mammalian gene in the recombinant DNA must code for) a product of some interest, such as a hormone. To find and isolate that gene from the entire mammalian genome is a formidable task, particularly when the DNA contains numerous introns. One way is to determine the amino acid sequence in the protein, deduce the sequence of DNA codons responsible for that amino acid sequence, and then synthesize DNA of the proper sequence, one base at a time.

This last step is difficult, and the method is only practical for short polypeptides. It is easier to synthesize a copy DNA (cDNA) from mRNA. In this procedure, RNA must be extracted from the cells, and the mRNA must be separated from the much more plentiful rRNA by some method such as affinity chromatography, taking advantage of the poly-*A* tail on the mature mRNA. In some instances the desired mRNA is present in great abundance and can be purified. The mixture of mRNAs is fractionated according to size (molecular weight) by a technique called gel electrophoresis, the fractions are recovered, and activity is tested in a cell-free translation system. The cDNA is then prepared from the mRNA that mediates synthesis of the desired protein by the action of reverse transcriptase (Figure 38-19). The second strand of DNA is synthesized by DNA polymerase using the first cDNA strand as a template. Thus a molecule of DNA is prepared for insertion into a plasmid and cloning in bacteria.

In many cases, however, the desired mRNA is present in small quantities, and the cDNA prepared from the partially purified RNA is impure. A mixture of DNA fragments containing many different genes is "shotgunned" into bacteria,

mines which n
cess that passe

When an
and its effects
condition can
ervoir of muta
Inbreeding enc
ity of recessive

Most mu
in which muta
conditions and
there could be
environment h
and mutations

Frequency o

Although whe
prevail at diffe
others, and ind
pairs) is more
possible to es
traits.

Relativel
Drosophila the
of 0.01% per l
per 100,000 lo
then a single nc
is mutated. Ho
son carries app
duced contains

Since mo
ful. Fortunatel
until by chance
produced. At t
rying the muta

All anim
pools. In huma
hidden mutatic
mutant individ
type.

——MOLI

The fundamen
manner (**neopl**
normal cells ha
rapidly, invadi
cells that lose
degree. Thus tl
cells of the tum

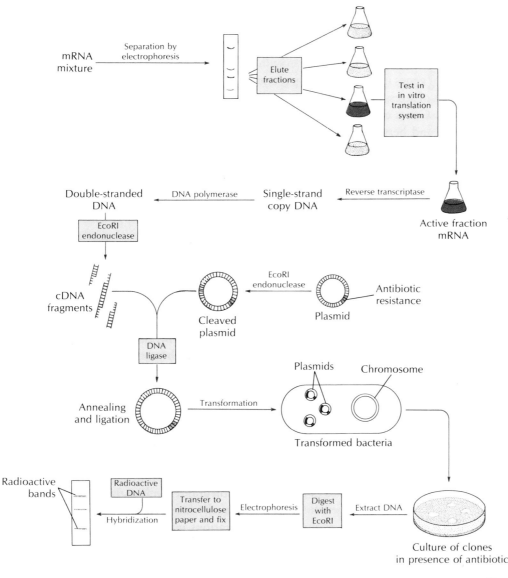

Figure 38-19

Steps in preparing DNA for a mammalian gene for cloning in bacteria and for determining that the gene has been cloned.

and their colonies are screened for the desired product by one of several methods, such as antibody binding.

When colonies of bacterial clones have been identified, their DNA can be extracted, cleaved by the endonuclease, and fractionated by gel electrophoresis, and the cDNA is used as a probe to confirm whether the mammalian gene was being replicated in the bacteria. In this assay the DNA fractions on the electrophoresis plate are heated to separate them into single strands (denaturation), then transferred and fixed to a nitrocellulose filter paper by a kind of blotting technique. A solution containing radioactive cDNA is spread over the plates, and they are washed. Wherever radioactive spots remain on the plates, cDNA has hybridized with the denatured DNA and reformed the double strands according to the rules of base pairing, indicating the identity of the two (Figure 38-19).

To obtain a useful product from the bacterial colonies containing the mammalian genes, the mammalian genes must be transcribed and translated by the bacteria. This can be accomplished by inserting the proper messages into the plasmid DNA, causing the little chemical factories of the bacterial colonies to turn in many cancerous cells, perhaps all, has a genetic basis, and investigation of the genetic damage in cancer is now a major thrust of cancer research.

___ Oncogenes and Proto-oncogenes

Some 40 genes have been identified in cells that, under certain circumstances of activation, are associated with neoplastic growth. They have been named **oncogenes** and are derived from normal cellular genes. In their benign, inactivated state these genes are often referred to as **proto-oncogenes.** In a number of cases the proteins encoded by oncogenes are known, but just how these proteins may account for the neoplastic phenotype remains unclear.

Transformation of proto-oncogenes to oncogenes

Several mechanisms are now known that explain how proto-oncogenes can become oncogenes. For example, cellular genes can become oncogenes through the action of certain **retroviruses.** Retroviruses are viruses in which the genome is single-stranded RNA, and this is transcribed into DNA in the host cell by reverse transcriptase. The viral DNA is then integrated into the host DNA, and the cell machinery is taken over to produce viral gene products. Insertion of viral DNA into host DNA is mutagenic and may damage cellular genes, as well as bring cellular genes under the influence of powerful regulatory elements of the viral genome. These events may enhance the production of gene product by release of the proto-oncogene from its normal controls or by increase in transcription of the proto-oncogene by a viral enhancer.

Another mechanism for transformation of proto-oncogenes is chromosomal translocation: portions of two different chromosomes are exchanged, one for the other. Thus the proto-oncogene may be released from its normal control on its original chromosome, or it may be placed near an element that promotes transcription on its new chromosome.

Mutations have been implicated in the conversion of some proto-oncogenes to oncogenes. At least one case is known in which a change in a single base (a guanine is replaced by a thymine) in a 5000 nucleotide gene is responsible for the transformation. This mechanism can help account for the fact that certain chemicals and ionizing radiation, which are known to cause mutations in cells, are also powerful carcinogens (cancer-causing agents).

Gene amplification (p. 833) is not normally a feature of mammalian chromosomes, but it has been found in a number of tumors. The amplified DNA includes some proto-oncogenes, but amplification has been found only in cells that have taken some steps toward neoplastic growth. It is not known whether amplification is a cause or an effect of the basic genetic damage.

Gene products and actions of oncogenes

Because each somatic cell in vertebrates carries the entire genome, all the potential oncogenes are present in all cells. The same oncogene can be activated in founder cells of different tissue types, and these will result in different types of cancers. However, the actions to trigger the neoplastic phenotype in the wide variety of cancers may be much smaller. At the present time, only four biochemical functions are known for the numerous proteins encoded by the oncogenes described so far: (1) protein phosphorylation, which is a common mechanism in cells to regulate the activity of enzymes (see p. 94); (2) metabolic regulation by proteins that bind

---

Of the many ways that cellular DNA can sustain damage, the three most important are ionizing radiation, ultraviolet radiation, and chemical mutagens. The high energy of ionizing radiation (x rays and gamma rays) causes electrons to be ejected from the atoms it encounters, resulting in ionized atoms with unpaired electrons (free radicals). The free radicals (principally from water) are highly reactive chemically, and they react with molecules in the cell, including DNA. Some damaged DNA is repaired, but if the repair is inaccurate, a mutation results. Ultraviolet radiation is of much lower energy than ionizing radiation and does not produce free radicals. It is absorbed by pyrimidines in DNA and causes formation of a double covalent bond between the adjacent pyrimidines. UV repair mechanisms can also be inaccurate. Chemical mutagens react with the DNA bases and cause mispairing during replication.

There is a
Bernard Sh
letter from
suggested
perfect chi
her beauty
declined th
that the ch
inherit her
Shaw was
parental ga
random, it

mines which new genes merit survival; the environment imposes a screening process that passes the fit and eliminates the unfit.

When an allele of a gene is mutated to the new allele, it tends to be recessive and its effects are normally masked by its partner allele. Only in the homozygous condition can such mutant genes be expressed. Thus a population carries a reservoir of mutant recessive genes, some of which are lethal but seldom expressed. Inbreeding encourages the formation of homozygotes and increases the probability of recessive mutants appearing.

Most mutations are destined for a brief existence. There are cases, however, in which mutations may be harmful or neutral under one set of environmental conditions and helpful under a different set. Should the environment change, there could be a new adaptation beneficial to the species. The earth's changing environment has provided numerous opportunities for new gene combinations and mutations, as evidenced by the great diversity of animal life today.

Frequency of mutations

Although whether or not a mutation occurs is random, different mutation rates prevail at different loci. Some *kinds* of mutations are more likely to occur than others, and individual genes differ considerably in length. A long gene (more base pairs) is more likely to have a mutation than a short gene. Nevertheless, it is possible to estimate average spontaneous rates for different organisms and traits.

Relatively speaking, genes are extremely stable. In the well-studied fruit fly *Drosophila* there is approximately one detectable mutation per 10,000 loci (rate of 0.01% per locus per generation). The rate for humans is one per 10,000 to one per 100,000 loci per generation. If we accept the latter, more conservative figure, then a single normal allele is expected to go through 100,000 generations before it is mutated. However, since human chromosomes contain 100,000 loci, every person carries approximately one new mutation. Similarly, each spermatozoon produced contains, on the average, one mutant allele.

Since most mutations are deleterious, these statistics are anything but cheerful. Fortunately, most genes are recessive and are not detected by natural selection until by chance they have increased enough in frequency for homozygotes to be produced. At that point most are eliminated, since the zygote or individual carrying the mutants does not reproduce.

All animals carry large numbers of lethal and semilethal alleles in their gene pools. In human populations this has been called the "genetic load," or load of hidden mutations. By genetic recombination these continue to surface to produce mutant individuals less "fit" (in the Darwinian sense) than the "wild-type" genotype.

___MOLECULAR GENETICS OF CANCER

The fundamental defect in cancer cells is that they proliferate in an unrestrained manner (**neoplastic** growth). The mechanism that controls the rate of division of normal cells has somehow broken down, and the cancer cells multiply much more rapidly, invading other tissues in the body. Cancer cells originate from normal cells that lose their constraint on division and become dedifferentiated in some degree. Thus there are many kinds of cancer, depending on the original founder cells of the tumor. In recent years mounting evidence has indicated that the change

Inbreeding has surfaced as a serious problem in zoos holding small founder populations of rare mammals. Matings of close relatives tend to bring together genes from the same ancestor and make it more likely that two copies of a deleterious gene will come together. The result is "inbreeding depression." One management solution is to enlarge genetic diversity by bringing together captive animals from different zoos, or by introducing new stock from the wild when this is possible. Paradoxically, where founder populations are extremely small and no wild stock can be obtained, deliberate inbreeding is recommended. This selects for genes that tolerate inbreeding; deleterious genes disappear because they kill the animals bearing them.

in many cancerous cells, perhaps all, has a genetic basis, and investigation of the genetic damage in cancer is now a major thrust of cancer research.

___ Oncogenes and Proto-oncogenes

Some 40 genes have been identified in cells that, under certain circumstances of activation, are associated with neoplastic growth. They have been named **oncogenes** and are derived from normal cellular genes. In their benign, inactivated state these genes are often referred to as **proto-oncogenes.** In a number of cases the proteins encoded by oncogenes are known, but just how these proteins may account for the neoplastic phenotype remains unclear.

Transformation of proto-oncogenes to oncogenes

Several mechanisms are now known that explain how proto-oncogenes can become oncogenes. For example, cellular genes can become oncogenes through the action of certain **retroviruses.** Retroviruses are viruses in which the genome is single-stranded RNA, and this is transcribed into DNA in the host cell by reverse transcriptase. The viral DNA is then integrated into the host DNA, and the cell machinery is taken over to produce viral gene products. Insertion of viral DNA into host DNA is mutagenic and may damage cellular genes, as well as bring cellular genes under the influence of powerful regulatory elements of the viral genome. These events may enhance the production of gene product by release of the proto-oncogene from its normal controls or by increase in transcription of the proto-oncogene by a viral enhancer.

Another mechanism for transformation of proto-oncogenes is chromosomal translocation: portions of two different chromosomes are exchanged, one for the other. Thus the proto-oncogene may be released from its normal control on its original chromosome, or it may be placed near an element that promotes transcription on its new chromosome.

Mutations have been implicated in the conversion of some proto-oncogenes to oncogenes. At least one case is known in which a change in a single base (a guanine is replaced by a thymine) in a 5000 nucleotide gene is responsible for the transformation. This mechanism can help account for the fact that certain chemicals and ionizing radiation, which are known to cause mutations in cells, are also powerful carcinogens (cancer-causing agents).

Gene amplification (p. 833) is not normally a feature of mammalian chromosomes, but it has been found in a number of tumors. The amplified DNA includes some proto-oncogenes, but amplification has been found only in cells that have taken some steps toward neoplastic growth. It is not known whether amplification is a cause or an effect of the basic genetic damage.

Gene products and actions of oncogenes

Because each somatic cell in vertebrates carries the entire genome, all the potential oncogenes are present in all cells. The same oncogene can be activated in founder cells of different tissue types, and these will result in different types of cancers. However, the actions to trigger the neoplastic phenotype in the wide variety of cancers may be much smaller. At the present time, only four biochemical functions are known for the numerous proteins encoded by the oncogenes described so far: (1) protein phosphorylation, which is a common mechanism in cells to regulate the activity of enzymes (see p. 94); (2) metabolic regulation by proteins that bind

Of the many ways that cellular DNA can sustain damage, the three most important are ionizing radiation, ultraviolet radiation, and chemical mutagens. The high energy of ionizing radiation (x rays and gamma rays) causes electrons to be ejected from the atoms it encounters, resulting in ionized atoms with unpaired electrons (free radicals). The free radicals (principally from water) are highly reactive chemically, and they react with molecules in the cell, including DNA. Some damaged DNA is repaired, but if the repair is inaccurate, a mutation results. Ultraviolet radiation is of much lower energy than ionizing radiation and does not produce free radicals. It is absorbed by pyrimidines in DNA and causes formation of a double covalent bond between the adjacent pyrimidines. UV repair mechanisms can also be inaccurate. Chemical mutagens react with the DNA bases and cause mispairing during replication.

GTP (guanosine triphosphate); (3) influencing mRNA production to control gene expression; and (4) influencing DNA replication. It is yet to be shown how any of these functions could lead to the many characteristics shown by neoplastic cells.

___ SUMMARY

The nucleic acids in the cell are DNA and RNA, which are large polymers of nucleotides, each composed of a nitrogenous base, pentose sugar, and phosphate group. The nitrogenous bases in DNA are adenine (A), guanine (G), thymine (T), and cytosine (C); and those in RNA are the same except that uracil (U) is substituted for thymine. DNA is a double-stranded, helical molecule in which the bases extend toward each other from the sugar-phosphate backbone: A always pairs with T, and G always pairs with C.

The genetic material in prokaryotes is carried in the DNA of their single, circular chromosome, which never condenses into the rodlike bodies we see in mitosis of eukaryotes. Most bacteria also have plasmids, which are small circles of DNA carrying relatively few, but often important, parts of the bacterial genome. Viruses also carry genetic information in DNA and sometimes in RNA. Neither viruses nor plasmids can replicate outside the host cell.

Genetic information in eukaryotes is found in the DNA of their mitochondria, plastids, and nuclear chromosomes. The chromosomes carry far more DNA than bacterial chromosomes or the other organelles mentioned. During interphase the eukaryotic chromosomes are dispersed into long strands of chromatin. The chromatin is comprised of DNA molecules organized with basic proteins called histones in repeating units known as nucleosomes. During mitosis the chromosomes are tightly contracted into rodlike bodies that avoid entanglement of the genetic material during its division between two daughter cells.

The position of the centromere is characteristic of a given chromosome. The locations of heterochromatin and euchromatin indicate regions where genes are inactive and active, respectively. Evidence of gene activity can be observed with the light microscope in the puffs of the polytene chromosomes of insects.

One gene most commonly controls the production of one protein or polypeptide (one gene–one polypeptide hypothesis), but the various types of RNA are also encoded on the genes.

The strands of DNA are antiparallel and complementary, being held in place by hydrogen bonds between the paired bases. In DNA replication the strands part, and the enzyme DNA polymerase synthesizes a new strand along each parent strand, using the parent strand as a template.

The sequence of the bases in DNA is a code for the amino acid sequence in the ultimate product protein. Each three bases (triplet) make up a codon specifying a particular amino acid. Cells have considerable ability to repair damaged DNA strands.

Proteins are synthesized by transcription of the information coded into DNA into the base sequence of a molecule of messenger RNA (mRNA), which functions in concert with ribosomes (containing ribosomal RNA [rRNA] and protein) and transfer RNA (tRNA). Ribosomes attach to the strand of mRNA and move along it, assembling the amino acid sequence of the protein. Each amino acid is brought into position for assembly by a molecule of tRNA, which itself bears a base sequence (anticodon) complementary to the respective codons of the mRNA. In eukaryotes the sequences of bases in DNA coding for amino acids in a protein (exons) are interrupted by intervening sequences (introns). The introns are spliced out of the primary mRNA before it leaves the nucleus, and the protein is synthesized in the cytoplasm. Some introns can self-catalyze their own cutting and splicing, and some RNAs can catalyze the assembly of other RNAs, thus qualifying them as enzymes.

On a molecular level, a gene is a portion of DNA bearing the information necessary to synthesize a functional product, whether mRNA (and therefore ultimately a protein), rRNA, or tRNA. Genes, and the synthesis of the products for which they are responsible, must be regulated: turned on or off in response to varying environmental conditions or cell differentiation. The operon model describes an important mode of gene regulation in prokaryotes. Gene regulation in eukaryotes is more complex, and a number of mechanisms are known. One inactivation mechanism is the methylation of cytosine residues that lie next to guanine residues in the DNA strand, thus preventing transcription. In some cells genes are activated by the direct action of a hormone in the nucleus.

Modern methods in molecular genetics have made spectacular advances possible. Restriction endonucleases cleave DNA at specific base sequences, and such cleaved DNA from different sources can be rejoined to form recombinant DNA. Combining mammalian with plasmid or viral DNA, a mammalian gene can be introduced into bacterial cells, which then multiply and express the mammalian gene, for example, the gene for synthesis of insulin or interferon.

A mutation is a physicochemical alteration in the bases of the DNA that changes the effect of the gene. Although rare and usually detrimental to the survival and reproduction of the organism, mutations are occasionally beneficial and provide new genetic material on which natural selection can work.

Many types of cancer (neoplastic growth) are associated with genetic damage in which certain normal genes (proto-oncogenes) become abnormally activated to oncogenes. Mechanisms for transformation of proto-oncogenes into oncogenes include the action of retroviruses, chromosomal translocation, mutation, and gene amplification. Only four biochemical actions are known for the products of the 40 oncogenes described so far.

Review questions

1. Name the purines and pyrimidines in DNA and tell which ones pair with each other in the double helix. What are the purines and pyrimidines in RNA and to what are they complementary in DNA?
2. Explain the essential differences in the state of DNA in each of the following: prokaryote chromosome, plasmid, mitochondrion, eukaryote chromosome.
3. Define each of the following: chromatin, nucleosomes, histones, centromere, acrocentric, telocentric, heterochromatin, euchromatin, polytene chromosomes.
4. What is the "one gene–one enzyme hypothesis"?
5. Give a modern definition of a gene.
6. Explain the following with regard to DNA: base pairing, antiparallel, complementary.
7. Explain how DNA is replicated.
8. Why is it not possible for a codon to consist of only 2 bases?
9. Explain the transcription and processing of mRNA in the nucleus.
10. Explain the role of mRNA, tRNA, and rRNA in protein synthesis.
11. Explain four ways that genes can be regulated in eukaryotes.
12. In modern molecular genetics, what is recombinant DNA, and how is it prepared?
13. Name three sources of phenotypic variation.
14. Distinguish between proto-oncogene and oncogene. What are three mechanisms whereby a proto-oncogene can become an oncogene?

Selected references

Anderson, W.F. 1984. Prospects for human gene therapy. Science **226**: 401-408. *Genetic engineering offers hope to correct some genetic disorders of humans.*

Bishop, J.M. 1987. The molecular genetics of cancer. Science **235**:305-311. *A concise summary of the evidence to the present on oncogenes and their products.*

Brown, D.D. 1981. Gene expression in eukaryotes. Science **211**:667-674. *A review of knowledge concerning gene regulation in eukaryotes.*

Cech, T.R. 1986. RNA as an enzyme. Sci. Am. **255**:64-75 (Nov.). *Evidence is accumulating that RNA can catalyze its own replication. The first nucleic acids may have duplicated themselves with the assistance of proteins.*

Croce, C.M., and G. Klein. 1985. Chromosome translocations and human cancer. Sci. Am. **252**:54-60 (Mar.). *How a proto-oncogene can become an oncogene when it is translocated to another chromosome near genetic sequences that enhance its activity.*

Darnell, J.E., Jr. 1983. The processing of RNA. Sci. Am. **249**:90-100 (Oct.). *Describes removal of introns, capping, and polyadenylation of RNA.*

Kornberg, R.D., and A. Klug. 1981. The nucleosome. Sci. Am. **244**:52-64 (Feb.). *The nucleosome is an important part of the organization of chromosomes in eukaryotes.*

Koshland, D.E., Jr. 1987. Frontiers in recombinant DNA. Science **236**:1157. *This is a lead editorial in an issue of* Science *that contains a number of articles on exciting applications of recombinant DNA research.*

Motulsky, A.G. 1983. Impact of genetic manipulation on society and medicine. Science **219**:135-140. *A summary of ethical considerations raised by genetic engineering. We must improve biological education for everyone so that we can make informed decisions.*

Watson, J.D. 1969. The double helix. New York, Atheneum Publishers. *An enlightening and entertaining history of events leading to an understanding of the genetic code.*

CHAPTER 39

ORGANIC EVOLUTION

The singularity of the earth is not that it harbors life, since life must surely exist elsewhere in the universe. It is that life on earth is so enormously diverse and pervasive. Even from space the ubiquity of life is evident in the kaleidoscopic patterns of grasslands, forests, lakes, and croplands stretching across the earth's surface, their colors changing with the seasons.

Each ecosystem is biologically and physically distinct. Each is occupied by a great variety of organisms living interdependently and manifesting every conceivable kind of adaptation to their surroundings. Nutrients are withdrawn and again released; energy captured by plants flows through the system bringing order out of disorder in apparent defiance of the second law of thermodynamics; organisms die and are replaced by offspring, a renewal that threads its descent faithfully through generations of ancestors. It is a drama of incalculable complexity that has been and is now unfolding before our eyes.

Yet despite the apparent permanence of the natural world, change characterizes all things on earth and in the universe. Countless kinds of animals and plants have flourished and disappeared, leaving behind an imperfect fossil record of their existence.

The earth itself bears its own record of change, transformed as it is by processes that have occurred over a vast span of time and still are at work today. These changes are irreversible, and for life at least they appear in the broad sense to be directional and progressive, as primitive organisms of the young earth have yielded to the advanced, complex creatures of the present. We call this historical change in the set of instructions for building an organism **organic evolution.** Because every feature of life as we know it today is a product of the evolutionary process, biologists consider organic evolution to be the keystone of all biological knowledge.

The great contribution of Charles Robert Darwin and Alfred Russel Wallace (Figure 39-1) was that they simultaneously provided the first credible explanation for evolutionary change, the **principle of natural selection.** The principal alternative explanation to account for the diversity of life on earth is that of **special creation,** which holds that all life forms were divinely created much as we find them today. However, special creation rests on premises that are not verifiable and thus must be accepted on faith. The study of biology, on the other hand, must account for the physical evidence that life has evolved and changed over time. Students should be aware that the extensive evidence for evolution and acceptance of the fact that evolution has occurred are not necessarily incompatible with religious concepts of creation. Indeed, evolution can be viewed as a creative process operating over long periods of time. How matter and life may have appeared in our universe is not at issue in evolutionary theory. It is regrettable therefore that the recent reemergence of rigid antievolutionary views, in the guise of "creation

A   B

science," should portray creation by a supernatural being and evolution as irreconcilable.

In our treatment of the principle of organic evolution in this chapter, we consider Darwin's theory of natural selection and the evidence for evolution, the concept of species and how they change with time, and the question of how new species arise. The chapter closes with a brief synopsis of human evolution.

DEVELOPMENT OF THE IDEA OF ORGANIC EVOLUTION

Evolution is no longer a subject for debate among biologists; as a process it is known and accepted, even though the forces that determine the course of evolution are not fully understood. Yet before the eighteenth century, much of the speculation on the origin of species rested on myth and superstition rather than on observation. Nevertheless, long before this time, there were those who were thinking about and attempting to interpret the order of nature.

Some of the early Greek philosophers, notably Xenophanes, Empedocles, and Aristotle, developed the germ of the idea of change and natural selection within the restrictions of belief in spontaneous generation. They recognized fossils as evidence of a former life that they believed had been destroyed by some natural catastrophe. Despite their spirit of intellectual inquiry, the Greeks failed to establish an evolutionary concept, probably because of their limited experience with the natural world.

With the gradual decline of ancient science, beginning well before the rise of Christianity, debate on evolution nearly ended. The opportunity for fresh thinking became even more restricted as the biblical account of the earth's creation became accepted as a tenet of faith. The year 4004 BC was fixed by Archbishop James Ussher (mid-seventeenth century) as the time of true creation of life. In this atmosphere evolutionary views were considered rebellious and heretical.

Still, some speculation continued. The French naturalist Buffon (1707-1788)

Figure 39-1

Founders of theory of natural selection. **A,** Charles Robert Darwin (1809-1882), as he appeared in 1881, the year before his death. **B,** Alfred Russel Wallace (1823-1913) in 1895. Darwin and Wallace independently arrived at the same conclusion. A letter and essay from Wallace written to Darwin in 1858 spurred Darwin to write *The Origin of Species,* published in 1859.
A, Courtesy American Museum of Natural History; **B,** courtesy British Museum (Natural History), London.

stressed the influence of environment on the modifications of animal type. He also extended the age of earth to 70,000 years from 6000 years.

Lamarckism: the First Scientific Explanation of Evolution

The first complete explanation of evolution was authored by Jean Baptiste de Lamarck (1744-1829) (Figure 39-2), a pioneering French biologist. Lamarck, who published his concept in 1809, the year of Darwin's birth, was the first biologist to make a convincing case for the idea that fossils were the remains of extinct animals and not, as some argued, stones molded in chance imitations of life. Lamarck's idea of **inheritance of acquired characteristics** was engagingly simple: organisms, by striving to adapt to their environment, acquire adaptations during their lives that are passed on by heredity to their offspring. The giraffe, according to Lamarck, developed its long neck by constantly stretching for food. Genetic variation then arose preferentially in this direction, and the trait of long necks became common in the population. Snakes lost their legs, Lamarck believed, because legs were a handicap in moving through dense vegetation. A man who developed his muscles by exercise was supposed to pass his strength on to his sons. Lamarck's attitude toward adaptation was straightforward: an organism could get any adaptation it needed. Despite the uncluttered appeal and efficiency of Lamarck's concept, it was effectively disproved in this century by geneticists who could find no mechanism that would make such a process possible. The unveiling of the genetic code showed that DNA transcription and translation are a one-way process with no known way for the environment to feed information back into the cell's DNA.

Charles Lyell and the Doctrine of Uniformitarianism

While eighteenth- and early nineteenth-century zoologists wrestled with conflicting concepts of organic evolution, geologists were marshaling sound evidence for the physical evolution of the earth's crust. The geologist Sir Charles Lyell (1797-1875) (Figure 39-3), in his *Principles of Geology* (1830-1833), stated that the changes in the earth's surface were the result of ongoing natural forces operating with uniformity over great periods of time. The idea that processes occurring now are the same as those in the past came to be called the doctrine of uniformitarianism. As one writer of the time stated, "no vestige of a beginning—no prospect of an end."

Lyell was able to show that such forces, acting over long periods of time, could account for all observed changes, including the formation of fossil-bearing rocks. His familiarity with fossils, the natural history of contemporary marine and freshwater animals, and sedimentary rocks led him to conclude that the earth's age must be reckoned in millions of years rather than in thousands.

Charles Darwin (1809-1882) was thus not the first to propose the idea of evolution nor even to suggest a mechanism for its action. His predecessors had already established an intellectual climate that made a hypothesis of evolutionary mechanism possible, if not inevitable. Darwin himself later acknowledged that the theory of natural selection had been suggested by others before him. But so forcefully did Darwin present his ideas and his array of carefully collected scientific data that no one since has been able to challenge his preeminence in this field. Today the fact of organic evolution can be denied only by abandoning reason. As

Figure 39-2

Jean Baptiste de Lamarck (1744-1829), French naturalist who developed the first reasoned account of evolution. The scientific establishment of his time was contemptuous of his "use-and-disuse" explanation of evolution. Yet, in France today, where Darwinism has never been fully accepted, Lamarck is regarded as the father of evolution. This engraving was made in 1821. Courtesy British Museum (Natural History), London.

the noted English biologist Sir Julian Huxley wrote, "Charles Darwin effected the greatest of all revolutions in human thought, greater than Einstein's or Freud's or even Newton's, by simultaneously establishing the fact and discovering the mechanism of organic evolution."*

——DARWIN'S GREAT VOYAGE OF DISCOVERY

"After having been twice driven back by heavy south-western gales, Her Majesty's ship *Beagle,* a ten-gun brig, under the command of Captain Fitz Roy, R.N., sailed from Devonport on the 27th of December, 1831." Thus began Charles Darwin's account of the historic 5-year voyage of the *Beagle* around the world (Figure 39-4). Darwin, not quite 23 years old, had been invited to serve as naturalist without pay on the *Beagle,* a small vessel only 90 feet in length, which was about to depart on a second extensive surveying voyage to South America and the Pacific. It was the beginning of one of the most important voyages of the nineteenth century.

During the voyage (1831-1836) (Figure 39-5), Darwin endured almost constant seasickness and the erratic companionship of the authoritarian Captain Fitz Roy. But his youthful physical strength and early training as a naturalist equipped him for his work. The *Beagle* made many stops along the harbors and coasts of South America and adjacent regions. Darwin made extensive collections and observations on the fauna and flora of these regions. He unearthed numerous fossils of animals long since extinct and noted the resemblance between fossils of the South American pampas and the known fossils of North America. In the Andes he encountered seashells embedded in rocks at 13,000 feet. He experienced a severe earthquake and watched the mountain torrents that relentlessly wore

*Huxley, J. In Bowman, R.I. 1966. The Galápagos. Berkeley, University of California Press.

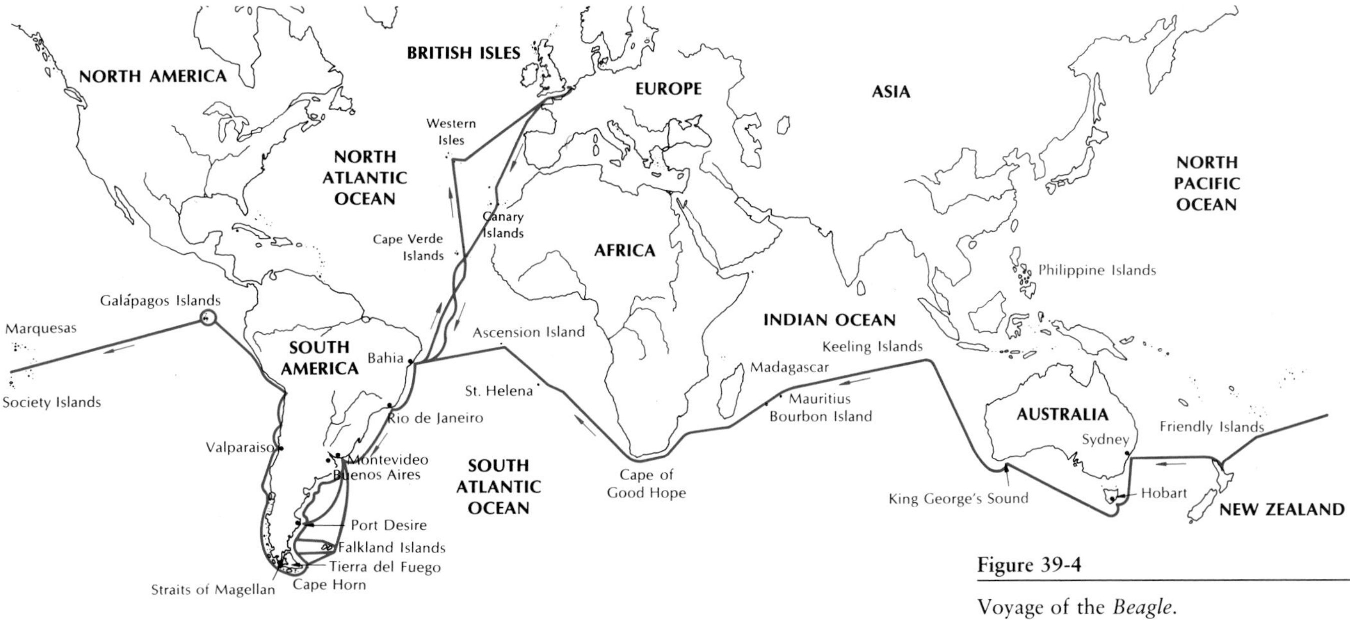

Figure 39-3

Sir Charles Lyell (1797-1875), English geologist and friend of Darwin. His book *Principles of Geology* greatly influenced Darwin during Darwin's formative period. The photograph was made about 1856.
Courtesy British Museum (Natural History), London.

Figure 39-4

Voyage of the *Beagle.*
Modified from Moorhead, A. 1969. Darwin and the *Beagle.* New York, Harper & Row, Publishers, Inc.

A   **B**

Figure 39-5

Charles Darwin and H.M.S. *Beagle*. **A**, Darwin in 1840, 4 years after the *Beagle* returned to England. **B**, H.M.S. *Beagle* in the Straits of Magellan.

A, From a watercolor by George Richmond; **B**, courtesy American Museum of Natural History.

away the earth—observations that strengthened his conviction that natural forces were responsible for the geological features of the earth. Finally he realized that he was witnessing the results of evolution.

In mid-September of 1835, the *Beagle* arrived at the Galápagos Islands, a volcanic archipelago straddling the equator 600 miles west of the coast of Ecuador (Figure 39-6). The fame of the islands stems from their infinite strangeness. They are unlike any other islands on earth. Some visitors today are struck with awe and wonder; others with a sense of depression and dejection. Circled by capricious currents, surrounded by shores of twisted lava, bearing skeletal brushwood baked by the equatorial sun, almost devoid of vegetation, inhabited by strange reptiles and by convicts stranded by the Ecuadorian government, the islands indeed had few admirers among mariners. By the middle of the seventeenth century, the islands were already known to the Spaniards as "Las Islas Galápagos"—the tortoise islands. The giant tortoises, used for food first by buccaneers and later by American and British whalers, sealers, and ships of war, were the islands' principal attraction. At the time of Darwin's visit, the tortoises were already being heavily exploited.

During the *Beagle*'s 5-week visit to the Galápagos, Darwin began to develop his views of the evolution of life on earth. His original observations of the giant tortoises, marine iguanas that feed on seaweed, and drab ground finches, a family that had evolved into several species, each with distinct beak and feeding adaptations, all contributed to the turning point in Darwin's thinking.

Darwin was struck by the fact that, although the Galápagos Islands and the Cape Verde Islands (visited earlier in this voyage of the *Beagle*) were similar in climate and topography, their fauna and flora were altogether different. He recognized that Galápagos plants and animals were related to those of the South American mainland, yet differed from them sometimes in curious ways. Each island often contained a unique species of a particular group of animals that was nonetheless related to forms on other islands. In short, Galápagos life must have originated in continental South America and then undergone modification in the various environmental conditions of the separate islands. He concluded that living forms were neither divinely created nor immutable; they were, in fact, the products of evolution. Although Darwin devoted only a few pages to Galápagos animals and plants in his monumental *Origin of Species,* published more than two decades after visiting the islands, his observations on the unique character of the animals and plants were, in his own words, the "origin of all my views."

On October 2, 1836, the *Beagle* landed in England. Most of Darwin's extensive collections had long since preceded him to England, as had most of his notebooks and diaries kept during the cruise. Three years after the *Beagle*'s return to England, Darwin's journal was published. It was an instant success and required two additional printings within the first year. In later versions, Darwin made extensive changes and abbreviated the original ponderous title typical of nineteenth-century books to simply *The Voyage of the Beagle.* The fascinating account of his observations written in a simple, appealing style has made the book one of the most lasting and popular travel books of all time.

Curiously, the main product of Darwin's voyage, his theory of evolution, did not appear in print for more than 20 years after the *Beagle*'s return. In 1838 he "happened to read for amusement" an essay on populations by T.R. Malthus (1766-1834), who stated that animal and plant populations, including human populations, tend to increase beyond the capacity of the environment to support them. Darwin had already been gathering information on the artificial selection of animals under domestication by humans. After reading Malthus's article, Darwin realized that a process of selection in nature, a "struggle for existence" because of overpopulation, could be a powerful force for evolution of wild species.

He allowed the idea to develop in his own mind until it was presented in an essay in 1844—still unpublished. Finally in 1856 he began to pull together his voluminous data into a work on the origin of species. He expected to write four volumes, a "very big" book, "as perfect as I can make it." However, his plans were to take an unexpected turn.

In 1858, he received a manuscript from Alfred Russel Wallace (1823-1913),

Why was Darwin so successful? Although a famous scientist in his time, Darwin remained a mysterious figure to all but his family and a close circle of friends. From these sources emerges the portrait of a man possessing many abilities. He was a patient thinker with the capacity to weave together a theoretical structure from a mountain of empirical data. He kept meticulous records, enjoyed experimentation, and maintained a set and ordered routine in his life. Certainly he possessed a streak of competitiveness and the desire to excel. Perhaps most important of all, his intellectual development was strongly influenced by his scientific friends and, especially, his devoted family, an unfailing wellspring of love, respect, and support.

Figure 39-6

The Galápagos Islands.
Photograph by C.P. Hickman, Jr.

an English naturalist in Malaya with whom he had been corresponding. Darwin was stunned to find that in a few pages, Wallace summarized the main points of the natural selection theory that Darwin had been working on for two decades. Rather than withhold his own work in favor of Wallace as he was inclined to do, Darwin was persuaded by two close friends, the geologist Lyell and the botanist Hooker, to publish his views in a brief statement that would appear together with Wallace's paper in the *Journal of the Linnaean Society*. Portions of both papers were read before an unimpressed audience on July 1, 1858.

For the next year, Darwin worked urgently to prepare an "abstract" of the planned four-volume work. This was published in November 1859, with the title *On the Origin of Species by Means of Natural Selection, or the Preservation of Favoured Races in the Struggle for Life*. The 1250 copies of the first printing were sold the first day! The book instantly generated a storm that has never completely abated. Darwin's views were to have extraordinary consequences on scientific and religious beliefs and remain among the greatest intellectual achievements of all time.

Once Darwin's excessive caution had been swept away by the publication of *The Origin of Species*, he entered an incredibly productive period of evolutionary thinking for the next 23 years, producing book after book. He died April 19, 1882, and was buried in Westminster Abbey. The little *Beagle* had already disappeared, having been retired in 1870 and presumably broken up for scrap.

EVIDENCES FOR EVOLUTION
Reconstructing the Past

Fossils and their formation

The strongest and most direct evidence for evolution, and the evidence that first persuaded Darwin to think seriously about evolution, is the fossil record of the past, even though fossils and the sediments containing them provide dim and imperfect views of an ancient life (Figure 39-7). The complete record of the past is always beyond our reach since so many organisms left no fossils. Yet incomplete as the record is, biologists rely on the discoveries of new fossils and the continued study of existing fossils to interpret phylogeny and relationships of both plant and animal life. It would be difficult indeed to make sense out of the evolutionary patterns or classification of organisms without the support of the fossil record. The documentary evidence of evolution as a general process—progressive changes in life from one geological era to another, past distribution of lands and seas, and environmental conditions of the past (paleoecology)—all depend on what fossils teach us.

A fossil may be defined as the remains of past life uncovered from the crust of the earth. It refers not only to complete remains (mammoths and amber

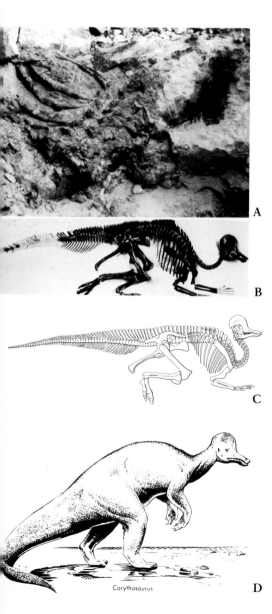

Figure 39-7

Reconstruction of the appearance of *Corythosaurus*, a dinosaur from the upper Cretaceous period of North America, approximately 90 million years ago. **A**, Portion of skeleton as discovered and partially worked out of rocks in New Mexico quarry. **B**, Skeleton, 30 feet in length on slab. **C**, Drawing prepared from skeleton. **D**, Artist's reconstruction of living animal.

A and B, Courtesy American Museum of Natural History; C and D, from Colbert, E.H. 1969. Evolution of the vertebrates, ed. 2. New York. Copyright © 1969. Reprinted by permission of John Wiley & Sons, Inc.

Figure 39-8

This bit of Green River shale from Farson, Wyoming, bears the impression of a "double-armored herring" *(Knightia)*, approximately 4 inches long, which swam there during the Eocene epoch, approximately 55 million years ago.

Fossil remains may on rare occasions include soft tissues preserved so well that recognizable cell organelles can be viewed with the electron microscope! Insects are frequently found entombed in amber, the fossilized resin of trees. One study of a fly entombed in 40 million year old amber revealed structures corresponding to muscle fibers, nuclei, ribosomes, lipid droplets, endoplasmic reticulum, and mitochondria (Figure 39-9). This extreme case of mummification probably occurred because chemicals in the plant sap diffused into the embalmed insect's tissues.

insects), actual hard parts (teeth and bones), and petrified skeletal parts that are infiltrated with silica or other minerals by water seepage (ostracoderms and molluscs), but also to molds, casts, impressions, and fossil excrement (coprolites).

The fossil record is biased because preservation is selective. Vertebrate skeletal parts and invertebrates with shells and other hard structures have left the best record (Figure 39-8). It is unlikely that jellyfishes, worms, caterpillars, and such fossilize, but now and then a rare chance discovery such as the Burgess shale deposits of British Columbia or the Precambrian fossil bed of south Australia reveals an enormous amount of information about soft-bodied organisms. Certain regions have apparently provided ideal conditions for fossil formation—for example, the tar pits of Rancho La Brea in Hancock Park, Los Angeles; the great dinosaur beds of Alberta, Canada, and Jensen, Utah; and the Olduvai Gorge of Tanzania.

A common method of fossil formation is the quick burial of animals under waterborne sediments. Rapid burial is usually important because it slows or prevents decomposition by oxidation, solution, and bacterial action.

Most fossils are laid down in deposits that become stratified. If left undisturbed, which is rare, the older strata are the deeper ones; however, the layers are usually tilted or folded or show faults (cracks). Often old deposits exposed by erosion are later covered with new deposits in a different plane. Sedimentary rock such as limestone may be exposed to tremendous pressures or heat during mountain building and may be metamorphosed into rocks such as crystalline quartzite, slate, or marble. Fossils are often destroyed during these processes.

Since various fossils are correlated with certain strata, they often serve as a means of identifying the strata of different regions. Certain widespread marine invertebrate fossils such as various foraminiferans and echinoderms are such good indicators of specific geological periods that they are called "index," or "guide," fossils.

Geological time

Long before the age of the earth was known, geologists began dividing its history into a table of succeeding events, using as a basis the accessible deposits and correlations from sedimentary rock. Time was divided into eons, eras, periods,

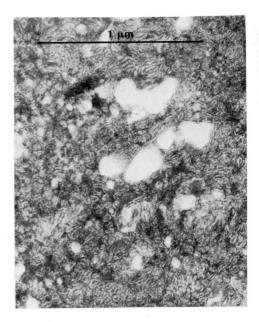

Figure 39-9

Electron micrograph of tissue from a 40-million-year-old fossilized fly. Membranous arrays characteristic of endoplasmic reticulum cisternae are visible. The scale bar is 1 μm.
Courtesy G.O. Poinar, Jr., University of California, Berkeley.

The more well-known carbon-14 (^{14}C) dating method is of little help in estimating the age of geological formations because the short half-life of ^{14}C restricts its use to quite recent events (less than about 40,000 years). It is especially useful, however, for archeological studies. This method is based on the production of radioactive ^{14}C (half-life of approximately 5570 years) in the upper atmosphere by bombardment of nitrogen-14 (^{14}N) with cosmic radiation. The radioactive ^{14}C enters the tissue of living animals and plants, and an equilibrium is established between atmospheric ^{14}C and ^{14}N in the organism. At death, ^{14}C exchange with the atmosphere stops. In 5570 years only half of the original ^{14}C remains in the preserved fossil. Its age is found by comparing the ^{14}C content of the fossil with that of living organisms.

and epochs. These are shown on the endpaper inside the back cover of this book. Time during the last eon (Phanerozoic) is expressed in eras (for example, Cenozoic), periods (for example, Tertiary), epochs (for example, Paleocene), and sometimes smaller divisions of an epoch.

Until recently scientists had no way to measure the absolute passage of time. Before about 1950, the "law of stratigraphy" was applied, a method of relative dating that relied on the sequence of layers of rock: the oldest at the bottom and the youngest at the top. In the late 1940s, radiometric dating methods were developed for determining the absolute age in years of rock formations. Several independent methods are now used, all based on the radioactive decay of naturally occurring elements into other elements. These "radioactive clocks" proceed independent of pressure and temperature changes and therefore are not affected by often violent earth-building activities.

One method, potassium-argon dating, depends on the decay of potassium-40 (^{40}K) to argon-40 (^{40}A) (12%) and calcium-40 (^{40}Ca) (88%).

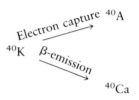

The half-life of potassium-40 is 1.3 billion years. This means that half the original atoms will decay in 1.3 billion years, and half the remaining atoms will be gone at the end of the next 1.3 billion years. This continues until all the radioactive potassium-40 atoms are gone. To measure the age of the rock, one calculates the ratio of remaining potassium-40 atoms to the amount of potassium-40 originally there (the remaining potassium-40 atoms plus the argon-40 into which they have decayed). Several such isotopes exist for dating purposes, some for dating the age of the earth itself. One of the most useful radioactive clocks depends on the decay of uranium into lead. With this method, rocks over 2 billion years old can be dated with a probable error of less than 1%.

As far as the fossil record of macroscopic organisms is concerned, the recorded history of life begins near the base of the Cambrian period of the Paleozoic era approximately 600 million years BP. The period before the Cambrian is called the Precambrian era. Although the Precambrian era occupies 85% of all geological time, it is a puzzling era because of the lack of macrofossils. It has received much less attention than later eras, partly because of the absence of fossils and partly because oil, which provides the commercial incentive for much geological work, seldom exists in Precambrian formations. There are, however, evidences for life in the Precambrian era: well-preserved bacteria and algae, as well as casts of jellyfishes, sponge spicules, soft corals, segmented flatworms, and worm trails. Most, but not all, are microfossils.

Major extinctions

After the Cambrian explosion of biological diversity, the subsequent history of evolution is punctuated by five major extinctions over the past 600 million years, followed by radiations of the survivors. The most cataclysmic of these extinction episodes happened about 225 million years ago, when at least half of the families of shallow-water marine invertebrates, and fully 90% of marine invertebrate species, disappeared within a few million years. This was the **Permian extinction.** The

Cretaceous extinction, which occurred about 65 million years ago, marked the end of the dinosaurs, as well as numerous marine invertebrates and many small reptile groups. The many terrestrial niches opened at this time allowed the mammals to flourish and radiate.

What caused these extinctions? There is increasing evidence suggesting that the Cretaceous extinction, and probably others, coincided with the impact on earth of a huge asteroid or comet (p. 518). Other extinctions, the Permian for example, may have been caused by drastic climatic alterations and changes in sea level.

_____ Classification and Homology

Biologists long before Darwin discovered that, despite the great diversity of life, it was possible to arrange living things into a logical system of classification. The "natural system of classification" that originated with Aristotle and later was formalized by Linnaeus (p. 117) is based on the degree of similarity of morphological characters.

For example, the domestic cat is clearly related to the wild cats and jungle cats of the Old World, and all are included in the same genus, _Felis_. This genus shares an obvious relationship with the "big cats" (lions, tigers, leopards, and jaguars of the genus _Panthera_), as well as lynxes and bobcats (genus _Lynx_), so all are grouped into a taxon of higher rank, the family Felidae. All members of this family share a number of traits that set them apart from all other families, such as a round head, 30 teeth, and digitigrade feet with retractile claws. At the same time, the cat family is clearly related as a group to a number of other families (for example, the dog, bear, raccoon, weasel and otter, and hyena families). All of these bear a distinctive anatomy and way of life that characterize them as flesh eaters; they are grouped together in the order Carnivora.

Thus animals can be grouped with respect to certain combinations of traits. Linnaeus had introduced his classification system partly as a means for cataloging nature and partly to reveal the Creator's purpose since he believed that each group carried a set of similar features that corresponded to God's plan. Darwin, however, recognized an alternative explanation: the marvelous similarities that allowed plants and animals to be classified into neat hierarchies existed because of a "blood" relationship, caused by descent and divergence from a common ancestor. This became one of the strongest arguments for his theory.

One of the most important tools used by Linnaeus and his successors to search out natural relationships was the concept of homology, discussed in Chapter 6 (p. 114). Homology is the term used to describe structures that, because of common ancestry, share a fundamental structure and relative position. The classic example we cited on p. 114 was the vertebrate pentadactyl limb. The human arm, front leg of a frog, wing of a bat or bird, and flipper of a whale are homologous because they are built of the same bones, even though they look different and function differently (Figure 39-10). Each forelimb has about the same number of bones arranged in a similar way. Common structures, constructed of the same building blocks, must reflect common descent from an ancestor having these structures.

Embryological homologies remind us of our kinship with distant vertebrate ancestors. Gill slits, which in fish develop into respiratory organs, appear in the embryos of all vertebrates (Figure 36-16, p. 791). In the adults of terrestrial vertebrates the gill arches disappear altogether or become modified beyond recognition. Thus, the history in our embryonic development is extraordinarily persis-

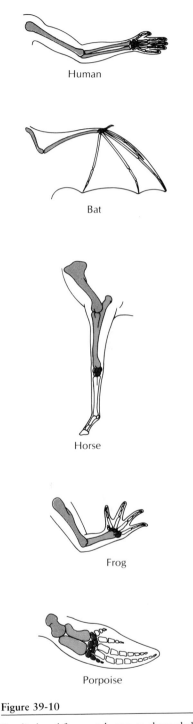

Human

Bat

Horse

Frog

Porpoise

Figure 39-10

Forelimbs of five vertebrates to show skeletal homologies. _Pink,_ humerus; _gray,_ radius and ulna; _red,_ wrist; _white,_ phalanges. Most generalized or primitive limb is that of humans—the feature that has been a primary factor in human evolution because of its wide adaptability. Various types of limbs have been structurally modified for adaptations to particular functions.

tant. Early embryological stages are especially unyielding to major changes since they control the induction and differentiation of all later embryological stages. Embryological similarities, like anatomical homologies, attest to evolution.

As the evolutionary theory gained acceptance, the unveiling of homologies received new impetus because it was seen as a key to understanding the course of evolution. The example of the pentadactyl limb tells us that all these vertebrates are related by common ancestry and should be grouped together as four-footed, five-toed vertebrates. But if we want to know if bats are more closely related to frogs or humans, the pentadactyl limb cannot help us because all three forms in question have it. We would discover, however, that bats and humans are warm blooded, have hair, suckle their young, and possess other homologous features that relate them, whereas the frog has none of these things. The frog possesses its own set of characteristics that relate it to salamanders and other amphibians. Working in this way, looking for more and more restricted homologies, we can characterize groups and subgroups, arriving at a classification that represents natural and true ancestry.

Biochemical homologies

Common ancestry and homology can now be demonstrated even more forcefully for proteins than for anatomical structures, mainly because protein chains have many more recognizable units (100 or more amino acids) than do anatomical structures, which seldom contain more than a dozen or so parts. By comparing amino acid chains of homologous proteins (hemoglobin, for example) from different animals, we can derive a molecular genealogical tree in much the same way that phylogenetic trees are constructed from fossils.

It has been possible to trace the evolution of at least one protein, cytochrome c, to the origin of eukaryotic cells more than 1 billion years ago. Cytochrome c is in part a small enzyme composed of 104 amino acid units. As different mutations accumulated, the amino acid sequences of the molecule in descendant species became different. Just how much any two species differ corresponds with the distance that they are separated on the phylogenetic tree. Elaborate family trees constructed entirely from sequence diversion agree remarkably well with those obtained by classical morphology and embryology. Molecular evolution is an active field of research today. Unlike anatomical structures, which arise from the complex interactions of genes over lengthy periods of embryological development, protein molecules are the *immediate* products of DNA, which are the fundamental molecules of evolutionary change. Such studies carry us closer to an understanding of the genetic variation that underlies all evolutionary change.

▬ NATURAL SELECTION
▬ Darwin's Theory of Natural Selection

The concept of natural selection, often expressed in the phrase "survival of the fittest," was the centerpiece of Darwin's theory of evolution. It was the first plausible alternative to Lamarck's concept of evolution by inheritance of acquired characteristics. Darwin's theory of natural selection was nearly as simple as Lamarck's (although less obvious) but was based on observable facts and supported by an abundance of evidence.

1. **All organisms show variation.** No two individuals are exactly alike. They differ in size, color, physiology, behavior, and other ways. Darwin realized that

The use of the word "theory" to describe evolutionary science has caused many people to believe that the principle carries little more weight than an idea or a conjecture, since we often use the word in common speech (for example, "I have a theory about what really happened there"). When creationists charge that evolution is "just a theory," this may be what they have in mind. Scientists, on the other hand, use "theory" to refer to a body of generalizations and principles developed within a field of inquiry, such as mathematics or physics or biology (p. 6). Thus we have the theory of numbers, theory of gravitation, and theory of evolution. Such theories are built from sets of facts, not conjectures and speculations. It is also important to emphasize that evolution as a process is a fact, and the only argument among scientists today is with the details of how it operates, not whether it operates.

much of this variation is inherited, even though he did not understand how. The mechanism of genetic mutation was to be discovered many years later. Darwin pointed out that humans have taken advantage of inherited variation to produce useful new breeds and races of livestock and plants. He believed that, if selection is possible under human control, it can also be produced by agencies operating in nature. He reasoned that *natural* selection can have the same effect as *artificial* selection.

2. **In nature all organisms produce more offspring than can survive.** In every generation the young are more numerous than the parents. Darwin calculated that, even in a slow-breeding species such as the elephant, a single pair breeding from age 30 to 90 and having only six young in this span could give rise to 19 million elephants in 750 years. Why, he asked, are there not more elephants?

3. **Accordingly, there is a struggle for survival.** If more individuals are born than can possibly survive, there must be a severe struggle for existence among them. Darwin wrote in *The Origin of Species* that "it is the doctrine of Malthus applied with manifold force to the whole animal and vegetable kingdoms." Competition for food, shelter, and space becomes increasingly severe as overpopulation develops.

4. **Some individuals of a species have a better chance for survival than others.** Because individuals of a species differ, some are favored in the struggle for survival; others are handicapped and eliminated.

5. **The result is natural selection.** Out of the struggle for existence results the survival of the fittest. Under natural selection, individuals bearing favorable variations survive and have a chance to breed and transmit their characteristics to their offspring. The less fit die without reproducing. Natural selection is simply the differential survival or reproduction of favored variants. The process continues with succeeding generations so that organisms gradually become better adapted to their environment. Should the environment change, there must also be a change in the characteristics that have survival value, or the species is eliminated. Reproduction is what really counts in natural selection.

6. **Through natural selection new species originate.** The differential reproduction of variants can gradually transform species and result in the long-term "improvement" of types. According to Darwin, when different parts of an animal or plant population are confronted with slightly different environments, each diverges from the other. In time they differ enough from each other to form separate species. In this way two or more species may arise from a single ancestral species. Through adaptation to a changed environment, a group of animals may also diverge enough from their ancestors to become a different species.

Appraisal of Darwin's Theory

Darwin supported his theory of evolution by an overwhelming amount of evidence that had never before been accumulated. He convincingly demonstrated evolution and then provided a logical explanation for it. Through the well-chosen body of evidence so lucidly expounded in *The Origin of Species*, he made evolution understandable. Nevertheless, despite its elegance, it is hardly surprising that parts of his theory have been modernized because of more recent biological knowledge. When Darwin wrote *The Origin of Species*, nothing was known about the inheritance of variation. It was not even known that sexual reproduction involved the combination of a single sperm with a single egg. One always marvels that Darwin's deductions and conclusions were so sound in view of the scientific ignorance that prevailed at the time.

The popular phrase "survival of the fittest" was not originated by Darwin but was coined a few years earlier by the British philosopher Herbert Spencer, who anticipated some of Darwin's principles of evolution. Unfortunately the phrase later came to be coupled with unbridled aggression and violence in a bloody, competitive world. In fact, natural selection operates through many other characteristics of living things. The fittest animal may be the most helpful, the most caring, or the most loving. Fighting prowess is only one of several means toward successful reproductive advantage.

"If I lived twenty more years and was able to work, how I should have to modify the Origin, and how much the views on all points will have to be modified! Well it is a beginning and that is something . . ."
Charles Darwin, in a letter to Joseph Hooker, 1869

Figure 39-11

Thomas Henry Huxley (1825-1895), one of England's greatest zoologists, on first reading the convincing evidence for natural selection in Darwin's *Origin of Species*, is said to have exclaimed, "How extremely stupid not to have thought of that!" He became Darwin's foremost advocate and engaged in often bitter debates with Darwin's critics. Darwin, who disliked publicly defending his own work, was glad to leave such encounters to his "bulldog," as Huxley called himself.
Courtesy British Museum (Natural History), London.

Gregor Mendel published his theories on inheritance in 1868, 14 years before Darwin died. But Darwin presumably never read the work and if others knew about it, they failed to bring it to his attention. Had Darwin been aware of Mendel's work, he may still have not appreciated its importance. It took other scientists years to establish the relationship of genetics to Darwin's theory of natural selection.

The most serious weakness in Darwin's theory was his failure to identify correctly the mechanism of inheritance. Darwin saw heredity as a blending phenomenon in which the characteristics of the parents melded together in the offspring. We now know that many of the kinds of variations that Darwin thought were inheritable are not. Eventually, Mendelian genetics provided the particulate kind of inheritance that Darwin's theory of natural selection required. Only variations arising from changes in the genes (mutations) are inherited and furnish the material on which natural selection can act.

Thus the first step in modernizing Darwin's theory was the demonstration that the operative units in inheritance are self-reproducing, nonblending genes. Next was the discovery of chromosomes in which genes were found to be precisely located in linkage groups. Biologists could then understand how the ultimate source of variation—mutations—could be inherited in the usual Mendelian fashion.

Ironically, when Mendel's work was rediscovered in 1900, it was viewed as antagonistic to Darwin's theory of natural selection. During the early part of this century, the popularity of Darwin's theory continued to ebb. When mutations were discovered, most geneticists accepted the idea that "mutation pressure"—continuously recurring mutations—was the basis of evolutionary change. Natural selection was relegated to the role of executioner, a negative force that merely eliminated the obviously unfit.

Emergence of Modern Darwinism: the Synthetic Theory

In the 1930s a new breed of geneticists began to reevaluate Darwin's theory from a different perspective. These were population geneticists, scientists who studied variation in natural populations of animals and plants and who had a sound knowledge of statistics and mathematics, disciplines that earlier geneticists woefully lacked. Gradually, a new comprehensive theory emerged that brought together population genetics, paleontology, biogeography, embryology, systematics, and, later, animal behavior.

Population genetics, which combines Darwinian natural selection with modern principles of heredity, shows that evolution is a change in the genetic composition of populations. Such changes occur when the population is exposed to an environmental challenge—some change in the physical or biotic environment. Interactions between population and environment have become the dominant theme of the modern synthetic theory of evolution.

We can recognize three ways in which populations respond to an environmental challenge. First, they may become extinct. This is the fate suffered by most species that have been confronted with a drastic environmental change.

Following a second course, a population may somehow successfully meet all environmental challenges to which they are exposed, undergoing only minor changes in their genetic makeup and persisting through time relatively unaltered. Examples include the tuatara of New Zealand (p. 531), the modern lungfishes, and the modern oysters—all forms that closely resemble their ancestors of millions of years ago. Internal changes that we cannot detect may take place, but for the most part the stability of these populations is real.

The third response to environmental challenge is the evolution of adaptations in a population that enable it to invade a new ecological habitat. Populations with this response are the winners in an evolutionary sense since they generate new species and enrich the earth's biota with the great diversity we find today.

We should not leave the impression that all debate ended once the modern synthetic theory (often called neo-Darwinism) emerged. Some of the major areas of controversy today will be discussed in the pages that follow.

MICROEVOLUTION: GENETIC CHANGE WITHIN POPULATIONS AND SPECIES

Microevolution is the change of gene frequencies within single, natural populations. These are the small evolutionary events that are going on all the time. The evolutionary process is much too slow to be observable in a single lifetime. Nevertheless, within individual populations it is often possible to observe shifts in gene frequencies that produce new adaptive traits. Occasionally we can witness the emergence of new species. We have stressed that evolution is the result of interaction between populations and their environments. How these interactions cause evolution is the subject of this section.

Genetic Variation within Species

Biological variation exists within all natural populations. We can verify this fact simply by measuring a single characteristic in several dozen specimens selected from a local population: tail length in mice, length of a leg segment in grasshoppers, number of gill rakers in sunfishes, number of peas in pods, and height of adult males of the human species. When the values are graphed with respect to frequency distribution, they tend to approximate a normal, or bell-shaped, probability curve (Figure 39-12). Most individuals fall near the mean, fewer fall above or below the mean, and the extremes make up the "tails" of the frequency curve with increasing rarity. The larger the population sample, the closer the frequencies resemble a normal curve. Sometimes significant deviations from normality are discovered in populations, and such instances may reveal important facts about variations within the population.

As pointed out in Chapter 38, the ultimate source of all genetic variation is the mutation of genes. The *immediate* source of genetic variation, however, is not mutations because a mutation is a rare event. Rather it is the recombination of existing alleles by sexual reproduction.

All of the alleles of all genes possessed by members of a population are referred to collectively as the **gene pool.** The interbreeding population is the visible manifestation of the gene pool, which thus has continuity through successive generations. The gene pool of large populations must be enormous because at observed mutation rates many mutant alleles can be expected at all gene loci. Let us suppose that two alleles are present at a chromosomal locus: *A* and *a*. Two kinds of sex cells would be produced, and among individuals in the population there are three possible genotypes: *A/A, A/a,* and *a/a*. When two pairs of alleles are involved, 2^2 or four kinds of sex cells can be produced (see the Punnett square on p. 807), and within the population there are nine possible genotypes. With larger numbers of loci and pairs of alleles, the potential number of different kinds of sex cells rises rapidly, that is, 2^3, 2^4, 2^5, and so on.

Most animals have thousands of gene loci. If the number of gene loci in an individual were 10,000 (about average for many mammal species), and 10% (1000) of these loci were heterozygous, the individual could produce a staggering 2^{1000} genetically different kinds of sex cells, a number larger than all the atoms in

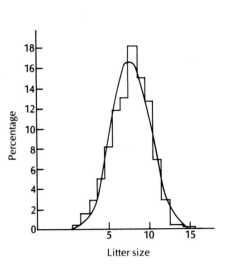

Figure 39-12

Frequency distribution of litter size in mice, showing how variation in litter size tends to conform to a bell-shaped, or normal, curve. Adapted from Falconer, D.S. 1981. Introduction to quantitative genetics, ed. 2, London, Longman Group Ltd.

In fact, the amount of potential variation is even greater than our example suggests. After protein electrophoresis (a method used to distinguish different forms of the same enzyme [allozymes]) was introduced in the 1960s, geneticists were surprised to discover that at least 30% of the roughly 10,000 loci in *Drosophila* are polymorphic, with two to six alleles at each locus. These alleles appear at such high frequencies that the average fly is likely to be heterozygous at about 12% of its loci. But this is not the whole story because genes exert different influences in the presence of other genes. Gene A may act differently in the presence of gene B from what it does in the presence of gene C. The diversity produced by this interaction and the addition of new mutations now and then add to the complexity of population genetics.

the universe! So with just two alleles existing for 10% of the genome, there is obviously copious material on which natural selection can act. Even if no new mutations occurred, the shuffling of existing genes would provide enough variation to fuel natural selection for a long time.

Variation and natural selection

What does variation signify for evolution? As already stated, genetic variation produced in whatever manner furnishes the material on which natural selection works to produce evolution. Natural selection does this by favoring beneficial variations and eliminating those harmful to the organism. Selective advantages of this type represent a slow process, but on a geological time scale they can bring about striking evolutionary changes represented by the various taxonomic units (species, genera, and so on), adaptive radiation of groups, and the various kinds of adaptations.

The fact must be stressed, however, that natural selection works on the whole animal and not on a single hereditary characteristic. The organism that possesses the superior "hand of cards," that is, the most beneficial combination of characteristics, is going to be selected over one not so favored. This concept helps explain certain puzzling instances in which an animal may bear characteristics that confer no advantage, or may even be harmful, but in the overall combination, is successful. Thus in a population, pools of variations are created on which natural selection can work to produce evolutionary change.

___ Genetic Equilibrium

In the human population, brown eyes are dominant to blue, curly hair is dominant to straight, and a Roman nose is dominant to a straight nose. Why has the dominant gene not gradually supplanted the recessive one in each instance so that we are all brown eyed, curly haired, and Roman nosed? It is a common belief that a characteristic dependent on a dominant gene increases in proportion because of its dominance. This is not the case because there is a tendency in *large* populations for genes to remain in equilibrium generation after generation. Strange as it may seem, a dominant gene does not change in frequency with respect to its recessive allele(s). Dominance describes the *phenotypic effect* of an allele, not its abundance.

This principle is based on a theorem that is the foundation of the entire genetic basis of evolution, the **Hardy-Weinberg equilibrium** (see box). According to this theorem, gene frequencies and genotype ratios in large biparental populations reach an equilibrium in one generation and *remain constant* thereafter *unless* disturbed by new mutations, natural selection, or genetic drift (chance). Such disturbances are the source materials of evolution.

A rare gene, according to this principle, does not disappear merely because it is rare. That is why certain rare traits, such as albinism and cystic fibrosis, persist for endless generations. Variation is retained even though evolutionary processes are not in active operation. Whatever changes occur in a population—gene flow from other populations, mutations, natural selection, and migration—involve the establishment of a new equilibrium with respect to the gene pool, and this new balance is maintained until upset by disturbing factors.

For example, albinism in humans is caused by a recessive allele *a.* Only one person in 20,000 is an albino, and this individual must be homozygous *(a/a)* for

Hardy-Weinberg Equilibrium: Why Gene Frequencies Do Not Change

The Hardy-Weinberg law is a logical consequence of Mendel's first law of segregation and expresses the tendency toward equilibrium inherent in Mendelian heredity.

Let us select for our example a population having a single locus bearing just two alleles, T and t. The phenotypic expression of this gene might be, for example, the ability to taste a chemical compound called phenylthiocarbamide. Individuals in the population will be of three genotypes for this locus: T/T, T/t (both tasters), and t/t (nontasters). In a sample of 100 individuals, let us suppose we have determined that there are 20 of T/T genotype, 40 of T/t genotype, and 40 of t/t genotype. We could then set up a table showing the allelic frequencies as follows (remember that every individual has two alleles for each gene):

Genotype	Number of individuals	Number of T alleles	Number of t alleles
T/T	20	40	
T/t	40	40	40
t/t	40		80
TOTAL	100	80	120

Of the 200 alleles, the proportion of the T allele is $80/200 = 0.4$ (40%), and the proportion of the t allele is $120/200 = 0.6$ (60%). It is customary in presenting this equilibrium to use p and q to represent the two allele frequencies. The dominant gene is represented by p, the recessive by q. Thus:

$$p = \text{Frequency of } T = 0.4$$
$$q = \text{Frequency of } t = 0.6$$
$$\text{Therefore} \quad p + q = 1$$

Having calculated allele frequencies in the sample, let us determine whether these frequencies will change spontaneously in a new generation of a population. Assuming the mating is random (and this is important; all mating combinations of genotypes must be equally probable), each individual will contribute an equal number of gametes to the "common pool" from which the next generation is formed. This being the case, the frequencies of gametes in the "pool" will be proportional to the allele frequencies in the sample. That is, 40% of the gametes will be T, and 60% will be t (ratio of $0.4:0.6$). Both eggs and sperm will, of course, show the same frequencies. The next generation is formed as follows:

Sperm ＼ Eggs	Probability of $T = 0.4$	Probability of $t = 0.6$
Probability of $T = 0.4$	Probability of T/T individual = $0.4 \times 0.4 = 0.16$	Probability of T/t individual = $0.4 \times 0.6 = 0.24$
Probability of $t = 0.6$	Probability of T/t individual = $0.4 \times 0.6 = 0.24$	Probability of t/t individual = $0.6 \times 0.6 = 0.36$

Collecting the genotypes, we have:

$$\text{Frequency of } T/T = 0.16$$
$$\text{Frequency of } T/t = 0.48$$
$$\text{Frequency of } t/t = 0.36$$

Next, we determine the values of p and q from the randomly mated population:

$$T\ (p) = \frac{0.16 + \frac{1}{2}\ (0.48)}{1} = 0.4$$

$$t\ (q) = \frac{0.36 + \frac{1}{2}\ (0.48)}{1} = 0.6$$

The new generation bears exactly the same genotype frequencies as the parent population! There has been no increase in the frequency of the dominant gene T. Thus, *in a freely interbreeding, sexually reproducing population, the frequency of each allele remains constant generation after generation.* The more mathematically minded reader will recognize that the genotype frequencies. T/T, T/t, and t/t are actually a binomial expansion of $(p + q)^2$:

$$(p + q)^2 = p^2 + 2\ pq + q^2 = 1$$

the recessive allele. Obviously there are many carriers in the population with normal pigmentation, that is, people who are heterozygous *(A/a)* for albinism. What is their frequency? A convenient way to calculate the frequencies of a gene in a population is with the binomial expansion of $(p + q)^2$ (see box). We will let p represent the frequency of the dominant allele A and q the frequency of the recessive allele a for albinism.

Assuming that mating is random (a questionable assumption for albinos, but one that we will accept for our example), genotype distribution will be $p^2 = A/A$, $2 pq = A/a$, and $q^2 = a/a$. Only the frequency of genotype a/a is known with certainty, 1/20,000; therefore:

$$q^2 = 1/20,000$$

$$q = \sqrt{1/20,000} = \frac{1}{141}$$

$$p = 1 - q \qquad = \frac{140}{141}$$

The frequency of carriers is as follows:

$$A/a = 2\ pq = 2 \times \frac{140}{141} \times \frac{1}{141} = \frac{1}{70}$$

One person in every 70 is a carrier! Even though a recessive trait may be rare, it is amazing how common a recessive allele may be in a population. There is a message here for anyone proposing a eugenics program designed to free a population of a "bad" gene. It is practically impossible. Since only the homozygous recessives reveal the phenotype to be artificially selected against (by sterilization, for example), the gene continues to surface from the heterozygous carriers. For a recessive allele present in two of every 100 persons (but homozygous in only one in 10,000 persons), it would require 50 generations of complete selection against the homozygotes just to reduce its frequency to one in 100 persons.

___ Processes of Evolution: How Genetic Equilibrium Is Upset

The Hardy-Weinberg law shows that populations do not change—and thus do not evolve—as long as certain conditions are met: (1) large population, (2) random mating, (3) no mutations, (4) no selection, and (5) no migration. We already know that these conditions cannot all prevail indefinitely and that the Hardy-Weinberg equilibrium can be disturbed by mutations, genetic drift, natural selection, and migration. Earlier (p. 836) we discussed mutations, the ultimate source of variability in all populations. We will now look at other factors that cause gene frequencies to change and thus lead to evolution.

Genetic drift and neutral mutations

Genetic drift refers to accidental changes in gene frequency that may occur when a few individuals at random become isolated from a large population. This could happen when a small number of individuals (called **founders**) migrate to a remote habitat. These few will carry with them only an incomplete sample of the gene pool of the parent population; alleles not carried will be lost. Thus the founders are certain to be different from the parent population. The smaller the number of colonizers, the greater are random changes in gene frequencies that can lead to new species.

However, small populations that descend from founders may be less able to cope with a new environment because the loss by genetic drift of some genes present in the ancestral gene pool has reduced their "fitness," or capacity to adapt. This may be one reason why small populations on islands often become extinct. Despite the limited genetic diversity of founder populations, however, they are sometimes phenomenally successful. All of the billions of starlings in North America are the descendants of a few birds introduced into New York City in 1890 (p. 557). Most of the (unfortunately) successful introductions of birds and mammals into North America, New Zealand, and Australia began from a handful of individuals.

Whatever the importance of genetic drift, it must be considered a rare event. Once a founder population becomes established, it is then guided in its evolution by natural selection.

Neutral mutations are those that do not confer differences in fitness and are therefore not affected by natural selection. If a mutation is truly neutral, that is, is neither beneficial nor harmful, it will remain in the population at some frequency determined largely by accidents of genetic drift. The mutation may by chance be lost or may increase in frequency (again purely by chance) and become fixed in the gene pool.

Over the last 10 years or so, geneticists have discovered far more variation in protein structure—mostly enzymes—than could ever have been predicted. The presence of different forms of the same protein in a population is called **protein polymorphism.** Most of these protein variants appear to be neutral because they seem to be unaffected by Darwinian selection.

The discovery of neutral mutations has given rise to the "neutralist school," which maintains that most if not all biochemical variation in gene pools is neutral with respect to adaptation and natural selection. If the advocates of the neutralist school are correct, molecular evolution (for example, protein structure) occurs by genetic drift. This has led to a lively debate between "selectionists," who believe that natural selection is the dominant force in evolution, and "neutralists," who maintain that most evolution is determined by drift and is therefore "non-Darwinian."

Much biochemical variation appears to be trivial and neutral with respect to adaptation. For example, there have been many attempts to discover adaptive properties in the ABO blood group system of humans. The blood groups of this system are distributed unevenly among human populations in different parts of the world (Figure 39-13). Some correlation has been found between the frequency of certain blood groups and the incidence of particular diseases, but most geneticists now believe that the blood groups are neutral in humans. Genetic drift appears to be responsible for differences in frequencies of the blood groups among different human populations.

Nevertheless, not all "invisible" protein polymorphism is neutral. Experiments have shown that many enzymatic differences are adaptive. For example, animals living in cold climates have enzymes that are more active at lower temperatures, whereas populations of the same species living in warmer climates possess enzymes that are more active at higher temperatures. Discoveries of this kind suggest to selectionists that much protein polymorphism is adaptive in ways not yet understood.

During the 1960s, genetic drift was deemphasized as a factor of much importance in evolution because it could operate only in small populations. However, evolutionists now realize that most breeding populations of animals *are* small. A natural barrier such as a stream or a ravine between two hilltops may effectively separate a breeding population into two separate and independent units of selection.

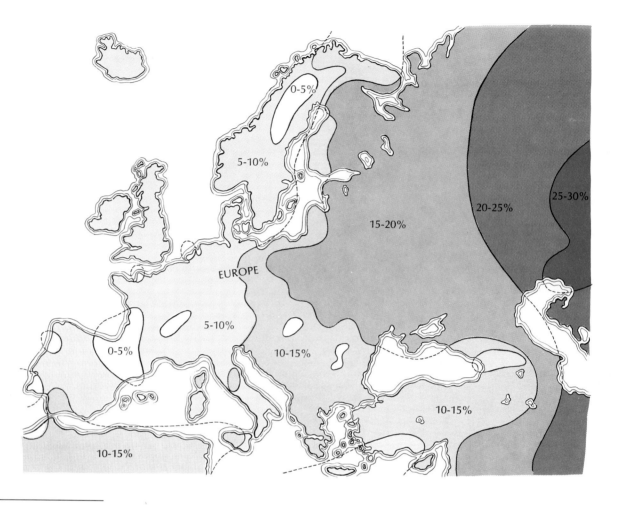

Figure 39-13

Frequencies of the blood-type allele B among humans in Europe. The allele is common in the east and rare in the west. The allele may have arisen as a new mutant gene in a small population of people in Asia that were, or became, isolated from their parent population eastward where the allele did not exist. By random genetic drift, the new allele may have spread through the founding population. The spread of the allele westward began when the founding population lost its isolation and intermixed with people in the west.
Modified from Mourant, A.E. 1954. The distribution of human blood. Toronto, Ryerson Press.

Natural selection

We have already stated that natural selection is the principal guiding force in evolution, and we have alluded to its mechanism. We will now more firmly define what it does. *Natural selection guides evolution by sorting out new adaptive combinations from a population gene pool derived from mutations, recombination, and other sources of genetic variation acting over many generations.*

Let us examine this definition. Sexually reproducing populations are composed of many individuals, and each individual has a different genotype. These differences have arisen by mutations, genetic recombinations, and perhaps genetic drift, nonrandom mating, or migration. Suppose that one of these genotypes is better than any of the others; that is, the animal bearing it is *better adapted* to the environment than are other animals having different genotypes. What basis have we for concluding that a particular genotype is better, that is, has greater **adaptive value** than others? The basis is reproductive success. Most organisms contribute progeny to the next generation. The genotype that contributes the most progeny, and therefore the most genotypes to the next generation, has the greatest adaptive value and chance of success. If this differential reproduction continues systematically generation after generation, the genotype will increase in frequency while others decline in frequency. Natural selection, then, produces differences in gene pools from one generation to the next.

We have stressed, and will stress once more, that natural selection works on the whole organism—the entire phenotype—and not on individual genes. The

Figure 39-14

Light and melanic forms of the peppered moth *Biston betularia* on, **A**, a lichen-covered tree in unpolluted countryside and, **B**, a soot-covered oak near industrial Birmingham, England. The dark melanic coloration that appeared in 1850 is controlled by a dominant allele.
Photographs by H.B.D. Kettlewell.

winning genotype may well contain genes that have detrimental effects on the individual but are preserved because other genes are strongly beneficial.

Natural selection at work

Despite a current debate over the importance of natural selection as the principal guiding force for evolution, natural selection remains the centerpiece of evolutionary biology. Differential survival and reproduction of different genotypes change the genetic composition of populations. Changing gene frequency therefore is the whole basis of evolution.

There are many examples that show how natural selection alters populations in nature. Sometimes selection can proceed rapidly as, for example, in the development of high resistance to insecticides by insects, especially flies and mosquitoes. Doses that at first killed almost all pests later were ineffective in controlling them. As a result of selection, mutations bestowing high resistance, but previously rare in the population, increased in frequency.

Perhaps the most famous instance of rapid selection is that of **industrial melanism** (dark pigmentation) in the peppered moth of England (Figure 39-14). Before 1850 the peppered moth was always white with black speckling in the wings and body. In 1849 a mutant black form of the species appeared. It became increasingly common, reaching frequencies of 98% in Manchester and other heavily industrialized areas by 1900. The peppered moth, like most moths, is active at night. It rests during the day in exposed places, depending on its cryptic coloration for protection. The mottled pattern of the normal white form blends perfectly with the lichen-covered tree trunks. With increasing industrialization, the soot and grime from thousands of chimneys killed the lichens and darkened the bark of trees for miles around centers such as Manchester. (This part of England was known as the "Black Country.") Against a dark background the white moth was conspicuous to predatory birds, whereas the mutant black form was camouflaged. The result was rapid natural selection: the easier-to-see white form was preferentially selected by birds, whereas the melanic form was subject to far less predation. Selection pressure thus tended to eliminate the white form while favoring the genes that contributed to black wings and body.

Another way to describe the selective quality of the genes that determine color is in terms of their **fitness.** Fitness, or adaptive value, describes the way individuals differ in their reproductive success because of differences in the genes they carry. In this case the single dominant mutation for the black form of the peppered moth increased its fitness in England's "Black Country." This same trait, however, confers low fitness in nonpolluted countryside areas. On the other hand, genes for the white form confer high fitness in nonpolluted countryside areas but low fitness near industrial centers.

Of special interest is the rapidity of the change. Rather than requiring thousands of years, the change occurred in only about 50 years. White moths still survived in nonpolluted areas of England and with the recent institution of pollution-control programs, the white forms are reappearing in the woods around cities. Industrial melanism is a dramatic example of rapid directional selection caused by a changing environment and leading to shifting gene frequencies.

EVOLUTION OF NEW SPECIES

We have described how populations change over periods of time because of genetic variation. Mutations are the fundamental source of genetic variation, further enhanced by sexual interchange and recombination. Natural selection acting on these variations is the force that produces evolutionary change and maintains the adaptive well-being of populations. In this discussion we consider how new species arise. The process of species multiplication, that is, the division of a parent species into several daughter species, is called **speciation.**

Species Concept

Although a definition of species is needed before we can theorize about species formation, biologists do not agree on a single rigid definition that applies in all cases. A traditional but now obsolete concept of species is as static and immutable units subject only to minor and unimportant variations. According to this view, a species is a group within which the individuals closely resemble one another but are clearly distinguishable from those of any other group. The flaw in this definition is the fact of biological variation. Many species are made up of individuals that can be arranged into completely intergrading series. Sometimes the gaps between species can be detected only with the greatest difficulty, and it becomes almost impossible to decide where two species are to be separated. Yet species are realities in nature. We observe and collect them, study them, sample their populations, and describe their behavior.

The difficulty with the traditional system is that classification is based on appearance alone (size, color, length of various parts, and so on). Other criteria can be used: behavioral differences, reproductive characteristics, genetic composition, and ecological niche. The single property that maintains the integrity of a species more than any other property, especially among biparental organisms, is *interbreeding*. The members of a species can interbreed freely with each other, produce fertile offspring, and share a common gene pool. Interbreeding of different species usually is either physically or behaviorally impossible, or it produces sterile offspring. There are of course exceptions, but they are rare. Species are thus usually considered genetically closed systems, whereas **races** within a species are open systems and can exchange genes. A species therefore may be defined as *a group of organisms of interbreeding natural populations that is reproductively isolated from other such groups and that shares a common gene pool* (Dobzhansky, 1951; Mayr, 1970*†).

How Species Become Reproductively Isolated

The definition of a species states that gene pools of different species are **reproductively isolated** from each other. With the exception of occasional hybridization, gene flow between species does not occur. Reproductive isolation, then, ensures the independence of a new species. No new species could maintain its

---

*Dobzhansky, T. 1951. Genetics and the origin of species, ed. 3. New York, Columbia University Press.

†Mayr, E. 1970. Population, species and evolution. Cambridge, Mass., Harvard University Press.

integrity if it freely interbred with other species. If it did interbreed with another species, and the offspring from these matings were just as viable and fertile as those of matings within the species, the two species would simply amalgamate. Reproductive isolation, then, is the *result* of successful speciation. How does it happen?

There appears to be only one certain way for barriers to be built between animal species: **geographical isolation.** Unless some individuals of a population can be segregated for many generations from the parent population, new adaptive variations that may arise may become lost through interbreeding with the parent population. Geographical isolation permits a unique sample of the population gene pool to be "pinched off," or segregated, so that diversification can occur.

___ Allopatric Speciation

Allopatric ("in another land") populations of a species are those that occupy different geographical areas and do not interbreed. Speciation that occurs as a result of geographical isolation is known as allopatric speciation.

Let us imagine a single interbreeding population of mammals that is split into two isolated populations by some geographical change—the uplifting of a mountain barrier, the sinking and flooding of a geological fault, or a climatic change that creates a hostile ecological barrier such as a desert. Gene flow between the two isolated populations is no longer possible. The populations are almost certainly different from each other even when first isolated, since just by chance alone one population contains alleles not present in the other.

With the passage of time, genetic recombination accentuates the difference, especially if the populations are small. Mutations that appear in the separated populations are certain to be different. Furthermore, the climatic conditions of the separated regions are usually different; one may be warm and moist, the other cool and dry. Natural selection acting on the isolated gene pools favors those mutations and recombinations that best adapt the populations to their respective environments, unique food supplies, and new predator-prey relationships.

The two populations continue to diverge morphologically, physiologically, and behaviorally until distinct geographical races are formed. After some indefinite time span, perhaps only a few thousand years, but possibly a much longer period of time, evolutionary diversification progresses until the two races are reproductively isolated. If the barrier separating them fails (erosion of mountains, shifting of river flow, reestablishment of a hospitable ecological bridge), the two populations again intermix. However, they may no longer interbreed because fertility between them is impaired by their accumulated gene differences. They are now distinct species.

Adaptive radiation on islands

Many times in the earth's history, a single parental population has given rise not just to one or two new species, but to an entire family of species. The rapid multiplication of related species, each with its unique specializations that fit it for a particular ecological niche, is called **adaptive radiation.**

Young islands are especially ideal habitats for rapid evolution. If they are formed by volcanoes that rose from a platform on the ocean floor, as were the Galápagos Islands, they are at first devoid of life. In time they are colonized by plants and animals from the continent or from other islands. These newcomers encounter an especially productive situation for evolution because of the abundance of new opportunities.

Geographical isolation is not the same thing as reproductive isolation. Geographical isolation refers to the spatial separation of two populations. It prevents gene exchange and is a precondition for speciation. Reproductive isolation is the result of speciation, and refers to the various physical, physiological, ecological, and behavioral barriers that prevent interbreeding with any other species.

On the crowded mainland almost every ecological niche is already occupied. (An ecological niche is an animal's unique way of life in an environment; see Chapter 41.) Every animal is specialized and adapted for a particular way of life, and all food sources are exploited by one species or another. Although variations continue to appear within a population, those offspring in each generation most like their parents are most likely to survive because they are best suited for that particular set of environmental conditions. Competition between different species is too keen on the mainland (where evolution has been proceeding for a long time) for many new evolutionary experiments to succeed.

On a young island, however, new arrivals find new ecological opportunities and no competitors. Those surviving the sea or air voyage and the landfall may be able to become established, multiply, and spread out. They are in a new land that in all probability differs ecologically from their original home. New variations may serve some of the descendants of the colonizers in establishing themselves in new niches. Offspring that differ slightly from their parents may find that these differences enable them to exploit alternate food sources as yet unused by other animals. They thus flourish and produce offspring, some of which bear similar characteristics. With each passing generation these animals become increasingly successful at this alternative way of life, at the same time becoming increasingly different from their ancestors. The outcome is a new genetic blueprint, in short, a new species. Equally well, immigrants to an island may fail to become established if the limited variety that is provided by a small group of colonizers does not happen to fit the new environment.

On archipelagoes, such as the Galápagos Islands, isolation plays an extremely important evolutionary role. Not only is the entire archipelago isolated from the continent, but each island is geographically isolated from the others by the sea, the most inhospitable environment for land animals. Moreover, each island is different from every other to a greater or lesser extent in its physical, climatic, and biotic characteristics. As colonizing animals find their way to the different islands, they are presented with an environmental challenge that is unique for each island. Moreover, each new arrival carries with it only a small fraction, a biased sample, of the population's gene pool. This further stimulates diversification, already encouraged by isolation, new ecological opportunities, and lack of competition. Archipelagoes, more than any other place on the earth's surface, offer the raw materials and opportunities for rapid evolutionary changes of great magnitude.

Darwin's finches

Let us consider the evolution of the family of 13 famous Galápagos finch species (Figure 39-15). Their fame rests not on their beauty—they are, in fact, inconspicuous, dull colored, and rather unmusical in song—but on the enormous impact they had in molding Darwin's theory of natural selection. Darwin noticed that the Galápagos finches (the name "Darwin's finches" was popularized in the 1940s by the British ornithologist David Lack) are clearly related to each other, but that each species differs from the others in several respects, especially in size and shape of the beak and in feeding habits. Darwin reasoned that, if the finches were specially created, it would require the strangest kind of coincidence for 13 similar finches to be created on the Galápagos Islands and nowhere else.

Darwin's finches are an excellent example of how new forms may originate from a single colonizing ancestor. Lacking competition from other land birds, they underwent adaptive radiation, evolving along various lines to occupy available niches that on the mainland would have been denied to them. They thus assumed

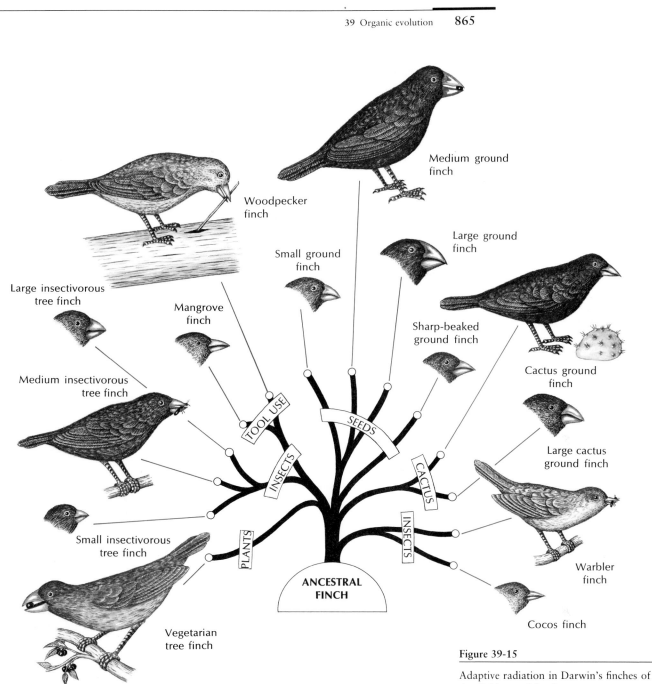

Figure 39-15

Adaptive radiation in Darwin's finches of the Galápagos Islands.

the characteristics of mainland families as diverse and unfinchlike as warblers and woodpeckers.

A fourteenth Darwin's finch is found on isolated Cocos Island, far to the north of the Galápagos archipelago. It is similar in appearance to the Galápagos finches and almost certainly has descended from the same ancestral stock. Since Cocos Island is a single small island with no opportunities for isolation of finch populations, there has been no species multiplication.

___ Sympatric Speciation

The Cocos Island example emphasizes the fundamental requirement for geographical isolation in speciation. In its absence, there is no other reasonable way for gene pools of animals to become separated long enough for genomic differences to accumulate. Is there any mechanism for sympatric ("in the same land") speciation in animals to occur? Most biologists think not, while acknowledging that it may occur under special circumstances and as an isolated event.

Examples of sudden sympatric speciation are common among the higher plants. The best evidence suggests that from 30% to 47% of the 235,000 species of flowering plants have evolved by polyploidy, mostly through the fusion of unreduced gametes. But in animals, polyploidy is an exceptional event and probably a rare process in animal evolution. There is some evidence that sympatric speciation might occur (although it is impossible to prove) in certain insects and in animal parasites.

MACROEVOLUTION: MAJOR EVOLUTIONARY EVENTS

We have seen that the gradual accumulation of small changes in gene frequency within a population, leading to the formation of new species by classical Darwinian natural selection, is called microevolution. **Macroevolution** is the term used to describe large-scale events in organic evolution operating over millions of years. The appearance of major groups (for example, molluscs, land vertebrates, and birds) represents completely new designs for living. Can historical trends of this magnitude be explained by shifting gene frequencies within populations (that is, by microevolution) operating over vast time scales? The orthodox answer is yes, and the supporting evidence comes especially from molecular and physiological studies showing that organisms share homologies at many levels (DNA, RNA, blood proteins, cellular organization, metabolic pathways, and so on).

Nevertheless, many evolutionists are troubled by the idea that evolution has always progressed at a stately pace, with new lineages appearing gradually over vast periods of time. More often than not, the fossil record suggests that evolution moves in leaps. Many species remain virtually unchanged for millions of years, then suddenly disappear to be replaced by a quite different, but related, form. Moreover, most major groups of animals appear abruptly in the fossil record, fully formed, and with no fossils yet discovered that form a transition from their parent groups. Thus, it has seldom been possible to piece together ancestor-dependent sequences from the fossil record that show gradual, smooth transitions between species.

There are two principal explanations for the absence of transitional forms in the fossil record. The traditional explanation is that evolution has always progressed at the same steady pace, and that "gaps" in the record are imperfections because transitions occur by microevolution in restricted localities and in small populations, under circumstances not favorable to fossil formation. In this view, fossils appear in the record only after a major group has become widely established by migration. The alternative view, which appears to be gaining favor among paleontologists, is that the abruptness of the fossil record is a true record of macroevolutionary "spurts."

One emerging explanation for rapid speciation events is that of **punctuated equilibria** (Eldredge and Gould, 1972). In this view, new species emerge rather suddenly, then remain mostly unchanged until they become extinct. In other words, long periods of equilibrium are abruptly "punctuated" by the extinction of a species, and its replacement by a new descendant species. It has been observed that most species survive 5 to 10 million years before becoming extinct, and during this time they apparently change very little. A speciation event that occurs in a small, isolated population over perhaps 50,000 years represents 1% or less of the species' life span. Fifty thousand years is a "geological instant" in the earth's history but represents thousands of generations in the evolving species. We know that gradual Darwinian selection can accomplish dramatic genetic changes much more rapidly than this. Thus punctuated equilibrium does not require a new explanation for speciation because 50,000 to 100,000 years is sufficient time for a new species to evolve by Darwinian selection. Eldredge and Gould have repeatedly emphasized that the long periods of stasis following the rapid transition are the more interesting phenomenon. Figure 39-16 compares the punctuated equilibrium model with the conventional Darwinian (gradualistic) model of evolution.

Evolutionists who have long lamented the imperfect state of the fossil record were treated in 1981 to the opening of an uncensored page of fossil history in Africa. Peter Williamson, a British paleontologist working in fossil beds 400 m deep near Lake Turkana, was able to document a remarkably clear record of speciation events in freshwater molluscs. The geology of the Lake Turkana basin reveals a history of instability. Earthquakes, volcanic eruptions, and climatic changes caused the waters of the lake to rise and fall periodically, sometimes by hundreds of feet. The 13 lineages of snails collected by Williamson show long periods of stability interrupted at intervals by relatively brief periods of rapid change in shell shape when the waters of the lake receded, isolating the snail populations from each other. Then the newly evolved species remained unchanged through thick deposits before becoming extinct and being replaced by new fossils resembling (but clearly distinct from) the ancestral form. The transitions took place within 5000 to 50,000 years. In the few meters of sediment where the speciation event occurred, all the transition forms were visible.

Williamson's study is a remarkably fine-grained record of evolution at work. It conforms well to the punctuated equilibrium model of Eldredge and Gould.

HUMAN EVOLUTION

Although Darwin did not include the human species among the evidences he used in *The Origin of Species* to support his theory of natural selection, few missed the implication: humans have evolved slowly from lower life forms. Such an idea was repugnant to the Victorian world, which responded with predictable outrage—especially to the notion that humans were related to apes. In fact, there was at the time virtually no fossil evidence linking humans with apelike creatures. Darwin and his advocates, especially the spirited and brilliant Thomas H. Huxley (Figure 39-11), rested their case mostly on cogent anatomical comparisons between humans and apes. To these zoologists, the close resemblances between apes and humans could only be explained by assuming a common origin.

The search for fossils, especially for a "missing link" that would prove a connection between apes and people, began slowly but was spurred by the unearthing of two excellent skeletons of the Neanderthal man in the 1880s and then by the discovery of the famous Java man *(Homo erectus)* by Eugene Dubois in 1891. The most spectacular discoveries, however, have been made in the last three decades in Africa, especially in the period 1967 to 1977, which American paleoanthropologist Donald C. Johanson refers to as the "golden decade."

We are no longer searching for a mythical "missing link." Living apes and humans are descended from a common ancestor in the past and, although closely related much as cousin is to cousin, they have evolved along independent lineages.

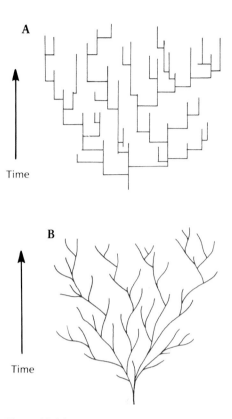

Figure 39-16

Punctuation and gradualism compared. **A,** The punctuation model sees evolutionary change as being concentrated in relatively rapid bursts *(straight lateral lines)* followed by prolonged periods of little or no change *(straight vertical lines)*. **B,** The gradualistic (Darwinian) model views evolution as proceeding more or less steadily through time.

In *The Origin of Species* (1859), Darwin did not offer evidences of human evolution, only suggesting in the conclusion that in the future, "Much light will be thrown on the origin of man and his history." But 12 years later he applied his evolutionary speculations to humans in his *Descent of Man* (1871). This book dealt in large part with sexual selection, attempting to show that human racial and sexual differences are caused by this mechanism. But he also clearly asserted that humans respond to the forces of natural selection, as do all other animals.

Figure 39-17

Common tree shrew *Tupaia glis* of tropical rain forests and montane forests of Southeast Asia.
Photograph by Ron Garrison, San Diego Zoo.

Figure 39-18

Mindanao tarsier *Tarsius syrichta carbonarius* of Mindanao island in the Philippines.
Courtesy of San Diego Zoo.

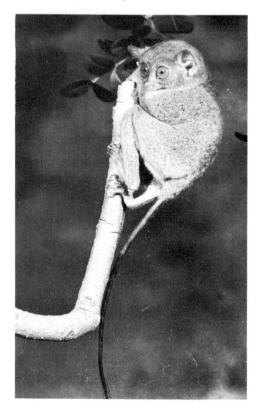

___ Our Closest Relatives, the Primates

Humans are primates, a fact that even the nonevolutionist Linneaus recognized. All primates share certain significant characteristics: grasping fingers on all four limbs, flat fingernails instead of claws, and forward-pointing eyes with binocular vision and excellent depth perception. The more advanced primates have large brains and high intelligence.

The earliest primate is believed to have been a small, nocturnal animal similar to **tree shrews** (Figure 39-17). These gave rise to **prosimians,** a primitive primate group that includes **lemurs, tarsiers** (Figure 39-18), and **lorises.** All were arboreal forms, a life-style that was important in guiding the evolution of increased intelligence. Flexible limbs are essential for active animals moving through trees. Grasping hands and feet, in contrast to the clawed feet of squirrels and rodents, enable the primates to grip limbs, hang from branches, seize food and manipulate it, and, most significantly, use tools. Highly developed sense organs, especially good vision, and proper coordination of limb and finger muscles are essential for an active arboreal life and were selected for. Of course, sense organs are no better than the brain that processes sensory information. Precise timing, judgment of distance, and alertness were necessary concomitants of an arboreal life and required a large cerebral cortex.

The earliest fossils of anthropoid monkeys and apes appeared in Africa in late Eocene deposits, some 40 million years ago. For some reason, these primates became day-active rather than nocturnal. Vision became the dominant special sense, now enhanced by color vision. Early in their evolution some monkeys migrated, perhaps by rafting on islands of vegetation, to South America, where their isolated descendants gave rise to the New World monkeys (ceboids), such as the howler monkey (Figure 39-19), spider monkey, and the tamarin. The Old World monkeys (cercopithecoids), such as baboons (Figure 39-20), mandrills, and

Figure 39-19

Red howler monkey *Alouatta seniculus,* a New World monkey of tropical rain forests and deciduous forests throughout South America. New World monkeys have prehensile tails, are completely arboreal, show little sexual dimorphism, and lack the ischial callosities (sitting pads) of Old World monkeys.
Photograph by Ron Garrison, San Diego Zoo.

colobus monkeys, differ from the New World monkeys in lacking a prehensile tail, and having close-set nostrils, more opposable, grasping thumbs, and more advanced teeth.

Ancestors of Apes and Humans

Humans evolved not from monkeys but from some apelike ancestor that evolved independently from the prosimians. About 25 million years ago a group of large, arboreal, fruit-eating apes called the **dryopithecines** appeared. At this time the woodland savannas were appearing over vast areas of Africa and Europe, as well as North America. Perhaps motivated by the greater abundance of food on the ground, these apes began to come down out of the trees and enter the savanna as pioneers. From these descended the modern **pongids,** or great (anthropoid) apes, represented by the gibbons, orangutan (Figure 39-21), chimpanzees, and gorillas.

Despite the obvious anatomical differences that separate humans and apes, there is abundant evidence for a common ancestry. New genetic techniques reveal that humans and chimpanzees differ in only 2.5% of their genes! Humans and chimpanzees are in fact more closely related to each other than chimpanzees are to gorillas. Given the genetic similarity, why are chimpanzees so different biologically from people? Geneticists point out that by affecting the *expression* of other genes, even a single gene mutation can profoundly influence embryonic timing and development and lead to significant phenotypic change. Such **regulatory mutations** were responsible, it is believed, for the rapid evolution of primates in general and for the remarkable speciation of humans in particular.

The first hominids

During the Miocene and Pliocene epochs (about 4 to 20 million years ago) the replacement of forests with grasslands provided an impetus for the apes to move out onto the savannas. Because of the benefits of standing upright—better view of

Figure 39-20

Olive baboon *Papio anubis* and baby, an Old World monkey of central Africa. Old World monkeys lack prehensile tails, may be either arboreal or ground dwelling, exhibit distinct sexual dimorphism, and possess ischial callosities.
Photograph by L. L. Rue, III.

Figure 39-21

Bornean orangutan *Pongo pygmaeus,* an anthropoid (pongid) ape of the rain forests of Borneo.
Courtesy San Diego Zoo.

Figure 39-22

Reconstructed fossil skull of the hominid *Australopithecus afarensis,* a near human believed to represent a direct ancestor of true humans *(Homo).*
Courtesy The Cleveland Museum of Natural History.

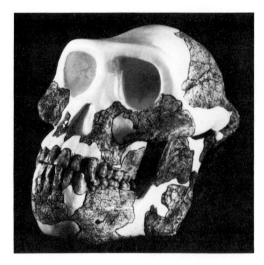

Figure 39-23

Lucy *(Australopithecus afarensis),* the most nearly complete skeleton of an early hominid ever found. Lucy is dated at 2.9 million years old.
Courtesy The Cleveland Museum of Natural History.

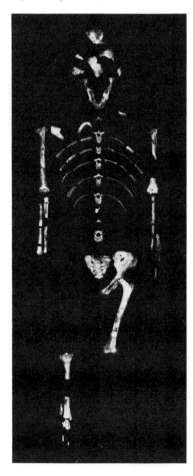

predators, freeing of hands for using tools and gathering food—emerging hominids slowly evolved toward upright posture. This important transition was really an enormous leap because it required extensive redesigning of the skeleton and muscle attachments.

The evidence concerning the earliest hominids of this period is remarkably sparse. Although several different early hominids have been identified, they all disappeared virtually without a trace of their descendants. Not until about 4 million years ago, after a lengthy fossil gap, do the first "near humans" appear. One was the recently discovered *Australopithecus afarensis,* a short, bipedal hominid with a face and brain size resembling those of a chimpanzee (Figure 39-22). Numerous fossils of this species have now been unearthed, the most celebrated of which was the 40% complete skeleton of a female discovered in 1974 by Donald Johanson and named "Lucy" (Figure 39-23 and 39-24). *A. afarensis* is considered by many paleoanthropologists to be ancestral to all human and humanlike forms that followed (Figure 39-25).

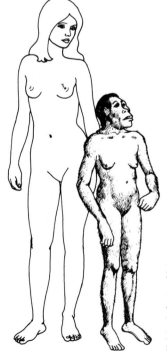

Figure 39-24

Lucy beside a modern human being *(Homo sapiens).* Lucy was only 1.07 m (3½ feet) tall and probably weighed about 27 kg (60 pounds). She was about 25 years old when she died.

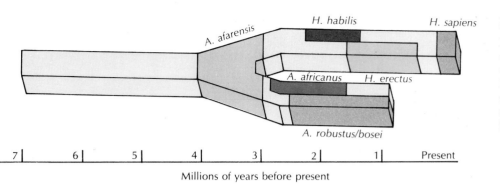

Millions of years before present

Figure 39-25

Proposed phylogeny of the hominids according to D. Johanson, discoverer of Lucy. Other schemes have been proposed. Richard Leakey, discoverer of *Homo habilis*, believes that *A. afarensis* is ancestral only to the australopithecines rather than to all later hominids.

___ Emergence of *Homo,* the True Human

Between 3 and 4 million years ago two quite separate hominid lines emerged that coexisted for at least 2 million years. One was the bipedal species *Australopithecus africanus*, the "southern African ape," with a brain size only about one third as large as that of modern humans. Several different lines of australopithecines appeared, some large and robust (Figure 39-26) and probably approaching the size of a gorilla.

Until they became extinct between 1.75 and 1 million years ago, the australopithecines shared the countryside with an advanced, fully erect hominid, *Homo habilis*, the first true human (Figure 39-27). *Homo habilis*, meaning "able man," was more lightly built but larger brained than the australopithecines and unquestionably used stone and bone tools to kill animals. This species appeared about 2 million years ago and survived for perhaps one-half million years.

About 1.5 million years ago *Homo erectus* appeared, probably as a descendant of *Homo habilis*. *Homo erectus* was a large hominid standing 150 to 170 cm (5 to 5.5 feet) tall. It had a low but distinct forehead, strong browridges, and a brain capacity of about 1000 cc (about intermediate between the brain capacity of *Homo habilis* and modern humans) (Figure 39-28). *Homo erectus* was a social species living in tribes of 20 to 50 people. They successfully hunted large animals that they butchered and cooked at campsites that often were caves. Fire was probably used for warmth. All in all, *Homo erectus* had a successful and complex

Figure 39-26

Intact cranium of *Australopithecus boisei*, a robust hominid that coexisted with early *Homo* for perhaps 2 million years.
Photograph by Peter Kain. Copyright Richard Leakey.

Figure 39-27

Skull of *Homo habilis*, earliest member of the true human genus *Homo*, discovered by Richard Leakey in 1972. This fossil is dated at 1.6 to 1.8 million years old.
Photograph by Peter Kain. Copyright Richard Leakey.

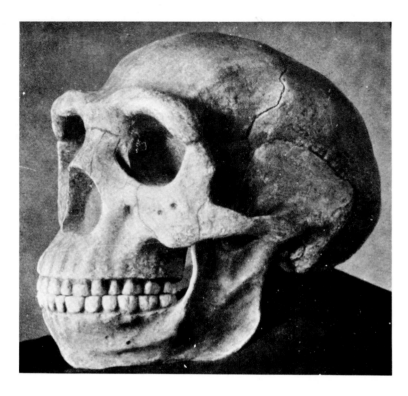

Figure 39-28

Skull of *Homo erectus,* discovered near Peking, China, and known as the "Peking man." This unusually well preserved skull has not been accurately dated but is estimated to be 0.5 to 0.7 million years old. Other more recent finds in Africa have been reliably dated as old as 1.5 million years.
From Dillion, L.S. 1978. Evolution, ed. 2. St. Louis, The C.V. Mosby Co.

culture and became widespread throughout the tropical and temperate Old World.

Homo sapiens: modern hominids

After the disappearance of *Homo erectus* about 300,000 years ago, subsequent human evolution and the establishment of *Homo sapiens* ("wise man") threaded a complex course. From among the many early subcultures of *Homo sapiens,* modern humans' immediate predecessors, the **Neanderthals,** emerged about 130,000 years ago (Figure 39-29). With a brain capacity well within the range of modern humans, the Neanderthals were proficient hunters and tool-users. The Neanderthals were not a homogeneous species but varied from place to place in response to

Figure 39-29

Neanderthal skull *(Homo sapiens),* discovered in France.
Courtesy Musée de l'Homme; from Dillion, L.S., 1978. Evolution, ed. 2. St. Louis, The C.V. Mosby Co.

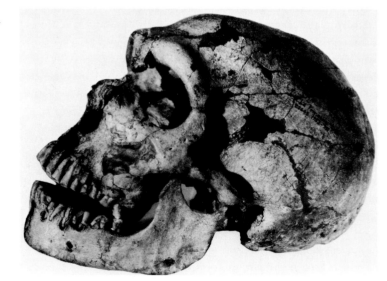

Figure 39-30

Restoration of prehistoric hominids. A Cro-Magnon *(right)* is compared with a Neanderthal *(center)* and a Java man *(Homo erectus) (left)*.
Courtesy J.M. McGregor.

local conditions and the isolation of populations from one another. Nevertheless, they dominated the Old World in the late Pleistocene epoch.

About 30,000 years ago the Neanderthals were rather abruptly replaced and quite possibly exterminated by the Cro-Magnons (Figure 39-30). The origin of the Cro-Magnons is obscure, although there is some evidence they came from Africa. They were tall people with a culture far superior to that of the Neanderthals. Implement crafting improved rapidly, and human culture became enriched with aesthetics, artistry, and sophisticated language.

The Unique Human Position

Despite our remarkable symbolic language, capacity for conceptual thought, and ability to manipulate our environment to suit ourselves, we are the outcome of the same forces that have directed the evolution of every organism from the time life originated on earth. Mutation, isolation, genetic drift, and natural selection have operated for us as for other animals. Yet we have what no other animal has, a nongenetic cultural evolution that involves a constant feedback between past and future experience. Finally, we owe much of our success to our arboreal ancestry that bequeathed us with binocular vision, superb visuotactile discrimination, and manipulative skills in the use of our hands. If the horse (with one toe instead of five fingers) had human intellect and culture, could it have accomplished what humans have?

SUMMARY

Organic evolution explains the diversity of living organisms, their characteristics, and their distribution as the historical outcomes of gradual, continuous change from previously existing forms. Evolutionary theory is strongly identified with Charles Robert Darwin who, with Alfred Russel Wallace, presented the first credible explanation for evolutionary change, the principle of natural selection. An alternative nineteenth-century scientific theory of evolution, Jean Baptiste de Lamarck's concept of inheritance of acquired characteristics, was disproved early in this century by geneticists. Darwin derived most of the material later used to construct his theory when, as a young man, he made a 5-year voyage around the world on the H.M.S. *Beagle.*

Evidences of organic evolution come from several sources. Fossils, the preserved remains of long-dead organisms, form an incomplete record of the progression of ancient life through geological time. The age of fossils can be reliably estimated because the geological formations in which they lie can be dated by measuring the radioactive decay of naturally occurring elements.

Following the Cambrian explosion some 600 million years ago (when most multicellular animal phyla originated), animal evolution has been punctuated by periodic mass extinctions followed by rapid radiations of the survivors.

The similarities that allow organisms, both living and extinct, to be classified into neat hierarchies is also evidence of evolution because of the presence of homologies. Homologies are structures that, because of common ancestry, share a fundamental structural plan and relative position. Common ancestry is also revealed by molecular evolution, which involves comparing homologous proteins and other molecules in different species.

Natural selection is the guiding force of evolution. The principle is founded on the observed facts that no two organisms are exactly alike, that at least some variations are inheritable, that all species tend to overproduce their kind, and that more individuals are produced than can survive. Therefore survival and reproduction of some variants are favored over others; that is, organisms possessing those adaptations making them better adapted to the environment will survive and produce more offspring than those with less favorable variations. Thus their inheritable characteristics will appear in greater proportion in the next generation. The essence of natural selection is reproductive success.

Darwin's theory of natural selection was gradually modified in this century when the mechanisms of heredity were established by geneticists. The modern synthesis, a new view of evolution that emerged in the 1930s, emphasized natural selection by gradual shifting of gene frequencies within populations. This is also known, especially today, as microevolution.

Mutations are the ultimate source of all new variation. However, sexual reproduction, in which the genotypes of two parents are mixed in the offspring, is the immediate source of variation among individuals in a population.

In large, interbreeding populations, gene frequencies tend to remain stable as predicted by the Hardy-Weinberg equilibrium. Such populations will not evolve unless disturbed by mutations, genetic drift, natural selection, or migration.

Genetic drift is the change in gene frequencies in small populations as a result of chance alone, in the absence of natural selection. Once considered to be an unimportant force in evolution, genetic drift has received new attention from advocates of the neutralist school of evolution, who argue that most evolutionary changes represent the fixation of neutral mutations by random drift.

Natural selection, on the other hand, works on the products of the genetic code, and it guides evolution by sorting out better-adapted organisms from a population composed of individuals with different genotypes. The spread of the mutant melanic form of the peppered moth in industrial England is an example of natural selection in action.

A species is defined as a natural, interbreeding population of organisms sharing a common gene pool that is reproductively isolated from other groups. The evolution of new species (speciation) may occur when some individuals become geographically isolated from the parent population. This is called allopatric speciation. Young islands, with their new ecological opportunities and absence of competitors for colonizing forms, are ideal habitats for adaptive radiation: multiplication of related species. The 13 finch species of the Galápagos Islands illustrate the crucial importance of isolation in speciation.

Sympatric speciation, the origin of new species in the absence of geograph-

Galápagos tortoise.
Photograph by C.P. Hickman, Jr.

ical isolation, often occurs in plants but has never been convincingly demonstrated in animals.

Large-scale events in life's history and the origin of new evolutionary trends are called macroevolution. Some scientists believe macroevolution can be explained by long-term extrapolation of microevolutionary mechanisms, that is, gradual changes in gene frequencies. This is the traditional view. Others more recently have viewed evolution as making sudden leaps, with new designs arising rapidly. The punctuated equilibrium hypothesis asserts that new species arise almost instantaneously in geological time, then remain almost unchanged during their subsequent history on earth. The fossil record appears to support this view.

Humans are primates, a generalized group having an arboreal ancestry and characterized by having grasping fingers and forward-facing eyes capable of binocular vision. Modern primates descended from a shrewlike ancestor and radiated over a period of about 80 million years into two major lines of descent: the prosimians (lemurs, lorises, and tarsiers) and the anthropoids (monkeys, apes, and hominids). About 25 million years ago a group of fruit-eating anthropoids called the dryopithecines emerged and began to venture from the woods onto the savanna.

The earliest hominids appeared about 4 million years ago; these were the bipedal australopithecines. These gave rise to, and coexisted with, the species *Homo habilis,* the first user of stone tools. *Homo erectus* appeared about 1.5 million years ago and was eventually replaced by *Homo sapiens* some 300,000 years ago.

Review questions

1. Briefly summarize Lamarck's concept of the mechanism of evolution. What is wrong with this concept?
2. What is the "doctrine of uniformitarianism" and what was its importance?
3. Explain why the *Beagle's* visit to the Galápagos Islands in 1835 was so important to Darwin's thinking.
4. What was the key idea contained in Malthus's essay on populations that was to help Darwin formulate his theory of evolution?
5. Compare the roles of Darwin and Alfred Russel Wallace in developing the theory of natural selection.
6. Explain how each of the following serves as evidence that evolution has occurred: fossils; different ages of sequential layers of rock as determined by radioactive clocks; natural system of classification; biochemical homologies. How might a creationist explain these observations?
7. Provide a summary of Darwin's theory of natural selection. How does natural selection work?
8. Creationists deride the idea of evolution, saying that it is "merely a theory." How would you answer this charge?
9. If Darwin were alive today, what changes do you think he would make in his theory?
10. Explain why the following statement is or is not true: "Random mutations in genes are more likely to be favorable when the environment changes."
11. Distinguish between immediate and ultimate sources of variation in sexually reproducing populations.
12. Can natural selection act on a population if no new mutations occur? Why?
13. It is a common belief that because some alleles are dominant and others are recessive, the dominants will eventually replace (drive out) all the recessives. How does the Hardy-Weinberg equilibrium answer this notion?

14. Assume you are sampling a trait in animal populations that is controlled by a single gene pair *A* and *a*, and that you can distinguish all three phenotypes *AA*, *Aa*, and *aa* (intermediate inheritance). Here are the results:

POPULATION	*AA*	*Aa*	*aa*	TOTAL
I	300	500	200	1000
II	400	400	200	1000

Calculate the *expected* distribution of phenotypes in each population. Is population I in equilibrium? Is population II in equilibrium?

15. If after studying a population for a trait determined by a single gene pair you find that the population is not in equilibrium, what possible reasons could there be for the lack of equilibrium?
16. Explain why genetic drift is more likely to occur in small populations.
17. What is the difference between genetic drift and natural selection? Between natural selection and adaptation?
18. In a sexually reproducing population, what is the basis for concluding that a particular genotype makes an animal better adapted to its environment than different genotypes in other animals?
19. Explain how industrial melanism of the peppered moth has evolved under the influence of natural selection.
20. What is the difference between natural selection and speciation?
21. Offer a biological definition of species.
22. What is the difference between a "race" and a "species"?
23. Distinguish between reproductive isolation and geographical isolation.
24. How does allopatric speciation occur?
25. What is the evolutionary lesson provided by the Darwin's finches on the Galápagos Islands?
26. Distinguish between microevolution and macroevolution.
27. What are the two principal explanations for the presence of "gaps" (absence of transitional forms) in the fossil record?
28. What anatomical characteristics set the primates apart from all other mammals?
29. What species was "Lucy" and where should this species be placed on the hominid family tree?
30. The genus *Australopithecus* coexisted with the genus *Homo* for at least 2 million years. In what ways did the two genera differ?
31. When approximately did the different species of *Homo* appear and how did they differ socially?
32. Is human evolution subject to the same pressures as the evolution of other animals? Why or why not?

Selected references

The literature on evolution is particularly extensive. The following references have been, of necessity, selected somewhat arbitrarily but are points of departure for the interested student. The September 1978 issue of *Scientific American*, devoted exclusively to evolution, should also be consulted.

Darlington, P.J., Jr. 1980. Evolution for naturalists: the simple principles and complex reality. New York, John Wiley & Sons. *A delightfully written view of evolution by a scientist with a deep understanding of the natural world.*

Darwin, C. 1859. On the origin of species by means of natural selection, or the preservation of favoured races in the struggle for life. London, John Murray. *There were five subsequent editions by the author.*

Eldredge, N., and S.J. Gould. 1972. Punctuated equilibria: an alternative to phyletic gradualism. In T.J.M. Schopf (ed.). Models in paleobiology, San Francisco, Freeman, Cooper & Co.

Gould, S.J. 1977. Ever since Darwin: reflections in natural history. New York, W.W. Norton. *Fascinating collection of essays written with humor and cool logic.*

Irvine, W. 1955. Apes, angels, and Victorians: the story of Darwin, Huxley, and evolution. New York, McGraw-Hill Book Co. *Masterly description of two central figures in the nineteenth century Darwinian revolution, told with skill and wit.*

Leakey, R.E. 1981. The making of mankind. New York, E.P. Dutton. *Beautifully illustrated and clearly written nonpolemical review of human and primate origins.*

Lowenstein, J.M. 1985. Molecular approaches to the identification of species. Amer. Sci. **73:**541-547. *How newly developed immunological methods are used to unravel taxonomic relationships and dates of evolutionary divergence.*

McMenamin, M.A.S. 1987. The emergence of animals. Sci. Am. **256:**94-102 (April). *The sudden diversification of animal life during the Cambrian period is known as the "Cambrian explosion." Alternative explanations are explored.*

Miller, J., and B. Van Loon. 1982. Darwin for beginners. New York, Pantheon Books, Inc. *Cleverly illustrated and witty presentation of the major concepts of evolution, Darwin's life and thought, and the controversy Darwin generated.*

Patterson, C. 1978. Evolution. London, Routledge & Kegan Paul, Ltd. *Lucid explanation of evolutionary theory. Highly recommended.*

Pilbeam, D. 1984. The descent of hominoids and hominids. Sci. Am. **250:**84-96 (Mar.) *How recent fossil finds have influenced interpretations of primate evolution.*

Stanley, S.M. 1981. The new evolutionary timetable: fossils, genes, and origin of species. New York, Basic Books. *Well-written explanation of the punctuated equilibrium model; semipopular treatment.*

Taylor, G.R. 1983. The great evolution mystery. New York, Harper & Row. *Provocative book, pointing to what the author believes are deficiencies in modern evolutionary theory and arguing that mechanisms other than natural selection, as yet unidentified, have influenced evolution.*

PART FIVE

THE ANIMAL AND ITS ENVIRONMENT

C.P. Hickman, Jr.

CHAPTER 40

THE BIOSPHERE AND ANIMAL DISTRIBUTION

BIOSPHERE

All life is confined to a thin veneer of the earth called the **biosphere.** From the first remarkable photographs of earth taken from the Apollo spacecraft, revealing a beautiful blue and white globe lying against the limitless backdrop of space, viewers were struck and perhaps humbled by our isolation and insignificance in the enormity of the universe. The phrase "spaceship earth" became a part of the vocabulary, and the realization evolved that all the resources we will ever have for sustaining life are restricted to a thin layer of land and sea and a narrow veil of atmosphere above it. We could better appreciate just how thin the biosphere is if we could shrink the earth and all of its dimensions to a 1 m sphere. We would no longer perceive vertical dimensions on the earth's surface. The highest mountains would fail to penetrate a thin coat of paint applied to our shrunken earth; a fingernail's scratch on the surface would exceed the depth of the ocean's deepest trenches.

Fitness of the Earth's Environment

Astronomers have estimated conservatively that in the vastness of the universe there are millions of stars, similar to our sun, that have planetary systems. The astronomer Harlow Shapley believes that in our galaxy alone there are *at least* 100,000 planets having conditions suitable for life. If he is correct, we can no longer maintain that only the earth harbors life. Nevertheless, there are so many requirements for life that only a small number of planets can fulfill the special conditions that would permit evolution of life at all similar to that on earth.

First, a planet suitable for life must receive a steady supply of light and heat from its sun for many billions of years. This means that its orbit must be nearly circular.

Second, water must be present on the planet to permit the evolution of complex biochemical systems based on carbon. Earth is indeed a watery planet, with water covering 71% of its surface. Viewed from space, water, not land, is earth's most conspicuous surface feature.

Third, the temperature must be suitable, meaning practically within the range of $-50°$ to $+100°$ C, as it is over most of the earth's surface. Life at temperatures above $100°$ C is impossible because biopolymers based on carbon and water would be rapidly hydrolyzed. Temperatures much below the freezing point of water prevent growth of organisms by slowing chemical processes.

Fourth, all life requires a suitable array of major and minor elements. Oxygen, carbon, hydrogen, and nitrogen form 95% of living matter and thus dominate the composition of life on earth. These four are supported by seven other major elements (phosphorus, calcium, potassium, sulfur, sodium, chlorine, and magnesium) and a large number of minor or "trace" elements. Perhaps 46 elements in all are found in living matter; many are essential for life, whereas others are present only because they exist in the environment with which the organism interacts. Not all planets possess elements in the same proportions, even in our solar system.

There are many other properties of earth that make it an especially fit environment for life. It is large enough to have a surface density that permits molecules to collect and align properly. Matter in cells is in a colloidal state, an intermediate between conditions too solid to allow change and conditions too fluid or gaseous to permit molecular organization. The earth's gravity is strong enough to hold an extensive gaseous atmosphere but not so strong that more than a trace of free hydrogen remains. Another consideration, especially important for life on earth today, is the oxygen-ozone atmospheric screen that absorbs lethal ultraviolet radiation from the sun. So effective is this absorption that rays with wavelengths shorter than 283 nm (nanometer = 10^{-12}m) fail to reach the earth's surface.

In his classic book *The Fitness of the Environment*, published in 1913, the biochemist L.J. Henderson maintained that earth possesses "the best of all possible environments for life." We may agree; however, we realize that the surface of any body that is the size and age of earth in a similar orbit revolving about a similar sun and having a similar elemental composition should also have an excellent environment for life.

The organism and its environment share a reciprocal relationship. The environment is fit for the organism, and the organism is fitted to the environment and adapts to its changes. As an open system, an animal is forever receiving and giving off materials and energy. The building materials for life are obtained from the physical environment, either directly by producers such as green plants or indirectly by consumers that return inorganic substances to the environment by excretion or by the decay and disintegration of their bodies.

The living form is a transient link that is built up out of environmental materials, which are then returned to the environment to be used again in the re-creation of new life. Life, death, decay, and re-creation have been the cycle of existence since life began.

In this continuous interchange between organism and environment, both are altered in the process, and a favorable relationship is preserved. The environment of earth, with its living and nonliving components, is not a static entity but has undergone an evolution in every way as dramatic as the evolution of the animal kingdom. It is still changing today, more rapidly than ever before, under the heavy-handed influence of humans.

The primitive earth of 3.5 billion years ago, barren, stormy, and volcanic with a reducing atmosphere of ammonia, methane, and water, was wonderfully fit for the prebiotic syntheses that led to life's beginnings. Yet, it was totally unsuited, indeed lethal, for the kinds of living organisms that inhabit the earth today, just as early forms of life could not survive in our present environment. The appearance of free oxygen in the atmosphere, produced largely if not almost entirely by life, is an example of the reciprocity between organism and environment. Although oxygen was at first poisonous to early forms of life, its gradual accumulation over the ages from photosynthesis forced protective biochemical alterations to appear that led eventually to complete dependence on oxygen for life.

Earth's biosphere and the organisms in it have evolved together. Again, as living organisms adapt and evolve, they act on and produce changes in their environment; in so doing they must themselves change.

DISTRIBUTION OF LIFE ON EARTH
Ecosphere and Its Subdivisions

The **ecosphere** includes all life on earth and the physical environments in which it lives and with which it interacts. It includes the biosphere, which is supported by three more subdivisions of the ecosphere: lithosphere, hydrosphere, and atmosphere.

The **lithosphere** is the rocky material of the earth's outer shell and is the ultimate source of all mineral elements required by living organisms. The **hydrosphere** is the water on or near the earth's surface, and it extends into the lithosphere and the atmosphere. Water is distributed over the earth by a global hydrological cycle of evaporation, precipitation, and runoff. Some five sixths of the evaporation is from the ocean, and more water is evaporated from the ocean than is returned to it by precipitation. Ocean evaporation therefore provides much of the rainfall that supports life on land. The gaseous component of the ecosphere, the **atmosphere,** extends to some 3500 km above the surface of the earth, but all life is confined to the lowest 8 to 15 km (troposphere). The oxygen-ozone atmospheric screen layer, mentioned previously, is concentrated mostly between 20 and 25 km. The main gases present in the troposphere are (by volume) nitrogen, 78%; oxygen, 21%; argon, 0.93%; carbon dioxide, 0.03%; and variable amounts of water vapor.

Atmospheric oxygen has originated almost entirely from plant photosynthesis. As discussed in Chapter 3, the primitive earth contained a reducing atmosphere devoid of oxygen. When oxygen-evolving photosynthesis appeared about 3 billion years ago, oxygen gradually began to accumulate in the atmosphere. It is believed that by the mid-Paleozoic era, some 400 million years BP, the oxygen concentration had reached its present level of about 21%. Since then oxygen consumption by animals and plants has approximately equaled oxygen production. The present surplus of free oxygen in the atmosphere resulted from the fossilization of plants before they could decay or be consumed by animals. As these vast stores of fossil fuels are burned by our industrialized civilization, the oxygen surplus that accumulated over the ages could conceivably be depleted. Fortunately, this is unlikely for two reasons: (1) most of the total fossilized carbon is deposited in noncombustible shales and rocks, and (2) the oxygen reserves in the atmosphere and in the oceans are so enormous that the supply could last thousands of years, even if all photosynthetic replenishment suddenly were to cease.

There is concern, however, that the rapid input of carbon dioxide into the atmosphere from the burning of fossil fuels may significantly affect the earth's heat budget. Much of the sun's short-wave light energy absorbed by the earth's surface is reradiated as longer-wave infrared heat energy (Figure 40-1). Materials in the atmosphere, especially carbon dioxide and water vapor, impede this heat loss and allow the atmosphere to warm up. The accumulation of carbon dioxide could lead to an increase in the temperature of the biosphere as a whole. This is called the "greenhouse effect," since the atmosphere acts to trap reradiated heat from the earth in much the same way the glass of a greenhouse traps the heat reradiated by the plants and soil inside.

The concern over the long-range effects of increasing atmospheric carbon dioxide, primarily from burning of fossil fuel, stems not from mere conjecture. Atmospheric carbon dioxide increased from about 280 parts per million (ppm) before the Industrial Revolution to an average 335 ppm today and is increasing at a rate of 1.3 ppm per year. It is expected to exceed 600 ppm in the next century. In the past century the global temperature has increased 0.4° C, and it will have increased 3° C when the carbon dioxide has doubled in the next century, a warming of unprecedented magnitude. This is expected to shift climatic zones profoundly, creating drought conditions in the American wheatlands and wetter conditions along the coasts. Melting of the earth's ice sheets is another possibility, with flooding of coastal lowlands throughout the world. Whatever the outcome, a fascinating, unplanned global experiment is underway.

___ Terrestrial Environment: Biomes

A biome is a major biotic unit bearing a characteristic and easily recognized array of plant life. Botanists long ago recognized that the terrestrial environment of the earth could be divided into large units having a distinctive vegetation: forests, prairies, deserts, and so on. Animal distribution has always been more difficult to map, even after naturalists accumulated the basic information about animal ranges and centers of distribution from around the world. Plant and animal distributions do not coincide. Eventually zoogeographers accepted plant distribution as the basic biotic units and recognized the biomes as distinctive combinations of plants and animals. A biome is therefore identified by its dominant plant formation (Figure 40-2), but, since animals depend on plants, each biome supports a characteristic fauna.

Each biome is distinctive, but its borders are not. Anyone who has traveled across the United States knows that plant communities grade into one another over broad areas. The moist deciduous forests of the Appalachians give way gradually to the drier oak forests of the upper Mississippi states, and then to oak woodlands with grassy understory. This yields to tall and mixed prairie (now corn and wheatlands), then to desert grasslands, and finally to desert shrublands. The indistinct boundaries where the dominant plants of adjacent biomes are mixed together form an almost continuous gradient called an **ecocline.** The main factors determining an ecocline are temperature and precipitation.

From what we have said, it is obvious that biomes are mainly a convenient way to organize our concepts about different life communities. Nevertheless, anyone can distinguish a grassland, deciduous forest, coniferous forest, or shrub desert by the dominant plants in each. And we can make reasonable assumptions about the kinds of animals that live in each biome.

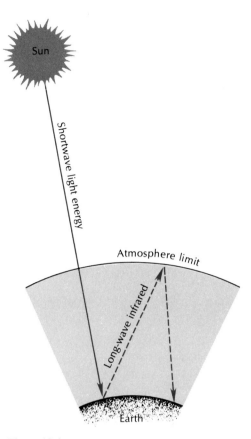

Figure 40-1

"Greenhouse effect." Carbon dioxide and water vapor in the atmosphere are transparent to sunlight but absorb heat energy reradiated from the earth, leading to warming of atmospheric air.

Figure 40-2

Some of the major biomes of North America.

The principal terrestrial biomes are temperate deciduous forest, temperate coniferous forest, tropical forest, grassland, tundra, and desert. In this brief survey, we will refer especially to the biomes of North America and will consider predominant features of each.

Temperate deciduous forest

The temperate deciduous forest, best developed in eastern North America, encompasses several forest types that change gradually from the northeast to the south. Deciduous, broad-leaved trees such as oak, maple, and beech that shed their leaves in winter predominate. Seasonal aspects are better defined in this biome than in any other. The deciduous habit is an adaptation of dormancy for low energy levels from the sun in winter and freezing winter temperatures. In summer, the relatively dense forests form a closed canopy that creates deep shade below. Consequently there has been a selection for understory plants that grow rapidly in the spring and flower early before the canopy develops. The mean annual precipitation is relatively high (75 to 125 cm, or 30 to 50 inches) and rain falls periodically throughout the year.

Animal communities in deciduous forests respond to seasonal change in various ways. Some, such as the insect-eating warblers, migrate. Others, such as the woodchuck, hibernate away the winter months. Others that are unable to escape survive by using the available food (for example, deer) or stored food supplies (for example, squirrels). Hunting and habitat loss have eliminated virtually all the large carnivores that once roamed the eastern forests, such as mountain lions, bobcats, and wolves. Deer, on the other hand, thrive in the second-growth eastern forests under the protection of strict hunting management. Insect and invertebrate communities are abundant in deciduous forests because decaying logs and forest floor litter provide excellent shelter.

Coniferous forest

In North America the coniferous forests form a broad, continuous, continent-wide belt stretching across Canada and Alaska, and south through the Rocky

The heavy exploitation of the deciduous forests of North America began in the seventeenth century and reached a peak in the nineteenth century. Logging removed nearly all of the once-magnificent stands of temperate hardwoods. With the opening of the prairie for agriculture, many eastern farms were abandoned and allowed to return gradually to deciduous forests.

Figure 40-3

Moose browsing on dwarf birch in the coniferous forest biome.
Photograph by C.P. Hickman, Jr.

Mountains into Mexico. It is dominated by evergreens—pine, fir, spruce, and cedar—which are adapted to withstand freezing and take full advantage of short summer growing seasons. The conical trees with their flexible branches shed snow loads easily. The northern area is the **boreal** (northern) **forest,** often referred to as **taiga** (a Russian word, pronounced "tie-ga"). The taiga is dominated by white and black spruce, balsam, subalpine fir, larch, and birch. In the central region of North America, the taiga merges into **lake forest,** dominated by white pine, red pine, and eastern hemlock. However, most of this forest was destroyed by logging and is replaced by shrubby wasteland that characterizes much of Michigan, Wisconsin, and Minnesota today. The large **southern pine forests** occupy much of the southeastern United States.

Mammals of the boreal and lake coniferous forests are deer, moose (Figure 40-3), elk, snowshoe hare, several furbearers, and a variety of rodents. They are adapted physiologically or behaviorally for long, cold, snowy winters. Common birds are chickadees, nuthatches, warblers, and jays. One bird, the red crossbill, has a beak specialized for picking seeds from cones. Mosquitoes and flies are pests to both animals and humans in this biome. Southern coniferous forests lack many mammals found in the north, but they have more snakes, lizards, and amphibians.

Tropical forest

The worldwide equatorial belt of tropical forests is an area of high rainfall, high humidity, relatively high and constant temperatures, and little seasonal variation in day length. These conditions have nurtured luxurious, uninterrupted growth that reaches its greatest intensity in the rain forests. In sharp contrast to temperate deciduous forests, dominated as they are by a relatively few tree species, tropical forests contain thousands of species, none of which is dominant. A single hectare typically contains 50 to 70 tree species as compared with 10 to 20 tree species in an equivalent area in the eastern United States. Climbing plants and epiphytes are common among the trunks and limbs (Figure 40-4). A distinctive feature of tropical forests is the stratification of life into six, and occasionally as many as eight, feeding strata (Figure 40-5).

Insectivorous birds and bats occupy the air above the canopy; below it birds, fruit bats, and mammals feed on leaves and fruit. In the middle zones are arboreal

Figure 40-4

Air plants (bromeliads) in a tropical forest. The pineapple-like bromeliads and other epiphytes (plants that live on the surface of other plants) are common in tropical America. They tolerate lack of soil and a precarious water supply. Large bromeliads are often little ecosystems in themselves, supporting a variety of small, interdependent plant and animal life.
Photograph by L.L. Rue, III.

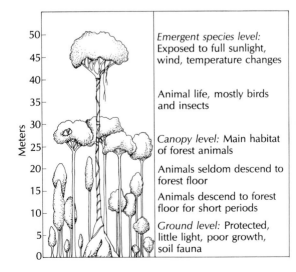

Figure 40-5

Profile of tropical rain forest, showing stratification of animal life. The animal biomass is small compared with the biomass of the trees.

Each month an area of undisturbed tropical forest the size of El Salvador is being converted to other uses. Unlike the forest clearing in the temperate zones, which made possible sustained, productive agriculture, tropical soils quickly become depleted, forcing farmers to move on to clear more forest. Another pressure on tropical forests is multinational timber companies that own and cut large tracts of timber to make furniture for developed countries. Cattle ranchers also clear huge forest tracts to raise beef that is later sold to North America's fast-food chains. As one specialist remarked, "Everyone's hand is on the chain saw."

mammals (such as monkeys and tree sloths), numerous birds, insectivorous bats, insects, and amphibians. A distinct group of animals ranges up and down the trunks, feeding from all strata. On the ground are large mammals lacking climbing ability, such as the large rodents of South America (for example, capabara, paca, and agouti). Finally, a mixed group of small insectivorous, carnivorous, and herbivorous animals searches the litter and lower tree trunks for food. No other biome can match the tropical forest in incredible variety of animal species. Food webs are intricate and notoriously difficult for ecologists to unravel.

The tropical forests, especially the enormous expanse centered in the Amazon Basin, are the most seriously threatened of forest ecosystems. Large areas are being cleared for agriculture by "slash-and-burn" methods, but, because of low fertility, farms are soon abandoned. It may seem paradoxical that a biome as luxuriant as the tropical forest should have poor soil. This occurs because nutrients released by decomposition are rapidly absorbed by plants, leaving no reservoir of humus. Once the plants are removed, the soil rapidly becomes a hard, bricklike crust called laterite. Tropical plants cannot recolonize such areas.

Grassland

One of the most extensive prairie biomes in the world is in North America, but the original plant and animal components have been almost completely destroyed by humans. Virtually all the major native grasses have been replaced by cultivated grains in croplands and by alien species of grasses in grazing lands. Rainfall is too scant to support trees but is sufficient to prevent the formation of deserts. Very few of the once dominant herbivore, bison, survive, but jackrabbits, prairie dogs, ground squirrels, and antelope remain. Mammalian predators include coyotes, ferrets, and badgers, although, of these, only coyotes are common.

Tundra

The tundra is characteristic of severe, cold climatic regions, especially the treeless Arctic regions and high mountaintops. Plant life must adapt itself to a short growing season of about 60 days and to a soil that remains frozen for most of the year. Most tundra regions are covered with bogs, marshes, ponds, and a spongy mat of decayed vegetation, although high tundras may be covered only with lichens and grasses. Despite the thin soil and short growing season, the vegetation of dwarf woody plants, grasses, sedges, and lichens may be quite profuse. The plants of the alpine tundra of high mountains, such as the Rockies and Sierra Nevadas, may differ from the Arctic tundra in some respects. Characteristic animals of the Arctic tundra are the lemming, caribou, musk-ox, arctic fox, arctic hare, ptarmigan, and (during the summer) many migratory birds.

Desert

Deserts are extremely arid regions where rainfall is low (less than 25 cm a year), and water evaporation is high. Desert plants, such as various shrubs and cacti, have reduced foliage and other adaptations for conserving water (Figure 40-6). Many large desert animals have developed remarkable anatomical and physiological adaptations for keeping cool and conserving water (p. 655). Most smaller animals avoid the most severe conditions by living in burrows or developing nocturnal habits. Mammals found there include the white-tailed deer, peccary, cottontail, jackrabbit, kangaroo rat, and ground squirrel. Typical birds are the road-

Figure 40-6

Sonoran desert of Arizona and central Mexico has a diversified vegetation. Conspicuous here are giant saguaro cactus, creosote bushes, and bur sage.
Photograph by L.L. Rue, III.

runner, cactus wren, turkey vulture, and burrowing owl. Reptiles are numerous, and a few species of toads are common. Arthropods include a great variety of insects and arachnids.

Aquatic Environments

Inland waters

Whereas freshwater habitats make up only a small fraction of the total water on earth, they are an essential component of the global water cycle. Inland water habitats are also indispensable to the many land animals that depend on them as their source of drinking water or for reproduction. The freshwater habitat presents several challenges to the numerous aquatic animals that live there: (1) it is changeable, lacking the stability of either land or the oceans; (2) seasonal changes are often pronounced; (3) lakes and ponds gradually fill in, grow smaller, and are eventually replaced by land; (4) aquatic organisms require special adaptations for conserving salt and eliminating excess water since fresh water contains little salt; (5) animals living in streams must bear special adaptations, such as organs of attachment or streamlined shapes, to avoid being swept away in the unremitting current; and (6) many freshwater habitats have been severely damaged by human pollution such as the dumping of toxic industrial wastes and enormous quantities of sewage.

Inland waters are divided broadly into running-water, or **lotic** (L., *lotus*, action of washing) habitats, and standing-water, or **lentic** (L., *lentus*, slow) habitats. Lotic habitats follow a gradient from mountain brooks to streams and rivers. Brooks and streams with a high velocity of water flow are high in dissolved oxygen as a result of their turbulence. Energy input is chiefly in the form of organic detritus washed in from adjacent terrestrial areas. More slowly moving rivers have

Acid rain, caused mainly by the industrial emissions of sulfur and nitrogen oxides from the burning of fossil fuels, has led to serious deterioration of water quality in lakes, ponds, and rivers of southern Scandinavia and eastern North America. These emissions interact with atmospheric moisture to yield high concentrations of hydrogen, sulfate, and nitrate ions. Acid rains having a pH below 4 are not uncommon. Most sensitive to acid rain are recently glaciated areas having low buffering capacity. The acidity of Maine's lakes increased eightfold between 1937 and 1974; hundreds of lakes in the Adirondacks are now sterile, or are approaching sterility, and are lost to practical and recreational purposes.

less dissolved oxygen and more floating algae and other plants. Their fauna is tolerant of lower oxygen concentration.

Lentic habitats, such as ponds and lakes, tend to have still lower oxygen concentration, particularly in the deeper areas. Animals living on the bottom or on submerged vegetation (**benthos**) include snails and mussels, crustaceans, and a wide variety of insects. Many swimming forms, called **nekton,** are found in lakes and larger ponds. Depending on the nutrients available, there may be a large contingent of small floating or weakly swimming plants and animals (**plankton**).

Oceans

By almost any measure, the oceans represent by far the largest portion of the earth's biosphere. They cover 71% of the earth's surface to an average depth of 3.75 km (2.3 miles), with their greatest depths reaching to more than 11.5 km (7.2 miles) below sea level. The marine world is relatively uniform as compared with land, and in many respects it is less demanding on life forms. However, the evident monotony of the ocean's surface belies the variety of life below. The oceans are the cradle of life, and this is reflected in the variety of organisms living there—more than 200,000 species of plants and animals. The vast majority of these forms, about 98%, live on the sea bed (**benthic**); only 2% live freely in the open ocean (**pelagic**). Of the benthic forms, most occur in the intertidal zone or shallow depths of the oceans. Less than 1% live in the deep ocean below 2000 m.

The most productive areas are concentrated along continental margins and a few areas where the waters are enriched by organic nutrients and debris lifted by upwelling currents into the sunlit, or **photic,** zone, where photosynthetic activity occurs. With certain notable exceptions (see box p. 902), all life below the photic zone must be supported by the light "rain" of organic particles from above.

The life of the ocean is divided into regions, or provinces, each with its own distinctive life forms (Figure 40-7). The **littoral,** or intertidal, zone, where sea and land meet, is paradoxically both the harshest and the richest of all marine environments. It, as well as the animals living there, is subjected to pounding surf, sun, wind, rain, extreme temperature fluctuations, erosion, and sedimentation. Yet because of the diversity of available habitats and the bounteous supply of nutrients, animals such as barnacles, snails, chitons, limpets, mussels, sea urchins, and many others flourish there. Below the littoral zone is the **sublittoral** zone, which is always submerged. It also supports a rich variety of animal life, as well as forests of brown algae.

Figure 40-7

Major marine zones.

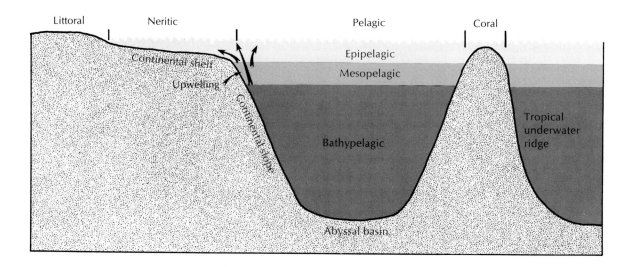

An **estuary** is a semienclosed transition zone where fresh water flows into the sea. Despite an unstable salinity caused by the variable entry of fresh water, the estuary is a nutrient-rich habitat that supports a diverse fauna.

The **neritic,** or shallow water, zone surrounds the continents and extends out to a depth of 200 m. This zone is more productive than the open ocean because it benefits from nutrients delivered by rivers and by upwelling at the edge of the continental shelf. Plant growth is prolific, which in turn supports a diverse animal life, including most of the world's fisheries.

Areas of **upwelling,** although small and restricted to a few regions, are vital sources of nutrient renewal for the surface photic zone. Some of the world's most productive fisheries are—or were—centered on upwelling regions. Before its collapse in 1972, the Peruvian anchovy fishery that depended on the Peru Current provided 22% of all the fishes caught in the world! Earlier, the California sardine fishery and the Japanese herring fishery, both fisheries of upwelling regions, were intensively harvested to the point of collapse and have never recovered.

The **pelagic** zone is the vast open ocean. Despite its size (comprising 90% of the total oceanic area), the pelagic zone is relatively impoverished biologically because, as organisms die, they sink out of the photic zone, carrying nutrients into the bathypelagic zone where they are immobilized. However, in areas of upwelling and where oceanic currents converge, nutrients are replenished and productivity may be high. The enormously productive polar seas are an example. Before their populations were overexploited by humans, the baleen whales are estimated to have consumed 77 million tons of Antarctic krill (a shrimplike animal) per year, far more than the entire catch of all the fishes, crustaceans, and molluscs taken by all the world's fishing fleet in any one year! The enormous krill population was sustained by phytoplankton, the base of the food chain, which in turn flourished because of the abundance of nutrients in the Antarctic sea.

In the great ocean depths exists the **abyssal** habitat, characterized by enormous pressure, perpetual darkness, and a constant temperature near 0° C. It has remained a world unknown to humans until recently, when baited cameras, bathyscaphs, and deep-water trawls have been lowered to view and sample the ocean bottom. Benthic forms of the sea floor survive on that meager portion of the gentle rain of organic debris from above that escapes consumption by organisms in the water column. Despite the unusual properties of the habitat, sea anemones, sea urchins, crustaceans, and polychaete worms—indeed nearly all major invertebrate groups—exist here, as well as fishes (Figure 40-8). Recently self-contained benthic communities of animals that are completely independent of solar energy and the rain of organic debris from above were discovered adjacent to vents of hot water issuing from rifts in the ocean floor (see box, p. 902).

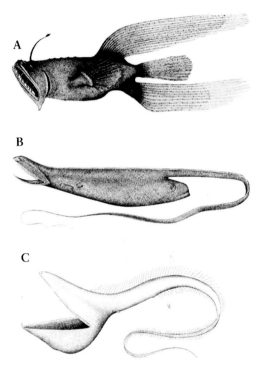

Figure 40-8

Fishes of the abyssal habitat. Because their entire lives are spent in absolute darkness, their eyes are degenerate and often functionless. Huge mouths and capacious stomachs are adaptations for large but very infrequent meals, the organic debris drifting down from above. **A,** *Caulophryne* sp. **B,** *Saccopharynx ampullaceus.* **C,** *Eurypharynx pelecanoides.*
Courtesy Smithsonian Institution.

ANIMAL DISTRIBUTION (ZOOGEOGRAPHY)

The study of zoogeography tries to explain why animals are found where they are, their patterns of dispersal, and the factors responsible for their dispersal. With the exception of the human species, which is able to live nearly anywhere on earth, and creatures such as the house mouse and the cockroach, which share human habitations, the different kinds of animals occupy limited habitats on earth. It is not always easy to explain why animals are distributed as they are, since similar habitats on separate continents may be occupied by quite different kinds of animals. A particular species may be absent from a region that supports similar animals for any of three reasons: (1) there may be barriers that prevent it from getting there; (2) having gotten there, it may be unable to adapt to the new habitat or compete successfully with resident species and may become extinct; or (3) having once arrived and adapted, it has subsequently evolved into a distinct species.

Thus there are always good reasons why animals are found where they are (or are not found where one thinks they ought to be), and it is the task of the zoogeographer to discover what these reasons are. Usually this means going back into the past. The fossil record plainly shows that animals once flourished in regions from which they are now absent. Extinction has played a major role, but many groups left descendants that migrated to other regions and survived. For example, camels originated in North America, where their fossils are found, but spread to Eurasia by way of Alaska (true camels) and to South America (llamas) during the Pleistocene epoch. Then they became extinct in North America at the close of the Ice Age. Thus the history of an animal species or its ancestor must be known before one can understand why it is where it is. The earth's surface is undergoing constant change. Many areas that are now land were once covered with seas; fertile plains may be claimed by advancing desert; impassable mountain barriers may arise where none existed before; or inhospitable ice fields may retreat before a warmer climate to be replaced by forests. Geological change has been responsible for much of the alteration in animal (and plant) distribution and has been a powerful influence in shaping organic evolution.

Major Faunal Realms

The nineteenth-century naturalists recognized that it was possible to divide the terrestrial world into several distinct realms of animal distribution, separated from one another by topographical and climatic barriers. Philip L. Sclater (1858) first proposed a region system based on bird families. Later (1876), Alfred Russel Wallace, codiscoverer with Charles Darwin of the principle of natural selection, modified Sclater's pattern and applied it to vertebrates in general. Wallace recognized six land realms based on animal ranges. Although later scientists slightly modified his biogeographical boundaries, the Wallacean regions are the ones we recognize today (Figure 40-9). Great oceanic barriers serve as longitudinal dividers between the Nearctic and Palearctic realms in the north and the Neotropic and Australian realms in the south. The great northern land mass is divided by a subtropic, warm temperate dry belt into Palearctic, Ethiopian, and Oriental realms. These realms are better recognized by the differences between them than by faunal similarities within them. As with the biomes described earlier, the

Figure 40-9

The six major faunal realms of the earth.

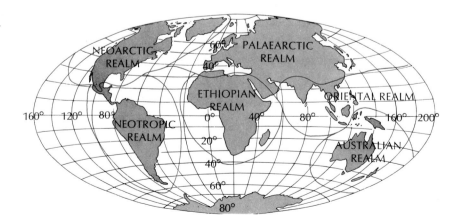

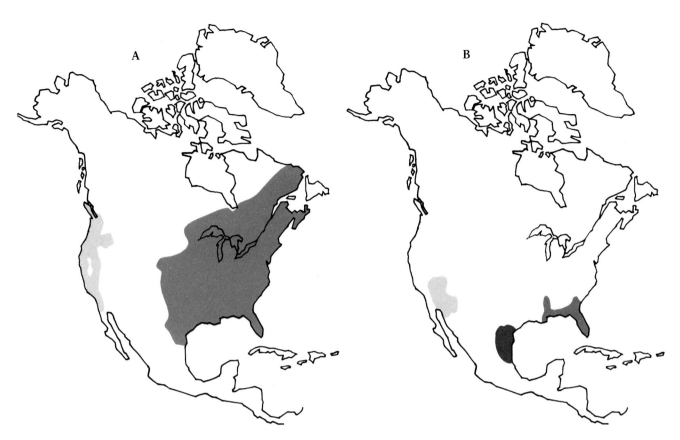

Figure 40-10

Disjunct distributions in North America. **A,** Moles of the family Talpidae probably entered North America across the Bering land bridge that once joined North America and Asia during the Tertiary period. Eastern and western populations are now separated by the Rocky Mountains. **B,** Gopher turtles of the genus *Gopherus* are now separated into three fully isolated populations.

A, Redrawn from Burt, W.H.; **B,** redrawn from Blair, W.F. Both in C.L. Hubbs (ed.). 1958. Zoogeography, Washington, D.C., AAAS Pub. No. 51. Copyright 1958 by AAAS.

boundaries between realms are transition zones that are often poorly defined, especially if the barrier between them is largely climatic. Or they may be clear and distinct if the separation is a mountain range or salt water; to land animals, even a narrow saltwater strait is an effective barrier.

Distributional Processes

The association of animals in ways that enable zoogeographers to recognize faunal realms suggests that the major realms have been separated from each other for long periods of time. Of particular interest are the numerous instances of **discontinuous,** or **disjunct, distributions:** the same or closely related organisms or group living in widely separated areas of a continent, or even the world (Figure 40-10). How could an animal group become separated in this way? In some cases they may be **evolutionary relicts,** the scattered remnants of a once widespread and successful group that can no longer compete (or can compete only marginally) with modern forms. The three surviving genera of lungfishes—one each in South America, Africa, and Australia—are good examples (Figure 23-16, p. 478). But this explanation cannot apply to many species of modern animals—mammals, for instance—that compete successfully in their habitats, yet may be widely but unevenly distributed around the world. Nor can it explain the appearance of new animal species, which has occurred millions of times over the history of the earth. Any convincing explanation of animal distribution must also explain speciation.

There are only two ways for disjunct distributions to arise. Either an organism moves to a new location (**dispersal**), or the location moves, carrying the organism passively with it (**vicariance**).

Distribution by dispersal

By dispersal, animals spread into new localities from their place of origin. Changes in environment may force animals to shift their domiciles, although most great groups have spread to gain favorable conditions, not to avoid unfavorable ones. The reproductive rate of animals is so great that there is continuous pressure on individuals to move into new areas, reducing the competition for food, shelter, and space to live and breed. This **population pressure** expands the range of the species and maintains healthy populations. Survival of the species may depend on successful dispersal to new areas when the old habitat ceases to be habitable because of geographical and ecological changes through time. Dispersal followed by isolation is also a biological prerequisite for the formation of new species because, as we pointed out previously (p. 863), geographical isolation of a founding colony from the parent population permits new adaptive variations to arise.

Dispersal movements involve emigration from one region and immigration into another. Dispersal is a *one-way,* outward movement that must be distinguished from migration, a *periodic* movement back and forth between two localities, such as the seasonal migration of birds. Dispersing animals may move **actively** under their own power, whether they are creepers, climbers, runners, hoppers, or flyers; or they may be **passively** dispersed by wind and air currents, by floating or rafting on rivers, lakes, or the sea, or by hitching rides on other animals. The chief deterrents to the spread of animals are barriers such as mountain chains, cliffs, canyons, sand dunes, lava flows, rivers, and, most effective of all, oceans. However, what constitutes a barrier to one animal may serve as an avenue of dispersal for another. A river may obstruct the dispersal of a gray squirrel but facilitate the movement of a beaver. In addition to the physical barriers listed are climatic barriers, such as deserts, and biotic barriers, such as competition between species, predators, changes in vegetation, and absence of suitable food. Deserts are especially important barriers to most animals because they have an inhospitable climate, too hot and dry for most species, and are lacking in vegetation for cover, food, and suitable places to live. Deserts are of course the chosen habitats for a variety of animals bearing appropriate desert adaptations; to these animals, non-desert habitats are barriers to dispersal.

Distribution by vicariance

Disjunct distributions of animals may be created by the splitting of ancestral populations by changing geography—separation of areas that were once joined—rather than by dispersal. The study of fragmentation of biotas in this manner is called **vicariance** (L., *vicarius,* a substitute) **biogeography.** Vicariance distribution can occur when a species is separated by some new barrier, such as a mountain range or a desert, or when landmasses become fragmented, resulting in the separation of their fauna and flora. Many biogeographers use the word "vicariance" as a synonym for "allopatry," which, as we have seen, is simply a way for new species to arise in geographically separated areas. Vicariance biogeography, however, has recently attracted a school of enthusiastic adherents who claim that vicariance is a primary principle of biogeography, to the virtual exclusion of dispersal. In their view, dispersal can only lead to competition, and often to extinction, as species expand into already occupied habitats, whereas vicariance is responsible for the multiplication of species. The vicariance explanation fits in well with the accepted idea of allopatric speciation: biotas must be geographically

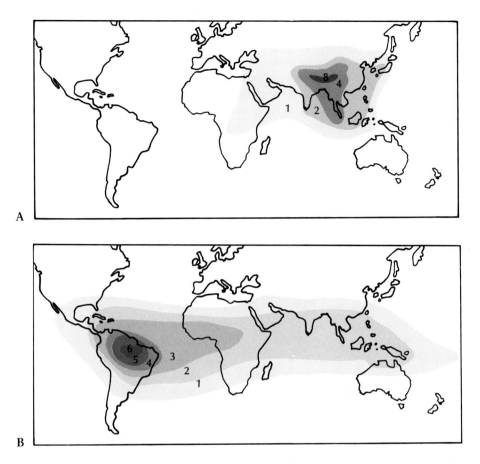

A

B

Figure 40-11

Centers of origin of, **A**, pheasants of the subfamily Phasianinae and, **B**, crocodiles of the order Crocodilia. The figures indicate numbers of species or genera within each contour. The center of origin is considered to be the area of highest taxon density.
Redrawn from Pielou, E.C. 1979. Biogeography. New York, John Wiley & Sons, Inc.

separated before speciation can occur (p. 863). It does not, however, explain the common observation that many taxonomic groups have centers of origin (Figure 40-11). The **center of origin**, the place where a group is believed to have originated, can be located by drawing a contour map composed of lines of equal species (or genus) richness. The center of origin is outlined by the contour surrounding the highest species (or genus) density for the group (Figure 40-11). The problem for the vicariance advocate is to explain how species multiplication can occur within a restricted area, as center-of-origin maps imply, if virtually all speciation occurs allopatrically, that is, by geographical splitting of ancestral populations. Various models have been offered as solutions to this problem, but the explanations are beyond the scope of this book. The reasonable conclusion is that both dispersal and vicariance occur, separately and in combination. It may be true that most speciation happens by vicariance (allopatry), but this places no restriction on dispersal after (or even before) speciation occurs.

Continental drift theory

It is no accident that the current enthusiasm for vicariance biogeography coincides with the recent acceptance of the continental drift theory by geologists. The continental drift theory is not new (it was proposed in 1912 by the German meteorologist Alfred Wegener), but it remained controversial and largely rejected until the newly proposed theory of **plate tectonics** provided a mechanism to account for drifting continents. According to the theory of plate tectonics (tectonics means "deforming movement"), the earth's surface is composed of 6 to 10 rocky plates, about 100 km thick, that shift about on a more malleable underlying layer. Wegener proposed that the earth's continents had been drifting like rafts follow-

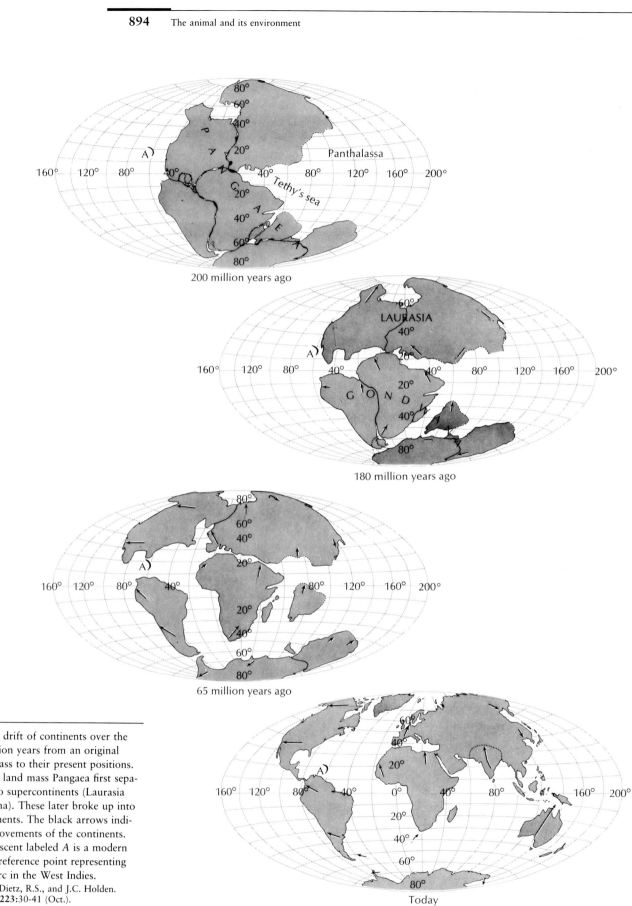

Figure 40-12

Hypothesized drift of continents over the past 200 million years from an original single land mass to their present positions. The universal land mass Pangaea first separated into two supercontinents (Laurasia and Gondwana). These later broke up into smaller continents. The black arrows indicate vector movements of the continents. The black crescent labeled *A* is a modern geographical reference point representing the Antilles arc in the West Indies.

Modified from Dietz, R.S., and J.C. Holden. 1970. Sci. Am. **223**:30-41 (Oct.).

ing the breakup of a single great landmass called Pangaea ("all land"). According to recent workers who have considerably revised Wegener's dating, this occurred some 200 million years ago. Two great supercontinents were formed: a northern Laurasia and a southern Gondwana, separated from each other by the Tethys Sea (Figure 40-12). At the end of the Jurassic period, some 135 million years ago, the supercontinents began to fragment and drift apart. Laurasia split into North America, most of Eurasia, and Greenland. Gondwana split into South America, Africa, Madagascar, Arabia, India, Australia, and Antarctica. This theory is supported by the appearance of fit between the continents, by recent airborne paleomagnetic surveys, by seismographic studies, by the presence of midoceanic ridges where the tectonic plates were born, and by a wealth of biological data.

Continental drift explains several otherwise puzzling distributions of animals, such as the similarity of invertebrate fossils in Africa and South America, as well as certain similarities in present-day fauna at the same latitudes on the two continents. However, the continents have been separated for all of the Cenozoic era and probably for much of the Mesozoic era as well, much too long to explain the dispersal of modern organisms such as the placental mammals. Continental drift theory is, nevertheless, enormously useful in explaining interconnections between flora and fauna of the past.

The present distribution of the marsupial mammals is an excellent example of the influence of continental breakup. The marsupials appeared in the Middle Cretaceous period, about 100 million years BP, probably in South America. Because South America was at that time connected to Australia through Antarctica (which was then much warmer than it is today), the marsupials spread through all three continents. They also moved into North America, but there they met the placental mammals, which had dispersed to that continent from Asia. The marsupials evidently could not compete with placentals, and so became extinct in North America. (North American marsupials today, that is, the opossums, are relatively recent arrivals from South America.) The placentals followed the marsupials into South America, but by that time the marsupials had expanded and were too firmly established to be driven into extinction. In the meantime, about 50 million years BP, Australia drifted apart from Antarctica, barring entrance to the placentals. Australia remained in isolation, allowing the marsupials to diversify into the present rich and varied fauna.

Temporary land bridges have also been important pathways of dispersal. An important and well-established land bridge that no longer exists connected Asia and North America across the Bering Strait. It was across this corridor that the placentals moved from Asia into North America.

Today a land bridge connects North and South America at the Isthmus of Panama, but from the mid-Eocene epoch (50 million years BP) to the end of the Pliocene epoch (3 million years BP), the two continents were completely separated by a water barrier. During this long period, all of the major groups of placental mammals were evolving in distinctive directions on each continent. When the land bridge was reestablished at the end of the Pliocene epoch, a tide of mammals began to flow in both directions. This has been called the "Great American Interchange." For a period both continents gained in mammal diversity, but direct competition for similar ecological niches soon led to the extinction of large numbers of mammals on both continents. North American invaders such as deer, foxes, bears, raccoons, tapirs, llamas, mastodons, and antelopes displaced many South American residents occupying similar niches, and today nearly half of the South American mammals are descendants of recent North American invaders

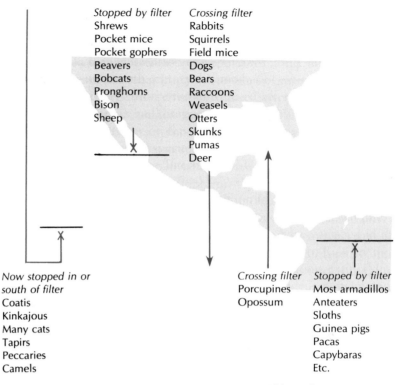

Old Northerners

Stopped by filter
Shrews
Pocket mice
Pocket gophers
Beavers
Bobcats
Pronghorns
Bison
Sheep

Crossing filter
Rabbits
Squirrels
Field mice
Dogs
Bears
Raccoons
Weasels
Otters
Skunks
Pumas
Deer

Now stopped in or south of filter
Coatis
Kinkajous
Many cats
Tapirs
Peccaries
Camels

Crossing filter
Porcupines
Opossum

Stopped by filter
Most armadillos
Anteaters
Sloths
Guinea pigs
Pacas
Capybaras
Etc.

Old Southerners

Figure 40-13

Central America as a filter zone. The Isthmus of Panama emerged 3 to 4 million years ago, permitting the extensive interchange of many groups of mammals but preventing the passage of others. "Old Northerners" and "Old Southerners" are extant mammals of northern and southern ancestry, respectively.

Modified from Simpson, G.G. 1980. Splendid isolation. New Haven, Conn., Yale University Press.

(Figure 40-13). Only a few South American invaders survived in North America: porcupines, armadillos, and opossums. Several other South American groups, including the giant ground sloths and the giant armadillos, entered North America but subsequently died out.

SUMMARY

The biosphere is the thin blanket of life surrounding the earth. The presence of life on earth is possible because numerous conditions for life are fulfilled on this planet. These include a steady supply of energy from the sun, the presence of water, a suitable range of temperatures, the correct proportion of major and minor elements, and the screening of lethal ultraviolet radiation by atmospheric ozone.

The earth environment and living organisms have evolved together, each deeply marking the other.

The ecosphere is composed of the biosphere, the thin film of life on the earth's surface; the lithosphere, the earth's rocky shell; the hydrosphere, the global distribution of water; and the atmosphere, the blanket of gas surrounding the earth.

The earth's terrestrial environment is composed of biomes that bear a distinctive array of plant life and associated animal life. The eastern deciduous forest is characterized by distinct seasons and autumn leaf fall. North of the deciduous forest is the coniferous forest, which in its northern range is called taiga, an area dominated by needle-leafed trees adapted for heavy snowfall. Animals of the taiga are adapted for long, snowy winters.

The tropical forest is the richest biome, characterized in part by a great diversity of plant species and the vertical stratification of animal habitats. Tropical forest soils rapidly deteriorate when the forest is removed.

The most modified biome is the grassland, or prairie, which has been largely converted to agriculture and grazing. The tundra biome of the far north and the desert biome are both severe environments for animal life, but they are populated nevertheless with organisms that have evolved appropriate adaptations.

Freshwater habitats include rivers and streams (lotic habitats) and ponds and lakes (lentic habitats). All are ephemeral habitats that are strongly influenced by nutrient input.

The oceans occupy 71% of the earth's surface. The photic, or sunlit, zone supports photosynthetic activity by phytoplankton. The rain of nutrients from the photic zone supports the great diversity of life below on the sea bed (benthos). The littoral, or tidal, zone is biologically rich but physically harsh. The neritic, or shallow-water, zone overlying the continental shelf is the locus of the world's great fisheries, which are especially productive in areas of upwelling where nutrients are constantly renewed. The open ocean, or pelagic zone, occupies most of the ocean's area but has low biological productivity.

Zoogeography is the study of animal distribution on earth. Several great faunal realms are recognized by distinctive animal associations. These are the Nearctic, Neotropical, Palearctic, Ethiopian, Oriental, and Australian realms.

Animals have become distributed around the earth by dispersal, the spread of populations from the centers of origin, and by vicariance, the separation of populations by barriers or continental drift. Continental drift, now strongly supported by plate tectonic theory, helps explain how animal groups become geographically separated so that speciation can occur. It also explains how certain groups, such as the marsupial mammals, can become isolated from competing groups. Temporary land bridges have also served as important pathways for animal dispersal.

Review questions

1. What are the special conditions on earth that make this planet especially fit for life?
2. What is the justification for saying that the earth and the life forms on it have evolved together and that each has deeply influenced the other?
3. What is the ecosphere? How would you distinguish between the following subdivisions of the ecosphere: lithosphere, hydrosphere, atmosphere?
4. What is the origin of oxygen on earth? What would happen to the earth's supply of oxygen if photosynthesis were suddenly to cease?
5. What effect is the burning of fossil fuels having on the atmosphere?
6. What is a biome? Briefly describe six examples of biomes.
7. What are some kinds of adaptations expected of animals living in freshwater habitats?
8. What are the most productive marine evironments, and why are they so productive?
9. What is the source of nutrients for animals living in the abyssal habitat?
10. What are some reasons why a species may be absent from a habitat or region to which it should adapt well?
11. What geographical boundaries or barriers divide the six faunal realms recognized by Alfred Russel Wallace?
12. Define and distinguish between the alternative explanations for disjunct distributions among animals: disperal and vicariance.
13. Who first proposed the continental drift theory and what finally convinced geologists that the theory was correct?
14. In what way does the continental drift theory help explain the present distribution of marsupial mammals on earth?
15. What was the Great American Interchange, when did it occur, and what were the results?

Selected references

Brown, J.H., and A.C. Gibson. 1983. Biogeography. St. Louis, The C.V. Mosby Co. *A wealth of information in this balanced treatment.*

Cox, C.B., and P.D. Moore. 1985. Biogeography: an ecological and evolutionary approach, ed. 4. Boston, Blackwell Scientific Publications. *Highly readable account with a strong ecological emphasis.*

Dietz, R.S., and J.C. Holden. 1970. The breakup of Pangaea. Sci. Am. **223**:30-41 (Oct.). *The sequence of continental drift since the early Mesozoic era is mapped.*

Henderson, L.J. 1913. The fitness of the environment. New York, Macmillan, Inc. *Explains how conditions on our planet made life possible.*

Marshall, L.G., S.D. Webb, J.J. Sepkoski, Jr., and D.M. Raup. 1982. Mammalian evolution and the great American interchange. Science **215**:1351-1357. *North American mammalian immigrants to South America over the Panama land bridge were remarkably successful. The reasons are explored.*

Pielou, E.C. 1979. Biogeography. New York, John Wiley & Sons, Inc. *A landmark publication in the field with a superb analysis of contemporary thinking in biogeography. Highly recommended.*

Revelle, R. 1982. Carbon dioxide and world climate. Sci. Am. **247**:35-43 (Aug.). *Atmospheric carbon dioxide levels are increasing. What will be the effect?*

Stehli, F.G., and S.D. Webb (eds.). 1985. The Great American Biotic Interchange: Topics in geobiology, vol. 4. New York, Plenum Press. *Contributed chapters on the faunal interchange between North and South America. Advanced.*

CHAPTER 41

ANIMAL ECOLOGY

Life is confined largely to the interfaces between land, air, and water. It does not penetrate very far below the land surface, very high into the atmosphere, or very abundantly into the depths of the ocean. In the preceding chapter we defined where life does occur in the biosphere. Living things within the biosphere may be examined at several different levels of organization.

The most inclusive level of organization below the biosphere is the **ecosystem** (Figure 41-1). An ecosystem is a complex, self-sustaining natural system of which living organisms are a part, together with the nonliving components. All the interactions that bind the living (**biotic**) and nonliving (**abiotic**) components together are included in the ecosystem. A particular ecosystem is usually defined by the investigator studying it, and it may be quite large, such as a grassland, a forest, a lake, or even an ocean, or it may be more restricted, such as a river bank or treehole. Whatever the size or the biological structures it contains, certain characteristics of any ecosystem can usually be described. The sun's energy is fixed by plants and then transferred to consumers and decomposers. Nutrients are cycled and recycled through the various living components of the ecosystem. No ecosystem is ever completely closed, however. There is always some flow of resources and organisms into and out of an ecosystem.

The next level of organization is the **community**, an assemblage of living organisms sharing the same environment and having a certain distinctive unity (Figure 41-1). Communities comprise the living elements of an ecosystem. Like ecosystems, communities may be large or small, ranging from the coniferous forest community that may span a continent to the inhabitants of a rotting log community or the community of microorganisms living in a human's large intestine. The elements of a community are closely interdependent.

The **population**, the next lower level of organization, is an interbreeding group of organisms of the *same species* sharing a particular space. Every community is composed of several populations, including those of plants, animals, and microorganisms. Energy and nutrients flow through a population. Its size is regulated by its relationships to other populations in the community and by the abiotic characteristics of the ecosystem in which it is found. Because members of a population are interbreeding, they share a **gene pool** and thus are a distinct genetic unit.

The lowest level of organization within the biosphere, in ecological terms, is the **organism** itself. It is the living expression of the species. Each organism responds to its environment. To understand why animals are distributed as they are, ecologists must examine the varied mechanisms that animals use to compensate for environmental stresses and alterations.

Ecologists have, in fact, become increasingly interested in the physiological

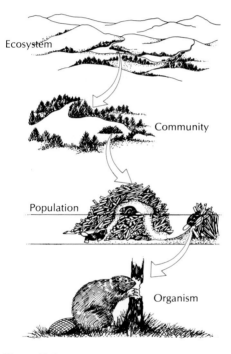

Figure 41-1

Relationships between ecosystem, community, population, and organism.

and behavioral mechanisms of animals. Both are intrinsic to the animal-habitat interrelationship. For example, the success of certain warm-blooded species under extreme temperature conditions such as in the Arctic or in a desert depends on near-perfect balance between heat production and heat loss and between appropriate insulation and special heat exchangers. Other species succeed in these situations by escaping the most extreme conditions by migration, hibernation, or torpidity. Insects, fishes, and other poikilotherms (animals having variable body temperature) compensate for temperature change by altering biochemical and cellular processes involving enzymes, lipid organization, and the neuroendocrine system. Thus, the physiological capacities with which the animal is endowed permit it to live under changing and often adverse environmental conditions. Physiological studies are necessary to answer the "how" questions of ecology.

The animal's behavioral responses are also part of the animal-habitat interaction and of interest to the ecologist. Behavior is involved in obtaining food, finding shelter, escaping enemies and unfavorable environments, finding a mate, courting, and caring for the young. Genetically determined mechanisms such as behavioral repertoires are acted on by natural selection. Those that improve adaptability to the environment assist in survival and the evolution of the species.

The term **ecology,** coined in the last century by the German zoologist Ernst Haeckel, is derived from the Greek *oikos,* meaning "house" or "place to live." Haeckel called ecology the "relation of the animal to its organic as well as inorganic environment." Although we no longer restrict ecology to animals alone, Haeckel's definition is still basically sound.

The term **environment** is often used in reference to the organism's immediate surroundings, but it is not always clear whether it is meant to include the living, as well as the nonliving, surroundings. To the ecologist it certainly includes both. Ultimately the environment consists of everything in the universe external to the organism. Since ecology is really biology of the environment, it obviously is a broad and far-flung field of study.

The ecologist may choose to focus on any level of organization within the biosphere. **Ecosystem analysis** is largely interdisciplinary, incorporating physics, chemistry, and other sciences to assist in the comprehension of the role of the environment in determining the distribution and abundance of organisms. **Community ecology** is similar but of more restricted scope; it is possible to focus on the interactions of a few species and to study energy transfers in detail. **Population biology** stresses genetics, evolution, seasonal changes, and other phenomena affecting population dynamics. Some ecologists study the organism itself (**organismic biology**) to see how it responds to the environment, hour by hour, day by day; such studies have become physiological and behavioral. All contribute to ecological understanding. None stands alone.

In the last several years the word "ecology" has come into popular misuse as a synonym for environment, which often makes biologists wince. As people concerned about the environment, we can be environmentalists; a person engaged in the scientific study of the relationship of organisms and their environment is an ecologist. He or she is usually an environmentalist too, but environment is not the same as ecology.

ECOSYSTEM ECOLOGY

The abiotic component of an ecosystem can be characterized by its physical parameters such as temperature, moisture, light, and altitude and by its chemical features, which include various essential nutrients. These characteristics determine the basic nature of the ecosystem.

The biotic component—the populations of plants, animals, and microorganisms that form the communities of the ecosystem—may be categorized into producers, consumers, and decomposers. The **producers,** mostly green plants, are **autotrophs,** which use the energy of the sun to synthesize sugars from carbon dioxide by photosynthesis. This energy is made available to the **consumers** and **decomposers.** They are **heterotrophs,** which exploit the self-nourishing autotrophs by converting organic compounds of plants into compounds required for their own growth and activity. In this discussion we consider the flow of energy through the ecosystem, which involves the concepts of productivity and the food chain, trophic levels, biogeochemical cycling, and limiting factors within the environment.

Solar Radiation and Photosynthesis

Virtually all life depends on the energy of the sun. The sun releases energy produced by the nuclear transmutation of hydrogen to helium. Solar radiation is received at the earth's surface in wavelengths of approximately 280 to 13,500 nm. Ultraviolet radiation with wavelengths of less than 280 nm is cut off sharply by the ozone layer in the upper atmosphere (Figure 41-2). Radiation with wavelengths greater than 760 nm is long-wave infrared radiation that heats the atmosphere, warms the earth, and produces currents of air and water. The most important part of solar radiation is the wavelengths between 310 and 760 nm; this is the portion that we call **visible light** because of its effect on the human retina. It is also the range that drives all important photobiological processes, including photosynthesis, photochemical effects, phototropism (orientation of plants toward light), and animal vision.

The flow of energy through the ecosystem begins with **photosynthesis.** Light striking a green plant is absorbed by chlorophyll, sending low-energy electrons into a higher energy level. These excited electrons drop back to a ground state in approximately 10^{-7} seconds, but in this brief interval their energy is channeled into a sequence of energy-yielding reactions. Part of the energy is used to synthesize ATP; the remainder causes the reduction of pyridine nucleotides (NADP). Both ATP and reduced NADP are then used to synthesize sugars from carbon

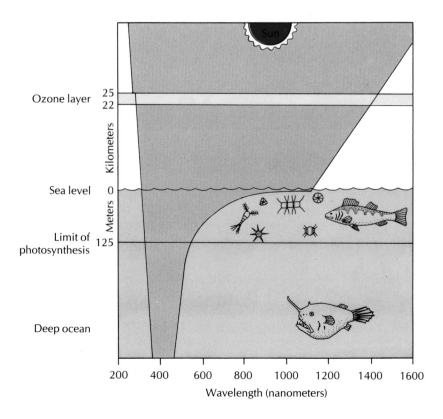

Figure 41-2

Narrowing of spectrum of sunlight by atmospheric absorption and by absorption in seawater.

dioxide and water. Photosynthesis in the individual leaf begins at low light intensity and increases at first linearly; that is, the rate of photosynthesis increases as light intensity increases until it reaches a maximum. The leaf achieves its highest rate of photosynthesis at only approximately one-tenth the intensity of full sunlight.

The amount of solar energy that reaches the earth's atmosphere is estimated at 15.3×10^8 gram calories per square meter each year (g cal/m^2/yr). Much of this energy is dissipated by dust particles or consumed in the evaporation of water. Only a very small fraction is used in the photosynthetic conversion of carbon dioxide to carbohydrates. Calculated on an annual or growing season basis, the photosynthetic efficiency of land area is approximately 0.3% and that of the ocean approximately 0.13%. These estimates are low because they are based on the total energy available for the year rather than on the growing season alone. During brief intervals of very active growth, plants may store a maximum of 19% of the available light energy.

Life without the sun

Until recently it was believed that all animals depended directly or indirectly on primary production from solar energy. However, in 1977 and 1979, dense communities of animals were discovered living on the sea floor adjacent to vents of hot water issuing from rifts (Galápagos Rift and East Pacific Rise) where tectonic plates on the sea floor are slowly spreading apart. These communities (see figure) included several species of molluscs, some crabs, polychaete worms, enteropneusts (acorn worms), and giant pogonophoran worms. The seawater above and immediately around vents is 7° to 23° C where it is heated by basaltic intrusions, whereas the surrounding normal seawater is 2° C.

It was discovered that the producers in the vent communities are chemoautotrophic bacteria that derive energy from the oxidation of the large amounts of hydrogen sulfide in the vent water and fix carbon dioxide into organic carbon. Some of the animals in the vent communities, for example, the bivalve molluscs, are filter feeders that ingest the bacteria. Others, such as the giant pogonophoran tubeworms (see p. 400), which lack mouths and digestive tracts, harbor colonies of symbiotic bacteria in their tissues and use the organic carbon that these bacteria synthesize.

Thus the deep-sea vent communities are self-contained, closed systems that depend entirely on energy issuing from the earth's interior. Each is a miniature ecosystem, altogether separate from other known ecosystems, all of which depend on solar energy and photosynthesis.

A

B

Animal communities near a Galápagos Rift thermal vent, photographed at 2800 m (about 9000 feet) from the deep submersible *Alvin*. **A,** "Spaghetti worms" (acorn worm of the phylum Hemichordata) lie in a tangled web across rocks. **B,** Giant pogonophoran worms, *Riftia pachyptila,* grow in dense profusion among rocks covered with mussels, limpets, polychaete worms, and crabs. An unknown fish swims by.

A, Photograph by J. Childress, University of California, Santa Barbara; B, photograph by J. Edmond, Massachusetts Institute of Technology.

Production and the Food Chain

The energy accumulated by plants in photosynthesis is called **production.** Because it is the first step in the input of energy into the ecosystem, the *rate* of energy storage (that is, energy storage per unit of time) by plants is known as **primary productivity.** The total rate of energy storage, the **gross productivity,** is not entirely available for growth because plants also use energy for maintenance and reproduction. When this energy consumption, or plant **respiration,** is subtracted from the gross productivity, the **net primary productivity** remains. Plant growth results in the accumulation of plant **biomass.** Biomass is expressed as the *weight* of dry organic matter per unit of area and thus differs from productivity, which is the *rate* at which organic matter is formed by photosynthesis.

The level of productivity of different ecosystems depends on the availability of nutrients and the limitations of temperature levels and moisture availability. Highly productive ecosystems are intertidal communities, floodplain forests, swamps and marshes, estuaries, coral reefs, and certain crop ecosystems (for example, rice and sugar cane). In such systems net production can exceed 14,000 $g/m^2/yr$ of dry organic matter (intertidal kelp and sea palm communities). Less productive (1000 to 2000 $g/m^2/yr$) ecosystems are most temperate forests, most agricultural crops, lakes and streams, and grasslands. Least productive (70 to 200 $g/m^2/yr$) ecosystems are tundra and alpine regions, deserts, and the open ocean. Extreme desert, rock, and ice regions have virtually zero productivity.

The net productivity of plants is the energy that supports almost all the rest of life on earth. Plants are eaten by consumers, which are themselves consumed by other consumers, and so on in a series of steps called the **food chain.** Food chains are descriptions of the way energy flows through the ecosystem. A diagram of a food chain shows arrows leading from one species to another, meaning that the first species is food for the second. But the first may be food for several other organisms as well. Seldom, in fact, does one organism live exclusively on another. Although some food chains are simple and short, for example, the one in which the whale feeds mainly on plankton, it is more common for several food chains to be interwoven into a complex **food web** (Figure 41-3).

Figure 41-3

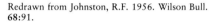

Midwinter food web in *Salicornia* salt marsh of San Francisco Bay area. Although they are not shown, parasites are also important members at all levels of the food web.
Redrawn from Johnston, R.F. 1956. Wilson Bull. **68:**91.

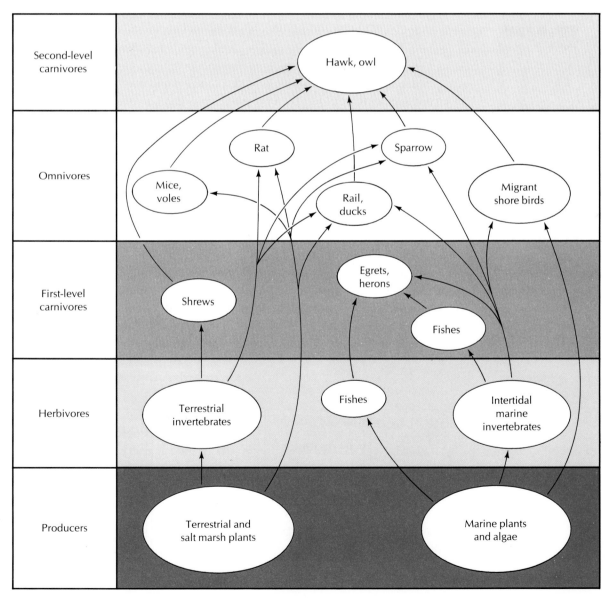

Despite their complexity, food chains tend to follow a pattern. Green plants, the base of the food chain, are eaten by grazing **herbivores,** which can convert the energy stored in plants into animal tissue. This is the base of the **grazing food chain.** Herbivores may be eaten by small **carnivores** and these by large carnivores. There may be two or three, sometimes even four, levels of carnivores. At the end of the chain are the **top carnivores,** which, lacking predators, die and decompose, replenishing the soil with nutrients for plants that start the chain. At each level, beginning with the plants, are parasites, which qualify as herbivores or carnivores depending on whether they are parasites of plants or animals. They usually deprive their hosts of only insignificant amounts of energy, but other detrimental effects may be significant. Sometimes the parasites have parasites.

There are numerous examples of food chains. In the forest, for instance, many small insects (primary consumers) feed on plants (producers). A smaller number of spiders and carnivorous insects (secondary consumers) prey on small insects; still fewer birds (tertiary consumers) live on the spiders and carnivorous insects; and finally one or two hawks (quaternary or top consumers) prey on the birds.

The decomposers have traditionally been considered the final step in the herbivore-carnivore food chain, since they reduce organic matter into nutrients that become available again to the producers. Ecologists now recognize that the decomposers comprise their own distinct **detritus food chain,** consisting of **detritus feeders,** such as earthworms, mites, millipedes, crabs, aquatic worms, and molluscs, and **microorganisms,** such as bacteria and fungi. Dead organic matter, such as fallen leaves or dead animals, is decomposed and used by fungi, bacteria, and protozoa. Detritus feeders then eat the microorganisms as well as much of the dead organic matter directly. They are in turn eaten by small carnivores; the detritus food chain thus leads up into grazing food chains and it is important in recycling nutrients into grazing food chains.

____ Trophic Levels

Each step in the food chain may be considered a **trophic level** (Gr. *trophe,* food). At each transfer within the food chain, 80% to 90% of the available energy is lost as heat. This limits the number of steps in the chain to four or five. Therefore the number of top consumers that can be supported by a given biomass of plants depends on the length of the chain.

Humans, who occupy a position at the end of the chain, may eat the grain that fixes the sun's energy; this very short chain represents an efficient use of the potential energy. Humans also may eat beef from animals that eat grass that fixes the sun's energy; the addition of a trophic level decreases the available energy by

"R.F.D.2."

Steve Stinson and Roanoke Times and World-News

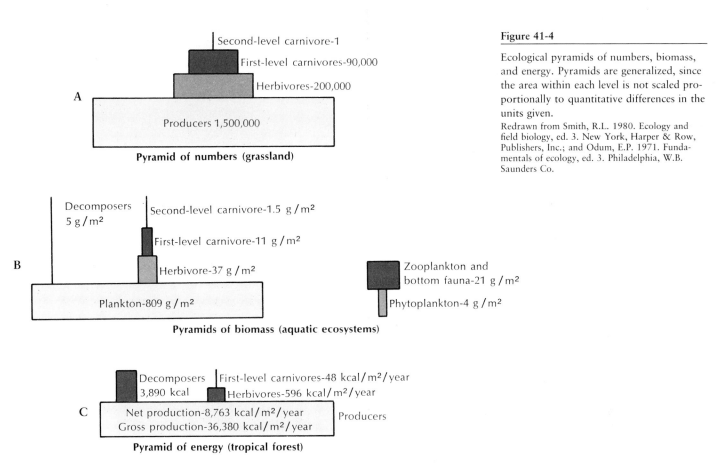

Figure 41-4

Ecological pyramids of numbers, biomass, and energy. Pyramids are generalized, since the area within each level is not scaled proportionally to quantitative differences in the units given.
Redrawn from Smith, R.L. 1980. Ecology and field biology, ed. 3. New York, Harper & Row, Publishers, Inc.; and Odum, E.P. 1971. Fundamentals of ecology, ed. 3. Philadelphia, W.B. Saunders Co.

an order of 10. In other words, it requires 10 times as much plant biomass to feed humans as meat eaters as to feed humans as grain eaters. Let us consider the person who eats the bass that eats the sunfish that eats the zooplankton that eats the phytoplankton that fixes the sun's energy. The 10-fold loss of energy occurring at each trophic level in this five-step chain explains why bass do not form a very large part of the human diet. In this food chain, for a person to gain a pound by eating bass, the pond must produce 5 tons of phytoplankton biomass.

These figures must be considered as we look to the sea for food. The productivity of the oceans is very low and limited largely to regions of upwelling where nutrients are brought up and made available to the phytoplankton producers and to estuaries, marshes, and reefs. Such areas occupy only a small part of the ocean. The rest is a watery desert.

Ocean fisheries supply 18% of the world's protein, but most of this is used to supplement livestock and poultry feed. If we remember the rule of 10-to-1 loss in energy with each transfer of material between trophic levels, then the use of fish as food for livestock rather than as food for humans is a poor use of a valuable resource in a protein-deficient world. Of the fishes that we do eat, the preference is for species such as flounder, tuna, and halibut, which are three or four steps up the food chain. Every 125 g of tuna requires 1 metric ton of phytoplankton food to produce. If humans are to derive greater benefit from the oceans as a food source in the future, we must eat more of the fishes that are at lower trophic levels.

When we examine the food chain in terms of biomass at each level, it is apparent that we can construct **ecological pyramids** of numbers or of biomass. A pyramid of numbers (Figure 41-4, *A*), also known as an **Eltonian pyramid** (after the British ecologist Charles Elton, who first devised the scheme), depicts the numbers of individual organisms that are transferred between each trophic level. Although providing a vivid impression of the great difference in numbers of organisms involved in each step of the chain, a pyramid of numbers does not indicate the actual weight of organisms at each level.

More instructive are pyramids of biomass (Figure 41-4, *B*), which depict the total bulk of organisms. Such pyramids usually slope upward because mass and energy are lost at each transfer. However, in some aquatic ecosystems in which the producers are the algae, which have short life spans and rapid turnover rates, the pyramid is inverted. This happens because the algae can tolerate heavy exploitation by the zooplankton consumers. Therefore the base of the pyramid is smaller than the biomass it supports.

A third type of pyramid is the pyramid of energy, which shows rate of energy flow between levels (Figure 41-4, *C*). An energy pyramid is never inverted because less energy is transferred from each level than was put into it. A pyramid of energy gives the best overall picture of community structure because each level reveals its true importance in the community regardless of its biomass.

___ Nutrient Cycles

All of the elements essential for life are derived from the environment, where they are present in the air, soil, rocks, and water. When plants and animals die and their bodies decay, or when organic substances are burned or oxidized, the elements and inorganic compounds essential for life processes (nutrients) are released and returned to the environment. Decomposers fulfill an essential role in this process by feeding on plant and animal remains and on fecal material. The result is that nutrients flow in a perpetual cycle between the biotic and abiotic components of the ecosystem.

Nutrient cycles are often called **biogeochemical cycles** because they involve exchanges between living organisms (bio-) and the rocks, air, and water of the earth's crust (geo-). Geochemistry is the discipline that deals with the chemical composition of the earth and the exchange of elements therein.

Nutrient and energy cycles are closely interrelated, since both influence the

Figure 41-5

Nutrient cycles and energy flow in a terrestrial ecosystem. Note that nutrients are recycled, whereas energy flow *(red)* is one way.

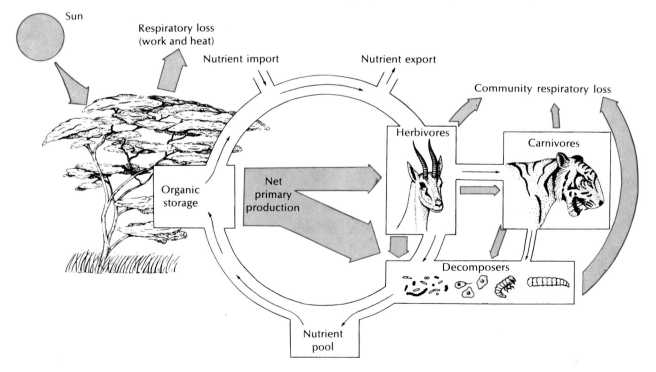

abundance of organisms in an ecosystem. However, unlike nutrients, which recirculate, energy flow follows one direction; it does not follow a cycle because it is lost as heat as it is used. The continuous input of energy from the sun keeps nutrients flowing and the ecosystem functioning. This interrelationship is depicted in Figure 41-5. Among the most important biogeochemical cycles are those of carbon and nitrogen.

Carbon cycle

Carbon is the basic constituent of organic compounds and living tissue and is required by plants for the fixation of energy by photosynthesis. It is hardly necessary to emphasize the total dependence of life on the availability of this element. Carbon circulates between carbon dioxide (CO_2) gas in the atmosphere and living organisms through assimilation and respiration; it is also withdrawn into long-term reserves of fossil fuel deposits (humus and peat and finally coal and oil) (Figure 41-6).

The cycling of carbon parallels and is linked to the flow of energy that begins with the fixation of energy during photosynthetic production. Plants synthesize glucose, a 6-carbon compound, from CO_2 that is withdrawn from the atmosphere; they then use this sugar to build higher carbohydrates, especially cellulose, the main structural carbohydrate of plants. Plants require 1.6 kg of CO_2 from the atmosphere for each kilogram of cellulose produced. The concentration of CO_2 in the atmosphere today is only 0.03%, in contrast to the relative abundance of atmospheric oxygen, 21% of air.

Two aspects of the low concentration of CO_2 require emphasis. The first is that CO_2 availability limits energy fixation by plants. Physiologists have shown that, if the atmospheric CO_2 is increased by 10%, plant photosynthesis increases 5% to 8%.

The second point is that there has been a tremendous increase in the consumption of fossil fuels by humans in the last two decades. More than 10 billion tons of CO_2 enters the atmosphere each year from industrial and agricultural activities; of this 75% comes from burning fossil fuels and 25% from the release of CO_2 from soil because of frequent plowing (a surprising and unappreciated effect of agricultural practice). Some authorities argue that the input of CO_2 into the atmosphere by human activities is not really chemical pollution, since the increase is removed by photosynthesis or by precipitation as carbonates in the ocean. There is concern, however, that even small increases in CO_2, which traps radiated heat, may increase the temperature of the earth's biosphere (p. 882).

Nitrogen cycle

Like carbon, nitrogen is a basic and essential constituent of living material, particularly in proteins and nucleic acids. Despite the high concentration of nitrogen in the atmosphere (78% of air), it is almost totally unavailable to life forms in its gaseous state (N_2). The nitrogen cycle converts N_2 into a chemical form that living organisms can use. This conversion is called **nitrogen fixation.** Some atmospheric N_2 is fixed by lightning, which produces ammonia and nitrates that are carried to earth by rain and snow. But at least 10 times as much N_2 is biologically fixed by bacteria and cyanobacteria (Figure 41-7).

Most important in terrestrial systems are bacteria associated with legumes (members of the pea family), whose nitrogenous contribution to soil enrichment is well known. The old agricultural practice of allowing fields to lie fallow (without crops) for a season every few years enabled N_2 fixers to replenish the element in a

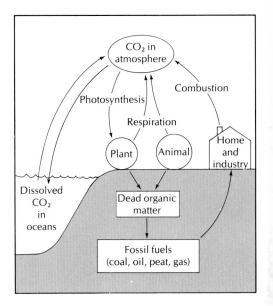

Figure 41-6

Carbon cycle showing circulation of carbon as CO_2 gas between living components of environment and long-term storage as fossil fuels.

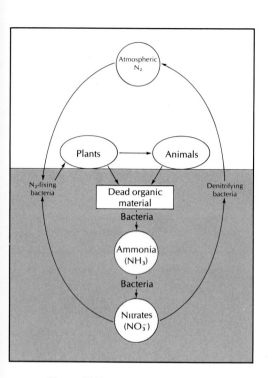

Figure 41-7

Nitrogen cycle showing circulation of nitrogen between organisms and through environment *(red)*.

usable form. Aerobic rhizobia bacteria produce nodules on the roots of legumes in which molecular N_2 is converted to ammonia and nitrates, which plants can use to build protein. Plant proteins are transferred to consumers, which build their own proteins from the amino acids supplied. Plants' and animals' waste products (urea and excreta) and their ultimate decomposition provide for the return of organic nitrogen to the substrate. Then decomposers break down proteins, freeing ammonia, which is used by bacteria in a series of steps as shown in Figure 41-7. Finally nitrate (NO_3^-) is produced, which may be used once again by plants or may be degraded to inorganic nitrogen by denitrifying bacteria and returned to the atmosphere. Nitrates also are carried by runoff to streams and lakes and eventually to the sea. A cycle similar to this terrestrial cycle occurs in aquatic ecosystems except that there is a steady loss of nitrogen to deep-sea sediments.

The nitrogen cycle is a near-perfect, self-regulating cycle in which losses in one phase are balanced by gains in another. There is little absolute change in nitrogen in the biosphere as a whole. However, human activities have caused steady losses of soil nitrogen by slowing natural addition of organic nitrogen through current agricultural practice and by the harvesting of timber. The latter causes an especially heavy outflow, from both timber removal and soil disturbance.

COMMUNITIES

Communities represent the most tangible concept in ecology. We are all familiar with the differences between forest, grassland, desert, and salt marsh, and we have little trouble picturing, at least in a general way, the kinds of plant and animal communities associated with each of these ecosystems. Communities comprise the *biotic* portion of the ecosystem; each consists of a certain combination of species that forms a functional unit. Although communities are sometimes difficult to define because the assemblages of species within similar communities are not always the same, communities do exist, and they all possess a number of attributes. It is beyond the scope of this book to discuss all the aspects of community form and function. In this discussion we examine two important principles that operate in community organization.

Ecological Dominance

Biological communities are typically dominated by a single species or limited group of species that greatly influences the nature of the local environment. All other species in that community must adapt to conditions created by the dominants. Communities often are named for the dominant organisms, for example, black spruce forest, beech-maple forest, oyster community, and coral reef. It is not always easy to specify just what constitutes a dominant organism. It may be the most numerous, the largest, or the most productive, or it may in some other manner exert the greatest influence on the rest of the community. In a woodland a tree obviously means more to the community than a poison ivy plant (even though a casual visitor may carry away a more lasting impression of the poison ivy).

Dominant species are often so because they occupy space that might otherwise be occupied by other species. When a dominant species is eliminated for some reason, the community changes. The American chestnut once dominated large regions of the eastern United States, where it made up more than 40% of the overstory trees in climax deciduous forests. After the invasion in 1900 of the chestnut blight (a fungus), which was apparently brought into New York City on

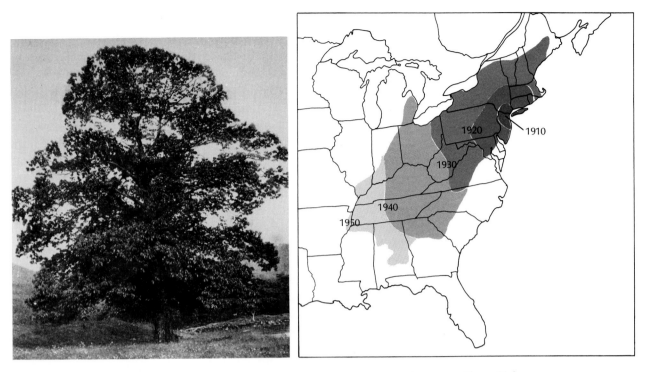

Figure 41-8

American chestnut tree *Castanea dentata* (*left*) and map (*right*) showing the spread of chestnut blight following its accidental introduction into New York City in 1900. Map courtesy Agricultural Research Service.

nursery stock from Asia, the chestnut was eliminated from its entire range within a few years (Figure 41-8). The chestnut position was filled by oak, hickory, beech, and red maple, and chestnut-oak forests became oak and oak-hickory forests. Tree squirrel populations that relied on chestnuts as their major food decreased to a small fraction of their former abundance.

The appearance of the forests changed. To the subdominant species, whose specialized environmental requirements were attuned to the particular conditions of a chestnut forest, the change required adjustments that some could not meet. The shift in dominance produced changes throughout the community.

Succession

The removal of the American chestnut and changes in community structure that followed are an example of succession. **Succession** is a change or series of changes in species and structure of a community leading finally to a relatively stable, self-maintaining **climax community.** If the succession occurs in previously unoccupied habitats, such as bare rock or sand, it is known as **primary succession.** Whereas, when the succession occurs because of a disturbance, either natural or by humans, it is **secondary succession.**

A familiar example of secondary succession occurs in old farm fields. The first plants to colonize the field after cultivation ceases are quickly growing annual plants. As years go by, more slowly growing perennial grasses and other plants succeed the annuals, then finally woody plants and trees. Ultimately the climax community is reestablished. Such successions are especially noticeable in the northeast United States on farms abandoned after the Midwest was opened to agriculture.

Ecological Niche Concept

An animal's position in the environment is characterized by more than the habitat in which we expect to find it; it also has a "profession," a special role in life that distinguishes it from all other species. This is its **niche,** which Elton defined as the animal's place in the biotic environment, that is, what it does and its relation to its food and its enemies. The niche concept has now been broadened to include an animal's position in time and space as well. Moreover, an animal's niche is defined by and is a property of its community. Two populations of the same species in different communities will have different niches because they are interacting with different assemblages of other animals in the two communities.

It is an accepted rule that no two species can occupy the same niche at the same time and place (principle of competitive exclusion; see Principle 28, p. 15). If they did, they would be in direct competition for exactly the same food and space. Should this happen, one species would have to diverge into a different niche, move to a different habitat, or face extinction.

Related populations that live in the same geographical area are called **sympatric** (meaning literally "same country") species, as opposed to **allopatric** ("different country") species that occupy different locations. Sympatric species might be expected to have similar niches and be in danger of direct competition. This can be avoided in a number of ways, as the following examples illustrate.

In parts of eastern and southern Africa the two species of rhinoceros, the black and the white, are sympatric: they share the same habitat. However, the black rhinoceros is a browser, feeding on leaves and woody plants, whereas the white rhinoceros is a grazer, eating grasses and herbs. They do not compete for the same food, and consequently they occupy distinct ecological niches.

The three species of boobies of the Galápagos Islands—the white, blue-footed, and red-footed—offer a second example. All three are plunge divers that feed on the same kinds of marine fish frequenting the ocean around the islands. The blue-footed booby, however, always fishes close to the shore; the white booby flies a mile or two from land to fish; and the red-footed booby makes long hunting forays many miles from shore. Again, although sympatric, they do not compete for the same food. The three species have divided up the food resources so that a different portion of the sea becomes the undisputed hunting territory of each species. This is an example of **niche diversification.**

The niche concept is a fundamental one to the biologist because it explains how animals avoid endless struggles with other animals. A species and its niche are reflections of the same thing: a unique way of life. A species is master of its own niche and thus is not in direct competition with similar species, even though competition was used to decide the boundaries of the niche in the first place. This concept is discussed in Chapter 39, p. 864.

POPULATIONS

A population is a group of organisms belonging to the same species that shares a particular space. Whether the population is gray squirrels in an eastern woods or bluegill sunfish in a farm fishpond, it bears a number of attributes unique to the group. A population shares a common gene pool; it has a certain density, birth rate, death rate, age ratio, and reproductive potential; and it grows and differentiates much like the individual organisms of which it is composed.

Population Interactions

Populations in a community may affect each other in various ways, although in certain cases there may be no apparent interaction (**neutral** effect). One interaction, **competition,** has already been mentioned. Although it is true that two populations cannot occupy the same niche at the same time and place, some degree of competition often occurs in natural populations. Because natural selection favors decreased competition, this often leads to niche diversification, such as the three species of boobies in the Galápagos Islands. Another adaptation may be that the two populations use the same resource but at different times of the year. One population may be excluded by a superior competitor from a habitat it could otherwise occupy. When two such populations occupy adjacent habitats, competition may be intense along the border between them.

Another interaction is the **predator-prey** relationship, in which individuals of one population kill and feed on individuals in the prey population. Often a population is preyed on by more than one species of predator, and predators commonly have more than one prey species. Predators are large relative to their prey, and they usually consume several to many prey animals in their lifetime. A conceptually related interaction is the **parasite-host** relationship. Both parasites and predators feed at the expense of their host-prey. However, parasites are small relative to their hosts, and parasites have only one host—at least at a given stage in their life cycle. In most cases, parasites do not kill their hosts because, if the host dies, the parasite perishes along with it. Most parasites are **symbiotic;** that is, they live in (**endoparasites**) or on (**ectoparasites**) the body of their host. Organisms such as mosquitoes and biting flies are intermediate between parasites and predators as we have defined them; they are sometimes referred to as **micropredators** or **intermittent parasites.**

Another type of interaction that is sometimes symbiotic is **commensalism,** in which the commensal enjoys some benefit from the relationship but does not harm the host. A classic example of commensalism is the association of pilot fishes and remoras with sharks. These fish get the "crumbs" left over when the host makes its kill, but we now know that some remoras feed on ectoparasites of the sharks. This is an example of commensalism grading into **mutualism,** an interaction in which both species benefit (Figure 41-9). An example of mutualism is that of the termite and the protozoa living in its gut. The protozoa can digest the wood eaten by the termite, although the termite cannot, and the termite lives on the waste products of the protozoan metabolism. In return, the protozoa gain a congenial place to live and a food supply. Mutualism has been recognized in recent years as much more prevalent in the animal kingdom than previously believed. Many cases of mutualistic interactions are symbiotic, but many are not. Some investigators prefer to distinguish **protocooperation** from mutualism. Protocooperation describes mutually beneficial interactions that are not physiologically necessary to the survival of the partners.

Population Growth

Living organisms have reproductive potential beyond that required for replacement of their numbers, some far in excess of such requirement. Some female insects lay thousands of eggs, field mice can produce as many as 17 litters of four to seven young each year, and a single female codfish may spawn 6 million eggs per season. It has been calculated that a bacterium dividing three times per hour would produce a colony a foot deep over the entire earth in 1½ days, and the colony would be over our heads 1 hour later. The growth potential of a population, the so-called biotic potential rate, never proceeds unchecked indefinitely because of the limitations of the environment. Most populations reach a certain level and then fluctuate about the level. What keeps populations in check? How is the "balance of nature" explained?

Most people are aware that lions, tigers, and wolves are predators, but the world of invertebrates also includes predaceous animals. These range from protozoa, jellyfish and their relatives, and various worms to predaceous insects, sea stars, and many others.

Figure 41-9

Mutualism (or protocooperation) between African Cape buffalo and cattle egret. Egrets benefit by feeding on parasites embedded in buffaloes' skin and on insects disturbed by buffaloes as they graze. Buffaloes in turn are rid of pests and warned of approaching danger when watchful egrets take to the air in noisy flight.
Photograph by L.L. Rue, III.

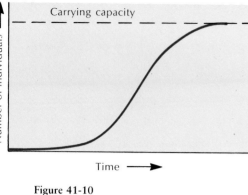

Number of individuals →

Carrying capacity

Time ⟶

Figure 41-10

Population growth curve.

If numbers of individuals are plotted on a graph against units of time, a growth curve of the population may be derived. The growth curve of a population with limited resources will be sigmoid in shape (Figure 41-10). In the beginning, with ample space and abundant food, the starting population breeds and grows as fast as its reproductive potential allows. As births exceed deaths, and more animals join the reproductive population, the increase is exponential. But, as the environmental resistance increases, commonly because of a depletion of food and space, the growth rate decreases as the upper population limit is approached. The simple sigmoid curve can be demonstrated easily with populations in the laboratory (bacteria, protozoa, and some insects), but the situation in nature is usually more complex. Annual plankton blooms in ponds and lakes, eruptions of insect pests, and growth of weeds in old fields may demonstrate the sigmoid population growth curve. The mathematical form of the growth curve, described by the **logistic equation,** is explained in the box below.

Exponential and logistic growth

The sigmoid growth curve (Figure 41-10) can be described by a simple model called the **logistic equation.** The slope at any point on the growth curve is the *growth rate,* that is, how rapidly the population size is changing with time. If **N** represents the number of organisms and **t** the time, we can, in the language of calculus, express growth as an *instantaneous rate:*

$$dN/dt = \text{the rate of change in the number of organisms per time at a particular instant}$$

When populations are growing in an environment of unlimited resources (unlimited food and space, and no competition from other organisms), growth is limited only by the inherent capacity of the population to reproduce itself. Under these ideal conditions growth is expressed by the symbol **r,** which is defined as the instantaneous rate of population growth per capita. The index **r** is actually the difference between the birth rate and death rate per individual in the population at any instant. The growth rate of the population as a whole is then:

$$\frac{dN}{dt} = rN$$

This expression describes the rapid, **exponential growth** illustrated by the early upward-curving portion of the sigmoid growth curve (Figure 41-10). But growth rate for populations in the real world slows as the upper population limit is approached, and eventually stops altogether. At this point **N** has reached its maximum density; that is, the space being studied has become "saturated" with animals. This limit is called the **carrying capacity** of the environment and is expressed by the symbol **K.** The sigmoid population growth curve can now be described by the **logistic equation** which is written as follows:

$$\frac{dN}{dt} = rN\left(\frac{K\text{-}N}{K}\right)$$

This equation states that the rate of population increase per unit of time (dN/dt) = rate of population growth per capita (**r**) × population size (**N**) × unutilized freedom for population growth ([**K-N**]/**K**). One can see from the equation that when the population approaches the carrying capacity, that is, when **K-N** approaches 0, dN/dt also approaches 0 and the curve will flatten. It is not uncommon for populations to overshoot the carrying capacity of the environment so that **N** exceeds **K.** When this happens, the population runs out of some resource (usually food or shelter). The rate of population growth term dN/dt then becomes negative and the population must decline.

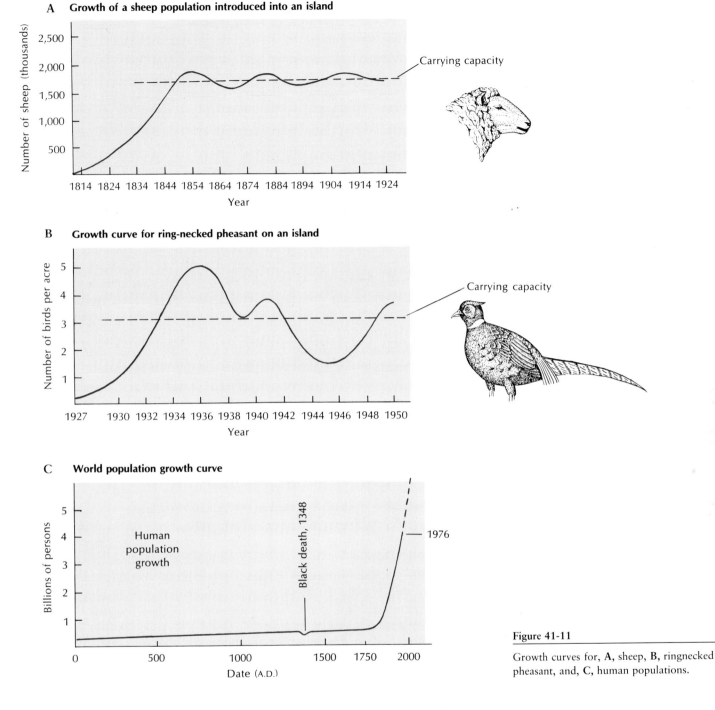

Figure 41-11

Growth curves for, **A**, sheep, **B**, ringnecked pheasant, and, **C**, human populations.

    When populations reach the carrying capacity of the environment, they may remain at this level for a time, or they may fluctuate above and below the upper limit. For example, when sheep were introduced on the island of Tasmania around 1800, their growth was represented by a sigmoid curve with a small overshoot, followed by mild oscillations around a final population size of 1,700,000 sheep (Figure 41-11, *A*). A similar pattern but with larger fluctuations was recorded for a population of ring-necked pheasants introduced on an island in Ontario, Canada (Figure 41-11, *B*). Some populations oscillate cyclically with large year-to-year changes in abundance. One of the best-known examples is the cyclic oscillation in the abundance of lynx and snowshoe hare in which peaks of abundance alternate with spectacular "crashes" (Figure 27-22, p. 582).

The growth of the human population was slow for a very long time. For most of their evolutionary history, humans were hunters and gatherers who depended on and were limited by the natural productivity of the environment. With the development of agriculture, the carrying capacity of the environment increased, and the population grew steadily from 5 million around 8000 BC, when agriculture was introduced, to 16 million around 4000 BC. Despite the toll taken by terrible famines, disease, and war, the population reached 500 million by 1650. With the coming of the Industrial Revolution in Europe and England in the eighteenth century, followed by a medical revolution, discovery of new lands for colonization, and better agriculture practices, the carrying capacity of the earth for humans increased dramatically. The population doubled to 1 billion around 1850. It doubled again to 2 billion by 1930, to 4 billion in 1976, and passed 5 billion in 1986. Thus, the growth has been exponential and remains high (Figure 41-11, C).

However, recent surveys provide hope that the world population growth is slackening. Between 1970 and 1987 the annual growth rate decreased from 1.9% to 1.7%. At 1.7%, it will take 41 years for the world population to double rather than 36 years at the higher annual growth rate figure. The decrease is credited to better family planning programs, but increasing deaths from starvation and disease also are contributing factors.

Unlike other animals, which can do little to increase the carrying capacity of their environment, humans have done so repeatedly. Unfortunately, when the carrying capacity is increased, humans respond as would any animal by increasing their population. Because the earth's resources are finite, the time will inevitably come when the carrying capacity can be extended no further. The question is whether humans will be able to anticipate this limit and check population growth in time or whether we will overshoot the resources and experience a crash. Indeed, for millions of Third World people time has already run out. The rapid population growth is not something that thinking people view with equanimity.

___ How Population Growth Is Curbed

What determines the number of animals in a natural population? For a laboratory culture of animals the answer seems fairly clear. They grow until they reach the carrying capacity of the environment, at which point growth is suppressed by competition for the limited resources of space and food. Thus, the forces that limit growth arise from within the population; that is, they are **density-dependent** mechanisms caused by crowding. Food and space limitations are commonly observed density-dependent factors, but others, such as disease and predation, may operate as well.

Transmission of infectious **disease** may be much higher under crowded conditions. Epidemics may sweep through crowded populations as did the terrible bubonic plague (black death) through the crowded, dirty cities of Europe in the fourteenth century. Parasitism also causes more mortality than would prevail at lower population densities. **Predation** increases when prey becomes abundant and easy to catch, and larger populations of prey are often followed by an increase in the predator populations. **Shelter** for suitable nest sites and as refuge from bad weather becomes less available with crowding.

The continuous competition for food and space to live brings forth yet another density-dependent force: **stress.** Although the physiology of stress is not thoroughly understood, there is ample evidence that, when certain natural populations become crowded, such as populations of lemmings, voles, and snowshoe

Table 41-1 Comparison of *K*- and *r*-Selected Animals

Strategy	*K*-Selected species	*r*-Selected species
Population	At carrying capacity	Below capacity
Environment	Constant or predictable	Unstable
Intraspecific competition	Keen	Lax
Development	Slow	Fast
Body size	Large	Small
Investment per offspring	Large	Small
Number of offspring	Few	Many
Niche	Specialists	Generalists

From Gould, J.L. 1982. Ethology. New York, W.W. Norton & Co.

hares, a neuroendocrine imbalance involving the pituitary and adrenal glands appears. Growth is suppressed. Reproduction fails, and individuals become irritable and aggressive. Under such conditions snowshoe hares suffer from a lethal "shock disease" that results in a population crash.

Overcrowding may also cause **emigration** away from the birth area. Overcrowded mice increase their locomotor activity and begin to explore new areas. In the case of lemmings, overpopulation produces the famous mass "marches" recorded at intervals in Scandinavia. When songbird populations become overcrowded, the less successful become "vagabond" birds that are forced out into less preferred habitats. Emigration is one major force resulting in the colonization of new habitats. The rapid dispersal of the European starling throughout the United States and southern Canada, following its introduction into New York City in 1890, is an excellent example (p. 557).

Not all of the forces that limit growth are density dependent. Extreme changes in the weather or unusually cold, hot, wet, or dry weather is a **density-independent** hazard of varying severity to animal populations. Local insect populations are sometimes pushed to the point of extinction by a severe winter. Hurricanes and volcanic eruptions may destroy entire populations. Hailstorms have been known to kill most of the young of wading-bird populations. Prairie grass and forest fires kill everything unable to escape.

Ecologists have distinguished between organisms whose populations are controlled more by density-dependent factors and those controlled by density-independent factors. Organisms with a range of adaptations to survive density-dependent control are called K-selected (from the K term in the logistic equation), and those with adaptations for density-independent control are r-selected (from the r term) (Table 41-1). In natural populations of various species, there is a continuum of highly r-selected to highly K-selected, and the terms are relative; that is, species A may be r-selected with respect to species B and K-selected compared with species C. Nevertheless, certain characteristics of each can be recognized. Organisms that are r-selected generally have very high reproductive rates, high (often catastrophic) mortality, short life, and population sizes variable in time, usually well below the carrying capacity of the environment, whereas K-selected organisms have the opposite characteristics. K-selected populations tend to be composed of larger, long-lived individuals able to compete in crowded circumstances (N close to K), not making a heavy commitment to reproduction because that would reduce the chance of individual survival. Populations that are

K-selected species are often known as K-strategists, and r-selected organisms as r-strategists. For many r-strategists the carrying capacity of the environment for the adults is rarely reached because the mortality of the progeny is so high. This is true for many marine organisms that produce enormous numbers of offspring. Many parasites are r-strategists; unless the hosts are crowded, transmission to a new host may be hazardous, and the reproductive capacity of the parasite is high to offset this loss. Examples of r-strategists among plants would include the annual weeds that quickly colonize disturbed ground, reproduce, and disperse their seed before perennial species can mature, shading and crowding out the annuals, thus altering and stabilizing the environment.

r-selected can rapidly exploit an unstable environment before better competitors arrive, producing large numbers of young that disperse quickly and increase the chances that at least a few will find another favorable place to grow. All such characteristics are genetically determined and are the products of relentless natural selection over the millenia.

Thus, population numbers are influenced by factors generated within the population and by forces from without, and usually no single mechanism can account fully for how growth is curbed in a given population. Natural populations are controlled by density-dependent and density-independent forces. Humans are the greatest force of all. By altering animal habitats we change the balances that set the old limits to animal numbers. Our activities can increase or exterminate whole populations of animals.

▬ SUMMARY

The most inclusive level of environmental organization below the biosphere is the ecosystem, which includes interacting biotic and abiotic components. The biotic components are communities of various species of organisms, each species being comprised of a population, and each population being the sum of the individual organisms that share a gene pool.

Energy flows into the ecosystem as sunlight. Nearly all life depends on the conversion of that energy into organic compounds by the fixation of carbon dioxide in green plants. The only animal communities known that are completely independent of solar energy are recently discovered communities associated with deep-sea thermal vents.

Food chains of producers (plants) and consumers (herbivores and carnivores) may be recognized, but in nature, complex food webs exist. Ecological pyramids can be constructed to show the decrease in numbers of individuals, biomass, and energy at each level of the food chain. Nutrients, such as carbon, nitrogen, and other elements, are cycled and recycled in the ecosystem. Each community is usually dominated by one or a few species that greatly influence the nature of the local environment. Each species has its own niche, or the role it plays in the ecosystem. Populations in the community may have no interaction (neutral), or they may interact as competitors, predators and prey, parasites and hosts, commensals, mutuals, or protocooperators.

Populations introduced into an environment where there is an upper limit to resources usually grow in a manner that can be described by a sigmoid curve; at first, when food and space are not limiting, the growth is exponential, and then the curve levels off as the carrying capacity of the environment is reached. The curve can be expressed mathematically as the logistic equation. This equation takes into account the intrinsic rate of increase of the population (r) and the carrying capacity of the environment (K). Species of organisms can be categorized as to their adaptations to survive their mechanisms of population growth regulation. Species whose populations approach K and are regulated by density-dependent factors are K-selected, and species that generally do not approach K and are regulated by density-independent factors are r-selected. Populations that are r-selected are opportunists, quickly moving into an unstable environment and producing large numbers of progeny before better competitors arrive. Density-dependent mechanisms of population control include competition, disease, and predation; density-independent mechanisms include extreme weather, abrupt environmental change, and hazardous conditions for the progeny with accompanying high mortality.

Review questions

1. What is the distinction between *ecology* and *environment?* Ask several nonbiologist friends how they would define *ecology.*
2. Discuss the distinction between ecosystem, community, and population.
3. What happens to solar energy reaching the earth?
4. Recently discovered animal communities surrounding deep-sea thermal vents are said to exist in total independence of solar energy. How can this be possible?
5. Define *production* as the word is used in ecology. What is the distinction between production, primary productivity, gross productivity, and net primary productivity.
6. What is a food chain? How does a food chain differ from a food web?
7. Distinguish between producers, herbivores, carnivores, and detritus feeders in a food chain.
8. What is a trophic level, and how does it relate to a food chain?
9. How is it possible to have an inverted pyramid of biomass in which the consumers have a greater biomass than the producers? Can you think of an example of an inverted pyramid of *numbers* in which there are, for example, more herbivores than plants on which they feed?
10. The pyramid of energy has been offered as an example of the second law of thermodynamics (p. 74). Why?
11. What pools of CO_2 are involved in the carbon cycle? What is the significance of the carbon cycle?
12. What biological and nonbiological processes are involved in nitrogen fixation?
13. Species dominance among plants has sometimes been called "space capture." Is this an acceptable description?
14. Define the niche concept. How is direct competition avoided in sympatric species?
15. Define predation. How does the predator-prey relationship differ from the parasite-host relationship?
16. What are some types of symbiotic interactions among animals? Is mutualism always a symbiotic relationship?
17. Contrast exponential and logistic growth of a population. Under what conditions might you expect a population to exhibit exponential growth? Why cannot exponential growth be perpetuated indefinitely?
18. Population growth may be hindered by either density-dependent or density-independent mechanisms. Define and contrast these two mechanisms. Offer examples of how the human population growth might be curbed by either agent.
19. What are the characteristics of a K-selected species? Of an r-selected species? Are humans K-strategists or r-strategists? (Refer to marginal note on p. 915.)

Selected references

Childress, J.J., H. Selbeck, and G. Somero. 1987. Symbiosis in the deep sea. Sci. Am. **256:**114-120 (May). *Describes the biology of the communities of life surrounding hydrothermal vents on the ocean floor.*

Colinvaux, P. 1986. Ecology. New York, John Wiley & Sons. *Comprehensive college textbook.*

Ehrlich, P.R. and J. Roughgarden. 1987. The science of ecology. New York, Macmillan Publishing Company. *Comprehensive, but with more emphasis on evolution and behavior ecology than is usually found in introductory treatments.*

Krebs, C.J. 1985. Ecology: the experimental analysis of distribution and abundance, ed. 3. New York, Harper & Row, Publishers. *Important treatment of population ecology.*

Lederer, R.J. 1984. Ecology and field biology. Menlo Park, Ca., The Benjamin/Cummings Publishing Co., Inc. *Nontechnical treatment that emphasizes organismic adaptations.*

Odum, E.P. 1983. Basic ecology. Philadelphia, Saunders College Publishing. *This the author's successful* Fundamentals of Ecology *trimmed and updated.*

Smith, R.L. 1986. Elements of ecology, ed. 2. New York, Harper & Row, Publishers. *Clearly written, well-illustrated general ecology text.*

CHAPTER 42

ANIMALS IN THE HUMAN ENVIRONMENT

In 1846, 23-year-old Francis Parkman, guided by a desire to study American Indians and haunted by images of a western wilderness yet unseen, began a journey westward on the Oregon Trail. Near the Platte River in what is now Nebraska, he described the prairie scene that unfolded before him while he rode back to camp after a furious buffalo chase.*

> The face of the country was dotted far and wide with countless hundreds of buffalo. They trooped along in files and columns, bulls, cows, and calves, on the green faces of the declivities in front. They scrambled away over the hills to the right and left; and far off, the pale swells in the extreme distance were dotted with innumerable specks. Sometimes I surprised shaggy old bulls grazing alone, or sleeping behind the ridges I ascended. They would leap up at my approach, stare stupidly at me through their tangled manes, and then gallop heavily away. The antelope were very numerous, and as they are always bold when in the neighborhood of buffalo, they would approach to look at me, gaze intently with their great round eyes, then suddenly leap aside, and stretch lightly away over the prairie, as swiftly as a race horse. Squalid, ruffian-like wolves sneaked through the hollows and sandy ravines. Several times I passed through villages of prairie dogs, who sat, each at the mouth of his burrow, holding his paws before him in a supplicating attitude, and yelping away most vehemently, whisking his little tail with every squeaking cry he uttered. Prairie dogs are not fastidious in their choice of companions; various long checkered snakes were sunning themselves in the midst of the village, and demure little gray owls, with a large white ring around each eye, were perched side by side with the rightful inhabitants. The prairie teemed with life.

When the first British settlements were founded in Virginia and Massachusetts, there could not have been fewer than 60 million bison, inaccurately called buffalo *(Bison bison)*, roaming the American wilderness. The two recognized subspecies, *B.b. bison*, the plains bison, and the somewhat larger and darker wood bison, *B.b. athabascae*, extended from northwestern Canada to Mexico and from the Rocky Mountains to the Atlantic seaboard from New York to Florida (Figure 42-1). On the great plains, where their domain was most secure, the bison made enormous annual migrations in vast herds numbering millions of animals. In the spring they moved northward to graze on new grass that had sprung up behind the retreating snow; in autumn they returned southward. Late eighteenth- and early nineteenth-century observers recorded moving herds of bison covering areas of several square miles so densely that nowhere was the ground visible. Because they were nomadic, bison did not overgraze the land despite their incredible numbers. There was also ample herbage to feed the numerous antelopes and prairie dogs

---

*Parkman, F. 1931. The Oregon Trail. New York, Farrar & Rinehart.

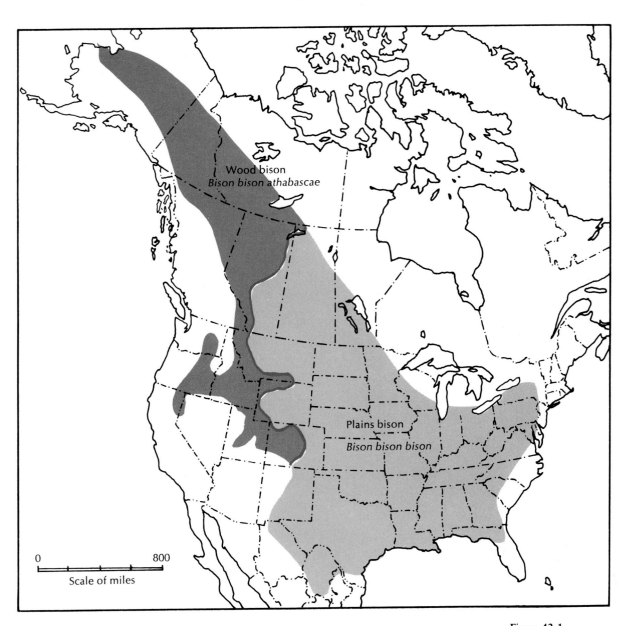

Wood bison
Bison bison athabascae

Plains bison

Bison bison bison

0   800
Scale of miles

Figure 42-1

Distribution of American bison before the arrival of Europeans in North America. From Hall, E.R., and K.R. Kelson. 1959. The mammals of North America. New York, The Ronald Press Co.

observed by Parkman. The bison population easily sustained losses inflicted from predation by their natural enemies—the wolves, grizzly bears, and pumas—and even the hunting pressure by Indians had little effect on their numbers.

To the Plains Indians the bison were providers of virtually every necessity of life. The high-quality meat was eaten fresh or was dried, pounded, and stored as pemmican for future use. The skins provided homes, clothing, moccasins, and warm robes. The rawhide was turned into saddles, ropes, and quivers, and the sinews became bowstrings. Spoons and other utensils were carved out of the horns; hooves were reduced to glue; and the ribs became scrapers and sled runners.

The bison culture of the Plains Indians reached full flower in the seventeenth and eighteenth centuries after the Indians acquired horses, which had been introduced into North America by the Spaniards at the end of the sixteenth century. With their new mobility, hunting became easier, and the tribes could follow the bison migrations, carrying all their belongings with them. When rifles were introduced to the Indians by white traders along the Missouri, Platte, and Arkansas

Figure 42-2

The Still Hunt by James Henry Moser. This painting was commissioned for the Smithsonian Institution and was first displayed at the 1888 Cincinnati Exhibition as part of a crusade to save the bison from extinction. Although vast herds like this no longer existed in 1888—the bison was already an endangered species—many who had crossed the prairies in the first half of the nineteenth century attested to the painting's authenticity. Market hunters such as the one portrayed commonly shot more than 100 bison from a single position before moving on. Courtesy Mammoth Museum, Yellowstone National Park.

rivers, hunting pressure increased substantially, but even this made few inroads on the total bison population because the Indian population was low.

In eastern America the rapid westward advance of pioneers was accompanied by decimation of the bison. By 1810 no bison remained east of the Mississippi. Their extinction clearly portended the ultimate fate of the plains bison. In 1846, when Parkman traveled the Oregon Trail, the bison were restricted to the western plains and mountains, although there were still probably 30 million left. The final act in the great bison slaughter began in the mid-nineteenth century. Congress in 1864 authorized the building of the Union Pacific Railroad across the heart of the continent, directly through the bison range. Thousands of hunters moved over the plains, killing bison to feed the huge construction gangs (Figure 42-2). As the railroad penetrated westward, "buffalo" robes and tongues, the latter considered a delicacy in the East, were collected and shipped east by rail. The carcasses were left to rot on the ground. Hunting pressure reached its zenith in the early 1870s when some 4 million bison were killed each year.

Contributing to the slaughter were thousands of soldiers instructed to kill bison in order to starve out the Plains Indians. By 1883, the annihilation was virtually complete, except for a single herd of about 10,000 animals in North Dakota. In September of that year, hunters were sent to North Dakota, and by guarding all available water holes, they destroyed the last free herd of bison (Figure 42-3).

Those who protested the destruction were ignored by politicians, who believed it to be a justifiable means to deprive the Plains Indians of a way of life.

Fortunately, small groups of semidomesticated bison still existed on a few private estates. A census taken in 1889 estimated the total remnant population throughout the United States and Canada at 541 animals. Following the senseless North Dakota slaughter of 1883, strong voices of protest from a few influential individuals aroused public interest in the bison's plight. In Yellowstone National Park, a herd of 20 bison that was constantly threatened by poachers was given full protection in 1890. Some farmers succeeded in building up small breeding herds from remnant groups they had been protecting.

By the turn of the century the bison population was beginning to recover. The American Bison Society, formed in 1905, received both federal government backing and powerful private support in its mission to reestablish the American bison as a viable species secure from extinction. The National Bison Range was established in Montana in 1908, and other reserves in both the United States and Canada followed. Shortly after the turn of the century a large herd of wood bison was discovered near Lake Athabasca in western Canada. The 17,300-square-mile Wood Buffalo National Park in Alberta and Northwest Territories was established in 1922. By 1967, there were 23,000 bison in North America, including plains bison, wood bison, hybrids of the two subspecies, and zoo captives. Most exist today as wards of humans.

The destruction of the American bison was a tragic episode in American history that becomes only slightly more comprehensible when viewed in the context of the exploitive spirit of the times. Within a few decades 60 million large animals of a single species were destroyed, a holocaust of unprecedented magnitude. The herds were vast, probably the largest aggregation of land animals on earth, and few people could foresee their fate. The same mentality resulted in the annihilation of the passenger pigeon. Fortunately, unlike the passenger pigeon, the bison were able to flourish in captivity and could recover from remnant populations. It was inevitable that the bison would disappear from the plains, since roaming herds of half-ton animals were incompatible with a West given over to agriculture. We cannot deny that the human species must exploit nature to live, but the destruction of the bison was cataclysmic in scope, senseless in direction, and erosive of human values. Their fate is sobering commentary on the turbulent relationship between wildlife and advancing human civilization.

It is of interest to compare the great prairies that "teemed with life" when Francis Parkman saw them in 1846 with their appearance at the turn of the cen-

Figure 42-3

Huge stacks of skulls and other bones representing approximately 25,000 bison await shipment beside a Saskatchewan railway siding in 1890. For a brief period following the bison slaughter, bones collected from the prairies were shipped east to be ground into fertilizer.

From Hewitt, C.G. 1921. The conservation of wildlife of Canada. New York, The Scribner Book Companies, Inc.

tury just a few years after the arrival of homesteaders. Bison were replaced by longhorn cattle and sheep, and the ranges were overstocked. Sheep, with their close-cropping teeth, were especially destructive; within a few years the highly palatable prairie grasses, legumes, and forbs disappeared. These species were replaced by wheat grasses, bromegrasses, and annual weeds—rapidly converting the prairie into low-grade pasture. With continued grazing, mesquite appeared, and the topsoil thinned. Boomtime "sodbusting" for growing wheat accelerated soil deterioration. No longer bound by grass roots, the soil was lifted by wind and washed away by rain. The American naturalist and conservationist William Temple Hornaby wrote in 1908 of the Montana prairie, where "the awful sheep herds have gone over it, like swarms of locusts, and now the earth looks scalped and bald and lifeless. Today it is almost as bare of cattle as of buffaloes, and it will be years in recovering from the fatal passage of sheep."

Conditions were ideal for the creation of the Dust Bowl, requiring only an extended drought, which occurred in the 1930s. From Texas to Canada the "black blizzards" swept across an exhausted land, joining economic depression in bringing misery to prairie farmers (Figure 42-4). A single storm on May 12, 1934, lifted an estimated 300 million tons of soil from Texas, Oklahoma, Colorado, and Kansas, and deposited it eastward in amounts up to 100 tons per square mile.

Today the American prairie is a changed land. Soil conservation practices introduced in the late 1930s eventually transformed the prairies into the most productive agricultural region of the world, dominated by a monoculture of cereal grains (Figure 42-5). These grains comprise more than one half of the world's grain exports and are the single most important export factor in the U.S. balance of trade. Except for pests, few animals exist in the uniformly cultivated fields, but certain areas are still important refuges for wildlife. One of these is the broad Platte River valley in Nebraska, through which Francis Parkman traveled in 1846. In south-central Nebraska, where the Platte River curves eastward through an

Figure 42-4

"Black blizzards" of the 1930s moved millions of tons of soil off the prairies, darkening the skies to the Atlantic seaboard. Thousands of "dusted out" families streamed westward to California, a migration later portrayed in John Steinbeck's novel *The Grapes of Wrath*. This photograph was taken in 1936 in Gregory County, South Dakota.
Courtesy U.S. Department of Agriculture, Soil Conservation Service, Washington, D.C.

Figure 42-5

Harvesting wheat near Regina, Saskatchewan, in the North American grain belt. Sound soil conservation practices instituted after the Dust Bowl have transformed the prairies into the world's most productive cropland. Today, the average North American farmer, using irrigation, fertilizers, machines, and a considerable input of fossil fuel energy, produces enough to feed himself and 70 others.
Photograph by L.L. Rue, III.

enormous bordering marshy area, migratory birds are attracted in great numbers: hundreds of thousands of ducks, countless songbirds, and most of the mid-continent population of sandhill cranes and white-fronted geese. Over the years, however, the marshy wetlands have been gradually drained for cropland until only 20% of the original area remains for wildlife.

The Platte River itself is seriously threatened by an unremitting demand for water for irrigation and other human needs. The yearly flow today is only 16% of what it was before the Nebraska Territory became a state. If the additional water developments that are planned for the Platte River are all carried out, the river will be dry within 35 years. Attempts by the U.S. Fish and Wildlife Service to set aside a portion of the wetlands as a permanent refuge for wildlife have up to this time been defeated, primarily by suspicious landowners who fear that they will be displaced from their land. Meanwhile, cattle raisers, farmers, and industry vigorously compete for the remaining water, a dispute in which the wildlife has no voice.

Little remains of the vast climax grasslands that occupied the heart of North America only 150 years ago. No other ecosystem on this continent has been as thoroughly transformed from its natural state. It emphasizes our capacity as humans to reshape the face of nature and manipulate the environment to suit our own needs. This capacity is both the outcome of our biological success as a species and the origin of our ecological problems.

ENVIRONMENT AND HUMAN EVOLUTION
Humans as Hunters

Hunting is an ancient human activity. For most of the period that true humans (genus *Homo*) have walked the earth, their survival depended on the success of the hunt. Only in the last evolutionary instant of human existence on earth have humans of necessity replaced hunting with work. However, the ancient hunting urge, basically a male pursuit, is abundantly manifested today in big-game hunting, fox hunting, game-bird hunting, angling, and numerous other expressions of sport hunting that have survived cultural suppression.

Whereas humans today stand fully revealed as predators, we have our origins in an apelike lineage called the dryopithecines, which thrived on a diet of fruit rather than meat. Human evolution, treated briefly in Chapter 39, was a sequence of fundamental changes in anatomy, behavior, and intelligence from which humans emerged as hunters. Our early dryopithecine ancestors were creatures of the forest. However, climatic changes in the late Miocene and early Pliocene epochs (about 8 million years BP) produced major environmental transformations in North America and Asia. Tropical and subtropical forests yielded to temperate forests and great expanses of grasslands. The early hominids gradually left the dwindling forests altogether to become creatures of the open terrain. As they did so, they consumed more meat and fewer plant foods because plant foods were much scarcer on the savanna than in the forests. Manual dexterity and excellent eye-hand coordination, together with superior intelligence and upright posture, were advantages that enabled the hominid lineage to surpass the rest of the wildlife. They became increasingly dependent on animal food, ranging from insects, tortoises, lizards, and rodents to carrion abandoned by predators on the savanna.

We can speculate that carrion feeding and the first use of tools—sharpened pebbles used to tear away parts of hoofed animals killed by predators—appeared together. Weapons, too, doubtless were needed, since lions and hyenas could not be expected to relinquish their kills voluntarily. Hominids were now aggressors. Using tools and weapons of their own invention, they became hunters, fully equipped for a predatory existence.

Emergence of the family unit

These changes occurred slowly since some 10 million years separate the first true hominid from *Homo erectus,* the first indisputable big-game hunter. The beginnings of social organization, as well as the use of tools, appeared among the Pleistocene australopithecines that lived in open habitats 2 to 3 million years BP. A new social order, especially the family unit, developed to maximize food-foraging potentials. Division of labor between the sexes was required: men ranged widely to hunt game and collect kills abandoned by lions, while women cared for the young, gathered plant foods, and guarded the home site. Family coherence was strengthened by the length of human childhood and associated necessary parental care. The use of language, which presumably existed in at least rudimentary form some 3 million years ago, also reinforced the family bond and facilitated intergroup communication. With individuals united into families and families into tribes and clans, cooperative hunting behavior was employed, making early humans far more formidable adversaries of wildlife. The stage was readied for humans to use their intelligence to exploit a new way of life on the open savanna. Gradually, early hominids acquired the biblical dominion "over the fish of the sea, and over the fowl of the air, and over the cattle, and over all the earth, and over every creeping thing that creepeth upon the earth" (Genesis 1).

Impact of human hunting

There is no evidence that the small-brained australopithecines used any but the smallest stone tools. Nor had they discovered the use of fire. The large-brained *Homo erectus,* who shared much of the Pleistocene epoch with the australopithecines, took a much larger step toward environmental conquest. Wooden clubs

The effectiveness of early tools was demonstrated by the late Louis Leakey, who skinned and butchered an antelope in less than 20 minutes using a pebble tool found at Olduvai Gorge in Tanzania.

and spears entered the arsenal of hunting weapons, and there is evidence that bolas were employed about 400,000 years ago to capture antelopes. Even more significant *Homo erectus* learned the use of fire, perhaps as early as 800,000 years ago. With fire they warmed themselves, smoked animals from caves, lighted the darkness to ward off real and imagined terrors of the night, and cooked and tenderized their meat. There is also evidence that Pleistocene humans set fire to the vegetation to concentrate their prey, perhaps stampeding them over cliffs or into swamps where they would be trapped. They also must have discovered that the new growth that sprang up following grass and forest fires attracted wildlife and eased the task of finding it. Humans were thus beginning to modify the environment for their own needs. They were also conducting cooperative big-game drives that were having an increasing impact on wildlife.

Modern humans' immediate predecessors, the Neanderthals, which appeared about 130,000 years ago, left a relatively rich array of fossils, artifacts, and evidences of cultural activities. Already the world's most formidable predators, they depended entirely on their skill as hunters for their survival. The refinement of mass harvesting techniques coincided with a human population explosion that brought the total world population to perhaps 4 million about 20,000 years ago. Large mammal herds could not sustain their numbers against mass slaughter techniques employed by nomadic peoples who followed the herds wherever they went. Animal populations went into serious decline. Magic hunting rituals and polychrome cave art of primitive hunters are thought to be attempts to restore animals to their former abundance, as well as to achieve spiritual domination over the quarry.

For countless millennia humans had drawn all their sustenance from the animal kingdom. If the hunter outwitted the quarry, the family could eat; if he failed, they starved. Animals that shared the hunter's world occupied his thoughts and plans and were the center of all his activities. Although he killed the animals ruthlessly, he nevertheless respected them, felt a kinship with them, and believed himself closely related to them. But as successful hunters, humans were putting themselves out of the hunting business.

Transition from Hunting to Agriculture

Agriculture, the domestication and cultivation of plants, especially cereal crops, arose independently in the Near East and Central America as the hunter-gatherer economies began to fail. The first cultivated cereals, perhaps wheat and barley, were probably selected from natural hybrids of wild species. Sites along the Nile Valley in Egypt provide evidence that milled grain was used for food as early as 15,000 years ago. The domestication of animals soon followed. Sheep were domesticated in Iraq at least 11,000 years ago, and evidence of domesticated goats was discovered in 10,000-year-old Palestinian sites. Cattle (8000 years BP) and pigs (5000 to 6000 years BP) soon followed. Cattle became perhaps the earliest beasts of burden, used for pulling carts, turning waterwheels, and carrying people, as well as providing milk and meat.

The advent of agriculture was nothing less than a revolution. No subsequent human event—not the great plagues nor the Industrial Revolution nor the devastating wars of humankind—has had a greater permanent impact on the course of human history. The hunter-gatherer economy that had persisted for more than 2 million years was brought rapidly to a close, although certain hunting populations persisted, some to the present century (such as Australian aborigines, pygmies of the Congo rain forest, and North American Eskimos). A nomadic economy was replaced by more or less permanent villages and communities centered in agricul-

Mass slaughters of animals by early humans were not uncommon. At one site at the foot of a high rock cliff in southern France were discovered the bones of at least 40,000 wild horses in a deposit up to 245 cm (8 feet) thick. These were presumably stampeded over the cliff by Neanderthal hunters who set grass fires.

tural areas. Civilization, defined as a state of social culture that is organized around cities and that practices agriculture, was born. The most important result of the new agricultural technology was a vast increase in the carrying capacity of the human environment. A given area of agricultural land could support hundreds of times more people than it could during the early hunter economy. The human population began to soar. Agricultural production continued to diversify, and more and more land was turned to the production of crops and domesticated animals.

The human population, estimated to be 5 million at the beginning of the agricultural age 10,000 years ago, reached an estimated 40 million by 3500 BC and 100 million by 2000 BC. It probably stood at nearly 200 million at the time of the birth of Christ.

As limits to growth were reached, new lands were conquered and exploited. The human species now emerged as a powerful reshaper of the earth. Forests were cut or burned, irrigation systems were built, hillsides were terraced, and grasslands were converted to barren wastelands by overgrazing. Landscapes have suffered the greatest ravages where civilization has existed the longest. The vegetation of the Mediterranean region was largely destroyed 2500 years ago. Most of the pastureland of Spain, Italy, Greece, and Yugoslavia was severely damaged by erosion, which was caused by grazing of domestic animals, especially goats. Before the birth of Christ large sections of northern Africa and the Middle East were so heavily overcropped that they became deserts or wastelands, largely abandoned by both humans and animals. Great civilizations that once flourished in Mesopotamia (Iraq), Palestine, Greece, Italy, and China fell victim to exploitive overgrazing.

Despite the calamities that had befallen the Mediterranean area, until about 300 years ago most of the earth's biosphere was virtually unmarked by human activity. In the Americas, the forests, lakes, prairies, and rivers lay wild and natural and rich in native fauna and flora. Most of northern Europe and Asia was dominated by virgin forests, and the soil remained untouched. The environments of Australia, New Zealand, the East Indies, and the Pacific islands were yet to feel the violent impact of civilization's arrival. The seas were clean and abounded in marine life undisturbed by humans.

This was the scene that the Swedish zoologist-conservationist K. Curray-Lindahl calls "the last idyll." It was a heritage of incalculable richness for future generations. At that point, 300 years ago, history provided abundant warnings of the bitter harvest to be reaped from unbridled human exploitation of nature. Tragically, they were lessons unheeded. Human population was about to begin an unprecedented growth. Stimulated by the Industrial Revolution, the discovery of abundant fossil fuels for energy, the colonization of new lands, improved agricultural technology, and medical advances, the human population doubled between the years 1650 and 1850, reaching 1 billion people. It doubled again to 2 billion people by 1930; in 1976 it stood at 4 billion people. Now, with each passing day, the problem of feeding so many people intensifies. The population of the human species has reached plague proportions.

Today little of the earth's biosphere resembles its original pristine state. Numerous animal species have been destroyed by human activities in the last three centuries, and many others are threatened with extinction. According to the American biologist W.H. Murdy* (1975), the present ecological crisis is a "crisis in human evolution." We must, he argues, participate in our own evolution, using our knowledge to ensure our survival and the enhancement of human cultural values in an environment of which all living things are a part. We alone have the enormous responsibility of preserving that which is not yet destroyed by countering the destructive forces within us. Our options are rapidly diminishing.

---

*Murdy, W.H. 1975. Anthropocentrism: a modern version. Science, **187**:1168-1172.

WILDLIFE AND PEOPLE
Pitting Wildlife against Progress

In June 1978, the U.S. Supreme Court ruled that construction on the $120 million Tellico dam project nearing completion in Tennessee must be stopped to save from certain extinction an endangered species of fish, the snail darter. The snail darter, discovered only in 1973, requires shallow, fast-flowing water for spawning, a critical habitat that could be destroyed by the dam's reservoir.

The ruling did nothing to dissipate polarization of public opinion that had already developed over the dam issue. It clearly brought into focus the continuing confrontation between human beings and their environment—a conflict that has no easy resolution and from which no winners can emerge. It pits wildlife against "progress," intangible, long-term benefits of nature against the tangible, short-term benefits of public works projects, principles against practicality. That a small fish could quite literally bring to a halt an enormously expensive construction project astounded Congress, delighted preservationists, and left the average citizen incredulous.

The cause of the snail darter's sudden rise to fame was the Endangered Species Act, a surprisingly powerful piece of legislation passed by Congress in 1966 and expanded and reinforced in 1973. Congress declared in 1966 that "one of the unfortunate consequences of growth and development . . . has been the extermination of some native fish and wildlife . . . and serious losses in other species of native wild animals." It pledged through the act to conserve and protect native fish and wildlife threatened with extinction. The most significant section of the act prohibits federal agencies from jeopardizing endangered species or habitats that have been designated as critical. This legislation was a landmark in wildlife protection because it was the first time that a U.S. domestic law was not enacted mainly to promote human, rather than wildlife, interests. In the past, wildlife reserves or seasons closed to hunting, trapping, and fishing were usually established to perpetuate animals for sport, future market use, or some other nonaltruistic motive.

The snail darter's reprieve from extinction turned out to be short lived. The Endangered Species Act was amended by Congress in 1978 to create a committee empowered to grant exemptions to the act when the benefits of a project "clearly outweigh" the benefits of species preservation. The committee was soon dubbed the God Committee because its decisions could save a species or send it to extinction. It was becoming apparent that the Tellico dam project was a boondoggle and that even with the dam mostly completed, it would be profitable to abandon construction and leave the valley unflooded for agriculture and recreation. A new study showed that the dam's electricity would actually be produced at an economic loss. The God Committee then ruled in favor of the snail darter. But in 1979 the dam's proponents fashioned an amendment to a large water projects appropriation that exempted the Tellico dam from *all* laws that stood in its way. Congress, disgruntled with the whole matter, passed it without debate and it was signed into law in September 1979. Dam construction immediately resumed, and within a year the reservoir began to fill.

Extinction

The Endangered Species Act grew out of rapidly rising public concern for the destructive effects of civilization on wildlife. One measure of human impact is the extinction rate of animals. Since 1600, when reasonably accurate record keeping

The International Union for the Conservation of Nature and Natural Resources of Lausanne, Switzerland, periodically publishes a *Red Data Book* listing threatened wildlife. The recognized categories are (1) critically endangered species that are in immediate danger of extinction, such as the California condor, the whooping crane, and the blue whale; (2) vulnerable or threatened species that are still abundant but are depleted and are under threat, for example, the grizzly bear, the California bighorn, and the prairie falcon; and (3) rare species, which comprise small populations and, although not threatened, are subject to risk, for example, the Galápagos tortoise and the Komodo dragon. This and other such lists published by federal agencies of different nations are unfortunately replete with statistical confusion arising from differences of opinion as to when species or subspecies should be added or removed from the lists, and to which category each belongs.

began, approximately 225 species of vertebrate animals are known to have become extinct (the true extinction total is far higher but includes numerous invertebrates and tropical animals unknown to biologists). The rate of extinction increased significantly between 1600 and 1900: 21 extinctions in the seventeenth century, 36 in the eighteenth century, and 84 in the nineteenth century. Toward the end of the nineteenth century the number of extinctions increased alarmingly, and although conservation efforts instituted after 1920 stabilized the rate somewhat, 85 more species had disappeared forever by 1974. Furthermore, numerous animal species are today on the verge of extinction; these are the approximately 185 critically endangered species and subspecies recognized by the U.S. Fish and Wildlife Service and by the International Union for the Conservation of Nature and Natural Resources.

In North America the rapid transition from wilderness to a highly industrialized society has had an especially severe impact on native wildlife. Sixteen mammals, eight birds, and nine fishes are known to have vanished in the continental United States alone. If animals from Hawaii and Puerto Rico are included, the list rises to 62 birds, mammals, and fishes that have become extinct since 1600, most of them in the twentieth century. The victims include, among mammals, the great sea mink (last seen in 1860), Merriam's elk (1906), and six subspecies of North American wolf. Extinct birds include the great auk (1833), the Labrador duck (1875), the passenger pigeon (1914), the Carolina parakeet (1914), and the heath hen (1932).

But these figures provide no perception of the enormous loss of tropical species that is taking place now and will continue as the tropical forests rapidly disappear. It is conservatively estimated that the tropical forests contain at least 10 million animal species—mostly arthropods—of which less than 1 in 10 has been identified. We appear to be on the threshold of perhaps the greatest mass extinction the earth has known.

Extinction is, of course, a natural event in organic evolution, as we have discussed previously (especially in Chapter 39). Sometimes individual species disappear gradually as their populations wane in competition with better-adapted forms. At other times whole taxa become extinct within a relatively brief geological period, as for example the complete demise of the dinosaurs at the end of the Cretaceous period (70 million years BP) or the concurrent extinction of many large mammals and ground-nesting birds at the end of the Pleistocene epoch of the Cenozoic era (11,000 years BP). Extinctions may be associated with global or regional climatic changes that adversely affect vegetation and habitat or with the invasion of exotic animals that are able to outcompete the endemic fauna; in the case of recent extinctions, they may have been the result of human predation.

Although we cannot fully assess the impact of early humans on animal extinction, the spread of civilization across the earth has unmistakably been linked to the demise of many animal species. Historically, waves of extinction have attended the arrival of explorers and colonizers to each new continent, island, or archipelago. At first, most extinctions were caused by direct hunting, usually for food but often for commercial products such as furs and ornamental feathers. The development of the musket was responsible for the sharp rise in bird extinctions during the late eighteenth and early nineteenth centuries. Later, various indirect causes of extinction assumed prominence: habitat destruction, exotic animal introductions, and pest and predator control.

Causes of extinction today

Habitat alteration

The destruction of animal habitats is today the major threat to survival of endangered species. When habitats are destroyed—whether by logging, drainage of marshes, damming of rivers, urban development, or agriculture—animals must move and adapt to other habitats or die. Although it has been difficult to educate the public about the importance of habitat preservation—many still believe that conservation is primarily the protection of animals against wanton killing—nearly everyone is aware that the destruction of an ecosystem by strip mining, cleancut logging, or subdivision development adversely affects animals.

Habitat encroachment by humans often has subtle and far-flung effects that are not so readily apparent. Development and land exploitation more frequently than not cause habitat fragmentation. As patches of habitat become smaller and more isolated from each other, many animals that require large home ranges find themselves restricted to areas too small for their food, shelter, and reproductive requirements. Animals in isolated habitats of restricted size become much more vulnerable to nearby human activities, such as pesticide use, air pollution, noise, and human intrusion. Sometimes the exploitation of a small area has widespread ramifications to wildlife in undisturbed areas. A pond or stream may be the only source of water to wildlife for miles around; if it is destroyed by pollution or diverted for irrigation, large numbers of animals are adversely affected.

Migratory animals are especially vulnerable to habitat alteration. The habitat of ducks, geese, and other migratory waterfowl embraces the northern breeding grounds, southern wintering grounds, and all the areas they visit in between. All portions are necessary for their survival. Migratory fishes, such as salmon and steelhead trout, if barred from their spawning grounds by a dam or polluted stretch of river, will be destroyed just as effectively as if they were decimated by overfishing.

Yet another dimension of habitat alteration is successional changes that follow the clearing or burning of a mature forest. Species requiring stable, mature forest wilderness (such as grizzly bears, wolverines, bighorn sheep, caribou, and ivory-billed woodpeckers) decrease in numbers following a severe forest fire, whereas the luxurious new plant growth that follows a fire, fertilized by nutrients in the ash, encourages successional species such as deer, elk, rabbits, grouse, and quail (Figure 42-6). Not all forest fires are bad, despite the admonitions of Smokey the Bear. Periodic forest fires stimulate forest diversity, recirculate nutrients, and control diseases that tend to flourish in overmature forest stands. Some species such as Kirtland's warbler depend on fires for their survival. This endangered species requires 8- to 20-year-old jack pines for nesting. Jack pine cones remain tightly sealed by resins, opening only during a fire to release their seeds. Fire suppression in the Kirtland warbler's Michigan habitat resulted in uniform stands of trees too old for the birds' narrow habitat requirement. In 1964, when the cause of the warbler's demise was discovered, the U.S. Forest Service instituted periodic controlled burns to encourage the species' recovery.

The rapid destruction of the tropical forests has alarming implications for both humans and animals. Some 40% of the tropics of the world has been destroyed during the past 150 years, and virtually all the rest may be cleared by the year 2000. Some 400 million people make their living by slash-and-burn techniques, growing crops for a year or two, then moving on. Within 25 years, when the

Figure 42-6

Regrowth after a forest fire in Oregon (Tillamook Burn) provides excellent habitat for game species such as deer, elk, bear, grouse, and quail. Controlled fires are now considered an important forest management tool for removing diseased trees and opening land to new growth.
Photograph by L.L. Rue, III.

human population is expected to have at least doubled in size, there will be little tropical forest left.

Conservationist Norman Myers (1984) points out that everyone is involved in the decline of the rain forest. The human fondness for tropical hardwoods, used in furniture and luxury applications in nearly all homes throughout the affluent world, has expanded twentyfold in just 30 years. But the paramount reason for the destruction of Latin American forests is cattle raising. Because of rising beef prices at home, the United States has steadily increased beef imports, 17% of which comes from Latin America (most of the rest comes from Australia and New Zealand). Latin American beef is raised on grass rather than grain, and so it is very lean, suitable only for the fast-food trade. Because tropical beef is cheaper than home-raised, its infusion into the fast-food trade shaves a nickel off the price of a hamburger and saves American consumers about $500 million each year. But in responding to a lucrative market, Latin American cattle ranchers are clearing at least 20,000 square kilometers of forest each year. Beef exports from Central America represent that area's most dynamic export trade.

Most tropical countries, beset as they are by overwhelming human problems arising from explosive population growth, have neither the resources nor the motivation to study and protect the natural communities on which they depend. Nevertheless, several Latin American countries have plans to create reserves and parks in large portions of rain forest that still remain intact. Developed countries should be offering every possible encouragement and economic assistance to help these countries implement their plans. Areas already under protection have not fared well in Latin America because of haphazard protection and strong social and economic pressures to use the reserves for other purposes. Reserves in Latin America may slow the loss of natural habitat, but sweeping social changes will be required to prevent its eventual destruction.

Hunting

Until relatively recently, uncontrolled hunting for meat and oil and for commercial animal products (furs, hides, ornamental feathers, horns, and tusks) was the most serious menace to wildlife. Historically, the marine mammals, especially the furbearers (fur seals, harp seals, and sea otters) were subjected to egregious mass slaughters wherever they were discovered around the globe. The sea otter suffered 170 years of unrelenting persecution along northern Pacific coasts until, in 1910, it was considered extinct. Fortunately, some had survived and now, under strict protection, the species is slowly recovering.

The northern fur seal was nearly exterminated by a century and a half of sealing on the Pribilof Islands by Russian, American, Canadian, and British sealers. After the United States purchased Alaska and the Pribilofs from Russia in 1867, it attempted to limit harvesting, but Canadian and British sealers continued indiscriminate and wasteful shooting in the open sea. Finally in 1911, the United States arranged an international treaty that stopped open-sea hunting. From a low of about 125,000 seals, the population is approaching its former size of an estimated 2 to 3 million animals. Some 30,000 fur seals are collected annually by licensed sealers, operating in a rational, well-managed industry. The baby harp seal industry along the coast of Labrador and in the Gulf of St. Lawrence is also well managed, but the bloody method of harvesting has sparked an outcry from animal lovers and protectionists that may well bring an end to the persecution. It must be remembered that seal harvesting is supported by human demand for fur coats, not by the sealers' putative desire for butchery.

The vigorous public denunciation of harp seal hunting stems in part from the

Costa Rica has established enormous areas of unbroken rain forest as national parks and plans to extend the parks until they occupy 10% of the nation's territory—proportionately more than any other Latin American country. Brazil has set aside 175,000 square kilometers (68,000 square miles) of Amazonia as parks and reserves and has ambitious plans to expand this about tenfold. Peru, Venezuela, and Colombia are also establishing new parks, some of which embrace huge areas of rain forest. Unfortunately, most of these are little more than "paper parks," poorly understood by politicians and subject to constant invasion by people for farming, logging, and other disruptive activities.

"Bambi factor"—our separation of animals into "good" and "bad" based on human illusions of imaginary animal characteristics. Baby harp seals, with their big, dark, liquid eyes, look lovable and innocent; it seems inconceivable that anyone could club them to death by the thousands and skin them for their fur—or, for that matter, any other reason. Yet snakes, coyotes, and weasels, which are just as innocent (although perhaps not so lovable), seldom benefit from public outcries over their persecution. As long as an irrational public is concerned only about furry creatures with warm brown eyes, conservation organizations will use them as a selling point. The conservation movement, however, is firmly committed to the preservation of the entire pyramid of life.

Present-day game hunting and angling are, as we mentioned earlier, atavistic displays of an ancient hunting urge. "There is in the passion for hunting," wrote Charles Dickens, "something deeply implanted in the human breast." Sport hunting is a well-managed industry in the United States and Canada, probably the best managed in the world. None of the 74 bird species and 35 mammal species hunted legally in the United States is either threatened or endangered. Nevertheless, much of the urban public associates hunting with the wanton and repugnant killing of beautiful forest creatures. The confrontation between the nation's 21 million hunters and the numerous antihunting organizations is not new, but the polarization of viewpoints is more intense than ever before. However one views hunting emotionally, the morality of killing animals and the arguments about cruelty are side issues to conservation, often obscuring more important issues.

Two inescapable facts are that (1) hunting interests in North America have been a principal force in acquiring land for wildlife refuges and migratory bird sanctuaries and (2) until humans are prepared to reestablish wolves, coyotes, mountain lions, lynxes, and other natural predators in animal habitats, controlled hunting is necessary to manage game populations (Figure 42-7). It should be added, however, that because hunters killed the carnivores and raptors in the first place to keep them from feeding on the hunters' game, hunters themselves are responsible for the present predator-prey imbalance. It is unlikely that the hunter-nonhunter controversy will disappear.

In other parts of the world, game hunting still is a major threat to numerous

Figure 42-7

Starving white-tailed deer. Hunting is necessary to crop deer populations that otherwise quickly proliferate beyond the capacity of the land to support them in winter. Humans have replaced natural predators.
Photograph by L.L. Rue, III.

Poaching of African elephants for the illegal ivory trade has reached massive proportions. The price of ivory reached $34 per pound in 1980, up from $2.30 per pound in 1970. It was estimated in 1980 that 50,000 to 150,000 elephants are killed illegally each year from a total African population of 1.3 million elephants.

Rhinoceros horns (actually the "horn" is a compacted mass of hairs) are in great demand by Indians and Chinese who consider them to be a powerful aphrodisiac (in part because of their phallic shape) and remedy for a variety of ailments. Daggers with intricately carved rhinoceros horn handles are traditional gifts at puberty rites in the Middle East. Carvers in North Yemen prepared more than 2000 horns in 1975 and 1976 alone.

Figure 42-8

Elephant tusks taken from poachers by Kenya conservation authorities. These are only a few from an enormous warehouse in Mombasa containing thousands of huge tusks, representing, of course, the slaughter of thousands of elephants.
Photograph by F.M. Hickman.

species such as Bengal tigers, leopards, cheetahs, rhinoceroses, crocodiles, and elephants (Figure 42-8). Animal poaching in game preserves is increasingly prevalent as the rising human population of Africa and Asia places enormous strains on available food supplies. How long can national parks and game preserves be protected for the enjoyment of tourists, while natives adjacent to the preserves starve? Here, as elsewhere, it is human problems that conflict with wildlife.

Feral domestic and introduced animals

Humans have introduced numerous species into new habitats all over the world, where they affect the indigenous fauna by preying on them (cats, dogs, mongooses) or competing with them for food and habitat (rabbits, rats, mice, cattle, goats, sheep, pigs). Many introductions are deliberate, others are accidental, and both are facilitated by the mobility of humans. In 1929 a noted tropical entomologist, while on a rice ship from Trinidad to Manila, amused himself by listing every kind of animal on board, from cockroaches and fleas to rats and pet animals. He found no less than 41 species, mostly insects. When he opened his luggage in Manila, out walked several beetles.

Of course, not every animal introduced to a new land manages to establish itself, but of those that do, nearly all change the balance of the ecosystem in some way. Many have inflicted an enormous cost to agriculture—the fire ant, the boll weevil, and the gypsy moth are a few latter-day examples in the United States— and are responsible for much, if not most, of the heavy use of pesticides on crops and forests. The demise of both the American elm and the chestnut (p. 909) was caused by accidental introduction of diseases carried by insect vectors.

The natural balance of island communities is especially vulnerable to damage by introduced animals. In the absence of natural checks that keep their populations regulated on the continent, introduced species often flourish at the expense of the native island fauna, many of which have evolved in the absence of mammalian predators and consequently lack protective avoidance behavior. Flightless birds are highly vulnerable. A classic example is the dodo, a very large flightless member of the pigeon family, which inhabited the island of Mauritius.

Its fate was sealed by the first Dutch colonists who arrived in 1644 with pigs and dogs. Both the colonists and the introduced animals preyed on the dodos, annihilating the last of them in 1693. The dodo thus bears the distinction of being the first extinct animal whose demise could be fully documented.

No archipelago has suffered more disfigurement of the native fauna and flora than the Hawaiian Islands, where the whole of the native lowland forest community has been destroyed by introduced plants and animals. Sixteen species of birds became extinct between 1837 and 1945, half of them members of the extraordinary honeycreeper family. Only 14 species of honeycreepers survive, nearly all endangered. Although some of the original forest remains in Hawaii, especially in upland habitats where cultivation is impractical, nearly all is overrun by feral domestic animals, rats, and a host of other introduced vertebrate and invertebrate pests. It is only a matter of time before what remains of the marvelous indigenous fauna and flora of Hawaii has either vanished forever or is disfigured beyond recognition.

The unique Pleistocene fauna of New Zealand, which still flourished when the Polynesians arrived in the ninth century AD, was dominated by large ratite

Rabbits, introduced into New Zealand in the 1850s, quickly reached epidemic population densities that bared the ground of pasture, created deserts where lush tussock country had flourished before, and forced cattle and sheep ranchers into bankruptcy. Their incredible rate of reproduction created what farmers called "rabbit arithmetic"—that 2 times 3 equals 9 million (the progeny of two rabbits in 3 years).

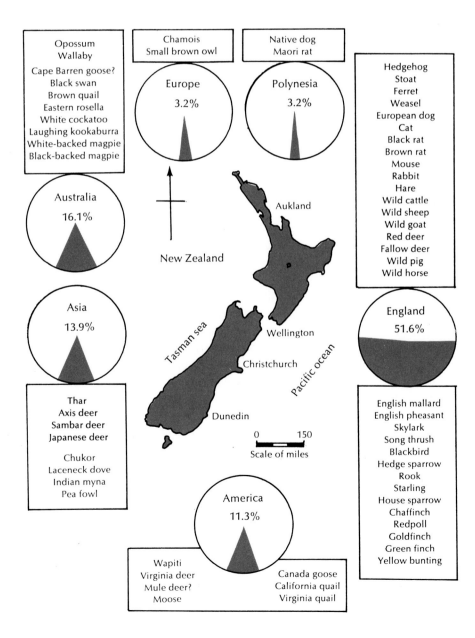

Figure 42-9

Introduced birds and mammals that have established populations in New Zealand, with countries of origin. The colored areas of each diagram represent the proportion of all introduced animals that have come from that region.

From Elton, C.S. 1958. The ecology of invasions by animals and plants. New York, John Wiley & Sons, Inc.; after Wodzicki, K.A. 1950. Bull. D.S.I.R., New Zealand 98:1-255.

(flightless) birds (21 species of moas and four species of kiwis) and had a rich representation of other birds, reptiles, and amphibians. The only land mammals were two bats and perhaps a rodent. Today, largely as a result of the depredations of introduced animals, all 21 moa species are extinct. Three kiwi species survive under strict protection. Nearly one third of the original fauna has vanished in a thousand years of human habitation. Approximately 60 species of birds and mammals have been introduced (Figure 42-9), of which about half have become problem animals. Although conservation programs are vigorously supported in New Zealand, it will never be possible to reconstitute the original fauna and flora.

Predator and pest control

Predators and pests that compete with humans for the same food have long been relegated to the category of "bad" living things. Among mammalian predators, wolves, coyotes, bobcats, grizzly bears, and cougars kill livestock and game and in doing so have engendered the wrath of sheepherders and hunters. In the United States and Canada, millions of taxpayer dollars have been spent to shoot, poison, and trap predators to benefit a relatively small but vociferous group. As a result, cougars, grizzly bears, and wolves have virtually disappeared from western grazing lands. The adaptable coyote, however, has been able to maintain a relatively healthy population despite vigorous control programs. Biologists had long argued that elimination of all predators by mass poisoning programs was poor management, since a large part of the coyotes' diet consists of rodents, which compete with sheep for forage. In 1972, yielding finally to citizen protests over mass poisoning programs—stimulated in part by such polemics as Jack Olsen's book *Slaughter the Animals, Poison the Earth*—the U.S. Fish and Wildlife Service banned all poisoning on public lands. Contrary to fears expressed by the sheep industry, the ban neither resulted in a dramatic increase in coyotes nor was responsible for decline in the sheep industry, which was already waning for economic reasons.

Animal traffic

The pet trade is big business in North America, England, and Europe. Between 1967 and 1974, American imports of exotic mammals, birds, and reptiles increased sixfold, reaching nearly 4 million annually in the mid-1970s. More than 100 million fishes, mostly tropical, are imported each year for the 20 million Americans who own aquariums. An astonishing variety of endangered species are sold illegally as pets, especially in the United States. The list includes monkeys of various species, cheetahs, leopards of three species, ocelots, numerous species of endangered and threatened birds, lizards, snakes, and fishes. The result is that the exotic pet trade, a curious phenomenon that seems to be associated with an affluent public's avidity for new kinds of gratification, is rapidly assuming a prominent position among the causes of extinction.

The collection and shipment of animals from Southeast Asia and Central and South America by careless hunters and callous dealers are extraordinarily wasteful—only about one animal in 10 survives the numerous hazards and gross mistreatment that occur between capture and purchase by the ultimate owner. Most collecting is done by indigenous peasant hunters who know little about live-capture techniques. Mortality is high at the time of capture. Many species are collected by shooting the mother to capture the baby; not infrequently several females are killed to secure a single uninjured infant. The rarer the animal, the greater is the price it will fetch, and the more enterprising are its human pursuers. Thus, an animal's rarity often promotes those elements that were responsible for its demise in the first place. Of the wild animals that survive the perilous trip to American homes, most die within a few weeks from improper care. Most of the rest are disposed of as people become bored with them or discover they cannot cope with them.

Some 60% of all wild mammals imported into the United States are destined for biomedical research, and of these, the great majority are primates. They are used for the preparation of vaccines, for physiological and behavioral experiments, and for defense-related research. The global demand for primates for research is of such proportions, estimated at 200,000 animals per year, that many species of both New World and Old World monkeys have been pushed close to extinction. Of the 149 species of primates, 54 (36%) are severely threatened today. The demand by zoos and circuses for primates, especially for the pongids (chimpanzees, orangutans, gorillas, and gibbons) also contributes to the drain, although in general, zoos are far less damaging to wildlife than the general public imagines. For example, in 1974 zoos imported into the United States only 4% of the wild mammals and 2% of the wild birds; the pet trade and biomedical research consumed the rest. Furthermore, nearly all of the larger zoos breed animals in captivity; zoos trade animals with each other and in some instances have been able to rebuild stocks of severely endangered species.

To relieve the stress that biomedical research places on native primate populations, several American users have begun breeding programs for rhesus monkeys and other primates. Although demand is certain to outpace the production of home-breeding programs for some years to come, biomedical researchers can only welcome any move that will reduce their contribution to the world wildlife crisis.

ENVIRONMENTAL MOVEMENT

In 1854, Chief Seattle of the Dunwanish tribe in Puget Sound admonished president Franklin Pierce in a moving and visionary letter:

> The great Chief in Washington sends word that he wishes to buy our land. How can you buy or sell the sky—the warmth of the land? The idea is strange to us. Yet we do not own the freshness of the air or the sparkle of the water. How can you buy them from us? Every part of this earth is sacred to my people. Every shiny pine needle, every sandy shore, every mist in the dark woods, every clearing and humming insect is holy in the memory and experience of my people.
>
> We know that the white man does not understand our ways. One portion of the land is the same to him as the next, for he is a stranger who comes in the night and takes from the land whatever he needs. The earth is not his brother but his enemy, and when he has conquered it he moves on. He leaves his father's graves, and his children's birthright is forgotten.

The American Indians, harmoniously and inextricably linked with the creatures around them and in spiritual kinship with nature, could not comprehend the white settlers' reckless and indifferent exploitation of land. The Indians venerated nature; the white settlers abused it. We may never fully understand why the ethics of these two people diverged as far as they did, but it is possible to reconstruct historical events that led to the white settlers' attitude—an attitude that threatens our very survival today.

We have already traced the beginnings of human society. Early humans were merely a part of wildlife's ecological web. People exploited nature for food and shelter as did the animals around them, but as long as people were fundamentally hunters and gatherers, their environmental requirements were small, and they remained in harmony with their surroundings. As civilization developed, people began to draw away from nature, spinning an increasingly artificial world around themselves. To satisfy their needs, new and more effective ways were developed to obtain them from the environment. When this caused habitat destruction and diminishing returns, the change was regarded as an inevitable and acceptable consequence of a chosen way of life. Agriculture, as we have already seen, vastly accelerated both habitat alteration and wildlife extermination, since wild animals now became competitors with humans for their food. Changes often occurred so slowly that people were unaware of how altered their environment had become.

At the time the first settlers came to America, much of the wildlife of England and Europe belonged to the privileged classes. Only the nobility hunted deer and grouse; the peasants hunted rabbits. Such nonegalitarian restraints were in large part responsible for the presence of wild boar and other large mammals in densely settled parts of Europe for hundreds of years.

In America, natural resources and wildlife appeared inexhaustible. From the first, settlers encountered deer, turkey, salmon, and sturgeon in abundance. No aristocracy restricted their right to hunt them; wildlife belonged to everyone. Thus began the "great American experiment," founded on the philosophy that every individual was entitled to any resources that could be gathered through personal initiative. It was an exploitive and destructive ethic, but no one viewed it that way. Living was hard for the settler despite the plenitude of natural resources. It was the pioneer family pitted against trees, wild animals, and hostile Indians—not conditions expected to foster a conservation ethic.

Even before the pioneers had reached the Mississippi River, attitudes were changing along the eastern seaboard. As the wildlife disappeared, eastern colonists began enacting laws to protect game and fish. Poorly enforced and relatively ineffective at first, such laws nevertheless reflected changing public attitudes toward wildlife. Gradually and almost grudgingly, the federal government began to assume leadership in environmental legislation.

The establishment of Yellowstone National Park in 1872 marked the real beginning of the environmental movement in the United States. Environmental wounds were becoming obvious to everyone, and by the turn of the century, an increasing number of determined Americans began to prod a reluctant Congress to act on environmental legislation. Public sentiment for animals in jeopardy had been quickened by the loss of the passenger pigeon and the near miss with the bison, the latter becoming a symbol of vanishing America.

At a time when pollution, energy and economic crises, and burgeoning social problems besiege the nation's conscience, most people are apt to respond indifferently to the passing of a few remote species of animals. If one has never seen a whooping crane or a blue whale—and never expects to see one or any other animal on the endangered species list—of what possible concern is their extinction? And is not the sacrifice of a few insignificant animals and plants necessary for progress? Such myopic views are affirmations of a philosophy at least as old as civilization: that the value of plants and animals is determined by their benefit to humans. This is an anthropocentric position that sees living things merely as instruments to human survival. It ignores the interrelatedness of all living components of the environment while subscribing to the notion that, because humans are the highest animal, they alone can make correct decisions about the exploitation of nature. According to this view, some animals and plants are more valuable to humans than others. Such a narrowly conceived attitude is rejected by modern ecological theory, which reaffirms the value of all living things in nature because all are parts of an interrelated whole.

Extinction therefore is an extremely serious result of environmental abuse because it is irreversible; an extinct animal or plant can never be re-created. Extinction removes a functional part from the web of life, and both diversity and stability of the ecosystem are reduced. Vanishing wildlife is an "early warning signal" of environmental malaise, and when we see the list of endangered animals lengthen year after year, we cannot remain indifferent.

Many people have not remained indifferent. With the establishment of Yellowstone National Park the conservation movement almost at once began to gather momentum. In 1886 Canada created Glacier National Park in British Columbia, followed in 1887 by the magnificent Banff National Park in the Canadian Rockies. Australia and New Zealand followed suit with their own national parks in the next decade; soon national parks were springing up all over the world. Admittedly, during the early decades of the environmental movement, national parks were created to preserve scenic beauty; the preservation of wildlife was incidental. But as the surrounding land became emptied of wildlife, parks acquired status as sanctuaries. And while Americans and Canadians pondered the new idea of establishing parks just for animals, large game preserves were already appearing in Africa. Today, the 386 national wildlife refuges scattered among all but one state in the United States and the hundreds of national parks and game preserves all over the world are manifestations of an environmental movement that has matured far beyond the most fanciful visions of the founders of Yellowstone National Park.

Conservation is much more than trying to fence off beautiful national parks and wildlife sanctuaries from an otherwise deteriorating biosphere. Most of all, it requires a common understanding among all humanity that our fate is inextricably linked to the living world at large. It is a worldwide problem of enormous complexity. In the final analysis, anything we do now or in the future becomes futile unless the growth of the human population can be stemmed. That is the ultimate problem.

Many believe that we are on a threshold of a new awakening, that we will be able to bridle our self-indulgence, our extraordinary propensity to isolate ourselves from nature, and then reaffirm both the meaningfulness of the human experience and our kinship with all living things.

SUMMARY

This final chapter concerns one aspect of the humanization of the earth, the human impact on animal extinction. We begin by documenting the great bison slaughter of the nineteenth century, the most cataclysmic known destruction of an animal species by humans. Approximately 60 million large mammals were reduced within a few decades to remnant herds totaling less than 600 animals. The principal domain of the bison, the North American grasslands, was then given over to highly productive croplands.

Humans have always been animal hunters, but the small populations of early hominids could not have had a significant impact on animal numbers or habitat. With the evolution of modern humans *(Homo sapiens)* some 130,000 years ago, more sophisticated hunting techniques appeared. Humans soon proved themselves capable of eliminating species through overhunting and habitat modification with the use of fire.

Habitat disruption increased very rapidly after the introduction of agriculture some 15,000 years ago. Agriculture increased the carrying capacity of the human environment, permitting the human population to increase much more quickly than ever before. Humans soon emerged as powerful reshapers of the earth's natural habitats.

Although extinction is a natural process in organic evolution, the extinction rate of animals has increased measurably, if not alarmingly, with the spread of human civilization. The Endangered Species Act, passed in 1966 by the U.S. Congress and reinforced in 1973, was the first legislation enacted that was designed to slow the extinction of animals. It is a surprisingly tough and uncompromising law, largely altruistic in purpose, that recognizes the importance of preserving the habitat of an endangered species.

Of the major causes of extinction today, habitat alteration is the greatest threat to species survival because it disrupts or destroys entire communities of organisms on which animals depend. Uncontrolled hunting was until recently a major cause of extinction, but in developed countries hunting and angling today are so closely managed that no legally hunted species is threatened or endangered. In developing countries, however, enormous hunting pressures on endemic animal species continue.

The introduction of nonendemic species into new habitats all over the world has changed natural balances that have evolved in various ecosystems and has led to numerous documented animal extinctions. Predator and pest control on croplands and grazing lands has also endangered several mammalian carnivores. The pet trade has recently emerged as a serious threat to wildlife, largely because of wasteful collecting techniques.

People have become increasingly aware that the endless quest for progress has left deep environmental wounds and forced many species into extinction. Much of the past indifference toward animal extinction stems from the widely held view that animals and plants exist solely as instruments to human survival. This myopic view is rejected by modern ecological theory. Today there are encouraging signs that a new ethic—a conservation strategy—is emerging that cherishes natural systems and recognizes the interdependence of all life forms.

Selected references

The environmental movement has spawned an avalanche of literature, some of which has been given over to emotional condemnation of just about every aspect of human interaction with wildlife, and some of which is concerned principally with pollution. We have not included these kinds of treatment in the following selection.

Bates, M. 1960. The forest and the sea. New York, Random House, Inc. *A lucid, thoughtful treatment of nature's economy and human intervention.*

Druett, J. 1983. Exotic intruders: the introduction of plants and animals into New Zealand. Auckland, Heinemann Publishers. *The author informatively documents the history of introductions into New Zealand.*

Ehrlich, P., and A. Ehrlich. 1981. Extinction: the causes and consequences of the disappearance of species. New York, Random House, Inc. *Lively and well-documented argument that survival of all species is essential for our own intrinsic good.*

Eno, A.S., R.L. Di Silvestro, and W.J. Chandler (eds.). 1986. Audubon Wildlife Report 1986. New York, The National Audubon Society. *Valuable compendium of information on how federal conservation systems work: how they are organized, funded, operated, and how they respond to controversial issues.*

Myers, N. 1979. The sinking ark: a new look at the problem of disappearing species. New York, Pergamon Press. *Outlines broad-scale conservation measures we can take to stop the destruction of species.*

Myers, N. 1984. The primary source: tropical forests and our future. New York, W.W. Norton & Company. *Why we cannot afford to lose our tropical forests is convincingly explained.*

Norton, B.G. (ed.). 1986. The preservation of species: the value of biological diversity. Princeton, NJ, Princeton University Press. *Contributors consider the dimensions of the endangered species problem, the ethics of maintaining diversity, and suggestions for management.*

Odell, R. 1980. Environmental awakening: the new revolution to protect the earth. Cambridge, Mass., Ballinger Publishing Co. *Thoughtful and carefully researched discussion of the key issues in the environmental movement.*

Rolston, H., III. 1986. Philosophy gone wild: essays in environmental ethics. Buffalo, NY, Prometheus Books. *Fifteen thought-provoking essays that argue for the adoption of environmental ethics.*

ORIGINS OF BASIC CONCEPTS AND KEY DISCOVERIES IN ZOOLOGY

384-322 BC: *Aristotle. The foundation of zoology as a science.*

Although this great pioneer zoologist and philosopher cannot be appraised by modern standards, there is scarcely a major subdivision of zoology to which he did not make some contribution. He was a true scientist, for he emphasized observational and experimental methods. Despite his lack of scientific background, he was one of the greatest scientists of all time.

○ **130-200 AD:** *Galen. Development of anatomy and physiology.*

This Roman investigator has been praised for his clear concept of scientific methods and blamed for passing down to others certain glaring errors that persisted for centuries. His influence was so great that for centuries students considered him the final authority on anatomical and physiological subjects.

1347: *William of Occam. Occam's razor.*

This principle of science has received its name from the fact that it is supposed to cut out unnecessary and irrelevant hypotheses in the explanation of phenomena. The gist of the principle is that, of several possible explanations, the one that is simplest, has the fewest assumptions, and is most consistent with the data at hand is the most probable.

○ **1543:** *Vesalius, Andreas. First modern interpretation of anatomical structures.*

With his insight into fundamental structure, Vesalius ushered in the dawn of modern biological investigation. Many aspects of his interpretation of anatomy are just now beginning to be appreciated.

○ **1616-1628:** *Harvey, William. First accurate description of blood circulation.*

Harvey's classic demonstration of blood circulation was the key experiment that laid the foundation of modern physiology. He explained bodily processes in physical terms, cleared away much of the mystical interpretation, and gave an auspicious start to experimental physiology.

○ **1627:** *Aselli, G. First demonstration of lacteal vessels.*

This discovery, coming at the same time as Harvey's great work, supplemented the discovery of circulation.

○ **1649:** *Descartes, René. Early concept of reflex action.*

Descartes postulated that impulses originating at the receptors of the body were carried to the central nervous system where they activated muscles and glands by what he called "reflection."

✳ **1651:** *Harvey, William. Aphorism of Harvey: Omne vivum ex ovo (all life from the egg).*

Although Harvey's work as an embryologist is overshadowed by his demonstration of blood circulation, his *De Generatione Animalium,* published in 1651, contains many sound observations on embryological processes.

○ **1652:** *Bartholin, Thomas. Discovery of lymphatic system.*

The significance of the thoracic duct in its relation to the circulation was determined in this investigation.

○ **1658:** *Swammerdam, Jan. Description of red blood corpuscles.*

This discovery, together with Swammerdam's observations on the valves of the lymphatics and the alterations in shape of muscles during contraction, represented early advancements in the microscopical study of bodily structures.

○ **1660:** *Malpighi, Marcello. Demonstration of capillary circulation.*

By demonstrating the capillaries in the lung of a frog, Malpighi was able to complete the scheme of blood circulation. Harvey never saw capillaries and thus never included them in his description.

△ **1665:** *Hooke, Robert. Discovery of the cell.*

Hooke's investigations were made with cork, and the term "cell" fits cork much better than it does animal cells, but by tradition the misnomer has stuck.

△ **1672:** *de Graaf, R. Description of ovarian follicles.*

De Graaf's name is given to the mature ovarian follicle, but he believed that the follicles were the actual ova, an error later corrected by von Baer.

△□ **1675-1680:** *van Leeuwenhoek, Anthony. Discovery of protozoa.*

The discoveries of this eccentric Dutch microscopist revealed a whole new world of biology.

□ **1693:** *Ray, J. Concept of species.*

Although Ray's work on classification was later overshadowed by that of Linnaeus, Ray was really the first to apply the species concept to a particular kind of organism and to point out the variations that exist among the members of a species.

○ **1733:** *Hales, Stephen. First measurement of blood pressure.*

This was further proof that the bodily processes could be measured quantitatively—more than a century after Harvey's momentous demonstration.

✳ **1745:** *Bonnet, Charles. Discovery of natural parthenogenesis.*

Although somewhat unusual in nature, this phenomenon has yielded much information about meiosis and other cytological problems.

□ **1758:** *Linnaeus, Carolus. Development of binomial nomenclature system of taxonomy.*

So important is this work in taxonomy that 1758 is regarded as the starting point in the determination of the generic and specific names of animals. Besides the value of his binomial system, Linnaeus gave taxonomists a valuable working model of conciseness and clearness that has never been surpassed.

✳ **1759:** *Wolff, Kaspar F. Embryological theory of epigenesis.*

This embryologist, the greatest before von Baer, did much to overthrow the preformation theory then in vogue and, despite many shortcomings, laid the basis for the modern interpretation of embryology.

○ **1760:** *Hunter, John. Development of comparative investigations of animal structure.*

This vigorous eighteenth-century anatomist gave a powerful impetus not only to anatomical observations but also to the establishment of natural history museums.

□ **1763:** *Adanson, M. Concept of empirical taxonomy.*

This botanist proposed a scheme of classification that grouped individuals into taxa according to shared characteristics. A species would have the maximum number of shared characteristics according to this scheme. The concept has been revived recently by exponents of numerical taxonomy. It lacks the evaluation of the evolutionary concept and has been criticized on this account.

‡ **1763:** *Kölreuter, J.G. Discovery of quantitative inheritance (multiple genes).*

Kölreuter, a pioneer in plant hybridization, found that certain plant hybrids had characteristics more or less intermediate between the parents in the F_1 generation, but in the F_2 there were many gradations from one extreme to the other. An explanation was not forthcoming until after Mendel's laws were discovered.

□ **1768-1779:** *Cook, James. Influence of geographical exploration on development of biology.*

This famous sea captain made possible a greater range of biological knowledge because of the able naturalists whom he took on his voyages of discovery.

△ **1772:** *Priestley, J., and J. Ingenhousz. Concept of photosynthesis.*

These investigators first pointed out some of the major aspects of this important phenomenon, such as the use of light energy for converting carbon dioxide and water into released oxygen and retained carbon.

△ **1774:** *Priestley, J. Discovery of oxygen.*

The discovery of this element is of great biological interest because it helped in determining the nature of oxidation and the exact role of respiration in organisms.

△ **1778:** *Lavoisier, Antoine L. Nature of animal respiration demonstrated.*

A basis for the chemical interpretation of the life process was given a great impetus by the careful quantitative studies of the changes during breathing made by this great investigator. His work also meant the final overthrow of the mystical phlogiston hypothesis that had held sway for so long.

☐ 1781: *Abildgaard, P. First experimental life cycle of a tapeworm.*

Life cycles of parasites may be very complicated, involving several hosts. That a parasitic worm could require more than one host was a revolutionary concept in Abildgaard's day and was widely disbelieved until the work of Küchenmeister over 60 years later.

✕ 1791: *Smith, William. Correlation between fossils and geological strata.*

By observing that certain types of fossils were peculiar to particular strata, Smith was able to work out a method for estimating geological age. He laid the basis of stratigraphic geology.

○ 1792: *Galvani, L. Animal electricity.*

The lively controversy between Galvani and Volta over the twitching of frog legs led to extensive investigation of precise methods of measuring the various electrical phenomena of animals.

✕ 1796: *Cuvier, Georges. Development of vertebrate paleontology.*

Cuvier compared the structure of fossil forms with that of living ones and concluded that there had been a succession of organisms that had become extinct and were succeeded by the creation of new ones. To account for this extinction, Cuvier held to the theory of catastrophism, or the simultaneous extinction of animal populations by natural cataclysms.

✕ 1801: *de Lamarck, J.B. Evolutionary concept of use and disuse.*

De Lamarck gave the first clear-cut expression of a hypothesis to account for organic evolution. His assumption was that acquired characteristics were inherited; modern evolutionists have refuted this part of the hypothesis.

○ 1802: *Young, T. Hypothesis of trichromatic color vision.*

Young's hypothesis suggested that the retina contained three kinds of light-sensitive substances, each having a maximum sensitivity in a different region of the spectrum and each being transmitted separately to the brain. The three substances combined produced the colors of the environment. The three pigments responsible are located in three kinds of cones. Young's explanation has been modified in certain details by other investigators.

○ 1811: *Bell, C., and F. Magendie. Discovery of the functions of dorsal and ventral roots of spinal nerves.*

This demonstration was a starting point for an anatomical and functional investigation of the most complex system in the body.

✳ 1817: *Pander, C.H. First description of three germ layers.*

The description of the three germ layers was first made on the chick, and later the concept was extended by von Baer to include all vertebrates.

✳ 1827: *von Baer, Karl. Discovery of mammalian ovum.*

The very tiny ova of mammals escaped de Graaf's eyes, but von Baer brought mammalian reproduction into line with that of other animals by detecting the ova and their true relation to the follicles.

△ 1828: *Brown, Robert. Brownian movement first described.*

This interesting phenomenon is characteristic of living matter and sheds some light on the structure of cells.

☐ 1828: *Thompson, J.V. Nature of plankton.*

Thompson's collections of these small forms with a tow net, together with his published descriptions, are the first records of the vast community of planktonic animals. He was also the first to work out the true nature of barnacles.

△ 1828: *Wöhler, F. First to synthesize an organic compound.*

Wöhler succeeded in making urea (a compound formed in the body) from the inorganic substance ammonium cyanate; thus, $NH_4OCN \rightarrow NH_2CONH_2$. This success in producing an organic substance synthetically was the stimulus that resulted in the preparation of thousands of compounds by others.

✳✕ 1830: *von Baer, Karl. Biogenetic law formulated.*

Von Baer's conception of this law was conservative and sounder in its implication than has been the case with many other biologists (Haeckel, for instance). Von Baer stated that embryos of higher and lower forms resemble each other more the earlier they are compared in their development, and *not* that the embryos of higher forms resemble the adults of lower organisms.

✕ 1830: *Lyell, C. Modern concept of geology.*

The influence of this concept not only did away with the catastrophic theory but also gave a logical interpretation of fossil life and the correlation between the formation of rock strata and the animal life that existed at the time these formations were laid down.

△ 1831: *Brown, Robert. First description of cell nucleus.*

Others had seen nuclei, but Brown was the first to name the structure and to regard the nucleus as a general phenomenon. This description was an important preliminary to the formulation of the cell theory a few years later, for Schleiden acknowledged the importance of the nucleus in the development of the cell concept.

○ 1833: *Hall, Marshall. Concept of reflex action.*

Hall described the method by which a stimulus can produce a response independently of sensation or volition and coined the term "reflex action." It remained for the outstanding work of the great Sir Charles Sherrington in the twentieth century to explain much of the complex nature of reflexes.

†○ 1835: *Bassi, Agostino. First demonstration of a microorganism as an infective agent.*

Bassi's discovery that a certain disease of silkworms was caused by a small fungus represents the beginning of the germ theory of disease that was to prove so fruitful in the hands of Pasteur and other able investigators.

△ 1835: *Dujardin, Felix. Description of living matter (protoplasm).*

Dujardin associated the jellylike substance that he found in protozoa and that he called "sarcode" with the life process. This substance was later called "protoplasm," and the sarcode idea may be considered a significant landmark in the development of the protoplasm concept.

†○ 1835: *Owen, Richard. Discovery of Trichinella.*

This versatile investigator is chiefly remembered for his researches in anatomy, but his discovery of this very common parasite in humans is an important landmark in the history of parasitology.

△○ 1838: *von Liebig, Justus. Foundation of biochemistry.*

The idea that vital activity could be explained by chemicophysical factors was an important concept for biological investigators studying the nature of life.

△ 1838-1839: *Schleiden, M.J., and T. Schwann. Formulation of the cell theory.*

The cell doctrine, with its basic idea that all plants and animals were made up of similar units, represents one of the truly great landmarks in biological progress. Many biologists believe the work of Schleiden and Schwann has been rated much higher than it deserves in light of what others had done to develop the cell concept. R. Dutrochet (1824) and H. von Mohl (1831) described all tissues as being composed of cells.

△ 1838: *Mulder, G.J. Concept of the nature of proteins.*

Mulder proposed the name "protéine," later "protein," for the basic constituents of protoplasmic materials.

△ 1839-1846: *Purkinje, J.E., and Hugo von Mohl. Concept of protoplasm established.*

Purkinje proposed the name "protoplasm" for living matter, and von Mohl did extensive work on its nature, but it remained for Max Schultze (1861) to give a clear-cut concept of the relations of protoplasm to cells and its essential unity in all organisms.

☐ 1839: *Verhulst, P.F. Logistical theory of population growth.*

According to this theory, animal populations have a slow initial growth rate that gradually speeds up until it reaches a maximum and then slows down to a state of equilibrium. By plotting the logarithm of the total number of individuals against time, an S-shaped curve results that is somewhat similar for all populations.

☐○ 1840: *von Liebig, J., and F.F. Blackman. The law of minimum requirements.*

Von Liebig interpreted plant growth as being dependent on essential requirements that are present in minimum quantity; some substances, even in minute quantities, were needed for normal growth. Later Blackman discovered that the rate of photosynthesis is restricted by the factor that operates at a limiting intensity (for example, light or temperature). The concept has been modified in various ways since it was formulated. We now know that a factor interaction other than the minimum ones plays a part.

□ **1840:** *Müller, J. Theory of specific nerve energies.*

This theory states that the kind of sensation experienced depends on the nature of the sense organ with which the stimulated nerve is connected. The optic nerve, for instance, conveys the impression of vision however it is stimulated.

○ **1842:** *Bowman, William. Histological structure of the nephron (kidney unit).*

Bowman's accurate description of the nephron afforded physiologists an opportunity to attack the problem of how the kidney separates waste from the blood.

✳ **1842:** *Steenstrup, Johann J.S. Alternation of generations described.*

Metagenesis, or alternation of sexual and asexual reproduction in the life cycle, exists in many animals and plants. The concept had been introduced before this time by L.A. de Chamisso (1819).

× **1843:** *Owen, Richard. Concepts of homology and analogy.*

Homology, as commonly understood, refers to similarity in embryonic origin and development, whereas analogy is the likeness between two organs in their functioning. Owen's concept of homology was merely that of the same organ in different animals under all varieties of form and function.

○ **1844:** *Ludwig, C. Filtration theory of renal excretion.*

About the same time that Bowman described the renal corpuscle, Ludwig showed that the corpuscle functions as a passive filter and that the filtrate that passes through it from the blood carries into the urinary tubules the waste products that are concentrated by the resorption of water as the filtrate moves down the tubules. Other investigators, Cushny, Starling, and Richards, have confirmed and extended his observations.

○△ **1845:** *von Helmholtz, H., and J.R. von Mayer. The formulation of the law of conservation of energy.*

This landmark in human thinking showed that in any system, living or nonliving, the power to perform mechanical work always tends to decrease unless energy is added from without. Physiological investigation could now advance on the hypothesis that the living organism is an energy machine and obeys the laws of energetics.

‡ **1848:** *Hofmeister, W. Discovery of chromosomes.*

This investigator made sketches of bodies, later known to be chromosomes, in the nuclei of pollen mother cells (*Tradescantia*). Schneider further described these elements in 1873, and Waldeyer named them in 1888.

□ **1848:** *von Siebold, Carl T.E. Establishment of the status of protozoa.*

Von Siebold emphasized the unicellular nature of protozoa, fitted them into the recently developed cell theory, and established them as the basic phylum of the animal kingdom.

○ **1851:** *Bernard, Claude. Discovery of the vasomotor system.*

Bernard showed how the amount of blood distributed to the various tissues by the small arterioles was regulated by vasomotor nerves of the sympathetic nervous system, as he demonstrated in the ear of a white rabbit.

✳ **1851:** *Waller, A.V. Importance of the nucleus in regeneration.*

When nerve fibers are cut, the parts of the fibers peripheral to the cut degenerate in a characteristic fashion. This wallerian degeneration enables one to trace the course of fibers through the nervous system. In the regeneration process the nerve cell body (with its nucleus) is necessary for the downgrowth of the fiber from the proximal segment.

○ **1852:** *von Helmholtz, Herman. Determination of rate of nervous impulse.*

This landmark in nerve physiology showed that the phenomena of nervous activity could be measured experimentally and expressed numerically.

○△ **1852:** *von Kölliker, Albrecht. Establishment of histology as a science.*

Many histological structures were described with marvelous insight by this great investigator, starting with the publication of the first text in histology (*Handbuch der Gewebelehre*).

○ **1852:** *Stannius, H.F. Stannius's experiment on the heart.*

By tying a ligature as a constriction between the sinus venosus and the atrium in the frog and also one around the atrioventricular groove, Stannius was able to demonstrate that the muscle tissues of the atria and ventricles have independent and spontaneous rhythms. His observations also indicated that the sinus is the pacemaker of the heartbeat.

†□ **1852:** *Küchenmeister, G.F.H. Relationship of "bladder worms" to adult tapeworms.*

This was the first demonstration that "bladder worms" (cysticerci) were the juveniles of taeniid tapeworms. It was followed rapidly by numerous other lifecycle studies of cestodes.

✳ **1854:** *Newport, G. Discovery of the entrance of spermatozoon into a frog's egg.*

This was a significant step in cellular embryology, although its real meaning was not revealed until the concept of fertilization as the union of two pronuclei was formulated about 20 years later (Hertwig, 1875).

†○ **1855:** *Addison, T. Discovery of adrenal disease.*

The importance of the adrenals in maintaining normal body functions was shown when Addison described a syndrome of general disorders associated with the pathology of this gland.

× **1856:** *Discovery of Neanderthal fossil hominid* (Homo sapiens).

Many specimens of this type of fossil hominid have been discovered (Europe, Asia, Africa) since the first one was found near Düsseldorf, Germany. Their culture was Mousterian (100,000 to 40,000 years BP), and although they were of short stature, their cranial capacity was as large as or larger than that of modern humans.

○ **1857:** *Bernard, Claude. Discovery of formation of glycogen by the liver.*

Bernard's demonstration that the liver forms glycogen from substances brought to it by the blood showed that the body can build up complex substances as well as tear them down.

□ **1858:** *Sclater, P.L. Distribution of animals on basis of zoological regions.*

This was the first serious attempt to study the geographical distribution of organisms, a study that eventually led to the present science of zoogeography. A.R. Wallace worked along similar lines.

✳△ **1858:** *Virchow, Rudolf. Aphorism of Virchow:* Omnis cellula e cellula *(every cell from a cell).*

△ **1858:** *Virchow, Rudolf. Formulation of the concept of disease from the viewpoint of cell structure.*

He laid the basis of modern pathology by stressing the role of the cell in diseased tissue.

× **1859:** *Darwin, Charles. The concept of natural selection as a factor in evolution.*

Although Darwin did not originate the concept of organic evolution, no one has been more influential in the development of evolutionary thought. The publication of *The Origin of Species* represents the greatest single landmark in the history of biology.

○ **1860:** *Bernard, Claude. Concept of the constancy of the internal environment (homeostasis).*

Bernard's realization that there are mechanisms in living organisms that tend to maintain internal conditions within relatively narrow limits, despite external change, is one of the most valuable generalizations in physiology.

✳ **1860:** *Pasteur, Louis. Aphorism of Pasteur:* Omne vivum e vivo *(every living thing from the living).*

✳ **1860:** *Pasteur, Louis. Refutation of spontaneous generation.*

Pasteur's experiment with the open S-shaped flask proved conclusively that fermentation or putrefaction resulted from microbes, thus ending the long-standing controversy regarding spontaneous generation.

□ **1860:** *Wallace, A.R. The Wallace line of faunal delimitation.*

As originally proposed by Wallace, there was a sharp boundary between the Australian and Oriental faunal regions so that a geographical line drawn between certain islands of the Malay Archipelago, through the Makassar Strait between Borneo and Celebes, and between the Philippines and the Sanghir Islands separated two distinct and contrasting zoological regions. On one side Australian forms predominated; on the other, Oriental ones. The validity of this division has been questioned by zoogeographers in the light of more extensive knowledge of the faunas of the two regions.

△ **1861:** *Graham, Thomas. Colloidal states of matter.*

This work on colloidal solutions has resulted in one of the most fruitful concepts regarding protoplasmic systems.

□ **1862:** *Bates, H.W. Concept of mimicry.*

Mimicry refers to the advantageous resemblance of one species to another for protection against predators. It involves palatable or edible species that imitate dangerous or inedible species that employ warning coloration against potential enemies.

□ **1864:** *Haeckel, Ernst. Modern zoological classification.*

The broad features of zoological classification as we know it today were outlined by Haeckel and others, especially R.R. Lankester, in the third quarter of the nineteenth century. B. Hatschek is another zoologist who deserves much credit for the modern scheme, which is constantly undergoing revision. Other schemes of classification in the early nineteenth century did much to resolve the difficult problem of classification, such as those of Cuvier, de Lamarck, Leuckart, Ehrenberg, Vogt, Gegenbauer, and Schimkevitch.

△ **1864:** *Schultze, Max. Protoplasmic bridges between cells.*

The connection of one cell to another by means of protoplasmic bridges has been demonstrated in both plants and animals. Modern techniques, including electron microscopy, have shown that many cells communicate by gap junctions, which have channels capable of passing molecules of up to 1000 daltons in molecular weight.

‡ **1866:** *Haeckel, Ernst. Nuclear control of inheritance.*

At the time the hypothesis was formed, the view that the nucleus transmitted the inheritance of an animal had no evidence to substantiate it.

□ **1866:** *Kovalevski, A. Taxonomic position of tunicates.*

The great Russian embryologist showed in the early stages of development the similarity between amphioxus and the tunicates and how the latter could be considered a branch of the phylum Chordata.

‡ **1866:** *Mendel, Gregor. Formulation of the first two laws of heredity.*

These first clear-cut statements of inheritance made possible an analysis of hereditary patterns with mathematical precision. Rediscovered in 1900, his paper confirmed the experimental data geneticists had found at that time. Since that time Mendelian genetics has been considered the chief cornerstone of hereditary investigation.

○ **1866:** *Schultze, Max. Histological analysis of the retina.*

Schultze's fundamental discovery that the retina contained two types of visual cells—rods and cones—helped to explain the physiological differences of vision in high and low intensities of light.

×✳ **1867:** *Kovalevski, A. Germ layers of invertebrates.*

The concept of the primary germ layers laid down by Pander and von Baer was extended to invertebrates by this investigator. He found that the same three germ layers arose in the same fashion as those in vertebrates. Thus, an important embryological unity was established for the animal kingdom.

○ **1869:** *Langerhans, P. Discovery of islet cells in pancreas.*

These islets were first found in the rabbit, and since their discovery they have been one of the most investigated tissues in animals. The hypothesis that they have come from the transformation of pancreatic acinus cells has now given way to the belief that they originated from the embryonic tubules of the pancreatic duct.

△ **1869:** *Miescher, F. Isolation of nucleoprotein.*

From pus cells, a pathological product, Miescher was able to demonstrate in the nuclei certain phosphorus-rich substances, nucleic acids, which are bound to proteins to form nucleoproteins. In recent years these complex molecules have been the focal point of significant biochemical investigations on the chemical properties of genes, with the wider implications of a better understanding of growth, heredity, and evolution.

1871: *Quetelet, Lambert A. Foundations of biometry.*

The applications of statistics to biological problems is called biometry. By working out the distribution curve of the height of soldiers, Quetelet showed biologists how the systematic study of the relationships of numerical data could become a powerful tool for analyzing data in evolution, genetics, and other biological fields.

□ **1872-1876:** Challenger *expedition.*

Not only did this expedition establish the science of oceanography, but a vast amount of material was collected that greatly extended the knowledge of the variety and range of animal life.

□ **1872:** *Dohrn, Anton. Establishment of Naples Biological Station.*

The establishment of this famous station marked the first of the great biological stations, and it rapidly attained international importance as a center of biological investigation.

○ **1872:** *Ludwig, C., and E.F.W. Pflüger. Gas exchange of the blood.*

By means of the mercurial blood pump, these investigators separated the gases from the blood and thereby threw much light on the nature of gas exchange and the place where oxidation occurs (in the tissues).

×□ **1874:** *Haeckel, Ernst H. Taxonomic position of phylum Chordata.*

The great German evolutionist based many of his conclusions on the work of the Russian embryologist Kovalevski, who in 1866 showed that the tunicates as well as amphioxus had vertebrate affinities.

× **1874:** *Haeckel, Ernst H. Gastrea hypothesis of metazoan ancestry.*

According to this hypothesis, the hypothetical ancestor of all the Metazoa consisted of two layers (ectoderm and endoderm) similar to the gastrula stage in embryonic development, and the endoderm arose as an invagination of the blastula. Thus, the diploblastic stage of ontogeny was to be considered as the repetition of this ancestral form. This concept in modified form has had wide acceptance.

‡✳ **1875:** *Hertwig, Oskar. Concept of fertilization as the conjugation of two sex cells.*

The fusion of the pronuclei of the two gametes in the process of fertilization paved the way for the concept that the nuclei contained the hereditary factors and that both maternal and paternal factors are brought together in the zygote.

△ **1875:** *Strasburger, Eduard. Description of "indirect" (mitotic) cell division.*

The accurate description of the processes of cell division that Strasburger made in plants represents a great pioneer work in the rapid development of cytology during the last quarter of the nineteenth century.

△ **1876:** *Pasteur, Louis. The Pasteur effect.*

In his studies of fermentation processes in wine making, Pasteur discovered that cells consumed much more glucose and produced much more lactic acid in the absence of oxygen than in its presence. It is now known that the effect occurs in many different kinds of cells and is a consequence of the lower energy yield of glycolysis compared with oxidative phosphorylation.

○ **1877:** *Pfeffer, Wilhelm. Concept of osmosis and osmotic pressure.*

Pfeffer's experiments on osmotic pressure and the determination of the pressure in different concentrations laid the foundation for an understanding of a general phenomenon in all organisms.

○ **1878:** *Balfour, Francis M. Relationship of the adrenal medulla to the sympathetic nervous system.*

By showing that the adrenal medulla has the same origin as the sympathetic nervous system, Balfour really laid the foundation for the interesting concept of the similarity in the action of epinephrine and sympathetic nervous mediation.

○△ **1878:** *Kühne, Willy. Nature of enzymes.*

The study of the action of chemical catalysts has steadily increased with biological advancement and now represents one of the most fundamental aspects of biochemistry.

△ **1879:** *Flemming, Walther. Chromatin described and named.*

Flemming shares with a few others (especially Strasburger, 1875) the description of the details of mitotic cell division. The part of the nucleus that stains deeply he called **chromatin** (colored), which gives rise to the chromosomes.

✳ **1879:** *Fol, Hermann. Penetration of ovum by a spermatozoon described.*

Fol was the first to describe a thin conelike body extending outward from the egg to meet the sperm. Compare Dan's acrosome reaction (1954).

△ **1879:** *Kossel, Albrecht. Isolation of nucleoprotein.*

Nucleoproteins were isolated in the heads of fish sperm; they make up the major part of chromatin. They are combinations of proteins with nucleic acids, and this study was one of the first of the investigations of a topic that interests biochemists at the present time.

Nobel Laureate (1910).

† **1880:** *Laveran, C.L.A. Protozoa as pathogenic agents.*

This French investigator first demonstrated that the causative organism for malaria is a protozoan. This discovery led to other investigations that revealed the role the protozoa play in causing diseases such as sleeping sickness and kala-azar. It remained for Sir Ronald Ross to discover the role of mosquitoes in spreading malaria.

Nobel Laureate (1907).

○ **1880:** *Ringer, S. Influence of blood ions on heart contraction.*

The pioneer work of this investigator determined the inorganic ions necessary for contraction of frog hearts and made possible an evaluation of heart metabolism and the replacement of body fluids.

△ **1882:** *Flemming, Walther. First accurate counts of nuclear filaments (chromosomes) made.*

△ **1882:** *Flemming, W. Mitosis and spireme named.*

†△ **1882:** *Metchnikoff, Élie. Role of phagocytosis in immunity.*

The theory that microbes are ingested and destroyed by certain white corpuscles (phagocytes) shares with the theory of chemical bodies (antibodies) the chief explanation for the body's natural immunity. Metchnikoff first studied phagocytosis in the cells of starfish and water fleas *(Daphnia)*.

Nobel Laureate (1908).

△ **1882:** *Strasburger, Eduard. Cytoplasm and nucleoplasm named.*

□ **1882-1924:** *The* Albatross *of the Fish Commission.*

The *Albatross*, under the direction of the U.S. Fish Commission, was second only to the famed *Challenger* in advancing scientific knowledge about oceanography.

○△ **1883:** *Golgi, Camillo, and R. Cajal. Silver nitrate technique for nervous elements.*

The development and refinement of this technique gave a completely new picture of the intricate relationships of neurons. Modifications of this method have given valuable information concerning the cellular element—the Golgi apparatus.

Nobel Laureates (1906).

✳ **1883:** *Hertwig, Oskar. Origin of term "mesenchyme."*

This important tissue may arise embryonically from all three germ layers but arises chiefly from the mesoderm. It is a protoplasmic network whose meshes are filled with a fluid intercellular substance. Mesenchyme gives rise to a great variety of tissues, and both its cells and intercellular substance may be variously modified; its most common derivative is connective tissue.

†□ **1883:** *Leuckart, Rudolph, and A.P. Thomas. Life history of sheep liver flukes.*

This discovery represents the first time a complete life cycle was worked out for a digenetic trematode. Working independently, these investigators found that the trematode used a snail intermediate host.

‡ **1883:** *Roux, Wilhelm. Allocation of hereditary functions to chromosomes.*

This hypothesis could not have been much more than a guess when Roux made it, but how fruitful was the idea in the light of the enormous amount of evidence since accumulated!

△ **1884:** *Flemming, W., E. Strasburger, and E. van Beneden. Demonstration that nuclear filaments (chromosomes) double in number by longitudinal division.*

This concept represented a further step in understanding the precise process in indirect cell division.

○ **1884:** *Rubner, Max. Quantitative determinations of the energy value of foods.*

Although Liebig and others had estimated calorie values of foods, the investigations of Rubner put their determinations on a sound basis. His work made possible a scientific explanation for metabolism and a basis for the study of comparative nutrition.

‡ **1885:** *Hertwig, Oskar, and E. Strasburger. Concept of the nucleus as the basis of heredity.*

The development of this idea occurred before Mendel's laws of heredity were rediscovered in 1900, but it anticipated the important role the nucleus with its chromosomes was to assume in hereditary transmission.

‡ **1885:** *Rabl, Karl. Concept of the individuality of the chromosomes.*

The view that the chromosomes retain their individuality through all stages of the cell cycle is accepted by all cytologists, and much evidence has accumulated in support of the hypothesis. However, it has been virtually impossible to demonstrate the chromosome's individuality through all stages.

✳ **1885:** *Roux, Wilhelm. Mosaic theory of development.*

In the early development of the frog's egg, Roux showed that the determinants for differentiation were segregated in the early cleavage stages and that each cell or groups of cells would form only certain parts of the developing embryo (mosaic or determinate development). Later, other investigators showed that in many forms blastomeres, separated early, would give rise to whole embryos (indeterminate development).

✳‡ **1885:** *Weismann, August. Formulation of germ plasm theory.*

Weismann's great theory of the germ plasm stresses the idea that there are two types of protoplasm—germ plasm, which gives rise to the reproductive cells or gametes, and somatoplasm, which furnishes all the other cells. Germ plasm, according to the theory, is continuous from generation to generation, whereas the somatoplasm dies with each generation and does not influence the germ plasm.

1886: *Establishment of Woods Hole Biological Station in Massachusetts.*

This station is by all odds the greatest center of its kind in the world. Many of the biologists in America have studied there, and the station has increasingly attracted investigators from many other countries as well. The influence of Woods Hole on the progress of biology cannot be overestimated.

△ **1887:** *Fischer, Emil. Structural patterns of proteins.*

The importance of proteins in biological systems has made their study the central theme of much modern biochemical work.

Nobel Laureate (1902).

✳ **1887:** *Haeckel, Ernst H. Concept of organic form and symmetry.*

Symmetry refers to the spatial relations and arrangements of parts in such a way as to form geometrical designs. Although many others before Haeckel's time had studied and described types of animal form, Haeckel gave us our present concepts of organic symmetry, as revealed in his monograph on radiolarians collected on the *Challenger* expedition.

‡ **1889:** *Hertwig, R., and E. Maupas. True nature of conjugation in paramecium.*

The process of conjugation was described (even by van Leeuwenhoek) many times and its sexual significance interpreted, but these two investigators independently showed the details of the divisions before conjugation and the mutual exchange of the micronuclei during the process.

†○ **1889:** *von Mering, J., and O. Minkowski. Effect of pancreatectomy.*

The classic experiment of removing the pancreas stimulated research that led to the isolation of the pancreatic hormone insulin by Banting (1922).

† **1890:** *Smith, Theobald. Role of an arthropod in disease transmission.*

The transmission of the sporozoan *Babesia*, which is the active agent in causing Texas cattle fever, by the tick *Boophilus*, represented the first demonstration of the role of an arthropod as a vector of disease.

∗ **1891:** *Driesch, Hans. Discovery of totipotent cleavage.*

The discovery that each of the first several blastomeres, if separated from each other in the early cleavage of the sea urchin embryo, would develop into a complete embryo stimulated investigation on totipotent and other types of development.

× **1890-1891:** *Dubois, Eugene. Discovery of the fossil human* Pithecanthropus erectus *(now* Homo erectus*).*

Although not the first fossil human to be found, the Java man represents one of the first significant primitive humans that have been discovered.

× **1893:** *Dollo, I. Concept of the irreversibility of evolution.*

In general, the overall evolutionary process is one-way and irreversible insofar as a whole complex genetic system is concerned, although back mutations and restricted variations may occur over and over.

○ **1893:** *His, Wilhelm. Anatomy and physiology of the atrioventricular node and bundle.*

The specialized conducting tissue of the heart has given rise to many investigations, not the least of which were those of His, whose name was given to the intricate system of branches that are reflected over the inner surface of the ventricles.

∗ **1894:** *Driesch, Hans. Constancy of nuclear potentiality.*

Driesch's view was that all nuclei of an organism were equipotential but that the activity of nuclei varied as tissues differentiated. More recent studies have emphasized the crucial importance of the cytoplasm of the embryo in determining expression of the nuclear genome.

□ **1894:** *Merriam, C.H. Concept of life zones in North America.*

This scheme is based on temperature criteria and the importance of temperature in the distribution of plants and animals. According to this concept, animals and plants are restricted in their northward distribution by the total quantity of heat during the season of growth and reproduction, and their southward distribution is restricted by the mean temperature during the hottest part of the year.

†□ **1895:** *Bruce, D. Life cycle of protozoan blood parasite* (Trypanosoma).

The relation of this parasite to the tsetse fly and to wild and domestic animal infection in Africa is an early demonstration of the role of arthropods as vectors of disease.

†○ **1895:** *Roentgen, W. Discovery of x-rays.*

This great discovery was quickly followed by its application in the interpretation of bodily structures and processes and represents one of the greatest tools in biological research.

Nobel Laureate (1901).

× **1896:** *Baldwin, J.M. Baldwin evolutionary effect.*

It is the belief that genetic selection of genotypes will be channeled or canalized in the same direction as the adaptive modifications that were formerly nonhereditary. It is now understood that the *capacity* to respond to environmental conditions is itself hereditary.

× **1896:** *Russian Hydrographic Survey. Biology of Lake Baikal.*

This lake in Siberia is more than 800 km long and 80 km wide and has an extreme depth of more than a mile (the deepest lake in the world). Its unique fauna is a striking example of evolutionary results from long-continued isolation. Almost all of the species in certain groups are endemic (found nowhere else). This remarkable fauna represents the survival of ancient freshwater animals that have become extinct in surrounding areas.

○ **1897:** *Abel, J.J., and A.C. Crawford. Isolation of the first hormone (epinephrine).*

The purification and isolation of one of the active principles of the suprarenal medulla led to discovery of its chemical nature and naming by J. Takamine (1901) and its synthesis by F. Stolz (1904).

△○ **1897:** *Buchner, E. Discovery of zymase.*

Buchner's discovery that an enzyme (a nonliving substance) manufactured by yeast cells was responsible for fermentation resolved many problems that had baffled Pasteur and other investigators. Zymase is now known to consist of a number of enzymes.

Nobel Laureate (1907).

× **1897:** *Canadian Geological Survey. Dinosaur fauna of Alberta, Canada.*

The discovery and exploration of the Upper Cretaceous fossil beds along the Red Deer River in Alberta revealed one of the richest dinosaur faunal finds known.

†○ **1897:** *Eijkman, Christian. Discovery of the cause of a dietary deficiency disease.*

Eijkman's pioneer work on the causes of beriberi led to the isolation of the antineuritic vitamin (thiamine). This work may be called the key discovery that resulted in the development of the important vitamin concept.

Nobel Laureate (1929).

†□ **1897:** *Ross, Ronald. Life history of malarial parasite* (Plasmodium).

This notable achievement represents a great landmark in the field of parasitology and the climax of the work of many investigators on the problem.

Nobel Laureate (1902).

△○ **1897:** *Sherrington, C.S. Concept of the synapse in the nervous system.*

If the nervous system is composed of discrete units, or neurons, functional connections must exist between these units. Sherrington showed how individual nerve cells could exert integrative influences on other nerve cells by graded excitatory or inhibitory synaptic actions. The electron microscope has in recent years added much to our knowledge of synaptic structure.

Nobel Laureate (1932).

△ **1898:** *Benda, C., and C. Golgi. Discovery of mitochondria and the Golgi apparatus.*

These interesting cytoplasmic inclusions were actually seen by various observers before this date, but they were both named in 1898, and the real study of them began at this time. W. Flemming and R. Altmann first demonstrated mitochondria. The Golgi apparatus was demonstrated by V. St. George (1867) and G. Platner (1855), but Camillo Golgi, with his silver nitrate impregnation method, gave the first clear description of the apparatus in nerve cells. Mitochondria are now known to play an important role in cell respiration and energy production.

× **1898:** *Osborn, Henry F. Concept of adaptive radiation in evolution.*

This concept states that, starting from a common ancestral type, many different forms of evolutionary adaptations may occur. In this way evolutionary divergence can take place, and the occupation of a variety of ecological niches is made possible, according to the adaptive nature of the invading species.

× **1900:** *Chamberlain, T.C. Hypothesis of the freshwater origin of vertebrates.*

The evidence for this hypothesis as first proposed was based mainly on the discovery of early vertebrate fossils in sediments thought to be of freshwater origin, such as Old Red Sandstone. The fossil evidence has been reevaluated by recent investigators, who interpret it to support a marine origin of the vertebrates.

‡ **1900:** *Correns, Karl E., Erick Tschermak von Seysenegg, and Hugo De Vries. Rediscovery of Mendel's laws of heredity.*

In their genetic experiments on plants these three investigators independently obtained results similar to Mendel's, and in their survey of the literature they found that Mendel had published his now-famous laws in 1866. A few years later, W. Bateson and others found that the same laws applied to animals.

× **1900:** *Andrews, C.F. Discovery of fossil beds in the Fayum Depression region of Egypt.*

Here Andrews and others discovered numerous Eocene and Oligocene fossils of protomonkeys believed to be ancestral to Old World monkeys and thus in the lineage leading to hominids.

†○ **1900:** *Landsteiner, Karl. Discovery of blood groups.*

This fundamental discovery made possible successful blood transfusions and initiated intense work on the biochemistry of blood.

Nobel Laureate (1930).

∗ **1900:** *Loeb, Jacques. Discovery of artificial parthenogenesis.*

The possibility of getting eggs that normally undergo fertilization to develop by chemical and mechanical methods has been accomplished in a number of different animals, from sea urchin and frog eggs (Loeb, 1900) to the rabbit egg (Pincus, 1936). The phenomenon has been a useful tool in studying the biology of fertilization.

‡ **1901:** *Montgomery, T.H. Homologous pairing of maternal and paternal chromosomes in the zygote.*

Montgomery showed that in synapsis before reduction division, each pair is made up of a maternal and a paternal chromosome. This phenomenon, verified a year later by W.S. Sutton, is of fundamental importance in the segregation of hereditary factors (genes).

×‡ **1901:** *De Vries, Hugo. Mutation theory of evolution.*

De Vries concluded from his study of the evening primrose *Oenothera lamarckiana* that new characters appear suddenly and are inheritable. Although the variations in *Oenothera* were probably not mutations at all, since many of them represented hybrid combinations, the evidence that mutations are the ultimate source of all new variation has steadily mounted.

‡ **1902:** *McClung, Clarence E. Discovery of sex chromosomes.*

The discovery in the grasshopper that a certain chromosome (X) had a synaptic mate (Y) different in appearance or else lacked a mate altogether gave rise to the theory that certain chromosomes determined sex. H. Henking actually discovered the X chromosome in 1891.

○ **1903:** *Bayliss, William M., and Ernest H. Starling. Action of the hormone secretin.*

The demonstration of the action of secretin, a hormone released from the mucosa of the stomach, marked the real birth of the science of endocrinology. It was the first unequivocal proof that physiological functions could be chemically integrated without participation of the nervous system.

‡ **1903:** *Boveri, Theodor, and W.S. Sutton. Parallelism between chromosome behavior and Mendelian segregation.*

This concept states that synaptic mates in meiosis correspond to the Mendelian alternative characters and that the formula of character inheritance of Mendel could be explained by the behavior of the chromosomes during maturation. This is therefore a cytological demonstration of Mendelian genetics.

‡ **1903:** *Sutton, W.S. Constitution of the diploid group of chromosomes.*

Sutton related Mendelian transmission of inherited characters to cytology by showing that the diploid group of chromosomes is made up of two chromosomes of each recognizable size, one member of which is paternal and the other maternal in origin.

○ **1904:** *Cannon, Walter B. Mechanics of digestion observed by means of x-ray films.*

The clever application of x-rays to a study of the movements and other aspects of the digestive system has revealed an enormous amount of information on the physiology of the alimentary canal. Cannon first used this technique in 1898.

○ **1904:** *Carlson, Anton J. Pacemaking activity of neurogenic hearts.*

Heartbeats may originate in muscle (myogenic hearts) or in ganglion cells (neurogenic hearts). Carlson showed that in the arthropod *Limulus* the pacemaker was located in certain ganglia on the dorsal surface of the heart and that experimental alterations of these ganglia by temperature or other means altered the heart rate. Neurogenic hearts are restricted to certain invertebrates.

□ **1904:** *Jennings, Herbert S. Behavior patterns in protozoa.*

The careful investigations of this life-long student of the behavior of these lower organisms led to concepts such as the trial-and-error behavior and many of our important beliefs about the various forms of tropisms and taxes.

□ **1904:** *Nuttall, George H.F. Serological relationships of animals.*

This method of determining animal relationships is striking evidence of evolution. It has been used in recent years to help establish the taxonomic position of animals.

○ **1905:** *Haldane, John B.S., and John G. Priestley. Role of carbon dioxide in the regulation of breathing.*

By their clever technique of obtaining samples of air from the lung alveoli, these investigators showed how the constancy of carbon dioxide concentration in the alveoli and its relation to the concentration in the blood were the chief regulators of the mechanism of respiration.

‡ **1906:** *Bateson, William, and Reginald C. Punnett. Discovery of linkage of hereditary units.*

Although Bateson and Punnett first discovered linkage in sweet peas, it was Morgan and associates who correctly interpreted this great genetic concept. All seven pairs of Mendel's alternative characteristics were on separate chromosomes, which simplified the problem.

○ **1906:** *Einthoven, W. Mechanism of the electrocardiogram.*

The invention of the string galvanometer (1903) by Einthoven supplied a precise tool for measuring the bioelectrical activity of the heart and quickly led to the electrocardiogram, which gives accurate information about disturbances of the heart's rhythm.

Nobel Laureate (1924).

○ **1906:** *Hopkins, Frederick G. Analysis of dietary deficiency.*

Hopkins tried to explain dietary deficiency by a biochemical investigation of the lack of essential amino acids in the diet, an approach that has led to many important investigations in nutritional requirements.

Nobel Laureate (1929).

△ **1906:** *Tswett, M. Principle of chromatography.*

This is the separation of chemical components in a mixture by differential migration of materials according to structural properties within a special porous sorptive medium. It was refined and became a widely used method 30 or 40 years later (Martin and Synge, 1941). The

technique was first used by Tswett to separate pigments of plants.

† **1906-1913:** *Gorgas, William C. Control of malaria and yellow fever.*

By intelligent and industrious application of the knowledge that mosquitoes carry malaria and yellow fever, Gorgas saved 71,000 lives and made possible the construction of the Panama Canal.

✳ **1907:** *Boveri, Theodor. Qualitative differences of chromosomes.*

Boveri showed in his classic experiment with sea urchin eggs that chromosomes have qualitatively different effects on development. He found that only those cells that had one of each kind of chromosome developed into larvae; those cells that did not have representatives of each kind of chromosome failed to develop.

○△ **1907:** *Harrison, Ross. Tissue culture technique.*

The culturing of living tissues in vitro, that is, outside the body, has given biologists an important tool for studying tissue structure and growth. Harrison overcame great technical difficulties to successfully culture living tissues in suitable media.

× **1907:** *Discovery of the Heidelberg fossil human* (Homo heidelbergensis *now* H. erectus heidelbergensis).

This fossil consisted of a lower jaw with all its teeth. The jaw shows many simian characteristics, but the teeth show patterns of primitive humans. This type is supposed to have existed during mid-Pleistocene times and is more or less intermediate between Java man and modern humans.

○ **1907:** *Hopkins, Frederick G. Relationship of lactic acid to muscular contraction.*

Hopkins showed that, after being formed in muscular contraction, a part of the lactic acid is oxidized to furnish energy for the resynthesis of the remaining lactic acid into glycogen. This discovery did much to clarify part of the metabolic reactions involved in the complicated process of muscular contraction.

Nobel Laureate (1929).

✳ **1907:** *Wilson, H.V. Reorganization of sponge cells.*

In this classic experiment, Wilson showed that the disaggregation of sponges, by squeezing them through fine silk bolting cloth so that they are separated into minute cell clumps, resulted in the surviving cells coming together and organizing themselves into small sponges when they were in seawater.

‡ **1908:** *Garrod, A.E. Discovery that gene products are proteins.*

Certain hereditary disorders are caused by enzyme deficiencies wherein the mutations of a single gene may be responsible for controlling the specificity of a particular enzyme in certain disorders. Garrod's work was largely ignored until 1940.

×‡ **1908:** *Hardy, Godfrey H., and W. Weinberg. Hardy-Weinberg population formula.*

This important theorem states that, in the absence of factors (such as mutation and selection) causing change in gene frequency, the frequency of a particular gene in any large population will reach an equilibrium in one generation and thereafter will remain stable regardless of whether the genes are dominant or recessive. Its mathematical expression forms the basis for the calculations of population genetics.

○△ 1909: *Arrhenius, S., and S.P.L. Sörensen. Dissociation theory.*

The sensitivity of most biological systems to acid and alkaline conditions has made pH values of the utmost importance in biological research.

Arrhenius, Nobel Laureate (1903).

‡ 1909: *Castle, William E., and J.C. Philips. The inviolability of germ cells to somatic cell influences.*

That germ cells are relatively free from somatic cell influences was shown by the substitution of a black guinea pig ovary in a white guinea pig, which gave rise to black offspring when mated with a black male.

× 1909: *Johannsen, W.L. Limitations of natural selection on pure lines.*

This investigator found that when a hereditary group of characteristics becomes genetically homogeneous, natural selection cannot change the genetic constitution with regard to these characteristics. He showed that selection could not itself create new genotypes but was effective only in isolating genotypes already present in the group; it therefore could not effect evolutionary changes directly.

†□ 1909: *Nicolle, Charles J.H. Body louse as vector of typhus fever.*

The demonstration that typhus fever was transmitted from patient to patient by the bite of the body louse paved the way for the control of this dreaded epidemic disorder by using DDT to delouse populations.

Nobel Laureate (1928).

○ 1910: *Dale, Henry H. Nature of histamine.*

Dale and colleagues found that an extract from ergot had the properties of histamine (β-imidazolyl ethylamine), which can be produced synthetically by splitting off carbon dioxide from the amino acid histidine. It is also a constituent of all tissue cells, from which it may be released by injuries or other causes. The pronounced effect of histamine in dilating small blood vessels, contracting smooth muscle, and stimulating glands has caused it to be associated with many physiological phenomena, such as anaphylaxis, shock, and allergies.

Nobel Laureate (1936).

† 1910: *Ehrlich, Paul. Chemotherapy in the treatment of disease.*

The discovery of salvarsan as a cure for syphilis represents the first great discovery in this field. Another was the dye sulfanilamide, discovered by Domagk in 1935.

Nobel Laureate (1908).

□ 1910: *Heinroth, O. Concept of imprinting as a type of behavior.*

This is a special type of learning that is demonstrated by birds (and possibly other animals). It is based on the fact that in many species of birds the hatchlings are attracted to the first large object they see and thereafter will follow that object to the exclusion of all others.

‡ 1910: *Morgan, Thomas H. Discovery of sex linkage.*

Morgan and colleagues discovered that the results of a cross between a white-eyed male and a red-eyed female in *Drosophila* were different from those obtained from the reciprocal cross of a red-eyed male with a white-eyed female. This was a crucial experiment because it showed for the first time that how a trait behaved in heredity depended on the sex of the parent, in contrast to most Mendelian characters, which behave genetically the same way whether introduced by a male or female parent.

‡ 1910-1920: *Morgan, Thomas H. Establishment of the theory of the gene.*

The extensive work of Morgan and associates on the localization of hereditary factors (by genetic experiments) on the chromosomes of the fruit fly *(Drosophila)* represents some of the most significant work ever performed in the field of heredity.

Nobel Laureate (1933).

□ 1910: *Murray, J., and J. Hjort. Deep-sea expedition of the* Michael Sars.

Of the many expeditions for exploring the depths of the oceans, the *Michael Sars* expedition, made in the North Atlantic regions, must rank among the foremost. The expedition yielded an immense amount of information about deep-sea animals, as well as important data regarding the ecological pattern of animal distribution in the sea.

□ 1910: *Pavlov, Ivan P. Concept of the conditioned reflex.*

The idea that acquired reflexes play an important role in the nervous reaction patterns of animals has greatly influenced the development of modern psychology.

Nobel Laureate (1904).

✳ 1911: *Child, Charles M. Axial gradient hypothesis.*

This hypothesis attempts to explain the pattern of metabolism from the standpoint of localized regional differences along the axes of organisms. The differences in the metabolic rate of different areas have made possible an understanding of certain aspects of regeneration, development, and growth.

× 1911: *Cuénot, L. The preadaptation concept.*

Preadaptation refers to a morphological or physiological characteristic that has been selected for (arisen) in one environment, but that is coincidentally adaptive in a new environment. This favorable conjunction of characteristics and suitable environment is considered to be an important factor in progressive evolution (opportunistic evolution).

† 1911: *Funk, Casimer. Vitamin hypothesis.*

Vitamin deficiency diseases are those in which effects can be definitely traced to the lack of some essential constituent of the diet. Thus, beriberi is caused by an insufficient amount of thiamine, scurvy by a lack of vitamin C, and so on.

✳ 1911: *Harvey, E.B. Cortical changes in the egg during fertilization.*

In the mature egg, cortical granules gather at the surface of the egg, but on activation of the egg these granules, beginning at the point of sperm contact, disappear in a wavelike manner around the egg. These granules are supposed to release material during their breakdown that helps form the ensuing fertilization membrane.

△ 1911: *Rutherford, Ernest. Concept of the atomic nucleus.*

This cornerstone of modern physics must be of equal interest to the biologist in the light of the rapid advances of molecular biology. Rutherford also discovered (1920) the proton, one of the charged particles in the atomic nucleus.

Nobel Laureate (1908).

× 1911: *Walcott, Charles D. Discovery of Burgess shale fossils.*

The discovery of a great assemblage of beautifully preserved invertebrates in the Burgess shale of British Columbia and their careful study by an American paleontologist represent a landmark in the fossil record of invertebrates. These fossils date from the middle Cambrian age and include the striking *Aysheaia*, which has a resemblance to the extant *Peripatus*.

○✳ 1912: *Gudernatsch, J.F. Role of the thyroid gland in the metamorphosis of frogs.*

This investigator found that the removal of the thyroid gland of tadpoles prevented metamorphosis into frogs, and also that the feeding of thyroid extracts to tadpoles induced precocious metamorphosis. In 1919 W.W. Swingle showed that the presence and absence of inorganic iodine would produce the same results.

□ 1912: *Wegener, Alfred L. Concept of continental drift.*

This theory postulates that the continents were originally joined together in one or two large masses that gradually broke up during geological time, and the fragments drifted apart to form the current landmasses. The theory has been revived recently on the basis of surveys of paleomagnetism, seismographic studies, tectonic plate studies, and a wealth of biological evidence, for example, the finding of an amphibian fossil in Antarctic regions.

△ 1913: *Michaelis, L., and M. Menten. Enzyme-substrate complex.*

On theoretical and mathematical grounds these investigators showed that an enzyme forms an intermediate compound with its substrate (the enzyme-substrate complex) that subsequently decomposes to release the free enzyme and the reaction products. D. Keilin of Cambridge University and B. Chance of the University of Pennsylvania later demonstrated the presence of such a complex by color changes and measurements of its rate of formation and breakdown that agree with theoretical predictions.

× **1913:** *Reck, H. Discovery of Olduvai Gorge fossil deposits.*

This region in East Africa has yielded an immense number of early mammalian fossils as well as the tools of Stone Age humans, such as stone axes. Among the interesting fossils discovered were elephants with lower jaw tusks, horses with three toes, the odd ungulate (chalicothere) with claws on the toes, and, more recently, several hominid fossils (see Leakey, Mary D., 1959).

□ **1913:** *Shelford, Victor E. Law of ecological tolerance.*

This law states that the potential success of an organism in a specific environment depends on how it can adjust within the range of its toleration to the various factors to which the organism is exposed.

‡ **1913:** *Sturtevant, A.H. Formation of first chromosome map.*

By the method of crossover percentages, it has been possible to locate the genes in their relative positions on chromosomes—one of the most fruitful discoveries in genetics, for it led to the extensive mapping of the chromosomes in *Drosophila.*

○ **1913:** *Tashiro, Shiro. Metabolic activity of propagated nerve impulse.*

The detection of slight increases in carbon dioxide production in stimulated nerves, as compared with inactive ones, was evidence that conduction in nerves is a chemical change. Later (1926), A.V. Hill was able to measure the heat given off during the passage of an impulse. Increased oxygen consumption and other metabolic changes have also been measured in excited nerves.

○ **1914:** *Kendall, Edward C. Isolation of thyroxine.*

The isolation of thyroxine in crystalline form was a landmark in endocrinology. It was artificially synthesized by Harington in 1927.

✳ **1914:** *Lillie, Frank R. Role of fertilizin in fertilization.*

According to this theory, the jelly coat of eggs contains a substance, fertilizin, now known to be an acid mucopolysaccharide, which combines with the antifertilizin on the surface of sperm and causes the sperm to clump together.

‡ **1914:** *Shull, George H. Concept of heterosis.*

When two standardized strains or races are crossed, the resulting hybrid generation may be noticeably superior to both parents as shown by greater vigor, vitality, and resistance to unfavorable environmental conditions. First worked out in corn (maize), such hybrid vigor may also be manifested by other kinds of hybrids. Although its exact nature is still obscure, the phenomenon may be caused by the bringing together in the hybrid of many dominant genes of growth and vigor that were scattered among the two inbred parents, or it may be caused by the complementary reinforcing action of genes when brought together.

‡ **1916:** *Bridges, Calvin B. Discovery of nondisjunction.*

Bridges explained an aberrant genetic result by a suggested formula that later he was able to confirm by cytological examination. A pair of chromosomes failed to disjoin at the reduction division so that both chromosomes passed into the same cell. It was definite proof that genes are located on the chromosomes.

□ **1917:** *Grinnell, Joseph. Concept of the ecological niche.*

The niche is the role of an animal in relation to all of the resources in its environment, both physical and biological. C. Elton has done much to develop the concept.

○ **1918:** *Krogh, August. Regulation of the motor mechanism of capillaries.*

The mechanism of the differential distribution of blood to the various tissues has posed many problems from the days of Harvey and Malpighi, but Krogh showed that capillaries were not merely passive in this distribution but that they had the power to contract or dilate actively, according to the needs of the tissues. Both nervous and chemical controls are involved.

Nobel Laureate (1920).

□ **1918:** *Loeb, Jacques. Forced movements, tropisms, and animal conduct.*

Regarded as the founder of a mechanist "school," Loeb opposed the generally anthropomorphic and teleological interpretations of animal behavior. While Loeb carried his mechanistic doctrine to an extreme, he did inspire more rigorous, experimental approaches in animal behavior studies.

○ **1918:** *Starling, Ernest H. The law of the heart.*

Within physiological limits, the more the ventricles are filled with incoming blood, the greater is the force of their contraction at systolé. This is an adaptive mechanism for supplying more blood to tissues when it is needed. It is now recognized that only the hearts of lower vertebrates obey Starling's law; the intact heart of mammals is regulated principally by nervous and hormonal controls.

1919: *Aston, Francis W. Discovery of isotopes.*

Radioactive isotopes have proved of the utmost value in biological research because of the possibility of tracing the course of various elements in living organisms.

Nobel Laureate (1922).

△ **1920:** *Herzog, R.O., and W. Jancke. Development of x-ray diffractometry.*

When a parallel beam of x-rays is passed through crystallized biological material, the rays are spread and an image of the diffracted pattern is recorded (by rings, spots, and the like). By measurements and mathematical calculations information about structure can be obtained.

△ **1921:** *Hopkins, Frederick G. Isolation of glutathione.*

The discovery of this sulfur-containing compound gave a great impetus to the study of the complicated nature of cellular oxidation and metabolism.

Nobel Laureate (1929).

○ **1921:** *Langley, John N. Concept of the functional autonomic nervous system.*

Langley's concept of the functional aspects of the autonomic system dealt mainly with the mammalian type, but the fundamental principles of the system have been applied with modification to other groups as well. Two divisions—sympathetic and parasympathetic—are recognized in the functional interpretation of excitation and inhibition in the antagonistic nature of the two divisions.

○ **1921:** *Loewi, Otto, and H.H. Dale. Isolation of acetylcholine.*

This key demonstration led to the neurohumoral concept of the transmission of nerve impulses to muscles.

Nobel Laureates (1936).

○ **1921:** *Richards, A.N. Collection and analysis of glomerular filtrate of the kidney.*

This experiment was direct evidence of the role of the glomeruli as mechanical filters of cell-free and protein-free fluid from the blood and was striking confirmation of the Ludwig-Cushny theory of kidney excretion.

✳ **1921:** *Spemann, Hans. Organizer concept in embryology.*

Certain parts of the developing embryo known as organizers induce specific developmental patterns in responding tissue. While many aspects of induction remain obscure, Spemann's work was important in showing that development is a sequence of programmed stages and that each stage is necessary for the next.

Nobel Laureate (1935).

†○ **1921, 1922:** *Banting, Frederick, C.H. Best, and J.J.R. Macleod. Extraction of insulin.*

The great success of this hormone in relieving a distressful disease, diabetes mellitus, and the dramatic way in which active extracts were obtained have made the isolation of this hormone the best known in the field of endocrinology.

Banting and Macleod, Nobel Laureates (1923).

○ **1922:** *Erlanger, Joseph, and H.S. Gasser. Differential conduction of nerve impulses.*

By using the cathode-ray oscillograph, these investigators found that there were several different types of mammalian nerve fibers that could be distinguished structurally and that had different rates of conducting nervous impulses according to the thickness of the nerve sheaths (most rapid in the thicker ones).

Nobel Laureates (1944).

○ **1922:** *Kopeč, S. Concept of neurosecretory systems.*

This concept has emerged from the work of many investigators. R. Cajalin (1899) discovered that the nerve fibers to the neurohypophysis are extensions of the neurons of the hypothalamus. Kopeč found that the substance responsible for metamorphosis in the larva of moths originated in the brain where certain cerebral ganglia served as glands of internal secretion. Kopec's pioneering studies were largely ignored by vertebrate endocrinologists until the 1950s, when through the studies of B. Scharrer, E. Scharrer, B. Hanström, G.W. Harris, and others it was recognized that vertebrates possess a neurosecretory system (hypothalamo-hypophyseal) analogous to invertebrate systems.

□ 1922: *Schjelderup-Ebbe, T. Social dominance-subordinance hierarchies.*

This observer found certain types of social hierarchies among birds in which higher-ranking individuals could peck those of lower rank without being pecked in return. Those of the first rank dominated those of the second rank, who dominated those of the third, and so on. Such an organization, once formed, may be permanent.

□ 1922: *Schmidt, Johann. Life history of the freshwater eel.*

The long, patient work of this oceanographer in solving the mystery of eel migration from the freshwater streams of Europe to their spawning grounds in the Sargasso Sea near Bermuda represents one of the most romantic achievements in natural history.

△ 1923: *de Hevesy, George. First isotopic tracer method.*

Tracer methodology has proved especially useful in biochemistry and physiology. For instance, it has been possible by the use of these labeled units to determine the fate of the particular molecule in all steps of a metabolic process and the nature of many enzymatic reactions. The exact locations of many elements in the body have been traced by this method.

Nobel Laureate (1943).

△ 1923: *Warburg, Otto. Manometric methods for studying metabolism of living cells.*

The Warburg apparatus has been useful in measuring gas exchange and other metabolic processes of living tissues. It proved of great value in the study of enzymatic reactions in living systems and was a standard tool in many biochemical laboratories for many years.

Nobel Laureate (1931).

○ 1924: *Cleveland, L.R. Symbiotic relationships between termites and intestinal flagellates.*

This study was made on one of the most remarkable examples of mutualism known in the animal kingdom. Equally important were the observations this investigator and others made of the symbiosis between the wood-roach (*Cryptocercus*) and its intestinal protozoa.

○ 1924: *Houssay, Bernardo A. Role of the pituitary gland in regulation of carbohydrate metabolism.*

This investigator showed that, when a dog had been made diabetic by pancreatectomy, the resulting hyperglycemia and glucosuria could be abolished by removing the anterior pituitary gland.

Nobel Laureate (1947).

× 1925: *Dart, Raymond. Discovery of Australopithecus africanus.*

This important fossil, once referred to as the "missing link," bore many human characteristics but had a brain capacity only slightly greater than the higher apes. Most authorities place this hominid on or near the main branch of human ancestry.

△ 1925: *Mast, S.O. Nature of ameboid movement.*

By studying the reversible sol-gel transformation in the protoplasm of an ameba, Mast not only gave a logical interpretation of ameboid movement but also initiated many fruitful concepts about the contractile nature of cytoplasmic gel systems, such as the furrowing movements (cytokinesis) in cell divisions.

†○ 1925: *Minot, G.R., M.W.P. Murphy, and G.H. Whipple. Liver treatment of pernicious anemia.*

These investigators discovered that the feeding of raw liver had a pronounced effect in the treatment of pernicious anemia (a serious blood disorder). Much later investigation by numerous workers has led to some understandings of the antipernicious factor, vitamin B_{12}, whose chemical name is cyanocobalamin. This complex vitamin contains, among other components, porphyrin that has a cobalt atom instead of iron or magnesium at its center.

Nobel Laureates (1934).

□ 1925: *Rowan, William. Photoperiodism hypothesis of bird migration.*

Rowan demonstrated that birds subjected to artificially increased day length in winter increased the size of their gonads and showed a striking tendency to migrate out of season. Other workers, stimulated by Rowan's experiments, showed that naturally increasing day length in spring acts through the brain neurosecretory system to set in motion the migratory disposition in birds.

△ 1926: *Sumner, James B. Isolation of enzyme urease.*

The isolation of the first enzyme in crystalline form was a key discovery and was followed by others that have helped unravel the complex nature of these important biological substances.

Nobel Laureate (1946).

○ 1927: *Bozler, E. Analysis of nerve net components.*

Bozler's demonstration that the nerve net of cnidarians was made up of separate cells and contained synaptic junctions resolved the old problem of whether or not the plexus in this group of animals was an actual network.

○ 1927: *Eggleton, P., G.P. Eggleton, C.H. Fiske, and Y. Subbarow. Role of phosphagen (phosphocreatine) in muscular contraction.*

The demonstration that phosphagen is broken down during muscular contraction and then is resynthesized during recovery gave an entirely new concept of the initial energy necessary for the contraction process. Confirmation of this discovery received a great impetus from the discovery of E. Lungsgaard (1930) that

muscles poisoned with monoiodoacetic acid, which inhibits the production of lactic acid from glycogen, would still contract and that the amount of phosphocreatine broken down was proportional to the energy liberated. It is now known that phosphocreatine is a high-energy phosphate storage compound, readily donating its phosphate group to ADP to form ATP.

○ 1927: *Heymans, Corneille. Role of carotid and aortic reflexes in respiratory control.*

The carotid sinus and aortic areas contain pressoreceptors and chemoreceptors, the former responding to mechanical stimulation, such as blood pressure, and the latter to oxygen lack. When the pressoreceptors are stimulated, respiration is inhibited; when the chemoreceptors are stimulated, the respiratory rate is increased. (A much more potent stimulator of respiration, however, is the effect of carbon dioxide on the respiratory center of the brain.)

Nobel Laureate (1938).

‡ 1927: *Muller, Hermann J. Artificial induction of mutations.*

By subjecting fruit flies (*Drosophila*) to mild doses of x-rays, Muller found that the rate of mutation could be increased 150 times over the normal rate.

Nobel Laureate (1946).

× 1927: *Stensiö, E.A. Appraisal of the Cephalaspida (ostracoderm) fish fossil.*

The replacement of amphioxus as a prototype of vertebrate ancestry by the ammocoete lamprey larva, currently of great interest, has to a great extent been the result of this careful fossil reconstruction. It is generally believed that living Agnatha (lampreys and hagfishes) are descended from these ancient forms.

× 1928: *Garstang, Walter. Hypothesis of the ascidian ancestry of chordates.*

According to this hypothesis, primitive chordates were sessile, filter-feeding marine organisms very similar to present-day ascidians. The actively swimming prevertebrate was considered a later stage in chordate evolution. The tadpole ascidian larva, with its basic organization of a vertebrate, had evolved within the group by progressive evolution and by neoteny became sexually mature, ceased to metamorphose into a sessile, mature ascidian, and became the true vertebrate.

‡△ 1928: *Griffith, F. Discovery of the transforming principle (DNA) in bacteria (genetic transduction).*

By injecting living nonencapsulated bacteria and dead encapsulated bacteria of the *Pneumococcus* strain into mice, it was found that the former acquired the ability to grow a capsule and that this ability was transmitted to succeeding generations. This active agent or transforming principle was isolated from the encapsulated type by other workers later and was found to consist of DNA.

F. Sanfelice (1893) had actually found the same principle when he discovered that nonpathogenic bacilli grown in a culture medium containing the metabolic products of true tetanus bacilli would also produce toxins and would do so for many generations.

△○ **1928:** *Weiland, H., and A. Windaus. Structure of the cholesterol molecule.*

Sterol chemistry has been a focal point in the investigation of such biological products as vitamins, sex hormones, and cortisone. Cholesterol is the precursor of bile acids and steroid hormones in animals.

Nobel Laureates (1927, 1928).

○ **1929:** *Berger, Hans. Demonstration of brain waves.*

The science of electroencephalography, or the electrical recording of brain activity, has revealed much about both the healthy and the diseased brain.

○ **1929:** *Butenandt, A., and E.A. Doisy. Isolation of estrone.*

This discovery was the first isolation of a sex hormone and was arrived at independently by these two investigators. Estrone was found to be the urinary and transformed product of estradiol, the actual hormone. The male hormone testosterone was synthesized by Butenandt and L. Ruzicka in 1931. The second female hormone, progesterone, was isolated from the corpora lutea of sow ovaries in 1934.

Doisy, Nobel Laureate (1943).

† **1929:** *Fleming, Alexander. Discovery of penicillin.*

The chance discovery of this drug from molds and its development by H. Florey a few years later gave us the first of a notable line of antibiotics that have revolutionized medicine.

Nobel Laureate (1945).

○ **1929:** *Heymans, Corneille. Discovery of the role of the carotid sinus and bodies in regulating the respiratory center and arterial blood pressure.*

Heymans found that the carotid sinus contains pressure-sensitive receptors (baroreceptors) that, acting through a nervous reflex to the medulla, help to keep the blood pressure stabilized. He also discovered oxygen-sensitive chemoreceptors in the carotid and aortic bodies. These are the origin of a reflex that responds to a drop in blood oxygen tension by increasing respiratory rate and blood pressure.

Nobel Laureate (1938).

△ **1929:** *Lohmann, K., C. Fiske, and Y. Subbarow. Discovery of ATP.*

The discovery of ATP (adenosine triphosphate) was the culmination of a long search for the direct source of energy in biochemical reactions of many varieties, such as muscular contraction, vitamin action, and many enzymatic systems.

× **1930:** *Fisher, R.A. Statistical analysis of evolutionary variations.*

With Sewall Wright and J.B.S. Haldane, Fisher analyzed mathematically the interrelationships of the factors of mutation rates, population sizes, selection values, and others in the evolutionary process. Although many of their theories are in the empirical stage, evolutionists

in general agree that they have great significance in evolutionary interpretation.

‡ **1931:** *Stern, C., H. Creighton, and B. McClintock. Cytological demonstration of crossing over.*

Proof that crossing over in genes is correlated with exchange of material by homologous chromosomes was independently obtained by Stern in *Drosophila* and by Creighton and McClintock in corn. By using crosses of strains that had homologous chromosomes distinguishable individually, they definitely demonstrated cytologically that genetic crossing over was accompanied by chromosomal exchange.

○□ **1932:** *Bethe, A. Concept of the ectohormone (pheromone).*

An organism secretes these substances to the outside of the body, where they affect the physiology or behavior of another individual of the same species. They have been demonstrated in insects, where they may function in trail making, sex attraction, development control, and the like.

× **1932:** *Danish scientific expedition. Discovery of fossil amphibians (ichthyostegids).*

These fossils were found in the upper Devonian sediments in eastern Greenland and appear to be intermediate between advanced crossopterygians *(Osteolepis)* and early amphibians. They are the oldest known forms that can be considered amphibians. Many of their characters show primitive amphibian conditions.

× **1932:** *Lewis, G. Edward. Discovery of Ramapithecus fossil.*

Ramapithecus brevirostris was a small humanlike ape of the Miocene and the earliest known hominid. Since Lewis's initial find in India, other fossil fragments of this genus have been discovered in East Africa, Greece, Turkey, and Hungary.

×‡ **1932:** *Wright, Sewall. Genetic drift as a factor in evolution.*

In small populations the Hardy-Weinberg formula of gene frequency may not apply because of "sampling error." If some genotypes are by chance underrepresented in matings in a small population, gene frequencies in the population may change.

○ **1933:** *Goldblatt, M., and U.S. von Euler. Discovery of prostaglandins.*

These fatty acid derivative compounds have been isolated from many mammalian tissues (such as seminal plasma, pancreas, seminal vesicle, brain, and kidney). They have a variety of hormonelike and pharmacological actions, such as stimulating smooth muscle contractions and relaxation, lowering blood pressure, and inhibiting enzymes and hormones.

× **1933, 1938:** *Haldane, John B.S., and A.I. Oparin. Heterotroph hypothesis of the origin of life.*

This hypothesis is based on the idea that life was generated from nonliving matter under the conditions that existed before the appearance of life and that have not been duplicated since. The hypothesis stresses the idea that living sys

tems at present make it impossible for any incipient life to gain a foothold as primordial life was able to do.

‡ **1933:** *Painter, T.S., E. Heitz, and H. Bauer. Rediscovery of giant salivary chromosomes.*

These interesting chromosomes were first described by Balbiani in 1881, but their true significance was not realized until these investigators rediscovered them. It has been possible in a large measure to establish the chromosome theory of inheritance by comparing the actual cytological chromosome maps of salivary chromosomes with the linkage maps obtained by genetic experimentation.

○ **1933:** *Wald, George. Discovery of vitamin A in the retina.*

The discovery that vitamin A is a part of the visual purple molecule of the rods not only gave a better understanding of an important vitamin but also showed how night blindness can occur when there is a deficiency of this vitamin in the diet.

Nobel Laureate (1967).

○ **1934:** *Dam, Carl P.H., and E.A. Doisy. Identification of vitamin K.*

The isolation and synthesis of this vitamin are important not merely because of the vitamin's practical value in certain forms of hemorrhage but also because of the light they throw on the physiological mechanism of blood clotting.

Nobel Laureates (1943).

△ **1934-1935:** *Danielli, J.F., and H. Davson. Concept of the cell (plasma) membrane.*

Danielli proposed a hypothesis that cell membranes consist of two layers of lipid molecules surrounded on the inner and outer surfaces by a layer of protein molecules. The electron microscope reveals the plasma membrane of a thickness of 7.5 to 10 nm, consisting of two dark (protein) membranes (2.5 to 3 nm thick) separated by a light (lipid) interval membrane of 2.5 to 3 nm thickness. Recent studies confirm the Danielli-Davson model, while demonstrating a greater dynamic role for the peripheral proteins.

○ **1934:** *Wigglesworth, V.B. Role of the corpus allatum gland in insect metamorphosis.*

This small gland lies close to the brain of an insect, and it has been shown that during the larval stage this gland secretes a juvenile hormone that causes the larval characteristics to be retained. Metamorphosis and maturation occur as the gland secretes less and less of the hormone. Removal of the gland causes the larva to undergo precocious metamorphosis; grafting the gland into a mature larva will cause the latter to grow into a giant larval form. The gland was first described by A. Nabert in 1913.

○ **1935:** *Hanström, B. Discovery of the X-organ in crustaceans.*

This organ, together with the related sinus gland, constitutes an anatomical complex that has proved of great interest in understanding crustacean endocrinology. The neurosecretory cells in the X-organ (part of the brain) produce a molt-inhibiting hormone that is stored in the sinus gland of the eyestalk, while a molting hormone is produced in the Y-organ. The interrelations of these two hormones control the molting process.

○ **1935-1936:** *Kendall, Edward C., and P.S. Hench. Discovery of cortisone.*

Kendall had first isolated this substance, which he called compound E, from the adrenal glands. Its final stages were prepared by Hench later and involved a long tedious chemical process. A hormone that controls the synthesis and release of cortisone and similar steroids was isolated in 1943 from the pituitary and was called adrenocorticotropic hormone (ACTH).

Nobel Laureates (1950).

△ **1935:** *Stanley, Wendell M. Isolation of a virus in crystalline form.*

This achievement of isolating a virus (tobacco mosaic) is noteworthy not only in giving information about these small agents responsible for many diseases but also in affording much speculation on the differences between the living and the nonliving.

Nobel Laureate (1946).

○ **1936:** *Young, J.Z. Demonstration of giant fibers in squid.*

These giant fibers are formed by the fusion of the axons of many neurons whose cell bodies are found in a ganglion near the head. Each fiber is really a tube, up to 1 mm in diameter, consisting of an external sheath filled with liquid axoplasm. Because of their size, they have been a very valuable system in neurophysiology.

△ **1937:** *Findlay, G.W.M., and F.O. MacCullum. Discovery of interferon.*

These protein substances, produced by cells in response to viruses and other foreign substances, selectively inhibit virus replication and are one of the body's important defenses against viral infection. They also may play a role in regulation of the immune response.

△ **1937:** *Krebs, Hans A. Citric acid (tricarboxylic acid) cycle.*

In a brilliant example of reasoning and deduction, Krebs showed the existence of a cycle of reactions central to aerobic energy production. In the cycle, two carbon fragments from carbohydrate, fatty acids, or amino acids are oxidized to carbon dioxide. The electrons derived from the oxidation reactions in the cycle are passed through a further series of reactions (electron transport system), in which energy is captured in ATP, and finally to oxygen as the final electron acceptor to produce water.

Nobel Laureate (1953).

× **1937:** *Sonneborn, T.M. Discovery of mating types in paramecium.*

Sonneborn's discovery that only individuals of complementary physiological classes (mating types) would conjugate opened up a new era of protozoan investigation that promises to shed new light on the problems of species concept and evolution.

△○ **1937:** *König, P., and A. Tiselius. Development of electrophoresis.*

Electrophoresis is an extremely valuable method of separating proteins in a solution and, after numerous refinements, is currently in wide use. It depends on the differential migration of proteins on an inert carrier when an electrical field is applied.

□ **1938:** *Remane, A. Discovery of the new phylum Gnathostomulida.*

This marine phylum was first described by P. Ax in 1956. The members of the phylum are small wormlike forms (about 0.5 mm long) and show great diversity among the different species. One of their major characteristics is their complicated jaws provided with teeth. They live in sandy substrata and are worldwide in distribution.

○ **1938:** *Schoenheimer, R. Use of radioactive isotopes to demonstrate synthesis of bodily constituents.*

By labeling amino acids, fats, carbohydrates, and the like with radioactive isotopes, it was possible to show how these were incorporated into the various constituents of the body. Such experiments demonstrated that pairs of the cell were constantly being synthesized and broken down and that the body must be considered a dynamic equilibrium.

□ **1938:** *Skinner, B.F. Measurement of motivation in animal behavior.*

Skinner worked out a technique for measuring the rewarding effect of a stimulus, or the effects of learning on voluntary behavior. His experimental animals (rats) were placed in a special box (Skinner's box) containing a lever that the animal could manipulate. When the rat pressed the lever, small pellets of food would or would not be released, according to the experimental conditions.

○△ **1938:** *Svedberg, Theodor. Development of ultracentrifuge.*

In biological and medical investigation this instrument has been widely used for the purification of substances, the determination of particle sizes in colloidal systems, the relative densities of materials in living cells, the production of abnormal development, and the study of many problems concerned with electrolytes.

Nobel Laureate (1926).

○ **1939:** *Brown, Frank A., Jr., and O. Cunningham. Demonstration of molt-inhibiting hormone in eyestalk of crustaceans.*

Although C. Zeleny (1905) and others had shown that eyestalk removal shortened the intermolt period in crustaceans, Brown and Cunningham were the first to present evidence to explain the effect as being caused by a molt-inhibiting hormone present in the sinus gland.

× **1939:** *Discovery of coelacanth fishes.*

The collection of a living specimen of this ancient fish *(Latimeria),* followed later by other specimens, has brought about a complete reappraisal of this "living fossil" with reference to its ancestry of the amphibians and land forms.

✳ **1939:** *Hörstadius, S. Analysis of the basic pattern of regulative and mosaic eggs in development.*

The masterful work of this investigator has done much to resolve the differences in the early development of regulative eggs (in which each of the early blastomeres can give rise to a whole embryo) and mosaic eggs (in which isolated blastomeres produce only fragments of an embryo). Regulative eggs were shown to have two kinds of substances, both of which were necessary in proper ratios to produce normal embryos. Each of the early blastomeres has this proper ratio and thus can develop into a complete embryo; in mosaic eggs the regulative power is restricted to a much earlier time scale in development (before cleavage), and thus each isolated blastomere will give rise only to a fragment.

×□ **1939:** *Huxley, Julian. Concept of the cline in evolutionary variation.*

This concept refers to the gradual and continuous variation in character over an extensive area because of adjustments to changing conditions. This idea of character gradients has proved a very fruitful one in the analysis of the mechanism of evolutionary processes, for such a variability helps to explain the initial stages in the transformation of species.

†‡ **1940:** *Landsteiner, Karl, and A.S. Wiener. Discovery of Rh blood factor.*

Not only was a knowledge of the Rh factor of importance in solving a fatal disease of infancy, but it has also yielded a great deal of information about relationships between human races.

Landsteiner, Nobel Laureate (1930).

△‡ **1941:** *Beadle, George W., and E.L. Tatum. Biochemical mutation.*

By subjecting the bread mold *Neurospora* to x-ray irradiation, they found that genes responsible for the synthesis of certain vitamins and amino acids were inactivated (mutated) so that a strain of this mold carrying the mutant genes could no longer grow unless these particular vitamins and amino acids were added to the medium on which the mold was growing. This discovery—that alterations in various genes resulted in loss of a corresponding biosynthetic enzyme—first revealed the precise way in which genes and enzymes are related.

Nobel Laureates (1958).

○ **1941:** *Cori, Carl F., and Gerty T. Cori. Lactic acid metabolic cycle.*

The regeneration of muscle glycogen reserves in mammals involves the passage of lactic acid from the muscles through the blood to the liver, the conversion of lactic acid there to glycogen, the production of blood glucose from the liver glycogen, and the synthesis of muscle glycogen from the blood glucose.

Nobel Laureates (1947).

△○ **1941:** *Martin, A.J.P., and R.L.M. Synge. Development of partition chromatography.*

This is a powerful method for separation of chemically similar substances in mixtures. It depends on differential solubility of a compound in stationary and moving phases of solvents. Variants of the method include column, paper, and thin-layer chromatography.

○ **1941:** *Szent-Györgyi, Albert. Role of ATP in muscular contraction.*

The demonstration showing that muscles get their energy for contraction from ATP (adenosine triphosphate) has done much to explain many aspects of the puzzling problem of muscle physiology.

Nobel Laureate (1937).

✳ **1942:** *McClean, D., and I.M. Rowlands. Discovery of the enzyme hyaluronidase in mammalian sperm.*

This enzyme dissolves the cement substance of the follicle cells that surround the mammalian egg and facilitates the passage of the sperm to the egg. This discovery not only aided in resolving some of the difficult problems of the fertilization process but also offers a logical explanation of cases of infertility in which too few sperm may not carry enough of the enzyme to afford a passage through the inhibiting follicle cells.

△ **1943:** *Claude, A. Isolation of cell constituents.*

By differential centrifugation, Claude found it possible to separate, in relatively pure form, particulate components, such as mitochondria, microsomes, and nuclei. These investigations led immediately to a more precise knowledge of the chemical nature of these cell constituents and aided the elucidation of the structure and physiology of the mitochondria—one of the great triumphs in the biochemistry of the cell.

Nobel Laureate (1974).

✳ **1943:** *Holtfreter, J. Tissue synthesis from dissociated cells.*

By dissociating the cells of embryonic tissues of amphibians (by dissolving with enzymes or other agents the intercellular cement that holds the cells together) and heaping them in a mass, he found that the cells in time coalesced and formed the type of tissue from which they had come. This is an application to vertebrates of Wilson's discovery with sponge cells. It showed that, by some mechanism, cells could recognize other cells of the same type.

‡ **1943:** *Sonneborn, T.M. Extranuclear inheritance.*

The view that in paramecia cytoplasmic determiners (plasmagenes) that are self-reproducing and capable of mutation can produce genetic variability has thrown additional light on the role of the cytoplasm in hereditary patterns.

△‡ **1944:** *Avery, O.T.C., C.M. MacLeod, and M. McCarty. Agent responsible for bacterial transformation.*

These workers were able to show that the bacterial transformation of nonencapsulated bacteria to encapsulated cells was really attributable to the DNA fraction of debris from disrupted encapsulated cells to which the nonencapsulated cells were exposed. This key demonstration showed for the first time that nucleic acids and not proteins were the basic material of heredity, and initiated the study of molecular genetics. However, the profound significance of Avery's conclusions was only gradually recognized.

△ **1945:** *Cori, Carl F. Hormone influence on enzyme activity.*

The delicate balance that insulin and the diabetogenic hormone of the pituitary exercise over the activity of the enzyme hexokinase in carbohydrate metabolism has opened up a whole new field of the regulative action of hormones on enzymes.

Nobel Laureate (1947).

□ **1945:** *Griffin, D., and R. Galambos. Development of the concept of echolocation.*

Echolocation refers to a type of perception of objects at a distance by which echoes of sound are reflected back from obstacles and detected acoustically. These investigators found that bats generated their own ultrasonic sounds that were reflected back to their own ears so that they were able to avoid obstacles in their flight without the aid of vision. Their work climaxed an interesting series of experiments inaugurated as early as 1793 by Spallanzani, who believed that bats avoided obstacles in the dark by reflection of sound waves to their ears. Others who laid the groundwork for the novel concept were C. Jurine (1794), who proved that ears were the all-important organs in perception; H.S. Maxim (1912), who advanced the idea that the bat made use of low-frequency sounds inaudible to human ears; and H. Hartridge (1920), who proposed the hypothesis that bats emitted sounds of high frequencies and short wavelengths (ultrasonic sounds).

△ **1945:** *Lipmann, Fritz A., and Hans Krebs. Discovery of coenzyme A.*

The discovery of this important coenzyme made possible a better understanding of the breaking down of fatty acid chains and furnished important further knowledge of the reactions in the tricarboxylic acid cycle.

Nobel Laureate (1953).

△ **1945:** *Porter, Keith R. Description of the endoplasmic reticulum.*

The endoplasmic reticulum is a very complex cytoplasmic structure consisting of a lacelike network of irregular anastomosing tubules and vesicular expansions within the cytoplasmic matrix. Associated with the reticulum complex are small, dense granules of ribonucleoprotein and other granules known as microsomes that are fragments of the endoplasmic reticulum. The reticulum complex plays an important role in the synthesis of proteins.

‡ **1946:** *Lederberg, J., and E.L. Tatum. Sexual recombination in bacteria.*

These investigators found that two different strains of bacteria (*Escherichia coli*) could undergo conjugation and exchange genetic material, thereby producing a hereditary strain with characteristics of the two parent strains. W. Hayes (1952) found that recombination still occurred after one parent strain was killed.

Nobel Laureates (1958).

✕ **1946:** *Libby, Willard F. Radiocarbon dating of fossils.*

The radiocarbon age determination is based on the fact that carbon 14 in the dead organism disintegrates at the rate of one half in 5560 years, one half of the remainder in the next 5560 years, and so on. This is based on the assumption that the isotope is mixed equally through all living matter and that the cosmic rays (which form the isotopes) have not varied much in periods of many thousands of years. The limitation of the method is around 40,000 years.

Nobel Laureate (1960).

✕ **1946:** *White, E.I. Discovery of the primitive chordate fossil* Jamoytius.

The discovery of this fossil in the freshwater deposits of Silurian rock in Scotland throws some light on the early ancestry of vertebrates, for this form seems to be intermediate between amphioxus or ammocoete larva (of lampreys) and the oldest known vertebrates, the ostracoderms. Morphologically, *Jamoytius* represents the most primitive vertebrate yet discovered. It could well serve as an ancestor of such forms as amphioxus and the jawless ostracoderms.

○ **1947:** *Holtz, P. Discovery of norepinephrine (noradrenaline).*

This hormone (vasoconstriction effects) has since been found in most vertebrates and shares with epinephrine (metabolic effects) the functions of the chromaffin part of the adrenal gland.

✕ **1947:** *Sprigg, R.C. Discovery of Precambrian fossil bed.*

The discovery of a rich deposit of Precambrian fossils in the Ediacara Hills of South Australia has been of particular interest because the scarcity of such fossils in the past has given rise to vague and uncertain explanations about Precambrian life. It was all the more remarkable that the fossils discovered were those of soft-bodied forms, such as jellyfish, soft corals, and segmented worms, including the amazing *Spriggina*, which shows relationship to the trilobites. The fossils are prearthropod but not preannelid.

△ **1947:** *Szent-Györgyi, Albert. Concept of the contractile substance actomyosin.*

This protein complex consists of the two components, actin and myosin, and is considered the source of muscular contraction when triggered by ATP. Neither actin nor myosin will contract singly.

Nobel Laureate (1937).

‡ **1947:** *McClintock, Barbara. Concept of mobile genetic elements.*

Studying the pattern of certain mutations in maize, McClintock deduced that it must be caused by movement of genetic elements ("jumping genes") within the genome. Her observations were ignored or discounted for years until mobile genetic elements were discovered in bacteria in the 1960s, then in many kinds of eukaryotes in the 1970s. We now know that transposable genes help to account for the huge diversity of antibodies in vertebrates.

Nobel Laureate (1983).

□ **1948:** *von Frisch, Karl. Communication mechanisms of honeybees.*

After many years of patient work, von Frisch deciphered the meaning of bee "dances," behavior patterns of individual bees returning to the hive. The dances communicate information to the other bees regarding location of food and water, location and suitability of sites for a new hive, and other information.

Nobel Laureate (1973).

□ **1948:** *Hess, Walter R. Localization of instinctive impulse patterns in the brain.*

By inserting electrodes through the skull, fixing them in position, and allowing such holders to heal in place, it was possible to study the brain of an animal in its ordinary activities. When the rat could automatically and at will stimulate itself by pressing a lever, it did so frequently when the electrode was inserted in the hypothalamus region of the brain, indicating a pleasure center. In this way, by placing electrodes at different centers, rats can be made to gratify such drives as thirst, sex, and hunger.

Nobel Laureate (1949).

△ **1948:** *Hogeboom, G.H., W.C. Schneider, and G.E. Palade. Separation of mitochondria from the cell.*

This was an important discovery in unraveling the amazing enzymatic activity of the mitochondria in the tricarboxylic acid cycle. The role mitochondria play in the energy transfer of the cell has earned for these rod-shaped bodies the appellation "powerhouse" of the cell.

Palade, Nobel Laureate (1974).

△ **1949:** *de Duve, C. Discovery of the lysosome organelle.*

These small particles were first identified chemically, and later (1955) morphologically with the electron microscope. They contain the enzymes for digestion of materials taken into the cell by pinocytosis and phagocytosis and for the digestion of the cell's own cytoplasm. When the cell dies, their enzymes are also released and digest the cell (autolysis).

Nobel Laureate (1974).

† **1949:** *Enders, J.F., F.C. Robbins, and T.H. Weller. Cell culture of animal viruses.*

These investigators found that poliomyelitis virus could be grown in ordinary tissue cultures of nonnervous tissue instead of being restricted to host systems of laboratory animals or embryonated chick eggs.

Nobel Laureates (1954)

○ **1949:** *von Euler, U.S. Role of norepinephrine as transmitter.*

Von Euler isolated and identified norepinephrine as the neurotransmitter in the sympathetic nervous system (1940). Later he isolated and characterized norepinephrine storage granules in nerves and described how the substance was taken up, stored, and released by them.

Nobel Laureate (1970).

‡ **1949:** *Pauling, Linus. Genic control of protein structure.*

Pauling and his colleagues demonstrated a direct connection between specific chemical differences in protein molecules and alterations in genotypes. Making use of the hemoglobin of patients with sickle cell anemia (which is caused by a homozygous condition of an abnormal gene), Pauling was able by the method of electrophoresis to show a considerable difference in the behavior of this hemoglobin in an electric field compared with that from a heterozygote or from a normal person.

Nobel Laureate (1954).

○ **1949:** *Selye, Hans. Concept of the stress syndrome.*

In 1937 Selye began his experiments, which led to what he called the "general adaptation syndrome." This involved the chain reactions of many hormones, such as cortisone and ACTH, in meeting stress conditions faced by an organism. Whenever the stress experience exceeds the limitations of these body defenses, serious degenerative disorders may result.

△ **1950:** *Caspersson, T., and J. Brachet. Biosynthesis of proteins.*

Investigations to solve the vital problem of protein synthesis from free amino acids have been under way since Emil Fischer's (1887) outstanding work on protein structure. Many competent research workers, such as Bergmann, Lipmann, and Schoenheimer, have made contributions to an understanding of protein synthesis, but Caspersson and Brachet were the first to point out the significant role of ribonucleic acid (RNA) in the process—and most biochemical investigations of the problem since that time have been directed along this line.

△ **1950:** *Chargaff, Erwin. Base composition of DNA.*

The discovery that the amount of purines was equal to the amount of pyrimidines in DNA, the amount of adenine was equal to that of thymine, and the amount of cytosine was equal to that of guanine paved the way for the DNA model of Watson and Crick. The two major functions of DNA are replication and hereditary information storage.

□ **1950:** *Lorenz, Konrad. Development of ethology.*

Lorenz is regarded as the founder of ethology, a biological approach to analyzing behavior, emphasizing a comparative method and stressing innate factors in behavior development. His classic studies on the greylag goose first came to the attention of English-speaking scientists during the 1950s, though the work itself commenced in the 1930s.

Nobel Laureate (1973).

□ **1950:** *von Frisch, Karl. Discovery of polarized light perception by honeybees.*

Honeybees are able to detect, and navigate to, polarized sunlight reaching the earth's surface, using specialized retinular cells in a specific but small area of the compound eyes. Subsequently it has been found that some crustaceans, cephalopod molluscs, and teleost fishes, as well as many other terrestrial arthropods, can orient by polarized light.

✳ **1952:** *Briggs, Robert, and Thomas J. King. Demonstration of possible differentiated nuclear genotypes.*

The belief that all cells of a particular organism have the same genetic endowment has been questioned as the result of the work of these investigators who transplanted nuclei of different ages and sources from blastulas and early gastrulas into enucleated zygotes and observed variable and abnormal development.

□ **1952:** *Kramer, G. Orientation of birds to positional changes of sun.*

Kramer showed that birds (starlings and pigeons) can be trained to find food in accordance with the position of the sun. He found that the general orientation of the birds shifted at a rate (when exposed to a constant artificial sun) that could be predicted on the basis of the birds' correcting for the normal rotation of the earth. Birds were able to orient themselves in a definite direction with references to the sun, whether the light of the sun reached them directly or was reflected by mirrors. They were capable also of finding food at any time of day, indicating an ability to compensate for the sun's motion across the sky.

△ **1952:** *Palade, G.E. Analysis of the fine structure of the mitochondrion.*

The important role of mitochondria in the enzymatic systems and cellular metabolism has focused much investigation on the structure of these cytoplasmic inclusions. Each mitochondrion is bounded by two membranes; the outer is smooth and the inner is thrown into small folds or cristae that project into a homogeneous matrix in the interior. Some modifications of this pattern are found.

Nobel Laureate (1974).

△ **1952:** *Zinder, N., and J. Lederberg. Discovery of the transduction principle.*

Transduction is the transfer of DNA from one bacterial cell to another by means of a phage. It occurs when an infective phage picks up from its disintegrated host a small fragment of the host's DNA and carries it to a new host where it becomes a part of the genetic equipment of the new bacterial cell.

Lederberg, Nobel Laureate (1958).

‡△ **1953:** *Crick, Francis H.C., James D. Watson, and Maurice H.F. Wilkins. Chemical structure of DNA.*

Based on knowledge of the chemical nature of DNA and studies of x-ray diffraction by Wilkins, Crick and Watson formulated the hypothesis that the DNA molecule was made up of two chains twisted around each other in a helical structure with pairs of nitrogenous bases projecting toward each other—adenine opposite thymine and guanine opposite cytosine. Genes are considered to be segments of these molecules, with the sequence of bases coding for the amino acids in a protein. Each of the complementary strands acts as a model or template to form a new strand before cell division. The hypothesis has been widely accepted and is now amply confirmed.

Nobel Laureates (1962).

△ **1953:** *Palade, G.E. Description of cytoplasmic ribosomes.*

The description of ribonucleic acid—rich granules (usually on the endoplasmic reticulum) represents one of the key links in the unraveling of the mechanism of protein synthesis. The ribosomes (about 25 nm in diameter) serve as the site of protein synthesis; the number of ribosomes (polysomes) involved in the synthesis of a protein depends on the length of messenger RNA and the protein being synthesized.

Nobel Laureate (1974).

× **1953:** *Urey, Harold, and Stanley Miller. Demonstration of the possible primordia of life.*

By exposing a mixture of water vapor, ammonia, methane, and hydrogen gas to electrical discharge (to simulate lightning) for several days, these investigators found that several complex organic substances, such as the amino acids glycine and alanine, were formed when the water vapor was condensed into water. This demonstration offered a very plausible hypothesis to explain how the early beginnings of life substances could have started by the formation of organic substances from inorganic ones.

○† **1953:** *Sperry, R.W. Independence of brain hemispheric function.*

By severing the bundle of nerve fibers separating the two cortical hemispheres, Sperry developed the "split-brain" technique, which he used to show that the two brain halves govern two sets of activities. He also contributed much to developmental neurobiology.

Nobel Laureate (1981).

△ **1954:** *Del Castillo, J., and B. Katz. Impulse transmission at nerve junction.*

These researchers demonstrated that the chemical transmitter substance is released from the presynaptic terminal in discrete multimolecular packets, or quanta, each containing several thousand molecules.

Katz, Nobel Laureate (1970).

✳ **1954:** *Dan, J.C. Acrosome reaction.*

In echinoderms, annelids, and molluscs Dan showed that the acrosome region of the spermatozoon forms a filament and releases an unknown substance at the time of fertilization. Evidence suggests that the filament is associated with the formation of the fertilization cone. Other observers had described similar filaments before Dan made his detailed descriptions. The filament (1 to 75 μm long) may play an important role in the entrance of the sperm into the cytoplasm.

○ **1954:** *Du Vigneaud, V. Synthesis of pituitary hormones.*

This investigator isolated the posterior pituitary hormones oxytocin and vasopressin. Both were found to be polypeptides, and oxytocin was the first polypeptide hormone to be produced artificially. Oxytocin contracts the uterus during childbirth and releases the mother's milk; vasopressin raises blood pressure and decreases urine production.

○ **1954:** *Huxley, H.E., A.F. Huxley, and J. Hanson. Sliding filament model of muscular contraction.*

By means of electron microscopic studies and x-ray diffraction studies, these investigators showed that the proteins actin and myosin were found as separate filaments that apparently produced contraction by a sliding reaction in the presence of ATP. The concept is widely accepted.

A.F. Huxley, Nobel Laureate (1963).

△ **1954:** *Sanger, Frederick. Structure of the insulin molecule.*

Insulin is the important hormone used in the treatment of diabetes. It was the first protein for which a complete amino acid sequence was known. The molecule was found to be made up of 17 different amino acids in 51 amino acid units. Although one of the smallest proteins, its formula contains 777 atoms.

Nobel Laureate (1958).

△ **1955-1957:** *Kornberg, Arthur, and S. Ochoa. Synthesis of nucleic acids outside of cells (in vitro).*

By mixing the enzyme polymerase, extracted from the bacterium *Escherichia coli*, with a mixture of nucleotides and a tiny amount of DNA, Kornberg was able to produce synthetic DNA. Ochoa obtained the synthesis of RNA in a similar manner by using the enzyme polynucleotide phosphorylase from the bacterium *Azotobacter vinelandii*. This significant work reveals more insight into the mechanism of nucleic acid duplication in the cell.

Nobel Laureates (1959).

✳ **1955:** *Kettlewell, H.B.D. Natural selection in action: industrial melanism in moths.*

Using field techniques, Kettlewell was able to demonstrate a selective advantage of dark (melanic) mutants in a barklike cryptic moth species in industrial areas of Great Britain, where the tree trunks had been darkened by soot. This demonstration provided an evolutionary explanation for the increasing incidence of melanism in cryptic moths that has been noted throughout the industrialized regions of the world and constitutes a classic example of the effects of a strong selective pressure.

○ **1956:** *von Békésy, Georg. The traveling wave hypothesis of hearing.*

Helmholtz (1868) had proposed the resonance hypothesis of hearing on the basis that each cross fiber of the basilar membrane, which increases in width from the base to the apex of the cochlea, resonates at a different frequency; von Békésy showed that a traveling wave of vibration is set up in the basilar membrane and reaches a maximal vibration in the part of the membrane appropriate for that frequency.

Nobel Laureate (1961).

△ **1956:** *Borsook, H., and P.C. Zamecnik. Site of protein synthesis.*

By injecting radioactive amino acids into an animal, they found that the ribosome of the endoplasmic reticulum is the place where proteins are formed.

‡ **1956:** *Ingram, V.M. Nature of a mutation.*

By tracing the change in one amino acid unit out of more than 300 units that make up the protein hemoglobin, Ingram was able to pinpoint the difference between normal hemoglobin and the mutant form of hemoglobin that causes sickle cell anemia.

△ **1956:** *Sutherland, Earl W., and T.W. Rall. Discovery of cyclic AMP.*

An intracellular mediating agent (cyclic adenosine-3′, 5′-monophosphate, or cyclic AMP) is found in all living animal tissues, and changes in cyclic AMP levels (by effects of hormones) cause the hormones to produce different target effects depending on the type of cell in which they are found. Cyclic AMP is formed by a metabolic reaction in which the enzyme adenyl cyclase converts ATP to cyclic AMP.

Sutherland, Nobel Laureate (1971).

‡ **1956:** *Tjio, J.H., and A. Levan. Revision of human chromosome count.*

The time-honored number of chromosomes in humans, 48 (diploid), was found by careful cytological technique to be 46 instead.

△ **1957:** *Calvin, Melvin. Chemical pathways in photosynthesis.*

By using radioactive carbon 14, Calvin and colleagues were able to analyze step by step the incorporation of carbon dioxide and the identity of each intermediate product involved in the formation of carbohydrates and proteins by plants.

Nobel Laureate (1961).

△ **1957:** *Holley, Robert W. The role of transfer RNA in protein synthesis.*

Nucleotides of transfer RNAs differ from each other only in their bases. Holley also devised methods that precisely established the transfer RNAs used in the transfer of certain amino acids to the site of protein synthesis.

Nobel Laureate (1968).

□ **1957:** *Ivanov, A.V. Analysis of phylum Pogonophora (beard worms).*

Specimens of this phylum were collected in 1900 and represent one of the most recent phyla to be discovered and evaluated in the animal kingdom. Collections have been made in the waters of Indonesia, the Okhotsk Sea, the Bering Sea, and the Pacific Ocean. They are found mostly in the abyssal depths. Thought originally to belong to the Deuterostomia division, more recent evidence indicates that they are protostomes. At present 80 species divided into two orders have been described.

△ **1957:** *Perutz, Max F., and J.C. Kendrew. Structure of hemoglobin.*

The mapping of a complex globular protein molecule of 600 amino acids and 10,000 atoms arranged in a three-dimensional pattern represented one of the great triumphs in biochemistry. Myoglobin of muscle, which acts as a storehouse for oxygen and contains only one heme group instead of four (hemoglobin), was found to contain 150 amino acids.

Nobel Laureates (1962).

□ **1957:** *Sauer, E. Celestial navigation by birds.*

By subjecting Old World warblers to various synthetic night skies of star settings in a planetarium, Sauer was able to demonstrate that the birds made use of the stars to guide them in their migrations.

○ **1958:** *Lerner, A.B. Discovery of melatonin in the pineal gland.*

Lerner and associates discovered that the production of melatonin in the pineal gland is increased in darkness and decreased in the light. Activity of the hormone appears important in regulation of gonadal functions affected by photoperiod, and its discovery represented a breakthrough in understanding the function of the pineal gland.

△ **1958:** *Meselson, M., and F.W. Stahl. Confirmation "in vivo" of the duplicating mechanism in DNA.*

This was really confirmation of the self-copying of DNA in accordance with Watson and Crick's scheme of the structure of DNA. These investigators found that, after producing a culture of bacterial cells labeled with heavy nitrogen 15 and then transferring these bacteria to a cultural medium of light nitrogen 14, the resulting bacteria had a DNA density intermediate between heavy and light, as would be expected on the basis of the Watson-Crick hypothesis.

○ **1959:** *Burnet, F. Macfarlane. Clonal selection hypothesis of immunity.*

One of the most puzzling aspects of the acquired immune response has been how to account genetically for the ability to generate specific antibodies against the enormous variety of potential antigens. N.K. Jerne suggested that the information necessary to generate the antibodies was present in the host before exposure to the antigens. Burnet proposed that we have a large variety of antibody-producing cells, each present in such a small number that its product cannot be detected. Upon exposure to a particular antigen, a clone of cells that can make antibody to that antigen is "selected" and multiplies rapidly, thus increasing antibody to a detectable level. A corollary to the theory is that antigens present when the organism is embryonic or very young are recognized as "self," and the immune system does not respond. This was confirmed by P. Medawar.

Burnet and Medawar, Nobel Laureates (1960); Jerne, Nobel Laureate (1984).

‡ **1959:** *Ford, C.E., P.A. Jacobs, and J.H. Tjio. Chromosomal basis of sex determination in humans.*

By discovering that certain genetic defects were associated with an abnormal somatic chromosomal constitution, it was possible to determine that male-determining genes in humans were located on the Y chromosome. Thus, a combination of XXY (47 instead of the normal 46 diploid number) produced sterile males (Klinefelter's syndrome), and those with XO combinations (45 diploid number) gave rise to Turner's syndrome or immature females.

× **1959:** *Leakey, Mary D. Discovery of Australopithecus (Zinjanthropus) boisei fossil hominid.*

This fossil is a robust form within the small-brained, large-jawed genus *Australopithecus*, first discovered by Dart (1925). Its status is at present controversial: some believe it to be a species distinct from *A. africanus*, whereas others believe that *A. africanus* and *A. boisei* are females and males within a single, polytypic species.

○ **1959:** *Butenandt, A.F.J. Chemical identification of a pheromone.*

Butenandt and associates chemically analyzed the first pheromone—the sex attractant substance of silk moths *(Bombyx mori)*. Named bombykol, it is a doubly unsaturated fatty alcohol with 16 carbon atoms. The chemical nature of numerous other pheromones has since been determined.

△† **1950s-present:** *Bennacerraf, B., J. Dausset, and G. Snell. Genetics and function of the major histocompatibility complex (MHC).*

This complex of genes codes for cell surface antigens. These antigens are critically important in various interactions between cells, such as rejection of transplanted tissues and control of the immune response. Much work remains to be done before we completely understand the role of the MHC in immunity.

Nobel Laureates (1980).

△* **1950s-1960s:** *Levi-Montalcini, Rita, and Stanley Cohen. Isolation and characterization of nerve growth factor (NGF) and epidermal growth factor (EGF).*

Growth factors are proteins necessary for the growth, development, and maintenance of their target cells. Since the discovery of NGF and EGF, other growth factors have been identified that act on many cell types. Some oncogenes are derived from genes for growth factors; therefore, some cancers may have their origins in a malfunction of genes encoding factors that regulate normal cell growth.

Nobel Laureates (1986).

○ **1959-1960:** *Yalow, Rosalyn, and S.A. Berson. Development of radioimmunoassay.*

The development of this technique made it possible for the first time to measure minute quantities of a hormone directly in a complex mixture of proteins such as serum. A radioactively labeled hormone is mixed with a specific antibody to the hormone, together with a sample of the unknown. The labeled compound competes with the unlabeled in the formation of the antigen-antibody complex, and the amount of unknown is obtained by comparison with a standard curve prepared with known amounts of unlabeled hormone. The technique was originally described for insulin, but it has been rapidly extended to many other hormones and has revolutionized endocrinology.

Yalow, Nobel Laureate (1977).

□ **1960:** *Tinbergen, Niko. Development of the concept of specific searching images in predators.*

Tinbergen's studies of songbird predation in Dutch pinewoods helped to lay the foundation for the science of ethology.

Nobel Laureate (1973).

△ **1960:** *Hurwitz, J., A. Stevens, and S. Weiss. Enzymatic synthesis of messenger RNA.*

The exciting knowledge of the coding system of DNA and its translation to protein synthesis (ribosomes) was further elucidated when it was discovered that an enzyme, RNA polymerase, was responsible for the synthesis of RNA from a template pattern of DNA.

△ **1960:** *Jacob, François, and Jacques Monod. The operon hypothesis.*

The operon hypothesis is a postulated model of how enzyme synthesis is regulated in prokaryotes. The model proposes that refinement in regulation involves an inducible system for allowing structural genes to synthesize needed enzymes and a repressible system that cuts off the synthesis of unneeded enzymes.

Nobel Laureates (1965).

△ **1960:** *Strell, M., and R.B. Woodward. Synthesis of chlorophyll a.*

Strell and Woodward with the aid of many co-workers finally solved this problem, which had been the goal of organic chemists for generations.

Woodward, Nobel Laureate (1965).

△ **1961:** *Hurwitz, J., A. Stevens, and S.B. Weiss. Confirmation of messenger RNA.*

Messenger RNA transcribes directly the genetic message of the nuclear DNA and moves to the cytoplasm, where it becomes associated with a number of ribosomes or submicroscopic particles containing protein and nonspecific structural RNA. Here the messenger RNA molecules serve as templates against which amino acids are arranged in the sequence corresponding to the coded instructions carried by messenger RNA.

‡△ **1961:** *Jacob, François, and Jacques Monod. The role of messenger RNA in the genetic code.*

The transmission of information from the DNA code in the genes to the ribosomes represents an important step in the unraveling of the genetic code, and these investigators proposed certain deductions for confirmation of the hypothesis, which in general has been established by many researchers.

Nobel Laureates (1965).

○ **1961:** *Miller, J.F.A. Function of the thymus gland.*

Long known as a transitory organ that persists during the early growth period of animals, the thymus is now recognized as a vital processing center in the development of certain lymphocytes (T cells). These cells are very important in the immune response of vertebrates.

‡△ **1961:** *Nirenberg, Marshall W., and J.H. Matthaei. Deciphering the genetic code.*

By adding a synthetic RNA composed entirely of uracil nucleotides to a mixture of amino acids, these investigators obtained a polypeptide made up solely of a single amino acid, phenylalanine. On the basis of a triplet code, it was concluded that the RNA code word for phenylalanine was UUU and its DNA complement was AAA. This was the beginning of decoding.

Nirenberg, Nobel Laureate (1968).

△ **1961:** *Mitchell, Peter. The chemiosmotic-coupling hypothesis of ATP formation.*

Although it is known that the oxidation of food molecules in the cell results in net synthesis of ATP from ADP and inorganic phosphate, the molecular mechanism to drive the reaction is still unclear. Current evidence supports the chemiosmotic hypothesis, which suggests that the energy derived from electron transport serves to pump hydrogen ions across the inner mitochondrial membrane, actively setting up an electrochemical gradient. The gradient of hydrogen ions is then coupled with an ATPase complex in the inner membrane, providing the free energy to form the high-energy phosphate bond.

○ **1962:** *Copp, Harold. Discovery of calcitonin.*

This peptide hormone is produced by the ultimobranchial glands (lower vertebrates) or their embryological derivatives, the parafollicular cells of the thyroid gland (mammals). It lowers the blood calcium and increases calcium uptake by bone cells, thus antagonizing the action of parathormone.

△‡ **1962-1972:** *Arber, W., H.O. Smith, and D. Nathans. Discovery, isolation, and characterization of restriction endonucleases.*

Use of these bacterial enzymes has been essential in the recent explosion of knowledge in molecular genetics and recombinant DNA technology. Arber is given credit for predicting the existence of such enzymes, Smith for isolating the first restriction endonuclease and describing its reaction, and Nathans for applying the enzymes in studies of gene regulation and organization.

Nobel Laureates (1978).

× **1964:** *Hamilton, W.D. Concepts of kin selection and inclusive fitness.*

The problem of neuter castes in social insects had destroyed Lamarckian and baffled Darwinian evolutionary explanations. Hamilton, however, demonstrated that the peculiar mode of reproduction in most social insects having neuter castes (haplodiploidy, or males haploid and females diploid) resulted in a situation in which sisters would share more common genes with one another than with their own progeny. Hence, helping to rear "kin" (sisters) could be more adaptive from the point of view of one's genes (inclusive fitness) than rearing one's own young (individual fitness). These ideas have since been extended to social behaviors of many

species—vertebrate as well as invertebrate— and provide a fundamental core of the emerging field of sociobiology.

× **1964:** *Hoyer, B.H., B.J. McCarthy, and E.T. Bolton. Phylogeny and DNA sequence.*

These investigators presented evidence that certain homologies exist among the polynucleotide sequences in the DNA of such different forms as fishes and humans. Such sequences may represent genes that have been retained with little change throughout vertebrate history. Possible phenotypic expressions may be bilateral symmetry, notochord, hemoglobin, and so on.

△‡ **1966:** *Khorana, H.G. Proof of code assignments in the genetic code.*

By using alternating codons (CUC and UCU) in an artificial RNA chain, Khorana was able to synthesize a polypeptide of alternating amino acids (leucine and serine) for which these codons respectively stood.

Nobel Laureate (1968).

△ **1967:** *Katz, Bernard, and R. Miledi. Calcium entry at nervous synapses.*

These investigators proposed that the arrival of an action potential at a presynaptic terminal causes calcium influx that facilitates binding of synaptic vesicles with the presynaptic membrane.

△‡ **1967:** *Ptashne, M. Isolation of first repressor.*

Repressors are protein substances supposedly formed by regulatory genes and function by preventing a structural gene from making its product when not needed by the cells.

□ **1967:** *MacArthur, Robert H., and E.O. Wilson. Theoretical ecology.*

Mathematical models in conjunction with field studies form the basis of theoretical ecology. This is well shown in the *Theory of Island Biogeography* by MacArthur and Wilson. By the methods found in their book and the investigations of many other researchers, it has been possible to determine the equilibrium of species, the number of extinctions, and other factors on islands. This viewpoint has been a revelation in ecological studies.

□ **1968:** *Goodall, Jane. Behavior of free-living chimpanzees.*

A long-term complete study of primate social behavior was done in the field, providing an impetus and stimulus for a large number of similar studies on other primates.

○ **1960s-1970s:** *Hubel, D.H., and T.N. Wiesel. Understanding of stereoscopic vision.*

Analysis of processing of images in eyes and brain have opened a new field in vision physiology.

Nobel Laureates (1981).

△† **1960s-1970s:** *Bergström, S., B. Samuelsson, and J. Vane. Characterization of prostaglandins.*

Prostaglandins are chemical transmitters of intracellular and intercellular signals. They are involved in a variety of physiological and

pathological functions. Bergström is credited with isolating prostaglandins and determining their structures; Samuelsson determined their biosynthesis and metabolism; and Vane found that vascular endothelium produces a prostaglandin (prostacyclin) that inhibits platelet aggregation.

Nobel Laureates (1982).

△ **1970:** *Temin, H.M., and D. Baltimore. Demonstration of DNA synthesis from an RNA template.*

Many viruses carry RNA rather than DNA as a genetic material, and it was not known how the RNA virus could enter the host cell and induce the host to synthesize more RNA virus particles. Temin and Baltimore independently found an RNA-dependent DNA polymerase in RNA viruses; thus, the RNA virus reproduces by entering a host cell and producing a DNA copy from the RNA template; then the host cell machinery makes new virus particles from the DNA template.

Nobel Laureates (1975).

○△ **1970:** *Edelman, Gerald M., and R.R. Porter. The structure of gamma globulin.*

By using myeloma tumors (which contain pure immunoglobulin proteins) the investigators were able to work out after many years a complete analysis of the large gamma globulin molecule, which is made up of 1320 amino acids and 19,996 atoms, with a molecular weight of 150,000.

Nobel Laureates (1972).

△○ **1971:** *Cheung, W.Y. Discovery of calmodulin.*

The calcium-binding protein interacts reversibly with intracellular calcium to form a complex that regulates a broad spectrum of cellular activities.

△ **1971:** *Berg, P., D. Jackson, and R. Symons. Recombinant DNA.*

A DNA molecule from both a bacterial virus and an animal tumor virus were opened with a restriction endonuclease, then spliced together. This was the first recombinant DNA from two different organisms.

Berg, Nobel Laureate (1980).

† **1971-1973:** *Cormack, A.M., and G.N. Hounsfield. Invention of computer-assisted tomography (CAT scanner).*

This x-ray diagnostic technique produces clear images of internal body structures. A basic problem was how to obtain accurate measurements of the x-ray attenuation coefficient for all points in the x-ray section being examined. Cormack and Hounsfield each developed mathematical solutions to the problem.

Nobel Laureates (1979).

△ **1972:** *Singer, S.J., and G.L. Nicolson. Fluid-mosaic model of biological membranes.*

The plasma membrane of cells contains a bimolecular leaflet of lipids with the surface interrupted by proteins. Some proteins, called extrinsic proteins, are attached to the lipid surface, whereas others, called intrinsic proteins, penetrate the bilayer and may completely span the membrane. This model remains the most widely accepted model of membrane structure.

× **1972:** *Gould, Stephen J., and J. Eldridge. Concept of punctuated equilibrium.*

This hypothesis proposes that species remain virtually unchanged for long periods of time until a sudden spurt of rapid evolution produces a distinct species. Since the hypothesis is an alternative to more orthodox Darwinian gradualism, it has produced lively debate among evolutionists.

○ **1972:** *Woodward, R.B., and A. Eschenmoser. The synthesis of vitamin B$_{12}$.*

Vitamin B$_{12}$ is the last vitamin to be synthesized. This complex molecule is not made up of polymers; new methods of organic chemistry were required before it could be synthesized. It contains the metal ion cobalt.

○△ **1972:** *Jerne, Niels K. Hypothesis of anti-idiotype regulation of the immune system.*

After the discovery by J. Oudin and H. Kunkel that antibodies themselves bear sites (idiotypes) that stimulate production of other antibodies against them (anti-idiotypes), Jerne proposed that regulation of the immune system was mediated by a complex network of antibodies specific for other antibodies. Binding of the antibodies to each other could be stimulatory or suppressive. There is now considerable evidence to support this concept.

Nobel Laureate (1984).

□ **1973:** *The Endangered Species Act.*

This omnibus legislation, created in 1966 and greatly strengthened in 1973, is the first United States domestic law concerned exclusively with wildlife and enacted for purely altruistic motives. Its strength resides in that portion of the act specifically prohibiting federally funded projects from jeopardizing endangered species or their habitats. Some 400 species of plants and animals have been placed on the endangered list.

△ **1974:** *Brown, M., and J. Goldstein. Discovery of the LDL receptor on cell surfaces.*

Cholesterol, which is an essential component of cell membranes, is taken up by body cells in association with low density lipoprotein (LDL). Discovery that the presence of a receptor for LDL on the cell surface was essential for this process had important implications for understanding of endocytosis of other macromolecules. Brown and Goldstein also described the intracellular regulation of cholesterol metabolism, and they cloned the gene for the LDL receptor and thus were able to work out the molecular structure of the receptor.

Nobel Laureates (1985).

△† **1975:** *Milstein, Cesar, and G. Köhler. Hybridoma technique and monoclonal antibodies.*

Hybridomas are cells formed by hybridizing myeloma cells (from immune system tumors) with lymphocytes. Each hybridoma produces a clone of identical daughter cells, all manufacturing identical antibodies. Monoclonal antibodies are of great value in research, are useful diagnostic tools, and have great potential in disease therapy.

Nobel Laureates (1984).

△ **1975:** *Miller, J., and P. Lu. Alteration of amino acid sequence.*

By inserting known amino acids to replace naturally occurring amino acids in the lac repressor protein, these investigators have been able to determine which amino acids are necessary for the various functions of the lac repressor.

△ **1975:** *Sanger, J.W. Chromosome migration on spindle fibers.*

The proteins (actin and myosin) that contract muscles were found to be involved in the migration of chromosomes along the spindle fibers. Actin bundles are present on chromosomal spindle fibers.

△ **1975-1977:** *Gilbert, W., A. Maxam, and F. Sanger. Methods for determining base sequence in DNA.*

Practical methods for determining the base sequence in DNA have great implications for molecular genetics and DNA technology.

Gilbert and Sanger, Nobel Laureates (1980).

† **1977:** *Eradication of smallpox.*

The global eradication of variola major, the virulent form of smallpox, was achieved by making preventive immunization available to underdeveloped areas through a World Health Organization campaign.

‡△ **1977:** *Discovery of discontinuous genes in eukaryotes.*

Evidence from several laboratories led to the amazing conclusion that the DNA in many genes of eukaryotes coded for RNA sequences that do not become part of the final product (introns). They are clipped out, and the remaining sequences (exons) are spliced together before the RNA is exported to the cytoplasm.

△ **1981:** *Complete sequencing of human mitochondrial DNA.*

This landmark research by a group at England's Medical Research Council revealed that the human mitochondrial genome contains 16,569 base pairs. Packed into the genome with remarkable economy are the genes for making two ribosomal RNAs, 22 transfer RNAs, and 13 assorted proteins.

BOOKS AND PUBLICATIONS THAT HAVE GREATLY INFLUENCED DEVELOPMENT OF ZOOLOGY

Aristotle. 336-323 BC. De anima, Historia animalium, De partibus animalium, *and* De generatione animalium. *These biological works of the Greek thinker have exerted an enormous influence on biological thinking for centuries.*

Vesalius, Andreas. 1543. De fabrica corporis humani. *This work is the foundation of modern anatomy and represents a break with the Galen tradition. His representations of some anatomical subjects, such as muscles, have never been surpassed. Moreover, he treated anatomy as a living whole, a viewpoint adopted by most present-day anatomists.*

Fabricius of Aquapendente. 1600-1621. De formato foetu *and* De formatione ovi pulli. *This was the first illustrated work on embryology and may be said to be the beginning of the modern study of development.*

Harvey, William. 1628. Essay on the motion of the heart and the blood. *This great work represents one of the first accurate explanations in physical terms of an important physiological process. It initiated an experimental method of observation that gave an impetus to research in all fields of biology.*

Descartes, René. 1637. Discourse on method. *This philosophical essay gave a great stimulus to a mechanistic interpretation of biological phenomena.*

Buffon, Georges. 1749-1804. Histoire naturelle. *This extensive work of many volumes collected together natural history facts in a popular and pleasing style. It had a great influence in stimulating a study of nature. Many eminent biological thinkers, such as Erasmus Darwin and Lamarck, were influenced by its generalizations.*

Linnaeus, Carolus. 1758. Systema naturae. *This work lays the basis for the classification of animals and plants. With few modifications, the taxonomic principles outlined therein have been universally adopted by biologists.*

Wolff, Caspar Friedrich. 1759. Theoria generationis. *The theory of epigenesis was here set forth for the first time in opposition to the preformation theory of development so widely held before Wolff's work.*

von Haller, Albrecht. 1760. Elementa physiologiae. *An extensive summary of various aspects of physiology that greatly influenced physiological thinking for many years. Some of the basic concepts laid down therein are still considered valid, especially those on the nervous system.*

Malthus, Thomas R. 1798. Essay on population. *This work stimulated evolutionary thinking among such people as Darwin and Wallace.*

de Lamarck, Jean Baptiste. 1809. Philosophie zoologique. *This publication was of great importance in focusing the attention of biologists on the problem of the role of the environment as a factor in evolution. Lamarck's belief that all species came from other species represented one of the first clear-cut statements on the mutability of species, even though his theory of use and disuse has not been accepted by most biologists.*

Cuvier, Georges. 1817. Le règne animal. *A comprehensive biological work that dealt with classification and a comparative study of animal structures. Its plates are still of value, but the general plan of the work was marred by a disbelief in evolution and a faith in the doctrine of geological catastrophes. The book, however, exerted an enormous influence on contemporary zoological thought.*

von Baer, Karl Ernst. 1828-1837. Entwickelungsgeschichte der Thiere. *In this important work are laid down the fundamental principles of germ layer formation and the similarity of corresponding stages in the development of embryos that have proved to be the foundation studies of modern embryology.*

Audubon, John J. 1828-1838. The birds of America. *The greatest of all ornithological works, it has served as the model for all monographs dealing with a specific group of animals. The plates, the work of a master artist, have seldom been surpassed.*

Lyell, Charles. 1830-1833. Principles of geology. *This great work exerted a profound influence on biological thinking, for it did away with the theory of catastrophism and prepared the way for an evolutionary interpretation of fossils and the forms that arose from them.*

Beaumont, William. 1833. Experiments and observations on the gastric juice and the physiology of digestion. *Beaumont's observations on various functions of the stomach and digestion were accurate and thorough. This classic work paved the way for the brilliant investigations of Pavlov, Cannon, and Carlson of later generations.*

Müller, Johannes. 1834-1840. Handbook of physiology. *The principles set down in this work by perhaps the greatest of all physiologists have set the pattern for the development of the science of physiology.*

Darwin, Charles. 1839. Journal of researches (Voyage of the *Beagle*). *This book reveals the training and development of the naturalist and the material that led to the formulation of Darwin's concept of organic evolution.*

Schwann, Theodor. 1839. Mikroskopische Untersuchungen über die Uebereinstimmung in der Struktur und dem Wachstum der Thiere und Pflanzen. *The basic principles concerning the cell doctrine are laid down in this classic work.*

Kölliker, Albrecht. 1852. Mikroskopische Anatomie. *This was the first textbook in histology and contains contributions of the greatest importance in this field. Many of the* histological descriptions Kölliker made have never needed correction. In many of his biological views he was far ahead of his time.

Maury, Matthew F. 1855. The physical geography of the sea. *This work has often been called the first textbook on oceanography. This pioneer treatise stressed the integration of such knowledge as was then available about tides, winds, currents, depths, circulation, and such matters. Maury's work represents a real starting point in the fascinating study of the oceans and has had a great influence in stimulating investigations in this field.*

Virchow, Rudolf. 1858. Die Cellularpathologie. *In this work Virchow made the first clear distinction between normal and diseased tissues and demonstrated the real nature of pathological cells. The work also represents the death knell of the old humoral pathology, which had held sway for so long.*

Darwin, Charles. 1859. On the origin of species. *The most influential book ever published in biology. Although built around the theme that natural selection is the most important factor in evolution, the enormous influence of the book can be attributed to the extensive array of evolutionary evidence it presented. It also stimulated constructive thinking on a subject that had been vague and confusing before Darwin's time.*

Marsh, George P. 1864. Man and nature: physical geography as modified by human action. *A work that had an early and important influence on the conservation movement in America.*

Mendel, Gregor. 1866. Versuche über Pflanzenhybriden. *Careful, controlled pollination technique and statistical analysis gave a scientific explanation that has influenced all geneticists after the "rediscovery" in 1900 of Mendel's classic paper on the two basic laws of inheritance.*

Owen, Richard. 1866. Anatomy and physiology of the vertebrates. *This work contains an enormous amount of personal observation on the structure and physiology of animals, and some of the basic concepts of structure and function, such as homologue and analogue, are here defined for the first time.*

Brehm, Alfred E. 1869. Tierleben. *The many editions of this work over many years have indicated its importance as a general natural history.*

Bronn, Heinrich G. (ed.). 1873 to present. Klassen and Ordnungen des Tier-Reichs. *This great work is made up of exhaustive treatises on the various groups of animals by numerous authorities. Its growth extends over many years, and it is one of the most valuable works ever published in zoology.*

Balfour, Francis M. 1880. Comparative embryology. *This is a comprehensive summary of* embryological work on both vertebrates and invertebrates up to the time it was published. It is often considered the beginning of modern embryology.

Semper, Karl. 1881. Animal life as affected by the natural conditions of existence. *This work first pointed out the modern ecological point of view and laid the basis for many ecological concepts of existence that have proved important in the further development of this field of study.*

Bütschli, Otto. 1889. Protozoen (Bronn's Klassen und Ordnungen des Tier-Reichs). *This monograph has been of the utmost importance to students of protozoa. No other work on a like scale has ever been produced in this field of study.*

von Hertwig, Richard. 1892. Lehrbuch der Zoologie. *This text has proved to be an invaluable source of material for many generations of zoologists. Its illustrations have been widely used in other textbooks.*

Weismann, August. 1892. Das Keimplasma. *Weismann predicted from purely theoretical considerations the necessity of meiosis or reduction of the chromosomes in the germ cell cycle, a postulate that was quickly confirmed cytologically by others.*

Hertwig, Oskar. 1893. Zelle und Gewebe. *In this work a clear distinction is made between histology as the science of tissues and cytology as the science of cell structure and function. Cytology as a study in its own right dates from this time.*

Korschelt, E., and K. Heider. 1893. Lehrbuch der vergleichende Entwicklungsgeschichte der wirbellosen Thiere, 4 vols. *A treatise that has been a valuable tool for all workers in the difficult field of invertebrate embryology.*

Wilson, Edmund B. 1896. The cell in development and heredity. *This and subsequent editions represented the most outstanding work of its kind in the English language. Its influence in directing the development of cytogenetics cannot be overestimated, and in summarizing the many investigations in cytology, the book has served as one of the most useful tools in the field.*

Pavlov, Ivan. 1897. Le travail des glandes digestives. *This work, a landmark in the study of the digestive system, describes many of Pavlov's now classic experiments, such as the gastric pouch technique and the rate of gastric secretions.*

De Vries, Hugo. 1901. Die Mutationstheorie. *The belief that evolution is the result of sudden changes or mutations is advanced by one who is commonly credited with the initiation of this line of investigation into the causes of evolution.*

Sherrington, Charles. 1906. The integrative action of the nervous system. *The basic concepts of neurophysiology elaborated in this book, especially the concept of the integrative action of the nervous system, remain the basis of modern neurophysiology.*

Garrod, Archibald. 1909. Inborn errors of metabolism. *This pioneer book showed that certain congenital diseases were caused by defective genes that failed to produce the proper enzymes for normal functioning. It laid the basis for biochemical genetics, which later received a great impetus from the work of Beadle and Tatum.*

Henderson, Lawrence J. 1913. The fitness of the environment. *This book pointed out in a specific way the reciprocity that exists between living and nonliving nature and how organic matter is fitted to the inorganic environment. It has exerted a considerable influence on the study of ecological aspects of adaptation.*

Shelford, Victor E. 1913. Animal communities in temperate America. *This work was a pioneer in the field of biotic community ecology.*

Bayliss, William M. 1915. Principles of general physiology. *If a classic book must meet the requirements of masterly analysis and synthesis of what is known in a particular discipline, then this great work must be called one.*

Wegener, Alfred. 1915. The origin of continents and oceans. *In the first (German) edition of this book, Wegener developed the idea of continental drift. Wegener's interpretation of continental history was out of favor for many decades but was substantiated by geophysical work (plate tectonics) in the 1960s and 1970s. Biogeographers now invoke Wegenerian hypotheses to explain distribution of many groups of plants and animals.*

Matthew, W.D. 1915. Climate and evolution. *Matthew, in contrast to Wegener, viewed the positions of continents as permanent. He explained distribution of plants and animals by dispersal between continents across land bridges, such as the Bering bridge between Asia and Alaska and the Panamanian bridge between Central and South America. Matthew's ideas dominated biogeographical thinking until the revival of Wegener's hypotheses in the 1960s and 1970s.*

Morgan, Thomas H., A.H. Sturtevant, C.B. Bridges, and H.J. Muller. 1915. The mechanism of Mendelian heredity. *This book gave an analysis and synthesis of Mendelian inheritance as formulated from the epoch-making investigations of the authors. This classic work stands as a cornerstone of our modern interpretation of heredity.*

Doflein, F. 1916. Lehrbuch der Protozoenkunde, ed. 6 (revised by E. Reichenow, 1949). *A standard treatise on protozoa. Its many editions have proved helpful to all workers in this field.*

Thompson, D'arcy W. 1917. Growth and form. *In this pioneering work the author attempted to reduce the great diversity of life into common themes and designs.*

Kukenthal, W., and T. Krumbach. 1923. Handbuch der Zoologie. *An extensive treatise on zoology that covers all phyla. The work has been an invaluable tool for all zoologists who are interested in the study of a particular group.*

Fisher, Ronald A. 1930. Genetical basis of natural selection. *This work has exerted an enormous influence on the newer synthesis of evolutionary mechanisms that emerged in the 1930s.*

Dobzhansky, Theodosius. 1937. Genetics and the origin of species. *The vast change in the explanation of the mechanism of evolution, which emerged about 1930, is well analyzed in this work by a master evolutionist. Other syntheses of this new biological approach to the evolutionary problems have appeared since this work was published, but none of them has surpassed the clarity and fine integration of Dobzhansky's work.*

Spemann, Hans. 1938. Embryonic development and induction. *In this work the author summarizes his pioneer investigations that have proved so fruitful in experimental embryology.*

Hyman, Libbie H. 1940. The invertebrates: Protozoa through Ctenophora. 1951. Platyhelminthes and Rhynchocoela, The acoelomate Bilateria. 1951. Acanthocephala, Aschelminthes, and Entoprocta, the pseudocoelomate Bilateria. 1955. Echinodermata, the coelomate Bilateria. 1959. Smaller coelomate groups: Chaetognatha, Hemichordata, Pogonophora, Phoronida, Ectoprocta, Brachiopoda, Sipunculida, The coelomate Bilateria. 1967. Mollusca I. *This series of volumes is a monumental work by a single author and is the only such coverage of the invertebrates in the English language. It is a benchmark in completeness, accuracy, and incisive analysis. It is an absolutely essential reference for all teachers and investigators in invertebrate zoology.*

Schrödinger, Erwin. 1945. What is life? *The emphasis placed on the physical explanation of life and the popularization of the idea of a chemical genetic "codescript" provided a new point of view of biological phenomena so well expressed in the current molecular biological revolution.*

Lack, David. 1947. Darwin's finches. *This classic, written several years after Lack's study visit to the Galápagos Islands, introduced competition theory into animal ecology, stressed the importance of ecological isolation in speciation, and provided a cogent model for adaptive radiation.*

Grassé, P.P. (ed.). 1948. Traité de zoologie. *This is a series of many treatises by various specialists on both invertebrates and vertebrates. Since it is relatively recent and comprehensive, the work is valuable to all students who desire detailed information on the various groups.*

Allee, W.C., A.E. Emerson, O. Park, T. Park, and K.P. Schmidt. 1949. Principles of animal ecology. *The basic ecological principles laid down in this comprehensive work are a landmark in the field.*

Tinbergen, Niko. 1951. The study of instinct. *The classic statement of the approaches and theoretical interpretations of early ethologists, especially K. Lorenz, is presented. This work brought ethology to the attention of American behavioral scientists, generating both controversy and collaboration.*

Blum, H.F. 1951. Time's arrow and evolution. *In this thoughtful book, Blum explores the relation between the second law of thermodynamics ("time's arrow") and organic evolution and recognizes that mutation and natural selection have been restricted to certain channels in accordance with the law, even though these two factors appear to controvert the principle of pointing the direction of events in time.*

Simpson, George G. 1953. The major features of evolution. *In a comprehensive synthesis of modern evolutionary theory, the author draws on evidence from paleontology, population genetics, and systematics. This book, along with Simpson's earlier* Tempo and mode in evolution, *have greatly influenced our thinking about the mechanisms of evolution within the framework of natural selection.*

Crick, Francis H.C., and James D. Watson. 1953. Genetic implications of the structure of deoxyribonucleic acid. *The solution of the structure of the DNA molecule has served as the cornerstone for an explanation of genetic replication and control of the cell's attributes and functions.*

Burnet, F. MacFarlane 1959. The clonal selection theory of acquired immunity. *This was an important new theory to explain the specificity of the acquired immune response and has to a great extent replaced the template theory formulated by Haurowitz in 1930.*

Carson, Rachel L. 1962. Silent spring. *Though a polemic addressed to a lay audience, this book was highly valuable in increasing public awareness of the danger to the ecosystem of indiscriminate use of pesticides.*

Mayr, Ernst. 1963. Animal species and evolution. *This lucid and thoroughly documented landmark in evolutionary study helped to clarify and integrate the emerging synthetic theory of evolution.*

Watson, James D. 1965. Molecular biology of the gene. *With perhaps the clearest statement of what molecular biology is, this important book introduced the field to the new generation of investigators in molecular genetics.*

Williams, George C. 1966. Adaptation and natural selection. *This scholarly and insightful work, with its well-reasoned explanation of adaptation and its convincing defense of natural selection, was highly influential in clarifying the role of natural selection in evolutionary change.*

MacArthur, R.H. 1972. Geographical ecology. *One of the most important theoretical ecology books to have appeared in the 1970s.*

Gurdon, John B. 1974. The control of gene expression in animal development. *Gurdon developed an experimental system, extended from the pioneering nuclear transplantation experiments of Briggs and King, that enabled him to transfer genetic information from one animal to another. His studies, summarized in this influential work, are interpreted to mean that nuclei of embryonic cells do not undergo alterations or loss of genetic material during development.*

Wilson, Edward O. 1975. Sociobiology: the new synthesis. *The rapidly developing fields of behavioral biology, behavioral ecology, and population genetics are reviewed. Wilson attempts to provide the basis for a new biological science, utilizing concepts developed in these heretofore disparate fields, devoted to analyzing social behaviors at all phylogenetic levels.*

GLOSSARY

aboral (ab-oʹrəl) (L. *ab*, from + *os*, mouth). A region opposite the mouth.

acanthor (ə-kanʹthor) (Gr. *akantha*, spine or thorn, + *or*). First larval form of acanthocephalans in the intermediate host.

acclimatization (ə-klīʹmə-də-zāʹshən) (L. *ad*, to, + Gr. *klima*, climate). Gradual physiological adaptation in response to relatively long-lasting environmental changes.

acetabulum (asʹə-tabʹū-lum) (L. a little saucer for vinegar). True sucker, especially in flukes and leeches; the socket in the hip bone that receives the thigh bone.

acinus (asʹə-nəs), pl. **acini** (asʹə-ni) (L. grape). A small lobe of a compound gland or a saclike cavity at the termination of a passage.

acoelomate (a-sēlʹə-māt') (Gr. *a*, not, + *koilōma*, cavity). Without a coelom, as in flatworms and proboscis worms.

acontium (ə-känʹchē-əm), pl. **acontia** (Gr. *akontion*, dart). Threadlike structure bearing nematocysts located on mesentery of sea anemone.

acrocentric (akʹrō-senʹtrək) (gr. *akros*, tip, + *kentron*, center). Chromosome with centromere near the end.

actin (Gr. *aktis*, ray). A protein in the contractile tissue that forms the thin myofilaments of striated muscle.

adaptation (L. *adaptatus*, fitted). An anatomical structure, physiological process, or behavioral trait that improves an animal's fitness for existence.

adaptive value. Degree to which a characteristic helps an organism to survive and reproduce; lends greater fitness in environment; selective advantage.

adductor (ə-dukʹtər) (L. *ad*, to, + *ducere*, to lead). A muscle that draws a part toward a median axis, or a muscle that draws the two valves of a mollusc shell together.

adenine (adʹnēn, adʹə-nēn) (Gr. *adēn*, gland, *ine*, suffix). A purine base; component of nucleotides and nucleic acids.

adenosine (ə-denʹə-sen) **di-, tri-) phosphate** (ADP and ATP). A nucleotide composed of adenine, ribose sugar, and two (ADP) or three (ATP) phosphate units; ATP is an energy-rich compound that, with ADP, serves as a phosphate bond—energy transfer system in cells.

adipose (adʹə-pōs) (L. *adeps*, fat). Fatty tissue; fatty.

adrenaline (ə-drenʹə-lən) (L. *ad*, to, + *renalis*, pertaining to kidneys). A hormone produced by the adrenal, or suprarenal, gland; epinephrine.

adsorption (ad-sorpʹshən) (L. *ad*, to, + *sorbere*, to suck in). The adhesion of molecules to solid bodies.

aerobic (a-rōʹbik) (Gr. *aēr*, air, + *bios*, life). Oxygen-dependent form of respiration.

afferent (afʹə-rənt) (L. *ad*, to, + *ferre*, to bear). Adjective meaning leading or bearing toward some organ, for example, nerves conducting impulses toward the brain or blood vessels carrying blood toward an organ; opposed to efferent.

aggression (L. *aggredi*, attack). A primary instinct usually associated with emotional states; an offensive behavior action.

agonistic behavior (Gr. *agōnistēs*, combatant). An offensive action or threat directed toward another organism.

alate (āʹlāt) (L. *alatus*, wing). Winged.

allantois (ə-lanʹtois) (Gr. *allas*, sausage, + *eidos*, form). One of the extraembryonic membranes of the amniotes that functions in respiration and excretion in birds and reptiles and plays an important role in the development of the placenta in most mammals.

allele (ə-lēl') (Gr. *allēlōn*, of one another). Alternative forms of genes coding for the same trait; situated at the same locus in homologous chromosomes.

allograft (aʹlō-graft) (Gr. *allos*, other, + graft). A piece of tissue or an organ transferred from one individual to another individual of the same species, not identical twins; homograft.

allometry (ə-lomʹə-trē) (Gr. *allos*, other, + *metry*, measure). Relative growth of a part in relation to the whole organism.

allopatric (Gr. *allos*, other, + *patra*, native land). In separate and mutually exclusive geographical regions.

alpha helix (Gr. *alpha*, first, + L. *helix*, spiral). Literally the first spiral arrangement of the genetic DNA molecule; regular coiled arrangement of polypeptide chain in proteins; secondary structure of proteins.

altricial (al-triʹshəl) (L. *altrices*, nourishers). Referring to young animals (especially birds) having the young hatched in an immature, dependent condition.

alula (alʹyə-lə) (L. dim. of *ala*, wing). The first digit or thumb of a bird's wing, much reduced in size.

alveolus (al-vēʹə-ləs) (L. dim. of *alveus*, cavity, hollow). A small cavity or pit, such as a microscopic air sac of the lungs, terminal part of an alveolar gland, or bony socket of a tooth.

ambulacra (amʹbyə-lakʹrə) (L. *ambulare*, to walk). In echinoderms, radiating grooves where podia of water-vascular system characteristically project to outside.

amebocyte (ə-mēʹbə-sīt) (Gr. *amoibē*, change, + *kytos*, hollow vessel). Any free body cell capable of movement by pseudopodia; certain types of blood cells and tissue cells.

ameboid (ə-mēʹboid) (Gr. *amoibē*, change, + *oid*, like). Ameba-like in putting forth pseudopodia.

amictic (ə-mikʹtic) (Gr. *a*, without, + *miktos*, mixed or blended). Pertaining to female rotifers, which produce only diploid eggs that cannot be fertilized, or to the eggs produced by such females.

amino acid (ə-mēʹnō) (amine, an organic compound). An organic acid with an amino group (—NH₂). Makes up the structure of proteins.

amitosis (āʹmī-tōʹsəs) (Gr. *a*, not, + *mitos*, thread). A form of cell division in which mitotic nuclear changes do not occur; cleavage without separation of daughter chromosomes.

amniocentesis (amʹnē-ō-sin-teʹsəs) (Gr. *amnion*, membrane around the fetus, + *centes*, puncture). Procedure for withdrawing a sample of fluid around the developing embryo for examination of chromosomes in the embryonic cells and other tests.

amnion (amʹnē-än) (Gr. *amnion*, membrane around the fetus). The innermost of the extraembryonic membranes forming a fluid-filled sac around the embryo in amniotes.

amniote (amʹnē-ōt). Having an amnion; as a noun, an animal that develops an amnion in embryonic life, that is, reptiles, birds, and mammals.

amphiblastula (amʹfə-blasʹchə-lə) (Gr. *amphi*, on both sides, + *blastos*, germ, + L. *ula*, small). Free-swimming larval stage of certain marine sponges; blastula-like but with only the cells of the animal pole flagellated; those of the vegetal pole unflagellated.

amphid (amʹfəd) (Gr. *amphidea*, anything that is bound around). One of a pair of anterior sense organs in certain nematodes.

amphipathic (am-fi-paʹthək) (Gr. *amphi*, on both sides, + *pathos*, suffering, passion). Adjective to describe a molecule with one part soluble in water (polar) and another part insoluble in water (nonpolar).

amplexus (am-plekʹsəs) (L. embrace). The copulatory embrace of frogs or toads.

ampulla (am-pūlʹə) (L. flask). Membranous vesicle; dilation at one end of each semicircular canal containing sensory epithelium; muscular vesicle above tube foot in water-vascular system of echinoderms.

amylase (amʹə-lās') (L. *amylum*, starch + *ase*, suffix meaning enzyme). An enzyme that breaks down starch into smaller units.

anadromous (an-adʹrə-məs) (Gr. *anadromos*, running upward). Refers to fishes that migrate up streams from the sea to spawn.

---

bat / āpe / ärmadillo/ herring / fēmale / finch / līce / crocodile / crōw / duck / ūnicorn / ə indicates unaccented vowel sound "uh" as in mammal, fishes, cardinal, heron, vulture / stress as in bi-olʹo-gy, biʹo-logʹi-cal

G-1

anaerobic (an´ə-rō´bik) (Gr. *an*, not, + *aēr*, air, + *bios*, life). Not dependent on oxygen for respiration.

analogy (L. *analogus*, ratio). Similarity of function but not of origin.

anastomosis (ə-nas´tə-mō´səs) (Gr. *ana*, again, + *stoma*, mouth). A union of two or more blood vessels, fibers, or other structures to form a branching network.

androgen (an´drə-jən) (Gr. *anēr, andros*, man, + *genēs*, born). Any of a group of vertebrate male sex hormones. Compare estrogen.

androgenic gland (an´drō-jen´ək) (Gr. *anēr*, male, + *gennaein*, to produce). Gland in Crustacea that causes development of male characteristics.

aneuploidy (an´ū-ploid´ē) (Gr. *an*, without, not, + *eu*, good, well, + *ploid*, multiple of). Loss or gain of a chromosome, cells of the organism have one fewer than normal chromosome number, or one extra chromosome, for example, trisomy 21 (Down syndrome).

angiotensin (an´jē-o-ten´sən) (Gr. *angeion*, vessel, + L. *tensio*, to stretch). Blood protein formed from the interaction of renin and a liver protein, causing increased blood pressure and stimulating release of aldosterone and ADH.

Angstrom (after Ångström, Swedish physicist). A unit of one ten millionth of a millimeter (one ten thousandth of a micrometer); it is represented by the symbol Å.

anhydrase (an-hī´drās) (Gr. *an*, not, + *hydōr*, water, + *ase*, enzyme suffix). An enzyme involved in the removal of water from a compound. Carbonic anhydrase promotes the conversion of carbonic acid into water and carbon dioxide.

anisogametes (an´īs-ō-gam´ēts) (Gr. *anisos*, unequal, + *gametēs*, spouse). Gametes of a species that differ in form or size.

anlage (än´lä-gə) (Ger. laying out, foundation). Rudimentary form; primordium.

annulus (an´yəl-əs) (L. ring). Any ringlike structure, such as superficial rings on leeches.

antenna (L. sail yard). A sensory appendage on the head of arthropods, or the second pair of the two such pairs of structures in crustaceans.

antennal gland. Excretory gland of Crustacea located in the antennal metamere.

anterior (L. comparative of *ante*, before). The head end of an organism, or (as an adjective) toward that end.

antibody (an´tē-bod´ē). A protein, usually circulating and dissolved in the blood, that is capable of combining with the antigen that stimulated its production.

anticodon (an´tī-kō´don). A sequence of three nucleotides in transfer RNA that is complementary to a codon in messenger RNA.

antigen (an´-ti-jən). Any substance capable of stimulating an immune response, most often a protein.

aperture (ap´ər-chər) (L. *apertura* from *aperire*, to uncover). An opening; the opening into the first whorl of a gastropod shell.

apical (ā´pə-kl) (L. *apex*, tip). Pertaining to the tip or apex.

apical complex. A certain combination of organelles found in the protozoan phylum Apicomplexa.

apopyle (ap´-ə-pīl) (Gr. *apo*, away from, + *pylē*, gate). In sponges, opening of the radial canal into the spongocoel.

appendicular (L. *ad*, to, + *pendere*, to hang). Pertaining to appendages; pertaining to vermiform appendix.

arboreal (är-bōr´ē-al) (L. *arbor*, tree). Living in trees.

archenteron (ärk-en´tə-rän) (Gr. *archē*, beginning, + *enteron*, gut). The main cavity of an embryo in the gastrula stage; it is lined with endoderm and represents the future digestive cavity.

archeocytes (ärk´ē-ō-sites) (Gr. *archaios*, beginning, + *kytos*, hollow vessel). Ameboid cells of varied function in sponges.

areolar (a-rē´ə-ler) (L. *areola*, small space). A small area, such as spaces between fibers of connective tissue.

arginine phosphate. Phosphate storage compound (phosphagen) found in many invertebrates and used to regenerate stores of ATP.

Aristotle's lantern. Masticating apparatus of some sea urchins.

artiodactyl (är´ti-o-dak´təl) (Gr. *artios*, even, + *daktylos*, toe). One of an order or suborder of mammals with two or four digits on each foot.

asconoid (Gr. *askos*, bladder). Simplest form of sponges, with canals leading directly from the outside to the interior.

asexual. Without distinct sexual organs; not involving formation of gametes.

assimilation (L. *assimilatio*, bringing into conformity). Absorption and building up of digested nutriments into complex organic protoplasmic materials.

atoke (ā´tōk) (Gr. *a*, without, + *tokos*, offspring). Anterior, nonreproductive part of a marine polychaete, as distinct from the posterior, reproductive part (epitoke) during the breeding season.

ATP. Adenosine triphosphate. In biochemistry, an ester of adenosine and triphosphoric acid.

atrium (ā´trē-əm) (L. *atrium*, vestibule). One of the chambers of the heart; also, the tympanic cavity of the ear; also, the large cavity containing the pharynx in tunicates and cephalochordates.

auricle (aw´ri-kl) (L. *auricula*, dim. of *auris*, ear). One of the less muscular chambers of the heart; atrium; the external ear, or pinna; any earlike lobe or process.

auricularia (ə-rik´u-lar´e-ə) (L. *auricula*, a small ear). A type of larva found in Holothuroidea.

autogamy (aw-täg´ə-me) (Gr. *autos*, self, + *gamos*, marriage). Condition in which the gametic nuclei produced by meiosis fuse within the same organism that produced them to restore the diploid number.

autosome (aw´-tō-sōm) (Gr. *autos*, self, + *sōma*, body). Any chromosome that is not a sex chromosome.

autotomy (aw-täd´ə-mē) (Gr. *autos*, self, + *tomos*, a cutting). The breaking off of a part of the body by the organism itself.

autotroph (aw´tō-trōf) (Gr. *autos*, self, + *trophos*, feeder). An organism that makes its organic nutrients from inorganic raw materials.

autotrophic nutrition (Gr. *autos*, self, + *trophia*, denoting nutrition). Nutrition characterized by the ability to use simple inorganic substances for the synthesis of more complex organic compounds, as in green plants and some bacteria.

avicularium (L. *avicula*, small bird, + *aria*, like or connected with). Modified zooid that is attached to the surface of the major zooid in Ectoprocta and resembles a bird's beak.

axial (L. *axis*, axle). Relating to the axis, or stem; on or along the axis.

axolotl (ak´sə-lot´l) (Nahuatl, *atl*, water, + *xolotl*, doll, servant, spirit). Larval stage of any of several species of the genus *Ambystoma* (such as *Ambystoma tigrinum*) exhibiting neotenic reproduction.

axoneme (aks´ə-nēm) (L. *axis*, axle, + Gr. *nēma*, thread). The microtubules in a cilium or flagellum, usually arranged as a circlet of nine pairs enclosing one central pair; also, the microtubules of an axopodium.

axopodium (ak´sə-pō´di-um) (Gr. *axon*, an axis, + *podion*, small foot). Long, slender, more or less permanent pseudopodium found in certain sarcodine protozoa. (Also **axopod**.)

B cell. A type of lymphocyte that is most important in the humoral immune response.

basal body. Also known as kinetosome and blepharoplast, a cylinder of nine triplets of microtubules found basal to a flagellum or cilium; same structure as a centriole.

basis, basipodite (bā´səs, bā-si´pə-dīt) (Gr. *basis*, base, + *pous*, podos, foot). The distal or second joint of the protopod of a crustacean appendage.

bathypelagic (bath´ə-pe-laj´ik) (Gr. *bathys*, deep + *pelagos*, open sea). Relating to or inhabiting the deep sea.

benthos (ben´thäs) (Gr. depth of the sea). Organisms that live along the bottom of seas and lakes; adj., **benthic**. Also, the bottom itself.

biogenesis (bī´ō-jen´ə-səs) (Gr. *bios*, life, + *genesis*, birth). The doctrine that life originates only from preexisting life.

bioluminescence. Method of light production by living organisms in which usually certain proteins (luciferins), in the presence of oxygen and an enzyme (luciferase), are converted to oxyluciferins with the liberation of light.

biomass (Gr. *bios*, life, + *maza*, lump or mass). The weight of total living organisms or of a species population per unit of area.

biome (bī´ōm) (Gr., *bios*, life, + *ōma*, abstract group suffix). Complex of plant and animal communities characterized by climatic and soil conditions; the largest ecological unit.

biosphere (Gr. *bios*, life, + *sphaira*, globe). That part of earth containing living organisms.

bipinnaria (L. *bi*, double, + *pinna*, wing, + *aria*, like or connected with). Free-swimming, ciliated, bilateral larva of the asteroid echinoderms; develops into the brachiolaria larva.

biramous (bī-rām´əs) (L. *bi*, double, + *ramus*, branch). Adjective describing appendages with two distinct branches, contrasted with uniramous, unbranched.

bat / āpe / ärmadillo/ herring / fēmale / finch / līce / crocodile / crōw / duck / ūnicorn / ə indicates unaccented vowel sound "uh" as in mammal, fishes, cardinal, heron, vulture / stress as in bi-ol´o-gy, bi´o-log´i-cal

bivalent (bī-vāl′ənt) (L. *bi*, double + *valen*, strength, worth). The pairs of homologous chromosomes at synapsis in the first meiotic division, a tetrad.

blastocoel (blas′tō-sēl) (Gr. *blastos*, germ, + *koilos*, hollow). Cavity of the blastula.

blastomere (Gr. *blastos*, germ, + *meros*, part). An early cleavage cell.

blastopore (Gr. *blastos*, germ, + *poros*, passage, pore). External opening of the archenteron in the gastrula.

blastula (Gr. *blastos*, germ, + L. *ula*, dim.). Early embryological stage of many animals; consists of a hollow mass of cells.

blepharoplast (blə-fä′rə-plast) (Gr. *blepharon*, eyelid, + *plastos*, formed). See basal body.

blood type. Characteristic of human blood given by the particular antigens on the membranes of the erythrocytes, genetically determined, causing agglutination when incompatible groups are mixed; the blood types are designated A, B, O, AB, Rh negative, Rh positive, and others.

Bohr effect. A characteristic of hemoglobin that causes it to dissociate from oxygen in greater degree at higher concentrations of carbon dioxide.

BP. Before the present.

brachial (brak′ē-əl) (L. *brachium*, forearm). Referring to the arm.

brachiolaria (brak′ē-ō-lār′ē-ə) (L. *brachiola*, little arm, + *aria*, pertaining to). This asteroid larva develops from the bipinnaria larva and has three preoral holdfast processes.

branchial (brank′ē-əl) (Gr. *branchia*, gills). Referring to gills.

brown fat. Mitochondria-rich, heat-generating adipose tissue of endothermic vertebrates.

buccal (buk′əl) (L. *bucca*, cheek). Referring to the mouth cavity.

buffer. Any substance or chemical compound that tends to keep pH levels constant when acids or bases are added.

calorie (kal′ō-ry) (L. *calere*, to be warm). Unit of heat defined as the amount of heat required to heat 1 g of water from 14.5 to 15.5 C; 1 cal = 4.184 joules in the International System of Units.

capitulum (ka-pi′tə-ləm) (L. small head). Term applied to small, headlike structures of various organisms, including projection from body of ticks and mites carrying mouthparts.

carapace (kar′ə-pās) (F. from Sp. *carapacho*, shell). Shieldlike plate covering the cephalothorax of certain crustaceans; dorsal part of the shell of a turtle.

carbohydrate (L. *carbo*, charcoal, + Gr. *hydōr*, water). Compounds of carbon, hydrogen, and oxygen having the generalized formula $(CH_2O)_n$; aldehyde or ketone derivatives of polyhydric alcohols, with hydrogen and oxygen atoms attached in a 2:1 ratio.

carboxyl (kär-bäk′səl) (carbon + oxygen + yl, chemical radical suffix). The acid group of organic molecules —COOH.

carinate (kar′ə-nāt) (L. *carina*, keel) Having a keel, in particular the flying birds with a keeled sternum for the insertion of flight muscles.

carnivore (kar′nə-vōr′) (L. *carnivorous*, flesh eating). One of the flesh-eating mammals of the order Carnivora. Also, any organism that eats animals. Adj., **carnivorous.**

carotene (kär′ə-tēn) (L. *carota*, carrot, + *ene*, unsaturated straight-chain hydrocarbons). A red, orange, or yellow pigment belonging to the group of carotenoids; precursor of vitamin A.

cartilage (L. *cartilago*; akin to L. *cratis*, wickerwork). A translucent elastic tissue that makes up most of the skeleton of embryos, very young vertebrates, and adult cartilaginous fishes, such as sharks and rays; in higher forms much of it is converted into bone.

caste (kast) (L. *castus*, pure, separated). One of the polymorphic forms within an insect society, each caste having its specific duties, as queen, worker, soldier, and so on.

catadromous (kə-tad′rə-məs) (Gr. *kata*, down, + *dromos*, a running). Refers to fishes that migrate from fresh water to the ocean to spawn.

catalyst (kad′ə-ləst) (Gr. *kata*, down, + *lysis*, a loosening). A substance that accelerates a chemical reaction but does not become a part of the end product.

cecum, caecum (sē′kəm) (L. *caecus*, blind). A blind pouch at the beginning of the large intestine; any similar pouch.

cellulose (sel′ū-lōs) (L. *cella*, small room). Chief polysaccharide constituent of the cell wall of green plants and some fungi; an insoluble carbohydrate $(C_6H_{10}O_5)_n$ that is converted into glucose by hydrolysis.

centriole (sen′trē-ol) (Gr. *kentron*, center of a circle, + L. *ola*, small). A minute cytoplasmic organelle usually found in the centrosome and considered to be the active division center of the animal cell; organizes spindle fibers during mitosis and meiosis. Same structure as basal body or kinetosome.

centrolecithal (sen′tro-les′ə-thəl) (Gr. *kentron*, center, + *lekithos*, yolk, + Eng. *al*, adjective). Pertaining to an insect egg with the yolk concentrated in the center.

centromere (sen′trə-mir) (Gr. *kentron*, center, + *meros*, part). A small body or constriction on the chromosome to which a spindle fiber attaches during mitosis or meiosis.

centrosome (sen′trə-sōm) (Gr. *kentron*, center, + *sōma*, body). Minute body in cytoplasm of many plant and animal cells that contains one or two centrioles and is the center of dynamic activity in mitosis.

cephalization (sef′ə-li-zā-shən) (Gr. *kephale*, head). The process by which specialization, particularly of the sensory organs and appendages, became localized in the head end of animals.

cephalothorax (sef′ə-lä-thō′raks) (Gr. *kephale*, head, + *thorax*). A body division found in many Arachnida and higher Crustacea, in which the head is fused with some or all of the thoracic segments.

cercaria (ser-kär′ē-ə) (Gr. *kerkos*, tail, + L. *aria*, like or connected with). Tadpolelike larva of trematodes (flukes).

chelicera (kə-lis′ə-rə), pl. **chelicerae** (Gr. *chēlē*, claw, + *keras*, horn). One of a pair of the most anterior head appendages on the members of the subphylum Chelicerata.

chelipeds (kēl′ə-peds) (Gr. *chēlē*, claw, + L. *pes*, foot). Pincerlike first pair of legs in most decapod crustaceans; specialized for seizing and crushing.

chiasma (kī-az′mə), pl. **chiasmata** (Gr. cross). An intersection or crossing, as of nerves; a connection point between homologous chromatids where crossing over has occurred at synapsis.

chitin (kī′tən) (Fr. *chitine*, from Gr. *chitōn*, tunic). A horny substance that forms part of the cuticle of arthropods and is found sparingly in certain other invertebrates; a nitrogenous polysaccharide insoluble in water, alcohol, dilute acids, and digestive juices of most animals.

chlorogogue cells (klōr′ə-gog) (Gr. *chlōros*, light green, + *agōgos*, a leading, a guide). Modified peritoneal cells, greenish or brownish, clustered around the digestive tract of certain annelids; apparently they aid in elimination of nitrogenous wastes and in food transport.

chlorocruorin (klō′rō-kroo′ə-rən) (Gr. *chlōros*, light green, + L. *cruor*, blood). A greenish iron-containing respiratory pigment dissolved in the blood plasma of certain marine polychaetes.

chlorophyll (klō′rō-fil) (Gr. *chlōros*, light green, + *phyllōn*, leaf). Green pigment found in plants and in some animals; necessary for photosynthesis.

chloroplast (klō′rō-plast) (Gr. *chlōros*, light green, + *plastos*, molded). A plastid containing chlorophyll and usually other pigments, found in cytoplasm of plant cells.

choanocyte (kō-an′ō-sīt) (Gr. *choanē*, funnel, + *kytos*, hollow vessel). One of the flagellate collar cells that line cavities and canals of sponges.

cholinergic (kōl′-i-nər′jik) (Gr. *chōle*, bile + *ergon*, work). Type of nerve fiber that releases acetylcholine from axon terminal.

chorion (kō′rē-on) (Gr. *chorion*, skin). The outer of the double membrane that surrounds the embryo of reptiles, birds, and mammals; in mammals it contributes to the placenta.

choroid plexus (kōr′oid plek′sus) (Gr. *chorion*, skin, + *eidos*, form; L. *plexus*, interwoven). Vascular network in the brain ventricles that regulates secretion and absorption of cerebrospinal fluid.

chromatid (krō′mə-tid) (Gr. *chromato*, from *chrōma*, color, + L. *id*, feminine stem for particle of specified kind). A replicated chromosome joined to its sister chromatid by the centromere; separates and becomes daughter chromosome at anaphase of mitosis or anaphase of the second meiotic division.

chromatin (krō′mə-tin) (Gr. *chrōma*, color). The nucleoprotein material of a chromosome; the hereditary material containing DNA.

chromatophore (krō-mat′ə-fōr) (Gr. *chrōma*, color, + *pherein*, to bear). Pigment cell, usually in the dermis, in which usually the pigment can be dispersed or concentrated.

chromomere (krō′mō-mir) (Gr. *chrōma*, color, + *meros*, part). One of the chromatin granules of characteristic size on the chromosome; may be identical with a gene or a cluster of genes.

chromonema (krō-mə-nē′mə) (Gr. *chrōma*, color, + *nēma*, thread). A convoluted thread in prophase of mitosis or the central thread in a chromosome.

chromoplast (krō′mō-plast) (Gr. *chrōma*, color, + *plastos*, molded). A plastid containing pigment.

chromosome (krō′mə-sōm) (Gr. *chrōma*, color, + *sōma*, body). A complex body, spherical or rod shaped, that arises from the nuclear network during mitosis, splits longitudinally, and carries a part of the organism's genetic information as genes composed of DNA.

chrysalis (kris′ə-lis) (L. from Gr. *chrysos*, gold). The pupal stage of a butterfly.

cilium (sil′i-əm), pl. **cilia** (L. eyelid). A hairlike, vibratile organelle process found on many animal cells. Cilia may be used in moving particles along the cell surface or, in ciliate protozoans, for locomotion.

cinclides (sing′klid-əs), sing. **cinclis** (sing′kləs) (Gr. *kinklis*, latticed gate or partition). Small pores in the external body wall of sea anemones for extrusion of acontia.

circadian (sər′kə-dē′-ən) (L. *circa*, around, + *dies*, day). Occurring at a period of approximately 24 hours.

cirrus (sir′əs) (L. curl). A hairlike tuft on an insect appendage; locomotor organelle of fused cilia; male copulatory organ of some invertebrates.

cisternae (sis-ter′nē) (L. *cista*, box). Space between membranes of the endoplasmic reticulum within cells.

cistron (sis′trən) (L. *cista*, box). A series of codons in DNA that code for an entire polypeptide chain.

cladistics (klad-is′-təks) (Gr. *cladus*, branch, sprout). A system of arranging taxa by analysis of primitive and derived characteristics so that the arrangement will reflect phylogenetic relationships.

cleavage (O.E. *cleofan*, to cut). Process of nuclear and cell division in animal zygote.

climax (klī′maks) (Gr. *klimax*, ladder). Stage of relative stability attained by a community of organisms, often the culminating development of a natural succession. Also, orgasm.

cline (klīn) (Gr. *klinein*, slope, bend). A pattern of gradual genetic change in a population according to its geographical range.

clitellum (klī-tel′əm) (L. *clitellae*, packsaddle). Thickened saddlelike portion of certain midbody segments of many oligochaetes and leeches.

cloaca (klō-ā′kə) (L. sewer). Posterior chamber of digestive tract in many vertebrates, receiving feces and urinogenital products. In certain invertebrates, a terminal portion of digestive tract that serves also as respiratory, excretory, or reproductive duct.

clone (klōn) (Gr. *klōn*, twig). All descendants derived by asexual reproduction from a single individual.

cnidocil (nī′dō-sil) (Gr. *knidē*, nettle, + L. *cilium*, hair). Triggerlike spine on nematocyst.

cnidocyte (nī′dō-sīt) (Gr. *knidē*, nettle, + *kytos*, hollow vessel). Modified interstitial cell that holds the nematocyst; during development of the nematocyst, the cnidocyte is a cnidoblast.

coacervate (kō′ə-sər′vət) (L. *coacervatus*, to heap up). An aggregate of colloidal droplets held together by electrostatic forces.

cochlea (kōk′lēə) (L. snail, from Gr. *kochlos*, a shellfish). A tubular cavity of the inner ear containing the essential organs of hearing;

occurs in crocodiles, birds, and mammals; spirally coiled in mammals.

cocoon (kə-kun′) (Fr. *cocon*, shell). Protective covering of a resting or developmental stage, sometimes used to refer to both the covering and its contents; for example, the cocoon of a moth or the protective covering for the developing embryos in some annelids.

codon (kō′dän) (L. code, + on). A sequence of three adjacent nucleotides that codes for one amino acid.

coelenteron (sē-len′tər-on) (Gr. *koilos*, hollow, + *enteron*, intestine). Internal cavity of a cnidarian; gastrovascular cavity; archenteron.

coelom (sē′lōm) (Gr. *koilōma*, cavity). The body cavity in triploblastic animals, lined with mesodermal peritoneum.

coelomocyte (sē′lō′mə-sīt) (Gr. *koilōma*, cavity, + *kytos*, hollow vessel). Another name for amebocyte; primitive or undifferentiated cell of the coelom and the water-vascular system.

coelomoduct (sē-lō′mə-dukt) (Gr. *koilos*, hollow, + L. *ductus*, a leading). A duct that carries gametes or excretory products (or both) from the coelom to the exterior.

coenecium, coenoecium (sə-nēs[h]′ē-um) (Gr. *koinos*, common, + *oikion*, house). The common secreted investment of an ectoproct colony; may be chitinous, gelatinous, or calcareous.

coenocytic (sē-nə-sit′ik) (Gr. *koinos*, common, + *kytos*, hollow vessel). A tissue in which the nuclei are not separated by cell membranes; syncytial.

coenzyme (kō-en′zīm) (L. prefix, *co*, with, + Gr. *enzymos*, leavened, from *en*, in, + *zymē*, leaven). A required substance in the activation of an enzyme; a prosthetic or nonprotein constituent of an enzyme.

collagen (käl′ə-jən) (Gr. *kolla*, glue, + *genos*, descent). A tough, fibrous protein occurring in vertebrates as the chief constituent of collagenous connective tissue; also occurs in invertebrates, for example, the cuticle of nematodes.

collenchyme (käl′ən-kīm) (Gr. *kolla*, glue, + *enchyma*, infusion). A gelatinous mesenchyme containing undifferentiated cells; found in cnidarians and ctenophores.

collencyte (käl′lən-sīt) (Gr. *kolla*, glue, + *en*, in, + *kytos*, hollow vessel). A type of cell in sponges that is star shaped and apparently contractile.

colloblast (käl′ə-blast) (Gr. *kolla*, glue + *blastos*, germ). A glue-secreting cell on the tentacles of ctenophores.

colloid (kä′loid) (Gr. *kolla*, glue, + *eidos*, form). A two-phase system in which particles of one phase are suspended in the second phase.

columella (kä′lə-mel′ə) (L. small column). Central pillar in gastropod shells.

comb plate. One of the plates of fused cilia that are arranged in rows for ctenophore locomotion.

commensalism (kə-men′səl-iz′əm) (L. *cum*, together with, + *mensa*, table). A relationship in which one individual lives close to or on another and benefits, and the host is unaffected; often symbiotic.

community (L. *communitas*, community, fellowship). An assemblage of organisms that are associated in a common environment and interact with each other in a self-sustaining and self-regulating relation.

competition. Some degree of overlap in ecological niches of two populations in the same community, such that they both depend on the same food source, shelter, or other resources.

complement. Collective name for a series of enzymes and activators in the blood, some of which may bind to antibody and may lead to rupture of a foreign cell.

conjugation (kon′ju-gā′shən) (L. *conjugare*, to yoke together). Temporary union of two ciliate protozoa while they are exchanging chromatin material and undergoing nuclear phenomena resulting in binary fission. Also, formation of cytoplasmic bridges between bacteria for transfer of plasmids.

conspecific (L. *com*, together, + *species*). A member of the same species.

contractile vacuole. A clear fluid-filled cell vacuole in protozoa and a few lower metazoa; takes up water and releases it to the outside in a cyclical manner, for osmoregulation and some excretion.

copulation (Fr. from L. *copulare*, to couple). Sexual union to facilitate the reception of sperm by the female.

copy DNA (cDNA). DNA "copied" from an mRNA molecule using the viral enzyme reverse transcriptase.

corium (kō′re-um) (L. *corium*, leather). The deep layer of the skin; dermis.

cornea (kor′nē-ə) (L. *corneus*, horny). The outer transparent coat of the eye.

corneum (kor′nē-əm) (L. *corneus*, horny). Epithelial layer of dead, keratinized cells. Stratum corneum.

cornified (kor′nə-fīd) (L. *corneus*, horny). Adjective for conversion of epithelial cells into nonliving, keratinized cells.

corona (kə-rō′nə) (L. crown). Head or upper portion of a structure; ciliated disc on anterior end of rotifers.

corpora allata (kor′pə-rə əl-la′tə) (L. *corpus*, body + *allatum*, aided). Endocrine glands in insects that produce juvenile hormone.

cortex (kor′teks) (L. bark). The outer layer of a structure.

coxa, coxopodite (kox′ə, kəx-ä′pə-dīt) (L. *coxa*, hip, + Gr. *pous, podos*, foot). The proximal joint of an insect or arachnid leg; in crustaceans, the proximal joint of the protopod.

creatine phosphate. High-energy phosphate compound found in the muscle of vertebrates and some invertebrates, used to regenerate stores of ATP.

crista (kris′ta), pl. **cristae** (L. *crista*, crest). A crest or ridge on a body organ or organelle; a platelike projection formed by the inner membrane of mitochondrion.

crossing over. Exchange of parts of nonsister chromatids at synapsis in the first meiotic division.

cryptobiotic (Gr. *kryptos*, hidden, + *biōticus*, pertaining to life). Living in concealment; refers to insects and other animals that live in secluded situations, such as underground or in wood; also tardigrades and some nematodes, rotifers, and others that survive harsh environmental conditions by assuming for a time a state of very low metabolism.

ctenoid scales (ten′oid) (Gr. *kteis, ktenos*, comb). Thin, overlapping dermal scales of the more advanced fishes; exposed posterior margins have fine, toothlike spines.

cuticle (kū′ti-kəl) (L. *cutis*, skin). A protective, noncellular, organic layer secreted by the external epithelium (hypodermis) of many invertebrates. In higher animals the term refers to the epidermis or outer skin.

cyanobacteria (sī-an-ō-bak-ter′ē-ə) (Gr. *kyanos*, a dark-blue substance, + *bakterion*, dim. of baktron, a staff). Photosynthetic prokaryotes, also called blue-green algae, cyanophytes.

cyanophyte (sī-an′ō-fīt) (Gr. *kyanos*, a dark-blue substance, + *phyton*, plant). A cyanobacterium, blue-green alga.

cycloid scales (sī′-kloid) (Gr. *kyklos*, circle). Thin, overlapping dermal scales of the more primitive fishes; posterior margins are smooth.

cydippid larva (sī-dip′pid) (Gr. *kydippe*, mythological Athenian maiden). Free-swimming larva of most ctenophores; superficially similar to the adult.

cyrtocyte (ser′tō-sīt) (Gr. *kyrtē*, a fish basket, cage, + *kytos*, hollow vessel). A protonephridial cell with a single flagellum enclosed in a cylinder of cytoplasmic rods.

cysticercoid (sis′tə-ser′coəd) (Gr. *kystis*, bladder, + *kerkos*, tail, + *eidos*, form). A type of juvenile tapeworm composed of a solid-bodied cyst containing an invaginated scolex; contrast with cysticercus.

cysticercus (sis′tə-ser′kəs) (Gr. *kystis*, bladder, + *kerkos*, tail). A type of juvenile tapeworm in which an invaginated and introverted scolex is contained in a fluid-filled bladder; contrast with cysticercoid.

cystid (sis′tid) (Gr. *kystis*, bladder). In an ectoproct, the dead secreted outer parts plus the adherent underlying living layers.

cytochrome (sī′tə-krōm) (Gr. *kytos*, hollow vessel, + *chrōma*, color). Several iron-containing pigments that serve as electron carriers in aerobic respiration.

cytokinesis (sī′tə-kin-ē′sis) (Gr. *kytos*, hollow vessel, + *kinesis*, movement). Division of the cytoplasm of a cell.

cytopharynx (Gr. *kytos*, hollow vessel, + *pharynx*, throat). Short tubular gullet in ciliate protozoa.

cytoplasm (sı′tə-plasm) (Gr. *kytos*, hollow vessel, + *plasma*, mold). The living matter of the cell, excluding the nucleus.

cytoproct (sī′tə-prokt) (Gr. *kytos*, hollow vessel, + *prōktos*, anus). Site on a protozoan where undigestible matter is expelled.

cytosol (sī′tə-sol) (Gr. *kytos*, hollow vessel, + L. *sol*, from *solutus*, to loosen). Unstructured portion of the cytoplasm in which the organelles are bathed.

cytosome (sī′tə-sōm) (Gr. *kytos*, hollow vessel, + *sōma*, body). The cell body inside the plasma membrane.

cytostome (sī′tə-stōm) (Gr. *kytos*, hollow vessel, + *stoma*, mouth). The cell mouth in many protozoa.

Darwinism. Theory of evolution by natural selection.

definitive host. The host in which sexual reproduction of a symbiont takes place; if no sexual reproduction, then the host in which the symbiont becomes mature and reproduces; contrast intermediate host.

deme (dēm) (Gr. populace). A local population of closely related animals.

demography (də-mäg′grə-fē) (Gr. *demos*, people, + *graphy*). The properties of the rate of growth and the age structure of populations.

deoxyribonucleic acid (DNA). The genetic material of all organisms, characteristically organized into linear sequences of genes.

deoxyribose (dē-ok′sē-rī′bōs) (L. *deoxy*, loss of oxygen, + ribose, a pentose sugar). A 5-carbon sugar having 1 oxygen atom less than ribose; a component of deoxyribonucleic acid (DNA).

dermal (Gr. *derma*, skin). Pertaining to the skin; cutaneous.

dermis. The inner, sensitive mesodermal layer of skin; corium.

desmosome (dez′mə-sōm) (Gr. *desmos*, bond, + *sōma*, body). Buttonlike plaque serving as an intercellular connection.

determinate cleavage. The type of cleavage, usually spiral, in which the fate of the blastomeres is determined very early in development; mosaic cleavage.

detritus (də-trī′tus) (L. that which is rubbed or worn away). Any fine particulate debris of organic or inorganic origin.

Deuterostomia (dū′də-rō-stō′mē-ə) (Gr. *deuteros*, second, secondary, + *stoma*, mouth). A group of higher phyla in which cleavage is indeterminate (regulative) and primitively radial. The endomesoderm is enterocoelous, and the mouth is derived away from the blastopore. Includes Echinodermata, Chordata, and a number of minor phyla. Compare with Protostomia.

dextral (dex′trəl) (L. *dexter*, right-handed). Pertaining to the right; in gastropods, shell is dextral if opening is to right of columella when held with spire up and facing observer.

diapause (dī′ə-pawz) (Gr. *diapausis*, pause). A period of arrested development in the life cycle of insects and certain other animals in which physiological activity is very low and the animal is highly resistant to unfavorable external conditions.

diffusion (L. *diffusus*, dispersion). The movement of particles or molecules from area of high concentration of the particles or molecules to area of lower concentration.

digitigrade (dij′ə-də-grād) (L. *digitus*, finger, toe, + *gradus*, step, degree). Walking on the digits with the posterior part of the foot raised; compare plantigrade.

dihybrid (dī-hī′brəd) (Gr. *dis*, twice, + L., *hibrida*, mixed offspring). A hybrid whose parents differ in two distinct characters; an offspring having different alleles at two different loci, for example, A/a B/b.

dimorphism (dī-mor′fizm) (Gr. *di*, two, + *morphē*, form). Existence within a species of two distinct forms according to color, sex, size, organ structure, and so on. Occurrence of two kinds of zooids in a colonial organism.

dioecious (dī-ē′shəs) (Gr. *di*, two, + *oikos*, house). Having male and female organs in separate individuals.

diphycercal (dif′i-ser′kəl) (Gr. *diphyēs*, twofold, + *kerkos*, tail). A tail that tapers to a point, as in lungfishes; vertical column extends to tip without upturning.

diphyodont (dī′fi-ə-dänt) (Gr. *diphyēs*, twofold, + *odous*, tooth). Having deciduous and permanent sets of teeth successively.

dipleurula (dī-ploor′ū-lə) (Gr. *di*, two, + *pleura*, rib, side, + L. *ula*, small). A hypothetical, simple ancestral form of echinoderms; elongated and bilaterally symmetrical, with three pairs of coelomic sacs.

diploblastic (di′plə-blas′tək) (Gr. *diploos*, double, + *blastos*, bud). Organism with two germ layers, endoderm and ectoderm.

diploid (dip′loid) (Gr. *diploos*, double, + *eidos*, form). Having the somatic (double, or 2n) number of chromosomes or twice the number characteristic of a gamete of a given species.

DNA. See deoxyribonucleic acid.

dominance hierarchy. A social ranking, formed through agonistic behavior, in which individuals are associated with each other so that some have greater access to resources than do others.

dorsal (dor′səl) (L. *dorsum*, back). Toward the back, or upper surface, of an animal.

Down syndrome. A congenital syndrome including mental retardation, caused by the cells in a person's body having an extra chromosome 21; also called trisomy 21.

dyad (dī′əd) (Gr. *dyas*, two). One of the groups of two chromosomes formed by the division of a tetrad during the first meiotic division.

ecdysiotropin (ek-dē-zē-o-tro′pən) (Gr. *ekdysis*, to strip off, escape, + *tropos*, a turn, change). Hormone secreted in brain of insects that stimulates prothoracic gland to secrete molting hormone. Prothoracicotropic hormone.

ecdysis (ek′də-sis) (Gr. *ekdysis*, to strip off, escape). Shedding of outer cuticular layer; molting, as in insects or crustaceans.

ecdysone (ek-dī′sōn) (Gr. *ekdysis*, to strip off). Molting hormone of arthropods, stimulates growth and ecdysis, produced by prothoracic glands in insects and Y organs in crustaceans.

ecology (Gr. *oikos*, house, + *logos*, discourse). Part of biology that deals with the relationship between organisms and their environment.

ecosystem (ek′ō-sis-təm) (eco[logy] from Gr. *oikos*, house, + system). An ecological unit consisting of both the biotic communities and the nonliving (abiotic) environment, which interact to produce a stable system.

ecotone (ek′ō-tōn) (eco[logy] from Gr. *oikos*, home, + *tonos*, stress). The transition zone between two adjacent communities.

ectoderm (ek′tō-derm) (Gr. *ektos*, outside, + *derma*, skin). Outer layer of cells of an early embryo (gastrula stage); one of the germ layers, also sometimes used to include tissues derived from ectoderm.

ectolecithal (ek′tō-les′ə-thəl) (Gr. *ektos*, outside, + *lekithos*, yolk). Yolk for nutrition of the embryo contributed by cells that are separate from the egg cell and are combined with the zygote by envelopment within the eggshell.

ectoplasm (ec′tō-plazm) (Gr. *ektos*, outside, + *plasma*, form). The cortex of a cell or that part of cytoplasm just under the cell surface; contrasts with endoplasm.

ectothermic (ek′tō-therm′ic) (Gr. *ektos*, outside, + *thermē*, heat). Having a variable body temperature derived from heat acquired from the environment; contrasts with **endothermic**.

effector (L. *efficere*, bring to pass). An organ, tissue, or cell that becomes active in response to stimulation.

efferent (ef′ə-rənt) (L. *ex*, out, + *ferre*, to bear). Leading or conveying away from some organ, for example, nerve impulses conducted away from the brain, or blood conveyed away from an organ; contrasts with **afferent**.

eleocyte (el'ē-ə-sīt) (Gr. *elaion*, oil, + *kytos*, hollow vessel). Fat-containing cells in annelids that originate from the chlorogogue tissue.

elephantiasis (el-ə-fən-tī'ə-səs). Disfiguring condition caused by chronic infection with filarial worms *Wuchereria bancrofti* and *Brugia malayi*.

embryogenesis (em'brē-ō-jen'ə-səs) (Gr. *embryon*, embryo, + *genesis*, origin). The origin and development of the embryo; embryogeny.

emigrate (L. *emigrare*, to move out). To move *from* one area to another to take up residence; contrasts with immigrate.

emulsion (ə-məl'shən) (L. *emulsus*, milked out). A colloidal system in which both phases are liquids.

endemic (en-dem'ik) (Gr. *en*, in, + *demos*, populace). Peculiar to a certain region or country; native to a restricted area; not introduced.

endergonic (en-dər-gän'ik) (Gr. *endon*, within, + *ergon*, work). Used in reference to a chemical reaction that requires energy; energy absorbing.

endite (en'dīt) (Gr. *endon*, within). Medial process on an arthropod limb.

endocrine (en'də-krən) (Gr. *endon*, within, + *krinein*, to separate). Refers to a gland that is without a duct and that releases its product directly into the blood or lymph.

endocytosis (en'dō-sī-tō'sis) (Gr. *endon*, within, + *kytos*, hollow vessel). The engulfment of matter by phagocytosis and of macromolecules by pinocytosis.

endoderm (en'də-dərm) (Gr. *endon*, within, + *derma*, skin). Innermost germ layer of an embryo, forming the primitive gut; also may refer to tissues derived from endoderm.

endolecithal (en'də-les'ə-thəl) (Gr. *endon*, within, + *lekithos*, yolk). Yolk for nutrition of the embryo incorporated into the egg cell itself.

endolymph (en'do-limf) (Gr. *endon*, within, + *lympha*, water). Fluid that fills most of the membranous labyrinth of the vertebrate ear.

endometrium (en'də-mē'trē-əm) (Gr. *endon*, within, + *mētra*, womb). The mucous membrane lining the uterus.

endoplasm (en'də-pla-zm) (Gr. *endon*, within, + *plasma*, mold or form). The portion of cytoplasm that immediately surrounds the nucleus.

endoplasmic reticulum. A complex of membranes within a cell; may bear ribosomes (rough) or not (smooth).

endopod, endopodite (en'də-päd, en-dop'ə-dīt) (Gr. *endon*, within, + *pous, podos*, foot). Medial branch of a biramous crustacean appendage.

endopterygote (en'dəp-ter'i-gōt) (Gr. *endon*, within, + *pteron*, feather, wing). Insect in which the wing buds develop internally; has holometabolous metamorphosis.

endorphin (en-dor'fin) (contraction of endogenous morphine). Group of opiate-like brain neuropeptides that modulate pain perception and are implicated in many other functions.

endoskeleton (Gr. *endon*, within, + *skeletos*, hard). A skeleton or supporting framework within the living tissues of an organism; contrasts with exoskeleton.

endosome (en'də-sōm) (Gr. *endon*, within, + *sōma*, body). Nucleolus in nucleus of some protozoa that retains its identity through mitosis.

endostyle (en'də-stīl) (Gr. *endon*, within, + *stylos*, a pillar). Ciliated groove(s) in the floor of the pharynx of tunicates, cephalochordates, and larval cyclostomes, useful for accumulating and moving food particles to the stomach.

endothermic (en'də-therm'ic) (Gr. *endon*, within, + *thermē*, heat). Having a body temperature determined by heat derived from the animal's own oxidative metabolism; contrasts with ectothermic.

enkephalin (en-kef'lin) (Gr. *endon*, within, + *kephale*, head). Group of small brain neuropeptides with opiate-like qualities.

enterocoel (en'tər-ō-sēl') (Gr. *enteron*, gut, + *koilos*, hollow). A type of coelom formed by the outpouching of a mesodermal sac from the endoderm of the primitive gut.

enterocoelic mesoderm formation. Embryonic formation of mesoderm by a pouchlike outfolding from the archenteron, which then expands and obliterates the blastocoel, thus forming a large cavity, the coelom, lined with mesoderm.

enterocoelomate (en'ter-ō-sēl'ō-māte) (Gr. *enteron*, gut, + *koilōma*, cavity, + Eng. *ate*, state of). An animal having an enterocoel, such as an echinoderm or a vertebrate.

enteron (en'tə-rän) (Gr. intestine). The digestive cavity.

entozoic (en-tə-zō'ic) (Gr. *entos*, within, + *zōon*, animal). Living within another animal; internally parasitic (chiefly parasitic worms).

entropy (en'trə-pē) (Gr. *en*, in, on, + *tropos*, turn, change in manner). A quantity that is the measure of energy in a system not available for doing work.

enzyme (en'zīm) (Gr. *enzymos*, leavened, from *en*, in, + *zyme*, leaven). A protein substance, produced by living cells, that is capable of speeding up specific chemical transformations, such as hydrolysis, oxidation, or reduction, but is unaltered itself in the process; a biological catalyst.

ephyra (ef'ə-rə) (Gr. *Ephyra*, Greek city). Refers to castlelike appearance. Medusa bud from a scyphozoan polyp.

epidermis (ep'ə-dər'məs) (Gr. *epi*, on, upon, + *derma*, skin). The outer, nonvascular layer of skin of ectodermal origin; in invertebrates, a single layer of ectodermal epithelium.

epididymis (ep'ə-did'ə-məs) (Gr. *epi*, on, upon, + *didymos*, testicle). Part of the sperm duct that is coiled and lying near the testis.

epigenesis (ep'ə-jen'ə-sis) (Gr. *epi*, on, upon, + *genesis*, birth). The embryological (and generally accepted) view that an embryo is a new creation that develops and differentiates step by step from an initial stage; the progressive production of new parts that were nonexistent as such in the original zygote.

epigenetics (ep'ə-jə-net'iks) (Gr. *epi*, on, upon, + *genesis*, birth). Study of mechanisms by which the genes produce phenotypic effects.

epipod, epipodite (ep'ē-päd, e-pip'ə-dīt) (Gr. *epi*, on, upon, + *pous, podos*, foot). A lateral process on the protopod of a crustacean appendage, often modified as a gill.

epithelium (ep'i-thē'lē-um) (Gr. *epi*, on, upon, + *thele*, nipple). A cellular tissue covering a free surface or lining a tube or cavity.

epitoke (ep'i-tōk) (Gr. *epitokos*, fruitful). Posterior part of a marine polychaete when swollen with developing gonads during the breeding season; contrast with atoke.

erythroblastosis fetalis (ə-rith'-rō-blas-tō'səs fə-tal'əs) (Gr. *erythros*, red, + *blastos*, germ, + *osis*, a disease; L. *fetalis*, relating to a fetus). A disease of newborn infants caused when Rh-negative mothers develop antibodies against the Rh-positive blood of the fetus. See blood type.

erythrocyte (ə-rith'rō-sīt) (Gr. *erythros*, red, + *kytos*, hollow vessel). Red blood cell; has hemoglobin to carry oxygen from lungs or gills to tissues; during formation in mammals, erythrocytes lose their nuclei, those of other vertebrates retain the nuclei.

estrus (es'trəs) (L. *oestrus*, gadfly, frenzy). The period of heat, or rut, especially of the female during ovulation of the egg. Associated with maximum sexual receptivity.

estuary (es'chə-we'rē) (L. *aestuarium*, estuary). An arm of the sea where the tide meets the current of a freshwater stream.

ethology (e-thäl'-ə-jē) (Gr. *ethos*, character, + *logos*, discourse). The study of animal behavior in natural environments.

euchromatin (ū'krō-mə-tən) (Gr. *eu*, good, well, + *chrōma*, color). Part of the chromatin that takes up stain less than heterochromatin, contains active genes.

eukaryotic, eucaryotic (ū'ka-rē-ot'ik) (Gr. *eu*, good, true, + *karyon*, nut, kernel). Organisms whose cells characteristically contain a membrane-bound nucleus or nuclei; contrasts with prokaryotic.

euploidy (ū'ploid'ē) (Gr. *eu*, good, well, + *ploid*, multiple of). Change in chromosome number from one generation to the next in which there is an addition or deletion of a complete set of chromosomes in the progeny; the most common type is polyploidy.

euryhaline (ū'-rə-hā'līn) (Gr. *eurys*, broad, + *hals*, salt). Able to tolerate wide ranges of saltwater concentrations.

euryphagous (yə-rif'ə-gəs) (Gr. *eurys*, broad, + *phagein*, to eat). Eating a large variety of foods.

eurytopic (ū-rə-täp'ik) (Gr. *eurys*, broad, + *topos*, place). Refers to an organism with a wide distribution range.

eutely (u'te-lē) (Gr. *euteia*, thrift). Condition of a body composed of a constant number of cells or nuclei in all adult members of a species, as in rotifers, acanthocephalans, and nematodes.

evagination (ē-vaj'ə-nā'shən) (L. *e*, out, + *vagina*, sheath). An outpocketing from a hollow structure.

evolution (L. *evolvere*, to unfold). Organic evolution is any genetic change in organisms, or more strictly a change in gene frequency from generation to generation.

excision repair. Means by which cells are able to repair certain kinds of damage (dimerized pyrimidines) in their DNA.

exergonic (ek'sər-gän'ik) (Gr. *exō*, outside of, + *ergon*, work). An energy-yielding reaction.

exite (ex′īt) (Gr. *exō*, outside). Process from lateral side of an arthropod limb.

exocrine (ek′sə-krən) (Gr. *exō*, outside, + *krinein*, to separate). A type of gland that releases its secretion through a duct; contrasts with **endocrine**.

exon (ex′ən) (Gr. *exō*, outside). Part of the mRNA as transcribed from the DNA that contains a portion of the information necessary for final gene product.

exopod, exopodite (ex′ə-päd, ex-äp′ə-dīt) (Gr. *exō*, outside, + *pous, podos*, foot). Lateral branch of a biramous crustacean appendage.

exopterygote (ek′səp-ter′i-gōt) (Gr. *exō*, without, + *pteron*, feather, wing). Insect in which the wing buds develop externally during nymphal instars; has hemimetabolous metamorphosis.

exoskeleton (ek′sō-skel′ə-tən) (Gr. *exō*, outside, + *skeletos*, hard). A supporting structure secreted by ectoderm or epidermis; external, not enveloped by living tissue, as opposed to endoskeleton.

exteroceptor (ek′stər-ō-sep′tər) (L. *exter*, outward, + *capere*, to take). A sense organ excited by stimuli from the external world.

facilitated transport. Mediated transport of a substance into a cell in the direction of a concentration gradient, contrast with active transport.

FAD. Abbreviation for flavine adenine dinucleotide, an electron acceptor in the respiratory chain.

fatty acid. Any of a series of saturated organic acids having the general formula $C_nH_{2n}O_2$, occurs in natural fats of animals and plants.

fermentation (L. *fermentum*, ferment). Enzymatic transformation, without oxygen, of organic substrates, especially carbohydrates, yielding products such as alcohols, acids, and carbon dioxide.

fiber, fibril (L. *fibra*, thread). These two terms are often confused. Fiber is a fiberlike cell or a strand of protoplasmic material produced or secreted by a cell and lying outside the cell. Fibril is a strand of protoplasm produced by a cell and lying within the cell.

filipodium (fı′li-pō′de-əm) (L. *filum*, thread, + Gr. *pous, podos*, a foot). A type of pseudopodium that is very slender and may branch but does not rejoin to form a mesh.

filter feeding. Any feeding process by which particulate food is filtered from water in which it is suspended.

fission (L. *fissio*, a splitting). Asexual reproduction by a division of the body into two or more parts.

fitness. Degree of adjustment and suitability for a particular environment. Genetic fitness is relative contribution of one genetically distinct organism to the next generation; organisms with high genetic fitness are naturally selected and become prevalent in a population.

flagellum (flə-jel′əm) (L. a whip). Whiplike organelle of locomotion.

flame bulb. Specialized hollow excretory or osmoregulatory structure of one or several small cells containing a tuft of cilia (the "flame") and situated at the end of a minute tubule; connected tubules ultimately open to the outside. See **solenocyte, protonephridium**.

fluke (O.E. *flōc*, flatfish). A member of class Trematoda or class Monogenea. Also, certain of the flatfishes (order Pleuronectiformes).

FMN. Abbreviation for flavin mononucleotide, the prosthetic group of a protein (flavoprotein) and a carrier in the electron transport chain in respiration.

food vacuole. A digestive organelle in the cell.

fouling. Contamination of feeding or respiratory areas of an organism by excrement, sediment, or other matter. Also, accumulation of sessile marine organisms on the hull of a boat or ship so as to impede its progress through the water.

fovea (fō′vē-ə) (L. small pit). A small pit or depression; especially the fovea centralis, a small rodless pit in the retina of some vertebrates, a point of acute vision.

free energy. The energy available for doing work in a chemical system.

gamete (ga′mēt, gə-mēt′) (Gr. *gamos*, marriage). A mature haploid sex cell; usually, male and female gametes can be distinguished. An egg or a sperm.

gametic meiosis. Meiosis that occurs during formation of the gametes, as in humans and other higher animals.

gametocyte (gə-mēt′ə-sīt) (Gr. *gametēs*, spouse, + *kytos*, hollow vessel). The mother cell of a gamete, that is, immature gamete.

ganoid scales (ga′noid) (Gr. *ganos*, brightness). Thick, bony, rhombic scales of some primitive bony fishes; not overlapping.

gap junction. An area of tiny canals communicating the cytoplasm between two cells.

gastrodermis (gas′tro-dər′mis) (Gr. *gastēr*, stomach, + *derma*, skin). Lining of the digestive cavity of cnidarians.

gastrolith (gas′trə-lith) (Gr. *gastēr*, stomach, + *lithos*, stone). Calcareous body in the wall of the cardiac stomach of crayfish and other Malacostraca, preceding the molt.

gastrovascular cavity (Gr. *gastēr*, stomach, + L. *vasculum*, small vessel). Body cavity in certain lower invertebrates that functions in both digestion and circulation and has a single opening serving as both mouth and anus.

gastrula (gas′trə-lə) (Gr. *gastēr*, stomach, + L. *ula*, dim.). Embryonic stage, usually cap or sac shaped, with walls of two layers of cells surrounding a cavity (archenteron) with one opening (blastopore).

gastrulation (gas′trə-lā′shən) (Gr. *gastēr*, stomach). Process by which an early metazoan embryo becomes a gastrula, acquiring first two and then three layers of cells.

gel (jel) (from gelatin, from L. *gelare*, to freeze). That state of a colloidal system in which the solid particles form the continuous phase and the fluid medium the discontinuous phase.

gemmule (je′mūl) (L. *gemma*, bud, + *ula*, dim.). Asexual, cystlike reproductive unit in freshwater sponges; formed in summer or autumn and capable of overwintering.

gene (Gr. *genos*, descent). The part of a chromosome that is the hereditary determiner and is transmitted from one generation to another. It is a nucleic acid sequence (usually DNA) that encodes a functional polypeptide or RNA sequence.

gene pool. A collection of all of the alleles of all of the genes in a population.

genetic drift. Change in gene frequencies by chance processes in the evolutionary process of animals. In small populations, one allele may drift to fixation, becoming the only representative of that gene locus.

genome (jē′nōm) (Gr. *genos*, offspring, + L. *ōma*, abstract group). All the genes in a haploid set of chromosomes.

genotype (jēn′ō-tīp) (Gr. *genos*, offspring, + *typos*, form). The genetic constitution, expressed and latent, of an organism; the total set of genes present in the cells of an organism; contrasts with **phenotype**.

genus (jē-nus), pl. **genera** (L. race). A group of related species with taxonomic rank between family and species.

germ layer. In the animal embryo, one of three basic layers (ectoderm, endoderm, mesoderm) from which the various organs and tissues arise in the multicellular animal.

germ plasm. The germ cells of an organism, as distinct from the somatoplasm; the hereditary material (genes) of the germ cells.

gestation (je-stā′shən) (L. *gestare*, to bear). The period in which offspring are carried in the uterus.

glochidium (glō-kid′e-əm) (Gr. *glochis*, point, + *idion*, dim.). Bivalved larval stage of freshwater mussels.

glomerulus (glä-mer′u-ləs) (L. *glomus*, ball). A tuft of capillaries projecting into a renal corpuscle in a kidney. Also, a small spongy mass of tissue in the proboscis of hemichordates, presumed to have an excretory function. Also, a concentration of nerve fibers situated in the olfactory bulb.

glycogen (glī′kə-jən) (Gr. *glykys*, sweet, + *genēs*, produced). A polysaccharide constituting the principal form in which carbohydrate is stored in animals; animal starch.

glycolysis (glī-kol′i-sis) (Gr. *glykys*, sweet, + *lysis*, a loosening). Enzymatic breakdown of glucose (especially) or glycogen into phosphate derivatives with release of energy.

gnathobase (nāth′ə-bās′) (Gr. *gnathos*, jaw, + base). A median basic process on certain appendages in some arthropods, usually for biting or crushing food.

Golgi complex (gōl′jē) (after Golgi, Italian histologist). An organelle in cells that serves as a collecting and packaging center for secretory products.

gonad (gō′nad) (N.L. *gonas*, a primary sex organ). An organ that produces gametes (ovary in the female and testis in the male).

gonangium (gō-nan′jē-əm) (N.L. *gonas*, primary sex organ + *angeion*, dim. of vessel). Reproductive zooid of hydroid colony (Cnidaria).

gonoduct (Gr. *gonos*, seed, progeny, + duct). Duct leading from a gonad to the exterior.

gonopore (gän′ə-pōr) (Gr. *gonos*, seed, progeny, + *poros*, an opening). A genital pore found in many invertebrates.

green gland. Excretory gland of certain Crustacea; the antennal gland.

gregarious (L. *grex*, herd). Living in groups or flocks.

guanine (gwä′nēn) (Sp. from Quechura, *huanu*, dung). A white crystalline purine base, $C_5H_5N_5O$, occurring in various animal tissues and in guano and other animal excrements.

gynandromorph (ji-nan′drə-mawrf) (Gr. *gyn*, female, + *andr*, male + *morphē*, form). A bisexual form with the characteristics of both sexes; bisexual mosaic.

gynocophoric canal (gī′nə-kə-fōr′ik) (Gr. *gynē*, woman, + *pherein*, to carry). Groove in male schistosomes (certain trematodes) that carries the female.

habitat (L. *habitare*, to dwell). The place where an organism normally lives or where individuals of a population live.

habituation. A kind of learning in which continued exposure to the same stimulus produces diminishing responses.

halter (hal′tər), pl. **halteres** (hal-ti′rēz) (Gr. leap). In Diptera, small club-shaped structure on each side of the metathorax representing the hindwings; believed to be sense organs for balancing; also called balancer.

haploid (Gr. *haploos*, single). The reduced, or n, number of chromosomes, typical of gametes, as opposed to the diploid, or 2n, number found in somatic cells. In certain lower phyla, some mature animals have a haploid number of chromosomes.

hectocotylus (hek-tə-kät′ə-ləs) (Gr. *hekaton*, hundred, + *kotylē*, cup). Specialized, and sometimes autonomous, arm that serves as a male copulatory organ in cephalopods.

hemal system (hē′məl) (Gr. *haima*, blood). System of small vessels in echinoderms; function unknown.

hemerythrin (hē′mə-rith′rin) (Gr. *haima*, blood, + *erythros*, red). A red, iron-containing respiratory pigment found in the blood of some polychaetes, sipunculids, priapulids, and brachiopods.

hemimetabolous (he′mi-mə-ta′bə-ləs) (Gr. *hēmi*, half, + *metabolē*, change). Refers to gradual metamorphosis during development of insects, without a pupal stage.

hemoglobin (Gr. *haima*, blood, + L. *globulus*, globule). An iron-containing respiratory pigment occurring in vertebrate red blood cells and in blood plasma of many invertebrates; a compound of an iron porphyrin heme and a protein globin.

hemolymph (hē′mə-limf) (Gr. *haima*, blood, + L. *lympha*, water). Fluid in the coelom or hemocoel of some invertebrates that represents the blood and lymph of higher forms.

hemozoin (hē-mə-zo′ən) (Gr. *haima*, blood, + *zōon*, an animal). Insoluble digestion product of malaria parasites produced from hemoglobin.

hepatic (hə-pat′ic) (Gr. *hēpatikos*, of the liver). Pertaining to the liver.

herbivore ([h]ərb′ə-vōr′) (L. *herba*, green crop, + *vorare*, to devour). Any organism subsisting on plants. Adj., **herbivorous.**

hermaphrodite (hə[r]-maf′rə-dīt) (Gr. *hermaphroditos*, containing both sexes; from Greek mythology, Hermaphroditos, son of Hermes and Aphrodite). An organism with both male and female functional reproductive organs. **Hermaphroditism** may refer to an aberration in unisexual animals; **monoecism** implies that this is the normal condition for the species.

hermatypic (hər-mə-ti′pik) (Gr. *herma*, reef, + *typos*, pattern). Relating to reef-forming corrals.

heterocercal (het′ər-o-sər′kəl) (Gr. *heteros*, different, + *kerkos*, tail). In some fishes, a tail with the upper lobe larger than the lower, and the end of the vertebral column somewhat upturned in the upper lobe, as in sharks.

heterochromatin (he′tə-rō-krōm′ə-tən) (Gr. *heteros*, different, + *chrōma*, color). Chromatin that stains intensely and appears to represent inactive genetic areas.

heterochrony (hed′ə-rō-krōn-y) (Gr. *heteros*, different, + *chronos*, time). Change in the relative time of appearance and rate of development of characteristics already present in ancestors.

heterodont (hed′ə-ro-dänt) (Gr. *heteros*, different, + *odous*, tooth). Having teeth differentiated into incisors, canines, and molars for different purposes.

heterotroph (hat′ə-rō-träf) (Gr. *heteros*, different, + *trophos*, feeder). An organism that obtains both organic and inorganic raw materials from the environment in order to live; includes most animals and those plants that do not carry on photosynthesis.

heterozygote (het′ə-rō-zī′gōt) (Gr. *heteros*, different, + *zygōtos*, yoked). An organism in which the pair of alleles for a trait is composed of different genes (usually dominant and recessive); derived from a zygote formed by the union of gametes of dissimilar genetic constitution.

hexamerous (hek-sam′ər-əs) (Gr. *hex*, six, + *meros*, part). Six parts, specifically, symmetry based on six or multiples thereof.

hibernation (L. *hibernus*, wintry). Condition, especially of mammals, of passing the winter in a torpid state in which the body temperature drops nearly to freezing and the metabolism drops close to zero.

histogenesis (his-tō-jen′ə-sis) (Gr. *histos*, tissue, + *genesis*, descent). Formation and development of tissue.

histone (hi′stōn) (Gr. *histos*, tissue). Any of several simple proteins found in cell nuclei and complexed at one time or another with DNA. Histones yield a high proportion of basic amino acids on hydrolysis; characteristic of eukaryotes.

holoblastic cleavage (Gr. *holo*, whole, + *blastos*, germ). Complete and approximately equal division of cells in early embryo. Found in mammals, amphioxus, and many aquatic invertebrates that have eggs with a small amount of yolk.

holometabolous (hō′lō-mə-ta′bə-ləs) (Gr. *holo*, complete, + *metabolē*, change). Complete metamorphosis during development.

holophytic nutrition (hōl′ō-fit′ik) (Gr. *holo*, whole, + *phyt*, plant). Occurs in green plants and certain protozoa and involves synthesis of carbohydrates from carbon dioxide and water in the presence of light, chlorophyll, and certain enzymes.

holozoic nutrition (hōl′ō-zō′ik) (Gr. *holo*, whole, + *zoikos*, of animals). Type of nutrition involving ingestion of liquid or solid organic food particles.

home range. The area over which an animal ranges in its activities. Unlike territories, home ranges are not defended.

homeostasis (hō′mē-ō-stā′sis) (Gr. *homeo*, similar, + *stasis*, state or standing). Maintenance of an internal steady state by means of self-regulation.

homeothermic (hō′mē-ō-thər′mik) (Gr. *homeo*, alike, + *thermē*, heat). Having a nearly uniform body temperature, regulated independent of the environmental temperature; "warm blooded."

hominid (häm′ə-nid) (L. *homo, hominis*, man). A member of the family Hominidae, now represented by one living species, *Homo sapiens.*

hominoid (häm′ə-noid). Relating to the Hominoidea, a superfamily of primates to which the great apes and humans are assigned.

homocercal (hō′mə-ser′kal) (Gr. *homos*, same, common, + *kerkos*, tail). A tail with the upper and lower lobes symmetrical and the vertebral column ending near the middle of the base, as in most teleost fishes.

homodont (hō′mō-dänt) (Gr. *homos*, same, + *odous*, tooth). Having all teeth similar in form.

homograft. See allograft.

homology (hō-mäl′ə-jē) (Gr. *homologos*, agreeing). Similarity of parts or organs of different organisms caused by similar embryonic origin and evolutionary development from a corresponding part in some remote ancestor. May also refer to a matching pair of chromosomes. Serial homology is the correspondence in the same individual of repeated structures having the same origin and development, such as the appendages of arthropods. Adj., **homologous.**

homozygote (hō-mə-zī′gōt) (Gr. *homos*, same, + *zygotos*, yoked). An organism in which the pair of alleles for a trait is composed of the same genes (either dominant or recessive but not both). Adj., **homozygous.**

humoral (hū′mər-əl) (L. *humor*, a fluid). Pertaining to an endocrine secretion.

hyaline (hī′ə-lən) (Gr. *hyalos*, glass). Adj., glassy, translucent. Noun, a clear, glassy, structureless material occurring, for example, in cartilage, vitreous body, mucin, and glycogen.

hybridoma (hī-brid-ō′mah) (contraction of hybrid + myeloma). Fused product of a normal and a myeloma (cancer) cell, which has some of the characteristics of the normal cell.

hydatid cyst (hī-da′təd) (Gr. *hydatis*, watery vesicle). A type of cyst formed by juveniles of certain tapeworms (*Echinococcus*) in their vertebrate hosts.

hydranth (hī′dranth) (Gr. *hydōr*, water, + *anthos*, flower). Nutritive zooid of hydroid colony.

hydroid. The polyp form of a cnidarian as distinguished from the medusa form. Any cnidarian of the class Hydrozoa, order Hydroida.

hydrolysis (Gr. *hydōr*, water, + *lysis*, a loosening). The decomposition of a chemical compound by the addition of water; the splitting of a molecule into its groupings so that the split products acquire hydrogen and hydroxyl groups.

hydrosphere (Gr. *hydōr*, water, + *sphaira*, ball, sphere). Aqueous envelope of the earth.

hydrostatic skeleton. A mass of fluid or plastic parenchyma enclosed within a muscular wall to provide the support necessary for antagonistic muscle action; for example, parenchyma in acoelomates and perivisceral fluids in pseudocoelomates serve as hydrostatic skeletons.

bat / āpe / ärmadillo/ herring / fēmale / finch / līce / crocodile / crōw / duck / ūnicorn / ə indicates unaccented vowel sound "uh" as in mammal, fishes, cardinal, heron, vulture / stress as in bi-ol′o-gy, bi′o-log′i-cal

hydroxyl (hydrogen + oxygen, + yl). Containing an OH⁻ group, a negatively charged ion formed by alkalies in water.

hyperosmotic (Gr. *hyper*, over, + *ōsmos*, impulse). Refers to a solution whose osmotic pressure is greater than that of another solution to which it is compared; contains a greater concentration of dissolved particles and gains water through a semipermeable membrane from a solution containing fewer particles; contrasts with **hypoosmotic**.

hyperparasitism (hī′pər-par′ə-sid-iz-əm) (Gr. *hyper*, over, + *para*, beside, + *sitos*, food). Parasitism of a parasite by another parasite.

hypertrophy (hī-pər′trə-fē) (Gr. *hyper*, over, + *trophē*, nourishment). Abnormal increase in size of a part or organ.

hypodermis (hī′pə-dər′mis) (Gr. *hypo*, under, + L. *dermis*, skin). The cellular layer lying beneath and secreting the cuticle of annelids, arthropods, and certain other invertebrates.

hypoosmotic (Gr. *hypo*, under, + *ōsmos*, impulse). Refers to a solution whose osmotic pressure is less than that of another solution with which it is compared or taken as a standard; contains a lesser concentration of dissolved particles and loses water during osmosis; contrasts with **hyperosmotic**.

hypophysis (hī-pof′ə-sis) (Gr. *hypo*, under, + *physis*, growth). Pituitary body.

hypostome (hī′pə-stōm) (Gr. *hypo*, under, + *stoma*, mouth). Name applied to structure in various invertebrates (such as mites and ticks), located at posterior or ventral area of mouth.

hypothalamus (hī-pō-thal′ə-mis) (Gr. *hypo*, under, + *thalamos*, inner chamber). A ventral part of the forebrain beneath the thalamus; one of the centers of the autonomic nervous system.

imago (ə-mā′gō). The adult and sexually mature insect.

immunoglobulin (im′yə-nə-glä′byə-lən) (L. *immunis*, free, + *globus*, globe). Any of a group of plasma proteins, produced by plasma cells, that participates in the immune response by combining with the antigen that stimulated its production. Antibody.

imprinting (im′print-ing) (L. *imprimere*, to impress, imprint). Rapid and usually stable learning pattern appearing early in the life of a member of a social species and involving recognition of its own species; may involve attraction to the first moving object seen.

indeterminate cleavage. A type of embryonic development in which the fate of the blastomeres is not determined very early as to tissues or organs, for example, in echinoderms and vertebrates; regulative cleavage.

indigenous (ən-dij′ə-nəs) (L. *indigena*, native). Pertains to organisms that are native to a particular region; not introduced.

inductor (in-duk′ter) (L. *inducere*, to introduce, lead in). In embryology, a tissue or organ that causes the differentiation of another tissue or organ.

inflammation (in′fləm-mā′shən) (L. *inflammare*, from *flamma*, flame). The complicated physiological process in mobilization of body defenses against foreign substances and infectious agents and repair of damage from such agents.

infraciliature (in-frə-sil′e-ə-tər) (L. *infra*, below, + *cilia*, eyelashes). The organelles just below the cilia in ciliate protozoa.

infundibulum (in′fun-dib′u-ləm) (L. funnel). Stalk of the neurohypophysis linking the pituitary to the diencephalon.

innate (i-nāt′) (L. *innatus*, inborn). A characteristic based partly or wholly on gene differences.

instar (inz′tär) (L. form). Stage in the life of an insect or other arthropod between molts.

instinct (L. *instinctus*, impelled). Stereotyped, predictable, genetically programmed behavior. Learning may or may not be involved.

integument (ən-teg′ū-mənt) (L. *integumentum*, covering). An external covering or enveloping layer.

intermediary meiosis. Meiosis that occurs neither during gamete formation nor immediately after zygote formation, resulting in both haploid and diploid generations, such as in foraminiferan protozoa.

intermediate host. A host in which some development of a symbiont occurs, but in which maturation and sexual reproduction do not take place (contrasts with **definitive host**).

interstitial (in-tər-sti′shəl) (L. *inter*, among, + *sistere*, to stand). Situated in the interstices or spaces between structures such as cells, organs, or grains of sand.

intron (in′trän) (L. *intra*, within). Portion of mRNA as transcribed from DNA that will not form part of mature mRNA, that is, that does not include part of message for gene product.

introvert (L. *intro*, inward, + *vertere*, to turn). The anterior narrow portion that can be withdrawn (introverted) into the trunk of a sipunculid worm.

invagination (in-vaj′ə-nā′shən) (L. *in*, in, + *vagina*, sheath). An infolding of a layer of tissue to form a saclike structure.

inversion (L. *invertere*, to turn upside down). A turning inward or inside out, as in embryogenesis of sponges; also, reversal in order of genes or reversal of a chromosome segment.

irritability (L. *irritare*, to provoke). A general property of all organisms involving the ability to respond to stimuli or changes in the environment.

isogametes (īs′o-gam′ēts) (Gr. *isos*, equal, + *gametēs*, spouse). Gametes of a species in which both sexes are alike in size and appearance.

isolecithal (ī′sə-les′ə-thəl) (Gr. *isos*, equal, + *lekithos*, yolk, + *al*). Pertaining to a zygote (or ovum) with yolk evenly distributed. Homolecithal.

isotonic (Gr. *isos*, equal, + *tonikos*, tension). Said of solutions having the same or equal osmotic pressure; isosmotic.

isotope (Gr. *isos*, equal, + *topos*, place). One of several different forms (species) of a chemical element, differing from each other in atomic mass but not in atomic number.

juvenile hormone. Hormone produced by the corpora allata of insects; among its effects are maintenance of larval or nymphal characteristics during development.

juxtaglomerular apparatus (jək′stə-glä-mer′yə-lər) (L. *juxta*, close to, + *glomus*, ball). Complex of sensory cells located in the afferent arteriole adjacent to the glomerulus and a loop of the distal tubule, which produces the enzyme renin.

keratin (ker′ə-tən) (Gr. *kera*, horn, + *in*, suffix of proteins). A scleroprotein found in epidermal tissues and modified into hard structures such as horns, hair, and nails.

kinesis. (kə-nē′səs) (Gr. *kinēsis*, movement). Movements by an organism in random directions in response to stimulus.

kinetodesma (kə-nē′tə-dez′mə), pl. **kinetodesmata** (Gr. *kinein*, to move + *desma*, bond). Fibril arising from the kinetosome of a cilium in a ciliate protozoan, and passing along the kinetosomes of cilia in that same row.

kinetosome (kin-et′ə-sōm) (Gr. *kinētos*, moving, + *sōma*, body). The self-duplicating granule at the base of the flagellum or cilium; similar to centriole, also called basal body or blepharoplast.

kinety (kə-nē′tē) (Gr. *kinein*, to move). All the kinetosomes and kinetodesmata of a row of cilia.

kinin (kī′nin) (Gr. *kinein*, to move, + *in*, suffix of hormones). A type of local hormone that is released near its site of origin; also called parahormone or tissue hormone.

K-selection (from the K term in the logistic equation). Natural selection under conditions that favor survival when populations are controlled primarily by density-dependent factors.

kwashiorkor (kwash-ē-or′kər) (from Ghana). Malnutrition caused by diet high in carbohydrate and extremely low in protein.

labium (lā′bē-əm) (L. a lip). The lower lip of the insect formed by fusion of the second pair of maxillae.

labrum (lā′brəm) (L. a lip). The upper lip of insects and crustaceans situated above or in front of the mandibles; also refers to the outer lip of a gastropod shell.

labyrinthodont (lab′ə-rin′thə-dänt) (Gr. *labyrinthos*, labyrinth, + *odous*, *odontos*, tooth). A group of fossil-stem amphibians from which most amphibians later arose. They date from the late Paleozoic era.

lacteal (lak′te-əl) (L. *lacteus*, of milk). Noun, one of the lymph vessels in the villus of the intestine. Adj., relating to milk.

lacuna (lə-kū′nə), pl. **lacunae** (L. pit, cavity). A sinus; a space between cells; a cavity in cartilage or bone.

lagena (lə-jē′nə) (L. large flask). Portion of the primitive ear in which sound is translated into nerve impulses; evolutionary beginning of cochlea.

Lamarckism. Hypothesis, as expounded by Jean Baptiste de Lamarck, of evolution by the acquisition during an organism's lifetime of characteristics that are transmitted directly to offspring.

lamella (lə-mel′ə) (L. dim. of *lamina*, plate). One of the two plates forming a gill in a bivalve mollusc. One of the thin layers of bone laid concentrically around an osteon (Haversian canal). Any thin, platelike structure.

larva (lar′və), pl. **larvae** (L. a ghost). An immature stage that is quite different from the adult.

lek (lek) (Sw. play, game). An area where animals assemble for communal courtship display and mating.

leminscus (lem-nis′kəs) (L. ribbon). One of a pair of internal projections of the epidermis from the neck region of Acanthocephala, which functions in fluid control in the protrusion and invagination of the proboscis.

lentic (len′tik) (L. *lentus*, slow). Of or relating to standing water such as swamp, pond, or lake.

leukemism (lū´kə-mi-zəm) (Gr. *leukos*, white, + *ismos*, condition of). Presence of white pelage or plumage in animals with normally pigmented eyes and skin.

leukocyte (lū´kə-sīt´) (Gr. *leukos*, white, + *kytos*, hollow vessel). A blood cell with a nucleus and ameboid properties; a white blood cell. Has no hemoglobin.

limax form (lī´məx) (L. *limax*, slug). Form of pseudopodial movement in which entire organism moves without extending a discrete pseudopodium.

lipase (lī´pās) (Gr. *lipos*, fat, + *ase*, enzyme suffix). An enzyme that accelerates the hydrolysis or synthesis of fats.

lipid, lipoid (li´pid) (Gr. *lipos*, fat). Certain fatlike substances, often containing other groups such as phosphoric acid; lipids combine with proteins and carbohydrates to form principal structural components of cells.

lithosphere (lith´ə-sfir) (Gr. *lithos*, rock, + *sphaira*, ball). The rocky component of the earth's surface layers.

littoral (lit´ə-rəl) (L. *litoralis*, seashore). Adj., pertaining to the shore. Noun, that portion of the sea floor between the extent of high and low tides, intertidal; in lakes, the shallow part from the shore to the lakeward limit of aquatic plants.

lobopodium (lō´bə-pō´de-əm) (Gr. *lobos*, lobe, + *pous, podos*, foot). Blunt, lobelike pseudopodium.

locus (lō´kəs), pl. **loci** (lō´sī) (L. place). Position of a gene in a chromosome.

logistic equation. A mathematical expression describing an idealized sigmoid curve of population growth.

lophocyte (lō´fə-sīt) (Gr. *lophos*, crest, + *kytos*, hollow vessel). Type of sponge amebocyte that secretes bundles of fibrils.

lophophore (lōf´ə-fōr) (Gr. *lophos*, crest, + *phoros*, bearing). Tentacle-bearing ridge or arm within which is an extension of the coelomic cavity in lophophorate animals (ectoprocts, brachiopods, and phoronids).

lorica (lo´rə-kə) (L. corselet). Protective external case found in some protozoa, rotifers, and others.

lotic (lō´tik) (L. *lotus*, action of washing or bathing). Of or pertaining to running water, such as a brook or river.

lumen (lū´mən) (L. light). The cavity of a tube or organ.

lymphocyte (lim´fō-sīt) (L. *lympha*, water, goddess of water, + Gr. *kytos*, hollow vessel). Cell in blood and lymph that has central role in immune responses. See **T cells** and **B cells.**

lysosome (lī´sə-sōm) (Gr. *lysis*, loosing, + *sōma*, body). Intracellular organelle consisting of a membrane enclosing several digestive enzymes that are released when the lysosome ruptures.

macromolecule. A very large molecule, such as a protein, polysaccharide, or nucleic acid.

macronucleus (ma´krō-nū´klē-əs) (Gr. *makros*, long, large, + *nucleus*, kernel). The larger of the two kinds of nuclei in ciliate protozoa; controls all cell functions except reproduction.

madreporite (ma´drə-pōr´īt) (Fr. *madrépore*, reef-building coral, + *ite*, suffix for some body parts). Sievelike structure that is the intake for the water-vascular system of echinoderms.

malacostracan (mal´ə-käs´trə-kən) (Gr. *malako*, soft, + *ostracon*, shell). Any member of the crustacean subclass Malacostraca, which includes both aquatic and terrestrial forms of crabs, lobsters, shrimps, pill bugs, sand fleas, and others.

malpighian tubules (mal-pig´ē-ən) (Marcello Malpighi, Italian anatomist, 1628-1694). Blind tubules opening into the hindgut of nearly all insects and some myriapods and arachnids, and functioning primarily as excretory organs.

mantle. Soft extension of the body wall in certain invertebrates, for example, brachiopods and molluscs, which usually secretes a shell; thin body wall of tunicates.

marasmus (mə-raz´məs) (Gr. *marasmos*, to waste away). Malnutrition, especially of infants, caused by a diet deficient in both calories and protein.

marsupial (mär-sū´pē-əl) (Gr. *marsypion*, little pouch). One of the pouched mammals of the subclass Metatheria.

mastax (mas´təx) (Gr. jaws). Pharyngeal mill of rotifers.

matrix (mā´triks) (L. *mater*, mother). The intercellular substance of a tissue, or that part of a tissue into which an organ or process is set.

maturation (L. *maturus*, ripe). The process of ripening; the final stages in the preparation of gametes for fertilization.

maxilla (mak-sil´ə) (L. dim. of *mala*, jaw). One of the upper jawbones in vertebrates; one of the head appendages in arthropods.

maxilliped (mak-sil´ə-ped) (L. *maxilla*, jaw, + *pes*, foot). One of the pairs of head appendages located just posterior to the maxilla in crustaceans, a thoracic appendage that has become incorporated into the feeding mouthparts.

mediated transport. Transport of a substance across a cell membrane mediated by a carrier molecule in the membrane.

medulla (mə-dul´ə) (L. marrow). The inner portion of an organ in contrast to the cortex or outer portion. Also, hindbrain.

medusa (mə-dū´sə) (Gr. mythology, female monster with snake-entwined hair). A jellyfish, or the free-swimming stage in the life cycle of cnidarians.

Mehlis' gland (me´ləs). Glands of uncertain function surrounding the ootype of trematodes and cestodes.

meiofauna (mī´ō-faə-nə) (Gr. *meion*, smaller, + L. *faunus*, god of the woods). Small invertebrates found in the interstices between sand grains.

meiosis (mī-ō´səs) (Gr. from *meioun*, to make small). The nuclear changes by means of which the chromosomes are reduced from the diploid to the haploid number; in animals, usually occurs in the last two divisions in the formation of the mature egg or sperm.

melanin (mel´ə-nin) (Gr. *melas*, black). Black or dark-brown pigment found in plant or animal structures.

meninges (mə-nin´jez), sing. **meninx** (Gr. *mēninx*, membrane). Any of three membranes (arachnoid, dura mater, pia mater) that envelope the vertebrate brain and spinal cord. Also, solid connective tissue sheath enclosing the central nervous system of lower vertebrates.

menopause (men´ō-pawz) (Gr. *men*, month, + *pauein*, to cease). In the human female, that time of life when ovulation ceases; cessation of the menstrual cycle.

menstruation (men´stroo-ā´shən) (L. *menstrua*, the menses, from *mensis*, month). The discharge of blood and uterine tissue from the vagina at the end of a menstrual cycle.

meroblastic (mer-ə-blas´tik) (Gr. *meros*, part, + *blastos*, germ). Partial cleavage occurring in zygotes having a large amount of yolk at the vegetal pole; cleavage restricted to a small area on the surface of the egg.

merozoite (me´rə-zō´īt) (Gr. *meros*, part, + *zōon*, animal). A very small trophozoite at the stage just after cytokinesis has been completed in multiple fission of a protozoan.

mesenchyme (me´zn-kīm) (Gr. *mesos*, middle, + *enchyma*, infusion). Embryonic connective tissue; irregular or amebocytic cells often embedded in gelatinous matrix.

mesocoel (mez´ō-sēl) (Gr. *mesos*, middle, + *koilos*, hollow). Middle body coelomic compartment in some deuterostomes, anterior in lophophorates, corresponds to hydrocoel in echinoderms.

mesoderm (me´zə-dərm) (Gr. *mesos*, middle, + *derma*, skin). The third germ layer, formed in the gastrula between the ectoderm and endoderm; gives rise to connective tissues, muscle, urogenital and vascular systems, and the peritoneum.

mesoglea (mez´ō-glē´ə) (Gr. *mesos*, middle, + *glia*, glue). The layer of jellylike or cement material between the epidermis and gastrodermis in cnidarians and ctenophores; also may refer to jellylike matrix between epithelial layers in sponges.

mesohyl (me´sə-hil) (Gr. *mesos*, middle, + *hylē*, a wood). Gelatinous matrix surrounding sponge cells; mesoglea, mesenchyme.

mesonephros (me-zō-nef´rōs) (Gr. *mesos*, middle, + *nephros*, kidney). The middle of three pairs of embryonic renal organs in vertebrates. Functional kidney of fishes and amphibians; its collecting duct is a wolffian duct. Adj., **mesonephric.**

messenger RNA (mRNA). A form of ribonucleic acid that carries genetic information from the gene to the ribosome, where it determines the order of amino acids as a polypeptide is formed.

metabolism (Gr. *metabolē*, change). A group of processes that includes digestion, production of energy (respiration), and synthesis of molecules and structures by organisms; the sum of the constructive (anabolic) and destructive (catabolic) processes.

metacentric (me´tə-sen´trək) (Gr. *meta*, among, + *kentron*, center). Chromosome with centromere at or near the middle.

metacercaria (me´tə-sər-ka´rē-ə) (Gr. *meta*, after, + *kerkos*, tail, + L. *aria*, connected with). Fluke juvenile (cercaria) that has lost its tail and has become encysted.

metacoel (met´ə-sēl) (Gr. *meta*, after, + *koilos*, hollow). Posterior coelomic compartment in some deuterostomes and lophophorates; corresponds to somatocoel in echinoderms.

---

bat / āpe / ärmadillo / herring / fēmale / finch / līce / crocodile / crōw / duck / ūnicorn / ə indicates unaccented vowel sound "uh" as in mammal, fishes, cardinal, heron, vulture / stress as in bi-ol´o-gy, bi´o-log´i-cal

metamere (met′ə-mēr) (Gr. *meta*, after, + *meros*, part). A repeated body unit along the longitudinal axis of an animal; a somite, or segment.

metamerism (mə-ta′mə-ri′zəm) (Gr. *meta*, between, after, + *meros*, part). Condition of being made up of serially repeated parts (metameres); serial segmentation.

metamorphosis (Gr. *meta*, after, + *morphē*, form, + *osis*, state of). Sharp change in form during postembryonic development, for example, tadpole to frog or larval insect to adult.

metanephridium (me′tə-nə-fri′di-əm) (Gr. *meta*, after, + *nephros*, kidney). A type of tubular nephridium with the inner open end draining the coelom and the outer open end discharging to the exterior.

metanephros (me′tə-ne′fräs) (Gr. *meta*, between, after, + *nephros*, kidney). Embryonic renal organs of vertebrates arising behind the mesonephros; the functional kidney of reptiles, birds, and mammals. It is drained from a ureter.

microfilament (mī′krō-fil′ə-mənt) (Gr. *mikros*, small, + L. *filum*, a thread). A thin, linear structure in cells; of actin in muscle cells and others.

micron (μ) (mī′krän) (Gr. neuter of *mikros*, small). One one thousandth of a millimeter; about 1/25,000 of an inch. Now largely replaced by micrometer (μm).

microneme (mī′krə-nēm) (Gr. *mikros*, small, + *nēma*, thread). One of the types of structures comprising the apical complex in the phylum Apicomplexa, slender and elongate, leading to the anterior and thought to function in host cell penetration.

micronucleus. A small nucleus found in ciliate protozoa; controls the reproductive functions of these organisms.

microthrix (mī′krə-thrix), pl. **microtriches** (Gr. *mikros*, small, + *thrix*, hair). A small microvillus-like structure on the surface of a tapeworm's tegument.

microtubule (Gr. *mikros*, small, + L. *tubule*, pipe). A long, tubular cytoskeletal element with an outside diameter of 20 to 27 μm. Microtubules influence cell shape and play important roles during cell division.

microvillus (Gr. *mikros*, small, + L. *villus*, shaggy hair). Narrow, cylindrical cytoplasmic projection from epithelial cells; microvilli form the brush border of several types of epithelial cells.

mictic (mik′tik) (Gr. *miktos*, mixed or blended). Pertaining to haploid egg of rotifers or the females that lay such eggs.

miracidium (mīr′ə-sid′ē-əm) (Gr. *meirakidion*, youthful person). A minute ciliated larval stage in the life of flukes.

mitochondrion (mīd′ə-kän′drē-ən) (Gr. *mitos*, a thread, + *chondrion*, dim. of *chondros*, corn, grain). An organelle in the cell in which aerobic metabolism takes place.

mitosis (mī-tō′səs) (Gr. *mitos*, thread, + *osis*, state of). Nuclear division in which there is an equal qualitative and quantitative division of the chromosomal material between the two resulting nuclei; ordinary cell division (indirect).

monocyte (mon′ə-sīt) (Gr. *monos*, single, + *kytos*, hollow vessel). A type of leukocyte that becomes a phagocytic cell (macrophage) after moving into tissues.

monoecious (mə-nē′shəs) (Gr. *monos*, single, + *oikos*, house). Having both male and female gonads in the same organism; hermaphroditic.

monohybrid (Gr. *monos*, single, + L. *hybrida*, mongrel). A hybrid offspring of parents different in one specified character.

monomer (mä′nə-mər) (Gr. *monos*, single, + *meros*, part). A molecule of simple structure, but capable of linking with others to form polymers.

monophyletic (mä′nə-phī-le′tik) (Gr. *monos*, single, + *phyletikos*, pertaining to a phylum). Referring to a taxon whose units all evolved from a single parent stock; contrasts with **polyphyletic**.

monosaccharide (mä′nə-sa′kə-rīd) (Gr. *monos*, one, + *sakcharon*, sugar, from Sanskrit *sarkarā*, gravel, sugar). A simple sugar that cannot be decomposed into smaller sugar molecules; the most common are pentoses (such as ribose) and hexoses (such as glucose).

monozoic (mo′nə-zō′ik) (Gr. *monos*, single, + *zōon*, animal). Tapeworms with a single proglottid, do not undergo strobilation to form chain of proglottids.

morphogenesis (mor′fə-je′nə-səs) (Gr. *morphē*, form, + *genesis*, origin). Development of the architectural features of organisms; formation and differentiation of tissues and organs.

morphology (Gr. *morphē*, form, + L. *logia*, study, from Gr. *logos*, work). The science of structure. Includes cytology, the study of cell structure; histology, the study of tissue structure; and anatomy, the study of gross structure.

morula (mär′u-lə) (L. *morum*, mulberry, + *ula*, dim.). Solid ball of cells in early stage of embryonic development.

mosaic cleavage. Type characterized by independent differentiation of each part of the embryo; determinate cleavage.

Müller's larva. Free-swimming ciliated larva that resembles a modified ctenophore, characteristic of certain marine polyclad turbellarians.

mutation (mū-tā′shən) (L. *mutare*, to change). A stable and abrupt change of a gene; the heritable modification of a characteristic.

mutualism (mū′chə-wə-li′zəm) (L. *mutuus*, lent, borrowed, reciprocal). A type of interaction in which two different species derive benefit from the association and in which the association is necessary to both; often symbiotic.

myofibril (Gr. *mys*, muscle, + L. dim. of *fibra*, fiber). A contractile filament within muscle or muscle fiber.

myogenic (mī′o-jen′ik) (Gr. *mys*, muscle, + N.L. *genic*, giving rise to). Originating in muscle, such as heart beat arising in vertebrate cardiac muscle because of inherent rhythmical properties of muscle rather than because of neural stimuli.

myosin (mī′ə-sin) (Gr. *mys*, muscle, + *in*, suffix, belonging to). A large protein of contractile tissue that forms the thick myofilaments of striated muscle. During contraction it combines with actin to form actomyosin.

myotome (mī′ə-tōm′) (Gr. *mys*, muscle + *tomos*, cutting). A voluntary muscle segment in cephalochordates and vertebrates: that part of a somite destined to form muscles; the muscle group innervated by a single spinal nerve.

nacre (nā′kər) (F. mother-of-pearl). Innermost lustrous layer of mollusc shell, secreted by mantle epithelium. Adj., **nacreous**.

NAD. Abbreviation of nicotinamide adenine dinucleotide, an electron acceptor or donor in many metabolic reactions.

naiad (nā′əd) (Gr. *naias*, a water nymph). An aquatic, gill-breathing immature insect (nymph).

nares (na′rēz), sing. **naris** (L. nostrils). Openings into the nasal cavity, both internally and externally, in the head of a vertebrate.

natural selection. A nonrandom reproduction of genotypes that results in the survival of those best adapted to their environment and elimination of those less well adapted; leads to evolutionary change.

nauplius (naw′plē-əs) (L. a kind of shellfish). A free-swimming microscopic larval stage of certain crustaceans, with three pairs of appendages (antennules, antennae, and mandibles) and median eye. Characteristic of ostracods, copepods, barnacles, and some others.

nekton (nek′tən) (Gr. neuter of *nēktos*, swimming). Term for actively swimming organisms, essentially independent of wave and current action. Compare with plankton.

nematocyst (ne-mad′ə-sist′) (Gr. *nēma*, thread, + *kystis*, bladder). Stinging organelle of cnidarians.

neoteny (nē′ə-tē′nē, nē-ot′ə-nē) (Gr. *neos*, new, + *teinein*, to extend). The attainment of sexual maturity in the larval condition. Also, the retention of larval characters into adulthood.

nephridium (nə-frid′ē-əm) (Gr. *nephridios*, of the kidney). One of the segmentally arranged, paired excretory tubules of many invertebrates, notably the annelids. In a broad sense, any tubule specialized for excretion and/or osmoregulation; with an external opening and with or without an internal opening.

nephron (ne′frän) (Gr. *nephros*, kidney). Functional unit of kidney structure of vertebrates, consisting of a Bowman's capsule, an enclosed glomerulus, and the attached uriniferous tubule.

neritic (nə-rid′ik) (Gr. *nērites*, a mussel). Portion of the sea overlying the continental shelf, specifically from the subtidal zone to a depth of 200 m.

neurogenic (nū-rä-jen′ik) (Gr. *neuron*, nerve, + N.L. *genic*, give rise to). Originating in nervous tissue, as does the rhythmical beat of some arthropod hearts.

neuroglia (nū-räg′le-ə) (Gr. *neuron*, nerve, + *glia*, glue). Tissue supporting and filling the spaces between the nerve cells of the central nervous system.

neurolemma (nū-rə-lem′ə) (Gr. *neuron*, nerve, + *lemma*, skin). Delicate nucleated outer sheath of a nerve cell; sheath of Schwann.

neuron (Gr. nerve). A nerve cell.

neuropodium (nū′rə-pō′de-əm) (Gr. *neuron*, nerve, + *pous*, *podos*, foot). Lobe of parapodium nearer the ventral side in polychaete annelids.

neurosecretory cell (nu′rō-sə-krēd′ə-rē). Any cell (neuron) of the nervous system that produces a hormone.

niche. The role of an organism in an ecological community; its unique way of life and its relationship to other biotic and abiotic factors.

nitrogen fixation (Gr. *nitron*, soda, + *gen*, producing). Reduction of molecular nitrogen to ammonia by some bacteria and cyanobacteria, often followed by **nitrification**, the oxidation of ammonia to nitrites and nitrates by other bacteria.

notochord (nōd′ə-kord′) (Gr. *nōtos*, back + *chorda*, cord). An elongated cellular cord, enclosed in a sheath, which forms the primitive axial skeleton of chordate embryos and adult cephalochordates.

notopodium (nō′tə-pō′de-əm) (Gr. *nōtos*, back, + *pous*, *podos*, foot). Lobe of parapodium nearer the dorsal side in polychaete annelids.

nucleic acid (nu′klē′ik) (L. *nucleus*, kernel). One of a class of molecules composed of joined nucleotides; chief types are deoxyribonucleic acid (DNA), found in cell nuclei (chromosomes and mitochondria), and ribonucleic acid (RNA), found both in cell nuclei (chromosomes and nucleoli) and in cytoplasmic ribosomes.

nucleoid (nu′klē-oid) (L. *nucleus*, kernel, + *oid*, like). The region in a prokaryotic cell where the chromosome is found.

nucleolus (nu-klē′ə-ləs) (dim. of L. *nucleus*, kernel). A deeply staining body within the nucleus of a cell and containing RNA; nucleoli are specialized portions of certain chromosomes that carry multiple copies of the information to synthesize ribosomal RNA.

nucleoplasm (nu′klē-ə-plazm′) (L. *nucleus*, kernel, + Gr. *plasma*, mold). Protoplasm of nucleus, as distinguished from cytoplasm.

nucleoprotein. A molecule composed of nucleic acid and protein; occurs in the nucleus and cytoplasm of all cells.

nucleosome (nu′klē-ə-som) (L. *nucleus*, kernel, + *sōma*, body). A repeating subunit of chromatin in which one and three quarters turns of the double-helical DNA are wound around eight molecules of histones.

nucleotide (nu′klē-ə-tīd). A molecule consisting of phosphate, 5-carbon sugar (ribose or deoxyribose), and a purine or a pyrimidine; the purines are adenine and guanine, and the pyrimidines are cytosine, thymine, and uracil.

nuptial flight (nup′shəl). The mating flight of insects, especially that of the queen with male or males.

nymph (L. *nympha*, nymph, bride). An immature stage (following hatching) of a hemimetabolous insect that lacks a pupal stage.

ocellus (ō-sel′əs) (L. dim. of *oculus*, eye). A simple eye or eyespot in many types of invertebrates.

octomerous (ok-tom′ər-əs) (Gr. *oct*, eight, + *meros*, part). Eight parts, specifically, symmetry based on eight.

ommatidium (ä′mə-tid′e-əm) (Gr. *omma*, eye, + *idium*, small). One of the optical units of the compound eye of arthropods.

oncomiracidium (än′kō-mīr′ə-sid′e-əm) (Gr. *onkos*, barb, hook, + *meirakidion*, youthful person). A ciliated larva of a monogenetic trematode.

ontogeny (än-tä′jə-ne) (Gr. *ontos*, being, + *geneia*, act of being born, from *genēs*, born). The course of development of an individual from egg to senescence.

oocyst (ō′ə-sist) (Gr. *oion*, egg, + *kystis*, bladder). Cyst formed around zygote of malaria and related organisms.

ooecium (ō-ēs′e-əm) (Gr. *oion*, egg, + *oikos*, house, + L. *ium*, from Ger. *ion*, dim.). Brood pouch; compartment for developing embryos in ectoprocts.

oogonium (ō′ə-gōn′e-əm) (Gr. *oion*, egg, + *gonos*, offspring). A cell that, by continued division, gives rise to oocytes; an ovum in a primary follicle immediately before the beginning of maturation.

ookinete (ō-ə-kī′nēt) (Gr. *oion*, egg, + *kinein*, to move). The motile zygote of malaria organisms.

ootype (ō′ə-tīp) (Gr. *oion*, egg, + *typos*, mold). Part of oviduct in flatworms that receives ducts from vitelline glands and Mehlis′ gland.

operculum (ō-per′kū-ləm) (L. cover). The gill cover in bony fishes; horny plate in some snails.

ophthalmic (äf-thal′mik) (Gr. *ophthalmos*, an eye). Pertaining to the eye.

opisthaptor (ä′pəs-thap′tər) (Gr. *opisthen*, behind, + *haptein*, to fasten). Posterior attachment organ of a monogenetic trematode.

opisthosoma (ō-pis′thə-sō′mə) (Gr. *opisthe*, behind, + *sōma*, body). Posterior body region in arachnids and pogonophorans.

opsonization (op′sən-i-zā′shən) (Gr. *opsonein*, to buy victuals, to cater). The facilitation of phagocytosis of foreign particles by phagocytes in the blood or tissues, mediated by antibody bound to the particles.

organelle (Gr. *organon*, tool, organ, + L. *ella*, dim.). Specialized part of a cell; literally, a small organ that performs functions analogous to organs of multicellular animals.

organizer (or′gan-ī-zer) (Gr. *organos*, fashioning). Area of an embryo that directs subsequent development of other parts.

osmole. Molecular weight of a solute, in grams, divided by the number of ions or particles into which it dissociates in solution. Adj., **osmolar.**

osmoregulation. Maintenance of proper internal salt and water concentrations in a cell or in the body of a living organism, active regulation of internal osmotic pressure.

osmosis (oz-mō′sis) (Gr. *ōsmos*, act of pushing, impulse). The flow of solvent (usually water) through a semipermeable membrane.

osmotic potential. Osmotic pressure.

osphradium (äs-frā′de-əm) (Gr. *osphradion*, small bouquet, dim. of *osphra*, smell). A sense organ in aquatic snails and bivalves that tests incoming water.

ossicles (L. *ossiculum*, small bone). Small separate pieces of echinoderm endoskeleton. Also, tiny bones of the middle ear of vertebrates.

osteoblast (os′tē-ō-blast) (Gr. *osteon*, bone, + *blastos*, bud). A bone-forming cell.

osteoclast (os′tē-ō-clast) (Gr. *osteon*, bone, + *klan*, to break). A large, multinucleate cell that functions in bone dissolution.

osteon (äs′tē-än) (Gr. bone). Unit of bone structure; Haversian system.

ostium (L. door). Opening.

otolith (ōd′əl-ith′) (Gr. *ous*, *otos*, ear, + *lithos*, stone). Calcareous concretions in the membranous labyrinth of the inner ear of lower vertebrates, or in the auditory organ of certain invertebrates.

oviger (ō′vi-jər) (L. *ovum*, egg, + *gerere*, to bear). Leg that carries eggs in pycnogonids.

oviparity (ō′və-pa′rəd-ē) (L. *ovum*, egg, + *parere*, to bring forth). Reproduction in which eggs are released by the female; development of offspring occurs outside the maternal body. Adj., **oviparous** (ō-vip′ə-rəs).

ovipositor (ō′və-päz′əd-ər) (L. *ovum*, egg, + *positor*, builder, placer, + *or*, suffix denoting agent or doer). In many female insects a structure at the posterior end of the abdomen for laying eggs.

ovoviviparity (ō′vo-vī-və-par′ə-dē) (L. *ovum*, egg, + *vivere*, to live, + *parere*, to bring forth). Reproduction in which eggs develop within the maternal body without additional nourishment from the parent and hatch within the parent or immediately after laying. Adj., **ovoviviparous** (ō′vo-vī-vip′ə-rəs).

oxidation (äk′sə-dā′shən) (Fr. *oxider*, to oxidize, from Gr. *oxys*, sharp, + *ation*). The loss of an electron by an atom or molecule; sometimes addition of oxygen chemically to a substance. Opposite of reduction, in which an electron is accepted by an atom or molecule.

oxidative phosphorylation (äk′sə-dād′iv fäs′fər-i-lā′shən). The conversion of inorganic phosphate to energy-rich phosphate of ATP, involving electron transport through a respiratory chain to molecular oxygen.

paedogenesis (pē-dō-jen′ə-sis) (Gr. *pais*, child, + *genēs*, born). Reproduction by immature or larval animals caused by acceleration of maturation. Progenesis.

paedomorphosis (pē-dō-mor′fə-səs) (Gr. *pais*, child, + *morphē*, form). Displacement of ancestral juvenile features to later stages of the ontogeny of descendants.

pair bond. An affiliation between an adult male and an adult female for reproduction. Characteristic of monogamous species.

pallium (pal′e-əm) (L. mantle). Mantle of a mollusc or brachiopod.

pangenesis (pan-jen′ə-sis) (Gr. *pan*, all, + *genesis*, descent). Darwin′s hypothesis that hereditary characteristics are carried by individual body cells that produce particles that collect in the germ cells.

papilla (pə-pil′ə) (L. nipple). A small nipplelike projection. A vascular process that nourishes the root of a hair, feather, or developing tooth.

papula (pa′pū-lə) (L. pimple). Respiratory processes on skin of sea stars; also, pustules on skin.

parabiosis (pa′rə-bī-ō′sis) (Gr. *para*, beside, + *biosis*, mode of life). The fusion of two individuals, resulting in mutual physiological intimacy.

parapodium (pa′rə-pō′de-əm) (Gr. *para*, beside, + *pous*, *podos*, foot). One of the paired lateral processes on each side of most segments in polychaete annelids; variously modified for locomotion, respiration, or feeding.

parasitism (par′ə-sīd′izəm) (Gr. *parasitos*, from *para*, beside, + *sitos*, food). The condition of an organism living in or on another organism (host) at whose expense the parasite is maintained; destructive symbiosis.

parasympathetic (par′ə-sim-pə-thed′ik) (Gr. *para*, beside, + *sympathes*, sympathetic, from *syn*, with, + *pathos*, feeling). One of the subdivisions of the autonomic nervous system, whose fibers originate in the brain and in anterior and posterior parts of the spinal cord.

parenchyma (pə-ren′kə-mə) (Gr. anything poured in beside). In lower animals, a spongy mass of vacuolated mesenchyme cells filling spaces between viscera, muscles, or epithelia; in some, the cells are cell bodies of muscle cells. Also, the specialized tissue of an organ as distinguished from the supporting connective tissue.

parenchymula (pa′rən-kī′mū-lə) (Gr. *para*, beside, + *enchyma*, infusion). Flagellated, solid-bodied larva of some sponges.

parthenogenesis (pär′thə-nō-gen′ə-sis) (Gr. *parthenos*, virgin, + L. from Gr. *genesis*, origin). Unisexual reproduction involving the production of young not fertilized by males; common in rotifers, cladocerans, aphids, bees, ants, and wasps. A parthenogenetic egg may be diploid or haploid.

pathogenic (path′ə-jen′ik) (Gr. *pathos*, disease, + N.L. *genic*, giving rise to). Producing or capable of producing disease.

peck order. A hierarchy of social privilege in a flock of birds.

pecten (L. comb). Any of several types of comblike structures on various organisms, for example, a pigmented, vascular, and comblike process that projects into the vitreous humor from the retina at a point of entrance of the optic nerve in the eyes of all birds and many reptiles.

pectines (pek′tīnz) (L. comb, pl. of *pecten*). Sensory appendage on abdomen of scorpions.

pectoral (pek′tə-rəl) (L. *pectoralis*, from *pectus*, the breast). Of or pertaining to the breast or chest; to the pectoral girdle; or to a pair of horny shields of the plastron of certain turtles.

pedal laceration. Asexual reproduction found in sea anemones, a form of fission.

pedicel (ped′ə-sel) (L. *pediculus*, little foot). A small or short stalk or stem. In insects, the second segment of an antenna or the waist of an ant.

pedicellaria (ped′ə-sə-lar′ē-ə) (L. *pediculus*, little foot, + *aria*, like or connected with). One of many minute pincerlike organs on the surface of certain echinoderms.

pedipalps (ped′ə-palps′) (L. *pes, pedis*, foot, + *palpus*, stroking, caress). Second pair of appendages of arachnids.

pedogenesis. See paedogenesis.

peduncle (pē′dun-kəl) (L. *pedunculus*, dim. of *pes*, foot). A stalk. Also, a band of white matter joining different parts of the brain.

pelage (pel′ij) (Fr. fur). Hairy covering of mammals.

pelagic (pə-laj′ik) (Gr. *pelagos*, the open sea). Pertaining to the open ocean.

pellicle (pel′ə-kəl) (L. *pellicula*, dim. of *pellis*, skin). Thin, translucent, secreted envelope covering many protozoa.

pentadactyl (pen-tə-dak′təl) (Gr. *pente*, five, + *daktylos*, finger). With five digits, or five fingerlike parts, to the hand or foot.

peptidase (pep′tə-dās) (Gr. *peptein*, to digest, + *ase*, enzyme suffix). An enzyme that breaks down simple peptides, releasing amino acids.

peptide bond. A bond that binds amino acids together into a polypeptide chain, formed by removing an OH from the carboxy group of one amino acid and an H from the amino group of another to form an amide group—CO—NH—.

periostracum (pe-rē-äs′trə-kəm) (Gr. *peri*, around, + *ostrakon*, shell). Outer horny layer of a mollusc shell.

periproct (per′ə-präkt) (Gr. *peri*, around, + *prōktos*, anus). Region of aboral plates around the anus of echinoids.

perisarc (per′ə-särk) (Gr. *peri*, around, + *sarx*, flesh). Sheath covering the stalk and branches of a hydroid.

perissodactyl (pə-ris′ə-dak′təl) (Gr. *perissos*, odd, + *daktylos*, finger, toe). Pertaining to an order of ungulate mammals with an odd number of digits.

peristalsis (per′ə-stal′səs) (Gr. *peristaltikos*, compressing around). The series of alternate relaxations and contractions that serve to force food through the alimentary canal.

peristomium (per′ə-stō′mē-əm) (Gr. *peri*, around, + *stoma*, mouth). Foremost true segment of an annelid; it bears the mouth.

peritoneum (per′ə-tə-nē′əm) (Gr. *peritonaios*, stretched around). The membrane that lines the coelom and covers the coelomic viscera.

petaloids (pe′tə-loids) (Gr. *petalon*, leaf, + *eidos*, form). Describes flowerlike arrangement of respiratory podia in irregular sea urchins.

pH (potential of *h*ydrogen). A symbol referring to the relative concentration of hydrogen ions in a solution; pH values are from 0 to 14, and the lower the value, the more acid or hydrogen ions in the solution. Equal to the negative logarithm of the hydrogen ion concentration.

phagocyte (fag′ə-sīt) (Gr. *phagein*, to eat, + *kytos*, hollow vessel). Any cell that engulfs and devours microorganisms or other particles.

phagocytosis (fag′ə-sī-tō-səs) (Gr. *phagein*, to eat, + *kytos*, hollow vessel). The engulfment of a particle by a phagocyte or a protozoan.

phagosome (fa′gə-sōm) (Gr. *phagein*, to eat, + *sōma*, body). Membrane-bound vesicle in cytoplasm containing food material engulfed by phagocytosis.

pharynx (far′inks), pl. **pharynges** (Gr. *pharynx*, gullet). The part of the digestive tract between the mouth cavity and the esophagus that, in vertebrates, is common to both digestive and respiratory tracts. In cephalochordates the gill slits span the pharynx.

phasmid (faz′mid) (Gr. *phasma*, apparition, phantom, + *id*). One of a pair of glands or sensory structures found in the posterior end of certain nematodes.

phenotype (fē′nə-tīp) (Gr. *phainein*, to show). The visible or expressed characteristics of an organism, controlled by the genotype, but not all genes in the genotype are expressed.

pheromone (fer′ə-mōn) (Gr. *pherein*, to carry, + *hormōn*, exciting, stirring up). Chemical substance released by one organism that influences the behavior or physiological processes of another organism.

phosphagen (fäs′fə-jən) (phosphate + gen). A term for creatine phosphate and arginine phosphate, which store and may be sources of high-energy phosphate bonds.

phosphatide (fäs′fə-tīd′) (phosphate + ide). A lipid with phosphorus, such as lecithin. A complex phosphoric ester lipid, such as lecithin, found in all cells. Phospholipid.

phosphorylation (fäs′fə-rə-lā′shən). The addition of a phosphate group, such as H_2PO_3, to a compound.

photosynthesis (fōd′ō-sin′thə-sis) (Gr. *phōs*, light, + *synthesis*, action or putting together). The synthesis of carbohydrates from carbon dioxide and water in chlorophyll-containing cells exposed to light.

phototaxis (fō-to-tak′sis) (Gr. *phōs*, light, + *taxis*, arranging, order). A taxis in which light is the orienting stimulus. An involuntary tendency for an organism to turn toward (positive) or away from (negative) light.

phyllopodium (fī′lə-pō′dē-əm) (Gr. *phyllon*, leaf, + *pous, podos*, foot). Leaflike swimming appendage of branchiopod crustaceans.

phylogeny (fī′läj′ə-nē) (Gr. *phylon*, tribe, race, + *geneia*, origin). The origin and development of any taxon, or the evolutionary history of its development.

phylum (fī′ləm), pl. **phyla** (N.L. from Gr. *phylon*, race, tribe). A chief category, between kingdom and class, of taxonomic classifications into which are grouped organisms of common descent that share a fundamental pattern of organization.

pilidium (pī-lid′ē-əm) (Gr. *pilidion*, dim. of *pilos*, felt cap). Free-swimming, hat-shaped larva of nemertine worms.

pinacocyte (pin′ə-kō-sīt′) (Gr. *pinax*, tablet, + *kytos*, hollow vessel). Flattened cells comprising dermal epithelium in sponges.

pinna (pin′ə) (L. feather, sharp point). The external ear. Also a feather, wing, or fin or similar part.

pinocytosis (pin′o-sī-tō′sis, pīn′o-sī-to′sis) (Gr. *pinein*, to drink, + *kytos*, hollow vessel, + *osis*, condition). Taking up of fluid by endocytosis; cell drinking.

placenta (plə-sen′tə) (L. flat cake). The vascular structure, embryonic and maternal, through which the embryo and fetus are nourished while in the uterus.

placoid scale (pla′koid) (Gr. *plax, plakos*, tablet, plate). Type of scale found in cartilaginous fishes, with basal plate of dentin embedded in the skin and a backward-pointing spine tipped with enamel.

plankton (plank′tən) (Gr. neuter of *planktos*, wandering). The passively floating animal and plant life of a body of water; compares to **nekton.**

plantigrade (plan′tə-grād′) (L. *planta*, sole, + *gradus*, step, degree). Pertaining to animals that walk on the whole surface of the foot (for example, humans and bears); compares to **digitigrade.**

planula (plan′yə-lə) (N.L. dim. from L. *planus*, flat). Free-swimming, ciliated larval type of cnidarians; usually flattened and ovoid, with an outer layer of ectodermal cells and an inner mass of endodermal cells.

planuloid ancestor (plan′yə-loid) (L. *planus*, flat, + Gr. *eidos*, form). Hypothetical form representing ancestor of Cnidaria and Platyhelminthes.

plasma cell (plaz′mə) (Gr. *plasma*, a form, mold). A descendant cell of a B cell, functions to secrete antibodies.

plasma membrane (plaz′mə) (Gr. *plasma*, a form, mold). A living, external, limiting, protoplasmic structure that functions to regulate exchange of nutrients across the cell surface.

plasmid (plaz′məd) (Gr. *plasma*, a form, mold). A small circle of DNA carried by a bacterium in addition to its large chromosome.

plasmodium (plaz-mō′dē-əm) (Gr. *plasma*, a form, mold, + *eidos*, form). Multinucleate ameboid mass, syncytial.

plastid (plas′təd) (Gr. *plast*, formed, molded, + L. *id*, feminine stem for particle of specified kind). A membranous organelle in plant cells functioning in photosynthesis and/or nutrient storage, for example, chloroplast.

platelet (plāt′lət) (Gr. dim. of *platys*, flat). A tiny, incomplete cell in the blood that releases substances initiating blood clotting.

pleiotropic (plī-ə-trō′pic) (Gr. *pleiōn*, more, + *tropos*, to turn). Pertaining to a gene producing more than one effect; affecting multiple phenotypic characteristics.

pleopod (plē′ə-päd) (Gr. *plein*, to sail, + *pous*, *podos*, foot). One of the swimming appendages on the abdomen of a crustacean.

pleura (plu′rə) (Gr. side, rib). The membrane that lines each half of the thorax and covers the lungs.

plexus (plek′səs) (L. network, braid). A network, especially of nerves or blood vessels.

pluteus (plū′-dē-əs), pl. **plutei** (L. *pluteus*, movable shed, reading desk). Echinoid or ophiuroid larva with elongated processes like the supports of a desk; originally called "painter's easel larva."

podium (pō′de-əm) (Gr. *pous*, *podos*, foot). A footlike structure, for example, the tube foot of echinoderms.

poikilothermic (poi-ki′lə-thər′mik) (Gr. *poikilos*, variable, + thermal). Pertaining to animals whose body temperature is variable and fluctuates with that of the environment; cold blooded; compares to **ectothermic.**

polarization (L. *polaris*, polar, + Gr. *iz*, make). The arrangement of positive electrical charges on one side of a surface membrane and negative electrical charges on the other side (in nerves and muscles).

Polian vesicles (pō′le-ən) (from G.S. Poli, Italian naturalist). Vesicles opening into ring canal in most asteroids and holothuroids.

polyandry (pol′y-an′drē) (Gr. *polys*, many, + *anēr*, man). Condition of having more than one male mate at one time.

polygamy (pə-lig′ə-mē) (Gr. *polys*, many, + *gamos*, marriage). Condition of having more than one mate at a time.

polygyny (pə-lij′ə-nē) (Gr. *polys*, many, + *gynē*, woman). Condition of having more than one female mate at one time.

polymer (pä′lə-mər) (Gr. *polys*, many, + *meros*, part). A chemical compound composed of repeated structural units called monomers.

polymerization (pə-lim′ər-ə-zā′shən). The process of forming a polymer or polymeric compound.

polymorphism (pä′lē-mor′fi-zəm) (Gr. *polys*, many, + *morphē*, form). The presence in a species of more than one structural type of individual.

polynucleotide (poly + nucleotide). A nucleotide of many mononucleotides combined.

polyp (päl′əp) (Fr. *polype*, octopus, from L. *polypus*, many footed). The sessile stage in the life cycle of cnidarians.

polypeptide (pä-lē-pep′tīd) (Gr. *polys*, many, + *peptein*, to digest). A molecule consisting of many joined amino acids, not as complex as a protein.

polyphyletic (pä′lē-fī-led′ik) (Gr. *polys*, many, + *phylon*, tribe). Derived from more than one ancestral source; opposed to monophyletic.

polyphyodont (pä′lē-fī′ə-dänt) (Gr. *polyphyes*, manifold, + *odous*, tooth). Having several sets of teeth in succession.

polypide (pä′li-pīd) (L. *polypus*, polyp). An individual or zooid in a colony, specifically in ectoprocts, which has a lophophore, digestive tract, muscles, and nerve centers.

polyploid (pä′lə-ploid′) (Gr. *polys*, many, + *ploidy*, number of chromosomes). Characterized by a chromosome number that is greater than two full sets of homologous chromosomes.

polysaccharide (pä′lē-sak′ə-rid, -rīd) (Gr. *polys*, many, + *sakcharon*, sugar, from Sanskrit *sarkarā*, gravel, sugar). A carbohydrate composed of many monosaccharide units, for example, glycogen, starch, and cellulose.

polysome (polyribosome) (Gr. *polys*, many, + *sōma*, body). Two or more ribosomes connected by a molecule of messenger RNA.

polytene chromosomes (pä′li-tēn) (Gr. *polys*, many, + *tainia*, band). Chromosomes in the somatic cells of some insects in which the chromatin replicates repeatedly without undergoing mitosis.

polyzoic (pä′lē-zō′ik) (Gr. *polys*, many, + *zoon*, animal). A tapeworm forming a strobila of several to many proglottids; also, a colony of many zooids.

pongid (pän′jəd) (L. *Pongo*, type genus of orangutan). Of or relating to the primate family Pongidae, comprising the anthropoid apes (gorillas, chimpanzees, gibbons, orangutans).

population (L. *populus*, people). A group of organisms of the same species inhabiting a specific geographical locality.

portal system (L. *porta*, gate). System of large veins beginning and ending with a bed of capillaries; for example, hepatic portal and renal portal systems in vertebrates.

preadaptation. The possession of a condition that is selected for in an ancestral environment and that coincidentally predisposes an organism for survival in some other environment.

prebiotic synthesis. The chemical synthesis that occurred before the emergence of life.

precocial (prē-kō′shəl) (L. *praecoquere*, to ripen beforehand). Referring (especially) to birds whose young are covered with down and are able to run about when newly hatched.

predaceous, predacious (prē-dā′shəs) (L. *praeda*, prey). Living by killing and consuming other animals; predatory.

predator (pred′ə-tər) (L. *praeda*, prey). An organism that preys on other organisms for its food.

prehensile (prē-hen′səl) (L. *prehendere*, to seize). Adapted for grasping.

primate (prī′māt) (L. *primus*, first). Any mammal of the order Primates, which includes the tarsiers, lemurs, marmosets, monkeys, apes, and humans.

primitive (L. *primus*, first). Primordial; ancient; little evolved; said of species closely approximating their early ancestral types.

proboscis (prō-bäs′əs) (Gr. *pro*, before, + *boskein*, feed). A snout or trunk. Also, tubular sucking or feeding organ with the mouth at the end as in planarians, leeches, and insects. Also, the sensory and defensive organ at the anterior end of certain invertebrates.

producers (L. *producere*, to bring forth). Organisms, such as plants, able to produce their own food from inorganic substances.

progesterone (prō-jes′tə-rōn′) (L. *pro*, before, + *gestare*, to carry). Hormone secreted by the corpus luteum and the placenta; prepares the uterus for the fertilized egg and maintains the capacity of the uterus to hold the embryo and fetus.

proglottid (prō-gläd′əd) (Gr. *proglōttis*, tongue tip, from *pro*, before, + *glotta*, tongue, + *id*, suffix). Portion of a tapeworm containing a set of reproductive organs; usually corresponds to a segment.

prokaryotic, procaryotic (pro-kar′ē-ät′ik) (Gr. *pro*, before, + *karyon*, kernel, nut). Not having a membrane-bound nucleus or nuclei. Prokaryotic cells are more primitive than eukaryotic cells and persist today in the bacteria and cyanobacteria.

promoter. A region of DNA to which the RNA polymerase must have access for transcription of a structural gene to begin.

pronephros (prō-nef′rəs) (Gr. *pro*, before, + *nephros*, kidney). Most anterior of three pairs of embryonic renal organs of vertebrates; functional only in adult hagfishes and larval fishes and amphibians; vestigial in mammalian embryos. Adj., **pronephric.**

proprioceptor (prō′prē-ə-sep′tər) (L. *proprius*, own, particular, + receptor). Sensory receptor located deep within the tissues, especially muscles, tendons, and joints, that is responsive to changes in muscle stretch, body position, and movement.

prosimian (prō-sim′ē-ən) (Gr. *pro*, before, + L. *simia*, ape). Any member of a group of primitive, arboreal primates: lemurs, tarsiers, lorises, and so on.

prosoma (prō-sōm′ə) (Gr. *pro*, before, + *sōma*, body). Anterior part of an invertebrate in which primitive segmentation is not visible; fused head and thorax of arthropod; cephalothorax.

prosopyle (präs′ə-pīl) (Gr. *prosō*, forward, + *pyle*, gate). Connections between the incurrent and radial canals in some sponges.

prostaglandins (präs′tə-glan′dəns). A family of fatty acid tissue hormones, originally discovered in semen, known to have powerful effects on smooth muscle, nerves, circulation, and reproductive organs.

prostomium (prō-stō′mē-əm) (Gr. *pro*, before, + *stoma*, mouth). In most annelids and some molluscs, that part of the head located in front of the mouth.

protease (prō′tē-ās), (Gr. *protein*, + *ase*, enzyme). An enzyme that digests proteins; includes proteinases and peptidases.

protein (prō′tēn, prō′tē-ən) (Gr. *protein*, from *proteios*, primary). A macromolecule of carbon, hydrogen, oxygen, and nitrogen and sometimes sulfur and phosphorus; composed of chains of amino acids joined by peptide bonds; present in all cells.

---

bat / āpe / ärmadillo/ herring / fēmale / finch / līce / crocodile / crōw / duck / ūnicorn / ə indicates unaccented vowel sound "uh" as in mammal, fishes, cardinal, heron, vulture / stress as in bi-ol′o-gy, bi′o-log′i-cal

prothoracic glands. Glands in the prothorax of insects that secrete the hormone ecdysone.

prothoracicotropic hormone. See ecdysiotropin.

prothrombin (pro-thräm′bən) (Gr. *pro*, before, + *thrombos*, clot). A constituent of blood plasma that is changed to thrombin by a catalytic sequence that includes thromboplastin, calcium, and plasma globulins; involved in blood clotting.

protist (prō′tist) (Gr. *protos*, first). A member of the kingdom Protista, generally considered to include the unicellular eukaryotic organisms (protozoa and eukaryotic algae).

protocoel (prō′tə-sēl) (Gr. *protos*, first, + *koilos*, hollow). The anterior coelomic compartment in some deuterostomes, corresponds to the axocoel in echinoderms.

protocooperation. A mutually beneficial interaction between organisms in which the interaction is not physiologically necessary to the survival of either.

protonephridium (prō′tō-nə-frid′ē-əm) (Gr. *protos*, first, + *nephros*, kidney). Primitive osmoregulatory or excretory organ consisting of a tubule terminating internally with flame bulb or solenocyte; the unit of a flame bulb system.

protoplasm (prō′tə-plazm) (Gr. *protos*, first, + *plasma*, form). Organized living substance; cytoplasm and nucleoplasm of the cell.

protopod, protopodite (prō′tə-päd, prō-top′ə-dīt) (Gr. *protos*, first, + *pous, podos*, foot). Basal portion of crustacean appendage, containing coxa and basis.

Protostomia (prō′də-stō′mē-ə) (Gr. *protos*, first, + *stoma*, mouth). A group of phyla in which cleavage is determinate, the coelom (in coelomate forms) is formed by proliferation of mesodermal bands (schizocoelic formation), the mesoderm is formed from a particular blastomere (called 4d), and the mouth is derived from or near the blastopore. Includes the Annelida, Arthropoda, Mollusca, and a number of minor phyla. Compares with **Deuterostomia.**

proventriculus (pro′ven-trik′ū-ləs) (L. *pro*, before, + *ventriculum*, ventricle). In birds the glandular stomach between the crop and gizzard. In insects, a muscular dilation of foregut armed internally with chitinous teeth.

proximal (L. *proximus*, nearest). Situated toward or near the point of attachment; opposite of distal, distant.

pseudocoel (sū′do-sēl) (Gr. *pseudēs*, false, + *koilōma*, cavity). A body cavity not lined with peritoneum and not a part of the blood or digestive systems, embryonically derived from the blastocoel.

pseudopodium (sū′də-pō′dē-əm) (Gr. *pseudēs*, false, + *podion*, small foot, + *eidos*, form). A temporary cytoplasmic protrusion extended out from a protozoan or ameboid cell, and serving for locomotion or for taking up food.

puff. Strands of DNA spread apart at certain locations on giant chromosomes of some flies where that DNA is being transcribed.

pupa (pū′pə) (L. girl, doll, puppet). Inactive quiescent stage of the holometabolous insects. It follows the larval stages and precedes the adult stage.

purine (pū′rēn) (L. *purus*, pure, + *urina*, urine). Organic base with carbon and nitrogen atoms in two interlocking rings. The parent substance of adenine, guanine, and other naturally occurring bases.

pyrimidine (pī-rim′ə-dēn) (alter. of pyridine, from Gr. *pyr*, fire, + *id*, adj. suffix, + *ine*). An organic base composed of a single ring of carbon and nitrogen atoms; parent substance of several bases found in nucleic acids.

queen. In entomology, the single fully developed female in a colony of social insects such as bees, ants, and termites, distinguished from workers, nonreproductive females, and soldiers.

radial cleavage. Type in which early cleavage planes are symmetrical to the polar axis, each blastomere of one tier lying directly above the corresponding blastomere of the next layer; indeterminate cleavage.

radial symmetry. A morphological condition in which the parts of an animal are arranged concentrically around an oral-aboral axis, and more than one imaginary plane through this axis yields halves that are mirror images of each other.

radula (re′jə-lə) (L. scraper). Rasping tongue found in most molluscs.

ratite (ra′tīt) (L. *ratis*, raft). Having an unkeeled sternum; compares with **carinate.**

recapitulation. Summing up or repeating; hypothesis that an individual repeats its phylogenetic history in its development.

recombinant DNA. DNA from two different species, such as a virus and a mammal, combined into a single molecule.

redia (rē′dē-ə), pl. **rediae** (rē′dē-ē) (from Redi, Italian biologist). A larval stage in the life cycle of flukes; it is produced by a sporocyst larva, and in turn gives rise to many cercariae.

regulative development. Progressive determination and restriction of initially totipotent embryonic material.

releaser (L. *relaxare*, to unloose). Simple stimulus that elicits an innate behavior pattern.

renin (rē′nin) (L. *ren*, kidney). An enzyme produced by the kidney juxtaglomerular apparatus that initiates changes leading to increased blood pressure and increased sodium reabsorption.

replication (L. *replicatio*, a folding back). In genetics, the duplication of one or more DNA molecules from the preexisting molecule.

respiration (L. *respiratio*, breathing). Gaseous interchange between an organism and its surrounding medium. In the cell, the release of energy by the oxidation of food molecules.

restriction endonuclease. An enzyme that cleaves a DNA molecule at a particular base sequence.

rete mirabile (rē′tē mə-rab′ə-lē) (L. wonderful net). A network of small blood vessels so arranged that the incoming blood runs countercurrent to the outgoing blood and thus makes possible efficient exchange between the two bloodstreams. Such a mechanism serves to maintain the high concentration of gases in the fish swim bladder.

reticular (rə-tīk′ū-lər) (L. *reticulum*, small net). Resembling a net in appearance or structure.

reticuloendothelial system (rə-tic′ū-lō-en-dō-thēl′i-əl) (L. *reticulum*, dim. of net, + Gr. *endon*, within, + *thele*, nipple). The fixed phagocytic cells in the tissues, especially the liver, lymph nodes, spleen, and others; also called RE system.

rhabdite (rab′dīu) (Gr. *rhabdos*, rod). Rodlike structures in the cells of the epidermis or underlying parenchyma in certain turbellarians. They are discharged in mucous secretions.

rheoreceptor (rē′ə-rē-cep′tər) (Gr. *rheos*, a flowing, + receptor). A sensory organ of aquatic animals that responds to water current.

rhopalium (rō-pā′lē-əm) (N.L. from Gr. *rhopalon*, a club). One of the marginal, club-shaped sense organs of certain jellyfishes; tentaculocyst.

rhoptries (rōp′trēz) (Gr. *rhopalon*, club, + *tryō*, to rub, wear out). Club-shaped bodies in Apicomplexa comprising one of the structures of the apical complex; open at anterior and apparently functioning in penetration of host cell.

rhynchocoel (ring′kō-sēl) (Gr. *rhynchos*, snout, + *koilos*, hollow). In nemerteans, the dorsal tubular cavity that contains the inverted proboscis. It has no opening to the outside.

ribosome (rī′bə-sōm). A small organelle composed of protein and ribonucleic acid. May be free in the cytoplasm or attached to the membranes of the endoplasmic reticulum; functions in protein synthesis.

ritualization. In ethology, the evolutionary modification, usually intensification, of a behavior pattern to serve communication.

RNA. Ribonucleic acid, of which there are several different kinds, such as messenger RNA, ribosomal RNA, and transfer RNA (mRNA, rRNA, tRNA).

rostellum (räs-tel′ləm) (L. small beak). Projecting structure on scolex of tapeworm, often with hooks.

rostrum (räs′trəm) (L. ship's beak). A snoutlike projection on the head.

r-selection (from the r term in the logistic equation). Natural selection under conditions that favor survival when populations are controlled primarily by density-independent factors; contrast with **K-selection.**

saccule (sa′kūl) (L. *sacculus*, small bag). Small chamber of the membranous labyrinth of the inner ear.

sagittal (saj′ə-dəl) (L. *sagitta*, arrow). Pertaining to the median anteroposterior plane that divides a bilaterally symmetrical organism into right and left halves.

saprophagous (sə-praf′ə-gəs) (Gr. *sapros*, rotten, + *phagos*, from *phagein*, to eat). Feeding on decaying matter; saprobic; saprozoic.

saprophyte (sap′rə-fīt) (Gr. *sapros*, rotten, + *phyton*, plant). A plant living on dead or decaying organic matter.

saprozoic nutrition (sap-rə-zō′ik) (Gr. *sapros*, rotten, + *zōon*, animal). Animal nutrition by absorption of dissolved salts and simple organic nutrients from surrounding medium; also refers to feeding on decaying matter.

sarcolemma (sär′kə-lem′ə) (Gr. *sarx*, flesh + *lemma*, rind). The thin, noncellular sheath that encloses a striated muscle fiber.

sarcomere (sär′kə-mir) (Gr. *sarx*, flesh, + *meros*, part). Transverse segment of striated muscle believed to be the fundamental contractile unit.

sacroplasm (sär′kə-plaz′-əm) (Gr. *sarx*, flesh, + *plasma*, mold). Clear, semifluid substance between fibrils or muscle tissue.

schizocoel (skiz′ə-sēl) (Gr. *schizo*, from *schizein*, to split, + *koilōma*, cavity). A coelom formed by the splitting of embryonic mesoderm. Noun, **schizocoelomate**, an animal with a schizocoel, such as an arthropod or mollusc. Adj., **schizocoelous.**

schizocoelous mesoderm formation (skiz′ō-sēl-ləs). Embryonic formation of the mesoderm as cords of cells between ectoderm and endoderm; splitting of these cords results in the coelomic space.

schizogony (skə-zä′gə-nē) (Gr. *schizein*, to split, + *gonos*, seed). Multiple asexual fission.

sclerite (skle′rit) (Gr. *skleros*, hard). A hard chitinous or calcareous plate or spicule; one of the plates making up the exoskeleton of arthropods, especially insects.

scleroblast (skler′ə-blast) (Gr. *skleros*, hard, + *blastos*, germ). An amebocyte specialized to secrete a spicule, found in sponges.

sclerotic (skle-räd′ik) (Gr. *skleros*, hard). Pertaining to the tough outer coat of the eyeball.

sclerotization (skle′rə-tə-zā′shən). Process of hardening of the cuticle of arthropods by the formation of stabilizing cross linkages between peptide chains of adjacent protein molecules.

scolex (skō′leks) (Gr. *skōlex*, worm, grub). The holdfast, or so-called head, of a tapeworm; bears suckers and, in some, hooks, and posterior to it new proglottids are differentiated.

scrotum (skrō′təm) (L. bag). The pouch that contains the testes in most mammals.

scyphistoma (sī-fis′tə-mə) (Gr. *skyphos*, cup, + *stoma*, mouth). A stage in the development of scyphozoan jellyfish just after the larva becomes attached, the polyp form of a scyphozoan.

sedentary (sed′ən-ter-ē). Stationary, sitting, inactive; staying in one place.

seminiferous (sem-ə-nif′rəs) (L. *semen*, semen, + *ferre*, to bear). Pertains to the tubules that produce or carry semen in the testes.

semipermeable (L. *semi*, half, + *permeabilis*, capable of being passed through). Permeable to small particles, such as water and certain inorganic ions, but not to larger molecules.

septum, pl. **septa** (L. fence). A wall between two cavities.

serial homology. See **homology.**

serosa (sə-rō′sə) (N.L. from L. *serum*, serum). The outer embryonic membrane of birds and reptiles; chorion. Also, the peritoneal lining of the body cavity.

serotonin (sir′ə-tōn′ən) (L. *serum*, serum). A phenolic amine, found in the serum of clotted blood and in many other tissues, that possesses several poorly understood metabolic, vascular, and neural functions; 5-hydroxytryptamine.

serous (sir′əs) (L. *serum*, serum). Watery, resembling serum; applied to glands, tissue, cells, fluid.

serum (sir′əm) (L. whey, serum). The liquid that separates from the blood after coagulation; blood plasma from which fibrinogen has been removed. Also, the clear portion of a biological fluid separated from its particulate elements.

sessile (ses′əl) (L. *sessilis*, low, dwarf). Attached at the base; fixed to one spot, not able to move about.

seta (sīd′ə), pl. **setae** (sē′tē) (L. bristle). A needlelike chitinous structure of the integument of annelids, arthropods, and others.

siliceous (sə-li′shəs) (L. *silex*, flint). Containing silica.

simian (sim′ē-ən) (L. *simia*, ape). Pertaining to monkeys or apes.

sinistral (si′nə-strəl, sə-ni′strəl) (L. *sinister*, left). Pertaining to the left; in gastropods, shell is sinistral if opening is to left of columella when held with spire up and facing observer.

sinus (sī′nəs) (L. curve). A cavity or space in tissues or in bone.

siphonoglyph (sī′fän′ə-glif′) (Gr. *siphōn*, reed, tube, siphon, + *glyphē*, carving). Ciliated furrow in the gullet of sea anemones.

siphuncle (sī′fun-kl) (L. *siphunculus*, small tube). Cord of tissue running through shell of nautiloid, connecting all chambers with body of animal.

solenocyte (sō-len′ə-sīt) (Gr. *sōlēn*, pipe, + *kytos*, hollow vessel). Special type of flame bulb in which the bulb bears a flagellum instead of a tuft of cilia. See **flame bulb, protonephridium.**

soma (sō′mə) (Gr. body). The whole of an organism except the germ cells (germ plasm).

somatic (sō-mat′ik) (Gr. *sōma*, body). Refers to the body, for example, somatic cells in contrast germ cells.

somatocoel (sə-ma′tə-sel) (Gr. *sōma*, body + *koilos*, hollow). One of the pair of sacs that form the main body coelom of echinoderms.

somatoplasm (sō′mə-də-pla′zm) (Gr. *sōma*, body, + *plasma*, anything formed). The living matter that makes up the mass of the body as distinguished from germ plasm, which makes up the reproductive cells. The protoplasm of body cells.

somite (sō′mīt) (Gr. *soma*, body). One of the blocklike masses of mesoderm arranged segmentally (metamerically) in a longitudinal series beside the neural tube of the embryo; metamere.

speciation (spē′sē-ā′shən) (L. *species*, kind). The evolutionary process by which new species arise; the process by which variations become fixed.

species (spē′shez, spē′sēz) sing. and pl. (L. particular kind). A group of interbreeding individuals of common ancestry that are reproductively isolated from all other such groups; a taxonomic unit ranking below a genus and designated by a binomen consisting of its genus and the species name.

spermatheca (spər′mə-thē′kə) (Gr. *sperma*, seed, + *thēkē*, case). A sac in the female reproductive organs for the reception and storage of sperm.

spermatogonium (spər′mad-ə-gō′nē-əm) (Gr. *sperma*, seed + *gonē*, offspring). Precursor of mature male reproductive cell; gives rise directly to a spermatocyte.

spermatophore (spər-mad′ə-fōr′) (Gr. *sperma*, *spermatos*, seed, + *pherein*, to bear). Capsule or packet enclosing sperm, produced by males of several invertebrate groups and a few vertebrates.

sphincter (sfingk′tər) (Gr. *sphinkter*, band, sphincter, from *sphingein*, to bind tight). A ring-shaped muscle capable of closing a tubular opening by constriction.

spicule (spi′kyul) (L. dim. *spica*, point). One of the minute calcareous or siliceous skeletal bodies found in sponges, radiolarians, soft corals, and sea cucumbers.

spiracle (spi′rə-kəl) (L. *spiraculum*, from *spirare*, to breathe). External opening of a trachea in arthropods. One of a pair of openings on the head of elasmobranchs for passage of water. Exhalent aperture of tadpole gill chamber.

spiral cleavage. A type of early embryonic cleavage in which cleavage planes are diagonal to the polar axis and unequal cells are produced by the alternate clockwise and counterclockwise cleavage around the axis of polarity; determinate cleavage.

spongin (spun′jin) (L. *spongia*, sponge). Fibrous, collagenous material making up the skeletal network of horny sponges.

spongioblast (spun′je-o-blast) (Gr. *spongos*, sponge, + *blastos*, bud). Cell in a sponge that secrets spongin, a protein.

spongocoel (spun′jō-sēl) (Gr. *spongos*, sponge, + *koilos*, hollow). Central cavity in sponges.

sporocyst (spō′rə-sist) (Gr. *sporos*, seed, + *kystis*, pouch). A larval stage in the life cycle of flukes; it originates from a miracidium.

sporogony (spor-äg′ə-ne) (Gr. *sporos*, seed, + *gonos*, birth). Multiple fission to produce sporozoites after zygote formation.

sporozoite (spō′rə-zō′īt) (Gr. *sporos*, seed, + *zōon*, animal, + *ite*, suffix for body part). A stage in the life history of many sporozoan protozoa; released from oocyts.

statoblast (stad′ə-blast) (Gr. *statos*, standing, fixed, + *blastos*, germ). Biconvex capsule containing germinative cells and produced by most freshwater ectoprocts by asexual budding. Under favorable conditions it germinates to give rise to new zooid.

statocyst (Gr. *statos*, standing, + *kystis*, bladder). Sense organ of equilibrium; a fluid-filled cellular cyst containing one or more granules (statoliths) used to sense direction of gravity.

statolith (Gr. *statos*, standing, + *lithos*, stone). Small calcareous body resting on tufts of cilia in the statocyst.

stenohaline (sten-ə-hā′līn, -lən) (Gr. *stenos*, narrow, + *hals*, salt). Pertaining to aquatic organisms that have restricted tolerance to changes in environmental saltwater concentration.

stenophagous (stə-näf′ə-gəs) (Gr. *stenos*, narrow, + *phagein*, to eat). Eating few kinds of foods.

stenotopic (sten-ə-tä′pik) (Gr. *stenos*, narrow, + *topos*, place). Refers to an organism with a narrow range of adaptability to environmental change; having a restricted geographical distribution.

stereogastrula (ste′rē-ə-gas′trə-lə) (Gr. *stereos*, solid, + *gastēr*, stomach, + L. *ula*, dim.). A solid type of gastrula, such as the planula of cnidarians.

sternum (ster′nəm) (L. breastbone). Ventral plate of an arthropod body segment; breastbone of vertebrates.

sterol (ste′rōl), **steroid** (ste′roid) (Gr. *stereos*, solid, + L. *ol*, [from *oleum*, oil]). One of a class of organic compounds containing a molecular skeleton of four fused carbon rings; it includes cholesterol, sex hormones, adrenocortical hormones, and vitamin D.

stigma (Gr. *stigma*, mark, tattoo mark). Eyespot in certain protozoa. Spiracle of certain terrestrial arthropods.

stolon (stō′lən) (L. *stolō, stolonis*, a shoot, or sucker of a plant). A rootlike extension of the body wall that gives rise to buds that may develop into new zooids, thus forming a compound animal in which the zooids remain united by the stolon. Found in some colonial anthozoans, hydrozoans, ectoprocts, and ascidians.

stoma (stō′mə) (Gr. mouth). A mouthlike opening.

stomochord (stō′mə-kord) (Gr. *stoma*, mouth, + *chordē*, cord). Anterior evagination of the dorsal wall of the buccal cavity into the proboscis of hemichordates; the buccal diverticulum.

strobila (strō′bə-lə) (Gr. *strobilē*, lint plug like a pine cone [*strobilos*]). A stage in the development of the scyphozoan jellyfish. Also, the chain of proglottids of a tapeworm.

stroma (strō′mə) (Gr. *strōma*, bedding). Supporting connective tissue framework of an animal organ; filmy framework of red blood corpuscles and certain cells.

structural gene. A gene carrying the information to construct a protein.

sycon (sī′kon) (Gr. *sykon*, fig). A type of canal system in certain sponges. Sometimes called syconoid.

symbiosis (sim′bī-ōs′əs, sim′bē-ōs′əs) (Gr. *syn*, with, + *bios*, life). The living together of two different species in an intimate relationship. Symbiont always benefits; host may benefit, may be unaffected, or may be harmed (mutualism, commensalism, and parasitism).

sympatric (sim′pa′-trik) (Gr. *syn*, with, + *patra*, native land). Having the same or overlapping regions of distribution. Noun, **sympatry.**

synapse (si′naps, si-naps′) (Gr. *synapsis*, contact, union). The place at which a nerve impulse passes between neuron processes, typically from an axon of one nerve cell to a dendrite of another nerve cell.

synapsis (si-nap′səs) (Gr. *synapsis*, contact, union). The time when the pairs of homologous chromosomes lie alongside each other in the first meiotic division.

syncytium (sin-sish′e-əm) (Gr. *syn*, with, + *kytos*, hollow vessel). A mass of protoplasm containing many nuclei and not divided into cells.

syndrome (sin′drōm) (Gr. *syn*, with + *dramein*, to run). A group of symptoms characteristic of a particular disease or abnormality.

syngamy (sin′gə-mē) (Gr. *syn*, with, + *gamos*, marriage). Fertilization of one gamete with another individual gamete to form a zygote, found in most animals with sexual reproduction.

synkaryon (sin-ker′e-on) (Gr. *syn*, with, + *karyon*, nucleus). Zygote nucleus resulting from fusion of pronuclei.

syrinx (sir′inks) (Gr. shepherd's pipe). The vocal organ of birds located at the base of the trachea.

systematics (sis-tə-mad′iks). Science of classification and evolutionary biology.

T cell. A type of lymphocyte most important in cellular immune response and in regulation of most immune responses.

tactile (tak′til) (L. *tactilis*, able to be touched, from *tangere*, to touch). Pertaining to touch.

tagma, pl. **tagmata** (Gr. *tagma*, arrangement, order, row). A compound body section of an arthropod resulting from embryonic fusion of two or more segments; for example, head, thorax, abdomen.

tagmatization, tagmosis. Organization of the arthropod body into tagmata.

taiga (tī′gä) (Russ.). Habitat zone characterized by large tracts of coniferous forests, long, cold winters, and short summers; most typical in Canada and Siberia.

taxis (tak′sis), pl. **taxes** (Gr. *taxis*, arrangement). An orientation movement by a (usually) simple organism in response to an environmental stimulus.

taxon (tak′son), pl. **taxa** (Gr. *taxis*, arrangement). Any taxonomic group or entity.

taxonomy (tak-sän′ə-mi) (Gr. *taxis*, arrangement, + *nomos*, law). Study of the principles of scientific classification; systematic ordering and naming of organisms.

tectum (tek′təm) (L. roof). A rooflike structure, for example, dorsal part of capitulum in ticks and mites.

tegument (teg′ū-ment) (L. *tegumentum*, from *tegere*, to cover). An integument: specifically external covering in cestodes and trematodes, formerly believed to be a cuticle.

telencephalon (tel′en-sef′ə-lon) (Gr. *telos*, end, + *encephalon*, brain). The most anterior vesicle of the brain; the anterior-most subdivision of the prosencephalon that becomes the cerebrum and associated structures.

teleology (tel′ē-äl′ə-jē) (Gr. *telos*, end, + L. *logia*, study of, from Gr. *logos*, word). The philosophical view that natural events are goal directed and are preordained, as opposed to the scientific view of mechanical determinism.

telocentric (tē′lō-sen′trək) (Gr. *telos*, end, + *kentron*, center). Chromosome with centromere at the end.

telolecithal (te-lō-les′ə-thəl) (Gr. *telos*, end, + *lekithos*, yolk, + *al*). Having the yolk concentrated at one end of an egg.

telson (tel′sən) (Gr. *telson*, extremity). Posterior projection of the last body segment in many crustaceans.

template (tem′plət). A pattern or mold guiding the formation of a duplicate; often used with reference to gene duplication.

tentaculocyst (ten-tak′u-lō-sist) (L. *tentaculum*, feeler, + Gr. *kystis*, pouch). One of the sense organs along the margin of medusae; a rhopalium.

tergum (ter′gəm) (L. back). Dorsal part of an arthropod body segment.

territory (L. *territorium*, from *terra*, earth). A restricted area preempted by an animal or pair of animals, usually for breeding purposes, and guarded from other individuals of the same species.

test (L. *testa*, shell). A shell or hardened outer covering.

tetrad (te′trad) (Gr. *tetras*, four). Group of two pairs of chromatids at synapsis and resulting from the replication of paired homologous chromosomes; the bivalent.

tetrapods (te′trə-päds) (Gr. *tetras*, four + *pous, podos*, foot). Four-footed vertebrates; the

group includes amphibians, reptiles, birds, and mammals.

therapsid (thə-rap′sid) (Gr. *theraps*, an attendant). Extinct Mesozoic mammal-like reptile from which true mammals evolved.

thermocline (thər′mō-klīn) (Gr. *thermē*, heat, + *klinein*, to swerve). Layer of water separating upper warmer and lighter water from lower colder and heavier water in a lake or sea; a stratum of abrupt change in water temperature.

Tiedemann's bodies (tēd′ə-mənz) (from F. Tiedemann, German anatomist). Four or five pairs of pouchlike bodies attached to the ring canal of sea stars, apparently functioning in production of coelomocytes.

tight junction. Region of actual fusion of cell membranes between two adjacent cells.

tornaria (tor-na′rē-ə) (L. *tornare*, to turn). A free-swimming larva of enteropneusts that rotates as it swims; resembles somewhat the bipinnaria larva of echinoderms.

torsion (L. *torquere*, to twist). A twisting phenomenon in gastropod development that alters the position of the visceral and pallial organs by 180 degrees.

toxicyst (tox′i-sist) (Gr. *toxikon*, poison, + *kystis*, bladder). Structures possessed by predatory ciliate protozoa, which on stimulation expel a poison to subdue the prey.

trachea (trā′kē-ə) (M.L. windpipe). The windpipe. Also, any of the air tubes of insects.

transcription. Formation of messenger RNA from the coded DNA.

transduction. Condition in which bacterial DNA (and the genetic characteristics it bears) is transferred from one bacterium to another by the agent of viral infection.

transfer RNA (tRNA). A form of RNA of about 70 to 80 nucleotides, which are adapter molecules in the synthesis of proteins. A specific amino acid molecule is carried by transfer RNA to a ribosome–messenger RNA complex for incorporation into a polypeptide.

transformation. Condition in which DNA in the environment of bacteria somehow penetrates them and is incorporated into their genetic complement, so that their progeny inherit the genetic characters so acquired.

translation (L. a transferring). The process in which the genetic information present in messenger RNA is used to direct the order of specific amino acids during protein synthesis.

trichinosis (trik-ən-o′səs). Disease caused by infection with the nematode *Trichinella spiralis.*

trichocyst (trik′ə-sist) (Gr. *thrix*, hair, + *kystis*, bladder). Saclike protrusible organelle in the ectoplasm of ciliates, which discharges as a threadlike weapon of defense.

triglyceride (trī-glis′ə-rīd) (Gr. *tria*, three, + *glykys*, sweet, + *ide*, suffix denoting compound). A triester of glycerol with one, two, or three acids.

triploblastic (trip′lō-blas′tik) (Gr. *triploos*, triple, + *blastos*, germ). Pertaining to metazoa in which the embryo has three primary germ layers—ectoderm, mesoderm, and endoderm.

trisomy 21. See **Down syndrome.**

trochophore (trōk′ə-fōr) (Gr. *trochos*, wheel, + *pherein*, to bear). A free-swimming ciliated marine larva characteristic of most molluscs and certain ectoprocts, brachiopods, and marine worms; an ovoid or pyriform body with preoral circlet of cilia and sometimes a secondary circlet behind the mouth.

trophallaxis (trŏf′ə-lak′səs) (Gr. *trophē*, food, + *allaxis*, barter, exchange). Exchange of food between young and adults, especially certain social insects.

trophoblast (trŏf′ə-blast) (Gr. *trephein*, to nourish, + *blastos*, germ). Outer ectodermal nutritive layer of blastodermic vesicle; in mammals it is part of the chorion and attaches to the uterine wall.

trophozoite (trŏf′ə-zō′ĭt) (Gr. *trophē*, food, + *zōon*, animal). Adult stage in the life cycle of a protozoan in which it is actively absorbing nourishment.

tropomyosin (trŏp′ə-mī′ə-sən) (Gr. *tropos*, turn, + *mys*, muscle). Low—molecular weight protein surrounding the actin filaments of striated muscle.

troponin (trə-pōn′in). Complex of globular proteins positioned at intervals along the actin filament of skeletal muscle; thought to serve as a calcium-dependent switch in muscle contraction.

tube feet (podia). Numerous small, muscular, fluid-filled tubes projecting from body of echinoderms; part of water-vascular system; used in locomotion, clinging, food handling, and respiration.

tubulin (tū′bū-lən) (L. *tubulus*, small tube, + *in*, belonging to). Globular protein forming the hollow cylinder of microtubules.

tundra (tun′drə) (Russ. from Lapp, *tundar*, hill). Terrestrial habitat zone, located between taiga and polar regions; characterized by absence of trees, short growing season, and mostly frozen soil during much of the year.

tunic (L. *tunica*, tunic, coat). In tunicates, a cuticular, cellulose-containing covering of the body secreted by the underlying body wall.

turbellarian (tər′bə-lar′ə-an) (L. *turbellae*, a stir or tumult). Flatworm of phylum Platyhelminthes, class Turbellaria.

typhlosole (tif′lə-sōl′) (Gr. *typhlos*, blind, + *sōlen*, channel, pipe). A longitudinal fold projecting into the intestine in certain invertebrates such as the earthworm.

umbilical (L. *umbilicus*, navel). Refers to the navel, or umbilical cord.

bat / āpe / ärmadillo / herring / fēmale /finch / līce / crocodile / crōw / duck / ūnicorn / ə indicates unaccented vowel sound "uh" as in mammal, fishes, cardinal, heron, vulture / stress as in bi-ol′o-gy, bi′o-log′ical

umbo (um′bō), pl. **umbones** (əm-bō′nēz) (L. boss of a shield). One of the prominences on either side of the hinge region in a bivalve mollusc shell. Also, the "beak" of a brachiopod shell.

ungulate (un′gū-lət) (L. *ungula*, hoof). Hooved. Noun, any hooved mammal.

urethra (ū-rē′thrə) (Gr. *ourethra*, urethra). The tube from the urinary bladder to the exterior in both sexes.

utricle (ū′trə-kəl) (L. *utriculus*, little bag). That part of the inner ear containing the receptors for dynamic body balance; the semicircular canals lead from and to the utricle.

vacuole (vak′yə-wōl) (L. *vacuus*, empty, + Fr. ole, dim.). A membrane-bounded, fluid-filled space in a cell.

valence (vā′ləns) (L. *valere*, to have power). Degree of combining power of an element as expressed by the number of atoms of hydrogen (or its equivalent) that that element can hold (if negative) or displace in a reaction (if positive). The oxidation state of an element in a compound. The number of electrons gained, shared, or lost by an atom when forming a bond with one or more other atoms.

valve (L. *valva*, leaf of a double door). One of the two shells of a typical bivalve mollusc or brachiopod.

vector (L. a bearer, carrier, from *vehere, vectum*, to carry). Any agent that carries and transmits pathogenic microorganisms from one host to another host.

veliger (vēl′ə-jər, vel-) (L. *velum*, veil, covering). Larval form of certain molluscs; develops from the trochophore and has the beginning of a foot, mantle, shell, and so on.

velum (vē′ləm) (L. veil, covering). A membrane on the subumbrella surface of jellyfish of class Hydrozoa. Also, a ciliated swimming organ of the veliger larva.

vestige (ves′tij) (L. *vestigium*, footprint). A rudimentary organ that may have been well developed in some ancestor or in the embryo.

vicariance (vī-kar′ē-ənts) (L. *vicarius*, a substitute). Geographical separation of ancestral populations; allopatry.

villus (vil′əs), pl. **villi** (L. tuft of hair). A small fingerlike, vascular process on the wall of the small intestine. Also one of the branching, vascular processes on the embryonic portion of the placenta.

virus (vī′rəs) (L. slimy liquid, poison). A submicroscopic noncellular particle composed of a nucleoprotein core and a protein shell; parasitic; will grow and reproduce in a host cell.

viscera (vis′ər-ə) (L. pl. of *viscus*, internal organ). Internal organs in the body cavity.

vitalism (L. *vita*, life). The view that natural processes are controlled by supernatural forces and cannot be explained through the laws of physics and chemistry alone, as opposed to mechanism.

vitamin (L. *vita*, life, + *amine*, from former supposed chemical origin). An organic substance required in small amounts for normal metabolic function; must be supplied in the diet or by intestinal flora because the organism cannot synthesize it.

vitellaria (vī′təl-lar′e-ə) (L. *vitellus*, yolk of an egg). Structures in many flatworms that produce vitelline cells, that is, cells that provide eggshell material and nutrient for the embryo.

vitelline membrane (və-tel′ən, vī′təl-ən) (L. *vitellus*, yolk of an egg). The noncellular membrane that encloses the egg cell.

viviparity (vī′və-par′ə-dē) (L. *vivus*, alive, + *parere*, to bring forth). Reproduction in which eggs develop within the female body, with nutritional aid of maternal parent as in therian mammals, many reptiles, and some fishes; offspring are born as juveniles. Adj., **viviparous** (vī-vip′ə-rəs).

water-vascular system. System of fluid-filled closed tubes and ducts peculiar to echinoderms; used to move tentacles and tube feet that serve variously for clinging, food handling, locomotion, and respiration.

X-organ. Neurosecretory organ in eyestalk of crustaceans that secretes molt-inhibiting hormone.

Y-organ. Gland in the antennal or maxillary segment of some crustaceans that secretes molting hormone.

zoecium, zooecium (zō-ē′shē-əm) (Gr. *zōon*, animal, + *oikos*, house). Cuticular sheath or shell of Ectoprocta.

zoochlorella (zō′ə-klōr-el′ə) (Gr. *zōon*, life, + *Chlorella*). Any of various minute green algae (usually *Chlorella*) that live symbiotically within the cytoplasm of some protozoa and other invertebrates.

zooid (zō-oid) (Gr. *zōon*, life). An individual member of a colony of animals, such as colonial cnidarians and ectoprocts.

zygote (Gr. *zygōtos*, yoked). The fertilized egg.

zygotic meiosis. Meiosis that takes place within the first few divisions after zygote formation; thus, all stages in the life cycle other than the zygote are haploid.

INDEX

Page numbers in *italics* indicate illustrations. Page numbers
followed by *n* indicate marginal notes. Page numbers
followed by *t* indicate tables.

ORIGIN OF LIFE AND GEOLOGIC TIME TABLE

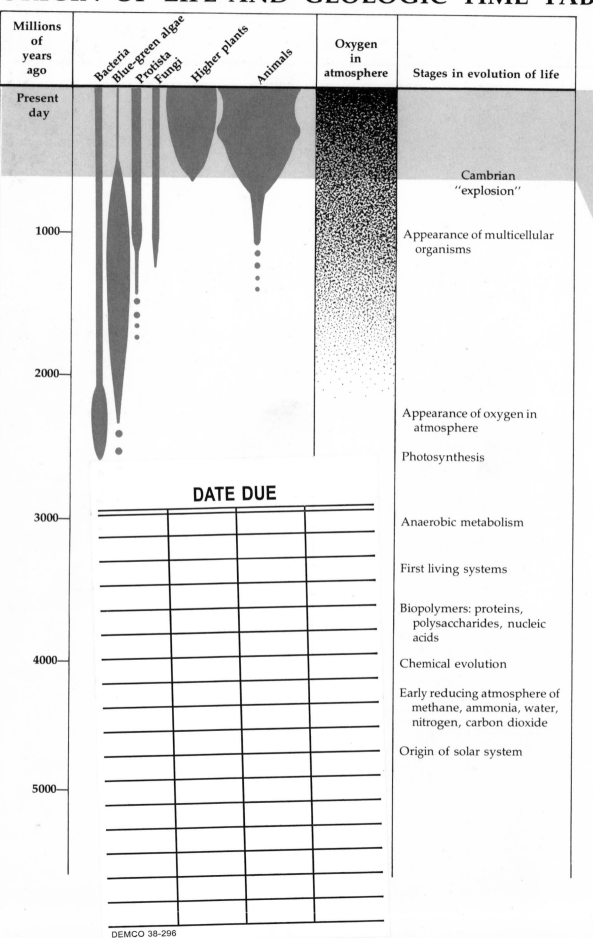

DEMCO 38-296